Flora of the
Sydney
Region

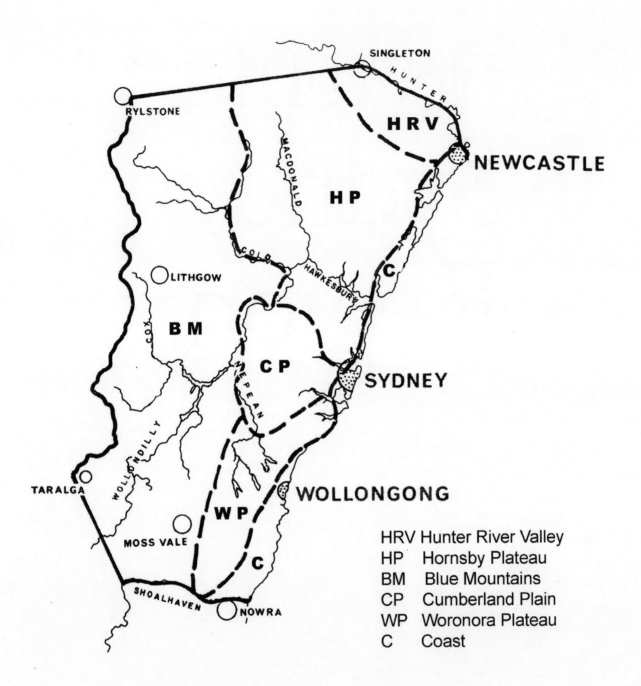

HRV Hunter River Valley
HP Hornsby Plateau
BM Blue Mountains
CP Cumberland Plain
WP Woronora Plateau
C   Coast

Flora of the Sydney Region

FIFTH EDITION

A Complete Revision by

Belinda J. Pellow

Murray J. Henwood

Roger C. Carolin

*The authors wish to acknowledge
the work of the previous authors*

Noel C. W. Beadle

Obed D. Evans

Mary D. Tindale

SYDNEY UNIVERSITY PRESS

Published 2009 by SYDNEY UNIVERSITY PRESS

University of Sydney Library

sydney.edu.au/sup

Sydney University Press
Fisher Library F03
University of Sydney NSW 2006 AUSTRALIA
Email: sup.info@sydney.edu.au

2011 printing

National Library of Australia Cataloguing-in-Publication entry

Author:	Pellow, Belinda J. (Belinda Jane)
Title:	Flora of the Sydney region : a complete revision / Belinda J. Pellow, Murray J. Henwood, Roger C. Carolin.
Edition:	5th ed.
ISBN:	9781920899301 (pbk.)
Notes:	Includes index. Bibliography.
Subjects:	Botany--New South Wales--Sydney Region.
Other Authors/Contributors:	Henwood, Murray J. (Murray James) Carolin, R. C. (Roger Charles)
Dewey Number:	581.99441

Cover design and layout by Miguel Yamin, the University Publishing Service

Printed in Australia

CONTENTS

PREFACE TO THE FIFTH EDITION

Since its publication as the *Handbook of the Vascular Plants of the Sydney District and Blue Mountains* in 1962, the Flora has maintained a tradition of improvement and refinement. Designed originally as a teaching aid for botany students at the University of Sydney, the Flora was rapidly adopted by professional and amateur botanists as an adjunct to the *Flora of New South Wales* (Moore and Betche 1893) and *Maiden's Forest Flora* (1903 – 1925).

From its inception, the *Flora of the Sydney Region* has been continuously updated and improved to reflect the outcomes from taxonomic research of the day. The region covered in the fifth edition is unaltered from the fourth edition but is now known to contain over 3000 native and naturalized species. Similarly, the Flora has always been the product of a collaboration between several taxonomists. The fifth edition of the *Flora of the Sydney Region* maintains this tradition, and represents a partnership between the John Ray Herbarium (The University of Sydney), the University of Sydney Library and the Janet Cosh Herbarium (University of Wollongong).

As with previous editions, this edition is very much a product of its time. In 2005, support from the University of Wollongong Educational Strategies Development Fund (ESDF) was provided to revise four of the larger angiosperm families in the Sydney region. Then, in 2006, the NSW Environment Trust provided funds to complete the current revision and to present it in an electronic format, the eFlora. In 2009 further support from University of Wollongong ESDF enabled the completion and publication of the print-on-demand format of the Flora. As a result, the fifth edition of the *Flora of the Sydney Region* will be the first to be available in two formats; electronically and in the more familiar print form.

The arrangement and content of the fifth edition reflects recent taxonomic research. Much of this progress has been provided by the routine use of nucleotide sequences to offer insights into the evolutionary relationships within and between plant families. In the flowering plants, for example, we have seen the Epacridaceae become a subfamily of Ericaceae, and genera like *Trachymene* and *Hydrocotyle* move from Apiaceae to Araliaceae. Where such changes have been endorsed by the taxonomic community we have rewritten keys and descriptions accordingly. Where the outcomes of taxonomic research are more equivocal, we have maintained traditional definitions of taxa.

The fifth edition of the Flora is completely revised. There are, however, two threads present in the current edition that have consistently run through the Flora since its origin. The first of those threads is the intention of the authors '*to provide complete keys to the identification of all species of vascular plants known to occur in this region*'. The second thread is the presence of Associate Professor Roger Carolin as co-author. As can be seen in the preface to the fourth edition Roger Carolin and Mary Tindale have maintained a 40-year connection with the *Flora of the Sydney Region*. It is a testament to their taxonomic skills that much of the Fifth Edition of the *Flora of the Sydney Region* is based on their keys from the fourth edition.

Preface to Fourth Edition (1994)

Thirty years ago the first edition of this Flora appeared with the title *Handbook of the Vascular Plants of the Sydney District and Blue Mountains*. The initiative for this came some ten years previously when the then Professor of Botany at the University of Sydney, Prof. N.A. Burges, suggested to members of the Botany School that an up-to-date method for identifying plants of the Sydney district was needed for classes held in the School. At the time a series of duplicated laboratory notes, prepared by Dr Patrick Brough and Dr John McLuckie, and based upon the *Flora of New South Wales* by Charles Moore and Ernst Betche published in 1893 was used for this purpose. *The Handbook* was eventually completed by Prof. N.C.W. Beadle, O.D. Evans and R.C. Carolin with sections written by Dr. M.D. Tindale in 1962.

In 1972 the area covered was extended to the Sydney basin, as defined by current geological research, and the publication renamed the *Flora of the Sydney Region*. The intention was, and still is, to provide complete keys to the identification of all species of vascular plants known to occur in this region. In terms of floristics, this region is one of the richest in Australia with over 2500 native species as well as over 500 exotic species to be found in the region. A revision of the *Flora* was produced in 1982 to bring it up-to-date with current research at the time.

Since that date, the pace of taxonomic research in Australia has substantially increased. Notably, thirteen volumes of the *Flora of Australia* have been published and, as this Flora goes to press, three volumes of the four volumes of the *Flora of New South Wales* have appeared with the fourth in press. In addition a large number of technical papers of importance to the flora of the region have appeared. The research towards these publications has changed the situation drastically with numerous new taxa recognized and published as well as taxa not previously found in the region but now known to occur there. In addition a number of name changes have been made necessary by the International Code of Botanical Nomenclature. The previous edition of the *Flora of the Sydney Region* is now very much out-of-date.

The present complete revision has offered the opportunity to change the format from an indented key to a bracketed key. There are advantages in both these formats. The saving in space and the simplification of typesetting were considerations in the change, but in addition, it has bought the alternatives in the key together so that immediate comparisons are easier.

The present publication is not meant as a substitute for the semi-monographic floras, such as *Flora of Australia* and the *Flora of New South Wales*. It covers a much smaller area than either of these and, in addition, the descriptions given only cover the main diagnostic characters which distinguish the species concerned. It does, however, include the most populated parts of New South Wales and the single volume is more convenient to use in the field.

ACKNOWLEDGEMENTS

We would like to acknowledge our debt to Noel Beadle, Obed Evans and Mary Tindale who constructed many of the keys used in the previous editions of this work. Although the format of the keys has changed, their contribution is still apparent in many of them. Funding for the project was provided by the Environment Trust of New South Wales, the University of Wollongong and the University of Sydney.

We would like to thank the management committee of the Janet Cosh Herbarium for supporting the project particularly in the early days when funding was being sought. The staff of the National Herbarium of New South Wales provided their expertise and use of their facilities. In particular Louisa Murray, Peter Weston, Darren Crayn, Barry Conn, Surrey Jacobs, Karen Wilson, Peter Wilson and Elizabeth Brown. Marco Duretto from the Tasmanian Herbarium assisted with aspects of the Rutaceae. Keven Theile from the Western Australian Herbarium assisted with *Viola*.

The support of Mary Peat, Su Hanfling and Rowan Brownlee from the University of Sydney and Kris French from the University of Wollongong has been essential for the execution of this project. David Keith, Janice Hughes, Elizabeth Rosser and Jean Clarke provided support and constructive comment.

HOW TO USE A FLORA

The Identification Process

Possibly everyone starts identifying plants by 'matching' the unknown plants with authentically identified specimens or pictures. The most important drawback to this method is probably the fact that many different, authentically identified specimens or images have to be compared before a particular specimen can be identified with accuracy. Over 3000 species occur in the Sydney Region and so this method is, at best, very time-consuming with such large numbers of possibilities to compare. In fact, this method is only useful when dealing with a small number of possibilities or when the user already has knowledge of the plant species which may reduce the number of choices available.

Keys, on the other hand, are devices that progressively eliminate numbers of possibilities. A key asks you to determine which one of a number of contrasting characteristics occurs in the unknown specimen. For example, how many stamens are present in the flower: 1, or 3, or 5, or 10 and so on. Having determined the answer to that question, a number of the initial set of potential species are rejected and you proceed to determine which one of another pair of contrasting characteristics occurs in the unknown specimen, and so on until the choice is narrowed to one and the plant is identified. A key, then, is like a tree; each branching point represents an alternative route through the key and is defined by contrasting characteristics. You follow the branches by selecting one of two contrasting characteristics until you arrive at the last branch that represents the identification – the name. In most keys there is only a pair of contrasting characteristics at each step (branching). We can draw up the key as in the diagram below in which the tree is presented starting at the top of the diagram (the **bold** words in the caption represent the identification path of a plant with five stamens, five petals and compound leaves). However, this is not an efficient use of space when such a tree is printed.

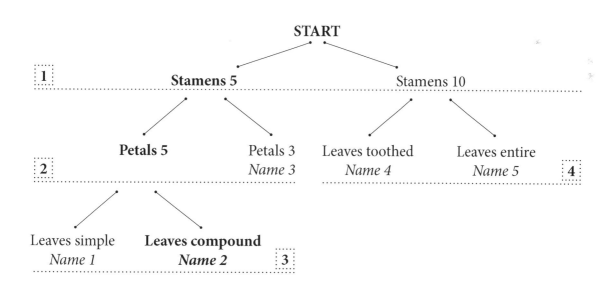

Bracketed Key

1 Stamens 5. . . . . **2**
1 Stamens 10 4
2 Petals 5 . . . . **3**
2 Petals 3 . . . . Name 3
3 Leaves simple. . . . . Name 1
3 Leaves compound . . . . **Name 2**
4 Leaves toothed . . . . Name 4
4 Leaves entire . . . . Name 5

In this edition of the *Flora of the Sydney Region* we use a bracketed key. We have opted to use the bracketed format here primarily because pairs of contrasting characters are brought together which removes the need to flick through a number of pages to locate and compare the characters and because this format also suits the presentation of the text electronically.

The keys in this book continue to species level. In some plant guides the keys only go to the family or generic level and then are replaced by a matching system. Since matching is a subjective method and relies very much on personal judgement, we consider the objective, analytical key method to be inherently more accurate than matching, provided the information in the key is correct.

The main problem with keys of any type is the communication between the writer and the user. The user must understand what the writer is saying. For this reason special words have evolved to describe the features of plants, and often ordinary vernacular English words have been assigned special meanings. These we have defined in the glossary and many of them are illustrated. DON'T BE AFRAID OF THESE WORDS. Just look them up when you encounter them in the text and eventually you will remember what they mean. Once you have mastered them you should be able to go to any English-speaking country, pick up the local plant key and identify the plants of that country. Furthermore, since so many of the words are common to botanical books written in other European languages, albeit with slight variations in spelling, you will be well on the way to understanding the keys written in these languages too! So you see, it is worth the effort to master botanical terminology if you are interested in plants and plant identification.

The Scientific Names

The scientific name of a species consists of the generic name and a species "epithet" which is a qualifier, often an adjective. Both are written in Latin or Greek form, eg

Cryptocarya **obovata** The Cryptocarya (with) obovate (shaped leaves). **Adjective**
Ranunculus **repens** The creeping Ranunculus. **Adjective**
Eucalyptus **smithii** Smith's Eucalyptus. **Genitive noun**
Acacia **clunies rossiae** The Acacia of Clunies-Ross. **Genitive noun**

(N.B. In these examples, the generic nouns are written in italic letters, specific epithets in bold italics, and the form of the epithet in bold letter).

A guide to the pronunciation of botanical names may be found in Hall and Johnson (1993).

Abbreviations of personnel names, or the names themselves, follow the generic specific and subspecific names. These refer to the persons responsible for the publication of the name. e.g.

(i) *Eucalyptus punctata* DC.: De Candolle (abbreviated to DC.) was the first person to publish this name and use it for this species and we have continued to do so.

(ii) *Coricarpia leptopetala* (F.Muell.) Domin: F. Mueller (abbreviated to F.Muell.) first published the epithet "*leptopetala*" for this species but placed it in a different genus. Domin decided that this species belonged to the genus *Coricarpia*, and was the first person to publish this combination of generic name and specific epithet together. It is this combination that we recognize as the name for this species.

(iii) *Pultenaea aristata* Sieber ex DC.: Sieber suggested the name (in this case by writing it on his preserved specimens of this species), but never published it. De Candolle (abbreviated DC.) was the first person to publish the name and use it for this species.

Since different authors, by mistake, might use the same combination of generic name and specific epithet for quite different species, the authorities for a name may assume some importance. Generally speaking they are not referred to all the time, but in publications they should be referenced, either by citing them or indicating a reference book that has been used to identify the material in question.

How to Use this Flora

Start at the Key to phyla and using the method outlined above, determine the phyla to which the unknown plant belongs. With practice you will soon be able to recognise the phylum a plant belongs to without using the Phyla Key.

Proceed to the appropriate phylum and using the key determine what family the plant belongs to. Once you have determined the family you can move to the next key by turning to the family indicated.

Working through the family key you will be able to determine the genus the plant belongs in. The genus will be numbered so turn to the appropriate name and number and then use the key provided to determine the species name for the plant.

It is always useful to check your identification by matching it against an authentically identified herbarium specimen or photograph.

The Arrangement of the Flora

Previous editions of the Flora have used the classification system of Cronquist (1988). Many changes in our understanding of the evolution of plants have occurred since this system was published. The Angiosperm Phylogeny Group publishes the latest additions to our knowledge of angiosperm phylogeny on their webpage and in most cases the numerical arrangement of families within this edition reflects their representation. However for simplicity Cronquist's major grouping of dicotyledons and monocotyledons has been maintained in this edition. Family circumscriptions considered by the authors to be in a state of flux at the time of writing have not been applied. The arrangement of genera within families and species within genera has not been substantially changed since the previous edition.

As far as possible, similar families and genera follow each other in the book, although opinion differs amongst botanists as to both the composition and the arrangement of many families and genera.

All known species of native or naturalized plants occurring in the geographic area covered by this flora have been included. Recent changes in nomenclature have been recorded in the index and synonyms may be accessed by referring to it.

Species listed under the NSW Threatened Species Act (1995) are indicated using the following terms: Vulnerable, Endangered, Critically Endangered and Extinct.

Authorities follow the Australian Plant Name Index and recent updates.

GLOSSARY

Abaxial: the side or face of a lateral organ away from the axis, such as the lower surface of a leaf.

Abortive: imperfectly developed; defective; barren.

Accumbent: (of cotyledons) those which have edges curved against the radicle.

Achene: a dry indehiscent fruit, formed from a superior ovary of one carpel and containing one seed which is free from the pericarp (Fig. 9 a).

Acicular: needle-shaped.

Acroscopic: pointing towards the apex.

Acrostichoid: resembling the fern genus *Acrostichum.*

Acrostichoid sori: densely covering the dorsal surface without having distinct sori.

Actinomorphic: (of flowers) having a radially symmetrical perianth.

Aculeate: prickly.

Acuminate: tapering to a point (Fig. 2).

Acute: pointed; sharp (Fig. 2).

Adaxial: the side or face turned towards the axis, such as the upper surface of the leaf; ventral.

Adhesion (hence adherent): the sticking together of floral parts of different whorls without organic fusion.

Adnate: (1) organically fused to another but different kind of part, e.g. stamens to petals; (2) (of anthers) fused to the filament by their whole length.

Adventitious: arising in irregular or anomalous positions.

Aerial roots: adventitious roots growing in the air.

Aestivation: the arrangement of the petals and sepals in the unexpanded bud (Fig. 5).

Aggregate fruit: the aggregation of fruits derived from an apocarpous gynoecium (Fig. 9 u). see Multiple fruit.

Alate: winged.

Alternate: arranged singly at different heights on the axis and in 2 rows longitudinally; commonly used also to include spiral arrangement.

Amphibious: growing on land or in water; growing with part of the plant in mud or water and with part in the air.

Anadromous venation: a type of venation in which the first set of veins is given off on the upper side of the midrib towards the apex as in *Polystichum* (Fig. 10).

Anastomose: to form a network.

Anatropous: (of ovules) having the micropyle facing the placenta; inverted.

Androecium: the stamens collectively.

Androgynophore: a stalk bearing male and/or female parts of flower.

Angiosperm: plant with seeds enclosed in an ovary.

Annual: plant completing its life cycle within one year after germination, and then dying.

Annular: in the form of a ring.

Annulus: the elastic ring of cells in the sporangium of a fern; a ring.

Anomalous: irregular; abnormal.

Anterior: towards or in the front; that side of a flower away from the main axis.

Anther: the part of the stamen which produces the pollen and consisting of the microsporangium and connective (Fig. 6).

Anthocarp: a false fruit consisting of true fruit which is surrounded by the base of the perianth eg. Nyctaginaceae.

Antrorse: bent towards apex.

Apiculate: with a small abrupt point (Fig. 2).

Apocarpous: applied to a gynoecium which consists of several free or slightly coherent or basally connate carpels.

Appendage: an attachment developed on, and projecting from, the surface of an organ.

Appressed: pressed up against another organ.

Aquatic: (plant) growing in water.

Arachnoid: cobweb-like; formed of tangled hairs or fibres.

Arborescent: growing to the size of a tree; resembling a tree in habit of growth.

Areola (pl. areolae) (hence areolate): a space in any reticulated surface, e.g. space between veins.

Aril (hence arillate): an expansion of the funicle growing partially or wholly over the testa of certain seeds (Fig. 39).

Aristate: bearing, or tapering to, a bristle (Fig. 2).

Article: (1) a joint; (2) a one-seeded part of a carpel into which some fruits break or segregate.

Articulate: having nodes; having joints where separation takes place; jointed.

Ascending: growing obliquely at first but finally upwards; rising or growing up.

Attenuated: gradually diminished in breadth towards either extremity.

Auricle (hence auriculate): an ear-like appendage at the base of a leaf or of a pinna or pinnule.

Autotrophs : none parasitic, ie. Obtains nutrients and charbohydrates independently.

Awn (hence awned): a straight or bent, bristle-like, branched or unbranched appendage.

Axil (hence axillary): the upper angle between the axis and any organ which arises from it, especially a leaf or bract.

Axile: on the axis; e.g. in axile placentation where the placentas are on the central axis of the ovary (Fig. 8).

Barbed: bearing sharp spine-like hooks which are bent backwards.

Barbellate: finely barbed.

Barren: sterile.

basal: attached or grouped at the base.

Basifixed: attached at or near the base.

Basipetal: developing in the direction of the base.

Basiscopic: pointing towards the base.

Beak: a pointed projection.

Bearded: having a tuft or tufts of hairs.

Berry: an indehiscent succulent fruit having the pericarp differentiated into the epicarp (skin) and pulp containing usually many (sometimes one) seeds (Fig. 9 s, t).

Bi: prefix signifying two or twice.

Bicolorous: having two colours.

Biennial: a plant which lives for more than one but less than two years.

Bifid: divided at one end into two parts (2-fid).

Bifoliate: two-leaved.

Bifurcate: divided into two; forked.

Bilateral: having two sides.

Bipartite: divided into two nearly to the base.

Bipinnate: twice pinnately divided (Fig. 3).

Bisexual: having both fertile male and female organs in the same flower; hermaphrodite.

Biternate: divided into three with each segment divided again into three (Fig. 3).

Bole: the trunk of a tree up to the first branch.

Bract: leaf-like structure or scale that subtends an inflorescence or flower.

Bracteate: having bracts.

Bracteole: a small bract situated on the pedicel below a flower but not subtending it, usually paired in dicotyledons and single in monocotyledons.

Bristle: a short stiff hair.

Bulb: a storage organ consisting of a short, underground stem surrounded by swollen leaf-bases and outer, dry, protective leaf-bases.

Bulbel: a bulb arising from a mother bulb.

Bulbiferous: bearing bulbs.

Bulbil: a small bulb in the axil of a scale or foliage leaf; a deciduous, axillary bud produced in the leaf axils of certain ferns.

Bulbous: bulb-shaped.

Bullate: applied to leaves when the spaces between the veins present convexities on one side and concavities on the other; bubble-like.

Burr: a rough or prickly compound structure developed from a seed or fruit and associated appendages (bracts, perianth, etc.) (Fig. 42).

Caducous: falling off early.

Caespitose: tufted.

Callous: hardened and abnormally thickened.

Callus: (1) an abnormally thickened part appearing as a knot or lump; (2) the hard basal projection (often sharply pointed) at the base of a floret or spikelet of some grasses.

Calyculus: a fringe or ring of tissue below the perianth of the flower of Loranthaceae.

Calyptra: the cap-like covering surmounting some flowers and consisting of connate perianth segments (see Operculum).

Calyx: the sepals collectively.

Campanulate: bell-shaped.

Campylotropous: applied to an ovule which is curved so that the axis is at right angles to the funicle.

Canopy: the cover of foliage of a plant community or layer of a plant community.

Capillary: hair-like.

Capitate: enlarged and globular at the tip.

Capsule: a dry, dehiscent fruit of two or more carpels (Fig. 9 i–9l).

Carpel: a unit of the gynoecium in which one or more ovules are enclosed. It is usually divided into stigma, style, and ovary.

Carpellodes: sterile carpels.

Carpophore: the persistent, filiform, central axis which supports the two mericarps of Apiaceae.

Cartilaginous: hard and tough and resembling cartilage.

Caruncle (hence carunculate): a warty protuberance near the hilum of a seed.

Caryopsis: a dry, indehiscent fruit formed from a unilocular, superior ovary (probable 3- or 2-carpellary) of the Poaceae and containing one seed closely fused to the pericarp (Fig. 9 b).

Catadromous venation: a type of venation in which the first set of veins is given off on the lower side of the midrib as in most species of *Lastreopsis* (Fig. 11).

Catkin: an inflorescence consisting of unisexual, apetalous flowers arranged in a compact spike.

Cauda: slender, elongated portion of perianth segments of some orchids.

Caudex: (1) a compact axis consisting of stem and sometimes root; (2) the trunk of a tree fern.

Caudicle: the thread supporting the pollen-masses of orchids.

Cauline: pertaining to the stem; attached to the stem.

Chalaza: the part of an ovule to which the extremity of the funicle is attached.

Chestnut: dark red-brown.

Ciliate: having the margin fringed with hairs.

Cilium (pl. cilia): a (short) hair.

Circinnate: coiled from the apex downward.

Circumscissile: opening or divided by a transverse line around the circumference so that the top comes off like a lid.

Cladode: a stem which takes on the function of a leaf and which bears scale-leaves.

Clathrate: in the form of a lattice.

Clavate: club-shaped.

Claw: (1) the narrowed base of a petal, sepal, bract or similar structure; (2) the bundles of stamens of *Melaleuca*.

Cleistogamous: applied to flowers which never open, but which are fertile and self-pollinated.

cm: centimetre.

Cochlear: a type of aestivation (Fig. 5).

Coenosori: sori which have united.

Cohesion (hence coherent): the sticking together of floral parts of the same whorl without organic fusion.

Collateral: lying one beside the other.

Columella: the persistent, central axis in some fruits and cones.

Column: an upgrowth of the axis above the ovary of orchids which incorporates styles, stigmas and stamens (Fig. 45). Or applies to the lower part of and awn which is differentiated into parts

Coma: a tuft of long hairs on the testa of some seeds.

Commissure: a joint or seam.

Community (of plants): any assemblage of plants.

Compound head: an inflorescence made up of a number of heads arranged in a large head.

Compound leaf: a leaf divided to the base or midvein into leaflets (Fig. 3).

Compound umbel: an inflorescence made up of a number of umbels arranged in a large umbel (Fig. 4).

Compressed: flattened laterally.

Conduplicate: (of cotyledons) those which are folded so that they almost surround the radical.

Cone: (1) a compact group of sporophylls (male or female) borne (usually spirally) on a central axis. (2) the woody multiple fruit of Casuarinaceae, made up of true fruits surrounded by woody bracts and bracteoles.

Congested: crowded.

Connate: organically fused to one or more members of the same whorl.

Connective: the part of the anther that connects the lobes.

Connivent: lying or standing side by side, but not touching or fused.

Contorted: twisted; convolute (Fig. 5).

Contracted: narrowed or shortened.

Convolute: rolled so that the margins overlap (Fig. 3).

Cordate: heart-shaped (often applied to bases only) (Fig. 2).

Coriaceous: leathery.

Corm: a short, swollen, upright, underground stem formed annually in the stem-base and in which food reserves are stored. It is surrounded by dry, protective leaf-bases.

Cormel: a small corm in the axil of a leaf of a larger corm.

Cormil: a small corm or corm-like bud in the axil of a foliage leaf or bract.

Corolla: the petals collectively.

Corona: a ring of tissue arising from the base of the corolla, perianth or filaments of a flower and standing between the perianth lobes and the stamens

Corymb (hence corymbose): a racemose inflorescence in which all the flowers, although they originate at different levels on the stem, are ultimately borne at about the same level (Fig. 4).

Costa: the mid vein of a pinna.

Costate: ribbed.

Costule: the mid-vein of a pinnule or segment of lesser order, except the central vein of an ultimate segment which is usually called the midrib or mid-vein.

Cotyledon: a seed-leaf of the embryo of a seed plant.

Crenate: having the margin cut regularly into rounded teeth or lobes (Fig. 3).

Crenulate: slightly crenate.

Crisped: curled, very wavy or crumpled.

Critically Endangered: listed under the New South Wales Threatened Species Conservation Act 1995.

Crown: the part of a tree or shrub above the first branching.

Cruciform: in the form of a cross.

Ctenitis-hair: a short, articulated, unbranched, redish hair.

Culm: aerial stem of grasses, sedges and rushes.

Cultrate: very narrow-oblong.

Cuneate: wedge-shaped (Fig. 2).

Cupular: cup-shaped.

Cyanogenetic: producing hydrogen cyanide (prussic acid) as a result of the hydrolysis of a cyanogenetic glycoside.

Cyathiform: cup-like.

Cyathium: inflorescence of *Euphorbia* and related genera (Fig. 26).

Cyme (hence cymose): an inflorescence in which the terminal flower terminates the growth of the main axis which is replaced by one or two lateral buds, the process being repeated throughout the inflorescence (Fig. 4 h,i).

Cypsela: the fruit of members of the family Asteraceae; i.e. an inferior nut (Fig. 41).

Deciduous: falling seasonally (e.g. leaves and bark of some trees, parts of the flower).

Decompound: a leaf having divisions that are themselves compound.

Decorticating: shedding the outer bark, usually in long strips.

Decumbent: reclining but with the summit ascending (applied to branches whose lower portions lie on or near the ground while the tips grow upwards).

Decurrent: extending downward beyond the place of insertion (applied to pinnae, leaves or petioles when their edges are continued downwards forming raised lines) (Fig. 2).

Decurved: bent downward in a curve.

Decussate: in pairs alternately at right angles.

Definite: (1) of a precise and constant number, as of stamens in a flower. (2) (of inflorescences) terminating in a flower, growth being resumed by lateral buds as in cymes.

Deflexed: bent sharply downwards.

Dehiscence: the manner in which the wall of a mature organ ruptures to allow the contents to escape (Fig. 9).

Dehiscent: opening or bursting at maturity.

Deltoid: of the shape of the Greek letter Δ (delta); triangular (Fig. 2).

Dendritic: branched like a tree.

Dentate: toothed (Fig. 3).

Denticulate: finely toothed (Fig. 3).

Depressed: flattened endwise or from above.

Diadelphous: applied to stamens when they are united by their filaments into two clusters.

Dichasium: a cyme in which two new, equal floral axes arise in turn opposite each other and beneath each flower after the latter has terminated growth (Fig. 4).

Dichopodial: when an axis repeatedly forks giving rise to a bifurcation where a bud is seated in the fork.

Dichotomous: divided into two C. equal branches.

Dicotyledons: a group of Angiosperms (Magnoliophyta) whose embryos have two cotyledons.

Dictyostele: a stele with large overlapping leaf-gaps.

Diffuse: spreading and much branched.

Digitate: branching from the axis (from a stem or a petiole) like fingers from a hand.

Dimidiate: applied to organs in which partial imperfections exist, e.g. in the pinnae of *Adiantum* spp. where the lamina is fully developed on one side of the midrib only.

Dimorphous (dimorphic): occurring in two forms.

Dioecism (hence dioecious): yhe condition of a species having unisexual flowers of which male and female are borne on different plants.

Disarticulate: to separate at a joint.

Disc: (1) a plate (almost circular) of tissue found sometimes between the whorls of floral segments; (2) the dilated stylar bases on the ovary and fruit of some Apiaceae (Fig. 27).

Disc flower: flower (usually tubular) borne on the central portion of the heads of some Asteraceae.

Dissected: divided into segments.

Dissepiment: vertical septum in the interior of an ovary dividing it wholly or partially into 2 or more parts.

Distal: remote from; at the end.

Distichous: arranged in two opposite rows in the same plane.

Divaricate: extremely divergent; straggling.

Domatia: small pocket, depression or group of hairs on the undersurface of a leaf usually at the point were the midvein and lateral veins converge.

Dorsal: pertaining to, or attached to, the back of an organ.

Dorsifixed: attached at or by the back.

Dorsiventral: having structurally different upper and lower surfaces.

Drupaceous: applied to fruit with a structure of a drupe, but derived from more than 1 carpel (Fig. 9 r).

Drupe: a succulent indehiscent fruit derived from a single carpel and having the pericarp differentiated into epicarp (skin), soft mesocarp and stony endocarp (Fig. 9 q).

Drupel: small drupe in an aggregate fruit (Fig. 9 u).

Dune: mound (usually of definite shape) of blown sand.

Echinate: covered with long and sharp almost prickly protuberances.

Echinulate: having tiny prickles.

Elater: one of the four hygroscopic bodies developing from the perispore.

Emarginate: notched at the summit or apex (Fig. 2).

Embryo: a young plant within the seed or within the archegonium.

Endangered: listed under the *New South Wales Threatened Species Conservation Act 1995.*

Endemic: peculiar to a particular geographic region.

Endocarp: the innermost layer of a layered pericarp, identified only when it is stony.

Endosperm: nutritive tissue developed in seeds; in Magnoliophyta (Angiosperms) it develops from the triple fusion nucleus; in Cycadophyta and Pinophyta it is the female gametophyte.

Entire: without division, incision, or separation.

Ephemeral: short-lived; annual.

Epicalyx: an involucre outside the true calyx and resembling a second calyx.

Epicarp: the outermost layer of the pericarp; the skin.

Epidermis: the outer layer of cells of all parts of plants.

Epigynous: referring to stamens (sometimes also to sepals and petals) which are attached to a floral tube which is fused to the ovary (Fig. 7).

Epipetalous: (of stamens) borne on the petals.

Epiphyte: a plant perched, but not parasitic on, another plant or other object.

Erecto-patent: between spreading and erect.

Ericoid: (of leaves) small and sharply pointed like those of the heaths.

Excurrent: where the vein runs through the apex of a segment and projects beyond it as a mucro.

Exfoliate: to come away in scales or flakes.

Exindusiate: without an indusium.

Exotic: introduced from abroad.

Exsert(ed): projecting beyond the surrounding parts.

Extinct: listed under the *New South Wales Threatened Species Conservation Act 1995.*

Extrorse: (of stamens) opening towards the circumference of the flower.

Facultative: occasional; not essential; incidental.

Falcate: sickle-shaped.

Family: a taxonomically related group of genera.

Farinaceous: consisting of starch; floury.

Fascicles (hence fascicled): bundles or clusters.

Ferruginous: of the colour of rusty iron.

Fertilization: the union of gametes.

Fibrillose: furnished with hair-like appendages.

-fid: cleft or branched, e.g. 2-fid, 3-fid.

Filament: (1) any thread-like body; (2) the stalk of a stamen (Fig. 6).

Filamentous (filiform): thread-like.

Fimbriate: fringed with fine hairs.

Fistular: hollow throughout its length.

Flabellate: fan-shaped.

Flaccid: limp; flabby.

Flexuous: bent alternately in opposite directions; zig zagged.

Floccose: having tufts of soft woolly hairs.

Flora: (1) the assemblage of plant taxa of any area; (2) a book or treatise dealing with the plant taxa of an area.

Floral: belonging to the flower or seated near the flower.

Floral leaves: leaves subtending flowers.

Floral tube: a usually tubular or cup-like structure present in some flowers and interpreted as the fused basal portions of the androecium and perianth or as an upgrowth of the receptacle. It may be free or adnate to the part or whole of the ovary (Fig. 7). The Hypanthium of some authors.

Floret: (1) the flower of Asteraceae; (2) the flower of grasses together with the lemma and palea.

Flower: the sexual reproductive structure of the Magnoliophyta (Angiosperms) usually consisting of gynoecium, androecium, and perianth (Fig. 6).

Foetid: having an offensive odour.

Foliaceous: leaf-like.

Foliolate: bearing leaflets, e.g. 3- (or tri-)foliolate, having 3 leaflets.

Follicle: a dry dehiscent fruit formed from 1 carpel and having 1 longitudinal line of dehiscence (Fig. 9 g).

Forest: a closed community dominated by trees which have flat-topped crowns and long boles.

Forest Rain (RF): a forest dominated by trees usually with mesomorphic leaves, few herbs, but lianas and epiphytes often abundant.

Forest, Dry Sclerophyll (DSF): a forest in which xeromorphic shrubs form a continuous or discontinuous layer below the trees.

Forest, Wet Sclerophyll (WSF): a forest in which mesomorphic shrubs form a usually continuous layer below the trees.

Free: not united with any other part.

Free-central placentation: placentation where the placentas are on the central axis of the ovary (Fig. 8).

Frond: that part of the fern plant that is analogous to the leaf.

Fruit: the seed bearing structure developed from the ovary of Magnoliophyta (Angiosperms) after fertilization (Fig. 9)

Funicle (funiculus): the stalk of an ovule (Fig. 39).

Fused: joined and growing together.

Fusiform: spindle-shaped; cigar-shaped.

Gene: an hereditary factor which, either alone or in conjunction with another gene or genes, produces a character in an organism.

Geniculate: abruptly bent so as to resemble a knee-joint.

Genus (pl. genera): a taxonomic group of closely related species.

Gibbous: more convex in one place than another; lobed with a short obtuse spur or swelling.

Glabrous: without hairs.

Gland: a structure, embedded or projecting, which secretes such substances as nectar or oil.

Glandular: having glands. eg. of hairs (Fig. 24)

Glaucous: dull green with a whitish blue lustre.

Glochidiate: having barbed bristles.

Glumaceous: resembling a glume.

Glume: (1) bract subtending the flower in the spikelet of Cyperaceae; (2) empty bract (usually 2, one on either side) at the base of the spikelets of grasses (Fig. 48).

Glutinous: covered with a sticky exudation.

Grain (Caryopsis): the fruit of grasses.

Grass: a member of the family Poaceae.

Grassland: plant community, either natural or induced, which is dominated by grasses.

Gymnosperm: seed plants whose ovules are not enclosed in an ovary (in this Flora divided into the Cycadophyta and Pinophyta).

Gynobasic: applied to a style which extends to the base of, or into a pit in the top of, a gynoecium between the carpels or articles (Fig. 7).

Gynoecium: the carpels of one flower collectively.

Gynophore: the stalk of a superior ovary (Fig. 27).

Gynostegium: formed when stamens and style are fused and form a central column. Found often in plants of the Apocynaceae.

Habit: the general appearance of a plant, including size, shape, growth-form and disposition of its various parts.

Habitat: the environment in which the plant lives; the natural abode of a plant.

Hair: an epidermal appendage consisting of one elongated cell or a number of cells (Fig. 24)

Halophyte: a plant which grows in and tolerates salty places or one which accumulates salt.

Hastate: triangular with spreading basal lobes (Fig. 2).

Haustoria: organ of parasitic plant which is attaches to the host plant and absorbs nutrients.

Head: a racemose inflorescence of sessile flowers crowded together on a receptacle and usually surrounded by an involucre (Fig. 4).

Heath: a plant community dominated by small shrubs (usually less than 2 metres high) which usually have ericoid leaves.

Hemi: half.

Herb: a plant which does not produce a woody stem.

Herbaceous: pertaining to herbs. When applied to the texture of a leaf or frond. midway between membranous and coriaceous, not succulent.

Hermaphrodite: bearing both male and female sex organs in the same flower; bisexual.

Hilum: the scar left on the testa of a seed at the spot where it was attached to the funicle or placenta.

Hip: the aggregate fruit of *Rosa* (Fig. 9 v).

Hirsute: covered with long spreading hairs.

Hispid: densely covered with short stiff hairs or bristles.

Hoary: covered with hairs so short as not readily to be distinguished by the naked eye and yet giving the surface a greyish hue.

Homosporous: producing only one morphologically distinct kind of spore from which develops a gametophyte producing both male and female gametes.

Hyaline: translucent and usually colourless.

Hyalo-membranous: hyaline and membranous.

Hybrid: (as used here) offspring of two plants of different taxa.

Hybrid swarm: the population of plants resulting from the backcrossing of a hybrid with one or both of the parents.

Hydathode: a water-excreting opening in the enlarged tip of a vein.

Hydrophyte: a plant growing submerged, or almost so, in water.

Hypanthium: see Floral tube.

Hypogynous: referring to stamens (sometimes also to petals and sepals) which are attached to the receptacle below the gynoecium. The floral tube is absent (Fig. 7).

Imbricate: with the edges overlapping.

Immersed: embedded in an organ.

Imparipinnate: pinnate with an odd terminal leaflet (Fig. 3).

Incumbent: (of cotyledons) with the back of one lying against the radicle.

Incurved: bending or curved inwards or upwards (Fig. 3).

Indefinite: (1) (of floral parts) too many to be counted easily; (2) (of inflorescences) capable of constant extension by means of the main axis.

Indehiscent: not opening at maturity.

Indumentum: a general term for the hairy covering of plants.

Induplicate: with the edges bent inward and the external face of these edges applied to each other without twisting.

Indusium: (1) any covering of a sorus whether a modified organ or merely the incurved margin of the pinna; (2) the pollen-cup of Goodeniaceae (Fig. 25).

Inferior ovary: one to which the sepals or floral tube are apparently attached; one to which the floral tube is fused (Fig. 7).

Inflexed: bent inwards; incurved.

Inflorescence: a group of flowers borne on one stem; the way in which flowers are arranged on a stem (Fig. 4).

Infra: below (eg infraterminal)

Integument: a covering; one of the outer layers of tissue on an ovule

Interjugary glands: the additional glands occurring along the leaf rhachis between the insertion of successive pairs of pinnae, occurring below the single and often slightly larger gland which is found at or just below the insertions of these pinnae.

Intermediate leaves: leaves which develop after the juvenile and before the mature leaves in plants which have dimorphic foliage.

Internode: the portion of the stem between successive nodes.

Interpetiolar: between the petioles of opposite leaves (applied to stipules, Fig. 29).

Interrupted: (of inflorescences) with the flowers in distinct clusters and with bare rhachis between the clusters (Fig. 47 i).

Intramarginal: situated within the margin and near the edge.

Introrse: turned inward; towards the axis.

Invaginated: turning in and enclosing e.g. the fig is an inflorescence turn inside out (Fig. 9 w).

Involucre: (1) a whorl or several whorls of bracts surrounding an inflorescence, a flower or a cone; (2) the indusium of a hymenophyllous fern.

Involute: folded inwards or backwards (Fig. 3).

Ironbark: a eucalypt tree whose bark is dark-coloured, hard, deeply furrowed, and often impregnated with kino.

Irregular flower: a flower which has one or more of its perianth segments very dissimilar in shape from the others of the whorl; a bilaterally symmetrical flower.

Isobilateral: having the same structure on both sides.

Isomerous: having segments of successive whorls equal in number.

Joint: (1) articulation; (2) node; (3) a segment of some cladodes.

Juvenile leaves: the first-formed leaves, especially when they differ from the mature leaves.

Keel (hence keeled): (1) a ridge, usually on the back, like the keel of a boat

Kino: reddish, astringent exudation from the bark or wood of some trees (Myrtaceae).

Labellum: one of the petals of orchids (and a few other plants) usually differentiated in form and size from the two lateral petals (Fig. 45).

Laciniate: cut into narrow slender teeth or lobes.

Lamina: the expanded portion of a leaf or fern frond; leaf blade.

Lanate: woolly.

Lanceolate: narrow and tapering at each end and more than 4 times as long as broad (Fig. 2).

Lateral: fixed on or near the sides of an organ; arising from a leaf axil.

Lax: loose, not compact.

Leaflet: a separate portion of a compound leaf.

Leaf-opposed: opposite to a stem leaf.

Legume: (1) a dry dehiscent fruit formed from a single carpel and having 2 longitudinal lines of dehiscence (Fig. 9 e); (2) (colloquial) a member of the Family Fabaceae.

Lemma: the lower of two bracts enclosing the flower of a grass (Fig. 48).

Lemma, staminate: a lemma enclosing a male flower.

Lemma, sterile (barren): an empty lemma.

Lenticular: disc shaped with convex sides

Liana (liane): a climbing or twining plant.

Lignotuber: a woody swelling, partly or wholly underground, at the base of the stem of certain plants and containing numerous cortical buds.

Ligulate: strap-shaped.

Ligule: (1) membranous or hairy outgrowth at the inner junction of the leaf sheath and blade of grasses (Fig. 47); (2) the limb of the corolla of a ray floret of members of Asteraceae; (3) membranous structure near the base of the leaf of some Lycopodiaceae.

Limb: the upper free and often spreading portion of a connate corolla.

Linear: long and narrow with parallel sides (Fig. 2).

Littoral: belonging to, or growing near, the sea.

Loculicidal dehiscence: dehiscence of a capsule in such a way that the openings occur in the carpel walls and not at the septa or repla (Fig. 9 k).

Loculus: a compartment within an ovary (Fig. 9 k).

Lodicule: one of two scales below the stamens and ovary of a grass and regarded as reduced perianth.

Lomentum: a dry fruit derived from 1 carpel, which breaks up transversely into (usually) 1-seeded articles at maturity (Fig. 9 f).

Lunate: crescent shaped.

Lyrate: pinnatifid with the terminal lobe much larger than the others; lyre shaped (Fig. 3).

m: metre.

Mallee: a shrub with many stems arising from a lignotuber (usually applied to eucalypts only).

Mangrove: a shrub or small tree growing in salt or brackish water and with aerial pneumatophores.

Marginal: placed upon or attached to the edge (Fig. 8).

Maritime: belonging to the sea; confined to the sea coast.

Marsh: a waterlogged area; swamp.

Massula: the hardened mucilage enclosing a cluster of microspores in *Azolla* and *Salvinia*.

Mealy: covered with coarse flour-like powder.

Megasporangium: the sporangium producing megaspores.

Megaspore: the larger (female) spore of heterosporous plants.

Membranous: thin and translucent.

Mericarp: a unicarpellary unit formed by the dehiscence of a schizocarp (Fig. 9 n,o).

-merous: a suffix used to refer to the number of parts, e.g. 5-merous meaning that each floral whorl contains 5 segments (possibly excepting the gynoecium).

Meristele: the portion of a stele received by each leaf.

Mesocarp: the middle layer, usually fleshy, of a 3-layered pericarp.

Mesomorphic: soft and with little fibrous tissue, but not succulent.

Microphyllous: (of leaves) small (usually less than 2cm long), and usually hard and narrow.

Microsporangium: a sporangium producing microspores.

Microspore: the smaller (male) sexual spore of a heterosporous plant.

Microsporocarp: a body containing the microsporangia.

Midrib (mid-vein): the principal vein which runs from the base to the apex of the leaf.

Minute: so small as to be difficult to see with the naked eye.

Mixed sporangia: sporangia of different ages.

mm: millimetre.

Monangial: used of a sorus containing one sporangium.

Moniliform: constricted and appearing bead-like.

Monochasium: a cymose inflorescence with a single branch at each node (Fig. 4).

Monocotyledon: a group of Magnoliophyta (Angiosperms) whose embryos have one cotyledon.

Monodelphous: stamens which have their filaments united to form one bundle.

Monoecious: having unisexual flowers of which both male and female are borne on the same plant.

Monolete: a spore with a single straight tetrad on the proximal face.

Monomorphic: having only one form.

Monopodial: referring to a stem with a single main axis which produces lateral organs of which the youngest is always at the tip.

Monotypic genus: a genus which has only one species.

Morphology: form and structure; the study of these.

Mucro: a short terminal point.

Mucronate: having a short terminal point (Fig. 2).

Mucronulate: having a very small mucro.

Multiple fruit: collection of fruits produced by more than one flower (Fig. 9 w).

Muricate: covered with short hard-pointed protuberances.

Naked: (1) without hairs; (2) without perianth segments.

Nectar: a (sweet) fluid secreted from a specialized gland.

Nectary: a specialized gland that secretes nectar (Fig. 27).

Net-veined: see reticulate.

Neuter: sterile.

Nodding: hanging downwards.

Node: the portion of the stem from which the leaf or bract arises.

Nut: a dry, indehiscent, one-seeded fruit formed from 2 or more carpels (Fig. 9 c).

Ob: a prefix signifying that the meaning of the simple word is reversed, e.g. obcordate – the reverse of cordate.

Obconic: conic but attached at the narrow end.

Obcordate: heart-shaped and with the notch at the apex (Fig. 2).

Oblanceolate: roughly lanceolate but with the distal end broader than the basal portion (Fig. 2).

Obovate: almost ovate but with the distal end broader (Fig. 2).

Obsolete: lacking or rudimentary.

Obtuse: blunt or rounded at the apex (Fig. 2).

Ochrea: a sheath formed from 2 stipules encircling the stem in most Polygonaceae (Fig. 29).

Operculum: the structure covering the stamens and style of a flower bud (e.g. *Eucalyptus*) formed from the fused perianth segments.(see Calyptra) (Fig. 36)

Opposite: arising at the same level but on opposite sides.

Orbicular: perfectly or nearly circular (Fig. 2).

Order: a taxonomic group of related families.

Osmophores: sent producing glands

Ovary: the basal portion of a carpel or group of fused carpels in which one or more ovules are enclosed, and which after fertilization develops into the fruit (Fig. 6).

Ovate: less than 4 times as long as broad with the broadest part in the lower third (Fig. 2).

Ovoid: the solid analogue of ovate.

Ovule: the megasporangium (nucellus) together with the integuments of a seed plant. After fertilization the ovule develops into the seed.

Palate: the projection in the throat of a 2-lipped corolla (Scrophulariaceae).

Palea: (1) the upper of two bracts enclosing the flower of a grass; (Fig. 48); (2) the chaffy scales on the petiole and rhachis of many ferns.

Paleaceous: clothed with chaffy scales.

Palmate: divided into 5 or more leaflets, the leaflets diverging from the same point (Fig. 2 and 3).

Palmatifid: divided into 5 or more distinct lobes almost to the petiole (Fig. 3).

Panicle (hence paniculate): a much branched racemose inflorescence (Fig. 4).

Papilionaceous: butterfly-like; shaped like the flower of a pea, and belonging to the same family, Fabaceae.

Papilla: a small elongated protuberance.

Pappus: the appendages (hairs, scales, etc.) at the top of a cypsela of the Asteraceae (Fig. 41).

Papyraceous: papery.

Paraphyses: sterile filaments occurring amongst the sporangia of Filicopsida.

Parapinnate: pinnate and with an equal number of leaflets (i.e. without a terminal leaflet) (Fig. 3).

Parasite: an organism living on or in, and deriving nourishment from, another organism (the host).

Parietal: attached to the sides or to a wall.

Parietal placentation: the ovules attached to parietal placentas (i.e. on the outside wall of the ovary) (Fig. 8).

Partite: subdivided into segments, the divisions extending nearly to the base.

Pectinato-pinnate: pinnate with narrow segments set like the teeth of a comb.

Pedate: ternate with the lateral leaflets cleft (Fig. 3).

Pedicel: the stalk of each single flower, or of a sporangium of a fern or of a spikelet of a grass. (Fig. 6).

Pedicellate: on a pedicel; having a pedicel.

Peduncle: (1) the stalk of an inflorescence; (2) of a solitary flower; (3) the stalk of a sporocarp.

Pedunculate: on a peduncle.

Pellucid: transparent.

Peltate: having the stalk (petiole) attached at the back and ± in the centre (Fig. 2).

Penniveined: pinnately veined (Fig. 2).

Penta-: five, used as a prefix.

Pentagonous: 5-sided.

Peppermint (Bark): a fibrous barked tree or the bark itself, the fibres being short and not removable in long strings. (Plate 000).

Perennial: living for more than 2 years.

Perfect flower: bisexual flower.

Perianth: the calyx and corolla collectively, especially when they are morphologically similar.

Pericarp: the wall of a fruit. developed from the ovary wall after fertilization.

Perigynous: referring to stamens (sometimes also to sepals and petals) which are attached to the rim of a lateral expansion of the receptacle (the expansion arising below the ovary, or attached to the rim of a floral tube which is not fused to the gynoecium (Fig. 7).

Perispore: the folded membrane which assists a spore to float in the air.

Persistent: remaining until the part that bears it is fully matured.

Petal: one of the usually conspicuous segments forming the inner whorl of the perianth (Fig. 6).

Petaloid: assuming the characters of petals.

Petiole (hence petiolate): the stalk of a leaf or the stipe of a pteridophyte.

Petiolule: the stalk of a pinna, pinnule, or ultimate segment.

Phyllode: a flat petiole of lamina-like appearance (Fig. 38).

Phyllopodium (pl. phyllopodia): a short, erect, scaly base of a stipe to which a frond is articulated

Pilose: sprinkled with rather long simple hairs.

Pinna: the primary segments of a divided leaf lamina or of a larger pinna.

Pinnate leaf: a compound leaf whose leaflets are arranged on opposite sides of a common rhachis (Fig. 3).

Pinnatifid: cut into lobes on both sides about half-way to the midrib (Fig. 3).

Pinnule: the ultimate segment of a divided pinna.

Pistil: a free carpel or group of fused carpels.

Pistillate: female.

Pitted: having numerous small depressions on the surface.

Placenta: the part of the ovary to which the ovules are attached.

Placentation: the arrangement of the placentas and of the ovules thereon (Fig. 8).

Plano-convex: flat on one side, convex on the other (Fig. 3).

Plicate: folded longitudinally.

Plumose: like a feather, i.e. with a central axis and finer hairs arising from it.

Plumule: the shoot of the embryo.

Pollen: the male fertilizing element of seed plants.

Pollen grain: microspore of seed plants, consisting of a spore coat and enclosed haploid nuclei.

Pollination: the transference of pollen from a microsporangium to the stigma of Magnoliophyta (Angiosperm) or the pollen chamber of Cycadophyta and Pinophyta.

Pollinium: a mass of pollen grains cohering by means of their waxy texture or fine threads.

Polygamous: having hermaphrodite and unisexual flowers mixed together.

Polyhedral: having many faces or planes.

Polymorphic: displaying many diversities of form.

Polypetalous: having free petals.

Posterior: next to or towards the main axis.

Prickle: a hard, pointed emergence arising from subepidermal tissue, but not containing vascular tissue.

Process: a projecting appendage.

Procumbent: trailing or spreading along the ground without putting forth roots.

Proliferous: having adventitious leaf-buds which produce roots and new plants.

Prophyll: bracteole.

Prostrate: lying close to or on the ground.

Protandrous: having the male organ maturing before the female (e.g. anthers before stigmas).

Protogynous: having the female organ maturing before the male (e.g. stigmas before anthers).

Protostelic: having a simple and primitive type of stele with a solid central vascular strand.

Pseudo-bulb: the swollen bulb-like internodes of the stems of many orchids (Fig. 39).

Pseudo-dichotomous: falsely dichotomous.

Pseudoglumes: the bracts subtending the spikelets in *Carex*, the true glumes are absent

Pubescent (hence pubescence): covered with short soft hairs.

Pulvinus: a swollen base of a leaf pinna, or pinnule, often capable of changing form which brings about movements of the leaf pinna, or pinnule.

Punctate: marked with dots, depressions or translucent glands.

Punctiform: in the form of a dot; reduced to a dot.

Pungent: sharp.

Pustule: pimple or blister.

Pyrene: the endocarp and the enclosed seed of a drupe or drupaceous fruit.

Pyriform: pear-shaped.

Quadrangular: four-angled (hence four-sided)

Quincuncial: a type of aestivation (Fig. 5).

Quinquangular: five-angled (hence five-sided).

Raceme (hence racemose): an inflorescence of stalked flowers whose growing point continues to add to the inflorescence so that the youngest flowers are nearest the apex (Fig. 4).

Rachilla: the axis of a grass spikelet above the glumes, or the axils of a sedge spikelet.

Radiate: spreading from a common centre.

Radical (of leaves): arising at the base of the stem and forming a rosette or tuft; basal.

Radicle: the rudimentary root in the embryo which develops into the tap-root.

Rainforest (RF): a community dominated by trees which have soft leaves and in which lianas and epiphytes are conspicuously present.

Ray: (1) the strap-like part of the corolla of a ray or ligulate floret of Asteraceae (Fig. 41); (2) a branch of an umbel.

Receptacle: (1) the part of the axis which bears the floral parts (Fig. 6); (2) (in ferns) the axis bearing the sporangia and, if present, paraphyses.

Recurved: curved backwards or downwards at a sharp angle (Fig. 3).

Regular: (of flowers) having a radially symmetrical perianth.

Reniform: resembling a kidney in shape (Fig. 2).

Replum: a thin wall formed by the ingrowth of a false septum from the parietal placentas.

Resupinate: turned through 180o; reversed.

Reticulate: forming a network (Fig. 2).

Retrorse: bent backwards.

Retuse: very obtuse with a slight depression at the apex.

Revolute: rolled backwards from the extremity or edge on to the under surface (Fig. 3).

Rhachilla: the axis within a spikelet of a grass.

Rhachis: the axis or axes of a compound leaf bearing pinnae or pinnules, or of an inflorescence.

Rhizome: an underground stem.

Rhizophore: leafy branch of *Selaginella* which eventually produces true roots.

Rhomboidal: diamond-shaped or C. so (Fig. 2).

Rootstock: a swollen root and part or all of a very short stem, sometimes partly underground.

Rostellum: an extension of the upper edge of the stigma of some orchids (Fig. 45).

Rostrate: beaked.

Rotate: having a regular, sympetalous corolla with a short tube and spreading limb; resembling a wheel.

Rugose: covered with wrinkles.

Rugulate: with small wrinkled ridges irregularly distributed.

Runner: a slender prostrate stem, having a bud at the end which sends out leaves and roots.

Rupestral: plants growing on walls or rocks.

Saccate: pouch-like.

Saggitate: shaped like the head of an arrow (Fig. 2).

Samara: a winged achene or winged nut (Fig. 9 d).

Saprophyte: an organism using non-living organic matter for foodstuffs.

Scaberulous: slightly rough to the touch.

Scabrous (scabroso): rough to the touch.

Scale: (1) any thin scarious body, usually a degenerate or rudimentary leaf; (2). a flat or bullate surface appendage more than one cell wide and usually one cell thick except sometimes towards the base.

Scape: a peduncle arising from near the ground and often solitary, leafless or with scale-leaves.

Scarious: dry and membranous.

Schizocarp: a dry fruit which on dehiscence, breaks up into individual carpels, each of which is called a mericarp (Fig. 9 m,o).

Schizocarp-capsule: a schizocarp whose mericarps dehisce (Fig. 9 p).

Sclerophyll (hence sclerophyllous): a plant with hard stiff leaves.

Scrub: a community dominated by shrubs.

Scurfy: scaly, the scales being bran-like.

Secondary thickening: the formation of additional vascular and supporting tissue through the activities of the vascular cambium.

Section: a subgroup of a genus used to identify closely related species.

Secund: with the parts directed to one side only.

Seed: the fertilized ovule of a plant, containing an embryo, and sometimes endosperm, which are covered by the seed coat.

Segment: (1) each subdivision of a divided or dissected leaf; (2) a part, e.g. a petal is a segment of the corolla.

Sepal: one of the usually green segments forming the outer of the two whorls of leaf-like structures of a flower, and collectively known as the calyx (Fig. 6).

Sepaloid: assuming the character of sepals.

Septate: divided by partitions.

Septicidal: (of capsules) dehiscing in the septa, i.e. along the lines of junction of the carpels (Fig. 9 l).

Septum: a partition.

Serrate: notched on the edge with assymetrical teeth which point forward (Fig. 3).

Serrulate: minutely serrate (Fig. 3).

Sessile: without a stalk (pedicel, peduncle, or petiole).

Seta: a stiff hair or bristle.

Setose: bristly.

Sheath: a long tubular structure, either entire or split longitudinally on one side.

Shrub: a woody plant less than 8 metres high and usually with many stems. In this book *tall shrub* is used to indicate a height of 2–8m; *small shrub* c. 1–2m; and *dwarf shrub* less than 1m.

Silicula: a dry dehiscent fruit, almost as broad as long or broader, formed from a superior ovary of two carpels and with 2 parietal placentas connected by a false-septum (Fig. 40).

Siliqua: a dry dehiscent fruit, longer than broad, formed from a superior ovary of two carpels and with 2 partietal placentas connected by a false-septum (Fig. 9 h, 40).

Simple: (of leaves) undivided.

Sinuate: with a deep wavy margin (Fig. 3).

Sinus: a depression or cleft between two adjacent lobes.

Slender: long and thin.

Solenostele (hence solenostelic): a tubular stele which has internal and external phloem.

Sorus: a group of sporangia on the fronds of ferns.

Spadix: an inflorescence which is a spike with a fleshy axis, the whole usually surrounded by a spathe (Fig. 26).

Spathe: a large bract situated at the base of a spadix which it encloses in a sheath (Fig. 26).

Spathulate: spoon-shaped; enlarged and rounded towards the summit (Fig. 2).

Species: a taxonomic unit of classification; the largest group of organisms potentially capable of interbreeding to produce fertile offspring for many generations.

Spicate: resembling a spike.

Spike: a racemose inflorescence of sessile flowers (or spikelets in the case of grasses) borne on a simple elongated axis (Fig. 4).

Spikelet: (1) a unit of the grass inflorescence usually consisting of two glumes with one of more florets; (2) a small spike in which each flower is subtended by a glume (Cyperaceae and some Restionaceae) (Fig. 46,48).

Spike-like: outwardly resembling a spike but with short branches which are concealed by the spikelets of grasses.

Spine: a hard pointed structure.

Spinescent: terminating in a spine.

Spinule: a small spine.

Spinulose: with small spines on the surface.

Sporangiophore: an organ bearing a sporangium.

Sporangium: an organ which produced spores.

Spore: a unicellular or few-celled asexual or sexual reproductive unit, not containing an embryo.

Sporocarp: the compact mass of sori enveloped in the indusium or sporophyll to form a hard subglobular or bean-shaped resting body – found only in water ferns.

Sporophyll: a leaf-like structure which bears one or more sporangia.

Spur: a conical or cylindrical projection from the base or side of one of the perianth whorls.

Squamule: a small scale.

Squarrose: with spreading or divergent scales.

Stamen: the structure in the flower which produces the pollen and consists of an anther and usually a filament (Fig. 6).

Staminate flowers: male flowers.

Staminode: a sterile stamen, usually modified morphologically.

Standard: the large upper petal of the Papilionaceous (Fabaceae) flower.

Stele: the usually cylindrical central vascular portion of the axis of a vascular plant.

Stellate: star-shaped, e.g. of hairs (Fig. 24).

Sterile: without reproductive organs.

Stigma: that part of the style adapted for the reception and germination of pollen grains (Fig. 6).

Stipe (hence stipitate): (1) a (small) stalk; (2) the stalk or petiole of a fern.

Stipella: a minute stipule at the base of a petiolule.

Stipule (hence stipulate): an appendage, pairs of which occur at the base of the petioles of some dicotyledons (Fig. 29).

Stock: (see caudex).

Stolon: a lateral stem, growing above the ground, and which roots at its nodes; a runner.

Stoloniferous: producing stolons; in the form of a stolon.

Stomate: an opening usually found in leaf epidermis which is bounded by two guard cells and allows the transfer of gasses.

Stomium: an opening in the annulus of a sporangium through which dehiscence occurs and the spores are released.

Stramineous: straw-like or straw-coloured.

Striate: marked with parallel longitudinal lines.

Stringybark: (1) a eucalypt bark characterized by long fibres which can be removed in long strings or ropes (Page 182); (2) a tree with such a bark.

Style: that part of the pistil situated above the ovary and bearing the stigma (Fig. 6).

Sub (as a prefix): (1) under or below; (2) almost or approaching.

Subfamily: a group of closely related genera within a family.

Subspecies (ssp.): a subgrouping within a species used to describe variants which are isolated by various means.

Subtend: to stand below or close to.

Subulate: narrow and gradually tapering to a fine point (Fig. 2).

Succulent: (1) juicy; fleshy – applied either to fruits, leaves or stems; (2) a plant with fleshy habit, hence leaf-succulent, stem-succulent.

Sucker: (1) a shoot of either subterranean or subaerial origin; (2) juvenile, when applied to the leaves of eucalypts.

Suffruticose: woody in lower part of stem, but with herbaceous branches.

Superficial: (as applied to sori) arising from the surface of a frond.

Superior ovary: one which lies above the point of insertion of the calyx; or, when a floral tube is present, one which is free from the floral tube (Fig. 7).

Suture: a seam or plane of junction between the carpels of a syncarpous ovary; (2) the putative fold (dorsal suture) and junction (ventral suture) of the carpel.

Syconium: a fig, the multiple fruit of the genus *Ficus* (Fig. 9 w).

Sympetalous (synpetalous): with the petals connate.

Sympodial: (i) a dichotomy where at each forking, one branch continues to develope and the other aborts; (2) where the apical bud ceases growth for one reason or another and growth is conitinued by a lateral bud

Syn (as a prefix): together.

Synangium: a composite sporangium with a series of loculi.

Syncarpous: having the carpels connate into a single unit.

Taproot: the main root of a plant when it descends perpendicularly into the soil; the mature radicle.

Taxon: a term used to describe any taxonomic category, e.g. subspecies, species, family, etc.

Taxonomy: classification.

Tendril: part of a plant modified into a slender elongated organ used in climbing.

Terete: cylindrical or nearly so.

Ternate: arranged in threes.

Terrestrial: growing on land.

Tesselated: divided into squares or squarish pieces.

Testa: the seed coat.

Tetragonous: having four angles.

Tetrahedral: having four sides.

Thorn: a reduced branch with a hard, sharp point.

Thyrse: a series of cymes arranged in a racemose manner and reducing to a single flower distally (Fig. 4).

Tomentose: covered with closely matted short hairs.

Translator: (of Apocynaceae) a horny elastic body to which the pollen masses are fixed.

Trapeziform: a plane shape of 4 unequal sides

Tree: a woody plant usually with a single stem (trunk) exceeding 8m in height.

Tribe: a group of related genera within a family or subfamily.

Trifid (3-fid): cleft into three to c. the middle.

Trifoliolate: having three leaflets (Fig. 3).

Trigonous: having three angles.

Trimerous: having three (or multiples of three) segments in each floral whorl.

Tripinnate: thrice pinnate.

Tripliveined: with three main veins.

Triquetrous: having three prominent angles or ridges.

Triternate: thrice ternate, i.e. divided into 3, three times giving 27 segments.

Triveined: having three C. longitudinal main veins.

Truncate: terminating abruptly as if cut off transversely (Fig. 2).

Trunk: the main stem of a tree.

Tuber: the swollen end of an underground stem containing food reserves. A *root tuber* is a swollen part of a root, containing food reserves.

Tuberculate: warted or warty; having swellings or nodules, e.g. of hairs (Fig. 24).

Tuberous: swollen – the term is usually applied to roots and not to true tubers.

Tufted: growing in (little) clumps.

Turbinate: shaped like a top.

Turgid: bloated; inflated; swollen.

Twiner: a climbing plant supporting itself by winding spirally round an object.

Umbel: a racemose inflorescence in which all the pedicels arise at the tip of the peduncle and the flower lie at the same level (Fig. 4).

Umbellate: in the form of an umbel.

Umbellule: an ultimate umbel of a compound umbel.

Umbo: an elevation in the centre.

Undulate: wavy, corrugated.

Uni: one.

Unilateral: one-sided.

Uniserial: in a single row or series.

Urceolate: urn-shaped.

Utricle: the membranous covering, possibly a modified glume, which surrounds the ovary of Cyperaceae, eg. *Carex*.

Valvate: (1) with the edges touching (Fig. 5); (2) opening by valves.

Valve: (1) a distinct portion into which some organs break; (2) the 3 inner perianth segments of some Polygonaceae.

Variety: a taxonomic subdivision of a polymorphic species.

Vascular bundle: the primary conducting system consisting principally of xylem and phloem.

Vegetation: the total aggregation of plant communities in an area.

Vein: the vascular bundles of a leaf.

Velamen: the absorbent multiple epidermis on the roots of some epiphytes, especially orchids.

Velutinous: densely covered with fine, short, soft, erect hairs.

Venation: the manner in which the veins of leaves are arranged (Fig. 2).

Ventral: the surface facing the main axis

Vernation: the manner in which unexpanded leaves are arranged in the leafbud.

Verrucose: warted.

Versatile: (of anthers) joined to the filament by the midpoint and swinging freely.

Verticillate: arranged in a whorl or in whorls.

Villous: covered with long weak hairs.

Virgate: twiggy.

Viscid: coated with a sticky substance.

Vitta: linear, longitudinal oil glands in the wall of the fruit of many Apiaceae.

Viviparous: germinating or sprouting from the seed or bud while still attached to the parent plant.

Vulnerable: listed under the *New South Wales Threatened Species Conservation Act 1995.*

Whorl: a group of three or more appendages arising from the axis at the same level.

Wing: (1) the membranous border of many seeds; (2) the two lateral petals of a Papilionaceous (Fabaceae) flowers.

Xeromorph: a plant which has the morphological characters of a xerophyte but which is not necessarily drought resistant.

Xerophyte: a drought resistant plant.

Zygomorphic: a flower which has one or more of its perianth segments very dissimilar in shape from the others of the whorl; a bilaterally symmetrical flower.

ABBREVIATIONS

Aust.	Australia
c.	about
cm	centimetre(s)
cosmop.	cosmopolitan
cv.	cultivar
diam.	diameter
DSF.	Dry Sclerophyll Forest
E.	East
Eur.	Europe
f.	form of
Fig.	Figure
Fl.	Flowering time
Introd.	Introduced
LGA.	Local Goverment Area
m	metre(s)
mm	millimetre(s)
μm	micrometer, micron
Mts	Mountains
N.	North
Pl.	Plain
R.	River
RF.	Rainforest
S.	South
Ss.	Sandstone(s)
ssp	subspecies
sp., spp	species (singular and plural respectively)
Subtrop.	Subtropical, Subtropics
Temp.	Temperate
Trop.	Tropical, Tropics
var.	variety
W.	West
WS,	Wianamatta Shales
WSF.	Wet Sclerophyll Forest
*	Introduced from abroad
±	more or less

ILLUSTRATIONS

FIGURE 2

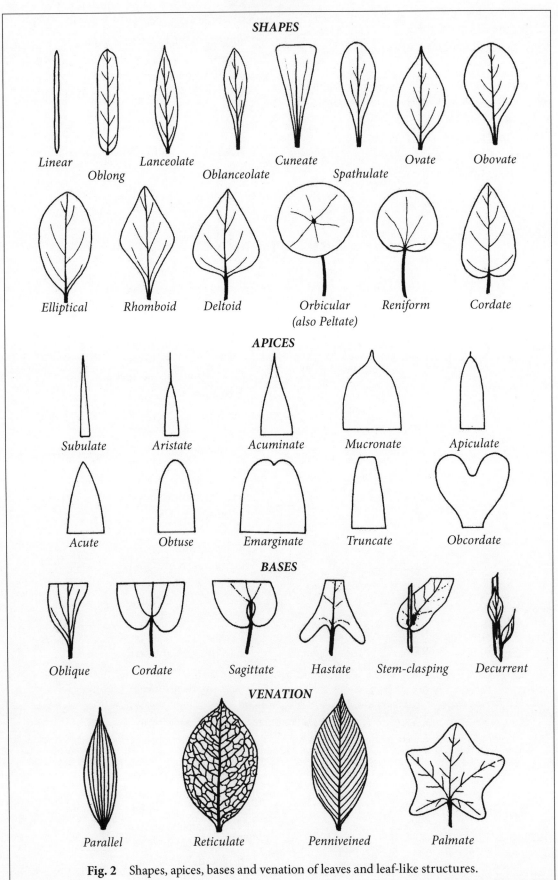

SHAPES

Linear   Oblong   Lanceolate   Oblanceolate   Cuneate   Spathulate   Ovate   Obovate

Elliptical   Rhomboid   Deltoid   Orbicular (also Peltate)   Reniform   Cordate

APICES

Subulate   Aristate   Acuminate   Mucronate   Apiculate

Acute   Obtuse   Emarginate   Truncate   Obcordate

BASES

Oblique   Cordate   Sagittate   Hastate   Stem-clasping   Decurrent

VENATION

Parallel   Reticulate   Penniveined   Palmate

Fig. 2   Shapes, apices, bases and venation of leaves and leaf-like structures.

FIGURE 3

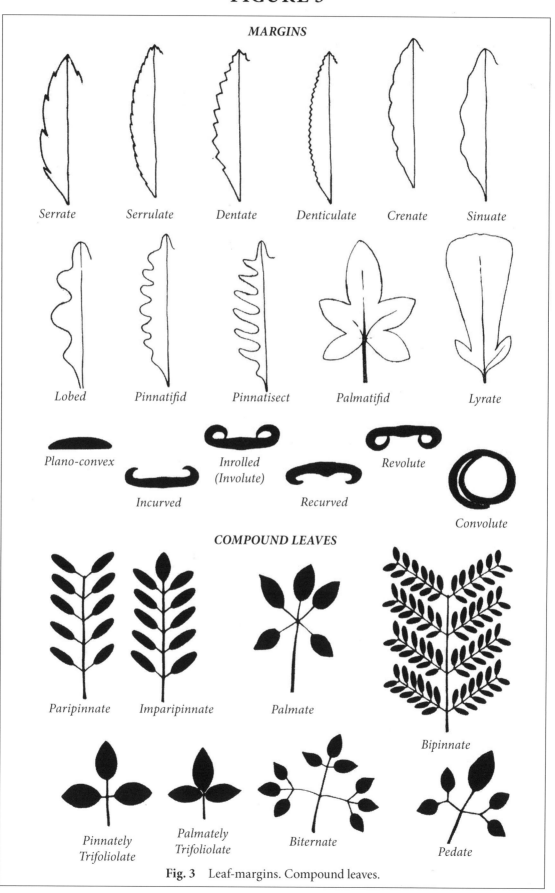

MARGINS

Serrate   Serrulate   Dentate   Denticulate   Crenate   Sinuate

Lobed   Pinnatifid   Pinnatisect   Palmatifid   Lyrate

Plano-convex

Incurved

Inrolled
(Involute)

Recurved

Revolute

Convolute

COMPOUND LEAVES

Paripinnate   Imparipinnate   Palmate   Bipinnate

Pinnately
Trifoliolate

Palmately
Trifoliolate

Biternate

Pedate

Fig. 3 Leaf-margins. Compound leaves.

FIGURE 4

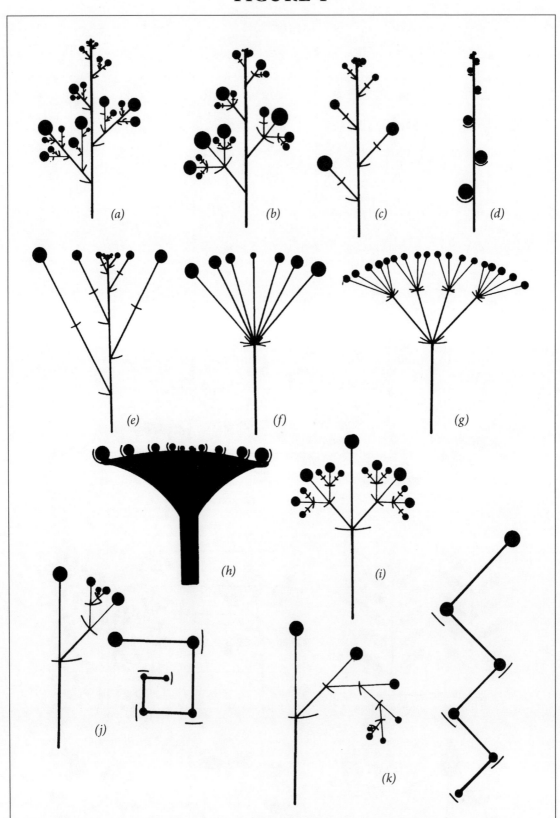

Fig. 4 Inflorescences: (a) Panicle (b) Thyrse (c) Raceme (d) Spike (e) Corymb (f) Simple Umbel (g) Compound Umbel (h) Capitulum (Head) (i) Dichasium (j) Monochasium-bostryx (k) Monochasium-cincinnus.

FIGURE 5 & 6

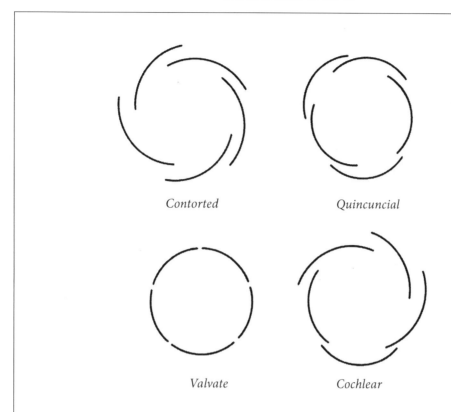

Contorted

Quincuncial

Valvate

Cochlear

Fig. 5　Aestivation.

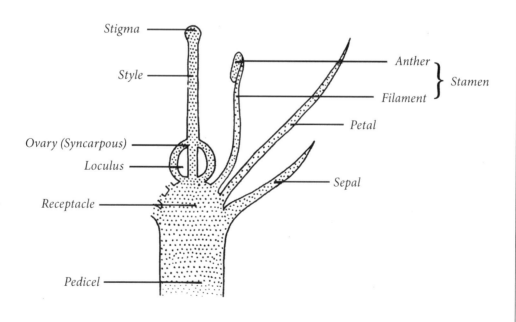

Stigma

Style

Ovary (Syncarpous)

Loculus

Receptacle

Pedicel

Anther

Filament

} Stamen

Petal

Sepal

Fig. 6　Parts of the flower.

FIGURE 7 & 8

Ovary superior
Stamens hypogynous
Carpels free

Ovary superior
Stamens perigynous
Carpels free

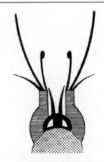

Ovary inferior
Stamens epigynous
carpels connate
Floral-tube extended
above ovary

Ovary inferior
Stamens epigynous
carpels connate

Style gynobasic
Disc hypogynous
Stamens hypogynous

Disc epigynous
Stamens epigynous

Fig. 7   Flower Types.

Marginal

Axile

Parietal

Free-central

Fig. 8   Placentation.

FIGURE 9

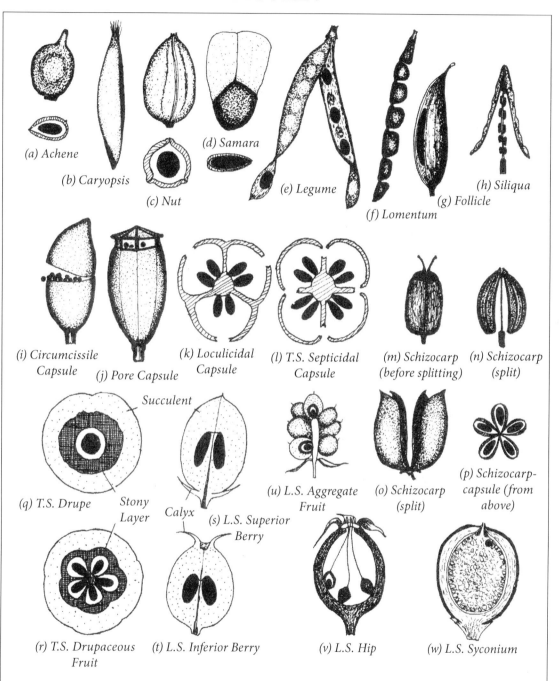

(a) Achene

(b) Caryopsis

(c) Nut

(d) Samara

(e) Legume

(f) Lomentum

(g) Follicle

(h) Siliqua

(i) Circumcissile Capsule

(j) Pore Capsule

(k) Loculicidal Capsule

(l) T.S. Septicidal Capsule

(m) Schizocarp (before splitting)

(n) Schizocarp (split)

Succulent

(q) T.S. Drupe

Stony Layer

Calyx

(s) L.S. Superior Berry

(u) L.S. Aggregate Fruit

(o) Schizocarp (split)

(p) Schizocarp-capsule (from above)

(r) T.S. Drupaceous Fruit

(t) L.S. Inferior Berry

(v) L.S. Hip

(w) L.S. Syconium

Fig. 9 Fruits: (a) achene (transverse section below); (b) caryopsis; (c) nut (transverse section below); (d) samara (transverse section below); (e) legume; (f) lomentum; (g) follicle; (h) siliqua; (i) circumcissile capsule; (j) pore capsule; (k) transverse section of a loculicidal capsule showing positions of dehiscence; (l) transverse section of a septicidal capsule showing positions of dehiscence; (m) schizocarp with 2 carpels, before splitting; (n) schizocarp with 2 carpels after splitting; (o) schizocarp with 4 carpels after splitting, (p) schizocarp capsule with 5 carpels after splitting and dehiscence (drawn from above); (q) transverse section of drupe; (r) transverse section of a drupaceous fruit with 5 carpels; (s) longitudinal section of a superior berry; (t) longitudinal section of an inferior berry; (u) longitudinal section through an aggregate fruit consisting of a cluster of drupels, (v) longitudinal section of a hip, the true fruits are achenes, one of which is sectioned; (w) vertical section through the syconium of a fig (multiple fruit). Shading: seeds exposed by dehiscence or sectioning are shown in solid black; ovary walls when sectioned are cross-hatched.

FIGURES 10 TO 13

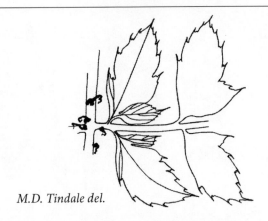

M.D. Tindale del.

Fig. 10 Dryopteridaceae: Pinnules of *Polystichum australiense* with anadromous veins and aristate margins. X⁵⁄₂.

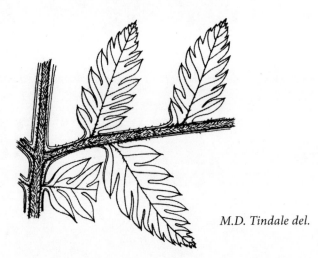

M.D. Tindale del.

Fig. 11 Dryopteridaceae: Pinnules of *Lastreopsis decomposita* with catadromous veins, ridges of the rhachises decurrent along the margin of the pinnules. X³⁄₂.

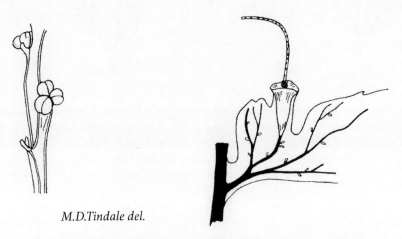

M.D.Tindale del.

Fig. 12 (left) PSILOTACEAE: Branch of *Psilotum nudum* with synangia X5.

Fig. 13 (right) HYMENOPHYLLACEAE: Tubular involucre of *Crepidomanes venosum* with an elongated receptacle. X6.

FIGURES 14 TO 18

M.D. Tindale del.

Fig. 14 DICKSONIACEAE: Leaf-segment of *Dicksonia antarctica* bearing marginal sori with 2-lipped indusia. X⅔.

Fig. 15 DENNSTAEDTIACEAE: *Histiopteris incisa*. Portion of a leaf-segment with marginal sori, indusia opening inwards, slightly anastomosing veins. X¹⁶⁄₃.

Fig. 16 LINDSAEACEAE: Pinna of *Lindsaea linearis* with an almost marginal linear sorus, the indusium opening outwards; junction of pinna with main rhachis hidden. X²⁰⁄₃.

Fig. 17 ADIANTACEAE: Pinnule of *Adiantum aethiopicum* with marginal sori and reniform indusia, the venation flabellate. X4.

Fig. 18 POLYPODIACEAE: *Microsorum scandens*. Portion of a leaf-segment with superficial sori and simple anastomosing veins. X¹⁶⁄₃.

FIGURES 19 TO 23

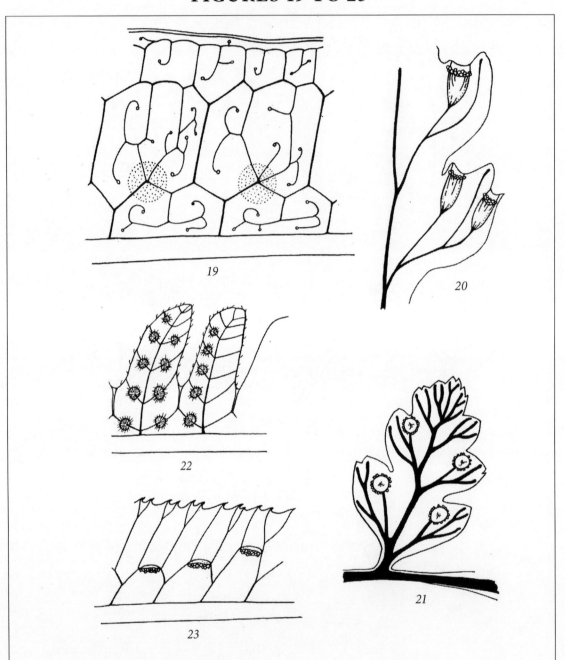

19

20

22

21

23

Fig. 19   POLYPODIACEAE: Portion of a leaf-segment of *Microsorum pustulatum* with superficial sori and complex anastomosing veins. $\times 10/3$.

Fig. 20   DAVALLIACEAE: Portion of a pinnule of *Davallia solida var. pyxidata* with marginal sori and cupuliform indusia. $\times 10$.

Fig. 21   DAVALLIACEAE: Pinnule of *Rumohra adiantiformis* with superficial sori, peltate indusia and free veins. $\times 8/3$.

Fig. 22   THELYPTERIDACEAE: Portion of a pinna of *Christella* sp. with excurrent veins between the segments $\times 8/3$.

Fig. 23   BLECHNACEAE: Portion of a leaf-segment of *Doodia aspera* with superficial sori attached on the inner side of veinlets which are parallel to the costa. $\times 16/3$.

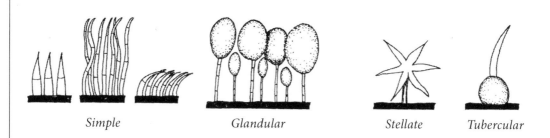

Simple   *Glandular*   *Stellate*   *Tubercular*

Fig. 24   Hair Types.

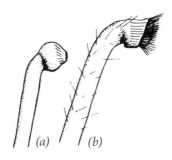

(a)   *(b)*

Fig. 25   GOODENIACEAE: *Indusia*: (a) *Dampiera stricta*, (b) *Goodenia stelligera*.

FIGURES 26 & 27

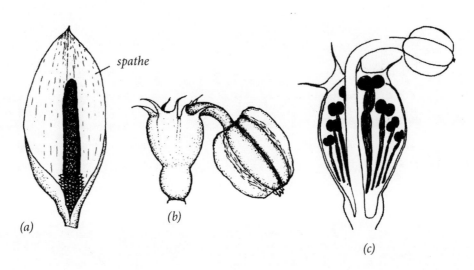

Fig.26 Flower: (a) complete spadix; (b) complete cyathium; (c) cyathium in longitudinal section (not median) showing 2 of the 5 groups of stamens and the female flower (ovary has not been cut).

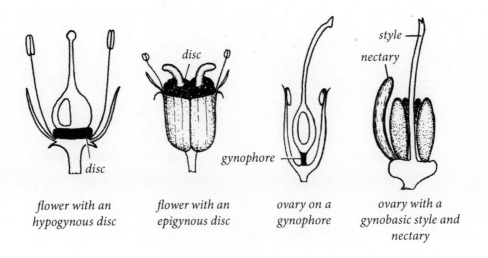

flower with an hypogynous disc   *flower with an epigynous disc*   *ovary on a gynophore*   *ovary with a gynobasic style and nectary*

Fig. 27 Floral disc, nectories and gynophore.

Key to the Phyla of Vascular Plants

1 Plants not producing seeds 2
1 Plants producing seed (Spermatophyta). 4

2 Stems jointed. Leaves scale-like, arranged in whorls at the nodes; lower portions of the leaves forming cylindrical sheaths . . . . **2 MONILOPHYTA**
2 Stems not jointed; either leafless; or, if leaves present, not in whorls and not fused to form sheaths at the nodes. 3

3 Sporangia solitary and born in the axils of sporophylls, often in strobili (cones) or born in fertile zones along the stem. Not fused into synangia. . . . . **1 LYCOPODIOPHYTA**
3 Sporangia not as above, either scattered over the lower surface of the lamina or aggregated into sori; or 2 – 3 fused into a synangia . . . . **2 MONILOPHYTA**

4 Ovules (and seeds) enclosed within carpels (fruits); the carpels bearing stigmas upon which the pollen lodges. Pollen produced in the anthers of stamens. Carpels and stamens usually surrounded by a perianth; the whole constituting a flower which may be uni- or bi-sexual (Angiosperms) . . . . **5 MAGNOLIOPHYTA**
4 Ovules (and seeds) borne naked on sporophylls (or on scales projecting from sporophylls, which are usually arranged in cones. Microsporophylls bearing the pollen sacs also arranged in cones. Pollen deposited directly on the ovule (Gymnosperms). 5

5 Leaves pinnate . . . . **3 CYCADOPHYTA**
5 Leaves simple . . . . **4 PINOPHYTA**

Key to the Lycopodiophyta and Monilophyta Families

Based on original key by Mary D. Tindale

Within the region these 2 phyla are represented by C. 130 species distributed amongst 27 families and 54 genera. Most of these ferns and fern allies grow in rainforests but others are found in open forests, heathland, swamps or lagoons, along creeks and rivers or at the edge of water-holes in fields. Some species are epiphytic on tree-trunks, occurring high in the branches, others are terrestrial on the forest floor or growing in rock-crevices.

Specimens should not be collected without the rhizome, as the hairs and/or scales of this organ are very important in the identification of these plants. In tree ferns the lower portion of the stipe with the basal tuft of scales or hairs is essential. Wherever possible, fertile fronds should be collected, as most keys are largely based on the characteristics of the sori and sporangia.

Key to the families

1 Plants growing in soil, mud or epiphytic 2
1 Whole plants floating with the roots or root-like submerged leaves suspended in the water 35

2 Plants aquatic, rhizome submerged in the mud but some of the laminas may be floating. Sporocarps hard and woody, borne at the base of or along the stipes, 3–9 mm long. . . **25 MARSILEACEAE** . . p38
2 Plants not aquatic. 3

3 Stems articulate with whorled, 1-veined leaves fused into a nodal sheath. . . . . **3 EQUISETACEAE** .p7
3 Stems not articulate with several-veined leaves or, if the leaves are 1-veined, then not whorled 4

4 Sporangia 2–3 (rarely 4), fused into synangia (Fig. 12), 2–6 mm in diam. or length, occurring in the axils of 2-lobed appendages. Sterile lateral appendages either minute and scale-like or leaf-like and 1–4 cm long. Stems simple or dichotomously branched. Rupestral or epiphytic plants **5 PSILOTACEAE** .. p7
4 Sporangia free or many fused laterally or, if 2–4, then borne dorsally in a rosette on the pinnae (Gleicheniaceae) 5

5 Sporangia axillary and solitary in the sporophylls, often in strobili (cones) or borne in fertile zones along the stems. Rupestral, epiphytic or terrestrial plants. 6
5 Sporangia on the margin of or on the lower surface of the lamina or fertile spike, usually grouped in sori, relatively small, indusiate or exindusiate. Terrestrial, epiphytic or rupestral plants 7

6 Sporangia and spores uniform. Ligules absent . . . . **2 LYCOPODIACEAE** . . p5
6 Sporangia and spores of two kinds. Ligules present . . . . **1 SELAGINELLACEAE** . . p5

7 Sporangia embedded in 2 rows in a simple spike or panicle borne on a fleshy glabrous stalk attached at or near the base of a sterile frond. Vernation straight or inclined. Terrestrial or rarely epiphytic plants with fleshy roots . . . . **4 OPHIOGLOSSACEAE** . . p8
7 Sporangia borne on the lower surface or margin of the lamina, usually in definite sori. Vernation circinnate 8

8 Sporangia either; non-soral, exindusiate, in oblong or oval patches which are later usually confluent; or else in comb-like lateral or apical sporangiophores 9
8 Sporangia in definite sori, never in comb-like sporangiophores 10

9 Sporangia non-soral, exindusiate, in oblong or oval patches which are later mostly confluent. Stipes with 2 basal flap-like expansions. Terrestrial, often arborescent ferns with short, hard, erect, non-scaly trunks or rhizomes. . . . . **19 OSMUNDACEAE** . . p31

9 Sporangia in comb-like, lateral or apical sporangiophores with protective flanges serving as false indusia. Stipes without flap-like expansions. Wiry or ribbon-like terrestrial plants under 15 cm high, with creeping or erect rhizomes, clothed with hairs or scales . . . . **22 SCHIZAEACEAE.** . p35

10 Fronds delicate, translucent, usually 1 cell thick, lacking stomates. Receptacle elongated and columnar. Epiphytic or rupestral ferns under 12 cm high . . . .**17 HYMENOPHYLLACEAE.** . p28
10 Plants without the above combination of characters 11

11 Arborescent ferns with a tall, erect trunk up to 3 m or more high. Stipes clothed at the base with hairs or scales. Sori round, exindusiate or with semi-circular, cup-shaped or globular indusia. Pedicel short or long 12
11 Non-arborescent ferns, either erect or straggling, terrestrial or rupestral ferns or small epiphytes . 13

12 Sori superficial on the veins or in their axils . Base of the stipes clothed with scales . . . .
. . . . **10 CYATHEACEAE.** . p17
12 Sori marginal, terminal on the minor veinlets (Fig.14). Base of the stipe clothed with hairs . . . .
. . . . **13 DICKSONIACEAE.** . p21

13 Fronds pseudodichotomously branched with dormant apical buds . . **15 GLEICHENIACEAE.** . p25
13 Fronds not branching in this manner 14

14 Rhizome or stem clothed with bristles or hairs or, if clothed with pleuricellular hairs (several cells in transverse section but narrowed at the base), then a reduced pair of stipule-like pinnae present at the base of each primary pinna. Large terrestrial ferns 0.3–2 m high. 15
14 Rhizome clothed with scales. Terrestrial, climbing or epiphytic ferns 17

15 Sori linear, marginal (Fig.15). Rhizome clothed with simple or pleuricellular hairs . . . .
. . . . **12 DENNSTAEDTIACEAE.** . p19
15 Sori circular, protected by recurved flap, true indusium small and/or membranous or indusium absent. Rhizome or stem clothed with simple hairs or bristles. 16

16 Indusium 2-lipped, the inner valve thin. Lamina ± leathery. Hairs on stipe base long . . . .
. . . . **13 DICKSONIACEAE.** . p21
16 Indusium with the 2 lips fused to form a cup with an almost even rim. Lamina soft. Hairs on stipe base short. . . . . **12 DENNSTAEDTIACEAE.** . p19

17 Sori apparently marginal or nearly so 18
17 Sori superficial or, if rather close to the margin, then borne dorsally on the veins and parallel to the midrib of the ultimate segments of the lamina 26

18 Stipes articulated to the rhizome. Scales of the rhizome ciliate or fimbriate, peltately affixed. Rhizome long-creeping, epiphytic on trees of climbing over rocks . . . . **11 DAVALLIACEAE.** . p18
18 Stipes not articulated to the rhizomes. Scales of the rhizome not peltately affixed 19

19 Indusium absent. Sori marginal, borne on the vein-tips, ± confluent. Stipes and rhachises black or dark red-brown . . . .**6 ADIANTACEAE.** . p9
19 Indusium present. Either true, or false and formed by the reflexed margin of the pinna 20

20 Indusium opening outwards towards the margin, attached at the base (Fig. 16). Pinnules cuneate, flabellate or dimidiate. . . . .**18 LINDSAEACEAE.** . p30
20 False indusium consisting of the reflexed margin of the pinna, opening inwards away from margin 21

21 False indusium covering and bearing the sorus, consisting of a sharply reflexed, reniform flap (Fig. 17). Sori ± reniform. Rhachises and stipes black or dark brown . . . .**6 ADIANTACEAE.** . p9
21 False indusium covering but not bearing the sorus. Sori slightly elongated or linear, continuous or interrupted. 22

22 Lamina simple, ribbon-like. Epiphytic ferns . . . .**24 VITTARIACEAE.** . p37
22 Lamina decompound, 1-pinnatifid to incompletely bipinnate. Terrestrial ferns 23

23 Sori short, interrupted, borne at the end of veins, may become confluent with age. Fronds at least 2-pinnate. . . . .**6 ADIANTACEAE.** . p9

23 Sori in longer linear bands along frond margins. Fronds 1–4 pinnate 24

24 Fronds 2 or more- pinnate and pinnae ± opposite. Basal pinnules of the pinnae often reduced and stipule-like . . . . **12 DENNSTAEDTIACEAE.** . p19

24 Fronds 1-4 pinnate. Pinnae without the above characters 25

25 Fronds 1-pinnate, leathery, sori in narrow continous bands along margin of fronds Sori borne on a vascular commissure connecting the vein-tip Stipe very wiry, glossy, black or dark brown. covered with scales . . . . **6 ADIANTACEAE.** . p9

25 Fronds1 to irregularly 4-pinnate, herbaceous. Sori in broad linear bands along margin of frond. Rhachis deeply grooved. **21 PTERIDACEAE.** . p34

26 Indusium absent 27

26 Indusium present 29

27 Stipes not articulated to the rhizome. Veins free (hidden in *Grammitis billardieri*). Small caespitose epiphytes under 15 cm high with hairy stipes. Scales of the rhizome non-peltate non-clathrate, with unicellular bristles along the margin. . . . . **16 GRAMMITIDIACEAE.** . p27

27 Stipes articulated to the rhizome. Scales of the rhizome peltate 28

28 Veins free and forked. Sori terminal on the veins (Fig. 20&21). Rhizome long-creeping . . . .
. . . .**11 DAVALLIACEAE.** . p18

28 Veins anastomosing (Fig. 18&19) . . . . **20 POLYPODIACEAE.** . . p32

29 Sori orbicular. Indusium peltate or reniform 30

29 Sori linear or oblong or cultrate or oval, with similarly shaped indusia or exindusiate in *Pleurosorus*, sometimes sori confluent with age 33

30 Scales of the rhizome peltately affixed . . . .**11 DAVALLIACEAE.** . p18

30 Scales of the rhizome not peltately affixed 31

31 Veins free. . . . . **14 DRYOPTERIDACEAE.** . p22

31 Veins reticulate 32

32 Areoles each with 1–3 free included veinlets. Excurrent veins absent. Indusia peltate. Lamina 1-pinnate. . . . . **14 DRYOPTERIDACEAE.** . p22

32 Areoles without free included veinlets. Excurrent veins present between the lobes of the primary pinnae (Fig. 22). Indusia reniform-orbicular . . . .**23 THELYPTERIDACEAE.** . p36

33 Sori borne on elongated vascular commissures parallel to the midvein or attached to the outer walls of the areolae but facing the midvein (Fig. 23) . . . .**9 BLECHNACEAE.** . p14

33 Sori borne obliquely to midvein on free veins never on anastomosing veins 34

34 Scales of the rhizome and axes non-clathrate. Indusia mostly single except for a few double indusia occurring on each primary pinna especially on the basal acroscopic veins **8 ATHYRIACEAE.** . p13

34 Scales of the rhizome and axes clathrate. Indusia single, absent in *Pleurosorus* **7 ASPLENIACEAE**..p11

35 Leaves minute, up to 1.5 mm long. Whole plants 0.8–6 cm long. Rhizome with numerous roots, bearing leaves in two alternate rows . . . . **26 AZOLLACEAE.** . p39

35 Leaves 1–3 cm long. Whole plants 5–20 cm long. Rhizome rootless, bearing leaves in whorls of three, the third leaf of each whorl being finely dissected and serving as a root . **27 SALVINIACEAE.** . p39

1 LYCOPODIOPHYTA

1 SELAGINELLACEAE

Sporophytes terrestrial or rarely epiphytic, annual or, in the 2 species in the region, perennial, sometimes xerophytic. Stems erect, scandent or creeping, with alternate or dichotomous branches usually in one plane, bearing rhizophores. Leaves simple, scale-like, usually small, uniform, borne in spirals or whorls or in 4 decussate rows, otherwise dimorphic and platystichous, i.e. borne in 2 planes; those of the upper plane vertical but flat; those of the lower plane lateral, horizontal. Sporophylls uniform or dimorphic, platystichous or tetragonous, arranged in strobili. Ligule minute, colourless, present at the base of each leaf and sporophyll at least in the younger portions of the plant. Sporangia unilocular, bivalvate, occurring singly in the axils of the sporophylls, borne in 2 types of sporangia, i.e. megasporangia containing 1–4 megaspores and microsporangia with numerous microspores. 1 gen., trop. to temp., a few arctic alpine.

1 Selaginella P.Beauv.
17 species in Aust. (11 native, 6 naturalized); all states and territories

Key to the species

1 Sterile leaves all similar or almost so, narrowly lanceolate to ovate, acute, spreading or slightly reflexed, 1.2–3 mm long. Aerial stems erect, 5–38 cm high, undivided or with rather rigid lateral branches. Strobili terminal on the branches. Widespread. Damp sandy areas in scrub forest, heathlands or near swampy land . . . . **S. uliginosa** (Labill.) Spring
1 Sterile leaves dimorphic; those on the upper surface appressed, narrowly ovate, attenuate, 1–2 mm long, with finely toothed margins; those on the lower surface spreading on the soil, broadly lanceolate, 2–4 mm long, with finely toothed margins. Aerial stems prostrate, creeping, much branched. Strobili lateral on the branches. Common garden weed and nursery escape, naturalized in places. Introd. from central and southern. Africa and the Azores . . . . ***S. kraussiana** (Kunze) A.Braun

2 LYCOPODIACEAE
Club Mosses

Plants terrestrial or epiphytic, perennial, usually herbaceous. Main stem cylindrical or polyhedral, short or elongated, not tuberous except in the genus *Phylloglossum*. Leaves simple, usually small, in spirals or whorls, arranged along simple or more usually branched stems or borne in a tuft at the base of a short, erect, unbranched stem. Sporophylls similar to the sterile leaves and arranged in fertile zones along the stem and branches or modified and clustered in strobili. Sporangia unilocular, reniform, bivalvate, occurring singly in the axils of the sporophylls. Spores uniform, globoso-tetrahedral. 4 gen., cosmop.

Key to the genera

1 Stems with branches equal in length throughout. Roots in a basal tuft. Main stems not long and indeterminate. Sterile leaves similar to the slightly smaller sporophylls of the fertile zones in *H. varia* . . . . **1 HUPERZIA**
1 Stems with unequal branches almost throughout. Roots at intervals on the lower surface of the main stems. Main stems indeterminate but the branchlet system determinate. Leaves dimorphic with the sporophylls modified into distinct strobili 2

2 Strobili either pendent and terminal on the branches of the much-branched upper portion of the aerial stems or the strobili erect and lateral (or sometimes also a terminal one) on simple or several times forked aerial stems . . . . **3 LYCOPODIELLA**
2 Strobili ± erect and borne at the tips of laterally arising branches from the aerial stems . . . .
. . . . **2 LYCOPODIUM**

1 Huperzia Bernh.

12 species native Aust.; Qld, NSW, Vic., Tas.

One species in the area

Erect terrestrial and pendent epiphytic plants; stems tufted, several times forked, 20–50 cm long. Sterile leaves dark green, thin, crowded, widely spreading to right angles, lanceolate or narrowly ovate, ± acute, not keeled, decurrent along the stem, 8–15 mm long, 1–2.5 mm wide; margins entire. Gradual transition from the sterile to the fertile zone. Strobili always terminal, usually forked, 8–15(-20) cm long, ± 4-sided; sporophylls ovate, ± acute, 3–5 mm long, 2 mm wide, ± keeled, decurrent along the stem; margins entire. Blue Mts, uncommon. RF. On rocks, terrestrial or epiphytes. . . . . **H. varia** (R.Br.) Trevis.

2 Lycopodium L.

4 species native Aust.; Qld, NSW, Vic., Tas., S.A.

One species in the area

Terrestrial plants up to 1 m high, with an elongated subterranean main stem and rigidly erect lateral stems. Sterile foliage dimorphic, both types sometimes occurring on the same plant; the commoner type of sterile leaves being lanceolate or narrowly ovate, densely imbricate, acuminate, 1–2.5 mm long, 0.2–0.75 mm wide, with scarious margins which are densely ciliate towards the base; the other type of leaves more spreading, entire, acuminate, 2.5–4 mm long, 0.2–0.5 mm wide. Strobili erect, sessile, terminal, 1–3 cm long; sporophylls overlapping, broadly ovate, acute, c. 2–3 mm long, 2 mm wide, with scarious fluted margins. Scrub, scrub-heathland, open forests, RF margins; on sandstone hillsides usually in sheltered moist positions or on the edge of swamps. **L. deuterodensum** Herter

3 Lycopodiella Holub

5 species native Aust.; all states and territories except NT

Terrestrial plants, branched anisotomously; main stems long, indeterminate, rooting at intervals on the lower surface; aerial shoots erect, simple or much divided (in *L. lateralis* with irregularly, laterally branched underground stems). Strobili either; erect and ending simple or forked branches which arise dorsally; or, sessile and nodding. Sporophylls modified. Sporangia subglobose or reniform. Spores rugulate.

Key to the species

1 Strobili erect, lateral on the branches, also a terminal strobilus in some cases. Sterile leaves linear-subulate, not arcuate-ascending, 4–7 mm long, 0.2–0.5 mm wide; the margins revolute, entire; the apex acuminate, with a few marginal teeth. Sporophylls broadly ovate, narrowing abruptly into a long, acuminate, toothed, apical portion; the margin lacerate or dentate. Plants erect, pendulous or straggling, simple to several times forked, 10–50(-75) cm high. Widespread. Swamps; heath; on sandstone cliff-faces, exposed hillsides or wet rocks. . . . . **L. lateralis** (R.Br.) B.Øllgaard

1 Strobili nodding at maturity, always terminal on the branches; lateral strobili absent. Sterile leaves linear-subulate, arcuate-ascending, 2–4.5 mm long, 0.1–0.3 mm wide; the margins entire; the apex cuspidate. Sporophylls broadly ovate, narrowing into a long cuspidate, toothed, apical portion; the margins ciliate-fimbriate. Plants usually subscandent, up to 2.5 m long, with a widely trailing rhizome rooting at intervals. Widespread. Moist cliff-faces; railway cuttings; RF . . . . **L. cernua** (L.) Pic. Serm.

2 MONILOPHYTA

3 EQUISETACEAE
1 gen., cosmop. except Australasia.

1 Equisetum L.
1 species naturalized Aust.; NSW

One species in the area

Perennial herb with pubescent, long-creeping rhizome. Stems erect, jointed, hollow, with 4–19 grooves and 4–19 small whorled leaves fused into a nodal sheath. Sterile stems up to 40 cm high, usually much branched; the branches simple, green. Fertile stems shorter, unbranched, light green with darker leaves. Strobili 1–4 cm long, terminal on the fertile stems, composed of elongate longitudinally dehiscing sporangia hanging from a peltate structure. Spores round, green, with 4 paddle shaped elaters. Recorded from Belrose. Introd. from Europe. *Common Horsetail* . . . . * **E. arvense** L.

4 PSILOTACEAE

Sporophytes terrestrial or epiphytic. Rhizomes creeping and branched, covered with fine brown rhizoids. Aerial axes erect or pendulous, rigid or flaccid, simple or branched dichotomously. Sterile lateral appendages undivided. Synangia composed of 2–3 (very rarely 4) fused sporangia, occurring singly in the axils of bifid or bifoliate appendages. Spores bilateral. 2 gen., trop. to temp.

Key to the genera

1 Sterile lateral appendages 0.5–2 mm long; the apex non-mucronate. Synangia composed of 3 (very rarely 2 or 4) sporangia (Fig. 12) . . . . **1 PSILOTUM**
1 Sterile lateral appendages 6–31 mm long; the apex mucronate. Synangia composed of 2 sporangia. . . . . **2 TMESIPTERIS**

1 Psilotum Sw.
2 species native Aust.; all states and territories except Tas.

One species in the area

Sporophytes 15–60 cm long, pendulous or erect. Stems and branches triquetrous with the leaves in 3 vertical rows; the ultimate branchlets 0.5–1.5 mm broad. Synangia numerous, borne on the angles of the branchlets. Widespread but uncommon away from the coast. Epiphytic in RF or open forests or more commonly in crevices of sandstone cliffs. *Skeleton Fork Fern.* . . . . **P. nudum** (L.) Beauv.

2 Tmesipteris Bernh.
7 species native Aust.; Qld, NSW, Vic., Tas.

The Australian species are small epiphytes growing on tree fern trunks especially on *Todea* and *Dicksonia* and occasionally on rocks. Rhizome long creeping, hairy. Lower portions of stems bearing a number of small, stiff, linear to narrowly lanceolate scale-leaves. Aerial axes pendulous, 1-pinnate, rarely the rhachis once forked. Sterile and fertile appendages occurring in irregular zones along the rhachis. Sterile lateral appendages green, subsessile, glabrous, simple, 1-veined, 4–31 mm long, 1–9 mm broad, obliquely very narrowly oblong to narrowly oblong, obliquely narrowly oblanceolate to lanceolate or elliptic. Fertile appendages green, with a comparatively long petiolule, bipartite almost to the base, bearing a single synangium just below the bifurcation; the lobes narrower than the sterile appendages. Synangia brownish, spindle-shaped, capsule-like, bivalvate with rounded or pointed ends, dehiscing longitudinally and around the middle. Spores bilateral.

Key to the species

1 Apices of the majority of the sterile lateral appendages on each plant acute or rounded. Synangia rounded. Aerial shoots 5–16.5 cm long including the stems. Sterile lateral appendages almost equally-sided at the base 2

1 Apices of sterile lateral appendages truncate or bilobed. Synangia pointed or rounded. Aerial shoots 8.5–86 cm long including the stems. Spores 85–90 µm diam. Sterile lateral appendages not dilated to very much dilated at the base 3

2 Apices of the sterile lateral appendages acute. Spores 60 µm diam. Widespread but not common. RF or tall open forests, epiphytic on tree ferns . . . . **T. parva** N.A.Wakef.

2 Apices of the majority of the sterile lateral appendages rounded. Spores 70–75µm diam. Widespread but not common. RF, epiphytic on tree ferns . . . . **T. ovata** N.A.Wakef.

3 Lower portions of the stems smooth except for the upper groove. Base of the sterile lateral appendages, mostly 3–9 mm broad; the upper side of the base much dilated and convex. Sterile lateral appendages light to yellow-green when dried. Synangia pointed. Widespread, usually in the mountains. RF, epiphytic on tree ferns . . . . **T. obliqua** Chinnock

3 Lower portions of the stems ridged on all sides. Base of the sterile lateral appendages 1–3 mm broad; the upper side of the base slightly dilated or not dilated. Sterile lateral appendages usually dark green when dried. Synangia pointed or more rarely rounded. Widespread. RF or sandstone gullies, on rocks or on trunks of *Todea* . . . . **T. truncata** R.Br.

5 OPHIOGLOSSACEAE

Sporophytes herbaceous, perennial, terrestrial or rarely epiphytic. Rhizome non-palaceous, mostly short, erect, unbranched, bearing thick, fleshy, hairless roots. Fronds erect or pendulous, 1–several; each composed of a common basal stalk, a sterile and a fertile segment; sterile lamina sessile or petiolate, simple, lobed, palmately or pinnately compound or decompound; veins free or anastomosing; fertile spike simple or paniculate, appearing to arise on the adaxial surface of the sterile blade but really lateral in origin; vernation straight or inclined, never circinnate. Sporangia in 2 marginal rows, free or coalescent, exindusiate, devoid of an annulus, globular, ovoid or hemispherical, sessile or with a very short pedicel; dehiscing horizontally or vertically into 2 valves. Spores very thick-walled, tetrahedral or bilateral, very numerous. 3 gen., mostly cosmop.

Key to the genera

1 Veins of the lamina free, dichotomous. Fertile spikes 1–3-pinnate; the sporangia distinct. Sterile lamina 1-pinnate to 3-pinnate-pinnatifid . . . . **1 BOTRYCHIUM**

1 Veins of the lamina anastomosing, with or without free vein-endings in the areolae. Fertile spikes simple; the sporangia coalescing. Sterile lamina simple or irregularly forked near the apex . . . .
. . . . **2 OPHIOGLOSSUM**

1 Botrychium Sw.

2 species native Aust.; Qld, NSW, Vic., Tas., S.A.

One species in the area

Herbs 7–50 cm high. Sterile frond tripartite, pale green, 3-pinnate to 3-pinnate-pinnatifid; the petiole 4–15 cm long. Fertile spike 2–3-pinnate, lanceolate-deltoid to broadly ovate-deltoid, borne on a fleshy stalk 6–40 cm long. Widespread. Sclerophyll forests or in grassy places on the edge of RF. Uncommon. *Parsley Fern* . . . . **B. australe** R.Br.

2 Ophioglossum L.

7 species native Aust.; all states and territories

One species in the area

Small herbs 1–13 cm high with 1–5 sterile leaves. Rootstock slightly tuberous with fleshy roots. Sterile lamina light green, rather fleshy, very narrowly elliptic to broadly elliptic or lanceolate to ovate, entire, 7–45 mm long, 2–9 mm broad; the apex acute or obtuse; the veins anastomosing, with free vein-endings in some of the areolae. Spike 5–15 mm long, 1.5–2 mm broad, borne on an elongated peduncle. Widespread. Open grassland, freshwater swamps or in peaty soil on heathland. *Austral Adder's Tongue* . . . .
. . . . **O. lusitanicum** L.

6 ADIANTACEAE

Small to medium-sized ferns, usually growing amongst rocks, often xerophytic, other species confined to RF. Rhizomes erect or creeping, dictyostelic or more usually solenostelic, clothed with bristle-like or woolly hairs or more usually with scales which range from fuscous or ferruginous to black with a paler border; the paleae intermixed with silky hairs in some genera. Stipes stramineous or castaneous or maroon or brown or black, often glossy, glabrous to densely scaly and/or hairs, grooved above or terete, not articulated to the rhizome. Main rhachis grooved above or terete. Fronds uniform or rarely dimorphic, simple, 1–5-pinnate or -pedate or sometimes palmate, thin to coriaceous, glabrous, hairy or scaly, covered with waxy powder in some genera. Ultimate segments equilateral or unilateral, dimidiate or trapeziform or cuneate-flabelliform or linear or lanceolate or ovate to roundish, articulated or non-articulated. Veins usually free, sometimes anastomosing, without free included veinlets in the areolae. Sori superficial or submarginal in some genera, but appearing to be marginal in the genera occurring in New South Wales, exindusiate or with a false indusium, rarely with a vascular commissure, borne on the distal vein-endings and sometimes between them, then either rounded or confluent into ± continuous, intramarginal, linear or oblong sori, otherwise spreading inwards along the veins and sometimes occurring in the spaces between them, rarely monangial. Indusium false, either; a reflexed, modified, marginal flap bearing and covering the sorus (*Adiantum*) (Fig. 17); or the reflexed, continuous veinless, ± modified margin of the ultimate segment or in other genera the margin flat and unmodified as a false indusium. Paraphyses present in some genera, sometimes clavate. Sporangia mixed, usually large; the pedicel short to subsessile or elongated, with 3–6-rows of cells; the annulus including 14–24 indurated cells. Spores globoso-tetrahedral or tetrahedral, rarely bilateral, smooth or reticulate-spinose or corrugated or otherwise ornamented, without a perispore. c. 28 gen. cosmop., often in dry country.

Key to the genera

1 False indusium covering and bearing the sorus, consisting of a sharply reflexed, flap formed by the lamina (Fig. 17). Rhachises and stipes black or dark brown. . . . . **1 ADIANTUM**
1 Inudium false or absent. Sori not borne on a flap formed by the lamina 2

2 False indusium continuous along front margin. Fronds 1-pinnate. Rhizome long, creeping **2 PELLAEA**
2 False indusium interrupted. Fronds decompound. Rhizome tufted or shortly creeping . . . .
. . . . **3 CHEILANTHES**

1 Adiantum L.

8 species in Aust. (2 endemic, 6 native); all states and territories

Terrestrial ferns. Rhizomes tufted to long-creeping, solenostelic or with a few meristeles, clothed with narrow, black or dark brown, entire or toothed, non-peltate scales which are sometimes paler towards the margin. Stipes not articulated to the rhizome, polished, black or dark brown, clothed near the base with scales, sometimes hairy in the upper parts. Main rhachis grooved above. Lamina 1–4-pinnate, decompound or digitate in the Australian species, glabrous or sometimes clothed with scattered hairs or setae. Ultimate segments dimidiate or flabellate-cuneate, mostly herbaceous, sometimes almost membranous or coriaceous. Veins free, fine, flabellate, branching dichotomously without any main veins. Sori marginal in shallow notches or deep grooves, borne on the lower surface of sharply reflexed marginal lobules. False indusium consisting of the ± reniform, reflexed margin of the frond-segment, opening inwards, bearing and covering the sori (Fig. 17). Sporangia borne along the veins which are continuous into the lobule; the pedicel long, narrow; the annulus including c. 18 thick-walled cells. Spores tetrahedral or sometimes bilateral, smooth, dark, lacking a perispore.

Key to the species

1 Each ultimate frond-segment attached by a comparatively long stalk coming from the middle or almost the middle of the base, not dimidiate. 2
1 Each ultimate frond-segment attached by a very short stalk at the lower corner of the segment, dimidiate 3

2 Scales on rhizomes entire, yellow. Rhizome creeping, stoloniferous. Plants 10–60 cm high. Widespread. Along the banks of rivers and creeks or in damp, open situations. *Common Maidenhair Fern* . . . . . .
. . . .**A. aethiopicum** L.
2 Scales on rhizomes with ciliate margins, dark brown. Rhizome erect or shortly creeping. Plants to 75 cm high. Along the banks of rivers and creeks . . . .**A. atroviride** Bostock

3 Sori in deep notches between the lobes of the ultimate leaf-segments. Rhizome usually short and tufted. 4
3 Sori in shallow grooves at the tips of the lobes of the ultimate frond-segments, not in the notches between the lobes, Rhizome long-creeping 5

4 Ultimate frond-segments sparingly setulose with minute, stiff, black or brown setae especially near the margin. Rhizome tufted; the rootlets with prominent dark bulbils. Plants 10–35 cm high. Widespread. RF. On earth banks or near waterfalls. . . . .**A. diaphanum** Blume
4 Ultimate frond-segments ± clothed with white, soft hairs. Rhizome tufted or rarely shortly creeping, lacking bulbils. Plants 10–65 cm high. Widespread. RF or open forests. On earth banks or amongst rocks. *Rough Maidenhair Fern* . . . .**A. hispidulum** Sw.

5 Terminal pinna of the frond very short in comparison with the length of the basal primary pinnae. Plants 60–120 cm high. Widespread. RF or open forests, Grassy slopes, alluvial flats or near creeks; usually in stands. *Giant Maidenhair Fern*. . . . . **A. formosum** R.Br.
5 Terminal pinna of the frond much attenuated, as long as the basal primary pinnae. Plants up to 80 cm high. Widespread. RF. Creek banks or on moist cliff-faces . . . .**A. silvaticum** Tindale

2 **Pellaea** Link
6 species native Aust.; Qld, NSW, Vic., Tas.

Small terrestrial ferns usually growing amongst rocks. Rhizome short and thick or shortly to long-creeping, solenostelic, densely clothed with narrow, non-peltate scales. Stipes wiry, often black or dark brown, mostly glossy, densely scaly in many species, not articulated to the rhizome. Lamina 1-pinnate in the Australian species, otherwise decompound, palmate or pedate, mostly glabrous, otherwise inconspicuously pubescent or with a few scales. Pinnae usually with persistent petiolules articulated to the segments, uniform, mostly coriaceous and entire. Veins hidden or inconspicuous, free and forked in the Australian species or sometimes anastomosing. Sori marginal, at first distinct, later forming a linear or oblong band, borne on the unconnected vein-tips, protected by a false indusium i.e. the continuous, reflexed margin. Paraphyses usually absent. Sporangia with a vertical, incomplete annulus composed of 14–20 thick-walled cells and c. 6–8 thin-walled cells. Spores globoso-tetrahedral, spiny in the Australian species, otherwise smooth or verrucose or reticulate-spiny.

Key to the species

1 Pinnae shortly stalked, with petiolules 1–5 mm long. Scales and hairs whitish to light brown. 2
1 Pinnae subsessile, with petiolules 0.2–1 mm long, 14–44 pairs, 0.6–5 cm long, 0.2–1.2 cm broad. Juvenile fronds pinnate composed of c. 5–10 pairs of pinnae 3

2 Pinnae symmetric, 14–30 mm long. Lateral pinnae reducing in size towards apex. Fronds to 50 cm high. Usually amongst rocks . . . . **P. calidirupium** Brownsey & Lovis

2 Pinnae asymmetric, 20–60 mm long. Lateral pinnae all more or less the same size. Juvenile fronds simple, suborbicular or ovate to broadly ovate or hastate; the base cordate. Fronds including the stipes up to 65 cm long. Widespread. RF. Usually amongst rocks. Uncommon . . . **P. paradoxa** (R.Br.) Hook.

3 Pinnae 0.6–2 cm long, mostly 0.2–0.6 cm broad. Fronds including the stipes mostly 13–40 cm long. Widespread. RF. Usually amongst rocks. *Sickle Fern* . . . . **P. nana** (Hook.) Bostock

3 Pinnae 2.5–6 cm long, 0.5–1.3 cm broad. Fronds including the stipes 15–65 cm long. Widespread. RF or open forests. Often on damp rocks or on loamy soil. *Sickle Fern* . . . .**P. falcata** (R.Br.) Fée

3 Cheilanthes Sw.
Rock Ferns
15 species in Aust. (9 endemic, 6 native); all states and territories

Small, terrestrial, mostly xerophytic ferns found on rocky hillsides. Rhizome tufted or shortly creeping, clothed with scales which are narrow, entire or slightly denticulate, non-peltate, concolorous, and non-sclerotic or with pale margins and a dark central sclerotic band. Stipes tufted, light chesnut-coloured or dark red-brown or dark brown or purplish black or black, polished, shallowly grooved above, paleaceous at the base; the upper and middle regions scaly or hairy or glabrous. Main rhachis chestnut-coloured to black, grooved above, terete or angled. Fronds uniform or dimorphic. Lamina small, narrow, often basiscopically produced, mostly herbaceous, glabrous or clothed with hairs and/or scales, bipinnatifid to decompound, with very small ultimate segments. Veins free. Sori marginal, borne on the tips of the veins, small, rounded, often spreading laterally and in contact but discontinuous. Indusium composed of the ± reflexed and slightly to moderately developed frond-margin, roundish, not continuous but later ± confluent, sometimes absent. Sporangia with an annulus of 14–24 thick-walled cells, several thin-walled cells and a stomium of numerous cells. Spores globoso-tetrahedral, warty or spiny or smooth, without a perispore.

Key to the species

1 Lamina tripinnate, deltoid. Widespread. Amongst rocks in open forest or on rocky hillsides . . . .**C. austrotenuifolia** H.M.Quirk & T.C.Chambers

1 Lamina bipinnate, lanceolate 2

2 Lamina glabrous. Indusium consisting of the slightly reflexed frond-margin. Widespread. Open forest; ti-tree scrub; rocky hillsides; pastures . . . .**C. sieberi** Kunze ssp. **sieberi**

2 Lamina clothed on the lower surface with lanceolate scales and on the upper surface with simple hairs. Indusium absent. Widespread. Open forest or woodland. Rocky hillsides . . . .**C. distans** (R.Br.) Mett.

7 ASPLENIACEAE

Terrestrial or epiphytic ferns. Rhizome shortly creeping or suberect, rarely long-creeping, dictyostelic, clothed with non-peltate, clathrate scales. Stipes mostly fascicled, not articulated to the rhizome. Fronds simple, lobed, 1-pinnate to pinnately decompound, membranous to coriaceous or almost cartilaginous. Veins free or reticulate, without free vein-endings in the areolae. Sori elongated along the veins with a similarly shaped indusium, rarely exindusiate. Sporangia with an annulus interrupted by the long, narrow pedicel which is composed mainly of 1 row of cells; the annulus including mostly 18–28 thick-walled cells. Spores bilateral, often spinulose, tuberculate or, with a perispore. 5 gen., cosmop.

Key to the genera

1 Indusia absent. Lamina densely hairy . . . . **2 PLEUROSORUS**

1 Indusia present. Lamina glabrous or scaly . . . . **1 ASPLENIUM**

1 Asplenium L.

Spleenworts

29 species native Aust.; all states and territories

Mostly terrestrial ferns but some species are huge RF epiphytes. Rhizome usually shortly creeping, clothed with thin, chestnut-coloured or grey or purplish, clathrate scales. Stipes with X-shaped vascular bundles in transverse section of the upper portion. Fronds simple or lobed to pinnately decompound, often viviparous by buds on the axes or pinnae, glabrous or clothed with small, clathrate scales. Veins free, forked, but sometimes with a few anastomoses near the margin or with an intramarginal vein present. Sori oblong to linear, borne on the basiscopic side of the veinlets. Indusium oblong to linear, elongated along the veins, usually with tapering ends, single in all the Australian species. Sporangia with an annulus including 20–28 thick-walled cells. Spores bilateral, smooth or spiny,

Key to the species

1 Sori mostly marginal, 1 to each incurved lobule of a pinna, affixed to the central veinlet of the lobule but with a prominent indusium projecting towards the upper margin of the lobule, so as to make the sorus appear marginal 2
1 Most of the sori remote from the margin of the lamina 3

2 Fertile fronds thickly herbaceous to membranous, soft, flaccid. Fronds 1-pinnate to 3-pinnate-pinnatifid. Plants up to 120 cm high. Widespread. RF. Terrestrial or epiphytic . . . . **A. bulbiferum** G.Forst. ssp. **gracillimum** (Colenso) Brownsey
2 Fertile fronds thick, coriaceous almost cartilaginous. Fronds 1–2-pinnate. Plants up to 70 cm high. Widespread. RF. Epiphytic or on rocks . . . . **A. flaccidum** G.Forst. ssp. **flaccidum**

3 Fronds simple or irregularly divided into segments at the base only. 4
3 Fronds once or more pinnate 6

4 Fronds sessile or subsessile, simple, entire, 60–180 cm long, 3–21 cm broad; the apex acute, subacute or emarginate. Sori 1 mm apart, often 4–6 cm long, very numerous. Midrib glabrous. Veins connected in an intramarginal vein. Large epiphyte with tufted fronds encircling a hollowed central region. Widespread. RF. On trees and rocks. *Bird's Nest Fern* . . . . **A. australasicum** (J.Sm.) Hook.
4 Fronds stipitate, simple or irregularly divided into segments at the base only, up to 45 cm long, up to 5.5 cm broad; the apex often proliferous and very attenuated except in juvenile fronds. Sori 3–4 mm apart, up to 1.5 cm long. Main rhachis or midrib ± scaly. Intramarginal veins absent. Rhizome short, tufted. Damp rocks in shady places or in closed gullies, occasionally epiphytic 5

5 Fronds lobed in lower third, or pinnate at base. Coast and adjacent plateaus. . . . . **A. attenuatum** R.Br. var. **attenuatum**
5 Fronds all simple. Lower Blue Mts, northwards . . . . . . **A. attenuatum** R.Br. var. **indivisum** F.Muell.

6 Fronds 2–3-pinnate or 1-pinnate and with 1–2-pinnatisected pinnae. Proliferous buds absent on the main rhachis 7
6 Fronds 1-pinnate 8

7 Stipes green, dull. Fronds 2–3-pinnate. Lamina pale green, cartilaginous. Indusium thick. Terrestrial ferns. Coast north from La Perouse. Rock crevices on headlands near the sea . . . . . **A. difforme** R.Br.
7 Stipes black, glossy. Fronds 1-pinnate with 1–2-pinnatisected pinnae. Lamina dark green, almost coriaceous. Indusium membranous. Coast and adjacent plateaus north from Mt. Keira, Culoul Range. RF or tall open forest. Epiphytic or growing on rocks, uncommon . . . **A. aethiopicum** (Burm.f.) Bech.

8 Apices of the pinnae finely acuminate. Epiphytic or rupestral ferns usually 30–150 cm high. Widespread. RF. Common . . . . **A. polyodon** G.Forst.
8 Apices of the pinnae acute to broadly rounded. 9

9 Apex of the frond attenuated, filiform and devoid of pinnae, sometimes proliferous at the tip. Stipes prostrate, flaccid, dark-coloured at the base only. Middle pinnae flabellate, rhomboid-cuneate or orbicular-reniform. Lamina membranous to subcoriaceous, flaccid, straggling; the margin not reinforced. Plants c. 10–75 cm high. Widespread. RF and open forests. On rocks or tree trunks. *Necklace Fern* . . . . **A. flabellifolium** Cav.

9 Apex of the frond never finely attenuated nor filiform nor proliferous at the tip. Stipes erect, rigid. Middle pinnae obliquely cultrate to very broadly oblong or very broadly ovate or obliquely narrowly lanceolate to lanceolate or a few of the basal pinnae rounded 10

10 Lamina thick and almost cartilaginous; the margin reinforced. Stipes green. Pinnae 2–8 (rarely up to 14) pairs, prominently stalked; the petioles not persistent, up to 6 cm long and 1.5 cm broad. Indusium firm to cartilaginous; the margin entire. Plants 10–45 cm high. Coast south from Sydney. Cliffs overlooking the sea. *Shore Spleenwort* . . . . . **A. obtusatum** G.Forst. ssp. **northlandicum** Brownsey

10 Lamina herbaceous to thinly coriaceous; the margin not reinforced. Stipes black, glossy. Pinnae 10–30 pairs, up to 1 cm long and 0.8 cm broad, subsessile or very minutely stalked, articulate on persistent petiolules. Jenolan Caves and other caves in the ranges. Limestone regions in crevices in rocks . . . . **A. trichomanes** L.

2 Pleurosorus Fée
2 species native Aust.; all states and territories

Rhizome short, tufted, densely clothed with narrow, clathrate, purplish-brown scales. Lamina 1-pinnate to bipinnate thinly coriaceous to coriaceous, 4–12 cm long, 1.5–4 cm broad, with 1-lobed, fan-shaped pinnules. Veins free, flabellate. Sori exindusiate, broadly to narrowly oblong, sometimes confluent in age and covering the lower surface, borne along divergent veinlets, 2–5 mm long, 1–1.5 mm broad. Spores bilateral, orbicular-oblong, black, with a reticulate, broadly winged perispore. On hillside in dry areas, usually under boulders or in rock crevices

Key to the species

1 Rhachis and lamina densely clothed with soft, short, ferruginous hairs each bearing a large glandular head. Blue Mts . . . . **P. subglandulosus** (Hook. & Grev.) Tindale

1 Rhachis and lamina densely clothed with soft, short, silver to reddish non-glandular hairs. Western parts of area. . . . . **P. rutifolius** (R.Br.) Fée

8 ATHYRIACEAE

Terrestrial ferns. Rhizome or caudex short, oblique, erect or shortly creeping, dictyostelic, clothed with non-clathrate scales which are entire or bearing marginal teeth composed of 2 cells which may separate at their apices. Stipes often robust, with 2 vascular strands joined higher up into a U-shaped bundle. Rhachises mostly papillate in the groove on the upper surface and scaly or hairy on the lower surface. Fronds uniform or dimorphic, mostly thin. Costae grooved. Lamina simple to decompound. Veins free or rarely anastomosing especially towards the margin, lacking free included veinlets. Sori dorsal, linear to oblong, usually elongated along the veins, sometimes ± curved and inflated, sometimes reniform or rarely small and round; the basal acroscopic sori often double (i.e. occurring on both sides of the veinlet). Indusium narrow, elongated along the veinlets, sometimes J-shaped to horseshoe-shaped, almost round or reniform, rarely absent. Sporangia globose; the annulus including 12–20 (mostly 16) thick-walled cells; pedicel slender. Spores bilateral, with perispore. c. 23 gen., cosmop.

Key to the genera

1 Ridges of the grooves on the upper surface of the rhachis and costae open at the junctions with the pinnae and pinnules. Multiseptate hairs absent on the upper surface of the lamina. . . **1 DIPLAZIUM**

1 Ridges of the grooves on the upper surface of the rhachis and costae not open at the junctions with the pinnae and pinnules. Multiseptate hairs present on the upper surface of the lamina . . . . **2 DEPARIA**

1 Diplazium Sw.

11 species in AUST, (8 native, 2 endemic, 1 naturalized); Qld, NSW, Vic., Tas.

One species in the area

Terrestrial ferns up to 1.6 m high. Rhizome tufted, often forming a short trunk; scales thick, acuminate, black with a brown margin. Fronds 3-pinnate at the base, with the tertiary segments often deeply lobed. Lamina bright green, submembranous to herbaceous, flaccid, with few or no scales on the lower surface. Veins free, pinnate; the lateral veinlets 4–6 per pinnule, simple or forked. Sori oblong, single or a few double sori on the basal acroscopic veins. Indusia thin, pale brown. Widespread. RF. . . . .**D. australe** (R.Br.) N.A.Wakef.

2 Deparia Hook. & Grev.

1 species native AUST,; Qld, NSW, Vic.

One species in the area

Terrestrial ferns 0.3–0.8m high. Rhizome long-creeping with distant fronds, clothed with thin, hair-pointed, light brown, comparatively broad scales. Fronds 1-pinnate-pinnatifid or almost bipinnate at the base. Lamina 15–30 cm long, herbaceous. Veins free, pinnate; the lateral veinlets 4–6 per pinnule, simple or forked. Sori dorsal, broadly linear or oblong, single or the lowest in a segment sometimes double. Indusia light brown; the margins erose-dentate. Coast and gullies in adjacent plateaus and lower Blue Mts Uncommon. RF and tall open forests on rocks. . . . . **D. petersenii** (Kunze) M.Kato ssp. **congrua** (Brack.) M.Kato

9 BLECHNACEAE

Terrestrial ferns. Rhizome mostly short and erect, sometimes with a short trunk up to 50 cm high, otherwise creeping or more rarely twining up trees, dictyostelic, clothed with non-peltate, non-clathrate, often black scales. Stipe not articulated to the rhizome, with several vascular bundles. Fronds usually pinnatifid or pinnate, sometimes simple or lobed, uniform or dimorphic; the fertile fronds often much narrower than the sterile ones. Veins free or anastomosing into a row or rarely several rows of areolae near the costae. Sori short and often later confluent, more usually elongated into linear coenosori which are borne on vascular commissures parallel to the costa. Indusium opening inwards towards the costules or costae, frequently linear or oblong, rarely absent. Sporangia large, with a longitudinally interrupted annulus which includes 14–28 thick-walled cells. Spores bilateral, monolete, with or without a perispore. 9 gen., cosmop., especially S. hemisphere.

Key to genera

1 Coenosori in a continuous or almost continuous line on each side of the midrib covering the vascular commissures, having a membranous indusium opening inwards towards the costa. Veins of the sterile fronds free. . . . . **1 BLECHNUM**
1 Sori in 1–2 (rarely 3) rows, rather short or sometimes later confluent, 1 in each areola attached on the inner side of the veinlet or veinlets parallel to the costa (Fig. 23). Veins areolate in the fertile fronds and most of the sterile fronds, rarely free in the latter . . . . .2 DOODIA

1 Blechnum L.

18 species in Aust. (11 endemic, 7 native); all states and territories

Terrestrial ferns. Rhizome usually erect or oblique, sometimes creeping or forming a short trunk up to 60 cm high, with a complex dictyostele, clothed with scales which are non-clathrate, usually thick, linear to ovate, black or dark brown or chestnut-coloured or with a dark centre and pale borders. Stipes not articulated to the rhizome, with several vascular strands. Fronds uniform or dimorphic; the fertile fronds usually much narrower than the sterile ones. Lamina simple, pinnatifid or pinnate in the Australian species, coriaceous or sometimes herbaceous, glabrous or clothed with a few scales; margin entire or serrate. Pinnae articulate to the rhachis in some species, often auriculate; the sterile flange beyond the sorus broad in some species but almost non-existent in others. Veins free in the sterile fronds, simple or once or twice forked, often prominent; a vascular commissure parallel to the costa in the fertile fronds. Coenosori linear, mostly continuous, rarely interrupted, superficial, one on each side of and parallel to the costa, close to the latter, often confluent with age and covering the whole of the lower surface of the fertile pinnae. Indusium linear, marginal or intramarginal, membranous, attached to the fertile commissure and opening towards the costa. Sporangia with an annulus including 14–28 thick-walled cells. Spores bilateral, reniform to almost globular, with or without a perispore.

Key to the species

1 Fertile and sterile fronds dissimilar 2
1 Fertile and sterile fronds similar; the fertile pinnae or frond-segments slightly narrower to one-third of the width of the sterile. 8

2 Sterile fronds simple, ± linear-elliptic or 1-pinnatifid into 2–18 lateral segments which are 0.7–2.7 cm broad and constricted at the base; the terminal segment much elongated. Main rhachis very broadly winged by the decurrent bases of the frond-segments; the wing 0.3–1.5 cm broad on each side. Fertile fronds simple or pinnatifid. Rhizome erect, covered with the broken bases of the stipes. Plants up to 65 cm high. Widespread. RF gullies, often near waterfalls and on the banks of creeks . . . .
. . . . **B. patersonii** (R.Br.) Mett. ssp. **patersonii**
2 Sterile fronds pinnate, pinnatisect or, if pinnatifid into numerous segments, then the segments not noticeably constricted at the base 3

3 Sterile frond-segments attached by their broad bases. Rhizome erect or long-creeping. 4
3 Sterile frond-segments (pinnae) stalked except a few of the upper pairs. Rhizome erect or shortly creeping. 6

4 Rhizome long-creeping, with the stipes scattered along the rhizome but often clustered at the apex of the rhizome. Plants 8–35 (rarely up to 45) cm high. Lamina firm, paler beneath, glabrous or almost so. Basal frond-segments widely spaced and much reduced in size. Middle sterile frond-segments 0.5–1.5 cm long. Blue Mts (Jenolan Caves). Cooler mountain regions usually amongst rocks and in boggy situations or amongst grass. *Alpine Water-fern*. . . . .
. . . . **B. penna-marina** (Poir.) Kuhn ssp. **alpina** (R.Br.) T.C.Chambers & P.A.Farrant
4 Rhizome erect 5

5 Lamina flaccid, slightly paler beneath, glabrous. Main rhachis of the fertile fronds dull and mostly pale stramineous. Basal frond-segments reduced to auricles. Middle sterile frond-segments usually 1.2–4.5 cm long. Plants up to 60 cm high. Illawarra ranges. RF; in caves near waterfalls . . . . **B. chambersii** Tindale
5 Lamina very rigid, usually much paler paler beneath, frequently glaucous or reddish, often clothed with chestnut-coloured scales when young, glabrous when old. Main rhachis of the fertile fronds glossy, black or atropurpureous. Basal frond-segments oblong or broadly oblong. Middle sterile frond-segments 3.5–11.5 cm long. Plants up to 60 cm high. Widespread. Common. RF; on the edges of creeks on Ss or basalt gullies or in peaty soil near swamps . . . . **B. nudum** (Labill.) Mett. ex Luerss.

6 Lower lateral pinnae not or almost imperceptibly reduced in size. Rhizome shortly creeping with the fronds at intervals. Scales of the rhizome glossy, dark brown with a paler border. Plants erect, up to 120 cm high. Widespread. RF. Terrestrial, growing in soil usually on the banks of creeks. *Hard Water Fern* . . . .
. . . . **B. wattsii** Tindale
6 Lower lateral sterile pinnae auricle-like and/or much reduced in size. Rhizome with an erect portion bearing a crown of fronds. 7

7 Sterile and fertile pinnae not auriculate at the base or with the lower pairs very slightly auriculate. Scales of the rhizome dull, mostly concolorous, rarely with a paler border. Stipes of mature plants stramineous, sometimes mottled with brown; the base dark brown, smooth or slightly tuberculate. Plants up to 90 cm high. Widespread. Usually on the tablelands, along the banks of creeks in open forest or in narrow mountain gorges, at the edges of waterfalls or in swampy creeks in cleared RF areas. *Soft Water Fern* . .
. . . .**B minus** (R.Br.) Ettingsh.

7 Sterile and fertile pinnae very prominently auriculate at the base. Scales of the rhizome glossy, with a dark central band. Stipes of mature plants dark red-brown or black, very tuberculate. Plants up to 120 cm high. Coast. Swampy low-lying land, near the sea . . . . **B. camfieldii** Tindale

8 Sterile frond-segments attached by their broad bases. Plants up to 110 cm high. Widespread. RF; margins of RF; open forests, on hillsides often amongst rocks. *Gristle Fern* . . . . **B. cartilagineum** Sw.

8 Sterile frond-segments (pinnae) very shortly and distinctly stalked in the greater part of the frond or in the lower third 9

9 Pinnae articulated to the main rhachis. Rhizome erect with a crown of fronds. Spores minutely papillate. Sterile pinnae very glossy, 6–13 cm long, 8–12 mm broad; the fertile pinnae 4–9 cm long, 4–8 mm broad. Plants up to 110 cm high. Coast and Woronora Plateau. Edge of brackish swamps or lakes, usually in sandy soil . . . . **B. indicum** Burm.f.

9 Pinnae not articulated to the main rhachis. Rhizome shortly creeping or long-creeping. Spores with an alate, either reticulate or tuberculate perispore. Pendulous plants up to 120 cm long 10

10 Fertile pinnae almost as broad as the sterile pinnae. Rhizome with very fleshy circinnate buds. Scales of the rhizome fawn, dull, almost as broad as long. Spores tuberculate, 56–90μm 44–86 μm, with a narrow wing 3.75–6μm, usually developed on one side of the spore. Pinnae pale green, shiny, rounded or slightly auriculate at the base. Mainly Blue Mts and Illawarra ranges near the coast. RF, usually on rock ledges, caves or near waterfalls, on wet cliff-faces or rarely epiphytic. . . . .**B. gregsonii** (Watts) Tindale

10 Fertile pinnae c. one third to a half of the width of the sterile pinnae. Rhizome with non-fleshy, circinnate buds. Scales of the rhizome light fawn or brown or ferruginous, sometimes slightly darker towards the centre when old, dull to slightly glossy, narrowly lanceolate to narrowly ovate. Spores with a reticulate perispore, 49–75μm X44-56μm; the wings 3.75–11μm diam, surrounding the spore. Pinnae light to medium green, dull, slightly auriculate at the base. Widespread. Uncommon. RF and tall open forests. Ss, usually on rock ledges under caves or near waterfalls or on wet cliff-faces . . . .
. . . . **B. ambiguum** (C.Presl) Kaulf. ex C.Chr.

2 **Doodia** R.Br.
Rasp Ferns
8 species in Aust. (4 endemic, 4 native); all states and territories

Terrestrial ferns, mostly 10–40 cm (sometimes 60cm) high, usually in RF or covering banks in open eucalypt forests. Rhizome mostly black, with a short, erect, tufted portion clothed with small, thick, dark, non-peltate scales and bearing numerous wiry, black roots; underground runners emerging from the tufted portion. Stipes borne in a crown, not articulated to the rhizome, dark brown or black, scaly especially near the base. Fronds 1-pinnatifid to 1-pinnate, harsh, submembranous to coriaceous, uniform to very dimorphic. Pinnae mostly 15–40 pairs, with a prickly margin. Veins prominent, areolate in the fertile fronds, sometimes simple or forked in the sterile fronds but mostly anastomosing with several rows of areolae parallel to the midrib. Sori short, superficial, distinct or confluent, cultrate to oblong, often slightly curved, 1–2 (rarely 3) rows parallel to the midrib, 1 to each areola, attached on the inner side of the veinlet or veinlets parallel to the costa (Fig. 23). Indusium cultrate to oblong, membranous, opening towards the midrib.Sporangia with an annulus of 14–16 thick-walled cells. Spores bilateral, brown to dark brown, smooth, reniform or sub-orbicular.

Key to the species

1 Pinnae attached by their bases except the lowermost pair sometimes shortly stalked and somewhat enlarged. Sori distinct (Fig. 23). Fronds harsh, coriaceous to subcoriaceous, uniform. Widespread. RF gullies or tall open forests. *Rasp Fern* . . . .**D. aspera** R.Br.

1 Pinnae in the lower third of the frond or almost all the pinnae very shortly stalked. Sori distinct or confluent. Fronds submembranous to coriaceous, uniform to very dimorphic 2

2 Plants erect and robust. Pinnae stalked towards the base of the fronds. Fronds harsh, coriaceous to subcoriaceous, uniform. Widespread. Exposed slopes in gullies; open forests or RF amongst rocks . . . . **D. australis** (Parris) Parris

2 Plants decumbent or subdecumbent. Pinnae stalked in the lower half of the frond. Fronds submembranous to subcoriaceous, often very dimorphic. 3

3 Fronds divided for the greater portion of the lamina but often with an elongated apical pinna, subdimorphic to very dimorphic. Widespread. RF and moist shady places on banks, often near creeks and waterfalls. . . . . **D. caudata** (Cav.) R.Br.

3 Fronds undivided for the greater part, with a few, short, broad, pinnae towards the base. Cumberland Plain and lower Blue Mts RF margins or open forests. Uncommon . . . . **D. linearis** J.Sm.

10 CYATHEACEAE

Tree ferns or large terrestrial ferns; the trunks thick or sometimes massive, erect, often covered with densely matted adventitious roots, dictyostelic. Fronds borne spirally on the trunk in tree ferns, 1–4-pinnate; the bases of the stipes clothed with scales, persistent or, if caducous leaving round or oval scars on the trunk; the dermal appendages hairs or scales or both. Upper surface of the pinna-rhachis and costae raised or grooved. Lamina membranous to coriaceous. Pinnules symmetrical or almost so; the fertile and sterile segments uniform. Veins free or anastomosing. Cubical cells present in association with the sclerenchyma. Sori superficial, round, indusiate or exindusiate, terminal on the veins. Indusium hemitelioid, cup-shaped, globular or absent. Receptacle variable, erect and knob-like or columnar or slightly raised. Paraphyses mostly present. Sporangia numerous, rather small, composed of 14–28 indurated cells, oblique, not interrupted by the pedicel; the stomium well developed; the pedicel short. Spores trilete, globose, without a perispore; the exine smooth or papillose. 1 gen., mainly trop. to subtrop., some south-temp.

1 Cyathea Sm.
16 species native Aust.; Qld, NSW, Vic., Tas., WA

Tree ferns with an erect caudex up to 20m high. Bases of the stipes persistent or deciduous, clothed with distinctive scales. Stipes fawn or yellow to dark brown or red-brown or purplish black or sometimes black, usually muricate or aculeate; the spines obtuse, acute or pungent, straight or curved. Lamina lanceolate to broadly ovate, bipinnate to decompound, membranous to thick and chartaceous but mostly coriaceous. Veins free; minor veinlets simple, once or twice forked or sometimes pinnate; costae and costules mostly clothed with distinctive scales and/or hairs. Sori on the axils of, or dorsal on the veinlets, exindusiate or indusiate; the indusium cupuliform or hemitelioid (semicircular) or consisting of a ring tuft of scales. Receptacle elevated, oval, globular or columnar. Sporangia pyriform or cuneate; the pedicel short and thick, consisting of 4 rows of cells; the annulus oblique; the stomium lateral, definite. Paraphyses usually present, filamentous, mostly fawn or reddish. Spores tetrahedral, trilete, smooth or papillose.

In Cyathea the base of a stipe with the basal tuft of hairs should be collected, also a portion of the lamina should be examined under the microscope for details of the costal scales or hairs.

Key to the species

1 Lower surface of the secondary rhachises purplish black or rarely bright red-brown, aculeate with short sharp prickles. Scales of the bases of the stipes pale stramineous or almost white. Scales of the costae never bullate; the margin setose with very numerous purplish spinules. Most of the old bases of the stipes persistent on the caudex. Tree ferns mostly 1–7 m high. Widespread. RF or along the coast not far from the ocean, on mountain slopes near creeks. *Prickly Tree fern.* . . . . . .**C. leichhardtiana** (F.Muell.) Copel.

1 Lower surface of the secondary rhachises fawn or yellow-brown or brown or light stramineous; the tubercles obtuse or coarsely pointed 2

2 Scales of the costae substellate, never bullate; the margin of the scales setose throughout their length with numerous red, brown or white spinules. Scales of the bases of the stipes of 2 types, i.e. broad, white or pale stramineous paleae and narrower, dark red scales. Old bases of the stipes not persistent on the caudex. Tree ferns mostly 2–10 m high. Coast and gullies in adjacent plateaus mainly south of Sydney. RF in shady gullies, along creeks . . . . **C. cooperi** (Hook. ex F.Muell.) Domin

2 Scales of the costae bullate; the margins of the scales entire or with long, white or fawn fimbriae and/or with 1–4 dark red spinules. Scales of the bases of the stipes bright brown, glossy, coriaceous. Old bases of the stipes persistent on the caudex. Tree ferns mostly 2.5–20 m high. Widespread, especially Blue Mts RF gullies or mountain slopes; open forests. *Rough Tree fern.* . . . . **C. australis** (R.Br.) Domin

11 DAVALLIACEAE

Epiphytic, climbing or terrestrial ferns. Rhizome mostly fleshy, dorsiventral and long-creeping or sometimes short and erect, stoloniferous in one genus, dictyostelic or solenostelic, mostly with a highly dissected solenostele, densely clothed with brown or chestnut or black, peltate-based, entire or ciliate, often hair-pointed or long-attenuated scales. Stipes mostly distant except in *Nephrolepis*, articulated or not articulated to the rhizome in a few non-Australian species. Lamina simple, pinnate to decompound, linear or oblong to broadly deltoid, herbaceous to coriaceous, glabrous or hairy. Pinnae not articulated or articulated to the main rhachis. Ultimate segments mostly unequally-sided at the base except in simple fronds. Veins free, rarely anastomosing. Sori indusiate or rarely exindusiate, rounded or rarely elongated along the margin, terminal or dorsal on the veins, marginal or intramarginal. Indusium peltate, reniform or reniform-orbicular or lunate or cup-shaped, either affixed by the sinus or opening outwards and attached at the base and sometimes at the sides. Sporangia with an annulus including 9–19 thick-walled cells; pedicel elongated, composed of 3 rows of cells. Spores bilateral, mostly oblong to reniform or oval, tuberculate or smooth, with or without a perispore. 9 gen., mainly trop. to warm temp.

Key to the genera

1 Stipes not articulated to the rhizome 2
1 Stipes articulated to the rhizome 3

2 Fronds at least 2-pinnate. Pinnae not articulated to the main rhachis. Indusium orbicular-peltate (Fig. 21). . . . . **3 RUMOHRA**
2 Fronds 1-pinnate. Pinnae articulated to the main rhachis. Epiphytic on trees or climbing over rocks . . . . **2 ARTHROPTERIS**

3 Fronds at least 2-pinnate. Scales of the rhizome ciliate or fimbriate, peltately affixed. Indusium a cylindrical tube opening towards the margin, affixed by the base and sides (Fig. 20). Epiphytic on trees or climbing over rocks . . . . **1 DAVALLIA**
3 Fronds 1-pinnate. Pinnae articulated to the main rhachis. Indusium reniform or lunate. Terrestrial. . . . . **4 NEPHROLEPIS**

1 Davallia Sm.

4 species native Aust.; Qld, NSW, Vic.

One species in the area

Rhizome long-creeping, 0.4–1.2 cm broad, clothed with brown or chestnut, acuminate, finely fimbriate scales each of which has a black peltate base. Stipes 10–20 cm long, 0.5–3 mm broad near the middle, deciduous, leaving 2 rows of round scars along the rhizome. Lamina 15–70 cm long, 4–25 cm broad, 3-pinnate to 3-pinnate-pinnatifid, thinly coriaceous to coriaceous, glossy. Veins free, simple or mostly once forked. Sori never confluent, 1–5 on each ultimate segment (Fig. 20). Indusium 1–1.2 mm long, 0.2–1 mm broad; the apex concave or convex. Widespread. RF. Epiphytic on trees or climbing over rocks in the drier open forests. *Hare's Foot Fern.* . . . . **D. solida** (G.Forst.) Sw. var. **pyxidata** (Cav.) Noot.

2 Arthropteris J.Sm. ex Hook.f.

4 species native Aust.; Qld, NSW

Epiphytes or climbing ferns. Rhizomes long creeping, solenostelic, clothed with peltate-based scales. Stipes distant, in 2 ranks, articulated to the rhizome by a joint which may be close to or remote from the rhizome. Fronds 1-pinnate. Pinnae articulated to the main rhachis, sometimes very deeply lobed, often with a row of lime-dots near the margin on the upper surface of the lamina. Veins free, forked. Sori superficial, uniserial, orbicular, terminal on the veins, exindusiate or with a reniform-orbicular or orbicular-peltate indusium. Sporangia mixed at maturity; the pedicels long, narrow; the annulus composed of 9–19 thick-walled cells, a 2-celled stomium and 5–8 thin-walled cells. Spores bilateral, oblong or globose, with a broadly winged perispore; the wing often deeply laciniate with curved spines

Key to the species

1 Indusium absent. Pinnae thinly coriaceous, usually 3–10 cm long. Upper primary pinnae acuminate; the lower ones acute or rounded. Fronds including the stipes 25–60 cm long. Widespread. RF. On trees or rocks . . . . **A. tenella** (G.Forst.) J.Sm.
1 Indusium rather persistent. Pinnae to 3 cm long, rarely longer. 2

2 Indusium pubescent, orbicular-peltate or reniform. Pinnae membranous or herbaceous, 0.2–1.7 cm long with numerous long hairs, obtusely rounded, margins crenate. Fronds including the stipes 10–20 cm long. Widespread. RF. On trees and rocks . . . . **A. beckleri** (Hook.) Mett.
2 Indusium glabrous, reniform. Pinnae usually 3 cm or less long, with scattered short hairs on the veins, obtusely rounded, margins entire to crenate. Fronds including the stipes 15–30 cm long. North from Illawarra escarpment. RF. On trees. Endangered . . . . **A. palisotii** (Desv.) Alston

3 Rumohra Raddi
1 species native Aust.; Qld, NSW, Vic., Tas.

One species in the area
Rhizome long-creeping, densely clothed with chestnut-coloured to dark brown, peltate scales. Stipes scattered along the rhizome. Fronds including the stipes 3–65 cm long, 3-pinnate except those of dwarf plants which are often 1-pinnate-pinnatifid to 2-pinnate, broadly deltoid or lanceolate to broadly ovate or slightly pentagonous, coriaceous. Ultimate segments with blunt crenate teeth. Sori orbicular, superficial, embedded in the lamina, pustulate on the upper surface (Fig. 21). Indusium rather persistent, Sporangia with an incomplete annulus including 12–15 thick-walled cells. Spores with a narrowly winged perispore. Widespread but less common on the coast. Mostly RF. On cliffs or rocks or climbing on tree trunks . . . . **R. adiantiformis** (G.Forst.) Ching

4 Nephrolepis Schott
c. 9 species native Aust.; Qld, NSW, Vic., WA, NT

One species in the area
Rhizome short, erect or oblique, bearing palaceous stolons with fleshy, scaly ovoid tubers; scales of the rhizome, stolons and tubers pale fawn to light chestnut-coloured, hair-pointed, peltate, coarsely fimbriate at the base. Stipes not articulated to the rhizome. Fronds 1-pinnate, submembranous to almost coriaceous, 10–75 cm long (including the stipes). Pinnae articulated to the rhachis, very deciduous, 35–70 pairs, 1–3.5 cm long, 4–9 mm wide, narrowly oblong to very broadly oblong, closely spaced, ± at right angles to the rhachis; the margin crenate; the upper base of each pinna auriculate; the lower base truncate; a row of lime-dots often on the upper surface of the pinna near the margin. Veins free once or more forked, not reaching the margin. Sori submarginal, uniserial. Indusium lunate or reniform, opening towards the pinna apex. Spores bilateral, oblong-reniform. Native to Queensland, Northern Territory and the north coast of New South Wales. Naturalized as a garden escape in urban bushland in the suburbs of Sydney. Terrestrial, rupestral or rarely epiphytic. *Fishbone Fern*. . . . . *N. cordifolia** (L.) C.Presl

12 DENNSTAEDTIACEAE
Terrestrial ferns with creeping or more rarely erect rhizomes which are clothed with simple hairs or more rarely with pleuricellular hairs (as in *Histiopteris*) or rarely with dark scales, solenostelic or dictyostelic. Fronds usually scattered along the rhizome or sometimes in a tuft at the apex of an erect rhizome. Lamina mostly decompound or sometimes 1–2-pinnate, coriaceous or herbaceous, mostly uniform or almost so. Ultimate segments usually oblique at the base, hairy or glabrous. Veins free or rarely anastomosing

without free vein endings in the areolae (Fig. 15). Sori round or linear, marginal or submarginal, terminal on the free veins or sometimes borne on a vascular commissure joining the vein-endings. Indusium in the New South Wales species cupuliform or pouch-shaped, 2-valved, rounded or linear (with an outer false indusium and an inner true indusium) or 1-valved and composed of a small, reflexed, marginal flap or the linear, reflexed, margin of the lamina. Paraphyses present or absent. Sporangia basipetal, gradate or mixed in origin; the annulus with 13–23 indurated cells, longitudinal and interrupted by the pedicel; the stomium definite; the pedicel long, slender, with 2 rows of cells. Spores tetrahedral or sometimes bilateral, without a perispore. c. 12 gen., trop. to temp.

Key to the genera

1 Sori orbicular, not confluent, terminal or almost so on the veins 2
1 Sori linear, ± continuous and elongated in a marginal line. False indusium consisting of the recurved margin of the lamina, with or without a thin, inner true indusium which faces the costa 3

2 Indusium absent or with a short, roundish, reflexed false Indusium. . . . .**2 HYPOLEPIS**
2 Indusium cup-shaped, outer false indusium fused to the inner indusium . . . . . **1 DENNSTAEDTIA**

3 Veins free below the sori. Indusium 2-valved; the outer false indusium coriaceous; the inner true indusium thin. Rhizome clothed with simple hairs. Stipule-like pinnules absent at the base of the primary pinnae. . . . . **3 PTERIDIUM**
3 Veins anastomising below the sori (Fig. 15). Indusium 1-valved; the false indusium scarious, consisting of the recurved margin of the lamina; the inner true indusium absent. Rhizome clothed with scales and/or pleuricellular hairs which are narrowed at the base. and Stipule-like pinnules present at the base of the primary pinnae . . . .**4 HISTIOPTERIS**

1 Dennstaedtia Bernh.
1 species endemic Aust.; Qld., NSW, Vic., WA

One species in the area
Terrestrial ferns. Rhizome widely creeping, 3–5 mm diam., densely clothed with red-brown bristly hairs. Stipes red-brown, clothed near the base with the same type of hairs as on the rhizome, not articulated to the rhizome. Fronds including the stipes usually 0.9–1.5 m long, mostly 0.3–0.5 m broad. Lamina dark green, soft, membranous to thinly herbaceous, 3-pinnate-pinnatifid. Veins free. Sori round, marginal, terminal on the veinlets, mostly in sinuses. Indusium double, cyathiform, formed by the fusion of the inferior indusium and ± modified lobe of the lamina. Receptacle short, columnar. Widespread. RF gullies, or tall open forests . . . . **D. davallioides** (R.Br.) T.Moore

2 Hypolepis Bernh.
Ground Ferns
8 species native Aust.; all states and territories except NT

Terrestrial ferns often forming extensive colonies. Rhizome slender, solenostelic, long-creeping, clothed with simple brown or reddish hairs. Stipes not articulated to the rhizome, clothed with hairs which are sometimes glandular-tipped or viscid. Lamina bipinnate to decompound; the apex characterized by intermittent growth, often clothed with short glandular-tipped or strigose hairs, sometimes glabrous, herbaceous, uniform. Veins free. Sori orbicular, usually terminal on the veinlets, never confluent, either marginal and indusiate or intramarginal and exindusiate, borne on punctiform receptacles. Indusium sometimes absent, mostly a small, modified, reflexed lobule of the margin, usually protected but never bearing the young sori, later often turned back. Sporangia with an annulus including 13–15 thick-walled cells; pedicel 2 rows of cells, long, narrow. Spores bilateral, monolete, without a perispore, mostly spinulose or tuberculate, rarely smooth.

Key to the species

1 Secondary rhachises, costae, costules and minor veinlets clothed often rather sparsely with stiff, acute, non-glandular, septate hairs. Main rhachis yellowish white or straw yellow or red-brown or light to dark brown, smooth or with scattered brown or orange-brown tubercles, usually almost glabrous. Rhizome 1–3 mm diam. Plants 30–90 cm high. Widespread. Along creeks or in swampy land in open forest, at edge of RF or in open pasture. . . . . **H. muelleri** N.A.Wakef.

1 Secondary rhachises, coastae, costules and minor veinlets densely clothed with crisped, septate hairs a number of which have glandular apical cells 2

2 Stipes dark red-brown or purplish-red. Main rhachis red-brown or purplish red or orange-red, with scattered, dark red tubercles. Hairs of the costae and costules with brownish lumina. Rhizome 1–2 mm diam. Plants 30–60 cm high. Blue Mts Mountain gullies near creeks or in swampy land. . . . .
. . . . **H. rugulosa** (Labill.) J.Sm.

2 Stipes brown or orange-brown or deep yellow but dark brown near the base. Main rhachis light brown or bright golden yellow or orange-red, with scattered, brown or orange-brown tubercles. Hairs of the costae and costules with opaque lumina. Rhizome 3–8 mm diam. Plants up to 3 m high. Widespread. Mountain RF and RF margins . . . . **H. glandulifera** Brownsey & Chinnock

3 Pteridium Gled. ex Scop.
3 species in Aust. (2 native, 1 naturalized); all states and territories

One species in the area

Large terrestrial ferns often occurring in dense stands. Rhizome extensively creeping, 2–10 mm diam., clothed with dark red or ferruginous, simple hairs. Stipes 0.3–1.5 m long, up to 1 cm broad at the middle. Fronds 0.6–3 m high including the stipes. Lamina 3–4-pinnate, farinaceous beneath. Ultimate segments often falcate, linear to very broadly oblong; the margin recurved, entire; the base decurrent; the upper surface glabrous or with a few reddish hairs near the margin; the lower margin glabrous or with a few reddish hairs; the lower surface pubescent with short, appressed, white hairs; the midrib glabrous or with a few dark red hairs. Rhachises, costae and costules with free lobes. Veins free below the sori, forked, oblique, without membranous wings. Sori linear, borne on a submarginal vascular commisure. Indusium with a green, coriaceous outer valve and a papery, brown or fawn, usually glabrous, rarely ciliate, entire or erose inner valve. Spores tetrahedral. Widespread. Open forests, damp sandy flats, sandstone gullies, pasture land, or sand dunes. *Bracken* . . . . **P. esculentum** (G.Forst.) Cockayne

4 Histiopteris (J.Agardh) J.Sm.
1 species native Aust.; all states and territories

One species in the area

Large terrestrial ferns sometimes forming dense stands. Rhizome long-creeping, 2–7 mm diam., clothed with entire, glossy-metallic, pleuricellular hairs which are not flattened at the base. Stipes yellow or orange or red-brown or brown, glabrous except for the pleuricellular hairs near the base. Fronds including the stipes usually 1–4 m long, 2-pinnate to 3-pinnate-pinnatifid; the sterile fronds usually less divided than the fertile, pale green, glaucous beneath; the basal pinnules stipule-like. Veins obliquely branched, repeatedly dichotomous in each lobe, some anastomosing but without free vein-endings in the areolae (Fig. 15). False indusium 0.5–1 mm broad, entire or slightly lobed. Ultimate segments sessile, opposite, narrowly oblong-deltoid to very broadly oblong-deltoid; the apex obtuse or broadly rounded; the margin entire, ± crenate or lobed. Widespread. Mostly growing in moist sheltered situations near the base of sandstone cliffs, in sandstone gullies or sometimes in mountain RF. *Bat's Wing Fern* . . . .
. . . . **H. incisa** (Thunb.) J.Sm.

13 DICKSONIACEAE

Tree ferns or large terrestrial ferns; trunks 1 mor more high, often massive and bearing adventitious roots as well as often persistent stipe-bases (as in *Dicksonia*) or with stout stipes from the creeping rhizome (as in *Calochlaena*). Stipes not articulated, flattened above or 1 or more times grooved on the upper surface, densely clothed with hairs; the grooves continuous. Lamina chartaceous or coriaceous or subcoriaceous, 2-pinnate to decompound, dimorphic or non-dimorphic, clothed with hairs (never scales). Ultimate segments mostly ± lanceolate, crenate to 2-pinnatifid. Veins free, forked or pinnate, close to the margin. Cubical cells present in association with the sclerenchyma. Sori round, borne on the vein-endings, almost marginal, protected by a 2-lipped indusium, an inner or true indusium and on outer false indusium consisting of a modified marginal laminal lobe. Paraphyses filiform. Sporangia gradate; the pedicels short to long; the annulus oblique, ± interrrupted or not interrupted by the pedicel; stomium definite, horizontal. Spores trilete, globose or globose-tetrahedral. 6 gen., north- and south-temp., trop.

Key to the genera

1 Tree ferns with thick or massive trunks. Pinnules symmetrical at the base, markedly dimorphic. . . . . **1 DICKSONIA**

1 Large terrestrial ferns with stipes 0.5–1 cm diam. Pinnules asymmetrical at the base, slightly dimorphic. . . . . **.2 CALOCHLAENA**

1 Dicksonia L'Hér.
3 species native Aust.; Qld, NSW, Vic., Tas., S.A.

One species in the area

Arborescent tree ferns with a crown of large fronds. Caudex 1–4.5 m high, 10–30 cm diam., clothed with the persistent bases of old stipes, densely matted with coarse fibrous roots. Stipes not winged, verruculose, densely clothed at the base with soft ferruginous acicular hairs which are mostly 2–4.5 cm long. Main rhachis fawn or light stramineous. Fronds 3-pinnate to 3-pinnate-pinnatifid, rather harsh, dark green above, paler beneath, 1.8–3.6 m long. Ultimate lobules rounded in the fertile segments; the margin serrate or aristate; more oblong in the sterile segments and the margins aristate. Sori 0.8–1.3 mm diam. Indusium with the 2 lips equal or almost so, persistent (Fig. 14). Widespread. Deep mountain gullies; RF. Usually along creeks. *Soft Tree-fern* . . . .**D. antarctica** Labill.

2 Calochlaena (Maxon) M.D. Turner & R.A. White
2 species native Aust.; Qld, NSW, Vic., Tas.

One species in the area

Large terrestrial ferns. Stem creeping, 0.8–2.5 cm diam., clothed with soft silky silvery hairs intermixed with long, ferruginous or red-brown hairs. Stipes yellow or brown, often mottled, black or dark brown towards the base which is long pilose. Fronds including the stipes 0.6–1.5 m high, usually 0.3–0.8 m broad. Lamina 3-pinnate-pinnatifid, subcoriaceous to herbaceous, pale green or yellowish-green. Sori round, 0.8–1.5 mm diam., up to 20 on each terminal segment. Indusium bivalvate; the outer valve formed by the recurved lobule of the lamina; the inner thin, dentate-ciliate, usually hidden at maturity. Spores tetrahedral. Widespread. Open forest, usually on poor soils in open places or sheltered gullies, often on sandstone cuttings. *Rainbow Fern* . . . . **C. dubia** (R.Br.) M.D.Turner & R.A.White

14 DRYOPTERIDACEAE

Terrestrial or climbing ferns. Rhizome creeping or erect, sometimes forming a short trunk, dictyostelic (often complex), dorsiventral in some genera, clothed with non-clathrate, mostly non-peltate scales which lack stiff unicellular hairs; the marginal teeth of these scales usually formed by 2 adjacent cells. Stipes mostly not articulated to the rhizome, with 3–7 vascular strands. Fronds uniform or dimorphic. Lamina simple to pinnately decompound or sometimes flabellate, often basiscopically produced, glabrous or clothed with multicellular hairs. Rhachises and costae glabrous, papillate or clothed with multicellular hairs. Costae grooved or raised. Pinnae articulated or not articulated to the rhachis. Veins free or anastomosing, with or without free veinlets in the areolae, anadromous or catadromous. Sori indusiate or rarely exindusiate, mostly dorsal or terminal on the veinlets, rarely marginal or projecting beyond the margin, usually orbicular and comparatively small but sometimes oblong or linear or aristichoid and spreading along the veins. Indusium peltate, reniform, oblong or linear. Sporangia with pedicels composed of 1 or 3 rows of cells; the annulus longitudinal, interrupted by the pedicel, including 10–40 thick-walled cells (usually under 20 in genera occurring in Australia). Spores bilateral, monolete, usually with a perispore. 29 gen., cosmop.

Key to the genera

1 Veins anastomising. Areoles each with 1–3 free included veinlets. Excurrent veins absent. Indusia peltate. Lamina 1-pinnate . . . . **1 CYRTOMIUM**
1 Veins free 2

2 Ridges of the upper surface of the main rhachis continuous with the thickened margin of the pinnae (Fig. 11). Channel of the main rhachis glanduloso-pubescent or clothed densely with articulate, reddish, unbranched hairs. Costae and costules with scattered, red or yellow, oblong or rounded, glandular hairs. Indusium mostly orbicular-reniform or rarely peltate . . . . **4 LASTREOPSIS**
2 Ridges of the upper surface of the main rhachis never continuous with margin of the frond. Channel of the main rhachis glabrous or clothed with scales. Costae and costules without oblong or rounded, glandular hairs 3

3 Indusia orbicular-reniform, attached by a deep sinus, rarely interspersed with a few peltate indusia. Basal acroscopic pinnule or lobe of the middle primary pinnule much closer to the main rhachis than the basal basiscopic pinnule or lobe. Lamina often basiscopically produced. Rhizomes long-creeping. Main rhachis without scaly buds . . . . **2 ARACHNIODES**
3 Indusia orbicular-peltate. Basal acroscopic pinnule or lobe of the middle primary pinnae not distinctly closer to (Fig. 10) or slightly more remote from the main rhachis than the basal basiscopic pinnule or lobe. Lamina not basiscopically produced. Rhizome short and erect. Main rhachis with scaly proliferous buds in some species . . . . **3 POLYSTICHUM**

1 **Cyrtomium** C.Presl
1 species naturalized Aust.; Qld, NSW, Vic., S.A.

One species in the area

Tufted terrestrial plants 30–60 cm high. Rhizome erect, short, densely clothed with broad, fimbriate, reddish-brown scales. Fronds 1-pinnate. Pinnae 3–7 pairs, with an entire or 3-cleft terminal pinna, coriaceous, glossy, falcate, lanceolate to ovate, shortly petiolulate, auriculate on the acroscopic side at the base, 6–10 cm long, 2–4 cm broad; the margin incised with sharp teeth; the apex acuminate. Main rhachis brown or red-brown, clothed with fibrillose scales. Veins anadromous, anastomosing to form large areolae each with 1–3 free veinlets in each areola. Sori orbicular, dorsal on the veins, uniformly covering the lower surface. Indusium peltate, rather persistent, glabrous, fawn; the margin erose. Sporangia each with a narrow pedicel; annulus including 14–16 thick-walled cells. Spores bilateral, monolete, brown, globoso-ellipsoidal, covered with balloon-like wings. Garden escape in coastal Sydney suburbs. In crevices of cliffs above high-water mark and on old tramway cuttings near the ocean. A cultivar derived from a native of Japan, Korea and China. *Holly Fern* . . . . ***C. falcatum** (L.f.) C.Presl

2 **Arachniodes** Blume
1 species native Aust.; Qld, NSW

One species in the area

Terrestrial ferns. Rhizome long-creeping, 0.6–1 cm diam., clothed with red-brown to dark brown, papery, entire or slightly denticulate scales. Stipes scattered along the rhizome, 15–60 cm long, often longer than the lamina, clothed towards the base with scales similar to those on the rhizome. Main rhachis without gemmiferous scaly buds; the ridges of the upper surface never continuous with the the frond-margin; the intervening channel glabrous or clothed with scales. Fronds including the stipes 45–120 cm long. Lamina 3–4-pinnate, glossy, subcoriaceous to coriaceous, 5-angular; the apex abruptly acuminate. Rhachises clothed with fibrillose, dark brown scales which are fimbriate at the base. Primary pinnae 18–25 pairs; the basal pair basiscopically produced; tertiary segments obliquely rhomboidal. Ultimate segments or lobes very unequally-sided at the base; the margin aristate; the apex mucronate. Veins not prominent. Sori round, terminal, on the anterior basal veinlet of each group. Indusium mostly orbicular-reniform but a few peltate, fairly persistent, black in the centre. Spores dark brown; the perispore reticulate. Coast. Uncommon. RF in stands, usually near creeks . . . . **A. aristata** (G.Forst.) Tindale

3 **Polystichum** Roth

7 species endemic Aust.; all states and territories except WA & NT

Terrestrial ferns. Rhizomes erect, short, thick, densely clothed with lacerate, non-peltate, non-clathrate scales which are often of several kinds. Stipes clustered, not articulated to the rhizome, often densely scaly. Main rhachis with proliferous scaly buds in some species; the ridges of the upper surface never continuous with the frond-margin; the intervening channel glabrous or scaly. Lamina coriaceous, anadromous, 1-pinnate to decompound, seldom dilated towards the base, not basiscopically produced, glandular hairs absent. Pinnae not articulated to the rhachis, unequally-sided, usually clothed with fibrillose scales having fimbriate bases; basal acroscopic pinnule or lobe of the middle primary pinnae either slightly more remote from or slightly closer to the main rhachis than the basal basiscopic pinnule or lobe (Fig. 10). Ultimate segments mostly with aristate marginal teeth or at least a mucronate apex. Veins free, anadromous (Fig. 10). Sori orbicular, superficial. Indusium orbicular-peltate, rarely absent. Sporangia with an annulus of 12–18 or more thick-walled cells and 6–9 thin-walled cells including a 2-celled stomium, with or without glandular hairs on the long, narrow pedicels. Spores bilateral, mostly globoso-ellipsoidal or oblong, with a perispore which is tuberculate or echinulate or covered with balloon-like wings or plicate and broadly winged.

Key to the species

1 Proliferous buds absent on the main rhachis. Fluffy, fawn, caducous squamules present on the stipes and rhachises. Scales of the rhizomes and bases of the stipes dull, narrowly lanceolate to ovate except for a few inconspicuous, cultrate scales 4–8 mm long. Spores black or dark brown, tuberculate. Plants 30–60 cm high. Mainly in the south of the region. Uncommon. RF on rocky cliffs, often near waterfalls and usually on mountain slopes . . . . **P. formosum** Tindale
1 Proliferous buds present near the apex of the main rhachis. Squamules absent on the stipes and rhachises. 2

2 Scales of the bases of the stipes glossy and mostly with a pale border. Distal lobes of the pinnules obtuse. Pedicels of the sporangia often with 1–2, stalked glands. Spores dark brown; the perispore covered with rounded balloon-like protuberances. Plants 30–120 cm high. Mainly on the ranges; common on the Blue Mts RF and open forests on hillsides, often forming large stands. *Mother Shield Fern* . . . . **P. proliferum** (R.Br.) C.Presl
2 Scales of the bases of the stipes dull and without a paler border. Distal lobes of the pinnules aristate (Fig. 10). Pedicels of the sporangia glandless. Spores with brown, reticulate, broadly winged perispores. Plants 30–110 cm high. Widespread. Open forests on rocky hillsides or on grassy slopes or in RF gullies . . . . **P. australiense** Tindale

4 **Lastreopsis** Ching

Shield Ferns
15 species native Aust.; Qld, NSW, Vic., Tas., S.A.

Terrestrial ferns mostly 0.3–1 m high. Rhizome long-creeping, shortly creeping or rarely erect; scales thin, narrowly lanceolate to narrowly ovate,never clathrate nor iridescent brown or rarely castaneous or almost black; the apex acute or acuminate; the cells thick-walled, rectangular or hexagonal, with red or yellow lumina; the margin entire or slightly to very denticulate or with a few fimbriate or glandular-headed processes . Fronds decompound, 5-angular, with the lowest pair of primary pinnae strongly basiscopically produced, catadromous throughout (Fig. 11) or more often anadromous in the upper segments, rarely anadromous throughout. Frond-margin thickened including where it is decurrent along the costae. Main rhachis bordered above by 2 prominent ridges which are continuous with the thickened frond-margin of the pinnae (Fig. 11); the broad, intervening shallow channel filled by a scarcely raised vein which is mostly clothed with short, articulated, unbranched, reddish hairs or rarely glanduloso-pubescent. Costae raised. Veins free; the minor veinlets simple or forked. Sori round, up to 2 mm diam., terminal or medial on the simple minor veinlets or their acroscopic branchlets, indusiate or rarely exindusiate. Indusium reniform-orbicular or rarely peltate, glabrous or villous; the margin crenate, entire or glandular-fimbriate. Sporangia with an annulus composed of 13–16 thick-walled cells and 8–9 thin-walled cells including a 2-celled stomium; the pedicel long, narrow, usually with 1 or more rarely 2, oblong, red or yellow, stalked glands. Spores bilateral, globoso-ellipsoidal, usually covered with balloon-like wings over the whole surface of the spore (i.e. sacco-rugulate) or with a crested perispore having a broken or more rarely an interrupted wing. Glandular hairs cylindrical or rarely rounded, yellow or orange or red, scattered over the lower surface and sometimes the upper surface of the lamina, costae, costules and sometimes on the indusia.

Key to the species

1 Main rhachis densely clothed on the lower surface with spreading, dark red or dark brown, bristle-like, bulbous-based scales; channel on the upper surface clothed with bristle-like scales and stiff, club-shaped, glandular-tipped, dark red-brown or whitish hairs, usually 2-celled with red articulations. Indusium orbicular-peltate or reniform. Rhizome long-creeping. Lamina submembranous to subcoriacecous. Sori near the margin of the segments. Plants 15–90 cm high. Blue Mts RF. Endangered. . . . .
. . . . **L. hispida** (Sw.) Tindale
1 Main rhachis clothed on the lower surface with short, soft hairs and often with dark brown, narrowly lanceolate to ovate, bullate or flattened scales; channel on the upper surface with non-club-shaped, non-glandular, acute or obtuse hairs of 3–5 cells with red or red-brown articulations. Indusium orbicular-reniform (or often orbicular in *L. acuminata*) 2

2 Rhizome 0.15–0.6 cm diam., long-creeping. Scales of the main rhachis dark brown, narrowly lanceolate to ovate. Indusium light brown, clothed with yellow, oblong, glandular hairs especially towards the centre. Plants 30–90 cm high. Widespread. RF and tall open forests . . . .
. . . . **L. microsora** (Endl.) Tindale ssp. **microsora**
2 Rhizome 0.8–4 cm diam., shortly creeping or short and erect or oblique with a dense crown of fronds. . 3

3 Upper pairs of primary pinnae with a basal secondary pinna arising from the main rhachis. Rhizome shortly creeping with closely spaced stipes. Sori almost marginal but never in a row around the margin. Indusium light brown, often with a dark centre; the margin glandular-fimbriate. Scales of the main rhachis dark brown, lanceolate to ovate, rarely absent. Plants 25–90 cm high. Widespread. RF or sometimes on hillsides in tall open forests . . . . **L. decomposita** (R.Br.) Tindale
3 Upper pairs of primary pinnae without secondary pinnae arising from the main rhachis. Rhizome short, erect or oblique, with a dense crown of fronds, covered with the broken bases of the old stipes. Sori halfway between the midrib and the margin of the segments. Indusium dark brown, almost entire, lacking glandular hairs. Scales absent on the main rhachis. Plants 25–90 cm high. Widespread. RF and tall open forests, usually near creeks. . . . . **L. acuminata** (Houlston) C.V.Morton

15 GLEICHENIACEAE

Erect or straggling terrestrial ferns which usually grow in dense clumps or stands. Rhizome long-creeping, branched, sparsely clothed with hairs or scales, mostly protostelic or rarely solenostelic. Fronds 1-pinnate or several times pinnate, often pseudo-dichotomously branched, with dormant terminal scaly or hairy leaf-apices; the lamina clothed with simple or branched hairs and/or fringed scales. Vernation circinnate. Ultimate segments flat or saccate, herbaceous or coriaceous; stomates present. Veins free. Sori exindusiate, rosette-like, superficial, borne on the lower surface of the lamina, often surrounded by floccose hairs. Sporangia few, 1–12 in each sorus, broadly pear-shaped; the annulus complete, transverse or oblique, dehiscing by a longitudinal slit; the pedicel short. Paraphyses present in some genera. Receptacle small, rounded. Spores bilateral or tetrahedral, without a perispore. 5 gen., trop to temp., mostly S. hemisphere.

Key to the genera

1 Ultimate segments usually 0.1–0.5 cm long, flat, recurved or pouch-like, each with a sorus. Pinnules of the ultimate branches of the frond divided into very numerous segments. Rhizome clothed with scales. Spores tetrahedral . . . .**1 GLEICHENIA**

1 Ultimate segments usually 1–6 cm long, flat; sori numerous on each segment, mostly in a single row on each side of the midrib. Pinnules of the ultimate branches of the frond undivided 2

2 Minor veinlets of the segments once forked, rarely the lowest pair twice or thrice forked. Rhizome clothed with scales. Sporangia 1–6 (mostly 4) in each sorus. Nodes devoid of accessory pinnae. Spores bilateral . . . .**2 STICHERUS**
2 Minor veinlets of the segments 2–10 times branched, rarely interspersed with a few once forked veinlets. Rhizome clothed with branched hairs. Sporangia 6–12 in each sorus. Nodes usually with a pair of divaricate accessory pinnae which are pectinato-pinnate. Spores tetrahedral in the Australian species . . . .**3 DICRANOPTERIS**

1 Gleichenia Sm.
Coral Ferns
7 species native Aust.; all states and territories

Terrestrial ferns, often of straggling habit, frequently growing in rock crevices. Rhizome long-creeping, clothed with scales. Fronds herbaceous or coriaceous, once forked or branched pseudo-dichotomously. Primary lateral axes mostly pectinato-pinnate, devoid of accessory pinnae, clothed often very densely with woolly-fimbriate or laciniate scales and/or with stellate hairs. Dormant leaf-apices in the forks of the axes, protected by imbricate fimbriate scales. Pinnae in divaricate pairs, bipinnate. Pinnules ± linear. usually spreading almost at right angles to the axes. Ultimate segments less than 5 mm long, flat, recurved or pouch-like, semiorbicular, oblong or ovate-triangular, glabrous or often densely woolly-tomentose with fimbriate or stellate scales and hairs, Sori solitary, often immersed, almost terminal on the basal acroscopic branch of the vein, composed of 2–4 (rarely 6) sporangia. Paraphyses present. Spores tetrahedral.

Key to the species

1 Ultimate segments of the mature fronds pouch-like on the lower surface, with a bar of green tissue across the base except sometimes in sterile fronds, opening with a comparatively narrow aperture. Sori mostly 2-sporangiate. Plants 0.3–2.8 m high. Widespread. Sandstone cliffs; swampy land; ti-tree scrub; RF margins; open forests . . . . **G. dicarpa** R.Br.
1 Ultimate segments of the mature fronds flat or slightly recurved on the lower surface. Sori mostly 3–4-sporangiate, more rarely 2-sporangiate 2

2 Lower surface of the lateral axes stellate-tomentose. Lower surface of the ultimate segments pale green. Plants 0.3–2.6 m high. Widespread. Sandstone cliffs; swampy land; ti-tree scrub; open forests . . . .
. . . . **G. microphylla** R.Br.
2 Lower surface of the lateral axes glabrous or almost so. Lower surface of the ultimate segments glaucous. Plants 0.3–2.6 m high. Widespread. Sandstone cliffs, often near the sea; open forests; RF margins; near waterfalls in the mountains. . . . . **G. rupestris** R.Br.

2 Sticherus C.Presl
5 species native Aust.; Qld, NSW, Vic., Tas.

Ferns of erect or straggling habit, forming dense thickets. Rhizome long-creeping, clothed with scales. Fronds uniform, pinnate or repeatedly pseudo-dichotomously branched; accessory branches at the nodes absent. Ultimate branches and some of the preceding axes pectinato-pinnate. Pinnules oblong or linear, often glaucous beneath. Dormant leaf-apices clothed with scales. Veins free, once forked, rarely the lowest pair twice or thrice branched. Sori numerous, rosette-like, mostly in a single row on either side of the midrib, composed of 1–6 (mostly 4) sporangia. Paraphyses present. Spores bilateral.

Key to the species

1 Primary axes without pinnules. Pinnules serrate almost to the base, acutely angled to the axes, pale green or slightly glaucous beneath, glabrous or clothed on the lower surface with stellately branched hairs and scales. Plants 0.3–1 m high. Widespread. In moist gullies; open forests on hillsides or along creeks at the edge of RF. *Umbrella Fern* . . . .**S. flabellatus** (R.Br.) St John var. **flabellatus**

1 Primary axes with simple or lobed pinnules. Pinnules entire or bluntly and crenately lobed, glaucous beneath 2

2 Lower surface of the pinnules always glabrous. Lower surface of the ultimate lateral axes (i.e. costae) clothed with flat, narrowly lanceolate to narrowly ovate, shortly fimbriate scales. Pinnules widely spreading at 60°–90° to the axes. Plants up to 2 m high. Widespread. Rocky hillsides; margins of RF; open forests . . . .**S. lobatus** N.A.Wakef.

2 Lower surface of the pinnules densely clothed with stellately branched hairs and scales. Lower surface of the ultimate lateral axes (i.e. costae) clothed with substellate scales (with a long terminal portion and several elongated processes towards the base) as well as a few, narrowly lanceolate, fimbriate scales. Pinnules acutely angled, usually spreading at 40°–70° to the axes. Widespread. Sheltered gullies; mountain slopes; along creeks; pendulous in crevices of sandstone cliff-faces; near waterfalls . . . . . . **S. urceolatus** M.Garrett & Kantvilas

3 Dicranopteris Bernh.
1 species native Aust.; Qld., NSW,A.W., NT

One species in the area

Terrestrial ferns often of straggling habit, forming dense thickets. Rhizome long-creeping, cylindrical, often ± glaucous, 1.5–3 mm diam., glabrescent, clothed at first with rather stiff, sharply acuminate, septate hairs which have several short branches at the base. Stipes up to 30 cm or more long, 1–3 mm broad near the middle, widely spaced on the rhizome, rigid, glossy, smooth. Fronds light to yellowish-green above, very glaucous on the lower surface, glabrous or with a few reddish stellate hairs. Accessory pinnae at each forking of the rhachis except at the ultimate forks. Ultimate segments sessile, 25–60 pairs, entire, obtuse or bifid, linear to very narrowly oblong or the smaller segments narrowly oblong. Sori 0.8–1 mm diam. 8–17 pairs on each segment, in a single subcostular or sometimes medial row. Spores tetrahedral, white, not reticulate. Plants 0.3–2 m high. Coast and adjacent plateaus north of Royal National Park. Uncommon. Banks of streams at the edge of RF; railway cuttings; open forests on the banks of roads or in clefts of sandstone cliffs . . . . **D. linearis** (Burm.f.) Underw. var. **linearis**

16 GRAMMITIDACEAE

Very small to medium-sized epiphytes or small ferns growing in rock crevices. Rhizome erect to ascending or shortly creeping or rarely long-creeping, with a dictyostele or a simple solenostele, clothed with glossy, concolorous, brown or chestnut-coloured, seldom peltate, entire, ciliate or setulose scales. Stipes usually crowded, borne in 2 rows, mostly not articulated to the rhizome, sometimes articulated or pseudo-articulated, with 1–2 vascular bundles, bearing unicellular or branched, reddish or pale-coloured hairs. Fronds uniform. Lamina simple, narrow, forked, pinnatifid or pinnate, with adnate pinnae, bearing unicellular or branched reddish or pale-coloured hairs or setae. Veins mostly free, simple or forked, more rarely casually anastomosing without free included veinlets. Sori exindusiate, orbicular or oblong or elliptic, sometimes elongate and coenosoric, superficial or immersed in the lamina in cavities or pouches or deep vertical grooves, terminal or dorsal on the veins. Sporangia glabrous or setulose; annulus interrupted, including 8–16 thick-walled cells; pedicel elongated, composed of 1 row of cells except at the top. Paraphyses mostly absent, rarely filamentous or with clavate glandular heads. Spores globoso-tetrahedral, trilete, without a perispore. 20 gen., mainly trop., extends to temp.

1 Grammitis Sw.
15 species in Aust. (6 endemic); Qld., NSW, Vic., Tas.

Small caespitose epiphytic or rupestral ferns. Rhizome tufted or shortly creepingFronds simple or slightly lobed, subcoriaceous or coriaceous, linear-oblanceolate; the margin entire or undulate. Lamina spathulate, lanceolate, elliptic or linear. Lateral veins 1-forked, free, hidden in the thick lamina except the prominent midrib. Sori in a closely crowded, oblique row on each side of, and very close to the costa.

Key to the species

1 Stipe absent, basal 5 mm of frond glabrous. Fronds 1.7 – 4.5 cm long, 1–4 mm wide, subcoriaceous, glabrous or hairy. Sori rounded to oblong. Rhizome up to 2.5 cm long. Growing on rocks in RF and WSF. Endangered . . . . **G. stenophylla** Parris

1 Stipes crowded, up to 2 cm long, densely clothed with long, simple, glandular-tipped hairs and a few forked or stellate hairs. (If stipes absent then hairs present on basal 5 mm of fronds). Fronds including the stipes 2–15 cm long, 3–10 mm wide, Sori elliptic. Rhizome up to 2.5 cm long. Widespread. RF or WSF. Moist rocks or sometimes on trees, near waterfalls, or often in crevices in cliffs. *Finger Fern*. . . . . **G. billardieri** Willd.

17 HYMENOPHYLLACEAE
Filmy Ferns

Small epiphytic or terrestrial ferns mostly found in rainforests under very moist conditions. Rhizome long-creeping or short and thick, often slender or filiform, clothed with hairs, sometimes without roots. Stipes unwinged or frequently winged almost to the base, mostly filiform or slender, glabrous or hairy. Lamina simple or bifid or palmatifid or 1-pinnate to pinnately decompound, usually one cell in thickness except the veins, membranous, almost transparent, lacking stomates. Pinnae with plain or pitted walls, glabrous or clothed with simple to elaborately branched hairs. Ultimate segments small, mostly 1-veined. Veins free except in a few cases. Sori marginal, terminal on the pinnules or on short lateral lobes. Involucre (indusium) bivalvate tubular or obconic; the lips often much distended. Receptacle columnar, included to long-exserted. Sporangia spherical to compressed, very shortly stalked or sessile, clustered along the receptacle, basipetal or maturing at the same time; the annulus complete, oblique or almost transverse, opening by a ± longitudinal slit. Spores tetrahedral or globoso-tetrahedral, without a perispore.

Key to the genera

1 Involucre bivalvate (deeply or to the base) or bivalvate in the upper portion but obconic at the base . . . . **1 HYMENOPHYLLUM**

1 Involucre tubular (Fig. 13) or obconic 2

2 Fronds 1-pinnate with the lower pinnae often again imperfectly pinnate. Ultimate segments pinnately veined. Rhizome 0.2–0.3 mm diam., densely tomentose with orange hairs. . . . . **2 CREPIDOMANES**

2 Fronds pinnately decompound. Ultimate segments 1-veined. Rhizome 1–10 mm diam., clothed with dark red, bristle-like hairs. . . . . **3 CEPHALOMANES**

1 Hymenophyllum Sm.
24 species native Aust.; Qld, NSW, Vic., Tas.

Small epiphytic or terrestrial ferns. Rhizomes creeping, narrow glabrous or clothed with sparse hairs. Fronds thin, simple and once or more rarely twice forked, or pinnately divided; margins entire or with minute teeth. Sori terminal on the ultimate segments or close to the main rhachis on short branches on the upper side of the primary pinnae. Involucre divided to the middle or base. Receptacle included or very slightly exserted. Sporangia large, sessile.

Key to the species

1 Margin of the lamina and the axes clothed with branched and often simple hairs. Rhizome widely creeping, c. 0.1–0.2 mm diam., sparsely pilose with simple or once forked hairs. Stipes not winged, 1–25 mm long, filiform. Lamina divided flabellately or digitately almost to the base of the frond into dichotomous segments, deltoid or broadly obovate or ± rounded in outline. Ultimate segments with a glabrous surface; the margin with teeth bearing simple and branched hairs; the apex rounded truncate or emarginate. Sori solitary or very rarely in pairs, deeply sunken in the apex of the ultimate segments. Involucre 1–1.5 mm long, 0.8–2.5 mm broad, the lips erose or toothed, each tooth bearing a reddish

forked hair. Receptacle included. Plants 1–5 cm high. Blue Mts; Illawarra ranges. RF, growing on rocks and trees. . . . . **H. lyallii** Hook.f.

1 Margin of the lamina and the axes not clothed with branched hairs .2

2 Margin of the lamina entire. 3
2 Margin of the lamina denticulate 6

3 Lamina simple, once or rarely thrice forked. Involucre and/or the lamina with a black thickened margin Tiny mat-forming ferns. Rhizome wiry, sparsely clothed with a few undivided hairs. Stipes 0.3–7 mm long, filiform. Lamina simple, once or more rarely twice forked, thin, light green, linear to narrowly oblong, 4–35 mm long, 1–2.5 mm broad, glabrous; margins entire, glabrous, composed of 1–2 rows of black cells. Midrib black, glossy, without any lateral branchlets, clothed with a few, reddish, tubular hairs. Sori solitary at the apex of the frond or its lobes. Involucre oval or orbicular, 2-valved to the base, 0.8–1.5 mm long, light green, with a glossy black border 2–3 cells in width. Spores tetrahedral-globose or globose. Widespread but uncommon. RF, often near waterfalls, on rocks, logs or wet cliff-faces. . . . .
. . . . **H. marginatum** Hook. & Grev.
3 Lamina pinnately divided. Involucre and lamina without a black thickened margin 4

4 Stipes and the lower part of the main rhachis not winged. Sori slightly immersed at the base. Fronds 2–3-pinnate-pinnatifid, 8–35 cm (rarely 4–8 cm) long including the stipes, 15–55 mm broad. Widespread. RF on tree ferns, rocks or trees, often in mountain gullies . . . . . .**H. flabellatum** Labill.
4 Main rhachis winged; the wing usually continuous down the stipe 5

5 Sori free, terminal on the ultimate segments, often in pairs. Stipes broadly winged with a crisped or flat wing. Fronds 1–3-pinnate-pinnatifid, 5–25 cm long including the stipes, 10–55 mm broad. Widespread, common in Blue Mts RF or tall open forests on rocks tree ferns or trees . . . .**H. australe** Willd.
5 Sori immersed in the apices of the ultimate segments. Stipes with a short flat wing or unwinged. Fronds 1-pinnate-pinnatifid, 1.2–14 cm long including the stipes, 8–20 mm broad. Widespread. Uncommon. RF on tree ferns, rocks or trees . . . . **H. rarum** R.Br.

6 Involucre bivalvate in the upper portion but obconic at the base Rhizome long-creeping, wiry, 0.3–0.8 mm diam., clothed sparsely especially at the nodes with hairs which are dark brown, 2–6 cells long and with a spur at the base. Stipes wingless, 3–11 cm long. Fronds 8–35 cm long including the stipes. Lamina 3-pinnate-pinnatifid, submembranous to herbaceous, deltoid or lanceolate to very broadly ovate; the apex usually deflexed. Ultimate segments flat, cultrate to narrowly oblong, obtuse, denticulate. Veins free, not reaching the apex of each segment. Sori terminal on the short ultimate segments, immersed at the base. Involucre 0.8–1.2 mm long, 0.5–1 mm broad, orbicular to oval, entire, obconic at the base, 2-valved almost to the base. Receptacle included or slightly exserted. Widespread. RF, terrestrial or epiphytic, often on boulders. . . . . **H. bivalve** G.Forst.
6 Involucre bivalvate deeply or to the base 7

7 Fronds lanceolate to ovate or oblanceolate to narrowly obovate, 1–2-pinnate-pinnatifid, 2.5–7 cm long including the stipes. Sori solitary on the upper side of each primary pinna on short branches close to the main rhachis. Involucre slightly immersed in a short stalk, suborbicular, with a narrow base; the margin denticulate. Widespread. RF, growing on rocks and trees; also sometimes in drier areas . . . . .
. . . . **H. cupressiforme** Labill.
7 Fronds rhomboidal or palmate, 1-pinnate or 1-pinnate-pinnatifid, 1–2.7 cm long including the stipes. Sori terminal on the ultimate segments. Involucre rounded or oblong, not immersed; the margin minutely spinulose-denticulate. Blue Mts; Illawarra ranges, pass above Kiama. RF. Uncommon . . . . .
. . . . **H. pumilum** C.Moore

2 Crepidomanes (C.Presl) C.Presl
14 species native Aust.; Qld, NSW, Vic., Tas.

One species in the area
Rhizome widely creeping, c. 0.2–0.3 mm diam., very densely clothed with orange hairs. Stipes not winged, 1–5.5 cm long, filiform, glabrous or with a few unbranched hairs. Rhachis broadly winged towards the

apex. Lamina 1-pinnate or with the lower pinnae often again imperfectly pinnate, membranous, pellucid, pale green, irregular in outline, 2.5–10.5 cm long (rarely up to 15cm) including the stipes, 0.5–6.5 cm broad. glabrous or with a few unbranched hairs. Primary pinnae 2–9 pairs, rhomboidal or flabellate; the apex obtuse, truncate or emarginate; margin crenate or sinuate, glabrous, not specialized. Ultimate segments ± oblong except the apical segment which is usually linear. Veins reticulate, very dark brown; the midrib flexuose; the ultimate segments pinnately veined. Sori 2–14 on each frond, usually solitary on each pinna, partly sunken in a short upper basal lobe, but in large fronds 2 sori each occur on the upper and lower margins of the pinna. Involucre (Fig. 13) 1.5–2 mm long, c. 0.3–0.5 mm broad, tubular; the mouth broadly dilated, entire or very slightly 2-lipped. Receptacle long-exserted. Widespread, common in Blue Mts and southern ranges. RF, forming densely matted patches on trees, logs or tree ferns . . . . **C. venosum** (R.Br.) Bostock

3 Cephalomanes C.Presl
5 species native Aust.; Qld, NSW, Vic., NT

One species in the area
Small ferns. Rhizome long-creeping, 1–1.5 mm diam., densely clothed with dark red bristle-like hairs. Stipes and main rhachis with 2 almost imperceptible ridges. Fronds 2–3-pinnate-pinnatifid, membranous, 4–21 cm long including the stipes, 1–4 cm (rarely up to 7 cm) broad. Veins dark, not reaching the margin. Sori 12–130 on each frond. Involucre tubular with the mouth truncate or slightly expanded, often 2-lipped, entire or slightly erose. Receptacle long-exserted. Widespread. RF, epiphytic on the trunks of trees or tree ferns . . . .**C. caudatum** (Brack.) Bostock

18 LINDSAEACEAE
Mostly terrestrial ferns, sometimes climbing or epiphytic, common in the tropics. Rhizome creeping or rarely climbing, dorsiventral, protostelic (lindsaeoid type) or solenostelic, clothed with brown, rather stiff, entire, non-peltate, mostly non-clathrate scales 2–4 cells broad at the base and intermixed with or grading into bristle-like scales. Stipes ± glabrous except near the base, not articulated to the rhizome, with 1–2 vascular strands. Lamina simple 1-pinnate to decompound, mostly uniform but ± dimorphic in several Australian species, anadromous, usually clothed with minute 2–3-celled simple hairs. Rhachises grooved above; the raised edges of the groove mostly continuous with those of the minor axes and decurrent along the basiscopic margin of the ultimate segments. Pinnae and pinnules not articulated (except in one Malayan genus), dimidiate or equilateral, cuneate or rhomboidal or trapeziform. Sori indusiate, almost marginal, terminal on simple veins or borne on a vascular commissure joining 2 or more vein-endings. Indusium true, opening towards the margin, attached at the base and/or the sides, linear or oblong or rather pouch-shaped. Veins free and forked or sometimes laxly anastomosing without free, included veinlets in the oblique areolae; a midrib usually absent. Paraphyses present in many species, hairlike, usually 2–3-celled. Sporangia mixed; the annulus straight or slightly oblique, composed of 8–22 thick-walled cells and 2–6 stomial cells; the pedicel long, with 3 rows of cells at least at the apex. Spores tetrahedral or bilateral, smooth or minutely warty, without a perispore. 6 gen., trop., temp.

1 Lindsaea Dryand. ex Sm.
Wedge Ferns
15 species native Aust.; all states and territories

Mostly small terrestrial ferns (as in the 4 species below) but some species epiphytic. Rhizome long- or short-creeping, usually protostelic (in a modified form), clothed with scales which are narrow, chestnut, composed of 2–4 rows of thick-walled cells and intermixed with bristle-like scales. Stipes not articulated to the rhizome. Fronds mostly 1-pinnate to pinnately decompound, rarely simple. Pinnae or pinnules glabrous, usually thin, flabellate or dimidiate or cuneate; the basiscopic edge much thickened. Veins free and dichotomously flabellate or laxly reticulate. Sori almost marginal (Fig. 16) or intramarginal, terminal on the veins, usually one to each ultimate segment but sometimes continuous along the margin of the frond, mostly on the upper and outer margins of the segments. Indusium attached at the base, opening towards the margin of the segment (Fig. 16). Sporangia usually 130–200 µm long, with narrow pedicels;

the annulus straight or slightly oblique, interrupted, including 9–14 thick-walled cells. Spores tetrahedral or bilateral, without a perispore.

Key to the species

1 Fronds 1-pinnate, linear, dimorphic; the fertile fronds narrower and taller than the sterile fronds . . . 2
1 Fronds 1-pinnate-pinnatifid at the base, 2-pinnate or 2-pinnate-pinnatifid 3

2 Stipe and rhachis dark purplish or red-brown or black. Primary pinnae flabellate (Fig. 16). Widespread. Open forests, heath or scrub (often *Leptospermum*). Sandy soil, near swamps or amongst grass in damp situations. *Screw Fern* . . . . **L. linearis** Sw.
2 Stipe (except near the base) and rhachis yellowish. Primary pinnae cuneate-flabellate to subdimidiate. Coast and Cumberland Plain. Heath; Open forests. Damp hillsides, in sandy or peaty soil often under dry grass. . . . . **L. dimorpha** F.M.Bailey

3 Fronds dimorphic, sometimes 1-pinnate-pinnatifid at the base. Stipe (except near the base) and rhachis yellowish. . . . . **L. dimorpha** (see above)
3 Fronds not dimorphic, 1-pinnate-pinnatifid or 2-pinnate to 2-pinnate-pinnatifid 4

4 Rhizome long-creeping. Fronds mostly subcoriaceous, rigid. Stipes mostly red-brown or black, very wiry. Ultimate segments narrowly obovate to very broadly obovate or oblong-cuneate, equilateral. Mainly Coast south of Sydney and lower Blue Mts Uncomon. RF ravines . **L. trichomanoides** Dryand.
4 Rhizome knotted, shortly creeping. Fronds submembranous or herbaceous , rather flaccid. Stipes yellow or yellow-brown, flexuose. Ultimate segments cuneate-truncate or sometimes flabellate, equilateral or somewhat unilateral. Widespread. Open forests, woodland and RF margins. Damp sandy soil or in crevices of Ss cliffs . . . . **L. microphylla** Sw.

19 OSMUNDACEAE

Terrestrial ferns with non-scaly, short, hard, erect trunks or rhizomes. Stipes with 2 stipular expansions at the base. Fronds uniform or dimorphic, pinnately compound (mostly bipinnate) with circinnate vernation. Veins free. Sporangia non-soral, exindusiate, large, subglobular, maturing simultaneously; the annulus represented by a group of thick-walled cells near the distal end of each sporangium, dehiscing by a ventral-distal slit; the pedicel short and thick. Spores tetrahedral, green, without a perispore. 3 gen., trop. to temp.

Key to the genera

1 Mature fronds coriaceous, non-transparent, light green, usually 0.6–2.4 m long including the stipes; stomates present. Sporangia in oval or oblong masses along the minor veinlets, later usually confluent and covering almost the whole of the lower surface of the fronds . . . . **1 TODEA**
1 Mature fronds membranous or herbaceous, transparent, dark green, usually 0.3–1.4 m long including the stipes; stomates absent. Sporangia clustered on the basal and middle regions of the minor veinlets in oblong or irregular patches, much less abundant than in *Todea* . . . . **2 LEPTOPTERIS**

1 **Todea** Willd. ex Bernh.
1 species native Aust.; Qld, NSW, Vic., Tas., S.A.

One species in the area

Arborescent fern up to 3 m high; caudex erect, covered with the broken bases of the old stipes. with 2–14 crowns of clustered and often stiffly erect fronds. Stipes not articulated to the caudex, pilose near the base with soft fawn glandular-tipped hairs, winged at the base with 2 broad stipular extensions. Fronds 2-pinnate (or 1-pinnate-pinnatifid in young plants), coriaceous, non-transparent, light green, usually 6–2.4 m long including the stipes; stomates present. Fertile pinnae borne towards the base of the lamina, not reduced. Veins free. Widespread. In crevices of sandstone cliffs or on creek banks in open forests; less commonly in RF. *King Fern* . . . . **T. barbara** (L.) T.Moore

2 Leptopteris C.Presl

2 species endemic Aust.; Qld, NSW

One species in the area

Subarborescent ferns usually under 1 m high, rarely up to 3 m high; caudex or rhizome 0.2–1 m high. Stipes not winged by an extension of the lamina, 20–45 cm long, 3–4 mm broad near the middle, often glaucous, tomentose, with light ferruginous, floccose hairs, winged at the base by 2 dark red-brown (later cartilaginous) stipular extensions. Main rhachis clothed with floccose hairs. Fronds 2-pinnate to 2-pinnate-pinnatifid, narrowly elliptic to elliptic; the juvenile fronds broadly ovate-deltoid to narrowly ovate-deltoid. Primary pinnae 8–30 pairs, opposite or almost so. Ultimate segments often bifid or trifid. Widespread. RF in rocky mountain gullies, usually in caves and often near waterfalls, always in very moist conditions . . . .**L. fraseri** (Hook. & Grev.) C.Presl

20 POLYPODIACEAE

Small to large epiphytic ferns (sometimes bracket-epiphytes) or rarely terrestrial ferns. Rhizome usually long-creeping, sometimes shortly creeping, clothed with hairs in 2 genera, otherwise bearing peltate-based, clathrate or non-clathrate scales which do not have stiff hairs, solenostelic or dictyostelic, often with black sclerenchymatous strands in the cortex. Stipes in 2 rows if the rhizome is creeping, generally articulated to the rhizome. Fronds uniform to very dimorphic; humus-collecting scale-leaves present in some genera. Lamina simple or lobed or pinnate or dichotomously branched, glabrous or often clothed with peltate scales or stellate hairs. Pinnae articulated or not articulated to the main rhachis. Veins mostly anastomosing with free vein-endings in the areolae (Fig. 18 & 19) or very rarely free, often ending in hydathodes. Sori exindusiate, round and superficial or immersed; borne either at the vein-junctions or on the free veinlets usually in the areolae; otherwise elongated and parallel to the main veins or to the margin; sometimes acrostichoid over the whole fertile surface or in portions specialized as fertile. Sporangia glabrous or rarely setulose; annulus interrupted or complete, composed of 12–14 (rarely up to 24) thick-walled cells, 5–10 thin-walled cells and a stomium which is 2-celled or rarely undifferentiated; pedicel with 2–4-rows of cells. Paraphyses absent or simple, more usually clathrate, umbrella-shaped, stellate or with swollen apices. Spores bilateral, monolete, mostly smooth or tuberculate, often hyaline, without a perispore. C. 30 gen., mainly trop. to subtrop.

Key to the genera

1 Large bracket-epiphytes with short thick rhizomes. Sori confluent, forming large patches 5–50 cm diam. on specialised areas of the fertile fronds. Fronds clothed with stellate hairs, markedly dimorphic; the outer scale-leaves sterile, horizontally spreading, orbicular-cordate or orbicular-reniform, eventually brown and papery; inner fronds fertile, erect or drooping, forked repeatedly, ± coriaceous . . . .
. . . . **1 PLATYCERIUM**
1 Epiphytic or rupestral ferns with long-creeping rhizomes. Sori round or, if elongated and confluent, then 2–6.5 cm long. Scale-fronds absent 2

2 Lamina densely tomentose with appressed, persistent, stellate hairs. Paraphyses stellate. Scales of the rhizome non-clathrate. . . . . **2 PYRROSIA**
2 Lamina glabrous or with a few clathrate scales along the veins. Paraphyses absent or with enlarged apices, never stellate. Scales of the rhizome clathrate at least towards the middle 3

3 Stipe >20 cm long. Lamina deeply lobed, arching. . . . . **5 PHLEBODIUM**
3 Stipe <20 cm long. Lamina simple or pinnatifid 4

4 Free included veinlets in the areolae of the lamina absent except in rare cases in the Australian species; the veins almost hidden. Lamina simple . . . . **3 DICTYMIA**
4 Free included veinlets present in the areolae of the lamina (Fig. 18). Lamina simple or pinnatifid . . . .
. . . .**4 MICROSORUM**

1 Platycerium Desv.

4 species native Aust.; Qld, NSW

One species in the area

Large bracket epiphytes. Rhizome shortly creeping, branched, clothed with broad, costate scales. Fronds markedly dimorphic. Soral patch covering all or the greater part of the ultimate segments of the fertile fronds. Fertile fronds often erect, 25–100 cm long or longer, distinctly stipitate, up to thrice dichotomously forked. Sterile nest-leaves sessile, rigid, 10–30 cm diam.; the margins entire, sinuate or obtusely lobed. Coast and gullies in adjacent plateaus. RF. Epiphytic on trees or growing on rocks. *Elkhorn* . . . .
. . . . **P. bifurcatum** (Cav.) C.Chr.

2 Pyrrosia Mirb.

Felt Ferns

4 species native Aust.; Qld, NSW

Epiphytes on trees or growing on rocks. Rhizome long-creeping or shortly creeping, dictyostelic, densely clothed with persistent, non-clathrate scales which are fimbriate or entire and usually have a dark central spot on the expanded, peltately affixed base. Stipes articulated to the rhizome; the prolongation clothed with the same type of scales as those of the rhizome; the stipes short or sometimes up to twice as long as the lamina, usually ± clothed with caducous, stellate hairs. Lamina usually simple, mostly entire, thick, coriaceous, uniform or more rarely dimorphic, linear or elliptic or ovoid or spathulate or hastate or oblanceolate, ± clothed with a dense, often interwoven tomentum of stellate-peltate hairs; the base of the lamina cuneate, sometimes decurrent on the stipes. Veins complex and reticulate with free vein-endings, mostly obscure and hidden in the thick lamina. Sori exindusiate, superficial or partly submerged, occurring in numerous closely packed rows or more rarely 1–2-seriate, rounded or rarely oblong, often confluent in age. Paraphyses stellate. Sporangia rounded or cuneate; the annulus including 14–24 thick-walled cells; the pedicel elongated, composed of 2–3 rows of cells. Spores bilateral, rugose or minutely tuberculate or smooth, yellow or hyaline. Hydathodes pitted or often punctate, numerous on the upper surface or near the margin, absent in some species.

Key to the species

1 Sori occupying up to three-quarters of the abaxial surface of the fertile fronds, irregularly but densely crowded in 1–4 rows on each side of the midrib, slightly confluent in age, but retaining their identity longer than the sori of *P. confluens*. Fronds markedly dimorphic; sterile fronds frequently almost orbicular, 1–2.5 cm (sometimes up to 8cm) long including the stipes, 0.3–1.4 cm broad, often shorter than the fertile fronds. Scales of the rhizome squarrose, entire, thin-textured, pale-ferruginous. Widespread. RF. Usually covering rocks or tree trunks. *Rock Felt Fern* . . . . **P. rupestris** (R.Br.) Ching
1 Sori usually restricted to the extreme distal portions of the fertile fronds as 2 oblong or oval coenosori, occasionally fusing to form a single round patch but rarely a single row of unfused sori on each side of the midrib. Fertile and sterile fronds rather similar, mostly 3–7 cm long including the stipes (but ranging from 1.5–18 cm long), 0.5–1.5 cm broad; sterile fronds never orbicular. Scales of the rhizome appressed, fimbriate, thick, dark chestnut-coloured in the young condition, later dark brown or dark grey-brown or black; border pale, hyaline. Hunter River Valley. RF. On trees and logs . . . .
. . . . **P. confluens** (R.Br.) Ching var. **confluens**

3 Dictymia J.Sm.

1 species native Aust.; Qld, NSW

One species in the area

Epiphytic or rupestral ferns. Rhizome widely creeping, 2–6 mm diam., densely clothed with rather persistent, peltate-based, clathrate, squarrose, long-acuminate scales. Fronds 4–60 cm long including the stipes, 5–20 mm broad. Lamina dark green, simple, uniform, very thick; the base gradually narrowing into the stipe; the margin undulate or entire. Veins almost hidden except the prominent midrib. Sori narrowly oblong to oval, rarely orbicular, half-way between the midrib and margin. North from Blue Mts;

occasionally Coast and Cumberland Plain (Wiseman's Ferry). RF. Growing on rocks and trees, mostly in the ranges . . . .**D. brownii** (Wilkstr.) Copel.

4 **Microsorum** Link

8 species native Aust.; all states and territories except S.A.

Epiphytes on trees or growing on rocks. Rhizomes long-creeping, clothed with scales which are persistent or caducous, peltate-based, entire or toothed, narrowly lanceolate or lanceolate to broadly ovate or suborbicular, clathrate throughout or towards the middle, sometimes with a light brown border. Stipes smooth, unwinged or winged almost to the base, articulated to the rhizome (except in a few non-Australian species); phyllopodia clothed with scales similar to those on the rhizome. Fronds simple and entire or 1-pinnatifid, sometimes with both types of fertile fronds on the same rhizome (as in the 2 Australian species listed here). Lamina membranous to coriaceous, sometimes translucent, glabrous or with a few peltate scales along the main veins. Segments of the fronds 1–20 pairs, often narrowed towards the base; a terminal segment also present; margin entire or sinuate, never serrate, often slightly thickened. Veins prominent to almost hidden, complex, anastomosing with free vein-endings in the numerous areolae; the included veinlets facing in all directions, ending in hydathodes. Sori exindusiate, deeply pustulate, slightly immersed or superficial, mostly round or oval, rarely slightly elongated along the veins, in 1–3 rows with the sori (in the 2 species listed here) in the row closest to the midrib each borne on a distinctly thickened acroscopic veinlet which proceeds from the costa at the base of or near the base of a main lateral vein (Fig. 18 & 19). Paraphyses with slightly thickened apices. Sporangia with an annulus including 14–16 thick-walled cells. Spores bilateral, monolete.

Key to the species

1 Scales of the rhizome squarrose, persistent, dark brown or purplish-brown, narrowly lanceolate to lanceolate. Rhizome (when living) tough, wiry. Lamina dark green, membranous to herbaceous, with a distinct musk scent when fresh or freshly dried. Segments of the pinnatifid fronds 1.2–10 cm long. Fronds 3–50 cm long including the stipes. Widespread W to Widden Valley. RF. Climbing to a considerable height up tree trunks or trailing on rocks. (Fig. 18). *Fragrant Fern* . . . . **M. scandens** (G.Forst) Tindale

1 Scales of the rhizome appressed, deciduous, black or purplish-brown or dark to light grey, with a brown or light brown border, narrowly lanceolate or ovate. Rhizome (when living) fleshy, often very glaucous. Lamina light green, herbaceous to coriaceous, unscented. Segments of the pinnatifid fronds 1.8–15 cm long. Fronds 3–50 cm long including the stipes. Widespread. RF. Scrambling over rocks or climbing tree trunks or tree-ferns. (Fig. 19). *Kangaroo Fern* . . . . **M. pustulatum** (G.Forst.) Large, Braggins & P.S.Green ssp. **pustulatum**

5 **Phlebodium** (R.Br.) J.Sm.

1 species naturalized Aust.; NSW

One species in the area

Rhizome thick, 8–15 (30) mm diam., densely covered with golden-brown scales. Lamina bright green to glaucous, arching deeply lobed c. 40 cm long. Sori circular, in rows either side of the midrib. Epiphytic in its natural habitat it is recorded as growing as a ground cover in the Vaucluse area. Introd. from Central and South America. *Rabbits-foot fern* . . . . ***P. aureum** (L.) J.Sm.

21 PTERIDACEAE

Terrestrial ferns usually in forests, often in rock-crevices. Rhizomes erect or creeping, dictyostelic or solenostelic (often with perforations) clothed with non-peltate, mostly non-clathrate scales which often have a thinner, paler border. Stipes not articulated to the rhizome. Main rhachis grooved above. Lamina uniform or subdimorphic, 1-pinnate to decompound (usually pinnately divided, rarely digitately), coriaceous to herbaceous or sometimes almost membranous. Pinnae not articulated to the rhachis. Sori with a false indusium or exindusiate, linear, submarginal, intramarginal or rarely superficial, borne on a ± continuous vascular commissure connecting the vein-endings, sometimes avoiding the apices and the sinuses of the segments. Indusium false, consisting of a reflexed, often scarious margin of the frond. Veins free or anastomosing as a costal row or over the whole surface; free vein-endings in the areolae absent. Paraphyses often present, usually a row of uniform cells. Sporangia mixed; the annulus vertical, interrupted, with 16–34 indurated cells; the pedicel usually long, with 3 rows of cells. Spores tetrahedral or sometimes bilateral, pale to almost black, smooth, sculptured or tuberculate, without a perispore. 5 gen., trop, warm temp. and south temp.

1 Pteris L.

14 species in Aust. (11 native, 3 naturalized); all states and territories

Terrestrial ferns. Rhizome thick, shortly creeping, with a dictyostele or a complex solenostele, clothed with scales which are non-peltate but attached by a broad base. Stipes grooved on the upper surface, yellow or red or bright brown in the Australian species. Lamina 1-pinnate to pinnately decompound, usually with the pinnae divided near the base, glabrous or sometimes with articulate hairs. Secondary rhachises of species with decompound fronds deeply grooved on the upper surface, with raised edges; the basiscopic margins of the latter being continuous with the edges of the groove of the main rhachis; short spines often present at the base of the costules. Veins free below the sori (as in the 3 species below) or anastomosing without free vein-endings in the areolae. Sori linear, continuous along the margin of the pinnae or pinnules except at the apices, borne on a vascular commissure connecting the ends of the veins. Indusium consisting of the recurved modified margin of the pinnae or pinnule, opening introrsely. Paraphyses present, often numerous. Sporangia having long pedicels with 3 rows of cells; the annulus including 16–34 thick-walled cells. Spores mostly tetrahedral or sometimes bilateral, smooth, light-coloured to almost black, covered with tubercles or often elaborately sculptured.

Key to the species

1 Fronds 1-pinnate or imperfectly bipinnate with some of the lower primary pinnae divided into a few, ribbon-like, sessile segments. Veins free, transverse to the midrib, usually simple or once forked. . . . 2

1 Fronds 2–4-pinnate; the pinnules never ribbon-like. Veins reticulate or free and once forked. 4

2 Fronds 1-pinnate. Primary pinnae decreasing in size towards the base of the frond, very shortly stalked, not decurrent on the main rhachis. Plants 40–75 cm high. Coast and adjacent plateaus. Hillsides on granite or limestone, often near streams. Uncommon. *Chinese Brake*. . . . . **P. vittata** L.

2 Fronds imperfectly bipinnate with some of the lower primary pinnae divided into a few ribbon-like sessile segments. Primary pinnae increasing in size towards the base of the frond; the upper and middle primary pinnae sessile, decurrent 3

3 Lowest primary pinnae 20–35 cm long. Fronds scarcely dimorphic; the sterile fronds very slightly broader; the sterile pinnules ribbon-like. Plants 0.6–2.6 m high. Widespread. RF, often on basalt, usually amongst rocks. *Jungle Brake* . . . . **P. umbrosa** R.Br.

3 Lowest primary pinnae 5–16 cm long. Fronds markedly dimorphic; the sterile fronds shorter and broader; the sterile pinnules oblong or ovate. Plants 20–30 cm high. Coast and adjacent plateaus especially near Sydney. In railway cuttings. Introd. from Qld. . . . . ***P. ensiformis** Burm.f.

4 Veins of the ultimate segments free and forked below the sori, not forming a reticulum. Plants 0.6–1.3 m high. Widespread. RF and open forests. Often in rocky gullies, mostly along the banks of creeks or under rock-ledges on shale, limestone or basalt . . . . **P. tremula** R.Br.

4 Veins of the ultimate segments reticulate. Plants 0.6–2.6 m high. RF and tall open forests. Blue Mts Often near creeks or waterfalls, on basalt or Ss . . . . **P. comans** G.Forst. sens. lat.

22 SCHIZAEACEAE

Terrestrial or climbing ferns. Rhizome erect or creeping, radial or dorsiventral, clothed with simple hairs or flattened scales, protostelic, solenostelic or dictyostelic. Stipes and rhachises slender, erect or twining, with repeated equal-forking or sympodial or dichopodial branching. Lamina simple, bipinnate, forked or flabellately branched with palmate or pinnate segments, mostly uniform, rarely dimorphous, clothed with simple hairs (sometimes with glandular apical cells) or with flattened scales; the vernation circinnate.

Veins free and forked or rarely coarsely reticulate. Sporangia large, pyriform or obovoid, curved or radial, sessile or shortly pedicellate; the annulus apical and transverse, composed of a few large cells with a definite stomium, with dehiscence by a vertical slit, borne in monangial sori on lateral or apical sporangiophores, in 2 or 4 rows in the Australian genera, marginal in origin but forced into a superficial position by the protective flanges which serve as a false indusium. Spores bilateral or tetrahedral, without a perispore. 4 gen., mainly trop. and southern warm temp.

1 Schizaea Sm.

Comb Ferns

5 species native Aust.; all states and territories

Small terrestrial ferns growing on poor sandy soils in open forests and swamps, sometimes found near waterfalls. Rhizome creeping or ascending, protostelic (an advanced type); the apex covered with brown hairs. Stipes wiry, scarcely distinguishable from the lamina. Lamina simple, forked or flabellately dichotomous, terete, angular or flattened, often grasslike. Segments 1-veined. Sporangiophores at the apex of the lamina or its segments. Veins free, each segment 1-veined. Sporangia brown, in 2 or 4 rows, sometimes mixed with hairs, protected when young by the reflexed margins of the lobes; the annulus complete, distal, operculiform. Spores bilateral.

Key to the species

1 Sterile fronds simple, flat, without revolute margins, smooth, glossy, 1–2.5 mm broad, 7–10 pairs of sporangia on a soriferous pinnule. Soriferous heads single, on simple flattened fronds. Coast south of Gosford and adjacent plateaus; Blue Mts Near waterfalls on damp rocks or in moist sandstone caves . . . . **S. rupestris** R.Br.

1 Sterile fronds divided or, if simple, then 0.3–1 mm broad and terete or slightly flattened with revolute margins 2

2 Sterile fronds simple, terete or slightly flattened towards the apex, 0.3–0.5 mm broad, 5–9 pairs of sporangia on a soriferous pinnule. Soriferous heads single, on simple smooth or asperous fronds. Coast. Uncommon. Wet peaty soil or on hillocks in bogs . . . . **S. fistulosa** Labill.

2 Sterile fronds forked to much divided (with up to 175 branchlets) or, if undivided, then 0.5–1 mm broad and slightly flattened with revolute margins. 10–20 pairs of sporangia on a soriferous pinnule 3

3 Fertile fronds dichotomously divided 3 – 6 times, with 12–60 heads in a broad fan-shaped arrangement, borne on slightly flattened, very asperous branches. Sterile fronds always branched above the middle, with 22–175 slightly flattened very scabrous branches. Coast and adjacent plateaus north of Nowra. Open forests, sandstone scrub or heath or open woodland, usually on sandy soil **S. dichotoma** (L.) Sm.

3 Fertile fronds simple or dichotomously divided 1 – 2 times, with 1–7 heads in a fan-shaped arrangement. Sterile fronds usually branched below the middle 4

4 Sterile fronds unbranched or absent. Fertile fronds unbranched or dichotomously divided once rarely twice, the forkings more irregular than in *S. dichotoma*. Stipe smooth to rough. Widespread. Open forests, sandstone scrub or heath, among rocks, often in sandy soil . . . .**S. bifida** Willd.

4 Sterile fronds dichotomously divided 1 – several times. Fertile fronds dichotomously divided 1 – 5 times. Stipe rough. Moist areas. . . . . **S. asperula** N.A.Wakef.

23 THELYPTERIDACEAE

Moderately large terrestrial ferns, rarely climbing and proliferous. Rhizome dictyostelic, usually creeping but sometimes short and erect, non-dorsiventral, sparsely clothed with non-peltate, non-clathrate scales which bear unicellular needle-like or more rarely branched hairs on the margins and sometimes on the surfaces. Stipes usually sparsely scaly near the base with the upper and middle portions hairy, not articulated to the rhizome, with 2 vascular bundles near the base but these uniting into a U-shaped bundle higher in the stipe. Fronds uniform or rarely subdimorphic. Lamina mostly 1-pinnate-pinnatifid, rarely bipinnate or more compound, oblong or elliptic, never broadly deltoid-quinquangular, often clothed with simple, whitish, unicellular or rarely 2-celled, needle-like hairs, frequently with round, glandular hairs, reduced pinnae often present towards the lower portion of the lamina, aerophores often occurring at the base of each pinna. Veins all free or with veinlets from adjacent lobes confluent into an excurrent vein which usually runs to a callous sinus (Fig. 22). Sori orbicular or rarely elongated, borne dorsally on the veins or rarely terminal. Indusium reniform-orbicular or rarely oblong, sometimes absent, mostly clothed with needle-like hairs and/or glandular hairs. Sporangia smooth or bearing stiff hairs. Paraphyses often present. Spores bilateral mostly with a perispore. c. 29 gen., mostly trop. and subtrop.

Key to the genera

1 Lower primary pinnae scarcely shorter or sometimes longer than the others . . . . . **1 CYCLOSORUS**
1 Lower primary pinnae gradually much shorter and more distant, often reduced to auricles . . . .
. . . . .2 CHRISTELLA

1 Cyclosorus Link
1 species native Aust.; Qld, NSW, WA, NT

One species in the area

Terrestrial ferns up to 1 m high. Fronds coriaceous, erect. Lower primary pinnae scarcely shorter or sometimes longer than the others. Lobes of the primary pinnae ovate or ovate-triangular; the apices acute or mucronate. Sori usually forming a crowded row close to the margin of each ultimate segment, confluent in age. Coast north from Royal National Park. Fresh water swamps. . . . .
. . . . .C. **interruptus** (Willd.)H.Itô

2 Christella H.Lév.
5 species native Aust.; all states and territories except Tas.

Terrestrial ferns often in swamps or along creeks in RF. Rhizomes usually erect or suberect or creeping, more rarely long-creeping, clothed with scales which are narrow with many superficial setae. Fronds usually with 1–5 pairs of pinnae usually reduced in size towards the lower part of the frond, mostly auriculate on the acroscopic base; pinnae ± lobed, clothed with erect, acicular hairs and sometimes orange-red or orange-yellow glandular hairs, at least the basal pair of veins from adjacent lobes anastomosing to form an excurrent vein which runs to the callous-sinus (Fig. 22). Sori round; indusium orbicular-reniform, mostly setiferous; sporangia without setae or glands close to the annulus but always bearing oblong glandular hairs on the sporangial stalk. Spores dark, tuberculate or ridged, without a perispore.

Key to the species

1 Capitate hairs present on the costules. Hairs on the lower surface of the costae and costules more than 0.2 mm long. Rhizome sometimes erect. Lower Hunter River Valley. Uncommon. RF or open forests, along watercourses, usually in shady situations . . . . C. **hispidula** (Decne.) Holttum
1 Capitate hairs absent from the costules. Hairs on the lower surface of the costae and costules 0.2 mm long or less. Rhizome shortly creeping. Coast and adjacent plateaus. RF or open forests, along watercourses or near waterfalls. . . . . C. **dentata** (Forssk.) Brownsey & Jermy

24 VITTARIACEAE

Epiphytic or rarely rupestral, very small to medium-sized ferns. Rhizome suberect or creeping, clothed with clathrate, often metallic scales and dark roots covered with hairs. Fronds simple, entire (in *Vittaria*), forked or pinnate. Stipes very short, mostly poorly developed. not articulated to the rhizome. Veins simple or forked and free, or pinnate and joined in a submarginal commissure (in *Vittaria*) or anastomosing. Epidermis with spicular cells. Sori mostly in simple or branched lines, often in grooves, otherwise round or scattered over the lamina, exindusiate but protected by a false indusium composed of the recurved laminal margin or the flap-like margins of the grooves. Sporangia with a vertical annulus interrupted by the pedicel; stomium 4-celled; pedicel with 1–2 rows of cells except at the apex. Spores smooth, mostly hyaline, tetrahedral or bilateral. 6–9 gen., mainly pantrop. to subtrop.

1 **Vittaria** Sm.

2 species native Aust.; Qld, NSW, NT

One species in the area

Rhizome creeping, densely covered with dark roots and clathrate, metallic scales. Fronds simple, linear, ribbon-like, coriaceous, 20–80 cm long, usually 3–6 mm wide; lateral veins very oblique. Sori linear, immersed in submarginal grooves, exindusiate. Paraphyses branched; apices resembling inverted cones. Spores bilateral. Watagan Mts RF. Epiphytic on boulders. Uncommon.. . . . .**V. elongata** Sw.

25 MARSILEACEAE

Subaquatic ferns. Rhizome long-creeping, filiform or thick and woody, clothed with hairs. Sterile fronds thread-like or with 2 or 4 leaflets at the apex of a long stalk; the vernation circinnate. Spores of 2 kinds; megaspores single in each megasporangium; microspores numerous in each microsporangium. Sporangia numerous in sessile or stalked, usually globose to oblong sporocarps. 3 gen., trop. to temp.

Key to the genera

1 Sterile lamina with 2 pairs of opposite leaflets, resembling a 4-leafed clover . . . . **1 MARSILEA**
1 Sterile lamina linear, filiform or grass-like . . . .**2 PILULARIA**

1 **Marsilea** L.

Nardoo

8 species native Aust.; all states and territories except Tas.

Growing in grassland on the edges of swamps, weirs, rivers or waterholes, floating on shallow water or rooting in the mud on the banks, usually occurring on land which is subject to periodic inundation. Sterile fronds borne at intervals along the rhizome on long stalks; the lamina composed of 2 pairs of opposite, obovate, glabrous or hairy leaflets. Veins fine, very numerous, radiating from the base of the leaflet, reticulate but without free vein-endings in the long narrow areolae. Sporocarps borne on stalks which arise at intervals along the rhizomes, solitary or in groups, hard, woody, usually densely hairy, often ribbed, globose or oblong-ovoid or ellipsoid or almost square, usually bearing 1–2 basal teeth which are conical, flat-topped or hooked, sometimes without teeth. Sori attached to a gelatinous ring which is later exserted when the sporocarp bursts.

Key to the species

1 Sporocarps without teeth. Stalks of the sporocarps usually branched once to thrice. Leaflets of the terrestrial forms glabrous or with a few hairs at the base, broadly obovate-cuneate to very broadly obovate-cuneate, 2.4–4.2 cm long, 2.3–4.3 cm broad in the floating forms but narrowly obovate-cuneate to very broadly obovate-cuneate, 0.4–1.9 cm long, 0.3–2.2 cm broad in the terrestrial forms. Widespread. Edges of ponds and lagoons, on river banks and in deep water. . . . . . . **M. mutica** Mett.
1 Sporocarps with 1 or more usually 2 teeth. Stalks of the sporocarps very rarely branched. Leaflets of the terrestrial forms often very densely hirsute. Stalks shorter than the sporocarps 2

2 Sporocarps with 1–2 unequal teeth; the upper tooth truncate; the lower often undeveloped. Leaflets usually narrowly oblanceolate to oblanceolate0.1–1 cm long, 0.07–0.45 cm broad. Cumberland Plain. Low-lying land subject to inundation . . . . **M. costulifera** D.L.Jones
2 Sporocarps with 2 teeth which are equal in length. Leaflets very broadly obovate-cuneate to narrowly obovate-cuneate, 0.8–2.1 cm long, 0.5–1.3 cm broad. Cumberland Plain. In swampy land, on the edge of pools and lagoons . . . . **M. hirsuta** R.Br.

2 Pilularia L.

1 species native Aust.; NSW, Vic., Tas., S.A., WA

One species in the area

Growing amongst grasses and small sedges in soft mud at the edges of marshes, lagoons or wayside pools. Rhizome filiform, creeping under the water, often very elongated, with roots at each node. Sterile fronds bright green, simple, subulate, linear, 0.9–8 cm long, 0.2–0.6 mm broad, usually several fronds clustered at each node; the young leaves clothed with a few, simple, fawn hairs. Sporocarps sessile or borne on short, erect, or recurved stalks rarely more than 8 mm long, spherical, hard and woody when ripe, 2–4 mm diam., rather densely pubescent with silky fawn hairs. Conceptacle 4-celled. Cumberland Plain. At the edge of ponds, in the margins of marshes and seasonally dry depressions. Endangered. *Austral Pillwort*. . . . . **P. novae-hollandiae** A.Braun

26 AZOLLACEAE

Very small, floating ferns. Rhizome horizontal, zigzag. Fronds in 2 alternate rows; the vernation circinnate. Sori enclosed by an indusium, borne on the lower lobes of the leaves. Megasporangia and microsporangia present. 1 gen., trop. to temp.

1 Azolla Lam.

2 species native Aust.; all states and territories

Aquatic ferns, often forming a carpet on the surface of slow-moving or stagnant water. Plants ovate or triangular, usually 0.8–6 cm long, 0.8–4 cm broad. Leaves in 2 rows, very numerous, up to 1.5 mm long, imbricate, often papillose, unequally 2-lobed; the upper lobe with a colony of the cyanobacteria *Anabaena* in the central cavity; the lower lobe thin, submerged, bearing in the axils 2–4 megasporocarps or microsporocarps or sometimes a sorus of each type. Microsporocarps ovoid or globular, up to 2 mm long, protected by a membrane-like indusium which encloses a cluster of 7–100 orbicular-headed microsporangia, each of which is borne on a long narrow pedicel. Microsporangia containing 4–10 massulae which bear few to many glochidiate or non-glochidiate, septate or non-septate processes on one or all sides. Microspores 32–64, occurring in groups within the massulae. Megasporocarps oblong, with a conical apex, protected by a membrane-like indusium. Megaspores solitary in the megasporocarps, consisting of a rounded basal portion which is smooth or pitted or with raised hexagonal markings, surmounted by 3 or 9 floats.

Key to the species

1 Roots feathery in mature plants, sometimes simple in young specimens. Plants regularly pinnate. Massulae of the microsporangia conical, with rounded bases; 1–6 weak, non-barbed, non-septate processes borne only on one side. Megasporangia each with 9 apical swimming floats. Coast and adjacent plateaus; Cumberland Plain. Floating on slow-moving or stagnant water . . . **A. pinnata** R.Br.
1 Roots simple. Plants irregularly pinnate. Massulae of the microsporangia spherical, with numerous barbed processes borne on all sides. Megasporangia each with 3 apical swimming floats. Widespread. Floating on slow-moving or stagnant water. . . . . **A. filiculoides** Lam.

27 SALVINIACEAE

Small floating ferns. Rhizome hairy, rootless, bearing leaves in whorls of 3; the "water leaves" simulating roots. Vernation straight. Sori enclosed by an indusium, borne on the submerged "water-leaves". Megasporangia and microsporangia present. 1 gen., trop. to warm temp.

1 Salvinia Ség.

1 species naturalized Aust.; all states and territories except Tas.

One species in the area

Aquatic fern often forming vast carpets on slow-moving water. Plants c. 5–20 cm long. Foliar leaves almost orbicular, deeply cordate at the base, 1.5–3 cm long, 1.5–3 cm broad, clothed on the upper surface with hairs surmounted by a basket-like structure of 4 incurved cells joined at the apex; the veins very close together. Coast and Cumberland Plain. Floating on the surface of the water in dams, ponds and lagoons. Introd. from S. America. . . . . *S. molesta D.S.Mitch.

3 CYCADOPHYTA

28 ZAMIACEAE
9 gen., trop. to temp.

1 Macrozamia Miq.
Burrawang
42 species endemic Aust.; Qld, NSW, WA, NT

Dioecious perennials with a stout fleshy stem which only rarely appears above the ground in local species. Leaves pinnately compound; pinnae linear, coriaceous. Male (pollen or microsprore) cones and female (ovulate or megaspore) cones borne severally on each plant at intervals of 2 or more years. Male cones with numerous spirally arranged imbricate microsporophylls each with numerous pollen sacs (microsporangia) on the undersurface. Female cones with much larger and fewer megasporopylls each with 2 ovules. After pollination, but before fertilization, the ovules enlarge and become red or orange or more rarely yellow. The cones fall to pieces while the embryo is immature and the seeds ripen slowly on the ground.

Key to the species

1 Pinnae progressively reduced in size and becoming spine-like towards the base of the leaf; bases of the pinnae never red. Leaves mostly 20–100 in number, 0.7–2.5 m long; rhachis not twisted. Stems mostly sunken, except in shallow soils where they are sometimes 1–2 m high. 2

1 Pinnae all about the same size or abruptly reduced towards the base. Leaves 2–12 in the crown, up to 100 cm long. Stems sunken in the ground. 3

2 Seed more than 3 cm long. Leaves 1.5–2.5 m long, dark green. Male cones cylindrical, 20–45 cm long at maturity, 6–12 cm diam. Female cones cylindrical, 20–50 cm long, 10–20 cm diam. Pollination Oct.; fertilization Jan.–Feb. Seeds yellow or orange Coast and adjacent plateaus and eastern slopes of Blue Mts DSF; WSF . . . . **M. communis** L.A.S.Johnson

2 Seed less than 3 cm long. Leaves to 1.5 m long, paler green than *M. communis*. Male cones cylindrical, 20–45 cm long at maturity, 6–12 cm diam. Female cones cylindrical, 20–50 cm long, 10–20 cm diam. Pollination Oct.; fertilization Jan.–Feb. Seeds yellow or orange Coast in ranges in north of region. DSF. Ss. and sands . . . . **M. reducta** K.D.Hill & D.L.Jones

3 Rhachis very twisted (360° or more). Pinnae ± spreading with a pale or reddish slightly callous base; petiole rounded or flattened. Male cones cylindrical to ellipsoid, 8–25 cm long, 4–6 cm diam. Female cones ovoid to cylindrical, 10–25 cm long, 7–9 cm diam. Seeds c. 3 cm long, red when ripe. Northern parts of coast; Hunter River Valley. Open forest . . . . **M. flexuosa** C.Moore

3 Rhachis straight or twisted through an angle of up to 180°. 4

4 Pinnae secund, rising almost vertically from the rhachis, sometimes glaucous, with pink or pale callous base; rhachis recurved at the tip; petiole concave above. Male cones cylindrical to ellipsoid, 15–20 cm long, 4–5 cm diam. Female cones cylindrical to ovoid, 15–25 cm long, 7–9 cm diam. Seeds c. 3 cm long, red when ripe. Blue Mts Open forests . . . . **M. secunda** C.Moore

4 Pinnae spreading, with a red orange or pink callous at the base; petiole rounded or flat above. Male cones cylindrical or ellipsoid, 15–20 cm long, 7–9 cm diam. Female cones ovoid, 12–25 cm long, 5–10 cm diam. Seeds c. 3 cm long, orange-red when ripe 5

5 Pinnae with a terminal spine. Near Mountain Lagoon, Blue Mts D.S.F. Ss.. . . . .
. . . . **M. elegans** K.D.Hill & D.L.Jones
5 Pinnae without terminal spine. Widespread. Open forest on laterite . . . . . . **M. spiralis** (Salisb.) Miq.

4 PINOPHYTA

Key to the families

1 Leaves on older branches different from leaves on the juvenile branches . . . . **29 ARAUCARIACEAE**
1 Leaves all ± the same 2

2 Leaves 4–30 cm long,. 3
2 Leaves less than 1 cm long, often reduced to decurrent scales. Seeds in cones 4

3 Leaves acicular, borne 2–3 together near the top of dwarf-branches. Seeds in woody cones . . . .
. . . . .31 PINACEAE
3 Leaves flat, alternate. Seeds solitary or paired on a fleshy receptacle. . . . . . . **32 PODOCARPACEAE**

4 Spreading shrubs. Seed-cones c. 3 mm long, scarcely woody. . . . . **32 PODOCARPACEAE**
4 Erect trees or shrubs. Seed-cones 2–3 cm diam., becoming woody . . . . .30 CUPRESSACEAE

29 ARAUCARIACEAE

3 gen. cosmop.

One genus in the area

1 Wollemia W.G.Jones

1 species endemic Aust.; NSW

Tall tree to 40 m high with regular whorled branches. Bark densely covered with nodules. Leaves on juvenile shoots dull with and obtuse apex and paler than the adult leaves which are dark green, ± glossy and have an acute apex. Under surface of leaves ± glaucous. Male cones to 11 cm long. Female cones spiny, to 8 cm diam. Seed winged. Wollomi Nat. Park. Sheltered gullies. Endangered. *Wollomi Pine.* . . . .
. . . . **W. nobilis** W.G.Jones, K.D.Hill & J.M.Allen

30 CUPRESSACEAE

Monoecious or dioecious trees or shrubs. Leaves opposite, alternate or whorled, linear, flattened, acicular or scale-like. Male cones axillary or terminal with 2–9 microsporangia per scale. Female cones terminal, 1–many fertile scales. Ovules 1–12 per scale. Seed winged or not. Embryo with usually 2 but can be up to 15 cotyledons. 29 gen. world-wide.

Key to genera

1 Adult leaves decussate . . . . **1 CUPRESSUS**
1 Adult leaves in whorls of 3, rarely 4 when leaves juvenile . . . . **2 CALLITRIS**

1 Cupressus L.

5 species naturalized Aust.; all states and territories

One species in the area

Monoecious tree to 20 m high. Juvenile leaves decussate and spreading. Adult leaves, decussate, decurrent and adnate to the stem. Male and female cones solitary, terminal. Male cone scale with 2–6 microsporangia. Female cone scales with 1–20 ovules. Seeds narrowly winged. Occasionally naturalized near plantings. *Mexican Cypress* . . . . ***C. lusitanica** Mill.

2 Callitris Vent.

13 species endemic Aust.; all states and territories

Tall shrubs or small trees, ± columnar in shape. Leaves in whorls of 3 (rarely 4 when leaves juvenile), in the young plant free and linear but becoming adherent to, or decurrent on, the internode of the stem or branchlet so that only the scale-like tip of the leaf remains free. Male cones usually terminal, solitary or several together, cylindrical ovoid or obovoid. Female cones on short thick lateral branches on young main stems, with 6 scales in 2 whorls; scales 2–3 mm long, each with several erect ovules at the base. Fruiting cones almost globular, 2–3 cm diam.; the scales hardened, woody. Seeds compressed, hard, winged, usually only a few fertile in each cone.

Key to the species

1 Dorsal surface of leaf rounded. Leaves (including decurrent part) 2–4 mm long. Fruiting cones subglobular, up to 3 cm diam.; scales tuberculate, with a small dorsal point, united at the base. Trees or stunted shrubs up to 8 m high. Wolgan Valley northwards. DSF . . . . **C. preissii** Miq.
1 Dorsal surface of the leaf keeled 2

2 Fruiting cone scales with a broad conical protuberance on the dorsal surface near the middle. Leaves (including decurrent part) 2–4 mm long. Tips of main shoots drooping. Small tree or shrub up to 6 m high. Widespread. DSF. *Port Jackson Pine*. . . . . **C. rhomboidea** R.Br. ex Rich.
2 Fruiting cone scales with a small dorsal point near the apex. Tips of branches not drooping 3

3 Leaves (including decurrent part) 4–10 mm long. Trees or shrubs with ascending fastigiate branches, up to 6 m high. Widespread. DSF . . . .**C. muelleri** (Parl.) F.Muell.
3 Leaves (including decurrent part) 2–4 mm long. Trees, sometimes glaucous, with spreading branches. Hunter River Valley and Northern Blue Mts DSF (Populations of *C. endlicheri* on the Woronora Plateau are listed as an 'Endangered population') *Black Cypress Pine*. . . . . . . **C. endlicheri** (Parl.) F.M.Bailey

31 PINACEAE

11 gen., widely distributed, especially N. hemisphere

1 Pinus L.

19 species naturalized Aust.; all states and territories except NT

Monoecious, evergreen trees producing new branches in regular whorls. Bark rough or scaly. Shoots of two kinds: ordinary long shoots bearing scale leaves without chlorophyll; short shoots borne in the axils of the scale-leaves and with a definite number (2, 3 or 5) of green needle-like leaves surrounded by sheathing scale-leaves at the base, the short shoot not growing further and finally falling off as a whole. Male cones replacing short shoots at the base of a new year's growth; sporophylls each with 2 microsporangia on the lower surface. Female cones replacing long shoots and taking 1–2 years to ripen; ovuliferous scales each producing 2 ovules on the upper surface. Ripe cones woody. Seed winged.

Key to the species

1 Leaves in pairs, 5–30 cm long, stout, rigid. Buds spindle-shaped, pointed. Young shoots yellowish-brown. Female cones ± symmetrical, up to 18 cm long; scales with a pyramidal umbo. Tall tree. Introd. from S. Europe. *Cluster Pine* . . . .*P. pinaster* Aiton
1 Leaves in groups of 3, 10–15 cm long, thin, flexible. Buds cylindrical; young shoots greenish. Female cones oblique, up to 17 cm long; scales with a rounded umbo and a minute dorsal prickle. Widespread. Tall tree. Introd. from California. *Insignis Pine, Radiata Pine* or *Monterey Pine*. . . .*P. radiata* D.Don.

(*P. pinea* (Umbrella Pine) *P. contorta* (Lodgepole Pine) and *P. nigra* (Austrian Pine) are sparingly naturalized in some areas)

32 PODOCARPACEAE

Dioecious trees or shrubs. Leaves flat, acicular or scale-like, usually alternate. Male cones cylindrical or ovoid, with spirally arranged sporophylls. Female cones with 1–few ovuliferous scales; each with 1 ovule. Seed cones variable, often becoming fleshy. Embryo with 2 cotyledons. 17 gen., mainly S. hemisphere.

Key to the genera

1 Leaves 10–150 mm long . . . . .1 PODOCARPUS
1 Leaves 3–4 mm long . . . . 2 PHEROSPHAERA

1 Podocarpus L'Hér.
7 species endemic Aust.; all states and territories

Dioecious trees or shrubs. Leaves linear to linear-oblong or linear-lanceolate, alternate, coriaceous, with a prominent mid-vein. Male cones borne in the axils of scale-leaves or leaves near the base of the current year's shoots; the sporophylls each with 2 pollen sacs. Ovules borne singly or twinned at the end of the very short specialized lateral branches or receptacles, which are borne in a similar position to the male cones but on separate plants. Seeds blue-black, ± glaucous, ovoid or globular, solitary, rarely twin, borne upon the enlarged, fleshy, blue-black, ± glaucous receptacle.

Key to the species

1 Shrub c. 1 m high with weak straggling branches. Leaves linear, rigid, pungent-pointed, 4–6 cm long. Male cones 4–6 mm long, sessile in axillary clusters. Seeds 8–12 mm diam. Coast in sheltered places; Blue Mts in gullies . . . .**P. spinulosus** (Sm.) R.Br. ex Mirb.
1 Tall or small tree with fibrous bark. Leaves oblong-linear to linear-lanceolate, 5–15 cm long, 6–12 mm wide. Male cones 2–3 cm long, several together, sessile, in axillary clusters. Seeds 8–12 mm diam.; the fleshy receptacle sometimes 18 mm diam. Coast, e.g. Illawarra district. RF and gullies. Grown for ornamental purposes. *Plum Pine*. . . . . **P. elatus** R.Br. ex Endl.

2 Pherosphaera W.Archer
2 species endemic Aust.; NSW, Tas.

One species in the area

Shrub with weak straggling branches and green, drooping branchlets. Leaves narrow, crowded, spreading, 3–4 mm long, whitish on the upper surface, green and shining on the undersurface, shortly decurrent, spirally arranged. Male cones ovate or globular, c. 6 mm long, with a thin axis and stipitate sporophylls. Female cones at the tips of branchlets, c. 3 mm long, with 4–8 thin ovuliferous scales (previously known as *Microstrobos*). Blue Mts A few usually south facing rock ledges in reach of spray. Endangered. *Dwarf Mountain Pine* . . . .**P. fitzgeraldii** (F.Muell.) F.Muell. ex Hook.f.

5 MAGNOLIOPHYTA

To count the loculi in an ovary, cut a transverse section of the ovary and then lightly squeeze the bottom half. The ovules should then pop out and they can be brushed away with the finger and the loculi exposed.

Key to the classes

1 Flowers usually 4- or 5-merous (rarely 3-merous, when the plants are usually branching trees, shrubs or parasitic twiners). Leaf venation usually reticulate. Embryo with 2 cotyledons. Trees, shrubs, herbs, climbers . . . . **MAGNOLIOPSIDA** (Dicotyledons). . p44
1 Flowers usually 3-merous. Leaf venation usually parallel. Embryo with 1 cotyledon Plants usually herbaceous rarely woody and then ± unbranched). . . . .**LILIOPSIDA** (Monocotyledons). . p58

MAGNOLIOPSIDA

Key to the families

1 Flowers and fruits arranged on an invaginated floral axis, i.e. a fig . . . . . . **90 MORACEAE.** . p291
1 Flowers and fruits not as above. 2

2 Flowers unisexual and arranged within a cup-shaped collection of connate bracts, a number of male flowers (single stamens) surrounding a solitary female flower, i.e. a cyathium (Fig. 26). . . . .
. . . . **78 EUPHORBIACEAE.** . p268
2 Flowers not arranged as above 3

3 Stems green, jointed, ridged; leaves reduced to a whorl of scale-like teeth at each joint . . . .
. . . . **76 CASUARINACEAE.** . p265
3 Stems and leaves not as above. 4

4 Carpels free, 2 or more, ...**GROUP 1**
4 Either carpel solitary or carpels connate ; in both cases there is a single ovary but there may be a number of styles 5

5 Either petals or perianth segments free; or perianth absent. 6
5 Petals or perianth segments connate 12

6 Ovary inferior or half-inferior . . . .**GROUP 2**
6 Ovary superior or perianth absent 7

7 Stamens numerous, many more than twice as many as petals or perianth . . . .**GROUP 3**
7 Either stamens up to twice as many as petals or perianth, very rarely three times as many and then arranged in three whorls; or perianth absent 8

8 Either, one perianth whorl only present; or perianth segments in 2 or 3 whorls of 3 members each; or perianth absent 9
8 Calyx and corolla present (occasionally one very deciduous). 10

9 Either, stamens adnate to perianth; or floral tube present; or perianth absent . . . .**GROUP 4**
9 Perianth present and stamens free from it; no floral tube present . . . .**GROUP 5**

10 Flowers irregular . . . .**GROUP 6**
10 Flowers regular 11

11 Stamens and conspicuous staminodes (when present) twice the number of sepals, rarely three times the number; flowers bisexual . . . .**GROUP 7**
11 Stamens and minute staminodes (when present) less than twice the number of sepals; or flowers unisexual . . . .**GROUP 8**

12 Ovary inferior or half-inferior. . . . .**GROUP 9**
12 Ovary superior 13

13 Leaves opposite or whorled 14
13 Leaves alternate or basal 15

14 Stamens fewer than the corolla or perianth lobes. . . . .**GROUP 10**
14 Either, stamens equal in number to corolla or perianth lobes; or, sometimes, more numerous. . . . . .
. . . .**GROUP 11**

15 Stamens free from each other and free from the corolla or perianth . . . .**GROUP 12**
15 Stamens adnate to the corolla or perianth tube or connate (very rarely free and then plants with scale-like leaves and angular stems) 16

16 Only one perianth whorl present or petals minute . . . .**GROUP 13**

16 Calyx and corolla present and conspicuous; petals as large as sepals, larger or only slightly smaller. . .
. . . . **GROUP 14**

Group 1

17 Trees with petaloid sepals (includes Sterculiaceae) . . . . **99 MALVACEAE**. . p316

Group 2

1 Plants with jointed succulent stems bearing spines . . . . **52 CACTACEAE**. . p140
1 Plants without jointed succulent stems 2

2 Either, inner or outer stamens petaloid and numerous; or petals more than 8 3
2 No stamens petaloid; perianth whorls up to 8-merous or fused into an operculum 5

3 Aquatic herbs with floating leaves . . . . **33 NYMPHAEACEAE**. . p93
3 Terrestrial plants 4

4 Prostrate succulent herbs; outer stamens numerous and petaloid . . . . **48 AIZOACEAE**. . p126
4 Rainforest shrubs and small trees; inner stamens numerous and petaloid **40 EUPOMATIACEAE**.. p99

5 Either, sepals 2; or sepals and petals fused into an operculum 6
5 Either, sepals more than 2; or one perianth whorl with more than 2 segments. 7

6 Sepals 2 . . . . **59 PORTULACACEAE**. . p155
6 Sepals and petals fused into an operculum . . . . **69 MYRTACEAE**. . p171

7 Either, stamens numerous; or 5–10 and the leaves with oil dots 8
7 Stamens up to twice as many as the petals or perianth members; leaves without oil dots. 12

8 Perianth with only one whorl . . . . **48 AIZOACEAE**. . p126
8 Calyx and corolla present 9

9 Styles more than 1, free 10
9 Style simple 11

10 Leaves with 3 leaflets . . . . **86 CUNONIACEAE**. . p284
10 Leaves simple or compound with more than 3 leaflets . . . . **92 ROSACEAE**. . p299

11 Flowers in compound leafless axillary spikes; leaves without oil dots **112 SYMPLOCACEAE**. . p357
11 Flowers not as above; leaves usually with oil dots. . . . . **69 MYRTACEAE**. . p171

12 Flowers unisexual; perianth 4-merous . . . . **65 HALORAGACEAE**. . p163
12 Either, flowers bisexual; or rarely unisexual with a 3- or 5-merous perianth 13

13 Stipules interpetiolar, sometimes deciduous leaving a scar between the leaf bases 14
13 Stipules not interpetiolar or absent 15

14 Climber or straggling shrub with simple leaves. . . . . **63 APHANOPETALACEAE**. . p161
14 Tree or shrub with 1 or 3 foliate leaves . . . . **86 CUNONIACEAE**. . p284

15 Stamens twice as many as the petals . . . . **70 ONAGRACEAE**. . p210
15 Either, stamens as many as the petals or perianth members; or fewer 16

16 Stamens opposite the petals or perianth members 17
16 Stamens alternating with the petals or perianth members. 21

17 Plant parasitic on branches 18
17 Terrestrial plants 19

18 Flowers bisexual. . . . . **61 LORANTHACEAE**. . p156
18 Flowers unisexual (includes Viscaceae) . . . . **62 SANTALACEAE**. . p158

19 Leaves reduced to scales . . . . **62 SANTALACEAE**. . p158
19 Leaves well developed 20

20 Petals or perianth segments 4–5 . . . . **91 RHAMNACEAE** . . p293
20 Perianth segments 6–8 . . . . **61 LORANTHACEAE** . . p156

21 Only one perianth whorl present . . . . **91 RHAMNACEAE** . . p293
21 Two perianth whorls (calyx and corolla) present 22

22 Flowers solitary in the leaf axils . . . . **72 CELASTRACEAE** . . p214
22 Flowers in umbels, panicles, heads or pairs 23

23 Herbs or small shrubs; epigynous disc usually swelling at the base of the styles on the fruit . . . .
. . . . **138 APIACEAE** . . p418
23 Trees, shrubs, woody climbers or creepers; epigynous disc usually not swollen on the fruit 24

24 Styles 2 . . . . **139 ARALIACEAE** . . p424
24 Style single, 3–5-lobed (previously Escalloniaceae) . . . . **137 QUINTINIACEAE** . . p418

Group 3

1 Leaves alternate, aromatic, with pellucid oil dots, ... **36 WINTERACEAE** . . p94
1 Leaves without oil dots; or opposite with oil dots present 2

2 Flower unisexual . . . . **78 EUPHORBIACEAE** . . p218
2 Flowers bisexual 3

3 Ovary on a long gynophore . . . . **95 CAPPARACEAE** . . p315
3 Ovary sessile 4

4 Sepals 2–3 or united into a calyptra 5
4 Sepals 4 or more and not united into a calyptra 7

5 Trees or shrubs. Sepals united into a calyptra shed as the flower opens . . **86 CUNONIACEAE** . . p284
5 Herbs 6

6 Leaves succulent . . . . **59 PORTULACACEAE** . . p155
6 Leaves not succulent . . . . **45 PAPAVERACEAE** . . p118

7 Stamens connate into a tube or column . . . . **99 MALVACEAE** . . p316
7 Stamens free, or sometimes connate into bundles but never forming a tube or column 8

8 Leaves opposite or whorled 9
8 Leaves alternate 12

9 Aquatic plants with whorled leaves . . . . **35 CERATOPHYLLACEAE** . . p94
9 Not aquatic plants 10

10 Stipules interpetiolar, sometimes deciduous leaving a scar between the leaves or absent an leaves
appearing 3-foliate . . . . **86 CUNONIACEAE** . . p284
10 Stipules absent. Leaves simple 11

11 Leaves opposite . . . . **80 HYPERICACEAE** . . p278
11 Leaves whorled (Mimosoideae) . . . . **74 FABACEAE** . . p217

12 Petals yellow 13
12 Petals white, pink or greenish 15

13 Herbs; sepals greenish . . . . **97 RESEDACEAE** . . p316
13 Shrubs or trees 14
14 Flowers c. 3cm diam . . . . **82 OCHNACEAE** . . p279
14 Flowers less than 1cm diam. (the flower heads may be larger) (Mimosoideae) **74 FABACEAE** . . p217

15 Petals and sepals insert on a tube which extends above the ovary base (perigynous). Plants often deciduous . . . . **92 ROSACEAE.** . p299

15 Petals and sepals not as above. Plants not deciduous 16

16 Leaves dentate . . . .**87 ELAEOCARPACEAE.** . p286

16 Leaves entire or distantly angular (includes Flacourtiaceae) . . . .**84 SALICACEAE.** . p280

Group 4

1 Leaves reduced to scales; rootless twining parasites (includes Cassythaceae) **38 LAURACEAE.** . p95

1 Leaves conspicuous. 2

2 Perianth absent 3

2 Perianth present 5

3 Trees or shrubs . . . . **84 SALICACEAE.** . p280

3 Herbs. 4

4 Plants aquatic or creeping on mud . . . .**129 CALLITRICHACEAE.** . p395

4 Plants not aquatic or creeping on mud. . . . .**41 PIPERACEAE.** . p99

5 Perianth segments 4, deciduous . . . . **42 PROTEACEAE.** . p99

5 Either perianth segments 5–6; or 4 and persistent into the fruiting stage. 6

6 Staminodes present and bearded below the middle . . . . **60 OLACACEAE.** . p159

6 Staminodes absent or, if present, then glabrous 7

7 Trees . . . . **38 LAURACEAE.** . p95

7 Herbs or climbers 8

8 Stipules ocreous, i.e. sheathing the stem. . . . . **58 POLYGONACEAE.** . p150

8 Stipules not sheathing the stem or absent 9

9 Climbers. . . . . **51 BASELLACEAE.** . p140

9 Herbs. 10

10 Leaves simple . . . .**53 CARYOPHYLLACEAE.** . p141

10 Leaves compound. . . . . **92 ROSACEAE.** . p299

Group 5

1 Latex present in the bark . . . . **90 MORACEAE.** . p291

1 Latex absent. 2

2 Climbers; flowers unisexual; plants dioecious . . . .**44 MENISPERMACEAE.** . p116

2 Not climbers, rarely scramblers 3

3 Perianth segments scarious; filaments often connate at the base. . **50 AMARANTHACEAE.** . p136

3 Perianth segments herbaceous or petaloid; filaments rarely connate at the base. 4

4 Either, style solitary, simple (filiform or capitate); or rarely present as a tuft of hairs on the summit of the ovary. 5

4 Either, styles several; or single and branched above 11

5 Either, herbs; or trees with leaves covered in stinging hairs. 6

5 Trees or shrubs without stinging hairs on leaves 7

6 Style capitate or obscurely lobed . . . . **94 BRASSICACEAE.** . p306

6 Style filiform . . . . **93 URTICACEAE.** . p304

7 Ovary and fruit lobed . . . . **104 SAPINDACEAE.** . p339
7 Ovary entire, or nearly so 8

8 Either, leaves all compound; or reduced to 3-fid thorns on main branches and simple on side branches
. . . . **43 BERBERIDACEAE.** . p116
8 Leaves all simple, not reduced to thorns. 9

9 Leaves reduced to scales . . . . **62 SANTALACEAE.** . p158
9 Leaves well developed 10

10 Flowers arranged in a panicle (previously Icacinaceae) . . . **142 CARDIOPTERIDACEAE.** . p431
10 Flowers arranged in a raceme . . . . **56 PHYTOLACCACEAE.** . p149

11 Stipules ocreous, i.e. sheathing the stem . . . . **58 POLYGONACEAE.** . p150
11 Stipules not sheathing the stem or absent 12

12 Leaves with domatia in the angles of secondary and mid veins (previously Icacinaceae) . . . .
. . . . **140 PENNANTIACEAE.** . p428
12 Leaves without domatia 13

13 Ovary 1-locular 14
13 Ovary multi-locular 18

14 Trees, or tall shrubs (includes *Celtis* & *Trema* previously Ulmaceae). . **89 CANNABACEAE.** . p290
14 Herbs or small shrubs 15

15 Stigmas a tuft of more than 4 hairs on top of the ovary . . . . **93 URTICACEAE.** . p304
15 Styles up to 4. 16

16 Styles or stylar branches 4 (rarely 3); leaves opposite . . . . **53 CARYOPHYLLACEAE.** . p141
16 Styles or stylar branches 2–3; leaves often alternate. 17

17 Leaves palmately compound, with 3–7 leaflets . . . . **89 CANNABACEAE.** . p290
17 Leaves not palmately compound . . . . **49 CHENOPODIACEAE.** . p129

18 Flowers unisexual. 19
18 Flowers bisexual. 20

19 Each locule of ovary with 1 ovule. **78 EUPHORBIACEAE.** . p268
19 Each locule of ovary with 2 ovules, but only one may develop in fruit (previously Euphorbiaceae) . . .
. . . . **79 PHYLLANTHACEAE.** . p274

20 Plants aquatic or creeping on mud . . . . **77 ELATINACEAE.** . p267
20 Plants not aquatic or creeping on mud. . . . . **56 PHYTOLACCACEAE.** . p149

Group 6

1 Sepals 2, (includes Fumariaceae) . . . . **45 PAPAVERACEAE.** . p118
1 Sepals 3, 4 or 5, sometimes 1 or 2 sepals petaloid. 2

2 1 or 2 sepals petaloid. 3
2 Sepals all ± similar in texture. 4

3 Two lateral sepals small and green . . . . **106 BALSAMINACEAE.** . p343
3 Two lateral sepals large and petaloid. . . . . **75 POLYGALACEAE.** . p264

4 Fertile stamens 5, no staminodes present . . . . **85 VIOLACEAE.** . p282
4 Either fertile stamens 10, or if less than this then staminodes complete the complement to 10; or fertile
stamens 8 no staminodes present 5

5 Calyx spurred and the spur free from the pedicel . . . . **98 TROPAEOLACEAE.** . p316

5 Calyx without a spur or with a spur adnate to the pedicel 6

6 Style 5-fid . . . . .67 **GERANIACEAE**. . p168
6 Style simple . . . . 74 **FABACEAE**. . p217

Group 7

1 Sepals 2–3 . . . . 59 **PORTULACACEAE**. . p155
1 Sepals more than 3 2

2 Leaves with oil dots; style ± gynobasic . . . . .103 **RUTACEAE**. . p328
2 Leaves without oil dots; style(s) usually terminal on the ovary 3

3 Style solitary, simple 4
3 Styles several or single and branched above 10

4 Stamens perigynous 5
4 Stamens hypogynous 7

5 Calyx modified into barbed prickles . . . . 92 **ROSACEAE**. . p299
5 Calyx not modified into barbed prickles 6

6 Petals pink to purplish, as long as or longer than the sepals . . . . 68 **LYTHRACEAE**. . p170
6 Petals greenish, minute, much shorter than the sepals . . . . 100 **THYMELAEACEAE**. . p324

7 Style capitate . . . . 102 **MELIACEAE**. . p327
7 Style filiform 8

8 Flowers solitary in leaf axils. Petals bright yellow . . . . 47 **DILLENIACEAE**. . p122
8 Flowers usually in racemes or panicles, if solitary then petals not bright yellow 9

9 Stamens opening through terminal pores (includes Tremandraceae) . . 87 **ELAEOCARPACEAE**. p286
9 Stamens opening through longitudinal slits . . . . 104 **SAPINDACEAE**. . p339

10 Trees or shrubs 11
10 Herbs 14

11 Leaves opposite with interpetiolar stipules . . . . 86 **CUNONIACEAE**. . p284
11 Leaves alternate 12

12 Leaves simple, sometimes deeply lobed (includes Sterculiaceae) . . . . . . 99 **MALVACEAE**. . p316
12 Leaves compound 13

13 Fruit dry, winged; ovary lobed. Leaflets with 2 lobes near the base which are terminated by a resin
      gland . . . . .105 **SIMAROUBACEAE**. . p342
13 Fruit drupaceous, not lobed. Leaflets not as in Simaroubaceae . . . 101 **ANACARDIACEAE**. . p326

14 Leaves 3-foliolate . . . . .88 **OXALIDACEAE**. . p289
14 Leaves simple or pinnate-compound 15

15 Corolla yellow 16
15 Corolla pink, white or blue 17

16 Prostrate herb; leaves compound . . . . .71 **ZYGOPHYLLACEAE**. . p213
16 Erect herbs; leaves simple, entire to pinnatisect . . . . 97 **RESEDACEAE**. . p316

17 Leaves simple, not deeply dissected; ovary usually 1-locular . . .53 **CARYOPHYLLACEAE**. . p141
17 Leaves deeply dissected or compound; ovary 5-locular . . . . 67 **GERANIACEAE**. . p168

Group 8

Group 9

Group 10

11 Seeds not compressed, Leaves opposite or alternate . . . . . . **132 SCROPHULARIACEAE**. . p397
11 Seeds usually compressed and discoid. Leaves opposite 12

12 Leaves with glandular hairs . . . . **125 MARTYNIACEAE**. . p391
12 Leaves without glandular hairs . . . . **120 ACANTHACEAE**. . p374

13 Plants with branched hairs. Corolla mauve, yellowish-green or bluish green **123 LAMIACEAE**. . p378
13 Plants may have hairs but the hairs are not branched. Corolla white, purplish, blue, orange to pink or
   red . . . . **133 VERBENACEAE**. . p405

Group 11

1 Calyx and corolla present 2
1 One perianth whorl only present (the involucral bracts of Nyctaginaceae may resemble a calyx) . . . 15

2 Anthers opening through terminal pores . . . . **108 ERICACEAE**. . p343
2 Anthers opening through longitudinal slits. 3

3 Stamens free from corolla; leaves with oil dots . . . . **103 RUTACEAE**. . p328
3 Either stamens adnate to corolla; or, if almost free, then leaves without oil dots 4

4 At least some stamens opposite the corolla lobes. 5
4 Stamens alternating with the corolla lobes 6

5 Stamens 4–5 (includes Primulaceae). . . . . **109 MYRSINACEAE**. . p355
5 Stamens 8 . . . . **64 CRASSULACEAE**. . p161

6 Leaves with several fine longitudinal parallel veins (includes Epacridaceae) **108 ERICACEAE**. . p343
6 Leaves with reticulate venation or a single vein. 7

7 Plants with white, yellow, orange or watery latex (N.B. latex exudes in a globule on top of the broken
   stem or stalk; ordinary sap does not form a globule) (includes Esclepiadaceae) . . . .
   . . . . **115 APOCYNACEAE**. . p361
7 Plants without latex 8

8 Mangrove plant with pheumatophores; leaves leathery more than 14mm wide . . . .
   . . . . **120 ACANTHACEAE** p374
8 Plants not as above 9

9 Flowers irregular 10
9 Flowers regular 11

10 Style with 2 lobes . . . . **123 LAMIACEAE**. . p378
10 Style simple . . . . **133 VERBENACEAE**. . p405

11 Climber (previously Loganiaceae) . . . . **116 GELSEMIACEAE**. . p366
11 Plants not climbers 12

12 At least undersurface of leaves with stellate hairs. . . . . **128 BUDDLEJACEAE**. . p395
12 Plants glabrous or hairs if present not stellate 13

13 Stipules present or opposing leaves connected at base by a raised line or tissue flap . . . .
   . . . . **118 LOGANIACEAE**. . p368
13 Stipules absent. 14

14 Herb . . . . **117 GENTIANACEAE**. . p366
14 Tree or shrub . . . . **133 VERBENACEAE**. . p405

15 Stamens free from perianth tube 16
15 Stamens adnate to perianth tube 18

Group 12

Group 13

Group 14

2 Pollen cup (indusium) present at top of style . . . . **145 GOODENIACEAE**. . p483
2 No pollen cup present 3

3 Flowers irregular 4
3 Flowers regular 5

4 Stamens 8. Lateral sepals enlarged and petal-like . . . .**75 POLYGALACEAE**. . p264
4 Stamens 10. All sepals green and similar . . . .**74 FABACEAE**. . p217

5 Leaves sessile, sheathing the stem (includes Epacridaceae) . . . . **108 ERICACEAE**. . p343
5 Leaves petiolate. 6

6 Mangrove shrubs; stamens 5 . . . . **109 MYRSINACEAE**. . p355
6 Not mangrove shrubs; stamens usually more than 5 7

7 Leaves compound . . . . **102 MELIACEAE**. . p327
7 Leaves simple 8

8 Herbs. . . . . **99 MALVACEAE**. . p316
8 Trees . . . .**112 SYMPLOCACEAE**. . p357

9 Flowers unisexual, plants dioecious . . . .**107 EBENACEAE**. . p343
9 Flowers bisexual 10

10 Leaves with several longitudinal parallel veins 11
10 Leaves with reticulate venation or a single midvein 12

11 Shrubs with cauline leaves (includes Epacridaceae) . . . . **108 ERICACEAE**. . p343
11 Herbs with basal leaves. . . . . **127 PLANTAGINACEAE**. . p393

12 Flowers in a one-sided coiled inflorescence. Style ± gynobasic; ovary lobed **114 BORAGINACEAE**.p358
12 Flowers not arranged as in Boraginaceae. Style terminal on the ovary; ovary not lobed. 13

13 Stamens opposite the corolla lobes 14
13 Stamens alternating with the corolla lobes 16

14 Herbs . . . .**57 PLUMBAGINACEAE**. . p150
14 Trees or shrubs 15

15 Bark with latex . . . . **111 SAPOTACEAE**. . p357
15 Bark without latex . . . . **109 MYRSINACEAE**. . p355

16 Stamens 5 17
16 Stamens 2–4 23

17 Aquatic plants, sometimes in sites which dry in the summer; corolla lobes with membranous wings or
     bearded. . . . . **146 MENYANTHACEAE**. . p488
17 Terrestrial plants; corolla lobes without membranous wings, not bearded 18

18 Leaves pinnate. . . . . **110 POLEMONIACEAE**. . p356
18 Leaves simple, sometimes deeply lobed 19

19 Climbers, twiners, scramblers or creepers forming mats 20
19 ± erect herbs, shrubs or trees 21

20 Anthers opening through terminal pores . . . .**135 SOLANACEAE**. . p410
20 Anthers opening through longitudinal slits. . . . . **134 CONVOLVULACEAE**. . p407

21 Ovules (seeds) several in each loculus . . . .**135 SOLANACEAE**. . p410
21 Ovules (seeds) solitary in each loculus 22

22 All or the majority of leaves entire, some maybe irregularly and obscurely toothed near apex . . . . . .
. . . . **131 MYOPORACEAE.** . p396

22 Leaves serrate (includes Ehretiaceae) . . . . **114 BORAGINACEAE.** . p358

23 Stamens 2 . . . . .124 **LENTIBULARIACEAE.** . p390

23 Stamens 4 24

24 Herbs 25

24 Shrubs or trees 26

25 Stamens ± equal. . . . . **135 SOLANACEAE.** . p410

25 Stamens unequal; 2 long and 2 short . . . . **132 SCROPHULARIACEAE.** . p397

26 Leaves linear . . . . **130 SELAGINACEAE.** . p395

26 Leaves broader than linear 27

27 Shrubs or tree with a hard bark. Seeds solitary in each loculus . . . .
. . . . **131 MYOPORACEAE.** . p396

27 Shrub or tree with corky bark. Seeds several in each loculus . . . . .135 **SOLANACEAE.** . p410

LILIOPSIDA

Key to the families

1 Ovary inferior . . . . .GROUP 1

1 Ovary superior 2

2 Either, at least one whorl of the perianth petaloid; or perianth segment solitary and petaloid . .**GROUP 2**

2 Either, both whorls of the perianth not petaloid; or one or both absent . . . .**GROUP 3**

Group 1

1 Partially or wholly submerged herbs, sometimes marine . . . . **157 HYDROCHARITACEAE.** . p497

1 Terrestrial plants 2

2 Stamens 6 3

2 Stamens 3 or fewer 9

3 Climbers or twiners with no tuft of basal leaves . . . . **180 DIOSCOREACEAE.** . p555

3 Erect herbs with a tuft of radical (basal) leaves 4

4 Coarse herbs more than 2 m high 5

4 Herbs less than 2 m high 6

5 Inflorescence a spreading panicle of white to yellow flowers . . . .**166 AGAVACEAE.** . p505

5 Inflorescence a large compact head-like panicle of red flowers . . . **172 DORYANTHACEAE.** . p512

6 Flower tube long and narrow, more than 2 times as long as lobes . . **188 HAEMODORACEAE.** . p559

6 Flower tube when present shorter than or to 1.5 times as long as lobes 7

7 Inflorescence a short raceme or flower solitary . . . . **174 HYPOXIDACEAE.** . p514

7 Inflorescence umbellate 8

8 Leaves petiolate . . . . **181 ALSTROEMERIACEAE.** . p555

8 Leaves without petioles . . . . **168 AMARYLLIDACEAE.** . p509

9 Stamens 3, free from the style 10

9 Fertile stamens less than 3 12

10 Ovary winged . . . . **179 BURMANNIACEAE.** . p555

10 Ovary not winged. 11

11 Style simple; capsule with 1–2 seeds per loculus . . . . **188 HAEMODORACEAE**. . p559
11 Style branched; capsules with more than 2 seeds per loculus . . . . **175 IRIDACEAE**. . p515

12 Neither stamens nor staminodes broad and petaloid. . . . . **176 ORCHIDACEAE**. . p521
12 At least one of the staminodes broad and petaloid 13

13 Stamen petaloid, with an anther loculus on one edge and surrounded by petaloid staminodes . . . . .
. . . . **202 CANNACEAE**. . p659
13 Stamen not petaloid, with an anther loculus on either side; one staminode petaloid, two linear . . . . .
. . . . **203 ZINGIBERACEAE**. . p659

Group 2

1 Carpels free from each other 2
1 Either carpel solitary or carpels connate; in both cases there is a single ovary but there may be a number
of syles 5

2 Plants completely submerged including the flowers . . . . **161 ZANNICHELLIACEAE**. . p502
2 Plants emergent, at least the flowers 3

3 Carpels more than 6 . . . . **153 ALISMATACEAE**. . p493
3 Carpels up to 6 4

4 Perianth segment solitary . . . . **154 APONOGETONACEAE**. . p494
4 Perianth segments 4 . . . . **160 POTAMOGETONACEAE**. . p501

5 Leaves undivided, spirally coiled at the tip . . . . **195 FLAGELLARIACEAE**. . p588
5 Leaves not spirally coiled at the tip but sometimes palmately divided with flexuose tips to the segments . 6

6 Stamen solitary . . . . **190 PHILYDRACEAE**. . p562
6 Stamens 3–6. 7

7 Tall trees or shrubs with annular leaf scars 8
7 Herbs or climbers, sometimes woody below 9

8 Leaves palmatifid or pinnate . . . . **187 ARECACEAE**. . p558
8 Leaves linear, undivided. . . . . **170 ASTELIACEAE**. . p511

9 Aquatic herbs with bulbous petioles . . . . **191 PONTEDERIACEAE**. . p562
9 Terrestrial or marsh plants; petioles not swollen or absent 10

10 Leaves reduced to scales, subtending leaf-like cladodes; . . . . **165 ASPARAGACEAE**. . p504
10 Plants without cladodes; true leaves usually well developed. 11

11 Climbers or scramblers. 12
11 Herbs, sometimes woody towards the base, or plants tree-like 14

12 Leaves with several prominent longitudinal veins; the connecting lateral veins obscure . . . .
. . . . **184 PHILESIACEAE**. . p557
12 Leaves with 3–5 longitudinal veins; some connecting lateral veins prominent 13

13 Flowers unisexual, in axillary umbels . . . . **186 SMILACACEAE**. . p558
13 Flowers bisexual in racemes or spikes . . . . **185 RIPOGONACEAE**. . p557

14 Flowers unisexual. . . . . **177 LOMANDRACEAE**. . p551
14 Flowers bisexual. 15

15 Flowers in umbels or heads, sometimes opening one at a time 16
15 Flowers in racemes, spikes, panicles or solitary. 21

16 Outer perianth segments green, herbaceous; leaf sheath closed (i.e., the margins fused) . . . .
. . . . **189 COMMELINACEAE. . p560**

16 Outer perianth segments not green; leaf sheath open (i.e., with overlapping margins or margins that do not meet) or absent 17

17 Inner perianth segments yellow. **201 XYRIDACEAE. . p658**
17 Inner perianth segments not yellow 18

18 Either styles 3 free; or style deeply 3-fid . . . . **182 COLCHICACEAE. . p555**
18 Style simple or shortly 3-lobed 19

19 Plants with bulbs; umbels with one large scariose bract (spathe) at base . . **163 ALLIACEAE. . p503**
19 Plants with rhizomes or corms or without underground stems; umbels or heads with several bracts at the base or none. 20

20 Flowers less than 2.5 cm long . . . . **167 ANTHERICACEAE. . p505**
20 Flowers more than 2.5 cm long. Robust rhizomatous herbs with fleshy distichous leaves. . . . .
. . . . **164 AGAPANTHACEAE. . p504**

21 Perianth segments fused into tube longer than the lobes 22
21 Perianth segments free or fused into a tube shorter than the lobes. 24

22 Flowers c. 10cm long . . . . **183 LILIACEAE. . p557**
22 Flowers less than 8cm long 23

23 Stamens fused to perianth tube . . . . **171 BLANDFORDIACEAE. . p511**
23 Stamens free from perianth tube . . . . **169 ASPHODELACEAE. . p510**

24 Either, styles 3 free; or style deeply 3-fid . . . . **182 COLCHICACEAE. . p555**
24 Style simple or shortly 3-lobed 25

25 Flowers in a large thick complex spike . . . . **178 XANTHORRHOEACEAE. . p553**
25 Flowers in racemes or panicles 26

26 Perianth deciduous after flowering . . . . **169 ASPHODELACEAE. . p510**
26 Perianth persistent after flowering 27

27 Leaves distichous (previously Phormiaceae). . . . . **173 HEMEROCALLIDACEAE. . p512**
27 Leaves not distichous. **167 ANTHERICACEAE. . p505**

Group 3

1 Flowers grouped into spikelets, each flower covered by a glumaceous bract 2
1 Flowers not grouped into spikelets and not covered by a glumaceous bract 6

2 Carpels 2 or more, arranged at different levels on a gynophore; perianth absent . . . .
. . . . **192 CENTROLEPIDACEAE. . p562**
2 Carpels sessile, connate and arranged at the same level on the receptacle; or free; or ovary 1-locular. . 3

3 Leaf sheath closed or rarely open and then the perianth reduced to bristles or minute scales . . . .
. . . .**193 CYPERACEAE. . p562**
3 Leaf sheath open 4

4 Leaves reduced to sheathing scales arranged along the stem sometimes with short laminas; flowers usually unisexual and dioecious . . . . **198 RESTIONACEAE. . p654**
4 Leaves seldom reduced to sheathing scales 5

5 Perianth segments 0–3, usually 2; flowers grouped into a characteristic spikelet (Fig. 48) . . . .
. . . .**197 POACEAE. . p593**
5 Perianth segments 4–6; flowers grouped into a head . . . . **194 ERIOCAULACEAE. . p588**

6 Carpels 3 or more, free. 7
6 Either carpels connate or solitary 8

9 Flowers unisexual 10
9 Flowers bisexual 19

10 Aquatic herbs 11
10 Terrestrial plants, sometimes woody at the base 17

11 Floating plants with or without roots but not attached to the substratum. 12
11 Plants attached to the substratum. 13

13 Submerged plants; flowers sometimes emergent 14
13 Plants not submerged. 16

FIGURE 28

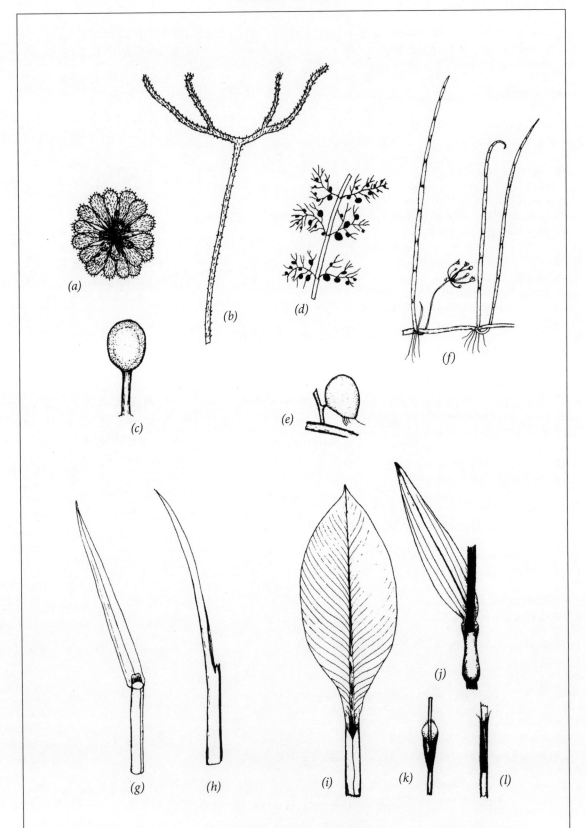

Fig. 28 (a) Plant of *Drosera spathulata* (nat. size); (b) single leaf of *Drosera binata* (about X½); (c) single glandular hair (enlarged); (d) *Utricularia flexuosa* with bladders (nat. size); (e) bladder enlarged; (f) *Lilaeopsis polyantha*; (g) leaf of a grass; (h) leaf of a *Juncus* sp.; (i) leaf of *Canna*; (j) leaf of *Commelina cyanea* (nat. size); (k) leaf of *Chordifex dimorphus* (nat. size); (I) leaf of *Hypolaena fastigiata* (nat. size).

FIGURE 29

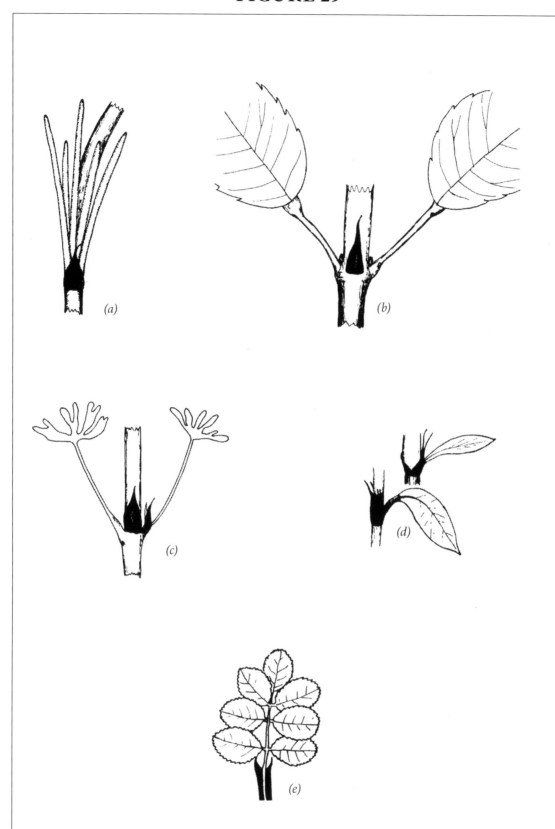

Fig. 29 (a) *Asparagus officinalis*, scale leaf subtending 5 leaf-like cladodes; (b) *Ceratopetalum*, with interpetiolar stipules; (c) interpetiolar stipules as in some *Geraniaceae*; (d) ochreous stipules as in *Polygonaceae* (upper-rear view) (lower-side view); (e) stipules adnate along the petiole as in *Rosa* spp.

FIGURE 30

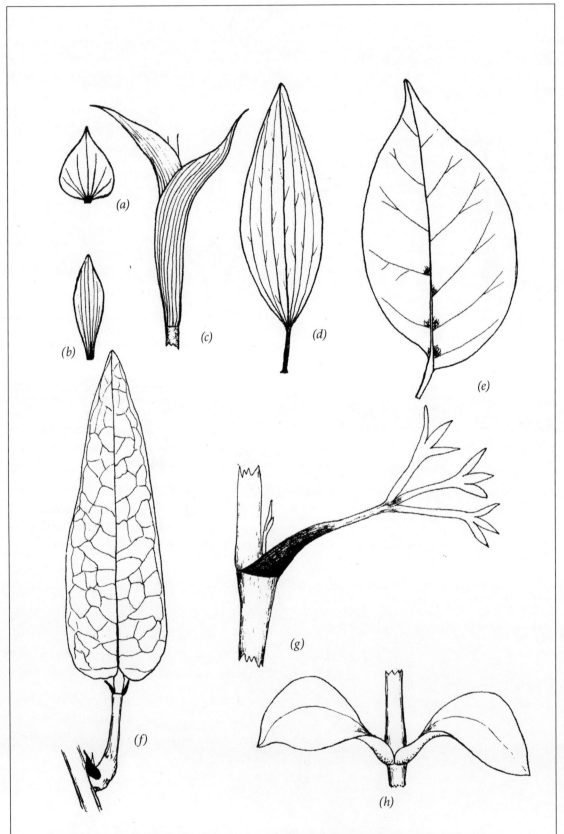

Fig. 30   (a) Leaf of *Epacris longiflora* (X2); (b) leaf of *Leucopogon parviflorus* (nat. size);
(c) part ofstem and two leaves of *Sprengelia incarnata* (nat. size); (d) leaf of *Trochocarpa
laurina* (nat. size); (e) under surface of pinna of *Toona ciliata* showing hairy glands (nat. size);
(f) leaf of *Hardenbergia violacea* showing reduced pinnae (nat. size); (g) dilated leaf base as in
many Apiaceae; (h) opposite leaves with bases touching as in Caryophyllaceae.

Vegetative Key to Magnoliophyta Families

This key may be used when no flowers are available. It will usually enable you to determine to which family a plant belongs. To identify specimens further it will be necessary to use a matching technique, preferably using an herbarium collection.

Easily identified families and groups

(The following 8 families and 10 groups are readily identified and are not keyed)

A. GRASSES. Herbs. Leaf bases modified to form an open sheath which encloses the stem; leaf blade with parallel venation and abruptly distinct from the sheath; LIGULE usually present at the top of the sheath (Fig. 28) . . . . **197 POACEAE**

B. PALMS. Leaves large, pinnately or palmately divided, the segments with parallel venation; leaf bases enclosing or clasping the stem. Young inflorescences enclosed by a large deciduous spathe. . . . .
. . . . **187 ARECACEAE**

C. STEMS SQUARE (or quadrangular), smooth or hairy but not scabrous; leaves opposite or rarely whorled, fragrant but usually without oil dots; fruit separating into 4 articles . . . . **123 LAMIACEAE**

D. HAIRS WITH SWOLLEN BASES present on leaves and stems. Herbs; whole plant scabrous and raspy to the touch; leaves alternate. Inflorescence cymose . . . . **114 BORAGINACEAE**

E. OCHREA (sheath) Present around the stem and petiole, formed from fused stipules (Fig. 29). Leaves alternate. Herbs . . . . **58 POLYGONACEAE**

F. Herbs with SWOLLEN NODES and opposite or rarely whorled leaves; leaves always simple with entire margins; the leaf bases often connate. Stipules mostly absent, sometimes membranous and connate . .
. . . . **53 CARYOPHYLLACEAE**

G. WOODY PLANT(usually shrubs) with leaves with PARALLEL VENATION (Fig. 30). Leaves alternate or spirally arranged, often crowded, rigid, ericoid, often sharp-pointed; the margins usually entire but sometimes denticulate; stipules absent. . . . . **108 ERICACEAE**

H. PHYLLODES present with a raised or depressed extra-floral nectary on the adaxial edge, hanging or standing with the nectary uppermost. (A few species of *Acacia* with exceptional additional structures are included in the key below) (Mimosoideae) . . . . **74 FABACEAE**

I. INSECTIVEROUS PLANTS with either sticky glands on the leaves (Fig. 28), or with bladders 1–2 mm long on submerged or subterranean leaves (Fig. 28). Very delicate herbs . . . . **GROUP 1**

J. FREE FLOATING WATER PLANTS; roots, when present, not rooted in the mud. (Note: there are 2 genera of ferns which are free floating, see *Azolla* and *Salvinia*) . . . . **GROUP 2**

K. Submerged plants rooted in mud. Leaves submerged or floating on the surface or sometimes the leaf blades held erect above the water surface. Freshwater or marine (Amphibious plants are not included in this group) . . . . **GROUP 3**

L. MANGROVES. Woody plants growing in saline mud near the coast . . . . **GROUP 4**

M. ROOTLESS EPIPHYTIC PARASITES. . . . . **GROUP 5**

N. TENDRILS present on stem or leaves; plants climbing or decumbent . . . . **GROUP 6**

O. LEAVES REDUCED TO SCALES ON THE AERIAL STEMS . . . . **GROUP 7**

P. EPIPHYTES on trees or rocks . . . . **GROUP 8**

Q. LATEX (milky white or yellow) present in stem and leaves . . . . **GROUP 9**

R. Either: leaves whorled; or leaves opposite and interpetiolar stipules present (Fig. 29). Herbs and woody plants . . . . **GROUP 10**

Key to the remaining groups

1 Plants climbing 2
1 Plants not climbing 3

2 Leaves opposite . . . . **GROUP 11**
2 Leaves alternate . . . . **GROUP 12**

3 Herbs with leaves with parallel venation; the leaves sometimes reduced to scales on a rhizome, often with a sheath and a blade (the blade penniveined in *Canna*). Terrestrial or amphibious. Many Liliopsida, a few *Magnoliopsida* 4
3 Leaves with reticulate venation or penniveined. Herbs and woody plants. Mainly Magnliopsida 6

4 Leaves borne on aerial stems, separated on the stem by distinct internodes . . . . **GROUP 13**
4 Leaves mainly basal or the leaf solitary, arising from a rhizome or underground upright stem or clustered at the base of upright aerial stems 5

5 Corms, bulbs or tubers present . . . . **GROUP 14**
5 Corms, bulbs and tubers absent; rhizomes present or the stems short and erect . . . . **GROUP 15**

6 Leaves compound with distinct leaflets 7
6 Leaves simple, entire or dissected (including pinnatisect) 8

7 Herbs . . . . **GROUP 16**
7 Woody plants . . . . **GROUP 17**

8 Leaves opposite (excluding the leaves subtending flowers) 9
8 Leaves alternate, alternate and basal, or all basal, or solitary (leaves subtending flowers sometimes opposite) 10

9 Herbs . . . . **GROUP 18**
9 Woody plants . . . . **GROUP 19**

10 Herbs 11
10 Woody plants 14

11 Stipules present (Note that some plants whose leaves are borne on underground stems have stipules . . . . **GROUP 20**
11 Stipules absent (Leaf bases sometimes dilated and sheathing) 12

12 Leaves all or mostly basal, or the leaf solitary or the leaves arising from the underground rhizomes . . . . **GROUP 21**
12 Leaves all or mostly cauline on aerial stems or stolons 13

13 Leaves (including the petioles) c. 25 mm long or less . . . . **GROUP 22**
13 Leaves (including the petioles) c. 30 mm long or longer . . . . **GROUP 23**

14 Thorns or prickles present on the stem or at the base of the petiole . . . . **GROUP 24**
14 Thorns or prickles absent 15

15 Either; leaves succulent or semi-succulent; or oil glands present . . . . **GROUP 25**
15 Leaves not succulent 16

16 Either; leaves terete (any length); or, leaves less than 30 mm long . . . . **GROUP 26**
16 Leaves more than 30 mm long, not terete 17

17 Stipules present. (Examine the youngest leaves before they have fully expanded and look for stipule scars at the base of the older leaves) . . . . **GROUP 27**

17 Stipules absent. 18

18 Adult leaves either with conspicuous hairs on one or both surfaces or along the midrib on the undersurface, or leaves white or yellow below. . . . .**GROUP 28**
18 Adult leaves glabrous or with inconspicuous hairs, often paler underneath but not white or yellow. 19

19 Leaf margins entire or undulate, sometimes with a few teeth near the apex . . . .**GROUP 29**
19 Leaf margins toothed or lobed or the leaves highly dissected . . . .**GROUP 30**

Group 1
Insectivorous plants

1 Leaves with glandular hairs, in a rosette and/or ascending and bifurcating (Fig. 28) . . . .
. . . . **54 DROSERACEAE**
1 Leaves (subterranean or in water) with bladders 1–2 mm long (Fig. 28). . **124 LENTIBULARIACEAE**

Group 2
Free-floating plants

1 Whole plant not exceeding 5 mm long . . . . **156 LEMNACEAE**
1 Plant up to 50 cm high. Petioles swollen and spongy . . . . **191 PONTEDERIACEAE**

Group 3
Submerged water plants

1 Leaves opposite or whorled on an elongated stem 2
1 Either; leaves alternate on an elongated stem (sometimes opposite when subtending flowers); or, leaves tufted on a short stem or arising from a rhizome. 8

2 Submerged leaves finely dissected into filiform segments. 3
2 Leaves entire or toothed or almost pinnatifid. 4

3 Leaves divided dichotomously . . . . **35 CERATOPHYLLACEAE**
3 Leaves divided pinnately (*Myriophyllum*). . . . . **65 HALORAGACEAE**

4 Broad stipules present. Leaves 2–7 cm long. Salt water plants (*Halophila*) **157 HYDROCHARITACEAE**
4 Stipules small or absent. Fresh or brackish water plants. 5

5 Stems creeping over the mud and rooting at the nodes; the plants forming mats. Leaves 3–15 mm long; minute stipules present . . . . **77 ELATINACEAE**
5 Stems free in water. (Plant may root at the nodes). Stipules absent. 6

6 Leaf bases broad and dilated or sheathing the stem (*Najas*) . . . . . . **157 HYDROCHARITACEAE**
6 Leaf bases neither broad and dilated nor sheathing the stem. 7

7 Leaves in whorls of 3–6 (*Egeria, Elodea*). . . . . **157 HYDROCHARITACEAE**
7 Leaves always opposite, rhomboidal spathulate or ovate, less than 10 mm long; stipules absent. . . . . . .
. . . . **129 CALLITRICHACEAE**

8 Leaves compound finely dissected or lobed (sometimes linear and stem-clasping) 9
8 Leaves entire or with 1–2 lobes or minutely toothed. 11

9 Leaves pinnatisect (sometimes linear) stem-clasping (*Cotula*) . . . .**143 ASTERACEAE**
9 Leaves pinnately divided 10

10 Leaf segments filiform (*Myriophyllum*) . . . . **65 HALORAGACEAE**
10 Leaf segments flat (*Rorippa*). . . . . **94 BRASSICACEAE**

11 Either; leaves lanceolate or broader; or, with long petioles and broader blades 12

11 Leaves linear, mostly ribbon-like, basal or cauline 19

12 Leaves sessile or the petioles c. as long as the blades 13
12 Petioles several times longer than the blades 14

13 Leaves with reticulate venation; 2 stipule-like vesicles present at the base of the petiole (*Ludwigia*) . . .
. . . . **70 ONAGRACEAE**
13 Leaves with parallel venation; vesicles absent . . . . **160 POTAMOGETONACEAE**

14 Leaf blades held erect, not floating on the surface of the water 15
14 Leaf blades floating on the surface of the water 17

15 Leaves (including petioles) c. 4 cm long (*Limosella*) . . . . **132 SCROPHULARIACEAE**
15 Leaves (including petioles) exceeding 10 cm long. 16

16 Leaf blades ovate to reniform, with a prominent midrib and conspicuous lateral veins (*Villarsia*). . . .
. . . . **146 MENYANTHACEAE**
16 Leaf blades with 3–9 ± equal longitudinal veins, often cordate at the base . . . **153 ALISMATACEAE**

17 Leaf blades ovate or elliptic, without lobes. . . . . **157 HYDROCHARITACEAE**
17 Leaf blades reniform or orbicular and deeply cordate 18

18 Leaf blades 2–8 cm across (*Nymphoides*). . . . . **146 MENYANTHACEAE**
18 Leaf blades mostly more than 8 cm across. **33 NYMPHAEACEAE**

19 Leaves c. 4 cm long, tufted on a rhizome (*Limosella*) . . . . **132 SCROPHULARIACEAE**
19 Either; leaves more than 4 cm long, tufted on a rhizome; or, leaves c. 4 cm long and alternate on a free
stem. 20

20 Fresh water plants. 21
20 Salt or brackish water plants. 25

21 Leaves all cauline, 4–6 cm long . . . . **160 POTAMOGETONACEAE**
21 Some or all leaves basal. 22

22 Several minute teeth present on both edges of the leaf near the tip (*Vallisneria*) . . . .
. . . . **157 HYDROCHARITACEAE**
22 Teeth absent from leaf margin 23

23 Plants slightly woolly-hairy. Leaves sword-shaped, 20–60 cm long . . . . . . **190 PHYLIDRACEAE**
23 Plants glabrous. Leaves linear 24

24 Leaves with a conspicuous midrib, broad, V-shaped in section particularly in the lower part . . . . . .
. . . . **199 SPARGANICEAE**
24 Leaves flat, without a conspicuous midrib. **158 JUNCAGINACEAE**

25 Leaves c. 10 mm wide, up to 1 m long . . . . **159 POSIDONIACEAE**
25 Leaves c. 5 mm wide or less 26

26 Stems erect. Leaves strictly basal . . . . **158 JUNCAGINACEAE**
26 Stems creeping 27

27 Leaves filiform, 1–2 mm wide (*Ruppia*) . . . . **160 POTAMOGETONACEAE**
27 Leaves 3–4 mm wide . . . . **162 ZOSTERACEAE**

Group 4
Mangroves

1 Leaves opposite . . . . **120 ACANTHACEAE**
1 Leaves alternate. . . . . **109 MYRSINACEAE**

Group 5
Rootless epiphytic parasites

1 Stems thread-like, leafless or almost so, forming tangled masses over the host and attached to the host by haustoria 2

1 Either; stems robust, erect or pendulous and with well developed leaves; or, stems flat, 3–9 mm wide and bearing scale-like leaves 3

2 Stems yellowish or reddish. Usually parasitic on herbs . . . . **134 CONVOLVULACEAE**

2 Stems green or with brown scale-leaves. Usually parasitic on shrubs or trees (*Cassytha*) . . . .
. . . . **38 LAURACEAE**

3 Leaves well developed . . . . **61 LORANTHACEAE**

3 Leaves reduced to scales (previously Viscaceae) . . . . **62 SANTALACEAE**

Group 6
Tendrils present

1 Tendrils (modified pinnae) at the end of a leaf, or the tip of the leaf prolonged into a coil 2

1 Tendrils on the stem (modified leaves or branches), or on the petiole (modified stipules) 4

2 Leaf tip prolonged into a coil. Leaves with parallel venation and a basal closed sheath. Plant bamboo-like . . . . **195 FLAGELLARIACEAE**

2 Leaves compound, the terminal pinnae modified into tendrils. 3

3 Trailing herbs with stems less than 1 m long. Leaflets 2–10 pairs (*Vicia*) . . . . **74 FABACEAE**

3 Stems several metres long, climbing. Leaflets 2–3 pairs . . . . **110 POLEMONIACEAE**

4 Tendrils on the petiole. . . . . **186 SMILACACEAE**

4 Tendrils on the vegetative stem, or on the peduncles 5

5 Leaves compound with 9 leaflets (*Cardiospermum*) . . . . **104 SAPINDACEAE**

5 Either; leaves compound with 5 leaflets; or, leaves simple. 6

6 Tendrils opposite leaves . . . . **.66 VITACEAE**

6 Tendrils in leaf axils, or on the stem near the bases of the petioles. 7

7 Tendrils in the leaf axils. Stipules present (sometimes small). Petioles usually with glands . . . .
. . . . **83 PASSIFLORACEAE**

7 Tendrils on the stem near the bases of and below the petioles. Stipules absent. Petioles without glands . .
. . . . **.73 CUCURBITACEAE**

Group 7
Leaves on aerial stems reduced to scales

1 Saprophytic herbs (one a climber) usually without chlorophyll . . . . **176 ORCHIDACEAE**

1 Chlorophyll present (some root parasites with a yellowish tinge in the cladodes) 2

2 Scale-leaves all bearing leaf-like cladodes in the axils . . . . **165 ASPARAGACEAE**

2 Leaf-like cladodes absent 3

3 Cladodes succulent. 4

3 Cladodes not succulent 5

4 Cladodes bearing numerous spines . . . . **52 CACTACEAE**

4 Cladodes without spines. Salt marsh plants (*Sarcocornia, Tecticornia*) . . . . **49 CHENOPODIACEAE**

5 Cladodes winged 6

5 Cladodes terete or flattened, not winged 7

6 Wings on the cladodes firm and fibrous (*Daviesia, Bossiaea*). . . . . **74 FABACEAE**
6 Wings on the cladodes herbaceous to membranous (*Tetratheca*). . . . . **87 ELAEOCARPACEAE**

7 Scale leaves in whorls of 4 or more, connate at the nodes . . . .**76 CASUARINACEAE**
7 Scale leaves alternate or spirally arranged. 8

8 Flowers in clusters at almost every node over the whole plant (dead flower parts persistent) (*Amperea*)
. . . .**78 EUPHORBIACEAE**
8 Flowers mainly at the ends of branches (dead flower parts persistent). 9

9 Total number of branches few (usually less than 12) on each plant (*Polygala*) . . .**75 POLYGALACEAE**
9 Shrubs with many branches. 10

10 Cladodes circular in section, not longitudinally striate when fresh (some becoming striate on drying)
. . . . 11
10 Cladodes angular, ± flattened especially near the tips, usually longitudinally striate when fresh . . 12

11 Scale leaves with hair-like tips, usually a few mm apart . . . . **62 SANTALACEAE**
11 Scale leaves with blunt tips, sometimes opposite. Internodes 1–several cm long (*Sphaerolobium,*
Spartium). . . . . **74 FABACEAE**

12 Cladodes often yellowish, or green but not glaucous . . . . **62 SANTALACEAE**
12 Cladodes usually glaucous (*Jacksonia*). . . . . **74 FABACEAE**

Group 8
Epiphytes

1 Leaves with parallel venation, or terete. Pseudobulbs and aerial roots with velamen often present
(Fig. 41) . . . .**176 ORCHIDACEAE**
1 Leaves with reticulate venation, or the venation obscure 2

2 Leaves alternate, 5–10 cm long. Plant finally rooting in the ground . . . . **137 QUINTINIACEAE**
2 Leaves opposite or whorled 3

3 Leaves sometimes whorled, succulent, c. 3 cm long or less . . . . **41 PIPERACEAE**
3 Leaves opposite; one leaf often smaller than the other in the pair; the larger 4–7 cm long . . . .
. . . . **122 GESNERIACEAE**

Group 9
Latex present in the leaves and young stems

1 Either; plants climbing or trailing; or, shrubs or trees. 2
1 Herbs. 8

2 Climbers or trailers 3
2 Shrubs or trees 6

3 Leaves alternate. 4
3 Leaves opposite 5

4 Stems herbaceous. Thorns absent. . . . . **134 CONVOLVULACEAE**
4 Stems woody. Thorns present (*Maclura*). . . . .**90 MORACEAE**

5 Either; two small glands present on the upper part of the petiole near the blade; or, leaf blades narrow or
rounded at the base . . . . **115 APOCYNACEAE**
5 Glands absent from the petiole but sometimes present on the midrib just above the petiole. Leaf
blades truncate or cordate at the base (if rounded then glands present on the midrib (previously
Asclepiadaceae). . . . . **115 APOCYNACEAE**

6 Leaves 1–2 cm wide (previously *Asclepiadaceae*) . . . . **115 APOCYNACEAE**
6 Leaves several cm wide 7

7 Leaves roughly triangular, truncate at the base (*Homalanthus*) . . . .**78 EUPHORBIACEAE**
7 Leaves not as above (*Ficus*) . . . .**90 MORACEAE**

8 Leaves all cauline, mostly less than 5 cm long 9
8 Either; leaves all basal or basal and cauline; or, all cauline and exceeding 5 cm long 11

9 Rhizomes and stolons absent. Plants erect or prostrate (*Euphorbia, Chamaesyce*). . . . .
. . . .**78 EUPHORBIACEAE**
9 Plants with rhizomes or stolons rooting at the nodes 10

10 Leaves mostly 1–2 mm wide, rarely 4 mm. Leaf margins usually entire . . . **144 CAMPANULACEAE**
10 Leaves mostly exceeding 5 mm wide. Leaf margins often dentate or lobed (previously *Lobeliaceae*). . .
. . . . **144 CAMPANULACEAE**

11 Leaves with bristly hairs on the petioles; or leaves blue-glaucous; or leaves mottled with white and less
than 4 cm wide (including the spines on the margins) . . . .**45 PAPAVERACEAE**
11 Leaves glabrous or with prickly hairs on the blades, c. 10 cm wide, the petioles not bristly-hairy nor the
leaves mottled with yellow (*Sylibum*) . . . . **143 ASTERACEAE**

Group 10
Leaves whorled, or interpetiolar stipules present

1 Leaves whorled, or stipules as large as leaves giving the appearance of 6–8 whorled leaves. 2
1 Leaves opposite; stipules interpetiolar. 12

2 Herbs. 3
2 Shrubs or trees 5

3 Stems not quadrangular. Leaves thick, broad, fleshy (*Peperomia*) . . . . **41 PIPERACEAE**
3 Stems angular, usually quadrangular, leaves narrow 4

4 Leaves in whorls of 4 or more (*Sheradia, Asperula, Galium*) . . . . **119 RUBIACEAE**
4 Leaves in whorls of 3 . . . . **68 LYTHRACEAE**

5 Leaves less than 25 mm long 6
5 Leaves mostly more than 30 mm long 9

6 Leaves 3–6 mm long, terete, grooved underneath, crowded on the stems. . . . . **108 ERICACEAE**
6 Leaves 6–25 mm long 7

7 Leaves (phyllodes) in whorls of 5–7 (*Acacia*) . . . . **74 FABACEAE**
7 Leaves in whorls of 3–4 rarely 5 8

8 Leaves in whorls of no more than 3 (*Mirbelia, Oxylobium, Aotus*). . . . . **74 FABACEAE**
8 Leaves in whorls of 3–5 often on the same shoot (*Tetratheca*) . . . . **87 ELAEOCARPACEAE**

9 Stipules present (*Caldcluvia*) . . . . **86 CUNONIACEAE**
9 Stipules absent 10

10 Leaves xeromorphic (*Helicia, Lambertia*) . . . . **42 PROTEACEAE**
10 Leaves mesomorphic 11

11 Leaves strongly lemon-scented . . . . **133 VERBENACEAE**
11 Leaves not lemon-scented (*Pittosporum*). . . . . **141 PITTOSPORACEAE**

12 Leaves simple, sometimes deeply dissected 13
12 Leaves compound. 17

13 Herbs 14
13 Trees, shrubs or climbers 15

14 Leaf margins entire or nearly so. . . . . **119 RUBIACEAE**
14 Leaf margins deeply dissected. . . . . **67 GERANIACEAE**

15 Leaf margins entire or nearly so. . . . . **119 RUBIACEAE**
15 Leaf margins dentate or serrate 16

16 Climber or straggling shrub . . . . **63 APHANOPETALACEAE**
16 Shrubs or trees. . . . . **86 CUNONIACEAE**

17 Herbs 18
17 Shrubs or trees. 19

18 Leaves cauline . . . . **71 ZYGOPHYLLACEAE**
18 Leaves mainly or all basal . . . . **67 GERANIACEAE**

19 Leaflet margins entire (*Eucryphia*) . . . . **86 CUNONIACEAE**
19 Leaflet margins serrate or dentate . . . . **86 CUNONIACEAE**

Group 11
Climbers. Leaves opposite

1 Prickles present on the stems 2
1 Prickles absent 3

2 Stems quadrangular (*Lantana*) . . . .**133 VERBENACEAE**
2 Stems rounded. Leaves with 3 main veins. . . . .**185 RIPOGONACEAE**

3 Leaves compound 4
3 Leaves simple or 1-foliolate 5

4 Leaflets 5–9 . . . . **121 BIGNONIACEAE**
4 Leaflets 3 (*Clematis*) . . . .**46 RANUNULACEAE**

5 Opposite leaves usually unequal in size. Leaves ovate to lanceolate, 4–7 cm long. Sometimes epiphytes
 . . . . **122 GESNERIACEAE**
5 Opposite leaves similar in size 6

6 Some or all the leaves hastate, semi-succulent, 1–3 (mostly c. 2) cm long, usually with glandular hairs
 (*Einadia*) . . . . **49 CHENOPODIACEAE**
6 Leaves not as above. 7

7 Two minute glands present on the upper part of the petiole close to the lamina. Stems trailing . . . . . .
 . . . . **115 APOCYNACEAE**
7 Glands absent from petiole 8

8 Numerous minute oil dots present in the leaves (*Palmeria*). . . . . **39 MONIMIACEAE**
8 Oil dots absent 9

9 Leaves xeromorphic, 2–5 cm long, with conspicuous reticulate venation (*Oxylobium*) . **74 FABACEAE**
9 Leaves soft. Leaf bases often joined by a ridge or flap of tissue 10

10 Juvenile leaves lobed or dissected. Young stems softly pubescent (*Lonicera*) . **150 CAPRIFOLIACEAE**
10 All leaves entire 11

11 Either; leaves c. 3 cm long, glabrous; or, under surface of leaves and young branches densely
 tomentose . . . . **1166 GELSEMIACEAE**
11 Leaves 5–12 cm long, glabrous (sometimes paler underneath) (*Parsonsia*) . . . **115 APOCYNACEAE**

Group 12
Climbers. Leaves alternate

1 Prickles or thorns present 2
1 Prickles and thorns absent 6

2 Axillary thorns present (*Maclura*) . . . . **90 MORACEAE**
2 Prickles present 3

3 Leaves compound 4
3 Leaves simple 5

4 Leaves 1-pinnate (*Rubus, Rosa*) . . . . **92 ROSACEAE**
4 Leaves 2-pinnate (Caesalpinioideae) . . . . **74 FABACEAE**

5 Leaves 3–5-lobed, rusty below (*Rubus*) . . . . **92 ROSACEAE**
5 Leaves entire, glabrous . . . . **186 SMILACACEAE**

6 Leaves compound 7
6 Leaves simple or 1-foliolate 8

7 Stipules present . . . . **74 FABACEAE**
7 Stipules absent (*Cephalaria*) . . . . **139 ARALIACEAE**

8 Stipules large, kidney-shaped. Leaves c. 6 cm long and 6 cm wide, 5–7-lobed . . . . **143 ASTERACEAE**
8 Stipules, if present, not as above 9

9 Leaves 2–3-pinnatisect (*Fumaria*) . . . . **45 PAPAVERACEAE**
9 Leaves entire or lobed but not pinnatisect 10

10 Stems producing tubers in the leaf axils . . . . **51 BASELLACEAE**
10 No such tubers formed 11

11 Leaves with parallel venation . . . . **184 PHILESIACEAE**
11 Leaves with palmate or reticulate venation 12

12 Leaves semi-succulent, ovate, with 5–7 principal veins (*Peperomia*) . . . . **41 PIPERACEAE**
12 Leaves not as above 13

13 Two stipelae present on the petiole near the base of the blade (Fig. 30). Small stipules present
(*Hardenbergia*) . . . . **74 FABACEAE**
13 No such leaflets present 14

14 Leaves linear to elliptic, c. 3 cm long or less, entire or undulate 15
14 Either; leaves more than 4 cm long (if c. 4 cm long then the bases cordate or sagittate); or, leaves
cordate, or orbicular, or peltate 16

15 Leaves usually undulate, silky or velvety or glabrous, broad-lanceolate to linear (*Billardiera*) . . . . . .
. . . . **141 PITTOSPORACEAE**
15 Leaves not undulate, few and distant, glabrous, linear to elliptic (*Comesperma*) . **58 POLYGALACEAE**

16 Leaves neither cordate nor sagittate at the base, ovate-lanceolate, definitely longer than wide 17
16 Leaves cordate or sagittate at the base, or c. as long as wide or deeply lobed 19

17 Leaves stem-clasping or petiolate and rounded at the base, 3–8 cm long. Young shoots and leaves
pubescent . . . . **47 DILLENIACEAE**
17 Leaves petiolate and not rounded at the base. Plants glabrous throughout 18

18 Petioles c. 6 mm long or less (*Celastrus*) . . . . **72 CELASTRACEAE**
18 Petioles more than 10 mm long (*Deeringia*) . . . . **50 AMARANTHACEAE**

19 Leaves leathery, almost semi-succulent, 3–5-lobed, dark green and glossy on the upper surface, light green underneath (*Hedera*) . . . . **139 ARALIACEAE**
19 Leaves not as above 20

20 Leaves peltate . . . . **44 MENISPERMACEAE**
20 Leaves not peltate 21

21 Midrib and 2 subsidiary veins extending from the top of the petiole to the tip of the leaf without interference from smaller veins. Underground tubers present . . . . **180 DIOSCOREACEAE**
21 Not as above 22

22 Leaf blades 5–20 cm wide, firm and tough, 3–5-veined at the base but not at the tip. . . . .
. . . . **44 MENISPERMACEAE**
22 Leaf blades usually less than 5 cm wide, soft; the veins not tough . . . . . **134 CONVOLVULACEAE**

Group 13
Liliopsida with cauline leaves

1 Leaves reduced to scarious sheaths on the aerial stems; green leaves absent except sometimes for green sheaths (Fig. 28) 2
1 Green leaves present on aerial stems 3

2 Sheaths open . . . . **198 RESTIONACEAE**
2 Sheaths closed (*Caustis*) . . . . **193 CYPERACEAE**

3 Leaves with a sheath and a blade; the sheath abruptly distinct from the blade 4
3 Leaves without a sheath, but often dilated into a broad base which is sometimes stem-clasping or stem-encircling, but not abruptly distinct from the blade 6

4 Sheath open. Leaves penniveined (Fig. 28i) . . . . **202 CANNACEAE**
4 Sheath closed 5

5 Leaf blades ovate-acuminate, herbaceous or semi-succulent, c. 2–6 times as long as broad . . . .
. . . . **189 COMMELINACEAE**
5 Leaf blades linear, 6 or more times as long as broad, soft or xeromorphic . . . . . **193 CYPERACEAE**

6 Stems erect, semi-woody. Leaves up to 40 cm long, 2–3 cm wide (*Cordyline*). . . . **170 ASTELIACEAE**
6 Otherwise 7

7 Leaves petiolate; the petioles 1–6 cm long; the blades 4–7 cm long, 10–15 mm wide . . . .
. . . . **181 ALSTROEMERIACEAE**
7 Leaves not petiolate as above (petioles absent or less than 1 cm long) 8

8 Bulb present. . . . . **183 LILIACEAE**
8 Bulb absent 9

9 Tubers present . . . . **176 ORCHIDACEAE**
9 Tubers absent 10

10 Leaves not twisted, flat or concave above or V-shaped in section or linear; bases green to pale green 11
10 Either; leaves flat, twisted; or, the bases green-grey or pale brown 12

11 Leaves tufted along the stem. . . . . **167 ANTHERICACEAE**
11 Leaves not tufted along the stem . . . . **173 HEMEROCALLIDACEAE**

12 Either; leaves with large irregular teeth at the tip; or, less than 50 cm long . . **177 LOMANDRACEAE**
12 Leaves without large irregular teeth at the tip, more than 50 cm long . **178 XANTHORRHOEACEAE**

Group 14
Liliopsida with basal leaves and corms bulbs or underground tubers present

1 Corms or bulbs present. Leaves usually mesomorphic .2

1 Corms or bulbs absent. 5

2 Corms present 3

2 Bulbs present 4

3 Leaves usually with an edge against (adaxial to) the stem. . . . . **175 IRIDACEAE**

3 Leaves with a surface against the stem (adaxial to) the stem . . . . **182 COLCHICACEAE**

4 Plants neither forming numerous hard bulbils around the main bulb nor with an odour of onions . . .
. . . .**168 AMARYLLIDACEAE**

4 Plants with an odour of onions or forming numerous hard bulbils around the main bulb . . . .
. . . . **163 ALLIACEAE**

5 Tubers hard (*Cyperus, Eleocharis*) . . . .**193 CYPERACEAE**

5 Tubers present, soft, succulent 6

6 Tubers usually globular, modified stems with buds. Leaves soft or semi-succulent, usually few, or the
leaf solitary . . . .**176 ORCHIDACEAE**

6 Tubers elongated, modified roots without buds . . . . **167 ANTHERICACEAE**

Group 15
Herbs (mostly Liliopsida). Leaves with parallel venation, arising in tufts or solitary.
Rhizomes present or the stem erect

1 Leaves (or green aerial stems) with transverse partitions, hollow, terete 2

1 Partitions in leaves or green aerial stems absent 4

2 Leaves or aerial stems c. 60 cm high or higher . . . .**193 CYPERACEAE**

2 Plants less than 40 cm high 3

3 Leaves with sheathing base or the leaves reduced to sheaths (Fig. 28 g–l). Both leaves and flowering
stems usually with partitions. Flowering stems longer than the leaves (*Juncus*) . . . .**196 JUNCACEAE**

3 Leaves without sheathing bases. Flowering stems without partitions and much shorter than the leaves
(*Lilaeopsis*) . . . . **138 APIACEAE**

4 Either leaves differentiated (abruptly) into a closed sheath which encircles the stem, and a blade; or,
leaves reduced to brown or colourless sheaths or scales on the rhizome (aerial parts are stems). 5

4 Green leaves present, with either a dilated base or an open sheath. 8

5 Green leaves present, with a closed sheath and a blade 6

5 Leaves reduced to sheaths or scales on the rhizome 7

6 Scattered hairs present on the leaves (*Luzula*) . . . .**196 JUNCACEAE**

6 Leaves glabrous or scabrous; the margins sometimes with cutting edges. . . . . . .**193 CYPERACEAE**

7 Aerial stems circular in section, without ridges or grooves and not increasing significantly in diameter
at the base near the rhizome (*Juncus*) . . . .**196 JUNCACEAE**

7 Aerial stems angular (often triangular), flat or ribbed and/or grooved; if circular in section the aerial
stems increasing to c. twice their diameter at the base near the rhizome . . . .**193 CYPERACEAE**

8 Leaves 8–15 cm wide. 9

8 Leaves c. or less than 2–3 cm wide 10

9 Leaves succulent . . . .**166 AGAVACEAE**

9 Leaves not succulent . . . .**172 DORYANTHACEAE**

10 Leaves 1–2m long 11
10 Leaves less than 1 m long 12

11 Leaves spongy; veins obscure. Plants amphibious . . . . **200 TYPHACEAE**
11 Leaves xeromorphic, with 12–15 parallel veins . . . . **155 ARACEAE**

12 Margins of dilated leaf bases coming together and fusing a few to several cm above the base; the leaf
  blade flat or terete. 13
12 Margins of the leaf bases not fused as above but the margins of the dilated base continuous individually
  with the margins of the blade 16

13 Margins fused for a few cm, then becoming free again; the leaf blade then V-shaped in section,
  becoming flat towards the middle and to . . . . **173 HEMEROCALLIDACEAE**
13 Margins fused together for the full length of the blade; the blade flat or terete 14

14 Lower part of the flowering stems round or flat and the same size and shape as the leaves . . . .
  . . . . **193 CYPERACEAE**
14 Lower part of the flowering stems round, or narrower than the leaves and different from the leaves in
  section 15

15 Leaf blades 1–3 mm wide, usually glabrous, rarely with a few hairs near or on the dilated base. Leaf
  blades mostly dark red-brown or blackish. . . . . **201 XYRIDACEAE**
15 Leaf blades 3–6 mm wide; if narrower, then the margins of the dilated base woolly hairy. Leaf bases
  not as dark as above. . . . . **175 IRIDACEAE**

16 Leaves less than 8 cm long, flat to filiform 17
16 Leaves more than 8 cm long 19

17 Leaves filiform. . . . . **192 CENTROLEPIDACEAE**
17 Leaves flat (wider than thick) 18

18 Leaves soft, herbaceous; veins conspicuous. Older leaves often transversely puckered. . . . .
  . . . . **194 ERIOCAULACEAE**
18 Leaves firm, rigid, midvein usually visible. **148 STYLIDIACEAE**

19 Leaves tapering from the middle downward; leaf bases sometimes dilated. *Magnoliopsida*. . . . .
  . . . . **See GROUP 21**
19 Leaves expanding downward from the tip to the base, sometimes the sides parallel, but never
  contracted below the middle. *Liliopsida* 20

20 Leaves herbaceous 21
20 Leaves xeromorphic 25

21 Leaf bases usually coloured, the stems deeply buried 22
21 Leaf bases usually brown, the stems not deeply buried. 23

22 Leaves terete, succulent . . . . **169 ASPHODELIACEAE**
22 Leaves flat . . . . **167 ANTHERICACEAE**

23 Leaves V-shaped in section. Aerial stems angular . . . . **193 CYPERACEAE**
23 Leaves flat or concave and rounded on the back 24

24 Leaf bases dilated to c. twice the width of the middle of the blade (*Juncus*). . . . . . **196 JUNCACEAE**
24 Leaf bases dilated to 3–6 times the width of the middle of the blade. . . . . **201 XYRIDACEAE**

25 Deeper underground parts orange to bright red . . . . **188 HAEMODORACEAE**
25 Otherwise 26

26 Leaves borne spirally on a massive ± woody caudex (sometimes above ground). Leaf bases ± woody. .
  . . . . **178 XANTHORRHOEACEAE**

26 Short rhizomes present. Leaf bases not woody 27

27 Either; leaves with a distinct midrib which is paler than the adjacent tissue; or, the leaf blade V-shaped in section in the lower part 28
27 Otherwise 29

28 Leaves with the adaxial surface folded and fused on either side of the midrib for a short length just above the sheath . . . . **173 HEMEROCALLIDACEAE**
28 Otherwise . . . . **171 BLANDFORDIACEAE**

29 Leaf bases pale (whitish cream or pale brown) sometimes with darker margins . . . .
. . . . **177 LOMANDRACEAE**
29 Leaf bases uniformly brown or red-brown . . . . **193 CYPERACEAE**

Group 16
Herbs with compound leaves

1 Leaflets 2–3 2
1 Leaflets more than 3 or the leaf pinnate bipinnate or pinnate-pinnatisect 7

2 Leaves with a very acid taste. Leaflets notched or forked. Bases of the petioles sometimes expanded and stipule-like . . . . **88 OXALIDACEAE**
2 Leaves not acid 3

3 Stipules present 4
3 Stipules absent 6

4 Leaflets 2–3 with entire margins or serrulate. Leaves mainly cauline; the stems sometimes rooting at the nodes (*Zornia, Trifolium, Meliolotus, Medicago, Canavalia, Kennedia, Glycine*) . . . . . **74 FABACEAE**
4 Leaflets conspicuously toothed or lobed; leaflets 3. Stolons or rhizomes present 5

5 Leaflets ± evenly toothed (*Duchesnea*) . . . . **92 ROSACEAE**
5 Leaflets deeply notched to c. half-way . . . . **138 APIACEAE**

6 Leaflets dissected. Leaves basal (*Ranunculus*) . . . . **46 RANUNCULACEAE**
6 Leaflets entire, toothed. Leaves cauline (*Bidens*) . . . . **143 ASTERACEAE**

7 Leaflets 5 or leaflets several to many and digitately arranged 8
7 Otherwise 9

8 Leaflets 5, palmately arranged . . . . **138 APIACEAE**
8 Leaflets 5, digitately arranged or leaflets several (*Lotus, Swainsona*) . . . . **74 FABACEAE**

9 Terminal leaflet or pinna absent; leaves pinnate or bipinnate 10
9 Terminal leaflet or segment always present 11

10 Leaves pinnate . . . . **71 ZYGOPHYLLACEAE**
10 Leaves bipinnate (*Neptunia*) . . . . **74 FABACEAE**

11 Stipules present 12
11 Stipules absent, or the leaf bases dilated into a sheath 13

12 Leaf margins entire (*Swainsona*) . . . . **74 FABACEAE**
12 Leaf margins toothed (*Geum, Potentilla, Acaena*) . . . . **92 ROSACEAE**

13 Leaf bases dilated into a sheath which clasps the stem; the sheath not bearing leaf segments . . . .
. . . . **138 APIACEAE**
13 Leaf bases not forming a sheath; if dilated green leaf segments usually present on the expanded base 14

14 Leaves opposite (*Bidens, Cosmos, Tagetes*) . . . . **143 ASTERACEAE**
14 Leaves alternate or alternate and basal 15

15 Leaflets or ultimate segments mostly 2–3 cm wide. Leaves 2–3-pinnate-pinnatisect (*Lycopersicon*) . . . .**135 SOLANACEAE**

15 Leaflets or ultimate segments filiform or less than 1 cm wide (rarely 2 cm wide and if so then the leaves once pinnate) . . . . **143 ASTERACEAE**

Group 17
Shrubs and trees. Leaves compound

1 Leaves opposite 2
1 Leaves alternate 4

2 Leaves with oil glands and a strong odour when crushed . . . .**103 RUTACEAE**
2 Leaves without oil glands 3

3 Leaves sessile; leaflets 3, entire or serrulate, up to 12 mm long (*Bauera*) . . . . . . **86 CUNONIACEAE**
3 Leaves petiolate; leaflets 3–11, serrate or coarsely toothed, 3–12 cm long (*Sambucus*) **149 ADOXACEAE**

4 Thorns or prickles present 5
4 Thorns and prickles absent 9

5 Thorns present (Caesalpinioideae) . . . . **74 FABACEAE**
5 Prickles present, or the stipules spinescent 6

6 Leaves bipinnate (*Caesalpinia*) . . . . **74 FABACEAE**
6 Leaves pinnate or with 3 leaflets 7

7 Stipules spinescent (*Robinia*) . . . . **74 FABACEAE**
7 Prickles on the stem 8

8 Leaflets usually 5 (*Geum*) . . . . **92 ROSACEAE**
8 Leaflets 3 (*Erythrina*) . . . . **74 FABACEAE**

9 Leaves consistently 3-foliolate . . . . **74 FABACEAE**
9 Leaves bipinnate or pinnate or pinnate-pinnatisect 10

10 Leaves bipinnate with distinct undivided leaflets 11
10 Either; leaves pinnate, with distinct undivided leaflets; or, pinnate-pinnatisect 13

11 Margins or leaflets entire; leaflets linear oblong or narrow-lanceolate (*Mimosoideae*) . . **74 FABACEAE**
11 Leaflets toothed, rarely entire and if so then broader than above 12

12 Young petioles and sometimes the leaflets with stellate hairs. Deciduous tree . . . . **102 MELIACEAE**
12 Hairs absent. Evergreen shrubs (*Polyscias*) . . . . **139 ARALIACEAE**

13 Leaves pinnate-pinnatisect . . . . **42 PROTEACEAE**
13 Leaves pinnate, with distinct undivided leaflets 14

14 Gland present at the junction of one or more lateral veins and the midvein of the leaflet on the undersurface; the gland either sunken and hairy or raised and ± triangular (Fig. 30e) 15
14 Glands absent 16

15 Leaflets 2–4 (rarely 5) pairs . . . . **101 ANACARDIACEAE**
15 Leaflets 3–12 pairs . . . . **102 MELIACEAE**

16 Fine spines c. 2 mm long, projecting from the leaflet margins. Leaves imparipinnate; leaflets mostly 3 pairs, c. 6 cm long and 3 cm wide . . . . **43 BERBERIDACEAE**
16 Spines absent 17

17 Leaves paripinnate. (Main rhachis sometimes slightly prolonged but the prolongation is shorter than the adjacent pinnae) 18
17 Leaves imparipinnate 19

18 Either; stipules (or 2 stipular scars) present; or, stems with thorns on the older parts. Leaflets opposite, mostly less than 3 cm long, flat concave or terete (*Caesalpinioideae*). . . . . **74 FABACEAE**

18 Stipules absent. Leaflets commonly more than 3 cm long, opposite subopposite or alternate, flat . . . .
. . . . **104 SAPINDACEAE**

19 Main rhachis of leaf jointed at the points of attachment of the leaflets. . . . . **139 ARALIACEAE**

19 Rhachis not jointed 20

20 Leaflets with serrate margins; the serrations 1–2 mm deep, regular and pointing towards the tips . . .
. . . . **92 ROSACEAE**

20 Margins entire or with smaller teeth than above 21

21 Leaves including the petioles) c. 10 cm long or less; stipules sometimes present (*Gompholobium, Indigofera, Tephrosia, Psoralea*) . . . . **74 FABACEAE**

21 Leaves more than 15 cm long; stipules absent 22

22 Leaflets c. 1 cm wide or less . . . .**101 ANACARDIACEAE**

22 Leaflets c. 2 cm wide or wider . . . .**105 SIMAROUBACEAE**

Group 18
Herbs. Leaves simple, opposite; venation reticulate

1 Leaves succulent, and/or the leaves with glandular hairs 2

1 Leaves not succulent. Hairs, if present, not glandular 5

2 Leaves cylindrical or triangular in cross-section. 3

2 Leaves flat 4

3 Leaves c. 1 cm long or less. . . . .**64 CRASSULACEAE**

3 Leaves 2–8 cm long. . . . . **48 AIZOACEAE**

4 Leaves glabrous. . . . .**64 CRASSULACEAE**

4 Leaves with hairs or lens-shaped vesicles . . . . **49 CHENOPODIACEAE**

5 Oil glands present in the leaves. . . . .**80 HYPERICACEAE**

5 Oil glands absent 6

6 Leaves with 2 glands on the upper surface of the petiole close to the blade. Stems trailing . . . .
. . . . **115 APOCYNACEAE**

6 Glands absent from the petiole. Stems not trailing, erect to decumbent 7

7 Leaves bullate (*Chloanthes*) . . . .**123 LAMIACEAE**

7 Leaves not bullate 8

8 Stems quadrangular especially near the tips and distinctly 4-ribbed 9

8 Stems not quadrangular. 13

9 Leaves with stinging hairs, dentate to serrate (*Urtica*). . . . .**93 URTICACEAE**

9 Stinging hairs absent. 10

10 Leaves serrate or toothed near the apex, often scabrous (*Phyla, Verbena*) . . . .
. . . .**133 VERBENACEAE**

10 Leaf margins entire or minutely denticulate. 11

11 Stems decumbent near the base, then ascending (*Anagallis*) . . . .**109 MYRSINACEAE**

11 Stems erect. 12

12 Some larger basal leaves usually present at the base of the plant which branches mainly in the lower portion . . . . **117 GENTIANACEAE**

12 Basal leaves absent. Plant branched top and bottom . . . . **68 LYTHRACEAE**

13 Leaves and stems scabrous; the leaf margins serrulate toothed or entire . . . . **65 HALORAGACEAE**
13 Leaves and stems not scabrous 14

14 Leaves petiolate 15
14 Leaves sessile 19

15 Leaf margins crenate toothed or lobe. 16
15 Leaf margins entire (sometimes wavy). 17

16 Plants erect, sometimes decumbent at the base . . . . **143 ASTERACEAE**
16 Plants with weak stems, often trailing and rooting at the nodes . . . . . **132 SCROPHULARIACEAE**

17 Large fleshy tap-root present. Stems prostrate. Paired leaves sometimes unequal in size . . . .
. . . . **55 NYCTAGINACEAE**
17 Otherwise 18

18 Rhizome present . . . . **120 ACANTHACEAE**
18 Rhizome absent. Chaffy or spiny floral remains often present in the leaf axils . . . .
. . . . **50 AMARANTHACEAE**

19 Leaves c. 1 cm long or less; the bases connate. Plants very fragile (*Mitrasacme*) . **118 LOGANIACEAE**
19 Leaves larger than above. 20

20 Leaves and stems with minute appressed hairs, or almost scabrous . . . . **143 ASTERACEAE**
20 Leaves glabrous or with soft hairs. 21

21 Leaves 1–3 cm wide; the bases of the opposite leaves sometimes connate . **132 SCROPHULARIACEAE**
21 Leaves 6–8 mm wide (*Epilobium*). . . . .**70 ONAGRACEAE**

Group 19
Trees and shrubs. Leaves simple, opposite

1 Leaves very scabrous on the upper surface (like sandpaper) 7–15 cm long, c. 5 cm wide. Tree or tall
shrub (*Ficus*) . . . .**90 MORACEAE**
1 Leaves not as above (sometimes scabrous but the leaves much smaller than above, when the plants are
less than 1 m high) 2

2 Leaves with oil glands (sometimes minute) and aromatic when crushed 3
2 Oil glands absent. 8

3 Leaf margins toothed 4
3 Leaf margins entire or rarely serrulate. 5

4 Leaves not aromatic. Fruit a drupe. . . . . **39 MONIMIACEAE**
4 Leaves aromatic. Fruit an achene. . . . . **37 ATHEROSPERMATACEAE**

5 Stems and petioles cinnamon brown, glabrous. Petiole 8–10 mm long; leaf blades 8–15 cm long, 3–4 cm
wide, with a distinct midrib. RF trees . . . . **38 LAURACEAE**
5 Otherwise 6

6 Leaves sessile, stem-clasping, thin, herbaceous, 5–10 cm long, 5 cm wide, obtuse at the tip . . . .
. . . .**80 HYPERICACEAE**
6 Otherwise 7

7 Plants with one of the following combinations of characters: (i) leaves and/or stems with stellate hairs;
or (ii) the leaf margins serrulate; or (iii) stems weak and the leaves 3–4 mm wide and soft; or (iv) leaf
blades articulate on the petiole . . . .**103 RUTACEAE**
7 Leaves and stems glabrous or with simple hairs; leaves petiolate or sessile, entire, sometimes 3-veined,
variable in size, sometimes xeromorphic and terete . . . .**69 MYRTACEAE**

8 Leaves semi-succulent or succulent. Young leaves usually with glandular hairs; older leaves mealy or covered with a felt-like layer and tasting of salt (*Einadia, Rhagodia, Chenopodium*) . . . . **49 CHENOPODIACEAE**
8 Otherwise 9

9 Stipules present. Leaves xeromorphic (*Chorizema, Oxylobium, Platylobium, Bossiaea*) . . **74 FABACEAE**
9 Stipules absent 10

10 Leaves scabrous; leaf margins mostly serrulate dentate or incised . . . . **65 HALORAGACEAE**
10 Leaves not scabrous. 11

11 Small shrubs with a bark with a strong fibre which is difficult to break and is readily stripped from the woody or leafy shoots when a specimen is picked. Leaf bases persistent on the stem as sharp protuberances . . . . **100 THYMELAEACEAE**
11 Otherwise 12

12 Leaves silvery white with copious hairs below, up to 10 cm long; the margins entire or toothed (*Celmisia*) . . . . **143 ASTERACEAE**
12 Otherwise 13

13 Leaf bases united or joined by a raised line or flap of tissue 14
13 Leaf bases not united or so joined. 15

14 Leaves acuminate or linear . . . . **118 LOGANIACEAE**
14 Leaves with blunt tips, sometimes with a mucro . . . . **150 CAPRIFOLIACEAE**

15 Leaves 1–3 mm wide (*Tetratheca*). **87 ELAEOCARPACEAE**
15 Leaves much wider 16

16 Leaf margins crenate dentate or serrate 17
16 Leaf margins entire 19

17 Leaf margins crenate; the teeth blunt (*Elaeodendron*) . . . . **72 CELASTRACEAE**
17 Leaf margins dentate or serrate; the teeth sharply pointed, almost spinescent 18

18 Leaves xeromorphic, with conspicuous reticulate fibrous venation. Adult leaves with entire margins (*Xylomelum*). . . . . **42 PROTEACEAE**
18 Leaves mesomorphic (RF species); venation reticulate but not strongly fibrous. . **136 POLYOSMACEAE**

19 Stems with an annular thickening c. 1 cm above each node (*Baloghia*) . . . . . **78 EUPHORBIACEAE**
19 Otherwise 20

20 Leaves (especially the young ones) tomentose underneath or on both surfaces (*Clerodendrum, Gmelina*). . . . . **123 LAMIACEAE**
20 Leaves glabrous on both surfaces 21

21 Leaves equally green on both surfaces (*Emmenosperma*) . . . . **91 RHAMNACEAE**
21 Leaves paler underneath 22

22 Petioles 2–4 cm long; leaf blades 10–16 cm long (*Xylomelum*). . . . . **42 PROTEACEAE**
22 Otherwise 23

23 Stems angular and ± ribbed. Leaf bases persisting after leaf-fall. Leaves 2–5 cm long, c. 8 mm wide. Petiole 1–3 mm long. Root parasite. . . . . **61 LORANTHACEAE**
23 Stems smooth, round or oval in section (sometimes slightly quadrangular in young stems). Leaves longer and/or broader than above and the petioles usually longer 24

24 Leaves glaucous above, flat; branches brittle and fragile, easily broken (*Santalum*) . . . . **62 SANTALACEAE**
24 Leaves glabrous above (except *Olea*). Branches tough, not brittle. . . . . **126 OLEACEAE**

Group 20

Herbs. Leaves alternate; stipules present. Magnoliopsida

1 Either; leaves linear or lanceolate; or, leaves c. 1 cm long or less 2

1 Leaves broad, reniform orbicular or hastate, entire or deeply incised or cordate-hastate.. 4

2 Leaves c. 1 cm long or less (*Poranthera*) . . . . **79 PHYLLANTHACEAE**

2 Leaves 1–5 cm long. 3

3 Stems tufted, arising from a rootstock. Leaves alternate, often crowded on the lower part of the stems (*Stackhousia*) . . . . **.72 CELASTRACEAE**

3 Stems solitary; upper leaves sometimes opposite (*Hybanthus*) . . . . **.85 VIOLACEAE**

4 Most or all the leaves arising from underground stems or at ground level from an underground stem. Stolons sometimes present 5

4 Plants erect or decumbent, sometimes rooting at the nodes, but all leaves arising from aerial stems . . 6

5 Leaves mainly basal, tufted on an underground stem (*Viola*). . . . . **.85 VIOLACEAE**

5 Leaves not tufted but arising remotely from creeping underground stems or from stolons (*Centella*, *Hydrocotyle*). . . . . **138 APIACEAE**

6 Hairs stellate, or stellate and simple with the stellate hairs on the veins underneath the leaf. . . . . **99 MALVACEAE**

6 Hairs simple or absent 7

7 Petioles c. as long as or shorter than the blades. . . . . **99 MALVACEAE**

7 Petioles 2–4 times as long as the blades 8

8 Leaves with simple hairs. . . . . **67 GERANIACEAE**

8 Leaves glabrous (*Viola*) . . . . **.85 VIOLACEAE**

Group 21

Herbs. Leaves basal, tufted or the leaf solitary; venation reticulate

1 Basal leaves with pungent pointed spines on the margins. 2

1 Pungent pointed spines absent 3

2 Leaves divided into ± linear lobes (*Eryngium*) . . . . **138 APIACEAE**

2 Leaves lobed but the lobes not linear (tribe *Cynareae*). . . . . **143 ASTERACEAE**

3 Basal leaves with entire margins, or lobed dentate cordate or sagittate 4

3 Basal leaves deeply lobed or pinnatisect 18

4 Leaf blades 10–45 cm long, ovate, cordate peltate or sagittate; petioles c. as long as or longer than the blades, soft, semi-succulent, grooved . . . . **155 ARACEAE**

4 Leaves not as above. 5

5 Either; leaves with several (up to 7, rarely 1) ± equal longitudinal veins; or, the leaf cordate and only one per plant 6

5 Either; leaves with a single main vein; or, the leaves succulent and the veins obscure 7

6 Underground tubers present . . . . **179 ORCHIDACEAE**

6 Tubers absent . . . . **127 PLANTAGINACEAE**

7 Either; leaves succulent; or, leaves thinly herbaceous and the whole plant readily crushed to pulp . . . 8

7 Leaves not succulent 9

8 Aerial stems ascending (often weak) leafy. Basal leaves 30–40 mm long, 10–20 mm wide (*Samolus*) . . . . **.113 THEOPHRASTACEAE**

8 Aerial stems decumbent, almost leafless. Basal leaves 15–25 mm long, linear-terete or oblong (*Calandrinia*) . . . . **59 PORTULACACEAE**

9 Basal leaves mostly more than 10 cm long, up to 10 cm wide 10
9 Basal leaves mostly c. 10 cm long or less (if longer the leaf linear or narrow-spathulate) 11

10 Petioles of the lowest leaves winged . . . . **135 SOLANACEAE**
10 Petioles not winged or short and thick . . . . **132 SCROPHULARIACEAE**

11 Plants of the estuarine coast. Leaves obovate-oblong, entire, 3–8 cm long, petiolate . . . .
. . . . **57 PLUMBAGINACEAE**
11 Plants found elsewhere 12

12 Leaves very hairy to woolly, or white on one or both surfaces 13
12 Leaves glabrous or with few hairs 15

13 Trailing or stoloniferous branches produced from the centre of the basal rosette of leaves (*Velleia*) . . .
. . . . **145 GOODENIACEAE**
13 Trailing branches absent 14

14 Petioles one quarter as long as the leaf blade, ± terete near the base (the base slightly dilated), grooved above. Leaves similar on both surfaces (*Brunonia*) . . . . **145 GOODENIACEAE**
14 Petioles absent or flat, shorter than above, usually winged. Leaves sometimes white on one or both surfaces . . . . **143 ASTERACEAE**

15 Bases of the leaves reddish-purple. Leaves linear; the margins often minutely serrulate . . . .
. . . . **148 STYLIDIACEAE**
15 Otherwise 16

16 Leaf blades reniform; petioles 3–5 times as long as the blades . . . .
. . . . **134 CONVOLVULACEAE**
16 Otherwise 17

17 Inflorescence a head . . . . **143 ASTERACEAE**
17 Inflorescence not a head . . . . **145 GOODENIACEAE**

18 Stems elongating from the basal rosette and becoming scrambling. Leaves finely dissected into flattish segments. Stems and leaves very soft and easily crushed to a pulp (*Fumaria*) . . . **45 PAPAVERACEAE**
18 Plants never scrambling 19

19 Plants with an odour such as aniseed carrots or parsnips. Leaf bases slightly sheathing; leaf segments filiform . . . . **138 APIACEAE**
19 Plant without such an odour 20

20 Petioles much longer than the blades 21
20 Petioles shorter than the blades or the leaves sessile 23

21 Leaves twice pinnatisect . . . . **45 PAPAVERACEAE**
21 Leaves once divided 22

22 Leaf blades palmately lobed or dissected into 3 (*Clematis*) . . . . **46 RANUNCULACEAE**
22 Leaf blades pinnately dissected into 3–7 lobes . . . . **143 ASTERACEAE**

23 Leaves with an odour of mustard or cabbage . . . . **94 BRASSICACEAE**
23 No such odour present . . . . **143 ASTERACEAE**

Group 22
Herbs. Leaves cauline, less than 25 mm long. Magnoliopsida

1 Leaves succulent 2

1 Leaves not succulent 7

2 Leaves bead-like, c. 5 mm long . . . . **.64 CRASSULACEAE**
2 Leaves not bead-like 3

3 Plants prostrate or decumbent 4
3 Plants erect 6

4 Leaves linear to narrow-oblong, c. 1 cm long or less (*Wilsonia*) . . . . **134 CONVOLVULACEAE**
4 Leaves spathulate to ovate 5

5 Clusters of leaves or inflorescences present in most or all of the leaf axils. Stems not red or brown
(*Rhagodia, Dysphania*). . . . . **49 CHENOPODIACEAE**
5 Otherwise. Leaves oblong-cuneate to obovate, 10–25 mm long. Stems often red or brown (*Portulaca*) . .
. . . . **59 PORTULACACEAE**

6 Plants with some spathulate basal leaves which are flat and glabrous (*Samolus*) . . . .
. . . . **113 THEOPHRASTACEAE**
6 Either; leaves terete; or, flat. With glandular hairs or a mealy surface (*Suaeda, Atriplex, Cheopodium*). .
. . . . **49 CHENOPODIACEAE**

7 Leaves and young stems with a faint odour of mustard or cabbage when crushed . . **94 BRASSICACEAE**
7 No such odour present. 8

8 Axillary buds forming small tufts of leaves in the leaf axils 9
8 No tufts of leaves present 10

9 Stems erect . . . . **70 ONAGRACEAE**
9 Stems prostrate . . . . **145 GOODENIACEAE**

10 Leaves glabrous 11
10 Leaves variously hairy 15

11 Stems prostrate 12
11 Stems erect or decumbent 13

12 Plants ± circular in outline; the branches radiating from a single central stem (*Soliva*) . . . .
. . . . **143 ASTERACEAE**
12 Stems irregularly placed on the soil surface, sometimes forming mats which are not circular in outline
(previously Lobeliaceae) . . . . **144 CAMPANULACEAE**

13 Leaves mostly 25 mm long (*Thesium*) . . . . **62 SANTALACEAE**
13 Leaves less than 20 mm long. 14

14 Leaves lobed or minutely toothed or with a minutely honeycombed surface . . . . **143 ASTERACEAE**
14 Leaves entire, not honeycombed . . . . **.81 LINACEAE**

15 Leaves triangular, truncate or cordate at the base, mostly less than 10 mm long . . . .
. . . . **132 SCROPHULARIACEAE**
15 Leaves otherwise . . . . **143 ASTERACEAE**

Group 23
Herbs. Leaves cauline, alternate, more than 25 mm long. Magnoliopsida

1 Leaves covered with water-filled vesicles . . . . **48 AIZOACEAE**
1 Vesicles absent 2

2 Plants growing above high-tide mark on sand or forming mats on coastal dunes 3
2 Otherwise 5

3 Plant growing on beach above high-tide mark. Leaves usually irregularly lobed (*Cakile*). . . . .
. . . . **94 BRASSICACEAE**
3 Plants forming mats on coastal dunes 4

4 Either; upper leaves stem-clasping or auriculate, green; or, leaves silvery white (*Senecio, Arctotis,*
Cryptostemma) . . . . **143 ASTERACEAE**
4 Upper leaves not stem-clasping or auriculate, green (*Scaevola*) . . . . **145 GOODENIACEAE**

5 Leaves with hairs especially when young 6
5 Leaves glabrous 14

6 Hairs glandular 7
6 Hairs not glandular 9

7 Leaves succulent or semi-succulent, sometimes hastate, often tasting of salt . . **49 CHENOPODIACEAE**
7 Otherwise 8

8 Leaves sessile. Stems slightly ribbed. Plant with a yellowish tinge . . . . **143 ASTERACEAE**
8 Leaves very shortly petiolate. Stems not ribbed. Plant green, sometimes the leaves purplish underneath
. . . . **132 SCROPHULARIACEAE**

9 Leaves deeply lobed (more than half-way) or pinnatisect 10
9 Leaves entire or toothed, or with shallow lobes, or the leaf hastate or cordate 11

10 Leaves mostly 3-lobed; the lobes sometimes symmetrically dissected into linear or spathulate lobes
(*Actinotus, Xanthosia*) . . . . **138 APIACEAE**
10 Otherwise (if 3-lobed, then axillary spines present) . . . . **143 ASTERACEAE**

11 Plants very fragile, with soft herbaceous stems and leaves which are readily crushed to a pulp. . . . . .
. . . .**93 URTICACEAE**
11 Otherwise 12

12 Hairs stellate . . . . **145 GOODENIACEAE**
12 Hairs simple, sometimes matted 13

13 Plants erect . . . . **143 ASTERACEAE**
13 Plants usually trailing or decumbent (*Goodenia, Coopernookia*) . . . . **145 GOODENIACEAE**

14 Leaf margins entire or slightly crenate, not reniform 15
14 Leaves toothed, lobed, dissected or reniform 18

15 Plant prostrate, rooting at the nodes. Leaves thick, shiny above, 2–8 cm long (*Selliera*) . . . .
. . . . **145 GOODENIACEAE**
15 Otherwise 16

16 Leaves peltate, with the odour and taste of mustard . . . . **98 TROPAEOLACEAE**
16 Otherwise 17

17 Leaf blades 5–15 cm long, semi-succulent. Stems erect, 1–2 m high . . . . . **56 PHYTOLACCACEAE**
17 Leaf blades smaller than *Phytolaccaceae*. Stems erect and shorter or decumbent. . . . .
. . . . **50 AMARANTHACEAE**

18 Plants erect 19
18 Plants ± prostrate or trailing, sometimes rooting at the nodes 21

19 Leaf bases not stem-clasping or decurrent . . . .**135 SOLANACEAE**
19 Leaf bases usually stem-clasping or decurrent 20

20 Leaf bases stem-clasping (Petiolate in some species of *Senecio*) . . . . **143 ASTERACEAE**
20 Leaf bases decurrent . . . . **145 GOODENIACEAE**

21 Petioles as long as the blades or longer 22
21 Petioles shorter than the blades 23

22 Leaves divided to half-way or deeper into usually 3 (or 5) lobes. Petioles c. twice as long as the blades
. . . . **46 RANUNCULACEAE**
22 Leaves shallowly 5-lobed. Petioles c. as long as the blades or longer . . . . **138 APIACEAE**

23 Plants with an unpleasant odour. Leaves not stem-clasping. . . . . **94 BRASSICACEAE**
23 Otherwise. Leaves stem-clasping . . . . **143 ASTERACEAE**

Group 24
Shrubs. Leaves alternate. Spinescent stipules, or prickles or thorns present on the stem

1 Spinescent stipules present (*Mimosoideae*) . . . . **74 FABACEAE**
1 Prickles or thorns present on the stem. 2

2 Prickles present on the stem 3
2 Thorns present on the stem 4

3 Stipules present . . . . **99 MALVACEAE**
3 Stipules absent . . . . **135 SOLANACEAE**

4 Leaves deeply and symmetrically lobed. Stipules leafy. Thorns in the axils and usually terminating
branches. . . . . **92 ROSACEAE**
4 Leaves not deeply lobed 5

5 Thorns present only below the nodes, apparently subtending leafy shoots . . . . **43 BERBERIDACEAE**
5 Otherwise 6

6 Leaves succulent, clustered at the nodes. Thorns mainly at the ends of the branches . **135 SOLANACEAE**
6 Leaves not succulent 7

7 Leaves dentate. Thorns axillary and terminal. 8
7 Leaves entire 9

8 Leaves up to 4 cm long, elliptic to linear or spathulate, tapering into a short petiole . .**85 VIOLACEAE**
8 Leaves c. 1 cm long, ovate to orbicular, almost sessile . . . . **141 PITTOSPORACEAE**

9 Leaves soft. Thorns axillary or near the nodes or terminal 10
9 Leaves xeromorphic. Thorns terminal only 11

10 Leaves 40–50 mm long, in clusters of c. 6 on short axillary shoots . . . . **92 ROSACEAE**
10 Leaves 10–25 mm long, not regularly clustered . . . . **141 PITTOSPORACEAE**

11 Either; leaves and tips of branches all pungent pointed; or, leaves not clustered . . . . **74 FABACEAE**
11 Tops of branches only pungent pointed and leaves clustered . . . . **91 RHAMNACEAE**

Group 25
Woody plants. Leaves alternate, either with oil dots or succulent

1 Leaves with oil dots, aromatic when crushed 2
1 Leaves succulent or semi-succulent, sometimes dotted with glands but not aromatic when crushed . . 6

2 Base of leaf blade truncate or auriculate. Leaves nearly sessile, tasting of pepper. . . **36 WINTERACEAE**
2 Otherwise 3

3 Leaves distichous, lying in one plane, mostly 12–15 cm long, 3–4 cm wide; petioles 6–8 mm long . . . .
. . . . **40 EUPOMATIACEAE**
3 Otherwise 4

4 Leaves mostly 5–15 cm long, 3–4 cm wide, with reticulate venation, distinctly dorsi-ventral; petioles 1–2 cm long. Mesomorphic plants from RF . . . . **38 LAURACEAE**

4 Leaves otherwise, much smaller than above; or, if as large or larger, then the leaves penniveined or isobilateral. Plants mostly not in RF 5

5 Either; oil glands often protruding as 'warts'; or, scurfy scales or stellate hairs present. Leaf blades sometimes auriculate on the petiole. Leaves usually soft and with little vascular tissue **103 RUTACEAE**

5 Leaves mostly xeromorphic, with tough veins. Larger leaves as in Lauraceae but isobilateral and penniveined; smaller leaves often terete or flat and less than 5 mm wide. Adult leaves usually glabrous
. . . . **69 MYRTACEAE**

6 Leaves succulent-leathery. Bark papery; inner layers red . . . . **42 PROTEACEAE**

6 Otherwise 7

7 Leaves terete, 1–2 mm diam., or flat and with glandular hairs, sometimes becoming mealy, often tasting of salt. . . . . **49 CHENOPODIACEAE**

7 Leaves flat, glabrous or with tufts of white hairs 8

8 Leaves glabrous. . . . . **69 MYRTACEAE**

8 Leaves with tufts of white hairs. . . . .**143 ASTERACEAE**

Group 26
Woody plants. Leaves alternate, terete (any length), or leaves not terete but less than 3 cm long

1 Leaves yellowish, distichous, 5–20 mm long. Shrub 1–2 m high, hemiparasitic . . . . **60 OLACACEAE**

1 Otherwise 2

2 Bark with a strong fibre which is difficult to break when a specimen is picked. Leaf bases persistent on the stem as hard protuberances. . . . . **100 THYMELAEACEAE**

2 Otherwise 3

3 Leaves terete, smooth, without ridges or grooves. 4

3 Leaves not as above. 5

4 Leaves 2–8 cm long, contracted at the base into a very short petiole or almost sessile **42 PROTEACEAE**

4 Either; leaves (sometimes regarded as cladodes) mostly 10–20 cm long; or, leaves 2–3 cm long and not contracted at the base . . . . **74 FABACEAE**

5 Either; stipules present; or, leaves 2–3 at each node on one side of the stem (not whorled) 6

5 Stipules absent 11

6 Leaves soft, with copious stellate and/or simple hairs underneath at least . . . . . **91 RHAMNACEAE**

6 Leaves not copiously hairy underneath 7

7 Leaves not harsh, soft or firm but not pungent. Stems woody near the base but the upper parts usually not woody (see also *Phyllanthaceae*). . . . .**78 EUPHORBIACEAE**

7 Leaves harsh and with much vascular tissue, terete and grooved, or flat. Stems woody or wiry throughout 8

8 Leaves not clustered, though often close together and overlapping . . . . **74 FABACEAE**

8 Leaves in clusters on short axillary branches or in threes at each node 9

9 Leaves clustered. Ends of the branches sometimes spinescent . . . . **91 RHAMNACEAE**

9 Leaves 2–3 at each node 10

10 Leaves 3 at each node, their bases touching or almost so . . . . **79 PHYLLANTHACEAE**

10 Leaves 2–3 at each node, their bases not touching . . . . **42 PROTEACEAE**

11 Young stems with glandular hairs . . . . **138 APIACEAE**

11 Hairs, if present, not glandular 12

12 Leaves with an odour resembling ants, 1–3 mm wide . . . . **143 ASTERACEAE**
12 No such odour present 13

13 Leaves tufted at the nodes as a result of the growth of axillary buds 14
13 Leaves not in tufts. 17

14 Leaves on the main stem less than 1 cm long 15
14 Leaves on the main stem more than 1 cm long 16

15 Leaves scabrous-hairy, with blunt tips . . . . **143 ASTERACEAE**
15 Leaves pungent pointed . . . . **.75 POLYGALACEAE**

16 Leaves xeromorphic, acuminate or obtuse. . . . . **42 PROTEACEAE**
16 Leaves soft, obtuse or emarginate, or terete and grooved . . . . **47 DILLENIACEAE**

17 Leaves minutely toothed at or near the tip. . . . . **42 PROTEACEAE**
17 Teeth absent 18

18 Leaves grooved on the upper surface. . . . . **74 FABACEAE**
18 Leaves grooved underneath, or with recurved margins or flat 19

19 Leaves decurrent on the stem, sessile, linear, c. 1 mm wide, mostly 2–3 cm long . **148 STYLIDIACEAE**
19 Leaves not decurrent 20

20 Plants slender, narrow, branching only in the upper part 21
20 Plants much branched both top and bottom 22

21 Leaf scars protruding; older stems therefore raspy . . . . **.75 POLYGALACEAE**
21 Leaf scars sunken; older stems therefore smooth . . . . **42 PROTEACEAE**

22 Leaves 2–3 cm long. 23
22 Leaves up to 2 cm long 25

23 Leaves pungent pointed . . . . **74 FABACEAE**
23 Leaves not pungent pointed, though sometimes finely pointed. 24

24 Leaves glabrous (see also *Phyllanthaceae*) . . . . **.78 EUPHORBIACEAE**
24 Leaves with some hairs. . . . . **42 PROTEACEAE**

25 Leaves pungent pointed 26
25 Leaves not pungent pointed 27

26 Leaves ± triangular, sometimes cordate at the base and always broader at the base . . **74 FABACEAE**
26 Leaves linear. . . . . **47 DILLENIACEAE**

27 Stems rigidly erect . . . . **74 FABACEAE**
27 Stems diffuse, spreading, decumbent or somewhat trailing. 28

28 Leaves slightly fragrant when crushed . . . . **141 PITTOSPORACEAE**
28 Leaves not fragrant . . . . **47 DILLENIACEAE**

Group 27
Woody plants. Leaves alternate, more than 3 cm long; stipules present

1 Leaves xeromorphic 2
1 Leaves soft. 3

2 Leaf margins with spines c. 2 mm long . . . . **.78 EUPHORBIACEAE**
2 Leaf margins entire . . . . **74 FABACEAE**

3 Stipules represented by 2 blunt projections (see also *Phyllanthaceae*) . . . . . . **.78 EUPHORBIACEAE**

3 Stipules otherwise 4

4 Leaves palmatifid or deeply 3-lobed 5
4 Leaves otherwise 6

5 Ridge of tissue encircling the stem at the nodes. Leaves palmatifid . . . . **78 EUPHORBIACEAE**
5 No ridge of tissue present encircling the nodes. Leaves palmatifid or deeply 3-lobed (previously *Sterculiaceae*) . . . . **99 MALVACEAE**

6 Hairs on the leaves stellate, sometimes mixed with simple hairs. 7
6 Hairs simple or absent 9

7 Leaves ovate to elliptic, usually with entire margins, never cordate or truncate at the base . . . .
. . . . **91 RHAMNACEAE**
7 Leaves cordate or truncate at the base; the margins crenate or dentate, often 3-veined at the base . . . 8

8 Tomentum usually brown or rusty (previously *Sterculiaceae*) . . . . **99 MALVACEAE**
8 Tomentum white or grey (*Howittia* sometimes brownish) . . . . **99 MALVACEAE**

9 Leaves 3-veined at the base (i.e. 2 lowest lateral veins joining the petiole . . . . . . **89 CANNABACEAE**
9 Otherwise 10

10 Leaf margins toothed or sinuate 11
10 Leaf margins entire. 13

11 Branches hanging vertically (*Weeping Willow*) . . . . **84 SALICACEAE**
11 Otherwise 12

12 2 or more glands (cup-like, stipule-like or protuberances) c. 1 mm long or diam. present at the top of the petiole or base of the midrib . . . . **78 EUPHORBIACEAE**
12 Glands absent. Young leaves pink-red . . . . **87 ELAEOCARPACEAE**

13 Leaves glabrous, not silvery white underneath 14
13 Leaves with hairs or silvery white underneath 15

14 Petiole c. two-thirds as long to as long as the leaf blade . . . . **99 MALVACEAE**
14 Petiole c. a quarter as long as the leaf blade or shorter (see also *Phyllanthaceae*) **78 EUPHORBIACEAE**

15 Leaves silvery white below but without hairs, finely dotted with glands on both surfaces. . . . .
. . . . **78 EUPHORBIACEAE**
15 Otherwise 16

16 Leaves glabrous or with simple long sometimes matted hairs. . . . . **92 ROSACEAE**
16 Leaves silvery or white (sometimes rusty) below with a dense tomentum of very short hairs. . . . .
. . . . **91 RHAMNACEAE**

Group 28
Woody plants. Leaves more than 3 cm long, alternate, hairy or white or yellow underneath

1 Stinging hairs present on the leaves . . . . **93 URTICACEAE**
1 Stinging hairs absent. 2

2 Leaves xeromorphic, with tough veins, entire lobed or pinnatisect 3
2 Leaves soft. 4

3 Leaves oblong to ovate; the tip and the base having the same shape, usually rusty below. Main lateral veins leaving the midrib at right angles . . . . **74 FABACEAE**
3 Otherwise . . . . **42 PROTEACEAE**

4 Leaves deeply lobed 5
4 Leaf margins entire or toothed 7

5 Leaves c. 5 cm long or less. . . . . **138 APIACEAE**
5 Leaves much longer than 5 cm 6

6 Leaves pinnatisect . . . . **143 ASTERACEAE**
6 Leaves 7–12-lobed, c. 20 cm wide. . . . . **139 ARALIACEAE**

7 Tomentum rusty (previously *Sterculiaceae*). . . . . **99 MALVACEAE**
7 Tomentum white, or the leaves whitish or yellowish underneath 8

8 Leaves yellowish or yellow-green underneath . . . .**107 EBENACEAE**
8 Otherwise. 9

9 Hairs stellate 10
9 Hairs not stellate 13

10 Young leaves from axillary buds semi-circular and resembling stipules. . . . . . .**135 SOLANACEAE**
10 Otherwise 11

11 Leaves up to 4 cm long, up to twice as long as broad, thick and leathery. . . . **145 GOODENIACEAE**
11 Leaves 4–15 cm long, several times longer than broad, herbaceous. 12

12 Leaf margins entire. . . . . **139 ARALIACEAE**
12 Leaf margins toothed or lobed . . . . **143 ASTERACEAE**

13 Leaves stem-clasping . . . . **47 DILLENIACEAE**
13 Leaves not stem-clasping but sometimes sessile. 14

14 Either; leaves with a strong scent of musk, or of ants or of marigolds when crushed; or, hairs in tufts. .
. . . .**143 ASTERACEAE**
14 Otherwise 15

15 Either; leaf margins toothed; or, leaves with glandular dots underneath; or, young leaves viscid. . . . .
. . . .**78 EUPHORBIACEAE**
15 Leaf margins entire, not gland dotted below, not viscid 16

16 Leaves densely hairy and white underneath. . . . . **91 RHAMNACEAE**
16 Leaves hairy mainly along the midrib underneath . . . . **141 PITTOSPORACEAE**

Group 29
Woody plants. Leaves alternate, more than 3 cm long, glabrous, entire

1 Branches hanging vertically (*Weeping Willow*). . . . . **84 SALICACEAE**
1 Otherwise 2

2 Leaves stem-clasping and/or auriculate . . . . **143 ASTERACEAE**
2 Leaves not stem-clasping 3

3 Young leaves (sometimes also the older ones) and the smallest stems viscid shining or varnished . . . 4
3 Leaves not viscid or varnished 5

4 Leaves tapering into a very short petiole; blades mostly less than 1 cm wide . . . . **104 SAPINDACEAE**
4 Petiole 1–2 cm long, distinct; blades 1–2 cm wide . . . . **143 ASTERACEAE**

5 Leaves xeromorphic (harsh and with tough veins or leathery). Plants usually in low fertility soils but a
  few *Proteaceae* in RF. 6
5 Leaves mesomorphic. Plants mostly in or near RF. 7

6 Leaves similar on both surfaces; veins strongly reticulate and very prominent. Buds without bud scales
. . . . **74 FABACEAE**

6 Either; leaf surfaces not similar and the leaf harsh and fibrous (if surfaces similar, then the buds covered with bud scales); or, leaves leathery, with obscure veins when fresh and the 2 surfaces similar . . . . . .
. . . . **42 PROTEACEAE**

7 Axillary buds forming tufts of leaves in the leaf axils . . . . **141 PITTOSPORACEAE**
7 Otherwise 8

8 Leaves c. 15 mm wide or less 9
8 Leaves 15 mm wide or wider 10

9 Leaves 10–15 mm wide . . . .**75 POLYGALACEAE**
9 Leaves mostly 5 mm wide . . . .**72 CELASTRACEAE**

10 Leaves not regularly spaced along the stem but often in clusters, sometimes also whorled 11
10 Leaves not clustered 12

11 Leaf blades 15–20 cm long . . . . **55 NYCTAGINACEAE**
11 Leaf blades 6–15 cm long . . . . **141 PITTOSPORACEAE**

12 Leaves sessile or almost so 13
12 Petioles distinct, c. 1 cm long or more 14

13 Stems reddish or brown . . . . **104 SAPINDACEAE**
13 Stems green . . . .**135 SOLANACEAE**

14 Petioles at least half as long as the blade . . . .**135 SOLANACEAE**
14 Petioles less than half as long as the blade 15

15 Leaves mostly 15–25 mm wide 16
15 Leaves 3–8 cm wide 17

16 Leaves soft, almost semi-succulent . . . .**135 SOLANACEAE**
16 Leaves firm, brittle . . . .**109 MYRSINACEAE**

17 Leaves somewhat rhomboidal, angular. Two, lowermost lateral veins prominent, giving a suggestion of 3 main veins (*Scolopia*) . . . . **84 SALICACEAE**
17 Otherwise 18

18 Hollow glands present on the undersurface of the leaves, either in the angle of the lateral veins and the midrib (gland triangular), or the glands round and remote from the midrib in the angles of the first branching of the lateral veins. Branchlets often wavy 19
18 No such glands present 20

19 Hollow glands pocket-like usually confined to axils along primary veins . . . .
. . . . **142 CARDIOPTERIDACEAE**
19 Hollow glands pit-like mostly in forks of secondary veins . . . .**142 PENNANTIACEAE**

20 Buds and youngest shoots with a rusty tomentum . . . .**109 MYRSINACEAE**
20 Otherwise 21

21 Midrib on the undersurface of the leaves brown or rusty . . . . **147 ROUSSEACEAE**
21 Midrib green. Latex present in the bark . . . . **111 SAPOTACEAE**

Group 30
Woody plants. Leaves alternate, more than 3 cm long, glabrous; margins toothed or lobed

1 Leaves pinnatisect or palmately lobed 2
1 Leaves serrate dentate denticulate or with shallow lobes 4

2 Leaves xeromorphic, with tough veins . . . . **42 PROTEACEAE**
2 Leaves soft 3

3 Leaves palmately lobed . . . . .78 **EUPHORBIACEAE**
3 Leaves pinnately lobed. . . . . .135 **SOLANACEAE**

4 Shrubs mostly 1–2 m high with soft herbaceous branches (sometimes the whole plant almost herbaceous). Leaves herbaceous, sometimes slightly yellowish, never viscid . . . . . **143 ASTERACEAE**
4 Otherwise 5

5 Leaves mostly more than 12 cm long 6
5 Leaves c. 10 cm long or less 9

6 Leaves xeromorphic . . . . **42 PROTEACEAE**
6 Leaves soft. 7

7 Leaf margins bluntly dentate or with broad shallow lobes . . . . .78 **EUPHORBIACEAE**
7 Leaf margins serrate 8

8 Serrations numerous and close together (2–5 per cm) (*Ehretia*) . . . . .114 **BORAGINACEAE**
8 Serrations few and distant (5–15 mm apart) . . . . **147 ROUSSEACEAE**

9 Leaves mostly lanceolate or narrower (ca. 15 mm wide or less). 10
9 Leaves mostly more than 15 mm wide. 12

10 Leaves herbaceous . . . . .135 **SOLANACEAE**
10 Leaves coriaceous or xeromorphic 11

11 Leaves 1–7 cm long . . . . .72 **CELASTRACEAE**
11 Leaves 5–20 cm long . . . . **42 PROTEACEAE**

12 Leaves mostly c. 10 cm long . . . . .112 **SYMPLOCACEAE**
12 Leaves 3–8 cm long 13

13 Leaf margins coarsely toothed; the teeth 2 mm deep or more and c. 1 cm apart 14
13 Leaf margins denticulate (teeth c. 1 mm long) 15

14 Leaves glossy on the upper surface; petioles 5–10 mm long . . . . .109 **MYRSINACEAE**
14 Leaves not glossy on the upper surface, almost sessile; the leaf bases sometimes swollen . . . .
. . . . **42 PROTEACEAE**

15 Young shoots viscid, and/or the leaves truncate at the base . . . . **145 GOODENIACEAE**
15 Young shoots not viscid 16

16 Leaves almost sessile . . . . .82 **OCHNACEAE**
16 Leaves with petioles 5–15 mm long. 17

17 Leaves slightly scabrous on the upper surface. .90 **MORACEAE**
17 Leaves smooth. . . . . .78 **EUPHORBIACEAE**

Descriptions of Magnoliophyta Families

33 NYMPHAEACEAE
8 gen., cosmop.

1 Nymphaea Zucc.
9 species in Aust. (3 introduced); Qld, NSW, Vic., NT, WA

Perennial, aquatic herbs with rhizomes corms or stolons. Leaves often large, usually on long petioles, floating, peltate or cordate. Flowers emergent, solitary, bisexual, regular. Sepals 3–5. Petals usually numerous and often grading into the stamens. Stamens numerous, epigynous; the connectives sometimes produced into an appendage. Ovary multi-locular, half-inferior; each loculus with numerous ovules scattered over the walls, filled with mucilage; stigmas as radiating ridges. Fruit a spongy berry breaking up irregularly when ripe.

Key to the species

1 Petals yellow, narrow-elliptic, up to 8 cm long. Stoloniferous plants. East Lakes and occasional in other fresh water lakes and ponds. Introd. from N. American *Yellow Waterlily*. . . . . ***N. mexicana** Zucc. and hybrids.
1 Petals blue or white, narrow-elliptic, up to 9 cm long. Plants with corms. Coast and Cumberland Plain. Fresh water lakes and ponds. Introd. from S. Africa. *Cape Waterlily* . . . .***N. caerulea** Savigny ssp. **zanzibarensis** (Casp.) S.W.L.Jacobs

34 CABOMBACEAE
Aquatic, rhizomatous herbs with entire floating leaves (which may be few or absent) and sometimes deeply dissected submerged leaves. The submerged parts covered with mucilage. Flowers emergent, bisexual, axillary, solitary. Perianth segments 6–8. Carpels free. 2 gen., trop. & warm temp.

Key to the genera

1 Leaves simple, floating. . . . . **1 BRASENIA**
1 Leaves mostly submerged, divided into linear segments . . . . **2 CABOMBA**

1 Brasenia Schreb.
1 species native Aust.; Qld, NSW, Vic.

Monotypic genus
Aquatic, rhizomatous herb. Stems and undersurface of leaves mucilaginous. Leaves alternate, peltate, orbicular to elliptic, 40–120 mm long, 30–90 mm wide, entire, floating. Flowers axillary, solitary, emergent. Perianth segments 6–8, dark red to purplish, narrow-oblong, up to 18 mm long. Stamens 12–40, free. Carpels 6–20, free, superior; ovules 2 per loculus. Fruit indehiscent, beaked. Picton Lakes; Colo River; Cattai River. *Watershield* . . . . **B. schreberi J.F.**Gmel

2 Cabomba Aubl.
1 species naturalized Aust.; Qld, NSW

One species in the area
Submerged, aquatic perennial up to 2 m long. Floating leaves few, up to 2 cm long but usually absent. Submerged leaves opposite or whorled, multi- or dichotomously branched into linear segments, up to 5 cm wide. Perianth white to cream. Carpels 2–4. Coast. Introd. from America. *Fanwort* . . . .
*C. caroliniana A.Gray

35 CERATOPHYLLACEAE
1 gen., widespread in fresh water

1 Ceratophyllum L.
1 species native Aust.; all states and territories except Tas.

One species in the area

Submerged, aquatic, monoecious, perennial herb. Leaves whorled and divided dichotomously into linear denticulate segments; stipules absent. Flowers unisexual, minute, solitary in leaf axils. Sepals 10–15, basally connate. Stamens 10–20 on a flat receptacle. Ovary 1-locular, superior, with a single pendulous ovule. Fruit a nut. Coast. Commonly grown in aquaria and rarely naturalized in the area. . . . .
. . . . ***C. demersum** L.

36 WINTERACEAE
9 gen., trop. and temp., mostly S. hemisphere

1 Tasmannia DC.
7 species in Aust.; Qld, NSW, Vic., Tas.

Shrubs. Leaves alternate, gland-dotted, aromatic, entire; without stipules. Flowers, regular, unisexual or polygamous, greenish white to white, apparently in terminal umbels. Sepals 2–3, connate into a cap which encloses bud. Petals usually 2. Stamens indefinite in number, with thick filaments. Carpels 1–5, free, with decurrent stigmas; placentas lateral or marginal. Fruit a berry.

Key to the species

1 Lamina of the leaf auriculate or truncate at the base with a distinct but very short petiole, elliptic to oblanceolate, 8–20 cm long, 15–40 mm wide, acuminate. Stamens 30–40 in male flower, with two carpellodes. Carpels solitary or rarely 2–3 in female flowers, scarcely stipitate. Berry purple, ovoid, c. 5 mm diam. Shrub up to 3 m high. Widespread. RF. Fl. summer. *Brush Pepperbush* . . . .
. . . . **T. insipida** R.Br. ex DC.
1 Lamina of the leaf tapering gradually towards the base, narrow-oblong-elliptic, 4–7 cm long, up to 1 cm wide, acute or obtuse, distinctly paler underneath, shortly petiolate. Carpels 1–3, almost sessile. Berry purple, globular, c. 5 mm diam. Shrub up to 2 m high. Higher Blue Mts Woodland. Fl. summer. *Mountain Pepperbush* . . . .**T. lanceolata** (Poir.) A.C.Sm.

37 ATHEROSPERMATACEAE

Trees or shrubs. Leaves opposite, often coarsely serrate, usually gland-dotted and aromatic; stipules absent. Flowers bisexual or unisexual, solitary or in racemes, cymes or panicles. Receptacle cup shaped. Perianth segments 4 or more in 2 or 3 rows. Stamens indefinite in number or a multiple of the perianth segments, perigynous; filaments with 2 appendages; anthers sessile or nearly so opening by basal valve. Carpels several or many, free, superior. Fruits several to many plumose nuts enclosed in the enlarged floral tube. 6–7 gen., trop. and subtrop.

Key to the genera

1 Leaves white or greyish on undersurface. Leaf hairs centrifixed. Flowers solitary, axillary, 15–20 mm diam.. . . . .**3 ATHEROSPERMA**
1 Leaves green and glabrous on both surfaces. Leaf hairs basifixed. 2

2 Flowers usually 3 together in a cyme, c. 2 cm diam.. . . . . **1 DORYPHORA**
2 Flowers numerous in a panicle, less than 2 cm diam. . . . . **2 DAPHNANDRA**

1 Doryphora Endl.

2 species endemic Aust.; Qld, NSW

Monotypic genus

Large tree, aromatic in all its parts. Young shoots and inflorescence silky hairy. Leaves elliptic to oblong-lanceolate, narrowed at the base, glabrous, ± coarsely serrate, 4–10 cm long; venation distinct on undersurface. Flowers bisexual, usually 3 together on a short axillary peduncle; the whole enclosed in the bud stage by 2 deciduous bracteoles. Perianth segments 6, white, lanceolate, c. 8 mm long, tapering to fine points. Fertile stamens 6; connective produced into a fine awn nearly as long as the perianth segments. Carpels several; style plumose. Floral tube enlarged in the fruit, ovoid, c. 12 mm long, with a narrow neck, splitting open and exposing the ripe carpels. Widespread. RF. Fl. early spring. *Sassafras* . . . .
. . . .**D. sassafras** Endl.

2 Daphnandra Benth.

6 species endemic Aust.; Qld, NSW

Small trees with young stems flattened at nodes. Leaves opposite, margins toothed. Flowers bisexual, in axillary panicles. Perianth segments 15–20 in 3 or 4 whorls. Stamens 4–7 with truncate anther appendage and filaments with glands. Staminodes 5–10. Carpels 4–12, tapering into a short style. Floral tube enlarged in the fruit to form a capsule in a similar manner to that of *Doryphora sassafras*. Seeds hairy.

Key to the species

1 Capsule glabrous, globose, 5–7 mm long. Leaves 6–12 cm long, 1.5–6 cm wide; ovate to elliptic, 7–12 cm long, acuminate, margins deeply serrate except in lower third; veins distinctly raised on upper surface. Illawarra region. RF. Endangered. Fl. spring. . . . .**D. johnsonii** Schodde
1 Capsule glabrous, urn-shaped, 8–20 mm long. Leaves 6–15 mm long, 1.5–4 mm wide; lanceolate; margins shallowly serrate; veins obscure on upper surface. Far northern parts of region. RF. Fl. spring. *Socket Wood*. . . . . **D. apatela** Schodde

3 Atherosperma Labill.

1 species endemic Aust.; NSW, Vic., Tas.

Monotypic genus

Small to medium sized tree, aromatic in all its parts. Young branches, undersurfaces of leaves and inflorescence brownish or greyish tomentose. Leaves very narrow-ovate to oblong, 3–10 cm long, entire or with a few teeth. Flowers unisexual, solitary, axillary, enclosed in the bud by 2 deciduous bracteoles. Perianth segments 8–10, in 2 rows, 6–10 mm long. Stamens 8–20; anthers short. Carpels numerous; the outer ones sterile; the inner ones tapering into a short style. Floral tube enlarged at fruiting stage, cup-shaped, enclosing numerous hairy carpels with persistent styles. Blue Mts RF and gullies. Fl. summer. *Black Sassafras* . . . .**A. moschatum** Labill.

38 LAURACEAE

Trees or large shrubs. Leaves alternate, rarely opposite, conspicuous or reduced to scales, entire, gland-dotted, usually aromatic; stipules absent. Flowers regular, bisexual or unisexual. Perianth segments 6 or 4. Stamens perigynous, usually in multiples of the perianth segments, in 2 rows, free, usually with glands at the base; anthers opening by valves from the base upwards. Ovary superior, enveloped in the floral tube, 1-carpellary, with a solitary basal ovule. Fruit a berry or drupe or the floral tube enlarging and ± enclosing the fruit and rarely adnate to it. 30 gen., mainly trop. and subtrop.

Key to the genera

1 Leaves reduced to scales. Rootless twinning hemiparasitic herb . . . .**1 CASSYTHA**
1 Leaves conspicuous. Trees or shrubs. 2

2 Ovary and fruit enclosed by the floral tube . . . .**2 CRYPTOCARYA**

2 Ovary and fruit enclosed by the floral tube near the base only **3**

3 Flowers unisexual and plants dioecious **4**
3 Flowers bisexual or unisexual and monoecious **5**

4 Leaves glaucous underneath, smooth and shining on the upper surface, acuminate . . .**3 NEOLITSEA**
4 Leaves green or slightly glaucous underneath, conspicuously reticulate on the upper surface, obtuse or scarcely acuminate . . . .**4 LITSEA**

5 Berry 18–25 mm diam. Fertile stamens 3. Leaves very finely reticulate on the upper surface . . . .
. . . .**5 ENDIANDRA**
5 Berry less than 12 mm diam. Fertile stamens 9. Leaves smooth . . . .**6 CINNAMOMUM**

1 Cassytha L.

Devil's Twine

14 species in Aust.; all states and territories

Twiners, parasitic on other plants. Leaves reduced to minute scales. Flowers small, bisexual, regular, white. Perianth segments 6, in 2 whorls, persistent. Stamens 9, perigynous, with glands at the base; anthers opening by valves. Ovary superior, 1-locular, 1-carpellary, with a solitary basal ovule. Fruit a berry enveloped in the enlarged floral tube but free from it.

Key to the species

1 Plants glabrous. Stems filiform. Flowers few in a globular head on a short peduncle. Floral tube and perianth together c. 2 mm long, glabrous. Floral tube at fruiting stage ovoid to ellipsoid, 4–6 mm long, reddish or yellowish. Coast to lower Blue Mts Heath. Fl. summer . . . .**C. glabella** R.Br. forma **glabella**
1 Plants pubescent at least when young. Stems comparatively stout. Flowers in short spikes or almost capitate. **2**

2 Stems pubescent, often tuberculate when older. Peduncles and perianth pubescent, ribbed. Floral tube globular at fruiting stage, greenish. Widespread. Heath and open forest. Fl. spring–summer . . . .
. . . . **C. pubescens** R.Br.
2 Stems glabrescent, more stout than in any other species. Peduncles pubescent. Floral tube with a blackish pubescence, globular at fruiting stage. Widespread. Fl. spring. . . . .**C. melantha** R.Br.

2 Cryptocarya R.Br.

46 species in Aust. (44 endemic, 2 native); Qld, NSW, NT, WA

Trees. Flowers bisexual, c. 2 mm long, in apparently terminal panicles. Perianth segments 6. Fertile stamens 9. Floral tube becoming enlarged and black, completely enclosing the fruit giving it the appearance of an inferior fruit.

Key to the species

1 Leaves green on both surfaces, oblanceolate to narrow-elliptic, 5–12 cm long, 2–3 cm wide, gradually tapering towards the base, ± acuminate, prominently reticulate underneath. Fruit globular, ± pointed, 12–16 mm diam. Small tree. Coast. RF. Fl. spring. *Murrogun* . . . . **C. microneura** Meisn.
1 Leaves glaucous underneath **2**

2 Leaves glabrous underneath, elliptic, 6–12 cm long, 3–5 cm wide, ± acuminate. Fruit globular, ± truncate, ± compressed, c. 20 mm diam. Tree. Widespread. RF. Fl. summer. *Brown Beech* or *Native Laurel* . . . . **C. glaucescens** R.Br.
2 Leaves pubescent underneath. **3**

3 Leaves obovate, 5–10 mm long, 2–4 mm wide. Fruit globular, c. 12 mm diam. Tree up to 40 m. Gosford area. RF. Fl. summer. *Pepperberry* . . . . **C. obovata** R.Br.

3 Leaves elliptic to ovate, 5–13 cm long, 2–5 cm wide. Fruit ellipsoid to ovoid, c. 20 mm long, pointed. Small tree or shrub. Brunkerville. RF. *Forest Maple* . . . . **C. rigida** Meisn.

3 Neolitsea Merr.

3 species endemic Aust.; Qld, NSW, NT

One species in the area

Small, dioecious trees. Mature leaves brown-pubescent underneath at least on the veins, whitish-glaucous. with 3 primary veins very prominent, ovate to oblong-elliptic 16–20 cm long. Petioles hairy. Flowers unisexual in sessile axillary clusters surrounded by several deciduous bracts. Perianth segments 4. Stamens 6. Anthers 4-celled; all stamens reduced to staminodes in female flowers. Drupe globular, surrounded by the floral tube at the base. North from Illawarra. RF & RF margins. *White Bolly Gum* . . .
. . . . **N. dealbata** (R.Br.) Merr.

4 Litsea Lam.

11 species in Aust. (10 endemic, 1 native); Qld, NSW, NT, WA

One species in the area

Dioecious tree. Leaves obovate to elliptic, 4–10 cm long, the apex rounded obtuse or scarcely acuminate, green or slightly glaucous underneath, conspicuously reticulate on the upper surface. Flowers unisexual, in short axillary racemes or clusters. Peduncles with 4 concave bracts c. 5 mm long near the top enclosing c. 6 flowers. Perianth segments 3–6 mm long. Fruit ovoid, black, c. 12 mm long, enclosed at the base in the cup-shaped floral tube c. one third the length of the fruit. Coast. RF. *Bolly Gum* or *She Beech* . . . . .
. . . . **L. reticulata** (Meisn.) F.Muell.

5 Endiandra R.Br.

38 species endemic Aust.; Qld, NSW, NT

Shrubs or trees. Flowers in axillary panicles shorter than the leaves. Floral tube short. Perianth segments 6, open. Stamens 3 alternating with 3 staminodes. Ovary ± enclosed in the floral tube but later enlarged. Fruit a black berry, not enclosed by the floral tube.

Key to the species

1 Leaves glaucous underneath with prominent domatia at the junction of the midvein and main laterals, elliptic-oblong, 6–12 cm long, up to 5 cm wide, obtuse, very shortly acuminate. Fruit ovoid, c. 30 mm long. Shrub to tall tree. Gosford district. RF. *Tick Wood* or *Rose Walnut* . . . . **E. discolor** Benth.
1 Leaves equally green on both surfaces, domatia absent, narrow-ovate to oblong-elliptic, 5–10 cm long, up to 3 cm wide, obtuse or obtusely acuminate. Fruit a globular berry, up to 25 mm long. Shrub or small tree. Coast and adjacent plateaus from Illawarra northwards. RF and gullies on Sandy soil. *Corkwood*.
. . . . **E. sieberi** Nees

6 Cinnamomum Schaeff

6 species in Aust. (5 endemic, 1 naturalized); Qld, NSW

Trees. Leaves opposite or alternate, glossy on the upper surface, ± glaucous underneath. Flowers small, in axillary panicles near the ends of the branches, ± unisexual. Female flowers usually rather larger and fewer in the panicle than the males with stamens slightly imperfect. Male flowers with a carpellode. Perianth segments 6, persistent. Fertile stamens 9. Fruit a globular or ovoid black berry seated upon the cup-like floral tube.

Key to the species

1 Leaves opposite, very narrow-ovate, tapering towards the base and apex, 8–16 cm long. Berry ovoid, c. 12 mm long. Perianth segments c. 4 mm long. Minnamurra Falls. RF. *Oliver's Ssssafras* . . . .
. . . . **C. oliveri** F.M.Bailey

1 Leaves alternate, ovate to elliptic, 7–8 cm long. Berry globular, 12 mm long. Perianth segments c. 2 mm long. Leaves and other parts with an odour of camphor when crushed. Naturalized in many places along the Coast. Introd. from China and Japan. Fl. spring. *Camphor Laurel* . . .*C.* **camphora** (L.) Nees

39 MONIMIACEAE

Trees shrubs or woody climbers. Branchlets with often flattened below nodes. Leaves opposite, often coarsely serrate, usually gland-dotted and aromatic; stipules absent. Flowers bisexual or unisexual. Receptacle ± expanded or globose to urceolate. Perianth segments 4 or more in 2 or 3 rows. Stamens indefinite in number or a multiple of the perianth segments, perigynous; anthers sessile or nearly so, opening by slits. Carpels several or many, free, superior. Fruits several to many drupes either resting upon the expanded receptacle or enclosed in the enlarged floral tube. 22 gen., trop. and subtrop.

Key to the genera

1 Woody climber or straggling shrub . . . . **3 PALMERIA**
1 Trees or shrubs 2

2 Fruits few, large and distinct; floral tube globular with a small orifice . . . . **1 WILKIEA**
2 Fruits numerous, small, crowded; floral tube hemspherical or flat . . . .**2 HEDYCARYA**

1 **Wilkiea** F.Muell.
6 species endemic Aust.; Qld, NSW

One species in the area

Shrub or small tree. Leaves ovate-elliptic to oblong-lanceolate, rigid, 4–16 cm long, sharply and irregularly toothed or almost entire; venation strongly reticulate on both surfaces but raised on undersurface. Flowers small, unisexual, in short axillary racemes and panicles. Floral tube nearly globular, c. 2 mm diam., with a small orifice. Female floral tube circumciss after flowering, enlarged to form a disc. Fruiting carpels ovoid, black, c. 12 mm long, seated upon the enlarged floral tube. Coast and adjacent plateaus. RF. Fl. summer. *Hard Wilkea* . . . . **W. huegeliana** (Tul.) A.DC.

2 **Hedycarya** J.R Frost & G.Forst.
2 species endemic Aust.; Qld, NSW, Vic., Tas.

One species in the area

Dioecious small tree or shrub, glabrous except the inflorescence and young shoots. Leaves ovate-elliptic to broad-lanceolate, 5–10 cm long, shortly acuminate, irregularly toothed to almost entire, thin, not rigid, on rather long petioles. Flowers unisexual, in short axillary raceme-like cymes. Floral tube hemispherical or flat, with 6–10 small inflexed perianth segments. Male flowers with numerous stamens covering the disc. Female flowers with numerous carpels. Aggregate fruit globular, 6–8 mm diam., with 10–20 drupes closely packed together. Widespread. In or near RF. Fl. spring. *Native Mulberry* **H. angustifolia** A.Cunn.

3 **Palmeria** F.Muell.
3 species endemic Aust.; Qld, NSW

One species in the area

Tall, woody climber or scrambling shrub. Young branches and leaves ± stellate hairy. Leaves broad-elliptic, 5–12 cm long. Flowers unisexual, in loose axillary racemes. Male floral tube hemispherical, 10 mm diam., with 5 connivent lobes. Anthers numerous, sessile or nearly so. Female floral tube ± globular, 3 mm diam., with a minute orifice. Carpels numerous, glabrous, with filiform styles slightly protruding through the orifice of the floral tube. Floral tube at fruiting stage much enlarged, fleshy, splitting open, 10–20 mm diam.; drupes globular, 5–7 mm diam. Coast and adjacent plateaus. RF. Fl. winter. *Anchor Vine* . . . . **P. scandens** F.Muell.

40 EUPOMATIACEAE

1 gen., N. Guinea and Aust.

1 Eupomatia R.Br.

2 species in Aust.; Qld, NSW, Vic.

One species in the area

Erect, glabrous shrub or small tree. Leaves alternate, elliptic to oblong-elliptic, shortly acuminate, 7–12 cm long, shortly petiolate, glossy, distichous, gland-dotted, aromatic; stipules absent. Flowers solitary, c. 25 mm diam., apparently axillary, on short peduncles. Perianth connate into a conical cap (calyptra), deciduous as the flower opens. Stamens spirally inserted on the cup-shaped receptacle; inner ones petaloid, obovate, sterile, folded over the carpels; outer ones fertile, linear-lanceolate, curved, spreading, with acuminate tips. Carpels numerous crowded into the cup-shaped receptacle; stigmas confluent on the flat areolate surface. Ovules several. Fruits 1–2-seeded, immersed in the enlarged receptacle; aggregate fruit 2 cm diam., berry-like. Seeds irregularly angular. Coast and adjacent plateaus. In or near RF. Fl. spring. *Bolwarra* . . . . **E. laurina** R.Br.

41 PIPERACEAE

Perennial herbs or tall climbers, sometimes succulent, often articulate at the nodes. Leaves alternate, opposite or whorled; stipules absent or minute. Flowers bisexual or unisexual, regular, usually closely packed in a spike. Perianth absent. Stamens 2-10, hypogynous. Ovary superior, 1-locular, with 1 basal ovule; stigma sessile. Fruit a small drupe. 8 gen., mostly trop.

Key to the genera

1 Tall climber. Leaves alternate . . . . .1 PIPER
1 Small succulent herbs. Leaves opposite or whorled . . . . **2 PEPEROMIA**

1 Piper L.

8 species in Aust. (6 endemic); Qld, NSW, NT

One species in the area

Tall glabrous climber. Leaves alternate, ovate, 7–10 cm long, shortly acuminate, entire, with 5–7 principal veins arising from the base or near it. Flowers unisexual, axillary; male spikes cylindrical, 12–20 mm long; female spikes ovoid, shorter. Drupes stipitate. Coast. RF. *Giant Pepper Vine* . . . .
. . . . **P. novae-hollandiae** Miq.

2 Peperomia Ruiz & Pav.

5 species in Aust. (3 endemic); Qld, NSW

Herbs. Leaves opposite or whorled, thick. Stems erect or decumbent, rooting. Flowers bisexual, crowded into terminal spikes. Drupes sessile.

Key to the species

1 Leaves opposite, ovate-elliptic, 1–3 cm long, shortly pubescent, thick, fleshy. Spikes mostly 5–8 cm long. Coast and adjacent plateaus. On rocks in WSF and RF . . . . **P. blanda** var. **floribunda** (Miq.) H.Huber
1 Leaves in whorls of 4, glabrous or nearly so, shining, ovate-rhomboidal, 5–15 mm long, thick, hard. Spikes 1–3 cm long. Woronora Plateau; Blue Mts On rocks and tree trunks in or near RF . . . .
. . . . **P. tetraphylla** (G.Forst) Hook. & Arn.

42 PROTEACEAE

Shrubs or trees. Leaves alternate, sometimes opposite or whorled, mostly coriaceous or xeromorphic, entire to pinnatisect, without stipules. Flowers in spikes, racemes or umbels, sometimes paired or rarely solitary, regular or irregular, bisexual. Perianth segments 4, in a single whorl, petaloid, valvate in bud, cohering when young and separating later (often slit on one side) or connate towards the base. Stamens 4, opposite the perianth segments and adnate to them; anthers often sessile. Hypogynous glands 3–4, separate or united or absent. Ovary superior, often stipitate, (ie. gynophore present (Fig. 27), 1-locular; placenta marginal or basal; style simple, often curved or hooked and protruding through the perianth before the stigma emerges; pollen from the anthers of the same flower often deposited upon a variously shaped pollen presenter below or around the stigma before the flower opens. Fruit an achene, drupe or follicle. 80 gen., temp. to trop., particularly Africa and Aust.

Key to the genera

1 Leaves opposite or whorled 2
1 Leaves alternate. 3

2 Leaves more than 10 cm long, flat. Fruit pear-shaped . . . .**13 XYLOMELUM**
2 Leaves less than 10 cm long, with recurved margins. Fruit horned . . . . **14 LAMBERTIA**

3 Flowers sessile 4
3 Flowers pedicellate 10

4 Leaves variously divided. 5
4 Leaves undivided 7

5 Flowers in loose, narrow spikes often grouped into panicles . . . .**3 SYMPHIONEMA**
5 Flowers in dense cone-like or globular spikes. 6

6 Spikes longer than broad, cone-shaped in fruit. Bracts persistent, woody, opening to release the fruits. Perianth tube splitting to the base . . . .**1 PETROPHILE**
6 Spikes globular when in flower, sometimes elongated later. Bracts falling with the fruits. Perianth tube not splitting to the base . . . .**2 ISOPOGON**

7 Flowers in dense many-flowered spikes. 8
7 Flowers in panicles or loose spikes. 9

8 Fruit a woody follicle. Flowers in pairs in each bract, Inflorescence usually ovoid or cylindrical >5 cm long. . . . . **6 BANKSIA**
8 Fruit a nut enclosed in the woody bracts. Flowers solitary in each bract. Inflorescence globular <5 cm long. . . . . **2 ISOPOGON**

9 Perianth white or bluish. Fruit a flat-topped hairy achene . . . . **4 CONOSPERMUM**
9 Perianth yellow. Fruit a drupe . . . . **5 PERSOONIA**

10 Flowers solitary or in pairs or clusters in the axils of leaves or bracts 11
10 Flowers in elongated or short and dense racemes, resembling heads or umbels 12

11 Flowers solitary, perianth segments equally spreading, yellow. Fruit a drupe. . . . . . **5 PERSOONIA**
11 Flowers in pairs or clusters, perianth segments rolled back to one side, usually whitish, sometimes cream. Fruit a woody follicle. . . . . **7 HAKEA**

12 Ovules 2 13
12 Ovules more than 2 14

13 Anthers on very short filaments, versatile. Fruit indehiscent, ± fleshy . . . .**8 HELICIA**
13 Anthers sessile not versatile. Fruit a follicle . . . . **9 GREVILLEA**

14 Flowers in large terminal pseudo-heads; surrounded by large red, pink or greenish (rarely white) bracts. Perianth red (rarely white). . . . . **12 TELOPEA**
14 Flowers in umbels or racemes. Bracts caducous or absent. Perianth white, cream or yellow 15

15 Flowers in umbels of 20 flowers or less. Hypogynous gland horseshoe shaped or absent . . . .
. . . . **11 STENOCARPUS**
15 Flowers in racemes. Hypogynous glands 3, , occasionally with a smaller 4th gland . . . **10 LOMATIA**

FIGURE 31 & 32

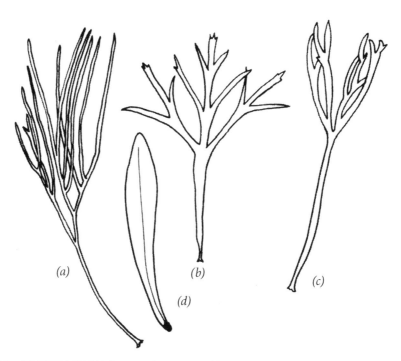

Fig. 31 PROTEACEAE, leaves of *Isopogon*: (a) *I. anethifolius*; (b) *I. anemonifolius*; (c) *I, dawsonii* ; (d) *I. fletcheri*. All X½.

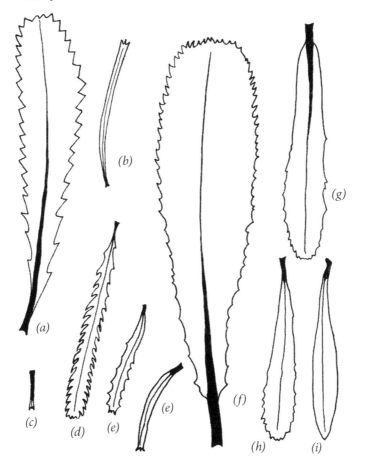

Fig. 32 PROTEACEAE, leaf-shapes of *Banksia*; (a) *B. serrata*, (b) *B. spinulosa* var. *spinulosa* (c) *B. ericifolia*, (d) *B. spinulosa var. collina* (e) *B. marginata*, (f) *B. robur*, (g) *B.oblongifolia* (h) *B. paludosa*. var. paludosa (i) *B. integrifolia*. All X1.

1 Petrophile R.Br. ex Knight
Conesticks
53 species endemic Aust.; Qld. NSW

Erect shrubs 1–3 m high. Leaves pinnately divided, the pinnae again divided pinnately or 2–3-chotomously; segments terete. Flowers in dense, terminal or axillary, cone-like spikes, each flower sessile within a bract. Bracts broad and woody after flowering, persistent, imbricate. Perianth whitish or yellow; segments equally spreading. Anthers sessile. Ovary sessile. Pollen presenter below stigma, club-shaped. Fruiting structure 'cones' ovoid to oblong. Fruit an achene, shorter than the bracts.

Key to the species

1 Cones axillary, 15–25 mm long, on short peduncles, 10–30 mm long. Perianth and bracts glabrous. Leaves up to 16 cm long; segments spreading but not pungent. Glabrous shrub up to 3 m high. Western Woronora to Robertson Plateau, Blue Mts Heath and DSF. Fl. summer. . . . . . . **P. pedunculata** R.Br.

1 Cones terminal (cones may appear axillary due to new growth above fruit), up to 50 mm long, sessile or nearly so. Bracts pubescent at least when young. Perianth silky-pubescent, 12–14 mm long. Leaves 3–10 cm long 2

2 Young shoots glabrous or nearly so. Leaf segments erect or ascending but not divaricate and not pungent. Shrub up to 3 m high. Widespread. Heath and DSF, usually on shallow sandy soils over Ss. Fl. spring–summer. Morphological intermediates between this and *P. sessilis* and *P. canescens* occur . . . .
. . . .**P. pulchella** (Schrad.) R.Br.

2 At least young shoots pubescent 3

3 Leaf segments spreading, very divaricate, rigid, ± pungent, glabrescent. Shrub up to 3 m high. Coast and adjacent plateaus in south of the region. DSF and heath. Ss . **P. sessilis** Sieber ex Schult. & Schult.f.

3 Leaf segments ascending, not pungent, greyish-pubescent. Blue Mts DSF and heath on deeper sands. Shrub up to 1.5 m high. Fl. spring . . . . **P. canescens** A.Cunn. ex R.Br.

2 Isopogon R.Br. ex Knight
Drumsticks
35 species endemic Aust.; Qld. NSW Vic. Tas. SA, WA

Erect shrubs 1–3 m high. Leaves pinnately or ternately divided and subdivided or simple and entire (Fig. 31). Flowers in dense terminal globular spikes, each flower sessile within a bract. Bracts tomentose outside, deciduous with the fruit. Perianth yellow, with a slender tube; segments equally spreading. Anthers sessile. Ovary sessile. Style filiform; pollen presenter below stigma, club-shaped. Fruiting structure 'cones', globose. Fruit an achene, shorter than the bracts, hirsute.

Key to the species

1 Leaves entire, oblanceolate, coriaceous, tapering gradually to an almost sessile base, 4–8 cm long, 5–20 mm wide. Stout, erect shrub up to 1.5 m high. Blue Mts, e.g. Blackheath. Sheltered places in DSF and heath, particularly on steep slopes. Vulnerable. Fl. spring–summer. . . . . **I. fletcheri** F.Muell.

1 Leaves much divided. 2

2 Perianth densely silky-hairy. Leaf segments flat, linear. Erect shrub 1–3 m high. Nepean River; Glenbrook Creek; Wolgan Valley. Fl. spring . . . . **I. dawsonii** R.T.Baker

2 Perianth sparingly and shortly hairy or almost glabrous at the tip 3

3 Leaf segments terete; leaves 4–16 cm long, pinnately divided and subdivided. Cones ovoid-globular, 12–25 mm diam. Erect shrub 1–3 m high, glabrous except the inflorescence. Widespread. DSF and heath. Ss. Fl. spring. . . . . **I. anethifolius** (Salisb.) Knight

3 Leaf segments flat, linear or 3-fid, pinnately or ternately divided and subdivided 4

4 Leaf segments 3–5 mm wide, smooth or obscurely ridged. Leaves 4–10 cm long. Perianth tip with distinct villous hairs. Cones subglobular, mostly 12–20 mm diam. Young shoots, and sometimes branches, pubescent. Shrub 1–2 m high. Widespread. DSF and heath. Ss. Fl. spring . . . . **I. anemonifolius** Knight

4 Leaf segments c. 2 mm wide; at least the ultimate segments ridged underneath at the margin and midrib. Leaves 4–10 cm long. Perianth almost glabrous at the tip or sparsely hairy. Cones subglobular, 12–20 mm diam. Young shoots and branches pubescent. Prostrate shrub or sometimes ascending to 1 m high. Blue Mts DSF and heath. Ss. Fl. spring–summer . . . . **I. prostratus** McGill.

3 Symphionema R.Br.
2 species endemic Aust.; NSW

Erect shrubs 30–100 cm high, glabrous or nearly so. Leaves alternate or opposite, divided into 3–9 or more narrow segments. Flowers small, in narrow spikes, 2–5 cm long, terminal or in the upper axils. Perianth pale yellow, 4–5 mm long; segments nearly equal, spreading. Anthers coherent in young flower, separating explosively. Style filiform, slightly bent, slightly longer than the stamens; pollen presenter absent. Fruit an oblong glabrous achene c. 2 mm long, usually with a single seed.

Key to species

1 Leaf segments flat, linear or linear-lanceolate, acute; leaves 2–4 cm long. Higher Blue Mts and Woronora Plateau. Wet or dry heath. Ss. Fl. spring. . . . . **S. montanum** R.Br.

1 Leaf segments almost terete, acute; leaves 1–3 cm long. Coast and adjacent plateaus. Wet heath. Ss. Fl. spring. . . . . **S. paludosum** R.Br.

4 Conospermum R.Br.
Coneseeds
53 species endemic Aust.; Qld. NSW, Vic., Tas., SA, WA

Small shrubs with few branches. Leaves entire, alternate or crowded. Flowers in short dense spikes which are arranged in a ± corymbose panicle, each flower sessile within a broad sheathing persistent bract. Perianth whitish or blue, irregular; tube straight; upper lobe usually broad, lower 3 narrower. Lower stamen reduced to a staminode; two lateral stamens with one fertile and 1 sterile loculus; upper stamen perfect; all coherent in the young flower, separating explosively. Style short, bent; pollen presenter absent. Fruit an achene, 2–3 mm long, with a flat or concave top, crowned with a ring of long hairs.

Key to the species

(There appears to be no fixed line of demarcation between *C. taxifolium* and *C. ericifolium* or *C. ellipticum*. Most of the plants on the higher Blue Mts are morphologically intermediate between *C. taxifolium* and *C. ericifolium*; they are often reduced to a single species as *C. taxifolium*. On the eastern slopes of the Blue Mts some intermediates (possibly hybrids) between *C. taxifolium* and *C. longifolium* var. *angustifolium* occur.)

1 Leaves 5–25 cm long. Peduncles 15–30 cm long, terminal or in the upper axils, few in number. Perianth bluish or white 2

1 Leaves mostly under 3 cm long. Peduncles 2–6 cm long, often numerous. Perianth white 5

2 Stems tangled, procumbent. Leaves linear, almost terete, channelled above, 8–16 cm long, ascending, curved inwards at the tip. Peduncles 18–50 cm long, slender. Perianth bluish to lilac. Shrub up to 1 m high. Widespread. Wet places. Ss. Fl. spring . . . . **C. tenuifolium** R.Br.

2 Stems erect. Leaves flat, almost erect, 1–20 mm wide. Perianth white. Erect shrubs up to 2 m high. . . 3

3 Leaves oblanceolate, more than 8 mm wide, tapering into a long petiole. Widespread, mostly north of Port Jackson. Heath and DSF. Ss. Fl. spring. . . . . **C. longifolium** Sm. ssp. **longifolium**

3 Leaves narrow-oblanceolate to linear, less than 8 mm wide 4

4 Widest leaves 4–8 mm wide. Blue Mts DSF. Ss. Fl. spring. . . . .

. . . . **C. longifolium** ssp. **mediale** L.A.S.Johnson & McGill.

4 Widest leaves less than 4 mm wide. Coast mostly south of Port Jackson. Heath and DSF. Ss. Fl. spring

. . . .**C. longifolium** ssp. **angustifolium** (Meisn.) L.A.S.Johnson & McGill.

5 Leaves elliptic to broad-lanceolate or oblong-cuneate. Coast and adjacent plateaus. Erect shrub up to 1 m high. Heath. Ss. and old dunes. Fl. spring . . . . **C. ellipticum** Sm.

5 Leaves lanceolate to linear or terete, crowded, erect or slightly spreading 6

6 Leaves flat, often twisted, lanceolate to linear, 1–3 cm long, the lower ones often longer. Erect shrub up to 1 m high. Widespread. Heath and DSF. Ss and old dunes. Fl. spring . . . . **C. taxifolium** Sm.

6 Leaves linear or terete, 5–15 mm long, c. 1 mm wide. Coast. Ss. and old dunes. Erect shrub up to 1 m high. Fl. spring . . . . **C. ericifolium** Sm.

5 **Persoonia** Sm.

Geebung

100 species endemic Aust.; all states and territories

Shrubs or small trees. Leaves alternate, entire. Flowers regular, solitary or sometimes paired in the leaf axils or in terminal racemes. Perianth yellow; segments equally spreading, recurved above, finally almost free. Filaments short, attached near the middle of the perianth segments. Ovary stipitate. Fruit a drupe. Fl. summer unless otherwise indicated.

Key to the species

1 Ovary villous . . . . .GROUP 1

1 Ovary glabrous 2

2 Flowers recurved (nodding) or spreading . . . . .GROUP 2

2 Flowers erect to spreading . . . . .GROUP 3

Group 1

Ovary villous

1 Leaves lanceolate to elliptic, flat or with thickened margins, 2–10 cm long 2

1 Leaves narrow-oblong to terete, less than 2 cm long. 5

2 Perianth densely pubescent with rust coloured hairs outside. Erect shrubs 50–150 cm high. Heath and DSF. Ss. Fl. spring–summer. 3

2 Perianth glabrous or sprinkled with a few hairs outside. 4

3 Mature leaves smooth, glabrous. Widespread from Sydney northwards . . **P. laurina** Pers.ssp. **laurina**

3 Mature leaves scabrous or sprinkled with a few hairs. Widespread from Sydney southwards . . . .

. . . .**P. laurina** ssp. **intermedia** L.A.S.Johnson & P.H.Weston

4 Pedicels hairy, 2–7 mm long. Ascending or decumbent branching shrub up to 60 cm high. Capertee. DSF. Ss. Vulnerable. . . . . **P. marginata** A.Cunn. ex R.Br.

4 Pedicels glabrous, 9–23 mm long. Erect to spreading shrub up to 2 m high. Blue Mts N of Glenbrook. Heath and DSF. Ss . . . . **P. oblongata** R.Br.

5 Perianth glabrous or sprinkled with a few appressed hairs, 8–10 mm long. Leaves glabrous, linear, 10–15 mm long. Prostrate shrub forming mats often exceeding 1 m diam. Higher Blue Mts and adjacent valleys; Woronora Plateau. Heath and DSF. Ss . . . . **P. chamaepitys** A.Cunn.

5 Perianth densely hirsute, c. 10 mm long. Leaves oblong to linear, 6–20 mm long. Spreading shrubs 50–100 cm high, hirsute or scabrous all over 6

6 Leaves linear to narrow-oblong, 1.5 mm or less wide; margins revolute. Gosford to Royal National Park. DSF. Endangered . . . . **P. hirsuta** Pers. ssp. **hirsuta**

6 Leaves spathulate to elliptic, 1.5 mm or more wide; margins recurved. Higher Blue Mts; Glen Davis; Putty; Hill Top. DSF . . . . **P. hirsuta** ssp. **evoluta** L.A.S.Johnson & P.H.Weston

Group 2
Ovary glabrous; flowers recurved (nodding) or spreading

1 Perianth segments each with a terminal or subterminal, subulate appendage 1–5 mm long 2
1 Perianth segments without conspicuous appendages or with a point less than 1 mm long 5

2 Leaves linear, 0.8 –30 mm long, 1–2 mm wide. Perianth and pedicels glabrous 3
2 Leaves lanceolate, 20 mm– 45 mm long, 5–30 mm wide. Pedicels pubescent, 2–4 mm long. Perianth slightly pubescent. Much branched, spreading shrubs to 2 m high. 4

3 Leaves 15–30 mm long. Pedicels 6–12 mm long, reflexed in fruiting stage. Perianth 8.5–11 mm long. Much branched shrub up to c. 1 m high. Cumberland Plain. DSF on laterite and sand. Endangered . . . . **P. nutans** R.Br.
3 Leaves 0.8–15 mm long. Pedicels 6–8 mm long. Perianth 6–8 mm long. Decumbent to prostrate shrub. Collected from Manly and Newport. Extinct. . . . . **P. laxa** L.A.S.Johnson & P.H.Weston

4 Appendages on perianth segments recurved, 1–2 mm long. Leaves 4–12 mm wide. Higher Blue Mts Heath and DSF. Ss . . . .**P. myrtilloides** Sieber ex Schult. & Schult.f. ssp. **myrtilloides**
4 Appendages on perianth segments reflexed, 2.5–5 mm long. Leaves 6–30 mm wide. Capertee. Heath and DSF. Ss . . . . **P. myrtilloides** ssp. **cunninghamii** (R.Br.) L.A.S.Johnson & P.H.Weston

5 Longest leaves up to 11 mm long, ovate to oblong-lanceolate, rigid. Shrub up to 1 m high, scabrous, pubescent or glabrous. Higher Blue Mts DSF and open ground . . . . **P. oxycoccoides** Sieber ex Spreng.
5 Longest leaves more than 11 mm long. 6

6 Leaves 2–6 cm long, 4–25 mm wide . . . . **P. oblongata** (see above)
6 Leaves up to 2 cm long. Prostrate to spreading shrubs. 7

7 Leaves acuminate, distinctly paler underneath, 3–9 mm wide, elliptic to ovate or obovate. Pedicels 3–6 mm long. Prostrate to spreading shrub. Higher Blue Mts Heath, WSF and DSF. Granite, basalt, etc., not Ss.. . . . . **P. acuminata** L.A.S.Johnson & P.H.Weston
7 Leaves acute, concolourous or slightly paler underneath, 1–4 mm wide, narrow-oblong to narrow-elliptic. Pedicels 2–4 mm long. Prostrate to spreading shrub. Higher Blue Mts DSF. Ss . . . . **P. recedens** Gand.

Group 3
Ovary glabrous; flower erect to spreading

1 Leaves mostly more than 2 mm wide 2
1 Leaves mostly 2 mm or less wide. Occasionally to a maximum of 6 mm but then plant with flaky bark
. . . . 13

2 Leaves up to 2.5 cm long. 3
2 Leaves more than 2.5 cm long 9

3 Leaves scabrous. 4
3 Leaves smooth or very slightly scabrous, 1–17 mm wide, elliptic, spreading away from stem, hairy at least when young 5

4 Leaves 15–50 mm long, 4–19 mm wide, oblanceolate, usually curved in towards the stem. Perianth often rust coloured. Erect to decumbent shrub. Blue Mts and Mt. Kembla. Open forests. **P. rigida** R.Br.
4 Leaves 3–9 mm long, 2–5 mm wide, elliptic to ovate. often curved downwards at tip. Prostrate to erect shrub. Higher southern Blue Mts Heath and DSF. Ss . . . . **P. microphylla** R.Br.

5 Prostrate to decumbent shrub. Nattai River; Bullio. DSF. Ss . . . .
. . . . **P. mollis** ssp. **revoluta** (Sieber ex Schult. & Schult.f.) S.Krauss & L.A.S.Johnson
5 Erect to ascending shrubs 6

6 Leaves less than 6 mm wide. Shoalhaven River and Illawarra Ranges. Heath and DSF. Ss . . . .
. . . . **P. mollis** ssp. **ledifolia** (A.Cunn. ex Meisn.) S.Krauss & L.A.S.Johnson
6 Leaves more than 6 mm wide 7

7 Flower buds coppery when fresh. Hornsby-Cowan area. WSF and DSF. Endangered . . . .
. . . .**P. mollis** ssp. **maxima** S.Krauss & L.A.S.Johnson
7 Flower buds silvery when fresh 8

8 Hairs on flowers c. 1 mm long. Blue Mts WSF and DSF . . . . **P. mollis** R.Br. ssp. **mollis**
8 Hairs on flowers c. 0.5 mm long. Oakdale to Hill Top and Illawarra Ranges. WSF and DSF . . . .
. . . . **P. mollis** ssp. **nectens** S.Krauss & L.A.S.Johnson

9 Bark rough and flaking, reddish where flaked. Leaves ovate to obovate or broad-elliptic, 13–80 mm
wide, mostly oblique or falcate, ± distinctly 3 veined. Pedicels 4–8 mm long. Shrub up to 5 m high.
Widespread. Heath and DSF on sandy soils . . . . **P. levis** (Cav.) Domin
9 Bark not easily flaking, greyish 10

10 Anthers white . . . . **P. laurina** ssp. **laurina** (see above)
10 Anthers yellow 11

11 Leaves paler on undersurface, revolute or distinctly recurved . . . .
. . . . **P. mollis** (see above dichotomies 5–8 for subspecies)
11 Leaves ± same colour on both surfaces 12

12 Leaves greyish at least when young, oblanceolate, 3–8 cm long, 4–18 mm wide. Erect shrub. Picton
southwards. DSF. Endangered . . . .**P. glaucescens** Sieber ex Spreng.
12 Leaves green to yellowish-green, oblanceolate to obovate, 3–10 cm long, 4–32 mm wide. Erect shrub.
Coast and adjacent ranges. Heath and DSF. Ss and deep sands . . . . **P. lanceolata** Andrews

13 Leaves terete or grooved 14
13 Leaves flat or slightly concave 16

14 Perianth segments without a terminal appendage, pubescent. Leaves 3–5 cm long, grooved on lower
surface. Floral leaves much smaller than foliage leaves. Shrub up to 4 m high. Widespread. DSF. Ss . . .
. . . .**P. pinifolia** R.Br.
14 Perianth segments each with a terminal or subterminal appendage, glabrous. Floral leaves similar to
foliage leaves. Shrubs, glabrous or nearly so, up to 3 m high 15

15 Leaves grooved on upper surface, 10–15 mm long, erect or incurved with a straight tip. Higher Blue
Mts DSF. Vulnerable . . . . **P. acerosa** Sieber ex Schult. & Schult.f.
15 Leaves grooved on undersurface, 15–25 mm long, ascending or spreading with incurved tip. Hornsby
Plateau. DSF . . . . **P. isophylla** L.A.S.Johnson & P.H.Weston

16 Pedicels hairy 17
16 Pedicels glabrous. Leaves linear to lanceolate, mostly less than 25 mm long 18

17 Leaves linear to narrow oblong, 2–8 cm long. Pedicels 2–8 mm long. Bark flaky. Fruit often with dark
stripes. Widespread. Forests on Ss and shale . . . . **P. linearis** Andrews
17 Leaves linear, 15–25 mm long, usually incurved, succulent. Pedicels 2.5–5 mm long. Bark smooth.
Newnes Plateau near Lithgow. DSF. Ss. Endangered . . . . . . **P. hindii** L.A.S.Johnson & P.H.Weston

18 Prostrate shrubs. Higher Blue Mts DSF . . . . **P. chamaepeuce** Lhotsky ex Meisn.
18 Erect to ascending shrub. Bargo area. DSF. Ss and laterite. Endangered . . . .
. . . . **P. bargoensis** L.A.S.Johnson & P.H.Weston

(Intermediate forms, hybrids or possible hybrids occur between several pairs of species, e.g. *P. levis* and *P.linearis* (= *P. lucida* R.Br.), *P. acerosa* and *P. angulata*, *P. myrtilloides* and *P. angulata*, *P. myrtilloides* and *P. levis*. The last mentioned occurs on the higher Blue Mts and on leaf characters alone, might be confused with *P. lanceolata*.)

6 Banksia L.f.

76 species in Aust. (75 endemic); all states and territories

Shrubs or trees. Leaves alternate, sometimes whorled or in pseudowhorls, xeromorphic (Fig. 32). Flowers sessile, in pairs around a thick rhachis of a dense terminal spike, each subtended by one thick bract and two similar bracteoles. Perianth tube slender. Anthers sessile in the concave tips of the perianth segments. Hypogynous glands 4. Ovary sessile. Style straight or hooked and protruding from a slit in the perianth tube but finally free; pollen presenter ± cone-shaped, just below the stigma. Fruit a follicle, opening in two hard woody valves set transversely on the rhachis. The bracts and bracteoles becoming consolidated with the fruits and the rhachis to form a thick woody cone to which the withered barren flowers adhere for a time.

Key to the species

1 Style permanently hooked at the tip 2
1 Style straight or gently curved from near the base 5

2 Leaves 10–15 mm long, linear, crowded, truncate or notched at the apex, with closely revolute margins. Spike up to 15 cm long. Perianth orange-red. Fruits scarcely protruding from the spike. Shrub or small tree, glabrous except the inflorescence. Widespread except higher Blue Mts Heath and DSF. Ss. Fl. winter–spring . . . . **B. ericifolia** L.f.
2 Leaves 3–8 cm long, linear to oblong. Perianth brownish-red to golden yellow 3

3 Undersurface of leaves brownish. Leaves linear to oblanceolate, 2–7 cm long, 2–5 mm wide; margins recurved, mostly entire except at the apex. Perianth 20–30 mm long, yellow to golden. Lignotuber absent. Shrub up to 5 m high, single-stemmed at the base. Widespread. DSF. Fl. winter . . . .
 . . . . **B. cunninghamii** Sieber ex Rchb. ssp. **cunninghamii**
3 Undersurface of leaves white or concealed by revolute margins. Lignotuber present. Shrubs up to 2 m high, many-stemmed at the base.. 4

4 Leaves 1–2 mm wide, revolute to the midrib, notched at the tip with a prominent point in the notch and several small teeth on each side towards the tip, sometimes fine serrations extending the whole length. Coast and adjacent plateaus south of Sydney. DSF. Ss. Fl. winter. . . . . **B. spinulosa** Sm. var. **spinulosa**
4 Leaves 2–8 mm wide; margins recurved showing white undersurface, usually coarsely and evenly serrate. Coast and adjacent plateaus north of Sydney and Blue Mts Fl. winter (Intermediates between these varieties occur between Sydney and the Hawkesbury River.) . . . .
 . . . .
 . . . . **B. spinulosa** var. **collina** (R.Br.) A.S.George

5 Leaves white underneath with a close tomentum, 2–10 cm long 6
5 Leaves not white underneath, 10–25 cm long. 13

6 Leaves 3–10 mm wide, 2–6 cm long, linear to oblong, retuse and emarginate or obtuse, serrate or entire; lateral veins underneath usually inconspicuous. Spike 4–9 cm long, ovoid to cylindrical. Perianth yellow. Styles spreading or reflexed with a small thin pollen presenter. Spreading shrub or small tree with very villous young branches. Widespread. Heath and DSF. Ss. Fl. summer–autumn . . . .
 . . . . **B. marginata** Cav.
6 Leaves 10–20 mm wide, occasionally some narrower; lateral veins ± conspicuous 7

7 Leaves in regular whorls of 3–5 or in pseudowhorls 8
7 Leaves mostly alternate 12

8 Trees up to 16 m high, rarely shrubby. Leaves mostly entire (leaves on young plants dentate) 9

8 Many-stemmed shrubs up to c. 2 m high, rarely an open shrub to 5 m tall usually with dentate leaves . 10

9 Leaves 4–10 cm long, usually dull on upper surface. Ovary hairy around the apex. Woodlands on sandy soils near the coast, particularly on stabilized dunes. Fl. summer–autumn. *Coastal Banksia, Silver Banksia* . . . . **B. integrifolia** L.f. ssp. **integrifolia**
9 Leaves 10–20 cm long, usually glossy on upper surface. Ovary hairy on upper half. Blue Mts, e.g. Mt Wilson. Woodlands and WSF, particularly on basalt. Fl. summer. . . . .
. . . . **B. integrifolia** ssp. **monticola** K.R.Thiele

10 Perianth 20–25 mm long yellowish-green to bluish green in bud. Leaves white tomentose and with rusty hairs on veins, elliptic to obovate, 4–12 cm long. Some of the bracts with a tuft of long brown hairs at the apex. Lignotuber absent. Blue Mts Cliffs, and rocky places. Fl. autumn–winter . . . .
. . . . **B. penicillata** (A.S.George) K.R.Thiele
10 Perianth up to 20 mm long, golden-brown in bud. Leaves white tomentose when young, usually glabrescent, obovate, 4–8 cm long. 11

11 Usually compact shrub to 2 m tall. Lignotuber present. Widespread but more common on coast and adjacent plateaus. Heaths, usually near swamps. Fl. summer . . . . . **B. paludosa** R.Br. ssp. **paludosa**
11 Open-branched shrub to 5 m tall. Lignotuber absent. Near Hilltop. DSF. Ss. Fl. winter . . . .
. . . . **B. paludosa** ssp. **austrolux** A.S.George

12 Stems glabrescent after first year . . . . **B. penicillata** (see above)
12 Young stems densely rusty tomentose, remaining hairy for several years. Leaves elliptic to oblong or obovate, 5–10 cm long, with rusty hairs when young, dentate to entire. Perianth yellowish-green. Widespread. Heath. Fl. summer . . . . **B. oblongifolia** Cav.

13 Mature leaves rusty brown-tomentose and strongly reticulate underneath, ovate-oblong, mostly 10–25 cm long, 2–8 cm wide, irregularly and coarsely serrate. Spikes 8–12 cm long, 7–8 cm diam., deep bluish green when young. Style spreading; pollen presenter very small. Shrub c. 1 m high with densely tomentose branches. Coast and adjacent plateaus. Swamps. Fl. summer (Hybrids between *B. robur* and *B. oblongifolia* occur) . . . . **B. robur** Cav.
13 Mature leaves with greyish or brownish hairs underneath or almost glabrous, green on both surfaces, with parallel transverse veins, margins coarsely and evenly serrate 14

14 Pollen presenter cylindrical, 2 mm long. Tree or tall shrub, usually with one main stem. Bark grey. Leaves elliptic-oblong to oblanceolate, 8–16 cm long, 2–4 cm wide. Spike 8–16 cm long, 8–10 cm diam. Fruits protruding, tomentose, rounded or ovate, 3 cm wide. Perianth dove-grey. Widespread. Heath and DSF. Ss and hind-dunes. Fl. summer–autumn. **B. serrata** L.f.
14 Pollen presenter ovoid, c. 1 mm long. Shrub up to 4 m high, with spreading gnarled habit, often branching near the base. Bark grey with yellowish patches. Leaves oblanceolate, 1–2 cm wide. Perianth whitish green. Coast and Cumberland Plain. Old dunes and sand banks. Fl. autumn . **B. aemula** R.Br.

7 **Hakea** Schrad. & J.C.Wendl.
150 species endemic Aust.; all states and territories

Shrubs or small trees. Leaves alternate, flat or terete. Flowers on short pedicels in axillary clusters. Perianth white to cream or pink; tube almost straight; segments rolled back to one side, usually becoming free. Anthers sessile in the concave tips of the perianth segments. Ovary glabrous, usually stipitate. Style longer than the perianth, curved and protruding from a slit in the perianth before being released; pollen presenter flat or conical, expanded around the stigma. Fruit a woody follicle splitting to the base into two solid valves. Seeds 2, each with a broad wing on one edge.

Key to the species

1 Leaves flat 2
1 Leaves terete, rigid, pungent pointed 6

2 Leaves with 3 or more prominent longitudinal veins. Fruit smooth or slightly rough, ovate or almost globular, 12–35 mm diam., with a very small straight beak. Flowers numerous in axillary clusters. . . . 3

2 Leaves smooth, without prominent veins or with the midvein only distinct. Fruit rough, with prominent tubercles or almost smooth, with a distinct beak. 4

3 Single-stemmed shrub or small tree to 5 m without a lignotuber. Branchlets glabrescent or with pale hairs at flowering. Leaves narrow-elliptic to linear, 4–13 cm long, 5–15 mm wide. Widespread. DSF, WSF. Fl. spring . . . . **H. dactyloides** (Gaertn.) Cav.

3 Multistemmed shrub to 3 m with a lignotuber. Branchlets brown-tomentose at flowering. Leaves linear to obovate, 2–13 cm long, 4–30 mm wide. Widespread. Heath. Fl. spring . . **H. laevipes** Gand. ssp. **laevipes**

4 Leaves more than 7 mm wide, lanceolate, 8–15 cm long. Small tree or shrub to 5m. Widespread. DSF. Ss. or junctions of Ss. and shales. Fl. spring. . . . . **H. salicifolia** (Vent.) B.L.Burtt. ssp. **salicifolia**

4 Leaves less than 7 mm wide. 5

5 Leaves narrow-lanceolate-oblong, 8–15 cm long. Small tree or shrub to 5 m. Coast and adjacent plateaus. Near creeks in Ss. Fl. spring. . . . . **H. salicifolia** ssp. **angustifolia** (A.A.Ham.) W.R.Barker

5 Leaves linear, 20–40 cm long, 1.8–2.0 mm wide. Shrub to 3 m. Restricted to ridges in Kanangra Boyd Nat. Park. DSF. Endangered . . . . **H. dohertyi** Haegi

6 Fruit narrow, acuminate, several times as long as broad, with several angular protuberances below the middle, 2–3 cm long. Leaves 2–5 cm long. Perianth 4–6 mm long, pubescent. Straggling shrub 1–3 m high. 7

6 Fruit ovoid-globular. Flowers glabrous or villous. 8

7 Pedicel and perianth with dense ± appressed silky hairs. Widespread. Wet heath. Ss. Fl. spring-summer
. . . . **H. teretifolia** (Salisb.) Britten ssp. **teretifolia**

7 Pedicels and perianth with dense ± raised tomentose hairs. Coast and adjacent plateaus. Wet heath. Ss.
. . . . **H. teretifolia** ssp. **hirsuta** (Endl.) R.M.Barker

8 Fruit less than 8 mm diam. when mature, slightly rugose to smooth, with a small protuberance near the top of each valve. Leaves up to 8 cm long. Perianth 4–5 mm long, whitish. Shrub 0.5–1 m high. Higher Blue Mts Swamps. Fl. spring–summer. . . . . **H. microcarpa** R.Br.

8 Fruit more than 1.5 cm diam. when mature; rugose, warted 9

9 Young stems and very young leaves hirsute to villous. Perianth 5–6 mm long, white or cream 10

9 Young stems and very young leaves glabrous or very slightly pubescent 11

10 Fruit distinctly beaked, 25–30 mm diam. Leaves 4–8 cm long. Young stems and leaves hirsute. Shrub 1–2 m high. Coast and adjacent plateaus. Heath. Ss. Fl. spring . . . . **H. gibbosa** (Sm.) Cav.

10 Fruit not distinctly beaked, c. 40 mm diam., rugose. Leaves 5–10 cm long. Young stems and leaves with a dense villous felt. Shrub up to 8 m high. Blue Mts, Bell to Mt Wilson. DSF. Ss. Fl. summer . . .
. . . . **H. constablei** L.A.S.Johnson

11 Perianth 8–10 mm long, pink. Pedicel glabrous. Flowers mostly borne on old stems below the leaves. Fruit 40–50 mm diam., rough, usually contracted into a short beak. Leaves 5–7 cm long. Young stems sparsely pubescent. Bushy shrub up to 2 m high. Coast and adjacent plateau. DSF. Ss. Fl. autumn– winter. . . . . **H. bakeriana** F.Muell. & Maiden

11 Perianth up to 7 mm long. Flowers borne on young stems amongst the leaves 12

12 Fruit usually less than 20 mm diam. Leaves patent, up to 6 cm long. Perianth 4–7 mm long, white or pinkish. Shrub spreading, up to 3 m high. Lignotuber present. Blue Mts DSF and heath. Sandy or rocky soils. Fl. winter–spring . . . . **H. decurrens** ssp. **physocarpa** W.R.Barker

12 Fruit 20 mm or more diam. Perianth <5 mm long. Lignotuber absent. 13

13 Fruit 20 mm diam., moderately rough, contracted into a short smooth beak. Perianth tube 2.5–5 mm long, white or cream. Leaves 2–6 cm long. Shrub spreading, up to 3 m high. Widespread. DSF and heath. Ss and WS. Fl. spring. . . . . **H. sericea** Schrad. & J.C.Wendl.

13 Fruit 25 mm diam., rough and often tuberculate, without a beak. Perianth to 2.2 mm long 14

14 Flowers white. Fruit mostly more than 3.5 cm long. Small compact tree or shrub to 5 m, single stemmed. Coast to lower Blue Mts DSF, WSF. Fl. winter. . . . . **H. propinqua** A.Cunn.

14 Flowers cream to yellow. Fruit mostly less than 3.5 cm long. Shrub compact to 2 m tall, single stemmed. Higher Blue Mts Ss. Fl. winter-spring . . . . **H. pachyphylla** Sieber ex Spreng.

8 Helicia Lour.
9 species in Aust.; (8 endemic). Qld. NSW NT

One species in the area

Shrub or small tree to 10 m high, glabrous. Leaves lanceolate to broad-lanceolate, coriaceous, 2–4 cm long, entire or sometimes toothed, alternate to irregularly opposite or whorled. Flowers c. 1 cm long, in axillary racemes 3–6 cm long. Perianth regular, pale cream, c. 8 mm long; segments ± equally revolute. Anthers with a short filament. Hypogynous glands ± connate. Ovary sessile. Pollen presenter conical. Fruit ± fleshy, with a hard endocarp, often bluish, c. 12 mm long. Minnamurra Falls. RF. Fl. spring . . . .

. . . . **H. glabriflora** F.Muell.

9 Grevillea R.Br. ex Knight
357 endemic Aust.; all states and territories.

Shrubs or trees. Leaves alternate. Flowers in racemes, often short and umbel-like; bracts deciduous. Perianth straight or curved; segments rolled back to one side. Hypogynous gland 1, horse-shoe shaped. Anthers sessile in the concave tips of the perianth segments. Style curved, protruding from a slit in the perianth before being finally released; pollen presenter flat to conical, extended around the stigma. Fruit a follicle, thin-walled and brittle when dry, usually oblique. Seeds 2, collateral, wingless or with an annular membranous wing.

Key to the species

1 Leaves deeply serrate or variously divided. Racemes usually one-sided . . . . .GROUP 1
1 Leaves entire. Racemes short and umbel-like. 2

2 Ovary hairy, sessile or shortly stipitate . . . . .GROUP 2
2 Ovary glabrous, stipitate. . . . . .GROUP 3

Group 1
Leaves deeply serrate or variously divided

1 Leaves, very coarsely serrate or pinnatifid, up to 25 cm long, linear to lanceolate. Perianth dark red . . 2
1 Leaves pinnate or ternate lobed, 3–12 cm long; the lobes sometimes again divided, oblong to ovate or obovate in outline 3

2 Undersurface of leaves covered with ± appressed silky usually brownish hairs. Branchlets angular, silky hairy. Leaves serrate or rarely entire, the teeth scarcely reaching halfway to the midrib, 10–25 cm long, 15–20 mm wide. Shrub 2–5 m high. Blue Mts; Cumberland Plain; Woronora Plateau. Banks of gullies or creeks. Fl. spring . . . . .G. longifolia Sieber ex Spreng.
2 Undersurface of leaves covered with tomentose crinkled brownish or white hairs. Branchlets tomentose. Leaves usually deeply serrate to pinnatifid, the teeth reaching almost to the midrib, very rarely entire. Shrub 2–5 m high. Blue Mts from Cox's River southwards. Woronora Catchment. DSF. Ss. ridges. Fl. winter–spring. . . . . .G. aspleniifolia Knight

3 Leaf segments terete, revolute to the midrib, c. 1 mm wide; leaves 1–25 cm long. Perianth red to orange, silky hairy outside. Shrub up to 5 m high. Capertee district; recorded from Illawarra ranges. DSF. Fl. spring . . . . **G. johnsonii** McGill.
3 Leaf segments flat, sometimes with recurved margins but never to the midrib, usually more than 1 mm wide 4

4 Undersurface of leaves glabrous 5

4 Undersurface of leaves hairy 6

5 Perianth glabrous, deep pink. Leaves bipinnate-partite, 2–6 pinnae with pungent tips. Ovary with a stalk c. 2 mm long. Straggling to spreading shrub up to 2 m high. Carrington Falls. Banks of creeks. Endangered. Fl. spring–autumn . . . . **G. rivularis** L.A.S.Johnson & McGill.
5 Perianth silky hairy, deep pink. Leaves pinnately divided almost to the midvein; lobes cuneate, 3–5-lobed, rigid, pungent-pointed. Straggling shrub up to 2 m high. Higher Blue Mts Wet places. Ss. Fl. most of the year. . . . . **G. acanthifolia** A.Cunn. ssp. **acanthifolia**

6 Style up to 10 mm long. Perianth whitish, c. 5 mm long. Leaf-lobes pungent-pointed. 7
6 Style more than 20 mm long. Perianth deep red. Leaf lobes not pungent pointed 9

7 Style with hairs. Leaf-lobes cuneate, 5–10 mm wide, tomentose with brownish hairs underneath. Spreading shrub up to 3 m high. Higher Blue Mts DSF. Fl. spring . . . . **G. ramosissima** Meisn. ssp. **ramossisima**
7 Style glabrous 8

8 Inflorescence ovoid dense. Style + ovary >5 mm long. Buds spreading. Leaf lobes narrow-linear, divided into 2–5 segments, silky hairy beneath. Straggling shrub up to 1.5 m high. Coast and Blue Mts in south of region. DSF. Fl. spring . . . . **G. raybrownii** Olde & Marriott
8 Inflorescence cylindrical. Style plus ovary 5 mm long. Buds appressed. Leaf-lobes linear, c. 2 mm wide, ternate or forked, with appressed brownish hairs underneath. Straggling shrub up to 1 m high. Illawarra district; Woronora Plateau; Blue Mts DSF. Fl. winter–spring. . . . .**G. triternata** R.Br.

9 Young shoots and undersurface of the leaves slightly pubescent with white hairs. Leaves pinnate-lobed; lobes oblong or ovate, rarely themselves lobed. Perianth dark pink. Small shrub. Higher Blue Mts Fl. spring–summer. (This plant is known to be a hybrid between *G. acanthifolia* and *G. laurifolia*; some forms closely resemble one or other of the parents) . . . . **G. x gaudichaudii** R.Br. ex Gaudich.
9 Branches, petioles and undersurface of leaves and inflorescences densely and softly hairy with spreading rusty hairs. Leaves very deeply pinnatifid with numerous oblong-linear lobes, obtuse or mucronate with recurved margins. Perianth dark red. Spreading shrub 2–4 m high. Terrey Hills to Mona Vale. DSF. Laterite. Endangered. Fl. summer . . . . **G. caleyi** R.Br.

Group 2
Leaves entire; ovary hairy

1 Ovary hairs only on lower ventral side . . . . **G. rosmarinifolia** ssp. **rosmarinifolia** (see below)
1 Ovary hairs not confined to lower half 2

2 Inside of perianth and style very densely covered with whitish reflexed hairs, the outside of the perianth densely rusty hairy. Perianth tube usually under 6 mm long, narrow or broad but not usually dilated below the middle. Racemes umbel-like. Perianth pinkish. 3
2 Inside of perianth with hairs similar in colour to those on the outside or glabrous. Perianth tube usually more than 6 mm long, very broad or dilated at the base. Racemes one sided or if umbel-like then flowers red to green 6

3 Stylar appendage obscure, less than 1 mm long. 4
3 Style produced into an appendage beyond the pollen presenter. Young shoots and branches densely covered with long rusty or dark hairs 5

4 Undersurface of leaf visible through the hairs. Pollen presenter orbicular. Perianth tube 2–3 mm long. Branchlets villous. Leaves elliptic 1–2 cm long. Erect shrub to 1.5 m high. between Wollombi and Putty. Heath and DSF. Fl. spring–autumn . . . .**G. buxifolia** ssp. **ecorniculata** Olde & Marriott
4 Undersurface of leaf obscured by hairs. Pollen presenter ovate. Perianth tube 2–3 mm long. Young shoots and branches and undersurface of leaves with closely appressed rusty or silvery hairs. Leaves linear to oblong-lanceolate or oblong-elliptic, 1–2 cm long. Erect shrub up to 1 m high. Smaller in all its parts than *G. buxifolia* excepting sometimes the leaves. Coast and adjacent plateaus and lower Blue Mts Heath and DSF. Ss. Fl. spring . . . . **G. sphacelata** R.Br.

5 Pollen presenter orbicular. Perianth tube 4–5 mm long. Leaves ovate oblong or lanceolate, 6–20 mm long. Shrub 1–2 m high. Coast and adjacent plateaus. Heath and DSF. Ss. Fl. spring . . . .
. . . . **G. buxifolia** (Sm.) R.Br. ssp. **buxifolia**

5 Pollen presenter oblong. Perianth tube 2–4 mm long. Leaves linear-lanceolate to lanceolate, 16–30 mm long. Shrub c. 1 m high. Wiseman's Ferry, Colo area and Blue Mts Heath and DSF. Ss. Fl. spring. . . . .
. . . .**G. phylicoides** R.Br.

6 Racemes one-sided 7
6 Racemes not one-sided, usually short, umbel-like 9

7 Prostrate trailing shrub forming a mat up to 3 m diam. Leaves broad-ovate to elliptic, 3–12 cm long, glabrous above, appressed silky hairy below. Perianth dark red. Blue Mts and adjacent valleys. DSF. Fl. spring–summer. . . . .**G. laurifolia** Sieber ex Spreng.
7 Erect shrubs 2–5 m high. 8

8 Under surface of leaves with ± appressed silky hairs. . . . . **G. longifolia** (see above)
8 Under surface of leaves tomentose with crinkled hairs . . . . **G. aspleniifolia** (see above)

9 Lobes of the perianth with an appendage more than 1 mm long, ovoid-acuminate in the bud 10
9 Lobes of the perianth obtuse, rounded in the bud or apiculate with a point less than 1 mm long. Perianth tube 7–10 mm long. 12

10 Leaves silky underneath, narrow-elliptic to linear, 1–2 occasionally 3 cm long, 1–7 mm wide. Inflorescences almost sessile with 1–4 flowers. Flowers red often green near base. Inside of perianth glabrous to sparsely hairy above beard. Shrub up to 2 m high. Hunter River Valley. DSF. Fl. spring. . .
. . . . **G. montana** R.Br.
10 Leaves pubescent to villous underneath, oblong-elliptic, 2–7.5 cm long, 3–15 mm wide. Inflorescences mostly pedunculate with 2–10 flowers. Flowers red, pink or orange often green or yellow at base. Inside of perianth densely to spreadingly hairy above beard. Shrub 1–3 m high. 11

11 Lower surface of leaf silky, gynoecium 22–27 mm long. Widespread in Blue Mts DSF. Fl. most of the year . . . . **G. arenaria** R.Br. ssp. **arenaria**
11 Lower surface of leaf velvety, gynoecium 26–35 mm long. Western Blue Mts DSF. Fl. most of the year
. . . . **G. arenaria** ssp. **canescens** (R.Br.) Olde & Marriott

12 Ovary densely hirsute with brown or reddish hairs 13
12 Ovary slightly pubescent with whitish hairs. 14

13 Perianth glabrous outside or nearly so, red but often becoming green towards the top. Leaves ovate to elliptic, up to 25 mm long, sessile, glabrous, minutely mucronate. Shrub c. 1 m high. Coast and adjacent plateaus in south of region. Open forests. Fl. spring–summer . . . . . . **G. baueri** ssp. **baueri**
13 Perianth pubescent outside, reddish or greenish or rusty-coloured from the hairs. Leaves broad-elliptic to obovate or ± orbicular, up to 4 cm long, shortly petiolate, with a mucro usually 1–2 mm long. Shrub up to c. 1 m high. Widespread. Open forests. Fl. most of the year . . . .**G. mucronulata** R.Br.

14 Leaves broad-elliptic to obovate or ± orbicular with a mucro 1–2 mm long. Not suckering. . . . .
. . . . **G. mucronulata** (see above)
14 Leaves narrow-elliptic to oblanceolate or linear, 1.5–6 cm long, 2–10 mm wide, with an inconspicuous mucro less than 1 mm long. Perianth red but greenish above, pubescent outside. Style red, with villous hairs. Suckering shrub up to 2 m high. Kedumba valley, Kiaramba ridge. DSF. Fl. spring–summer. . .
. . . . **G. kedumbensis** (McGill.) Olde & Marriott

Group 3
Leaves entire; ovary glabrous

1 Gynophore c. 15 mm long, nearly as long as the style. Hypogynous gland very large. Perianth pale violet to greenish, tinged with pale purple-brown. Leaves narrow-lanceolate, 8–16 cm long, acuminate,

3-veined. Tall shrub or small tree 2–8 m high. North side of the Hawkesbury River. Forest in gullies in Ss. Vulnerable. Fl. spring–summer. . . . . **G. shiressii** Blakely

1 Gynophore less than half as long as the style 2

2 Gynophore almost as thick as ovary, with a tuft of hairs. Leaves narrow elliptic to linear, 0.8–4 cm long, 0.8–3.0 mm wide, sometimes pungent, margins recurved to revolute. Racemes terminal, 4–12 flowered. Perianth glabrous outside, pink and cream, or whole pink, green, yellow, cream or orange. Erect to low spreading shrub to 2 m high. Kowmung and Hartley area. Open forest, woodland and riparian communities. Fl. winter–spring . . . .**G. rosmarinifolia** ssp. **rosmarinifolia** A.Cunn.

2 Gynophore glabrous and slender 3

3 Leaves very spreading, linear-subulate, rigid, pungent-pointed, with revolute margins, 1–2 cm long. Racemes very short and umbel-like, terminal, sessile. Perianth silky-pubescent outside, pale yellow and green or red; tube 10–12 mm long. Prickly divaricate shrub 1–2 m high. 4

3 Leaves not as above. Shrubs rarely prickly, sometimes tall. Young branches, undersurface of leaves and inflorescences ± silky hairy 5

4 Upper leaf surface with 3 prominent longitudinal veins. Low spreading shrub to 1 m. Leaves 0.6–0.8 mm wide. Cumberland Plain. Open forest and cleared areas on WS. Vulnerable Fl. spring–summer . . . .**G. juniperina** R.Br. ssp. **juniperina**

4 Upper leaf surface with midvein prominent. Erect shrub to 2 m high, rarely prostrate. Leaves 0.6–1.8 mm wide. On sandstone derived solis from Berrima to Tallong and west to Blue Mts Grows in DSF in moist areas beside creeks. Fl. Winter–spring . . . . .**G. juniperina** ssp. **sulphurea** (A.Cunn.) Makinson

5 Gynoecium more than 16 mm long, much longer than the perianth tube 6

5 Gynoecium less than 16 mm long, usually less than twice as long as the perianth tube. 8

6 Perianth pink, rarely white (near Heathcote), silky pubescent outside; tube 4–8 mm long. Style 10–15 mm long. Racemes dense, short, on short terminal peduncles. Leaves narrow-ovate to almost linear, 1–6 cm long, glabrous or slightly silky hairy above, closely silky tomentose below. Shrub 1–2 m high, branches often angular when young. Widespread. Heath and DSF. Ss. Fl. most of the year . . . . **G. sericea** (Sm.) R.Br.

6 Perianth red, very occasionally pale or even white. Style 20–25 mm long. 7

7 Leaves 1–5 cm long, ovate or elliptic to narrow-lanceolate. Racemes almost terminal on lateral branches; peduncles spreading or reflexed, 5–50 mm long. Perianth tube 8–10 mm long. Erect or bushy shrub 1–3 m high. Hornsby Plateau. Heath and DSF. Ss. Fl. most of the year. . . . . . **G. speciosa** (Knight) McGill.

7 Leaves 5–10 cm long, linear to lanceolate. Racemes mostly sessile in the axils of leaves or terminal on very short branches. Perianth tube 8–14 mm long. Erect shrub c. 2 m high. Woronora Plateau; Blue Mts Heath and DSF. Ss . . . . **G. oleoides** Sieber ex Schult. & Schult.f.

8 Perianth red to almost black or rarely greenish brown. Racemes dense and very short 9

8 Perianth whitish or pale to deep pink, silky pubescent outside. 14

9 Limb on late flower buds tomentose to villous, hairs ascending to spreading 10

9 Limb on late flower buds with silky appressed hairs. Never with rusty tomentum on bud limbs or leaves 3-veined 12

10 Leaves 3-veined and pungent pointed, lanceolate to oblong 10–35 mm long, 1.5–6 mm wide. Racemes held +/- erect. Glandular hairs present on outside of perianth. Shrub to 1.2 m high covered with loose shaggy hairs. Mangrove Mt. to Woy Woy. Heath and DSF. Ss. Fl. summer–autumn. . **G. oldei** McGill.

10 Leaves not or scarcely pungent pointed, glandular hairs not present on outside of perianth 11

11 Nectary protruding past receptacle rim. Limb of bud with rusty hairs. Leaves elliptic to obovate, mostly 2.5–6 cm long, 3–10 mm wide. Low dense shrub to 0.5 m high. North-western Blue Mts Heath and DSF, sometimes in swampy conditions. Vulnerable. Fl. winter–spring. . . .**G. evansiana** MacKee

11 Nectary scarcely protruding past receptacle rim. Limb of bud with pale hairs (occasionally light brown). Leaves narrow elliptic to lanceolate, 2–9 cm long, 2–8 mm wide. Decumbent shrub to 0.5 m

high. Grows in poorly drained areas and swamp margins. Restricted to southern edge of Sydney Basin. Fl. winter–summer . . . . **G. capitellata** Meisn.

12 Branchlets subterete to slightly angular with short hairs. Racemes on subterete peduncles to 10 mm long, scarcely pendent. Style with hairs confined to apex. Leaves elliptic 1.5–5.5 cm long, 2–4 mm wide. Low compact shrub to 0.5 m high. Woronora Plateau. Heath and DSF, Ss.. Fl. winter–summer . . . . **G. diffusa** Sieber ex Spreng. ssp. **diffusa**

12 Branchlets prominently angular, glabrous or with silky hairs between the ridges. Racemes on angular peduncles more than 10 mm long. Style with hairs covering top half 13

13 Leaves usually more than 7 cm long, 2–4 mm wide, margins bent in towards lower surface. Peduncles slender, 15–40 mm long. Perianth and style scarlet to crimson. Shrub to 1 m high. Calga to Mt. White. Heath and DSF, sometimes in swampy conditions. Fl. spring. . . . . **G. diffusa** ssp. **filipendula** McGill.

13 Leaves usually less than 7 cm long, 2.5–4 mm or occasionally 7 mm wide, margins recurved. Peduncles 10–15 mm long. Perianth and style dark red to dark burgundy. Shrub 1–2 m high. Waterfall, Helensburgh Georges River area. Often near creeks. Ss. DSF. Fl. winter–spring . . . . **G. diffusa** ssp. **costablei** Makinson

14 Leaves rigid with a pungent point, linear to narrow elliptic 2–5 cm long, 1–3 mm wide. Racemes many flowered. Perianth pale pink to dark pink or rarely white. Erect or low spreading shrub to 2 m high. Coastal Plain south form Heathcote. Heath and DSF. Ss. Fl. summer . . . . . . **G. patulifolia** Gand.

14 Leaves pliable not or scarcely pungent. 15

15 Gynoecium 4–7 mm long. Racemes with 6–14 flowers. Undersurface of leaf exposed or sometimes enclosed and 2-grooved 16

15 Gynoecium 7–13 mm long. Racemes 10 to many-flowered. 17

16 Major branches ascending to erect. Branchlets not or only occasionally arranged on one side of stem. Leaves 0.8–1.3 mm wide. Gynophore 1–2 mm long. Shrub ≤1 m high. West and south of Sydney. Heath and woodlands on shale in sandy or light clay soils. Vulnerable. Fl. winter –summer . . . . **G. parviflora** R.Br. ssp. **parvifolia**

16 Major branches ± spreading. Branchlets arranged on one side of stems. Leaves held upright 0.6–2 mm wide. Gynophore 0.5–0.6 mm long. NW of Sydney. Shrub ≤1 m high. Heath and woodland, on Ss. in skeletal sandy soils. Endangered. Fl. winter–spring . . . . . . **G. parvifolia** ssp. **supplicans** Makinson

17 Erect to spreading shrub to 1 m high. Leaves ascending, oblong elliptic 3–5 cm long, 1.8–2.5 mm wide, undersurface mostly exposed. Perianth pink occasionally white. Gynoecium 7–10 mm long. Gynophore 1–1.7 mm long. Lower Hunter area and possibly Putty. DSF and heathy woodland. Fl. winter–spring . . . . **G. humilis** Makinson ssp. **humilis**

17 Erect shrub 1–3 m high. Leaves spreading to loosely ascending, linear to narrow elliptic, 5–9 cm long, 1–3 mm wide, undersurface mostly enclosed. Perianth white. Gynoecium 7–13 mm long. Gynophore 1.3–1.5 mm long. Widespread. Heath and DSF. Ss. Fl. spring–summer . . **G. linearifolia** (Cav.) Druce

10 Lomatia R.Br.
9 species endemic Aust.; Qld., NSW, Vic., Tas.

Shrubs or small trees. Leaves alternate, entire to much divided. Flowers in axillary or terminal racemes. Perianth slit on the lower side in the bud; segments rolled back to one side. Hypogynous glands 3. Anthers sessile. Ovary on a long gynophore, glabrous. Style protruding from the perianth slit until released; pollen presenter expanded obliquely around the stigma. Fruit a follicle, opening almost flat. Seeds numerous in two rows, imbricate upwards with a broad terminal nearly straight wing and with a yellow powdery substance between the seeds.

Key to the species
(Intermediate forms, possibly hybrids, occur in some localities where these species grow together.)

1 Leaves mostly twice or thrice pinnately divided rarely simple pinnate, 10–30 cm long; segments sessile and decurrent, linear or lanceolate, usually deeply and sharply toothed. Flowers in a loose terminal

raceme or panicle much exceeding the leaves, 15 cm or more long. Perianth white. Shrub 50–150 cm high. Widespread. Heath and DSF and cleared areas. Fl. summer. *Crinkle Bush*. **L. silaifolia** (Sm.) R.Br.

1 Leaves undivided, entire to coarsely serrate. 2

2 Leaves ovate to lanceolate or oblong-elliptic, acute-obtuse at the base, with a distinct but short petiole, 10–18 cm long, 3–5 cm wide; veins distinctly reticulate on the upper surface; margins coarsely serrate. Raceme exceeding the leaves. Perianth yellowish, 10–12 mm long. Shrub up to 3 m high but usually shorter. Blue Mts Forests. Ss. Fl. summer . . . .**L. ilicifolia** R.Br.

2 Leaves linear to narrow-oblong-lanceolate, tapering very gradually into a very short petiole or sessile, 5–20 cm long, up to 2 cm wide, not distinctly reticulate on the upper surface; margins serrate. Racemes shorter than the leaves or scarcely exceeding them. Perianth greenish yellow, 8–10 mm long. Shrub to small tree up to 4 m high. Widespread. Along water courses or in forests. Ss. Fl. summer . . . .**L. myricoides** (C.F.Gaertn.) Domin

11 **Stenocarpus** R.Br.
9 species in Aust. (7 endemic); Qld. NSW, NT, WA

One species in the area

Tree or large shrub, glabrous. Leaves narrow- to ovate-lanceolate, alternate, entire, 3–10 cm long. Flowers 10–20 in an axillary umbel c. 3 cm diam.; peduncle 2–4 cm long. Perianth yellowish, slit on the lower side while in bud; limb nearly globular and rolled back to one side. Hypogynous gland 1 or absent. Anthers sessile. Ovary stipitate, slightly pubescent. Style long, protruding from the perianth slit until released; pollen presenter expanded obliquely around the stigma. Fruit a narrow coriaceous follicle. Seeds 5–8, flat, imbricate, winged at the lower end. Widespread. Sheltered gullies, RF. Fl. spring. *Scrub Beefwood*. . . . . **S. salignus** R.Br.

12 **Telopea** R.Br.
5 species endemic Aust.; NSW, Vic., Tas.

Nearly glabrous shrubs up to 4 m high. Leaves alternate. Flowers pedicellate in pairs, arranged in large terminal pseudo-heads with an involucre of coloured bracts. Perianth red, slit on the lower side when in bud; segments rolled back on one side. Hypogynous gland a short oblique almost complete ring. Anthers sessile. Ovary stipitate, glabrous. Style protruding from the perianth slit until released; pollen presenter expanded obliquely around the stigma. Fruit a coriaceous to woody follicle, recurved. Seeds several, in 2 rows, winged.

Key to the species

1 Leaves reticulate on the upper surface, narrow-oblong to obovate, 10–15 cm long, 2–4 cm wide, mostly dentate, sometimes divided. Involucral bracts bright red (rarely white), 5–8 cm long. Flowers in a dense compact pseudo-head, 8–15 cm diam. Shrub up to 4 m high. Widespread. DSF. Ss. Fl. spring on coast, summer on Blue Mts *Waratah* . . . . **T. speciosissima** (Sm.) R.Br.

1 Leaves not distinctly reticulate on the upper surface, very narrow-oblong to narrow-obovate, 8–10 cm long, 8–15 mm wide, entire or rarely with a few teeth towards the top. Involucral bracts ± inconspicuous, pinkish or similar to the leaves. Flowers in a loose pseudo-head c. 7 cm diam. Southern Blue Mts Shrub up to 6 m high. DSF. Fl. summer. *Monga Waratah*. **T. mongaensis** Cheel

13 **Xylomelum** Sm.
6 species endemic Aust.; Qld., NSW, WA

One species in the area

Small tree or tall shrub. Young shoots and inflorescence rusty villous; young leaves glabrous, reddish and cyanogenetic. Adult leaves lanceolate or ovate-lanceolate, glabrous, 10–20 cm long, usually entire on mature trees but sinuate and prickly toothed on younger plants, coriaceous, strongly veined, tapering to a long pointed apex. Flowers in dense opposite nearly sessile spikes 5–8 cm long. Perianth c. 10 mm long; segments white or cream, spreading, revolute. Anthers on short filaments. Hypogynous glands 4. Ovary

shortly stipitate. Style filiform; pollen presenter club-shaped, below the stigma. Fruit a woody pear-shaped follicle, 6–9 cm long, attached at the larger end, tardily dehiscent. Seeds 2, collateral, each with a large oblique brown wing. Widespread. DSF. Ss. Fl. spring. *Native Pear* or *Woody Pea*. . . . .
. . . .**X. pyriforme** (Gaertn.) Knight

14 Lambertia Sm.
10 species endemic Aust.; NSW, WA

One species in the area
Shrub 1–2 m high. Leaves linear to oblong, 3–5 cm long, mostly in whorls of 3, rigid, glossy above, mucronate, pale underneath with a prominent midvein; margins recurved. Inflorescence terminal, head-like, with 7 sessile flowers enclosed within an involucre of red imbricate bracts. Perianth red; tube 2–5 cm long; segments spirally and equally revolute. Anthers sessile. Style rigid, filiform, projecting 10–16 mm beyond the perianth; pollen presenter filiform, below the stigma. Hypogynous glands connate into a cup at the base of the ovary. Fruit a follicle, rugose, woody, with a short beak, and with a long horn on each valve. Widespread. Heath and DSF. Ss. Fl. most of the year. *Mountain Devil*. . . . . **L. formosa** Sm.

43 BERBERIDACEAE
Shrubs. Leaves alternate, simple or compound; stipules absent. Flowers bisexual, in cymo-panicles or racemes. Sepals and petals similar, in 2 or more whorls, free, imbricate. Stamens 6, hypogynous, free; anthers dehiscing by valves. Carpel solitary, with several ascending ovules. Fruit a berry. 13 gen., temp. America and Asia.

Key to the genera

1 Leaves of the main shoots reduced to spines; leaves of short shoots simple . . . .**1 BERBERIS**
1 Leaves of the main shoots compound, not reduced to spines . . . .**2 MAHONIA**

1 Berberis L.
Barberries
3 species naturalized Aust.; NSW, Vic.

One species in the area
Shrub up to 3 m high with trifid leaf spines. Leaves elliptic-oblong, up to 5 cm long, dentate with acuminate teeth, sessile or nearly so. Flowers in axillary racemes. Fruit bright red becoming purplish and glaucous. Naturalized at Yerranderie and Macquarie Pass. Introd. from Himalayas . . . ***B. aristata** DC.

2 Mahonia Nutt.
1 species naturalized Aust.; NSW

One species in the area
Shrub up to 2 m high. Leaves compound, pinnate with c. 11 leaflets; leaflets ovate, up to 9 cm long, stiff, serrato-dentate, with acuminate teeth. Flowers in terminal clusters of racemes. Fruit blue-black, glaucous. Naturalized at Mt. Wilson. Introd. from E. Asia . . . .***M. leschenaultii** (Wallich) Takeda

44 MENISPERMACEAE
Dioecious climbers. Leaves alternate, entire, palmately veined; stipules absent. Inflorescence axillary or lateral. Flowers unisexual, regular, small. Sepals 2–10, free. Petals 3–6, free, imbricate. Stamens 2–6, free or connate, hypogynous. Carpels free, superior. Fruit a drupe. 80 gen., mostly trop. and subtrop.

FIGURE 33

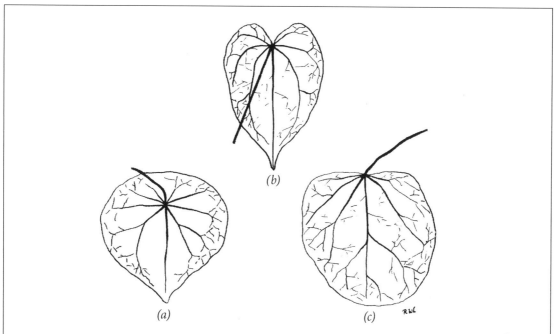

Fig. 33 MENISPERMACEAE, leaves (from below) of: (a) *Stephania japonica* var. *discolor*; (b) *Sarcopetalum harveyanum*; (c) *Legnephora moorei*. All X½.

Key to the genera

1 Leaves peltate (Fig. 33a). Flowers in compound umbels . . . . **1 STEPHANIA**
1 Leaves not peltate (very rarely slightly so). 2

2 Leaves broad-ovate, green, concolourous. Flowers in racemes . . . .**2 SARCOPETALUM**
2 Leaves nearly orbicular, whitish or hoary underneath. Flowers in dichotomous cymes. . . . .
. . . .**3 LEGNEPHORA**

1 Stephania Lour.

3 species in Aust.; Qld, NSW, NT

One species in the area

Slender climber. Leaves broad-ovate orbicular or ± triangular, up to 10 cm wide, peltate. Flowers in compound umbels. Sepals 6–10 in male flowers, 3–5 in female flowers. Petals 3–5. Stamens connate into a column. Carpel 1. Fruit a compressed drupe c. 5 mm diam. Coast. In or near RF. Near the sea. *Snake Vine* . . . . **S. japonica** var. **discolor** (Blume) Forman

2 Sarcopetalum F.Muell.

1 species in Aust.; Qld, NSW, Vic.

One species in the area

Woody climber. Leaves broad-ovate, deeply cordate at the base, acuminate, rarely slightly peltate, glabrous, mostly 5–10 cm wide. Flowers in simple racemes. Sepals 2–5, small. Petals 3–6, thick, gland-like. Stamens connate into a column. Carpels 3–6. Fruit a flattened drupe, 6–8 mm diam. In or near RF. Widespread. *Pearl Vine* . . . . **S. harveyanum** F.Muell.

3 Legnephora Miers
1 species endemic Aust.; Qld, NSW

One species in the area

Tall climber. Leaves nearly orbicular, sometimes slightly cordate at the base, 5–20 cm wide, whitish or hoary underneath. Flowers very small, in dichotomous cymes. Sepals, petals, and stamens 6. Carpels 3. Fruit a globular drupe, slightly flattened, red. Coast and adjacent plateaus. In or near RF . . . .
. . . . **L. moorei** (F.Muell.) Miers

45 PAPAVERACEAE

Herbs often with latex. Leaves basal and alternate, toothed, lobed or dissected; stipules absent. Flowers bisexual, regular, solitary. Sepals 2–3, sometimes connate into a calyptra, deciduous. Petals 4 or 6, imbricate, crumpled in the bud. Stamens numerous free. Ovary superior, with 2–12 parietal placentas; stigmas sessile. Fruit a capsule dehiscing by pores or valves. 40 gen., Trop. & temp. N. hemisphere.

Key to the genera

1 Sepals inconspicuous. Petals connivent corolla pouched or spurred. Stamens 6 . . . .**4 FUMARIA**
1 Sepals large, enclosing crumpled petals. Stamens numerous 2

2 Sepals connate into a calyptra . . . . **3 ESCHSCHOLZIA**
2 Sepals free or sometimes cohering 3

3 Sepals 2. Petals 4 . . . . **1 PAPAVER**
3 Sepals 3. Petals 6 . . . . **2 ARGEMONE**

1 Papaver L.
6 species naturalized Aust.; all states and territories

Bristly or glaucous herbs with latex. Flowers solitary on long peduncles. Sepals 2. Petals 4, crumpled in the bud. Stigmas sessile in many rows which radiate from the centre of a convex disc crowning the ovary. Fruit a capsule dehiscing by pores. Seeds numerous, minute.

Key to the species

1 Capsule covered with bristles, ovoid. Petals red, usually with a purple spot at the base. Leaves twice or thrice pinnatisect with narrow segments and lobes. Widespread. Introd. from Europe and W Asia. *Rough Poppy* . . . . ***P. hybridum** L.
1 Capsule glabrous 2

2 Plant covered with long spreading bristles. Leaves pinnatifid; lobes broadly toothed, each tooth ending in a conspicuous bristle. Petals red, c. 12 mm long. Capsule ovoid-oblong. Widespread but not common. Weed of cultivation, waste places and open communities, (may not be native). *Native Poppy* . . . .
. . . . **P. aculeatum** Thunb.
2 Plant glabrous or some parts pubescent with soft hairs or short bristles on the main veins of the leaves 3

3 Leaves pinnatisect, not stem-clasping. Petals orange to red, often darker towards the base, c. 2–5 cm long. Uncommon on waste ground. Introd. from Europe. . . . . ***P. dubium** L.
3 Leaves lobed or deeply toothed, stem-clasping, glaucous. Petals purplish pink, 2–4 cm long, with a dark spot at the base. Peduncles with short stiff hairs. Capsule obovoid-oblong to almost globular. Widespread. Weeds of cultivation and waste places. Introd. from the Mediterranean. *Varieties of the Opium Poppy* 4

4 Capsule indehiscent, globular, up to 5 cm diam . . . . ***P. somniferum** L. ssp. **somniferum**
4 Capsule dehiscing through pores, cylindrical to subglobular, up to 2 cm diam . . . .
. . . . ***P. somniferum** ssp. **setigerum** (DC.) Thell.

2 Argemone L.

2 species naturalized Aust.; Qld, NSW, Vic., NT, S.A.

Glaucous, glabrous annuals up to 1.5 m high. Leaves alternate up to 12 cm long, spiny-toothed or lobed, mottled with white. Flowers solitary, sessile or nearly so, on short branches. Sepals 3, horned. Petals 6, c. 25 mm long. Fruit a capsule, oblong to ovoid, up to 5 cm long.

Key to the species

1 Petals cream to pale yellow. Capsule with numerous spines. Widespread. Weed of cultivation and waste places. Introd. from North America. Fl. summer. *Mexican Poppy* ***A. ochroleuca** Sweet ssp. **ochroleuca**
1 Petals deep yellow. Capsule with few stout spines. Hunter River Valley. Weed of cultivated land. Introd. from N. America. Fl. summer. *American Poppy* . . . . . ***A. subfusiformis** Ownbey ssp. **subfusiformis**

3 Eschscholzia Cham.

1 species naturalized Aust.; NSW, Vic., S.A.

One species in the area

Annual herb, glabrous or glaucous. Leaves finely divided; basal leaves c. 20 cm long; cauline leaves smaller. Flowers solitary on long peduncles. Sepals connate into a calyptra, deciduous as flower opens. Petals 4, yellow, up to 3 cm long. Capsule ± cylindrical, c. 5 cm long, explosive when ripe. Widespread but not common. Garden escape. Introd. from California. *Californian Poppy* . . . . . ***E. californica** Cham.

4 Fumaria L.

7 species naturalized Aust.; all states and territories

Delicate, glabrous, ± glaucous annuals with weak angular trailing stems. Leaves twice or thrice pinnatisect, basal and alternate. Flowers irregular, arranged in bracteate racemes. Sepals 2, flat, scale-like. Petals connivent, 4, the upper one spurred or pouched. Stamens 6, united into 2 bundles. Ovary superior, 1-locular, with 2 placentas; style filiform, 2-fid. Fruit a nut.

Key to the species

1 Petals white with purple tips, 2–3 mm long; the upper petal laterally compressed, wings not concealing the keel. Inflorescence rather dense, c. 20-flowered. Pedicels of fruit recurved. Fruit globular. Coast and adjacent plateaus; Cumberland Plain. Weed of cultivation and waste places. Introd. from Europe. Fl. spring–summer. *White-flowered Fumitory* . . . .***F. capreolata** L. ssp. **capreolata**
1 Petals reddish or pink 2

2 Fruit with a large shallow pit at the apex or truncate, rugose when dry. Racemes 10–40-flowered usually c. 20. Bracts shorter than the pedicels. Leaf segments flat. Widespread. Weed of cultivation and waste places. Fl. spring–summer. Introd. from Europe . . . . * **F. officinalis** L. ssp **officinalis**
2 Fruit globular 3

3 Fruit smooth when dry, often with 2 small inconspicuous pits near the apex. Racemes usually c. 12-flowered. Petals 6–11 mm long, the lower with an erect margin. Pedicel of fruit straight, erect or spreading. Widespread. Weed of cultivation and waste places. Introd. from Europe. *Wall Fumitory* . . . . ***F. muralis** Sond. ex W.D.J. Koch. ssp. **muralis**
3 Fruit rugose when dry. Sepals serrate. Petals 9–11 mm long, the upper ones laterally compressed. Racemes 15–25-flowered. Widespread. Weed of cultivation and waste places. Fl. spring–summer. Introd. from Europe. *Bastard's Fumitory* . . . .***F. bastardii** Boreau

46 RANUNCULACEAE

Herbs with basal or alternate leaves, or climbers with opposite leaves; stipules absent. Flowers bisexual or unisexual, regular, solitary or in racemes or panicles. Sepals 4–5, imbricate or valvate. Petals mostly 5 or absent. Stamens usually numerous, hypogynous. Carpels free, superior. Fruit an achene. 70 gen., mostly temp. to cold.

Key to the genera

1 Climbers or scramblers with opposite leaves. Petals absent . . . . **1 CLEMATIS**
1 Herbs with basal or alternate leaves. Petals present 2

2 Petals yellow. Leaves sometimes divided into linear segments but then 1 mm wide or more and arranged in one plane only . . . . **2 RANUNCULUS**
2 Petals white. Leaves divided several times into narrow-linear or filiform segments less than 1 mm wide, and arranged in several planes . . . . **3 BATRACHIUM**

1 Clematis L.
Old Man's Beard
10 species in Aust. (1 naturalized); all states and territories except NT

Climbers or scramblers with woody stems. Leaves opposite, simple or compound with 3 leaflets; petioles twining. Flowers often unisexual, in axillary or terminal panicles. Sepals 4, white, petaloid, valvate. Petals absent. Stamens numerous. Carpels numerous; styles plumose, persistent, forming a long appendage on the fruit.

Key to the species

1 Anthers tipped with a subulate appendage nearly as long as the pollen sacs. Sepals 18–25 mm long. Leaflets rather thick, irregularly toothed, ovate-cordate to narrow-lanceolate, 2–7 cm long. Juvenile leaves simple, often purple with white streaks. Widespread. Forests. Fl. spring. **C. aristata** R.Br. ex DC.
1 Anther appendage short and obtuse. Sepals 12–18 mm long. Leaflets entire or with a single tooth on each side near the base, narrow-oblong to obovate 2–5 cm long, thinner than in *C. aristata*. Juvenile leaves as for *C. arsitata*. Widespread. Various situations. Fl. early spring . . . .
. . . . **C. glycinoides** DC. ssp. **glycinoides**

2 Ranunculus L.
Buttercups
47 species in Aust. (10 naturalized); all states and territories

Herbs with basal or alternate leaves, usually lobed or divided. Flowers bisexual. Sepals 5, imbricate. Petals usually 5 (sometimes more), usually with a nectary pit near the base. Stamens numerous. Carpels several to many, capitate; style persisting as a beak on the achene.

Key to the species

1 Leaves entire or toothed but not lobed or dissected, linear to lanceolate, c. 4–6 cm long. Flowers up to 20 mm diam. Petals obovate. Achenes pitted. Wingecarribee Swamp. Damp places. Introd from Europe. *Lesser Spearwort* . . . . ***R. flammula** L.
1 Leaves lobed or deeply dissected 2

2 Either; sepals reflexed, or fewer than 5. 3
2 Sepals not reflexed, 5 or more. 6

3 Petals conspicuous, flowers >5 mm diam. Receptacle hairy. 4
3 Petals inconspicuous flowers <5 mm diam. Receptacle glabrous 5

4 Fruiting head cylindrical with 70–100 achenes per head, achenes not flattened. Stems thick, hollow. Leaves divided into 3 or 5 obtusely toothed or lobed segments. Petals pale yellow, scarcely exceeding the sepals. Carpels very small, numerous. Glabrous, erect, much-branched, annual herbs. Coast and

Cumberland Plain. Wet places. Introd. from N. temp. regions. Fl. summer. *Celery-leaf Buttercup* . . .
. . . . ***R. sceleratus** L.

4 Fruiting head globose with less than 50 achenes per head, achenes flattened. Basal leaves ovate to deltoid in outline, 2–6 cm diam., divided into 3, each segment toothed or further divided. Sepals 5. Petals pale yellow, ovate, 6–11 mm long. Achenes almost circular, 2–3 mm diam.. Pubescent tufted herbs to 60cm high. annual herbs. Irrigated or swampy lowlands Introd. from Europe. *Pale Hairy Buttercup* . . . . . .
. . . .* **R. sardous** Crantz

5 Achenes tuberculate, hairy. Petals mostly 5. Blue Mts Moist places. Fl. spring . . . .
. . . . **R. pumilio** R.Br. ex DC. var. **pumilio**
5 Achenes smooth, glabrous. Petals mostly 3–4. Blue Mts, introduced to Coast. Moist places. Fl. spring. .
. . . .**R. pumilio** var. **politus** Melville

6 Achenes smooth or minutely rugose, pitted or with some obtuse ridges 7
6 Achenes covered with spiny tubercles 16

7 Stems decumbent and rooting at the nodes or stoloniferous 8
7 Stems not decumbent or rooting at the nodes or stoloniferous 14

8 Leaves glabrous or nearly so 9
8 Leaves pubescent to hirsute 13

9 Leaf segments filiform, 1–3 cm long, 1–2 mm wide. Flowers to 15 mm diam. Petals 5–12, narrow-elliptic to oblanceolate, with a narrow nectary. Achenes obscurely ridged. Weak perennial. Widespread. Aquatic or on damp ground. Fl. summer . . . . **R. inundatus** R.Br. ex DC.
9 Leaf segments narrow-oblong to lanceolate, up to 5 cm long, less than 4 mm wide 10

10 Petals more than 9 mm wide. Leaves usually with paler markings near the base. . . . .
. . . .***R. repens** (see below)
10 Petals less than 9 mm wide. Leaves without paler markings. 11

11 Nectary a thickened ridge near the base of the petal. Flowers 8–10 mm diam. Petals 5–9, narrow-elliptic to oblanceolate, up to 2 mm wide. Achenes obscurely rugose. Weak perennial. Higher Blue Mts Aquatic or on damp ground. Fl. summer . . . . **R. amphitrichus** Colenso
11 Nectary a distinct pocket near the base of the petal. 12

12 Leaves thick, 3 primary segments each with 3–5 lobes, rarely entire. Flowers 15–28 mm diam.. Petals c. 5, narrow-elliptic, up to 4 mm wide. Achenes smooth or wrinkled, 12–36, 2–4 mm long. Erect perennial up to 30 cm. Higher Blue Mts Damp places. Fl. summer . . . . **R. papulentus** Melville
12 Leaves thin, divided into many linear segments up to 2 mm wide. Petals 7–11. Achenes 30–85, 1–1.5 mm long. Perennial herb. Blue Mts Damp places. Fl. spring–summer. . . . .
. . . . **R. meristus** B.G.Briggs & R.O.Makinson

13 Petals obovate, c. 10 mm wide. Flowers up to 30 mm diam. Leaves ternate, with the central leaflet stalked; segments cuneate, dentate or lobed. Achenes smooth with a prominent border. Erect, stoloniferous herbs. Widespread. Damp ground. Introd. from Europe and Asia. Fl. spring. *Creeping Buttercup*. . . . .***R. repens** L.
13 Petals oblanceolate, c. 5 mm wide. Flowers up to 20 mm diam. Leaves ternate with the central lobe sessile; segments cuneate, toothed. Achenes with obscure ridges. Perennial herbs. Blue Mts Forests . .
. . . . **R. collinus** R.Br. ex DC.

14 Leaves, especially the basal ones, pinnate or pinnatisect, oblong in outline, up to 30 mm long. Flowers up to 12 mm diam. Petals obovate. Achenes smooth or obscurely rugose, with a short terete style at the top. Perennial herbs up to 20 cm high. Higher Blue Mts Wet places. Fl. spring–summer . . . .
. . . . **R. pimpinellifolius** Hook.
14 Leaves ternate or ternate-lobed, ovate to cuneate in outline 15

15 Flowers usually 20–40 mm diam. Sepals appressed. Petals obovate. Achenes smooth with a long recurved or divergent hooked style. Erect plant 10–60 cm high. Widespread. Open situations. Fl. spring–summer. *Common Buttercup*. . . . . **R. lappaceus** Smith

15 Flowers usually less than 12 mm diam. Sepals reflexed. Petals obovate. Achenes smooth with a short erect hooked style. Weak herbs up to 30 cm high. Widespread. In or near forests and shady places. Fl. summer. *Hairy Buttercup* . . . . **R. plebeius** R.Br. ex DC.

16 Slender, hairy annuals. Leaves 5–20 mm long, with slender petioles. Flowers almost sessile or on a slender peduncle opposite the leaf. Petals pale yellow, 2–3 mm long. Achene tuberculate, 2 mm long, with a short beak or with a short hooked spine and a curved beak. Coast and Cumberland Plain. Uncommon . . . . **R. sessiliflorus** R.Br. ex DC. var. **sessiliflorus**

16 Nearly glabrous annuals with stout stems, erect diffuse or prostrate and creeping and rooting at the nodes. Leaves 1–5 cm long; petioles with a sheathing base. Flowers 10–15 mm diam. (rarely 20 mm). Sepals reflexed. Petals yellow, slightly longer than the sepals. Achenes up to 5 mm long, with spiny tubercles on the sides and a stout recurved beak. Widespread. Weed in moist places and lawns. Introd. from the Mediterranean. Fl. spring. *Sharp Buttercup* or *Rough-seeded Buttercup*. . . . **R. muricatus** L.

3 Batrachium (DC.) Gray
1 species native/naturalized Aust.; NSW, Vic., Tas., S.A.

One species in the area
Petals white. Leaves divided several times into narrow-linear or filiform segments less than 1 mm wide, and arranged in several planes. Submerged, aquatic herb often rooting at the nodes, glabrous. Achenes smooth. Higher Blue Mts The native/naturalized status of this taxon is uncertain . . . .
. . . . ***R. trichophyllum** (Chaix) Bosch

47 DILLENIACEAE
Climbers, shrubs or small trees. Leaves alternate, simple, margins entire or sometimes toothed or lobed. Flowers solitary, in cymes or racemes. Sepals and petals 3 to many, usually 5. Petals yellow, often falling early. Stamens hypogynous, 3 to many, staminodes sometimes present. Carpels superior, free, 2 to 5, up to 20 in some genera. Styles free, simple. Fruit a cluster of follicles. 12 gen., trop. and subtrop., S.E. Asia, Aust.

Key to species

1 Filaments dilated, cohering into a short tube; stamens 10 in 2 whorls. . . . .**1 ADRASTAEA**
1 Filaments free or nearly so, ...**2 HIBBERTIA**

1 Adrastaea DC.
1 species endemic Aust.; Qld, NSW

One species in the area
Erect, rather large shrub with slender branches. Stems with loose reddish bark. Leaves linear-oblong, 15–40 mm long, 2–5 mm wide, villous underneath, margins recurved. Flowers sessile within clusters of leaves. Sepals 4–6 mm long. Petals 7–9 mm long. Carpels glabrous or sparsely hairy. Coastal heath and swamps. Not common. North from Royal National Park. Fl. spring–summer . . . . . . **A. salicifolia** DC.

2 Hibbertia Andrews
110 species in Aust.; all states and territories

Shrubs, sometimes almost herbaceous, or climbers. Leaves spirally arranged, entire, often with the midrib prominent underneath; stipules minute or absent. Flowers solitary, terminal or apparently axillary, sessile in a whorl of floral leaves or pedicellate. Sepals 5. Petals 5, yellow. Stamens few or numerous, free or nearly so, sometimes all on one side of the carpels, sometimes some reduced to staminodes, hypogynous; anthers erect, oblong ovate or orbicular, opening by longitudinal slits or terminal pores. Carpels (1-)

2–5 free or nearly so, superior, usually with 2–6 ovules in each. Styles filiform. Fruit a follicle, usually dehiscent at the top. Seeds reniform or nearly globular with an entire or divided aril.

Key to the species

1 Stamens all on one side of the carpels . . . . .GROUP 1
1 Stamens surrounding the carpels but sometimes more numerous on one side. Staminodes occasionally present. 2

2 Carpels glabrous . . . . .GROUP 2
2 Carpels tomentose or villous . . . . .GROUP 3

Group 1
stamens all on one side of the carpels

1 Carpels and sepals glabrous. 2
1 Carpels hairy. Sepals variously hairy, or occasionally glabrous 4

2 Peduncles 8–15 mm long. Stamens 4. Leaves narrow-oblong, 5–8 mm long, glabrous or nearly so; margins recurved. Trailing shrub with glossy reddish stems.Widespread. Heath. Ss. Fl. summer. . . . . .
. . . . **H. rufa** N.A.Wakef.
2 Flowers sessile. Stamens more than 4 3

3 Leaves incurved, linear, 4–6 mm long, 0.5 mm wide, clustered, without a mucro. Carpels 3. Stamens 8–12 . . . . .H. fasicularis (see below)
3 Leaves recurved, 5–10 mm long, 1 mm wide, not clustered, with a minute blunt yellowish mucro. Carpels 2. Stamens c. 6. Shrub up to 1 m high. Widespread. Heath. Ss. Fl. spring. . . . .
. . . .**H. cistiflora** N.A.Wakef. ssp. **cistiflora**

4 Leaves flat, or with slightly recurved margins, oblong, 1–2 cm long, shining or glabrous on upper surface. Flowers terminal or within a cluster of floral leaves on very short axillary branches. Petals broad, emarginate. Erect or diffuse shrubs c. 1 m high 5
4 Leaf margins strongly recurved to revolute. 6

5 Sepals glabrous. Stamens c. 11. Leaves shining. Widespread. Heath and DSF. Ss. Fl. spring–summer . .
. . . . **H. nitida** (R.Br. ex DC.) Benth.
5 Sepals with silky hairs. Stamens c. 16. Widespread. Heath and DSF. Fl. spring–summer. . . . .
. . . . **H. bracteata** (R.Br. ex DC.) Benth.

6 Calyx 4–6 mm long 7
6 Calyx more than 6 mm long 10

7 Leaves with long, usually pungent tips, linear, 5–10 mm long, upper surface glabrous or with minute hairs. Stems glabrous. Stamens usually 8. Erect, prostrate or diffuse shrub. Moist heath near the coast and on sandy, stony, sterile soils inland. Fl. spring–summer . . . .**H. acicularis** (Labill.) F.Muell.
7 Leaves without pungent tips, stellate hairs present at least on lower surface midvein. Stems with stellate hairs. 8

8 Peduncle usually more than 10 mm long. Leaves with some stellate hairs although those on the upper surface ± deciduous leaving tubercles; undersurface usually ± tomentose. Stamens usually 10–12. Ascending or erect shrub. Widespread. Ss and WS in various situations. Fl. late winter–summer (hybrids may occur with *H empetrifolia* ssp. *empetrifolia* where the two species occur together) . . . . .
. . . . **H. aspera** DC. ssp. **aspera**
8 Peduncles less than 10 mm long 9

9 Plants grey/green. Upper and lower leaf surfaces densely covered with stellate hairs. Sepals with short soft hairs, obtuse. Leaves 5–8 mm long. Shrub up to 70 cm high. Blue Mts Uncommon. Fl. spring–summer. . . . . **H. cistoidea** (Hook.) C.T.White

9 Plants green. Lower leaf surface sparsely hairy with hooked hairs, stellate hairs (if present), restricted to midvein. Stamens usually 4–8. Decumbent shrub with long sprawling branches. Widespread. Ss. Fl. spring–summer (hybrids may occur with *H aspera* ssp. *aspera* where the two species occur together) . . . . **H. empetrifolia** (DC.) Hoogland. ssp. **empetrifolia**

10 Hairs all simple but of varying lengths. 11
10 Hairs simple and/or stellate at least on stems 12

11 Under surface of leaves exposed between midvein and recurved margins. Leaves linear, 6–12 mm long, c. 1 mm wide, with dense short hairs, simple or stellate, overtopped by many long simple ones. Flowers, single, terminal on branches. Petals 5.5–6.5 mm long. Stamens 6–9. Low spreading shrub to 0.3 m high. Sandstone. Ridgetop woodland. Endangered. Fl. winter–spring . . . **H. superans** Toelken
11 Under surface of leaf not exposed between midvein and revolute margins. Leaves minutely scabrous to glabrescent, oblong-lanceolate, 1.5–7 mm long, 0.6–1.5 mm wide. Flowers single or in clusters, terminal on branches. Petals 6–8 mm long. Stamens 10–14. Ovary minutely pubescent. Shrub to 30 cm high. Ss. Coast to ranges. Endangered. Fl. spring . . . . **H. puberula** Toelken

12 Tuft of hairs present between petals and stamens. Upper margin of outer sepals ± recurved. Upper surface of leaves with stellate hairs overtopped by a few longer silky hairs. Leaves linear to linear-triangular, 3–20 mm long, 2–4 mm wide. Flowers in terminal clusters of 1 to 7. Petals 5.5–15 mm long. Stamens 10–14 in one cluster. Spreading shrub to 2 m high. Scrub to forest on ranges . . . . **H. incana** (Lind.) Toelken
12 Tuft of hairs absent between petals and stamens. Margins of sepals incurved. Upper surface of leaves scabrous. Sepals usually scabrous. 13

13 Low spreading shrub with dense short hairs overtopped by many longer simple hairs. Stamens 6–9 . . . .**H. superans** (see above)
13 Combination of characters not as above 14

14 Leaves 7–15 mm long, densely hairy, hairs not tuberculate based. Undersurface of leaf with a deep groove along either side of the midvein. Stamens 6–8. Branchlets pale orange-brown. Small upright shrub to c. 1 m high. Nowra & Menai. Endangered. Fl. winter–summer . . . . . . **H. sp. 'Menai'** (A.T. Fairley 2004)
14 Leaves 8–10 mm long, hairs moderately dense and with tuberculate bases. Midvein pressed up against revolute margins on undersurface of leaf. Stamens 6–16. Flowers solitary. A species with many forms differing in habit, foliage, size, time of flowering and indumentum. Shrub up to 60 cm high. Widespread. Heath. Ss. Fl. winter–summer . . . . **H. riparia** (R.Br. ex DC.) Hoogland.

Group 2
stamens surrounding the carpels; carpels glabrous

1 Climbers or prostrate shrubs 2
1 Erect or diffuse shrubs, never climbing. Stamens all fertile. 5

2 Leaves mostly less than 2 cm long. Flowers <3 cm diam. Stamens 20 or less 3
2 Leaves mostly 3–8 cm long. Flowers 3–5 cm diam. Stamens >30. 4

3 Sepals 6–8 mm long. Carpels mostly 3. Leaf length to width ratio <4:1 . . . .**H. diffusa** (see below)
3 Sepals 10–15 mm long. Carpels mostly 4. Leaf length to width ratio >4:1. Leaves 15–20 mm long c. 2 mm wide, glabrous and glaucous, apex acute. Flowers sessile, terminal. Stamens c. 20. Prostrate shrub. Mangrove Mt. Endangered. Fl. summer. . . . . **H. procumbens** DC.

4 Leaves tapering gradually at the base, often stem-clasping, obovate to lanceolate, entire, villous underneath. Staminodes absent. Stamens >30. Carpels 3–7, usually 5 (on seashore usually 3–4). Aril orange. Stems wiry. Widespread. Coastal sand dunes; open forests; margins of RF. Fl. most of the year. *Guinea Flower* . . . . **H. scandens** (Willd.) Dryand.
4 Leaves with a broad obtuse base, distinctly petiolate, oblong-elliptic to ovate; margins dentate. Staminodes numerous. Stamens >30. Carpels 3. Widespread. Forests on Ss. and shales. Fl. spring–summer. . . . . **H. dentata** R.Br. ex DC.

5 Leaves stem-clasping, narrow-oblong to narrow-lanceolate, 2–11 cm long, pubescent underneath, mostly with a short terminal point. Flowers 3–4 cm diam., sessile in a cluster of floral leaves. Sepals 12–16 mm long. Large shrub with softly pubescent young branches. Blue Mts; Cumberland Plain; Illawarra Ranges. Moist gullies. Fl. spring . . . .**H. saligna** R.Br. ex DC.

5 Leaves never stem-clasping, shortly petiolate or almost sessile, rarely more than 3 cm long 6

6 Leaves convex underneath; margins never recurved or revolute, sometimes incurved 7

6 Leaves flat or with recurved margins, sometimes folded 8

7 Leaves narrow-linear, 4–6 mm long, clustered, slightly covered with soft spreading hairs or finally glabrous. Bracts small. Sepals 4–6 mm long. Stamens usually 8–12. Small, erect or diffuse shrub. Widespread. Heath and DSF. Ss. and old dunes. Fl. winter–summer . . . . **H. fasciculata** R.Br. ex DC.

7 Leaves linear, 7–25 mm long, neither clustered nor crowded, stiff, glabrous. Bracts broad, pale brown, c. half as long as the calyx. Sepals c. 8 mm long. Erect or diffuse shrub with thin wiry branches. Coast and adjacent plateaus. Not common. Heath. Sandy soils. Fl. spring . . .**H. virgata** R.Br. ex DC. ssp. **virgata**

8 Leaves entire 9

8 Leaves usually apically 2–3-dentate, linear-cuneate to obovate, 5–15 mm long 11

9 Decumbent or prostrate shrubs, usually less than 30 cm high, with many weak, glabrous branches (pubescent when young) arising from a stout underground rootstock. Leaves obovate to linear-cuneate, length to width ratio <4:1, 5–15 mm long, tapering gradually into a short petiole, dark green, glabrous, smooth, slightly succulent. Margins with lateral teeth or lobes (coastal plants may have entire margins). Sepals c. 8 mm long. Petals c. 10 mm long. Stamens 20–25. Widespread except the higher Blue Mts Open forest and cleared land, chiefly on WS. Fl. spring–autumn . . . .**H. diffusa** R.Br. ex DC.

9 Erect or diffuse shrubs 0.5–2 m high (very variable in habit). 10

10 Stamens mostly 15–25. Plants glabrous except sometimes the young shoots. Leaves linear-oblong to obovate, length to width ratio >4:1mostly , 8–20 mm long, occasionally much longer. Sepals 5–6 mm long. Petals 8–10 mm long. Widespread. Heath and DSF. Ss. and old dunes. Fl. spring . . . .
. . . . **H. linearis** R.Br. ex DC.

10 Stamens 30–40. Plants ± covered with a crisped or shortly stellate greyish tomentum or sometimes pubescent. Leaves broad-oblong-spathulate to linear, very obtuse. Sepals 7–9 mm long. Petals 9–15 mm long. Widespread. Heath and DSF. Fl. spring . . . . **H. obtusifolia** DC.

11 Decumbent or prostrate shrubs, usually under 30 cm high. Leaves obovate to linear-cuneate, tapering to a short petiole, the two lateral teeth not exceeding the apex, not folded longitudinally. . . . .
. . . . **H. diffusa** (see above).

11 Erect shrubs to c. 60 cm high. Leaves 5–12 mm long including petiole, mostly linear-cuneate, narrowed at the base, folded longitudinally. Two lateral teeth often exceeding the truncate apex 12

12 Carpel solitary. Stamens 10–12. Widespread. DSF. Fl. spring–summer . . .**H. monogyna** R.Br. ex DC.

12 Carpels 3. Stamens 15–30. Widespread. DSF. Fl. spring–summer . . . .**H. circumdans** B.J.Conn

Group 3
Stamens surrounding the carpels; carpels hairy

1 Leaves 5–10 mm wide, 1–2 cm long, flat, obovate-oblong to cuneate. Flowers 15 mm diam. Stamens c. 15. Erect shrub with stellate and simple hairs. Bent's Basin; lower Blue Mts DSF. Fl. spring–summer. . .
. . . .**H. hermanniifolia** DC.

1 Leaves 1–4 mm wide, 3–10 mm long; margins recurved or revolute. 2

2 Flowers c. 12 mm diam on peduncles longer than the leaves. Stamens 15–20; several small staminodes usually present. Leaves linear, 3–6 mm long. Diffuse, prostrate or rarely erect shrubs, glabrous or with a few short spreading hairs. Widespread. Open forest and cleared areas on sandy soils and WS. Fl. spring
. . . . **H. pedunculata** R.Br. ex DC.

2 Flowers on peduncles shorter than the leaves or sessile 3

3 Stamens more than 30, several small staminodes usually present. Flowers nearly sessile, c. 2 cm diam. Sepals 8 mm long. Leaves linear, 6–8 mm long. Erect or decumbent, mostly hirsute shrub. Coast north of Botany Bay. Sandy soils. Fl. spring . . . . **H. vestita A.**Cunn. ex Benth.

3 Stamens usually 11–13, staminodes absent. Leaves linear to oblong-ovate, spathulate or cuneate, 3–6 mm long, mostly scabrous; margins revolute. Decumbent or prostrate shrubs, sometimes shortly hirsute. Widespread. Heath and DSF. Ss. Fl. late spring–summer . . . . . **H. serpyllifolia** R.Br. ex DC.

48 AIZOACEAE

Herbs or shrubs. Leaves alternate or opposite, entire, sometimes fleshy; stipules absent or minute. Flowers regular, bisexual. Perianth segments 4–5, connate at the base or tubular, persistent on the fruit. Stamens 4–5 or more, sometimes numerous when the outer ones are petaloid staminodes. Ovary superior or inferior, 3–multi-locular; placentas axile, parietal or basal. Fruit a capsule or berry-like. 140 gen., warm temp. and subtrop., particularly S. Africa.

Key to the genera

1 Perianth segments 5, connate at base or free (two whorls may be present in *Macarthuria*) 2
1 Perianth tubular (sometimes fused to ovary) 5

2 Leaves all alternate or basal 3
2 At least some leaves opposite or in whorls or pseudowhorls 4

3 Styles 5. Stamens 10 . . . . **2 GALENIA**
3 Styles 3. Stamens 8 . . . . **1 MACARTHURIA**

4 Seed with an appendage. Stamens 3–20 . . . . **3 GLINUS**
4 Seed without an appendage. Stamens 3–5. . . . . **4 MOLLUGO**

5 Ovary superior 6
5 Ovary inferior 7

6 Style solitary . . . . **12 TRIANTHEMA**
6 Styles 3. . . . . **11 SESUVIUM**

7 Petaloid staminodes absent. Leaves broad, flat . . . . **10 TETRAGONIA**
7 Numerous linear petaloid staminodes present outside the fertile stamens 8

8 Leaves broad, flat. . . . . **9 APTENIA**
8 Leaves ± triangular in cross section 9

9 Fruit berry-like. Leaves comparatively large, 10–15 mm wide near the base . . . . . **5 CARPOBROTUS**
9 Fruit a capsule. Leaves small, c. 5 mm diam. or less 10

10 Petaloid staminodes yellow-white. . . . . **8 DISPHYMA**
10 Petaloid staminodes purplish red to rose pink 11

11 Plants glabrous . . . . **6 LAMPRANTHUS**
11 Plants pubescent with coarse white hairs . . . . **7 DROSANTHEMUM**

1. Macarthuria Endl.

7 species endemic Aust.; Qld, NSW, NT, WA

One species in the area

Diffuse, herbaceous, glabrous annual. Leaves obovate to linear, up to 5 cm long and 1 cm wide but usually less, ± entire, obtuse, alternate, slightly succulent. Flowers arranged in loose cymo-panicles, c. 3 mm diam. Petals white to pinkish, c. 1 mm long. Stamens 8, connate at the base. Styles 3. Seeds arillate. Hunter River Valley near Hexham. Heath. Fl. summer. . . . . **M. neo-cambrica** F.Muell.

2. Galenia L.

1 species naturalized Aust.; NSW, Vic., S.A., WA

One species in the area

Grey-villous herb with prostrate stems. Leaves spathulate to obovate, 4–20 mm long. Petals 4–5, white, c. 3 mm long. Stamens 10 but sometimes reduced to staminodes. Styles 5. Hornsby Plateau; Hunter River Valley. Waste and cultivated land . . . .*G. pubescens (Eckl. & Zeyh.) Druce

3 Glinus L.

3 species in Aust.; all states and territories except Tas.

One species in the area

Prostrate annual herb. Glabrous. Leaves opposite or clustered, obovate to oblanceolate, 30 mm long. Perianth white or tinged with pink. Carpels 3. Styles 3. Fruit a dehiscing capsule. Seeds numerous, with an appendage. Ephemeral watercourses. Widespread but not common . . . . G. oppositifolius (L.) A.DC.

4 Mollugo L.

3 species in Aust.; Qld, NSW, Vic., NT, S.A., WA

One species in the area

Spreading annual herb. Leaves whorled up to 30 mm long, obovate to spathulate; upper leaves smaller and linear. Perianth 3 mm long, greenish, with membranous margins. Carpels 3. Styles 3. Fruit a dehiscent capsule. Seeds numerous, brown and ridged. Uncommon in the region. . . . . M. verticillata L.

5 Carpobrotus N.E.Br.

6 species in Aust. (4 native); all states and territories except NT

Glabrous perennial with prostrate stems rooting at the nodes. Leaves opposite, triangular in cross section, thick, firm, fleshy. Flowers large, showy, pedunculate or subsessile within the terminal pair of leaves. Perianth tube obconical, 2-keeled; the lobes very unequal. Petaloid staminodes numerous, linear, equal to or exceeding the 2 longest perianth segments. Ovary with 8 or more loculi; styles 8–11. Fruit obovoid, berry-like.

Key to the species

1 Staminodes yellow but becoming flesh-coloured with age. Leaves with all faces slightly concave, almost equilateral, 4–8 cm long, 4–7 times as long as wide; keel denticulate to serrulate at least above the middle. Perianth segments densely dotted almost to the membranous margin. Flowers 7–8.5 cm diam. Fruit often wider than long, yellowish. Near the sea and often in railway enclosures. Introd. from S. Africa. Fl. spring–summer. *Hottentot Fig* . . . .*C. edulis (L.) N.E.Br.
1 Staminodes purple to deep pink. Leaves with slightly convex to flat lateral faces; keel smooth or denticulate only at the apex. 2

2 Staminodes paler but not white towards the base. Leaves equilateral in the middle, 3.5–9 mm long, 6–12 times as long as wide. Perianth segments densely dotted, becoming more sparsely so towards the membranous margins. Flowers 3.5–8 cm diam. Fruit usually longer than wide, purplish. Frequently used to control erosion. Not common in the area. Introd. from S. America . . . .
. . . .*C. aequilaterus (Haw.) N.E.Br.
2 Staminodes white towards the base. Leaves ± equilateral in the middle, 3.5–10 cm long, 4–8 times as long as wide. Perianth segments with the membranous part dotted for c. half its width. Flower 4–6 cm diam. Fruit usually longer than wide, purplish. Near the sea. Fl. most of the year. *Pigface*. . . . .
. . . . C. glaucescens (Haw.) Schwantes

6 Lampranthus N.E.Br.

3 species naturalized Aust.; NSW, Vic., Tas., S.A., WA

Perennials with creeping stems rooting at the nodes and bearing dwarf upright shoots 2–8 cm high. Leaves opposite, succulent, ± triangular in cross section. Flowers solitary, terminal. Petaloid staminodes purplish, red, white or rose pink. Ovary 5-locular; styles 5. Fruit a capsule

Key to the species

1 Petaloid staminodes pink, c. 6 mm long. Leaves glaucous, up to 15 mm long. Pedicels 0.5–1 cm long. Prostrate plant with thin woody stems rooting at the nodes, often forming dense mats. Salt marshes on Parramatta and Lane Cove Rivers. Probably introd. from S. Africa . . . . . ***L. tegens** (F.Muell.) N.E.Br.
1 Petaloid staminodes white or purple, 1–1.5 cm long. Leaves glaucous, up to 4 cm long. Pedicels 1–2.5 cm long. Decumbent to ascending shrub. Garden escape. Introd. from S. Africa. Fl. spring–summer. *Ice Plant* . . . . ***L. multiradiatus** (Jacq.) N.E.Br.

7 Drosanthemum (Haw.) Schwantes

1 species naturalized Aust.; NSW, Vic., S.A., WA

One species in the area

Prostrate perennial rooting at the nodes, with short erect leafy branches. Leaves opposite, succulent, triangular in cross section, 5–8 mm long, pubescent with coarse hyaline vescicular hairs. Flowers solitary, terminal on the short branches. Petaloid staminodes linear, c. 6 mm long, rose pink. Styles 5. Fruit a soft capsule. Garden escape on sandy soils. Introd. from S. Africa . . . . ***D. candens** Schwantes

8 Disphyma N.E.Br.

1 species endemic Aust.; Qld, NSW, Vic., Tas., S.A., WA

One species in the area

Prostrate to decumbent, perennial shrub with erect or ascending leafy shoots up to 50 cm high. Leaves opposite, succulent, linear, 2–5 cm long, 2–3 mm wide, triangular in cross section, with concave faces. Flowers solitary, 2–3 cm diam. Perianth segments with scarious margins. Petaloid staminodes linear, yellow-white, as long as or exceeding the perianth segments. Styles 5. Fruit a capsule with 5 woody valves. Coastal headlands . . . . **D. crassifolium** (L.) L.Bolus ssp. **clavellatum** (Haw.) Chinnock

9 Aptenia N.E.Br.

1 species naturalized Aust.; NSW, S.A.

One species in the area

Perennial with prostrate stems. Leaves flat, broad-ovate, cordate, thick, opposite, 12–25 mm long. Flowers solitary, axillary or terminal. Perianth c. 15 mm long, with 2 large obtuse lobes at least as long as the tube and 2 small subulate ones. Petaloid staminodes purple. Ovary 4-locular; styles 4. Fruit a capsule with 4 valves. Occasionally naturalized in coastal areas. Introd. from S. Africa. *Heart-leaved Ice Plant* . . . . ***A. cordifolia** (L.f.) Schwantes

10 Tetragonia L.

8 species in Aust. (5 native); all states and territories except NT

Perennial or annual, prostrate herbs. Leaves alternate, flat, rather thick, papillose. Flowers solitary or several together, axillary, greenish or yellowish. Perianth with 3–5 valvate lobes. Stamens few to many; staminodes absent. Ovary 2–8-locular. Fruit a capsule enclosed in the floral tube which becomes bony on drying.

Key to the species

1 Pedicels more than 4 mm long. Fruit with 4 thick wings which are rounded in outline. Flowers often more than 2 in the axil. Leaves 2–4 cm long, elliptic to obovate; the apex broad and rounded; petioles

and young stems covered with loose hairs. Coast. Not common. Introd. from Africa. *African Spinach* .
. . . . ***T.decumbens** Miller

1 Flowers sessile or subsessile. 2

2 Styles 5–10. Fruit 10–12 mm diam., subglobular to turbinate or sometimes 3–4-angular or the angles produced upwards to form horns (very variable even on the same plant). Leaves triangular-ovate to lanceolate, 2–8 cm long. Flowers solitary or 2 together. Stamens 4–many. Petioles and young stems with vescicular hairs. Sea coasts and margins of salt marshes. *New Zealand Spinach* or *Warrigal Cabbage* . .
. . . .**T. tetragonioides** (Pall.) Kuntze

2 Styles 2–3. Fruit 2–4 mm long, papillose, truncate at top with 3–4 projections. Leaves ovate to circular 1–3 cm long. Flowers in groups of 2–5 . Stamens usually as many as perianth segments and alternating with them. Introd. from S. Africa. . . . . ***T. microptera** Fenzl

11 Sesuvium L.
1 species endemic Aust.; Qld, NSW, NT, WA

One species in the area
Succulent, glabrous, decumbent, perennial herb rooting at the nodes. Leaves opposite, linear to lanceolate, up to 7 cm long. Flowers solitary in leaf axils. Perianth 6–9 mm long, 5-lobed, pink with scarious margins, Stamens numerous. Ovary superior, 3-locular; styles 3. Capsule circumscissile. Coastal dunes and beaches in north of area. Fl. spring–summer . . . . **S. portulacastrum** L.

12 Trianthema L.
12 species in Aust. (10 endemic); Qld, NSW, NT, S.A., WA

Prostrate to ascending annuals or perennials. Leaves opposite, ± unequal in each pair, sheathing at the base. Flowers solitary or clustered, sessile. Perianth lobes 5, white to pink, unequal. Stamens adnate to perianth. Ovary superior, 1-locular; placentas basal; style solitary. Capsule circumscissile near the base.

Key to the species

1 Flowers solitary, fused to leaf sheath. Leaves elliptic to obovate, up to 55 mm long with a green sheath, not succulent. Stamens more than 5. Decumbent to ascending herb up to 50 cm high. Weed. Introd. from Trop. *Giant Pigweed*. . . . . ***T. portulacastrum** L.

1 Flowers clustered. 2

2 Stamens 5. Leaves terete to oblanceolate, up to 30 mm long, less than 4 mm wide, with a membranous sheath, succulent. Prostrate to ascending herb to 60 cm high. Coast. Usually on sandy soils . . . .
. . . . **T. triquetra** Willd.

2 Stamens 7–10. Leaves obovate to orbicular up to 5 mm long. Petiole base dilated and with scarious margins. Slender, prostrate, glabrous, herb. Collected once from Hawksebury River in 1839. Extinct . .
. . . .**T. cypseleoides** (Fenzl) Benth.

49 CHENOPODIACEAE
Annual or perennial herbs or small shrubs, often halophytic. Leaves usually alternate, often succulent, often mealy from collapsed bladder hairs; stipules absent. Flowers unisexual (plants monoecious or dioecious) or bisexual, usually regular, usually sessile or clustered in dense cymes. Perianth segments 5, herbaceous, usually connate, in 1 whorl, sometimes absent. Stamens as many as perianth segments and opposite them or 1, inserted on a disc, hypogynous. Ovary superior, 1-locular; ovule solitary; styles 1–3; stigmas 2–3. Fruit a nut, surrounded by the perianth which sometimes becomes succulent, or a 1-seeded berry. c. 100 gen., world wide, mostly in dry and/or salty areas.

Key to the genera

1 Plants apparently leafless with succulent jointed stems or branches. 2
1 Leaves conspicuous; stems or branches not succulent. 3

2 Flowers 5–9 in each flowering axil; herbs . . . . **1 SARCOCORNIA**
2 Flowers 3 in each axil; shrubs. . . . . **2 TECTICORNIA**

3 Leaves narrow, often ± terete; margins entire; hairs on leaves without swollen tips or leaves glabrous . **4**
3 Leaves flat, usually broad; margins frequently toothed, sinuate or hastate-sagittate; hairs on leaves with swollen glandular tips or leaves glabrous **8**

4 Fruiting perianth becoming succulent. . . . .**3 ENCHYLAENA**
4 Fruiting perianth dry **5**

5 Fruiting perianth without appendages (Fig. 34) . . . . **4 SUAEDA**

5 Fruiting perianth developing appendages (wings, lobes or spines). **6**

6 Appendages of spines (Fig. 34) . . . . **5 SCLEROLAENA**
6 Appendages of lobes or membranous wings (Fig. 34) **7**

7 Leaves not pungent pointed. . . . .**6 MAIREANA**
7 Leaves pungent pointed . . . . **7 SALSOLA**

8 Bracts enclosing female flowers greatly enlarged at fruiting (Fig. 34) . . . .**8 ATRIPLEX**
8 Neither bracts nor perianth greatly enlarged in fruit **9**

9 Ovary half-inferior. . . . . **12 BETA**
9 Ovary superior **10**

10 True fruit succulent. **11**
10 True fruit dry **12**

11 Prostrate to twining herb or shrubby perennials, rarely ascending. Stamens 1–3. . . . . .**10 EINADIA**
11 Spreading to erect, often much-branched shrubs. Stamens 5 . . . .**9 RHAGODIA**

12 Perianth usually 3-lobed; stamen solitary . . . . **13 DYSPHANIA**
12 Perianth 4–5-lobed **13**

13 Stamens 1–3 . . . .**10 EINADIA**
13 Stamens 4–5 . . . .**11 CHENOPODIUM**

1 Sarcocornia A.J.Scott.
3 species native Aust.; all states and territories except NT

One species in the area
Herb with succulent jointed stems up to 25 cm long, erect or decumbent. Stem sections terete, 10–15 mm long, green. Flowers bisexual, in terminal spikes 3–5 cm long; the clusters of flowers embedded on each side of the terete sections which are similar to, but shorter than, the vegetative sections; only the anthers and styles protruding. Perianth minute, 2–5-lobed, not enlarging after flowering. Stamens 2. Styles 2. Fruit embedded in the succulent stem. Salt marshes near the coast. Fl. spring–summer. *Samphire* or *Glasswort* . . . . **S. quinqueflora** (Bunge ex Ung. Sternb.) ssp. **quinqueflora**

2 Tecticornia Hook.f.
22 species endemic Aust.; all states and territories

One species in the area
Erect shrub up to 1 m high, with succulent jointed branches. Stem sections up to 4 cm long, ± glaucous. Flowers bisexual, in groups of 3 on each side of joints in terminal or intercalary spike-like inflorescences. Stamen solitary. Uncommon in salt marshes on Parramatta River. Fl. spring–summer . . . .
. . . . **T. pergranulata** (J.M.Black) K.A.Shep. & Paul G.Wilson ssp. **pergranulata**

FIGURE 34

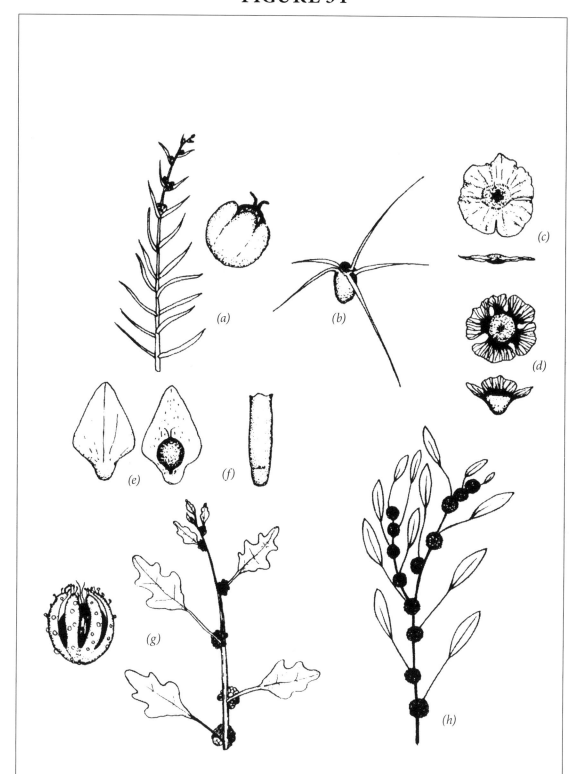

Fig. 34 CHENOPODIACEAE: (a) shoot (nat. size) of *Suaeda australis* and a mature fruit (enlarged) enclosed in the persistent perianth; (b) 'fruit' of *Sclerolaena muricata* (X2); (c) 'fruit' of *Maireana microphylla*, upper giving the surface view, lower the side view (X2); (d) 'fruit' of *Salsola tragus*, surface and side views (X2); (e) 'fruit' of *Atriplex cinerea*, left showing bracteoles from outside, right with one bracteole removed to expose the true fruit (X2); (f) 'fruit' of *Atriplex leptocarpa*; (g) shoot (nat. size) of *Chenopodium carinatum* and a mature fruit (enlarged) enclosed in the persistent perianth; (h) shoot of *Dysphania littoralis* (X2).

3 Enchylaena R.Br.

2 species endemic Aust.; all states and territories except Tas.

One species in the area

Diffuse or compact shrub up to 1 m high; stems usually tomentose, striate or ribbed. Leaves alternate, subterete, tomentose, 6–15 mm long. Flowers bisexual, axillary, solitary, sessile. Perianth enlarging after flowering, becoming succulent, red or yellow (black when dry), depressed-globular, 5–8 mm diam., glabrous. Fruit a nut enclosed in the perianth. Sea coasts. Fl. most of the year. *Ruby Saltbush*. . . . .
. . . . **E. tomentosa** R.Br.

4 Suaeda Forssk. ex Scop.

5 species in Aust. (2 native, 3 naturalized); all states and territories except Tas.

One species in the area

Glabrous herb up to 80 cm high. Leaves alternate, succulent, linear, 1–4 cm long (shorter towards the top of the plant), incurved. Flowers mostly bisexual, sessile, 1 mm diam., in a terminal panicle in the axils of reduced floral leaves. Perianth lobes 5, closing over the fruit; appendages absent. Stamens 5. Styles 2. Fruit enclosed in the almost globular perianth, 1.5–2 mm long. Sea coast and estuaries. Salt marshes. Fl. spring–summer *Seablite*. . . . . **S. australis** (R.Br.) Moq.

5 Sclerolaena R.Br.

62 species endemic Aust.; all states and territories except Tas.

Perennial or annual shrubs Leaves terete, linear or linear-lanceolate, ± succulent. Flowers bisexual, solitary, axillary. Perianth lobes 5, small, enclosing the fruit, with spines near the summit. Stamens 5. Styles 2.

Key to the species

1 Fruit with 2(3) erect spines to 5 mm long. Leaves linear, to 25 mm long, glabrous or hairy. Fruit c. 5 mm diam., hairy. Spreading shrub to 80 cm high, branches densely hairy. Introd. from western divisions of NSW. Uncommon . . . . ***S. bicornis** Lindl. var. **horrida** Domin
1 Fruit with 5(4) or more spreading spines 2

2 Fruit glabrous, c. 2.5 mm long, with 6(5) spreading spines 1–2 mm long. Leaves terete, to c. 15 mm long, glabrous or hairy. Shrub to 30 cm high, branches lightly hairy. Introd. from western divisions of NSW. Uncommon . . . . ***S. calcarata** (Ising) A.J.Scott
2 Fruit glabrous or hairy, with 5(4) spreading spines. 3

3 Leaves hairy on both surfaces, obovate to c.15 mm long, shortly petiolate. Fruit hairy 2–3 mm long. Spines 5(4) the longest 5–9 mm long, the shorter c. 4 mm long. Shrub to 1 m high with hairy branches. Introd. from western divisions of NSW. Uncommon . . . . ***S. birchii** (F.Muell.) Domin
3 Leaves dark green hairy on lower surface only. Fruit with longest spine 4–10 mm long; 2 shorter spines united near the base. Leaves linear or linear-lanceolate, 6–20 mm long, 1–2 mm wide. Shrub usually c. 50 cm high branches glabrous to hairy. Overgrazed pastures. Fl. most of the year. Introd. from western divisions of NSW. Uncommon. *Black Rolypoly* or *Electric Burr* 4

4 Plants glabrous or nearly so. Leaves linear to narrow oblong. . . . .
. . . . ***S. muricata** (Moq.) Domin var. **muricata**
4 Plants hairy 5

5 Branches with appressed hairs. Leaves obovate to oblanceolate . . . .
. . . . ***S. muricata** var. **semiglabra** (Ising) A.J.Scott
5 Branches woolly. Leaves linear to lanceolate . . . . ***S. muricata** var. **villosa** (Benth.) J.M.Black

6 Maireana Moq.

58 species endemic Aust.; all states and territories except Tas.

Shrubs or herbs. Leaves alternate or opposite, usually succulent. Flowers axillary, solitary or in pairs, sessile, bisexual or unisexual. Perianth 5-lobed, each lobe usually with a horizontal wing or appendage.

Key to the species

1 Fruiting perianth 5–6 mm diam., with 5 thickened scarcely succulent imbricate ± auriculate lobes; the tube vertically 5-ribbed, with 5 lesser alternating ribs. Leaves linear-lanceolate, mostly 6–15 mm long, villous. Undershrub up to 25 cm high, almost herbaceous. Singleton district . . . .
. . . . **M. enchylaenoides** (F.Muell.) Paul G.Wilson
1 Fruiting perianth winged, up to 10 mm diam. Leaves linear-terete, 2–4 mm long, glabrous. Undershrub up to 40 cm high. Cumberland Plain and Singleton district. Open communities. *Eastern Cotton bush* .
. . . . **M. microphylla** (Moq.) Paul G.Wilson

7 Salsola L.

1–2 species in Aust.; all states and territories

One species in the area

Annual herb mostly c. 50 cm high but up to 1 m, diffuse when young, becoming hemispherical at maturity. Leaves terete or linear, up to 3 cm long on younger parts, mostly c. 1 cm long on older parts, semi-succulent, pungent pointed. Flowers sessile, axillary, usually solitary. Perianth segments 5, closing over the fruit and finally each producing a horizontal wing; the whole perianth 4–7 mm diam. Widespread, often in saline areas. Fl. spring–summer. *Prickly Rolypoly* or *Saltwort* . . . . **S. tragus** L.

8 Atriplex L.

60 species in Aust. (3-4 naturalized); all states and territories

Monoecious or dioecious herbs or small shrubs. Leaves alternate or opposite, flat, with glandular hairs or glabrous. Flowers unisexual; male flowers usually terminal in a short spike, the flowers clustered in the axils of reduced floral leaves; female flowers few together, clustered in the axils. Male perianth 5-cleft; stamens 5. Female flowers enclosed by two large bracteoles which finally enclose the fruit; perianth absent; ovary 1-locular; styles 2.

Key to the species

1 Mature bracteoles twice as long as broad, tubular in the lower portion, free at the summit; the tube 2–3 mm diam.; (Fig. 34). Decumbent perennial up to 30 cm high, monoecious. Leaves linear to lanceolate, alternate, 2–4 cm long, mealy. Clay soils near stockyards. Probably introduced from western parts of NSW. Fl. most of the year. *Slender-fruit Saltbush* . . . . *A. leptocarpa** F.Muell.
1 Mature bracteoles as long as broad, rhomboidal in outline, closely appressed to each other. 2

2 Lower leaves truncate, hastate, or sagittate at base, 2–4 cm long, opposite or alternate, green, glabrous. Upper leaves triangular or ovate. Mature bracteoles roughly triangular, 2–4 mm long. Weak perennial with thin straggling branches mostly less than 50 cm long. Salty areas near the sea. Introd. from Europe and Asia. Fl. summer . . . . **A. prostrata** Boucher ex DC.
2 Lower leaves linear, lanceolate, elliptic or oblong, tapering to base 3

3 Mature bracteoles 6–14 mm long (Fig. 34) 4
3 Mature bracteoles 3–6 mm long, ± diamond-shaped, sometimes slightly or quite succulent. 5

4 Shrubs, mostly dioecious, up to 1 m high. Leaves oblong-lanceolate, obtuse, 2–6 cm long. Male flowers in dense globular clusters, sometimes with a few females intermixed. Whole plant mealy and grey. Salty areas near the sea. Fl. spring–summer. *Grey Saltbush* . . . . **A. cinerea** Poir.
4 Herbs, monoecious, up to 1 m high. Leaves narrow-elliptic, up to 7 cm long. Flowers clustered in spike-like inflorescences. Fl. spring–summer. Uncommon. Introd. from Europe. . . . . **A. patula** L.

5 Leaves 1–2 cm long, oblong or lanceolate, toothed lobed or entire, green or mealy or green above and mealy below. Mature bracteoles diamond-shaped, 4–6 mm long, sometimes red and fleshy. Prostrate perennial with stems up to 50 cm long. Drier parts of WS, often common near stockyards. Fl. most of the year. *Creeping Saltbush.* . . . . **A. semibaccata** R.Br.

5 Leaves 3–10 cm long, linear to lanceolate, green or mealy, entire or toothed; the lower ones hastate, petiolate. Flowers in terminal panicles. Mature bracteoles diamond-shaped to ovate, 5 mm long, fused below the middle. Annual up to 1 m high, often straggling and decumbent. Salty areas near the sea. Fl. summer . . . . **A. australasica** Moq.

9 Rhagodia R.Br.
11 species endemic Aust.; all states and territories

One species in the area

Shrub with straggling branches up to 1 m long. Leaves mostly alternate, ovate to broad-lanceolate, sometimes hastate, 1–3 cm long, paler underneath. Flowers bisexual in terminal, much branched, mealy inflorescences. Perianth lobes 5. Stamens 5. Styles 2. Fruit a 1-seeded berry. Sea coast. Fl. most of the year. *Coastal Saltbush* . . . . **R. candolleana** Moq. ssp. **candolleana**

10 Einadia Raf.
4 species in Aust.; all states and territories

Perennial herbs, sometimes shrubby, prostrate to ascending, sometimes twining. Leaves usually alternate, often sagittate or hastate. Flowers bisexual and female in same spike. Perianth segments 5. Stamens 1–3. Fruit succulent or dry.

Key to the species

1 Fruit red or yellow, succulent, 2–3 mm diam. Flowers in short terminal spikes 2
1 Fruit neither red nor yellow, dry 4

2 Leaves ovate to broadly hastate, up to 4 cm long, apex obtuse. Weak trailing perennial with fragile stems up to 1 m long. Widespread. Usually on clay soils . . . . **E. hastata** (R.Br.) A.J.Scott
2 Leaves linear-lanceolate to narrow-sagittate, up to 3 cm long, acute. Weak trailing perennial. Widespread on heavy soils. *Climbing Saltbush.* 3

3 Leaves linear . . . . **E. nutans** ssp. **linifolia** (R.Br.) A.J.Scott
3 Leaves sagittate or narrowly hastate . . . . **E. nutans** (R.Br.) A.J.Scott ssp. **nutans**

4 Lower leaves lanceolate to narrow-sagittate mostly up to 2 cm long and 5 mm wide; upper leaves lanceolate. Flowers in clusters in the upper axils or in terminal clusters on axillary branches. Prostrate or climbing herb. Widespread on heavy soils . . . . **E. polygonoides** (Murr) Paul G.Wilson
4 Lower leaves ovate to hastate, up to 5 cm long, 1–3 cm wide; apex acute upper leaves rarely lanceolate. Flowers in small clusters on terminal or axillary branches. Prostrate or climbing herb. Widespread on heavy soils. *Fishweed.* 5

5 Perianth segments broad enclosing most of the fruit . . . .
. . . . **E. trigonos** (Schul.) Paul G.Wilson ssp. **trigonos**
5 Perianth segments narrow at base exposing most of the fruit . . . .
. . . . **E. trigonos** ssp. **stellulata** (Benth.) Paul G.Wilson

11 Chenopodium L.
23 species in Aust. (9 naturalized); all states and territories

Herbs or shrubs. Flowers mostly bisexual, minute, sessile, in clusters. Perianth segments 5, imbricate, herbaceous, enclosing the fruit, usually hooded at the summit (Fig. 34). Stamens 5 or fewer. Styles 2–3. Fruit a minute nut enclosed in the persistent perianth.

Key to the species

1 Leaves glabrous. Perianth herbaceous but not fleshy, green. Leaves 3–8 cm long, rhomboidal; margins toothed. Flowers in a terminal leafy panicle. Herb up to 60 cm high. Widespread. Weed in waste places. Introd. from Europe, Asia and Afric. Fl. spring–summer. *Nettle-leaf Goosefoot* . . . . . . *C. **murale** L.

1 Leaves glandular-pubescent or mealy 2

2 Plants erect 3

2 Plants decumbent or with trailing stems 4

3 Leaves mealy-white, sinuately toothed, 3–6 cm long, ovate to rhomboidal (upper leaves lanceolate, entire). Stems angular, ribbed. Flowers in a terminal spreading panicle. Annual 1–2 m high. Widespread. Weed of cultivation and waste places. Introd. from temp. and trop. areas. Fl. chiefly winter. *Fat-hen* . . . . *C. **album** L.

3 Leaves green, glandular-pubescent, lanceolate, sinuate-toothed or almost entire. up to 8 cm long. Flowers in short terminal spikes. Annual or perennial up to 80 cm high with an odour of ants. Widespread. Weed. Introd. from America. Fl. summer. *Mexican Tea* . . . .*C. **ambrosioides** L.

4 Leaves densely mealy on undersurface (sometimes also, on upper surface), linear, hastate, triangular, up to 7 cm long, with entire margins or 2–3 lobes near the base. Habit variable; prostrate, 4–5 cm high and leaves mainly basal; or with trailing stems up to c. 30 cm long. Salty places. Fl. chiefly autumn. . . . . ?*C. **glaucum** L.

4 Leaves green on both surfaces or slightly mealy underneath 5

5 Leaves pinnatisect into ± linear segments, alternate, glandular-pubescent, green on both surfaces, 2–12 cm long. Flowers bisexual, in clusters of 2–5 subtended by a bract or small leaf; clusters in a leafy panicle. Perianth usually 3-lobed (sometimes 5-lobed). Stamens 5. Fruit membranous; perianth enclosing the fruit ovoid-oblong, 2 mm long. Seed reddish-brown. Prostrate or ascending, perennial herb with stems up to 50 cm long. Near Sydney. Waste places. Introd. from S. America. Fl. summer. *Scented Goosefoot* . . . .*C. **multifidum** L.

5 Leaves coarsely sinuately lobed, 1–4 cm long, green on both surfaces but with glandular hairs, oblong or ovate in outline. Flowers in short axillary clusters. Perianth less than 1 mm long. Annual, decumbent herbs 6

6 Perianth segments keeled, pubescent, dorsally broadening towards the apex and thus triangular in outline. Coast, Blue Mts, Hornsby Plateau. Weed of gardens and waste places. Fl. chiefly summer. *Green Crumbweed* . . . . C. **carinatum** R.Br.

6 Perianth segments concave but not keeled, with a few scattered hairs, curved in outline. Widespread. Weed of gardens, waste places, forests and woodlands. Fl. chiefly summer. *Small Crumbweed* . . . . .C. **pumilio** R.Br.

12 Beta L.
1 species naturalized Aust.; NSW, Vic., S.A.

One species in the area

Herbaceous biennial with erect or ascending stems up to 50 cm high. Lower leaves alternate, up to 10 cm long; petioles c. as long as or longer than the laminas; lamina semi-succulent, rhomboidal to ovate, sometimes cordate, glabrous. Upper leaves narrower with shorter petioles. Flowers bisexual, in clusters in the axils of small leaves which are usually alternate and remote from each other at the base of the spike-like terminal inflorescence. Perianth 5-lobed, becoming thickened and hardened, 5-ribbed at maturity. Stamens 5. Ovary half-inferior. Salty places, e.g. Cook's River. *Wild Beet*. . . . .*B. **vulgaris** L. ssp. **maritima** (L.) Threll.

13 Dysphania R.Br.

10 species endemic Aust.; all states and territories except Tas.

Annual or perennial aromatic herb with glandular hairs. Leaves alternate, simple, entire or dissected. Flowers unisexual and bisexual, mixed, clustered in the leaf axils (Fig. 34). Perianth segments 1–4, often inflated or spongy. Stamens and styles 1–2. Pericarp thin.

Key to the species

1 Terminal flowers male, perianth segments 3, inflated. Lateral flowers female, perianth segments 3–4, inflated. Prostrate to erect annual or perennial with simple or glandular multicellular hairs. Leaves elliptic 5–10 mm long, petiolate. Stamens 2, styles usually 1. Fruit subglobular. Coastal areas. FL. summer–autumn. . . . . **D. littoralis** R.Br.

1 Terminal flowers bisexual perianth segments 3, hooded. Lateral flowers female, perianth segments 1–2-hooded. Glabrous annual up to 10 cm high with ascending stems. Leaves obovate to oblong, usually 3–9 mm long; petioles c. twice as long. Flowers in groups, 2–3 mm diam. Stamens 1 or 2. Styles 1–2. Fruit ovoid. Coast and Cumberland Plain. Depressions, on clays or sands; sometimes common on sandy soils after fire. Fl. spring–summer. *Red Crumbweed* . . . . **D. glomulifera** (Nees) Paul G.Wilson ssp. **glomulifera**

50 AMARANTHACEAE

Herbs, small shrubs or climbers. Leaves alternate or opposite, entire, without stipules. Flowers usually small, bisexual or unisexual, regular, usually sessile within 2 scarious bracteoles, subtended by a scarious bract or floral leaf. Perianth segments 5, scarious or coloured, imbricate, persistent. Stamens 2–5, opposite the perianth segments. Ovary 1-locular, superior; styles 1–3. Fruit a capsule, nut or berry, 1-locular. Seeds 1–many. 65 gen., trop. to temp.

Key to the genera

1 Leaves alternate. 2
1 Leaves opposite 3

2 Erect or prostrate herbs. Fruit dry, membranous. . . . . **1 AMARANTHUS**
2 Glabrous shrub or climber up to 4 m high. Fruit a berry . . . . **2 DEERINGIA**

3 Perianth segments bracts and bracteoles not rigid 4
3 Perianth segments, bracts and bracteoles rigid, acute, often spinescent 7

4 Flowers in axillary clusters 5
4 Flowers in terminal heads or spikes 6

5 Perianth segments connate; inflorescence woolly . . . . **4 GUILLEMINEA**
5 Perianth segments free; inflorescence not woolly . . . .**3 ALTERNANTHERA**

6 Bracts white, acute; spikes sessile between uppermost leaves. . . . . **5 GOMPHRENA**
6 Bracts brown to dark grey, obtuse; spikes interrupted on long naked scapes . . . .**6 FROELICHIA**

7 Perianth segments 4 . . . . **8 NYSSANTHES**
7 Perianth segments 5 8

8 Flowers in axillary clusters . . . .**3 ALTERNANTHERA**
8 Flowers in terminal spikes 9

9 Flowers reflexed in fruit . . . .**7 ACHYRANTHES**
9 Flowers not reflexed in fruit. . . . .**6 FROELICHIA**

1 Amaranthus L.

20 species in Aust. (5 native); all states and territories

Erect to prostrate, usually monoecious, annual herbs. Leaves alternate, entire. Flowers unisexual, small, with a bract and 2 bracteoles at the base, in axillary or terminal clusters or spikes which often form a dense panicle. Perianth segments 3–5, green or scarious. Stamens 3–5, free. Styles 2–3. Fruit a nut or capsule , enclosed in the persistent perianth. Seed solitary, black, glossy.

Key to the species

1 Bracts with long prominent or rigid points, usually exceeding the styles. Fruit circumsciss.. 2
1 Bracts with short points or none, shorter or not much longer than the perianth. Perianth segments usually 3, if 5 then fruit with 3 swollen styles. Fruit indehiscent or bursting irregularly. 11

2 Bracts 10–15 mm long, spinescent, subtending the inflorescence branches. Bracteoles spinescent. Leaves ovate to lanceolate, 15–25 mm long occasionally more, paler and strongly reticulate underneath. Perianth segments 5. Coast. Weed. Hornsby Plateau. Introd. from trop. regions. *Needle Burr.* . . . . . .
. . . . *A. spinosus L.
2 Bracts and bracteoles spinescent or not, less than 10 mm long 3

3 Perianth segments 3 4
3 Perianth segments 5 5

4 Stems white, rigid. Spikes nearly all axillary, short. Bracts 3–4 mm long. Leaves obovate to lanceolate, 2–3 cm long, 5–10 mm wide, mucronate, petiole short reducing in size towards the apex of the flowering stems. Fruit c. as long as the perianth. Almost glabrous annual 20–80 cm high. Widespread. Uncommon weed. Introd. from Trop. America. *Tumble Weed* . . . . *A. albus L.
4 Spikes terminal. Bracts 5–6 mm long. Leaves 3–10 cm long, usually acute. Petiole long . . . . *
. . . . .A. powelli (see below)

5 Perianth segments obtuse or emarginate, often recurved. Stamens 5. Branches of the inflorescence short (10–15 cm long), thick. Hairy annual herb, base of plant and taproot red. Hunter River Valley. Introd. from America. *Red-root Amaranth.* . . . . *A. retroflexus L.
5 At least outer perianth segments acute to acuminate 6

6 Inflorescence deep red to maroon, 15–30 cm long, axillary and terminal. Bracts aristate, 2 mm long. Perianth spathulate, mucronate, 2 mm long. Stamens 3. Leaves 3–8 cm long, reddish. Erect annual herb with reddish stems. Uncommon garden escape. Introd. from S. America. *Love-lies-bleeding, Tassel Flower* . . . . *A. cordatus L.
6 Inflorescence not as above. 7

7 Stems and leaves glabrous 8
7 Stems and leaves hairy, may be glabrescent with age. 10

8 Apex of perianth segments recurved at least at maturity. Bracts spinescent, 3–5 mm long. Inflorescence terminal and dense. Perianth mucronate, 1.5–2.5 mm long. Stamens 5. Leaves up to 20 cm long. Erect annual. Uncommon garden escape. Introd. from S. America. *South American Amaranth* . . . .
. . . . *A. quitensis Kunth
8 Apex of perianth segments straight 9

9 Bracts 5–6 mm long. Styles thick at the base. Stamens and perianth segments 3–5. Bracts c. 5 mm long with a very prominent midrib. Inflorescence stiff, with few branches, 5–8 mm diam. Perianth segments straight. Widespread but uncommon. Weed. Introd. from N. America. . . . .*A. powellii S.Watson
9 Bracts 3–4 mm long. Styles slender at base, forming a narrow cleft between each other at the base. Inflorescence lax, usually with many branches. Perianth segments straight; inner ones oblong to lanceolate, acute-acuminate. Stamens 5. Widespread. Weed of cultivation and waste places. Introd. from Europe. *Slim Amaranth* . . . . *A. hybridus L.

10 Bract 3–4 mm long . . . . *A. hybridus (see above)

10 Bracts 2–3 mm long Styles slender at the base, forming a broad saddle between each other towards the base. Inflorescence lax, usually with many branches. Perianth ± recurved; inner ones oblanceolate, obtuse to mucronate. Stamens 5. Coast. Uncommon weed. Introd. from S. America. *Redshank* . . . . .
. . . .*A. cruentus** L.

11 Flowers in axillary clusters. 12
11 Flowers in elongated spikes, axillary or forming terminal panicles. 14

12 Perianth segments 5, 2–3 mm long; spathulate, mucronate. Styles 3, swollen. Fruit wrinkled and swollen below, slightly longer than perianth. Uncommon. Introd. from NW NSW. *Boggabri Weed* . . .
. . . . **A. mitchelli** Benth.
12 Perianth segments 3, 5 mm long, spathulate, acuminate or emarginate. Bracts and bracteoles shorter than the perianth Fruit oblong, much longer than the perianth, wrinkled, indehiscent, with 2–3 prominent styles 13

13 Fruit dark. Coast and Cumberland Plain. Weed near stockyards. Introd. from interior of Australia *Dwarf Amaranth* . . . . *A. macrocarpus** Benth. var. **macrocarpus**
13 Fruit straw-coloured. Coast. Weed near stockyards. Introd. from interior of Australia . . . .
. . . . *A. macrocarpus** var. **pallidus** Benth.

14 Prostrate to scrambling herb. Stems hairy. Fruit much longer than the perianth or bracts, nearly smooth, indehiscent. Bracts and bracteoles shorter or as long as he perianth. Leaves ovate to oblong-lanceolate, 1–3 cm long. Coast. Uncommon weed. Introd. from Europe. *Spreading Amaranth*. . . . . .
. . . .*A. deflexus** L.
14 Erect to spreading herbs. Stems glabrous 15

15 Fruit longer than perianth, striated. Bract shorter than perianth. Leaves elliptic up to 6 cm long. Stems erect to trailing. Perennial herb. Uncommon weed. Introd. from Europe. . . . . **A. blitum** L.
15 Fruit scarcely longer than the perianth, wrinkled. Bracts shorter than the perianth. Leaves ovate, up to 10 cm long. Stems spreading or ascending. Widespread. Annual herb. Common garden weed. Introd. from trop. regions. *Green Amaranth* . . . .*A. viridis** L.

2 Deeringia R.Br.
2 species native Aust.; Qld, NSW, WA

One species in the area
Woody, glabrous shrub or climber up to 4 m high or sometimes a small shrub. Leaves alternate, petiolate, ovate to ovate-lanceolate, 5–10 cm long, acuminate, entire. Flowers bisexual, in slender interrupted spikes 5–25 cm long in the upper axils or in a loose terminal panicle. Perianth segments 5, equal, scarcely 2 mm long. Stamens 5, connate at the base. Fruit a berry, ± globular, 4–5 mm diam., red, 3-furrowed. Widespread. Often near RF . . . .**D. amaranthoides** (Lam.) Merr.

3 Alternanthera Forssk.
8 or 9 species in Aust. (3 introduced); all states and territories

Herbs with prostrate or ascending stems. Leaves opposite. Flowers bisexual, in axillary spikes, heads or clusters. Perianth segments 5, scarious, whitish, hardened after flowering. Stamens 5, connate at the base, 2–3 often without anthers. Style very short or none; stigma capitate. Fruit a nut, usually compressed.

Key to the species

1 Plants glabrous except the young tips of the branches and the nodes. Leaves linear-lanceolate or narrow-oblong to oblanceolate and obovate, from less than 1 to 6 cm long, up to 6 mm wide. 2
1 Plants ± pubescent. Leaves usually more than 6 mm wide 3

2 Inner bracts 1.5–2.5 mm long. Perianth segments lanceolate, 2–3 mm long. Leaves 20–60 mm long. Flower clusters 4–8 mm diam., sessile. Fruits not much shorter than the perianth. Widespread. Margins of swamps and rivers. Weed of pastures and waste places. *Lesser Joyweed* . . . . . .**A. denticulata** R.Br.

2 Inner bracts up to 1.5 mm long. Perianth segments oblong, 2–3 mm long. Leaves 5–50 mm long. Flower clusters 4–8 mm wide. Prostrate perennial herb. Clay soils. *Plains Joyweed* . . . .
. . . . **A. sp. A** sensu Harden (1990)

3 Flower clusters on long peduncles up to 3 cm long, 8–10 mm diam. Perianth segments c. 4 mm long, without conspicuous points. Leaves obovate to narrow-elliptic, 3–6 cm long, 1–15 mm wide. Prostrate herb often forming dense mats and rooting at the nodes. Lower Hunter River Valley; Cumberland Plain. Ditches and wet places, often in slowly moving water. Introd. from S. America. *Alligator Weed* . . . . .
. . . . ***A. philoxeroides** Griseb.
3 Flower clusters sessile **4**

4 Two of the perianth lobes rigid, pungent pointed, 5 mm long; the others shorter. Stems prostrate, softly pubescent, rooting at the nodes, forming mats. Leaves obovate to nearly orbicular, 2–3 cm long. Spikes up to 15 mm long, 10 mm diam. Widespread. Weed; roadsides and waste land. Introd. from S. America. *Khaki Weed* . . . . ***A. pungens** Kunth
4 Perianth lobes ± equal, 2–3 mm long, acute or with a small point, thickened and hardened at the base when in fruit. Spikes globular to ovoid, 8–10 mm long, c. 6 mm wide. Ephemeral herb. Widespread but uncommon . . . . **A. nana** R.Br.

4 Guilleminea Kunth
1 species naturalized Aust.; Qld, NSW, S.A., WA, NT

One species in the area
Prostrate to ascending, hairy herbs. Leaves opposite, ovate to oblanceolate, up to 20 mm long. Flowers in spike-like woolly axillary clusters. Perianth segments 5, connate, c. 2 mm long. Stamens 5, connate. Style bifid. Fruit membranous. Weed, often in lawns. Introd. from America. *Small Matweed* . . . .
. . . . ***G. densa** (Humb. & Bonpl. & Schult.) Moq.

5 Gomphrena L.
19 species in Aust.; Qld, NSW, NT, S.A., WA

One species in the area
Woolly, branching, annual herb. Leaves opposite, spathulate, 2–5 cm long, narrowed into a very short petiole. Flowers bisexual, mostly in terminal globular shining silvery spikes 12 mm diam. which are sessile between the uppermost leaves. Coast. Weed. Introd. from S. America. *Gomphrena Weed* . . . . . .
. . . . ***G. celosioides** Mart.

6 Froelichia Moench
1 species naturalized Aust.; Qld, NSW

One species in the area
Erect or ascending herb up to 25 cm high. Leaves opposite, narrow-elliptic to narrow-oblong, 2–5 cm long, pubescent, acute, entire, almost sessile. Flowers bisexual, in interrupted spikes terminal on naked scapes. Bracts brown to dark grey, subtending the woolly flowers. Cumberland Plain. Introd. from S. America . . . . ***F. gracilis** (Hook.f.) Moq.

7 Achyranthes L.
1 species native Aust.; Qld, NSW, NT, WA

One species in the area
Erect or spreading, herbaceous annual or biennial up to 1 m high with a hard almost woody base; stems ± hairy. Leaves opposite, shortly petiolate, ovate or lanceolate to almost oblong, 2–6 cm long, softly pubescent underneath, less hairy on upper surface. Flowers numerous in long slender but rigid terminal spikes. Perianth segments 5, 3–4 mm long, reflexed after flowering. Coast. Weed in waste ground. *Chaff Flower* . . . . **A. aspera** L.

8 Nyssanthes R.Br.

2 species endemic Aust.; Qld, NSW

Much branched, stiff herbs. Leaves opposite. Perianth segments 4, green, rigid, very spreading or reflexed after flowering. Flowers in sessile head-like spikes or clusters in the upper forks of the branching system. Bracts and bracteoles spinescent, very spreading. Stamens 2 or 4, united at the base. Fruit membranous, 1-seeded, enclosed in the persistent perianth.

Key to the species

1 Stamens 4. Plants erect, ± pubescent, with soft suppressed hairs. Leaves elliptic-oblong to almost lanceolate, 2–7 cm long; the upper ones smaller. Coast to Blue Mts; Illawarra district; uncommon elsewhere. Weed of disturbed ground . . . . **N. erecta** R.Br.

1 Stamens usually 2. Plants usually diffuse, very similar to *N. erecta* but smaller in all its parts. Coast to Blue Mts; Illawarra district; uncommon elsewhere. Weed of disturbed ground, sometimes on RF margins. *Barbwire Weed.* . . . . **N. diffusa** R.Br.

51 BASELLACEAE

5 gen., trop. and subtrop. America and Asia

1 Anredera Juss.

1 species naturalized Aust.; NSW

One species in the area

Glabrous, herbaceous, perennial climber, producing tubers on roots and stems. Leaves thick, fleshy, alternate, cordate, 4–13 cm long; petioles compressed, 25 mm long; stipules absent. Flowers small, fragrant, numerous in axillary drooping racemes 7–13 cm long. Pedicels with a small subulate bract at the base and 2 small ovate bracteoles at the base of the flower. Perianth segments 5, white. Stamens 5. Stigmas 3. Ovule solitary. Coast and adjacent plateaus. Waste land in suburbs, gullies, and margins of RF. Introd. from S. America. *Lamb's Tails* or *Madeira Vine* . . . . ***A. cordifolia** (Ten.) Steenis

52 CACTACEAE

Herbs or shrubs. Stems modified into cladodes, succulent, ± spiny. Small leaves may be present on new growth. Areoles present from which spines, hairs, flowers and cladodes develop. Flowers bisexual, usually solitary, short lived. Hypanthuim elongated to cup-shaped. Perianth segments numerous and showy. Stamens numerous. Ovary inferior, 1-locule. Style simple, 3-many stigmas. Fruit a fleshy berry. 100 gen., temp. and subtrop. America

Key to the genera

1 Leaves alternate, broad, flat and ± persistent. Flowers pedunculate . . . .**1 PERESKIA**

1 Leaves absent or if present then reduced and terete. Flowers sessile . . . . **2 OPUNTIA**

1 Pereskia Mill.

1 species naturalized Aust.; Qld, NSW

One species in the area

Woody erect to spreading shrub, branches to 10 m long. Spines 1–5 cm in clusters along stem. Leaf axils with pairs of shorter curved spine. Leaves obovate to elliptic, 4–7 cm long, 20–30 mm wide. Flowers white, yellow or pink. Stamens numerous, shorter than the perianth. Ovary with leafy scales and sometimes spines. Fruit obovoid, 15–20 mm diam., fleshy, yellowish. Along river banks. Hawkesbury. Rarely naturalized. Introd. from America. *Barbados Gooseberry* . . . . ***P. aculeata** Mill.

2 **Opuntia** Mill.

28 species naturalized Aust., all states and territories except Tas.

Perennials with fleshy often flattened and jointed stems. Leaves small, caducous, with axillary areoles bearing spines and/or hairs. Sepals and petals numerous, imbricate. Stamens numerous. Ovary inferior, with numerous ovules scattered on the walls; style with several branches. Fruit a berry, reddish to purple when ripe.

Key to the species

1 Stem segments mostly less than 4 cm wide and 10 cm long. Areoles with white hair-like spines and 1–4 stout spines up to 3 cm long. Drier areas. Introd. from S. America. *Tiger Pear*. . .***O. aurantiaca** Lindl.
1 Stem segments mostly more than 5 cm wide and 10 cm long. 2

2 At least 3–8 spines per areole. Spines terete 2–4 cm long, darkening with age. Segments obovate and compressed, bright green. Leaves to 4 mm long red-tipped. Perianth yellow. Staminal filaments pink. Fruit obovoid, reddish. Not common. Introd. from S. America. . . . . ***O. elatior** Mill.
2 Areoles with 1 or 2 spines at most, rarely more. 3

3 Areoles with 1–2 long spines 1–4 cm long. Tree-like. Cumberland Plain. Weed, particularly near creeks. Introd. from S. America. *Tree Pear*. . . . . ***O.monocantha** Haw.
3 Areoles mostly with no long spines but sometimes those near the margin with a single spine. Coast and Cumberland Plain. Coastal sand dunes; open forests. Introd. from America. *Common Prickly Pear* . . .
. . . . ***O. stricta** Haw.

53 CARYOPHYLLACEAE

Herbs. Stems usually thickened and jointed at the nodes. Leaves opposite or apparently whorled, often connate basally; stipules present or absent. Flowers usually bisexual, regular. Sepals 4–5, free or connate, persistent, imbricate in bud. Petals 4–5, sometimes absent, imbricate in bud. Stamens up to 10. Ovary superior, usually 1-locular with basal or free central placentas, rarely multi-locular with axile placentas. Fruit a capsule or nut. 80 gen. cosmop. mainly N. hemisphere.

Key to the genera

1 Sepals free or nearly so. 2
1 Sepals connate 11

2 Stipules absent 3
2 Stipules present. 7

3 Styles 5. Capsule with 10 teeth . . . . **1 CERASTIUM**
3 Styles 3–4. Capsule with 6, 4 or 8 teeth or valves. 4

4 Styles 4. Capsule with 4 or 8 teeth or valves. 5
4 Styles 3. Capsule with 6 teeth 6

5 Leaves subulate. Capsule with 4 valves . . . . **5 SAGINA**
5 Leaves narrow-ovate to linear-lanceolate. Capsule with 8 teeth . . . . .3 **MOENCHIA**

6 Petals deeply notched or rarely absent. . . . . .2 **STELLARIA**
6 Petals present, entire. . . . . **4 ARENARIA**

7 Styles 5. Capsule with 5 teeth . . . . **6 SPERGULA**
7 Styles 2–3 8

8 Styles 2. . . . . **18 PARONYCHIA**
8 Styles 3. 9

9 Leaves obovate to spathulate . . . . .7 **POLYCARPON**

9 Leaves terete. Thui10

10 Bracts conspicuous . . . . **8 POLYCARPAEA**
10 Bracts inconspicuous . . . .**9 SPERGULARIA**

11 Styles 2. Fruit a 4-valved capsule or nut 12
11 Styles 3–5. Fruit a 5–6-valved capsule 17

12 Petals absent. . . . . **17 SCLERANTHUS**
12 Petals present 13

13 Flowers with large bracts at the base 14
13 Flowers without large bracts at the base 15

14 Bracts at base of flower herbaceous, green . . . . **11 DIANTHUS**
14 Bracts at base of flower scarious, straw-coloured . . . . **10 PETRORHAGIA**

15 Calyx up to 6 mm long. Petals constricted only near the base. . . . .**12 GYPSOPHILA**
15 Calyx more than 12 mm long. Petals constricted into a claw c. as long as the lamina or longer . . . 16

16 Petals with a corona . . . . **13 SAPONARIA**
16 Petals without a corona. . . . . **14 VACCARIA**

17 Styles 3. Capsule 6-toothed . . . . **15 SILENE**
17 Styles 5. Capsule 5-toothed . . . .**16 LYCHNIS**

1 Cerastium L.

7 species naturalized Aust.; all states and territories

Pubescent, annual herbs. Flowers in cymes. Sepals 5, free. Petals 5, white, deeply cleft. Stamens 5. Styles 5. Capsule splitting into twice as many teeth as styles.

Key to the species

1 Upper bracts entirely herbaceous. Lower leaves oblanceolate to obovate, 5–25 mm long. Pedicels shorter than the sepals. Sepals with hairs exceeding the apex. Widespread. Garden weed. Almost cosmop.. Fl. winter–summer. *Mouse-ear Chickweed* . . . . ***C. glomeratum** Thuill.
1 Upper bracts scarious on the tip or margins 2

2 Upper bracts scarious at least in the upper third. Lower leaves up to 20 mm long. Pedicels as long as or longer than the sepals. Hairs on sepals not exceeding the apex. Blue Mts Weed. Introd. from Europe. Fl. spring . . . . ***C. balearicum** F.Herm.
2 Upper bracts scarious on tip and margins only but green in centre of upper third. Leaves up to 30 mm long. Pedicels shorter than or equal to the sepals. Widespread. Weed. Introd. from Europe. Fl. spring . . . . ***C. vulgare** Hartm.

2 Stellaria L.

12 species in Aust. (3 naturalized); all states and territories except NT

Herbs with angular stems. Flowers in terminal leafy cymes or solitary and axillary. Sepals 5. Petals 5, usually white, deeply cleft, rarely absent. Styles 3. Capsule 6-valved, splitting almost to the base.

Key to the species

1 Stems with a line of hairs down one side of each node. Flowers in terminal leafy cymes, the peduncles at fruiting becoming lateral in the forks of branches. Annual, prostrate to ascending herbs. Cosmop. garden weeds. 2
1 Stems glabrous or with hairs not in lines 3

2 Sepals more than 4 mm long with glandular hairs. Petals present. Leaves green, ovate, mostly 12–18 mm long, upper ones sessile, lower ones petiolate. Capsule the same length as sepals. Widespread. Weed. Introd. from Europe. Fl. winter–summer. *Chickweed* . . . . ***S. media** Vill.

2 Sepals less than 4 mm long with simple hairs. Petals absent. Leaves yellowish-green, ovate, mostly less than 7 mm long, usually petiolate. Capsule exceeding the sepals. Plants rarely glabrous with age. Widespread. Weed. Introd. from Europe. Fl. spring. . . . . ***S. pallida** (Dumort.) Crep.

3 Leaves pungent-pointed, rigid, sessile, very narrow-ovate to ovate, mostly 6–8 mm long. Flowers on axillary peduncles usually exceeding the leaves. Sepals rigid, pungent pointed, c. 8 mm long. Petals equal in length to sepals. Much branched, decumbent or ascending perennial. Widespread. WSF, gullies and margins of RF. Fl. spring–summer. *Prickly Starwort* . . . . **S. pungens** Brongn.

3 Leaves not pungent-pointed, usually flaccid 4

4 Petals obvious. Flowers solitary, on axillary or terminal peduncles 5

4 Petals not obvious less than half length of sepals or absent. Inflorescence a terminal monochasium . . 6

5 Leaves linear, glabrous, sessile, 1–4 cm long. Branches long, weak, glabrous or scabrous. Peduncles usually longer than the leaves. Sepals 6–8 mm long. Capsule shorter than the sepals. Widespread. Swamps or wet places. Fl. spring–summer. *Swamp Starwort* . . . . **S. angustifolia** Hook.

5 Leaves ovate to very narrow-ovate, very acute, often with undulate margins, rarely exceeding 12 mm long; appearing petiolate. Stems glabrous or almost so except at the nodes. Sepals 4–5 mm long. Petals somewhat longer. Capsule ovoid, usually exceeding the sepals. Stems often very long. Widespread. Margins of RF. Fl. spring–summer. . . . . **S. flaccida** Hook.

6 Leaves mostly sessile. Sepals acute, 3–5 mm long. Petals 3–5 rarely absent. Stamens 3–10. Stems glabrous or occasionally with sparse hairs. Sepals 3–5 mm long. Capsule ovoid, equal to or longer than sepals. Annual with stems erect or ascending. Uncommon. Fl. spring–summer . . . . **S. multiflora** Hook.

6 Leaves petiolate. Sepals obtuse. Petals absent. Stamens 1–3. . . . . ***S. pallida** (see above)

3 Moenchia Ehrh.
1 species naturalized Aust.; NSW, Vic., Tas., S.A.

One species in the area
Annual, glabrous herb with slender stems c. 30 cm high. Leaves narrow-ovate to linear-lanceolate, 5–12 mm long; stipules absent. Flowers in axillary or terminal cymes. Sepals 4, free. Petals 4, opposite the sepals. Stamens 4–8. Styles 4. Capsule opening by 8 teeth. Blue Mts Weed in gardens, pastures and forests. Introd. from Europe. Fl. spring. *Erect Chickweed* . . . . ***M. erecta** (L.) G.Gaertn, B.Mey. & Scherb.

4 Arenaria L.
2 species naturalized in Aust.; NSW, Vic., Tas., S.A., WA

One species in the area
Branching annual 5–20 cm high, slightly pubescent. Leaves ovate, c. 5 mm long, acute. Flowers on slender pedicels in terminal cymes. Sepals 5, acute, c. 3 mm long. Petals much shorter, ovate, entire, white. Capsule c. as long as the calyx, opening by 6 narrow valves. Widespread. Weed in sandy soils. Almost cosmop.. Fl. spring–summer. *Thyme-leaved Sandwort.* . . . . ***A. leptoclados** (Rchb.) Guss.

5 Sagina L.
5 species in Aust. (2 native, 3 naturalized); NSW, Vic., Tas., S.A., WA

Small, matted or tufted herbs with subulate leaves; stipules absent. Flowers very small, usually solitary on slender terminal or axillary pedicels. Sepals, stamens and styles 4. Petals white, 4, entire, minute or absent. Capsule 4-valved, c. as long as the calyx.

Key to the species

1 Perennial herb with fibrous roots. Stems numerous up to c. 5 cm high, often rooting at the nodes, forming a mat; some stems without flowers at flowering stage. Flowers solitary, terminal; pedicels curved after flowering. Leaves 10–15 mm long, shortly awned. Widespread. Weed in gardens and pastures. Introd. from Europe. Fl. spring–summer. *Procumbent Pearlwort* . . . . *S. procumbens L.

1 Annual with slender taproot. Stems erect or ascending, up to 10 cm high, not rooting at the nodes; all stems with flowers at flowering stage. 2

2 Leaves 2–10 mm long, finely awned at apex, ciliate towards base. Flowers in a terminal cyme; often glandular hairy; pedicels straight after flowering. Fruiting sepals spreading. Coast and Cumberland Plain. Weed in moist gardens and pastures. Introd. from Europe. Fl. spring–summer. *Annual Pearlwort* . . . . *S. apetala Ard.

2 Leaves ± fleshy, 3–15 mm long, obtuse to acute or minutely mucronate; margins not ciliate. Inflorescence without glandular hairs. Fruiting sepals erect. Coastal areas. Introd. from Europe and Africa Fl. spring. *Sea Pearlwort* . . . . *S. maritima Don

6 Spergula L.
2 species naturalized Aust.; all states and territories except NT

One species in the area
Erect to ascending, glandular-pubescent or almost glabrous annual 20–40 cm high. Leaves narrow-linear, channeled beneath, 1–3 cm long, apparently whorled; stipules minute. Flowers on slender pedicels in terminal leafless cymes; pedicels spreading or reflexed in fruit. Sepals 5. Petals 5, entire, white, ovate, nearly as long as the sepals. Capsule ovoid, slightly exceeding the calyx. Seeds black, c. 1 mm diam. Widespread. Weed of cultivated land. Almost cosmop. Fl. spring–summer. *Corn Spurrey* . *S. arvensis L.

7 Polycarpon L.
1 species naturalized Aust.; all states and territories except NT

One species in the area
Spreading or prostrate, glabrous annual 6–12 cm high. Leaves obovate to spathulate, 5–12 mm long, petiolate, opposite or apparently in whorls of 4; stipules scarious. Flowers numerous in loose terminal cymes. Sepals 5, keeled, scarious on the margins, up to 2 mm long. Petals 5, entire, white, less than half as long as the sepals. Stamens usually 3. Styles 3. Capsule 3-valved. Widespread. Common weed in gardens and sheltered places. Introd. from Europe. Fl. spring–summer. *Four-leaf All-seed.* . . . . *P. tetraphyllum (L.) L.

8 Polycarpaea (L.) Lam.
12 species in Aust.; Qld, NSW, NT, S.A., WA

One species in the area
Annual herb up to 15 cm high, with pubescent stems. Leaves often apparently whorled, terete, 2–10 mm long; stipules prominent, hyaline. Flowers in terminal much branched cymes. Sepals 5, scarious, reddish-brown in centre, white towards the margin, acute, 2–3 mm long. Petals 5, entire shorter than the sepals. Stamens 5, adnate to petals in a small basal cup. Fruit a capsule. Sandy soils near Menangle Park. Uncommon in the area. Fl. summer.. . . . . P. corymbosa (L.) Lam. var. **corymbosa**

9 Spergularia (Pers.) J.Presl & C.Presl
12 species in Aust. (most naturalized, some possibly native); all states and territories except NT

Herbs sometimes woody at the base. Leaves linear, opposite; stipules hyaline. Flowers in cymes, 5-merous. Petals entire, pink. Stamens 1–10. Styles 3. Capsule 3-valved. Seeds sometimes winged.

Key to the species

1 Sepals glabrous, c. half as long as the capsule. Stipules triangular, scarcely connate. Cauline leaves up to 5 cm long. Stamens usually 10. Seeds with a fringed wing. Erect herbs with a branched woody stock. Ingleburn; Minto. Disturbed ground. Introd. from S. America. Fl. spring . . . . ***S. levis** Cambess.

1 Sepals pubescent, as long as the capsule or nearly so **2**

2 All seeds winged. Stipules not silvery. Sepals 4–7 mm long. Petals shorter than sepals, pink. Stamens 8–10. Capsule slightly longer than sepals 5–10 mm long. Saline areas. Probably native. Fl. spring–summer . . . . **S.** sp. **Butchers Gap** (P.Gibbons 234) L.G.Adams

2 Seeds not winged or occasionally a few winged **3**

3 Stamens mostly 4 or less **4**

3 Stamens 6–10 . . . . **5**

4 Stipules of younger branches connate for c. half their length, but sometimes splitting with age. Leaves 10–40 mm long, fleshy. Petals pink grading to white at the base, or entirely white. Capsule narrowly ovoid. Seed usually not winged. Coast. Saline wetland areas. Probably native. Fl. spring. . . . . **S. marina** (L.) Griseb.

4 Stipules free or slightly fused at base. Leaves 7–20 mm long, hardly fleshy. Petals pink. Capsule sub globose. Seed not winged. Annual often forming matt. Uncommon weed. Introd. from Europ. Fl. spring *Lesser Sand-spurrey* . . . . ***S. diandra** (Guss.) Heldr.

5 Stipules ± free, silvery, acuminate. Leaves with long awn, 1–4 cm long. Pedicels much longer than sepals and capsule. Petals 2.5–3.5 mm long, entirely pink. Stamens 10. Capsule c. 5 mm long. Widespread. Weed in gardens and waste places. Probably introd. from Europe. Fl. spring. *Sand-spurrey*. . . . . ***S. rubra** (L.) J.Presl & C.Presl

5 Stipules not silvery, or acuminate. Leaves with a short mucro or awn. Pedicels shorter than or equal to the sepals and capsule. Petals c. 2 mm long, white or pink inside. Stamens 2–5. Capsule c. 3 mm long. Weed of various soils, often in salt affected-areas. Introd. from Europe and Mediterranean. Fl. spring–summer. *Bocconi's Sand-spurrey*. . . . . ***S. bocconii** (Scheele) Graebn.

10 Petrorhagia (Ser. ex DC.) Link
2 species naturalized Aust.; all states and territories except NT

Annual herbs with stiff erect stems, 10–40 cm high. Leaves linear, 1–6 cm long. Flowers small, in terminal head-like clusters, 2–8 per cluster; calyces concealed by broad scarious bracts of which the innermost are 12–15 mm long. Petals pink. Calyx 5-toothed, 5-angled. Stamens 10. Styles 2. Capsule ovoid, 6–8 mm long, opening by 3 teeth. Seeds shield-shaped, hollowed on the inner face.

Key to the species

1 Stems glabrous or simple-hairy. Seeds 1.3–1.8 mm long, reticulate. Widespread. Weed in pastures and waste places. Introd. from the Mediterranean. Fl. spring–summer . . . . ***P. nanteuilii** (Burnat) P.W.Ball & Heywood

1 Stems glandular hairy. Flowers very shortly pedicellate. Seeds up to 1.3 mm long, papillose. Weed in pastures and waste places. Introd. from the Mediterranean. Fl. spring–summer . . . . ***P. velutina** (Guss.) P.W.Ball & Heywood

11 Dianthus L.
2 species naturalized Aust.; NSW, Vic., Tas.

One species in the area

Erect hairy annual up to 70 cm tall. Leaves linear, up to 7 cm long, entire. Flowers 3–6 in terminal clusters, with large herbaceous bracts at the base. Calyx up to 20 mm long, 5-toothed. Petals pink, toothed. Stamens 10. Styles 2. Capsule cylindrical, up to 12 mm long, opening by 4 teeth. Widespread but

not common. Weed in disturbed ground. Introd. from Europe. Fl. summer. *Deptford Pink* . . . .
. . . . *D. armeria* L.

12 Gypsophila L.

2 species naturalized Aust.; NSW, Vic., Tas., S.A.

One species in the area

Small, erect annual with sticky glandular hairs. Leaves linear, 5–15 mm long. Flowers in large dichasia on peduncles longer than the leaves. Calyx campanulate-tubular, 4–5 mm long, with 5 green ribs terminating in short teeth. Petals slightly longer than sepals, pink, notched at the apex. Stamens 10. Styles 2. Capsule ovoid-oblong, c. as long as the calyx, opening by 4 valves. Seeds minute. Coast and adjacent plateaus. Sandy soils. Introd. from Asia Minor. Fl. spring. *Annual Chalkwort* (previously known as *G. australis* which was considered native, but this now seems doubtful) . . . . . . *G. tubulosa* (Jaub. & Spach) Boiss.

13 Saponaria L.

2 species naturalized Aust.; all states and territories accept NT

One species in the area

Glabrous, perennial herb up to 1 m high. Leaves sessile or shortly stalked, elliptic to narrow-elliptic. Flowers in terminal dichasia. Calyx tube c. 20 mm long. Petals pink to white, clawed, with coronal scales. Styles 2. Widespread. Waste ground. Introd. from Europe. Fl. winter–spring. *Soapwort* . *S. officinalis* L.

14 Vaccaria Wolf

1 species naturalized Aust.; all states and territories

One species in the area

Glabrous, annual herb up to 60 cm high. Leaves sessile, opposite, ovate to lanceolate, 5–12 cm long, glaucous. Flowers in terminal compound dichasia; lower bracts leaf-like; upper bracts scarious. Calyx tube 12–18 mm long, inflated, with 5 green wings. Petals purplish, clawed. Styles 2. Casula. Waste land. Introd. from Europe. *Bladder Soapwort* . . . . *V. hispanica* (Mill.) Rauschert

15 Silene L.

14 species naturalized Aust.; all states and territories except NT

Erect or ascending, annual herbs. Flowers in monochasia. Calyx tubular, 5-toothed, 10–30-veined, sometimes inflated. Petals 5, exceeding the calyx, usually 2-fid, with a long narrow claw and usually 2 scales at the top of the claw. Stamens 10. Styles 3. Capsule usually enclosed within the calyx, opening by 6 teeth at the top. Seeds numerous, reniform.

Key to the species

1 Styles 5. Leaves lanceolate, 4–9 cm long. Calyx 15–30 mm long, 10–20-veined. Petals white. Capsule 10-valved. Dioecious, ± pubescent herb. Southern Blue Mts; Mt. Tomah. Grasslands. Introd. from Europe. Fl. summer. *White Campion* . . . . *S. latifolia* Poir. ssp. **alba** (Mill.) Greuter & Burdet
1 Styles 3. 2

2 Calyx 20-veined, inflated, 1–2 cm diam. Leaves elliptic to narrow-elliptic, 2–7 cm long. Petals white or pinkish. Glabrous, perennial herb. Widespread. Weed of cultivation and pastures. Introd. from Europe. Fl. spring–summer. *Bladder Campion* . . . . *S. vulgaris* (Moench) Garcke
2 Calyx 10-veined, not inflated, up to 5 mm diam 3

3 Plant not clammy or sticky. Calyx c. 10 mm long, narrowing slightly but not constricted towards the top; hairs not glandular. Lower leaves narrow-elliptic-obovate, up to 6 cm long. Petals pink. Annual up to 40 cm high. Coast and Cumberland Plain. Weed in waste places. Introd. from Europe. Fl. spring. . .
. . . . *S. nocturna* L.

3 Plant clammy and sticky above. Calyx c. 8 mm long, distinctly constricted at the base of the teeth; hairs glandular. Lower leaves oblong-spathulate, 2–6 cm long. Annual up to 30 cm high. Widespread. Weed in gardens and waste places. Introd. from Europe. Fl. spring–summer.. 4

4 Petal limb uniformly white or pink. *French Catchfly* . . . .
. . . .*S. gallica* var. **quinquevulnera** (L.) Mert. & W.D.J. Koch
4 Petal limb with crimson blotch. *Five-wounded Catchfly*. . . . . *S. gallica* L. var. **gallica**

16 Lychnis L.
2 species naturalized Aust.; Qld, NSW, Tas., Vic.

One species in the area
Erect perennial covered with a silky, silvery tomentum. Leaves oblong to lanceolate, 3–10 cm long. Flowers in cymes on long pedicels. Calyx 18 mm long, oblong, with 10 ribs. Petals red, entire, with a long claw and spreading lamina. Capsule opening by 5 teeth. Garden escape in some suburbs. Introd. from Europe and W. Asia. Fl. summer. *Rose Campion* . . . . *L. coronaria* (L.) Desr.

17 Scleranthus L.
9 species in Aust. (8 native, 1 naturalized); Qld, NSW, Vic., Tas., S.A.

Annual or perennial herb. Leaves opposite, linear to subulate, glabrous or hairy. Stipules absent. Flowers terminal or axillary. Sepals 4 or 5, connate into a tube, hardened at maturity, Petals absent. Styles 2. Ovary 1-locular. Nut membranous, enclosed within the calyx tube.

Key to the species

1 Flowers in pairs. Calyx lobes 4, less than 1 mm long; stamens 1. Ovary pyriform in shape. Leaves linear-subulate, c. 5 mm long, sometimes clustered. Small, densely branched, perennial herb, sometimes forming a mat or cushion. Blue Mts and Otford. Roadsides and grassy banks. Fl. spring–summer. *Twin-flowered Knawel* . . . . **S. biflorus** (R.J.Forst. & G.Forst.) Hook.f.
1 Flowers in clusters. Calyx lobes 5, more than 1 mm long; stamens 5–10. Ovary ovoid in shape. Leaves crowded, 10–14 mm long, hairy at base. Annual herb with decumbent or ascending stems. Introd. from Europe and Africa. Fl. summer. *Annual Knawel* . . . .*S. annuus* L.

18 Paronychia Mill.
3 species naturalized Aust.; Qld, NSW, Vic., S.A.

Small, perennial, decumbent or ascending herb, prostrate or diffuse; stems ± pubescent, becoming woody and knotted. Leaves opposite. stipules membranous, whitish, conspicuous. Flowers numerous, in axillary clusters; often concealed by silvery bracts. Petals 5 and minute or absent. Sepals 5, hooded. Stamens 2–5. Style bifid. Ovary 1-locular. Capsule 1-seeded, enclosed within the calyx.

Key to the species

1 Flower clusters and internodes visible. Sepals equal and deeply hooded, reddish, with an erect awn at the apex. Petals absent. Leaves linear to very narrow-ovate, mucronate, up to 8 mm long, crowded, finely hairy. Widespread. Weed of cultivation. Introd. from S. America. Fl. spring–summer. *Chilean Whitlow Wort*. . . . .*P. brasiliana* DC.
1 Flower clusters and internodes concealed by leaves and stipules. Sepals unequal and slightly hooded, yellowish-green, apex awned. Petals absent. Leaves elliptic to oblanceolate, mucronate up to 10 mm long, ciliate to appressed pubescent. Not common. Weed of cultivation. Introd. from Chile. Fl. summer. *Chile Nailwort*. . . . . *P. franciscana* Eastw.

54 DROSERACEAE

4 gen., cosmop.

1 Drosera L.

Sundews

54 species in Aust.; all states and territories

Perennial herbs. Leaves spirally arranged, covered above with glandular hairs and bordered with longer ones which are usually irritable, closing over small insects; stipules present or absent. Flowers regular, bisexual. Sepals and petals 5 or rarely 4, free, imbricate. Stamens as many as petals and alternating with them, hypogynous. Ovary superior, 1-locular, with parietal placentas. Fruit a small capsule (Fig. 28).

Key to the species

1 Leaves all basal. Roots fibrous, never tuberous 2
1 Leaves basal and cauline, peltate. Basal rosette of leaves often absent at time of flowering. Stems erect, occasionally bending over, 10–15 cm high, arising from a globular subterranean tuber. Flowers in terminal racemes. Petals white or pink 6

2 Leaves dichotomously divided into 2–8 segments, rarely undivided; segments linear, 5–10 cm long. Petiole long, glabrous. Flowers c. 25 mm diam., in a loose cyme with racemose branches. Petals white. Whole plant 10–60 cm high. Widespread in swamps and along creeks. Fl. spring–summer . . . . **D. binata** Labill.
2 Leaves undivided, usually in a rosette, reddish 3

3 Leaves with a petiole longer than lamina 4
3 Leaves with petiole shorter than lamina or sessile 5

4 Flowers solitary, minute, on filiform pedicels. Leaves orbicular, 1–2 mm diam. on slender petioles; stipules whitish, membranous, deeply incised, enclosing the central bud in resting stage in dry summers. Petals white. Whole plant c. 12 mm diam. Wet soil in open. Fl. spring–summer. Coast and adjacent plateaus and Cumberland Plain near coast . . . . **D. pygmaea** DC.
4 Flowers 7 or more per inflorescence. Leaves orbicular, 8–12 mm diam. Stipules entire attached to petioles. Petals orange or red. Swamps. Fl. winter–spring. . . . . **D. glandulifera** Lehm.

5 Styles 5, undivided. Leaves green, spathulate-cuneate, up to 2 cm long, 1–3 cm wide; stipules divided, membranous. Petals white. Cumberland Plain and Blue Mts Wet places. Fl. spring–summer . . . . **D. burmannii** Vahl
5 Styles 3–4, divided almost to the base. Leaves reddish, spathulate, 1–3 cm long; stipules membranous, laciniate. Flowers in a monochasia which resemble one-sided racemes. Sepals c. 3 mm long. Petals pink or white, c. as long as the sepals. Widespread. Wet places. Fl. spring–summer. . . . **D. spatulata** Labill.

6 Sepals ± pubescent, c. 4 mm long; margins ciliate-toothed. Widespread. Damp pastures and open forests. Fl. spring–summer . . . . **D. peltata** Thunb.
6 Sepals glabrous, c. 5 mm long; margins entire. Widespread. Damp places in gullies and forests. Fl. spring–summer . . . . **D. auriculata** Backh. ex Planch.

55 NYCTAGINACEAE

Herbs shrubs or trees. Leaves usually opposite. Flowers regular, bisexual or unisexual. Calyx tubular, often petaloid, with 5 (rarely 4) very short lobes; upper part deciduous; lower part persistent around the fruit, the whole referred to as an anthocarp. Petals absent. Stamens 1–many, hypogynous, involute in the bud. Ovary superior, 1-locular, with a solitary basal ovule; style simple. True fruit a nut. 30 gen., warm temp., particularly America.

Key to the genera

1 Tree with leaves mostly 10–20 cm long . . . . .3 PISONIA
1 Perennial herbs 2

2 Prostrate or ascending plants with slender stems. Perianth less than 1 cm long . . . . . **1 BOERHAVIA**
2 Erect plants. Perianth 2–3 cm long. . . . . **2 MIRABILIS**

1 **Boerhavia** L.
Tarvine
13 species in Aust. (8 native, 5 endemic); all states and territories except Tas.

Key to the species

Prostrate or ascending, glabrous or pubescent or sometimes glandular-viscid, perennial herb. Leaves opposite, petiolate. Inflorescence a cyme with flowers solitary, in pairs or clusters or in compound umbels. Flowers bisexual, 5-lobed, white to pink. Stamens 1–4. Fruit indehiscent and ribbed.

1 Flowers sessile or pedicellate, 2–6 in umbels on filiform, simple or branched axillary peduncles. Stems glabrous or with short glandular hairs only. Leaves ovate oblong or very narrow-ovate, cordate, 1–4 cm long, paler underneath; the margin often undulate. Perianth c. 5 mm long, constricted above the ovary; pale pink. Anthocarp 5-ribbed; 2–4 mm long. Clay and alluvial soils. . . **B. dominii** Meikle & Hewson
1 Flowers subsessile, 3–10 in small umbels; peduncles stout. Stems with long non glandular hairs and a few short glandular hairs. Leaves elliptic to ovate c. 5 mm long the margins undulate. Perianth pink. Stamens 3. Anthocarp 3–4 mm long 5-ribbed. Possibly introduced from western plains. . . . .
. . . . ***B. coccinea** Mill.

2 **Mirabilis** L.
1 species naturalized Aust.; Qld, NSW, Vic.

One species in the area

Erect, bushy, quick-growing, perennial herb up to 1 m high. Leaves opposite, ovate-lanceolate, acuminate, entire. Perianth funnel-shaped, with a long tube, white and varying shades of red and yellow, with a strong odour at night, opening in cloudy weather or late in the afternoon. Garden escape near habitation. Introd. from trop. America. Fl. autumn. *Four o'clock* or *Marvel of Peru*. . . . .***M. jalapa** L.

3 **Pisonia** L.
3 species in Aust.; Qld, NSW, NT, WA

One species in the area

Bushy small tree or shrub. Leaves alternate or in groups of 2–5, elliptic to obovate, mostly 10–20 cm long, 6–10 cm wide, entire, dark green, glossy. Flowers mostly bisexual, in terminal leafless cymo-panicles. Perianth funnel-shaped, 6 mm long. Anthocarp 5-ribbed, narrow, c. 25 mm long; the angles smooth, viscid. Minnamurra Falls. RF. Fl. summer–autumn. *Bird-lime Tree*. . . . .
. . . . **P. umbellifera** (J.R.Forst. & G.Forst.) Seem.

56 PHYTOLACCACEAE

Herbs, shrubs or trees. Leaves alternate, simple, entire; stipules absent or minute. Flowers bi- or unisexual, regular, in racemes or spikes. Sepals 4–5, free or connate basally. Petals absent. Stamens up to 8, free, hypogynous. Carpels 1–10, free or connate basally, superior, often becoming completely connate in the fruit. Fruit a berry. 18 gen., widespread, particularly in warmer parts of S. Africa and America.

Key to the genera

1 Stout herbaceous perennial with a fleshy rootstock. Carpels 5–10. Racemes erect, spike-like . . . .
. . . . **1 PHYTOLACCA**
1 Weak shrub. Carpel 1. Raceme slender, drooping in fruiting stage . . . . **2 RIVINA**

1 Phytolacca L.

3 species naturalized Aust.; all states and territories except Tas.

One species in the area

Herbaceous perennial 1–2 m high; stems woody towards the base. Leaves ovate-lanceolate, acute, tapering towards the base, entire, thin, 5–15 cm long. Flowers almost sessile in stout erect racemes. Petals white. Stamens 8. Carpels usually 8. Fruit a dark purple berry, depressed-globular, 8 mm diam., with 8 blunt ribs. Widespread. Weed of waste ground. Introd. from Trop. America. Fl. spring–summer. *Ink Weed* . . .
. . . .*P. octandra L.

2 Rivina L.

1 species naturalized Aust.; Qld, NSW

One species in the area

Weak shrub c. 1 m high with slender branches. Leaves ovate, acuminate, up to 9 cm long, on slender petioles. Flowers on slender pedicels in a raceme 7–12 cm long. Ovary subglobular, 1-locular. Fruit a small red berry. Minnamurra Falls. In or near RF. Introd. from S. America. Fl. spring–summer. *Coral Berry* . . . . *R. humilis L.

57 PLUMBAGINACEAE

10 gen., cosmop., on sea shores and in saline areas

1 Limonium Mill.

7 species in Aust.; (2 endemic, 5 naturalized); all states and territories

One species in the area

Perennial, glabrous herb. Leaves obovate-oblong, entire, 3–8 cm long, narrowed into a petiole; stipules absent. Scape 25–50 cm high, repeatedly forked to form a broad corymbose panicle. Flowers numerous, in short dense unilateral spikes, consisting of 2–3-flowered clusters. Calyx usually funnel-shaped, dry, membranous, ciliate, pale pink, 5-lobed, persistent. Petals 5, yellow, c. as long as the calyx. Stamens 5. Ovary unilocular, superior. Fruit a nut within the persistent calyx. Coast near salt marshes, e.g. near Kiama. Fl. summer. *Native Sea Lavender* . . . . **L. australe** (R.Br.) Kuntze

58 POLYGONACEAE

Herbs, shrubs or climbers. Leaves alternate, simple; stipules membranous, forming a sheath around the stem (ochreous). Flowers small, regular, bisexual or unisexual. Perianth segments 5–6, in 2 whorls, often greenish. Stamens 4–8, perigynous. Ovary superior, 1-locular, with 1 erect ovule; styles 2–3, sometimes united near the base. Fruit a small angular nut enclosed in the persistent perianth. 43 gen., cosmop., particularly N. hemisphere.

Key to the genera

1 Perianth segments or lobes 6, unequal. 2
1 Perianth segments 5, ± equal 5

2 Three outer perianth segments spiny; inner ones smaller .4 EMEX
2 Three outer perianth segments small; inner ones larger 3

3 Flowers unisexual. Inner fruiting perianth segments scarcely longer than nut. Leaf hastate, basal lobes narrow . . . . **3 ACETOSELLA**

3 Flowers bisexual. Inner fruiting perianth segments much longer than nut. Leaf shape variable, hastate to lanceolate. **4**

4 Leaf bases cordate, cuneate or rounded. Inner perianth segments leathery, often toothed, not inflated . . . . **1 RUMEX**
4 Leaf bases hastate or sagittate often acute. Inner perianth segments inflated and papery. . **2 ACETOSA**

5 Flowers unisexual. Calyx succulent or membranous at fruiting stage. . . . . . . **8 MUEHLENBECKIA**
5 Flowers mostly bisexual. Calyx enlarged but not otherwise changed at fruiting stage **6**

6 Nut much exceeding the perianth . . . . **9 FAGOPYRUM**
6 Nut enclosed by perianth **7**

7 Outer perianth segments winged. . . . . **7 FALLOPIA**
7 Outer perianth segments not winged **8**

8 Stipules entire when young, pale brown. . . . . **6 PERSICARIA**
8 Stipules deeply cut, silvery to white . . . . **5 POLYGONUM**

1 Rumex L.
13–15 species in Aust. (7 naturalized); all states and territories

Herbaceous perennials, rarely annuals, usually erect. Usually with taproot and basal rosette of leaves. Flowers small, on jointed pedicels in whorl-like clusters and arranged in long racemes or panicles. Perianth segments green or reddish, 6; the 3 inner (valves) enlarging and then enclosing the fruit. Stamens 6. Styles 3-fid, stigmas fringed. Nut triquetrous. All weeds of waste ground, pastures and gardens.

Key to the species

1 Inner perianth segments with long teeth, with or without a tubercle **2**
1 Inner perianth segments entire, tuberculate **6**

2 Flower clusters without floral leaves, except the lowest ones **3**
2 Flower clusters with linear leaves longer than the flowers. **5**

3 Flower clusters close together, many-flowered, without floral leaves or leafy only near the base of the long racemes. Perianth valves ovate-triangular, 4–5 mm long, obtuse, with c. 3 teeth on each side below the middle; only one of the valves with a large tubercle. Stems erect, stout. Basal leaves large, crenulate, cordate. Widespread. Damp places. Introduced from Europe and Asia. *Broad-leaf Dock* . . . . ***R. obtusifolius** L.ssp. **obtusifolius**
3 Flower clusters distant, several-flowered **4**

4 At least one of the valves with an oblong tubercle. Stem erect, with stiff upper branches spreading ± horizontally. Flower clusters 6–12-flowered. Perianth valves ovate-triangular, c. 5 mm long, reticulate, with 3–5 (rarely more) straight rigid teeth on each edge. Basal leaves oblong-cordate. Widespread. Weed of cultivation and pastures. Introd. from Europe and Asia. *Fiddle Dock* . . . ***R. pulcher** L. ssp. **pulcher**
4 Perianth valves without tubercles. Stems slender, erect. Flower clusters 5–16-flowered, distant in long racemes which are leafless except at the base. Perianth valves 4 mm long, with a long hooked point and 2–3 long hooked bristles on each side, reticulate, with a thickened midrib. Basal leaves oblong-cordate or almost hastate. Widespread. A native plant which has become a weed in gardens and pastures. *Swamp Dock* . . . . **R. brownii** Campd.

5 Flowers 2–8 in each whorl arranged in a spreading panicle. Perianth valves narrow-triangular, c. 2 mm wide with a tooth on each side at the base, acuminate. Basal leaves narrow-oblong. Cumberland Plain. Uncommon . . . . **R. stenoglottis** Rech.f.
5 Flowers numerous in each whorl arranged in an erect panicle. Perianth valves triangular, 2–3 mm long, acuminate, reticulate, with 2 straight rigid teeth on each side. Leaves linear to lanceolate. A native of inland NSW and recorded from Kurrajong. *Shiny Dock* . . . . **R. tenax** Rech.f.

6 Perianth valves oblong, obtuse, 2–3 mm long, each usually with a large oblong tubercle. Flower clusters rather distant, leafy at the summit, in a loose panicle with stiff spreading and ascending branches. Lower leaves oblong-lanceolate, crisped. Widespread. Weed of cultivation, waste places and pastures. Introd. from Europe and Asia. *Clustered Dock* . . . . ***R. conglomeratus** Murray

6 Perianth valves broad-ovate, 3–6 mm long, cordate, reticulate, often only 1 of the 3 bearing a well developed tubercle. Flower clusters many-flowered, close together, leafy only near the base of the stems, in a long dense panicle with erect branches. Lower leaves oblong, with crisped edges. Widespread. Weed of cultivation, waste places and pastures. Introd. from Europe and Asia. *Curled Dock* . . ***R. crispus** L.

2 Acetosa Mill.
2 species naturalized Aust.; all states and territories

Annual or perennial herbs or small shrubs; erect or climbing. Flowers small, in whorl-like clusters and arranged in long racemes or panicles; usually unisexual. Perianth segments 6, nor winged or toothed, the inner 3 membranous and inflated in fruit. Styles 3-fid, stigmas branched and fringed. Nut triquetrous.

Key to the species

1 Prostrate, diffuse or ascending and scrambling perennial herb with tubers. Leaves 3–5 cm long, hastate with spreading auricles. Flowers in a terminal panicle often large and much branched. Perianth valves orbicular, cordate, membranous, reticulate, often reddish, 4–7 mm long. Coast and adjacent plateaus; Cumberland Plain. Waste places. Introd. from S. Africa. *Turkey Rhubarb* or *Rambling Dock* . . . . ***A. sagittata** (Thunb.) L.A.S.Johnson & B.G.Briggs

1 Erect annual herb with fibrous roots. Leaves 5–10 cm long, triangular with a cordate to truncate base. Petiole longer than lamina on basal leaves. Flowers in clusters of 2 or 3 on a simple pedicel. Perianth valves 12–23 mm long; tinged pink to purple; emarginated at both ends. Occasionally naturalized. Introd. from N. Africa to Pakistan. *Bladder Dock* or *Rosy Dock*. . . . . ***A. vesicaria** (L.) A.Löve

3 Acetosella (Meisn.) Fourr.
1 species naturalized Aust.; Qld, NSW, Vic., Tas., S.A., WA

One species in the area
Erect dioecious herb 15–50 cm high. Flowers unisexual, very small, in slender leafless reddish panicles. Perianth valves entire, without tubercles. Leaves petiolate, hastate, with spreading auricles, acid in taste; venation obscure. Stipules silvery white. Widespread. Troublesome weed in gardens and pastures, spreading by rhizomes. Introd. from temperate regions of N. hemisphere. *Sheep Sorrel* . . . . ***A. vulgaris** Fourr.

4 Emex Neck. ex Campd.
2 species naturalized Aust.; all states and territories except Tas.

One species in the area
Glabrous, monoecious, prostrate annual. Leaves ovate, truncate or cordate at the base, mostly 3–6 cm long, on long petioles. Flowers unisexual, axillary, in whorl-like clusters. Female perianth enlarged and hardened at fruiting stage, 7–9 mm long, with a triangular tube and 6 lobes; outer lobes terminating in a rigid spreading spine; inner lobes smaller, ovate, closing over the fruit. Widespread. Weed of waste ground and pastures. Introd. from S. Africa. *Spiny Emex* or *Prickly Jacks* . . . . ***E. australis** Steinh.

5 Polygonum L.
5 species in Aust.; all states and territories

Prostrate to erect, annual or perennial herbs. Stipules deeply cut, silvery to white. Flowers small, usually bisexual, in axillary bracteate clusters up to 6-flowered, often forming panicles or sometimes solitary. Perianth segments usually 5, almost equal. Stamens 5–8. Styles 2–3, sometimes connate near the base. Nut enclosed in the persistent perianth.

Key to the species

1 Leaves linear-oblong, 3–15 mm long, only the midvein conspicuous underneath. Perianth c. 1 mm long with a very short tube. Stems usually less than 20 cm long. Widespread. Can be a weed in pastures and gardens. *Small Knotweed* . . . . **P. plebeium** R.Br.
1 Leaves oblong-lanceolate or elliptic, 1–4 cm long; lateral veins conspicuous underneath. Nut dull. Stems up to 1 m long. 2

2 Perianth tube at least half as long as the segments. Perianth 3–4 mm long. Leaves 5–20 mm long, ± uniform in size. Widespread. Waste land. Introd. from Europe. *Wire Weed* . . . *P. arenastrum Boreau
2 Perianth tube very short. Perianth 2–4 mm long. Leaves up to 50 mm long, those on the main stem much larger than those on the lateral stems. Widespread. Weed of cultivation and waste places. Fl. most of the year. Cosmop.. *Wire Weed* or *Hog Weed* . . . .*P. aviculare L.

6 Persicaria Mill.
15 species in Aust.; all states and territories

Herbs, annual or perennial, decumbent to erect. Stipules entire but breaking up with age, brownish. Flowers small, mostly bisexual, clustered, often forming spike-like inflorescences. Perianth segments usually 5. Stamens 5–8. Styles 2–3. Nut enclosed in the persistent perianth.

Key to the species

1 Flowers in dense hemispherical heads c. 1 cm diam.; perianth pink. Leaves ovate-lanceolate, 1–2 cm long with a dark brown zone across the upper surface, often tinged with red underneath. Trailing plant. Garden escape near habitation. Introd. from Himalayas. Fl. summer. . . . . *P. capitata (Buch.-Ham. ex D.Don) H.Gross
1 Flowers in spikes or spike-like racemes; if dense then more than 1 cm long 2

2 Stems prostrate or spreading. Perennial with appressed hairs. Perianth greenish or brown. Leaves lanceolate; stipular sheath sparsely hairy on the outside, ciliate on the margin. Flowers in spike-like axillary or terminal racemes 1–2 cm long. Widespread. In or near water. Fl. summer. *Creeping Knotweed* or *Trailing Knotweed* . . . .**P. prostrata** (R.Br.) Soják
2 Stems erect or ascending. Spikes often paniculate.. 3

3 Angles of branches and midveins of leaves ± lined with reversed bristles. Leaves ovate to lanceolate, bases hastate or sagittate, petiolate. Stems weak and straggling 4
3 Stems and leaves glabrous or ± softly hairy 5

4 Flowers arranged close together in ovate spikes on a branched peduncle. Leaves ovate, 3–7 cm long. Weak herb with stems up to 1.5 m long. Widespread. Wet places. Fl. summer. *Spotted Knotweed* . . . . **P. strigosa** (R.Br.) H.Gross
4 Flower clusters distant from each other and the spike thus interrupted. Leaves narrow-ovate, 3–8 cm long. Weak herb with stems up to 1.5 m long. Widespread. Margins of swamps, ponds, etc. Fl. summer . . . .**P. praetermissa** (Hook.f.) H.Hara

5 Stems and leaves glabrous or nearly so. 6
5 Stems and leaves ± hairy. 8

6 Stems stout, erect, over 1 m high. Leaves lanceolate or sometimes broader, 7–15 cm long, glandular-dotted underneath. Flowers in dense spikes 2–6 cm long forming a terminal panicle, the whole ± glandular-dotted. Perianth pink. Stipular sheaths ribbed, not ciliate. Annual herb. Widespread. In or near water. Fl. summer. *Pale Knotweed* . . . . **P. lapathifolia** (L.) Gray
6 Stems slender, erect or ascending up to 1 m high. Leaves lanceolate, acuminate, shortly petiolate. Stipules with long cilia on the margins 7

7 Perianth dotted with prominent glands, green whitish or reddish. Flowers rather distant or interrupted, in slender drooping spikes up to 10 cm long. Leaves 3–9 cm long. Weak herb up to 1 m high. Widespread.

Wet places. Possibly introd. from Europe. Fl. summer. *Water Pepper or Water Smartweed* . . . .
. . . . ?***P. hydropiper** (L.) Spach

7 Perianth without glands, pink. Flowers slightly crowded in rather slender spikes which are often paniculate and c. 5 cm long. Stems erect or procumbent, 30–50 cm high. Widespread. In or near water. Fl. summer. *Slender Knotweed* . . . . **P. decipiens** (R.Br.) K.L.Wilson

8 Stems with short glandular hairs. Leaves ovate to lanceolate, 3–15 cm long, sprinkled with glandular hairs. Perianth petaloid, glabrous, pink. Erect, branched herb up to 1 m high. Robertson. Damp places. Vulnerable. . . . . **P. elatior** (R.Br.) Soják

8 Stems with spreading or appressed simple hairs 9

9 Stipular sheaths pubescent on the outside; margin entire; all except the upper ones developing dilated, green collar like limbs; margins ciliate. Petiole more than 10 mm long. Leaves lanceolate to ovate, up to 13 cm long, softly pubescent especially underneath.. Perianth deep red to crimson. Spikes drooping, arranged in a terminal panicle. Erect, annual herb 1–2 m high. Garden escape. A variegated form is grown in gardens. Widespread. In or near water. Introd. from E. Asia. Fl. summer. *Prince's Feathers* . .
. . . . ***P. orientalis** (L.) Spach

9 Stipule sheath variously hairy; not developing a green limb. Margins with cilia 2–10 mm long 10

10 Erect, loosely branched perennial. Stems with appressed rather long hairs. Leaves lanceolate, up to 9 cm long, sprinkled with appressed hairs. Petiole c. 2 mm long. Stipular sheaths longer than the petioles, bordered with long cilia 4–10 mm long. Spikes 2–5 cm long. Perianth petaloid, white. Coast and Cumberland Plain. Wet places. Fl. summer . . . . **P. subsessilis** (R.Br.) K.L.Wilson

10 Erect herb to 60 cm high, ± glabrous. Leaves up to 8 cm long; antrorsely appressed hairs on main veins and margins; purple blotches often present in middle of upper surface. Petiole 0–7 mm long. Stipular sheath with cilia 2–3 mm long. Spikes 1–2.5 cm long, often interrupted at base. Perianth pink or paler. Weed of wasteland. Introd. from Europe. *Redshank.* . . . . ***P. maculosa** S.F.Gray

7 **Fallopia** Adans.
2 species naturalized Aust.; Qld, NSW, Vic., WA

One species in the area
Annual, twining herb. Perianth whitish. Leaves broad-sagittate, 2–5 cm long, acuminate, entire, with long petioles. Flowers in loose slender racemes. Fruit hard, 3-angled, 2–4 mm long. Widespread. Weed in gardens. Introd. from Europe. Fl. summer. *Black Bindweed* . . . . ***F. convolvulus** (L.) A.Löve

8 **Muehlenbeckia** Meisn.
14 species native Aust.; all states and territories

Glabrous climbers or shrubs. Flowers small, mostly unisexual, clustered within a sheathing bract and arranged in interrupted spike-like racemes or often forming a panicle. Perianth segments greenish or whitish, 5, enlarging with age. Stamens usually 8. Stylar branches 3. Nut triangular or nearly globular.

Key to the species

1 Slender, herbaceous, twining perennial. Leaves on slender petioles, ovate-cordate to broad-saggitate, acuminate, membranous; the margins crisped, mostly 2–5 cm long. Perianth succulent at fruiting stage. Care is necessary to avoid confusing this plant with *Fallopia convolvulus*. Coast. WSF; open forests on rivers; margins of RF. Fl. summer . . . . **M. gracillima** Meisn.

1 Robust climber, almost woody at the base, sometimes shrubby. Leaves broad-ovate-cordate, thick. Fruits ovoid, rugose. Flowers in large interrupted terminal panicles, 3–3.5 mm long. Blue Mts Creek banks, rocky places. . . . . **M. rhyticarya** F.Muell.

9 Fagopyrum Mill.

1 species naturalized Aust.; Qld, NSW, WA

One species in the area

Erect annual herb to 60 cm high. Stems tinged with red at maturity. Leaves hastate to cordate, 3–7 cm long, becoming sessile towards top of stem. Flowers in compact inflorescences; perianth tinged with pink, 2–3 mm long. Nut 5–6 mm long, smooth, triquetrous. Occasionally naturalized near areas of cultivation. Introd. from Asia. *Buckwheat.* . . . . **F. esculentum** Meonch

59 PORTULACACEAE

Herbs, usually fleshy, sometimes woody at the base. Leaves opposite or alternate, entire, thick; stipules, when present, scarious or split into hairs. Flowers bisexual, regular. Sepals 2. Petals 4–5, imbricate. Stamens 3–many. Ovary superior or half inferior, 1-locular, with basal or free central placentas; styles 3–6, free or basally connate. Fruit a capsule. c. 20 gen., warm temp., mostly America.

Key to the genera

1 Petals yellow; ovary half-inferior . . . . **1 PORTULACA**
1 Petals white, pink or purple; ovary superior 2

2 Petals connate at base . . . . **4 NEOPAXIA**
2 Petals free 3

3 Fruit red; roots tuberous. . . . . **3 TALINUM**
3 Fruit not red; roots not tuberous . . . . **2 CALANDRINIA**

1. Portulaca L.

20 species in Aust.; all states and territories

One species in the area

Prostrate, succulent annual. Stems often reddish or brownish. Leaves mostly alternate, oblong-cuneate to obovate, obtuse, 1–2 cm long, with or without minute stipular hairs. Flowers axillary, solitary or clustered. Sepals 5 mm long. Petals 4–6, yellow, scarcely exceeding the sepals. Stamens 8–15. Styles 4–6-lobed. Seeds black. Weed of cultivation. Cosmop. weed. *Purslane or Pigweed* . . . . ***P. oleracea** L.

2. Calandrinia Kunth

40 species in Aust.; all states and territories

Decumbent to ascending, annual or perennial, herbs. Leaves alternate or basal, usually thick and fleshy; stipules absent. Flowers in racemes. Sepals 2, persistent. Petals 4–5 sometimes more. Stamens various. Styles 3–4. Ovary 3–4-locular. Capsule 3-valved.

Key to the species

1 Sepals 7–10 mm long. Petals 5, 10–12 mm long. Leaves linear to narrow-obovate, up to 6 cm but usually c. 2 cm long. Flowers in a terminal leafy raceme up to 8 cm long. Decumbent or ascending, annual herb. Widespread. Introd. from N. America. Fl. spring . . . . ***C. menziesii** (Hook.) Torrey & A.Gray
1 Sepals up to 5 mm long 2

2 Fruiting pedicels reflexed or horizontal. Basal leaves terete to thickened-oblanceolate, up to 5 cm long. Petals 5, slightly exceeding the sepals. Taproot thin. Cumberland Plain; Blue Mts Open forests. Fl. spring . . . . **C. calyptrata** Hook.f.
2 Fruiting pedicels ascending. Basal leaves terete, linear to oblong, 15–30 mm long. Petals usually 4, slightly larger than the sepals. Taproot thick. Flowers in loose terminal racemes up to 8 cm long; bracts not leafy. Coast; Blue Mts Open forests. Ss. Fl. summer–autumn . . . . **C. pickeringii** A.Gray

3 **Talinum** Adans.

1 species naturalized Aust.; Qld, NSW

One species in the area

Erect, glabrous herb. Leaves alternate, fleshy, obovate to spathulate, 3–6 cm long, obtuse or acuminate at the apex, tapering to a petiole at the base; stipules absent. Flowers c. 4 mm diam., on filiform pedicels, in a loose cymose panicle 30–60 cm long. Petals pink. Fruit scarlet, globular, 3–4 mm diam. Garden escape near habitation. Introd. from S. America . . . .*T. paniculatum (Jacq.) Gaertn.

4 **Neopaxia** Ö.Nilsson

1 species in Aust.; NSW, Vic., Tas., S.A., WA

One species in the area

Stoloniferous herb with stems up to 30 cm long. Leaves alternate, linear to oblanceolate, up to 10 cm long, succulent, sometimes glaucous, with a membranous stem-clasping base. Petals 5, up to 10 mm long, white to pink. Stamens 5. Styles 3-fid. Wet places. Higher Blue Mts Damp places. Fl. summer . . . . .N. australasica (Hook.f.) Ö.Nilsson

60 OLACACEAE

15 gen., temp. to trop.

1 **Olax** L.

11 species endemic Aust.; all states and territories except Tas.

One species in the area

Erect, slender shrub 1–2 m high, yellowish-green, parasitic on the roots of neighbouring plants. Leaves narrow-oblong, mucronate, 5–20 mm long, distichous; stipules absent. Flowers solitary, axillary, regular, bisexual. Calyx cup-shaped, truncate, enlarged after flowering to enclose the fruit. Petals 5–6, yellowish, valvate, free or cohering, 4–6 mm long, spreading above. Functional stamens 3. Ovary superior, 1-locular. Fruit a drupe enclosed in the enlarged calyx but free from it. Widespread. Heath and DSF. Ss. Fl. most of the year . . . . .O. stricta R.Br.

61 LORANTHACEAE

Mistletoes

Shrubs, usually much-branched, epiphytic and parasitic on the branches of trees and shrubs or rarely (*Atkinsonia*) terrestrial and a root parasite. Branches usually brittle, often swollen at the nodes. Leaves opposite or rarely alternate, usually thick and leathery; stipules absent. Flowers regular, bisexual. Perianth segments usually 4–6, free or connate, valvate, petaloid, with a small fringe (the calyculus) below the perianth. Stamens as many as the perianth segments, opposite and adnate to them, epigynous. Ovary inferior, 1-locular. Fruit a 1-seeded berry; the inner layer of the floral tube forming a sticky pulp 65 gen., temp. to trop.

Key to the genera

1 Terrestrial shrub, parasitic on the roots of neighbouring trees or shrubs . . . . **1 ATKINSONIA**
1 Epiphytic parasites 2

2 Anthers versatile, dorsifixed; filaments attenuated at the summit. Flowers in a raceme of 3-flowered cymes, terminal on short lateral branches . . . . **6 MUELLERINA**
2 Anthers adnate; filaments not tapering at the summit but passing imperceptibly into the anthers . . . 3

3 Petals free or very shortly connate at the base. Flowers in an umbel of 2–3-flowered cymes or umbels. . . . . **2 AMYEMA**
3 Petals connate into a tube for the greater part of their length 4

4 Flowers in a simple raceme of 10–20 flowers . . . . **3 DENDROPHTHOE**

4 Flowers in a raceme or apparently an umbel of 3-flowered cymes or 2-flowered umbel 5

5 Flowers in a raceme or apparently an umbel of 3-flowered cymes. Corolla tube straight . . . . **4 AMYLOTHECA**

5 Flowers in 2-flowered umbel. Corolla tube curved . . . . **5 LYSIANA**

1 Atkinsonia F.Muell.
1 species endemic Aust.; NSW

Monotypic genus

Erect shrub 1–2 m high, parasitic on the roots of neighbouring trees or shrubs. Leaves opposite, lanceolate, 2–5 cm long, obtuse, narrowed into a petiole. Flowers on very short pedicels, in axillary racemes shorter than the leaves, sweetly scented. Bracteoles 2, close under the flower, with a third or subtending bract often a little lower than the pedicel. Petals usually 6, occasionally 7–8, free, linear, c. 6 mm long. Anthers dorsifixed, versatile. Fruit ovoid-oblong, scarlet. Blue Mts DSF. Ss. Fl. summer . . . . **A. ligustrina** (A.Cunn. ex F.Muell.) F.Muell.

2 Amyema Tiegh.
26 species in Aust. (32 endemic); all states and territories except Tas.

Epiphytic, parasitic shrubs. Leaves flat or terete, opposite alternate or spirally arranged. Flowers in axillary or lateral umbels of 2–3-flowered cymes or umbels. Petals free or very shortly connate at the base. Filaments compressed, passing imperceptibly into the adnate anthers.

Key to the species

1 Leaves terete, opposite, 3–10 cm long. Buds clavate, scurfy, 18–20 mm long. Hosts: *Casuarina* ssp. Hawkesbury and Nepean River districts. Fl. spring . . . . **A. cambagei** (Blakely) Danser

1 Leaves flat 2

2 Leaves 10–25 mm long (rarely up to 40 mm), linear-spathulate to linear-oblong. Branches rigid, erect or spreading. Buds c. 10 mm long. Fruit globular, 5 mm diam., red. Hosts: *Melaleuca* ssp. Widespread. Fl. summer . . . . **A. gaudichaudii** (DC.) Tiegh.

2 Leaves usually more than 50 mm long. Buds 20–30 mm long 3

3 Branches rigid, erect or spreading. Leaves elliptic-oblong to oblanceolate, rounded at the apex, 5–12 cm long, tapering at the base into a very short petiole. Fruit green. Many hosts but rarely on *Eucalyptus* ssp. Widespread. Fl. most of the year . . . . **A. congener** (Sieber ex Schult. & Schult.f.) Tiegh. ssp. **congener**

3 Branches pendulous 4

4 Leaves narrow-elliptic to broad-ovate, up to 12 cm long and 5 cm wide, greyish with a heavy close tomentum. Perianth segments green-grey, up to 2.5 mm long. Fruit ellipsoid to globular, 6–10 mm long, greyish. Hosts: *Acacia* spp. Hunter River Valley. Fl. winter–spring . . . . **A. quandang** (Lindl.) Tiegh. var. **quandang**

4 Leaves narrow-oblong to very narrow-ovate, often falcate, 7–24 cm long, green to yellow-green. Petioles often long and weak 5

5 All flowers in the cyme pedicellate. Perianth segments 5–7, red, up to 3 cm long. Fruit ovoid, up to 12 mm long. Hosts: chiefly *Eucalyptus* spp. Widespread. Fl. summer–autumn *Drooping Mistletoe*. . . . . **A. miquelii** (Lehm. ex Miq.) Tiegh.

5 Central flower of the ultimate cyme sessile. Perianth segments usually 5, red, up to 4 cm long. Fruit ovoid, up to 10 mm long. Hosts: chiefly *Eucalyptus* spp. Widespread. Fl. most of the year. *Drooping Mistletoe*. . . . . **A. pendulum** (Sieber ex Spreng.) Tiegh. ssp. **pendulum**

3 Dendrophthoe Mart.

6 species in Aust. (4 endemic); all states and territories except Tas. and SA

One species in the area

Epiphytic, parasitic shrub. Leaves mostly alternate, broad-ovate or narrow-ovate to narrow-lanceolate, obtuse, mostly 5–8 cm long, thick, ± glaucous, narrowed into a petiole. Flowers in short dense axillary racemes, all distinctly pedicellate. Perianth segments 5, orange-red, 3–4 cm long, connate for c. two-thirds of their length into a swollen tube; the lobes reflexed. Many hosts. Widespread. Fl. spring—summer . . . .**D. vitellina** (F.Muell.) Tiegh.

4 Amylotheca Tiegh.

2 species in Aust. (1 endemic); Qld, NSW

One species in the area

Epiphytic, parasitic shrub. Leaves mostly opposite, broad-obovate to orbicular or oblong-elliptic, obtuse, narrowed into a distinct petiole, up to 12 cm long; venation often strongly reticulate. Peduncles axillary, c. 1 cm long, each with 3–4 shorter 3-flowered branches. Perianth segments connate for c. three-quarters of their length into a scarcely dilated straight tube which is often split on the upper side. Parasitic on rainforest trees. Coast and adjacent plateaus. RF. Fl. summer . . . . . .**A. dictyophleba** (F.Muell.) Tiegh.

5 Lysiana Tiegh.

8 species endemic Aust.; all states and territories except Tas.

One species in the area

Epiphytic, parasitic shrub. Leaves opposite, linear, 1–3 mm wide, up to 14 cm long, narrowed into an obscure petiole; venation obscure. Flowers paired, axillary, almost sessile or with a very short peduncle (rarely 1–3-umbellate). Calyculus cylindrical. Petals 5, connate, 2.5–5 cm long, red, curved. Fruit ovoid, 6 mm diam., red or black. Usually on Casurinaceae species. Hunter River Valley Fl. summer. . . . . **L. exocarpi** ssp. **tenuis** (Blakely) Barlow

6 Muellerina Tiegh.

4 species endemic Aust.; Qld, NSW, Vic, SA

Epiphytic, parasitic shrub. Leaves opposite. Flowers in a raceme of 3-flowered cymes which is terminal on short lateral branches. Anthers versatile, dorsifixed; filaments attenuated at the summit. Fruits almost pear-shaped.

Key to the species

1 Leaves mostly 5–16 cm long, narrow- to broad-lanceolate or oblanceolate, 3-veined, light green on both surfaces. Branches drooping, 30–200 cm long. Fruit yellowish or green with a yellow apex. Hosts: chiefly *Eucalyptus* spp. Widespread. Fl. summer . . . . **M. eucalyptoides** (DC.) Barlow
1 Leaves 2–6 cm long, obovate to elliptic, penniveined, darker on the upper surface. Branches erect, divaricate, 30–100 cm long. Fruit pale pink or reddish on one side. Hosts: various but not *Eucalyptus* spp. Superficially rather like *Amyema congener*. Widespread. Fl. summer . . . . **M. celastroides** (Sieber ex Schult. & Schult.f.) Tiegh.

62 SANTALACEAE

Glabrous shrubs or small trees, usually parasitic on the branches and roots of other plants. Leaves entire, alternate or opposite, often reduced to scales; stipules absent. Flowers bisexual or unisexual, regular. Perianth segments 3–5, valvate, minute or petaloid. Stamens as many as the perianth segments and opposite them. Ovary inferior half-inferior or superior, 1-locular; ovule solitary, basal; style simple. Fruit a nut, drupaceous or a 1-seeded sticky berry,. 43 gen., temp. to trop.

Key to the genera

1 **Korthalsella** Tiegh.
6 species in Aust. (3 endemic); Qld, NSW, Vic., S.A.

One species in the area
Epiphytic parasite with opposite articulate green flattened branches 3–9 mm wide, and reduced leaves. Flowers minute, green or yellowish, clustered at the nodes, both sexes in the same cluster. Perianth lobes 3, persistent, crowning the very small globular berry. Hosts: *Doryphora sassafras* and other RF and WSF trees. Blue Mts; Illawarra. Fl. spring . . . . **K. rubra** (Tiegh.) Engl. ssp. **rubra**

2 **Viscum** L.
4 species in Aust. (2 endemic); Qld, NSW, Vic. , NT, WA

One species in the area
Epiphytic parasite with opposite angular green branches, and reduced leaves. Flowers minute in sessile clusters of 3–6 at the nodes, both sexes in the same cluster. Perianth segments usually 4, deciduous. Berry globular. Reported once from the area. **V. articulatum** Burm.f.

3 **Notothixos** Oliv.
4 species in Aust. (2 endemic); Qld, NSW, Vic.

Erect or pendent, epiphytic, parasitic, divaricate shrubs. Young parts of plant with dense tomentum. Inflorescence of 1 or more fan-like cymes with 3–13 flowers, each subtended by a pair of bracts. Flowers very small. Perianth segments 4, persistent. Inferior ovary barrel-shaped to cylindrical. Fruit with persistent perianth.

Key to the species

1 Leaf lamina up to 5 cm long. Bracts 0.5–0.8 mm long. Inflorescence with 1 terminal and 2 lateral fan-like cymes. Small shrub with a golden-yellow tomentum. Leaves elliptic or ovate to lanceolate; apex rounded, obtuse or acuminate; petiole 3–5 mm long. Berry globular, 5–6 mm diam. Hosts: epiphytic members of the Loranthaceae. Widespread. Fl. most of the year. . . . .**N. subaureus** Oliver

1 Leaf lamina 5–8 cm long. Bracts 1–1.5 mm long. Inflorescence of 4–12 pairs of fan-like cymes. Spreading to pendant shrub with white to tawny tomentum. Leaves elliptic to narrow-ovate, apex rounded, rarely acute; petiole 5–10 mm long. Berry ellipsoid 7 mm long. Host: members of the Sterculiaceae. RF and SF. Coast and ranges. Fl. spring–summer.. . . . . **N. cornifolius** Oliver

4 Exocarpos Labill.
10 species in Aust. (9 endemic); all states and territories

Tall shrubs or small trees, parasitic on the roots of other plants. Leaves alternate, reduced to minute scales or rudimentary. Branchlets angular-terete sometimes compressed, green. Flowers minute, bisexual or unisexual, in small axillary or terminal spikes or sessile clusters. Perianth divided to the base into 4–5 segments. Ovary superior. Fruit a globular nut resting upon the enlarged succulent pedicel.

Key to the species

1 Flowers in short cylindrical spikes 3–6 mm long including the peduncle. Perianth segments 5. Swollen pedicel bright red or yellowish at the fruiting stage, oblong-ovoid, 5–6 mm diam. Branchlets often drooping and graceful. Leaves scale-like triangular 0.5 mm long, (may be longer on new growth). Tall shrub or small tree. Widespread. Forests. Fl. spring–autumn. *Native Cherry* or *Cherry Ballart* . . . . **E. cupressiformis** Labill.

1 Flowers in small sessile axillary clusters. Perianth segments mostly 4. Swollen pedicel whitish at fruiting stage, 3–4 mm diam. Branchlets erect, prominently angular. Rudimentary leaves 1–3 mm long, present on young shoots only, subulate, caducous and leaving a minute triangular tooth-like base. Shrub. Widespread. Heath and DSF. Fl. most of the year. *Pale Ballart*. . . . . **E. strictus** R.Br.

5 Omphacomeria (Endl.) A.DC.
1 species endemic Aust.; NSW, Vic.

One species in the area

Wiry, dioecious shrub c. 1 m high, usually leafless, with terete striate branches. Flowers minute, unisexual, almost sessile; the males in axillary clusters of 3–5; females solitary. Perianth 4–5-lobed. Ovary inferior. Style very short; stigma 2-lobed. Fruit an ovoid drupe, 6–8 mm long, crowned by the persistent perianth lobes. Blue Mts Forests. Fl. spring–summer. *Leafless Sourbush*. . . . . **O. acerba** (R.Br.) A.DC.

6 Leptomeria R.Br.
17 species endemic Aust.; Qld, NSW, Vic., Tas., S.A., WA

One species in the area

Large shrub up to 3 m high with slender drooping green branches. Minute scale leaves falling off very early but leaving a decurrent notch at the node. Flowers bisexual, minute, yellowish-brown, in lateral spikes 15–20 mm long, each flower subtended by one small bract. Ovary inferior. Fruit drupaceous, 5 mm diam., greenish, succulent, very acid. Widespread. DSF and in more sheltered situations. Fl. summer. *Native Currant* . . . .**L. acida** R.Br.

7 Choretrum R.Br.
6 species endemic Aust.; Qld, NSW, Vic., S.A., WA

Erect shrubs. Leaves reduced to minute persistent scales. Branchlets green. Flowers bisexual, white, small, numerous, scattered singly along the branchlets apparently forming racemes or spikes of various lengths. Perianth 5-lobed, surrounded by a number of small bracts. Ovary inferior; style short. Fruit drupaceous, finally dry, crowned by the persistent perianth lobes.

Hybridization may occur between these species.

Key to the species

1 Branches terete smooth or slightly striate. Stamens with hair-like appendages. Flowering branches 1 mm or more wide, scarcely angled beneath the caducous scale leaves, rigid. Flowers arranged towards the ends of branches, solitary in a scale axil, almost sessile. Decumbent or ascending shrub less than 1 m high. DSF. Blue Mts Fl. spring–summer. *Dwarf Sourbush.* . . . . **C. pauciflorum** DC.

1 Branches angled, striate; stamens without hair-like appendages; Flowers very shortly pedicellate along the upper ends of the branchlets, each flower solitary within the axil of a minute persistent decurrent bract. 2

2 Flowering branchlets numerous, erect, wiry, c. 0.5 mm wide, acutely angled by the short lines decurrent from the caducous scale leaves. Leaves persisting on young branches Fruit dark when mature. Erect shrub up to 5 m high. Widespread. Heath and DSF. Fl. spring–summer . . . . . . **C. candollei** F.Muell.

2 Flowering branchlets strongly angular, broadly ridged below the caducous scale leaves. Leaves present only on young stems and growing tips. Fruit pale when mature. Shrub or small tree to 5 m high. WSF and DSF. Blue Mts Fl. summer. . . . . **C. sp. A** (sensu Harden 1992)

8 Santalum L.
6 species in Aust. (5 endemic); all states and territories except Tas.

One species in the area
Erect, glabrous shrub 1–2 m high. Leaves mostly opposite, 2–5 cm long, oblong-linear to lanceolate, obtuse, dark green on the upper surface, paler underneath; margins sometimes revolute. Flowers in short axillary cymes or racemes. Perianth c. 4 mm long; lobes triangular, shorter than the tube. Stamens inserted at the base of the perianth lobes. Ovary half-inferior. Fruit drupaceous, 6–8 mm diam., with a circular scar near the summit. Woronora Plateau; Blue Mts Fl. spring–summer. . . **S. obtusifolium** R.Br.

9 Thesium L.
1 species endemic Aust.; Qld, NSW, Vic., Tas.

One species in the area
Perennial herb c. 30 cm high, with slender wiry stems. Leaves alternate, linear, often more than 25 mm long. Flowers greenish yellow, solitary, on very short peduncles, adnate to the base of the subtending leaf. Perianth c. 2 mm long, cylindrical, 5-lobed; the lobes as long as the tube. Fruit an ovoid nut 2–3 mm long. Widespread. Blue Mts Vulnerable. Fl. spring–summer. *Toadflax* . . . . **T. australe** R.Br.

63 APHANOPETALACEAE
1 gen., Aust.

1 Aphanopetalum Endl.
2 species native Aust., Qld, NSW, Vic., WA

One species in the area
Straggling climber, glabrous. Leaves simple, glossy, thinly coriaceous, ovate- or elliptic-lanceolate to lanceolate, 4–10 cm long, obtuse or shortly acuminate, obtusely serrate. Flowers few in short cymes or sometimes a loose panicle. Sepals 4, soon enlarging, oblong to very narrow-ovate, c. 12 mm long. Petals minute or absent. Ovary 4-locular; stylar branches 4, ± united. Fruit a nut surrounded by the persistent sepals. Coast; Hunter River Valley. RF and gullies. Fl. spring . . . . **A. resinosum** Endl.

64 CRASSULACEAE
Perennial herbs. Leaves succulent. Flowers bisexual, regular, in terminal panicles or cymes. Sepals 3–6, usually free, sometimes connate into a tube. Petals 3–6, free or connate, often persistent. Stamens as many and alternating with or twice as many as sepals, hypogynous or ± perigynous. Carpels free, 3–6, often with a basal gland; placentas marginal. Fruit a follicle. c. 30 gen., cosmop.

Key to the genera

1 Stamens as many as petals. . . . . **1 CRASSULA**
1 Stamens twice as many as petals 2

2 Leaves alternate. . . . .**2 SEDUM**
2 Leaves opposite . . . .**3 BRYOPHYLLUM**

1 Crassula L.

14 species in Aust. (8 endemic); all states and territories

Small herbs often turning red. Leaves opposite, succulent, cylindrical, subcylindrical or globular; stipules absent. Sepals 4–5. Petals 4–5, white or reddish. Stamens 4–5, hypogynous or perigynous, alternating with petals. Carpels free or basally connate, equal and opposite to petals.

Key to the species

1 Leaves mostly 10–50 mm long. Inflorescences leafless. 2
1 Leaves less than 10 mm long. Inflorescences with leaf-like bracts 4

2 Leaves less than 10 mm wide, 10–30 mm long; linear, triangular or terete and curving upwards. Petals 5 cream to white, lobes 2–3 mm long. Erect perennial herb, up to 50 cm high. Naturalized near urban development. Introd. from S. Africa. Fl. summer . . . . . ***C. tetragona** ssp. **robusta** (Toelken) Toelken
2 Leaves more than 10 mm wide, ovate to broad elliptic. 3

3 Leaves obovate to broad-elliptic, with entire, often yellowish margins. . Petiole 5–25 mm long. Flowers in a terminal compound thyrse. Petals 4, reddish cream, 3–4 mm long. Ascending to decumbent, succulent, perennial herb up to 40 cm high. Naturalized near Sydney. Introd. from S. Africa. Fl. summer . . . . ***C. multicava** Lem.
3 Leaves ovate with toothed, often reddish margin. .Petiole to 3 mm long. Flowers in a terminal thyrse. Petals 5 , white or tinged with red, star shaped, lobes 4–8 mm long. Prostrate perennial herb, stems up to 80 cm long. Naturalized. Not common. DSF. Introd. from S. Africa . . . .
. . . . ***C. sarmentosa** Harv. var. **sarmentosa**

4 Flowers solitary, axillary, on pedicels rather shorter than the leaves. 5
4 Flowers in axillary clusters, the whole often resembling a spike; or paniculate. 6

5 Each carpel acuminate and subtended by a basal scale. Style c. half as long as ovary. Sepals and petals 1–2 mm long. Stems weak, floating and elongated when in water. Leaves linear to very narrow-ovate, mostly 2–6 mm long. Widespread. In ditches and on mud. Fl. winter–spring . . . .
. . . . **C. helmsii** (Kirk) Cockayne
5 Carpels obtuse, not subtended by a scale. Style c. one quarter the length of the ovary. Otherwise similar to *C. helmsii.* Cumberland Plain. Wet places, often on mud. Fl. winter–spring . . . .
. . . .**C. peduncularis** (Sm.) F.Meigen

6 Flowers in axillary clusters mixed with smaller leaves, the whole often forming a spike-like inflorescence. Sepals and petals c. 2 mm long. Ovary with 2 ovules. Succulent annual, flowering when very small but usually growing to a branching tuft up to 15 cm diam. Widespread in many situations . . . .
. . . .**C. sieberiana** (Schult. & Schult.f.) Druce
6 Flowers in a spreading, axillary, corymbose panicle. Sepals and petals c. 3 mm long. Ovary with 4 or more ovules. Erect, much branched annual up to 7 cm high. Widespread. Damp ground or on rocks. Fl. winter–spring. . . . .**C. decumbens** Thunb. var. **decumbens**

2 Sedum L.

Stonecrops

3 species naturalized Aust.; NSW, Vic., Tas.

Glabrous, succulent perennials. Leaves alternate. Flowers usually 5-merous, in small spreading cymes. Petals free or fused basally, yellow. Stamens twice as many as petals, hypogynous, shorter than petals. Follicles acute to acuminate.

Key to the species

1 Leaves less than 5 mm long, overlapping, ± gibbous at the base. Petals 6–8 mm long. Blue Mts, few records from Coast and adjacent plateaus. Rocky ground. Introd. from Europe. Fl. summer. ***S. acre** L.
1 Leaves more than 25 mm long, petiolate, not overlapping. Petals 6–8 mm long. Occasional along roadsides and in waste places. Coast and adjacent plateaus. Introd. from trop. America. Fl. winter–summer . . . . ***S. praealtum** A.DC.

3 Bryophyllum Salisb.

4 species naturalized Aust.; Qld, NSW

Erect, perennial, succulent herbs. Leaves opposite or whorled, often forming plantlets along the margin. Flowers pendent in terminal cymes, 4-merous. Sepals connate, shorter than petals. Petals connate. Stamens 8, epipetalous towards the base. Carpels with a basal scale.

Key to the species

1 Leaves ± cylindrical with a few acute teeth or lobes towards the apex, glaucous with purplish markings. Coast and adjacent plateaus. Cumberland Plain. Naturalized on rocky sites near habitation and along roadsides. Introd. from Madagascar. *Mother of Millions* . . . . . ***B. delagoense** (Eckl. & Zeyh.) Schinz
1 Leaves flat; lower ones simple; upper ones pinnate, green, sometimes with a reddish tinting. Coast and adjacent plateaus. Cumberland Plain. Roadsides and rubbish dumps. Introd. from S. Africa. *Resurrection Plant* . . . . ***B. pinnatum** (Lam.) Oken

65 HALORAGACEAE

Herbs or small shrubs, sometimes aquatic. Leaves simple or compound, opposite whorled or alternate; stipules absent. Flowers small, bisexual or unisexual. Floral tube cup-shaped, adnate to and sometimes continued above the ovary. Calyx 2–4-lobed or absent. Petals 2–4. Stamens 4–8. Ovary inferior, 1–4-locular; styles as many as the loculi, with papillose or plumose stigmas. Fruit small, indehiscent, 1–4-locular or a schizocarp separating into four 1-seeded mericarps. 8 gen., mainly southern hemisphere

Key to the genera

1 Aquatic plants or growing on mud. Petals absent in female flowers . . . .**4 MYRIOPHYLLUM**
1 Terrestrial plants, occasionally on mud. Petals present in female flowers. 2

2 Petals flat or almost so, twisted in the bud . . . .**3 HALORAGODENDRON**
2 Petals curved and hooded around the stamens, erect in the bud. 3

3 Flowers usually several in a dichasium in the axil of each bract. Fruit 2–4-locular, woody. . . . .
. . . .**1 HALORAGIS**
3 Flowers solitary in the axil of each bract. Fruit 1-locular, membranous. . . . .**2 GONOCARPUS**

1 Haloragis J.R.Forst. & G.Forst.

22 species in Aust.; all states and territories

Monoecious herbs, often scabrous. Leaves opposite or alternate, the upper ones gradually reduced to bracts. Flowers greenish or brownish, on short pedicels in a terminal raceme or panicle. Calyx lobes usually 4, persistent, short. Petals usually 4, small or often absent. Stamens usually 8. Ovary 4- (rarely 2-) locular. Fruit indehiscent.

Key to the species

1 Leaves palmately multifid or 3-fid, 2–5 cm long. Hairs on the stem scabrous, hooked at the tip. Erect perennial herb up to 60 cm high. Widespread. Damp places . . . . **H. heterophylla** Brongn.
1 Leaves simple, serrate to dentate or pinnately lobed 2

2 Leaves below the inflorescence mostly alternate towards the base of the plant, entire to dentate, lanceolate to ovate, 2–5 cm long. Hairs on the stem scabrous, hooked at the tip. Erect perennial up to 60 cm high. Hunter River Valley. Heavy soils prone to flooding. . . . . **H. aspera** Lindl.
2 Leaves below the inflorescence mostly opposite towards the base of the plant 3

3 Styles 2. Leaves sessile, lanceolate, 3–6 cm long, serrate, glabrous or with hairs on the margins and midrib. Bracts alternate. Erect perennial, sometimes shrubby at the base, up to 60 cm high. Blue Mts, e.g. Jenolan Caves. Open forests and grasslands . . . .**H. serra** Brongn.
3 Styles 4. Leaves sessile, lanceolate to oblong, 4–10 cm long, serrate. Shrubs up to 1.5 m high. Damp places. 4

4 Stems, leaves and ovary finely scabrous. Nepean River area. Vulnerable . . . . **H. exalata** F.Muell. var. **exalata**
4 Stems, leaves and ovary glabrous. Shoalhaven River area. Vulnerable. . . . .**H. exalata** var. **laevis** (Schindl.) Orchard

2 **Gonocarpus** Thunb.
36 species in Aust.; all states and territories

Monoecious herbs or small shrubs, often scabrous. Leaves opposite or alternate; the upper ones gradually reduced to bracts. Flowers small, greenish or brownish, in racemes. Calyx lobes usually 3–4, persistent. Petals 2–4. Stamens 4–8. Ovary 1-locular. Styles 3–4. Fruit indehiscent.

Key to the species

1 Bracts opposite, at least in the lower parts of the inflorescence. 2
1 Bracts all alternate 3

2 Leaves linear, almost terete, 6–15 mm long, fleshy, glabrous or with a few hairs. Stems slender, wiry, erect, up to 35 cm high with drooping tips. All bracts opposite. Widespread. Wet ground . . . . **G. salsoloides** Rchb. ex Spreng.
2 Leaves lanceolate to orbicular, serrulate, scabrous, 5–20 mm long. Upper bracts alternate. Stems erect, up to 50 cm high. Widespread. Open forest and heath . . . . **G. teucrioides** DC.

3 Stamens 4, often with 4 staminodes. Leaves petiolate, ovate, 1–2 cm long, serrate, pilose. Prostrate or decumbent herb. Mt. Wilson, Bent's Basin. Damp places . . . .**G. humilis** Orchard
3 Stamens 8 4

4 Leaves all alternate (occasionally the very lowest ones opposite), lanceolate, 1–3 cm long, pilose, denticulate, with recurved margins. Erect shrub up to 60 cm high. Rylstone. Open communities. . . . . **G. elatus** (A.Cunn. ex Fenzl) Orchard
4 Leaves mostly opposite below the inflorescence, occasionally in whorls of 3 5

5 Fruit and ovary covered with glistening papillae, 8-ribbed but not rugose. Petals glabrous. Leaves glabrous or slightly scabrous, serrulate, up to 2 cm long and 5 mm wide. Diffuse, decumbent herb with often ridged white stems up to 70 cm long. Hawkesbury River district. Wet places . . . . **G. chinensis** (Lour.) Orchard ssp. **verrucosus** (Maiden & Betche) Orchard
5 Fruit and ovary usually ribbed, ± rugose, often pubescent but never with glistening papillae 6

6 Leaves glabrous, often cordate 7

6 Leaves scabrous or pubescent **8**

7 Stems prostrate, rooting at the nodes but the scape often ascending. Inflorescence unbranched or rarely with secondary racemes. Widespread. Wet places . . . . **G. micranthus** Thunb. ssp. **micranthus**
7 Stems erect, up to 60 cm high. Inflorescence branched to 3rd or 4th degree. Widespread. Damp places . . . . **G. micranthus** ssp. **ramosissimus** Orchard

8 Herbs up to 35 cm high; stems not distinctly woody. Leaves lanceolate to narrow-elliptic, 6–20 mm long, serrate, scabrous, shortly petiolate. Widespread. Heath and grasslands. Fl. spring . . . .
. . . . **G. tetragynus** Labill.
8 Shrubs or undershrubs with a distinctly woody stem, usually more than 50 cm high. **9**

9 Leaves ovate to oblong, 1–4 cm long, densely softly pubescent, petiolate, serrate. Camden district. WSF and RF. . . . . **G. oreophilus** Orchard
9 Leaves linear-oblong, 1–3 cm long, pubescent with soft hairs or sometimes almost glabrous, serrate, shortly petiolate. Cumberland Plain and Woronora Plateau. Ss . . . . **G. longifolius** (Schindl.) Orchard

3 Haloragodendron Orchard
5 species endemic Aust.; NSW, Vic

Glabrous shrub up to 1 m high. Stems 4-angular. Leaves narrow-elliptic to oblanceolate, opposite, 25–40 mm long, dentate, coriaceous, sessile or nearly so. Flowers in terminal dichasia spikes or thyrses with leaf-like bracts. Petals 4, almost flat, twisted in the bud, 10–14 mm long. Stamens 8. Ovary glabrous; styles 4.

Key to the species

1 Fruit distinctly ribbed or winged. Between Hornsby and Gordon. Endangered . . . .
. . . . **H. lucasii** (Maiden & Betche) Orchard
1 Fruit not distinctly winged or ribbed. Uncommon. Blue Mountains . . . .
. . . .**H. gibsonii** Peter G.Wilson & M.L.Moody

4 Myriophyllum L.
Water Milfoil
36 species in Aust. (31 endemic, 5 naturalized); all states and territories

Monoecious or dioecious, glabrous, aquatic herbs or growing on mud or wet ground. Leaves alternate opposite or whorled; the submerged leaves usually finely pinnatisect. Flowers small, in the axils of the upper leaves; the upper ones usually male and the lower female. Male flowers with 4 calyx lobes and petals; stamens 2–8. Female flowers with minute or obsolete calyx; petals absent; ovary 2–4-locular; styles 2–4, short, stigmatic from the base, often plumose. Fruit a schizocarp, separating into 2–4 mericarps.

Key to the species

1 All leaves alternate. Emergent leaves linear to lanceolate, sometimes pinnatisect, usually less than 10 mm long. Flowers solitary, ± sessile. Male petals cream, c. 2 mm long **2**
1 At least submerged leaves whorled or opposite **3**

2 Leaves pinnately divided into usually 5 linear segments. Mericarps sparsely papillose. Coast and adjacent plateaus, Cumberland Plain. Shallow water or on mud . . . . . . **M. gracile** Benth. var. **gracile**
2 Leaves entire or shortly toothed. Mericarps densely papillose. Widespread. Usually in swamps but sometimes in still water . . . . **M. gracile** var. **lineare** Orchard

3 Leaves opposite, linear to filiform with terete segments, entire or 3-fid, 5–20 mm long. Mericarps smooth or rugose. Swampy areas in heath . . . .
. . . . **M. pedunculatum** Hook.f. ssp. **longibracteolatum** (Schindl.) Orchard
3 At least submerged leaves in whorls of 3–8 . . . . **4**

4 Emerged leaves entire or dentate **5**

4 Emerged leaves pinnately lobed, narrow-elliptic to ovate in outline, in whorls of 4–8 11

5 Emerged leaves thick, flat or compressed, lanceolate to narrow-elliptic or ovate to oblong. 6
5 Emerged leaves terete, entire or pinnatisect. 9

6 Emerged leaves lanceolate to narrow-elliptic, 15–45 mm long, dentate; submerged leaves finely dissected. Coast and adjacent plateaus; Cumberland Plain. Still water in ponds and lakes . **M. latifolium** F.Muell.
6 Emerged leaves ovate to oblong, entire or with a few short teeth towards the apex 7

7 Mericarps cylindrical, densely spinose, olive brown, c. 1.5 mm long. Emerged leaves glaucous, often red at tip, entire, 6–10 mm long, up to 3 mm wide. Widespread. Lakes and slow moving streams. *Cat-tail*. . . . . **M. caput-medusae** Orchard
7 Mericarps oblong to ovoid, smooth or rugose 8

8 Mericarps smooth or minutely papillose, yellowish-brown, usually 2–2.5 mm long. Emergent leaves ovate, 5–8 mm long, up to 4 mm wide, entire or with a few short teeth towards the apex. Blue Mts Lakes and slow moving streams . . . . **M. salsugineum** Orchard
8 Mericarps rugose, yellowish red or grey, c. 1 mm long. Emergent leaves ovate, 3–9 mm long, entire or toothed . . . . **M. verrucosum** (see below)

9 Emerged leaves mostly in whorls of 5–8, entire or pinnatifid with 6–8 segments, 5–20 mm long. Stems often with short curled hairs. Mericarps cylindrical, up to 1 mm long, yellowish-brown to red, papillose. Coast. Still water or on mud . . . . **M. crispatum** Orchard
9 Emerged leaves mostly in whorls of 3–6. Stems glabrous 10

10 Mericarps cylindrical, c. 1.5 mm long, yellowish-brown. Emerged leaves strictly whorled, usually in 5s, 8–15 mm long, mostly entire. Stigmas white. Widespread. Still or slow moving water . . . . **M. variifolium** Hook.f.
10 Mericarps ovoid, c. 1 mm long, reddish-purple. Emerged leaves often irregularly whorled, usually in 3s or 4s, 5–15 mm long, mostly entire. Stigmas red or pink. Coast and Cumberland Plain. Still water or on mud . . . . **M. simulans** Orchard

11 At least the upper emerged leaves not lobed to the midrib; the lobes not narrow-linear, 3–9 mm long. Bracteoles entire or denticulate. Mericarps rugose. Woronora Plateau and Blue Mts In water or on mud. . . . . **M. verrucosum** Lindl.
11 Emerged leaves all with fine linear lobes to the midrib, narrow-elliptic in outline, up to 25 mm long; submerged leaves more finely dissected. Bracteoles linear, divided. Mericarps not produced in Aust. Coast and Cumberland Plain. Still water. Introd. from S. America. *Possum Tail* or *Parrot's Feather*. **M. aquaticum** (Vell.) Verdc.

66 VITACEAE

Woody or herbaceous climbers with leaf-opposed tendrils; branches often articulate. Leaves alternate, simple or compound, usually articulate with the stem; stipules membranous. Flowers regular, bisexual, in leaf-opposed or sometimes apparently axillary cymo-panicles. Calyx small, entire or obscurely 4–5-lobed, persistent. Petals 4–5, valvate. Stamens 4–5, opposite the petals, inserted on the outside of a thick disc. Ovary 2–6-locular; placentas axile; style simple; stigma sometimes lobed. Fruit a berry. 12 gen., trop. to temp.

Key to the genera

1 Leaves pedately 5-foliolate (i.e. the leaves divided into 3, the 2 lateral leaflets divided into 2). Weak climber with herbaceous branches. . . . . **4 CAYRATIA**
1 Leaves simple or digitately 3–5-foliolate. 2

2 Stigma large, 4-lobed. Petal dorsally horned below the apex. Tall woody climbers . **2 TETRASTIGMA**
2 Stigma minute, not lobed 3

3 Tendrils mostly terminated by a dilated adhesion pad. Weak climber. . . . . . . **3 PARTHENOCISSUS**

3 Tendrils not dilated at the tip. Tall woody climbers . . . . **1 CISSUS**

1 Cissus L.
14 species in Aust. (7 endemic); Qld, NSW, Vic., NT

Tall or scrambling, woody climbers, Leaves simple or digitately 5-foliolate. Petals 4. Stamens 4. Disc obscurely 4-lobed. Style thin, simple. Fruit a berry.

Key to the species

1 Leaves simple, ovate to nearly cordate, denticulate or almost entire, 7–10 cm long, pubescent underneath, with glands in the axils of some of the main veins. Young shoots and inflorescence rusty-tomentose. Cymes dense, broadly corymbose, short. Berry globular, black, c. 1 cm diam. Tall, woody climber. Widespread. RF and in sheltered places. Fl. spring–autumn . . . . **C. antarctica** Vent.

1 Leaves compound. Leaflets 5, digitate, glabrous when mature, shortly acuminate, coriaceous. 2

2 Leaflets obtuse at the base, ovate to oblong-elliptic or very narrow-ovate, 5–8 cm long, entire or slightly toothed towards the apex, pale or glaucous underneath, finely reticulate; petiolules distinct, often more than 25 mm long. Cymes leaf-opposed. Flowers yellowish. Berry globular, bluish-black, 1–2 cm diam. Tall climber with rusty-tomentose young shoots. Widespread. RF and sheltered places. Fl. spring–summer . . . . **C. hypoglauca** A.Gray

2 Leaflets acute at the base. 3

3 Broadest leaflet of each leaf less than 2 cm wide, narrow-elliptic, often sinuate or dentate, ± paler on the lower surface; the reticulation not prominent. Cymes leaf-opposed. Scrambling climber with large tuber. Razorback Mt. Fl. summer . . . . **C. opaca** F.Muell.

3 Broadest leaflet of each leaf more than 3 cm wide, elliptic-oblong to slightly obovate, entire or dentate; the reticulation not prominent. Cymes apparently axillary. Berry ovoid. Tall climber. Coast and adjacent plateaus, e.g. Minnamurra Falls. RF. Fl. spring–summer . **C. sterculiifolia** (F.Muell. ex Benth.) Planch.

2 Tetrastigma (Miq.) Planch.
4 or 5 species in Aust. (4 endemic); Qld, NSW

One species in the area

Tall woody climber. Leaves palmately 3-foliolate; the leaflets ovate-elliptic, c. 10 cm long, dentate, acuminate, coriaceous. Flowers polygamo-dioecious, in apparently axillary inflorescences. Petals 4, horned. Style very short, surmounted by the swollen 4-lobed stigma. Fruit a berry. Coast N. of Gosford. RF. Fl. summer . . . . **T. nitens** (F.Muell.) Planch.

3 Parthenocissus Planch.
1 species naturalized Aust.;

One species in the area

Weak climber with branched leaf-opposed tendrils, most of the tendril branches terminating in a dilated adhesion pad. Leaves palmately 5-foliolate, deciduous; leaflets ovate-elliptic, c. 7mm long, serrate, acute or acuminate. Inflorescence terminal or distinctly leaf-opposed. Petals 4–5, not horned. Style short, thick, simple. Fruit a blackish berry. Garden escape near habitation. Introd. from N. America. Fl. spring–summer. *Virginia Creeper* . . . . ***P. quinquefolia** (L.) Planch.

4 Cayratia Juss.
8 species in Aust. (5 endemic); Qld, NSW

One species in the area

Weak climber with angular or striate branches. Leaves compound; leaflets 5, pedate, ovate, coarsely toothed or lobed, thin and glabrous or thick and pubescent, 3–6 cm long. Cymes rather dense, on long peduncles. Petals usually 4, green. Disc very prominent, entire. Style filiform. Berry depressed-globular,

5–6 mm diam. Widespread. RF margins and sheltered but often sunny places. Fl. summer, dying back in winter . . . .**C. clematidea** (F.Muell.) Domin

67 GERANIACEAE

Herbs, sometimes becoming woody and shrub-like. Leaves opposite or alternate, petiolate, entire or much dissected; stipules present. Flowers bisexual, regular or slightly irregular. Sepals 5, imbricate. Petals 5, imbricate. Stamens 10, in 2 rows; the outer row sometimes sterile. Ovary superior, 5-locular, 5-lobed, produced upwards into a beak consisting of 5 connate styles; stigmas 5, free. Fruit a schizocarp consisting of 5 mericarps which separate basally; the awns adhering to the beak of the ovary at the top or becoming free, coiled or spiraled. 7 gen., temp., especially Africa.

FIGURE 35

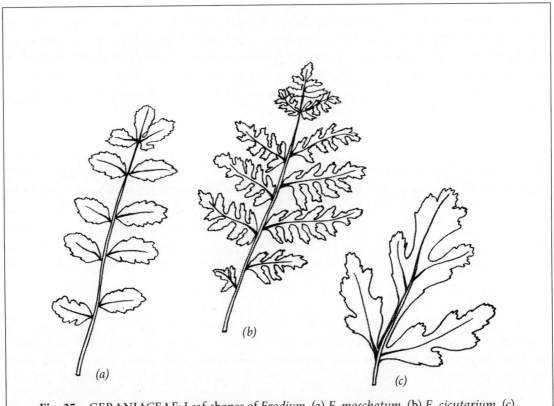

Fig. 35　GERANIACEAE: Leaf-shapes of *Erodium*, (a) *E. moschatum*, (b) *E. cicutarium*, (c) *E. crinitum*.

Key to the genera

1 Awns of the fruiting carpels glabrous on the inside surface. Fertile stamens 10 . . . .**1 GERANIUM**
1 Awns of the fruiting carpel hairy on the inside surface. Fertile stamens less than 10 2

2 Fertile stamens 5. Calyx without a spur . . . .**2 ERODIUM**
2 Fertile stamens usually more than 5. Calyx with a spur adnate to the pedicel (almost obsolete in *P. inodorum*) . . . .**3 PELARGONIUM**

1 Geranium L.

c. 35 species in Aust.(native and naturalized); all states and territories

Herbs. Leaves alternate, palmately divided or lobed, 1–4 cm wide. Flowers regular, up to 3 on axillary peduncles. Sepals hairy. Petals purple to pink or white. Stamens 10, fertile. Mericarp awn glabrous on inner surface, coiling upwards and ejecting the seed but remaining attached to the top of the beak or breaking off at the base and retaining the seed.

Key to the species

1 Flowers usually in pairs 2
1 Flowers solitary. 5

2 Mericarps wrinkled, glabrous or hirsute 3
2 Mericarps smooth, hirsute. Fruiting pedicels with spreading hairs 4

3 Leaves orbicular to reniform; 5–9 lobed the lobes broad and ± 3-fid. Petals 4–5 mm long, emarginate, mauve to pink, scarcely longer than the sepals. Tap root slender. Diffuse, softly villous annual. Widespread. Waste places and pastures. Introd. from Europe. Fl. spring-summer . . . .
. . . . *G. molle L. var. molle
3 Leaves ovate deeply divided with 3–9 pinnatisect primary lobes. Petals 9–12 mm long, obovate to spathulate, deep pink, longer than sepals. Tap root slender. Erect or decumbent herb to 40 cm high. Escape from cultivation. Higher Blue Mts Introd. from Europe. Fl. spring–summer. *Herb Robert* . . . .
. . . . *G. robertianum L.

4 Seeds black, covered with isobilateral areolae. Sepals ovate, hirsute, with no closely appressed hairs beneath the longer hairs. Tap root ± globular (napiform). Diffuse, hirsute perennial. Widespread. Grasslands and open forests. Fl. spring-summer . . . . .G. solanderi Carolin var. solanderi
4 Seeds dark brown, with ± elongated areolae. Sepals lanceolate to elliptic, with very short closely appressed hairs beneath the longer hairs. Leaves often purplish underneath. Tap root swollen, elongated (fusiform). Diffuse perennial, less hirsute than *G. solanderi*. Widespread. Margins of RF., WSF., creek banks. Fl. spring–summer. . . . . G. homeanum Turcz.

5 Petals white. Sepals ± reflexed. Seeds black, with elongated areolae. Diffuse perennial with minute soft spreading hairs. Tap root thick. Higher Blue Mts Open forest. Fl. spring–summer . . . .
. . . . G. graniticola Carolin
5 Petals pink. Leaves frequently purplish on undersurface 6

6 Hairs spreading. . . . . G. solanderi (see above)
6 Hairs ± appressed 7

7 Flowers 25–35 mm diam. Beak 15 mm long. Seeds brown, with minute ± isobilateral areolae. Tap root fleshy, ± branched. Weak, ascending to diffuse, perennial herb sparsely covered with appressed hairs. Higher Blue Mts Swamps and creek banks. Fl. spring—summer. . . . . G. neglectum Carolin
7 Flowers up to 15 mm diam. Beak 9–10 mm long. Diffuse perennial covered with appressed hairs. Tap root thick and branched 8

8 Seeds brown with shallow areolae. Widespread. Open forests, usually in damper places but not often in swamps. Fl. spring-summer. . . . . .G. potentilloides L'Hér. ex DC. var. potentilloides
8 Seed black with deep alreolae. Woodland or grassland at higher altitudes. Fl. summer. . . . .
. . . . G. potentilloides var. abditum Carolin

2 Erodium L'Hér. ex Aiton
9 species in Aust. (3 endemic, 6 naturalized); all states and territories

Annual or biennial herbs. Leaves deeply dissected, lobed or compound; cauline leaves opposite (Fig. 35). Flowers regular, 1–10 in axillary umbel-like cymes. Sepals hairy. Petals blue to purple or pink to white. Stamens 5, alternating with 5 staminodes. Mericarp hairy; awns hairy on the inner surface, becoming spirally twisted as they separate from the beak. Seed retained in the mericarp.

Key to the species

1 Leaves palmatifid or trifoliolate; the lobes incised and toothed. Petals blue. Beak 4–7cm long. Mericarp with a glabrous pit underlined by 1 groove on each side of the base of the awn. Cumberland Plain and Blue Mts Grasslands and woodlands. Fl. summer. *Crowfoot* . . . .**E. crinitum** Carolin

1 Leaves pinnate or deeply pinnatisect. Petals purple or pink to white 2

2 Leaflets deeply pinnately lobed (often cut to the main vein). Beak 2–4 cm long in fruit. Mericarp with a single pit on each side of the base of the awn. Widespread. Waste places, pastures and gardens. Fl. spring–summer. *Common Crowfoot* . . . . **E. cicutarium** Carolin

2 Leaflets toothed or lobed but not dissected more than half way to the main vein. Beak 3–4 cm long. Mericarp with a pit underlined by 1–2 grooves on each side of the base of the awn; the pit with some large glandular hairs. Widespread. Waste places and pastures. Introd. from the Mediterranean. Fl. summer. *Musky Crowfoot* . . . .***E. moschatum** (L.) L'Hér. ex Aiton

3 Pelargonium L'Hér. ex Aiton
11 species in Aust. (7 endemic, 4 naturalized); all states and territories

Herbs or shrubs. Leaves alternate or opposite, almost undivided to deeply lobed. Flowers in cymose umbels, irregular. Sepals basally connate; the posterior one with a nectiferous spur adnate to the pedicel. 2 posterior petals usually ± larger than the others and marked differently. Fertile stamens 3–8 with 2–7 staminodes. Mericarps hairy; awns with long hairs on the inner surface, detaching spirally from the beak.

Key to the species

1 Leaves deeply pinnately or palmately dissected, covered with harsh hairs, ovate in outline, up to 7 cm long and 6 cm wide; margins recurved; stipules ovate to triangular, c. 7 mm long. Flowers in umbels of 5–10. Calyx spur 2–4 mm long. Petals up to twice as long as the sepals, pink with darker markings. Fertile stamens 6. Shrubs up to 1 m high, fragrant. Escape from cultivation. Waste places. Fl. summer . . . .***P. asperum** Ehrh. ex Willd.

1 Leaves lobed or almost entire, covered with softly villous hairs or pubescent 2

2 Leaf lobes acute, ovate in outline, c. 7 mm long. Flowers in umbels of 5–10. Petals c. twice as long as sepals or longer. Spreading, often sprawling shrubs. Escape from cultivation, often on coastal sand dunes. Extremely variable. Fl. summer. *Show Geranium* . . . .***P. domesticum** L.

2 Leaf lobes obtuse. Petals at least one and a half times as long again as the sepals 3

3 Fertile stamens 3–5. Hairs on the calyx short, coarse, scattered; nectary spur almost obsolete. Petals only just exceeding the sepals, deep pink. Mericarp beak 6–8 mm long. Annual herb up to 40 cm high, with ascending basal leaves. Widespread. Shady damp places, usually on sandy soils. Fl. spring–summer. . . . .**P. inodorum** Willd.

3 Fertile stamens 6–8. Hairs on the calyx softly villous. Nectary spur 2–8 mm long 4

4 Bracts narrow-ovate to narrow-triangular. Flowers pedicellate, in umbels of 4–12. Leaves almost flat. Perennial herb with basal stock and some ascending basal leaves. Widespread. Sand dunes on coast and granite outcrops on Blue Mts Fl. summer. . . . .**P. australe** Willd.

4 Bracts ovate. Leaves distinctly undulate. Flowers sessile below the nectary spur. Shrubby perennial without basal stock and with cauline leaves. Sand dunes near the coast. Introd. from S. Africa. Fl. spring–summer. . . . .***P. capitatum** (L.) Aiton

68 LYTHRACEAE

Herb, shrub or tree. Leaves whorled or opposite sometime alternate, simple, margins entire; stipules minute or absent. Inflorescence axillary cymes, clusters of flowers or flowers solitary Flowers 4–6(7)-merous. Petals free, sometimes absent. Stamens up to twice as many as petals. Ovary 2–6-locular;

placentation axile; style simple usually capitate. Fruit a capsule, seeds numerous. 28 gen., temp to trop., particularly America.

Key to the genera

1 Flowers irregular; hypanthium spurred or swollen on one side at the bas . . . . **1 CUPHEA**
1 Flowers regular; hypanthium without swelling or spur at the base 2

2 Hypanthium narrow-tubular . . . . **2 LYTHRUM**
2 Hypanthium campanulate . . . . .3 ROTALA

1 Cuphea (Jacq.) J.F.Macbr.
1 species naturalized Aust.; Qld, NSW

One species in the area

Erect or spreading herb with glandular hairs. Leaves opposite, subsessile or with a short petiole; ovate to elliptic, 1–5.5 cm long, 0.5–2 cm wide, scabrous. Inflorescence a monochasial cyme or sometimes solitary in leaf axils. Hypanthium 11-veined and shortly spurred at base. Petals violet to purple. Stamens 11, not exceeding the petals. Fruit a capsule 2 mm diam. with 4–8 seeds. Possible collection from Cumberland DSF. Introd. from S. America. Fl. spring . . . .*C. carthagenensis* (Jacq.) J.F.Macbr.

2 Lythrum L.
5 species in Aust. (2 endemic); all states and territories

Herbs. Leaves opposite or alternate, sessile; stipules absent. Flowers regular, bisexual. Floral tube ribbed, enclosing the ovary. Sepals 4–6, very much shorter than the floral tube and alternating with 4–6 external appendages resembling sepals. Petals 4–6. Stamens as many, or twice as many, as the petals. Ovary superior, 2-locular; placentas axile; style capitate. Fruit an oblong capsule enclosed in the persistent floral tube, dehiscing at the summit.

Key to the species

1 Erect, often pubescent perennial 50–100 cm high. Leaves opposite, lanceolate, 2–5 cm long, slightly stem-clasping. Flowers 10–12 mm long, in a terminal leafy spike. Floral tube 5 mm long. Petals 5–6, c. 8 mm long, pink to purple or bluish. Stamens usually 12; 6 longer than the others, exserted. Widespread. Damp places. Fl. spring–summer. *Purple Loosestrife* . . . . **L. salicaria** L.
1 Decumbent or ascending, weak, glabrous annual. Leaves mostly alternate, linear to elliptic, 5–10 mm long. Flowers small, solitary, nearly sessile in most of the axils. Floral tube 2–3 mm long, increasing to c. 6 mm in fruit. Petals 1–4 mm long, pale purplish or bluish. Widespread. Wet places. Fl. spring–summer . . . . **L. hyssopifolia** L.

3 Rotala L.
6 species in Aust. (2 endemic, 3 native, 1 naturalized); Qld, NSW, NT, WA

One species in the area

Annual or perennial herb; stems creeping, erect or floating. Leaves opposite, decussate, sessile to subsessile, orbicular, 12 mm diam., lower surface pale green to reddish. Flowers numerous, sessile in terminal spikes. Petals 4, mauve to pink. Stamens 4, inserted at base of hypanthium. Ovary 2–4 locular. Capsule 2–4 mm long with 4 valves. Seed numerous. Grows in damp or submerged areas. Introd. from S. Asia. Fl. spring . . . . *R. rotundifolia* (Buch.-Ham. ex Roxb.) Koehne

69 MYRTACEAE

Shrubs, mallees or trees. Leaves without stipules, opposite or alternate, simple, usually entire, dotted with pellucid oil glands which may be obscure, usually aromatic. Flowers axillary, solitary or in cymes or umbels, or terminal in spikes, racemes or panicles. Floral tube adnate to the ovary and usually continued above the ovary summit. Sepals and petals 4–5, free above the floral tube or cohering to form a calyptra.

Stamens 5–numerous, free or united into 5 bundles opposite the petals. Ovary inferior, 1–multi-locular, with 1–many ovules per loculus; placentas axile, basal or rarely parietal; style simple, usually capitate. Fruit a capsule nut or berry, rarely a schizocarp. 150 gen., trop to warm temp., particularly S. America and Australia .

Key to the genera

1 Filaments connate . . . . .GROUP 1
1 Filaments free. 2

2 Leaves opposite. . . . . .GROUP 2
2 Leaves alternate or spirally arranged. . . . . .GROUP 3

Group 1

1 Flowers sessile in spikes or terminal heads . . . . 21 MELALEUCA
1 Flowers pedicellate in cymes 2

2 Petals white . . . . 12 LOPHOSTEMON
2 Petals yellow 3

3 Leaves opposite . . . . 13 TRISTANIA
3 Leaves alternate. . . . . 14 TRISTANIOPSIS

Group 2

1 Sepals with numerous linear lobes. . . . . 30 HOMORANTHUS
1 Sepals not lobed 2

2 Most leaves less than 25 mm long. Stamens usually up to twice as many as petals 3
2 Most leaves more than 25 mm long. Stamens numerous 13

3 Stamens 5, regularly opposite the petals. 28 MICROMYRTUS
3 Stamens 5-numerous, if 5 then not regularly opposite the sepals 4

4 Staminodes present alternating with the stamens . . . . 31 DARWINIA
4 Staminodes absent 5

5 Stamens more than twice as long as the petals 6
5 Stamens less than twice as long as the petals 7

6 Fruit hard and woody, persistent for several years . . . . .20 CALLISTEMON
6 Fruit not hard or woody or persistent . . . . 19 KUNZEA

7 Pedicels more than 4 mm long, fruit a berry . . . . .2 AUSTROMYRTUS
7 Pedicels less than 3 mm long, fruit a capsule 8

8 Ovary 2 locular. . . . . 26 BAECKEA
8 Ovary 3 locular. 9

9 Anther adnate to filament dehiscing by pore or short parallel slit 10
9 Anther versatile dehiscing by long parallel slit 11

10 Inflorescence usually with solitary flowers. Bracteoles 2. Ovules ≤13 per loculus . . .25 HARMOGIA
10 Inflorescence usually (2–)3 flowered. Bracteoles 4 to many. Ovules ≥14 per locule . 24 SANNANTHA

11 Stamens 3–10, some opposite centre of petals. Ovary with 3–4 ovules per locule 27 EURYOMYRTUS
11 No stamens opposite centre of petals. 12

12 Stamens 14–17. Ovary with 8–13 ovules per locule. . . . . 23 TRIPLARINA

12 Stamens 5–8. Ovary with 5–8 ovules per locule .22 OCHROSPERMA

13 Flowers in compact globular heads on long peduncles. 14
13 Flowers not in compact globular heads 15

14 Flowers united. Ovary 3-locular . . . . .11 SYNCARPIA
14 Flowers free. Ovary 2-locular . . . . .10 CHORICARPIA

15 Sepals greenish-white, petaloid, c. 9 mm long in fruiting stage. . . . . 9 BACKHOUSIA
15 Sepals less than 8 mm long in fruiting stage. 16

16 Fruit a capsule (almost always present under the tree or on it) 17
16 Fruit a berry. Leaves usually dark green 18

17 Leaves green; calyptra absent . . . . 15 ANGOPHORA
17 Leaves glaucous; calyptra present. . . . . .17 EUCALYPTUS

18 Ovary 4 or more-locular 19
18 Ovary 1–3-locular. 20

19 Lateral veins of the leaf bending towards the apex to become almost parallel with the leaf margin . . .
. . . . 8 RHODOMYRTUS
19 Leaf venation not as above. . . . . 4 DECASPERMUM

20 Leaves with 3 distinct longitudinal veins (i.e., intramarginal veins well away from the margins) . . . .
. . . . 1 RHODAMNIA
20 Leaf with one main longitudinal vein 21

21 Free part of floral tube deciduous before fruit. Fruit with circular scar on the top . . . . . 7 ACMENA
21 Free part of floral tube or sepals persistent on fruit. 22

22 Sepals deciduous, short tube present on top of fruit . . . . .6 WATERHOUSEA
22 Sepals persistent on fruit. 23

23 Fruit pink to blue . . . . 5 SYZYGIUM
23 Fruit whitish or black. 24

24 Leaves less than 10 mm wide . . . . .2 AUSTROMYRTUS
24 Leaves more than 10 mm wide . . . . 3 GOSSIA

Group 3

1 At least calyx and sometimes corolla fused into an operculum which is deciduous as the flower opens 2
1 Sepals and petals free 3

2 Lateral veins of leaf ± parallel making 50–60° angle with mid-vein, fruit ovoid to urceolate with disc
    deeply depressed, ≥10 mm long. . . . . 16 CORYMBIA
2 Lateral veins of leaf not as above. Fruit if urceolate then usually ≤10 mm long (longer in *E. robusta*). . .
    . . . . .17 EUCALYPTUS

3 Sepals awned . . . . .29 CALYTRIX
3 Sepals not awned 4

4 Stamens shorter or only slightly longer than the petals . . . . 18 LEPTOSPERMUM
4 Stamens more than twice as long as petals 5

5 Fruit not hard and woody, falling after the first year. Flowers arranged in leafy spikes or compact heads
. . . . 19 KUNZEA
5 Fruit hard and woody, persisting several years. Flowers in leafless compact spikes 20 CALLISTEMON

1. **Rhodamnia** Jack

13 species in Aust.; (12 endemic, 1 native) Qld, NSW

One species in the area

Tall shrub or small tree with pubescent branches. Leaves opposite, lanceolate to ovate-oblong, up to 12 cm long, acuminate, entire, distinctly 3-veined, laterale veins transverse. Flowers c. 1 cm diam, arranged in axillary cymes. Floral tube c. 3 mm long. Sepals and petals 4–5, white. Stamens numerous, free, arranged in several whorls. Ovary 1-locular with several ovules inserted on 2 parietal placentas; stigma scarcely peltate. Fruit a berry, 6 mm diam. Coast and adjacent plateaus. RF. Fl. summer. . . . .**R. rubescens** (Benth.) Miq.

2. **Austromyrtus** (Nied.) Burret

3 species endemic Aust.; Qld, NSW

One species in the area

Shrubs or small trees with smooth brown bark. Leaves opposite linear to linear-lanceolate, up to 4 cm long and 3 mm wide, with recurved margins, pubescent on lower surface, especially when young. Flowers with long pedicels arranged on short axillary branches or solitary, 5-merous. Corolla white. Stamens numerous, free, in several whorls. Ovary 2-locular with numerous ovules in each loculus. Fruit a berry white with dark spots. Coast and adjacent plateaus to Blue Mts Damp, sheltered places in forests on Ss. Fl. summer . . . . **A. tenuifolia** (Sm.) Burret

3 **Gossia** N.Snow & Guymer

16 species endemic Aust.; Qld, NSW

One species in the area

Shrubs or small trees. Bark smooth, brown mottled. Leaves lanceolate to elliptic, up to 8 cm long and 3.5 cm wide flat, glabrous. Petiole 2–4 mm long. Inflorescence 1–4 flowered. Sepals 5. Petals white and minutely ciliate. Fruit a berry, black. Illawarra ranges. RF. Fl. spring–summer. *Scrub Ironwood* . . . . **G. acmenoides** (F.Muell.) N.Snow & Guymer

4 **Decaspermum** J.R.Forst. & G.Forst.

2 species in Aust.; (1 endemic, 1 native). Qld, NSW

One species in the area

Shrubs or small trees. Leaves opposite, lanceolate to ovate, penniveined, closely dotted with oil glands. Flowers 4–5-merous, in axillary racemes. Floral tube scarcely produced above the ovary. Petals free. Stamens numerous, free. Fruit a red berry crowned by the persistent sepals and divided by a false-septum. North from Gosford area. RF. Fl. autumn. **D. humile** G.Don) A.J.Scott

5 **Syzygium** Gaertn.

52 species in Aust.; (47 endemic) QLD, NSW, NT, WA

Glabrous trees or shrubs with ± flaky bark. Leaves opposite, obovate to lanceolate, acuminate, entire. Flowers solitary or arranged in axillary cymes. Floral tube shortly continued above the ovary. Sepals 4, persistent or irregularly deciduous. Petals 4, persistent or deciduous, white-pink. Stamens numerous, arranged in several whorls. Ovary 2-locular with 1-several ovules inserted on axile placentas. Fruit a berry, ovate or urceolate, 10–15 mm long.

Key to the species

1 Branchlets 4-angular or 4-winged, with a pocket on either side just above each node. Leaves elliptic to obovate 3–10 cm long, 10–30 mm wide; oil glands usually faint. Fruit deep pink to red. Shrub to small tree. Coast and adjacent plateaus. RF. Fl. spring–summer. *Brush Cherry* . . . . **S. australe** (J.C.Wendl. ex Link) B.Hyland

1 Branchlets usually terete, without pockets above each node 2

2 Petals cohering, shedding as a cap as the flower opens. Leaves ovate to elliptic, 3–8 cm long, 15–35 mm wide; oil glands mostly faint. Fruit purplish blue. Tall tree. Gosford district. RF. Fl. spring–summer. *Giant Water Gum.* . . . . **S. francisii** (F.M.Bailey) L.A.S.Johnson

2 Petals free. Fruit magenta 3

3 Leaves with numerous large oil glands visible to naked eye, lanceolate to elliptic, 3–12 cm long, 10–40 mm wide. Fruit bluish purple or magenta. Shrub or small tree. Coast and adjacent plateaus north of Mt. Kembla. RF. Fl. summer. *Blue Lilly Pilly.* . . . . **S. oleosum** (F.Muell.) B.Hyland

3 Leaves with scattered faint oil dots mostly not visible to the naked eye, lanceolate to obovate, 4–10 cm long, 15–30 mm wide. Fruit magenta. Shrub or small tree. Coast. RF. Sandy soils. Vulnerable. Fl. summer . . . . **S. paniculatum** Gaertn.

6 Waterhousea B.Hyland
4 endemic species in Aust.; Qld, NSW

One species in the area

Tree with dark grey, fissured bark. Leaves opposite, lanceolate to elliptic, 5–16 cm long, 15–50 mm wide, with undulate margins; lower surface paler; oil glands numerous, small. Flowers in terminal panicles. Petals cohering, shed as a cap as the flower opens. Fruit with the short free part of the floral tube at top, fleshy, reddish green. Hunter River Valley. RF. Fl. spring. *Weeping Lilly Pilly* . . . .
. . . . **W. floribunda** (F.Muell.) B.Hyland

7 Acmena DC.
7 species in Aust.; (6 endemic) Qld, NSW, Vic., NT

One species in the area

Glabrous tree. Leaves opposite, narrow-ovate to ovate-oblong, up to 10 cm long, acuminate, entire. Flowers numerous, arranged in axillary cymose panicles. Floral tube continued and broadening above the ovary. Sepals 4, sometimes unequal, deciduous. Petals 4, white-pink. Stamens numerous, in more than one whorl, short. Ovary 2-locular with several ovules per loculus inserted on axile placentas. Fruit a nearly globular berry, 6–12 mm diam., white or purplish white. Widespread. RF. Fl. summer. *Lilly Pilly.*
. . . . **A. smithii** (Poir.) Merr. & L.M.Perry

8 Rhodomyrtus (DC.) Rchb.
13 species in Aust.; (12 endemic, 1 native) Qld, NSW

One species in the area

Shrub or small tree with pubescent young branches. Leaves opposite, glabrous, lanceolate to oblong, up to 12 cm long, shortly acuminate, entire. Flowers c. 10 mm diam., solitary or arranged in axillary cymes. Floral tube pubescent, not longer than the ovary. Sepals usually 5, ovate to deltoid, persistent. Petals usually 5, white to pale pink. Stamens numerous, free, arranged in several whorls. Ovary apparently multilocular due to the ingrowth of false-septa; ovules numerous, arranged in 6 rows on axile placentas. Fruit a berry, ovoid-globular. Coast from Gosford northwards. RF. Fl. spring–summer. *Native Guava.* . .
. . . . **R. psidioides** (G.Don) Benth.

9 Backhousia Hook. & Harv.
8 species endemic to Aust.; Qld, NSW

One species in the area

Trees or shrubs with pubescent young branches. Leaves ovate to lanceolate, opposite. Flowers arranged in cymes. Floral tube campanulate, continued well above the ovary. Sepals 4–5, c. 9 mm long at fruiting stage, petaloid, greenish or whitish, persistent. Petals ± persistent. Stamens numerous, free, arranged in several whorls. Ovary 2-locular. Fruit dry, often schizocarpic, crowned by the persistent calyx. Widespread. Margins of RF and gullies. Fl. summer. *Grey Myrtle.* . . . . . . **B. myrtifolia** Hook.f. & Harv.

10 Choricarpia Domin

2 species endemic to Aust., Qld, NSW

One species in the area

Small tree. Young stems 4-angled. Leaves opposite, broad-lanceolate to elliptic, entire or minutely dentate, acute or acuminate, up to 11 cm long, pale and scurfy underneath when young. Flowers arranged in compact globular heads on long peduncles. Floral tube usually pubescent, 2–3 mm long, not fused to each other. Sepals and petals 4–5, cream. Stamens numerous, free, arranged in 1 whorl. Ovary 2-locular, with 1 ovule per loculus. Fruit a capsule. Coast, e.g. Stanwell Park northwards. Margins of RF. Fl. spring–summer. *Brush Turpentine* . . . . **C. leptopetala** (F.Muell.) Domin

11 Syncarpia Ten.

3 species endemic to Aust.; Qld, NSW

One species in the area

Tall tree with fibrous bark. Leaves opposite, crowded into pseudo-whorls at ends of branches, lower surface hairy, elliptic to lanceolate, up to 12 cm long, acute or shortly acuminate, entire, with recurved margins. Flowers in globular clusters on long peduncles. Floral tubes united forming a compound structure, pubescent. Sepals 5, persistent. Petals 4, orbicular, white or cream. Stamens numerous, usually arranged in 2 whorls. Ovary 3-locular, with numerous ovules per loculus. Fruit a capsule united into a compound head 10–20 mm diam. Widespread. WSF; RF margins. Fl. spring–summer. *Turpentine* . . . . **S. glomulifera** (Sm.) Nied. **ssp. glomulifera**

12 Lophostemon Schott

4 species in Aust.; (3 endemic) Qld, NSW, NT, WA

One species in the area

Tall trees usually with a smooth pink trunk and branches, except for the rough base. Leaves alternate, crowded at the ends of branches, elliptic, up to 15 cm long. Sepals deciduous. Petals white. Stamens connate in bundles opposite the petals; the claws up to 15 mm long. Ovary 3-locular, with several ovules per loculus. Stigma capitate. Fruit a capsule, not exserteded beyond the floral tube. A native of RF from Bulladelah northwards but planted in streets and gardens and occasionally naturalized around Sydney. Fl. summer. *Brush Box* . . . . ***L. confertus** (R.Br.) Peter G.Wilson & J.T.Waterh.

13 Tristania R.Br.

1 species endemic to AUST;. NSW

One species in the area

Shrub or small tree. Leaves opposite, narrow-lanceolate to lanceolate, up to 9 cm long. Sepals persistent into fruiting. Petals yellow. Stamens connate in bundles opposite the petals; claw short and irregular. Ovary 3-locular with several ovules per loculus. Fruit a capsule, not exserted beyond the floral tube. Widespread. Usually in creek beds or on the banks. Ss. Fl. summer *Water Gum* **T. neriifolia** (Sims) R.Br.

14 Tristaniopsis Brongn. & Gris.

3 endemic species in Aust.; Qld, NSW, Vic.

Key to the species

Trees or shrubs. Leaves alternate, often crowded towards the ends of the branches. Sepals persistent into fruiting. Petals yellow. Stamens connate into bundles opposite the petals. Ovary 3-locular with several ovules in each loculus. Fruit a capsule, exserted beyond the floral tube.

1 Leaves pubescent underneath, oblanceolate, 6–12 cm long, with scattered oil glands. Petals obovate, 4–5 mm long. Stamen bundles shorter than petals. Tree to 30 m high. Widespread. Usually along streams. Fl. summer. *Water Gum* or *Kanuka* . . . . **T. laurina** (Sm.) Peter G.Wilson & J.T.Waterh.

1 Leaves glabrous underneath, elliptic, 4–10 cm long, with numerous dense oil glands. Petals orbicular, up to 3 mm diam. Stamen bundles c. as long as the petals. Shrub or tree up to c. 30 m high. Coast and adjacent plateaus; Illawarra ranges. RF and WSF, only rarely along streams. Fl. summer. *Mountain Water Gum* . . . . **T. collina** Peter G.Wilson & J.T.Waterh.

15 Angophora Cav.
15 species endemic to Aust.; Qld, NSW, Vic.

Trees or shrubs. Mature leaves opposite. Flowers in cymes arranged in terminal corymbs. Sepals 4 or 5, persistent. Petals 4 or 5, white-cream, spreading, free. Stamens numerous, free, arranged in several whorls. Ovary 3–4-locular. Fruit a usually ribbed capsule surmounted by the persistent calyx teeth. Seeds 1–few per loculus.

Key to the species

1 Bark smooth, pink to orange or greyish. Leaves without conspicuous oil glands 2
1 Bark rough 3

2 Capsule less than 15 mm diam. Leaves up to 20 cm long. Tall tree or rarely a shrub. Coast and adjacent plateaus; lower Blue Mts Heath, DSF and WSF. Fl. spring–early summer. *Smooth-barked Apple* or *Rusty Gum* . . . . **A. costata** (Gaertn.) Britten
2 Capsule more than 15 mm diam. Leaves up to 17 mm long. Tree up to 25 m high. Putty area and Judge Dowling Range. DSF. . . . . . . **A. euryphylla** (L.A.S.Johnson ex G.J.Leach) L.A.S.Johnson & K.D.Hill

3 Leaves cordate at base 4
3 Leaves tapering at base. 5

4 Capsule 14–20 mm diam. Young branches and inflorescences densely covered with reddish hairs. Leaves ovate to oblong, up to 10 cm long, obtuse. Shrub or small tree with flaky bark. DSF and scrubs. Coast and adjacent plateaus. Ss. Fl. summer. *Dwarf Apple* . . . . **A. hispida** (Sm.) Blaxell
4 Capsules 7–10 mm diam. Young branches and inflorescences with scattered reddish hairs. Leaves ovate-lanceolate to oblong, up to 12 cm long, mostly acute. Tree with rough bark. Coast and adjacent plateaus and lower Blue Mts Open forest, not on poorest soils. *Broad-leaved Apple* . . . . **A. subvelutina** F.Muell.

5 Adult leaves mostly more than 15 mm wide 6
5 Adult leaves mostly less than 15 mm wide. Small trees 7

6 Longest petiole less than 8 mm long. Leaves lanceolate to broad-lanceolate 4–11 cm long. Fruit 11–15 mm long. Tree often multi-stemmed. Restricted to Charmhaven–Wyee area. DSF. Vulnerable . . . .
. . . .
. . . . **A. inopina** K.D.Hill
6 Longest petiole more than 8 mm long. Leaves narrow-ovate to ovate, 8–12 cm long. Fruit 7–12 mm long. Tall, usually gnarled tree. Widespread. Usually on shales and alluvium. *Rough-barked Apple*. . . . . . .
. . . . **A. floribunda** (Sm.) Sweet

7 Pedicels mostly more than 10 mm long. Leaves lanceolate, 7–11 cm long, up to 15 mm wide. Buds 5–6 mm long. Fruit 10–14 mm long. Tree up to 15 m high. Kur-ing-gai Plateau . . . .
. . . . **A. crassifolia** (G.J.Leach) L.A.S.Johnson & K.D.Hill
7 Pedicels mostly less than 10 mm long. Leaves lanceolate, 6–10 cm long, up to 10 mm wide. Buds 4–5 mm long. Fruit 8–10 mm long. Tree up to 10 m high. Coast to lower Blue Mts DSF. Ss. *Narrow-leaved Apple* . . . . **A. bakeri** E.C.Hall

(Hybrid swarms occur between *A. floribunda* and the two species *A. bakeri* and *A. subvelutina*. Hybrids between other species may occur, but not in swarms. Some of these have been given specific names e.g., *A. hispida* X *A. bakeri* = *A.* X *clelandii* Maiden. *A. hispida* X *A. costata* = *A.* X *dichromophloia* Blakely)

16 Corymbia K.D.Hill & L.A.S.Johnson
5 species endemic to Aust.; All states and territories except Tas.

Tree or sometimes a mallee. Bark smooth and shedding in flakes or strips or rough and shortly fibrous persistent to small branches. Juvenile leaves opposite, adult leaves alternate. Lateral veins parallel and closely spaced. Inflorescence a corymbose panicle, compound, terminal or lateral,, umbels 3–7-flowered. Flowers sessile or pedicellate. Calyx and corolla fused into an operculum or calyx and corolla segments separately connate into two opercula. Opercula seated on the rim of the floral tube and usually leaving a scar when shed. Stamens numerous, anthers versatile and dehiscing by longitudinal slits. Ovary half inferior, usually 2-locular. Fruit a capsule enclosed by the floral tube except at the summit, opening by terminal valves which are enclosed by the free portion of the floral tube, disc depressed. *Leaf dimensions are length x width.*

Key to the species

1 Bark smooth, white or pink. 2
1 Bark rough 3

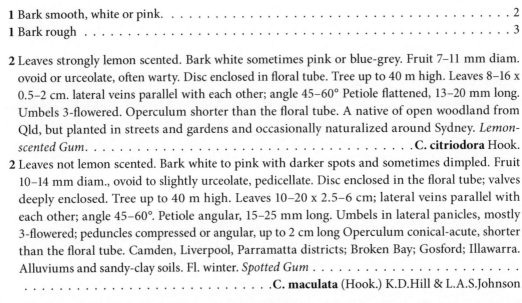

2 Leaves strongly lemon scented. Bark white sometimes pink or blue-grey. Fruit 7–11 mm diam. ovoid or urceolate, often warty. Disc enclosed in floral tube. Tree up to 40 m high. Leaves 8–16 x 0.5–2 cm. lateral veins parallel with each other; angle 45–60° Petiole flattened, 13–20 mm long. Umbels 3-flowered. Operculum shorter than the floral tube. A native of open woodland from Qld, but planted in streets and gardens and occasionally naturalized around Sydney. *Lemon-scented Gum.* . . . .**C. citriodora** Hook.

C. maculata

2 Leaves not lemon scented. Bark white to pink with darker spots and sometimes dimpled. Fruit 10–14 mm diam., ovoid to slightly urceolate, pedicellate. Disc enclosed in the floral tube; valves deeply enclosed. Tree up to 40 m high. Leaves 10–20 x 2.5–6 cm; lateral veins parallel with each other; angle 45–60°. Petiole angular, 15–25 mm long. Umbels in lateral panicles, mostly 3-flowered; peduncles compressed or angular, up to 2 cm long Operculum conical-acute, shorter than the floral tube. Camden, Liverpool, Parramatta districts; Broken Bay; Gosford; Illawarra. Alluviums and sandy-clay soils. Fl. winter. *Spotted Gum* . . . .
. . . .**C. maculata** (Hook.) K.D.Hill & L.A.S.Johnson

C. eximia

3 Leaves bluish, concolorous, bark yellow, rough on trunk and branches, flaky or scaly. Fruit sessile, ovoid-truncate or urceolate, 13–16 x 14–15 mm; disc enclosed in the floral tube; valves enclosed. Tree up to 20 m high, usually gnarled. Leaves 10–18 x 1–2.5 cm, sometimes paler underneath. Lateral veins parallel; angle 50–70°; intramarginal vein 1 mm from the margin. Umbels 7-flowered, in a terminal corymb; peduncles angular, almost terete, 10–25 mm long. Operculum hemispherical, shortly beaked, shorter than the floral tube. Lower Blue Mts; Woronora Plateau. Fl. spring. *Yellow Bloodwood* . . . . **C. eximia** (Schauer) K.D.Hill & L.A.S.Johnson

C. gummifera

3 Leaves green, discolorous, bark reddish or brown rough throughout (small branches and twigs smooth), fibrous-flaky, with longitudinal and transverse cracks. Fruit pedicellate, ovoid-truncate or urceolate, 12–20 x 10–18 mm; disc enclosed in the floral tube; valves enclosed. Tree usually c. 15 m high (up to 30 m), often with gnarled branches. Leaves 10–16 x 2–5 cm. Lateral veins parallel, close together; angle 60–70°; intramarginal vein c. 1 mm from the margin. Umbels 7-flowered, in a terminal corymb; peduncles compressed. Operculum hemispherical to conical, much shorter than the floral tube. Widespread. DSF. Mostly Ss, rarely shale. Fl. summer. *Red Bloodwood.* . . . .**C. gummifera** (Gaertn.) K.D.Hill & L.A.S.Johnson

17 Eucalyptus L'Hér.
>800 species worldwide, all but 16 endemic to Aust., all states and territories

Trees, tall shrubs or mallees. Bark either smooth throughout, or partly rough and partly smooth or rough throughout. Smooth-barked plants are referred to as gums but they may have some rough bark near the base or patches of rough (fibrous, flaky or ribbony) bark hanging loosely on the upper trunk or branches. Rough bark either with long matted fibres which can be removed in strings or ropes several centimetres or even metres long (stringybarks or mahoganies); or shortly fibrous, flaky or scaly (peppermints, boxes, scalybarks); or the fibres short and dark coloured or black and often impregnated with kino (ironbarks). Juvenile leaves (on seedlings and sometimes on mature branches and trunks which have been damaged) opposite, sometimes sessile or stem-clasping, differing from the adult leaves. Adult leaves usually alternate, petiolate, mostly with visible oil glands, usually isobilateral but sometimes paler underneath; venation usually prominent; midrib distinct and intramarginal veins usually present; lateral veins parallel or irregular and leaving the midrib at angles of 10–60° or forming a fine reticulation. Inflorescence usually a cymose umbel of 3, 7, 11, 15, 19, etc., up to 35 flowers or fewer by abortion, usually pedunculate in the leaf axils, or the umbels in axillary or terminal panicles. Flowers sessile or pedicellate. Calyx and corolla fused into a single structure, the operculum, or the calyx absent and a single petaline operculum present or calyx and corolla segments separately connate into two opercula. Operculua seated on the rim of the floral tube and usually leaving a scar when shed. Stamens numerous, in 2 or several rows. Anthers versatile or adnate; pollen sacs parallel or divergent and confluent at the apex, dehiscing by longitudinal slits or terminal pores; connective usually with a gland on the back or near the apex. Ovary 2–7-locular. Fruit a capsule enclosed by the floral tube except at the summit, opening by terminal valves which are either exserted or enclosed by the free portion of the floral tube. (Fig. 36). Floral tube usually smooth at maturity, rarely with longitudinal ribs, bearing at the summit the operculum scar or double scars, the scar of the filament bases, and within these a disc which is convex flat depressed or enclosed within the floral tube. (Fig. 36).

(Some species have sucker leaves that have neither adult nor juvenile characteristics. Be careful that you are observing the leaves of the correct type. You will often find capsules, and even buds, on the ground underneath the canopy of the tree that you are attempting to identify. Make sure that they have not fallen from an adjacent tree. Leaf dimensions are length x width.)

FIGURE 36

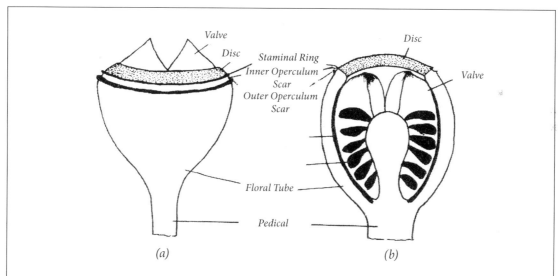

(a)   (b)

Fig. 36 MYRTACEAE, Eucalyptus; Diagram of two fruits illustrating the positions of the various structures used to identify the species. In (a) the fruit is drawn from the outside; the valves are exsert; the epigynous disc is prominent and remains fixed to both ovary and floral tube. In (b) a fruit with enclosed valves is drawn in longitudinal section; the epigynous disc is dragged away from the top of the ovary as the fruit matures and remains attached to and lines the inner part of the top of the floral tube.

FIGURE 37

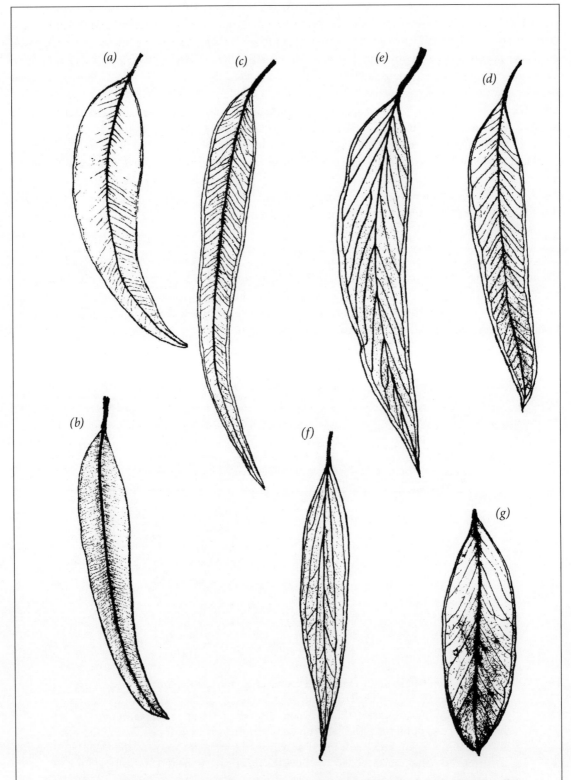

Fig. 37 MYRTACEAE, Eucalyptus; Venation types. (a) lateral veins more or less parallel; venation angle 50-60°; intermarginal vein about 1mm from margin. (Leaves paler on under surface); (b) lateral veins more or less parallel, with much reticulation; intermarginal vein 1mm or less from the margin. (Leaves paler on under surface-bloodwood type); (c–d) lateral veins somewhat to very irregular; venation angle 35-55°; intermarginal vein 2-3mm from margin. (Leaves usually isobilateral); (e) lateral veins irregular; venation angle 15-30° intramarginal vein 2-4mm from margin. (Leaves isobilateral); (f) lateral veins more or less parallel with the midrib. (Leaves isobilateral); (g) leaf of *E. camphora*.

Key to the species

Group 1
Leaves on adult plants opposite

1 Bark smooth, white, shed in ribbons. Straggling shrub or small tree. Leaves opposite throughout, sessile or stem-clasping, 4–6 x 3–4.5 cm, very glaucous, silver-blue. Umbels axillary, 3-flowered. Buds sessile, glaucous; operculum conical, slightly shorter than the floral tube. Fruit sessile, glaucous, usually hemispherical, 10 x 13 mm; disc flat; valves broad, slightly exserted. Upper Cox's River. Fl. early summer. *Silver-leaved Mountain Gum* . . . . **E. pulverulenta** Sims

E. pulverulenta

1 Bark fibrous on trunk and main branches, reddish-brown; smaller branches smooth and ribbony. Tree up to 12 m high. Juvenile and most of the adult leaves opposite, sessile or almost so, glaucous, silver-blue, mostly 3.5–4.5 x 3–5.5 cm; adult leaves sometimes alternate and narrower. Umbels axillary, 3-flowered. Buds sessile, glaucous; operculum conical, shorter than floral tube. Fruit sessile, globular to broadly pyriform, 6–10 x 7–10 mm; disc broad, convex; valves usually exserted. South-western part of area. Fl. early summer. *Argyle* or *Broad-leaved Apple, Mealy Stringybark, Blue Peppermint*. . . . .**E. cinerea** F.Muell. ex Benth.

E. cinerea

BARK TYPES

STRING BARK *Eucalyptus oblonga*

SMOOTH BARK *Eucalyptus grandis*

GREY GUM BARK *Eucalyptus punctata*

SCRIBBLY BARK *Eucalyptus haemastoma*

BARK TYPES

PEPPERMINT BARK *Eucalyptus piperita*

IRONBARK *Eucalyptus paniculata*

BOX BARK *Eucalyptus moloccana*

BLOODWOOD BARK *Corymbia gummifera*

Group 2
Umbels 3-flowered

E. longifolia

1 Pedicel as long as or longer than, the fruit. Tree up to 25 m high. Bark on trunk and larger branches dirty grey, fibrous-flaky, smaller branches with loose flakes of rough bark. Leaves 14–22 x 2–5 cm. Lateral veins slightly irregular, angle 45–65°; intramarginal vein 1–2 mm from margin. Peduncles terete, usually reflexed, 15–30 mm long. Operculum conical, usually beaked, as long as or longer than the floral tube. Fruit ovoid-truncate or campanulate, 15–30 x 10–12 mm; disc conspicuous, flat or convex; valves enclosed or the tips slightly exserted. Drier parts of Cumberland Plain; Illawarra. Usually on deep soils, often with a hardpan. Fl. spring–early summer. *Woollybutt* . . . . **E. longifolia** Link

1 Fruit sessile or the pedicel half as long as the fruit 2

E. baeuerlenii

2 Operculum much narrower than and c. as long as the floral tube, acutely conical. Mallee. Bark smooth. Leaves 12–20 x 1–2.5 cm, paler underneath. Lateral veins fine, irregular, forming a fine reticulum with the smaller veins; angle c. 40°; intramarginal vein within 1 mm of the margin. Peduncles angular, up to 4 mm long. Fruit campanulate, smooth or with a single rib, sessile, 7–9 x 8–11 mm; disc conspicuous, flat; valves exserted. Wentworth Falls, on terraces of south-facing Ss cliffs. Fl. late summer . . . . **E. baeuerlenii** Schauer

2 Operculum as wide as or wider than the floral tube. Trees 3

3 Fruit 10–20 mm diam 4

3 Fruit 7–9 mm diam. 6

E. bicostata

4 Leaves concolourous. Operculum warted. Tree up to 30 m high. Bark rough on lower part of trunk, the upper part and the branches smooth. Adult leaves 10–25 x 2.5–4 cm. Peduncles 5–7 mm long. Buds glaucous, sessile, 2-ribbed. Fruit sessile, glaucous, turbinate to globular, 2-ribbed, usually slightly warted, 14–17 x 14–20 mm; disc broad, smooth; valves exserted or just enclosed. Nullo Mt.; Jenolan Caves. *Eurabbie* . . . .**E. bicostata** Maiden, Blakely & Simmonds

4 Leaves discolourous. Operculum smooth. Trees up to 30 m high with a straight trunk. Bark rough throughout, with long fibres resembling a stringybark. Leaves 10–15 x 2–3 cm. Lateral veins ± parallel with each other; angle 50–60°. Peduncles compressed 8–25 mm long, 4–6 mm wide. Operculum c. as long as the floral tube. Fruit hemispherical-conical or campanulate, 8–18 x 12–20 mm, sessile or shortly pedicellate; disc prominent, flat; valves 4 usually exserted. Fl. summer. *Large-fruited Red Mahoganies* 5

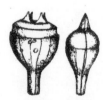

E. scias

5 Operculum with a prominent beak. Peduncles of the fruit c. 7 mm wide at the top. Hornsby Plateau and adjacent coast. WSF . . . . **E. scias** L.A.S.Johnson & K.D.Hill ssp. **scias**

5 Operculum apiculate. Peduncles of fruit c. 5 mm wide at the top. Woronora Plateau and adjacent coast. WSF. . . . . **E. scias** ssp. **callimastha** L.A.S.Johnson & K.D.Hill

E. viminalis

6 Juvenile and sucker leaves (usually procurable from shoots from damaged parts of mature trees) lanceolate, green, up to 3 cm wide. Tree up to 40nm high. Bark smooth or some rough bark present at base, usually with ribbons hanging on the trunk or caught in the branches. Adult leaves 11–18 x 1.5–2 cm. Lateral veins irregular; angle 30–50°; intramarginal vein 1–2 mm from margin. Twigs usually green or yellowish. Peduncles terete, 3–6 mm long. Operculum conical or hemispherical, c. as long as the floral tube. Fruit sessile or shortly pedicellate, ovoid to hemispherical, 5–6 x 7–8 mm; disc prominent, convex or domed; valves exserted. Nepean River, Camden to Menangle and higher elevations on better soils. WSF. Fl. most of the year. *Ribbon Gum* or *Manna Gum* . . . . **E. viminalis** Labill.

6 Juvenile and sucker leaves ovate to orbicular 7

E. dalrympleana

7 Juvenile and sucker leaves green or slightly glaucous. Buds and fresh fruits green. Tree up to 40m high. Bark mainly smooth, blotched. Adult leaves 10–20 x 1.5–2.5 cm. Lateral veins irregular; angle 30–40°; intramarginal vein 1–2 mm from the margin. Twigs usually reddish. Peduncles compressed, 4–7 mm long. Operculum conical, as long as or shorter than the floral tube. Fruit

usually ovoid-truncate, 7–8 x 8–9 mm, sessile or nearly so; disc broad, slightly convex to domed; valves exserted. Western Blue Mts Fl. summer–autumn. *Mountain Gum* . . . .
. . . . **E. dalrympleana** Maiden ssp. **dalrympleana**

7 Juvenile and sucker leaves grey or bluish. Buds and fresh fruits usually glaucous. Tree up to c. 30 m high. Bark smooth except near the base, whitish, blotched with pink or red. Adult leaves 10–25 x 2–3.5 cm. Lateral veins irregular; angle 30–50°. Peduncles almost terete, 5–10 mm long. Operculum conical, acute to obtuse, as long as or shorter than the floral tube. Fruit ovoid or ovoid-truncate, 5–6 x 6–7 mm, sessile or very shortly pedicellate; disc broad, domed; valves exserted. Higher Blue Mts; Megalong and Kanimbla Valleys. Fl. summer. *Candle-bark Gum* . . .
. . . . .E. **rubida** H.Deane Maiden ssp. **rubida**

E. rubida

Group 3
Ironbarks

1 Operculum c. twice as long as the floral tube 2
1 Operculum c. as long as or shorter than the floral tube 3

2 Adult leaves mostly 1–2 cm wide, 8–15 cm long, with a very fine point which is often recurved. Tree up to 45 m high. Umbels 3–7-flowered, arranged in panicles. Buds pedicellate. Fruits 6–8 x 5–7 mm, ± conical, slightly constricted at the orifice, often with one rib; disc small, oblique; valves enclosed or exserted. Hawkesbury and Colo Rivers and northward. Fl. July–Jan. *Northern Grey Ironbark* . . . . **E. siderophloia** Benth.

E. siderophloia

2 Adult leaves 2–4 cm wide, 8–18 cm long, without a fine point as above. Tree up to 30m high. Bark dark coloured to black, flaky, deeply furrowed; smaller branches smooth. Umbels 7-flowered, axillary. Fruit pyriform to ovoid-truncate, 6–9 x 7–8 mm, shortly pedicellate; disc conspicuous or narrow, flat or convex. Mainly drier parts of Cumberland Plain. WS. Fl. summer. *Broad-leaved Ironbark* . . . . .E. **fibrosa** F.Muell.

E. fibrosa

3 Leaves and buds glaucous. Tree up to 25 m high. Adult leaves 10–13 x 2.5–5 cm. Umbels axillary or in terminal panicles. Fruit pedicellate, usually pyriform, c. 10 x 9 mm; disc usually flat; valves deeply enclosed. Goulburn River Valley. Fl. mainly autumn–winter. *Caley's Ironbark* . . . .
. . . . **E. caleyi** Maiden ssp. **caleyi**
3 Leaves and buds not glaucous. 4

E. caleyi

4 Fruit 4-angled or 4-ribbed, pyriform, 6–10 mm long; disc flat; valves enclosed. Leaves 8–15 x 1.5–3 cm. Umbels 3–7-flowered, usually in a terminal panicle. Trees up to 25 m high 5
4 Fruit not 4-angled 6

E. fergusonii

5 Adult leaves with stomata on upper and lower surfaces. Buds 8–10 mm long. Hunter River Valley. WSF . . . . **E. fergusonii** R.T.Baker ssp. **fergusonii**
5 Adult leaves with stomata on lower surface only. Buds 10–13 mm long. Wollemi N.P. WSF . . . .
. . . . **E. fergusonii** ssp. **dorsiventralis** L.A.S.Johnson & K.D.Hill

6 Leaves paler underneath; lateral veins ± parallel with each other. Umbels in terminal panicles. Adult leaves 7–13 x 1.5–3 cm . Tree up to 40 m high. Bark grey to black. Umbels mostly 7-flowered. Fruit pyriform to conical, 6–10 x 5–10 mm; disc flat; valves enclosed.. 7
6 Leaves similar in colour on both surfaces 8

E. fergusonii dorsiventralis

7 Stomata on the lower surface of adult leaves except for a narrow band scattered along the midrib of the upper surface. Coast and adjacent plateaus; lower Blue Mts Usually on clay soils. Fl. winter–spring. *Grey Ironbark* . . . . .E. **paniculata** Sm. ssp. **paniculata**
7 Stomata on the upper surface of adult leaves scattered much less dense than on the lower surface. Port Stephen's to Newcastle. Usually on clay soils fl. winter–spring . . . .
. . . . **E. paniculata** ssp. **matutina** L.A.S.Johnson & K.D.Hill

8 Fruits 4–5 mm diam.. 9
8 Fruits more than 5 mm diam. 10

E. paniculata

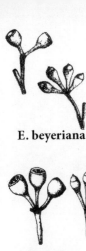

E. beyeriana

E. crebra

E. sieberi

E. nubila

E. placita

E. sideroxylon

E. pauciflora

E. moorei

9 Outer stamens sterile; anthers cuboid. Adult leaves 5–9 x 0.5–1.5 cm. Tree up to 30 m high. Umbels 3–7-flowered. Operculum conical to hemispherical. Fruit pyriform, c. 6 x 5 mm; disc small; valves usually enclosed. Cumberland Plain. Uncommon. Fl. spring . . . . **E. beyeriana** L.A.S.Johnson & K.D.Hill

9 Outer stamens fertile; anthers ± globular. Fruits 4–5 mm diam., 5–6 mm long; disc narrow, enclosed in the floral tube; valves enclosed; pedicels of same umbel often of different lengths. Tree up to 30 m high. Leaves 7–12 x 1–2 cm, with finely reticulate venation. Umbels 3–7-flowered, mostly in terminal panicles. Operculum conical or hemispherical. Drier parts of Cumberland Plain; foothills of Blue Mts; Hunter River Valley; Capertee. Usually clay soils. Fl. winter–early summer. *Narrow-leaved Ironbark* . . . . **E. crebra** F.Muell.

10 Operculum hemispherical, much shorter than the floral tube. Lateral veins irregular; angle as little as 15°. Tree up to 20 m high. Bark on trunk deeply furrowed, either dark greyish-brown and stringy or almost black and resembling an ironbark; branches smooth. Leaves green or subglaucous, 9–18 x 1.5–3 cm; petioles and young branches often reddish. Umbels 7–15-flowered, axillary. Fruit pyriform, 8–11 x 7–8 mm, tapering into the pedicel; disc conspicuous, flat or depressed; valves typically 3, enclosed. Widespread south of Gosford. Ss and laterites. Fl. summer. *Black Ash, Silvertop Ash* . . . . **E. sieberi** L.A.S.Johnson

10 Operculum conical, sometimes sharply beaked. Lateral veins ± parallel with each other, or irregular and the angle c. 30° or more **11**

11 All stamens fertile. Fruit 6 mm diam; valves exserted. Leaves 6–16 x 1.5–3 cm. Operculum c. as long as or slightly longer than, the floral tube. Tree up to 30 m high. Goulburn River Valley. Fl. early summer. *Dusky-leaved Ironbark* . . . . **E. nubila** Maiden & Blakely

11 Outer stamens infertile. Valves enclosed **12**

12 Annular scar remaining when operculum shed. Leaves ± glossy 7–15 x 1.5–3cm. Umbels 7-flowered. Fruit hemispherical, 6–9 mm long. Tree mostly up to 30 m high. Hunter River Valley. WSF . . . . **E. placita** L.A.S.Johnson & K.D.Hill

12 Annular scar not visible when operculum shed. Leaves dull green to glaucous. Leaves 6–11 x 1–2 cm. Umbels 7-flowered. Tree up to c. 30 m high. Cumberland Plain; Woronora Plateau; Hunter River Valley. WSF, usually on clay soils. Fl. spring–summer *Mugga*. . . . . **E. sideroxylon** A.Cunn. ex Woolls

Group 4
Smoothbarks with lateral veins ± parallel to the midrib

1 Operculum hemispherical, much shorter than the floral tube. Fruit mostly 6–8 mm diam., usually pyriform, tapering into a short pedicel; disc prominent, flat or depressed; valves enclosed **2**

1 Operculum conical, often beaked and very acute, c. as long as or longer than the floral tube. Fruit 4–6 mm diam. **3**

2 Tree up to c. 15 m high, usually with a short bole and a spreading crown. Bark smooth, white or mottled. Umbels 7–15-flowered; peduncles 10–20 mm long, terete or slightly flattened. Fruit 6–8 mm diam. Leaves 8–14 x 1.5–5 cm. High elevations in Blue Mts; Kanimbla and Hartley Valleys. Fl. early summer. *Snow Gum or White Sally* . . . . **E. pauciflora** Sieber ex Spreng.

2 Mallee 2–4m high. Umbels usually almost sessile. Fruit 7 x 6mm. Leaves 7–11 x 1–2.5cm. Clarence and Mt Wilson districts. *Mallee Snow Gum* . . . . . . **E. gregsoniana** L.A.S.Johnson & K.D.Hill

3 Leaves mostly 5–8 mm wide, 4–7 cm long; oil glands numerous, conspicuous. Small tree or mallee up to c. 8 m high. Bark smooth. Umbels 7–15-flowered, sessile or on peduncles up to 7 mm long. Fruit globular-truncate, 3–4 x 4–6 mm, sessile; disc flat or depressed; valves enclosed. Higher Blue Mts Wet soils on Ss. Fl. summer. *Narrow-leaved Sally* . . . . **E. moorei** Maiden & Cambage

3 Leaves 12–25 mm wide, 6–8 cm long. Tree up to c. 15 m high. Bark at the base rough and flaky; trunk and branches smooth, dull green or greyish-green. Umbels 7–23-flowered; peduncles almost terete, 3–5 mm long. Operculum hemispherical to acutely conical, almost as long as the floral tube. Fruit ovoid-truncate to pyriform or globular, 4–5 x 4–5 mm, sessile or very shortly pedicellate; disc narrow, depressed; valves enclosed. Higher Blue Mts; Kanimbla and Hartley Valleys. Usually in moist deep clay soils. Fl. autumn–spring. *Black Sally* . . . . **E. stellulata** Sieber ex DC.

E. stellulata

Group 5
Smoothbarks with wide, compressed peduncles

1 Disc flat or depressed; valves enclosed. Mallee up to 6 m high. Bark smooth, glaucous, sometimes with loose ribbons. Leaves coriaceous, green or glaucous, 14–17 x 2–4 cm. Lateral veins irregular; angle 30–40°; intramarginal vein 2–4 mm from margin. Umbels 7–11-flowered. Operculum acutely conical, as long as the floral tube. Fruit 10–12 x 10–13 mm, ribbed or wrinkled, cylindrical, tapering into a stout pedicel; disc broad, flat or depressed; valves enclosed. Helensburg to Gosford. Ss. Fl. early spring. *Yellow-top Ash*. . . . . . . **E. luehmanniana** F.Muell.
1 Disc domed, valves exserted 2

E. luehmanniana

2 Fruit 9 mm diam., shortly pedicellate, hemispherical. Mallee or small tree up to 6 m high. Leaves 6–12 x 1–3 cm. Operculum conical, longer than the floral tube. Upper Hunter Valley. Vulnerable. Fl. autumn. *Pokolbin Mallee* . . . . **E. pumila** Cambage
2 Fruit 10–15 mm diam 3

E. pumila

3 Fruit 13–15 mm diam., shortly pedicellate, 2-ribbed, conical to hemispherical. Tree up to 35 m high. Leaves 10–22 x 2–3.5 cm. Buds 2-ribbed or angular; operculum conical, c. as long as the floral tube. Uncommon in the northern parts of the region. Fl. summer. *Large-fruited Grey Gum* . . . . **E. canaliculata** Maiden
3 Fruit 10–12 mm diam., subsessile to pedicellate, turbinate, 1–2-ribbed. Tree up to 45 m high. Leaves glossy, green, up to 20 x 2.5cm. Buds glaucous; operculum hemispherical to conical, c. as long as the floral tube. Southern parts of the region at high altitudes. Fl. autumn–spring. *Maiden's Gum*. . . . . **E. maidenii** F.Muell.

E. canaliculata

E. maidenii

Group 6
Smoothbarks with leaves paler underneath

1 Valves strongly exserted, usually whitish; disc prominent, standing above the floral tube. Bark grey and roughish (but not fibrous) with patches of smooth usually pink bark 2
1 Valves enclosed or exserted and partly hidden by the floral tube; disc thin, inconspicuous, partly or wholly enclosed in the floral tube 5

2 Fruit 4–5 x 4–5mm, hemispherical. Tree up to 45 m high. Leaves 6–12 x 2 cm. Umbels 7–11-flowered; peduncles sometimes compressed. Operculum hemispherical to conical, as long as the floral tube. Northern parts of the region. Fl. summer. *Small-fruited Grey Gum* . . . . **E. propinqua** H.Deane & Maiden
2 Fruit 6–15 mm diam 3

E. propinqua

3 Fruit 13–15 mm diam . . . . **E. canaliculata** (see above)
3 Fruit 6–10 mm diam 4

4 Tree up to c. 25 m high, sometimes gnarled and shrubby in poor soils in exposed situations. Leaves 6–10 x 2–3 cm. Umbels 7–11-flowered; peduncles usually compressed. Fruit 6–12 x 6–10 mm, pyriform to almost globular, pedicellate or almost sessile. Operculum conical, longer than the floral tube. Widespread, usually on sandy soils and shales in wetter areas. Fl. summer. *Grey Gum*. . . . . **E. punctata** DC.
4 Mallee or small tree up to 6 m high. Hunter River Valley. . . . . **E. pumila** (see above)

E. punctata

5 Smallest branches and buds usually glaucous. Tree up to 45 m high. Bark smooth, deciduous, usually powdery. Leaves 13–20 x 2–3.5 cm, often undulate. Umbels mostly 3–11-flowered;

E. grandis

peduncles compressed. Operculum conical or shortly beaked. Fruits 7–8 x 6–8 mm, pyriform, thin, glaucous, pedicellate; valves usually incurved. Coast, northern parts of the region, usually on better soils. Fl. winter. *Flooded Gum* or *Rose Gum* . . . . **E. grandis** W.Hill ex Maiden

5 Smallest branches and buds green **6**

E. saligna

6 Juvenile leaves narrow-lanceolate to ovate, with undulate margins. Fruit conical or campanulate, 5–6 x 5–6 mm, sessile or shortly pedicellate; disc usually depressed, inconspicuous; valves enclosed or slightly exserted and often spreading. Tree up to c. 60 m high with a straight trunk. Bark usually smooth except near the butt; the trunk more fibrous south of Sydney where it hybridises with *E. botryoides*. Adult leaves 10–20 x 1.5–3 cm. Umbels 7–11-flowered; peduncles usually compressed or 2-angled, 8–12 mm long. Operculum hemispherical or beaked, c. as long as the floral tube. Coast and adjacent plateaus. Usually on better soils. Fl. late summer. *Sydney Blue Gum* . . . . **E. saligna** Sm.

E. deanei

6 Juvenile leaves broad-lanceolate to orbicular. Fruit campanulate or urceolate, 5 x 5mm, pedicellate; disc narrow, usually flat; valves broad-triangular, enclosed or slightly exserted. Tree up to 60 m high, with a straight trunk. Bark smooth to the base. Adult leaves 10–15 x 2–4 cm. Umbels 7–11-flowered; peduncles terete or slightly compressed, 10–13 mm long. Operculum hemispherical, often with a short beak, shorter than the floral tube. Blue Mts; Mooney Mooney Creek; Parson's Creek. Deep soils, usually in gullies. Fl. late summer. *Deane's Gum* . . . . **E. deanei** Maiden

Group 7
Smoothbarks with valves of the fruit well exserted from the orifice

E. camphora

1 Leaves with mucronate tips (Fig. 37), up to 14 x 3 cm. Fruit funnel-shaped, 5–6 x 6–7 mm. Tree up to 15 m high, sometimes mallee-like. Bark mostly smooth, dirty white, usually with ribbons. Umbels 7-flowered. Operculum conical acute or beaked, longer than the floral tube. Megalong Valley; Paddy's River. Swamps and creeks. Endangered. Fl. late summer . . . .
. . . .**E. camphora** R.T.Baker ssp. **camphora**

1 Leaves tapering into a point. Fruit not as in *E. camphora*. **2**

2 Mallee or small tree, usually with crooked and gnarled trunk **3**

2 Trees with upright habit and usually straight trunks **4**

E. dwyeri

3 Operculum more than 1.5 times as long as the floral tube. Bark creamy white or buff coloured, blotched with greenish or bluish patches. Leaves 10–13 x 1.5–2.5 cm. Umbels mostly 3- or 7-flowered. Buds usually reddish, never glaucous (see *E. dealbata* with which this species may be confused). Fruit pedicellate, campanulate to hemispherical or subcylindrical, 7–9 x 8–9 mm; disc prominent. Western parts of the area. Usually on stony ridges. Fl. winter–spring. *Dwyer's Mallee* . . . . **E. dwyeri** Maiden & Blakely

3 Operculum less than 1.5 times as long as the floral tube. Bark grey to greyish-brown. Leaves 5–10 x 2–4 cm, sometimes glaucous. Umbels 7-flowered. Buds 8 mm long. Fruit hemispherical 4–7 x 4–7 mm. Penrose. Swampy flats. Vulnerable. *Mountain Swamp Gum* . . . .
. . . .**E. aquatica** (Blakely) L.A.S.Johnson & K.D.Hill

4 Operculum much longer than the floral tube. **5**

4 Operculum c. as long as the floral tube or shorter **10**

E. blakelyi

5 Fruit 3–4 mm diam (sometimes up to 6 mm). Umbels 4–8-flowered. Tree up to 30 m high, usually with a short bole. Bark smooth, blotched. Juvenile leaves ovate to orbicular, 5–7 mm wide. Adult leaves 9–18 x 1–4 cm. Peduncles terete or very slightly compressed, 10–15 mm long. Operculum conical, 2–3 times as long as the floral tube. Western parts of the region. Fl. spring–summer. *Blakely's Red Gum* . . . . **E. blakelyi** Maiden

5 Fruit more than 4 mm diam **6**

6 Buds and usually smaller branches glaucous **7**

6 Buds and smaller branches not glaucous **8**

7 Fruit c. 8 mm diam. Tree up to 16m high. Bark slaty grey. Adult leaves 10–15 x 2–3 cm. Operculum 2–3 times as long as the floral tube. Coast in northern part of region. Vulnerable. *Slaty Red Gum* . . . . **E. glaucina** Blakely

7 Fruit 5–7 mm diam. Tree up to 15 m high, often crooked. Bark whitish, often rough near the base. Adult leaves 8–16 x 2–3 cm. Operculum usually 1 ¼ –2 ½ times as long as the floral tube. Western parts of the region. Fl. spring–summer. *Tumble-down Gum* . . . .
. . . . **E. dealbata** A.Cunn. ex Schauer

8 Operculum mostly 1.5–2 times as long as the floral tube, usually sharply pointed or beaked. Tree up to 20 m high with wide spreading branches and short bole. Bark dull white, often blotched with yellow or red. Leaves 12–20 x 0.8–1.5 cm. Fruit hemispherical to globular-truncate, 7–8 x 5–6 mm; disc sharp and domed; valves incurved. Along rivers. Hunter River Valley. Fl. mainly spring. Endangered population in Hunter Cathment . *River Red Gum* **E. camaldulensis** Dehnh.

8 Operculum 2–4 times as long as the floral tube 9

9 Umbels 7-flowered. Fruit 5–9 mm diam., pedicellate. Juvenile leaves elliptic to broad-lanceolate, 6–16 x 5–6 cm. Adult leaves 10–20 x 1.5–2.5 cm. Tree up to 40 m high. Bark bluish, usually blotched, with some deciduous ribbons. Widespread along the coast, Cumberland Plain; valleys of Blue Mts Usually open forests on fertile soils. (Plate 00) *Forest Red Gum* . **E. tereticornis** Sm.

9 Umbels more than 7-flowered. Fruit 4–6 mm diam., pedicellate. Juvenile leaves ovate to orbicular, 7–14 x 6–12 cm; stems bearing juvenile leaves square in section. Adult leaves 10–20 x 2.5–3.5 cm. Tree up to 30 m high, usually with a short bole. Bark blotched. Coast and Cumberland Plain. Usually swampy flats. WS. Fl. summer. *Cabbage Gum* . . .**E. amplifolia** Naudin ssp. **amplifolia**

10 Buds and smaller branches glaucous . . . . **E. dealbata** (see above)

10 Buds and branches not glaucous 11

11 Valves prominently exserted and sharply pointed. Fruit hemispherical, shortly pedicellate. Trees up to 20 m high, with drooping branches. Bark smooth, dirty white-grey, blotched. Juvenile leaves narrow-lanceolate, 15 mm wide. Adult leaves 12–16 x 1.5–2.5 cm. Umbels 7-flowered; peduncles terete or angular, 5–7 mm long. *Drooping Red Gums* 12

11 Valves obtuse, slightly exserted or just enclosed 13

12 Fruit less than 7 mm diam. Megalong Swamp; Cumberland Plain; Woronora Plateau. Fl. summer. Endangered population in some LGAs *Parramatta Red Gum* . . . .
. . . . **E. parramattensis** C.Hall ssp. **parramattensis**

12 Fruit more than 7 mm diam. Hunter River Valley. Vulnerable . . . .
. . . .**E. parramattensis** ssp. **decadens** L.A.S.Johnson & Blaxell

13 Bark smooth, usually with patches of flaky bark giving the tree a ragged appearance (bark sometimes completely smooth at some times of the year). Higher altitudes in poor shallow soils . . . . **E. mannifera** (2 subspecies Group 8)

13 Bark smooth, usually with loose ribbons. Operculum conical, much shorter than the floral tube. Tree up to 30m high. Juvenile leaves lanceolate to orbicular, usually glaucous. Adult leaves 8–14 x c. 1.5cm; lateral veins irregular 20–40°. Umbels 7-flowered; peduncles terete, 5–6 mm long. Fruit ovoid-truncate to campanulate, 4–5 x 4–5mm, pedicellate; disc flat or slightly convex; valves usually exserted. Kedumba Valley below Wentworth Falls; Nepean River. Vulnerable. Fl. summer. *Nepean River Gum* . . . . **E. benthamii** Maiden & Cambage

Group 8
Smoothbarks with valves of the fruit enclosed or flush with the orifice

1 Mallees, usually up to 6 m high, rarely trees 2

1 Trees 9

2 Leaves less than 8 mm wide 3

2 Leaves more than 8 mm wide 5

E. glaucina

E. dealbata

E. camaldulensis

E. tereticornis

E. amplifolia

E. dealbata

E. parramattensis

E. mannifera

E. benthamii

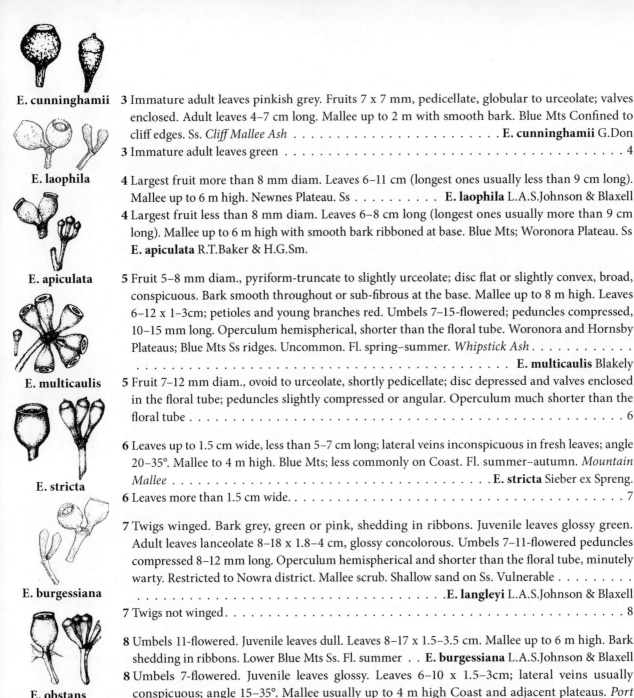

E. cunninghamii

3 Immature adult leaves pinkish grey. Fruits 7 x 7 mm, pedicellate, globular to urceolate; valves enclosed. Adult leaves 4–7 cm long. Mallee up to 2 m with smooth bark. Blue Mts Confined to cliff edges. Ss. *Cliff Mallee Ash* . . . . **E. cunninghamii** G.Don

3 Immature adult leaves green 4

E. laophila

4 Largest fruit more than 8 mm diam. Leaves 6–11 cm (longest ones usually less than 9 cm long). Mallee up to 6 m high. Newnes Plateau. Ss . . . . **E. laophila** L.A.S.Johnson & Blaxell

4 Largest fruit less than 8 mm diam. Leaves 6–8 cm long (longest ones usually more than 9 cm long). Mallee up to 6 m high with smooth bark ribboned at base. Blue Mts; Woronora Plateau. Ss
E. apiculata R.T.Baker & H.G.Sm.

E. apiculata

5 Fruit 5–8 mm diam., pyriform-truncate to slightly urceolate; disc flat or slightly convex, broad, conspicuous. Bark smooth throughout or sub-fibrous at the base. Mallee up to 8 m high. Leaves 6–12 x 1–3cm; petioles and young branches red. Umbels 7–15-flowered; peduncles compressed, 10–15 mm long. Operculum hemispherical, shorter than the floral tube. Woronora and Hornsby Plateaus; Blue Mts Ss ridges. Uncommon. Fl. spring–summer. *Whipstick Ash* . . . .
. . . . **E. multicaulis** Blakely

E. multicaulis

5 Fruit 7–12 mm diam., ovoid to urceolate, shortly pedicellate; disc depressed and valves enclosed in the floral tube; peduncles slightly compressed or angular. Operculum much shorter than the floral tube 6

6 Leaves up to 1.5 cm wide, less than 5–7 cm long; lateral veins inconspicuous in fresh leaves; angle 20–35°. Mallee to 4 m high. Blue Mts; less commonly on Coast. Fl. summer–autumn. *Mountain Mallee* . . . . **E. stricta** Sieber ex Spreng.

E. stricta

6 Leaves more than 1.5 cm wide. 7

7 Twigs winged. Bark grey, green or pink, shedding in ribbons. Juvenile leaves glossy green. Adult leaves lanceolate 8–18 x 1.8–4 cm, glossy concolorous. Umbels 7–11-flowered peduncles compressed 8–12 mm long. Operculum hemispherical and shorter than the floral tube, minutely warty. Restricted to Nowra district. Mallee scrub. Shallow sand on Ss. Vulnerable . . . .
. . . .**E. langleyi** L.A.S.Johnson & Blaxell

E. burgessiana

7 Twigs not winged. 8

8 Umbels 11-flowered. Juvenile leaves dull. Leaves 8–17 x 1.5–3.5 cm. Mallee up to 6 m high. Bark shedding in ribbons. Lower Blue Mts Ss. Fl. summer . . **E. burgessiana** L.A.S.Johnson & Blaxell

8 Umbels 7-flowered. Juvenile leaves glossy. Leaves 6–10 x 1.5–3cm; lateral veins usually conspicuous; angle 15–35°. Mallee usually up to 4 m high Coast and adjacent plateaus. *Port Jackson Mallee* (Intermediates between *E. stricta*, *E. obstans* and *E. apiculata* occur where their distributions meet) . . . .**E. obstans** L.A.S.Johnson & K.D.Hill

E. obstans

9 Fruit funnel-shaped, sessile or shortly pedicellate 10

9 Fruit otherwise, distinctly pedicellate 11

E. dawsonii

10 Fruit 5 mm diam., 5–6 mm long; disc very small, enclosed in the floral tube. Inflorescence terminal, paniculate. Tree up to 30 m high. Bark smooth-scaly. Leaves glaucous, 7–12 x 2–3.5 cm. Operculum hemispherical or conical, shorter than the floral tube. Western parts of the region. Fl. early summer *Slaty Gum*. . . . .**E. dawsonii** R.T.Baker

10 Fruit 6–8 mm diam, 5–9 mm long; disc flat, conspicuous; valves sometimes slightly exserted. Tree up to 20 m high. Umbels axillary. Bark rough at the base, smooth and whitish or yellowish and ribbony above. Leaves 9–12 x 2.5–3 cm, often glossy. Operculum conical to beaked, equal to or slightly longer than the floral tube. Higher altitudes. Usually in swampy ground. Fl. autumn–spring. *Swamp Gum* . . . . **E. ovata** Labill.

E. ovata

E. elata

11 Umbels 15–30-flowered. Leaves 1–1.5 cm wide, 10–20 cm long. Fruit 4–5 mm diam., ovoid-truncate; pedicel as long as or longer than the fruit; disc narrow, flat or depressed. Tree up to 40 m high. Bark hard and fibrous on lower trunk; upper trunk and branches smooth, pale. Operculum hemispherical, much shorter than the floral tube. Widespread. Gullies and River flats. *River Peppermint* . . . .**E. elata** Dehnh.

11 Umbels mostly 7–11-flowered (if more, then the leaves wider or the fruit larger than *E. elata*) 12

12 Leaves mostly 15–25 cm (rarely 10 cm) long, 2–3 cm wide; lateral veins irregular; angle 30–50°. Fruit usually longer than broad, 8–6 x 6–10 mm, pyriform, cylindrical or ovoid-truncate, pedicellate; disc narrow, depressed; valves just enclosed or their tips exserted. Tree up to 45 m high. Bark usually rough near the base but predominantly smooth and hanging in ribbons. Umbels 7-flowered; peduncles compressed, 7–20 mm long. Operculum conical, c. as long as the floral tube. Blue Mts Forests, chiefly in valleys. Fl. summer. *Monkey Gum* . . . .
. . . . **E. cypellocarpa** L.A.S.Johnson

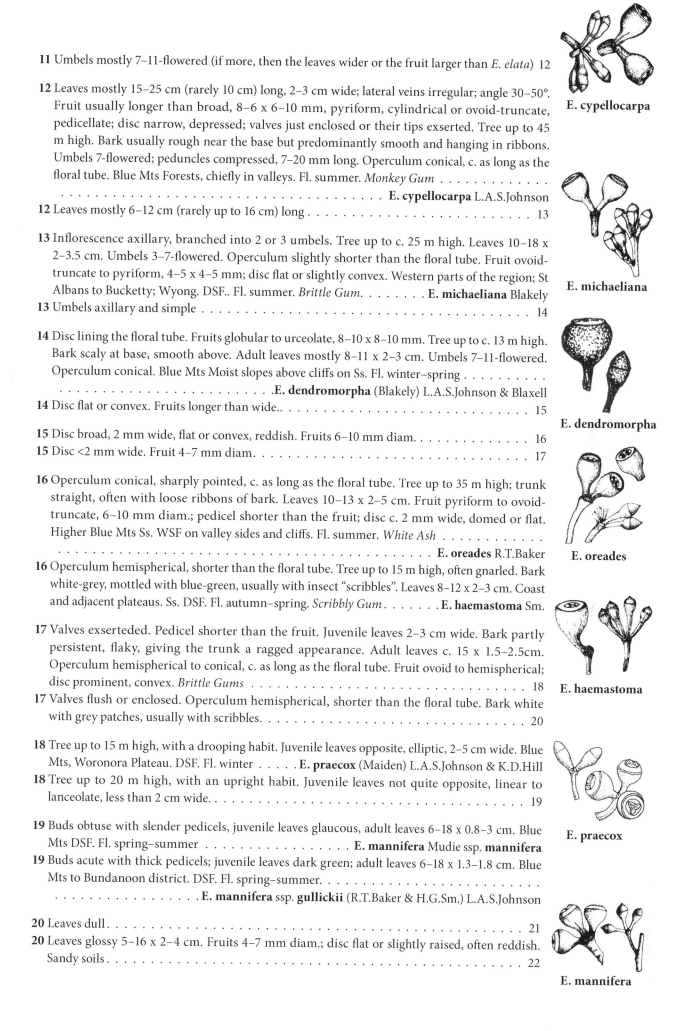

E. cypellocarpa

12 Leaves mostly 6–12 cm (rarely up to 16 cm) long 13

13 Inflorescence axillary, branched into 2 or 3 umbels. Tree up to c. 25 m high. Leaves 10–18 x 2–3.5 cm. Umbels 3–7-flowered. Operculum slightly shorter than the floral tube. Fruit ovoid-truncate to pyriform, 4–5 x 4–5 mm; disc flat or slightly convex. Western parts of the region; St Albans to Bucketty; Wyong. DSF.. Fl. summer. *Brittle Gum.* . . . . **E. michaeliana** Blakely

13 Umbels axillary and simple 14

E. michaeliana

14 Disc lining the floral tube. Fruits globular to urceolate, 8–10 x 8–10 mm. Tree up to c. 13 m high. Bark scaly at base, smooth above. Adult leaves mostly 8–11 x 2–3 cm. Umbels 7–11-flowered. Operculum conical. Blue Mts Moist slopes above cliffs on Ss. Fl. winter–spring . . . .
. . . . .E. dendromorpha (Blakely) L.A.S.Johnson & Blaxell

14 Disc flat or convex. Fruits longer than wide.. 15

E. dendromorpha

15 Disc broad, 2 mm wide, flat or convex, reddish. Fruits 6–10 mm diam. 16

15 Disc <2 mm wide. Fruit 4–7 mm diam. 17

16 Operculum conical, sharply pointed, c. as long as the floral tube. Tree up to 35 m high; trunk straight, often with loose ribbons of bark. Leaves 10–13 x 2–5 cm. Fruit pyriform to ovoid-truncate, 6–10 mm diam.; pedicel shorter than the fruit; disc c. 2 mm wide, domed or flat. Higher Blue Mts Ss. WSF on valley sides and cliffs. Fl. summer. *White Ash* . . . .
. . . . **E. oreades** R.T.Baker

E. oreades

16 Operculum hemispherical, shorter than the floral tube. Tree up to 15 m high, often gnarled. Bark white-grey, mottled with blue-green, usually with insect "scribbles". Leaves 8–12 x 2–3 cm. Coast and adjacent plateaus. Ss. DSF. Fl. autumn–spring. *Scribbly Gum.* . . . .**E. haemastoma** Sm.

17 Valves exserteded. Pedicel shorter than the fruit. Juvenile leaves 2–3 cm wide. Bark partly persistent, flaky, giving the trunk a ragged appearance. Adult leaves c. 15 x 1.5–2.5cm. Operculum hemispherical to conical, c. as long as the floral tube. Fruit ovoid to hemispherical; disc prominent, convex. *Brittle Gums* 18

E. haemastoma

17 Valves flush or enclosed. Operculum hemispherical, shorter than the floral tube. Bark white with grey patches, usually with scribbles. 20

18 Tree up to 15 m high, with a drooping habit. Juvenile leaves opposite, elliptic, 2–5 cm wide. Blue Mts, Woronora Plateau. DSF. Fl. winter . . . . .**E. praecox** (Maiden) L.A.S.Johnson & K.D.Hill

18 Tree up to 20 m high, with an upright habit. Juvenile leaves not quite opposite, linear to lanceolate, less than 2 cm wide. 19

19 Buds obtuse with slender pedicels, juvenile leaves glaucous, adult leaves 6–18 x 0.8–3 cm. Blue Mts DSF. Fl. spring–summer . . . . **E. mannifera** Mudie ssp. **mannifera**

E. praecox

19 Buds acute with thick pedicels; juvenile leaves dark green; adult leaves 6–18 x 1.3–1.8 cm. Blue Mts to Bundanoon district. DSF. Fl. spring–summer. . . . .
. . . . **E. mannifera** ssp. **gullickii** (R.T.Baker & H.G.Sm.) L.A.S.Johnson

20 Leaves dull. 21

20 Leaves glossy 5–16 x 2–4 cm. Fruits 4–7 mm diam.; disc flat or slightly raised, often reddish. Sandy soils. 22

E. mannifera

E. rossii

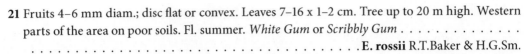

21 Fruits 4–6 mm diam.; disc flat or convex. Leaves 7–16 x 1–2 cm. Tree up to 20 m high. Western parts of the area on poor soils. Fl. summer. *White Gum* or *Scribbly Gum* . . . . **E. rossii** R.T.Baker & H.G.Sm.

21 Fruits 4–5 x 6–7 mm; disc usually convex, reddish. Leaves 7–14 x 1–1.5 cm. Tree up to 15 m high. Coast and adjacent plateaus. Sandy soils, rarely on clays. Fl. spring. *Snappy Gum* or *Narrow-leaved Scribbly Gum* . . . . **E. racemosa** Cav.

22 Valves at rim level. Tree up to 20 m high. Lower Blue Mts; Castlereagh; Hornsby Plateau. Fl. summer. *Hard-leaved Scribbly Gum* . . . . . **E. sclerophylla** (Blakely) L.A.S.Johnson & Blaxell

22 Valves enclosed. Tree up to 20 m high. Hunter River Valley, Morriset. Fl. spring. *Scribbly Gum* . . . . **E. signata** F.Muell.

Group 9
Stringybarks

(The Stringybark species are very difficult to identify because of the wide variations that occur in many of the characters. Identification is further complicated by the hybridization of the stringybarks amongst themselves and with peppermints and ashes.) (See also *E. bensonii* and *E. blaxlandii*.)

E. racemosa

E. sclerophylla

1 Adult leaves 2–6 x 1–1.7 cm; juvenile leaves oblong or elliptic to almost cordate. Shrub up to 4 m high (rarely a small tree). Bark stringy on main stems, flaky or rarely smooth on smaller branches. Umbels 7–15-flowered; peduncles compressed, 4–5 mm long. Operculum hemispherical, shorter than the floral tube. Fruit mostly c. 5 mm diam., hemispherical, sessile, crowded into globular heads; disc convex; valves enclosed. Highlands of Illawarra; Blue Mts Fl. early winter. *Privet-leaved Stringybark* . . . . **E. ligustrina** DC.

1 Adult leaves larger than those of *E. ligustrina* 2

E. ligustrina

2 Fruit pedicellate and clusters open 3
2 Fruit sessile or very shortly pedicellate and clusters crowded 9

3 Disc broad, domed; valves usually strongly exserted 4
3 Disc narrow, flat or depressed; valves enclosed or their tips just exserted. 7

E. cannonii

4 Fruit 15–17 mm diam. Buds 4-angled, 5–8 mm diam. Operculum beaked. Tree up to 10 m high, gnarled, spreading, with pendulous branches. Leaves 8–16 x 2–3 cm. Umbels 3–7-flowered; peduncles compressed, 15–20 mm long. Fruit turbinate, 15–17 x 15–17 mm; valves usually 3. Capertee valley. DSF. Vulnerable. Fl. spring . . . . **E. cannonii** R.T.Baker

4 Fruit 10–12 mm diam., hemispherical to globular. Buds 5 mm or less diam.. Operculum rounded or if beaked then without angles.. 5

E. macrorhyncha

5 Operculum longer than the floral tube, acutely conical or beaked. Tree up to 25 m high, usually with a straight trunk. Bark stringy throughout. Leaves 7–12 x 1.5–2 cm; lateral veins irregular; angle 20–40°; intramarginal vein mostly c. 2 mm from the margin. Umbels axillary or in a terminal panicle, 7–15-flowered; peduncles compressed or almost terete, 10–15 mm long. Western parts of the region. DSF. Fl. summer. *Red Stringybark* . . . . **E. macrorhyncha** F.Muell. ex Benth.

5 Operculum c. as long as or shorter than the floral tube, rounded. 6

E. laevopinea

6 Buds not angular. Leaves 10–16 x 1.5–3 cm. Umbels 7–11-flowered; peduncles almost terete, 7–15 mm long. Fruit globular or flat topped, 6–11 x 11–13 mm. Tall tree up to 40 m high. North-western Blue Mts, often on basalt. WSF. Fl. winter. *Silver-top Stringybark* . . . . **E. laevopinea** R.T.Baker

6 Buds angular. Leaves 7–15 x 1.4–3.2 cm. Umbels 7–11-flowered; peduncles angular, 7–14 mm long. Fruit globular to hemispherical, ± angular, 7–9 mm diam. Tree up to 25 m high. Bucketty region. DSF . . . . **E. prominula** L.A.S.Johnson & K.D.Hill

E. prominula

7 Operculum conical, acute, c. as long as the floral tube. Fruits almost sessile or shortly pedicellate, sometimes crowded into globular heads, globular-truncate, ovoid-truncate or hemispherical, 8–8 x 5–9 mm; disc convex, flat or depressed; valves enclosed or their tips exserted. Tree up to 26 m high. Juvenile leaves elliptic to broad-lanceolate. Adult leaves 10–15 x 1.2–3 cm; lateral veins irregular; angle 20–40°; intramarginal vein mostly 2–3 mm from the margin. Umbels 7–11-flowered; peduncles slightly compressed, 8–15 mm long. Widespread in drier areas. Fl. spring. *Thin-leaved Stringybark*. . . . . **E. eugenioides** Sieber ex Spreng.

E. eugenioides

7 Operculum hemispherical, much shorter than the floral tube, often with an umbo. 8

8 Disc sharp, thin, lining the floral tube. Tree up to 35 m high. Leaves 10–16 x 2–3 cm. Umbels mostly 7–15-flowered; peduncles compressed, 8–12 mm long. Fruit pyriform or ovoid-truncate, pedicellate, 8–9 x 8–9 mm; valves deeply enclosed. Higher Blue Mts Fl. summer–autumn. *Messmate* . . . . .E. obliqua L'Hér.

E. obliqua

8 Disc flat, depressed, lying below the rim of the fruit. Tree up to 30 m high. Leaves 10–15 x 2–3 cm. Umbels 7–11-flowered; peduncles slightly compressed or angular. Fruit globular-truncate, 8–10 x 8–10 mm; valves enclosed or their tips exserted. Illawarra Ranges; Colong district. Deep soils. Fl. summer–autumn *Yellow Stringybark* . . . . **E. muelleriana** A.W.Howitt

9 Adult leaves short, broad, less than 3 times as long as wide. 10

9 Adult leaves more than 3 times as long as wide. 11

E. muelleriana

10 Adult leaves similar on both surfaces, thick, glossy, often mucronate, 8–12 x 2.5–4 cm. Juvenile leaves cordate to orbicular. Mallee or small tree up to 8m high. Bark scaly-fibrous throughout. Umbels 7–15-flowered; peduncles angular-terete, up to 7 mm long, often obscured by the fruits when these are mature. Operculum hemispherical to conical, blunt, usually 2-angled, ± asymmetrical. Fruit sessile, hemipherical, 4–5 x 8–9 mm; disc thick, convex; valves usually enclosed. Royal Nat. Park to Gosford. DSF on laterite. Uncommon. Vulnerable. Fl. summer. *Heart-leaved Stringybark*. . . . . **E. camfieldii** Maiden

E. camfieldii

10 Leaves glossy above, ± duller below, 6–10 x 1.5–3 cm. Veins more conspicuous on lower surface; lateral veins irregular; angle 20–40°; intramarginal veins 2–4 mm from the margin. Juvenile leaves lanceolate, crinkled; margins and midribs stellate-hairy. Tree up to 30 m high. Umbels 7–15-flowered; peduncles compressed or angular, 5–8 mm long. Operculum conical-acute, c. as long as the floral tube. Fruits hemispherical or ovoid-truncate, 5–7 x 6–7 mm, crowded into dense heads; disc conspicuous, domed or flat; valves enclosed or their tips exserted. Coast and adjacent plateaus; lower Blue Mts; Megalong Valley. Common on the junction of Ss and WS. DSF. Fl. summer–autumn. *White Stringybark* . . . . **E. globoidea** Blakely

E. globoidea

11 Buds angular; operculum bluntly conical, rarely acute, much shorter than the floral tube. Fruits 9–12 mm diam., sessile, compressed-globular or globular-truncate, crowded into dense heads; disc flat or convex, broad, reddish; valves enclosed or their tips exserted. Tree up to 10 m high, frequently gnarled, sometimes shrubby. Juvenile leaves orbicular to broad-lanceolate, undulate, hispid. Adult leaves coriaceous, glossy, up to 17 x 3.5 cm; lateral veins irregular; angle 20–40°; intramarginal vein mostly 2–3 mm from the margin. Umbels 7–15-flowered; peduncles compressed or angular, 7–10 mm long. Coast and adjacent plateaus. Fl. summer. *Brown Stringybark* . . . . **E. capitellata** Sm.

E. capitellata

11 Buds sometimes compressed but without angles; operculum conical, acute or beaked, as long as or slightly longer than the floral tube. 12

12 Juvenile leaves narrow-lanceolate to linear. Adult leaves always 2 cm or less wide.. 13

12 Juvenile leaves lanceolate to elliptic or ovate. Adult leaves 1.2–3 cm wide. Fruits 5–10 mm diam. 15

E. tenella

13 Adult leaves mostly less than 6 cm long, 1–1.7 cm wide. Fruit with valves at rim level or just exserted. Umbels 7–11-flowered; peduncles terete, 3–6 mm long. Fruit globular to hemispherical, 5–7 mm diam. Tree up to 15 m high. Blue Mts DSF. Ss on shallow soils . . . . **E. tenella** L.A.S.Johnson & K.D.Hill

13 Adult leaves mostly more than 6 cm long;. Fruit with conspicuous disc and valve enclosed or to rim level 14

E. ralla

E. agglomerata

E. eugenioides

E. imitans

E. oblonga

E. scias

E. robusta

E. microcorys

14 Adult leaves linear, 6–15 x 0.6–2 cm. Umbels more than 7-flowered. Fruit globular, 5–7 mm diam., valves deeply enclosed; peduncles angular, 4–9 mm long. Tree up to 15 m high. Blue Mts, Nattai River. DSF. Ss, steep slopes. . . . .**E. ralla** L.A.S.Johnson & K.D.Hill

14 Adult leaves lanceolate, 8–12 cm x 1–2cm. Umbels more than 11-flowered. Fruit globular, 5–8 mm diam., valves enclosed to rim level. Tree to 20 m. Widespread. DSF. Shallow sandy soil . . .
. . . . **E. sparsifolia** DC.

15 Adult leaves bluish when fresh, 10–18 x 2–3 cm; lateral veins irregular; angle 20–40°; intramarginal vein up to 5 mm from the margin. Juvenile leaves lanceolate to ovate, softly tomentose. Fruit hemispherical or globular-truncate, compressed, 5–6 x 8 mm, usually sessile, crowded into dense heads, often becoming angular at maturity; disc broad, flat or convex; valves enclosed or their tips exserted. Umbels 7–15-flowered; peduncles usually compressed, 7–12 mm long. Operculum conical, beaked, c. as long as the floral tube. Tree up to c. 30 m high. Widespread. DSF on sandy soils. Fl. autumn–winter. *Blue-leaved Stringybark*. . . . . **E. agglomerata** Maiden

15 Adult leaves green. Operculum not drawn into a beak. 16

16 Leaves paler underneath, 10–14 x 2–3.3 cm. Umbels with 11 or more flowers. . . . .
. . . . **E. eugenioides** (see above)

16 Leaves similar in colour on both surfaces 17

17 Fruit greater than 7 mm diam. Adult leaves broad -lanceolate, 5–13 x 1.5–4 cm. Umbels more than 7-flowered. Fruit globular to hemispherical, 7–9 mm diam.; valves enclosed or their tips exserted. Tree up to 10 m high. Shoalhaven region. DSF. Ss . . . . . **E. imitans** L.A.S.Johnson & K.D.Hill

17 Fruit usually less than 7 mm diam. Adult leaves lanceolate, 8–15 x 0.8–2 cm. Umbels 7–11-flowered; peduncles angular or flattened, 6–11 mm long. Fruit 6–9 mm diam. valves to rim level or exerted. Tree up to 15 m high. Coast and adjacent plateaus; lower Blue Mts DSF. Ss, usually on skeletal soils. Fl. summer–autumn. . . . .**E. oblonga** Blakely

Group 10
Mahoganies except *E. globoidea*

1 Fruit 10–12 mm diam 2
1 Fruit less than 10 mm diam 3

2 Fruit c. as long as broad, 12–18 x 12–20 mm, hemispherical-conical or campanulate; disc prominent, flat; valves usually 4, usually exserted . . . .
. . . .**E. scias** L.A.S.Johnson & K.D.Hill (2 subspecies see Group 2)

2 Fruit slightly to very elongated, 12–15 x 10–12 mm, cylindrical to urceolate, pedicellate; disc enclosed in the floral tube; valves 3, enclosed. Tree up to 25 m high. Bark rough throughout, fibrous-flaky, furrowed. Leaves 10–18 x 4–8cm. Lateral veins parallel; angle 50–65°; intramarginal vein 1–2 mm from the margin, Umbels solitary in the axils or a few together in a panicle, 7–11-flowered; peduncles compressed, 2–3 cm long, 4–7 mm wide. Operculum conical, beaked, c. as long as the floral tube. Coast. DSF. Often in swamps. Fl. spring. *Swamp Mahogany*
. . . . **E. robusta** Sm.

3 Fruit usually longer than broad. 4
3 Fruit c. as long as broad, ± globular or depressed-globular or hemispherical. 5

4 Fruit ± conical, tapering into the pedicel, 7–10 x 5–6 mm; valves enclosed or exserted; disc obscure. Tree up to 30 m high. Bark subfibrous, brick red or rusty or yellowish. Leaves 8–10 x 2–3.5 cm. Lateral veins roughly parallel; angle 45–65°. Umbels 7-flowered; peduncles compressed. Operculum hemispherical, much shorter than the floral tube. Dora Creek. WSF on heavy soils. Fl. summer. *Tallow Wood* . . . . **E. microcorys** F.Muell.

4 Fruit cylindrical or barrel-shaped, 7–9 x 6–7 mm; disc narrow, obscure; valves enclosed or their tips just exserted. Floral tube of bud and sometimes of fruit with 1 or 2 ribs. Tree up to 25 m high. Bark as in *E. microcorys* but smaller branches smooth. Leaves 10–14 x 3–6 cm. Lateral veins regular or irregular, often parallel; angle 40–60°; intramarginal vein 1–2 mm from the margin. Umbels 7-flowered; peduncles very compressed, 7–10 mm long. Operculum hemispherical, obtuse or with a short beak, much shorter than the floral tube. Coastal. Deep, usually wet, and often saline soils. Fl. summer. *Bangalay* (Intermediates between *E. botryoides* and *E. saligna* occur extensively in the south of the region) . . . . **E. botryoides** Sm.

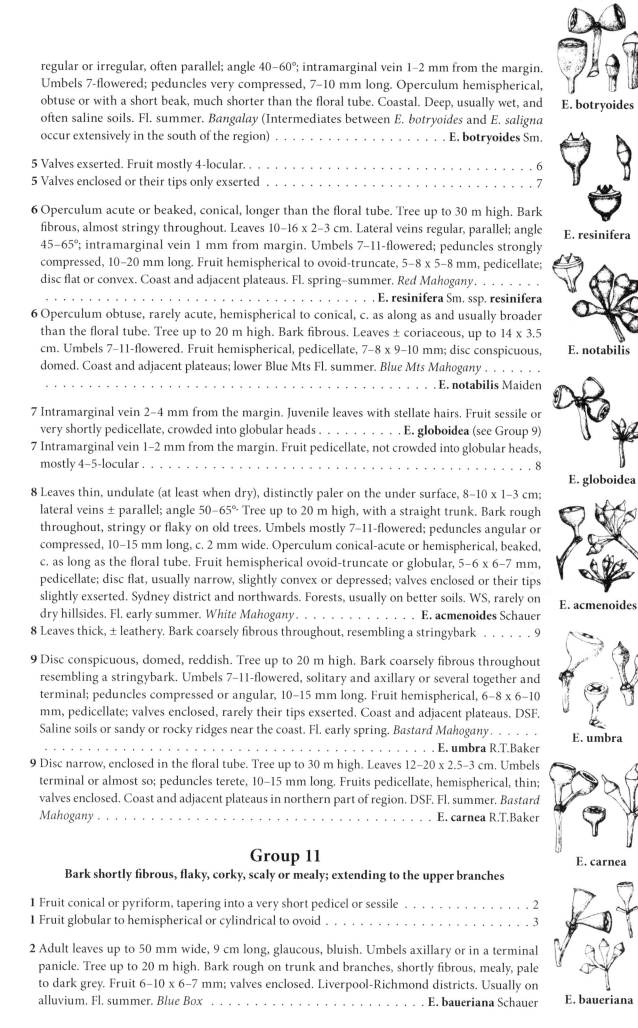

E. botryoides

5 Valves exserted. Fruit mostly 4-locular. 6
5 Valves enclosed or their tips only exserted 7

6 Operculum acute or beaked, conical, longer than the floral tube. Tree up to 30 m high. Bark fibrous, almost stringy throughout. Leaves 10–16 x 2–3 cm. Lateral veins regular, parallel; angle 45–65°; intramarginal vein 1 mm from margin. Umbels 7–11-flowered; peduncles strongly compressed, 10–20 mm long. Fruit hemispherical to ovoid-truncate, 5–8 x 5–8 mm, pedicellate; disc flat or convex. Coast and adjacent plateaus. Fl. spring–summer. *Red Mahogany*. . . . .
. . . . **E. resinifera** Sm. ssp. **resinifera**

E. resinifera

6 Operculum obtuse, rarely acute, hemispherical to conical, c. as along as and usually broader than the floral tube. Tree up to 20 m high. Bark fibrous. Leaves ± coriaceous, up to 14 x 3.5 cm. Umbels 7–11-flowered. Fruit hemispherical, pedicellate, 7–8 x 9–10 mm; disc conspicuous, domed. Coast and adjacent plateaus; lower Blue Mts Fl. summer. *Blue Mts Mahogany* . . . .
. . . . **E. notabilis** Maiden

E. notabilis

7 Intramarginal vein 2–4 mm from the margin. Juvenile leaves with stellate hairs. Fruit sessile or very shortly pedicellate, crowded into globular heads . . . . **E. globoidea** (see Group 9)
7 Intramarginal vein 1–2 mm from the margin. Fruit pedicellate, not crowded into globular heads, mostly 4–5-locular. 8

E. globoidea

8 Leaves thin, undulate (at least when dry), distinctly paler on the under surface, 8–10 x 1–3 cm; lateral veins ± parallel; angle 50–65°· Tree up to 20 m high, with a straight trunk. Bark rough throughout, stringy or flaky on old trees. Umbels mostly 7–11-flowered; peduncles angular or compressed, 10–15 mm long, c. 2 mm wide. Operculum conical-acute or hemispherical, beaked, c. as long as the floral tube. Fruit hemispherical ovoid-truncate or globular, 5–6 x 6–7 mm, pedicellate; disc flat, usually narrow, slightly convex or depressed; valves enclosed or their tips slightly exserted. Sydney district and northwards. Forests, usually on better soils. WS, rarely on dry hillsides. Fl. early summer. *White Mahogany*. . . . . **E. acmenoides** Schauer
8 Leaves thick, ± leathery. Bark coarsely fibrous throughout, resembling a stringybark 9

E. acmenoides

9 Disc conspicuous, domed, reddish. Tree up to 20 m high. Bark coarsely fibrous throughout resembling a stringybark. Umbels 7–11-flowered, solitary and axillary or several together and terminal; peduncles compressed or angular, 10–15 mm long. Fruit hemispherical, 6–8 x 6–10 mm, pedicellate; valves enclosed, rarely their tips exserted. Coast and adjacent plateaus. DSF. Saline soils or sandy or rocky ridges near the coast. Fl. early spring. *Bastard Mahogany*. . . . . .
. . . . **E. umbra** R.T.Baker
9 Disc narrow, enclosed in the floral tube. Tree up to 30 m high. Leaves 12–20 x 2.5–3 cm. Umbels terminal or almost so; peduncles terete, 10–15 mm long. Fruits pedicellate, hemispherical, thin; valves enclosed. Coast and adjacent plateaus in northern part of region. DSF. Fl. summer. *Bastard Mahogany* . . . . **E. carnea** R.T.Baker

E. umbra

E. carnea

Group 11
Bark shortly fibrous, flaky, corky, scaly or mealy; extending to the upper branches

1 Fruit conical or pyriform, tapering into a very short pedicel or sessile 2
1 Fruit globular to hemispherical or cylindrical to ovoid 3

2 Adult leaves up to 50 mm wide, 9 cm long, glaucous, bluish. Umbels axillary or in a terminal panicle. Tree up to 20 m high. Bark rough on trunk and branches, shortly fibrous, mealy, pale to dark grey. Fruit 6–10 x 6–7 mm; valves enclosed. Liverpool-Richmond districts. Usually on alluvium. Fl. summer. *Blue Box* . . . . **E. baueriana** Schauer

E. baueriana

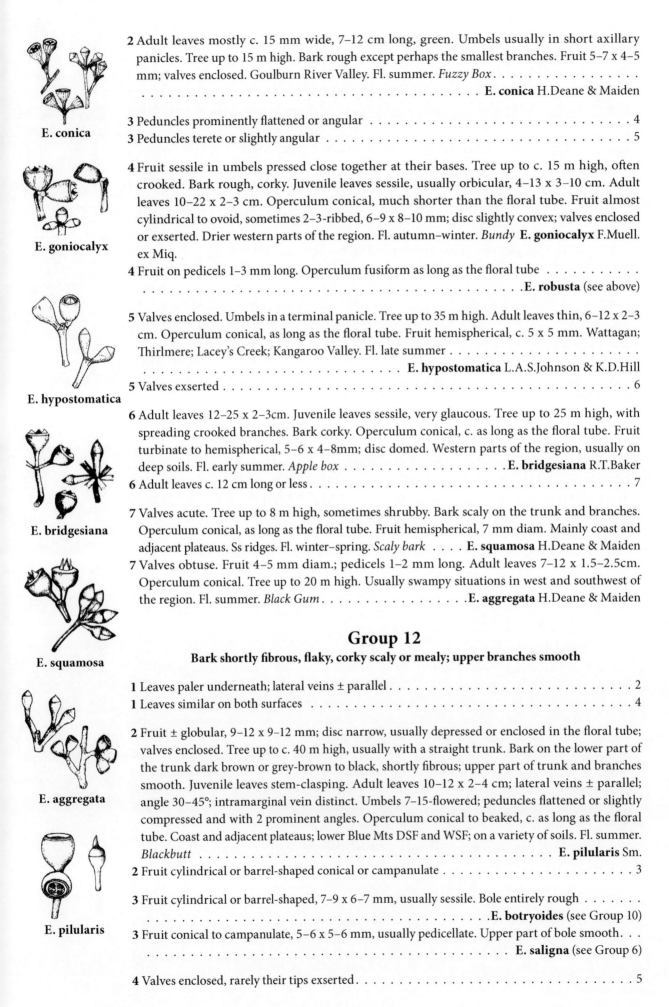

2 Adult leaves mostly c. 15 mm wide, 7–12 cm long, green. Umbels usually in short axillary panicles. Tree up to 15 m high. Bark rough except perhaps the smallest branches. Fruit 5–7 x 4–5 mm; valves enclosed. Goulburn River Valley. Fl. summer. *Fuzzy Box* . . . . **E. conica** H.Deane & Maiden

E. conica

3 Peduncles prominently flattened or angular 4
3 Peduncles terete or slightly angular 5

4 Fruit sessile in umbels pressed close together at their bases. Tree up to c. 15 m high, often crooked. Bark rough, corky. Juvenile leaves sessile, usually orbicular, 4–13 x 3–10 cm. Adult leaves 10–22 x 2–3 cm. Operculum conical, much shorter than the floral tube. Fruit almost cylindrical to ovoid, sometimes 2–3-ribbed, 6–9 x 8–10 mm; disc slightly convex; valves enclosed or exserted. Drier western parts of the region. Fl. autumn–winter. *Bundy* **E. goniocalyx** F.Muell. ex Miq.

E. goniocalyx

4 Fruit on pedicels 1–3 mm long. Operculum fusiform as long as the floral tube . . . . **E. robusta** (see above)

5 Valves enclosed. Umbels in a terminal panicle. Tree up to 35 m high. Adult leaves thin, 6–12 x 2–3 cm. Operculum conical, as long as the floral tube. Fruit hemispherical, c. 5 x 5 mm. Wattagan; Thirlmere; Lacey's Creek; Kangaroo Valley. Fl. late summer . . . . **E. hypostomatica** L.A.S.Johnson & K.D.Hill
5 Valves exserted 6

E. hypostomatica

6 Adult leaves 12–25 x 2–3cm. Juvenile leaves sessile, very glaucous. Tree up to 25 m high, with spreading crooked branches. Bark corky. Operculum conical, c. as long as the floral tube. Fruit turbinate to hemispherical, 5–6 x 4–8mm; disc domed. Western parts of the region, usually on deep soils. Fl. early summer. *Apple box* . . . . **E. bridgesiana** R.T.Baker
6 Adult leaves c. 12 cm long or less 7

E. bridgesiana

7 Valves acute. Tree up to 8 m high, sometimes shrubby. Bark scaly on the trunk and branches. Operculum conical, as long as the floral tube. Fruit hemispherical, 7 mm diam. Mainly coast and adjacent plateaus. Ss ridges. Fl. winter–spring. *Scaly bark* . . . . **E. squamosa** H.Deane & Maiden
7 Valves obtuse. Fruit 4–5 mm diam.; pedicels 1–2 mm long. Adult leaves 7–12 x 1.5–2.5cm. Operculum conical. Tree up to 20 m high. Usually swampy situations in west and southwest of the region. Fl. summer. *Black Gum* . . . . **E. aggregata** H.Deane & Maiden

E. squamosa

Group 12
Bark shortly fibrous, flaky, corky scaly or mealy; upper branches smooth

1 Leaves paler underneath; lateral veins ± parallel 2
1 Leaves similar on both surfaces 4

2 Fruit ± globular, 9–12 x 9–12 mm; disc narrow, usually depressed or enclosed in the floral tube; valves enclosed. Tree up to c. 40 m high, usually with a straight trunk. Bark on the lower part of the trunk dark brown or grey-brown to black, shortly fibrous; upper part of trunk and branches smooth. Juvenile leaves stem-clasping. Adult leaves 10–12 x 2–4 cm; lateral veins ± parallel; angle 30–45°; intramarginal vein distinct. Umbels 7–15-flowered; peduncles flattened or slightly compressed and with 2 prominent angles. Operculum conical to beaked, c. as long as the floral tube. Coast and adjacent plateaus; lower Blue Mts DSF and WSF; on a variety of soils. Fl. summer. *Blackbutt* . . . . **E. pilularis** Sm.

E. aggregata

2 Fruit cylindrical or barrel-shaped conical or campanulate 3

3 Fruit cylindrical or barrel-shaped, 7–9 x 6–7 mm, usually sessile. Bole entirely rough . . . . **E. botryoides** (see Group 10)
3 Fruit conical to campanulate, 5–6 x 5–6 mm, usually pedicellate. Upper part of bole smooth. . . . . **E. saligna** (see Group 6)

E. pilularis

4 Valves enclosed, rarely their tips exserted 5

4 Valves exserted 18

5 Umbels in a panicle which is usually terminal. 6
5 Umbels axillary, solitary (rarely in panicles in *E. bosistoana* and some forms of *E. piperita*) . . 10

6 Adult leaves with blunt apices, usually with a mucro, broad in comparison with length, 5–14 x 1.5–4.5 cm, sub-glaucous. Tree up to 20 m high. Rough bark scaly-fibrous. Operculum conical, shorter than the floral tube. Fruit pedicellate, hemispherical, ovoid to pyriform, 5–6 x 4–5mm. Capertee area. Fl. spring. *Red Box* . . . . **E. polyanthemos** Schauer ssp. **polyanthemos**
6 Adult leaves not as in *E. polyanthemos*. 7

7 Fruit c. 3 mm diam., 4 mm long, ovoid-pyriform to oblong, pedicellate; disc very thin. Tree up to 24 m high. Rough bark shortly fibrous, greyish; upper trunk and branches white. Leaves 5–12 x 0.8–2 cm. Operculum hemispherical, much shorter than the floral tube. Pokolbin State Forest. Fl. winter . . . . **E. largeana** Blakely & Beuzev.
7 Fruit more than 3 mm diam. 8

8 Fruit 8–10 mm diam., glaucous, sessile or shortly pedicellate, cylindrical pyriform or barrel-shaped, 9–12 mm long; disc enclosed. Tree up to 24 m high. Rough bark shortly fibrous, usually mealy-white. Leaves 10–15 x 2–3 cm, glaucous. Operculum conical, often angular, as long as the floral tube. Drier Western parts of the region. Forests on better soils. Fl. late summer–autumn. *White Box*. . . . . **E. albens** Benth.
8 Fruit 4–6 mm diam. 9

9 Fruit elongated, cylindrical to pyriform or urceolate, 7–8 x 4–5 mm. Tree up to 30 m high. Rough bark shortly fibrous, grey. Leaves 8–13 x 2.5–3.5 cm. Sometimes a few umbels axillary. Operculum conical, often beaked, c. as long as the floral tube. Common in drier parts of Cumberland Plain; Hunter River Valley. Forests on shales. Fl. summer. *Grey Box* . . . .
. . . . **E. moluccana** Roxb.
9 Fruit c. as broad as long, sub-globular to pyriform, 5–7 x 4–5 mm. Tree up to 15 m high. Rough bark shortly fibrous, greyish white. Leaves 7–12 x 1.5–2 cm. Operculum conical, as long as or longer than the floral tube. Mainly drier western parts of the region. Fl. summer–winter. *Green-leaved Box* . . . . **E. microcarpa** Maiden

10 Disc conspicuous, flat or convex 11
10 Disc enclosed in the floral tube or sharp and narrow. 16

11 Fruit 6–8 mm diam. 12
11 Fruit c. 5 mm diam.. 14

12 Pedicels slender and distinct from the fruit. Tree up to 20 m high. Bark on trunk and larger branches fibrous but not stringy, smooth on smaller branches. Leaves 10–15 x 2–5 cm; lateral veins irregular; angle 15–35°; intramarginal vein faint, 1–3 mm from the margin. Umbels 7–15-flowered; peduncles slightly compressed or angular. Operculum usually conical, shorter than the floral tube. Fruit pyriform or hemispherical, 6–8 x 5–7 mm, pedicellate; disc conspicuous, convex; valves enclosed or their tips exserted. Higher Blue Mts Ss. DSF. Fl. spring. *Broad-leaved Peppermint.* . . . . **E. dives** Schauer
12 Pedicel stout, merging into the fruit 13

13 Valves usually 4, enclosed or the tips slightly exserted. Rough bark shortly fibrous (peppermint). Tree up to c. 12 m high, with a short bole. Leaves 10–18 x 2–3 cm; lateral veins irregular; angle 15–35°; intramarginal vein up to 3 mm from the margin. Umbels 7–11-flowered; peduncles terete or slightly compressed, 7–12 mm long. Operculum hemispherical, shorter than the floral tube. Fruit pyriform, 6–8 x 6–8 mm; disc prominent, flat or depressed. Woronora Plateau; Blue Mts DSF. Ss. Uncommon. Fl. early spring. *Yertchuk* . . . . **E. consideniana** Maiden
13 Valves usually 3, enclosed. Rough bark hard and long-stringy or resembling an ironbark, deeply furrowed longitudinally. Twigs glaucous, reddish . . . . **E. sieberi** (see Group 3)

E. polyanthemos

E. largeana

E. albens

E. moluccana

E. microcarpa

E. dives

E. consideniana

E. sieberi

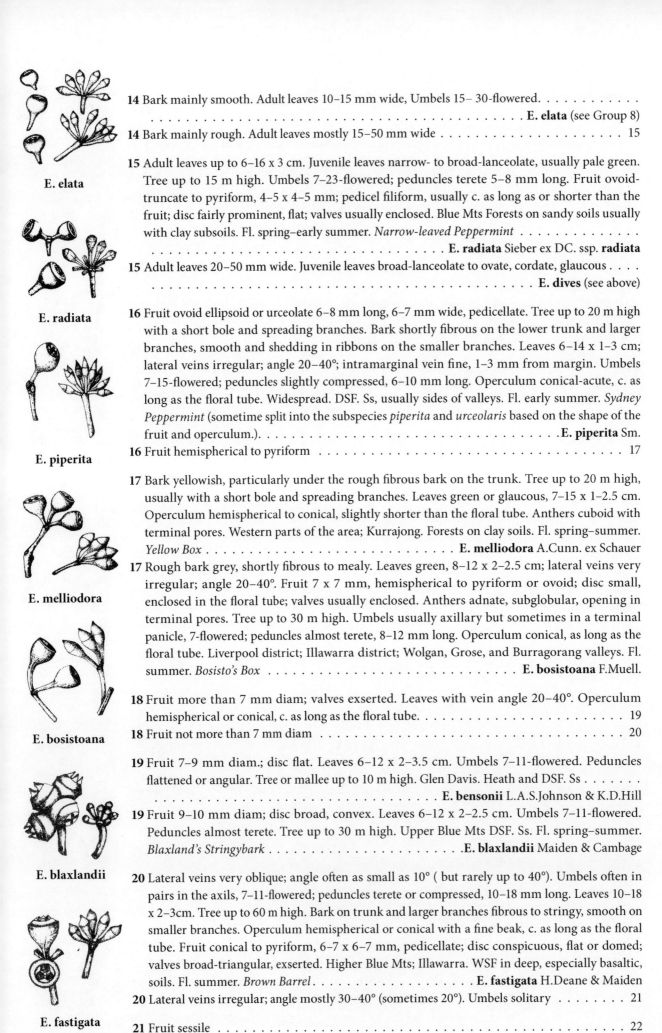

14 Bark mainly smooth. Adult leaves 10–15 mm wide, Umbels 15– 30-flowered. . . . .
. . . . **E. elata** (see Group 8)

14 Bark mainly rough. Adult leaves mostly 15–50 mm wide **15**

E. elata

15 Adult leaves up to 6–16 x 3 cm. Juvenile leaves narrow- to broad-lanceolate, usually pale green. Tree up to 15 m high. Umbels 7–23-flowered; peduncles terete 5–8 mm long. Fruit ovoid-truncate to pyriform, 4–5 x 4–5 mm; pedicel filiform, usually c. as long as or shorter than the fruit; disc fairly prominent, flat; valves usually enclosed. Blue Mts Forests on sandy soils usually with clay subsoils. Fl. spring–early summer. *Narrow-leaved Peppermint* . . . .
. . . . **E. radiata** Sieber ex DC. ssp. **radiata**

15 Adult leaves 20–50 mm wide. Juvenile leaves broad-lanceolate to ovate, cordate, glaucous . . . .
. . . . **E. dives** (see above)

E. radiata

16 Fruit ovoid ellipsoid or urceolate 6–8 mm long, 6–7 mm wide, pedicellate. Tree up to 20 m high with a short bole and spreading branches. Bark shortly fibrous on the lower trunk and larger branches, smooth and shedding in ribbons on the smaller branches. Leaves 6–14 x 1–3 cm; lateral veins irregular; angle 20–40°; intramarginal vein fine, 1–3 mm from margin. Umbels 7–15-flowered; peduncles slightly compressed, 6–10 mm long. Operculum conical-acute, c. as long as the floral tube. Widespread. DSF. Ss, usually sides of valleys. Fl. early summer. *Sydney Peppermint* (sometime split into the subspecies *piperita* and *urceolaris* based on the shape of the fruit and operculum.). . . . .**E. piperita** Sm.

16 Fruit hemispherical to pyriform **17**

E. piperita

17 Bark yellowish, particularly under the rough fibrous bark on the trunk. Tree up to 20 m high, usually with a short bole and spreading branches. Leaves green or glaucous, 7–15 x 1–2.5 cm. Operculum hemispherical to conical, slightly shorter than the floral tube. Anthers cuboid with terminal pores. Western parts of the area; Kurrajong. Forests on clay soils. Fl. spring–summer. *Yellow Box* . . . . **E. melliodora** A.Cunn. ex Schauer

17 Rough bark grey, shortly fibrous to mealy. Leaves green, 8–12 x 2–2.5 cm; lateral veins very irregular; angle 20–40°. Fruit 7 x 7 mm, hemispherical to pyriform or ovoid; disc small, enclosed in the floral tube; valves usually enclosed. Anthers adnate, subglobular, opening in terminal pores. Tree up to 30 m high. Umbels usually axillary but sometimes in a terminal panicle, 7-flowered; peduncles almost terete, 8–12 mm long. Operculum conical, as long as the floral tube. Liverpool district; Illawarra district; Wolgan, Grose, and Burragorang valleys. Fl. summer. *Bosisto's Box* . . . . **E. bosistoana** F.Muell.

E. melliodora

18 Fruit more than 7 mm diam; valves exserted. Leaves with vein angle 20–40°. Operculum hemispherical or conical, c. as long as the floral tube. **19**

18 Fruit not more than 7 mm diam **20**

E. bosistoana

19 Fruit 7–9 mm diam.; disc flat. Leaves 6–12 x 2–3.5 cm. Umbels 7–11-flowered. Peduncles flattened or angular. Tree or mallee up to 10 m high. Glen Davis. Heath and DSF. Ss . . . .
. . . . **E. bensonii** L.A.S.Johnson & K.D.Hill

19 Fruit 9–10 mm diam; disc broad, convex. Leaves 6–12 x 2–2.5 cm. Umbels 7–11-flowered. Peduncles almost terete. Tree up to 30 m high. Upper Blue Mts DSF. Ss. Fl. spring–summer. *Blaxland's Stringybark* . . . .**E. blaxlandii** Maiden & Cambage

E. blaxlandii

20 Lateral veins very oblique; angle often as small as 10° (but rarely up to 40°). Umbels often in pairs in the axils, 7–11-flowered; peduncles terete or compressed, 10–18 mm long. Leaves 10–18 x 2–3cm. Tree up to 60 m high. Bark on trunk and larger branches fibrous to stringy, smooth on smaller branches. Operculum hemispherical or conical with a fine beak, c. as long as the floral tube. Fruit conical to pyriform, 6–7 x 6–7 mm, pedicellate; disc conspicuous, flat or domed; valves broad-triangular, exserted. Higher Blue Mts; Illawarra. WSF in deep, especially basaltic, soils. Fl. summer. *Brown Barrel* . . . . **E. fastigata** H.Deane & Maiden

20 Lateral veins irregular; angle mostly 30–40° (sometimes 20°). Umbels solitary **21**

E. fastigata

21 Fruit sessile **22**

21 Fruit pedicellate 23

22 Fruit with acute valves, ovoid campanulate or conical, c. 6 x 5 mm; disc narrow, flat. Large branches usually rough. Leaves 9–20 x 1–2 cm, sinuate-dentate; lateral veins irregular; angle 30–40°; intramarginal vein 1–2 mm from margin. Umbels 7-flowered; peduncles terete or compressed, 5–8 mm long. Operculum hemispherical or conical, c. as long as the floral tube. Tree up to 50 m high. Illawarra. WSF on deep soils, usually in sheltered places. *Coastal White Box* . . . . **E. quadrangulata** H.Deane & Maiden

22 Fruit with obtuse valves, globular, 5 x 6 mm. Large branches rough. Leaves 9–13 x 1–1.5cm. Umbels 3–7-flowered; peduncles flattened or angular, 2–9 mm long. Operculum hemispherical or conical, shorter than the floral tube. Tree up to 40 m high. Boyd River; Moss Vale. Open forest on broad flats. Fl. spring–summer. *Paddy's River Box* or *Camden Woollybutt.* . . . . **E. macarthurii** H.Deane & Maiden

23 Large branches smooth; bark deeply furrowed, intermediate between an ironbark and a stringybark. Juvenile leaves pale green. Adult leaves 10–16 x 1–2 cm. Umbels 7–11-flowered; peduncles flattened or angular, 5–12 mm long. Operculum conical, c. as long as the floral tube. Mallee or tree up to 40 m high. Southern parts of the area. WSF on deep soils. Fl. summer. *Gully Ash* . . . . **E. smithii** R.T.Baker

23 Large branches rough, only the smaller ones smooth. Juvenile leaves greyish-green. Adult leaves 6–12 x 1–2 cm. Umbels 7-flowered. Operculum conical, c. as long as the floral tube. Fruit hemispherical to campanulate, 4–6 mm diam. Rylestone district. Tree up to 20 m high. DSF. Ss. Vulnerable. Fl. spring–summer . . . . **E. corticosa** L.A.S.Johnson

18 Leptospermum J.R. Forst. & G.Forst.
Tea Trees
77 species in Aust. (75 endemic); all states and territories

Shrubs or small trees. Leaves alternate, margins entire. Flowers solitary or arranged in axillary clusters. Floral tube shortly continued above the ovary. Sepals 5. Petals 5, pink to red or white. Stamens numerous, arranged in a single whorl, shorter or scarcely longer than the petals. Ovary 3–12-locular with numerous ovules per loculus. Fruit a capsule.

Key to the species

1 Capsule usually shed soon, persisting at most for a year, not hard and woody. Summit of the ovary pubescent . . . . **GROUP 1**
1 Capsule persisting for several years, hard and woody. Summit of the ovary glabrous 2

2 External surface of floral tube glabrous or occasionally very shortly pubescent. Sepals falling from mature capsule (except in *L. epacroideum* and *L. petraeum*) . . . . **GROUP 2**
2 External surface of floral tube with long, silky and/or dense hairs. Sepals persisting on capsule (except in *L. myrtifolium*) . . . . **GROUP 3**

Group 1

1 Ovary 8–12-locular. Capsule 7–8 mm diam., with a flat summit. Leaves obovate to narrow-elliptic, thick, 1–2 cm long, 5–7 mm wide, grey-green, obtuse or with a blunt mucro. Flowers solitary in the leaf axils, 12–15 mm diam., with two pairs of large caducous bracteoles and several smaller ones. Tall shrub or small tree with hard bark. Sand dunes near the sea. Fl. spring. *Victoria Tea-tree* or *Coast Tea-tree* . . . . **L. laevigatum** (Gaertn.) F.Muell.
1 Ovary 3–5-locular. Capsule ± domed 2

2 Floral tube glabrous or nearly so 3
2 Floral tube hairy 4

E. quadrangulata

E. macarthurii

E. smithii

E. corticosa

3 Leaves 2–4 mm wide, petiolate, up to 25 mm long. Flowers 5–6 mm diam. Petals white. Ovary mostly 5-locular. Capsule c. 3 mm diam. Shrub up to 5 m high with furrowed flakey bark. Widespread. Heath and DSF. Ss. Fl. spring–summer . . . . **L. polyanthum** Joy Thomps.

3 Leaves 1–2 mm wide, sessile, up to 15 mm long. Flowers 8–10 mm diam. Petals white. Ovary 4–5-locular. Capsule 3–4 mm diam. Shrub up to 5 m high; the bark shedding in strips. Lane Cove. Vulnerable. DSF. Fl. spring–summer . . . . **L. deanei** Joy Thomps.

4 Pedicel more than 2 mm long. 5
4 Flowers sessile or nearly so, pedicels up to 2 mm long. 6

5 Bark hard and smooth. Leaves narrow-obovate to oblong, 13–23 mm long, 2–3 mm wide. Flowers 6–10 mm diam. Capsule c. 4 mm diam. Shrub up to 4 m high, with drooping branches. Blue Mts Open forest, mostly on granite. Fl. spring–summer. . . . . **L. brevipes** F.Muell.

5 Bark shedding in fibrous strips. Leaves obovate to elliptic 4–10 mm long, up to 5 mm wide. Flowers c. 7 mm diam. Petals white to pink. Capsule up to 4 mm diam. Shrub up to c. 1 m high. Lithgow. Uncommon. Heath. Fl. summer . . . .**L. blakelyi** Joy Thomps.

6 Leaf tip obtuse, usually curving outwards. Leaves narrow-obovate to narrow-elliptic, up to 8 mm long, 1–3 mm wide. Flowers up to 18 mm diam. Capsule c. 3 mm diam. Shrub up to 2 m high, with hard bark. Widespread. DSF and heath. Fl. spring–summer . . . . **L. parvifolium** Sm.

6 Leaf tip acute or blunt, usually not curving outwards. Leaves obovate to almost linear, 10–20 mm long, 1–8 mm wide. Flowers up to 15 mm diam. Fruit 3–6 mm diam. Shrub or small tree up to 5 m high, with very flaky bark. Widespread. Heath and DSF. Ss and old sand dunes. Fl. variable . . **L. trinervium** (Sm.) Joy Thomps.

Group 2

1 Leaves obtuse or grooved and recurved or emarginate at the tip, rarely acute but never pungent 2
1 Leaves pungent-tipped, concave or involute, acute acuminate or acicular 11

2 Leaves distinctly emarginate 3
2 Leaves acute to obtuse 4

3 Capsule 4–5 mm diam. Leaves narrow-oblong to narrow-obovate, 15–35 mm long, 3–6 mm wide, with numerous oil dots. Flowers pedicellate, often clustered, c. 6 mm diam. Erect to drooping shrub up to 4 m high, with hard bark. Lower Blue Mts Creek banks. Fl. spring–summer. . . . .
. . . . **L. emarginatum** H.Wendl. ex Link

3 Capsule c. 7 mm diam. Leaves elliptic to obovate, 6–12 mm long, up to 6 mm wide, with fewer and larger oil dots than *L. emarginatum*. Flowers usually solitary, 8–10 mm diam. Erect to slightly drooping shrub up to 4 m high, with hard bark. Blue Mts Creek banks and river flats. Fl. summer . . . .
. . . . **L. obovatum** Sweet

4 Leaves orbicular to broad-elliptic 5
4 Leaves linear or oblong to obovate or elliptic. 6

5 Leaves 4–7 mm wide with a grooved recurved tip. Capsule 10 mm or more diam. sepals deciduous . . .
. . . . **L. rotundifolium** (see below)

5 Leaves c. 2 mm wide, obtuse, crowded together on the stem. Capsule c. 8 mm diam. sepals persistent. Flowers terminal on short lateral branches, c. 10 mm diam. Petals white to pale pink. Erect bushy shrub up to 2 m high. Coast in south of area. Endangered. Heath. Ss. Fl. late summer . . . .
. . . . **L. epacridoideum** Cheel

6 Leaves elliptic to obovate, 6–10 mm long, 3–6 mm wide. slightly grooved at the tip . . . .
. . . . **L. obovatum** (see above)

6 Leaves linear to narrow-elliptic or narrow-oblong or narrow-obovate to elliptic 7

7 Leaves heavily lemon-scented, narrow-obovate on side branches but often elliptic on main branches, c. 5 mm long, 1–2 mm wide. Flowers 10–12 mm diam. Petals pinkish white. Capsule 6–8 mm diam. Shrub up to 2 m high, with hard bark, generally with a pinkish cast to the foliage. Swamps in northern part of the Coast. Fl. summer . . . . **L. liversidgei** R.T.Baker & H.G.Sm.

7 Leaves not heavily lemon-scented, usually more than 2 mm wide 8

8 Leaves with 3 conspicuous veins, narrow-elliptic to oblanceolate, 15–35 mm long, 2–8 mm wide, silky but glabrescent, folded at the tip. Petals greenish to white. Capsule up to 10 mm diam. Shrub or small tree up to c. 5 m high. Widespread. DSF, on sandy soils. Fl. summer . . . . . **L. morrisonii** Joy Thomps.

8 Leaves without 3 conspicuous veins 9

9 Young stems almost glabrous. Leaves elliptic to oblanceolate, 10–20 mm long, 3–4 mm wide, with prominent oil dots and folded tip. Shrub up to 2 m high. Nowra district. Uncommon. DSF. Ss . . . . . .
. . . . **L. sejunctum** Joy Thomps.

9 Young stems pubescent. Leaves oblanceolate to elliptic or almost linear, 5–20 mm long, 1–5 mm wide, usually glabrous. Flowers up to 15 mm diam. Petals white. Capsule 5–10 mm diam. Shrubs or trees up to 7 m high with smooth soft bark 10

10 Leaves usually acute, pale green. Widespread. Heath and DSF on Ss. and deep sands. Fl. spring–summer . . . . **L. polygalifolium** Salisb. ssp. **polygalifolium**

10 Leaves usually obtuse, dull greyish-green. Coast and Hornsby Plateau N of Gosford . . . .
. . . . **L. polygalifolium** ssp. **cismontanum** Joy Thomps.

11 Sepals persisting on mature capsules 12

11 Sepals not persisting on mature capsule 13

12 Stamen ≤half the length of the petals. Leaves glossy, elliptic, mostly c. 15 mm long, c. 5 mm wide, thick, glabrescent. Flowers c. 22 mm diam. Petals white. Capsule 7–8 mm diam. Shrub up to 3 m high, with flaky bark. Kanangra Walls. Endangered. Fl. summer–autumn . . . . **L. petraeum** Joy Thomps.

12 Stamens >half the length of the petals . . . . **L. sphaerocarpum** (see Group 3)

13 Leaves held erect, linear to narrow-lanceolate, involute-acicular or concave-acuminate. Petals white 14

13 Leaves spreading, lanceolate to orbicular, concave, acute to acuminate 15

14 Flowers 6–10 mm diam. Leaves 3–15 mm long, 1–2 mm wide. Capsule 6–8 mm diam. Erect to spreading, slender shrub up to 3 m high, with hard bark. Widespread. Heath. Swamps and other damp sites. Fl. summer . . . . **L. juniperinum** Sm.

14 Flowers 10–12 mm diam. Leaves 10–20 mm long, 1–3 mm wide. Capsule 6–10 mm diam. Robust shrub up to 1 m high. Uncommon. Ss. on cliffs and escarpments. Fl. summer . . . . **L. rupicola** Joy Thomps.

15 Leaves orbicular, grooved and recurved at the tip, c. 6 mm wide. Flowers 12–30 mm diam. Capsule c. 10 mm diam. Erect or spreading shrub to 2 m high, with hard bark. Blue Mts; Woronora Plateau. Heath. Ss. Fl. spring–summer . . . . **L. rotundifolium** (Maiden & Betche) F.Rodway ex Cheel

15 Leaves neither orbicular nor grooved and recurved at tip. Erect shrubs 1–3 m high 16

16 Flowers c. 11 mm diam. Usually white; produced on new wood. Floral tube tapering towards the base. Capsule c. 9 mm diam. Ss. Wet places in DSF and heath. Fl. spring–summer. *Prickly Tea Tree* . . . . .
. . . . **L. continentale** Joy Thomps.

16 Flowers c. 16 mm diam., often pink; produced on short shoots from older branches. Floral tube obtuse at the base. Capsule c. 12 mm diam. Coast and adjacent plateaus. Ss. Heath and DSF. Fl. spring–summer . . . . **L. squarrosum** Gaertn.

Group 3

1 Leaves linear to linear-lanceolate, concave, acicular, 1–3 mm wide. Floral tube tomentose to villous. Flowers c. 8 mm diam. Rigid divaricate often decumbent shrub up to 2 m high. Widespread. Heath. Ss. Fl. spring–summer . . . . **L. arachnoides** Gaertn.

1 Leaves linear-lanceolate to oblong, flat to convex, acute to acuminate or mucronate 2

2 Leaves obtuse, obovate to elliptic, scarcely convex. Flowers 8–10 mm diam. Floral tube tapering towards the base. Capsule c. 6 mm diam. Sepals membranous, deciduous. Erect shrub up to 3 m high, with hard bark. Blue Mts Wet places. Fl. summer . . . . **L. myrtifolium** Sieber ex DC.

2 Leaves mucronate, oblong, often convex and paler underneath. Floral tube obtuse below. Capsule more than 7 mm diam. Sepals usually persistent on the capsule. Shrubs with hard bark 3

3 Stamens up to half as long as the petals 4
3 Stamens more than half as long as the petals 5

4 Bark smooth, compact. Leaves oblong to oblanceolate, 3–15 mm long, 2–5 mm wide, hairy at least underneath. Blue Mts and Hornsby Plateau. Swamps and along water courses. Fl. spring–summer. *Woolly Tea-tree* . . . . **L. lanigerum** (Aiton) Sm.

4 Bark shedding in patches and strips. Leaves oblong to obovate, 10–30 mm long, 3–8 mm wide, usually felted underneath. Blue Mts and Woronora Plateau. Swamps and along water courses. Fl. spring–summer . . . . **L. grandifolium** Sm.

5 Capsule broader than long, 15–20 mm diam. Leaves broad-elliptic, mostly 10–20 mm long, 5–10 mm wide. Flowers c. 15–30 mm diam. Petals white to dark red. Shrub up to 2 m high, with hard bark. Blue Mts DSF and heath. Fl. spring–summer. . . . . **L. macrocarpum** (Maiden & Betche) Joy Thomps.

5 Capsule as long as broad, less than 15 mm diam. 6

6 Leaves up to 20 mm long, 2–5 mm wide, lanceolate. Flowers 15–20 mm diam. Petals greenish white to pink. Capsule 7–10 mm diam. Blue Mts Heath and DSF. Ss. Fl. spring–summer . . . .
. . . .**L. sphaerocarpum** Cheel

6 Leaves more than 20 mm long, 3–5 mm wide, narrow-elliptic. Flowers 15– 30 mm diam. Petals dark red. Capsule 9–12 mm diam. Colo River. River banks. Ss. Fl. spring–summer. . . . .
. . . . **L. spectabile** Joy Thomps.

19 Kunzea Rchb.

36 species in Aust. (35 endemic); all states and territories

Shrubs. Leaves mostly alternate, linear to linear-lanceolate. Flowers sessile or sometimes pedunculate, usually arranged in short leafy spikes or terminal heads, sometimes with an involucre of scale-like bracts. Floral tube shortly continued above the ovary. Sepals 5. Petals 5, white to deep pink. Stamens numerous, free, longer than the petals, arranged in one or more whorls, sometimes with a small globular appendage on the anther. Ovary 2–5-locular, with 2 or more ovules per loculus. Fruit a thin-walled capsule.

Key to the species

1 Flowers crowded on leafy side branches or in the axils of the upper leaves. Petals and stamens white . 2
1 Flowers crowded into leafless terminal heads which may be displaced by further growth of the main axis. 3

2 Pedicels up to 2 mm long. Leaves usually spreading, concave but not appressed to the stem, linear to linear-oblong, up to 10 mm long. Floral tube pubescent or glabrous, c. 3 mm long. Shrub up to 3 m high. Coast and adjacent plateaus. Open forest, heathand in cleared areas. Ss and WS. Fl. summer. *Tick Bush* . . . . **K. ambigua** (Sm.) Druce

2 Pedicels more than 2 mm long. Leaves spreading, narrow-elliptic to oblanceolate, up to 25 mm long. Floral tube pubescent. Shrub up to 4 m high. Blue Mts; Hornsby Plateau. Heath and DSF. Along creek banks. Ss. Fl. summer . . . . **K. ericoides** (A.Rich.) Joy Thomps.

3 Ovary 2-locular. Petals white 4
3 Ovary 3-locular. Petals white, cream, pink or purple 5

4 Erect shrub up to 1.5 m high. Leaves narrow-oblong to oblanceolate, 6–14 mm long, concave towards the top, often convex towards the base. Sepals narrow-triangular, acuminate. Stamens 3–5 mm long. Floral tube villous. Petals shorter than sepals. Maroota to Kar-ing-gai Chase area. DSF. Ss. Vulnerable. Fl. summer. . . . .**K. rupestris** Blakely

4 Prostrate to ascending shrub up to 60 cm high. Leaves elliptic-oblong to obovate, 3–8 mm long, very slightly concave at the top and convex at the base or even flat. Sepals triangular. Petals ± longer than the sepals. Stamens up to 3 mm long. Floral tube silky hairy. Higher Blue Mts Wet mallee on Ss. Vulnerable. Fl. spring . . . . **K. cambagei** Maiden & Betche

5 Leaves distinctly 3-veined, oblanceolate to obovate, 3–9 mm long, 2–5 mm wide, concave. Sepals narrow-triangular, ± acuminate. Petals usually pink to purple but white in southern Blue Mts Floral tube villous at least at the base, longer than the ovary at flowering. Stamens 4–6 mm long. Capsule urceolate, 4 mm long, 2.5 mm wide. Erect or ascending shrub up to 1.5 m high. Widespread. Heath and DSF. Ss. Fl. spring–summer . . . . **K. capitata** (Sm.) Heynh.
5 Leaves 1-veined. 6

6 Leaves, linear to linear-oblanceolate, up to 4 mm long, 1–1.5 mm wide, channelled or concave ± appressed to stem. Sepals triangular to broad-triangular c. 2 mm long. Flowers pink to purple rarely white. Floral tube glabrous (sometimes hairy outside the region), only slightly longer than the ovary at flowering. Stamens 2–4 mm long. Capsule subglobular, 2–2.5 mm diam. Shrub up to 2 m high. Blue Mts and Woronora Plateau in south of region. Heath and DSF. Ss. Fl. summer . . **K. parvifolia** Schauer
6 Leaves broad-elliptic to elliptic 3–6 mm long, 2–2.5 mm wide. Sepals triangular c. 1 mm long. Flowers cream. Floral tube glabrous. Stamens 1.5–2 mm long. Capsule 3–3.5 mm diam. Ranges north of Yerranderie. DSF.. . . . . **K. sp. 'Mt. Cookem'** (Benson 4470)

20 Callistemon R.Br.

Bottlebrushes
36 species endemic Aust.; all states and territories

Shrubs or small trees with pubescent or tomentose young branches. Leaves terete or linear to lanceolate, acute or acuminate, sometimes pungent pointed. Flowers arranged in dense regular spikes usually with the growth of the main axis continuing beyond the spike. Floral tube shortly continued above the ovary. Sepals 5, deciduous. Petals 5, orbicular, green to yellow or white to pink or red. Stamens numerous, free, arranged in 2 or more whorls; filaments much longer than the petals and the same colour. Ovary 3–4-locular with numerous ovules per loculus. Fruit a woody capsule, usually remaining on the stem for several years.

Key to the species

1 Leaves 2 mm wide or less, rarely up to 3 mm wide, linear, incurved-concave or terete or almost flat . . 2
1 Leaves more than 3 mm wide, flat or ± concave, lanceolate to linear-lanceolate, not terete. 5

2 Filaments red 3
2 Filaments green-yellow, whitish or pinkish 4

3 Leaves 6–12 cm long, linear, concave, 1–2 mm wide, often pungent pointed. Spike 6–11 cm long. Filaments 20–25 mm long, pale red. Shrub 1–3 m high. Widespread. Damp places. Fl. spring–early summer. *Narrow-leaved Bottlebrush*. . . . . **C. linearis** (Schrad. & J.C.Wendl.) Sweet
3 Leaves mostly 2–4 cm long, linear to terete, 2–4 mm wide. Spike 4–8 cm long. Filaments 15–20 mm long, dark red. Shrub up to 1 m high. Widespread. Usually along creeks. Fl. summer . . . .
. . . . **C. subulatus** Cheel

4 Shrubs c. 1 m high. Young shoots greyish silvery. Leaves flat or with thickened margins, 1.5–2 cm long
. . . . **C. pityoides** (see below)
4 Shrub up to 2.5 m high. Leaves with sharply incurved margins, often terete, 4–8 cm long, often pungent pointed. Spikes 5–7 cm long. Filaments 10–15 mm long. Widespread. Usually in damp places. Fl. spring–early summer . . . . **C. pinifolius** (J.C.Wendl.) Sweet

5 Filaments red or pink 6
5 Filaments white to green or yellow. 11

6 Spike up to 5 cm long 7

6 Spike more than 5 cm long 8

7 Filaments pink . . . . **C. sieberi** (see below)
7 Filaments red . . . . **C. subulatus** (see above)

8 Aperture of mature capsule ± as wide as the middle of the capsule which is thus barrel-shaped. Vein reticulations between main and lateral veins usually clearly visible. Capsule 5–8 mm diam. Spike 6–12 cm long 9
8 Aperture of the mature capsule narrower than the middle of the capsule which is thus ± globular. Vein reticulations not usually clearly visible 10

9 Intramarginal veins set away from the margins. Leaves linear to linear-lanceolate, 9–14 cm long, 5–12 mm wide, acute. Filaments red, c. 20 mm long. Shrub up to c. 4 m high. Coast and adjacent plateaus. DSF. Vulnerable. Fl. spring–summer . . . . **C. linearifolius** (Link) DC.
9 Intramarginal veins forming a distinct ridge at the margins. Leaves linear, 6–10 cm long, 3–6 mm wide, acute-acuminate. Filaments red, 20–25 mm long. Stiff shrub 1–3 m high. Coast and adjacent plateaus. Damp places in heath and DSF. Fl. summer. *Stiff Bottlebrush* . . . . **C. rigidus** R.Br.

10 Leaves lanceolate 6–12 mm wide, 3–7 cm long, usually with a distinct rigid mucro. Spike 5–12 cm long. Filaments red to lilac, 20–25 mm long. Capsule 6–7 mm diam. Shrub 1–3 m high. Widespread. Damp places. Fl. spring–early summer. *Common Red Bottlebrush* or *Lemon Bottlebrush* . . . .
. . . . **C. citrinus** (Curtis) Skeels
10 Leaves linear to narrow-oblanceolate, 2–4 mm wide . . . . **C. subulatus** (see above)

11 Leaves narrow-elliptic, mostly 6–10 cm long, 5–10 mm wide, acute or obscurely acuminate with the midvein prominent beneath. Young shoots purplish silvery. Spikes 3–8 cm long. Filaments white to yellowish, 12–15 mm long. Bracts and bracteoles very deciduous. Small tree with papery bark. Widespread. Damp places. Fl. summer. *Willow Bottlebrush.* . . . . **C. salignus** (Sm.) Sweet
11 Leaves 1–6 cm long (sometimes longer, but then never more than 5 mm wide). 12

12 Shrubs c. 1 m high, sometimes up to 3 m. Young shoots greyish silvery. Rhachis of the spike pubescent. Leaves 1.5–2 cm long, 1–5 mm wide, acute. Spike 3–4 cm long; bracts and bracteoles often persistent until flowers open. Filaments white to yellowish or rarely pale pink, up to c. 9 mm long. Higher Blue Mts Bogs and swamps. *Alpine Bottlebrush* . . . . **C. pityoides** F.Muell.
12 Shrubs or trees 2–4 m high. Young shoots not silvery. Filaments white to yellow 13

13 Rhachis of the spike densely tomentose to pubescent. Leaves lanceolate, 2–3.5 cm long, 6–8 mm wide, acute or slightly acuminate; veins distinct underneath. Spike c. 3.5 cm long; bracts and bracteoles often persistent until the flowers open. Shrub 2–3 m high with papery bark. Woronora and Hornsby Plateaus. Uncommon. On shale and in RF gullies. Fl. spring . . . . **C. shiressii** Blakely
13 Rhachis of the spike glabrous or minutely pubescent, sometimes with some long villous hairs just below the floral tube 14

14 Leaves narrow-lanceolate to linear, mostly 4–6 cm long, 3–5 mm wide, with a distinct mucro. Spike c. 3.5 cm long. Young shoots flesh-coloured. Shrub or small tree. Woronora Plateau; Blue Mts Along and in creeks. Fl. summer. *River Bottlebrush* . . . . **C. sieberi** DC.
14 Leaves elliptic-oblong, 2.5–4 cm long, 6–9 mm wide, with a minute terminal point. Spike 3–8 cm long. Young shoots silvery yellow. Shrub or small tree up to 8 m high. Higher Blue Mts Creek banks and amongst wet rocks. Fl. spring–summer . . . . **C. pallidus** (Bonpl.) DC.

21 Melaleuca L.
215 species in Aust. (210 endemic); all states and territories

Shrubs or trees often with a papery bark. Leaves opposite or alternate. Flowers arranged in spikes or terminal heads; axis often continuing vegetative growth before flowering finishes. Floral tube shortly continued above the ovary. Petals white to yellow or red to purple. Stamens numerous; filaments connate into 5 bundles (claws) opposite the petals. Ovary 3-locular. Fruit a capsule.

Key to the species

1 Leaves opposite 2
1 Leaves mostly alternate or scattered 6

2 Stamens red or purplish 3
2 Stamens yellow to white 4

3 Flowers arranged in dense spikes 3–5 cm long. Capsule up to 10 mm diam., crowned by the persistent
   calyx. Stamens 18–30 mm long, red. Leaves lanceolate to oblong-elliptic, 13–40 mm long, 4–10 mm
   wide, with a prominent midrib. Shrub up to 6 m high, with corky bark. Widespread. Wet places, coastal
   headlands. Fl. spring–summer . . . . **M. hypericifolia** Sm.
3 Flowers arranged in few-flowered spikes up to 2 cm long. Capsule 3–4 mm diam. with spreading calyx
   teeth. Stamens 10–15 mm long, purplish. Leaves lanceolate to elliptic, 5–12 mm long, 1–3 mm wide;
   midrib not prominent. Shrub up to 2 m high with corky bark and often with a lignotuber. Widespread.
   Open forest and heath. Damp places and margins of swamps. Fl. summer . . . .**M. thymifolia** Sm.

4 Leaves equally 5–7-veined (sometimes obscurely so), lanceolate to suborbicular, 5–15 mm long, 4–5 mm
   wide. Flowers arranged in spikes 15–40 mm long. Stamens yellow; claws very short. Capsule c. 4 mm
   diam., with an undulate orifice. Shrub up to 2 m high, with corky bark. Coast; Woronora Plateau. Wet
   heath. Fl. summer . . . .**M. squarrosa** Donn ex Sm.
4 Leaves 1–3-veined, linear to lanceolate, more than 9 mm long, up to 4 mm wide; midvein more
   prominent than the laterals. Stamens white or pale yellow 5

5 Leaves concave, often keeled, 13–45 mm long, 1–3 mm wide. Staminal claw 8–15 mm long; filaments
   arranged along the side of the claw. Capsule c. 3 mm diam. Tree up to 10 m high, with papery bark.
   Coast and adjacent plateaus; lower Blue Mts Wet places on shales. Fl. spring–summer . . . .
   . . . . **M. linariifolia** Sm.
5 Leaves ± channelled above, presenting a convex surface on either side of the midrib, up to 18 mm long,
   c. 3 mm wide. Staminal claw less than 4 mm long; filaments attached terminally. Capsule c. 3 mm diam.
   Shrub or small tree with papery bark. Coast, e.g. Gosford district. Wet places. Vulnerable. Fl. summer
   . . . . **M. biconvexa** Byrnes

6 Stamens red to purple or pink 7
6 Stamens white to yellow 9

7 Leaves narrow-linear, up to 1 mm wide, with recurved-divergent tips, ± terete, mostly c. 10 mm long,
   with no prominent veins. Staminal claw 4–6 mm long; filaments attached towards the top. Flower
   spikes 2–4 cm long. Capsules with undulate orifices, c. 3 mm diam., in oblong clusters. Shrub up to 2 m
   high, with hard bark. Coast and adjacent plateaus. Open forests on WS. Fl. summer . . . .
   . . . .**M. diosmatifolia** Dum.Cours.
7 Leaves lanceolate to narrow elliptic 1.5-5 mm wide. 8

8 Flower spikes (or heads) 1–2 cm long. Leaves lanceolate, 1.5–3 mm wide, 5–12 mm long, apex acute,
   ± incurved, 3–5-veined; petiole to 1.5 mm long. Stamens pinkish-purple. Capsules with ± undulate
   orifices, 5–7 mm diam., arranged in short clusters. Shrub 1–2 m high, with corky bark. Woronora
   Plateau. Wet places. Ss. Fl. spring. . . . . **M. squamea** Labill.
8 Flower spikes >2 cm long. Leaves narrow elliptic 3.5–5 mm wide, 30–40 mm long, apex narrowly acute,
   petiole 2–2.5 mm long. Stamens pink, style red. Capsule 6 mm wide, narrowing towards theorifice.
   Shrub to 2.5 m high. Swampy areas. Megalong Valley. Vulnerable. Fl. Spring to early summer. . . . . .
   . . . . **M. sp.'Megalong Valley'**(Craven *et al* 10442)

9 Staminal claw scarcely exceeding the sepals 10
9 Staminal claw much exceeding the sepals. 16

10 Leaves linear to very narrow-oblong, up to 2 mm wide, sometimes pungent pointed 11
10 Leaves lanceolate to elliptic or narrow-elliptic, 2 mm wide or wider. 12

11 Capsules c. 2 mm diam., arranged in globular clusters. . . . . **M. nodosa** (see below)

11 Capsules 4–6 mm diam., arranged in loose clusters; aperture almost as wide as middle of capsule. Leaves 10–25 mm long. Flowers in short spikes or heads. Shrub up to 2 m high. Southern Blue Mts Heath. Ss. Fl. summer . . . . **M. capitata** Cheel

12 Leaves more than 30 mm long up to 25 mm wide, lanceolate-elliptic. Flower spikes 2–5 cm long. Stamens 8–12 mm long. Capsule c. 5 mm diam.; aperture almost as wide as middle of capsule. Tree up to 10 m high, with papery bark. Coast. Swamps and lake margins. Fl. autumn–winter. *Paperbark* or *Broad-leaved Tea-tree*. . . . . **M. quinquenervia** (Cav.) S.T.Blake

12 Leaves less than 25 mm long. 13

13 Leaves divergent from the stem but with incurved or erect tips. . . . . **M. squamea** (see above)

13 Leaves ascending or divergent, often with the edge turned towards the stem, acute or obtuse but not upcurved at the tip 14

14 Bark hard and rough. Capsule globose to urceolate 4–5 mm diam. Orifice narrower than middle of the capsule. Sepals ± persistent in the fruit. Flowers white. Leaves linear to narrow elliptic 5–15 mm long, 1–3 mm wide. Shrub or small tree. Capertee Valley. Open woodland, clayey soils, drier areas. Fl. Chiefly summer.. . . . . **M. lanceolata** Otto

14 Bark papery of fibrous 15

15 Capsule 2–5 mm diam., with an undulate orifice ± as wide as the middle of the capsule. Sepals ± persistent in the fruit. Flower spikes up to 4 cm long. Leaves lanceolate to narrow-elliptic, 4–12 mm long. Tall shrub or small tree, up to 5 m high. Coast. Swamps, often on deep sands. Fl. summer . . . . **M. sieberi** Schauer

15 Capsule 5–7 mm diam.; orifice distinctly narrower than middle of the capsule. Sepals not persistent in the fruit. Flower spikes 2–6 cm long. Leaves lanceolate-elliptic, 1–2.5 cm long. Shrub 1–3 m high Coast and adjacent plateaus. Wet heath. Ss. Vulnerable. Fl. summer . . . . **M. deanei** F.Muell.

16 Leaves with several longitudinal veins, lanceolate to ovate-acuminate, often twisted, 10–15 mm long, 3–6 mm wide, pungent pointed. Staminal claw 3–4 mm long. Capsules 3–4 mm diam., arranged in oblong clusters. Sepals triangular-acuminate. Shrub or small tree up to 6 m high, with papery bark. Widespread. In a variety of situations. Fl. summer. *Prickly-leaved Tea-tree* . . . . **M. styphelioides** Sm.

16 Leaves with 1–3 longitudinal veins or no conspicuous veins 17

17 Capsules arranged in dense globular clusters c. 2 cm diam. Leaves linear, 7–20 mm long, rigid, with a pungent point. Staminal claw c. 2 mm long. Shrub 1–3 m high, with corky bark. Widespread. Heath and DSF. Fl. spring–summer . . . . **M. nodosa** (Sol. ex Gaertn.) Sm.

17 Capsules arranged in oblong clusters. Leaves scarcely pungent. 18

18 Leaves narrow-linear, with no conspicuous veins, less than 1 mm wide, usually with an acute recurved or divergent tip 19

18 Leaves linear to lanceolate, with 1–3 conspicuous veins, 1 mm or more wide, with an acute or obtuse tip 20

19 Flower spikes 3–7 cm long. Staminal claw 5–6 mm long. Capsule c. 5 mm diam. Leaves 12–25 mm long. Tall shrub or small tree up to 5 m high, with hard bark. Coast. Damp places. Coastal headlands. Fl. Summer. **M. armillaris** (Sol. ex Gaertn.) Sm. ssp. **armillaris**

19 Flower spikes up to 25 mm long. Staminal claw 1.5–2 mm long. Capsules c. 3 mm diam. Leaves mostly c. 10 mm long (rarely 18mm). Shrub or small tree up to 6m high, with corky bark. Widespread. Wet places. Fl. summer . . . . **M. ericifolia** Sm.

20 Staminal claw less than 2 mm long. Leaves 2–3 mm wide. . . . . **M. sieberi** (see above)

20 Staminal claw c. 4 mm long. Leaves linear, 1–2 cm long, 1–2 mm wide, with 1 conspicuous vein. Capsule c. 2.5 mm diam. Flower spikes c. 3 cm long. Tall shrub or small tree up to 5 m high, with papery bark. Coast and Cumberland Plain. Clay soils. WS. Swamps. Fl. summer . . . . **M. decora** (Salisb.) Britten

22 Ochrosperma Trudgen

5 species endemic Aust.; Qld, NSW

One species in the area

Spreading shrub up to 50 cm high. Leaves opposite, elliptic to obovate, 2–6 mm long, 1–3 mm wide. Flowers up to 3 mm diam. Solitary or in pairs. Petals white, broad-obovate. Stamens 5, opposite the sepals. Capsule up to 3 mm diam. Upper Blue Mts DSF. Ss. Fl. spring–summer . . . . **O. oligomerum** (Radlk.) A.R.Bean

23 Triplarina Raf.

7 species endemic Aust.; Qld, NSW

One species in the area

Shrub up to 3.5 m high. Leaves obovate, 3.4–5 mm long, 1–1.7 mm wide, with slightly recurved truncate tips, margins entire, large oil glands in 2 small rows. Flowers paired in leaf axils c. 4.5 mm diam. White to deep pink. Stamens 15–17. Ovary 3 locular. Confined to Nowra district. Moist heath. Ss. Endangered. Fl. spring. . . . . **T. nowraensis** A.R.Bean

24 Sannantha Peter G.Wilson

15 species endemic Aust.; Qld, NSW, Vic.

One species in the area

Shrub to 4 m high. Leaves opposite, narrow-oblong to elliptic, 7–25 mm long, 1–6 mm wide, with scattered oil glands; margins recurved, entire. Flowers 5 merous, white, 2–3 or more on a common peduncle more than 4 mm long. Bracteoles 4–many. Flowers 5–8 mm diam. Stamens 5–15 in loose groups opposite petals. Ovary 3-locular; ≥14 ovules per loculus. Coast and adjacent plateaus; lower Blue Mts DSF. Fl. spring–summer. . . . .**S. pluriflora** (F.Muell.) Peter G.Wilson

25 Harmogia Schauer

1 species endemic Aust.; Qld, NSW

Monotypic genus

Shrub c. 1 m high. Leaves opposite, less than 1 mm wide, linear to terete 1.8–5 mm long; oil glands visible below. Flowers solitary, 6–7.5 mm diam. Bracteoles 2. Stamens 7–10, in loose groups opposite petals. Anthers fused to filaments. Ovary 3-locular with ≤13 ovules per loculus. Capsule 2 mm diam. Widespread. DSF and heath. Ss. Fl. Summer . . . .**H. densifolia** (Sm) Schauer

26 Baeckea L.

12 species endemic Aust.; Qld, NSW, Vic., Tas.

Shrubs mostly up to 1 m high, often heath-like. Leaves opposite. Flowers solitary or in small cymes in the axils of the upper leaves. Floral tube continued shortly above the ovary. Sepals usually persistent. Petals red to white. Stamens 5–c. 20, in a single whorl not opposite centre of petal. Anthers versatile. Ovary 2–locular. Fruit a capsule.

Key to the species

1 Leaves linear or almost terete >6 times long as wide. 2
1 Leaves orbicular to narrow-obovate or rarely linear, <6 times long as wide. 3

2 Leaves 5–20 mm long, c. 1 mm wide, very acute and almost pungent pointed. Flowers solitary in the axils of the upper leaves, c. 5 mm diam. Sepals entire. Flowers 5–6.5 mm diam, petals white. Stamens 8–15. Capsule c. 2 mm diam. Shrub up to 3 m high, with the tips of the branches drooping. Widespread. Gullies in damp places, often near waterfalls. Fl. summer . . . . **B. linifolia** Rudge

2 Leaves 3–6 mm long apex obtuse. Sepals reddish, triangular. Flowers 4 mm diam. Petals white. Stamens 8–10. Capsule 1–1.5 mm diam. Shrub spreading 1.5–2 m high. Heath. Ss. Kandos area. Endangered. Fl. Summer. . . . .**B. kandos** A.R.Bean

3 Leaves almost triangular in cross-section, c. 2 mm long. Stamens usually 12–15 sometimes more or fewer. Sepals entire. Flowers c. 4 mm diam. Petals white to pink. Undershrub up to 1 m high. Widespread. Heath. Ss. Fl. spring–summer. . . . .**B. brevifolia** (Rudge) DC
3 Leaves flattened or concave. Stamens up to 10 4

4 Leaves orbicular to broad-obovate, 2.5–5 mm long, 2.5–4 mm wide, often crenulate. Ovules 6–10 per loculus. Flowers 4–5 mm diam. Petals white or pinkish. Undershrub up to 1 m high. Coast and adjacent plateaus. Wet heath, swamp margins. Fl. spring–summer . . . . **B. imbricata** (Gaertn.) Druce
4 Leaves obovate to linear, up to 2 mm wide. Petals white 5

5 Sepals and young leaves ciliate-dentate. Ovules c. 4 per loculus. Flowers 4–6 mm diam. Leaves narrow-obovate, 2–6 mm long, c. 1 mm wide. Undershrub up to 1 m high. Widespread. Heath. Ss. Fl. spring . . . . **B. diosmifolia** Rudge
5 Sepals and young leaves entire. Ovules more than 10 per loculus. Flowers 4–6 mm diam. Leaves linear to narrow-oblong-elliptic, keeled, 4–9 mm long, c. 1 mm wide. Undershrub up to 1 m high. Higher Blue Mts Damp places. Fl. summer . . . . **B. utilis** F.Muell. ex Miq.

27 Euryomyrtus Schauer
7 species endemic Aust.; NSW, Vic., SA, WA

One species in the area
Diffuse or decumbent shrub up to c. 50 cm high. Leaves linear to terete, 4–7 mm long, up to 1 mm wide (rarely 2 mm wide), with scattered inconspicuous oil glands. Flowers 6–10 mm diam. Petals red to pale pink. Ovary 3 locular. Stamens 10, 5 of them opposite the centre of the petals. Capsule 4–5.5 mm diam. Widespread. DSF and heath. Ss. Fl. spring–summer. . . . . **E. ramosissima** (A.Cunn.) Trudgen ssp. **ramosissima**

28 Micromyrtus Benth.
22 species endemic Aust.; Qld, NSW, Vic., SA, WA

Glabrous shrubs up to 1.5 m high. Leaves opposite, 4-ranked. Flowers solitary in the axils of the upper leaves, sometimes crowded into heads. Floral tube short. Sepals 5, sometimes much reduced. Petals 5, white or pink. Stamens 5, opposite the petals. Ovary 1-locular with 2 or 4 ovules attached to a central strand. Fruit a nut.

Key to the species

1 Floral tube narrow-turbinate, c. 1 mm long. Ovules 2 per ovary. Flowers solitary or sometimes in small terminal heads, c. 2 mm diam. Leaves linear, c. 3 mm long, ciliate with long hairs. Spreading shrub up to 2 m high. Cumberland Plain. DSF. Lateritic river or lake sediments. Endangered. Fl. spring . . . . **M. minutiflora** (F.Muell.) Benth.
1 Floral tube ovate-turbinate, 2 mm long. Ovules 4 per ovary. Flowers in loose terminal heads 2

2 Leaves linear to oblong, with a prominent ciliate keel, 2–5 mm long. Flowers 6–7 mm diam. Petals pink. Undershrub up to 60 cm high. Areas near Hawkesbury River Heath. Vulnerable. Fl. summer . . . . **M. blakelyi** J.W.Green
2 Leaves linear to oblanceolate, without a prominent ciliate keel, 2–4 mm long. Petals white to pink. Spreading shrub up to 1.5 m high. Widespread. Heath, DSF, particularly on very shallow soils. Ss. Fl. spring–early summer . . . . **M. ciliata** (Sm.) Druce

29 Calytrix Labill.

75 species endemic Aust.; all states and territories

One species in the area

Shrub up to 2 m high. Leaves linear, 4–8 mm long, c. 0.5 mm wide, denticulate. Obtuse, triangular in cross-section, spirally arranged. Flowers arranged in terminal heads of up to 12. Bracts and bracteoles persistent, concave or keeled, obovate to elliptic. Floral tube much longer than the ovary, narrow below, abruptly dilated just below the insertion of sepals and petals. Sepals 5, broad-ovate, with long flexuous awns. Petals 5, white to pinkish, spreading, deciduous. Stamens numerous, arranged in several whorls; appendage attached to the connective. Ovary 1-locular, with 2 basally attached ovules; stigma capitate. Fruit a nut. Widespread. Heath. Ss. Fl. spring–summer . . . . **C. tetragona** Labill.

30 Homoranthus A.Cunn. ex Schauer

22 species endemic Aust.; Qld NSW, SA

One species in the area

Spreading shrub up to 2 m high. Leaves opposite, linear, terete, 2–12 mm long. Flowers paired, hanging down. Floral tube up to 6 mm long. Sepals with several narrow lobes. Petals yellow, c. 2 mm long. Style exceeding the petals. Putty to Rylestone. Uncommon. DSF. Ss. Fl. spring. . . . .
. . . . **H. cernuus** (R.T.Baker) Craven & S.R.Jones

31 Darwinia Rudge

45 species endemic Aust. NSW, Vic., SA, WA

Undershrubs or shrubs up to 2 m high with rigid or semi-rigid linear opposite leaves. Flowers arranged in terminal heads or pairs. Floral tube much longer than the ovary. Sepals usually 5, ± petaloid. Petals 5, white, incurved thus closing the floral tube except for a very short time as the style is extending. Stamens 10, alternating with 10 staminodes. Ovary 1-locular, with 2–3 basally attached ovules. Style long-exserted with a brush of hairs just below the stigma. Fruit a nut.

Key to the species

1 Leaves terete or nearly so; lower surface rounded. Flowers 2–20 per cluster. Bracteoles deciduous, 3–5 mm long. Floral tube 5–7 mm long; ovary c. 1 mm diam., with obscure ridges; style 12–18 mm long . 2
1 Leaves laterally compressed; lower surface keeled. Flowers usually 2–6 per cluster, rarely 8 3

2 Flowers usually >4 per cluster, apparently arranged irregularly. Erect or rarely decumbent shrubs, not rooting along the branches, up to 2 m high. Coast and adjacent plateaus. Heath. Ss. Fl. winter–spring. . . . . **D. fascicularis** Rudge ssp. **fascicularis**
2 Flowers usually 4 per cluster, arranged in pairs. Decumbent shrubs, often rooting along the branches, up to 60 cm high. Mt Banks to Wentworth Falls. Heath. Ss. Fl. spring. Endangered population in some LGAs. . . . . **D. fascicularis** ssp. **oligantha** B.G.Briggs

3 Bracteoles extending beyond the perianth, persisting after the flower opens. Style recurved. Leaves opposite, triangular in cross-section, 5–12 mm long. Flowers 2–4 or rarely 6 per cluster. 4
3 Bracteoles shorter than the perianth, deciduous or persistent 5

4 Style 6–12 mm long. Bracteoles 5–8 mm long. Floral tube 5–6 mm long. Erect to decumbent shrub 15–100 cm high. Blue Mts Heath. Ss Fl. spring–summer . . . . **D. taxifolia** A.Cunn. ssp. **taxifolia**
4 Style 15–24 mm long. Bracteoles 7–14 mm long. Floral tube 6–9 mm long. Erect to decumbent shrub. South Coast and Woronora Plateau. Heath. Ss. Fl. spring. . . . . **D. taxifolia** ssp. **macrolaena** B.G.Briggs

5 Sepals more than three-quarters as long as the petals 6
5 Sepals less then three-quarters as long as the petals 7

6 Floral tube 7–12 mm long. Style 12–20 mm long. Petals becoming dark red. Leaves opposite, triangular in cross-section, 8–18 mm long, 0.5–1 mm wide. Flowers 4–6 per cluster. Shrubs with main stems

prostrate and rooting; branches ascending to 50 cm high. Woronora Plateau. DSF. Fl. winter. . . . . . .
. . . .**D. grandiflora** (Benth.) R.T.Baker & H.G.Sm.

6 Floral tube 3–5 mm long. Style 4–9 mm long. Petals white to pink. Leaves decussate, triangular in cross-section, 6–11 mm long, 0.5–1 mm wide. Flowers 2–4 per cluster. Erect to ascending shrub up to 1.5 m high. Coast and adjacent plateaus. Heath and DSF. Ss. Fl. spring–summer . . . .
. . . . **D. diminuta** B.G Briggs

7 Flowers (incl. style) 4–9 mm long. Bracteoles yellow-green or yellow-brown. Young flowers white-cream 8
7 Flowers (incl. style) 9–25 mm long. Bracteoles dark brown or reddish-purple. Young flowers white or green. 10

8 Ovary c. 0.5 mm diam. Bracteoles deciduous, less than two-thirds as long as the floral tube. Floral tube 7–8 mm long. Flowers 2–4 per cluster. Leaves decussate, triangular in cross-section, 7–11 mm long, c. 0.5 mm wide. Erect shrub up to 80 cm high. Coast and Woronora Plateau. Heath. Ss. Fl. winter . . . . .
. . . . **D. leptantha** B.G.Briggs
8 Ovary 1–2 mm diam. 9

9 Floral tube glabrous between the rounded longitudinal ridges. Bracteoles persistent after flowers open, 3.5–6 mm long, more than two-thirds as long as the floral tube. Floral tube 3–6 mm long. Flowers 2–4 per cluster. Leaves decussate, triangular in cross-section, 6–17 mm long, 0.5–1 mm wide. Erect or spreading shrub up to 30 cm high. Coast in south of area and Woronora Plateau. Heath. Ss. Fl. winter–spring. . . . . **D. camptostylis** B.G.Briggs
9 Floral tube papillose between the 5 distinct ribs. Bracteoles 4–6 mm long, caducous. Floral tube 4.5–5 mm long. Flowers 1–6 per cluster. Leaves linear 11–18 mm long 0.8–1 mm wide. Erect shrub to 1.5 m high. Budawang Ranges Macquarie Pass. Heath and alloasuarina thickets. Gravelly or sandy soil. Fl. spring–winter. . . . . .D. briggsiae Craven & S.R.Jones

10 Main stems prostrate, up to 2.5 m long, with ascending branches up to 15 cm high. Leaves glaucous, decussate, triangular in cross-section, 8–17 mm long, 0.5–1 mm wide. Flowers 2–4 per cluster. Floral tube 7–8 mm long. Bracteoles reddish-brown, deciduous, 4–6 mm long. Hornsby Plateau, Hawkesbury River district. Heath. Ss. Vulnerable. Fl. winter–spring. (May hybridizes with *D. fascicularis* where they occur together.) . . . . **D. glaucophylla** B.G.Briggs
10 Erect to spreading shrubs with green leaves. 11

11 Flowers 4–8 per cluster. Stalk of flower cluster 1–2 mm long. Bracteoles persistent to flowering, reddish-brown to purplish, 8–11 mm long. Floral tube 5–8 mm long. Style 14–20 mm long. Leaves decussate, triangular in cross-section, 10–25 mm long, 0.5–1 mm wide. Erect shrub up to 3 m high. Hornsby Plateau, Gosford to Manly. Heath. Ss. Fl. winter–spring. . . . . .D. procera B.G.Briggs
11 Flowers up to 2 per cluster, sometimes 4 in *D. peduncularis*. 12

12 Stalk of flower cluster 4–7 mm long. Bracteoles deciduous, purplish red, 4–8 mm long. Floral tube 9–12 mm long. Style 6–10 mm long. Leaves decussate, triangular in cross-section. 7–12 mm long, 0.5–1 mm wide. Spreading shrub up to 1.5 m high. Hornsby Plateau, Hawkesbury river to Hornsby; Blue Mts DSF. Ss. Vulnerable. Fl. winter–spring . . . . **D. peduncularis** B.G.Briggs
12 Stalk of flower cluster up to c. 1 mm long. Bracteoles persistent after flower opens, purplish red, 4–8 mm long. Floral tube 5–8 mm long. Leaves decussate, 6–10 mm long, 0.5–1 mm wide. Erect shrub up to 80 cm high. Hornsby Plateau, Hawkesbury river to Port Jackson. Vulnerable. Fl. winter–spring . . .
. . . . **D. biflora** (Cheel) B.G.Briggs

70 ONAGRACEAE

Herbs, sometimes becoming woody. Leaves simple; stipules absent. Flowers regular, bisexual. Floral tube often extending above the ovary. Sepals 4–5, valvate. Petals 4–5, imbricate, twisted in the bud. Stamens usually twice as many as the petals. Ovary inferior, 4–5-locular; placentas axile; style capitate or lobed. Fruit usually a capsule, opening from the summit. Seeds few to numerous. 22 gen., temp. to subtrop., mostly America.

Key to the genera

1 Floral tube extending above the ovary. Sepals 4 . . . . **1 OENOTHERA**
1 Floral tube not extending above the ovary 2

2 Sepals persisting on to fruiting stage. Petals yellow. Aquatics or swamp plants . . . . . . **3 LUDWIGIA**
2 Sepals falling soon after flower opens. Terrestrial plants 3

3 Fruit a terete, indehiscent capsule; seeds numerous usually with a tuft of hairs. Petals pink white or
purple . . . .**2 EPILOBIUM**
3 Fruit ellipsoid, 4-angled; seeds 1–few, without a tuft of hairs; petals white, red to yellow with age . . . .
. . . . **4 GAURA**

1 Oenothera L.

13 species naturalized in Aust.; all states and territories

Herbs with alternate leaves. Flowers solitary in the axils of the upper leaves forming a leafy spike. Sepals 4, deflexed. Petals 4, obovate, spreading. Tubular free part of the floral tube extending far above the ovary and falling off with the sepals etc. Stamens 8. Ovary 4-locular. Capsule 4-valved, opening loculicidally from the summit.

Key to the species

1 Corolla whitish to deep pink 2
1 Corolla yellow sometimes tinged bronze-red with age 4

2 Flowers c. 1.5 cm diam. Free part of floral tube 5–6 mm long, minutely pubescent. Capsule obovoid;
upper fertile part 3–4 mm diam., 4-angled with 4 longitudinal ribs between the angles and passing
gradually into the lower sterile region. Leaves narrow-ovate to elliptic, c. 25 mm long, sinuate, sometimes
dentate, minutely pubescent. Erect herb. Cumberland Plain. Grasslands. Introd. from America. Fl.
spring–autumn . . . . ***O. rosea** L'Hér. ex Aiton
2 Flowers 2–5 cm diam. Free part of floral tube mostly more than 1 cm long 3

3 Hairs on the stem long, spreading. Fruit ribbed, 10–15 mm long; upper fertile part 6–8 mm diam.
Leaves ovate to lanceolate, 2–5 cm long, usually ± pinnatifid. Erect herb. Garden escape near habitation.
Introd. from trop. America. Fl. summer–autumn . . . . ***O. tetraptera** Cav.
3 Hairs on the stem short. Fruit ± 4-angled; upper fertile part 2–5 mm diam. Leaves narrow-ovate to
elliptic, c. 3 cm long, sinuate, obscurely dentate. Erect herb with underground runners. Cumberland
Plain. Waste land and grasslands. Introd. from America. Fl. spring–summer . . . . . ***O. speciosa** Nutt.

4 Flowers up to 1.5 cm diam 5
4 Flowers more than 2 cm diam 6

5 Hairs on the stem very short and all ± the same size. Leaves elliptic to oblong, up to 2 cm long, sinuate.
Capsule cylindrical, 15–20 mm long, pubescent. Erect herb up to 40 cm high, with a rhizome. Coast
and Cumberland Plain. Sandy soils, often on sand dunes. Introd. from S. America. Fl. spring–autumn
. . . . ***O. indecora** Cambess. ssp. **bonariensis** W.Dietr.
5 Some hairs on the stem 2 mm long or longer, others shorter. Leaves lanceolate to elliptic or oblong, up
to 4 cm long, sinuate. Capsule cylindrical, 20–30 mm long, villous-pubescent. Erect herb. Waste places,
usually on sandy soils. Coast. Introd. from America. Fl. winter–summer . . . . ***O. mollissima** L.

6 Hairs on the stem below the inflorescence with swollen red bases. Leaves narrow-ovate to narrow-
elliptic, up to 30 cm long at the base of the stem, pubescent. Flowers 5–7 cm diam. Capsule narrow-
ovoid, 20–25 mm long, c. 8 mm diam., angular. Seeds angular. Erect herb up to 60 cm high. Mostly
coast. On sandy soils. Introd. from America. Fl. summer–autumn . . . . ***O. glazioviana** Micheli
6 Hairs on the stem white to the base, scarcely swollen 7

7 Stems with a few long or short hairs or almost glabrous towards the base. Leaves linear to narrow-ovate, ± pubescent, 2–6 cm long, sinuate to almost flat at the margin, obscurely dentate to almost entire. Flowers 2–4 cm diam. Capsule cylindrical, tapering towards the base, 2–3 cm long. Seeds angular. Widespread. In disturbed ground. Introd. from S. America. Fl. most of the year (*O. stricta* hybridizes with *O. indecora* ssp. *bonariensis*) . . . .*O. stricta** Ledeb. ex Link ssp. **stricta**

7 Stems densely villous-pubescent 8

8 Cauline leaves obtuse or truncate at the base, quite sessile, up to 5 cm long, villous, obtuse or acute, dentate. Capsule cylindrical, up to 5 cm long, c. 4 mm wide, villous. Mostly coast. Weed of pastures and waste ground. Introd. from America. Fl. spring–summer . . . .*O. longiflora** L.

8 Cauline leaves narrow-ovate to ovate or narrow-oblong, up to 7 cm long, tapering towards the base. . 9

9 Capsule linear-obovoid, 20–25 cm long, villous. Flowers 3–4 cm diam. Leaves narrow-ovate to narrow-elliptic, up to 7 cm long and 15 mm wide, villous, tapering towards the apex, acute, dentate. Erect herb up to 50 cm high. Coast. Disturbed ground or on sand dunes. Introd. from S. America. Fl. spring. . . . . ***O. affinis** Cambess.

9 Capsule linear, 4 cm long, 4 mm diam., villous-pubescent. Flowers 3–4 cm diam. Leaves obovate to elliptic, up to 7 cm long and 15 mm wide, villous-pubescent, not tapering towards the apex as gradually or so acutely as *O. affinis*. Erect herb up to 50 cm high. Coastal sand dunes. Introd. from S. America. Fl. spring . . . . ***O. drummondii** Hook.

2 Epilobium L.

Willow Herbs

15 species in Aust. (5 endemic, 4 naturalized); all states and territories except NT

Perennial herbs, hoary to almost glabrous. Leaves sessile or subsessile. Sepals 4. Petals 4, pink, white or purple. Stamens 8. Floral tube slender, tubular, scarcely extending above the ovary. Ovary 4-locular; style filiform; stigma entire, clavate. Capsule long, terete, opening through 4 valves at the summit. Seeds numerous, minute, oblong, with a tuft of hairs at one end.

Key to the species

1 Stems with long (up to 0.5 mm) spreading simple hairs. Leaves alternate above, opposite below, hirsute, linear to lanceolate, 2–5 cm long, dentate. Flowers erect. Petals white to purple, 2–8 mm long. Capsule cylindrical, 3.5–6 mm long, pubescent. Stoloniferous herb up to 1.5 m high. Hornsby and Woronora Plateaus; Blue Mts Damp places. Fl. spring–summer . . . . **E. hirtigerum** A.Cunn.

1 Stems with short glandular and/or curved simple hairs. 2

2 Seeds with a pale incurved wing. Leaves narrow-ovate to oblong or elliptic, 1–6 cm long, glabrous or nearly so, dentate. Flowers erect. Petals 5–18 mm long. Capsule 3–7.5 cm long, puberulent. Stoloniferous herb. Blue Mts Damp places. Fl. summer . . . . **E. gunnianum** Hausskn.

2 Seeds without a pale incurved wing 3

3 Flowers nodding to one side of the inflorescence, up to 30 mm diam. Leaves lanceolate to narrow-oblong, 3–8 cm long, slightly glossy, dentate, glabrous except for the margins and midribs. Capsule up to 10 cm long. Rhizomatous herb up to 1 m high. Blue Mts Damp places. Fl. spring–summer. . . . . **E. pallidiflorum** Sol. ex A.Cunn.

3 Flowers erect 4

4 Seeds with lines of papillae on the surface. Petals mostly less than 5 mm long, deep pink to white. Leaves lanceolate, up to 8 cm long, minutely dentate. Capsule 45–85 mm long. Erect perennial herb usually up to c. 50 cm high. Widespread weed. Introd from N. America. Fl. most of the year . . . .*E. ciliatum** Raf.

4 Seeds with scattered papillae 5

5 Leaves grey-puberulent, narrow-oblong to linear, to 8 cm long, 1–5 mm wide; coarsely dentate (up to 8 teeth per side). Capsule up to 8 cm long. Stoloniferous herb up to 1 m high. Widespread. A variety of

sites, all mostly drier than the other ssp. Fl. spring–summer. . . . .
. . . . E. **billardiereanum** ssp. **cinereum** (Rich.) P.H.Raven & Engelhorn
5 Leaves glabrous except for the margins and veins lanceolate to ovate elliptic or oblong, up to 6 cm long,
± glossy, dentate. 6

6 Leaves narrow-ovate to ovate 6–18 mm wide, finely dentate (16–40 teeth per side). Capsule up to 8 cm
long. Stoloniferous herb up to 1 m high. Widespread south of Sydney. Damp places or sand dunes. Fl.
spring–summer. . . . . E. **billardiereanum** Ser. ssp. **billardiereanum**
6 Leaves lanceolate, 5–15 mm wide, dentate (4–12 teeth per side). Capsule up to 8 cm long. Stoloniferous
herb up to 1 m high. Inland damp places. Fl. spring–summer . . . .
. . . . E. **billardiereanum** ssp. **hydrophilum** P.H.Raven & Engelhorn

3 **Ludwigia** L.
8 species in Aust. (4 native, 3 naturalized, 1 doubtful); all states and territories except Tas.

Herbs or shrubby, often aquatic perennials. Leaves alternate, usually entire, sessile or petiolate. Flowers
solitary in the leaf axils. Petals 4–5, yellow. Stamens 4–10. Floral tube not extended above the ovary.
Ovary 4–5-locular. Capsule terete. Seeds numerous.

Key to the species

1 Prostrate, glabrous to villous herb, rooting at the nodes. Leaves ± petiolate with vesicular stipules,
lanceolate to obovate, mostly 3–6 cm long. Seeds embedded in the ovary wall. Coast and Cumberland
Plain. Wet situations or in ponds or creeks. Introd. from S. America. Fl. summer–autumn. *Water
Primrose* . . . . *L. **peploides** (Kunth) P.H.Raven ssp. **montevidensis** (Spreng.) P.H.Raven
1 Erect shrubs, seeds free from fruit 2

2 Hairy shrub up to 3 m high, sometimes with pneumatophores. Leaves sessile or nearly so, ovate to
lanceolate or elliptic, up to 12 cm long. Petals 10–25 mm long. Coast. Naturalized in swamps and lake
margins. Introd. from America. Fl. spring–autumn . . . . *L. **peruviana** (L.) H.Hara
2 Glabrous shrub to 3 m high. Stems 4-angled. Leaves linear to oblanceolate usually more than 12 cm
long. Petals 20–25 mm long. Garden escape. Introd. from S. America. Fl. summer–winter. . . . .
. . . . *L. **longifolia** (DC.) H.Hara

4 **Gaura** L.
2 species naturalized Aust.; NSW, S.A.

One species in the area
Branching perennial herb to 1.5 m high. Stems with curved hairs. Leaves narrow-elliptic to 9 cm long,
5–15 mm wide, margins toothed. Inflorescence up to 80 cm long, simple or branched. Flowers irregular.
Petals 4, 10–12 mm long, white at first becoming yellow or red with age. Fruit 6–8 mm long, 2–3 mm
wide, 4-angled. Garden escape, Camden area. Introd. from America. Fl. summer–autumn. . . . .
. . . . *G. **lindheimeri** Engelm. & A.Gray

71 ZYGOPHYLLACEAE
22 gen., Aust. & S. Africa

1 **Tribulus** L.
12 species in Aust. (9 endemic); all states and territories except Tas.

Prostrate herb. Leaves opposite, pinnate-compound with 8–16 oblong leaflets. Flowers regular, bisexual,
solitary. Petals yellow, 5, induplicate-valvate. Stamens 10, hypogynous. Ovary 5-locular, with axile
placentas, superior, surrounded at the base by a sinuate disc in the gland at the base of each stamen. Fruit
a muricate spiny schizocarp.

Key to the species

1 Style as long as stigma or shorter. Stigma elongated. Fruit with spines 3–8 mm long. Uncommon in the drier parts of the region. Probably introd. from the Mediterranean. Fl. spring–summer *Caltrop* . . . .?***T. terrestris** L.

1 Style longer than stigmas. Stigma hemispherical. Fruit with spines 0.5–2.5 mm long. Clay soils. Fl. spring–summer. *Yellow Vine* . . . . **T. micrococcus** Domin

72 CELASTRACEAE
(includes *Stackhousiaceae*)

Trees, shrubs, climbers or herbs. Leaves simple, entire, toothed or crenate, opposite or alternate; stipules small or absent, rarely prominent. Flowers, regular, bisexual or unisexual; in axillary or terminal spikes, racemes, panicles or cymes. Sepals 2–5, basally connate. Petals 4–5; free or fused at midpoint. Stamens usually as many as the petals and alternating with them. Ovary superior, occasionally appearing half-inferior when immersed in disc; 2–5-locular or; carpels 3–5 and connate along central axis. Style simple or branched. Fruit a capsule, shizocarp or drupaceous. 58 gen., widely distributed.

Key to the genera

1 Herbs, stems striated and sometimes woody. Leaves fleshy or leathery, petals free at base but connate at midpoint. Ovaries connate along central axis . . . . **1 STACKHOUSIA**
1 Trees, shrubs or climbers, petals and ovary free 2

2 Leaves opposite. Fruit drupaceous . . . . .5 ELAEODENDRON
2 Leaves alternate. Fruit a capsule 3

3 Woody climber or scrambling shrub. Ovary and fruit 3-locular. . . . . **2 CELASTRUS**
3 Erect shrubs. Ovary and fruit 2-locular 4

4 Leaves linear, ericoid, pungent; stipules prominent, lobed . . . . .3 APATOPHYLLUM
4 Leaves narrow-oblong to lanceolate, not pungent; stipules minute or absent. . . . . . . .4 MAYTENUS

1 Stackhousia Sm.
13 species in Aust. (12 endemic); all states and territories

Perennial herbs, usually glabrous. Leaves alternate, entire; stipules absent or minute and caducous. Flowers bisexual, regular, in terminal spikes or racemes, with a small bract and 2 bracteoles at the base of each. Calyx tubular, 5-lobed, small. Petals connate or cohering (except at the base) into a tubular corolla with 5 spreading lobes. Stamens 5, perigynous. Ovary 3–5-locular, superior; placentas axile. Fruit a schizocarp.

Key to the species

1 Spikes dense, flowers single along its axis. Corolla whitish to cream 2
1 Spikes often long and slender, usually interrupted with flowers in groups of 1–6 along its axis. Corolla greenish-yellow to yellow; lobes narrow. 3

2 Mature carpels each with 3 acute dorsal angles. Stems stout. Leaves thick, spathulate to obovate, 15–30 mm, long 8–15 mm wide, often obtuse, sometimes with short points. Corolla tube cream, 6–8 mm long, with obtuse lobes. Procumbent to ascending herb, stem often partly buried in sand. Coastal sand dunes. Fl. spring–summer. . . . . **S. spathulata** Sieber ex Spreng.
2 Carpels without angles, globular or obovoid, reticulate. Stems often slender. Leaves thin, usually linear (rarely narrow-ovate), mostly 1–2 cm long. Corolla white, drying cream; lobes obtuse; the tube 5–7 mm long. Spike elongating at fruiting stage. Erect herb up to 40 cm high. Widespread. Open forests. Fl. summer . . . .**S. monogyna** Labill.

3 Mature capsules pyriform with prominent round cavity at base, apex muricate. Leaves linear 1–2.5 mm wide. Corolla tube 2–6 mm long. Widespread. Woodland and grassland. Fl. Spring   **S. muricata** Lindl.

3 Mature capsules obovoid, cavity at base if present shallow and not round, apex not muricate. Leaves 1.5–8 mm wide or reduced to scales 4

4 Leaves reduced to minute distant scales. Branches slender, wiry, brownish, 30–60 cm high. Flowers in the axils of minute bracts, distant along the ends of the branches, shortly pedicellate. Calyx lobes narrow, acute. Corolla tube 2–3 mm long. Coast and adjacent plateaus. Damp sandy soils. Fl. summer . . . . **S. nuda** Lindl.

4 Leaves well developed on the basal portion of the plant but smaller on the long flowering stems, linear to obovate or elliptic, mostly 10–25 mm long, rather thick, corolla tube 4–6 mm long. Widespread. Heath and pasture; low forest on Ss. Fl. summer . . . .**S. viminea** Sm.

2 Celastrus L.
2 species endemic Aust.; Qld, NSW, Vic.

Woody climbers or scrambling shrubs. Leaves alternate, lanceolate to elliptic or ovate, usually coriaceous, dark green on the upper surface, yellowish underneath, entire or serrulate. Flowers in terminal panicles, 5-merous, dioecious, 3–4.5 mm diam., greenish; disc prominent. Ovary 3-locular. Fruit a coriaceous capsule opening loculicidally. Seeds enveloped in an orange or yellow-green aril

Key to the species

1 Fruit 2–5.5 mm long; inside surface of valves with red markings. Stipules filiform, 0.5–2 mm long. Leaves 3–8 cm long, 1–4 cm wide. Widespread. RF. Fl. summer–winter . . . . **C. australis** Harvey & F.Muell.

1 Fruit 7–10 mm long; inside surface of the valves pale brown lacking red markings. Stipules filiform, 1.5–4 mm long. Leaves 5–14 cm long, 2–7 cm wide. Illawarra region. RF. Fl. summer . . . . **C. subspicata** Hook.

3 Apatophyllum McGill.
2 species endemic Aust.; Qld, NSW

One species in the area
Small, spreading shrub up to 50 cm high, glabrous. Leaves linear, scattered, sessile, up to 20 mm long, pungent, with small lobed stipules. Flowers solitary in leaf axils, c. 2 mm diam, 5-merous. Ovary immersed in disc and appearing half-inferior, 2-locular, with 2 basal ovules per loculus. Fruit a capsule. Glen Davis. Ss. Endangered. Fl. spring. . . . .**A. constablei** McGill.

4 Maytenus Molina
Orangebarks
9 species endemic Aust.; Qld, NSW, NT, WA

One species in the area
Glabrous shrub. Leaves narrow-oblong to lanceolate, 1–7 cm long, entire or toothed, coriaceous. Flowers in short axillary or lateral racemes, greenish; pedicels slender, 4–6 mm long. Petals c. 2 mm long, broad-ovate. Capsule globular or ovoid, 2-locular, 4–6 mm diam., yellowish-brown when dry. Widespread. Open forests. Fl. spring–summer. **M. silvestris** Lander & L.A.S.Johnson

5 Elaeodendron Jacq.
2 species endemic Aust.; Qld, NSW, NT, WA

One species in the area
Shrub or small tree, usually glabrous. Leaves coriaceous, broad-ovate or obovate to elliptic or lanceolate, mostly 5–8 cm long, entire or crenate, usually opposite. Flowers greenish, in axillary or lateral cymes. Petals 2–3 mm long, ovate, often shortly 2-lobed. Ovary confluent with the disc in a conical mass,

2-locular. Fruit drupaceous, orange-red, ovoid or globular, c. 12 mm long. Coast and adjacent plateaus. Open forests and RF. Fl. spring. *Red-fruited Olive Plum* . . . .**E. australe** Vent.

73 CUCURBITACEAE

Herbs, usually monoecious, climbing by tendrils. Leaves alternate, usually palmately veined, without stipules. Flowers unisexual, regular, axillary. Calyx and corolla 5-lobed. Stamens 5, ± connate; anthers usually long, flexuous. Ovary inferior, 3–5-locular, with parietal placentas (rarely 1-locular with a basal ovule). Fruit usually indehiscent, berry-like sometimes called a 'pepo', rarely 1-seeded. 118 gen., mostly trop. and subtrop.

Key to the genera

1 Tendrils simple 2
1 Tendrils branched 3

2 Fruit mottled green, glabrous or with fine scabrous hairs. . . . . .2 CITRULLUS
2 Fruit with scattered hooked hairs and green strips; ripening yellow . . . . 3 CUCUMIS

3 Corolla white or greenish. Fruit 1-seeded . . . . .1 SICYOS
3 Corolla yellow. Fruit many-seeded 4

4 Leaves simple, lobed or toothed. Corolla c. 8 cm long. Fruit ribbed, flattened . . . . . . **4 CUCURBITA**
4 Leaves deeply palmately or pinnately lobed. Corolla up to 2 cm long. Fruit smooth, globose . . . .
. . . . .2 CITRULLUS

1 Sicyos L.

1 species native Aust.; Qld, NSW, Vic., Tas.

One species in the area

Slender climber with 3-branched tendrils, glabrous or sparingly pubescent. Leaves almost membranous, angularly 3- or 5-lobed, irregularly toothed, c. 4 cm long. Male flowers in short racemes on long peduncles; female flowers in clusters on short peduncles; corolla 3 mm long; both sexes in the same axils. Fruits almost ovate, acute, less than 12 mm long, densely covered with barbed bristles. Coast and gullies in adjacent plateaus. Near RF . . . . **S. australis** Endl.

2 Citrullus Schrad.

2 species naturalized Aust.; all states and territories except Tas.

One species in the area

Annual, trailing herb. Leaves palmate- or pinnate-lobed with up to 9 lobes, hairy. Corolla 1–2 cm long, yellow. Fruit globular, up to 30 cm diam., mottled green or yellowish. Cumberland Plain. Drier situations. Introd. from Asia. *Bitter Melon*. . . . . *C. lanatus (Thunb.) Matsum. & Nakai var. **lanatus**

3 Cucumis L.

6 species in Aust. (1 native, 5 naturalized); all states and territories

One species in the area

Monoecious annual with slender prostrate scabrous stems and simple tendrils. Leaves ovate-cordate in outline, 2–6 cm long, palmately divided into 5 obovate toothed lobes; the middle one longest and again often 3-lobed. Male flowers in clusters of 2–4. Female flowers solitary or twinned; ovary covered with rather distant bristles, marked with dark green longitudinal stripes. Corolla c. 4 mm long. Widespread. Weed, particularly in railway enclosures. Introd. from S. Africa. *Paddy Melon* . . . .
. . . . *C. myriocarpus Naudin ssp. **leptodermis** (Schweick.) C.Jeffery & P.Halliday

4 Cucurbita L.

1 species naturalized Aust.; NSW, ?Vic.

One species in the area

Monoecious, scabrous herb with trailing or climbing stems. Tendrils branched. Leaves ovate-cordate in outline, 20–30 cm diam., 5-lobed and toothed. Male flower peduncles 10–15 cm long; female flower peduncles 3 cm long. Corolla c. 8 cm long. Naturalized near habitation. Introd. from America. *Ironbark Pumpkin* . . . . *C. maxima* Lam.

74 FABACEAE

Herbs, climbers, shrubs or trees. Leaves alternate, opposite or sometimes whorled; compound, simple, reduced to scales or modified into phyllodes or tendrils. Stipules usually present. Inflorescence paniculate, racemose, cymose, umbellate or flowers solitary. Flowers usually bisexual, regular or irregular, 3 – 5 merous. Sepals and petals usually free sometimes connate. Stamens up to 10 or numerous, free or fused. Carpel solitary, ovary superior, style terminal. Fruit a legume (pod), 1–many-seeded. 730 gen. cosmop.

Key to the subfamilies

1 Flowers regular . . . . **2 MIMOSOIDEAE**
1 Flowers irregular 2

2 Posterior petal inside lateral ones . . . . **3 CAESALPINIOIDEAE**
2 Posterior petal outside lateral ones . . . . **1 FABOIDEAE**

1 FABOIDEAE

Herbs, shrubs, climbers or trees. Leaves usually alternate, simple or compound (3-foliolate or pinnate, sometimes with pinnae modified into tendrils); stipules often present. Flowers in spikes racemes or heads, bisexual, irregular. Sepals usually 5, usually basally connate into a tube. Petals 5, imbricate, the posterior one (standard) larger and outside the 2 lateral ones (wings); the two anterior petals connate or coherent along the lower margins forming the keel which envelopes the stamens and carpel. Stamens perigynous, usually 10; either all free, or the filaments all connate into a tube which is sometimes open on the upper side, or the upper stamen free. Ovary superior, 1-carpellate, 1-locular; placentas marginal; ovules 1–many; style simple. Fruit a legume, achene or lomentum. 500 gen., cosmop.

Key to the genera

(To determine the number of ovules in an ovary either count the seeds in the fruit (some of which may be small and ± aborted); or cut the ovary just below the base of the style and squeeze the ovary at the base; the ovules should pop out.)

1 Filaments all free . . . . .GROUP 1
1 Filaments variously connate 2

2 Leaves with more than 3 leaflets (the lower two sometimes close to the base and resembling stipules) . .
. . . . .GROUP 2
2 Leaves with 2–3 leaflets or simple 3

3 Herbs, trailers or twiners . . . . .GROUP 3
3 Trees or shrubs . . . . .GROUP 4

Group 1
Stamens free

1 Stipules fused within the leaf axil . . . . **7 PULTENAEA**
1 Stipules free from each other, sometimes minute 2

2 Ovary and fruit divided by a longitudinal partition . . . . **1 MIRBELIA**

2 Ovary and fruit without a longitudinal partition 3

3 Leaves compound . . . . **5 GOMPHOLOBIUM**
3 Leaves simple 4

4 Ovules 4 or more 5
4 Ovules 2 7

5 Leaves alternate, keel much shorter than other petals . . . .**2 CHORIZEMA**
5 Leaves mostly opposite or whorled, keel longer than wing petals 6

6 Floral bracts 3-lobed, hairs on stems simple . . . .**3 OXYLOBIUM**
6 Floral bracts simple, hairs in stems branched. . . . .**4 PODOLOBIUM**

7 Leaves well developed 8
7 Leaves on flowering stems mostly reduced to scales 14

8 Leaves modified to terete or filiform phyllodes more than 3 cm long . . . .**14 VIMINARIA**
8 Leaves not as above. 9

9 Leaves flat or revolute and pungent 10
9 Leaves revolute, recurved, incurved or involute but not pungent 11

10 Flowers in racemes or panicles . . . .**6 DAVIESIA**
10 Flowers in a head surrounded by leaves with enlarged bases . . . .**11 ALMALEEA**

11 Leaves recurved or revolute 12
11 Leaves incurved or involute 13

12 Bracteoles absent . . . .**9 AOTUS**
12 Bracteoles large, herbaceous. . . . .**8 PHYLLOTA**

13 Standard (excluding claw) much broader than long; flowers in racemes or spikes or heads not
surrounded by leaves with enlarged bases . . . .**10 DILLWYNIA**
13 Standard (excluding claw) slightly broader than long; flowers in heads surrounded by leaves with
enlarged bases . . . .**11 ALMALEEA**

14 Branchlets angular or winged 15
14 Branchlets terete 16

15 Tall shrub, greyish-green . . . .**12 JACKSONIA**
15 Shrub to 1 m high, dark green. . . . .**6 DAVIESIA**

16 Upper 2 calyx lobes united into an upper lip, longer than the lower 3. Small shrub c. 50 cm high . . . .
. . . .**13 SPHAEROLOBIUM**
16 Calyx lobes all free, short, nearly equal. Shrub often 2–3 m high . . . .**14 VIMINARIA**

Group 2
Stamens connate; leaves with more than 3 leaflets

1 Leaves palmate-compound . . . .**34 LUPINUS**
1 Leaves pinnate-compound 2

2 Leaves ending in a point or tendril. 3
2 Leaves ending in a leaflet 4

3 Stipules much smaller than the leaflets . . . .**42 VICIA**
3 Stipules equal to or much larger than the leaflets . . . .**40 PISUM**

4 Lower two leaflets close to the base of the petiole and resembling stipules; leaflets 5 or more 5
4 Lower two leaflets neither close to the base of the petiole nor resembling stipules; leaflets 5 or more . 6

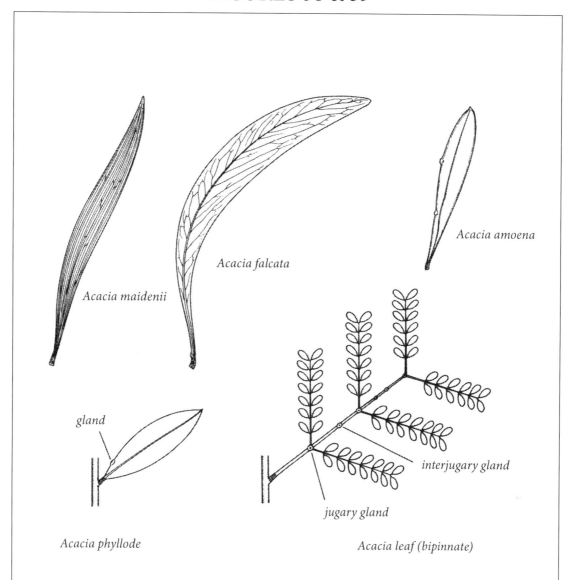

Fig. 38   Acacia: leaves and phyllodes

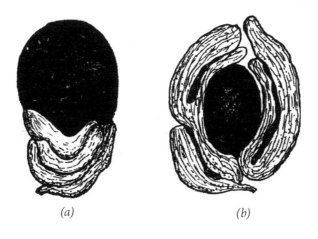

Fig. 39   Arils (funicles) of Acacia. (a) *A. implexa*; (b) *A. melanoxylon*.

5 Leaflets up to 5 . . . . **19 LOTUS**
5 Leaflets more than 7 . . . . **18 ORNITHOPUS**

6 Corolla white . . . .**39 ROBINIA**
6 Corolla pink to red or purple 7

7 Anthers tipped by a small gland. Hairs on leaves and stems T-shaped . . . .**35 INDIGOFERA**
7 Anthers not tipped by a small gland. Hairs not T-shaped 8

8 Leaves dotted with glands. . . . . **37 PSORALEA**
8 Leaves not dotted with glands 9

9 Plants ± pubescent . . . .**36 TEPHROSIA**
9 Plants glabrous . . . . **38 SWAINSONA**

Group 3
Herbs, trailers or climbers; stamens connate; leaves simple or with 2–3 leaflets

1 Leaves with 2 leaflets. 2
1 Leaves with 3 leaflets or simple 3

2 Leaves ending in a point or tendril. . . . . **41 LATHYRUS**
2 Leaves ending in a leaflet . . . . **17 ZORNIA**

3 Stipules adnate to the petiole; leaflets usually dentate 4
3 Stipules free from the petiole; leaves/leaflets usually entire 6

4 Legume enclosed in the calyx and persistent corolla; flowers usually in dense heads . .**20 TRIFOLIUM**
4 Legume exceeding calyx; corolla deciduous; flowers in racemes 5

5 Legume ovoid, straight; flowers in long racemes . . . .**21 MELILOTUS**
5 Legume curved or spirally twisted; flowers in short dense racemes . . . . **22 MEDICAGO**

6 Leaves simple. Corolla violet . . . . **45 HARDENBERGIA**
6 Leaves with 3 leaflets. 7

7 Rhachis of inflorescence nodose. Legume oblong, with a prominent rib on each valve near the suture. .
. . . . **43 CANAVALIA**
7 Rhachis of inflorescence not nodose. Legume not ribbed. 8

8 Style bearded above for more than half its length 9
8 Style glabrous or hairy above for much less than half its length 11

9 Wings very dark purple . . . . **48 MACROPTILIUM**
9 Wings yellow, white, pink or pale purplish 10

10 Style thickened above. . . . . **49 VIGNA**
10 Style not thickened above . . . .**50 DIPOGON**

11 Flowers 18–40 mm long; corolla red . . . . **44 KENNEDIA**
11 Flowers less than 18 mm long; corolla violet, pink, blue or white. 12

12 Legumes 5–9 cm long, covered with long, straight, rusty hairs. . . . . **46 PUERARIA**
12 Legumes up to 5 cm long, sparsely hairy if dense then hairs hooked. 13

13 Legume oblong to linear, splitting into 2 twisted valves . . . .**47 GLYCINE**
13 Legume constricted between seeds and breaking up at the constrictions . . . .**15 DESMODIUM**

Group 4
Trees or shrubs; stamens connate; leaves simple or with 2–3 leaflets

1 Branchlets spinous and pungent at the tips . . . . **29 ULEX**
1 Branches not spinous at the tips 2

2 Leaves with 3 leaflets 3
2 Leaves simple, sometimes scale-like 10

3 Corolla red . . . . **51 ERYTHRINA**
3 Corolla yellow or white or pink to purple 4

4 Corolla yellow or white 5
4 Corolla pink to purple 9

5 Corolla white . . . . **32 CHAMAECYTISUS**
5 Corolla yellow 6

6 Terminal leaflet with a distinct stalk . . . . **28 GOODIA**
6 Terminal leaflet sessile or with a very short stalk 7

7 Stamens fused into a tube but open on the upper side . . . . **25 CROTALARIA**
7 Stamens fused into a complete tube 8

8 Calyx teeth minute. Style coiled into a spiral or circle . . . . **33 CYTISUS**
8 Calyx teeth conspicuous. Style curved but not coiled . . . . **30 GENISTA**

9 Leaves without sunken black glands . . . . **16 LESPEDEZA**
9 Leaves with sunken black dots (glands) . . . . **37 PSORALEA**

10 Stamens united into a complete tube . . . . **31 SPARTIUM**
10 Stamen united into a tube but open on the upper side 11

11 Corolla blue, pink, purple or white . . . . **23 HOVEA**
11 Corolla yellow or orange 12

12 Leaves spirally arranged . . . . **24 TEMPLETONIA**
12 Leaves distichous 13

13 Leaves strongly reticulate, 2–5 cm long. Legume flat, winged, opening along one suture . . . .
   . . . . **26 PLATYLOBIUM**
13 Leaves not strongly reticulate, rarely exceeding 2 cm long, sometimes absent. Legume not winged,
   opening along both sutures . . . . **27 BOSSIAEA**

1 Mirbelia Sm.
20 species endemic Aust.; Qld, NSW, Vic., NT, WA

Small shrubs. Leaves simple, alternate opposite or in whorls of 3. Flowers sessile or shortly pedicellate, in small axillary or terminal clusters or short racemes. Bracts and bracteoles small or absent. Calyx lobes nearly equal in length. Corolla orange to yellow or yellow and red, or pink to reddish-purple or bluish. Ovary and legume divided longitudinally into 2 compartments by a false-septum projecting into the loculus from the lower suture. Ovules 2–many.

Key to the species

1 Corolla pink to purplish or bluish 2
1 Corolla orange to yellow or red and yellow, sometimes with purplish markings 4

2 Leaves mostly alternate, linear, rigid, pungent, up to 12 mm long. Branches pubescent. Erect or
   ascending shrub 30–50 cm high. Flowers nearly sessile, in axillary clusters. Calyx 4 mm long or

less. Corolla bluish purple or yellow and red with purple markings. Ovary and legume pubescent. Widespread. Heath. Uncommon. Fl. spring. . . . . **M. pungens** A.Cunn.ex G.Don

2 Leaves mostly in whorls of 3, narrow-lanceolate to linear, 5–25 mm long. Branches glabrous or slightly pubescent 3

3 Calyx c. 3 mm long. Diffuse shrub with slender branches up to 1 m long. Leaves up to 25 mm long, strongly reticulate, pungent, with recurved margins. Flowers several in axillary clusters. Corolla pink to purplish; standard c. 8 mm long. Ovules 2. Legume ovate, c. 5 mm long. Widespread. Common in heath. Sandy soils. Fl. spring . . . . **M. rubiifolia** (Andrews) G.Don

3 Calyx 6–8 mm long. Erect shrub up to 1 m high. Leaves narrow-linear, obtuse or almost pungent, scarcely reticulate; margins closely revolute. Corolla reddish-purple, drying pale purple or bluish. Flowers almost sessile in the upper axils, forming a terminal interrupted spike leafy at the base. Ovules c. 12. Legume thick, ovoid, 8 mm long. Loftus to Appin; north of Hawkesbury River; Blue Mts Heath and DSF. Ss. Fl. winter–spring . . . . **M. speciosa** Sieber ex DC. ssp. **speciosa**

4 Leaves ovate to lanceolate. Corolla orange to yellow, with a red centre 5

4 Leaves linear, up to 15 mm long, rigid, pungent. Corolla orange to red, usually with purplish markings or centre 6

5 Leaves up to 10 mm long, 1–3 mm wide, recurved at the margins and tip, pubescent underneath. Flowers up to 10 mm long. Decumbent to erect shrub up to 2 m high. Southern Tablelands. DSF. Fl. summer . . . . **M. oxylobioides** F.Muell.

5 Leaves 15–40 mm long, 10–15 mm wide, coriaceous, strongly reticulate and glossy on the upper surface, with recurved margins, silky hairy underneath. Flowers up to 18 mm long. Prostrate shrub. Blue Mts Heath. Ss. Fl. spring . . . . **M. platylobioides** (DC.) Joy Thomps.

6 Flowers 5–7 mm long, clustered . . . . **M. pungens** (see above)

6 Flowers 8–15 mm long, axillary, solitary. Legume ovoid, c. 6 mm long. Prostrate to ascending shrub sometimes forming mats. Higher Blue Mts Exposed heath. Ss. Fl. spring. . . . . **M. baueri** (Benth.) Joy Thomps.

2 Chorizema Labill.
18 species endemic to Aust.; Qld, NSW, WA.

One species in the area
Shrub 20–50 cm high, glabrous or nearly so. Branches slender and erect or ascending and angular. Leaves linear to oblong, with recurved margins, 10–25 mm long. Calyx 2–3 mm long. Corolla yellow with a red centre; standard 4–6 mm long. Legume turgid, c. 6 mm long. Coast and Cumberland Plain. Heavier soils. Grassland and heath. Fl. spring. Endangered population in some LGAs . *Eastern Flame Pea* . . . . **C. parviflorum** Benth.

3 Oxylobium Andrews
30 species endemic Aust.; Qld, NSW, Vic., Tas., WA.

Shrubs or climbers. Leaves simple, entire, mostly opposite or in whorls of 3, leaf margins distinctly recurved, ± pubescent underneath. Corolla yellow or orange and red. Ovary usually villous, sessile or nearly so. Ovules 4 or more. Legume ovoid or oblong, turgid.

Key to the species

1 Leaf apex recurved. Leaves alternate or opposite or whorled Flowers few in short terminal umbel-like racemes 2

1 Leaf apex mucronate with a straight pungent point. Leaves usually whorled. Flowers in axillary racemes or crowded towards the ends of branches 3

2 Leaves cordate, ovate, 2–10 mm long, 2–4 mm wide. Young branches with ± divergent silky hairs. Corolla orange-red; standard 6–8 mm long. Legume ovoid, acuminate, 8–10 mm long, pubescent.

Diffuse shrub c. 30 cm high. Coast and adjacent plateaus from Port Jackson northwards. Heath in damp soil. Ss. Fl. spring . . . .**O. cordifolium** Andrews

2 Leaves obtuse at the base, linear to elliptic or narrow-ovate, 5–10 mm long, 2–4 mm wide, usually scabrous above. Young branches with appressed villous hairs. Corolla orange-red; standard up to 10 mm long. Legume ovoid, acuminate, c. 10 mm long, pubescent. Low spreading or prostrate shrub. Hornsby Plateau. DSF and Swamp forest. Ss. Fl. spring . . . . **O. pulteneae** DC.

3 Leaves linear to oblong, mostly 3–5 cm long. Flowers in short axillary racemes. Calyx c. 5 mm long. Corolla yellow; standard 8–10 mm long. Legume ovoid, acuminate, 6–10 mm long, densely hairy. Tall shrub. Lower Blue Mts; Illawarra region. Forests. Fl. spring–summer. . . . . **O. arborescens** R.Br.

3 Leaves ovate to narrow-ovate, 1–2 cm long. Flowers crowded towards the end of the branches. Calyx c. 5 mm long. Corolla yellow; standard 8–10 mm long. Legume ovoid, acuminate, c. 8 mm long, silky hairy. Low shrub up to 2.5 m high. Higher Blue Mts Open forests. Fl. summer . . . **O. ellipticum** (Labill.) R.Br.

4 Podolobium R.Br.
6 species endemic Aust.; Qld, NSW, Vic.

Shrubs or climbers. Leaves simple, entire, toothed or lobed, mostly opposite or in whorls of 3, leaf margins ± thickened but not recurved. Corolla yellow or orange and red. Ovary usually villous, ovary stipitate. Ovules 4 or more. Legume ovoid or oblong.

Key to the species

1 Leaves with 3 or more lobes or teeth, reticulate, glabrous, glossy above, sometimes minutely pubescent below; lobes or teeth with pungent points. Flowers in short axillary or terminal racemes. Corolla yellow-orange; standard c. 8 mm long. Erect, very variable shrub up to 3 m high. Specimens from the Blue Mts have deeply 3-lobed leaves. Widespread. In a variety of situations. Fl. spring . . . .
. . . . **P. ilicifolium** (Andrews) Crisp & P.H.Weston
1 Leaves entire 2

2 Prostrate or climbing shrub. Leaves elliptic to lanceolate, reticulate, 2–5 cm long. Flowers mostly in terminal often umbel-like racemes. Calyx lobes longer than the tube. Corolla pale orange; standard c. 10 mm long. Coast to lower Blue Mts Woronora Plateau. Heavier soils. Fl. most of the year . . . .
. . . . **P. scandens** (Sm.) DC.
2 Erect or ascending shrub up to 2 m high. Leaves broad-ovate to narrow-ovate, 1–2.5 cm long, with a pungent acuminate tip; stipules spinous. Flowers in few-flowered axillary racemes. Calyx lobes shorter than the tube. Corolla orange-yellow; standard 5–6 mm long. Coast in north of region. WSF; margins of RF. Fl. spring–summer . . . . **P. aciculiferum** F.Muell.

5 Gompholobium Sm.
Golden Glory Peas and *Wedge Peas*
30 species endemic Aust.; Qld, NSW, Vic., Tas., WA

Shrubs. Leaves digitate- or pinnate-compound; leaflets narrow; stipules small or absent. Flowers terminal or in the upper axils, solitary or two or three together in a short raceme. Calyx deeply cleft. Corolla yellow-greenish or red. Ovary glabrous, shortly stipitate. Ovules 8 or more. Legume nearly globular, turgid.

Key to the species

1 Leaves with 3 leaflets. 2
1 Leaves with 4–30 leaflets, pinnately arranged. 9

2 Keel conspicuously ciliate. Pedicels usually longer than the calyx (sometimes shorter in *G. latifolium*) 3
2 Keel minutely ciliate or glabrous. Pedicels very short (sometimes long in *G. minus*). Leaflets linear to narrow-linear, usually with revolute margins. 4

3 Leaflets flat or the margins recurved, linear to linear-cuneate, mostly more than 25 mm long, 2–6 mm wide. Standard uniformly yellow, broad, 25–30 mm long but smaller on higher Blue Mts; lower petals nearly as long as the standard; keel densely fringed. Legumes mostly c. 15 mm long. Erect shrub up to 2 m high. Widespread. DSF. Sandy soils. Fl. spring. *Golden Glory Pea* . . . . **G. latifolium** Sm.

3 Leaflets with revolute margins, linear to broad-linear, 5–15 mm long or rarely longer. Standard pale greenish yellow, darker on the outside, 15–20 mm long. Legumes up to 15 mm long. Diffuse or much branched shrub 30–rarely 100 cm high. Blue Mts DSF and heath. Ss. Fl. spring . . . . **G. huegelii** Benth.

4 Leaves <12 mm long 5
4 Leaves >12 mm long 7

5 Old stems smooth or nearly so, glabrous or pubescent with spreading hairs. Buds ridged. Standard 10–15 mm long. Ovules 12–20. Leaflets linear to narrow-obovate, less than 10 mm long. Diffuse shrub 10–30 cm high. Widespread. In a variety of open situations. Fl. spring . . . . **G. minus** Sm.
5 Old stems tuberculate and glabrescent, slightly pubescent when young 6

6 Stipules absent. Corolla very deep orange to red; standard c. 8 mm long. Leaflets dark green to purplish, 7 mm long, 1 mm wide, linear to very narrow-ovate, margins recurved. Ovules c. 6. Pedicels up to 5 mm long. Diffuse shrub 10–20 cm high. Western Blue Mts DSF and heath. Ss and granites. Fl. summer
. . . . **G. uncinatum** A.Cunn. ex Benth.
6 Stipules small, conspicuously recurved. Corolla yellow. Leaflets grey-green, 3–10 mm long, 0.2–1 mm wide linear with slightly recuved margins. Ovules 12–20. Pedicels up to 5 mm long. Wiry erect shrub to 50 cm high. South from Blue Mts with records from Holsworthy DSF Fl. winter–spring. . . . .
. . . .**G. inconspicuum** Crisp

7 Leaflets acute, with a sharp point, 12–25 mm long. Ovules 6–15. Erect shrub up to c. 1 m high. Coast and adjacent plateaus south of Gosford; Blue Mts DSF and heath. Sandy soils. Fl. spring . . . .
. . . . **G. grandiflorum** Sm.
7 Leaflets obtuse, sometimes with a recurved tip. Corolla yellow; standard c. 18 mm long. Ovules 6–8. Erect shrubs up to c. 1 m high 8

8 Leaflets 15–30 mm long, usually c. 2 mm wide; apex usually with a ± recurved tip. Coast and adjacent plateaus north of Port Jackson. DSF and heath. Sandy soils. Fl. spring . . . .
. . . . **G. virgatum** Sieber ex DC. var. **virgatum**
8 Leaflets 10–13 mm long, c. 1 mm wide; apex usually quite obtuse or truncate. West Hornsby Plateau. DSF. Ss. Fl. spring . . . . **G. virgatum** var. **aspalathoides** (A.Cunn. ex Benth.) Benth.

9 Leaflets (3–)4–5(–7), usually less than 10 mm long, narrow-linear to linear, on a very short rhachis . Stems densely tuberculate. Corolla pale greenish yellow, darker on the outside; standard 8–10 mm long. Ovules 8–10. Shrub 30–50 cm high. Widespread. DSF and heath. Sandy soils. Fl. spring . . . .
. . . . **G. glabratum** Sieber ex DC.
9 Leaflets usually 20–30, 6–12 mm long, narrow-linear to narrow-elliptic, on an elongated rhachis 10–25 mm long. Stems not tuberculate. Calyx 5–6 mm long. Corolla yellow; standard slightly longer than the calyx. Ovules 8–10. Shrub 10–30 cm high, with thin branches. Coast and adj plateaus. Open forest and cleared areas. Sandy soils. Fl. spring–summer . . . . **G. pinnatum** Sm.

6 Daviesia Sm.
120 species endemic Aust.; all states and territories

Shrubs. Leaves modified into phyllodes, alternate, simple, entire, rigid, rarely reduced to minute scales; stipules minute or absent. Flowers solitary or in small usually axillary clusters or racemes or corymbs or umbels; bracts ± conspicuous. Bracteoles absent. Calyx lobes short, the upper 2 often connate. Corolla yellowish, usually with a dark red centre; standard broad, emarginate. Ovary glabrous, shortly stipitate. Ovules 2. Legume oblique, triangular.

Key to the species

1 Phyllodes reduced to minute scales on mature branches. Branches flat, winged or 3-angular, lax. Racemes 2–5-flowered. Calyx c. 4 mm long; the teeth as long as the tube. Straggling shrub c. 1 m high. Hawkesbury River; Blue Mts DSF and heath. Fl. spring . . . . **D. alata** Sm.

1 At least the lower phyllodes well developed 2

2 Phyllodes with revolute margins or terete, very rigid, pungent pointed. 3

2 Phyllodes flat or rarely recurved 4

3 Phyllodes linear, 15–30 mm long, with revolute margins. Flowers often solitary. Calyx 3–4 mm long; lobes c. as long as the tube. Shrub 30–50 cm high. Widespread except higher Blue Mts Open forest and cleared areas. Ss. Fl. spring . . . . **D. acicularis** Sm.

3 Phyllodes terete or vertically compressed, 10–25 mm long, almost patent. Racemes 2–6-flowered. Calyx 2–3 mm long; lobes much shorter than the tube. Erect shrub up to 60 cm high. Cumberland Plain; Blue Mts Fl. spring . . . .**D. genistifolia** A.Cunn. ex Benth.

4 Phyllodes not pungent-pointed, 3–10 cm long. Flowers in axillary or rarely terminal racemes or corymbs or umbels; bracts ± conspicuous; rhachis or peduncle ± elongated. Calyx with upper 2 lobes broad, truncate, connate almost to the top 5

4 Phyllodes pungent pointed 8

5 Venation prominent, reticulate 6

5 Venation obscure, parallel or pinnate 7

6 Flowers crowded on the end of the peduncle forming a corymbose raceme; pedicels long, filiform; bracts 2–3 mm long, obovate. Flowers 4–8 mm long. Phyllodes linear to narrow-ovate, 4–12 cm long, tapering towards the base; often broader on new growth after fire. Shrub 1–2 m high. Widespread. DSF. Ss. Fl. spring. . . . .**D. corymbosa** Sm.

6 Flowers in a raceme 2–5 cm long; bracts 2–4 mm long, ovate or oblong. Flowers 4–7 mm long. Phyllodes ovate-elliptic to ovate-lanceolate, mostly 2–15 cm long; tapering towards the base. Shrub up to 3 m high. Higher Blue Mts and valleys. DSF. Fl. spring. . . . .**D. latifolia** R.Br.

7 Phyllodes reduced on the upper part of the stems. Lower phyllodes rigid, narrow-obovate to linear, up to 9 cm long, obtuse to acute, thick. Racemes 1–2 cm long. Corolla c. 5 mm long. Shrub up to 1.5 m high. Higher Blue Mts DSF. Fl. spring–summer . . . .**D. leptophylla** A.Cunn. ex G.Don

7 Phyllodes not much reduced on upper part of stems, lanceolate or oblanceolate to narrow-oblanceolate, 2–20 cm long. Racemes usually less than 3 cm long, often with flowers in the upper half only. Standard 3–4 mm long. Shrub up to 2 m high. Widespread. DSF. Fl. spring–summer . . . . **D. mimosoides** R.Br. ssp. **mimosoides**

8 Flowers in small axillary or terminal umbel-like racemes on short peduncles. Phyllodes broad-ovate, cordate, less than 2 cm long, tapering into a fine pungent point. Shrub up to 1 m high. Coast and adjacent plateaus in scattered localities, e.g., Thirlmere, Newport, Barrenjoey. DSF and heath. Fl. spring–summer. . . . . **D. umbellulata** Sm.

8 Flowers solitary or rarely 2–4, on a very short peduncle, not in an umbel, axillary. Calyx c. 2 mm long. Standard 4–5 mm long 9

9 Weak, spreading shrub up to 1 m high with fewer slightly angular branches, not spinescent. Phyllodes sessile ovate, cordate, 6–10 mm long, tapering to a fine point, spreading or reflexed. Flowers solitary, on filiform pedicels 4–12 mm long. Upper calyx lobes united. Widespread but mainly coast. Open forests. Fl. spring . . . . **D. squarrosa** Sm.

9 Much-branched, erect shrub up to 2 m high, with short angular often spinescent branches. 10

10 Inflorescence an umbel with 2–5 flowers. Phyllodes lanceolate to narrow elliptic, 5–15 mm long, sessile or nearly so. Flowers on very short pedicels. Upper calyx lobes scarcely united at the base. Widespread. Common in many situations DSF. Fl. spring . . . . **D. ulicifolia** Andrews ssp. **ulicifolia**

10 Flowers solitary or in pairs in axils. Phyllodes linear, 9–20 mm long. Flowers on very short pedicels. Upper calyx lobes scarcely united at the base. Coastal in sandy soil. Fl. spring. . . . .
. . . . **D. ulicifolia** ssp. **stenophylla** G.Chandler & Crisp

7 **Pultenaea** Sm.

100 species endemic to Aust. all states and territories except NT

Erect spreading or prostrate shrubs. Leaves simple, mostly alternate. Stipules scarious, fused within the axil. Flowers either in terminal heads or axillary and then sometimes crowded towards the ends of the branches. Bracts frequently with broader more scarious stipules than the foliage leaves and sometimes with the lamina reduced or absent. Bracteoles persistent, ± adnate to the calyx, brown and membranous or green; with or without stipules which are often fused to the bracteoles making them appear to be lobed. Corolla yellow, often with a reddish centre. Standard nearly orbicular, longer than the lower petals, folded. Style straight or curved. Ovary sessile. Ovules 2. Legume ovate, small. Fl. spring except where otherwise stated.

Key to the species

1 Leaves with recurved margins; or, if flat, then paler or hoary underneath . . . .**GROUP 1**
1 Leaves with incurved margins; or with concave surface; or, if flat or convex, then darker underneath. If linear-terete then channelled on the upper surface 2

2 Ovary pubescent to the base . . . .**GROUP 2**
2 Ovary glabrous or with hairs on the upper half only, the base glabrous. 3

3 Bracteoles entire or 3-lobed at the apex only . . . .**GROUP 3**
3 Bracteoles deeply 3-lobed . . . .**GROUP 4**

Group 1

1 Unopened flower heads surrounded by conspicuously imbricate bract-stipules, the inner ones longer than the short pedicels. Keel dark coloured. 2
1 Bract-stipules, when present, not conspicuously imbricate. Flowers either in terminal heads or in the upper axils. Keel not always dark-coloured 8

2 Bract-stipules deciduous, usually falling by the time the flower opens, or, if remaining, easily dislodged
. 3
2 Bract-stipules persistent 4

3 Flower heads mostly dense, 2–3 cm diam. Calyx 5–6 mm long. Standard nearly twice as long as the calyx. Leaves cuneate-oblong, mostly 1–3 cm long, mucronate. Erect shrub up to 3 m high. Widespread. DSF. Ss. Fl. spring . . . . **P. daphnoides** J.C.Wendl.
3 Flower heads under 2 cm diam. Calyx c. 4 mm long. Leaves linear to linear-cuneate, obtuse or retuse, not mucronate, glabrous, 5–15 mm long. Erect shrub with glabrous usually angular branches, 0.5–2 m high, sometimes prostrate in exposed situations. Coast and adjacent plateaus; Hunter River Valley. DSF. Ss. Fl. spring–summer . . . . **P. retusa** Sm.

4 Calyx with appressed whitish hairs 5
4 Calyx with loose spreading greyish hairs.. 7

5 Bract-stipules mostly glabrous, brown, completely concealing the calyx; the inner ones 6–8 mm long. Calyx c. 6 mm long. Leaves 1–2 cm long. Prostrate to low spreading shrub with thin branches. Widespread. Heath, often in damp places. Ss. . . . . **P. paleacea** Willd.
5 Bract stipules mostly pubescent with whitish or silvery appressed hairs, mostly shorter than the calyx 6

6 Calyx c. 7 mm long. Leaves linear to obovate, 20–40 mm long. Bracteoles aristate. Hairs on upper portion of stems whitish, appressed. Weak shrub with branches up to 2 m long. Coast and adjacent Plateaus, mostly north of Sydney. Heath and DSF. Ss. Fl. spring . . . . **P. rosmarinifolia** Lindl.

6 Calyx 5–6 mm long. Leaves obovate to elliptic, 4–12 mm long, apex acute or obtuse and recurved. Bracteoles not atistate. Standard twice as long as the calyx or nearly so. Prostrate or decumbent-ascending shrub. Higher Blue Mts Damp ground. Ss . . . . **P. capitellata** Sieber ex DC.

7 Leaves with a minute recurved callous point, linear-cuneate, often emarginate, mostly 10–25 mm long. Undersurface of leaves and bract-stipules and upper portion of stems pubescent with mostly short grey hairs. Calyx c. 5 mm long. Standard at least twice as long as calyx. Erect shrub c. 1 m high Coast and adjacent Plateaus. DSF. Ss. Fl. spring. . . . . **P. linophylla** Schrad.
7 Leaves with a stiff straight or reflexed point Leaves 1–4 cm long, linear to narrow-elliptic or narrow-obovate, ± acuminate. Bract-stipules and younger stems villous with spreading soft greyish hairs. Calyx about 6 mm long. Standard less than twice as long as the calyx. Weak shrub with spreading or ascending branches up to 1 m long. Widespread south of the Hawkesbury River: Blackheath. DSF and heath . . . .**P. polifolia** A.Cunn.

8 Stipules up to 10 mm long, narrow, appressed, concealing the branches. Leaves linear or almost so, up to 4 cm long, with thickened margins. Flowers in dense heads c. 3 cm diam. Ovary glabrous in the basal half. Shrub up to 2 m high, with few erect branches. Widespread. DSF, usually in sheltered places. Ss . . . . **P. stipularis** Sm.
8 Stipules less than 5 mm long. Ovary pubescent all over 9

9 Flowers on filiform pedicels exceeding the leaves, solitary or paaired, terminal or in the upper axils. Calyx c. 3 mm long. Leaves linear-lanceolate, c. 5 mm long, rigid, acute. Leaf hairs not grooved. Prostrate, much-branched shrub forming a mat up to 1 m diam. WS. Endangered . . . .**P. pedunculata** Hook.
9 Flowers sessile or on pedicels less than 5 mm long. 10

10 At least the young stems, and undersurface of leaves with rather long spreading hairs. Leaves narrow- to broad-cuneate, up to 12 mm long, truncate emarginate or 2-lobed, scabrous above, tomentose below, with a mucronate reflexed point. Leaf hairs not grooved. Stipules spreading or recurved. Shrub up to 2 m high. Widespread. DSF and . . . . **P. scabra** R.Br.
10 Hairs appressed, whitish. Leaves linear-cuneate, 4–8 mm long, obtuse or truncate. Leaf hairs grooved. Stipules minute. Flowers in upper axils or 2–3 together at the ends of branches. Calyx 3–4 mm long. Standard c. twice as long as the calyx. Diffuse, much branched shrub, sometimes prostrate in exposed situations. Cumberland Plain; higher Blue Mts DSF. Usually on clay soils . . . . **P. microphylla** Sieber ex DC.

Group 2

1 Leaves terete, with a groove on the upper surface, 8–15 mm long. Flowers c. 8 mm long, almost sessile, 8 or fewer in a terminal head surrounded by leaves. Erect, graceful shrub up to 1 m high. Woronora Plateau; Blue Mts Wet ground. Ss . . . . **P. divaricata** H.B.Will.
1 Leaves not as above; margins incurved involute or flat 2

2 Hairs on calyx spreading or absent. Leaves alternate 3
2 Hairs on calyx silky, appressed. Leaves not always alternate 5

3 Leaves gradually tapering to a fine rather pungent point, lanceolate, 1–2 cm long, mostly 3–5-veined; stipules dark. Branches and leaf margins villous with soft hairs. Flowers nearly sessile, in the upper axils or clustered on short branches. Erect shrub 1–2 m high. Lower Blue Mts; Minnamurra. DSF. Sandy soils. Endangered population in some LGAs . . . . **P. villifera** Sieber ex DC.
3 Leaves obtuse or rather acute but not pungent pointed 4

4 Leaves 1–2 cm long, linear, glabrous or pubescent underneath. Stipules mostly with recurved points. Flowers shortly pedicellate, in heads c. 2 cm diam. Calyx villous, 5–6 mm long. Erect shrub. Widespread but not common. DSF . . . . **P. mollis** Lindl.
4 Leaves usually less than 1 cm long, oblong-linear, scabrous or hirsute or rust coloured. Flowers very few in a loose leafy head. Calyx 3–4 mm long. Standard c. twice as long as the calyx. In general appearance

resembles *P. villosa*. Coast and adjacent plateaus; southern Blue Mts Uncommon. DSF . . . .
. . . . **P. hispidula** R.Br. ex Benth.

5 Flowers in a compact head surrounded by conspicuous imbricate hirsute bract-stipules. Leaves linear to almost terete, less than 12 mm long; stipules up to 1 mm long. Erect to almost prostate shrub. Coast and adjacent plateaus south of Botany Bay. Uncommon. Wet heath . . . .**P. dentata** Labill.
5 Flowers arranged in short ± terminal racemes on lateral branches, each subtended by a bract similar to a foliage leaf. Leaves narrow-oblong, 2–3 cm long; stipules 1 mm long, (similar to *P. flexilis*). Leaves mostly opposite. Erect shrub up to 3 m high. Coast and adjacent plateaus. Open forests . . . .
. . . . **P. blakelyi** Joy Thomps.

Group 3

1 Leaves mostly opposite or in whorls of 3, flat, transverse-ovate to transverse-elliptic, up to 12 mm long, venation palmate, apex mucronate with a pungent point, glabrous, sometimes glaucous. Flowers in the leaf axils. Calyx 5–7 mm long. Bracteoles linear. Shrub up to 3 m high. Hunter River Valley; western Hornsby Plateau. DSF . . . . **P. spinosa** (DC.) H.B.Will.
1 Leaves mostly alternate, venation not appearing palmate 2

2 Flowers in ± leafy racemes with leaf-like bracts 3
2 Flowers in terminal compact heads; stipules of the bracts enlarged and the lamina much reduced or absent 9

3 Stipules less than 2 mm long, connate only at the base. Flowers distinctly pedicellate. Leaves flat to concave. Bracteoles simple 4
3 Stipules 2 mm long or longer, connate above the base, appressed 7

4 Leaves mucronate or with a distinct callous tip 5
4 Leaves obtuse to emarginate 6

5 Leaves oblanceolate to narrow-oblong, 1–3 cm long, paler above. Bracteoles linear. Calyx 4–5 mm long. Standard twice as long as the calyx. Erect shrub up to 4 m high, glabrous or young branches appressed-pubescent. Widespread. Open forest. Ss . . . .**P. flexilis** Sm.
5 Leaves narrow-linear to linear, up to 1.5 cm long. Bracteoles narrow-ovate. Calyx 4–7 mm long. Standard twice as long as the calyx. Erect to decumbent shrub up to 1.2 m high, glabrous to sparsely hairy. Northern Hornsby Plateau, e.g., Bucketty. DSF. Ss . . . . **P. setulosa** Benth.

6 Flowers 10–14 mm long, on pedicels 5–8 mm long. Bracteoles linear to lanceolate. Leaves narrow-obovate to narrow-oblong, 1–3 cm long; margins ± incurved. Erect shrub up to 2 m high. Northern part of the coast and lower Hunter River Valley. Open forests . . . . **P. euchila** DC.
6 Flowers up to 8 mm long on pedicels up to 2 mm long. Bracteoles ovate to triangular. Leaves narrow-obovate to linear, 5–8 mm long; margins ± incurved. Erect shrub. Higher Blue Mts, e.g. Jenolan Caves. DSF. . . . . **P. altissima** F.Muell.

7 Leaves concave, oblong-elliptic, 8–12 mm long, usually villous when young but becoming glabrous. Bracteoles membranous, 3-lobed at the apex. Flowers almost sessile. Standard c. 10 mm long. Erect shrub c. 1 m high. Widespread. DSF and heath. Ss . . . . **P. tuberculata** Pers.
7 Leaves with inrolled to involute margins, linear to linear-elliptic 8

8 Bracteoles simple . . . . **P. setulosa** (see above)
8 Bracteoles 3-fid at the apex . . . . **P. subspicata** (see below)

9 Leaves flat or with very slightly incurved margins or even slightly convex, up to 4 cm long. Stipules up to 1 cm long. . . . . **P. stipularis** (see Group 1)
9 Leaves with inrolled or involute margins 10

10 Leaves with a blunt callous tip, linear to linear-elliptic, up to 9 mm long, villous when young. Stipules 2–4 mm long. Calyx c. 6 mm long. Standard almost twice as long as the calyx. Low shrub up to c. 1 m high. Blue Mts, near Clarence. DSF. . . . .**P. subspicata** Benth.

10 Leaves with a distinct acuminate or aristate tip. 11

11 Leaves aristate, terminated by a bristle 2–3 mm long, linear, c. 12 mm long, tuberculate. Stipules 5–6 mm long. Calyx c. 9 mm long. Standard up to 13 mm long. Small shrub up to 1 m high. Woronora Plateau. DSF. Ss. Vulnerable . . . . **P. aristata** Sieber ex Benth.

11 Leaves acuminate with a short point less than 1 mm long. 12

12 Leaves villous when young, echinate when old, linear-terete, channelled above, mostly up to 12 mm long. Stipules 3 mm long. Calyx 4–5 mm long. Shrub up to 1.5 m high. Blue Mts, e.g. Mt. Victoria, King's Tableland. DSF and heath. Ss . . . . **P. echinula** Sieber ex DC.

12 Leaves glabrous, smooth, linear-terete, channelled above, 10–20 cm long. Stipules 4–5 mm long. Calyx 5 mm long. Erect shrub 1–2 m high. Blue Mts DSF. Ss. Vulnerable. . . . . **P. glabra** Benth.

Group 4

1 Margins of bracteoles or bracteole stipules beige in colour and fimbriate, large. Bracteoles deeply concave, concealing most of the calyx . . . . **P. tuberculata** (see Group 3)

1 Lateral lobes of bracteoles red-brown, margins not frayed; or lobes apparently absent. Bracteoles not concealing much of the calyx 2

2 Leaf apex shortly acuminate to aristate. Leaves concave, narrow-elliptic to narrow-obovate, c. 7 mm long, villous on the lower surface, almost glabrous on the upper surface. Calyx 5–7 mm long. Standard usually less than twice as long as the calyx. Decumbent or prostrate shrub. Cox's River Valley. DSF and heath . . . . **P. procumbens** A.Cunn.

2 Leaf apex acute to rounded lacking a point 3

3 Bracteoles lobed for up to half their length . . . . **P. subspicata** (see Group 3)

3 Bracteoles lobed to their base 4

4 Bracteoles <3 mm long attached ± at centre of calyx tube. Calyx glabrous on outer surface or lobes may have a few hairs. Flowers few, 5–7 mm long. Leaves 2–6 mm long, oblong-obovate, glabrous when mature. Erect shrub up to 1 m high. Cumberland Plain. Open forest on clay or lateritic soils. Endangered . . . . **P. parviflora** Sieber ex DC.

4 Bracteoles >3 mm long. Calyx moderately hairy 5

5 Leaves obovate, softly pubescent, 3–13 mm long, 3–5 mm wide. Stipules 3–6 mm long. Upper calyx lobes very much broader than lower ones, connate to the middle. Standard up to 12 mm long. Bracteoles attached on the calyx, 4–7 mm long. Erect shrubs up to 2 m high Hornsby Plateau, Lower Blue Mts to Gosford. Sheltered forest, DSF on Ss . . . . **P. ferruginea** Rudge

5 Leaves not as above. 6

6 Leaves narrow-oblong or slightly obovate, tuberculate or hairy underneath, 3–6 mm long, 1–2 mm wide. Bracteoles attached on the tube or at the base of the calyx, c. 3 mm long. Standard 5–8 mm long. Softly hairy shrub, spreading to erect, up to 2 m high, often with drooping tips to the branches. Widespread. Open forests. Common on heavy soils. . . . . **P. villosa** Willd.

6 Leaves crowded, narrow-oblong to slightly cuneate, silky hairy to glabrous underneath, up to 12 mm long. Flowers in dense terminal heads, sessile within the last leaves. Bracteoles attached at base of calyx c. 7 mm long, with 2 broad stipules, but not divided to the base. Stipules 4–5 mm long. Erect to prostrate shrub up to 1 m high. Higher Blue Mts Wet or damp places. Ss . . . . **P. canescens** A.Cunn.

8 Phyllota (DC.) Benth.
10 species endemic to Aust.; Qld, NSW, Vic., S.A., WA

Erect spreading or prostrate shrubs sometimes with underground rhizomes. At least young growth pubescent. Leaves alternate, simple, scattered, with revolute margins; stipules minute. Petals yellow to orange. Flowers in terminal heads or subterminal spikes; bracteoles herbaceous, longer than the calyx tube, attached at the base of the calyx. Ovules 2.

Key to the species

1 Leaves with a yellow recurved mucro. Flowers few in a subterminal spike. Corolla orange with deep orange to red keel. Rhizomatous, suckering shrub up to 75 cm high, often forming patches 1 mor more diam. Higher Blue Mts DSF and heath . . . .**P. squarrosa** (Sieber ex DC.) Benth.
1 Leaves obtuse or with a straight mucro. Non-suckering shrubs without rhizomes 2

2 Flowers 12–15 mm long. Young stems and calyx hirsute with long loose hairs. Leaves 6–20 mm long, 1.25–2.25 mm wide. Erect shrub up to 1 m high. Coast and adjacent plateaus. DSF. Ss. Fl. summer . . . . **P. grandiflora** Benth.
2 Flowers up to 10 mm long. Young stems and calyx with appressed ± villous hairs 3

3 Flowers in leafy spikes with few flowers or even solitary; bracteoles linear. Corolla deep orange-yellow to red. Leaves linear, 3–8 mm long, c. 0.5 mm wide. Procumbent shrub. Southern Blue Mts DSF. Fl. summer. Vulnerable . . . . **P. humifusa** Benth.
3 Flowers in compact leafy spikes, usually with numerous flowers; bracteoles lanceolate. Corolla yellow to orange. Leaves linear, 5–20 mm long, c. 1 mm wide. Erect or ascending shrub up to 2 m high. Widespread, but not in higher Blue Mts DSF and heath. Fl. spring–summer. . . . . **P. phylicoides** (Sieber ex DC.) Benth.

9 Aotus Sm.

15 species endemic Aust.; all states and territories except NT

Erect shrubs. Leaves alternate opposite or in whorls of 3, simple, with recurved or revolute margins; stipules absent. Flowers solitary or in clusters in the leaf axils; bracteoles absent. Corolla yellow with dark centre. Ovary villous. Ovules 2.

Key to the species

1 Standard yellow with a dark purple centre; keel purplish. Stems with appressed silky hairs. Leaves narrow-linear, 7–30 mm long, usually less than 1 mm wide; upper surface ± smooth. Much branched shrub up to c. 1 m high. Coast and adjacent plateaus north of George's River. DSF. Ss. Fl. spring . . . . **A. subglauca** var. **filiformis** Blakely & McKie
1 Standard yellow with an orange centre; keel yellow. Stems with close but not usually appressed, often rusty hairs. Leaves ovate to linear, 6–20 mm long, 1–5 mm wide; upper surface ± tuberculate. Shrub up to 2 m high. Widespread. DSF and heath. Fl. winter–spring . . . .**A. ericoides** (Vent.) G.Don

10 Dillwynia Sm.

Eggs and Bacon Peas or *Parrot Peas*
40 species endemic Aust.; all states and territories except NT

Erect spreading or prostrate shrubs. Leaves simple, alternate, narrow-linear, with incurved margins or terete and channelled above, sometimes twisted; stipules usually minute. Flowers in umbels, corymbs, racemes or axillary along the branches. Bracteoles distant from the calyx, deciduous. Calyx lobes shorter than or as long as the tube, 2 upper ones united higher up than the others. Corolla yellow or orange, often with a red or darker centre; standard reniform. Style hooked near the tip. Ovules 2. Legume ovate or rounded, turgid.

Key to the species

1 Calyx obtuse at the base (i.e. U-shaped in vertical section). Petals persistent, the legume surrounded by the withered petals; standard with a claw shorter than the calyx; lamina slightly broader than long . . 2
1 Calyx tapering towards the base (i.e. V-shaped in vertical section), 4–7 mm long. Petals deciduous; standard with a long claw; lamina more than twice as broad as long 4

2 Leaves appressed to the branches, 12–30 mm long, acute but not pungent pointed. Calyx lobes short and broad. Erect shrub 1–2 m high. Widespread. Open forests. Ss. Fl. spring. DSF . . . .**D. acicularis** Sieber ex DC.

2 Leaves spreading, rigid, 5–20 mm long, with fine pungent points. Calyx 3–5 mm long, pubescent . . . 3

3 Leaves sessile, 6–15 mm long. Inflorescence to 4.5 cm long. Calyx minutely ciliate; upper calyx lobes nearly or completely connate, all short and broad. Rigid, prickly shrub 1–2 m high with divaricate branches. Widespread. Open forests. Fl. spring . . . .**D. juniperina** Lodd.

3 Leaves 7–20 mm long petioles short, yellowish. Inflorescence to 2.5 cm long. Calyx lobes all short and broad. Erect prickly shrub to 2.5 m high. Widespread. DSF and woodland. Fl. winter–spring . . . . **D. sieberi** Lodd.

4 Stipules conspicuous, c. 1 mm long, subulate, persistent. Flowers crowded into leafy globular heads at the ends of the branches, vegetative growth sometimes continuing on before flowering has finished. Leaves 10–15 mm long, ± erect, straight, keeled, acute. Shrub 30–50 cm high, glabrous or young branches pubescent. Blue Mts, e.g. Clarence. Wet heath. Fl. spring–summer . . . **D. stipulifera** Blakely

4 Stipules minute or absent 5

5 Flowers almost sessile, in a dense terminal head with up to 9 flowers. Leaves 5–15 mm long, scabrous. Young branches and buds densely pubescent with whitish hairs. Calyx 5–6 mm long; lobes almost equal, nearly as long as the tube. Erect shrub 30–100 cm high. Higher Blue Mts DSF. Fl. spring–summer. . . . . **D. brunioides** Meisn.

5 Flowers not in dense terminal heads. 6

6 Flowers in sessile or pedunculate up to 9-flowered corymbs or umbels or racemes (or rarely solitary) which are terminal or axillary in the upper axils. Leaves 2–15 mm long 7

6 Flowers all axillary, mostly in pairs, often crowded towards the ends of the branches 14

7 Leaves ± conspicuously twisted. 8

7 Leaves straight or very occasionally slightly twisted. 11

8 Leaves branches and calyx usually scabrous with short rigid hairs. Leaves 3–8 mm long. Inflorescence sessile or sometimes pedunculate, mostly at the ends of the branches. Corolla orange-yellow with a darker centre. Prostrate to erect shrub up to 1.5 m high. Higher Blue Mts and adjacent valleys. DSF and heath. Fl. spring–summer. . . . .**D. phylicoides** A.Cunn.

8 Leaves glabrous. Stems hairy. Corolla yellow often with a red centre 9

9 Leaves 4–12 mm long. Erect or diffuse shrub 1–3 m high. Widespread. DSF and heath. Ss and dunes. Fl. spring . . . . **D. retorta** (C.J.Wendl.) Druce

9 Leaves up to 4 mm long.. 10

10 Flowers in terminal spikes which are sessile or on peduncles up to 3 mm long. Diffuse shrub up to 80 cm high. Cumberland Plain, e.g. Fairfield-Liverpool. WS. Open forest. Fl. spring . **D. parvifolia** R.Br.

10 Flowers in terminal umbels or corymbs with filiform peduncles more than 6 mm long. Low, spreading shrub up to c.a 80 cm high. Lower Blue Mts; northern part of coast. DSF. Ss, often with ironstones. Fl. spring–summer . . . . **D. 'trichopoda'** Maiden & Boorman 40290

11 Branches glabrous. 12

11 Branches pubescent. 13

12 Branches frequently terminating in spines. Leaves narrow-oblong and flat or terete, 3–5 mm long. Flowers solitary, terminal on short lateral branches or in the upper axils. Standard yellow with a dark centre. Divaricate shrub up to 150 cm high. Widespread but scattered. DSF. Ss. Fl. spring–summer . . . . **D. ramosissima** Benth.

12 Branches not spinescent. Leaves 5–15 mm long, glabrous, erect or slightly spreading with a recurved apex. Flowers in a terminal corymb, sessile or pedunculate. Standard yellow with purplish-brown lines in the centre. Weak spreading shrub 30–100 cm high. Widespread. Heath and open forest on deep sands and sandy alluvium, often in seasonally wet places. Fl. spring. . . . . **D. glaberrima** Sm.

13 Flowers solitary or rarely 2 on a short peduncle, terminal on short lateral branches, rarely in the upper axils. Leaves clustered, filiform, erect or ± spreading, 4–10 mm long, with very short recurved points. Branches minutely pubescent. Weak spreading shrub 60–100 cm high. Cumberland Plain; lower Blue Mts DSF on lateritized alluvium. Vulnerable. Fl. spring . . . .**D. tenuifolia** Sieber ex DC.

13 Flowers in terminal corymbs, sessile or pedunculate. . . . . **D. glaberrima** (see above)

14 Branches often terminating in spines. . . . . **D. ramosissima** (see above)

14 Branches not terminating in spines. 15

15 Branches, especially the young shoots, covered with silky white hairs. Leaves terete, 7–12 mm long, not keeled, not twisted, glabrous or slightly hairy. Flowers solitary or in pairs, sometimes appearing terminal but becoming infra-terminal by rapid growth of the stem apex. Calyx hairy or almost glabrous; upper 2 lobes connate more than half way, scarcely falcate. Erect shrub up to 1 m high. Higher Blue Mts DSF. Fl. spring. . . . . **D. sericea** A.Cunn.

15 Branches glabrous to hairy but not silky hairy (except rarely at the youngest tips). Calyx usually glabrous 16

16 Corolla with large dark red area in centre; standard orange and dark red to reddish-brown. Leaves scabrous with tubercles, hairy when very young. Calyx often tinged with red, glabrous or slightly hairy; upper lobes falcate. Shrub usually under 1 m high; branches villous when young, echinate when old. Dry heath. Ss. and laterites. Fl. spring–summer . . . . **D. rudis** Sieber ex DC.

16 Corolla yellow, often with a red centre. Branches not echinate 17

17 Leaves linear to scarcely terete, twisted, ± flat, 5–20 mm long, tuberculate-scabrous, erect, usually crowded. Branchlets with spreading to antrose hairs. Flowers with a minute peduncle or sessile. Shrub 1–2 m high. Widespread. Wet heath. Sandy soils usually in damp ground. Fl. spring . . . . **D. floribunda** Sm.

17 Leaves linear to ± terete, not or scarcely twisted. 18

18 Stems without decurrent leaf bases, appressed hairs on branchlets. Leaves 10–20 mm long, channelled above, almost smooth. Flowers, pedunculate, mostly 2 in leaf axils. Shrub to 2 m high. DSF and heath. Sandy soils. Fl. spring. . . . . **D. elegans** Endl.

18 Stems with prominent decurrent leaf bases. Leaves 6–15 mm long, ± trigonous. Stipules absent. Flowers mostly one in leaf axils. Peduncles absent. Shrub to 1.5 m high. Jamberoo area. Heath in damp situations. Ss. Fl. spring. . . . . **D. sp.'Barren Grounds'** (Chadwick s.n. NSW 39509)

11 Almaleea Crisp & P.H.Weston
5 species endemic Aust.; Qld, NSW, Vic., Tas.

Erect to procumbent shrubs up to 1 m high. Leaves simple, flat or concave, alternate, entire. Stipules absent or minute, free. Flowers yellow to orange with darker centre, arranged in terminal heads surrounded by leaves with enlarged bases. Bracts sometimes 3-lobed. Bracteoles small, attached to the pedicel below the flower. Standard orbicular to very broad-elliptic. Stamens free. Style hooked near the tip. Ovules 2. Legumes ovate, small. Fl. spring–summer.

Key to the species

1 Flowers 10 mm or more long. Bracts narrow, 5 mm long. Leaves concave, usually obtuse, with the tip incurved towards the stem, up to 10 mm long. Stipules minute. Blue Mts Wet places. Ss . . . . **A. incurvata** (A.Cunn.) Crisp & P.H.Weston

1 Flowers 6 mm long. Bracts leaf-like, ovate-linear, 3 mm long. Leaves very narrow-ovate, mostly flat, straight, up to 8 mm long. Stipules sometimes absent. Coast and adjacent plateaus. Wet sandy soil . . . .**A. paludosa** (Joy Thomps.) Crisp & P.H.Weston

12 Jacksonia R.Br. ex Sm.

40 species endemic Aust.; all states and territories except S.A..

One species in the area

Shrub up to c. 3 m high. Leaves reduced to minute scales. Branchlets angular, sometimes winged, often drooping, greyish in appearance due to fine hairs. Flowers in terminal racemes. Calyx deeply 5-lobed, valvate in bud. Petals 6–8 mm long, orange-yellow. Legume oblong, hairy, 8–12 mm long. Coast and adjacent plateaus; lower Blue Mts Usually in the open in DSF on lateritic sediments. Fl. summer. *Dogwood*. . . . . **J. scoparia** R.Br.

13 Sphaerolobium Sm.

12 species endemic Aust.; all states and territories except NT

Wiry shrub up to 1 m high; rush-like. Leaves mostly reduced to scales. Flowers in raceme. Calyx lobes imbricate in the bud; two upper ones large and united. Corolla yellow with orange centre, 3–5 mm long. Legume globular, stipitate.

Key to the species

1 Calyx and bracteoles punctate. Corolla wings larger than and enclosing keel. Style evenly curled, with a membranous flap inside the curve towards the top which is ± as broad as long and scarcely a ¼ the length of the style. Widespread. Wet heath on sandy or peat soils. Fl. spring. . . . . . **S. vimineum** Sm.
1 Calyx and bracteoles not punctate. Corolla wings equal to and exposing the keel. Style hooked near the base and straight above that, with a narrow membranous flap ¼ to ½ as long as the style. Widespread. Wet heath. Sandy soils. Fl. spring . . . . **S. minus** Labill.

14 Viminaria Sm.

1 species endemic Aust.; Qld, NSW, Vic., S.A., WA

Monotypic genus

Erect shrub 1–5 m high with long wiry often pendulous branchlets. Leaves modified into long cylindrical phyllodes, very rarely with 3 leaflets. Flowers in long terminal racemes. Calyx lobes nearly equal, short. Petals c. 8 mm long, yellow. Legume sessile, ovate. Coast and adjacent plateaus; lower Blue Mts Wet ground. Fl. spring. *Native Broom*. . . . .**V. juncea** (Schrad. & J.C.Wendl) Hoffmanns

15 Desmodium Desv.

Tick Trefoils

30 species in Aust., (7 sp. naturalized). all states and territories except Tas.

Weak shrubs or trailing or twining perennials. Leaves 3-foliolate or simple; stipules membranous, striate. Flowers small, usually in terminal racemes. Corolla pink or purple to blue or white. Stamens all united or upper one free. Fruit deeply indented on lower side, separating into 1-seeded articles, pubescent, with short appressed hairs.

Key to the species

1 Plants densely hairy 2
1 Plants glabrous or sparsely hairy 3

2 Plant stems rusty villous-tomentose. Leaflets orbicular to elliptic or ovate, 20–60 mm long, 8–35 mm wide. Corolla purple to pink. Pedicels or fruits as long as or longer than the calyx. Legume with 2–7 segments each 2–3 mm long. Prostrate trailing or twining herbs. Widespread. Forests. Fl. most of the year. . . . . **D. rhytidophyllum** F.Muell. ex Benth.
2 Plant stems covered with hooked hairs. Leaflets ovate to elliptic 2–10 cm long, 10–60 mm wide, upper surface with silver stripe down its middle. Corolla white, pink or blue. Legume with 3–10 segments each 5–7 mm long. Whale Beach. Fl. spring–summer. . . . .***D. uncinatum** (Jacq.) DC.

3 Corolla mauve to red or brownish. Leaflets ovate to elliptic, 15–70 mm long, all ± similar, lower surface with minute hooked hairs or glabrous. Pedicels shorter than calyx. Erect or ascending herb. Widespread. Forests. Fl. spring–autumn . . . .**D. brachypodum** A.Gray

3 Corolla pink to white. Leaflets, at least those of the lower leaves, rhombic, orbicular to obovate, up to 20 mm long. 4

4 Terminal leaflet with petiolule similar in length to lateral ones, less than or equal to 1 mm long. Leaflets rhombic usually truncate at apex. Leaf petioles 10–30 mm long; stipules 2–4 mm long. Flowers 1–1.5 mm long, pale pink to pale purple. Widespread. DSF. Fl. summer–autumn **D. gunnii** Benth. ex Hook.f.

4 Terminal leaflet with petiolule longer than lateral ones, 2–4.5 mm long. Leaflets orbicular to obovate. Leaf petiole 5–10 mm long; stipules 2–6 mm long. Flowers 4–5 mm long, pink to purple. Widespread. Grassland and forests. Fl. most of year . . . . **D. varians** (Labill.) G.Don

16 Lespedeza Michx.

2 species in Aust. (1 native, 1 naturalized) Qld, NSW, Vic.

One species in the area

Erect shrub with slender pubescent branches arising from a thick woody rootstock. Leaves pinnately 3-foliolate; leaflets linear-cuneate, 10–12 mm long, usually finely mucronate. Stipules small, subulate. Flowers c. 6 mm long, in clusters of 2–4. Calyx lobes rigid, very acute. Corolla pale pinkish purple. Legume sessile, nearly orbicular, 2–3 mm diam. indehiscent. Hartley; Minto; Richmond. Grasslands and open woodlands. Fl. summer–autumn. Endangered population in some LGAs . . . .**L. juncea** (L.f.) Pers. ssp. **sericea** (Thunb.) Steenis

17 Zornia J.F.Gmel.

19 species endemic Aust.; Qld, NSW, NT, WA

One species in the area

Decumbent or ascending perennial. Leaflets 2, ovate to linear, 4–25 mm long; stipules produced into a short auricle at the base. Flowers 6–8 mm long, almost enclosed in a pair of leafy bracts. Corolla yellow and red. Legume indented on the lower suture between the seeds, separating into 1-seeded articles. Widespread. Common amongst grass on heavier soils. Fl. spring–summer . . . .**Z. dyctiocarpa** DC. var. **dyctiocarpa**

18 Ornithopus L.

4 species naturalized Aust.; Qld, NSW, Vic., WA

One species in the area

Annual herb to 50 cm high. Leaves alternate, imparipinnate, pubescent. Stipules less than 1 mm long. 15–37 leaflets, oblong to elliptic 3–8 mm long, 2–4 mm wide. Inflorescence an axillary umbel with 2–5 flowers. Bracts leafy and pinnate. Calyx 3–5 mm long densely covered with hairs. Corolla 5–8 mm long, yellow. Legume curved, 20–5 mm long with a beak 7 mm long. Not common, naturalized in Sydney area. Introduced from Mediterranean. Fl. spring–summer. *Yellow Serradella* . . . . *O. compressus L.

19 Lotus L.

Trefoils

9 species in Aust.; (2 endemic, 7 naturalized) all states and territories

Herbs. Flowers subsessile, in umbels on axillary peduncles with 1–3 leafy bracts at the base of the umbel. Calyx 2-lipped; lobes almost equal. Stamens usually connate with the upper one free; filaments alternately long and short.

Key to the species

1 Corolla pink or white, 15 mm long or longer. Flowers 3–8 per umbel. Fruit 2–4 mm long. Perennial herbs up to 30 cm high. Widespread. Grasslands and open forests. Fl. summer. *Australian Trefoil* . . . . **L. australis** Andrews

1 Corolla yellow. Prostrate or ascending herbs 2

2 Flowers 10 mm long or longer 3
2 Flowers up to 8 mm long 4

3 Calyx teeth erect in the bud; 2 posterior ones with an obtuse sinus between them. Leaflets obovate to lanceolate or narrow-elliptic, 3–10 mm long. Stems ± solid. Decumbent to ascending glabrous to pubescent annual. Naturalized in a few places on the Blue Mts Grasslands. Introd. from Europe. Fl. spring–summer. *Bird'sfoot Trefoil* . . . . ***L. corniculatus** L. var. **corniculatus**
3 Calyx teeth spreading or recurved in the bud; 2 posterior ones with an acute sinus between them. Leaflets obovate to ovate, 10–20 mm long. Stems hollow. Naturalized in a few places on the Blue Mts Damp grassland. Introd. from Europe. Fl. spring–summer. *Bird'sfoot Trefoil* . .***L. uliginosus** Schkuhr

4 Fruit 6–12 mm long. Leaflets up to 20 mm long, narrow-obovate to lanceolate; peduncles usually longer than the leaves. Flowers (1–2) 3–4 per umbel 2. Villous, decumbent to ascending, perennial herb. Naturalized in a number of places. Introd. from Europe . . . . ***L. suaveolens** Pers.
4 Fruit 20–30 mm long. Leaves obovate to lanceolate, up to 20 mm long. Flowers 1–2 per umbel; peduncles usually shorter than the leaves. Villous annual. Naturalized in a number of places. Introd. from Europe. Fl. summer . . . . ***L. angustissimus** L.

20 Trifolium L.
Clovers
37 species naturalized Aust.; all states and territories except NT

Herbs. Leaves 3-foliolate; leaflets usually denticulate. Flowers small, usually numerous, arranged in heads with small persistent bracts. Calyx teeth 5, equal or unequal. Petals often persistent into the fruiting stage. Stamens connate with upper one free. Fruit small, dehiscing like a legume or circumcissile or indehiscent, usually enclosed in the persistent calyx. Seeds 1–2, rarely more.

Key to the species

1 Corolla yellow. Calyx 5-veined; throat open, glabrous inside. Flower heads on slender axillary peduncles, Fruit 1-seeded. Annuals 2
1 Corolla white to pink or red. Calyx 5- or more-veined or veining reticulate (except *T. hybridum*) . . . 3

2 Flowers 30–50 in a dense ovoid head c. 10 mm diam. Standard withering to brown then turned downwards over the divergent wings. Leaflets 5–12 mm long. Widespread. Common in waste places. Introd. from Europe. *Hop Clover*. . . . . ***T. campestre** Schreb.
2 Flowers 5–25 in a loose head c. 5 mm diam. Wings not divergent. Leaflets mostly 4–8 mm long. Widespread. Common in waste places. Introd. from Europe. *Yellow Suckling Clover*. ***T. dubium** Sibth.

3 Fertile flowers very few in the head with pink or white corolla; barren flowers without petals but with rigid calyx lobes enlarging after flowering. Peduncles turned downward to the ground at fruiting stage. Widespread. Pastures. Introd. from Europe. *Subterranean Clover*. . . . . ***T. subterraneum** L.
3 Flowers all fertile in the head. Heads sessile, or if pedunculate the peduncles not reflexed in fruiting stage 4

4 Calyx tube swollen and bladder-like in fruit, reticulately veined. 5
4 Calyx not swollen and bladder-like in fruit, with 5–12 longitudinal veins 7

5 Creeping perennial. Flowers nearly sessile, in dense ovate-globular heads 10–15 mm diam., with a basal involucre of several united bracts; peduncles mostly longer than the leaves. Fruiting head resembling a strawberry. Occasional. Grasslands. Introd. from Europe and Asia. *Strawberry Clover*. . . . . ***T. fragiferum** L.
5 Small annuals, glabrous except the calyx. Flower heads 5–8 mm diam., becoming enlarged in fruit. Flowers very small; pedicels twisted so that the standard is in an outer position 6

6 Calyx at fruiting stage with 2 short upper teeth. Fruiting head dense, globular, woolly tomentose. Occasional. Grasslands. Introd. from the Mediterranean. *Woolly Clover* . . . . *T. tomentosum** L.

6 Calyx at fruiting stage with 2 long upper teeth. Less tomentose but otherwise similar to *T. tomentosum*. Occasional in lawns, Introd. from the Mediterranean. *Shaftal Clover* . . . . *T. resupinatum** L.

7 Inflorescences sessile, obscured by broad membranous stipules. Standard shorter than the calyx. Flowers few in a head. Calyx teeth lanceolate, acuminate. Occasional only. Coast and Cumberland Plain. Introd. from W. Europe and the Mediterranean *Suffocated Clover* . . . . *T. suffocatum** L.

7 Inflorescences sessile or pedunculate but not obscured by stipules 8

8 Leaflets of upper leaves >5 times as long as wide 9

8 Leaflets of upper leaves <4 times as long as wide 10

9 Peduncles shorter than the leaves. Leaflets 30–50 mm long. Calyx teeth rigid, subulate, spreading, 5 mm long at fruiting. Stems rather stout, usually unbranched. Flower heads 4–8 cm long. Occasional weed. Introd. from the Mediterranean. *Narrow-leaved Clover* . . . . *T. angustifolium** L.

9 Peduncles longer than the leaves. Leaflets linear-oblong, 5–15 mm long. Calyx teeth plumose, much shorter than the tube. Stems slender, branching. Flowers in globular or cylindrical heads mostly 8–12 mm diam. Widespread. Common in vacant land. Introd. from Europe and W. Asia. *Hare's Foot Clover*
. . . . *T. arvense** L.

10 Flowers pedicellate 11

10 Flowers sessile 13

11 Inflorescences on peduncles shorter than the subtending leaf, rather loose. Flowers on very short pedicels. Standard notched at the summit, scarcely longer than the calyx. Occasional only. Introd. from Europe, W. Asia and Africa. *Nodding Clover* . . . .*T. cernuum** Brot.

11 Peduncle of inflorescence as long as or longer than the subtending leaf 12

12 Plants creeping, rooting at the nodes. Pedicels c. as long as the calyx tube. Corolla white; standard folded over the fruit after flowering. Widespread in pastures and as a weed. Introd. from Europe and W. Asia. Fl. spring–summer. *White* or *Dutch Clover* . . . .*T. repens** L.

12 Plants not rooting at the nodes. Inner flowers on pedicels 2–3 times as long as the calyx tube. Corolla usually pinkish; standard folded over the fruit after flowering. Coast. Pasture plant. Introd. from Europe and W. Asia. *Alsike clover*. . . . .*T. hybridum** L.

13 Leaf margins entire or nearly so. Leaflets elliptic to orbicular, more than 6 mm wide. Calyx with slender awn-like teeth, the lowest twice as long as the tube. Corolla partially fused. Widespread. Introd. from Europe and W. Asia. *Red Clover*. . . . .*T. pratense** L.

13 Leaf margins toothed. 14

14 Flowers red to purple inflorescence ovoid 20–60 mm long. Leaflets obovate; stipules oblong, almost free. Flower heads pedunculate, oblong. Calyx lobes subulate, hairy. Widespread but uncommon. Waste places, grasslands. Introd. from Europe. *Crimson Clover* . . . .*T. incarnatum** L.

14 Flowers pink or white <20 mm long 15

15 Inflorescences pedunculate peduncle longer than subtending leaf Standard yellowish to white. Leaflets oblong to lanceolate; stipules lanceolate-subulate, partly free. Flower heads sessile or pedunculate, ovoid, becoming oblong-conical in fruiting stage. Calyx lobes long and narrow. Tall erect or ascending, leafy annual. Widespread. Grasslands. Introd. from Syria and Egypt *Berseem Clover*. . . . .
. . . . *T. alexandrinum** L.

15 Inflorescence sessile or peduncle shorter than subtending leaf 16

16 Plants glabrous or with a few hairs on calyx and bases of leaflets. Standard longer than the calyx. Calyx-teeth equal, ovate, shortly awned. Widespread. Common in lawns and gardens. Introd. from Europe. *Clustered Clover* . . . .*T. glomeratum** L.

16 Plants hairy 17

17 Corolla white. Calyx teeth spreading at fruiting stage, almost pungent, curved outwards, the lowest tooth longer than the tube. Sydney. Introd. from Europe and W. Asia. *Rough Clover* . .*T. scabrum* L.

17 Corolla pink. Calyx teeth erect at fruiting stage, subulate, nearly equal. Heads partly enclosed in broad-ovate, acuminate stipules. Widespread but occasional. Introd. from Europe and W. Asia. *Knotted Clover*. . . . .*T. striatum* L.

21 Melilotus Mill.
4 species naturalized Aust.; all states and territories

Glabrous, erect herbs. Leaves 3-foliolate; leaflets denticulate. Flowers in long axillary pedunculate racemes. Calyx 5-toothed, nearly equal. Stamens connate with upper one free. Ovules 2–8. Fruit drooping, indehiscent, exceeding the calyx, 1–2-seeded.

Key to the species

1 Corolla white, 4–5 mm long. Fruit ovoid, reticulate. Racemes with 40–100 flowers, rather loose. Seeds 1 or rarely 2. Upper leaflets oblong. Widespread. Weed of cultivation and waste places. Introd. from Europe and W. Asia. Fl. summer. *Bokhara Clover* . . . . *M. albus* Medik.

1 Corolla yellow. 2

2 Fruit obtuse, reticulate, 1-seeded. Upper leaflets oblanceolate. Racemes with 10–60 flowers, rather dense. Flowers 2 mm long. Almost glabrous annual. Widespread. Weed of cultivation and waste places. Introd. from the Mediterranean and W. Asia. Fl. spring. *King Island Melilot* or *Hexham Scent* . . . . . .
. . . .*M. indicus* (L.) All.

2 Fruit acuminate, transversely ribbed, usually 1-seeded. Upper leaflets obovate. Racemes with 20–50 flowers. Flowers 5–6 mm long. Biennial. Blue Mts Weed of cultivation and waste places, uncommon. Introd. from Europe and Asia. *Common Melilot*. . . . .*M. officinalis* (L.) Pall.

22 Medicago L.
16 species naturalized Aust.; all states and territories

Herbs. Leaves 3-foliolate; leaflets toothed or lobed, with veins ending in a tooth; stipules usually toothed. Flowers in short axillary racemes with small persistent bracts; bracteoles absent. Peduncle often ending in an awn or bristle as long as the terminal flower. Calyx teeth 5, almost equal. Corolla often yellow. Stamens connate with the upper one free. Fruit short, at first straight but finally curved or coiled and exserted from the calyx, often with a row of spines or tubercles on each side of the dorsal suture or keel-like outer margin, 1–many-seeded. Seeds reniform.

Key to the species

1 Fruit without spines or tubercles, reniform or spirally coiled. 2

1 Fruit with spines or tubercles, spirally coiled. 4

2 Fruit reniform, c. 2 mm long, with concentric veins on the convex faces, 1-seeded, black when ripe. Flowers very small, numerous, in ovoid heads; peduncles slender, longer than the subtending leaf. Corolla yellow. Pubescent, procumbent annual or biennial. Widespread. Introd. from Europe and temp. Asia. *Black Medic*. . . . . *M. lupulina* L.

2 Fruit spirally coiled, with several seeds 3

3 Erect or ascending perennial. Flowers numerous in oblong racemes. Corolla violet. Legumes c. 5 mm diam., reticulate, with an opening in the centre of the spiral; coils 2–4, loose. Widespread. Probably a native of the Mediterranean and W. Asia. *Lucerne* or *Alfalfa*. . . . . *M. sativa* L.

3 Prostrate annual. Flowers few; peduncles awned. Corolla yellow. Fruit orbicular, c. 15 mm diam, depressed, thin, with 3–5 flat coils; central coil broadest. Widespread. Introd. from the Mediterranean. *Button Medic* . . . .*M. orbicularis* (L.) Bartal.

4 Fruit discoid to cylindrical, flat at the ends 5

4 Fruit ovoid or globular, spines present on ends. 7

5 Fruit with a row of tubercles on each side of the dorsal ridge. . . . . *M. polymorpha (see below)
5 Spines present on fruit. 6

6 Spines slender, divergent, c. as long as half the diam. of the fruit, almost hooked at the summit. Upper surface of leaflets glabrous. Stipules entire or toothed, but not deeply. Fruit discoid or almost cylindrical, 5–10 mm diam., flat and finely reticulate at both ends, with 1-½- 7 rather loose coils. Flowers 1–4. Peduncles awnless. Almost glabrous annual. Widespread. Introd. from the Mediterranean. *Burr Medic* *M. polymorpha L.
6 Spines rigid, divergent or ± appressed; fruit cylindrical or nearly so, 6–8 mm long, with 4–6 compact coils. Upper surface of leaflets hairy when young. Stipules deeply toothed, with a long entire point. Flowers 1–3; peduncles awned. Fruit cylindrical to truncate-conical, 5–8 mm diam., with 5–8 coils. Pubescent annual. Widespread. Introd. from the Mediterranean. *Barrel Medic* *M. truncatula Gaertn.

7 Stipules entire or slightly notched near the base. Leaflets small, very hairy on both surfaces. Fruit globular, 3–4 mm diam., with slender divergent hooked spines. Peduncles awned. Annual. Widespread. Introd. from W. Europe and W. Asia. *Small Woolly Burr Medic* . . . . *M. minima (L.) Bartal.
7 Stipules with long narrow teeth 8

8 Fruit 10–14 mm diam. with 8–11 coils, spines 3–6 mm long. Leaflets 15–20 mm long, toothed with dark patch at base. Flowers 3–7. Glabrous annual. Recorded once from Parramatta. Introd. from the Mediterranean. *Calvary Medic* . . . .*M. intertexta (L.) Mill.
8 Fruit less than 10 mm diam. 9

9 Leaflets obcordate-cuneate, toothed at the summit, with a reddish-brown spot in the centre. Fruit subglobular, slightly flattened at both ends. Flowers 1–5. Annual with a few spreading articulate hairs on stems and peduncles. Coast and Cumberland Plain. Introd. from Europe and W. Asia. *Spotted Medic*. . . . . *M. arabica (L.) Huds.
9 Leaflets without central spot, sometimes with dark flecks, deeply incised or toothed towards apex. . 10

10 Standard 4–6 mm long. Leaflets mostly deeply incised or lobed, sometimes with dark flecks. Fruit globular, 5–6 mm diam. with rigid hooked spines, 8–16 per coil, with up to 6 coils. Hairy annual. Coast. Introd. from the Mediterranean. *Cut-leaf Medic* . . . . *M. laciniata (L.) Mill.
10 Standard 2–3 mm long, Leaflets toothed towards top, without dark flecks. Fruit cylindrical or truncate and conical, 4–5 mm diam. 10–12 spines per coil, with up to 4 coils. Hairy annual. Uncommon. Introd. from Mediterranean rgion. *Small-leaved Burr Medic* . . . . *M. praecox DC.

23 Hovea R.Br.
38 species endemic Aust.; all states and territories

Shrubs. Indumentum of various simple hairs on most surfaces. Leaves alternate, simple, linear to oblong-elliptic, the lower ones (or all) sometimes ovate; margins recurved; stipules present but often caducous, rarely absent. Flowers in axillary clusters or very short racemes. Calyx densely hairy. Corolla blue to purplish. Stamens connate into a sheath. Ovules usually 2. Legumes turgid, globular or obliquely ovoid, 8–12 mm long.

Key to the species

1 Shrubs with few branches, erect to prostrate. Hairs straight and appressed. Lower surface of leaf usually partly visible through a sparse to moderately dense indumentum. Legume valves glabrous or sparsely hairy 2
1 Shrubs spreading to erect with many branches. Not all hairs straight and appressed. Lower surface of leaf with an indumentum of short curled hairs. Legume valves sparsely to densely hairy. 3

2 Leaves variable in shape, lower leaves boader than upper leaves, 1–8 cm long, 1.5–17 mm wide, upper surface glabrous. Calyx 3.5 mm long. Surface of legumes with short appressed hairs, sometimes glabrescent with age. Shrub with one to a few weak sub erect to prostrate stems. Widespread. DSF. Fl. spring. . . . . **H. heterophylla** A.Cunn. ex Hook.f.

2 Leaves uniformly linear to narrow linear, 3–11 cm long, 1–3.5 mm wide, upper surface glabrous. Calyx 5–6.5 mm long. Surface of legume glabrous. Erect shrub usually with a single main stem Nowra to Newcastle. DSF. Ss. Fl. winter–spring . . . . **H. linearis** (Sm.) R.Br.

3 Indumentum of long spreading hairs overtopping other shorter hairs by 0.8 mm. Stipule more than 3 mm long, less in *H. acutifolia* 4
3 Indumentum of mostly curled hairs, if long hairs present then overtopping others by less than 0.8 mm. Stipules less than 3 mm long 6

4 Leaves flat or weakly arched either side of midrib, upper surface glabrous. Apex of posterior calyx lobes usually acute . . . . **H. acutifolia** (see below)
4 Leaves strongly arched either side of midrib, sometimes weakly arched but then leaf midrib on upper surface hairy. Apex of posterior calyx lobes obtuse. 5

5 Upper surface of leaves glabrous; 2–5 mm long, 4–7 mm wide. Petiole 2–4 mm long. Inflorescence mostly 2-flowered, bracteoles 1.5–4 mm long. Shrub to 2.5 m high. South from Blue Mts DSF. Fl. winter–spring. . . . .**H. pannosa** A.Cunn. ex Hook.f.
5 Mid vein on upper surface of leaves with scattered hairs; 2–4.5 mm long, 3–6.5 mm wide. Petiole 2.5–3 mm long. Inflorescence mostly 3-flowered, bracteoles 3.5–6 mm long. Shrub to 3 m high. Widespread in the region. DSF. Ss. Fl. winter–spring . . . . **H. speciosa** I.Thomps.

6 Upper surface of leaf with hairs mainly on midrib. 7
6 Upper surface of leaf glabrous 9

7 Inflorescence on peduncle 5–20 mm long, 4–12-flowered, pedicels 1.5–6 mm long, standard 7–11 mm long. Leaves lanceolate to narrow elliptic, 4–9 mm long, 7–15 mm wide, lower surface with hairs tan to white greying with age. Shrub to 3 m high. West of Blue Mts, DSF. Ss or sandy soils. Fl. winter–spring . . . .**H. apiculata** A.Cunn. ex G.Don
7 Inflorescences sessile or on peduncle less than 3 mm long, mostly 2-flowered 8

8 Leaves narrow oblong, more than 4 mm wide, 1–7 cm long; margins recurved, apex obtuse to sub acute. Lower surface with cream to brown hairs. Petiole 2–6.5 mm long. Standard 11–7 mm long. Shrub to 3 m high. Tablelands. DSF. Fl. spring . . . . **H. purpurea** Sweet
8 Leaves linear less than 4 mm wide, 1–4 cm long; margins strongly recurved, apex truncate to obtuse. Lower surface with a dense covering of cream to brown hairs. Petiole 1–2 mm long. Standard 5–7 mm long. Shrub to 1.5 m high. Barrington Tops to Blue Mts DSF. Fl. winter . . **H. rosmarinifolia** A.Cunn.

9 Leaves narrow elliptic 7–20 mm wide, 4.5–7 cm long, tapering gradually at each end, lower surface covered with close, rusty orange hairs. Pedicels 1.5–4 mm long. Stipules 1.5–3 mm long. Shrub or small tree to 5 m. Springwood. Fl. winter–spring . . . . **H. acutifolia** A.Cunn. ex G.Don
9 Leaves linear 1–8 mm wide 10

10 Pedicels 3.5–6 mm long; leaves 2–8 cm long 2–8 mm wide, , lower surface covered with close, orange brown hairs. Stipules 1–1.5 mm long. Shrub to 3 m high. Widespread. SF along watercourses. Fl. winter–spring . . . .**H. longifolia** R.Br.
10 Pedicels 0.5–1.5 mm long; leaves 1–4 cm long, 1–4 mm wide . . . . **H. rosmarinifolia** (see above)

24 Templetonia R.Br.
11 species endemic Aust.; all states and territories except Tas.

One species in the area

Multistemmed undershrub up to 60 cm high, with a thick rootstock. Leaves linear to narrow-elliptic or lanceolate, up to 4 cm long, glabrous, concave or flat. Calyx almost equally 4-lobed. Standard up to 1 cm long, yellow to reddish-brown. Legume biconvex with hard leathery valves. Glen Davis. Open forest. Fl. spring . . . . **T. stenophylla** (F.Muell.) J.M.Black

25 Crotalaria L.

>40 species in Aust.; (18 endemic) all states and territories except Vic.

Shrubs. Leaves alternate, simple or palmately compound; stipules present or absent; stipellae absent. Flowers in terminal racemes. Calyx 5-toothed, sometimes 2-lipped. Petals yellow, often with brownish markings; keel curved to form a beak wrapped around the stamens. Stamens connate into a tube open on the upper side. Style with a line of hairs on the inside, pushing the pollen out of the beak of the keel when the wings are depressed. Legume inflated.

Key to the species

1 Leaves simple, ovate to elliptic, 6–15 cm long, glabrous on the upper surface, greyish hairy underneath. Wings c. 28 mm long. Legume up to 70 mm long. Shrub up to 5 m high. Hornsby Plateau. Garden escape. Introd. from Asia. Fl. winter–spring . . . .*C. lunata Bedd. ex Polhill
1 Leaves 3-foliolate 2

2 Ovary glabrous. Leaflets ovate to broad-elliptic, usually up to 5 cm long, glabrous. Racemes up to 40 cm long. Corolla greenish yellow; keel up to 50 mm long. Legume up to 10 cm long. Shrub up to 2 m high. Sydney. Naturalized near habitation. Introd. from Africa. Fl. summer–autumn. . . . .*C. agatiflora Schweinf. ssp. agatiflora
2 Ovary hairy 3

3 Leaflets broad-obovate to broad-elliptic, 10–50 mm wide, up to 5 cm long; stipules linear, up to 15 mm long. Corolla yellow with reddish lines; keel 6–8 mm long. Legume up to 45 mm long. Shrub up to 2 m high. Coast and adjacent plateaus; Cumberland Plain. Weed of cultivation or waste places. Introd. from Africa. Fl. summer. *Woolly Rattlepod* . . . .*C. incana ssp. purpurascens (Lam.) Milne-Redh.
3 Leaflets linear to narrow-elliptic, up to 10 mm wide, 3–4 cm long; stipules absent or minute. Corolla yellow with purplish-brown markings; keel c. 5 mm long. Legume up to 35 mm long. Annual up to 2 m high. Garden escape near Sydney. Introd. from Africa. Fl. autumn . . . .*C. lanceolata E.Mey. ssp. lanceolata

26 Platylobium Sm.

4 species endemic Aust.; Qld, NSW, Vic., Tas.

Shrubs up to 1 m high. Leaves simple, opposite, strongly reticulate, glabrous, ovate to ovate-lanceolate, cordate, mucronate, 2–5 cm long. Flowers axillary, solitary or 2 or rarely 4 together; peduncles with brown scarious imbricate bracts. Calyx hairy, 5-lobed; upper 2 lobes broad, c. twice as long as the narrow lower ones. Corolla yellow and red. Stamens connate into a sheath open above. Ovules c. 8. Legume flat, oblong, 2–4 cm long, stipitate, opening and rolling back from the lower suture; upper suture bordered by a narrow wing.

Key to the species

1 Calyx 8–10 mm long. Standard c. 18 mm long. Widespread. Sheltered situations on Ss and shale. Fl. early spring . . . . P. formosum Sm. ssp. formosum
1 Calyx 4–5 mm long. Standard 5–8 mm long. Hornsby Plateau. Uncommon. Sheltered situations. Fl. spring . . . . P. formosum ssp. parviflorum (Sm.) A.T.Lee

27 Bossiaea Vent.

50 species endemic to Aust.; all states and territories except NT

Shrubs, stems terete or broadly flattened, cladodes. Leaves simple, sometimes reduced to scales, entire, opposite or alternate, not strongly reticulate; stipules usually small. Flowers axillary, solitary or in clusters of 2–3, with imbricate bracts at the base of the peduncles. Calyx 5-lobed; upper 2 usually broader and larger than the others and ± united. Corolla yellow and red; standard usually reflexed, c. twice as long as the calyx. Stamens connate into a sheath open above at least at the base. Legume flat, not winged, opening along both sutures.

Key to the species

1 Leaves opposite 2
1 Leaves alternate. 3

2 Leaves orbicular, 3–7 mm diam. Straggling, glabrous shrub usually under 1 m high, with numerous slender wiry branches. Widespread. DSF. Ss. Fl. spring . . . . **B lenticularis** Sieber ex DC.
2 Leaves oblong or elliptic, obtuse, 1–2 cm long. Branches hoary. Fruit glabrous, stipitate. Prostrate to erect shrub up to 2 m high. Illawarra ranges, e.g. Upper Minamurra Falls. Open forests. Ss. Fl. spring . . . . **B. kiamensis** Benth.

3 Small prostrate or decumbent shrubs with weak branches. Flowers up to 8 mm long. 4
3 Erect shrubs. 7

4 Ovary densely hairy. Leaves elliptic to obovate, up to 20 mm long, obtuse, recurved or revolute. Rylestone. DSF. Fl. spring . . . . **B. scortechinii** F.Muell.
4 Ovary glabrous or hairy on top or margins 5

5 Upper calyx lobes hardly longer than the lower lobes, rounded or truncate. Leaves narrow-ovate to orbicular, symmetrical at the base, glabrous, distichous, 5–12 mm long; stipules as long as or shorter than the petioles. Rootstock thick, woody. Widespread. Various situations. Fl. spring . . . . **B. prostrata** R.Br.
5 Upper calyx lobes much larger than the lower ones. Leaves ovate, 3–10 mm long, ± distichous, pubescent underneath at least when young 6

6 Pedicels shorter than the calyx. Leaves with a subulate mucro, paler below. Flowers up to 9 mm long. Southern Blue Mts DSF. Fl. spring . . . . **B. neo-anglica** F.Muell.
6 Pedicels longer than the calyx. Leaves obtuse or with a short subulate mucro. Flowers up to 10 mm long. Widespread. Various situations. Fl. spring . . . . **B. buxifolia** A.Cunn.

7 Branchlets terete 8
7 Branchlets compressed or flattened 9

8 Branchlets spinescent. Leaves obovate to obcordate, often emarginate or with a small recurved point, 2–6 mm long. Flowers c. 8 mm long. Small shrub c. 50 cm high, rather rigid. Widespread. Ss and shales. Fl. spring . . . . **B. obcordata** (Vent.) Druce
8 Branchlets not spinescent. Leaves orbicular to broad-elliptic, up to 5 mm long, acute or apiculate. Flowers 8–11 mm long. Erect shrub to 2 m high. Yerranderie. DSF. Ss. Vulnerable. Fl. spring . . . . **B. oligosperma** A.T.Lee

9 Leaves well developed 10
9 Leaves reduced to minute scales except on seedlings or new growth after fire 12

10 Stipules large, 5–10 mm long, almost leaf-like, triangular, erect, appressed. Leaves elliptic to lanceolate, 5–25 mm long, mucronate, distichous. Flowers 8–10 mm long. Erect shrub up to 1 m high. Illawarra Ranges; Woronora and Hornsby plateaus. DSF and heath. Fl. spring . . . . . **B. stephensonii** F.Muell.
10 Stipules small 11

11 Leaves obovate to broad-rhomboid, 5–10 mm long, often mucronate, uniform in shape on one plant. Flowers c. 10 mm long. Seeds usually 2–4. Erect shrub to 1 m high. Widespread. DSF and heath. Ss. Fl. spring. . . . . **B. rhombifolia** Sieber ex DC. ssp. **rhombifolia**
11 Leaves obovate to almost linear, 5–25 mm long, often variable on one plant especially when some are broad. Flowers 8–15 mm long. Seeds usually 6 –8. Erect shrub to 1.5 m high. Very variable and very common. Widespread. DSF and heath. Ss. and old dunes. Fl. summer–autumn . **B. heterophylla** Vent.

12 Flowers less than 10 mm long. Calyx not usually more than 4 mm long. Fruit distinctly stipitate. Stems rather weak. Straggling shrub c. 50 cm high. Widespread. Heath. Fl. spring . . **B. ensata** Sieber ex DC.

12 Flowers more than 10 mm long. Calyx 5–6 mm long. Fruit almost sessile, c. 10 mm broad. Stems stout, erect, broader than those of *B. ensata*. Shrubs 75–150 cm high. Widespread. Heath and DSF. Sandy soils. Fl. spring . . . . **B. scolopendria** (Andrews) Sm.

28 Goodia Salisb.

1–3 species endemic Aust.; all states and territories except NT

One species in the area

Shrub up to 3 m high, glabrous or the young shoots pubescent or glaucous. Leaves 3-foliolate; leaflets ovate or cuneate-ovate, entire, 12–18 mm long. Flowers c. 1 cm long, in loose terminal or leaf-oppossed racemes. Calyx 5–6 mm long; upper lobes united into a 2-toothed lip; lower lobes narrow, c. as long as the tube. Corolla yellow and red. Stamens connate into a tube open on the upper side; anthers versatile, alternately long and short. Legume 2–3 cm long, c. 1 cm wide, on a slender stipe. Widespread in forests but not common. Fl. spring . . . . **G. lotifolia** Salisb.

29 Ulex L.

1 naturalized species Aust.; all states and territories except NT

One species in the area

Rigid, ± hairy shrub with ribbed branches ending in spines. Leaves narrow, rigid, spiny. Calyx c. 15 mm long, hairy, divided to the base into 2 lips; upper lip with 2 small teeth; lower lip with 3 teeth. Corolla bright yellow, slightly longer than the calyx. Widespread. Weed of open communities. Introd. from Europe. Fl. most of the year. *Gorse* or *Furze*. . . . . * **U. europaeus** L.

30 Genista L.

5 naturalized species Aust.; NSW, Vic., NT, S.A.

Shrubs up to c. 3 m high. Branches grooved or angular. Leaves 3-foliolate, entire, ± pubescent. Flowers c. 1 cm long, in short terminal racemes. Calyx 2-lipped; teeth conspicuous. Corolla yellow; standard broad, recurved. Stamens connate. Style incurved, short. Legume compressed, oblong.

Key to the species

1 Leaves sessile or nearly so. Leaflets linear to narrow-oblong, 10–15 mm long, glabrescent on the upper surface; margins recurved. Racemes terminal, many to few flowered. Calyx 7–8 mm long. Fruit with 2–3 seeds. Naturalized near Moss Vale. Introd. from the Mediterranean. Fl. spring. *Flax-leaf Broom* . . . . ***G. linifolia** L.

1 Leaves with a distinct but often short petiole usually more than 2 mm long. Leaflets obovate to oblong-elliptic or oblanceolate. 2

2 Leaflets obovate, 5–20 mm long, ± pubescent. Petiole 2–4 mm long. Flowers 3–9 per cluster, on short pedicels. Calyx 5–7 mm long. Fruit 15–25 mm long, with 4–6 seeds. Widespread. Roadsides and waste places. Introd. from the Mediterranean. Fl. spring–summer. *Montpelier Broom* . . . . ***G. monspessulana** (L.) L.A.S.Johnson

2 Leaflets elliptic to oblong or oblanceolate, 8–25 mm long, pubescent with closely appressed hairs underneath; petiole 6–15 mm long. Flowers 5–12 per cluster. Calyx c. 5 mm long. Fruit with 5–7 seeds. Garden escape, naturalized in a few places. Introd. from the Canary Islands. Fl. spring. *Madiera Broom* . . . . ***G. stenopetala** Webb. & Berthel.

31 Spartium L.

Monotypic genus

Shrub up to 2 m high, with terete striate glaucous branchlets. Leaves oblong to lanceolate, up to 25 mm long, glabrescent on the upper surface, nearly sessile. Flowers arranged in lax terminal racemes. Fruit flat; seeds up to 20. Introd. from the Mediterranean. Fl. summer. . . . . * **S. junceum** L.

32 Chamaecytisus Link

1 species naturalized Aust.; Qld, NSW, Vic., S.A., WA

One species in the area

Tall shrub or small tree with long drooping branches. Leaflets lanceolate, 1–2 cm long, glaucous above, pubescent below. Flowers in umbels mostly at the ends of short branchlets. Calyx tubular, 10–12 mm long, 2-lipped, densely pubescent; teeth conspicuous. Corolla white; standard c. twice as long as the calyx. Legume 4–5 cm long, c. 12 mm wide. Occasionally naturalized. Introd. from the Canary Islands (Teneriffe). Fl. winter–spring. *Tree Lucerne* or *Tagasaste* .*C. **palmensis** (H.Christ) Bisby & K.W.Nicholls

33 Cytisus Desf.

2 species naturalized Aust.; all states and territories except Qld & NT

One species in the area

Erect, glabrous shrub up to 3 m high, with green angular stems. Leaves 1–3-foliolate; leaflets 5–12 mm long, readily deciduous. Flowers axillary, c. 2 cm long. Calyx c. 5 mm long, glabrous, 2-lipped the lower with 2 minute teeth and the upper with three minute teeth. Corolla yellow. Style long, spirally coiled. Legume 25–40 mm long, black. Naturalized in a number of places particularly in Blue Mts Introd. from Europe. Fl. spring. *English Broom* . . . . *C. **scoparius** (L.) Link ssp. **scoparius**

34 Lupinus L.

Lupins

6 species naturalized Aust.; all states and territories except NT

Annual or perennial herbs. Leaves digitate, petiolate, stipulate. Flowers in terminal racemes. Calyx 2-lipped. Corolla with wings coherent at the apex. Stamens all connate. Fruit a legume, with 5–15 seeds.

Key to the species

1 Corolla yellow, c. 15 mm long. Leaflets oblanceolate to oblong, 3–6 mm long, villous. Fruit black, villous. Seeds dull black with whitish markings. Ascending, annual herb. Naturalized in various places particularly near the coast. Introd. from the Mediterranean. Fl. spring–summer.. . . . . . * L. **luteus** L.
1 Corolla blue sometimes with yellowish or purplish markings 2

2 Leaflets linear to linear-lanceolate, up to 5 mm wide, glabrescent above, ± hirsute below. Flowers mostly alternate. Fruit brownish. Seeds dull, pale to dark brown or greyish-brown with yellow markings. Naturalized in a few places on Cumberland Plain. Introd. from the Mediterranean. Fl. spring . . . . . .
. . . . *L. **angustifolius** L.
2 Leaflets obovate to narrow-oblong-elliptic, 6–12 mm wide, 2–6 cm long. Fruit dark brown, villous-pubescent 3

3 Hairs on the stem 2–4 mm long. Seeds purplish-brown with black lines around the hilum. On sand near the coast. Introd. from the Mediterranean. Fl. spring. . . . . * L. **pilosus** Murray
3 Hairs on the stem less than 2 mm long. Seeds greyish-brown with darker markings. Occasional near the coast. Introd. from the Mediterranean. Fl. spring. . . . . *L. **cosentinii** Guss.

35 Indigofera L.

40 species Aust. (8 naturalized, 32 endemic); all states except Tas.

Annual or perennial herb or shrub. Hairs 2-branched sometimes multicellular. Leaves alternate, pinnate; leaflets numerous. Inflorescence an axillary raceme. Corolla white, pink or purple. Stamens connate with the upper one free. Ovules 1, 2 or numerous. Legume narrow-cylindrical.

Key to the species

1 Shrub 100–150 cm high. Leaves 4–10 cm long. Leaflets 11–25, obovate to oblong, 12–40 mm long, glabrous. Stipules lanceolate up to 4 mm long. Flowers numerous, in racemes c. as long as the leaves.

Calyx with dark brown hairs 2 mm long, teeth shorter than the tube. Corolla mauve, pinkish or white; standard 6–8 mm long. Widespread. Open forests. Mostly on shales. Fl. spring . . . **I. australis** Willd.

1 Shrub 30–100 cm high, suckering from woody rhizome. Leaves 10–14 cm long. Leaflets 9–11 elliptic to ovate, 18–60 mm long. Stipules triangular, caducous. Flowers in racemes shorter or longer than the leaves. Calyx grey, teeth equal to tube. Corolla pink. Standard 10–15 mm long. Occasional garden escape. . . . . ***I. decora** Lindl.

36 Tephrosia Pers.

60 species in Aust. (9 naturalized, 51 endemic); Qld, NSW, NT, S.A.

Shrubs to 2 m high. Leaves alternate, imparipinnate, oblanceolate to oblong, up to 40 mm long. Flowers in pairs or clusters arranged in terminal racemes. Corolla purplish to pink. Stamens united; upper one free.

Key to the species

1 Stipules linear to lanceolate, up to 5 mm long. Legume up to 6 cm long, c. 5 mm wide. Occassional on the coast. Point Clare. Introd. from S. Africa. . . . . ***T. inandensis** H.M.L.Forbes

1 Stipules ovate, 6–10 mm long. Legume 5–8 cm long, 7–10 mm wide 2

2 Calyx with white hairs; teeth less than 2 mm long, triangular, acute. Occasional on Coast. Introd. from S. Africa . . . . ***T. grandiflora** (L'Hér. ex Aiton) Pers.

2 Calyx with rusty to golden hairs; teeth narrow-triangular, acuminate, more than 2 mm long. North from Wollongong. Introd. from S. Africa . . . . ***T. glomeruliflora** Meisn.

37 Psoralea L.

1 species naturalized in Aust.; NSW, Vic., S.A., WA

One species in the area

Shrub 2–3 m high. Leaves pinnate; leaflets 3–5 pairs, linear, acute, sprinkled with glands which may appear as black dots. Flowers 12 mm long, axillary, solitary, with a pair of bracts 1 mm long at the base of the peduncle. Corolla blue or purple. Ovule solitary. Fruit c. as long as the calyx, indehiscent. Introd. from S. Africa. Fl. summer. *African Scurf-pea* . . . . *** P. pinnata** L.

38 Swainsona Salisb.

84 endemic species in Aust.; all states and territories

Herbs or shrubs. Leaves imparipinnate. Stipules present. Flowers in axillary racemes. Calyx lobes ± equal or 2-lipped. Corolla purple, white or pink. Stamens united; upper one free. Style usually bearded.

Key to the species

1 Stems glabrous. Leaflets 5–10 pairs, obovate, 5–20 mm long, 3–5 mm wide. Legume ± compressed, 20–40 mm long. Shrub up to 1 m high. Blue Mts, not Ss. e.g. Cox's River Valley, Jenolan Caves. Open forests. Fl. summer. *Darling Pea* . . . . **S. galegifolia** (Andrews) R.Br.

1 Stems hairy. Leaflets up to 3 mm wide. 2

2 Hairs on the stem loosely appressed. Leaflets 11–25, oblong to elliptic, up to 15 mm long. Flowers up to 8 mm long. Legume 8–15 mm long. Widespread. Open forests and grasslands, not Ss . . . . **S. monticola** A.Cunn. ex Benth.

2 Hairs on the stem divergent. Leaflets 5–15, obovate, up to 15 mm long. Flowers up to 10 mm long. Legume 18–22 mm long. Widespread. Open forests and grasslands, not Ss . . . **S. reticulata** J.M.Black

39 Robinia L.

1 species naturalized in Aust.; Qld, NSW, Vic., S.A., WA

One species in the area

Tree, producing suckers. Leaves pinnate; leaflets 11–21, shortly stalked, 15–30 mm long; stipules 2 stout spines. Flowers numerous in axillary racemes. Calyx campanulate, lobes broad. Petals white, c. 18 mm long. Legume c. 8 cm long. Usually occurs where, at some time, it has been planted. Introd. from N. America. Fl. spring. *Black Locust* . . . . *R. pseudoacacia* L.

40 Pisum L.

1 species naturalized Aust.; NSW, NT, WA

One species in the area

Prostrate or climbing annual. Leaves imparipinnate, ending in a branched tendril. Stipules large ovate 1–10 cm long, 1–6 cm wide. Leaflets usually 6 sometime less, 1–7 cm long, 1–4 cm wide. Flowers solitary or sometimes 3 in an axillary raceme, purple, pink or white. Corolla 15 mm long, calyx 12 mm long. Legume to 12 cm long with 4–10 seeds. Escape from cultivation. Camden area. Introd. from Europe and Asia. Fl. spring. *Field Pea* . . . . *P. sativum* L. var. **arvense**

41 Lathyrus L.

7 species naturalized Aust.; all states and territories except NT

Annual or perennial climber. Leaves alternate, paripinnate, ending in a branched tendril. Flowers in axillary racemes or solitary. Calyx campanulate, regular or 2-lipped. Corolla with a broad standard which is longer than the keels. Stamens united into 2 groups. Ovary sessile or stipitate. Style compressed hairy on upper side.

Key to the species

1 Prostate or climbing perennial. Inflorescence 15–25 cm long with 3–15 flowers. Leaves 2-foliate. Stipules 1–3 cm long with a narrow basal lobe. Leaflets 2, ovate to elliptic, 3–8 cm long, 5–30 mm wide. Corolla pale purple to white, 2–3 cm long; calyx 7–10 mm long. Legume 5–11 cm long with 7–25 seeds. Occasionally naturalized. Native of Europe. Fl. spring–summer. *Everlasting Pea* . . . . . *L. latifolius* L.
1 Climbing annual herb. Inflorescence 5–10 cm long with mostly 1–3 flowers. Leaves 2-foliate. Stipules 1.2–2.5 cm long. leaflets lanceolate to ovate, 2–8 cm long, 3–20 mm wide. Corolla pale pink to purple, 2–3 cm long; calyx 8–12 mm long. Legume 6–10 cm long with 6–10 seeds. Occasionally naturalized. Native of Mediterranean. Fl. summer . . . . **L. tingitanus** L.

42 Vicia L.

8 species naturalized in Aust.; all states and territories except NT

Weak, annual herbs, sometimes climbing. Leaves pinnate, ending in a simple or branched tendril; leaflets mucronate; stipules present. Flowers axillary. Calyx 5-toothed. Stamens connate, with the upper one at least partially free. Style bent. Legumes linear or oblong; valves usually twisted spirally after dehiscence. Seeds globular or ovoid.

Key to the species

1 Flowers nearly sessile, 1–4 in the axils. Corolla purple and red. Leaflets 12–20, 1–2 cm long. 2
1 Flowers in racemes or solitary on long peduncles 3

2 Flowers 15–25 mm long; 1–4 in leaf axils. Fruit compressed, fawn, 5–7 cm long. Leaflets ovate or oblong, 6–10 mm wide, rounded truncate or notched at the ends. Widespread. Escaped from cultivation. Fl. winter–spring. *Common Vetch* . . . . *V. sativa* L. ssp. **sativa**
2 Flowers 10–15 mm long, 1 (rarely 2) in leaf axils. Fruit cylindrical, black when ripe, 3–5 cm long. Leaflets 2–4 mm wide. Widespread. Introd. from Europe. Fl. winter–spring. *Narrow-leaf Vetch* . . . .
. . . . *V. sativa* ssp. **nigra** (L.) Ehrh.

3 Flowers up to 30 per raceme. Leaflets oblong-elliptic 4

3 Flowers less than 10 per raceme. Leaflets linear to oblong 5

4 Plant stems villous-pubescent with dense spreading hairs. Stipules 2–5 mm wide. Pods glabrous. Lower calyx teeth as long or longer than the tube. Penshurst. Introd. from Europe. Fl. spring–summer. *Russian Vetch* . . . . ***V. villosa** Roth. ssp. **villosa**

4 Plant stems with appressed hairs or glabrous. Stipules less than 3 mm wide. Pod pubescent. Lower calyx lobes shorter than the tube.. Introd. from Europe. Fl. all year . . . .

. . . . **V. villosa** ssp. **eriocarpa** (Hausskn.) P.W.Ball

5 Legume with 2 seeds. Leaflets 12–20, linear-oblong, 5–15 mm long; tendrils branched; stipules toothed. Flowers 3–4 mm long, 3–8 together, pale coloured. Plants ± hairy. Widespread. Introd. from Europe. Fl. winter–spring. *Hairy Vetch* . . . . ***V. hirsuta** (L.) Gray

5 Legume with 4 seeds. Leaflets 4–10, linear, 5–15 mm long; tendrils simple or forked. Flowers 5–6 mm long, 1–3 together on filiform peduncles, pale coloured. Plants glabrous. Coast and Cumberland Plain. Introd. from Europe. Fl. winter–spring. *Slender Vetch* . . . . ***V. tetrasperma** (L.) Schreb.

43 Canavalia DC.

6 species in Aust. (5 native, 1 naturalized); Qld, NSW, NT, WA

One species in the area

Trailing to prostrate herb, glabrous except the young shoots. Leaves 3-foliolate; leaflets obovate to orbicular, mostly 4–8 cm long. Flowers in axillary racemes; peduncles 15–30 cm long. Calyx 2-lipped; upper lip nearly as long as the tube, with 2 broad rounded lobes. Corolla pale pink; standard orbicular, 18 mm diam. Stamens connate, upper one free. Legume 18–24 mm wide, with a rib on either side parallel to the upper suture. Coast near the sea, Shellharbour northwards. Fl. summer . . . . **C. rosea** (Sw.) DC.

44 Kennedia Vent.

16 species endemic Aust.; all states and territories

Twining or prostrate plants. Leaves 3-foliolate; stipules persistent, striate. Flowers axillary; bracts small. Calyx 2-lipped; upper lobes ± united. Stamens connate; upper one free. Ovules several. Legumes linear, compressed, with a pithy substance between the seeds.

Key to the species

1 Corolla 30–40 mm long, red; standard narrow-obovate. Leaflets usually ovate, 1–10 cm long, flat. Widespread. Coastal sand dunes; open forests; margins of RF. Fl. spring. . . . . . . **K. rubicunda** Vent.

1 Corolla up to 20 mm long 2

2 Leaflets orbicular to broad-obovate, less than 2.5 cm long, often with a wavy margin. Corolla scarlet; standard obovate, yellow near the base. Prostrate, mat-forming plant with silky white hairs. Widespread. Sandy soils. Fl. spring. *Running Postman* . . . . **K. prostrata** R.Br.

2 Leaflets broad-elliptic to orbicular, more than 2.5 cm long. Corolla pink to scarlet; standard orbicular with a white spot. Twining herb with white or rusty hairs. Northern Blue Mts Vulnerable. Fl. spring . .

. . . . **K. retrorsa** Hemsl.

45 Hardenbergia Benth.

2–3 species endemic Aust.; all states and territories except NT

One species in the area

Glabrous twiner, sometimes erect on Wianamatta Shale. Leaves broad-lanceolate to ovate, 3–9 cm long, reticulate. Flowers c. 10 mm long, numerous in each raceme. Calyx teeth short. Corolla violet with green centre. Legume flat, c. 4 cm long. Widespread. Heath to open forest. Fl. spring. *False Sarsparilla* . . . . . .

. . . . **H. violacea** (Schneev.) Stearn

46 Pueraria DC.

2 species naturalized Aust.; Qld, NSW

One species in the area

Twinning herb. Leaves alternate, 3 leaflets. Leaflets usually lobed, ovate to rhombic, 7–15 cm long, 5–13 cm wide, sparsely covered with appressed to spreading hairs; upper surface green lower surface grey-green. Stipules peltate, 8–16 mm long. Inflorescence axillary raceme, with 4–90 flowers. Calyx 5–12 mm long. Corolla 12–18 mm long, pink, purple or blue; standard with a yellow spot at base. Legume covered with rusty hairs, 5–9 mm long. Recorded in Picton area. Native of China and Japan. Noxious weed. Fl. summer. *Kudzu* . . . . *P. lobata (Willd.) Ohwi

47 Glycine Willd.

24 species native Aust.; all states and territories

Twining or prostrate perennials. Stems and branches with reflexed hairs. Leaves 3-foliolate. Flowers c. 8 mm long, in narrow axillary racemes. Corolla bluish, purple, mauve or pink. Stamens connate into a tube; upper stamen usually finally free. Legume linear, with a pithy substance between the seeds.

Key to the species

1 All 3 leaflets on short and equal petiolules. Occasionally central petiolule longer. Plants not stoloniferous. Stems with soft hairs. Leaflet with lateral veins obscure and at right angle to mid vein. Stipels not prominent. Leaflets of upper leaves linear oblong, lanceolate to narrow elliptic, the lower leaves usually broader and shorter. Leaflets sparse to moderately hairy on upper surface sometimes glabrous. Racemes to 21 cm long, flowers usually mauve. Widespread in many situations. Fl. most of the year . . . . **G. clandestina** J.C.Wendl.

1 Petiolule of median leaflet longer than those either side of it. Occasionally petiolules equal in length. Plants stoloniferous. Stipels obvious. 2

2 Leaflets with lateral veins prominent and at right angle to mid vein. Leaflets of upper leaves narrow lanceolate to lanceolate, upper surface with white hairs. Stems weakly hairy. Racemes to 5 cm long flowers usually pinkish. Widespread in open forests. Shales etc. Fl. most of the year . . . . **G. microphylla** (Benth.) Tindale

2 Leaflets with lateral veins at an acute angle to midrib. Leaflets of upper leaves ovate to linear, upper surface more or less glabrous. Stems with strigose hairs. Racemes to 14 cm long, flowers usually deep purple. Widespread in many communities. Fl. most of the year . . . . **G. tabacina** (Labill.) Benth.

48 Macroptilium (Benth.) Urb.

2 species naturalized Aust.; Qld, NSW, NT

One species in the area

Trailing or twining pubescent herb. Leaves alternate, with 3 ovate leaflets , 2–7 cm long, 18–50 mm wide, whitish below, stipellate; lateral leaflets lobed. Flowers in erect racemes. Petals very dark purple especially the wings which flutter in the wind. Upper stamen free. Legume reflexed, linear, up to 10 cm long. Coast. Pastures and waste places. Introd. from N. America. Fl. summer. *Siratro* . . . .*M. atropurpureum** (DC.) Urb.

49 Vigna Savi

12 species in Aust. (5 native, 7 naturalized); all states and territories except Tas.

One species in the area

Trailing, ± hairy herb. Leaves alternate, with 3 leaflets, stipellate; leaflets linear to ovate, 3–16 cm long, 5–15 mm wide. Flowers in erect few-flowered racemes. Corolla yellow or pink to purple. Upper stamen free. Legume erect, terete, up to 14 cm long. Coast and lower Blue Mts Pastures and waste places. Fl. spring–summer. **V. vexillata** ssp. **angustifolia** (Schumach. & Thonn.) Baker

50 Dipogon Liebm.
1 species naturalized Aust.; Qld, NSW, Vic., Tas., S.A., WA

One species in the area
Perennial twiner glabrous except the young tips, spreading actively to a distance of 3 m or more. Leaves 3-foliolate; leaflets triangular, 3–4 cm long, acute. Flowers in short racemes on axillary peduncles 5–10 cm long. Corolla pink or white; standard c. 15 mm long, reflexed. Legume straight. Garden escape near habitation. Introd. from Africa. Fl. winter–summer. *Dolichos Pea*. . . . .*D. lignosus (L.) Verdc.

51 Erythrina L.
5 species in Aust. (3 endemic, 2 naturalized); Qld, NSW, NT, S.A., WA

Deciduous tree with spines on the branches. Bark corky. Leaves 3-foliolate; stipules and stipellae small. Inflorescence a raceme, 20–40-flowered, axillary or terminal. Flowers large, appearing before the leaves, in dense short racemes on stout axillary peduncles. Standard longer than the wings, erect or recurved, red. Stamens connate into two groups. Ovary stipitate. Legume linear to oblong.

Key to the species
1 Leaflets flat, triangular to ovate, 7–12 cm long, 7–12 cm wide, articulate at the base; petioles 10–12 cm long, articulate on the stem. Calyx obliquely truncate or slightly toothed 11–15 mm long. Corolla red; standard scarcely recurved, 4–6 cm long; keel 2–3 cm long.. An horticultural hybrid widely naturalized. Fl. winter–spring. *Coral Tree* . . . .* **E. x sykesii** Barneby & Krukoff
1 Leaflets elliptic to ovate 3–6 cm long, 2–5 cm wide; petiole 5–10 cm long. Calyx 10 mm long. Corolla red; standard recurved, 4–5 cm long; keel 3–3.5 cm long. Naturalized in coastal areas often along streams. Fl. spring. *Cockspur Coral Tree* . . . .*E. crista-galli L.

2 MIMOSOIDEAE
Trees or shrubs. Leaves alternate, bipinnate or modified to phyllodes, often bearing one or more nectaries (glands) on the petiole or leaf-rhachis, with or without stipules. Flowers regular, usually bisexual, in spikes racemes or heads, usually small. Calyx tubular; lobes 4–5, valvate. Petals free or connate basally, 4–5, valvate. Stamens usually numerous, free or connate into a tube, hypogynous or perigynous. Ovary superior, 1-locular; placenta marginal. Fruit a legume. Seeds often with an aril which may be brightly coloured. 60 gen., trop to warm temp., particularly S. Africa and Aust.

Key to the genera
1 Filaments free. 2
1 Filaments connate basally into a tube 5

2 Stamens numerous, more than 10 3
2 Stamens 5 or 10. 4

3 Plants with bipinnate leaves AND spinose stipules . . . . **2 VACHELLIA**
3 Plants with or without bipinnate leaves. If leaves bipinnate then spinose stipule absent . . . **1 ACACIA**

4 Stamens 5 . . . . **3 NEPTUNIA**
4 Stamens 10 . . . . **4 LEUCAENA**

5 Pinnae 6–12 pairs per leaf; pinnules 20–40 pairs per pinna. . . . .**5 PARASERIANTHES**
5 Pinnae usually 1–2 pairs per leaf; pinnules 3–4 pairs per pinna . . . .**6 PARARCHIDENDRON**

1 Acacia Mill.
Wattles
959 species in Aust. (c. 945 endemic); all states and territories

Trees or shrubs. Leaves alternate, bipinnate or modified into phyllodes (Fig. 38); often with nectiferous glands on the petioles, rhachies or margin of the phyllodes; stipules present or absent. Flowers small, regular, yellow or cream, in globular heads or spikes. Sepals 4–5. Petals 4–5, valvate, sometimes coherent but usually free. Stamens numerous, free or nearly so. Legumes linear or oblong, flat to terete, straight, falcate or twisted, opening down both sutures. Funicle usually thickened under or around the seed into an aril.

(To count the number of flowers in a head most easily, select a head in which the flowers are just about to open and count the flower buds)

Key to the species

1 Leaves all bipinnate . . . . .GROUP 6
1 Leaves modified into phyllodes. 2

2 Flowers in cylindrical or oblong spikes . . . . .GROUP 1
2 Flowers in globular heads. 3

3 Flower heads solitary or in pairs or clusters, not in axillary racemes 4
3 Flower heads in axillary racemes. 5

4 Phyllodes with a distinct pungent point, or stipules spinescent . . . . .GROUP 2
4 Phyllodes without a pungent point (sometimes with a short rigid but not pungent point) spinescent
   stipules not present. . . . . .GROUP 3

5 Phyllodes with 2 or more prominent longitudinal veins. .GROUP 4
5 Phyllodes with 1 prominent longitudinal vein or none . . . . .GROUP 5

Group 1
Leaves modified into phyllodes; flowers in cylindrical or oblong spikes

1 Calyx and corolla mostly 4-merous 2
1 Calyx and corolla mostly 5-merous 10

2 Phyllodes rigid, pungent pointed, 12–25 mm long, narrow-lanceolate, scattered. Spikes often more than
   25 mm long. Rigid, spreading shrub 1–3 m high. Coast north of Sydney; Hornsby Plateau; Blue Mts
   DSF. Ss. Fl. spring . . . . **A. oxycedrus** Sieber ex DC.
2 Phyllodes not rigid or pungent pointed 3

3 Marginal gland adjacent to the pulvinus or absent. 4
3 Marginal gland distant from the pulvinus but often close to the base. 8

4 Phyllodes with a single prominent longitudinal vein or with the midvein much more prominent than
   any others 5
4 Phyllodes with several ± equally prominent longitudinal veins or the veins scarcely visible on the
   surface 6

5 Phyllodes 1–3 mm wide, 10–30 cm long, linear. Spikes slender, interrupted, 2–5 cm long, pale yellow.
   Shrub 2–4 m high. Coast and adjacent plateaus; Cumberland Plain. Fl. most of the year. . . . .
   . . . . **A. longissima** H.L.Wendl.
5 Phyllodes 5–15 mm wide, 5–15 cm long. Spikes dense, up to 25 mm long, deep yellow. Shrub or small
   tree up to 4 m high. Coast and adjacent plateaus; lower Blue Mts DSF. Fl. spring . . . .
   . . . . **A. subtilinervis** F.Muell.

6 Phyllodes mostly less than 10 mm wide, linear to narrow-elliptic, 5–12 cm long, thin, finely veined.
   Spikes pale yellow, interrupted. Large, bushy shrub or small tree. Widespread. Forests. Fl. Aug.–Sept. .
   . . . . **A. floribunda** (Vent.) Willd.
6 Phyllodes more than 10 mm wide 7

7 Shrub up to 3 m high. Phyllodes usually without a mucronate apex, coriaceous, often reddish when young; smaller veins uniformly prominent when dry. Flowers pale yellow, loosely and often irregularly placed in the spike. Widespread. Fl. Dec.–Feb. . . . . **A. obtusifolia** A.Cunn

7 Tree up to 6 m high. Legume much twisted and coiled. Phyllodes thin, very finely veined, 5–16 cm long, 8–15 mm wide, tapering towards each end, ± falcate. Flowers pale yellow. Small to medium tree. Widespread. Forests. Fl. Jan.–June . . . . **A. maidenii** F.Muell.

8 Tree up to 6 m high . . . .**A. maidenii** (see above)

8 Shrubs up to 4 m high. Flowers evenly and densely packed in the spike, yellow 9

9 Erect shrub 2–4 m high. Phyllodes oblong to narrow-elliptic, 8–22 cm long, 6–20 mm wide. Shrub 2–4 m high. Widespread. In many situations. Fl. June–Nov. *Sydney Golden Wattle* . . . .
. . . .**A. longifolia** (Andrews) Willd. var **longifolia**

9 Shrub ± prostrate. Phyllodes obovate-oblong, 5–9 cm long, 12–36 mm wide, with a mucronate apex at least when young. Legume very convex, often curved. Coastal sand dunes. Fl. Aug.–Oct. . . . .
. . . .**A. longifolia** var. **sophorae** (Labill.) Court

10 Phyllodes with one prominent vein, silvery pubescent, mucronate, 5–9 cm long, 1–2 cm wide, with thickened margins. Spikes short, dense, oblong or sometimes ovoid, bright yellow. Flowers 20–24 in each spike. Ovary densely tomentose. Legume tomentose, 2.5–4cm long, with 2–3 seeds. Blue Mts Fl. Sept. *Dorothy's Wattle* . . . . **A. dorothea** Maiden

10 Phyllodes with several longitudinal veins 11

11 Phyllodes densely silvery pubescent with minute hairs, falcate-lanceolate, 10–15 cm long, 5–20 mm wide. Spikes 2–6 cm long, bright yellow. Legume 6–8 cm long, 3–4 mm wide. Tree. Cumberland Plain; lower Blue Mts Hornsby Plateau. Fl. spring. *Coast Myall* . . . . . . **A. binervia** (J.C.Wendl.) J.F.Macbr.

11 Phyllodes glabrous or glaucous 12

12 Branchlets very acutely angular in section with sharp almost wing-like ridges. Phyllodes green . . 13

12 Branchlets terete to acute in section but not wing-like. Phyllodes ± glaucous. 14

13 Phyllodes with a network of minor veins visible on the surface, falcate-lanceolate, 7–10cm long. 10–20 mm wide. Spikes 2–6 cm long. Legume curved, biconvex, up to 10 cm long, 3–4 mm wide. Shrub up to 3 m high. Coast, between Palm Beach and Whale Beach. Probably introd. from some other part of the state. Fl. Sept. . . . . **A. leiocalyx** (Domin) Pedley ssp. **leiocalyx**

13 Phyllodes without a network of minor veins visible on the surface, very narrow-elliptic, ± falcate, mostly 6–19 cm long and 9–25 mm wide. Spikes up to 6 cm long, dense, bright yellow. Legume straight to twisted, up to 10 cm long and 4 mm wide. Shrub or small tree up to 15 m high. North-western Blue Mts; Hornsby Plateau. WSF. Fl. spring . . . . **A. matthewii** Tindale & S.J.Davies

14 Spikes pale yellow to almost white, up to 4 cm long, dense. Phyllodes elliptic, falcate, 5–15 cm long, 15–30 mm wide, with thickened margins. Legume straight or curved, up to 6 cm long, 2–3 mm wide. Shrub or tree up to 10 m high. Cox's River. Open forest . . . . **A. blakei** ssp. **diphylla** (Tindale) Pedley

14 Spikes deep or bright yellow 15

15 Ovary hairy. Phyllodes very narrow-elliptic, straight to falcate, 6–15 cm long, 9–20 mm wide. Spikes 3–6 cm long, dense. Legume ± straight, up to 15 cm long, 2–5 mm wide, biconvex. Tree up to 10 m high. Capertee-Glen Davis district. DSF. Fl. spring . . . . **A. cheelii** Blakely

15 Ovary glabrous. Phyllodes narrow-elliptic, scarcely falcate, up to 10cm long, mostly 7–20 mm wide. Spikes 3–5 cm long, dense. Legume straight or slightly curved, 3–10 cm long, up to 5 mm wide. Shrub or tree up to 8 m high. Open forests. Hunter River Valley. Fl. spring–summer . . . .
. . . . **A. bulgaensis** Tindale & S.J.Davies

Group 2
Leaves modified into phyllodes, pungent pointed; flowers in solitary, paired or clustered, globular heads

1 Phyllodes with 1 prominent vein on each surface or with none 2
1 Phyllodes with several prominent parallel veins on each surface, rigid 9

2 Phyllodes linear or linear-subulate or linear-tetragonous, spreading, tapering gradually to a point, sometimes very narrow-triangular but not oblique, less than 4 mm wide 3
2 Phyllodes obliquely ovate to elliptic or obliquely triangular to narrow lanccolate, 4 mm or more wide. 7

3 Flower heads almost sessile; peduncles up to 0.5 mm diam., less than half as long as the heads themselves. Flowers 20–30 per head. Petals mostly 5, without a midvein. Phyllodes linear-terete, 10–20 mm long, c. 2 mm wide, scarcely dilated at the gland. Erect shrub up to 2 m high, with rigid branches. Blue Mts DSF. Ss. Fl. spring . . . . **A. asparagoides** A.Cunn.
3 Flower heads on peduncles at least half as long as the phyllodes, frequently as long or longer 4

4 Young branches glabrous, angular. Petals mostly 4, with midrib absent or indistinct. Flowers cream, c. 20 per head; peduncles slender, up to 0.2 mm diam., clustered in the axils. Phyllodes linear, 16–30 mm long, 1.5–2 mm wide, tetragonous, scarcely dilated at the gland. Diffuse, straggling or erect shrub up to 2 m high. Southern Blue Mts DSF. Fl. winter–spring . . . . **A. genistifolia** Link
4 Young branches pubescent (although sometimes very finely so in *A. brownii*), terete or slightly tetragonous. Petals mostly 5, with a prominent midvein 5

5 Flowers cream, 20–50 per head; peduncles slender, less than 0.3 mm diam. Phyllodes 8–14 mm long, c. 1 mm wide, with a glandular dilation at the base which may not be very prominent when dry. Legume curved, 3–5 cm long, c. 3 mm wide, usually constricted between the seeds. Erect shrub with often drooping branches, up to 2 m high. Widespread. Heath and DSF. Sandy soils. Fl. spring. *Prickly Moses* . . . . **A. ulicifolia** (Salisb.) Court
5 Flowers bright yellow. Phyllodes widening very gradually or not at all towards the base and with scarcely any dilation at the gland. 6

6 Phyllodes terete (but often ± tetragonous when dry), 5–9 mm long, c. 1.5 mm wide. Peduncle up to 0.5 mm diam. Legume curved, 2–4 cm long, c. 5 mm wide, usually constricted between the seeds. Rigid, erect, divaricate shrub up to 2 m high. Widespread. Heath and DSF especially on exposed ridges. Ss. Fl. spring . . . . **A. echinula** DC.
6 Phyllodes compressed or tetragonous, mostly 13–25 mm long, c. 1 mm wide, Peduncles slender, up to 0.2 mm diam. Legume curved, 3–4 cm long, c. 3 mm wide, constricted between the seeds. Diffuse or sometimes prostrate shrub up to 1 m high. Widespread. Heath and DSF. Sandy soils and silts. Fl. spring–summer. . . . . **A. brownii** (Poir.) Steud.

7 Stipules spinescent, up to 12 mm long. Phyllodes obliquely ovate to narrow-lanceolate, 1–2 cm long. Legume straight or curved, 4–5 cm long, 3–4 mm wide. Large, straggling, spiny shrub up to 3 m high. Widespread, sometimes planted for hedges. DSF. Fl. spring . . . . **A. paradoxa** DC.
7 Stipules caducous or absent, not spinescent. 8

8 Phyllodes obliquely triangular or obliquely narrow-lanceolate, often less than 10 mm long, rigid, tapering to a fine pungent point; the principal vein near the lower margin. Flowers 20–30 per head. Legume 4–5 mm long, constricted between the seeds. Rigid shrub up to 2 m high. Western Blue Mts Ss. Fl. spring . . . . **A. gunnii** Benth.
8 Phyllodes oblong to narrow-obovate, not oblique, more than 10 mm long, obtuse with a pungent mucro; principal vein ± central. Flowers 30–50 per head. Legume 2–7 cm long, scarcely constricted between the seeds. Erect shrub up to 2 m high. Southern Blue Mts DSF. Fl. spring . . . . **A. aspera** Lindl.

9 Phyllodes terete or tetragonous, erect or slightly spreading, 2–6 cm long, abruptly narrowed into a fine point. Peduncles 5–12 mm long, Tall shrub. Blue Mts; Botany Bay, probably extinct on the coast. Fl. winter–spring. . . . . **A. quadrilateralis** DC.
9 Phyllodes flat, spreading. 10

10 Flower heads almost sessile; peduncles up to 3 mm long. 11

10 Flower heads on peduncles 10–20 mm long. Legume not striate 12

11 Phyllodes linear or linear-falcate, 1–2 mm wide, 1–4 cm long. Flowers 10–20 per head. Legumes 2–3cm long, pubescent striate. Low, bushy, pubescent shrub. Illawarra and west to the Mts Ss. Endangered. Fl. summer. . . . . **A. bynoeana** Benth.

11 Phyllodes oblong to narrow-elliptic, 3–8 mm wide, 2–5 cm long. Flowers 20–30 per head. Legume 2–5 cm long, pubescent, not striate. Bushy shrub up to 1 m high. Southern Blue Mts DSF. Fl. summer . . .

. . . . **A. lanigera** A.Cunn.

12 Phyllodes linear, 1–4 cm long, 1–3 mm wide, rigid. Legume straight or slightly curved, up to 10cm long, not constricted between the seeds on the suture lines, brown. Diffuse shrub 1–3 m high, with angular branches. Blue Mts; western Hornsby Plateau. WSF and DSF. Various soils. Fl. most of the year . . . . **A. trinervata** Sieber ex DC.

12 Phyllodes lanceolate to ovate, usually oblique, 6–18 mm long, 2–5 mm wide, rigid. Legume coiled or curved, up to 6cm long, constricted between the seeds on the suture lines, blackish. Spreading shrub up to 1.5 m high. North-western Hornsby Plateau; Hunter River Valley. Fl. winter–spring. . . . .

. . . . **A. amblygona** A.Cunn. ex Benth.

Group 3
Leaves modified into phyllodes with a soft point; flowers in solitary, paired or clustered globular heads

1 Phyllodes in regular or irregular whorls. 2

1 Phyllodes alternate or spirally arranged 5

2 Phyllodes elliptic to obliquely elliptic, 6–13 mm long, 1–2 mm wide. Flowers yellow, 18–35 per head. Flower heads solitary in the leaf axils; peduncle 6–12 mm long. Legume straight, flat, 3–7 cm long, c. 10 mm wide. Blue Mts; Hunter River Valley. DSF. Fl. winter–spring. *Golden-top Wattle.* **A. mariae** Pedley

2 Phyllodes linear-terete, 9–14 mm long, c. 1 mm wide, usually slightly reflexed at the tip. Flowers 20–25 per head; peduncles slender, up to 0.2 mm diam. Legume slightly curved, compressed, 2–4 cm long, c. 2 mm wide. 3

3 Stems and phyllodes glabrous, smooth. Undershrub, almost herbaceous, up to 50 cm high. Coast and adjacent plateaus. Usually in damp heath. Fl. spring–summer . . . . **A. baueri** Benth. ssp. **baueri**

3 Stems and phyllodes scabrous with the bases of harsh hairs, or pubescent 4

4 Flowers 10–15 per head. Legume dark brown, 1–2 cm long, 2–4 mm wide. Phyllodes scabrous. Blue Mts Heath. Vulnerable. Fl. spring–summer . . . . **A. baueri** ssp. **aspera** (Maiden & Betche) Pedley

4 Flowers c. 20 per head. Legume blue-black, 3–6 cm long, 10–15 mm wide, with reddish margins. Phyllodes pubescent . . . . **A.. gordonii** (see below)

5 Phyllodes less than 5 mm wide, linear. 6

5 Phyllodes 5 mm wide or wider, with 1–3 prominent longitudinal veins. 11

6 Phyllodes terete tetragonous or compressed, mostly up to 1 mm wide 7

6 Phyllodes flat, more than 1 mm wide 9

7 Phyllodes 10–15 mm long, curved at the tip, terete, pubescent, with no distinct longitudinal veins. Peduncles pubescent, 0.5 mm diam., c. as long as the subtending phyllodes. Flowers bright yellow, 20–30 per head. Legume flattened, straight, oblong, c. 4 cm long and 1cm wide. Straggling shrub c. 1 m high, with terete branchlets. Blue Mts Ss. Endangered. Fl. spring . . . . . **A..gordonii** (Tindale) Pedley

7 Phyllodes more than 5cm long, very slightly curved or straight at the tip 8

8 Peduncles glabrous, 3-12 mm long. Phyllodes linear-terete or obscurely tetragonous, mostly more than 6 cm long. Flower heads solitary or in clusters of 2–3, yellow, 15–20-flowered. Shrub up to 1 m high. Hornsby Plateau north of the Hawkesbury River; lower Blue Mts DSF. Ss. Fl. winter–spring . . . .

. . . . **A. juncifolia** Benth. ssp. **juncifolia**

8 Peduncles pubescent, 3-4 mm long. Phyllodes sub-tetragonous or compressed, 5–6 cm long, glabrescent, with several distinct longitudinal veins. Flowers pale yellow, 25–35 per head. Legume flattened, narrow-oblong, straight, c. 5cm long, 4–6 mm wide. Slender, erect shrub 1–2 m high, with angular branches. Blue Mts Wet ground. Ss. Fl. summer–autumn . . . .**A. ptychoclada** Maiden & Blakely

9 Phyllodes less than 15 mm long . . . . **A. mariae** (see above)
9 Phyllodes more than 20 mm long 10

10 Branchlets angular. Phyllodes 2–4 mm wide, 7–13 cm long, rigid, glabrous, with several prominent longitudinal veins. Peduncles pubescent, c. 0.5 mm diam. Flowers bright yellow, 20–35 per head. Legume straight, 6–8 cm long, c. 5 mm wide. Erect shrub up to 2 m high. Widespread. Usually in wet ground. Fl. spring. . . . . **A. elongata** Sieber ex DC.
10 Branchlets terete. Phyllodes 1–3 mm wide, 4–10 cm long, with 1–3 distinct longitudinal veins or none. Flowers yellow, 10–25 per head; peduncles 3–10 mm long. Legume straight, 3–10cm long, c. 3 mm wide. Erect shrub or tree up to 8 m high. Coast. Sandy soils. DSF. Fl. winter–spring . . . .
. . . . **A. cognata** Domin

11 Phyllodes viscid or varnished at least when young 12
11 Phyllodes neither viscid nor varnished. 14

12 Phyllodes 4–6 cm long, 4–10 mm wide, linear to elliptic, surface dotted with prominent glands. One longitudinal vein prominent laterals faint. Flowers bright yellow, 30–40 per head. Peduncles 5–8 mm long. Shrub up to 2 m high. Blue Mts, e.g. Capertee, Yerranderie. DSF. Ss. Fl. spring. *Varnish Wattle*. .
. . . . **A. verniciflua** A.Cunn.
12 Phyllodes mostly less than 4 cm long, not gland dotted 13

13 Flowers 30–40 per head; peduncles up to 5 mm long, densely covered with woolly hairs; heads solitary or 2–3 on a very short axis. Phyllodes with several prominent longitudinal veins, although the central one may be somewhat more prominent, elliptic to narrow-elliptic, 2–4 cm long Legume curved, 3–7 cm long, 2–3 mm wide, resinous and sparsely covered with hairs. Shrub up to 1 m high. Blue Mts, e.g. Capertee, Glen Davis. DSF. Fl. winter–spring. **A. ixiophylla** Benth.
13 Flowers 20–30 per head; peduncles 3–10 mm long, glabrous or hairy; heads in pairs in axil of phyllode. Phyllodes with 2 prominent longitudinal veins, oblanceolate, 1–4 cm long, 1.5–7 mm wide. Legume straight or curved, 1–5 cm long, 3–4 mm wide, woolly. Shrub to 3.5 m high. Glen Alice. Mallee. Fl. winter–spring.. . . . . **A. montana** Benth.

14 Phyllodes 1–3cm long, with thickened margins. 15
14 Phyllodes 4–15cm long, without thickened margins 18

15 Phyllodes oblong-falcate; the principal vein central or nearly so, scabroso-denticulate. Flowers c. 20 per head; peduncles shorter than or just as long as the phyllodes. Legume ovate or oblong, 10 mm wide, with 1–2 seeds. Shrub up to 1 m high. Widespread. Heath and DSF. Ss. Fl. most of the year . . .
. . . .**A. hispidula** (Sm.) Willd.
15 Phyllodes ovate to almost orbicular, ± oblique; margin undulate, smooth. 16

16 Phyllodes less than 1.6cm long, bright green. Flowers pale yellow 20–30 per head. Apex of branchlets twisted. Legume 3–9 cm long, straight or curved, flat, margins slightly undulate. Shrub open and drooping to 2 m high. Glen Davis and Yerranderie areas. Woodland at higher altitudes. Fl. all year. . .
. . . . **A. clandullensis** B.J.Conn & Tame
16 Phyllodes more than 1.6 mm long, grey-green.. 17

17 Flowers yellow, 25–75 per head. Branchlets usually pruinose. Phyllodes usually glabrous with some hairs on pulvinus and occasionally mid vein, margins thick, base truncate to cordate. Legume 3–9 cm long, straight or curved, flat, usually pruinose. Sparsely branching shrub to 2 m high. Widespread. DSF. Fl. all year.. . . . .**A. sertiformis** A.Cunn.
17 Flowers pale yellow 20–30 per head. Phyllodes usually hairy, base acute to obtuse. Legume 4–9 cm long, straight, flat. Open shrub to 3 m high. Bucketty to Megalong Valley. DSF. Fl. spring.. . . . .
. . . .**A. undulifolia** A.Cunn. ex G.Lodd.

18 Phyllodes obtuse, often mucronate, 5–12 cm long, 5–15 mm wide, narrow-oblong-elliptic. Flower heads yellow, in pairs or clusters; peduncles 2–3 mm long. Legume straight, narrow-oblong, 5 cm long. Erect shrub up to 3 m high, often gregarious. Widespread. In various situations. Fl. spring . . . . **A. stricta** (Andrews) Willd.

18 Phyllodes acute, 5–12 cm long, 4–20 mm wide, elliptic, with 1–3 prominent longitudinal veins and sometimes the lateral veins distinct. Flower heads yellow to pale yellow; peduncles 4–10 mm long. Shrub or tree up to 6 m high. Western Blue Mts DSF. Fl. spring . . . . **A. leprosa** Sieber ex DC.

Group 4
Leaves modified into phyllodes with 2 or more prominent longitudinal veins; flower heads in racemes

1 Phyllodes with usually 2 prominent longitudinal veins (very rarely 3 in *A. binervata*) 2
1 Phyllodes with 3 or more prominent longitudinal veins. 3

2 Flowers cream to pale yellow, mostly more than 20 per head. Phyllodes elliptic-falcate, 7–12 cm long, 10–30 mm wide, dark green; marginal gland below the middle, ± inconspicuous. Legume oblong, c. 12 mm long, flat, thin, straight. Tree. Coast and adjacent plateaus; near RF. Fl. spring. *Two-veined Hickory* . . . . **A. binervata** DC.

2 Flowers golden yellow, mostly less than 20 per head. Phyllodes narrow-elliptic-falcate, 5–15 cm long, 5–10 mm wide, ± glaucous; marginal gland near the base. Legume oblong, 30–80 mm long, 8–20 mm wide, flat, sometimes woody, straight to curved. Tree up to 15 m high, with drooping foliage. Western Blue Mts Usually on clay soils. Fl. summer–autumn. *Boree* or *Weeping Myall* . . . . **A. pendula** A.Cunn. ex G.Don

3 Phyllodes linear, 6–10 cm long, 3–4(5) mm wide, slightly viscid especially on the margins and veins; marginal gland basal, inconspicuous. Flowers 4–8 per head. Young branches angular, pubescent, viscid on the angles. Legume slightly curved or straight, up to 6cm long, 3–5 mm wide. Shrub up to 2 m high. Western Blue Mts DSF. Fl. spring . . . . **A. dawsonii** R.T.Baker

3 Phyllodes elliptic to elliptic-falcate, more than 8 mm wide. Flowers 30–50 per head 4

4 Phyllodes viscid and varnished at least when young. . . . . **A. ixiophylla** (see Group 3)
4 Phyllodes not viscid and varnished 5

5 Phyllodes mostly obtuse, thick, 7–11 cm long, 9–17 mm wide. Branchlets angled. Legume biconvex, curved or twisted, 8–10 cm long, 5–8 mm wide. Funicle twice encircling the seed (Fig. 39). Ribs on the young branches light-coloured. Peduncles 0.5–1 mm diam. Tree. Blue Mts; Illawarra ranges. Fl. spring–summer. *Blackwood*. . . . . **A. melanoxylon** R.Br.

5 Phyllodes mostly acute or acuminate, thin, 7–15 cm long or longer, up to 20 mm wide. Branchlets ± terete. Legume biconvex, c. 8 cm long, 4–5 mm wide, curved or twisted. Funicle folded under the seed (Fig. 39). Ribs on the young branches not conspicuous. Peduncles less than 0.5 mm diam. Tall shrub or small tree, often suckering. Widespread. WSF; margins of RF; gullies. Fl. most of the year. *Hickory* . . . . **A. implexa** Benth.

Group 5
Leaves modified into phyllodes, with 1 prominent longitudinal vein or none; flower heads in racemes.

1 Phyllodes with 2 or more marginal glands distant from the base of the phyllode 2
1 Phyllodes with 1 marginal gland; or marginal gland obsolete; or 2 marginal glands close to the base of the phyllode. 4

2 Phyllodes pubescent . . . . **A. dorothea** (see Group 1)
2 Phyllodes glabrous 3

3 Phyllodes 7–15cm long, 6–12 mm wide, narrow-elliptic, slightly falcate. Flowers golden yellow, 30–40 per head. Legume 10–13 cm long, 5–6 mm wide. Shrub up to 2 m high. Cambelltown; western Blue Mts DSF. Fl. spring–summer. *Sword-leaf Wattle* . . . . **A. gladiiformis** A.Cunn. ex Benth.

3 Phyllodes mostly 3–6 cm long, 5–12 mm wide, elliptic to oblanceolate, slightly falcate. Flowers up to 12 per head, yellow to pale cream. Legume straight, flat, up to c. 10 cm long, 4–5 mm wide. Erect shrub up to 3 m high. Blue Mts DSF. Fl. winter–spring. . . . . **A. amoena** H.L.Wendl.

4 Phyllodes with distinct lateral veins visible on the surface 5
4 Phyllodes without distinct lateral veins visible on the surface 21

5 Flowers less than 40 per head 6
5 Flowers more than 40 per head 20

6 Phyllodes fimbriate on the margins and sometimes also pubescent on the midrib, glabrous elsewhere, thin, usually with a prominent rounded marginal gland near the base, linear to narrow-oblong-elliptic, 2–4 cm long, 2–5 mm wide. Flowers golden yellow, 10–30 per head. Legume flat, straight, up to c. 7 cm long, 7 mm wide. Shrub or small tree up to 6 m high. Cumberland Plain; lower Blue Mts Open forests, often near streams. Fl. winter–spring. *Fringed Wattle*. . . . .**A. fimbriata** A.Cunn. ex G.Don
6 Phyllodes glabrous or with scattered hairs or hairy all over 7

7 Flowers up to 15 per head 8
7 Flowers mostly more than 15 per head 12

8 Flowers up to 8 per head, pale yellow to cream . . . .**A. myrtifolia** (see below)
8 Either; flowers more than 8 per head; or, if 8 or less, then golden yellow 9

9 Phyllodes 5–15 cm long, 8–25 mm wide, narrow-elliptic, reddish. Bipinnate leaves very often present. Flowers yellow to pale cream. Legume straight, or slightly curved, flat, up to c. 10 cm long, c. 5 mm wide. Shrub or tree up to c. 10 m high. Blue Mts Open forests. Fl. spring. *Red-leaf Wattle* . . . .
. . . . **A. rubida** A.Cunn.
9 Phyllodes mostly less than 5 cm long 10

10 Branchlets terete with appressed to spreading hairs. Phyllodes 3–10 mm wide, elliptic-oblong, almost glaucous. Flowers golden yellow. Legume straight, flat, up to 6 cm long, 8–15 mm wide. Seed placed transversely. Shrub up to 3 m high. Blue Mts Fl. winter–spring. . . **A. kybeanensis** Maiden & Blakely
10 Branchlets angled glabrous. 11

11 Heads with 3–5 flowers. Phyllodes narrow elliptic to oblanceolate, 1.5–3.5 mm long, 2.5–9 mm wide, lower margin ± straight upper margin curved. Marginal gland 2–6 mm above pulvinus, inconspicuous. Legume 5–8 mm long, 6–8 mm wide, seeds placed longitudinally. Shrub 1–3 m high. Hunter Valley south to Sydney. DSF. Ss. Fl. winter–spring. *Lunate-leaved Acacia*. . . . .**A. lunata** G.Lodd.
11 Heads with 5–10 flowers. Phyllodes straight, obovate to elliptic 1.5–3 cm long, 6–14 mm wide. Marginal gland 1–4 mm above pulvinus, inconspicuous. Legume 7–8 cm long, 6–8 mm wide, seeds placed longitudinally. Shrub 1–3 m high. Bowral–Wingello area. DSF. And heath. Fl. winter–spring . . . . . .
. . . . **A. leucolobia** Sweet

12 Marginal gland obscure and near the base of the phyllode or obsolete; when present, without a conspicuous vein leading to the midvein. 13
12 Marginal gland prominent and distant from the base of the phyllode; usually with a conspicuous vein leading to the midvein 16

13 Phyllodes glaucous 14
13 Phyllodes green 15

14 Phyllodes lanceolate-falcate, 12–40 mm wide, 7–18 cm long. Flower heads pale yellow, 2–4 mm diam., c. 20-flowered. Legume 5–10 cm long, c. 5–10 mm wide, slightly curved. Shrub or tree up to 5 m high. Coast and adjacent plateaus; Cumberland Plain. Open forests. Fl. winter . . . .**A. falcata** Willd.
14 Phyllodes linear to narrow-oblong, 3–7 mm wide, up to 5 cm long. Flower heads golden yellow, 15–30 per head. Legume rough, ± curved, flat, 5–10 cm long, 5–9 mm wide. Shrub up to 4 m high. Mostly western Blue Mts Open forests. Fl. winter–spring. *Western Golden Wattle* . . . . **A. decora** Rchb.

15 Phyllodes up to 4 cm long, 3–12 mm wide, lanceolate to elliptic or obovate, scarcely falcate, pale green to glaucous. Flowers golden yellow, c. 7–29 per head. Legume flat, straight, up to c. 8 cm long, c. 5 mm wide. Shrub up to 3 m high. Widespread but not on the coast. DSF and heath. Fl. winter–spring. *Box-leaf Wattle* . . . . **A. buxifolia** A.Cunn. ssp. **buxifolia**

15 Phyllodes more than 5 cm long, 15–40 mm wide, elliptic-falcate, green to yellowish-green. Flowers golden yellow, c. 20 per head. Legume straight or slightly curved, up to 15 cm long, 8–20 mm wide. Shrub or tree up to 15 m high. Blue Mts; Hunter River Valley. Open forests. Fl. spring . . . .
. . . . **A. obliquinervia** Tindale

16 Peduncles and young branches golden or silvery hairy. 17
16 Peduncles and young branches glabrous or nearly so 18

17 Phyllodes up to 4 cm long, 12–25 mm wide, ovate to obovate, hairy when young, glaucous. Flowers golden yellow. Racemes mostly longer than the subtending phyllodes. Legume straight or twisted, up to 12 cm long, 12–20 mm wide. Shrub or small tree up to 5 m high. Garden escape near habitation. Introd. from north coast of NSW and Queensland. *Queensland Silver Wattle* . . . .
. . . . ***A. podalyriifolia** A.Cunn. ex G.Don

17 Phyllodes 8–17 cm long, 17–40 mm wide, elliptic-falcate, glaucous. Flowers pale yellow to pale cream. Racemes shorter than the subtending phyllodes. Legume straight or slightly curved, up to 15 cm long, 12–25 mm wide. Shrub or tree up to 10 m high. Western Blue Mts DSF. Fl. winter–summer. *Broad-leaf Hickory* . . . . **A. falciformis** DC.

18 Phyllodes obliquely triangular to obliquely obovate, up to 3 cm long, 10–15 mm wide, marginal gland distant from the base and close to the widest part of the phyllode. Flowers golden yellow; flower heads often ± ovoid. Legume straight, flat, up to 7 cm long, 4–7 mm wide. Shrub up to 4 m high. Hunter River Valley; a garden escape near habitation on the coast. DSF and heath. Fl. winter–spring. *Knife-leaf Wattle* . . . . ***A. cultriformis** A.Cunn. ex G.Don
18 Phyllodes elliptic to oblanceolate, ± falcate, more than 5cm long. 19

19 Flowers pale yellow to almost white. Marginal gland mostly at least 10 mm from the base of the phyllode. Phyllodes oblanceolate, 5–10 cm long, 9–35 mm wide. Legume straight, flat, up to 20 cm long, 9–25 mm wide. Shrub or tree 3–8 m high. Widespread. DSF; WSF. Fl. most of the year. *Mountain Hickory* . . . . **A. penninervis** Sieber ex DC.
19 Flowers golden yellow. Marginal gland mostly less than 10 mm from the base of the phyllode. . . . . .
. . . . **A. obliquinervia** (see above)

20 Phyllodes 7–15 cm long, falcate-lanceolate, glabrous, green; marginal gland above the base and fairly distinct. Flower heads bright yellow; peduncles c. 0.5 mm diam. Shrub or tree up to 8 m high. Coast and adjacent plateaus. Garden escape into open forests. Introd. from inland parts of the State. Fl. spring. *Golden Wattle* . . . . ***A. pycnantha** Benth.
20 Phyllodes 15–30 cm long, linear-oblong to lanceolate-falcate, glaucous at least in juvenile stage, with a gland at the base. Racemes short, numerous. Flowers bright yellow; peduncles up to 0.2 mm diam. Shrub or tree up to 8 m high. Coast and adjacent plateaus. Introd. from WA. Fl. spring . . . .
. . . . ***A. saligna** (Labill.) H.L.Wendl.

21 Unexpanded racemes covered with large imbricate bracts. Phyllodes narrow-oblong to linear, up to 15 cm long, 2–10 mm wide, glabrous or glaucous. 22
21 Unexpanded racemes with no conspicuous imbricate bracts covering the flower heads. 23

22 Flowers 3–10 per head, pale yellow. Peduncles 1–5 mm long. 1 gland at base and 1 near apex of phyllode. Legume oblong, flat, glaucous, 2–4 cm long, 12–20 mm wide. Seeds placed transversely. Slender shrub 1–2 m high, with angular branches. Widespread. Heath and DSF. Sandy soils. Fl. autumn–winter. *Sweet Wattle* . . . . **A. suaveolens** (Sm.) Willd.
22 Flowers 12–17 per head, pale yellow. Peduncles 6–10 mm long. 1 gland to 10 mm above pulvinus. Legume 6–12 cm long, 5–12 mm wide, seeds placed longitudinally. Bushy shrub to 4 m high. Occasionally naturalized. Native of SA. Fl. autumn – winter.. . . . . . **A. iteaphylla** F.Muell. ex Benth.

23 Phyllodes hairy at least on the margins, sometimes glabrescent with age 24

23 Phyllodes glabrous, rarely with a few hairs near the base 26

24 Phyllodes with hairs on the margins and sometimes on the midvein, otherwise glabrous . . . .
. . . . **A. fimbriata** (see above)
24 Phyllodes with hairs on all surfaces 25

25 Phyllodes linear, 3–7 cm long, 3–4 mm wide, acute, often with a short mucro, pubescent with appressed
short hairs when young but glabrescent. Flowers c. 20 per head, golden yellow. Legume straight, up to
7 cm long and 7 mm wide. Shrub up to 3 m high. North-western Blue Mts DSF. Fl. winter–spring. *Blue
Bush* . . . . **A. caesiella** Maiden & Blakely
25 Phyllodes narrow-oblong-elliptic, 4–6 cm long, 4–8 mm wide, obtuse, with a short mucro, pubescent
with short hairs rarely glabrescent with age. Flowers up to 15 per head, golden yellow. Legume straight,
up to 6 cm long, 7–10 mm wide. Shrub up to 2 m high. Blue Mts DSF. Vulnerable. Fl. spring–summer
. . . . **A. clunies-rossiae** Maiden

26 Phyllodes mostly more than 5 mm wide 27
26 Phyllodes mostly less than 5 mm wide 36

27 Flowers up to 8 per head or sometimes up to 10 but then phyllodes without a prominent marginal
gland. 28
27 Flowers more than 8 per head 30

28 Flowers pale yellow 2–8 per head. Legume linear, thick, ± woody, curved, with thick margins, up to
11 cm long, 3–5 mm wide. Phyllodes narrow-elliptic to elliptic, 2–6 cm long, 5–30 mm wide, thick,
mucronate, with thickened margins, with a gland distant from the base of the phyllode but below the
middle. Glabrous shrub up to 2 m high, with angular often reddish branchlets. Widespread. Heath
and DSF. Fl. winter–spring. *Red-stemmed Wattle* . . . . **A. myrtifolia** (Sm.) Willd.
28 Flowers bright yellow 3–10 per head. Legume 5–8 cm long, 6–8 mm wide. 29

29 Heads with 3–5 flowers. Phyllodes narrow elliptic to oblanceolate, 1.5–3.5 cm long, 2.5–9 mm wide,
lower margin ± straight upper margin curved. . . . . **A. lunata** (see above)
29 Heads with 5–10 flowers. Phyllodes straight, obovate to elliptic 1.5–3 cm long, 6–14 mm wide. . . . . .
. . . . **A. leucolobia** (see above)

30 Phyllodes very acute, narrow-elliptic-oblong to elliptic-oblong, 5–12 cm long, 8–12 mm wide. Flowers
20–30 per head, pale yellow. Legume straight, 7–12 cm long, 13–18 mm wide. Shrub or tree up to 7
m high, with drooping branches and silvery to blue-grey bark. Northern Blue Mts; north-western
Hornsby Plateau. DSF; WSF. Ss. Fl. Feb.–Aug. . . . .**A. saliciformis** Tindale
30 Phyllodes ± obtuse, often with a short blunt mucro 31

31 Phyllodes mostly more than 7 cm long. 32
31 Phyllodes mostly less than 7 cm long. 33

32 Sepals free, although sometimes stuck together when young. Phyllodes oblanceolate to oblong, 6–9
mm wide, with a gland near the base, Flowers 15–30 per head, golden yellow. Shrub up to 2 m high.
Capertee, Glen Davis. Open forest. Fl. spring . . . . **A. hakeoides** A.Cunn. ex Benth.
32 Sepals connate; the lobes much shorter than the tube. Phyllodes oblanceolate, 8–15 mm wide, with
an obscure gland near the base. Flowers 15–30 per head, pale yellow to whitish. Legume straight or
slightly curved, thick, 5–12 cm long, 7–13 mm wide. Shrub or small tree up to 10 m high. Hunter River
Valley. Open forests. Fl. summer–winter. *Coobah* or *Native Willow* . . . . **A. salicina** Lindl.

33 Marginal gland inconspicuous, close to the base of the phyllode 34
33 Marginal gland conspicuous, either prominently exserted or in an angle in the margin of the phyllode
. . . . 35

34 Phyllodes thin, mostly more than 5 times as long as wide, 2–6 cm long, 4–8 mm wide. Flowers c. 20 per
head, golden yellow. Legume flat, 6–9 cm long, c. 7 mm wide. Wombeyan Caves. DSF on limestone. Fl.
summer . . . . **A. chalkeri** Maiden
34 Phyllodes thick, mostly less than 5 times as long as wide . . . . **A. buxifolia** ssp. **buxifolia**(see above)

35 Phyllode with an angle or indentation on the margin at the gland, elliptic to oblong 4–7 cm long. Flowers 15–25 per head, golden yellow. Legume straight, flat, up to 8 cm long, 6–12 mm wide. Shrub up to 3 m high. Western Blue Mts DSF. Fl. winter–spring. . . . . **A. obtusata** Sieber ex DC.

35 Phyllode not indented at the prominently exserted marginal gland, narrow-elliptic, 2–5 cm long, 8–12 mm wide. Flowers 8–15 per head, cream to golden yellow. Legume bluish, straight, flat, 3–8 cm long, 10–20 mm wide. Tall shrub or tree up to 18 m high. Coast and adjacent plateaus; Mulgoa. Open forests. Fl. spring. Endangered population in some LGAs. *Gosford Wattle* or *Sally Wattle* . . . .
. . . . **A. prominens** A.Cunn. ex G.Don

36 Phyllodes 6–10 cm long, 3–5 mm wide. Flowers 20–30 per head, bright yellow. Legume straight, narrow-oblong, up to 9 cm long, 5–8 mm wide. Slender shrub up to 3 m high. Blue Mts DSF. Ss. Vulnerable. Fl. summer . . . . **A. flocktoniae** Maiden

36 Phyllodes up to 5 cm long, rarely more and then less than 2 mm wide 37

37 Phyllodes elliptic to oblong, usually less than 8 times as long as broad 38

37 Phyllodes linear to sub terete usually more than 8 times as long as broad. 39

38 Heads with 3–5 flowers. . . . . **A. lunata** (see above)

38 Heads with 7–29 flowers . . . . **A. buxifolia** ssp. **buxifolia** (see above)

39 Flowers 9–20 per head. 40

39 Flowers up to 9 per head 41

40 Flowers 9–15 per head, golden yellow. Peduncle 2–4 mm long, glabrous. Phyllodes narrow-oblong to linear, 2–4 mm wide, 3–7 cm long, with thickened margins, mucronate. Legume oblong, flat, straight, c. 4 cm long and 10 mm wide. Shrub up to 3 m high. Blue Mts; Berowra. Heath and DSF. Fl. spring. *Hamilton's Wattle*. . . . .**A. hamiltoniana** Maiden

40 Flowers 12–20 per head pale to golden yellow. Peduncle 4–8 mm long, glabrous. Phyllodes terete to linear, 1–1.5 mm wide, 6–10 mm long, apex acute. Legume ± straight, 7–18 cm long, 4–8 mm wide. Erect spreading shrub to 4 m high. Capertee Valley. DSF. Fl. winter–summer.. . . **A. subulata** Bonpl.

41 Flowers mostly 5–9 per head, golden yellow to almost white. Peduncle 2–5 mm long, occasionally with hairs. Phyllodes 1–3 mm wide, 2–4 cm long. Legume oblong, flat, 4–8 cm long, 10–13 mm wide. Graceful shrub up to 4 m high with rather slender often drooping. Widespread. Heath and DSF. Fl. summer . . . . **A. linifolia** (Vent.) Willd.

41 Flowers mostly 4–8 per head, yellow to dark yellow. Peduncle 2–4 mm long, minutely hairy. Phyllodes 0.5–1.5 mm wide, mostly 1–6 cm long. Legume straight or slightly curved, 2.5–8.5 cm long, 4–7 mm wide. Straggling to erect shrub up to 2.5 m high, with erect or spreading branches. Blue Mts, Clarence district. DSF. Fl. winter–spring . . . . **A. meiantha** Tindale & Herscovitch

Group 6
Leaves bipinnate

1 Pinnules more than 2 mm wide, usually 4–12 mm wide. Prominent gland 1–4 mm long between the lowest pair of pinnae and the base of the petiole 2

1 Pinnules 0.5–1.5 mm wide. 1–several glands each 0.5–1 mm long often present between the lowest pair of pinnae and the base of the petiole. 10

2 Pinnules 4–8 pairs per pinna, light green. Pinnae 2–6 pairs. Flowers 14–20 per head, golden-yellow. Legume straight or nearly so, 5–11 cm long, 9–15 mm wide. Shrub or small tree up to 4 m high with ± glaucous bark. Rylestone and Glen Alice. DSF. Fl. July–Nov. *Mudgee Wattle*. . . . .
. . . . **A. spectabilis** A.Cunn. ex Benth.

2 Pinnules mostly more than 8 per pinna 3

3 Leaves 30–40 cm long. Mature pinnules 2.5–5 cm long; much paler beneath, 11–17 pairs; apex acuminate. Legume straight or slightly constricted, 12–14 cm long, c. 12 mm wide. Tree up to 20 m high. Widespread. WSF; RF (especially the margins); deep shady gullies; along watercourses. Coast; Cumberland Plain; Woronora Plateau; Blue Mts Fl. mainly late Dec.–March. *Mountain Cedar Wattle*. .
. . . .**A. elata** A.Cunn. ex Benth.

3 Leaves 3.5–20 cm long. Mature pinnules 7–20 mm long. not paler or much paler beneath; apex acute or obtuse **4**

4 Flowers 27–40 per head. Pinnules not paler beneath, 11–20 pairs. Pinnae 4–5 pairs. Legume constricted, 7.5–16 cm long, 5–10 mm wide. Tree up to 16 m high. Coast; Cumberland Plain; Hornsby Plateau; Hunter River Valley. WSF, in deep shady gullies usually near creeks. Ss and shales. Fl. late Nov.–Feb. . . . . **A. schinoides** Benth.

4 Flowers 6–15 per head. Pinnules much paler beneath **5**

5 Petioles 0.7–4 cm long, pinnules flat, 20 pairs. Pinnae 2–8 pairs. Legume not constricted, 3–11 cm long, 12–17 mm wide. Shrubs up to 2 m high or tree rarely up to 6 m high. (*Sunshine Wattles*) **6**

5 Leaves sessile or petiole less than 0.5 cm long, pinnules recurved. **9**

6 Branchlets usually hairy. **7**

6 Branchlets mostly glabrous. Flowers bright yellow to creamy white **8**

7 Peduncle 0.5–1 mm diam. 7–14 flowers per head. Inflorescence to 16 cm long. Coastal Sydney. DSF. Ss. Endangered. Fl. mostly March–Nov.. . . . . **A. terminalis** (Salisb.) J.F.Macbr. ssp. **terminalis**

7 Peduncle 0.3–0.5 mm diam. 5–9 flowers per head. Inflorescence to 33 cm long. Widespread. DSF. Ss. Fl. mostly March–Nov. . . . . **A. terminalis** ssp. **longiaxialis** Kodela & Tindale

8 Flowers pale yellow to creamy white, 5–13 per head. Gland on petiole 2–12 mm long. Widespread. DSF. Ss. Fl. mostly March–Nov.. . . . . **A. terminalis** ssp. **angustifolia** Tindale & Kodela

8 Flowers bright yellow occasionally paler, 5–7 per head. Gland on petiole 1.5–7 mm long. Widespread. DSF. Ss. Fl. mostly March–Nov. . . . . **A. terminalis** ssp. **aurea** Tindale & Kodela

9 Petioles ridged, hairy ± terete. Pinnae 3–13 pairs; pinnules 4–15 pairs. Axis of inflorescence not winged. Legume straight or curved flat, 6–13 cm long, 11–17 mm wide. Slender to spreading shrub or small tree. Bucketty to Mangrove Mt. DSF on Ss. Ridges and upper slopes. Fl. Mar–Sept. . . . . **A. kulnurensis** Tindale & Kodela

9 Petioles winged, usually glabrous, ± 4 sided. Pinnae 1–10 pairs; pinnules 7–17 pairs. Axis of inflorescence winged. Legume usually curved, slightly raised over seeds, 6–12 cm long, 11–13 mm wide. Slender shrub or small tree. Howes Mt. DSF on ridges and upper slopes. Fl. Dec–May. . . . . **A. alaticaulis** Kodela & Tindale

10 Interjugary glands never present on the rhachis **11**

10 Interjugary glands present on the rhachis between 1 or more pairs of pinnae **17**

11 Pinnules clothed with hairs **12**

11 Pinnules glabrous or with several marginal cilia, without an apical tuft of hairs **13**

12 Foliage silvery. Ridges of the stem smooth, densely pubescent. Legume blue, glabrous. Shrub or small tree up to 30 m high. Blue Mts, a garden escape on coast. Fl. Aug.–Oct. *Silver Wattle* . . . .**A. dealbata** ssp. **dealbata** Link

12 Foliage green. Ridges of the stem scabrous, with tubercles each bearing a tuft of minute hairs. Legumes black, scabrous, clothed at first with short, pale yellow or white, appressed hairs. Shapely tree 5–12 m (rarely up to 30m) high. Widespread. DSF and RF margins, usually near watercourses. Mainly Ss. Fl. Nov.–early Jan. . . . . **A. irrorata** Sieber ex Spreng. ssp. **irrorata**

13 Branchlets glabrous or clothed with hairs c. 0.1–0.2 mm long. Legumes brown or black, pubescent or glabrous **14**

13 Branchlets pilose or hispid; the hairs 0.8–2 mm long. Legume blue, glabrous **16**

14 Flowers 8–15 per head. Leaves ± sessile. Stems with insignificant ridges (0.1–0.2 mm high) which are densely pubescent with grey hairs or rarely almost glabrous. Straggly shrub 0.3–3 m high. Southern ranges, Bargo to Goulburn. DSF, often on dry stony ridges. Ss. Fl. Sept.–Oct., rarely Jan.. . . . . **A. jonesii** F.Muell. & Maiden

14 Flowers 20–30 per head **15**

15 Stems with small insignificant ridges usually 0.1–0.2 mm high. Pinnae 6–14 (usually 8–12) pairs. Pinnules 20–33 (rarely up to 40) pairs, 2.5–7 (mainly 3–5) mm long. Legume black or brown, 6–11 cm long, 3.5–7.5 mm wide. Shrub or tree 2–7 m (–12 m) high. . . . .**A. parramattensis** (see below)

15 Stems with broad wing-like ridges usually 0.6–2 mm high. Pinnae 4–12 pairs. Pinnules 15–35 pairs, 5–14 mm long. Legumes brown or dark brown, 4.5–10.5 cm long, 4–7 mm wide. Tree 4–14 m high. Widespread. DSF or in open undulating country. Shales. Fl. July–early Sept. *Green Wattle* . . . .
. . . .**A. decurrens** Willd.

16 Pinnae bluish, glaucous. Legume 8–14 mm wide, straight or almost so. Tree usually 5–10 m high. Garden escape. Introd. from inland areas of the state. Fl. Aug.–Sept. at low altitudes, Oct.–Jan. at high altitudes. *Cootamundra Wattle* . . . .***A. baileyana** F.Muell.

16 Pinnae green, non-glaucous. Legume 4–5 mm wide, ± constricted between the seeds. Bushy shrub 0.9–2.4 m high. Cumberland Plain and lower Blue Mts DSF, on gravelly clay ridges and in *Melaleuca* scrubs. Shales. Vulnerable. Fl. Aug.–Oct. . . . .**A. pubescens** (Vent.) R.Br.

17 Pinnules narrowly lanceolate to ovate or narrowly lanceolate-oblong; the lower surface villous with silvery hairs; the apices markedly acute to sub-acuminate. Legume submoniliform, almost straight-sided, densely lanate with grey or chestnut-coloured hairs. Tree 1.5–10 m high. Hornsby Plateau. DSF. Ss and basalt. Fl. April–June. . . . .**A. fulva** Tindale

17 Pinnules linear to oblong; the apices obtuse or subacute. Legume straight or submoniliform, glabrous or pubescent or tomentose but never lanate 18

18 Leaves sessile, 8–15 flowers per head . . . . **A. jonesii** (see above)

18 Leaves with petiole 0.5 – 2.5 cm long; 14–40 flowers per head 19

19 Flowers usually 14–18 per head (occasionally 20). Corolla 1–1.2 mm long. Ovary glabrous. Pinnules 3–5 mm long. Legumes blue-brown or blue-black, 5–11 cm long, 5–9 mm wide, submoniliform; when mature glabrous or very sparsely clothed with short, white, appressed hairs. Pinnae 14–30 (rarely more) pairs. Shrub or tree 2.5–10 m high; the trunk very silvery in young trees. Cumberland Plain; Hornsby Plateau; Blue Mts; Hunter River Valley. DSF. Ss, shales and laterite. Fl. Sept.–early Dec. sometimes also April–July in the more northerly regions . . . .**A. parvipinnula** Tindale

19 Flowers 20 or more per head. 20

20 Pinnules distinctly paler on lower surface. 21

20 Pinnules ± the same colour on upper and lower surface. 22

21 Branchlets rarely pruinose; pinnules hairy. Stems velvety pubescent. Pinnules broadly rounded, 1.2–3 mm long, glabrous on the upper surface, pubescent underneath, with cilia along the margins, 30–44 pairs. Pinnae 9–20 pairs. Flowers pale yellow to cream, 20–40 per head. Legumes black, submoniliform, 4–9 cm long, 4.5–8 mm wide. Glands of the leaf rhachises and peduncles of the inflorescences densely tomentose. Tree 5–15 m high. Woronora Plateau; Blue Mts; rarely a garden escape on the Cumberland Plain. DSF and coastal scrub. Ss, shales and slates. Fl. Oct.–Dec., mostly Nov. *Black Wattle* . . . .
. . . .**A. mearnsii** De Wild.

21 Branchlets pruinose; pinnules glabrous to ± hairy. Stems minutely hairy to glabrous. Pinnules oblong, 1–6 mm long, hairy to glabrous, 11–45 pairs, light green to silvery. Pinnae 5–18 pairs. Flowers bright yellow, 20–26 per head. Legume ±pruinose, flat straight to curved, 3–12 cm long, 4.5–12 mm wide. Shrub or tree to 13 m high. Widespread. DSF. Fl. July–Oct.. . . **A. leucoclada** Tindale ssp. **leucoclada**

22 Leaves light green. . . . .**A. leucoclada** (see above)

22 Leaves dark green. 23

23 Gland on petiole small, usually only 1. Flowers 25–50 per head, pale to bright yellow. Stems glabrous or sparsely clothed with short appressed hairs. Pinnules broadly rounded or subacute, 2.5–7 (mainly 3–5) mm long, glabrous except sparsely ciliate along the margins, 20–33 (rarely up to 40) pairs, dark green. Pinnae 6–14 (usually 8–12) pairs. Legume black or brown, submoniliform, 6–11 cm long, 3.5–7.5 mm wide. Glands of the leaf rhachises and peduncles of the inflorescences glabrous or slightly pubescent. Shrub or tree 2–7 (14 m) high. Woronora Plateau; Cumberland Plain; lower slopes and valleys of the

Blue Mts DSF. Shales, breccias or more rarely Ss. Fl. late Nov.–early Feb., rarely until April . . . .
. . . . **A. parramattensis** Tindale
23 Glands on petiole prominent, 1–5. Flowers usually 22–30 per head, bright yellow. Stems densley and minutely hairy, glabrescent with age. Corolla 1.5–1.8 mm long. Ovary clothed with long weak white hairs. Pinnules usually 6–10 (rarely 3–5) mm long, mostly 33–55 (rarely 23–68) pairs, dark green. Legume blue or purplish blue, glaucous, straight-sided, 4–12 cm long, 7–18 mm wide. Pinnae 5–14 pairs. Tree up to 14 m high; the trunk brown or grey or greyish-green in young trees. Woronora Plateau; Hornsby Plateau; Northern Blue Mts; Hunter River Valley. DSF on rocky hillsides and alluvial flats. Ss and alluvium. Fl. Aug.–Sept., rarely Oct. *Fern-leaf Wattle* . . **A. filicifolia** Cheel & M.B.Welch

2 Vachellia Wight & Arn.
9 species in Aust. (7 endemic, 2 naturalized); Qld, NSW, S.A., WA, NT

One species in the area
Multistemmed shrub to 4 m high. Leaves. Flower heads 1–3 in the axils of the leaves. Leaves bipinnate, pubescent with 1 gland between the lowest pair of pinnae and the base of the petiole. Interjugary glands absent. Pinnules 8–20 pairs. Pinnae 3–7 pairs. Legume turgid, indehiscent, black. Broke; Wondabyne. Probably introd. from trop. America. *Mimosa Bush* . . . . ***V. farnesiana** (L.) Wight & Arn.

3 Neptunia Lour.
5 species endemic Aust.; Qld, NSW, NT, S.A., WA

One species in the area
Decumbent to ascending, weak shrub with stems up to 2 m long, but usually much shorter. Leaves bipinnate with up to 3 pairs of pinnae, each pinnae with 8–22 pinnules; pinnules asymmetrically narrow-ovate, c. 8 mm long, glabrous. Flowers in small globular heads c. 7 mm diam. on long peduncles. Petals 5. Legume c. 2 cm long. Recorded once from Tuggerah Lakes, also occurs at Denman in the Goulburn River valley. . . . . **N. gracilis** Benth. forma **gracilis**

4 Leucaena Benth.
1 species naturalized Aust.; Qld, NSW, NT

One species in the area
Small tree up to 10 m high. Leaves bipinnate, with up to 6 pairs of pinnae; pinnules up to 20 pairs on each pinna, oblong to lanceolate, up to 15 mm long. Flowers in axillary globular heads, greenish to cream, 5-merous; heads up to 20mm diam. Stamens 10. Occasional in Hunter River valley. Weed in vacant land and pastures. Introd. from trop. America. ***L. leucocephala** (Lam.) de Wit

5 Paraserianthes I.C.Neilsen
2 species endemic Aust.; Qld, WA (naturalized in other states)

One species in the area
Small tree with downy branches. Leaves bipinnate, up to 20 cm long, with 6–12 pairs of pinnae; pinnules 20–40 per pinna, linear-oblong, 6–8mm long, with the principal vein near the edge, glabrous above, silky hairy below. Flowers in dense cylindrical axillary spikes 4–8 cm long, 1–3 spikes per axil. Filaments greenish yellow, c. 1 cm long, connate. Legume c. 10 cm long, 8mm wide. Seeds 8–11, placed transversely; funicle red. Slender shrub. Naturalized on the coast. Introd. from W. Aust. *Crested Wattle* . . . .
. . . . ***P. lophantha** (Willd.) I.C.Nielsen ssp. **lophantha**

6 Pararchidendron I.C.Neilsen
1 species native Aust.; Qld, NSW

One species in the area
Tree with young branches slightly rusty-pubescent. Leaves bipinnate, usually with 1–2 pairs of pinnae; pinnules 3–4 pairs per pinna, irregularly alternate, 2–6 cm long, mostly broad-oblong or rhomboidal and acuminate. Flowers shortly pedicellate, in globular umbels. Corolla greenish, c. 4 mm long. Filaments c.

1 cm long, connate. Legume twisted in a circle, smooth and reddish inside. Scattered along the coast and Hunter River Valley. RF. *Snow Wood* or *Stink Wood*.   **P. pruinosum** (Benth.) I.C.Nielsen var. **pruinosum**

3 CAESALPINIOIDEAE

Trees or shrubs, rarely herbs. Leaves bipinnate or pinnate, rarely simple, alternate; stipules usually present, caducous. Flowers usually bisexual, regular, in racemes or spikes. Sepals 5, usually free. Petals 5, free; the posterior one imbricate inside the two adjacent ones. Stamens usually 10, some often reduced to staminodes, often dehiscing by terminal pores, ± perigynous. Ovary superior, 1-locular; placenta marginal; ovules several; style simple. Fruit a legume, often with transverse partitions. 150 gen., trop. to warm temp.

Key to the genera

1 Prickles or thorns absent 2
1 Prickles or thorns present 3

2 Petals yellow . . . . **1 SENNA**
2 Petals reddish . . . . **3 CERATONIA**

3 Scrambling shrub covered with recurved prickles . . . . **2 CAESALPINIA**
3 Erect trees with straight branched thorns . . . . **4 GLEDITSIA**

1 **Senna** Mill.

46 species Aust. (33 endemic); all states and territories except Tas.

Shrubs. Leaves pinnate; stipules small, caducous except *S. didymobotrya*. Sepals distinctly imbricate. Petals spreading, yellow; lower outer petal much larger than others. Perfect stamens 4–10. Ovary incurved.

Key to the species

1 Stamens all perfect 2
1 Perfect stamens 4–7; staminodes 3–6. Leaflets flat 5

2 Leaflets 1–2 pairs per leaf, narrow-oblong to linear, up to 5 mm wide, with a small gland between the lowest pair. Spreading shrub up to 2 m high. Glen Davis. DSF. Fl. spring . . . .
. . . . **S. artemisioides** (Gaudich. ex DC.) Randell ssp. **zygopyhylla**
2 Leaflets more than 2 pairs per leaf, with a gland between all leaflet pairs 3

3 Leaflets 5–7 pairs per leaf, flat, elliptic, with 1–4 glands, up to 5 cm long. Shrub up to 3 m high. Coast. RF. Endangered. Fl. spring–summer . . . . **S. acclinis** (F.Muell.) Randell
3 Leaflets with recurved or revolute margins 4

4 Leaflets oblong-lanceolate to linear, 1–2.5 cm long, with recurved margins, paler underneath, with a narrow stipitate gland between each pair. Legume flat. Shrub up to 3 m high. Widespread. Uncommon. WSF; margins of RF. Fl. spring–summer . . . . **S. odorata** (R.Morris) Randell
4 Leaflets linear, 1.5–3 cm long, with revolute margins, often pungent pointed, with a narrow stipitate gland between each pair. Legume compressed. Spreading shrub c. 1.5 m high. Blue Mts Uncommon. Open forests. Fl. spring–summer . . . . **S. aciphylla** (Benth.) Randell

5 Glands on the leaf rhachis absent or reduced to a number of minute bristles. Perfect stamens 7. Bracts brown, broad-ovate, up to 2 cm long, covering the buds, deciduous. Leaves up to 25 cm long, pubescent; leaflets 10–16 pairs per leaf, elliptic, up to 5 cm long and 15mm wide. Legume flat, 15mm wide. Stout, divaricate shrub up to 2 m high. Garden escape near habitation. Introd. from Ethiopia. Fl. most of the year . . . . **S. didymobotrya** (Fresen.) H.S.Irwin & Barneby
5 Glands present on the leaf rhachis or petiole. Bracts narrow 6

6 Glands at base of petiole only 7

6 Glands between at least the basal pair of leaflets **8**

7 Gland conical. Leaves 10 cm long; leaflets 4–6 pairs per leaf, 2–4.5 cm long, up to 1 cm wide. Perfect stamens 6. Legume cylindrical, curved, c. 4 cm long. Erect, herbaceous perennial sometimes woody towards the base, up to 1 m high. Coast. Uncommon. WSF; margins of RF. Fl. summer . . . .
. . . . **S. clavigera** (Domin) Randell

7 Gland large, flattened, dark. Leaves up to 20 cm long; leaflets narrow-ovate, up to 5 cm long, 5–7 mm wide, 4–10 pairs per leaf. Perfect stamens 7. Legume thick, c. 5 mm wide. Diffuse shrub c. 1 m high. Widespread. Various situations on heavier soils. Fl. spring–summer . . . **S. barclayana** (Sweet) Randell

8 Plants densely pubescent with brownish yellow hairs. Leaves 5–10 cm long; leaflets 5–8 pairs per leaf, oblong-elliptic, 2–3 cm long, up to 1 cm wide. Glands c. 1 mm high, between most leaflet pairs. Perfect stamens 7. Legume ± compressed, 8–15 mm long. Shrub 2– 3 m high. Blue Mts Garden escape. Introd. from S. America. Fl. spring–summer . . . . ***S. multiglandulosa** (Jacq.) H.S.Irwin & Barneby

8 Plants glabrous or nearly so **9**

9 Narrow gland between each pair of leaflets. Leaflets acuminate, broad-lanceolate, 5–10 cm long. Leaves up to 20 cm long. Stout shrub up to 2 m high. Coast and adjacent plateaus and lower Blue Mts Margins of RF; near habitation. Introd. from S. America. Fl. spring–summer . . . .
. . . . ***S. septemtrionalis** (Viv.) H.S.Irwin & Barneby

9 Gland between the lowest pair of leaflets. Leaflets 4–6 pairs per leaf, obovate to elliptic, acute to obtuse, up to 10 cm long. Divaricate shrub up to 3 m high. Naturalized along the coast north of Sydney. Introd. from S. America. Fl. most of the year . . . . . . ***S. pendula** var. **glabrata** (Vogel) H.S.Irwin & Barneby

2 Caesalpinia L.
12 species Aust. (3 endemic, 2 naturalized); Qld, NSW, NT, WA

Shrubs, deciduous. Leaves bipinnate, alternate, stipules present. Flowers in terminal or axillary racemes, yellow. Sepal and petals 5. Stamens 10, in 2 whorls. Stigma lobed.

Key to the species

1 Woody, scrambling shrub; branches and racemes ± tomentose or pubescent, with scattered recurved prickles. Pinnae 6–10 pairs per leaf; leaflets 8–12 pairs per pinna, oblong, rarely more than 12mm long. Flowers numerous in axillary and terminal racemes 12–15 cm long. Corolla yellow. Filaments less than 2 cm long. Dundas; Wollongong. Introd. from Queensland and Indonesia. *Thorny Poinciana* . . . . . .
. . . . ***C. decapetala** (Roth.) Alston

1 Straggling shrub; branches glandular pubescent, without prickles. Pinnae 7–14 pairs per leaf; leaflets 7–11 pairs per pinna, elliptic to oblong, 4–10 mm long. Flowers in terminal racemes up to 15 cm long. Corolla yellow. Fillaments more than 5 cm long. Occasionally naturalized. *Bird-of-paradise Flower* . . .
. . . . ***C. gillesii** (Hook.) O.Dietr.

3 Ceratonia L.
1 naturalized species Aust.; Qld. NSW

One species in the area
Spreading tree c. 10 m high. Leaves pinnate-compound, with c. 6 leaflets; leaflets elliptic, 4–7 mm long, 3–4 mm wide, entire, glabrous, emarginate. Flowers unisexual, in lateral racemes c. 9 cm long. Sepals reddish, c. 2 mm long. Petals absent. Stamens 5. Legume 10–25 cm long, thick, compressed. Naturalized near Joorilands on the Wollondilly River. Hillsides. Introd. from the Mediterranean. Fl. summer–autumn. *Carob.* . . . .***C. siliqua** L.

4 Gleditsia L.

1 naturalized species Aust.; Qld, NSW, WA

One species in the area

Tall, spreading tree c. 10 m high, with branched thorns. Leaves bipinnate; pinnules c. 20 per leaflet, elliptic-ovate, 15–20 mm long, 6–8 mm wide, obscurely dentate, ciliate. Flowers in long narrow axillary racemes. Legumes 30–40 cm long, compressed, falcate, ± twisted. Naturalized along creeks in Picton area and sometimes near old habitations. Introd. from N. America. Fl. spring. *Honey Locust* . . . . ***G. triacanthos** L.

75 POLYGALACEAE

Shrubs herbs or twiners. Leaves alternate, entire, sometimes reduced to minute scales; stipules absent. Flowers irregular, bisexual. Sepals 5, free, much imbricate; the two inner wings usually larger and petal-like. Lower 3 petals connate forming a concave keel which encloses the stamens and ovary; two upper petals free. Stamens 8; filaments connate to above the middle in a sheath open on the upper side, epipetalous; anthers opening by apical pores. Ovary superior, 2-locular; ovules pendulous. Fruit a compressed capsule. 18 gen., temp. to trop.

Key to the genera

1 Sepals all yellowish-green, none petaloid. Leaves pungent pointed . . . . **3 MURALTIA**
1 Inner sepals (wings) large and petaloid. Leaves never pungent pointed 2

2 Keel with a filamentous crest . . . . **1 POLYGALA**
2 Keel not crested . . . . **2 COMESPERMA**

1 Polygala L.

17 species in Aust. (12 endemic, 5 naturalized); Qld, NSW, Vic., Tas., S.A., WA

Shrubs. Flowers in terminal or lateral racemes. 2 inner sepals large and petaloid. Anterior petal with a crest on the back, adnate to staminal sheath at the base; 2 lateral sepals much smaller. Ovary 2-locular; ovules 1 per loculus.

Key to the species

1 Herb to 50 cm high. Flowers less than 8 mm long 2
1 Shrubs usually more than 1 m high. Flowers more than 8 mm long 3

2 Flowers c. 2 mm long, in many flowered racemes. Leaves linear to narrow elliptic 2–20 mm long. Outer sepal c. 1 mm long. Petals pink to purple. Occasionally naturalized. Warragamba area. Introd. from S. America. Fl. spring-summer . . . . ***P. paniculata** L.
2 Flowers 3–6 mm long few to several in short lateral racemes. Herb up to 15 cm high, often woody at the base. Leaves elliptic to ovate, 5–15 mm long, conspicuously reticulate. Outer sepals broad-elliptic, c. 5 mm long. Petals purplish. Coast and Cumberland Plain. Grassland. Uncommon. Fl. spring–summer . . . . **P. japonica** Houtt.

3 Leaves linear or very narrow-elliptic, 2–6 cm long. Flowers in long terminal leafless racemes usually more than 10 cm long. Outer sepals obovate to broad-elliptic, c. 10 mm long. Petals pale purplish. Erect shrubs with slender twiggy branches. Coast and adjacent plateaus, e.g. Kurnell, Pymble. Waste places, especially sand dunes. Fl. spring-summer. Introd. from S. Africa . . . . ***P. virgata** Thunb.
3 Leaves elliptic to obovate or narrow-obovate, 2–3 cm long, not conspicuously reticulate. Flowers in short terminal racemes. Outer sepals ovate, 12–15 mm long. Petals deep pink. Shrub 1–2 m high. Coast. Waste land. Introd. from S. Africa. Fl. spring-summer . . . . ***P. myrtifolia** L.

2 **Comesperma** Labill.

24 species endemic Aust.; all states and territories

Slender shrubs or twiners. Leaves sometimes reduced to scales. Flowers in racemes. 2 inner sepals large and petaloid. Petals all ± the same length; keel not crested; 2 lateral petals separately attached to staminal tube. Capsule much narrowed at the base, rarely nearly orbicular.

Key to the species

1 Stems twining. Leaves few, distant, linear to elliptic, 5–30 mm long. Flowers blue. Racemes terminal or axillary, 3–5 cm long. Widespread. Heath and DSF. Ss. Fl. spring . . . . **C. volubile** Labill.
1 Stems erect 2

2 Petals purplish pink or rarely white. Leaves conspicuously developed, 5–12 mm long. Slender erect shrubs 1–1.5 m high 3
2 Petals blue. Leaves reduced to scales on upper part of stem or absent except towards the base. 4

3 Leaves with revolute or recurved margins, linear to elliptic. Flowers in terminal elongated racemes mostly several together forming a multiple raceme. Outer sepals 1–1.5 mm long. Capsule truncate or rounded at the top. Widespread. Heath and DSF. Ss. Fl. spring–summer. . . . .**C. ericinum** DC.
3 Leaves flat, obtuse, thick, oblong to elliptic. Racemes short, dense. Outer sepals c. 2 mm long. Capsule emarginate. Blue Mts Uncommon on coast. Fl. spring–summer . . . .**C. retusum** Labill.

4 Capsule narrowed into a basal stipe. Outer sepals almost as long as the petaloid ones. Filaments free above the middle. Plant with woody base and erect almost leafless stems up to 60 cm high. Widespread. Wet sandy soil. Fl. summer . . . . **C. defoliatum** F.Muell.
4 Capsule nearly orbicular, not narrowed into a basal stipe. Outer sepals half to two-thirds as long as petaloid ones. Filaments free above the middle. Plant usually with a woody stock and lax straggling almost leafless stems up to 20 cm long. Widespread. Heath and DSF. Ss. Fl. summer . . . .
. . . . **C. sphaerocarpum** Steetz

3 **Muraltia** DC.

1 species naturalized Aust.; NSW, Vic, S.A.

One species in the area

Erect or divaricate shrub 50–100 cm high, with rigid pubescent branches. Leaves clustered, linear-subulate, pungent pointed, 5–8 mm long. Flowers axillary, solitary or 2 together, c. as long as the leaves. Petals purplish. Coast near Sydney. Introd. from S. America. Fl. spring . . . .**M. heisteria** (L.) DC.

76 CASUARINACEAE

Dioecious or monoecious shrubs or trees with wiry articulate branchlets. Leaves reduced to ridges on the stems projecting at each node into whorls of teeth which are connate below. Flowers unisexual. Male flowers in whorls of cylindrical spikes, each flower with 2 bracteoles placed laterally; perianth segments 1–2, oblong, deciduous, usually hooded over the single stamen. Female flowers in globular or ovoid spikes, terminating very short lateral branches, each flower with 2 lateral bracteoles; perianth absent; ovary 1-locular, superior. Fruit a winged nut held between the woody bracteoles until shed. The whole fruiting structure known as a cone. 4 gen. Aust., Pacific and Malaysia.

Two genera are recognized in the area. However, the features that easily distinguish them are found only on the female plants of this largely dioecious family. The species are distinguished here within the one key mainly on vegetative characters which are found in both sexes.

N.B. To count the number of teeth in each whorl most easily, break branchlet at a node leaving the teeth as a crown on the broken stump.

Key to the genera

1 Teeth on the branchlets in whorls of 8–20. Bracteoles of cone thinly woody and without a dorsal protuberance; extending well beyond the cone body giving the cone a "spiky" appearance even before opening. Mature nut grey or yellowish-brown, dull, matt. . . . . **1 CASUARINA**

1 Teeth on the branchlets in whorls of 4–14. Bracteoles of cone thickly woody, convex and with a dorsal protuberance; not extending well beyond the cone body. Mature nut reddish-brown to black, glossy . . . . **2 ALLOCASUARINA**

1 Casuarina L.
6 species in Aust.; all states and territories except Tas.

2 Allocasuarina L.A.S.Johnson
58 species endemic Aust.; all states and territories

Key to the species of Casuarina and Allocasuarina

1 Surface of branchlet ridges uneven. Trees. 2

1 Surface of branchlet ridges smooth, may be pubescent with tiny hairs in *A. distyla* and *A. paludosa*. Trees or shrubs 3

2 Teeth on branchlets 10–14 in each whorl, erect and slightly overlapping. Branchlets erect. Cone 8–14 mm diam. broader than long. Nut 5 mm long, red-brown. Tree to 15 m high. Dioecious. western Blue Mts Woodland. *Bulloak* . . . . **A. luehmannii** (R.T.Baker) L.A.S.Johnson

2 Teeth on branchlets 9–13, spreading, not overlapping. Branchlets drooping. Cones 17–30 mm diam. longer than broad. Nut 7–12 mm long, dark brown. Tree to 10 m high. Dioecious. Coastal shale. *Drooping Sheoak* . . . . **A. verticillata** (Lam.) L.A.S.Johnson

3 Teeth on branchlets 4–5 in each whorl (rarely 6 in *A. nana*) 4

3 Teeth on branchlets 6 or more in each whorl 6

4 Furrows of branchlets glabrous. . . . . **A. glareicola** (see below)

4 Furrows of branchlets pubescent 5

5 Teeth on branchlets 4 (rarely 5). Tree with slender drooping angular branchlets C. 0.5 mm diam. Bark corky. Cones on peduncles >8 mm long, nearly globular, flat-topped, 18–25 mm diam. Nut 7–10 mm long. Male spikes slender, 2–4 cm long. Dioecious. Widespread. Forest, often on hillsides. *Forest Oak*. . . . . **A. torulosa** (Aiton) L.A.S.Johnson

5 Teeth on branchlets 5 (rarely 6) in each whorl. Shrub c. 1 m high with erect wiry branchlets 0.5–0.8 mm diam. Bark smooth to finely fissured. Cones ± sessile, cylindrical 12–15 mm diam.; nut 4–6 mm long. Male spikes 5–15 mm long, Usually dioecious. Higher parts of Blue Mts; Woronora Plateau. Heath. Ss . . . . **A. nana** (Sieber ex Spreng.) L.A.S.Johnson

6 Ridges on branchlets with a groove down the middle; teeth 6–8. Furrows of branchlets pubescent. Cones usually up to 2 cm long and 12 mm diam. Nut 3.5–5 mm long. Male spikes usually less than 25 mm long. Dioecious or monoecious shrub up to 2 m high but usually much less. Bark smooth. Coast and adjacent plateaus. Heath and DSF margins near swamps . . . . **A. paludosa** (Sieber ex Spreng.) L.A.S.Johnson

6 Ridges on branchlets without a groove down the middle 7

7 Furrows of branchlets pubescent 8

7 Furrows of branchlets glabrous. 10

8 Teeth on branchlets not overlapping at their base. Cone valves prominent, angular. Cones truncate, 1–3 cm long, 1–2 cm diam. Nut 4–10 mm long. Male spikes slender and weak. Small, often monoecious tree, with mostly erect branchlets Bark fissured. Widespread. Heath and open forests . . . . **A. littoralis** (Salisb.) L.A.S.Johnson

8 Teeth on branchlets overlapping at their base, at least on the young branchlets. Bark smooth but may be fissured at the base in very large trees 9

9 Teeth on branchlets flat; ridges flat to slightly rounded Cone 9–12 mm diam. Nut 5–7 mm long. Dioecious shrub or small tree up to 5 m high. Glen Davis. DSF. Ss . . .**A. gymnanthera** L.A.S.Johnson
9 Teeth on branchlets concave to ridged; ridges strongly convex to angular. Cones usually more than 16 mm diam., mostly 2–4 cm long, not truncate. Nut 4–8 mm long. Male spikes up to 8 cm long, erect and firm. Usually dioecious shrub, often large. Widespread. Common in moister parts of the area. Heath and DSF. Sandy soils . . . . **A. distyla** (Vent.) L.A.S.Johnson

10 Ridges of branchlets flat to slightly rounded 11
10 Ridges of branchlets angular or strongly convex 13

11 Teeth on branchlets 12–16 in each whorl. Cone cylindrical, often pubescent, mostly c. 12 mm diam. Branchlets erect or ascending; 0.9–1.2 mm diam. Tree. Coast; Cumberland Plain; Hunter River Valley. Near salt water estuaries; along sluggish creeks; occasionally on rising ground. *Swamp Oak* (Intermediate forms between *C. cunninghamiana* and *C. glauca* occur). . . . . **Casuarina glauca** Sieber ex Spreng.
11 Teeth on branchlets less than 12 in each whorl 12

12 Cone 9–12 mm diam., 14–40 mm long. . . . . **A. gymnanthera** (see above)
12 Cones up to 8 mm diam. and 14 mm long. Nut 3 mm long. Dioecious or monoecious shrub up to 2 m high. Castlereagh. Open forest on lateritic soil. Endangered . . . . **A. glareicola** L.A.S.Johnson

13 Teeth on branchlets withering, very broad-triangular. Cone 12–15 mm long, 8–10 mm diam. Dioecious shrub. Neilson Park. Heath. Ss. Endangered . . . . **A. portuensis** L.A.S.Johnson
13 Teeth on branchlets not withering or if they do then they are narrow-triangular 14

14 Internodes of branchlets same diam. throughout their length 15
14 Internodes of branchlets wider near their top than elsewhere. Cone 5–20 mm long, 5–12 mm diam. Dioecious or monoecious shrubs up to 2.5 m high 16

15 Teeth on branchlets 8–10 in each whorl. Branchlets 0.4–4.6 mm diam. Cones globular, mostly c. 8 mm diam. Small or large trees. Banks of fresh water streams. *River Oak* . . . . **Casuarina cunninghamiana** Miq. ssp. **cunninghamiana**
15 Teeth on branchlets up to 7 in each whorl . . . . **A. glareicola** (see above)

16 Teeth on branchlets broad to narrow-triangular, with straight or slightly convex margins, 6–7 in each whorl. Capertee. DSF. Ss . . . . **A. diminuta** L.A.S.Johnson ssp. **diminuta**
16 Teeth on branchlets broad-triangular, with convex margins, 6–10 in each whorl. Kingsford; Little Bay; Heathcote; Upper Blue Mts DSF. Ss. . . . . **A. diminuta** ssp **mimica** L.A.S.Johnson

77 ELATINACEAE
2 gen., cosmop.

1 Elatine L.
6 species in Aust.; all states and territories

One species in the area

Small, glabrous herbs, aquatic or creeping on mud. Leaves opposite or whorled, thin, obovate to very narrow-ovate or linear, 3–15 mm long; stipules small, caducous. Flowers minute, bisexual, regular, usually solitary, in one axil only of each pair of leaves. Sepals 3, membranous, obtuse. Petals 3, free. Stamens 3, hypogynous. Ovary superior, 3-locular; placentas axile. Fruit a capsule, membranous, depressed-globular, c. 1mm diam. Seeds cylindrical, slightly curved, marked with longitudinal and transverse lines. Widespread. Ponds and ditches. . . . . **E. gratioloides** A.Cunn.

78 EUPHORBIACEAE

Monoecious or dioecious trees, shrubs or herbs. Leaves simple, alternate or opposite, rarely very reduced; stipules present or absent. Inflorescence various. Flowers unisexual, regular. Sepals and petals 4–6, or petals and/or sepals absent. Stamens 1 to many. Ovary superior, 3- or rarely 1–2-locular; 1 ovule per loculus, pendulous; placentas axile; styles as many as loculi, free or connate towards the base, often branched. Fruit a schizocarpic capsule or succulent in *Claoxylon*, splitting septicidally into 2-valved mericarps and leaving a persistent central axis; mericarps usually dehiscing ventrally. 218 gen., cosmop.

Key to the genera

1 White latex present. 2
1 Latex absent or red latex present 5

2 Styles 2. . . . . **17 HOMALANTHUS**
2 Styles 3. 3

3 Leaves deeply palmately lobed . . . . **.19 MANIHOT**
3 Leaves simple 4

4 Stipules minute or absent . . . . **1 EUPHORBIA**
4 Stipules present, distinct, interpetiolar . . . . **.2 CHAMAESYCE**

5 Leaves with superficial red glands on lower surface . . . . **13 MALLOTUS**
5 Leaves without superficial red glands 6

6 Leaves with 2–5 glands at base of lamina on petiole or on margin just above the base 7
6 Leaves without glands at base of lamina. 11

7 Leaf veins palmate 8
7 Leaf veins pinnate 9

8 Hairs simple. Ovary 3–7-locular . . . . **8 VERNICIA**
8 Hairs stellate. Ovary 1–2-locular. . . . . **18 ALEURITES**

9 Glands on margin of lamina just above the base. Red latex present . . . . **15 BALOGHIA**
9 Glands on the petiole at base of lamina. Red latex absent. 10

10 3–5 glands present at base of lamina . . . . **.9 CLAOXYLON**
10 2 glands present at base of lamina . . . . **14 CROTON**

11 Leaves peltate . . . . **10 RICINUS**
11 Leaves not peltate 12

12 Petals present 13
12 Petals absent. 15

13 Stamens up to 6 . . . . **.6 MONOTAXIS**
13 Stamens more than 6 14

14 Anthers free . . . . **3 BEYERIA**
14 Anthers connate into a column . . . . **4 RICINOCARPOS**

15 Leaves crenate 16
15 Leaves entire. 18

16 Inflorescence terminal . . . . **16 ADRIANA**
16 Inflorescences axillary 17

17 Teeth on leaf margins with spines . . . . **.12 ALCHORNEA**
17 Leaves crenate or teeth blunt . . . . **11 ACALYPHA**

18 Leaves reduced to scales. Styles entire or slightly lobed .7 **AMPEREA**

18 Leaves not scale like. Styles branched . . . . 5 **BERTYA**

1 Euphorbia L.
C. 35 species in Aust.; all states and territories

Erect annual or biennial herbs containing milky latex. Leaves alternate on lower stems, opposite or spirally arranged above; bases symmetric; stipules minute or absent. Flowers unisexual, grouped in a terminal inflorescence (cyathium) (Fig. 26) consisting of 1 naked female flower surrounded by 8–15 male flowers or the female flower absent, the whole surrounded by an involucre of 5 connate bracts. Cyathia usually arranged in cymes. Lobes of the involucre short, alternating with 4–5 spreading glands which often secrete nectar and are entire or horned. Male flowers each with 1 stamen; filament articulate; female flower with a stipitate 3-locular ovary. Ovary and fruit hanging out of the involucre over the space where the 5th gland is usually absent. Seed usually have a caruncle.

Key to the species

1 Floral leaves red or white towards base. Cyathium with a solitary gland. Annual up to 70 cm high. Weed near the coast. Fl. summer. *Painted Spurge* . . . . *E.* **cyathophora** Murray

1 Floral leaves green or yellowish. 2

2 Most leaves usually more than 4 cm long 3

2 Leaves usually less than 4 cm long, obovate to obovate-cuneate, never decussate 4

3 Stem leaf margins entire. Leaves up to 16 cm long, oblong-lanceolate, sessile, decussate. Umbels with 2–5 rays; floral leaves large, ovate-lanceolate. Involucral glands with short blunt horns. Stout, glaucous biennial. Widespread. Weed. Introd. from Europe and Asia. Fl. summer. *Caper Spurge* . *E.* **lathyrus** L.

3 Stem leaf margins finely toothed. Leaves up to 60 mm long; linear to oblong-elliptic; alternate. Floral leaves ovate 8–20 mm long. Perennial with 2 or more stems up to 80 cm high. Occasionally naturalized. Introd. from Africa. . . . . **E. depauperata** A.Rich. var. **pubescens** Pax

4 Cauline leaves oblong to elliptic, sessile, fleshy, different in shape and colour from leaves on fertile stems. Leaves crowded 5–20 mm long. Fertile stem leaves circular to rhombic. Cyathia solitary. Involucral glands 4 each with 2 horn-like projections. Erect, glabrous, perennial herb to 70 cm high. Occasionally naturalized. Introd. from Europe. *Sea spurge* . . . .**E. paralias** L.

4 Cauline leaves petiolate, not markedly different from leaves on fertile stems. 5

5 Involucral glands with horns at each end, crescent shaped. Umbels of 2–3 repeatedly forked rays. Stems glabrous. Leaves obovate, 7–15 mm long. Herb up to 60 cm high. Widespread. Common weed in gardens and waste ground. Introd. from Europe and Asia. Fl. most of the year. *Petty Spurge* . . . . *E.* **peplus** L.

5 Involucral glands entire, slightly curved. Umbel of 5 long rays which are at first 3-forked and then 2-forked. Stems hairy. Leaves obovate-cuneate, 1–4 cm long. Erect annual up to 60 cm high. Coast and adjacent plateaus; Cumberland Plain. Waste ground. Introd. from Europe and Asia. Fl. summer. *Sun Spurge* . . . . *E.* **helioscopia** L.

2 Chamaesyce Gray
40 species in Aust.; all states and territories except Tas.

Prostrate or decumbent annual or perennial herbs containing milky latex. Leaves opposite; bases asymmetric, stipules present, distinct, interpetiolar. Flowers unisexual, grouped in a terminal inflorescence (cyathium) consisting of 1 naked female flower surrounded by 8–15 male flowers or the female flower absent, the whole surrounded by an involucre of 5 connate bracts. Cyathia usually arranged in cymes. Lobes of the involucre short, alternating with 4–5 spreading glands which often secrete nectar and usually have a petaloid appendage. Male flowers each with 1 stamen; filament articulate; female flower with a stipitate 3-locular ovary. Ovary and fruit hanging out of the involucre over the space where the 5th gland is usually absent. Seed without a caruncle.

Key to the species

1 Leaves 4–8 mm long, elliptic, ovate or ovate-oblong, opposite, in 2 ranks, often bluish green. Plants mostly less than 30 cm across. Cyathia solitary, axillary. Involucral glands red with a pink or white appendage. Fruit 2 mm long 2

1 Leaves 1–4 cm long, ovate to oblong 5

2 Stems glabrous. Fruit glabrous. Weeds in gardens and waste places 3

2 Stems pubescent. Fruits variously pubescent 4

3 Capsule with a fringed appendage at the base. Seeds smooth. Widespread. Open communities; pastures. Fl. spring–summer . . . . **C. dallachyana** (Baill.) Hassall

3 Capsule without a fringed appendage at the base. Seeds wrinkled. Widespread. Open communities. Fl. spring–summer. *Caustic Weed* . . . . **C. drummondii** (Boiss.) Hassall

4 Stem with a single longitudinal band of hairs. Leaves mostly c. 3 mm wide. Fruit fringed with hairs on the angle. Seeds with many prominent transverse wrinkles. Coast and adjacent plateaus. Weed in gardens. Introd. from trop. America. Fl. spring–summer. *Red Caustic Weed* . . . .
. . . . *C. prostrata** (Aiton) Small

4 Stem with 2 longitudinal bands of hairs. Leaves mostly c. 1 mm wide. Fruit pubescent. Seeds with a number of undulations. Coast and adjacent plateaus. Weed in gardens. Introd. from Trop. America. Fl. summer . . . . **C. maculata** L.

5 Plants densely hairy. Cyathia in head-like clusters, with red to whitish glands. Fl. spring–summer. Coast and Cumberland Plain. Weed. Introd. from the Tropics. Fl. summer . . . . **C. hirta** (L.) Millsp.

5 Plants glabrous or glabrescent or with 1–2 bands of hairs on stem. Leaves up to 30 mm long, dentate . 6

6 Glabrous, prostrate perennial, stems sometimes buried in sand and appearing erect. Leaves ovate. Plants forming a mat up to 1 m diam. Sand dunes near the sea. Fl. spring–summer . . . .
. . . . **C. psammogeton** (P.S.Green) P.I.Forst. & R.J.F.Hend.

6 Decumbent to ascending annual up to 60 cm high usually with 1–2 bands of hairs on stems. Leaves elliptic to oblong. Widespread. Weed. Introd. from America. Fl. summer . . **C. hyssopifolia** (L.) Small

3 Beyeria Miq.
15 species endemic Aust.; all states and territories

Shrubs up to 1.5 m high. Leaves entire, spirally arranged, often viscid, flat or slightly recurved, pale beneath; stipules absent. Flowers yellowish, axillary; the males often in clusters on recurved peduncles; females solitary. Calyx slightly enlarged in fruit. Stamens numerous; filaments short, connate. Petals usually 5, alternating with glands. Ovary 3-locular, with a single ovule per loculus; stigma broad, sessile.

Key to the species

1 Capsule and ovary glabrous or nearly so. Leaves mostly up to 5 cm long, 5–15 mm wide. Flowers c. 7 mm diam. Cumberland Plain; gullies in lower Blue Mts; Wolgan and Capertee Valleys. DSF. Fl. summer **B. viscosa** (Labill.) Miq.

1 Capsule and ovary hyaline-hirsute. Leaves up to 8 cm long, 1–1.8 mm wide. Flowers c. 8 mm diam. Jenolan Caves. WSF. Fl. summer . . . . **B. lasiocarpa** (F.Muell.) Müll.Arg.

4 Ricinocarpos Desf.
15 species endemic Aust.; Qld, NSW, Vic., Tas., WA

Erect to spreading, monoecious or dioecious shrubs. Leaves opposite or alternate; margins revolute or recurved; stipules absent. Flowers in terminal clusters, pedicellate; males and females often in same clusters. Calyx 4-6-lobed. Petals 4-6, white. Male flowers with numerous stamens connate into a column. Female flowers with 3-locular ovary. Fruit a capsule.

Key to the species

1 Sepals pubescent with stellate hairs. Leaves linear to oblong, 5–20 mm long, revolute, scabrous. Petals 10–15 mm long. Fruit globular, villous. Hornsby Plateau north of Hawkesbury River. DSF. Fl. winter–spring . . . . **R. bowmanii** F.Muell.

1 Sepals glabrous but for the ciliate margins. Leaves linear, 2–4 cm long, revolute, smooth. Petals 10–15 mm long. Fruit globular, densely muricate, c. 12 mm diam. Widespread. Heath and DSF. Sandy soils. Fl. winter–spring. *Wedding Bush* . . . . **R. pinifolius** Desf.

5 Bertya Planch.
25 species endemic Aust.; Qld, NSW, Vic.

Monoecious shrubs. Leaves spirally arranged; stipules absent. Flowers solitary or few together, axillary, with 3–8 small bracts on the slender pedicels. Perianth segments 5, petaloid in male flowers, smaller and narrower in females. Male flowers with filaments connate. Female flowers with 3-locular ovary; ovule solitary in each loculus; styles 3, 2–4-fid. Fruit usually with 1 seed.

Key to the species

1 Perianth of female flowers minutely stellate tomentose 2
1 Perianth of female flowers glabrous. Ovary glabrous or villous 3

2 Ovary densely tomentose but soon becoming glabrous. Capsule glabrous. Leaves with a loose stellate tomentum, 1–5 cm long, 5–15 mm wide; margins flat or slightly recurved. Flowers pedunculate; bracts 4 or 5. Erect shrub up to 3 m high. Ku-ring-gai Chase near Mt. Colah. WSF and RF. Fl. spring–summer . . . . **B. brownii** S.Moore

2 Ovary and capsule densely tomentose. Leaves 1–6 cm long, 3–7 mm wide; lower surface white-tomentose: margins recurved to revolute. Flowers sessile; bracts 8 to 10. Erect shrub up to 2 m high. Blue Mts DSF on Ss. Fl. spring–summer. . . . . **B. oleifolia** Planch.

3 Ovary and capsule densely hairy. Branches scurfy-tomentose. Ovary villous. Bracts 6, thick and almost equal. Leaves linear, up to 2.5 cm long, 0.5–1.5 mm wide, with recurved margins, tomentose underneath. Shrub up to 80 cm high. Upper Cox's River. Open forest. Fl. spring–summer . . . .
. . . . **B. rosmarinifolia** (Cunn.) Planch.

3 Ovary sparsely hairy to glabrous. Capsule glabrous 4

4 Perianth of female flower enlarging and enclosing capsule. Bracts 5–8. Leaves c. 2 mm wide; margins revolute to midrib; upper surface scabrous. Branches tomentose. Flowers on peduncles up to 15 mm long. Perianth segments 2–3 mm long. Fruit acute, 10 mm long. Shrub up to 1 m high. Woronora Plateau; Glenbrook. DSF. Fl. spring–summer. . . . . **B. pomaderroides** F.Muell.

4 Perianth of female flower not enlarging. Bracts usually 4. Leaves 3–12 mm wide; margins flat to recurved; upper surface glabrous. Branches scabrous to glabrous. Flowers ± sessile. Perianth segments 4 mm long. Fruit up to 12 mm long. Shrub up to 2 m high. Woodland on slopes. Northern Blue Mts Fl. spring . . . . **B. gummifera** Planch.

6 Monotaxis Brongn.
10 species endemic Aust.; Qld, NSW, WA

One species in the area
Small, glabrous, monoecious shrub 15–30 cm high, with a thick woody base and numerous thin wiry ascending or diffuse green stems. Leaves opposite or whorled, distant, not numerous, variable, lanceolate to linear or cuneate, 8–18 mm long; stipules very small. Flowers minute, white, in head-like cymes each with c. 12 male flowers and 1 female. Stamens usually twice as many as petals. Ovary 3-locular. Fruit globular. Widespread. Wet sandy soils. Fl. spring–summer. . . . . **M. linifolia** Benth.

7 Amperea A.Juss.

6 species endemic Aust.; Qld, NSW, Vic., Tas., WA

Erect, monoecious shrub 30–60 cm high with angular branches. Leaves cuneate to oblong; upper leaves scale-like spirally arranged; lower leaves alternate, 5–25 mm long. Stipules fringed. Flowers small, in axillary clusters surrounded by small brown bracts. Stamens up to 10. Ovary 3-locular. Capsule ovoid.

Key to the species

1 Most female flowers with a pedicel 1–13 mm long. Lower leaves 5–10 mm long; upper leaves less than 10 mm long. East Sydney (1892). Extinct . . . . **A. xiphoclada** var. **pedicellata** R.J.F.Hend.
1 All flowers with pedicels less than 1 mm long or sessile. 2

2 Leaves up to 25 mm long, both surfaces smooth. Widespread. Heath. Sandy soils. . . . . **A. xiphoclada** (Spreng.) Druce var. **xiphoclada**
2 Leaves up to 7 mm long, upper surface and lower midrib papillose. . . . . **A. xiphoclada** var. **papillata** R.J.F.Hend.

8 Vernicia Lour.

1 species naturalized Aust.; Qld, NSW

One species in the area

Tree 3–9 m high. Leaves spirally arranged, ovate to broad-ovate, 15–20 cm long, c. 18 cm wide, acuminate, sometimes cordate. Flowers monoecious in terminal panicles. Male flowers: petals pink; disc prominently lobed; stamens indefinite in number. Female flowers similar but gland less lobed; styles deeply 2-fid. Fruit ± fleshy, c. 2 cm diam. Naturalized at Central Colo. Introd. from Asia. *Tung Oil Tree* . . . . ***V. fordii** (Hemsley) Airy Shaw

9 Claoxylon A.Juss.

4 species in Aust.; Qld, NSW

One species in the area

Monoecious, glabrous or pubescent, small tree or tall shrub; small branches pale and brittle. Leaves spirally arranged, obovate elliptic or oblong, crenate, very thin, 9–18 cm long; petioles 2–4 cm long with 3–5 glands at the top; stipules minute. Flowers in axillary racemes shorter than the leaves. Perianth segments 3, 2 mm long. Stamens numerous on a central receptacle or disc. Ovary 3-locular, globular, pubescent; styles 3, free. Fruit a globular berry, 6 mm diam. Widespread. RF. Fl. spring **C. australe** Baill.

10 Ricinus L.

1 species naturalized Aust.; all states and territories except Tas.

One species in the area

Tall, branching, weak, monoecious shrub with a rank odour. Leaves spirally arranged, 10–60 cm diam., palmately veined, peltate, divided into 5–9 palmate serrate lobes; petioles long; stipules absent. Flowers in loose racemes on thick peduncles, the upper ones female. Male flowers with numerous stamens; filaments connate, branching. Female flowers: ovary 3-locular; styles 3, red, 2-fid. Fruit dry, 15 mm long, ovoid, spiny. Seeds carunculate, mottled. Coast and adjacent plateaus; Cumberland Plain. Waste ground. Introd. from Asia or Africa. Fl. summer. *Castor Oil Plant* . . . . ***R. communis** L.

11 Acalypha L.

8 species in Aust. (5 endemic, 2 naturalized); Qld, NSW

Monoecious or dioecious shrubs or herbs. Hairs simple. Leaves alternate, margins toothed. Flowers in axillary inflorescence, male flowers in spikes; perianth segments 4; stamens 8, free. Female flowers in clusters of 1–4; perianth segments 3–5; styles branched; ovary 2–3 locules with one ovule each. Fruit a capsule within the persistent bract.

Key to the species

1 Villous-pubescent shrub up to 2 m high, rarely prostrate near the sea. Leaves spirally arranged, ovate to narrow-ovate or oblong, 4–6 mm long, serrate-crenate; stipules small. Flowers monoecious. Male flowers in axillary spikes up to 10 cm long, c. 1 mm diam.; Female flowers in separate short spikes or below the males; styles divided into narrow branches. In or near RF. Coast and adjacent plateaus . . . . **A. nemorum** F.Muell. ex Müll.Arg.

1 Herb 10–50 cm tall. Leaves sparsely pubescent, ovate, 2–4 cm long, 5–15 mm wide. Petiole as long as the lamina or longer. Male inflorescences 1–2 cm long; female flowers concealed within bract. Garden escape. Introd. from Asia . . . . ***A. australis** L.

12 Alchornea Sw.
3 species in Aust.; Qld, NSW

One species in the area

Tall, dioecious shrub. Leaves spirally arranged, coriaceous, ovate or rhomboid, 3–9 cm long, sinuate-toothed with prickly points; stipules minute. Perianth usually 4-lobed. Male flowers: perianth lobes valvate; stamens 8, free. Female flowers: ovary 3-locular with 1 ovule per loculus; styles 3, broad, flat. In or near RF. Coast and adjacent plateaus . . . . **A. ilicifolia** (J.Sm.) Müll.Arg.

13 Mallotus Lour.
12 species in Aust.; Qld, NSW, NT, WA

One species in the area

Tree up to 20 m high. Leaves ovate to oblong, mostly up to 12 cm long, entire, glabrous, with 2 glands on the upper surface, grey-hairy with red glands on the lower surface. Flowers in racemes. Perianth segments 3–5. Illawarra ranges, e.g. Mt Keira. RF. *Red Kamala* . . . . **M. philippensis** (Lam.) Müll.Arg.

14 Croton L.
12 species in Aust.; Qld, NSW, NT

Tall shrubs or small trees, usually monoecious. Leaves spirally arranged, petiolate, usually with 2 stipitate glands near the base of the lamina; stipules minute. Flowers small, in clusters along the rhachis of a terminal raceme with the males chiefly or entirely in the upper portion. Sepals usually 5. Petals as many as sepals and c. as long. Male flowers: glands alternating with petals; stamens c. 11, free, inflexed in bud, inserted on an expanded disc. Female flowers: ovary usually 3-locular, with 1 ovule per loculus; styles usually 3, 2–4-fid. Fruit globose.

Key to the species

1 Leaves silvery-white beneath, with a close scaly tomentum, ovate to lanceolate, 5–8 cm long, entire or sinuate. Racemes 6–10 cm long. Fruit 6 mm diam. Tall shrubs. Widespread. Lower Blue Mts, e.g. Kurrajong. In or near RF. Fl. summer. *Queensland Cascarilla* . . . . **C. insularis** Baill.

1 Leaves green on both surfaces, glabrous or nearly so, ovate to elliptic or lanceolate, 4–12 cm long. Racemes mostly 3–5 cm long. Flowers few in each cluster, Petals fringed with woolly hairs. Small tree or tall shrub. Coast and adjacent plateaus, e.g. Bulli; lower Blue Mts, e.g. Kurrajong. In or near RF. Fl. summer. *Green Cascarilla* . . . . **C. verreauxii** Baill.

15 Baloghia Endl.
3 species in Aust.; Qld, NSW

One species in the area

Tall shrub or small tree, glabrous, monoecious, with red latex in the bark. Leaves opposite, very shortly petiolate, oblong to elliptic, 5–15 cm long, obtuse or obtusely acuminate, coriaceous, glossy; stipules absent. Flowers in short terminal racemes, male and female usually on different branches. Calyx 4–5-lobed. Petals as many as calyx lobes. Male flowers: stamens indefinite, shortly connate at the base, inserted upon or inside an irregularly lobed disc. Female flowers: ovary 3-locular, with 1 ovule per loculus; styles 3, almost free, 2-fid. Fruit hard, globular, 12–15 mm diam. Coast and adjacent plateaus. RF. Fl. summer. *Brush Bloodwood* . . . . **B. inophylla** (G.Forst.) P.S.Green

16 Adriana Gaudich.

2 species endemic Aust.; all states and territories except Tas.

One species in the area

Variable, dioecious shrub with stellate tomentum. Leaves spirally arranged, ovate to lanceolate, 10–15 cm long, coarsely and irregularly toothed, sometimes 3-lobed, paler underneath; stipules present as 2 blunt projections. Flowers in terminal spikes. Male spikes 5–25 cm long; female spikes shorter, dense. Male flowers: perianth segments 4–5, valvate, 3 mm long; stamens numerous on a slightly raised central receptacle; filaments short, free or basally connate. Female flowers: perianth segments 6–8, 4–6 mm long; ovary 3-locular with 1 ovule per loculus; styles 3, almost free, 2-fid, densely fringed, red. Fruit 8–12 mm diam. Widespread. Along creeks. Fl. spring. *Bitterbush* . . . . . . **A. tomentosa** Gaudich. var. **tomentosa**

17 Homalanthus A.Juss.

3 species native Aust.; Qld, NSW, NT

Small trees or shrubs, glabrous, monoecious, with milky latex. Leaves spirally arranged, broad-ovate, acuminate, thin, paler and prominently penniveined beneath, often turning red with age. Flowers small in terminal narrow racemes often exceeding the leaves. Perianth irregularly truncate or 2-lobed. Male flowers 2–3 mm diam., numerous, clustered; stamens few, free; disc absent. Female flowers few, pedicellate, at the base of the raceme; ovary 2-locular, with 1 ovule per loculus; styles 2, linear, entire. Fruit glaucous, compressed, 8–10 mm wide.

Key to the species

1 Capsule smooth. Bracts with 2 large glands. Leaves mostly 5–12 cm long. Seed half enveloped in a fleshy aril. Widespread. RF margins or in gullies in DSF. Fl. spring–summer. *Bleeding Hearts* . . . . **H. populifolius** Graham

1 Capsule with 2–6 conical projections. Leaves mostly up to 5 cm long. Seed with a short fleshy aril. Blue Mts; Cumberland Plain. Often on rocky hillsides . . . . **H. stillingiifolius** F.Muell.

18 Aleurites J.R.Forst. & G.Forst.

1 species native Aust.; Qld, NSW

One species in the area

Monoecious tree. Leaves alternate, entire or lobed, 10–20 mm long, 8–15 mm wide; two glands at apex of petiole. Flowers in terminal; panicles; male and female flowers mixed. Petals 8 mm long, white, turning red when dry. Fruit drupaceous, ridged, 5–6 cm diam. with 1 or 2 seeds. Garden escape. Native to N. Qld. *Candle-nut Tree* . . . . ***A. moluccana** Willd.

19 Manihot Mill.

2 species naturalized Aust.; Qld, NSW

One species in the area

Monoecious shrub or tree with milky latex, up to 5 m high. Leaves alternate, glabrous, lower surface ± glaucous, deeply 3–13-lobed. Lobes 5–10 cm long, 10–20 mm wide. Petiole 5–20 cm long. Flowers in racemes, perianth 8–13 mm long, green and purple. Fruit a capsule 1.5–2 cm diam. Occas. naturalized. Introd. from America . . . . ***M. grahamii** Hook.

79 PHYLLANTHACEAE

Monoecious or dioecious trees, shrubs or herbs. Leaves simple, alternate or opposite, rarely very reduced; stipules present or absent. Inflorescence various. Flowers unisexual, regular. Sepals and petals 4–6, or petals and/or sepals absent. Stamens 1 to many. Ovary superior, 3- or rarely 1–2-locular; ovules 2 per loculus, pendulous; placentas axile; styles as many as loculi, free or connate towards the base, often branched. Fruit a schizocarpic capsule or succulent *Breynia*, splitting septicidally into 2-valved mericarps and leaving a persistent central axis; mericarps usually dehiscing ventrally. 83 gen., cosmop.

Key to the genera

1 Petals present . . . . **1 PORANTHERA**
1 Petals absent 2

2 Styles branched . . . . **4 PHYLLANTHUS**
2 Styles entire or slightly lobed 3

3 Leaves in groups of 3 . . . . **2 MICRANTHEUM**
3 Leaves attached to stem singly 4

4 Leaves mostly less than 15 mm long . . . . **3 PSEUDANTHUS**
4 Leaves more than 15 mm long 5

5 Petioles 1–2 cm long . . . . **7 ACTEPHILA**
5 Petioles <1 cm long 6

6 Leaves glaucous beneath usually <30 mm long. Fruit <5 mm diam . . . . **6 BREYNIA**
6 Leaves not glaucous beneath, usually >40 mm long. Fruit >8 mm diam . . . . **5 GLOCHIDION**

1 Poranthera Rudge
8 species endemic Aust.; all states and territories except S.A.

Monoecious herbs or small shrubs. Leaves spirally arranged or sometimes opposite, entire; stipules present. Male and female flowers intermingled in very short dense racemes almost contracted into heads and forming a leafy terminal corymb. Sepals 5. Petals 5, white. Male flower with 5 stamens alternating with petals, and a rudimentary ovary. Female flower without stamens; ovary 3-locular, 6-lobed, with 2 ovules per loculus; styles 3, 2-fid almost to base.

Key to the species

1 Diffuse, glabrous, annual herb 8–15 cm high, with ascending branches. Leaves linear-spathulate to obovate, flat or nearly so, mostly 2–5 mm long; stipules entire. Flowers minute, in leafy corymbs. Widespread. Shady places. Fl. spring–summer . . . . **P. microphylla** Brongn.
1 Small, perennial shrubs 2

2 Stipules jagged. Plants 15–30 cm high, often minutely scabro-pubescent, with spreading branches. Leaves crowded, linear, with revolute margins, 6–12 mm long. Corymbs dense. Widespread. Heath and DSF. Sandy soils. Fl. spring . . . . **P. ericifolia** Rudge
2 Stipules entire. Plants 30–100 cm high, with few branches. Leaves linear to linear-lanceolate, flat or with slightly revolute margins, 1–5 cm long. Corymbs large, with long lateral branches. Widespread. DSF. Sandy soils. Fl. spring . . . . **P. corymbosa** Brongn.

2 Micrantheum Desf.
3 species endemic Aust.; Qld, NSW, Vic., Tas., S.A.

Monoecious, heath-like shrubs with many branches. Leaves linear to broad-lanceolate or oblong, sessile, in groups of 3, the groups alternate. Flowers very small, solitary or few together, axillary. Male flowers with 3–9 free stamens, and rudimentary ovary. Female flower with 3-locular ovary; styles 3, simple.

Key to the species

1 Stamens 3. Low erect or spreading shrub. Leaves usually linear, 4–8 mm long, spreading. Flowers 2 mm diam. Peduncles filiform, 2 mm long. Capsule elliptic-oblong, 6–7 mm long. Widespread. Heath and DSF. Sandy soils. Fl. spring . . . . **M. ericoides** Desf.

1 Stamens 6–9. Erect, often tall shrub. Leaves linear to oblong, 10–15 mm long, less spreading than *M. ericoides*. Perianth segments 4 mm long. Coast and adjacent plateaus. River banks, e.g. Upper George's River to Berrima. Fl. spring . . . .**M. hexandrum** Hook.f.

3 Pseudanthus Sieber ex Spreng.
8 species endemic Aust.; Qld, NSW, Vic., Tas., S.A.

Small, monoecious, heath-like shrubs. Leaves mostly spirally arranged, small, coriaceous, with thick margins; midvein prominent underneath; stipules present. Flowers in the axils of the upper leaves; males often clustered and on short peduncles; females solitary. Perianth segments 6, often larger in male than female flowers. Male flowers with 6 free stamens, and rudimentary ovary. Female flowers with a 3-locular 3-lobed ovary. Fruit 1-locular with 1 seed.

Key to the species

1 Perianth segments of male flower 8–12 mm long, linear, white. Flowers clustered at the ends of the branches. Female flowers few and inconspicuous. Leaves narrow-lanceolate to linear, 8–12 mm long. Erect, glabrous shrub up to 1 m high. Widespread. Heath and open forests. Sandy soils. Fl. spring. . . . . **P. pimeleoides** Sieber ex Spreng.

1 Perianth segments 1–2 mm long. Low, rigid shrubs. Fruit narrow-oblong, c. 4 mm long 2

2 Leaves oblong-linear to linear-spathulate, 4–8 mm long, glabrous. Old dunes near Sydney. Fl. autumn . . . .**P. orientalis** (Baill.) F.Muell.

2 Leaves ovate to orbicular, 2–4 mm long. Blue Mts Heath in rocky places. Fl. spring . . . .**P. divaricatissimus** Benth.

4 Phyllanthus L.
51 species in Aust. (40 endemic, 3 naturalized); all states and territories

Usually monoecious shrubs. Leaves entire, often distichous; stipules present. Flowers small, axillary; males often clustered; females solitary. Perianth segments 6, sepaloid, divided to the base in both sexes. Male flowers with 3 coherent or free stamens and 6 glands at the base of perianth segments. Female flowers with a 3-locular, 3-lobed ovary; styles 3, free, ± 2-lobed.

Key to the species

1 Main branch leaves mostly reduced to scale-like structures 2
1 Main branches with laminate leaves sometimes reduced but not scale like. 4

2 Perennial shrubs or plants at least with a woody base. Flowers with 6 sepals and 3 stamens. Styles deeply bifid. Leaves obovate to oblong, up to 20 mm long. Undershrub up to 60 cm high with reddish stems. Seed surface smooth. Coast and adjacent plateaus. Creek sides, RF. Fl. summer. . . **P. similis** Müll.Arg.

2 Annual herbs. Flowers with 5 sepals and 3 stamens or, 6 sepals and 4–6 stamens 3

3 Branchlets flattened. Styles shortly bifid. Leaves elliptic to oblanceolate, 4–12 mm long, 2–5 mm wide. Stamens 3, connate. Seed surface striated. Herb to 60 cm high. Naturalized in Sydney area. Introd. from the Americas. Fl. all year . . . .*P. amarus** K.Schum.

3 Branchlets rounded. Styles deeply bifid. Leaves elliptic to obovate, up to 25 mm long. Perianth c. 1 mm long. Stamens 4–6, free. Seed surface granular. Widespread but uncommon. near RF. Introd. from Madagascar. Fl. summer. . . . .*P. tenellus** Roxb.

4 Softly hairy shrub up to 50 cm high. Leaves obovate to narrow-obovate, up to 6 mm long. Stamens 3, free. Female perianth larger than male. Styles deeply bifid. Widespread. Heath and DSF. Ss. and WS. Fl. spring . . . . **P. hirtellus** F.Muell. ex Müll.Arg.

4 Glabrous or scabrous shrubs or herbs. Styles entire 5

5 Erect, glabrous shrub up to 2 m high. Stamens sometimes with prominent glands. Leaves broad-ovate to orbicular, up to 20 mm long, 4–12 mm wide; apex notched or obtuse. Capsule 3–4 mm diam. seed surface longitudinally ridged. Widespread. DSF. Ss. Fl. spring–summer . . . . **P. gunnii** Hook.f.

5 Small glabrous or minutely scabro-pubescent plant 5–30 cm high, with woody base and numerous erect to decumbent slender branches. Leaves greyish, oblong to narrow-elliptic, 10–20 mm long, 2–4 mm wide, flat or slightly concave but the margin ± thickened. Female perianth narrower than male. Capsule 1.5–3 mm diam. seed surface tuberculate. Cumberland Plain; northern Hornsby Plateau. WS or shale lenses in Ss. Fl. spring–summer . . . . **P. virgatus** G.Forst.

5 Glochidion J.R.Forst. & G.Forst.
15 species in Aust.; Qld, NSW, NT, WA

One species in the area

Small, monoecious tree. Leaves distichous, elliptic or ovate-lanceolate, ± acuminate, 5–10 cm long; stipules present. Flowers axillary. Perianth segments 6, sepaloid. Male flowers c. 3 mm diam., several together on slender peduncles 4–6 mm long; stamens 3, with connate filaments; glands absent. Female flowers solitary on short erect pedicels; ovary 5–7-locular; styles 5–7, connate basally, thick, erect. Fruit much depressed in the centre, greenish, 12–19 mm diam. Coast and adjacent plateaus. In and near RF; sheltered gullies. *Cheese Tree.*

Key to varieties

1 Leaves and fruit glabrous. Widespread. . . . .**G. ferdinandi** (Müll.Arg.) Bailey var. **ferdinandi**

1 Leaves and fruit pubescent. Uncommon. . . . .**G. ferdinandi** var. **pubens** Maiden ex Airy Shaw

6 Breynia J.R.Forst. & G.Forst.
4 species in Aust.; Qld, NSW, NT

One species in the area

Erect, monoecious shrub up to 2 m high. Leaves distichous, elliptic ovate or broad-oblong, 2–4 cm long, thin, glabrous, glaucous underneath; stipules present. Flowers axillary, small, solitary or several together, green; pedicels short, thin, drooping. Male flowers: perianth turbinate, sepaloid, 6-lobed; stamens 3 with connate filaments. Female flowers: ovary 3-locular with 2 ovules per loculus; styles 3, free, entire. Fruit a globular berry, c. 6 mm diam. Widespread. Near RF; sheltered places. Fl. summer. *Coffee Bush* . . . . **B. oblongifolia** Müll.Arg.

7 Actephila Blume
5 species in Aust.; Qld, NSW

One species in the area

Glabrous, monoecious, tall shrub or tree. Leaves spirally arranged, lanceolate to broad-lanceolate, 6–10 cm long, entire or irregularly sinuate; stipules caducous. Flowers in axillary clusters. Peduncles of male flowers up to 1 mm long, those of the female 6 mm long. Sepals 5–6, spreading, imbricate. Petals very small or absent. Stamens 3–6. Ovary 3-locular; styles 3, shortly 2-fid. Fruit hard, globular, c. 12 mm diam. Illawarra district. RF . . . .**A. lindleyi** (Steud.) Airy Shaw

80 HYPERICACEAE

8 gen., mainly temp. N. hemisphere

1 Hypericum L.

10 species in Aust. (1 native, 1 endemic, 8 naturalized); all states and territories

Herbs or shrubs. Leaves opposite, entire, sessile or nearly so; stipules absent. Flowers regular, bisexual. Sepals 5, imbricate. Petals 5, free, imbricate or contorted, yellow or orange. Stamens numerous, hypogynous, usually connate into a number of bundles; anthers subglobular, with a small gland at the summit. Ovary superior, 1-locular; placentas 3 (rarely 2 or 4), parietal, intruding; styles usually 3. Fruit a 3-valved septicidal capsule or berry.

Key to the species

1 Fruit a berry, purple-black when ripe. Leaves broad-ovate, 5–10 cm long, obtuse, with very small translucent glandular dots. Flowers c. 2 cm diam., few in a terminal cyme. Stamens 50–120, connate at the base into 5 bundles. Low shrub. Higher Blue Mts Damp forests. Introd. from Europe, N. Africa and Asia Minor. Fl. summer. *Tutsan* . . . . ***H. androsaemum** L.
1 Fruit a capsule 2

2 Semi aquatic herb 10–30 cm high with densely hairy leaves. Stems erect to creeping with roots occurring at nodes. Leaves ovate to orbicular, 10–30 mm long, 5–20 mm wide, bases stem clasping. Sepal margins with glandular hairs. Stamens connate in 5 bundles. Capsule 10 mm long. Introd. from Europe. Nepean River Fl. summer. . . . . ***H. elodes** L.
2 Glabrous herbs or shrubs 3

3 Styles 5. Capsule 5-valved, red; 10–15 mm long. Leaves ovate, 15–35 mm long. Flowers few in cymes, or solitary; up to 5cm diam.. Stamens in 5 bundles when united. Shrub to 1 m high; branches becoming decumbent. Disturbed areas. Blue Mts Introd. from Europe. Fl. . summer . . . ***H. kouytchense** H.Lév.
3 Styles 3. Capsule 3-valved 4

4 Rather stout, erect perennial 30–90 cm high, with creeping rhizomes. Leaves elliptic or oblong, 1–2 cm long, with numerous translucent oil dots. Flowers rather numerous, c. 2 cm diam., in a broad terminal leafy panicle. Petals twice as long as the sepals, with black dots along the margin. Stamens 50–100, in 3 bundles. Capsule twice as long as the calyx, striate. Widespread. Weed of pastures and waste places. Introd. from Europe and W. Asia. Fl. summer. *St John's Wort* . . . . ***H. perforatum** L.
4 Slender perennials with erect ascending or weak procumbent stems. Flowers few, in a terminal cyme or sometimes solitary. Stamens 20–50. Capsule smooth 5

5 Leaves cordate, ± stem-clasping, lanceolate to linear, 6–15 mm long. Prostrate to decumbent herb. Widespread. Open communities. Fl. summer . . . . **H. gramineum** G.Forst.
5 Leaves obtuse or ± cordate at the base, shortly petiolate, not stem-clasping, 6–20 mm long. Erect herb. Widespread. Open communities. Fl. summer . . . . **H. japonicum** Thunb.

81 LINACEAE

12 gen., mostly temp.

1 Linum L.

6 species in Aust. (2 native, 4 naturalized); all states and territories except NT

Herbs. Leaves alternate, sessile, entire; stipules absent. Flowers regular, bisexual, in cymes. Sepals and petals 5, free. Stamens 5, sometimes alternating with 5 staminodes, hypogynous. Ovary superior, 5-locular, with 2 ovules per loculus. Fruit a subglobular capsule.

Key to the species

1 Petals yellow. Slender glabrous annual. Leaves linear-lanceolate. Flowers small, on slender pedicels forming a loose corymbose panicle. Widespread. Garden weed. Introd. from the Mediterranean. Fl. spring–summer. . . . . ***L. trigynum** L.

1 Petals blue. 2

2 Styles united above the middle. Capsules not or scarcely exceeding the calyx. Leaves linear-lanceolate, 5–20 mm long. Flowers in loose terminal corymbs. Petals 8–12 mm long, deciduous. Glabrous, perennial herb with slender erect stems. Widespread. Grasslands and open forests on WS. Fl. spring. *Native Flax*
. . . . **L. marginale** A.Cunn. ex Planch.

2 Styles free. Capsule longer than the calyx. Larger and coarser than *L. marginale* in all its parts. Annual Widespread. Often along railways. Country of origin unknown. Fl. most of the year. *Cultivated Flax or Linseed* . . . . ***L. usitatissimum** L.

82 OCHNACEAE
30 gen., trop. and subtrop.

1 Ochna L.
1 species naturalized Aust.; NSW

Shrub up to 2 m high. Leaves spirally arranged, simple, narrow-oblong to narrow-elliptic, 2.5–6 cm long, 8–15 mm wide, serrato-dentate, obtuse to acute, sessile. Flowers c. 3 cm diam. Sepals 5. Petals 5, free, yellow. Stamens numerous, hypogynous, free. Ovary superior, c. 4-lobed, with 1 ascending ovule per loculus. Fruit a black berry, usually 4-lobed, on the swollen red receptacle; sepals red in fruiting stage. Garden escape near habitation. Introd. from S. Africa. Fl. summer . . . . . ***O. serrulata** (Hochst.) Walp.

83 PASSIFLORACEAE
10 gen., trop. to warm temp., particularly America.

1 Passiflora L.
Passion-flowers
12 species in Aust. (3 native); all states and territories except S.A.

Climbers with axillary tendrils. Leaves palmately lobed or divided or sometimes entire; stipules often small, caducous. Flowers bisexual, regular. Petals and calyx lobes 5, often similarly coloured. One or several rings of coloured filaments or appendages forming a corona around the petals. Stamens as many as calyx lobes, so united with the long stalk of the ovary as to appear inserted at or near its summit. Ovary superior, on a gynophore, 1-locular; placentas usually 3, parietal; styles 3, with large capitate stigmas. Fruit a berry; seeds enclosed in an aril.

Key to the species

1 Leaves minutely pubescent, up to 12 cm wide, cordate or truncate at the base, with 3 broad-triangular almost acute lobes, or some leaves ovate without lobes. Petiole with 2 glands very near the summit. Flowers pale orange-yellow or greenish. Calyx lobes c. 3 cm long. Fruit green, rather dry, 4–5 cm long. Coast and adjacent plateaus; lower Blue Mts Forests and partly cleared areas. Fl. summer. . . . .
. . . . **P. herbertiana** Ker Gawl.

1 Leaves glabrous. 2

2 Petals and upper surface of sepals crimson; petals more than twice as long as the sepals. Leaves c. 10 cm wide. Widespread. Uncommon. Fl. summer. *Red Passion-flower* . . . . **P. cinnabarina** Lindl.

2 Flowers white; or green white and purple. 3

3 Stipules broad, leafy, persistent, with an auriculate base. Leaves up to 8 cm wide, paler underneath; margins entire. Flowers wholly white or creamy white. Coast. In or near RF. Introd. from S. America. . . . . ***P. subpeltata** Ortega

3 Stipules narrow, linear, caducous. 4

4 Leaves deeply 5-lobed, up to 12 cm wide; lobes lanceolate to elliptic; margins entire. Stems slender. Flowers green and white; rays of the corona blue at the tip, white in the middle, purple at the base. Coast. In or near abandoned gardens. Introd. from Brazil. Fl. summer. . . . . ***P. caerulea** L.

4 Leaves deeply 3-lobed, margins serrate 5

5 Flowers pink. Rarely naturalized near RF., e.g. Mt Keira. Introd. from S. America. *Banana Passion-fruit* . . . . ***P. tarminiana** Coppens & V.Barney

5 Flowers green, white and often purple. Coast and adjacent plateaus. Moist gullies, RF. Introd. from Brazil. *Common Passion-fruit* . . . . ***P. edulis** Sims

84 SALICACEAE

Trees or shrubs usually dioecious. Leaves alternate, entire or serrate, with small glands often along margins; stipules present. Inflorescence axillary, dense spike like catkins, solitary or in clusters,. Flowers bisexual or unisexual with many often spreading stamens; nectary disc or glands present. Ovary superior with parietal placentation; styles free or fused 1–10. Fruit a capsule or berry. Capsule seeds with tuft of hairs. 55 gen., trop. to temp., mostly Africa and Asia.

Key to genera

1 Petals and/or sepals present, flowers solitary or in few flowered clusters. Fruit a berry 2

1 Petals and/or sepals absent, inflorescence a catkin. Fruit a capsule 3

2 Leaf margins entire or angular, lamina 3-veined at base . . . . **1 SCOLOPIA**

2 Leaf margins entire, lamina with 3 or more veins arising from above base. . . . . **2 DOVYALIS**

3 Leaves broad-elliptic, ovate, triangular or rhombic; usually more than 5 cm wide. Flowers with disc shaped nectary . . . . **3 POPULUS**

3 Leaves linear to obovate; usually less than 5 cm wide. Flowers with 1 or 2 elongated nectaries   **4 SALIX**

1 Scolopia Schreb.
1 species endemic Aust.; Qld, NSW

One species in the area

Glabrous shrub or small tree. Leaves alternate, ovate to lanceolate or rhomboidal 3–7 cm long, entire or distantly angular, obscurely 3-veined, smooth, shining. Flowers bisexual, very small, in short axillary racemes, sometimes forming a terminal panicle. Calyx 4-lobed. Petals 4, white, deciduous. Stamens many with slender filaments. Ovary superior, 1-locular; placentas parietal. Fruit a globular berry, c. 12 mm diam. Coast and gullies in adjacent plateaus. In or near RF. Fl. spring. *Flintwood* . . . . **S. braunii** (Klotzsch) Sleumer

2 Dovyalis E.Mey. ex Arn.
1 species naturalized Aust.; NSW

One species in the area

Dioecious tree with spines to 6 cm long. Leaves entire, glabrous, clustered on short lateral shoots, may be alternate on young branches; elliptic to ovate, 2–5 cm long, 1–3 cm wide. Flowers usually unisexual, inconspicuous. Fruit a yellow globular berry 20–40 mm diam. Occasionally naturalized near habitation. Introd. from Africa. *Kei Apple* . . . . ***D. caffra** (Hook.f. & Har.) Kook.f.

3 Populus L.
Poplars

2 species naturalized Aust.; Qld, NSW, Vic.

Deciduous trees which sucker from their roots. Leaves broad elliptic to triangular or rhombic sometimes cordate at the base; margins entire to toothed; petiole half as long as lamina or longer. Flowers circled by a coarsely toothed bract. Male flowers with 4 to many stamens and anthers reddish-purple. Ovary of female flower with 2 carpels and 2 entire or lobed stigmas. Seed numerous, hairy.

Key to the species

1 Leaves 3–9 cm long densely hairy below, 3–5-lobed or coarsely toothed. Catkins 3.5–6 cm long. Capsule c. 3 mm long. Spreading by suckers. Introd. from Eurasia and N. Africa . . . .*P. alba L.
1 Leaves 4–8 cm long, glabrous, closely toothed. Catkins 5–7 cm long. Capsule not seen. Spreading from suckers. Introd. from Eurasia . . . .*P. nigra L.

4 Salix L.
12–14 species naturalized Aust.; Qld, NSW, Vic., Tas., S.A.

Dioecious trees or shrubs. Leaves alternate, stipulate, linear, narrow-elliptic to oblong, up to 15 cm long, serrate to almost entire. Flowers unisexual, naked, in compact spikes (catkins). Female flowers: ovary solitary, sessile, with 2 parietal placentas; style 2-fid. Male plants uncommon. Fruit a capsule. Usually in damp areas and along river banks.

Key to the species

1 Trees with a narrow column shape. Partially evergreen with 30% of leaves retained over winter. Leaves linear-lanceolate 9–15 cm long, 6–10 mm wide, margins with glandular teeth. Male catkins terminal 3–10 cm long. Introd. from S. America. *Pencil Willow* . . . . . *S. humboldtiana Willd. 'Pyramidalis'
1 Trees or shrubs not with a column shape; completely deciduous. 2

2 Multi-stemmed trees or shrubs to 9 m high. Leaves ovate less than 3 times as long as broad. Catkins emerging long before leaves. Undersurface of leaves with grey hairs, some rusty hairs may be present. Stipules broad. Introd. from Eurasia and N. Africa. *Pussy Willow*. . . . .*S. cinerea L.
2 Trees with one to several trunks 10–20 m high. Leaves linear to oblanceolate more than 3 times as long as broad. Catkins emerging with leaves 3

3 Trees with long pendulous branches. Naturalized along the banks of some rivers in the region. Introd. from Asia. *Weeping Willow*. . . . . *S. babylonica L.
3 Trees with spreading or ascending branches 4

4 Twigs fragile at junctions, breaking with a snap. Leaves glossy green above, becoming glabrous on both surfaces 5
4 Twigs not breaking with a snap at the junctions 6

5 Leaf margins with course uneven glandular teeth. Leaves 6–15 cm long, glabrous from first; dark green above. Petiole with 2 glands towards the top. Male catkins 4–6 cm long, 10 mm wide. Rootlets red. Tree up to 20 m high. Widespread. Near fresh water. Introd. from Europe. Fl. spring. *Crack Willow.* . . . . .
. . . .*S. fragilis L. var. fragilis
5 Leaf margins with many small even glandular teeth. Leaves silky but becoming glabrous on upper surface. Male catkins 6–8 cm long, 8 mm wide. Spreading tree to 16 m high. Widespread in riverbeds. Introd. from Europe. (hybrid between *S. fragilis* var. *fragilis* and *S. alba* var. *vitellina*) . . . .
. . . .*S. X rubens Schrank

6 Branches yellow to orange. Leaves dull or slightly glossy above, becoming glabrescent; lanceolate, 5–10 cm long, serrulate. Rootlets white. Flowers crowded on the catkin rachis. Stamens 2. Tree up to 20 m

high. Bark grey-brown. Widespread. Near fresh water. Introd. from Europe. Fl. spring *White Willow*. .
. . . . ***S. alba** var. **vitellina** (L.) Stokes

6 Branches not yellow 7

7 Young branches in full sunlight dark reddish-brown. Leaves green on both surfaces slightly paler beneath, linear-lanceolate . Flowers not crowded on the catkin rachis. Stamens 4–7. Tree or shrub spreading up to 20 m high. Bark grey. Introd. from N. America. *Black Willow* . . . . ***S. nigra** Marshall
7 Young branches brown or olive green. Leaves dull or slightly glossy above, persistently silky, scarious below; lanceolate, 5–10 cm long, serrulate;. Rootlets white. Flowers crowded on the catkin rachis. Stamens 2. Tree up to 20 m high. Bark grey-brown. Widespread. Near fresh water. Introd. from Europe. Fl. spring. *Golden Willow* . . . . ***S. alba** L. var. **alba**

85 VIOLACEAE

Herbs or shrubs. Leaves basal, alternate or alternate and opposite, simple; stipules present or absent. Flowers irregular to regular, bisexual. Sepals and petals 5, imbricate. Stamens 5, subsessile; anthers connivent around the gynoecium, with a terminal membranous appendage. Ovary superior, 1-locular, usually with 3 parietal placentas; style simple, with a terminal stigma. Fruit a loculicidal capsule or berry. 23 gen., trop. to temp.

Key to the genera

1 Shrubs 1–2 m high. Flower regular. . . . . **3 MELICYTUS**
1 Herbs sometimes woody at the base. Flower irregular. 2

2 Lowest petal scarcely longer than the others . . . . **1 VIOLA**
2 Lowest petal much longer than the others. . . . . **2 HYBANTHUS**

1 Viola L.

11 species in Aust. (3 naturalized); Qld, NSW, Vic., Tas., S.A.

Small herbs. Leaves alternate or basal; stipules present. Flowers axillary on long peduncles with 2 bracteoles near or below the middle. Sepals produced into small appendages below their insertion. Lowest petal spurred or pouched at the base. Fruit a capsule opening in 3 valves.

Key to the species

1 Stipules large to 4 cm long, green, deeply lobed. Petals 5–10 mm long, white to partially mauve, yellow to orange at base. Lower petal broadly cuneate. Leaves ovate to elliptic 10–35 mm long, tapering at base; margins toothed to crenate. Annual or biennial herb. Naturalized on ranges. Introd. from Europe, Africa, Asia. Fl. summer. *Field Pansy* . . . . ***V. arvensis** Murray
1 Stipules small, translucent. Lower petal not broadly cuneate. 2

2 Stems erect or decumbent but not stoloniferous. Lower petal spurred. Flowers without sent 3
2 Stoloniferous plants. Lower petal not or only slightly spurred. Flowers with or without sent 5

3 Stems short, erect. Leaves basal, broad-lanceolate to almost ovate, cordate or truncate at the base, 2–7 cm long; stipules attached along the petiole. Petals purple or violet, 10–15 mm long; lateral ones bearded. Widespread. Open forest. Fl. summer . . . . **V. betonicifolia** Sm.
3 Stems elongated, decumbent or erect, up to 30 cm long. Leaves ovate to circular, scattered along stem. 4

4 Stipules entire or toothed, free. Leaves cauline, ovate to nearly orbicular, cordate, 16–30 mm long;. Petals white, c. 1 cm long; the spur short and broad. Widespread. Damp places. Fl. summer. *Swamp Violet* . . . . **V. caleyana** G.Don
4 Stipules fimbriate, fused to petiole at base. Leaves 10–45 mm long. Petals violet, 1–1.4 cm long, lower petal with a notched spur c. 5 mm long. Naturalized in the Katoomba and Mt Wilson area. Fl. summer. Introd. from Europe and Africa. *Common Dog-violet* . . . . ***V. riviniana** Rchb.

5 Petioles long with short retrorse hairs. Capsule hairy. Style hooked at the apex. Leaves ovate-orbicular, 1.5–6 cm diam., crenate-serrate, on a short erect rhizome. Flowers sweet-scented. Petals c. 1.5 cm long, deep violet to white. Stoloniferous, perennial herbs. Illawarra region and Hornsby Plateau. Shady places. Introd. from Europe. Fl. spring. *Sweet Violet* . . . . ***V. odorata** L.

5 Petioles glabrous or hairs if present not retrorse. Capsule usually glabrous. 6

6 Flowers concolorous 7

6 Flowers discolorous 9

7 Flowers 2–3 mm long, dark purple inconspicuous on scapes c. 3 cm long, shorter than the leaves. Lateral petals bearded inside. Leaves broad-ovate to rhombic, 5–7 mm long. Damp sites at high altitudes. Fl. summer . . . . **V. fuscoviolacea** (L.G.Adams) T.A.James

7 Flowers more than 5 mm long, pale blue 8

8 Mature leaves twice as wide as long or wider. Leaf lamina broad often flat topped, finely toothed; 4–15 mm long. Flowers with a narrow oblong anterior petal with small green patch at base, middle section whitish, apex pale violet; 6–7 mm long. Lateral petals with beard present or absent. Stems contracted into rosette or elongated and scrambling. Scattered on sandy soils in seasonally moist areas. Ss. Blue Mts Fl. spring–summer . . . . **V. silicestris** K.R.Thiele & Prober

8 Mature leaves not twice as wide as long. Leaf lamina longer than broad, spathulate with a distinct terminal tooth longer then lateral ones. Flower scapes usually exceeding leaves, 3–8 cm long, Petals 3–6 mm long, whitish to pale violet at apex. Lateral petals glabrous inside or nearly so. The whole plant often less than 25 mm high. Ss. Widespread. Moist ground in more exposed situations. Fl. spring–summer. . .

. . . . **V. sieberiana** Spreng.

9 Anterior petal ± rectangular. Robust plants with pale and dark violet flowers. Leaves reniform. Moist sandstone sites particularly on waterfalls and wet soakage areas from Bundanoon to Blue Mts . . . . . .

. . . . **V. sp. nov. B.** K.R.Thiele & Prober

9 Anterior petal ovate, circular or obovate-obcuneate. 10

10 Anterior petal elliptic to circular, broadest in middle third; distinctly three-nerved, central nerves scarcely anastomising; 8–10 mm long with a large green patch at base, middle section deep violet distinctly bordering white tip. Lateral petals bearded. Leaves orbicular with deeply cordate bases; margins with prominent teeth. Seed glossy black. Perennial stoloniferous herb. Stems contracted, leaves forming rosettes. Widespread. Coastal in moist areas. Headlands, swamps, RF and WSF margins. Fl. spring–summer . . . . **V. banksii** K.R.Thiele & Prober

10 Anterior petal obovate, broadest in the top third; obscurely three-nerved, central nerves anastomising; 8–10 mm long with a small green patch at base, middle section violet indistinctly separated from white tip. Lateral petals bearded inside or beard absent. Mature leaves ± semicircular; margins with obscure teeth. Seed dull brown and cream. Perennial stoloniferous herb. Stems contracted, leaves forming rosettes. Widespread. Coast and ranges. Dry to moist sites. Fl. spring–summer. *Ivy-leaved Violet* . . .

. . . . **V. hederacea** Labill.

2 Hybanthus Jacq.

10 species in Aust.; all states and territories

Herbaceous perennials sometimes flowering in the first year, 15–50 cm high (sometimes higher when growing amongst shrubs), with slender rather wiry stems. Leaves alternate or alternate and opposite; stipules present. Petals blue or orange; lowest petal much longer than the others, slightly pouched at the base. Fruit a capsule opening in 3 valves.

Key to the species

1 Petals orange; lower petal 8–15 mm long. Flowers solitary in the leaf axils with 2 minute bracteoles on the pedicel. Leaves linear to lanceolate, pubescent on the margins. Hornsby Plateau. DSF. Fl. summer. *Spade Flower* . . . . **H. stellarioides** (Domin) P.I.Forst.

1 Petals blue 2

2 Flowers in a slender raceme, each subtended by a minute bract. Lower petal 5–20 mm long. Leaves narrow-linear, 1–6 cm long; lower ones alternate; upper ones opposite. Widespread. Heath. Ss and old dunes. Fl. summer . . . . **H. monopetalus** (Schultes) Domin

2 Flowers on short pedicels in the axils of the upper leaves which are sometimes smaller than the other leaves. Lower petal 6–13 mm long. Leaves all alternate, linear to narrow-ovate, rarely more than 25 mm long. Coast and adjacent plateaus. Heath. Ss. Fl. spring–summer 3

3 Stems and leaves glabrous . . . . **H. vernonii** (F.Muell.) F.Muell. ssp. **vernonii**

3 Stems and leaves scabrous . . . .**H. vernonii** ssp. **scaber** E.Bennett

3 Melicytus J.R.Forst & G.Forst.
1 endemic species Aust.; NSW, Vic., Tas., S.A.

One species in the area

Glabrous, rigid, much-branched shrub up to 3 m high; the side branches often converted into thorns. Leaves alternate, elliptic to linear, 1–4 cm long, entire to irregularly toothed, coriaceous; stipules absent. Flowers regular or almost so, sometimes unisexual, axillary, solitary or 2 together on short slender peduncles. Sepals ovate to orbicular. Petals yellow to greenish, c. 4 mm long, with recurved tips. Fruit a berry, purplish, c. 6 mm diam. Widespread. River banks and near RF. Fl. spring. *Tree Violet* . . . .
. . . . **M. dentatus** (R Br. ex DC.) Molloy & Mabb.

86 CUNONIACEAE

Trees or shrubs. Leaves simple or compound, opposite or apparently whorled, often thick; stipules interpetiolar (Fig. 29) and often enclosing the apex. Flowers bisexual, regular. Floral tube sometimes present. Sepals 4–6, sometimes basally connate. Petals 4–6 or absent. Stamens 8–12. Ovary superior or inferior, 2–4-locular; placentas axile. Fruit a capsule nut or drupaceous. 26 gen., mostly trop. to warm temp.

Key to the genera

1 Petals absent or small, laciniate. Mostly tall shrubs or trees 2

1 Petals at least as long as the sepals, entire. Small shrubs or trees 4

2 Flowers in dense globular heads . . . . 1 CALLICOMA

2 Flowers in cymes or panicles or solitary 3

3 Leaves articulate on the petiole, 1- or 3-foliolate . . . . 2 CERATOPETALUM

3 Leaves not articulate on the petiole, simple . . . . 3 SCHIZOMERIA

4 Leaves simple . . . . .4 ACROPHYLLUM

4 Leaves compound 5

5 Leaves with 3 sessile leaflets less than 3 mm wide (may appear simple and whorled) . . . . .6 BAUERA

5 Leaves with 5–7 leaflets occasionally 3, more than 3 mm wide 6

6 Flowers with 10 stamens. Leaflet margins serrate . . . . 5 CALDCLUVIA

6 Flowers with numerous stamens. Leaflet margins entire . . . . 7 EUCRYPHIA

1 Callicoma Andrews
1 species endemic Aust.; Qld, NSW

One species in the area

Tall shrub or small tree. Leaves opposite, simple, elliptic-oblong or broad-lanceolate, acuminate, coarsely but regularly serrate, 5–10 cm long, glabrous and glossy above or with soft hairs, whitish or rusty tomentose below, with the parallel pinnate veins prominent. Flowers yellowish, numerous in dense globular heads 12–19 mm diam.; the heads on peduncles 1–3 cm long, several usually together on a short axillary common peduncle or in a terminal cluster. Sepals 4–5, free. Petals absent. Stamens much longer than the sepals. Widespread. In damp soil; often near creeks in rocky gullies; margins of RF. Ss. *Black Wattle* . . . . **C. serratifolia** Andrews

2 Ceratopetalum Sm.
6 species in Aust. (5 endemic); Qld, NSW

Tall shrubs or trees, glabrous. Leaves 1–3-foliolate, thick, shining, articulate on the petiole. Flowers small, in terminal panicles or corymbose cymes. Floral tube short, adnate to the base of the ovary. Sepals 4–5, valvate, persistent, enlarging and coloured reddish after flowering. Petal 4 or 5, or absent. Ovary half-inferior, 2-locular; ovules 4 per loculus; stylar branches 2, united at the base. Fruit a nut surrounded by the enlarged coloured sepals and thus resembling a flower but finally becoming dry.

Key to the species

1 Leaves 3-foliolate; leaflets lanceolate to ovate, 3–7 cm long, obtuse or acuminate, obtusely serrulate. Flowers in loosely trichotomous cymes or panicles; the common peduncle shorter than the leaves. Sepals 2 mm long, enlarging to 12 mm in fruit. Petals laciniate. Tall shrub or small tree. Widespread. Open forests, gullies or heath. Sandy soils. Fl. Oct.–Nov. (fruiting Dec.-Jan.). *Christmas Bush* . . . . **C. gummiferum** Sm.

1 Leaves 1-foliolate (apparently simple) elliptic to very narrow-ovate, mostly 8–16 cm long, obtusely serrate, acuminate, narrowed at the base. Flowers in rather dense corymbose cymes. Petals absent. Enlarged sepals c. 7 mm long. Tree with a smooth greyish bark fragrant when broken. Widespread. Common in RF and in gullies. Fl. spring–summer; fruit summer. *Coachwood* . . . . **C. apetalum** D.Don

3 Schizomeria D.Don
2 species in Aust.; Qld, NSW

One species in the area
Medium to tall tree with finely furrowed bark. Leaves opposite, simple, ovate or elliptic-lanceolate, 8–18 cm long, obtuse or acuminate, coarsely irregularly and obtusely serrate or nearly entire, coriaceous, reticulate, not articulate on the petiole. Petals small, laciniate. Ovary 2-locular, superior. Styles 2, free, short, recurved. Fruit drupaceous, ovoid to globular, whitish, up to 12 mm diam. Widespread. RF sheltered gullies. Fl. spring. *White Cherry* or *Crab Apple.* . . . . **S. ovata** D.Don

4 Acrophyllum Benth.
1 species endemic in Aust.; NSW

One species in the area
Small, erect glabrous shrub. Leaves simple, opposite or in whorls of 3, ovate or narrow-ovate, 2–6 cm long, nearly sessile, rigid, strongly reticulate, regularly and acutely serrate. Flowers numerous in dense axillary cymes near the ends of branches. Sepals 4–6, 2 mm long. Petals pink, very narrow, exceeding the sepals. Stamens 3–4 mm long. Ovary 2-locular, superior; styles 2, persistent, divaricate on the valves of the capsule. Springwood, Linden, Woodford and Lawson. In crevices of damp rocks on Ss. Fl. summer. . . . . **A. australe** (Cunn.) Hoogl.

5 Caldcluvia (F.Muell.) Hoogl.
2 species in Aust.; Qld, NSW

One species in the area
Small to medium-sized tree with thick corky bark. Leaves compound, with 5–7 leaflets; leaflets ovate to elliptic, up to 6 cm long. Petals 5, c. 1 mm long, slightly longer than sepals. Capsule sub-globular, 2–3 mm long; styles deciduous. Ourimbah. RF. Fl. summer. *Soft Corkwood* or *Rose-leaf Marara* . . . . **C. paniculosa** (F.Muell.) Hoogl.

6 **Bauera** Banks ex Andrews

4 species endemic Aust.; Qld, NSW, Vic., Tas., S.A.

Spreading shrubs; branches thin, wiry. Leaves opposite, sessile, 3-foliolate; appearing whorled; stipules absent. Leaflet margins entire lobed or toothed. Flowers regular, axillary, sometimes crowded at the ends of branches. Sepals 4–10, spreading, persistent. Petals 4–10, pink rarely white. Stamens indefinite. Ovary superior or partially inferior, 2-locular; placentas axile; styles 2, free, recurved. Fruit a 2-valved capsule.

Key to the species

1 Flowers sessile or nearly so, crowded at the ends of branches forming small leafy heads. Leaflets 5–6 mm long, obtuse, usually with one prominent lobe on each side. Sepals more distinctly 3-fid than leaflets, 4 mm long. Small, diffuse shrub. Coast and adjacent plateaus from Port Hacking northwards. Wet sandy heath. Fl. summer. . . . .**B. capitata** Sér ex DC.
1 Flowers on slender pedicels which are mostly longer than the leaves 2

2 Flowers 12–18 mm diam.; petals mostly pink. Leaflets oblong to lanceolate, 4–12 mm long, often serrate. Petals and sepals with many hairs. Sepals 3–4 mm long. Scrambling shrub up to 2 m high. Widespread. Usually in wet places on Ss. Common. Fl. spring–summer . . . .**B. rubioides** Andrews
2 Flowers 5 mm diam.; petals mostly white. Petals and sepals glabrous or with a few hairs. Sepals 2–3 mm long. Smaller in all its parts than *B. rubioides*. Wet heath. Sandy soils. Fl. spring–summer . . . .
. . . . **B. microphylla** Sieber ex DC.

7 **Eucryphia** Cav.

5 species endemic Aust.; Qld, NSW, Vic., Tas.

One species in the area

Tree or tall shrub. Leaves pinnate; leaflets usually 5–13 but fewer on flowering branches, oblong to ovate, 2–6 cm long, entire, coriaceous, dark green, nearly glabrous on upper surface, white-tomentose underneath; stipules small. Flowers regular, 25–30 mm diam., solitary or several together in upper axils. Sepals 4, broad, thin, cohering into a calyptra, falling as the flower opens. Petals 4, white, broad, much imbricate. Stamens very numerous. Ovary superior, 5–12-locular, silky tomentose. Fruit a capsule, oblong or ovoid, 8–16 mm long, hard, opening septicidally. Illawarra ranges. RF and WSF. Fl. summer–autumn. *Pinkwood* . . . . **E. moorei** F.Muell.

87 ELAEOCARPACEAE

Small shrubs or small to large trees. Leaves simple, alternate, opposite or whorled; stipules small and caducous or absent. Flowers bisexual, regular. Sepals 4 or 5-merous, valvate. Petals as many as sepals, usually imbricate. Stamens twice as many as petals and hypogynous; or numerous, mostly free, inserted around or within a perigynous disc. Ovary superior, 2–5-locular, with 1 or several ovules per loculus; style simple. Fruit a capsule or drupaceous. 8 gen., trop. to warm temp., except Africa.

Key to the genera

1 Small shrub. Flowers 4(rarely 5)-merous with 8(10) stamens.. . . . .**1 TETRATHECA**
1 Small to large trees. Flowers 5(rarely 4)-merous with numerous stamens. 2

2 Leaf lamina expanding just above the base and thus cordate. Fruit a capsule covered with soft bristles .
. . . .**2 SLOANEA**
2 Leaf lamina tapering gradually towards the base. Fruit drupaceous. . . . .**3 ELAEOCARPUS**

1 **Tetratheca** Sm.

Black-eyed Susan's

40 species endemic Aust.; all states and territories except NT

Small shrubs with terete, angled or winged branches. Leaves linear to ovate-elliptic or reduced to scales, usually opposite or whorled; stipules absent. Flowers regular, bisexual, solitary, on filiform pedicels in the axils of the upper leaves, 4-merous. Petals dark purple to pink or rarely white, involute around paired stamens in the bud. Stamens 8, hypogynous; anthers erect, quadrangular, dehiscing by terminal pores. Ovary superior, compressed, 2-locular, with axile placentas; ovules 1–2 per loculus, pendulous. Fruit a compressed capsule opening loculicidally.

Key to the species

1 Stems with 2–3 narrow wings or 2–3 angles in section except near the base. Leaves often reduced to scales but up to 2 cm long, linear-elliptic or rarely oblanceolate. Sepals deciduous. Petals 7–11 mm long. Ovary glabrous. Decumbent to sprawling undershrub. Previously in the St. George area of Sydney but now probably extinct there. Coast and adjacent plateau near Lake Macquarie. DSF. Ss. Vulnerable. Fl. spring . . . .**T. juncea** Sm.
1 Stems terete or ridged not winged 2

2 Sepals persistent around the fruit. Pedicels and sepals glandular-hirsute. Leaves mostly opposite, sub-opposite or rarely whorled linear, scabrous, with revolute margins. Petals 5–11 mm long. Ovary glandular-pubescent and with some simple hairs. Ovules 1 per loculus. Undershrub up to 50 cm high. Hornsby Plateau. Heath and DSF. Ss. Vulnerable. Fl. spring . . . .**T. glandulosa** Sm.
2 Sepals deciduous. Pedicels not glandular-hirsute. Leaves usually whorled, sometimes alternate (opposite on some branches of *T. shiressii*). 3

3 Branches dimorphic, i.e. some glabrous with linear opposite leaves, others pubescent with broader whorled leaves. Pedicels glabrous or nearly so. Sepals glabrous outside or nearly so. Petals 10–12 mm long. Ovary hirsute; ovules 2 per loculus. Ascending to decumbent undershrub. Woronora and Hornsby plateaus. Heath and DSF. Ss. Fl. spring . . . . **T. shiressii** Blakely
3 Branches not dimorphic. 4

4 Leaves linear, with revolute margins 5
4 Leaves broader than linear. 9

5 Leaves mostly alternate, linear, up to 15 mm long, with stiff greyish hairs especially towards the revolute margin. Branchlets and pedicels covered with greyish antrorse hairs. Sepals glabrous or with some scattered simple hairs. Petals 6–15 mm long. Ovary glabrous or pubescent; ovules 1 per loculus. Erect or ascending undershrub. Blue Mts; Hornsby Plateau. DSF. Ss. Fl. spring–summer . **T. decora** Joy Thomps.
5 Leaves mostly whorled. 6

6 Leaves obtuse or sometimes minutely mucronate, up to 10 mm long. Hairs on stem antrorse. Pedicels slender, glabrous. Sepals usually glandular-hirsute but rarely glabrous. Petals 5–13 mm long. Ovary glabrous or with a few simple hairs; ovules 2 per loculus. Erect undershrub. Coast and adjacent plateaus; lower Blue Mts Heath and DSF. Ss. Fl. spring–summer . . . . **T. ericifolia** Sm.
6 Leaves acute or acuminate. Hairs on stem retrorse or irregularly spreading or absent or in one species loosely antrorse. Sepals glabrous or simple-pubescent. 7

7 Hairs on the stem mostly very short, retrorse. Leaves up to 15 mm long. Pedicels and sepals glabrous or pubescent with very short hairs. Petals 5–11 mm long. Ovary minutely pubescent or glabrous; ovule 1 per loculus. Erect undershrub. Blue Mts Heath and DSF. Ss. Fl. summer . . . . . . **T. rubioides** A.Cunn.
7 Hairs on stem mostly not retrorse 8

8 Ovule 1 per loculus. Ovary pubescent with minute hairs or glabrous. Petals 5–10 mm long. Erect undershrub. Coast and adjacent plateaus in south of area. DSF and heath. Fl. spring-summer. . . . . . .
 . . . . **T. neglecta** Joy Thomps.

8 Ovules 2 per loculus. Ovary pubescent with shining white hairs. Petals 6–15 mm long. Erect undershrub. Blue Mts Heath and DSF. Ss. . . . . **T. rupicola** Joy Thomps.

9 Pedicels and sepals mostly pubescent, very rarely glabrous. Pedicels ± stout, erect or slightly spreading. Leaves narrow-elliptic to broad-elliptic or orbicular, up to 20 mm long; margins recurved. Petals 6–15 mm long. Ovules 2 per loculus. Erect undershrub. Hornsby Plateau and Blue Mts DSF and WSF. Fl. spring–summer. . . . . **T. thymifolia** Sm.

9 Pedicels and sepals glabrous or nearly so. Pedicels slender, usually hooked at the top. Leaves narrow-elliptic to obovate, up to 10 mm long; margins recurved. Petals 6–10 mm long. Ovules 2 per loculus. Erect or ascending undershrub. Higher Blue Mts DSF. Fl. spring–summer . . . . **T. bauerifolia** F.Muell. ex Schuchardt

2 Sloanea L.

5 species in Aust.; Qld, NSW

One species in the area

Tree. Leaves obovate-oblong, 10–30 cm long, narrowed towards the base but expanded just above, cordate, coriaceous, ± sinuate-toothed. Flowers cream, 20–25 mm diam., solitary or in short racemes, pendulous, axillary, on slender peduncles. Sepals pubescent, obovate-oblong, c. 8 mm long. Capsule opening by 4 woody valves, c. 12 mm long, densely covered with short soft bristles. Coast. RF. Fl. spring–summer. *Maiden's Blush* . . . .*S. australis* (Benth.) F.Muell.

3 Elaeocarpus L.

21 species in Aust.; Qld, NSW

Small or large trees. Leaves usually serrate, strongly reticulate. Flowers regular, bisexual, in axillary often one-sided racemes. Sepals 4–5, usually valvate. Petals as many as the sepals, white or pink, fringed lobed or entire, inserted around the base of the disc. Stamens numerous, inserted within the disc. Ovary superior, 2–5-locular. Fruit drupaceous.

Key to the species

1 Racemes tomentose, few-flowered. Sepals 4–5 mm long. Petals entire or slightly crenulate, 6 mm long. Anthers obtuse. Fruit black, subovoid, c. 8 mm long. Leaves glabrous on upper surface, usually tomentose on undersurface, lanceolate or oblanceolate to almost obovate, serrate, 3–7 cm long. Domatia absent. Petiole 3–10 mm long. Tree. Blue Mts Gullies. Fl. summer. *Black Olive Berry*. . . . .*E. holopetalus* F.Muell.

1 Racemes glabrous, flowers often rather numerous. Sepals 6–8 mm long. Petals fringed or lobed, as long as or slightly longer than the sepals. Domatia usually present on leaf 2

2 Sepals 4–5 mm long. Anthers obtuse. Leaves narrow-elliptic-oblong, up to 7 mm long and 2 mm wide, usually acuminate, obscurely serrate, tapering gradually towards the base, coriaceous, reticulate but not so distinctly as *E. reticulatus*; petioles 1–2 cm long. Racemes 2–6 cm long. Fruit dark blue, c. 6 mm diam. Tree. Coast from Wyong northwards. Gullies and sheltered places. Fl. spring. *Hard Quandong*. . . . . *E. obovatus* G.Don

2 Sepals 6–8 mm long. Anthers acute 3

3 Leaves elliptic-oblong to lanceolate or rarely narrower, mostly 7–12 cm long, acuminate, ± serrate, acute at the base, coriaceous; petioles 1–2 cm long. Racemes 2–8 cm long. Fruit dark blue, 8–10 mm diam. Small tree or large shrub. Widespread. Gullies and sheltered places. Fl. spring–summer. *Blueberry Ash* . . . . **E. reticulatus** Sm.

3 Leaves lanceolate or narrow-oblong, 10–20 cm long, serrate, coriaceous, obtuse at the base; petioles 3–5 cm long. Large tree. Coast and adjacent plateaus. RF. Fl. summer. *Pigeon Berry Ash* or *White Quandong* . . . .*E. kirtonii* F.Muell. ex F.M.Bailey

88 OXALIDACEAE

6 gen., mainly trop. and subtrop., particularly America

1 Oxalis L.

20 species in Aust. (6 native, c. 12 naturalized); all states and territories

Herbs with compound, usually 3-foliolate, leaves; leaflets palmate at the top of the petiole and folded downwards in repose; stipules absent, Flowers solitary or in umbels on axillary or basal peduncles, bisexual, regular. Pedicels with 2 bracts at the base. Sepals 5, imbricate. Petals 5, free, contorted. Stamens 10, hypogynous; filaments different lengths. Ovary 5-locular, superior; placentas axile. Fruit a loculicidal capsule; valves persistent on the axis. Seeds with a fleshy aril which shrinks on drying and expels the seed explosively. Stems and leaves with an acid taste.

Key to the species

1 Petals pink to purplish or mauve or white. 2
1 Petals yellow 9

2 Leaves and peduncles basal, stems reduced to a short stock 3
2 Leaves attached to aerial stems, often whorled. Peduncles axillary on aerial stems 8

3 Flowers solitary, on a peduncle which scarcely exceeds the leaves. Petals 25–30 mm long, pink or white with yellow base. Plants usually with long whitish hairs except on the upper surface of the leaflets. Leaflets suborbicular, 1–4 cm across. Garden escape near habitation. Introd. from S. Africa. Fl. winter-spring. *Large-flowered Wood Sorrel* . . . . *O. purpurea* L.
3 Flowers in umbels on peduncles exceeding the leaves 4

4 Leaflets roughly triangular, 2–6 cm wide, sometimes only half as long at the midvein, almost glabrous. Petals purplish, less than 10 mm long. Widespread. Troublesome garden weed. Introd. from Mexico. Fl. summer–winter. . . . . *O. latifolia* Kunth
4 Leaflets obcordate, not tending to a pronounced triangular shape. 5

5 Numerous small bulbils produced around parent bulb. Leaves and peduncles arising directly from the bulb. 6
5 No small bulbils produced. Leaves on a very short stem at ground level or a rhizome 7

6 Petals pink, c. 15 mm long. Inflorescences 8–15 flowered. Bulbs light brown; new bulbils sessile. Leaflets mostly 1–4 cm long, ± sprinkled with long whitish hairs. Widespread. Common and troublesome garden weed. Introd. from trop. America. Fl. most of the year. *Large-leaved Wood Sorrel* . . . .
. . . . **O. debilis** Kunth var. **corymbosa** (DC.) Lourteig
6 Petals pink, mauve or purple, c. 20 mm long. Inflorescence 1–4-flowered. Bulbs whitish; new bulbils on stems. Leaflets mostly 1–4 cm long. Occasional along roadsides and creeks. Introd. from trop. America. Fl. winter–summer. . . . . *O. brasiliensis* Lodd.

7 Bulb present. Stem succulent. Leaves on a short stem. Leaflets c. 5 cm long, ± sprinkled with glandular hairs. Petals white or pink with a yellow base, c. 12 mm long. Widespread. Garden escape. Introd. from S. Africa. Fl. winter–spring . . . . *O. bowiei* Herb. ex Lindl.
7 Bulb not present. Leaves produced on a rhizome. Leaflets up to 4 cm long, pubescent. Petals pinkish, mauve or white, c. 7 mm long. Garden escape naturalized in Blue Mts Introd. from S. America. Fl. most of the year . . . . *O. articulata* Savigny

8 Leaflets obcordate, pubescent underneath. Petals pink to pale lilac, c. 7 mm long. Bulb present. Garden escape near Sydney. Fl. spring–summer. . . . . *O. incarnata* L.
8 Leaflets linear, forked. Petals mauve, c. 10 mm long. Stems pubescent. Cambletown Cemetery. Introd. from S. Africa. Fl. autumn . . . . *O. bifurca* Lodd.

9 Rootstock swollen into a fleshy white tuber with numerous bulbils. Petals 15–25 mm long. Flowers drooping, 3–16 per umbel on a long erect peduncle. Leaves obcordate, 1–4 cm wide. Coast. Troublesome weed. Introd. from S. Africa. Fl. spring. *Soursob* . . . . *O. pes-caprae* L.

9 Rootstock not producing bulbils 10

10 Hairs on the stem mostly antrorse. Stipules conspicuous 11

10 Hairs on the stem spreading or retrorse. Stipules inconspicuous or conspicuous and truncate. . . . 14

11 Tape root stout. Fruit long and slender and held erect. Seed 1.5–1.8 mm long with v-shaped grooves and obtuse ridges. Leaflet 2–15 mm long. Widespread on heavy soils. Fl. winter–summer . . . .
. . . . **O. perennans** Haw.

11 Tape root absent or slightly developed. Fruit not slender 12

12 Leaflets bluish or purplish green, 2–15 mm wide. Capsule 10–20 mm long. Seed with u-shaped grooves and rounded ridges. Stoloniferous herb. Coast and Cumberland Plain. Sandy places, e.g. dunes, river sand levees. Fl. most of the year . . . . **O. rubens** Haw.

12 Leaflets green, up to 6 mm wide. Capsule less than 12 mm long 13

13 Seed strongly ribbed with v-shaped grooves; 1–1.4 mm long. Fruit 5–6 mm long. Stoloniferous herb. Widespread. Moist places in grasslands and open forests. Fl. most of the year . . . . **O. exilis** A.Cunn.

13 Seed smooth to shallowly ribbed. Fruit more than 6 mm long. Herb with erect to prostrate stems. Disturbed areas. Fl. winter–summer . . . . **O. thompsoniae** B.J.Conn & P.G.Richards

14 Stipules conspicuous, truncate, usually glabrous except for the ciliate margins. Capsule usually more than 10 mm long. Very variable stoloniferous herb with or without a swollen taproot. Widespread. Weed in gardens and pastures. Probably introd. from Europe. Fl. spring–autumn . *O. corniculata* L.

14 Stipules inconspicuous to obsolete, acute or obtuse 15

15 Plants with a thickened taproot. Leaves with sprinkled hairs. Capsule more than 10 mm long. Sprawling herb. Widespread. Sandy soils in a variety of communities. Fl. spring–summer. . . . .
. . . . **O. radicosa** A.Rich.

15 Plant without a thickened taproot. Leaves densely hairy. Capsule less than 10 mm long. Stoloniferous herb. Widespread. Forests but not usually on Ss. Fl. spring–summer . . . .**O. chnoodes** Lourteig

89 CANNABACEAE

Herbs, shrubs or trees with a characteristic tough bark on the twigs. Leaves opposite or alternate, simple, 3–5-veined from the base, or palmately compound; entire or serrate, often oblique; stipules small, caducous or persistent. Flowers uni- or bisexual, in axillary or lateral cymes. Perianth segments usually 5, basally connate. Stamens usually 5, opposite the perianth segments, hypogynous. Ovary superior, 1-locular; ovule solitary, pendulous; styles 2. Fruit drupaceous. 9 gen., worldwide.

Key to the genera

1 Leaves palmately compound. Leaflets 3–7 . . . . **1 CANNABIS**

1 Leaves not palmately compound 2

2 Leaves scabrous . . . . **2 TREMA**

2 At least the undersurface of the leaves smooth or hairy . . . .**3 CELTIS**

1 Cannabis L.
1 species naturalized Aust.; Qld, NSW, S.A.

One species in the area

Erect branching herb, dioecious, often woody at base 1–4 m high. Stems with glandular and non glandular hairs. Leaves palmate. Leaflets linear to lanceolate, 3–10 mm long, prominently veined; margins toothed; petiole 3–8 cm long. Fruit 4 mm long enclosed in glandular hairy bracts. Occasionally naturalized. Introd. from Eurasia. Prohibited plant. Fl. summer. *Indian Hemp, Marijuana* . . . . **C. sativa** L.

2 Trema Lour.

2 species native Aust.; Qld, NSW, Vic., NT, WA

One species in the area

Tall shrub with pubescent young branches, monoecious or polygamous. Leaves ovate-lanceolate, 3–8 cm long, scabrous, serrate, dark green on both surfaces or tomentose underneath. Flowers very small in axillary cymes, polygamous with male predominating. Fruit c. 3 mm diam. Widespread. RF and sheltered spots in WS; sometimes a weed in gardens. *Native Peach* . . . . **T. tomentosa** (Roxb.) H.Hara var. **aspera** (Brongn.) Hewson

3 Celtis L.

5 species in Aust. (2 native, 5 naturalized); Qld, NSW, NT, WA

Shrubs or small trees, monoecious or polygamous, up to 4m high. Leaves alternate, dentate or entire. Flowers polygamous, 5–4-merous; male flowers clustered; female flowers often solitary in leaf axils. Ovary with 2 divergent styles. Fruit a succulent drupe.

Key to the species

1 Leaves entire, elliptic to lanceolate, 5–8 cm long, acuminate, with translucent dots, scarcely oblique; both surfaces glabrous. Inflorescence many-flowered. Fruit ovoid or globular, 4–6 mm diam., blue-black. Coast and adjacent plateaus. RF and gullies. . . . . **C. paniculata** (Endl.) Planch.
1 Leaves dentate. Inflorescence 1–few-flowered. Upper surface of leaves scabrous, lamina ovate, 5–12 cm long, acuminate, margins coarsely toothed; oblique at the base; glabrous or pubescent underneath. Fruit globular, c. 8 mm diam., orange-red. Naturalized in places along the Nepean River. Introd. from N. America. *Hackberry* . . . . ***C. occidentalis** L.

90 MORACEAE

Dioecious or monoecious trees or shrubs with characteristic tough bark on the twigs, often with latex. Leaves alternate or opposite, simple, entire or dentate, often coriaceous; stipules present, often caducous leaving a scar. Flowers very small, unisexual, regular, in heads or spikes or enclosed within the enlarged invaginated floral axis. Perianth segments usually 4, free or connate basally. Stamens as many as the perianth segments and opposite them or 1–2. Ovary 1-locular, superior; ovule 1, pendulous; style usually 2-fid. Fruit a nut or drupaceous (in *Ficus* within the enlarged fleshy floral axis). 53 gen., trop. to warm temp.

Key to the genera

1 Flowers attached to the inner wall of the invaginated floral axis (fig or syconium) . . . . **1 FICUS**
1 Flowers in heads or spikes or racemes 2

2 Branches with rigid spines . . . . **2 MACLURA**
2 Branches without spines 3

3 Adult leaves entire. Plants straggling or climbing . . . . **3 TROPHIS**
3 Adult leaves denticulate or dentate 4

4 Leaves not strongly 3-veined from the base. Fruit solitary; fruiting perianth not fleshy . . **4 STREBLUS**
4 Leaves strongly 3-veined from the base. Fruits grouped together into an aggregate fruit; fruiting perianth black to red, fleshy . . . . **5 MORUS**

1 Ficus L.

42 species in Aust. (40 native); all states and territories

Trees or shrubs, usually with milky latex. Leaves alternate or opposite, entire; stipules large, caducous, enveloping the terminal bud. Flowers unisexual, minute, enclosed in the invaginated globular to pyriform floral axis (fig or syconium) which has a minute orifice at the apex. Male flowers in pairs or solitary near the mouth of the fig or mixed with the female. Figs at maturity usually enlarged and closed; nuts surrounded by the membranous or succulent perianth.

Key to the species

1 Climbing plant with adventitious roots. stipules petioles and young stems with soft hairs. Leaves on stems ovate, 1–3 cm long; petiole short; base asymmetric. Leaves on branches bearing fruit larger with 3 veins; petioles 1–2 cm long. Fruit solitary, 40–50 cm long, greyish. Occasionally naturalized. Introd. from Asia . . . .***F. pumilo** L.

1 Shrub or tree, sometimes epiphytic and strangling in early stages 2

2 Leaves very scabrous on the upper surface. Aerial roots absent 3

2 Leaves smooth on the upper surface. Aerial shoots present. Epiphytes or on rocks in early stages ("Strangler Figs") 4

3 Figs glabrous or slightly rough. Mature figs up to 2 cm diam. Leaves obovate to elliptic, 6–10 cm long. Erect shrub or small tree. Blackbutt Reserve, Newcastle. . . . . **F. fraseri** Miq.

3 Figs pubescent hispid, 8–12 mm diam. when mature. Leaves oblong-elliptic, 7–15 cm long. Shrub or small tree erect or scrambling. Coast and adjacent plateaus; valleys of the Blue Mts In or near RF. *Sandpaper Fig*. . . . . **F. coronata** Spin & Colla

4 Leaves green and glabrous on both surfaces 5

4 Leaves ± rust-coloured underneath, glabrous on the upper surface 7

5 Mature figs solitary on peduncles 4–6 mm long, globular 25–30 mm diam. Leaves ovate to oblong-elliptic, 7–14 cm long, thinly coriaceous, ± deciduous; lateral veins rather distant with a few short ones in between; reticulations rather conspicuous; petioles 20–50 mm long, jointed at apex. Tree. Coast and adjacent plateaus. RF; rocky places. *Deciduous Fig*. . . . . **F. superba** Miq. var. **henneana** (Miq.) Corner

5 Mature figs sessile or shortly stalked, mostly in pairs, globular, 6–10 mm diam. 6

6 Figs on short stalk 1–4 mm long; red to orange. Stipules more than 1 cm long. Leaves oblong-lanceolate to elliptic, 4–8 cm long; lateral veins lopping at margins, parallel; reticulations very small, not conspicuous; petioles 6–12 mm long, ± flattened. Tree. Coast and adjacent plateaus. In or near RF. *Small-leaved Fig* . . . . **F. obliqua** G.Forst.

6 Figs sessile; white to light brown. Stipules less than 1 cm long. Leaves oblong-elliptic to lanceolate, 5–20 cm long; lateral veins conspicuous; petioles 2.5–6 cm long, terete. Tree. RF. Minnamurra Falls. *White Fig* . . . . **F. virens** Aiton var. **sublanceolata** (Miq.) Corner

7 Leaves obovate to ovate or elliptic, 7–10 cm long, 5–6 cm wide, ± pubescent as well as rusty below; petioles mostly 1–3 cm long. Mature figs mostly 4–10 mm diam., usually marked with prominent warts, yellowish; peduncles 2–5 mm long. Spreading tree or shrub. Widespread. Rocky sides of inlets and gullies, not usually in dense RF. *Port Jackson Fig*. or *Rusty Fig* . . . . **F. rubiginosa** Desf. ex Vent.

7 Leaves ovate-elliptic to oblong, 10–25 cm long, 7–10 cm wide, glabrous on both surfaces; petioles mostly 5–10 cm long. Mature figs nearly globular or slightly pyriform, purplish, 18–25 mm diam.; peduncles 8–12 mm long. Large spreading tree. Coast; Woronora Plateau; Illawarra ranges. RF. Planted in many places. *Moreton Bay Fig* . . . . **F. macrophylla** Desf. ex Pers. ssp. **macrophylla**

2 **Maclura** Nutt.
2 species in Aust.; Qld, NSW

Dioecious shrubs, climbers or trees with stout axillary spines. Latex usually present although sometimes sparse. Leaves alternate, entire. Flowers in globular axillary heads. Fruits compound, globular, composed of the enlarged ± fleshy perianths and receptacles within which the small nuts are enclosed.

Key to the species

1 Woody climber or straggling shrub. Leaves oblong to elliptic, mostly 3–8 mm long. Compound fruit 1–2 cm diam., yellow-orange. Male flower heads 6–8 mm diam. Coast and adjacent plateaus. In or near RF. *Cockspur Thorn* . . . .**M. cochinchinensis** (Lour.) Corner

1 Trees or tall shrubs. Leaves ovate, mostly 5–15 cm long. Compound fruit 10–15 cm diam., yellowish-green. Male flower heads 15–20 mm diam. Occasionally naturalized. Introd. from N. America. *Osage Orange* . . . .*M. pomifera** (Raf.) G.K.Schneider

3 Trophis P.Browne
1 species in Aust.; Qld, NSW, NT, WA

One species in the area

Dioecious, straggling or climbing shrub, glabrous or the young shoots slightly pubescent. Leaves alternate, oblong-elliptic, 3–10 cm long, entire, acuminate to very obtuse, coriaceous. Male flowers in oblong cylindrical spikes 4–12 mm long; female flowers in globular heads 3–4 mm diam., slightly enlarged in fruiting stage. Coast and adjacent plateaus. In or near RF. *Crow Ash* . . . .**T. scandens** (Lour.) Hook & Arn. ssp. **scandens**

4 Streblus Lour.
2 species in Aust.; Qld, NSW

One species in the area

Tall monoecious shrub or tree, glabrous or nearly so. Leaves alternate, elliptic to ovate or lanceolate, mostly 1–7 cm long, mostly acuminate, denticulate, sometimes scabrous. Flowers axillary; male flowers in spikes 15–40 mm long; female flowers in very short spikes, usually containing only 3–4 flowers. Perianth segments c. 1 mm long. Ovary glabrous; style 2-fid. Fruit drupaceous, 6 mm diam., crowned by the stylar branches. Coastal areas; Woronora Plateau; Mt Wilson. RF. Fl. summer. *Whalebone Tree* . . . .**S. brunonianus** (Endl.) F.Muell.

5 Morus L.
1 species naturalized Aust.; NSW

One species in the area

Tree or shrub. Leaves alternate, ovate, 4–15 cm long, dentate, sometimes lobed, ± glabrous. Male flowers in drooping spikes 20–30 mm long; female flowers in shorter spikes. Fruit a drupe surrounded by the succulent usually black to red perianth; aggregate fruit up to 3 cm long. Weed in reserves in Sydney. Introd. from Asia. *White Mulberry* . . . .*M. alba** L.

91 RHAMNACEAE

Shrubs or trees. Leaves simple, alternate or opposite; stipules usually present. Flowers bisexual, regular. Disc present, perigynous or epigynous. Floral tube usually present. Sepals 4–5, valvate. Petals 4–5, hooded or concave, minute or absent. Stamens 4–5, perigynous or epigynous, opposite the petals or alternating with the single perianth whorl. Ovary superior or inferior, 2–4-locular; style short, simple or branched. Fruit a capsule or drupaceous. 60 gen., cosmop.

Key to the genera

1 Leaves mostly opposite, sometimes slightly displaced 2
1 Leaves alternate 3

2 Tree with well developed leaves . . . . **3 EMMENOSPERMA**
2 Spiny shrub with small leaves that fall early . . . .**6 DISCARIA**

3 Leaves glabrous 4
3 Leaves hairy at least underneath 5

4 Leaves more than 10 mm wide . . . . **1 RHAMNUS**
4 Leaves less than 10 mm wide . . . .**4 CRYPTANDRA**

5 Petals ± concave, not enclosing the stamens; or absent . . . . **5 POMADERRIS**
5 Petals hooded around the stamens. 6

6 Tree. Fruit more than 5 mm diam.. . . . . **2 ALPHITONIA**
6 Small shrubs. Fruit less than 5 mm diam.. 7

7 Inflorescence a single flower subtended by several to many bracts. Stipules hardened and connate around base of petiole . . . .**4 CRYPTANDRA**
7 Inflorescence a loose or dense compound cyme. Stipules membranous, free from each other and positioned between the petiole and stem . . . . **7 SPYRIDIUM**

1 Rhamnus L.
2 species in Aust. (1 native, 1 naturalized); Qld, Vic., NSW, S.A.

One species in the area

Spreading, evergreen shrub or small tree up to 6 m high. Leaves alternate, elliptic, 2.5–4 cm long, 10–25 mm wide, dentate, glabrous, shining on the upper surface. Flowers in short compact racemes. Petals 5. Fruit drupaceous, blue-black. Introd. from Europe. Fl. spring. *Buckthorn* . . . . ***R. alaternus** L.

2 Alphitonia Endl.
6 species in Aust. (4 endemic); Qld, NSW, NT, WA

One species in the area

Tall tree with rusty tomentose young branches and inflorescence. Leaves alternate, ovate or elliptic to very narrow-ovate, 7–12 cm long, obtuse or acuminate, coriaceous, entire, penniveined, white or rusty tomentose underneath; stipules small, caducous. Flowers 4–6 mm diam., numerous, in terminal or axillary cymes. Calyx 5-lobed, spreading. Petals 5, involute, enclosing the stamens. Ovary 2- or rarely 3-locular, immersed in the broad disc. Fruit drupaceous, blackish, globular to ovoid, usually 6–10 mm diam., with a small rim or annular line towards the base. Widespread. In or near RF. Fl. spring. *Red Ash* . . . . **A. excelsa** (Fenzl) Reisseck ex Benth.

3 Emmenosperma F.Muell.
2 species endemic Aust.; Qld, NSW, WA

One species in the area

Glabrous tree. Leaves opposite or nearly so, dark green on both surfaces, shining above, ovate, 5–8 cm long, acuminate, entire, penniveined; stipules minute, scarious. Flowers in terminal or axillary cymes or panicles. Calyx lobes 1–2 mm long, almost petaloid. Fruit drupaceous, orange-yellow, globular, 4–6 mm diam. In or near RF. Coast e.g., Royal National Park; lower Blue Mts, e.g. Kurrajong. Fl. spring. *Yellow Ash* . . . . **E. alphitonioides** F.Muell.

4 Cryptandra Sm.
40 species endemic Aust.; all states and territories except NT

Small shrubs. Leaves alternate, linear to ovate or elliptic; stipules hardened. Flowers solitary ± crowded at the ends of the branches, surrounded by persistent brown bracts. Sepals 5, persistent. Floral tube ± adnate to the ovary and produced beyond it. Petals 5, hooded around the stamens. Ovary 3-locular. Fruit a capsule enclosed in the persistent floral tube.

Key to the species

1 Sepals at least as long as the floral tube 2
1 Sepals shorter than the floral tube 3

2 Sepals 5–7 mm long, very silky hairy; the tube almost covered by bracts. Leaves linear or rarely ovate, with recurved margins, 2–6 mm long, not spinose. Divaricate shrub up to 1 m high. Coast and adjacent plateaus. Heath and DSF. Fl. winter–spring . . . . **C. propinqua** A.Cunn. ex Fenzl

2 Sepals 2–4 mm long, white tomentose. Leaves elliptic to oblanceolate, 2–5 mm long. Often spiny shrub up to 1 m high. Widespread. Heath and DSF. Fl. autumn–spring . . . .**C. amara** Sm. var. **amara**

3 Bracts acuminate, ciliate, often half as long as the floral tube. Leaves mostly linear-terete or with a slightly prominent midrib, 4–8 mm long, often clustered. Flowers in small terminal heads surrounded by leafy bracts. Floral tube narrow-campanulate, c. 4 mm long, silky hairy outside. Sepals narrow, spreading. Shrub up to 60 cm high. Widespread. Heath. Fl. autumn . . . . **C. ericoides** Sm.

3 Bracts broad and obtuse. 4

4 Floral tube glabrous at the base, tomentose above, 3–4 mm long. Spiny undershrub to 1 m high. Coast and adjacent plateaus. Open forest and cleared areas. WS. Fl. spring . . . . **C. spinescens** Sieber ex DC.

4 Floral tube hairy to base. Shrubs up to 1 m high 5

5 Leaves elliptic to obovate, 2–5 mm long, flat to recurved. Floral tube not constricted in lower half. Widespread. Open forest. Fl. winter . . . .**C. amara** var. **longiflora** F.Muell. ex Maiden & Betche

5 Leaves linear, 2–6 mm long, terete, revolute. Floral tube constricted in lower half at fruiting. Blue Mts Open forest. Fl. winter–spring . . . . **C. amara** var. **floribunda** Maiden & Betche

5 Pomaderris Labill.

Dogwoods

ca. 65 species in Aust.; all states and territories except NT

Shrubs to small trees. Undersurface of leaves and inflorescence covered with a close stellate tomentum often mixed with or concealed by longer simple hairs. Leaves simple, alternate; stipules brown, scarious, usually caducous. Flowers cream to yellow, in small cymes usually grouped into panicles or corymbs. Disc entirely adnate to ovary. Sepals 5. Petals 5, concave but not enveloping the stamens, or absent. Stamens 5. Ovary inferior, usually 3-locular; style 3-cleft or rarely almost entire. Capsule protruding above the disc, separating into 3 carpels.

Key to the species

1 Petals absent . . . . .GROUP 1

1 Petals present . . . . .GROUP 2

Group 1
petals absent

1 Undersurface of leaves with stellate hairs only 2

1 Undersurface of leaves with distinct simple hairs longer than the stellate ones; upper surface glabrous or pubescent (Simple hairs are longer than the stellate hairs and usually fairly conspicuous under a hand lens) 8

2 Upper surface of leaves glabrous; undersurface with prominent rusty stellate hairs. Leaves elliptic to obovate or oblong, 10–20 mm long, 4–7 mm wide. Flowers in compact terminal panicles. Floral tube and sepals villous, together c. 3 mm long. Young branches closely pubescent; the hairs only slightly brownish. Spreading, sometimes diffuse shrub up to 1 m high. Cox's River, often in damp places. Fl. spring . . . . **P. betulina** Hook. ssp. **betulina**

2 Upper surface of leaves with scattered scabrous stellate or simple hairs. 3

3 Leaves narrow-ovate to ovate, 6–13 cm long, irregularly crenate, wrinkled on the upper surface; lower surface with ± evenly spaced stellate hairs and prominent principal veins. Stems and inflorescence densely stellate pubescent. Flowers numerous in large terminal panicles often leafy at the base. Floral tube and sepals together c. 2 mm long. Shrub 1–2 m high. Widespread. WSF and gullies. Fl. spring . . .

. . . . **P. aspera** Sieber ex DC.

3 Leaves less than 4 cm long. **4**

4 Capsule and ovary hairy. Flowers in numerous small axillary cymes shorter than the leaves, the upper forming a leafy panicle. Floral tube and sepals c. 2 mm long, densely covered with simple and stellate hairs. Erect, small shrub usually with densely hairy branches **5**

4 Capsule and ovary glabrous. **6**

5 Leaf margin recurved to obscure the undersurface. Leaves linear, 5–10 mm long, less than 1.5 mm wide; with scattered short stiff erect hairs. Lower surface of leaf scarcely visible. Lodden Creek, Mt Wilson, Mt Irvine . . . . **P. phylicifolia** ssp. **ericoides** (Maiden & Betche) N.G.Walsh & F.Coates

5 Leaf margin slightly recurved not obscuring the undersurface. Leaves narrow oblong, more than 1.5 mm wide. Lower surface of leaf visible through the hairs. Sydney area and south. . . . . **P. phylicoides** Lodd. ex Link ssp. **phylicoides**

6 Leaves narrow-elliptic to narrow-oblong, 8–15 mm long, c. 2 mm wide . . . **P. angustifolia** (see below)

6 Leaves oblong or ovate to broad-elliptic or almost orbicular **7**

7 Upper surface of leaves with scattered few-armed stellate hairs, often with some simple hairs; lower surface with a close grey tomentum. Leaves broad-elliptic, 16–25 mm long, 10–18 mm wide, often emarginate. Flowers in terminal panicles. Floral tube and sepals villous, together c. 3 mm long. Erect shrub up to 2 m high. Wingello. Open forest. Endangered . . . . **P. cotoneaster** N.A.Wakef.

7 Upper surface of leaves hispid with simple hairs; lower surface rusty tomentose with some conspicuous brown stellate hairs. Leaves ovate to oblong, mostly 2–4 cm long, wrinkled. Flowers in terminal panicles. Floral tube and sepals villous, together c. 3 mm long. Young branches rusty tomentose. Shrub up to 2 m high. Cumberland Plain near Parramatta and Bankstown; Blue Mts, e.g Mt Caley. Open forest. Endangered population in some LGAs . . . . **P. prunifolia** Fenzl var. **prunifolia**

8 Leaves broad-elliptic, up to 14 mm long and 10 mm wide, very convex between the main lateral veins, yellow-grey tomentose underneath. Flowers in short compact terminal spikes. Floral tube and sepals together c. 3 mm long. Spreading shrub up to 3 m high. Higher Blue Mts Open forest . . . . **P. eriocephala** N.A.Wakef.

8 Leaves ovate to oblong, ± flat between the veins **9**

9 Leaves with a shining almost golden-brown tomentum underneath, narrow-elliptic to narrow-oblong, 8–20 mm long, 3–6 mm wide, obtuse. Flowers in small ± compact terminal panicles. Floral tube and sepals golden-villous, together c. 3 mm long. Erect shrub up to 2 m high. Berrima. Endangered . . . . **P. sericea** N.A.Wakef.

9 Leaves with rusty or grey tomentum underneath **10**

10 Hairs of the undersurface of the leaves forming a very close greyish tomentum with some scattered longer hairs also. Leaves ovate to narrow-ovate, 2–8 cm long, glabrous on the upper surface; margins slightly recurved. Flowers numerous in terminal panicles which are often leafy at the base. Floral tube and sepals together c. 3 mm long; the lower part silky hairy; the lobes sparingly hairy. Petals sometimes present. Shrub 1–3 m high. Widespread. Open forests. Fl. spring. . **P. discolor** (Vent.) Poir.

10 Hairs of the undersurface of the leaves numerous, long, silky, rusty to greyish. **11**

11 Lateral leaf veins terminating in a tuft of hairs. Leaves elliptic, c. 25 mm long, glabrous on the upper surface, densely grey-brown tomentose underneath. Flowers in dense panicles. Floral tube and sepals villous, together c. 3 mm long. Shrub up to 2 m high. Hawkesbury River district. Vulnerable . . . . **P. brunnea** N.A.Wakef.

11 Lateral leaf veins not terminating in a tuft of hairs **12**

12 Secondary veins on lower surface obscure. Leaves 1–2 cm long, 1.5–5 mm wide, margins recurved, occasionally revolute, glabrous and smooth on upper surface, lowers surface villous with simple and stellate hairs. Flowers cream, in terminal panicles. Capsule and ovary pubescent with simple hairs. Shrub to 3 m, young shoots with grey to rusty simple hairs over stellate hairs. Heath and shrubland. Hawkesbury River to Bulli. . . . . **P. mediora** N.G.Walsh & Coates

12 Secondary veins on lower surface distinct. Leaves lanceolate rarely ovate, 2–8 cm long, 10–20 mm wide, glabrous on the upper surface, densely tomentose underneath. Flowers yellow, in loose terminal panicles. Floral tube and sepals villous, together c. 2 mm long. Shrub up to 2 m high, with rusty-tomentose young shoots. Widespread. Open forest on Ss and shales. Fl. spring . . . .

. . . . **P. ligustrina** Sieber ex DC. ssp. **ligustrina**

Group 2
petals present

1 Leaves pubescent on both surfaces. 2
1 Leaves glabrous on the upper surface, closely or loosely tomentose underneath 6

2 Leaves with scattered hispid stellate hairs on the upper surface, narrow-elliptic to narrow-oblong, 8–15 mm long, c. 2 mm wide, brownish white tomentose underneath. Flowers in compact terminal spike-like panicles. Floral tube and sepals together c. 3 mm long. Erect shrub up to c. 2 m high. Cox's River. Open forest. . . . .**P. angustifolia** N.A.Wakef.
2 Leaves softly pubescent or hispid with long or short simple sometimes velutinous hairs on the upper surface. 3

3 Leaves broad-elliptic to almost orbicular, 2–8 cm long, 1.5–5 cm wide, often emarginate; upper surface densely shortly pubescent, ± bluish; undersurface rusty tomentose. Flowers in compact terminal panicles. Floral tube and sepals together c. 5 mm long. Erect shrub up to 2 m high. Leumeah; upper George's River; Hunter River Valley. Open forest . . . .**P. vellea** N.A.Wakef.
3 Leaves oblong to narrow-ovate, 1–10 cm long, entire or nearly so 4

4 Leaves pubescent and slightly hispid on the upper surface; softly and usually rusty tomentose underneath, 4–10 cm long. Flowers in terminal often large corymbose panicles. Floral tube and sepals densely hairy, together c. 4 mm long. Erect shrub 1–2 m high; branches petioles and inflorescence densely rusty tomentose. Widespread. Open forest. Ss . . . . **P. lanigera** (Andrews) Sims
4 Leaves velutinous on the upper surface, undersurface with a dense greyish tomentum, sometimes with long rusty hairs on the veins and margins 5

5 Leaves mostly 2–3 cm long, with scattered long hairs underneath. Flowers in loose terminal panicles. Floral tube and sepals villous, together c. 3 mm long. Erect shrub up to 3 m high. Burragorang. Creek banks. . . . . **P. velutina** J.H.Willis
5 Leaves up to 2 cm long, white-tomentose on lower surface. Floral tube and sepals with long white hairs. Shrub up to 2 m high. Blue Mts, e.g. Colong Caves . . . . **P. subcapitata** N.A.Wakef.

6 Hairs on the undersurface of the leaves very close, greyish, not silky, longer simple hairs mainly confined to veins or absent. 7
6 Simple hairs on the undersurface of the leaves silky, wavy or curled, whitish or rusty, not confined to the veins 10

7 Petals obovate tapering more evenly to the base, pale yellow to cream (rarely golden). 8
7 Petals auriculate, bright yellow or golden. Leaves elliptic to lanceolate, mostly 5–7 cm long, entire or slightly sinuate. Flowers in terminal usually corymbose panicles. Floral tube and sepals together 3–4 mm long. Erect shrubs 1–2 m high, with a close brown tomentum on the younger branches 9

8 Stamen filament adnate to base of petal. Leaves 1.5–3 cm long, 3–8 mm wide; margins recurved; veins on undersurface with grey to rusty simple hairs. Shrub to 2 m high. DSF. Bulli. Fl. spring . . . .
. . . . **P. adnata** N.G.Walsh & Coates
8 Stamen filament not adnate to base of petal. Leaves 3–6 cm long . . . . **P. discolor** (see Group 1)

9 Disc and undersurface of the leaves with simple hairs. Widespread. Ss. Fl. spring. **P. intermedia** Sieber
9 Disc, floral tube and undersurface of the leaves without simple hairs. Widespread. Ss. Fl. spring . . . . .
. . . . **P. elliptica** Labill. var. **elliptica**

10 Leaves 2–4 mm wide; narrow-oblong to lanceolate rarely elliptic; lateral veins not visible on the under surface of the leaves. . Floral tube and sepals together c. 3 mm long, villous. Carpels in fruit opening marginally. Small, erect shrub with white tomentose young branches very soon shedding the hairs. Higher Blue Mts; Appin; Mittagong. Often on rocky hillsides . . . .**P. ledifolia** A.Cunn.

10 Leaves more than 4 mm wide; lateral veins visible on under surface of leaves 11

11 Stamen filament adnate to base of petal. Under surface of leaves with loose appressed silky hairs; rusty on midviens. Leave narrow ovate, 4–6 cm long, 14–20 mm wide. Stipules narrow triangular, 2–6 mm long. Inflorescence c. 4–7 cm diam. Flowers cream to yellow. Shrub or small tree to 3 m high. Riparian. Kangaroo River. Critically Endangered. Fl. winter . . . . **P. walshii** Millott & K.L.McDougall

11 Stamen filament not adnate to base of petal 12

12 Inflorescence 5–10 cm diam. Flowers cream to dull yellow. Leaves lanceolate to ovate or oblong-elliptic, 4–10 cm long, 10–35 mm wide; undersurface soft with loose, wavy, rusty hairs. Shrub 1–3 m high; young branches velutinous . . . .**P. ferruginea** Sieber ex Fenzl

12 Inflorescence 1–4 cm diam. Flowers yellow to bright yellow. Leaves elliptic to oblong-elliptic or lanceolate, 1–5 cm long, 10–15 mm wide; yellowish or rusty underneath. Carpels in the fruit opening through a round aperture below the middle. Small, erect shrub with rusty, hairy young branches . 13

13 Leaves 1–5 cm long, 3–15 mm wide; nerves on under surface with straight appressed simple hairs. Inflorescence bracts usually falling before flowers open. Coast and adjacent plateaus; lower Blue Mts Open forest. Ss. Fl. spring . . . .**P. andromedifolia** A.Cunn. ssp. **andromedifolia**

13 Leaves 1.5–4 cm long, 10–15 mm wide; nerves on under surface with spreading curved to flexuous simple hairs. Inflorescence bracts persisting until flowers open. Campbelltown. Fl. spring. . . . . **P. andromedifolia** ssp. **confusa** N.G.Walsh & Coates

6 Discaria Hook.
2 species in Aust.; Qld, NSW, Vic., Tas.

One species in the area

Spreading, spiny, rigid shrub to 2 m high. Spines and stems green. Leaves small 3–15 mm long, falling early. Stipules red-brown, 1.5–3 mm long. Flowers white, in clusters of 10–50 at base of spines. Pedicels 5–10 mm long. Ovary inferior; disc prominent. Carpels 3; style minute. Stamens subequal to petals. Capsule prominently 3-lobed; 4–5 mm diam. Rocky situations in DSF. Blue Mts Fl. spring–summer. *Australian Anchor Plant* . . . .**D. pubescens** (Brongn.) Druce

7 Spyridium Fenzl
30 species endemic Aust.; NSW, Vic., Tas., S.A., WA

Shrubs, young branches white or rusty tomentose. Leaves alternate, undersurface greyish with stellate hairs; stipules , membranous. Flowers sessile, arranged in loose or dense compound cymes, white, hairy. Sepals 5. Petals 5, hooded. Floral disc glabrous, fused to the floral tube forming a notched projection adjacent to the staminal filaments. Stamens 5. Ovary inferior. Style entire or shortly 3-lobed. Capsule enclosed in the floral tube with sepals persistent.

Key to the species

1 Leaves ovate to elliptic, 8–25 mm long, 4–9 mm wide, usually tomentose below, with recurved margins; apex acute. Stipules pale brown. Floral leaves not whitish. Floral tube pubescent to the base, c. 4 mm long. Spreading shrub up to c. 50 cm high. Howes Valley; Hunter River Valley. DSF Ss. Fl. winter–spring . . . . **S. buxifolium** (Fenzl) K.R.Thiele

1 Leaves oblong to narrow oblong, 15–25 mm long, 3–6 mm wide, undersurface densely pubescent; margins recurved to revolute; apex obtuse. Stipules dark reddish-brown to black. Floral leaves whitish. Floral tube with hairs restricted to the top third, c. 2 mm long. Diffuse shrub up to 1.5 m high. Wollondilly and Nattai River valleys. DSF. Ss. Fl. winter . . . . **S. burragorang** K.R.Thiele

92 ROSACEAE

Herbs, shrubs or trees. Leaves simple or compound, alternate, usually with stipules which are sometimes adnate to the petiole. Inflorescences terminal, racemose or cymose. Flowers usually bisexual and regular. Floral tube present or the receptacle expanded to form a shallow cup. Sepals 4–5, usually imbricate, sometimes with an epicalyx. Petals 4–5, sometimes absent. Stamens few to numerous, perigynous or epigynous. Carpels 1–several, free or connate, superior or inferior. Fruit an achene, capsule, drupe or pome (inferior berry). c. 45 gen., mostly temp.

Members of the genus *Prunus* (Plums, Cherries etc) may be found around old building sites, but none appear to be naturalized.

Key to the genera

1 Herbs. 2
1 Woody plants, erect climbing or trailing 7

2 Leaves dissected, not compound 3
2 Leaves compound 4

3 Style not hooked in fruit. . . . . **4 APHANES**
3 Style hooked in fruit . . . . **1 GEUM**

4 Creeping stoloniferous perennials. Leaves with 3 leaflets . . . . **2 DUCHESNEA**
4 Erect perennials. Leaves with more than 3 leaflets 5

5 Leaves palmate . . . . **3 POTENTILLA**
5 Leaves pinnate or pinnate-lobed 6

6 Calyx without barbed prickles . . . . **1 GEUM**
6 Calyx with barbed prickles . . . . **5 ACAENA**

7 Ovary superior 8
7 Ovary inferior 9

8 Stipules narrow, fused to petiole at its base. Receptacle elongated. Hypanthium not enclosing achenes; an aggregate fruit formed . . . . **6 RUBUS**
8 Stipules broad, fused to petiole for most of its length. Hypanthium fleshy enclosing achenes to form a hip . . . . **7 ROSA**

9 Plants with thorns 10
9 Plants without thorns 11

10 Leaves lobed; the lobes dentate . . . . **8 CRATAEGUS**
10 Leaves entire or serrate, not lobed . . . . **9 PYRACANTHA**

11 Leaves compound. . . . . **13 SORBUS**
11 Leaves simple 12

12 Leaves hairy on lower surface 13
12 Leaves glabrous 15

13 Flowers more than 20 mm diam. Fruit more than 5 cm diam. Stipules caducous . . . . . . **14 MALUS**
13 Flowers less than 15 mm diam. Fruit less than 5 cm diam. Stipules persistent 14

14 Leaves entire, up to 8 cm long. . . . . **10 COTONEASTER**
14 Leaves dentate, more than 10 cm long . . . . **12 ERIOBOTRYA**

15 Leaves 3–7 cm long, 0.5–3 cm wide. . . . . **11 RHAPHIOLEPIS**
15 Leaves 8–14 cm long, 2–7 cm wide . . . . **15 PHOTINIA**

1 Geum L.

2 species in Aust. (1 endemic, 1 naturalized): NSW, Vic., Tas.

One species in the area

Erect, perennial herbs up to 1 m high. Leaves mostly radical, ovate in outline, pinnate or pinnate-lobed; leaflets irregularly deeply serrate; stipules leafy. Flowers in a terminal cyme. Petals 5, yellow. Fruiting achenes with a persistent hooked style surrounded by the persistent calyx. Naturalized on the Coast in the south of the area and Woronora Plateau. Introd. from Europe . . . . ***G. urbanum** L.

2 Duchesnea Sm.

1 species naturalized Aust.; Qld, NSW, Vic.

One species in the area

Creeping, stoloniferous, perennial herbs. Leaves trifoliolate, mostly radical; leaflets obovate, c. 3 cm long. Epicalyx and sepals green, c. 1 cm long. Petals yellow, shorter than the sepals. Fruit an achene. Receptacle at fruiting stage bright red, succulent, 12–15 mm diam.. Naturalized along the coast. Introd. from S. Asia. *Indian Strawberry* . . . . ***D. indica** (Andrews) Focke

3 Potentilla L.

4 species naturalized Aust.; NSW, Vic., Tas.

One species in the area

Erect, perennial herb with thick root stock. Leaves 5–7-palmate, pubescent; leaflets oblong, 2–10 cm long, 1–2 cm wide, serrate, ± truncate. Flowers in terminal much branched cymes, 20–25 mm diam. Sepals 5, glandular pubescent and villous. Petals 5, yellow. Carpels rugose. Receptacle ± swollen. Naturalized in the Moss Vale district. Introd. from Europe . . . . ***P. recta** L.

4 Aphanes L.

3 species in Aust. (1 endemic, 2 naturalized); NSW, Vic., Tas., S.A.

One species in the area

Annual herb up to 10 cm high. Leaves deeply dissected, up to 10 mm long, petiolate; stipules mostly lobed, often stem-clasping. Flowers in leaf-opposed clusters. Sepals persistent. Petals absent. Fruiting hypanthium urn-shaped, brown. Blue Mts Forests, on rocks . . . . **A. australiana** (Rothm.) Rothm.

5 Acaena L.

7 species native Aust.; all states and territories except NT

Perennial herbs. Leaves pinnate, usually silky hairy all over; stipules adnate to the petiole, sheathing at the base. Flowers small. Sepals 4–5, valvate in the bud. Petals absent. Stamens 2–10. Carpel solitary, rarely 2; style exserted. Fruit an achene enclosed in the hardened floral tube which is almost closed at the top with barbed prickles.

Key to the species

1 Flowers in dense globular heads on long terminal peduncles. Calyx prickles 4, 6–10 mm long, nearly equal. Floral tube 4-angled. Leaflets orbicular to oblong, dentate. Stoloniferous herb. Widespread. Grasslands, forests and cleared areas . . . . **A. novae-zelandiae** Kirk

1 Flowers in long interrupted spikes or in heads with further flowers on the peduncle below the head. Tufted herbs without stolons (except sometimes in *A.* x *anserovina*) 2

2 Flowers in heads with c. 4 flowers scattered on the peduncle below the head. Fruits with 4–6 slender spines at the apex and smaller ones below. Leaflets 11–17 per leaf, obovate, 8–14 mm long, green and glabrous above, glaucous and ± pilose below. Stamens 2–5, cream red or purple. Sometimes stoloniferous. Hybrid between *A. novae-zelandiae* and *A. ovina*. Coast in south of area but liable to be found wherever *A. novae-zelandiae* and *A. ovina* occur together . . . . **A. X anserovina** Orchard

2 Flowers in long interrupted spikes. Plants never stoloniferous **3**

3 Leaflets glabrous on upper surface or nearly so, with some hairs on main veins underneath, ovate to oblong, 4–23 mm long, serrate. Stamens 2–8. Calyx prickles several with 3–8 longer than the rest, thickened basally and forming longitudinal ridges. Blue Mts Grasslands and forests . . . .
. . . . **A. echinata** ssp. **subglabricalyx** (Bitter) Orchard
3 Leaflets densely evenly villous underneath, less so on upper surface **4**

4 Calyx prickles unequal, 3–6 longer than others, scarcely thickened basally. Leaflets obovate, 8–20 mm long, serrate. Stamens 3–5. Erect or ascending herb up to 10–50 cm high. Widespread. Grasslands . . .
. . . . **A. ovina** A.Cunn.
4 Calyx prickles ± equal, slender. Leaflets ovate to oblong, (5)8–25 mm long, serrate. Stamens 3–7. Erect or ascending herb c. 50 cm high. Widespread. Grasslands and open forest. . . . . . . **A. agnipila** Gand.

6 Rubus L.
20 species in Aust. (7 native); all states and territories except NT

Trailing climbing or sometimes erect, prickly shrubs. Canes usually do not bear flowers in the first year, these are referred to as *primocanes*. Leaves compound or simple. Flowers mostly in terminal or axillary racemes or panicles. Sepals 5, persistent. Petals 5, obovate. Stamens numerous. Carpels numerous, superior. Fruits: drupels clustered on the dry or spongy conical or oblong receptacle into an aggregate fruit.

Key to the species

(*R. parvifolius* and *R. moluccanus* ssp. *trilobus* hybridize in the area. The resulting hybrid is *R.* x *novus* Kuntze)

1 Leaves simple, ovate to orbicular-cordate, usually shortly and broadly 3–5-lobed, rusty tomentose below; margins serrate. Petals red or white. Sepals acuminate, 8–10 mm long, usually very silky hairy. Aggregate fruit red, ± globular, c. 12 mm diam. Tall, scrambling shrub or climber. Widespread. Margins of RF.; WSF. *Molucca Bramble*. **R. moluccanus** ssp. **trilobus** A.R.Bean
1 Leaves compound **2**

2 Leaves pinnate, some leaves may only have 3 leaflets **3**
2 Leaves palmate or with 3 leaflets; no leaves pinnate **5**

3 Petals red to pink. Hairs on stem and leaves simple. Leaves grey or white underneath; leaflets 3–5, ovate, up to 4 cm long, dentate. Aggregate fruit red, succulent, usually with few large drupels. Scrambling shrub. Widespread. Chiefly WSF. Fl. spring. *Native raspberry*. . . . . **R. parvifolius** L.
3 Petals white. Hairs on stem and leaves glandular. Leaves green on both surfaces; leaflets 5–7, elliptic to lanceolate, up to 8 cm long, dentate. Fruit red, dryish, with numerous drupels. Scrambling shrubs . . **4**

4 Flowers with 5 petals. Widespread on coast and ranges. Fl. spring. *Rose-leaf Bramble* . . . .
. . . . **R. rosifolius** Sm. ssp. **rosifolius**
4 Flowers with 9–13 petals. Widespread on coast. Fl. spring. *Rose-leaf Bramble* . . . .
. . . . **R. rosifolius** ssp. **commersonii** (Poir.) Tirveng

5 Canes with bristly red hairs. Fruit yellow. Leaves 3-foliate, leaflets broadly elliptic, 2–6 cm long, rounded at base and apex, upper surface glabrous lower surface with dense grey tomentum. Climber or shrub with long canes. Not common. Introd. from India. Fl. spring *Yellow Raspberry*. . . . ***R. ellipticus** Sm.
5 Canes with or without hairs, if present then not red and bristly. Fruit black or red **6**

6 Tall climber, bark orange. Prickles falcate, numerous on petioles, petiolules, lower midviens, base of sepals and pedicels. Flowers in loose leafless axillary panicles or racemes in the axils of the leaves. Sepals obtuse. Aggregate fruit dark red. Leaflets lanceolate to elliptic, often cordate at the base, 7–12 cm long, acuminate, glabrous or rarely tomentose underneath. Widespread. RF. Fl. spring. *Bush Lawyer* . .
. . . . **R. nebulosus** A.R.Bean

6 Scrambling or trailing shrubs. Flowers in terminal panicles on leafy stems arising from primocanes. Aggregate fruit black. 7

7 Canes with mostly glandular hairs. Leaflets distinctly paler underneath, elliptic to obovate, up to 8 cm long, acuminate. Flowers up to 3 cm diam. Sepals reflexed. Petals white or bright pink. Filaments longer than the styles. Upper Blue Mts and Illawarra region. Introd. from Europe . . . .
. . . . ***R. leightonii** Lees ex F.M.Leight.
7 Canes with simple hairs, or with some glandular hairs mixed with simple ones, or glabrous 8

8 Flowers on short lateral flowering canes arising in the axils of leaves on previous years primocanes. Petals white, c. 18 mm long. Leaflets almost concolourous. Filaments longer than the styles. Aggregate fruits up to 2 cm diam. Occasionally naturalized in urban areas. Introd. from N. America. *Dewberry*. .
. . . . ***R. roribaccus** Rydb.
8 Flowers on long lateral shoots arising in the axils of leaves on the current year's growth. The following species are referred to as *Blackberries* 9

9 Primocanes pubescent. Leaflets distinctly paler underneath, ovate to orbicular. Sepals usually with prickles. Petals pink or white, 10–14 mm long. Filaments longer than the styles. Blue Mts Introd. from Europe. . . . . ***R. vestitus** Weihe & Nees
9 Primocanes glabrous or with a few scattered hairs. 10

10 Glandular hairs mixed with simple ones on young branches and inflorescences. Leaflets ± paler on undersurface, obovate to elliptic. Flowers c. 2 cm diam. Petals pink. Filaments longer than the styles. Nepean Dam. Introd. from Europe. . . . . ***R. radula** Weihe ex Boenn.
10 Glandular hairs absent from young branches and inflorescences or nearly so 11

11 Secondary and tertiary veins of leaflets prominent underneath and impressed on the upper surface. Leaflets thicker than in *R. discolor*. Primocanes usually deep purplish on one side. Petals pink, 8–9 mm long. Stamens shorter than the styles or equalling them. Inflorescence cylindrical. Widespread. Introd. from Europe . . . . ***R. ulmifolius** Schott
11 Secondary and tertiary veins not prominent (particularly when mature). Primocanes a paler purple. Petals pinkish, 12–15 mm long. Inflorescence ± pyramidal in outline. Widespread. Introd. from Europe . . . . ***R. anglocandicans** A.Newton

(Several other species of *Rubus* may occasionally be found but they are doubtfully naturalized.)

7 **Rosa** L.

8 species naturalized in Aust.; Qld, NSW, Vic., Tas., S.A.

Prickly shrubs with imparipinnate leaves; stipules leafy, adnate to the petiole. Flowers solitary or in terminal corymbs. Floral tube deeply concave, almost closed at the mouth. Sepals and petals 5, imbricate. Stamens numerous. Carpels numerous, free, hairy, superior. Fruit a number of achenes enclosed in the fleshy floral tube; this false fruit is commonly called a "hip".

Key to the species

1 Floral tube and hip densely hairy. Climbing or scrambling shrubs with stout hooked prickles. Leaflets not glandular hairy, bright green on upper surface, ± shining and almost glabrous underneath. Flowers solitary or few together. Sepals and receptacle tomentose. Planted for hedges in the past, now naturalized. Introd. from China and Taiwan. *Macartney Rose*. . . . . ***R. bracteata** J.C.Wendl.
1 Floral tube and hip glabrous but sometimes with fine prickles. 2

2 Leaflets glandular hairy underneath, fragrant when rubbed. Peduncles with glandular hairs and prickles. Petals pink. Hip ovoid or cylindrical, often prickly. Erect or scrambling shrub to 3 mm high. Widespread. Weed in vacant land. Introd. from Europe. *Sweet Briar* . . . . ***R. rubiginosa** L.
2 Leaflets without glandular hairs underneath. Peduncles without prickles. Petals white to pinkish. Hip globose without prickles. Erect shrub to 3 m high. Not widespread. Introd. from Europe. *Dog Rose* . . .
. . . . ***R. canina** L.

8 Crataegus L.
2 species naturalized in Aust.; NSW, Vic., S.A.

One species in the area

Deciduous shrub or small tree with spinous branches. Leaves ovate in outline, deeply 3–5-lobed, 1–4.5 cm long; lobes conspicuously dentate. Flowers in umbel-like corymbs. Sepals 5. Petals white to red, 5 or sometimes more. Ovary inferior; style mostly solitary. Fruit a pome, deep red, 8–12 mm diam., with a single bony carpel. Southern Blue Mts; Illawarra Ranges. Weed in grasslands and forests. Introd. from Europe. *Hawthorn* . . . . *C. monogyna** Jacq.

9 Pyracantha M.Roem.
4 species naturalized Aust.; NSW, Vic.

Sprawling, usually ± spiny shrubs. Leaves simple, entire or serrate, obtuse with a minute mucro. Flowers in condensed corymbs. Sepals 5. Petals 5. Ovary inferior, 5-locular; styles 5, free. Fruit a pome, orange, usually with 5 bony carpels.

Key to the species

1 Leaves entire or very obscurely sinuous, pubescent underneath, narrow-oblong to narrow-ovate, up to 4 cm long and 1 cm wide. Fruit depressed-globular, c. 8 mm diam. Naturalized in sandstone communities. Introd. from Asia . . . . *P. angustifolia** (Franch.) C.K.Schneid.
1 Leaves serrate, glossy on the upper surface, lighter underneath, glabrous on both surfaces, elliptic to narrow-oblong, up to 6 cm long. Fruit depressed-globular, 4–5 mm diam.. Occasionally naturalized. Introd. from Himalayas . . . . *P. crenulata** (Roxb.) Roem.

10 Cotoneaster Medik.
6 species naturalized in Aust.; NSW, Vic., S.A.

Shrubs to small trees without prickles or thorns. Leaves simple, entire. Flowers in cymes or panicles terminating some of the lateral branches. Sepals 5. Petals 5, white. Fruit a pome, orange or red, with 2 bony carpels.

Key to the species

1 Leaves more than 30 mm long and 15–40 mm wide, elliptic, pubescent below but often glabrescent and then glaucous. Fruit orange, 4–5 mm diam. Naturalized in a number of places. Introd. from E. Asia . . . . *C glaucophyllus** Franch.
1 Leaves less than 30 mm long and 15 mm or less wide 2

2 Leaves elliptic to obovate, mostly up to 2 cm but sometimes 3 cm long, upper surface dull, glabrous, lower surface pubescent with longer coarser hairs than *C. glaucophyllus*. Veins inconspicuous on upper surface. Fruit dull red, 4–5 mm diam. Naturalized in a number of places. Introd. from E. Asia. . . . .*C. pannosus** Franch.
2 Leaves elliptic to ovate, up to 35 mm long, upper surface glossy, glabrescent with age, lower surface with yellow to grey tomentum. Veins conspicuous on upper surface. Fruit orange-red, 8–10 mm diam. Rarely naturalized. Introd. from China . . . . *C. franchetii** Bois

11 Rhaphiolepis Lindl.
2 species naturalized in Aust.; NSW

One species in the area

Shrub up to 2 m high with slender spreading branches. Leaves elliptic to obovate, 4–5 cm long, 2–3.5 cm wide, glabrous, stiff, dentate, strongly reticulate. Flowers in terminal cymes. Sepals 5. Petals 5, white or pinkish. Styles 2–3. Fruit a pome, purplish or bluish, with 2–3 papery-walled carpels. A garden escape naturalized in urban areas. Introd. from S. China. *Indian Hawthorn* . . . . *R. indica** (L.) Lindl.

12 **Eriobotrya** Lindl.

1 species naturalized Aust.; NSW

One species in the area

Small tree up to 7 m high. Leaves obovate, 10–24 cm long, 5–9 cm wide, dentate, pubescent underneath, almost sessile, glossy. Flowers in rusty-tomentose cymo-panicles. Sepals 5. Petals 5, white. Styles 2–5. Fruit a pear-shaped pome c. 4 cm long. Garden escape naturalized in a number of places. Usually in gullies. Introd. from E. Asia. *Loquat.* . . . . *E. japonica* (Thunb.) Lindl.

13 **Sorbus** L.

1 species naturalized Aust.; NSW

One species in the area

Tall, spreading tree c. 8 m high, with rough scaly bark. Leaves pinnate, with 7–9 leaflets; leaflets oblong to elliptic, 2–6 cm long, 1–3 cm wide, serrate, glabrescent beneath. Flowers in complex ± flat-topped cymo-panicles, 16–18 mm diam. Sepals 5. Petals 5, white. Styles 5. Fruit a pome, ovoid to globular, green or brown, with prominent lenticels, c. 2.5 cm diam.. Naturalized in lower Blue Mts Introd. from S. Europe. *Service Tree* . . . . *S. domestica* L.

14 **Malus** Mill.

1 species naturalized Aust.; NSW, Vic.

One species in the area

Small to medium deciduous tree. Leaves simple, ovate to elliptic, 4–3 cm long, 3–6 cm wide, margins toothed, upper surface sparsely hairy, lower surface densely hairy. Flowers 2–4 cm diam. white, pink or red, in few flowered racemes. Pome more than 5 cm diam. green to red. Cultivated plant occasionally naturalized around buildings and old habitations. Introd. from Europe and Asia. Fl. spring. *Apple* . . . . *M. pumila* Mill.

15 **Photinia** Lindl.

Shrub or tree to 12 m high. Young shoots and leaves maroon. Leaves oblong to elliptic, 8–14 cm long, 2–7 cm wide, margins with spiny teeth, upper surface dark green, lower surface yellow-green. Stipules with glandular margins. Inflorescence a terminal corym. Flowers white, 8 mm diam. Sepals triangular. Fruit globose to obovoid, 5–6 mm diam red. Cultivated, rarely naturalized. Kuring-gai Chase. Introd. from China. Fl. spring. *Chinese Photinia.* . . . . *P. serratifolia* (Desf.) Kalkman

93 URTICACEAE

Herbs shrubs or trees. Leaves simple, alternate or opposite, sometimes with stinging hairs; stipules caducous. Flowers regular, uni- or bisexual, in clusters or panicles. Perianth 4–5, free or connate, green. Stamens 4–5, opposite perianth segments, involute. Ovary superior, 1-locular; ovule 1, basal; style simple. Fruit a nut or drupaceous. 52 gen., mostly trop., a few temp.

Key to the genera

1 Trees or large shrubs. . . . . 1 DENDROCNIDE
1 Herbs. 2

2 Leaves alternate. 3
2 Leaves opposite. 5

3 Leaves very asymmetric at the base, penniveined, distichous . . . . 2 ELATOSTEMA
3 Leaves symmetric at the base or nearly so, with 3–5 principal veins from the base 4

4 Leaves entire . . . . 3 PARIETARIA
4 Leaves crenate. . . . . 4 AUSTRALINA

5 Leaves entire or sinuate, without stinging hairs . . . . **5 PILEA**
5 Leaves dentate, with stinging hairs. . . . . **6 URTICA**

1 Dendrocnide Miq.
6 species in Aust.; Qld, NSW

Dioecious or monoecious trees or shrubs with soft wood. Leaves alternate, covered or sprinkled with rigid stinging hairs. Flowers clustered in short axillary panicles. Fruit a nut, 2–3 mm diam., usually enclosed in the enlarged fleshy perianth and pedicel.

Key to the species

1 Leaves cordate at the base, orbicular-ovate, often more than 30 cm long, pubescent underneath. Pedicel at fruiting stage enlarged into an incurved fleshy mass. Tree. RF. *Giant Stinging Tree* or *Giant Nettle Tree*. . . . .**D. excelsa** (Wedd.) Chew
1 Leaves not cordate at the base, usually ovate, 10–15 cm long, shining, nearly glabrous but sometimes with a few stinging hairs. Pedicels not always enlarged. Tree or shrub. North from Richmond. RF. Uncommon. *Small-leaved Nettle Tree* . . . . **D. photinophylla** (Kunth) Chew

2 Elatostema J.R.Forst. & G.Forst.
2 species in Aust., Qld, NSW

One species in the area

Coarse, straggling herb with fleshy stems 30–60 cm high, often rooting at the base. Leaves distichous or alternate, broad-lanceolate to elliptic, 7–15 cm long, often very oblique at the base, slightly falcate, serrate; venation strongly reticulate. Flowers small, unisexual, in dense heads; male heads 12–25 mm diam., white, on peduncles up to 7 cm long; female heads rarely more than 8 mm diam., nearly sessile. Widespread. RF, usually near streams or on wet rocks. . . . .**E. reticulatum** Wedd.

3 Parietaria L.
4 species in Aust. (3 native, 1 introduced); all states and territories

Herbs. Leaves alternate, entire; stipules absent. Flowers mostly bisexual, axillary, clustered or cymose, with 1–3 green bracts. Perianth segments 4, connate, finally enlarged and enclosing the fruit.

Key to the species

1 Weak, diffuse, slightly pubescent annual. Leaves ovate, 7–30 mm long, thin, 3-veined from the base; petioles slender. Bracts linear to ovate-cordate. Perianth cylindrical at fruiting stage, deeply 4-lobed, c. 2 mm long. Widespread. Rocky places but not on Ss. *Native Pellitory* . . . . **P. debilis** G.Forst.
1 Perennial 30–60 cm high, with crisped hairs on stems and leaves and a woody rootstock. Leaves elliptic-lanceolate to ovate, mostly 1–4 cm long, 3- or 5-veined. Weed in Sydney suburbs. Introd. from Europe. *Pellitory* . . . .*****P. judaica** L.

4 Australina Gaudich.
1 or 2 species native Aust.; Qld, NSW, Vic., Tas.

One species in the area

Weak, diffuse, slightly pubescent perennial. Leaves alternate or subopposite, ovate to nearly orbicular, 5–40 mm long, crenate, 3-veined from the base, on slender petioles. Flowers small, unisexual, few in axillary clusters. Perianth ovoid-tubular at fruiting stage, enclosing the nut. Widespread. Moist shaded places in forests . . . . **A. pusilla** Gaudich.

5 Pilea Labill.

1 species naturalized Aust.; Qld, NSW

One species in the area

Dioecious, annual or perennial herb. Leaves opposite, elliptic-oblong, up to 5 mm long, entire to sinuate, petiolate. Flowers very small in axillary clusters; male flowers 3–4-merous; female flowers 3-merous, with 3 staminodes. Fruit a compressed nut within the persistent perianth segments. Naturalized near Swansea. Introd. from trop. America. *Artillery Plant* . . . . ***P. microphylla** (L.) Liebm.

6 Urtica L.

3 species in Aust. (1 native, 2 naturalized); all states and territories

Erect herbs covered with rigid stinging hairs. Leaves opposite, serrate, petiolate. Flowers unisexual, small, green or white. Perianth deeply 4-lobed; the lobes equal in male flowers but the 2 inner ones enlarged in the female. Stamens 4. Fruit a small nut enclosed in the perianth.

Key to the species

1 Hairy between the stinging hairs. Leaves lanceolate to ovate, 3–12 cm long, serrate. Dioecious perennial. Coast; Cumberland Plain. Introd. from Europe. *Stinging Nettle* . . . . ***U. dioica** L.
1 Glabrous between the stinging hairs 2

2 Leaves ovate to elliptic, mostly less than 5 cm long. Male and female flowers mixed in short axillary racemes. Monoecious annual. Widespread. Weed of waste and cultivated land. Introd. from Europe. *Small Nettle* . . . . ***U. urens** L.
2 Leaves lanceolate to broad-ovate, mostly more than 5 cm long. Dioecious or monoecious perennial with male and female flowers in separate axillary racemes or panicles. Widespread. Forests; weed in waste places. *Scrub Nettle* . . . . **U. incisa** Poir.

94 BRASSICACEAE

Herbs. Leaves basal or alternate, without stipules. Flowers usually in terminal racemes, regular, bisexual; bracts absent. Sepals 4, free, imbricate. Petals 4, clawed, often spreading in the form of a cross, rarely absent. Stamens usually 6, sometimes only 4; the 4 inner longer than the outer pair. Ovary superior, usually divided into 2 loculi by a membranous septum connecting the 2 parietal placentas. Ovules 2–many; stigmas sessile or almost so. Fruit a 2-locular capsule, either long (*siliqua*) or short (*silicula*), opening from the base by 2 valves; or the fruit separating transversely into 2 or more 1-seeded articles. A siliqua may be beaked, i.e., with a distinct sterile region above the valves. Seeds without endosperm; embryo curved; the radicle either bent along the edges of the 2 cotyledons (*accumbent*) or the radicle lying against the back of one of the cotyledons (*incumbent*) or folded so that they almost surround the radicle (*conduplicate*). 350 gen., mainly N. hemisphere, mostly cooler regions.

Hesperis matronalis L. (Damask Violet, Sweet Rocket) has been recorded as naturalizing in the Jenolan Caves area

Key to the genera

1 Fruit separating at maturity into 1-seeded articles . . . . **GROUP 1**
1 Fruit dehiscent from the base . . . . **GROUP 2**

Group 1

1 Fruit as broad as long . . . . **21 CORONOPUS**
1 Fruit longer than broad 2

2 Fruit 3–8-seeded, with a long conical beak, ± constricted around each seed (moniliform) separating into more than 2 articles (Fig. 40 j). . . . . **18 RAPHANUS**

2 Fruit with 2 superposed 1-seeded articles (Fig. 40 k,l) 3

3 Upper article with a short stout blunt beak. Cotyledons accumbent. . . . . **.19 CAKILE**
3 Upper article with a short thin subulate beak. Cotyledons conduplicate . . . . **20 RAPISTRUM**

Group 2

1 Fruit at least twice as long as broad (usually much longer) cylindrical or ± compressed (rarely
quadrangular) 2
1 Fruit usually less than twice as long as broad, triangular or ovate to orbicular. 12

2 Fruit almost quadrangular; valves 1-veined. Seeds in 1 row in each loculus 3
2 Fruit quite cylindrical or compressed 4

3 Fruit beaked. All leaves entire and stem-clasping . . . . **15 CONRINGIA**
3 Fruit not beaked. Lower leaves pinnatisect, not stem-clasping . . . . **1 BARBAREA**

4 Fruit cylindrical, usually 3-veined. Seeds in 1 row in each loculus. Cotyledons incumbent . . . .
. . . . **2 SISYMBRIUM**
4 Fruit sub-cylindrical or ± compressed. Cotyledons conduplicate or accumbent. 5

5 Beak of the fruit more than 4 mm long. Fruit subcylindrical. Cotyledons conduplicate 6
5 Beak of the fruit usually less than 4 mm long. Fruit compressed. Cotyledons accumbent 8

6 Each dry fruit valve with a strong central vein and faint lateral ones; beak slender. Sepals almost erect
. . . . **.3 BRASSICA**
6 Each fruit valve 3-veined (or obscurely so when ripe); beak usually thick 7

7 Sepals spreading. Beak of the fruit 10–15 mm long . . . . **5 SINAPIS**
7 Sepals almost erect. Beak of the fruit 4–8 mm long . . . . **6 HIRSCHFELDIA**

8 Petals pale yellow. Seeds in 2 rows in each loculus 9
8 Petals white or purplish 10

9 Fruit with a distinct beak; pedicel spreading . . . . **4 DIPLOTAXIS**
9 Fruit without a beak but with a persistent style; pedicels appressed to the stem . . . . . . **7 TURRITIS**

10 Valves flat. Seeds in 1 row in each loculus . . . . **10 CARDAMINE**
10 Valves ± convex. Seeds in 1–2 rows in each loculus. 11

11 Basal leaves lobed, stem leaves may be simple . . . . **8 RORIPPA**
11 Basal leaves absent or simple, stem leaves simple . . . . **11 IRENEPHARSUS**

12 Each loculus with 5 or more seeds 13
12 Each loculus with 1 seed 15

13 Fruit compressed at right angles to the septum . . . . **14 CAPSELLA**
13 Fruit compressed parallel to the septum or nearly globular 14

14 Petals white or pink, 2-fid . . . . **9 EROPHILA**
14 Petals yellow, obtuse or slightly notched . . . . **12 CAMELINA**

15 Fruit notched at the apex (Fig. 40 i) . . . . **16 LEPIDIUM**
15 Fruit not notched at the apex 16

16 Fruit compressed at right angles to the septum . . . . **17 CARDARIA**
16 Fruit compressed parallel to the septum . . . . **13 LOBULARIA**

FIGURE 40

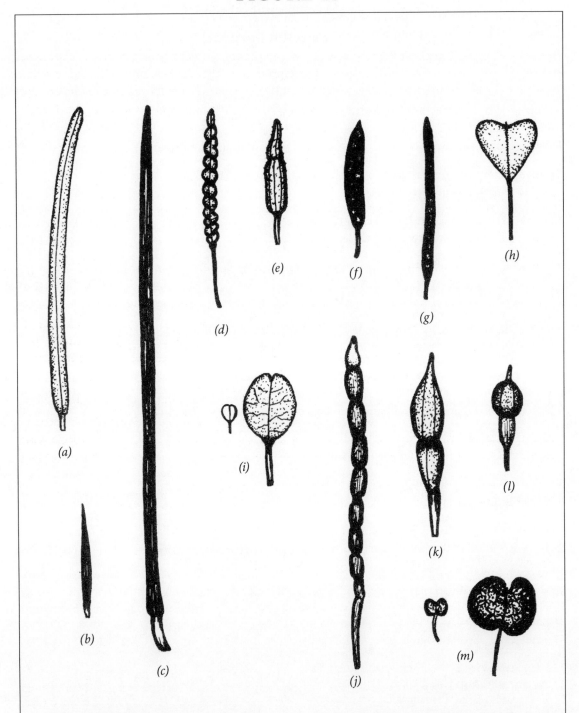

Fig. 40 BRASSICACEAE, fruits; (a) *Barbarea verna*; (b) *Sisymbrium officinale*; (c) *Sisymbrium orientale*; (d) *Brassica fruticulosa*; (e) *Sinapis arvensis*; (f) *Rorippa nasturtium-aquaticum*; (g) *Cardamine hirsuta*; (h) *Capsella bursa-pastoris*; (i) *Lepidium africanum*; (j) *Raphanus raphanistrum*; (k) *Cakile edentula*; (l) *Rapistrum rugosum*; (m) *Coronopus didymus*. All about natural size except the enlargements of i and m.

1 Barbarea Aiton

4 species in Aust. (2 endemic, 2 naturalized); NSW, Vic., Tas., S.A.

One species in the area

Biennial herb up to 50 cm high with a yellow tap-root and an erect, usually branching stem. Lower leaves pinnatisect, with 6–10 pairs of lateral lobes and a terminal one. Flowers 7–10 mm diam. Petals yellow. Racemes loose at fruiting stage. Fruits terete or bluntly 4-angled, 4–6 cm long, 2 mm diam., on short thick pedicels (Fig. 40 c). Leaves sometimes used for salads. Apparently a garden escape. Introd. from W. Europe. *Early Wintercress* . . . . *B. verna* (Mill.) Asch.

2 Sisymbrium L.

6 species naturalized Aust.; all states and territories

Erect, annual or biennial herbs, ± pubescent. Leaves pinnatisect. Flowers small, in terminal multiple racemes. Petals yellow. Fruits nearly cylindrical or tapering; valves with ± prominent midribs. Cotyledons incumbent.

Key to the species

1 Fruits 10–15 mm long, 2 mm diam., closely appressed against the axis of the raceme, subulate, almost sessile, stout at the base, tapering above. Branches of mature inflorescence rigid, spreading at right angles to the main axis. Widespread. Weed of waste ground; roadsides. Introd. from temp. Eurasia. *Hedge Mustard* . . . . *S. officinale* (L.) Scop.
1 Fruits 3–10 cm long, spreading 2

2 Plants grey pubescent. Fruits 6–10 cm long, 1–2 mm diam., rigid, cylindrical, on thick pedicels 3–4 mm long. Branches of the mature inflorescence ± erect but not rigid. Widespread. Weed of waste ground; roadsides. Introd. from the Mediterranean. *Wild Mustard* . . . . *S. orientale* L.
2 Plants glabrous or nearly so. Fruits 3–4 cm long, 1 mm diam., on pedicels 6–8 mm long. Coast and Cumberland Plain. Weed of waste ground; roadsides. Introd. from temp. Eurasia. *London Rocket* . . . .
. . . . *S. irio* L.

3 Brassica L.

9 species naturalized Aust.; all states and territories

Erect or ascending, annual or biennial herbs, glaucous or with scattered bristly hairs. Lower leaves lyrate. Flowers in racemes arranged in a terminal panicle. Petals usually yellow. Valves of the fruit convex, with a prominent midvein when dry; the lateral veins faint or reticulate; beak with few or no seeds. Seeds almost spherical, in 1 row in each loculus. Cotyledons conduplicate.

Key to the species

1 Plants hispid with large stiff scattered hairs on the leaves and stems. Lowers leaves regularly pinnatisect with 6–10 pairs of lobes. Sepals 3–4 mm long. Fruit 3–7 cm long. Stems not darkening on drying. Weed in sandy places near the sea. Introd. from the Mediterranean. *Mediterranean Turnip* . . . .
. . . . *B. tournefortii* Gouan
1 Plants glaucous, occasionally with scattered bristly hairs 2

2 Uppermost leaves stem-clasping, auriculate. Sepals 5–6 mm long. Fruit 4–7 cm long, with a slender ± conical seedless beak. Seeds large. Widespread. Weed of cultivation and waste places. Introd. from Europe. *Bird Rape* or *Wild Turnip* . . . . *B. rapa* L. ssp. **campestris** (L.) A.R.Chapham
2 Uppermost leaves never stem-clasping. Sepals 3–4 mm long. Plants glabrous throughout or setose-hispid on lowest part of stem 3

3 Seeds spaced 3–4 per cm in each loculus. Fruit 25–30 mm long. Coast and Cumberland Plain. Weed of cultivation and waste places. Introd. from Asia. *Indian Mustard* . . . . *B. juncea* (L.) Czernj

3 Seeds spaced 5–8 per cm in each loculus. Fruit 15–25 cm long. Widespread. Weed of vacant and cultivated land. Introd. from Europe. *Twiggy Turnip* . . . . ***B. fruticulosa** Cirillo

4 Diplotaxis DC.
2 species naturalized Aust.; NSW, Vic., Tas., S.A., WA

One species in the area
Annual, erect herb up to 50 cm high, with stiff hairs towards the base. Leaves mostly basal, toothed or lobed. Petals yellow, up to 10 mm long. Fruit 3–4 cm long, with a beak c. 2 mm long. Widespread. Weed of cultivation and waste places. Introd. from Europe. *Wall Rocket.* . . . . *D. muralis* (L.) DC.

5 Sinapis L.
2 species naturalized Aust.; all states and territories except NT

Annual, erect herbs. Basal leaves lyrato-pinnatisect. Flowers in terminal corymbs. Petals yellow. Fruit cylindrical, up to 60 mm long, constricted between the seeds; the beak often containing a seed.

Key to the species

1 Upper leaves sessile, simple, toothed. Fruit 1–3 mm wide; beak scarcely compressed, shorter than the valves. Seeds more than 8, reddish-brown. Widespread. Weed of cultivation and waste places. Introd. from Europe. *Charlock.* . . . . *S. arvensis* L.
1 Upper leaves petiolate, pinnate-lobed. Fruit 3–4 mm wide; beak compressed, at least as long as the valves. Seeds less than 8, pale brown. Coast and Cumberland Plain. Weed of cultivation and waste places. Introd. from Eurasia. *White Mustard`* . . . .*S. alba* L.

6 Hirschfeldia (L.) Larg.-Foss.
1 species naturalized Aust.; Qld, NSW, Vic., Tas., S.A.

One species in the area
Annual or perennial herb up to 1 m high, hirsute towards the base. Lower leaves pinnatifid; terminal lobe ovate, dentate. Upper leaves simple. Flowers c. 5 mm diam., in racemes up to 30 cm long. Petals yellow, sometimes with darker veins. Fruit 8–15 mm long, erect, ± appressed; style 4–8 mm long, swollen basally. Widespread. Weed of vacant and cultivated land. Introd. from Europe. *Buchan Weed or Hairy Brassica* .
. . . . *H. incana* (L.) Lagr.-Foss.

7 Turritis L.
1 species naturalized Aust.; NSW, Vic.

One species in the area
Erect, greyish-green annual or biennial 60–120 cm high, pubescent towards the base. Lower leaves narrow-oblong to narrow-ovate, sinuate-dentate to lyrate. Upper leaves saggitate, stem-clasping, glaucous. Petals yellow, c. 5 mm long. Fruit erect, 40–70 cm long; valves concave, ± raised in the midline by the prominent vein. Jenolan Caves. Introd. from Europe. *Tower Mustard.* . . . . *T. glabra* L.

8 Rorippa Scop.
7 species in Aust. (3 naturalized, 4 endemic); all states and territories

Annual, biennial or perennial herbs with erect or ascending stems, glabrous or with simple hairs. Leaves pinnately divided. Inner sepals saccate. Petals yellow or white. Fruit almost cylindrical or oblong; valves convex, without veins or with 1 inconspicuous vein. Seeds in 1–2 irregular rows in each loculus. Cotyledons accumbent.

Key to the species

1 Fruit less than 25 mm long 2
1 Fruit 25–50 mm long. 4

2 Erect or ascending biennial. Leaf segments toothed. Flowers in short racemes forming a terminal panicle. Sepals c. 1–2 mm long. Petals yellow, 1–2 mm long. Fruit oblong, 4–9 mm long, 2 mm wide, veinless, spreading. Widespread. Damp ground. Almost cosmop.. *Yellow Cress* or *Marsh Cress*. . . . . **R. palustris** (L.) Besser

2 Perennials. Stems procumbent, hollow, rooting below then ascending or floating in water. Leaf segments with undulate margins. Flowers in short terminal racemes. Petals white, twice as long as the sepals. Pedicels and fruits spreading or curved upwards. Fruit valves with an inconspicuous vein 3

3 Seeds in 2 distinct rows in each loculus, with c. 25 depressions on each face of the testa. Fruit 10–18 mm long. Widespread. Usually in water. Introd. from Europe and W. Asia. *Water Cress* . . . . ***R. nasturtium-aquaticum** (L.) Hayek

3 Seeds ± in 1 row in each loculus, with c. 100 depressions on each face of the testa. Fruit 15–22 mm long. Widespread. In or near water. Introd. from Europe and W. Asia. . . . . ***R. microphylla** (Boenn. ex Reichb.) Hyl.

4 Cauline leaves sagittate, sessile; basal leaves elliptic to narrow-ovate, petiolate, denticulate or sometimes entire. Style elongate. Fruits on short pedicels, at right angles to the axis. Seeds reticulate. Erect herbs up to 120 cm high. Widespread. Open forests . . . . **R. gigantea** (Hook.f.) Garn.-Jones

4 Cauline leaves never sagittate. 5

5 Seeds brown, smooth. Fruit 25–40 mm long, the lowest near the base of the stem; style c. 1 mm long. Petals 4–15 mm long. Lower leaves pinnatisect. Widespread . . . . **R. laciniata** (F.Muell.) L.A.S.Johnson

5 Seeds black, rugose, reticulate. Fruits up to 50 mm long; style c. 3 mm long. Petals 7–9 mm long. Lower leaves pinnatisect. Higher Blue Mts, e.g. Jenolan Caves. Open forests. . . . . **R. dictyosperma** (Hook.) L.A.S.Johnson

9 Erophila (L.) Chevall

1 species naturalized Aust.; NSW, Vic., Tas., S.A.

Annual hairy herb up to 20 cm high. Leaves basal, oblanceolate, up to 15 mm long, entire or with a few teeth. Petals pink or white, bifid, up to 4 mm long. Widespread. Damp places. Introd. from Europe. *Whitlow Grass.*

Key to subspecies

1 All hairs branched or stellate. Fruit compressed, narrow-elliptic to orbicular, 6–10 mm long; more than 2.5 times as long as broad . . . . ***E. verna** (L.) Chevall ssp. **verna**

1 Hairs a mix of simple, branched and stellate. Fruit ovate to circular, 3–6 mm long; less than 3 times as long as broad. In northwest of region . . . . ***E. verna** ssp. **praecox** (Steven) Walters

10 Cardamine L.

Bittercress

13 species in Aust. (11 native, 2 naturalized); all states and territories

Slender, almost glabrous, annual or perennial herbs with pinnately divided leaves. Petals white or lilac. Fruit linear, slightly diverging; valves flattened with inconspicuous veins. Seeds in 1 row in each loculus. Cotyledons accumbent.

Key to the species

1 Upper surface of leaves with scattered hairs 2
1 Upper surface of leaves glabrous 3

2 Stamens 4. Sepals 2–3 mm long, always with a few hairs. Petals white. Leaf segments 3–7 pairs, ovate, stalked, almost entire. Erect annual up to 40 cm high. Coast and adjacent plateaus; Cumberland Plain. Garden weed. Introd. from N. temp. regions . . . . ***C. hirsuta** L.

2 Stamens 6. Upper surface of leaves with scattered hairs. Leaf segments 3–7 pairs, ovate. Petals white, stalked. Annual herb up to 50 cm high. Widespread weed. Introd. from Europe. *Wood Bittercress.* . . . . * **C. flexuosa** With.

3 Margins of most pinnae with several minute cilia. Terminal pinnae mostly with 5–9 lobes or teeth. Basal leaves few to many to 8 cm long, simple or pinnate. Cauline leaves 2–5 cm long, pinnate. Petals white 3–4 mm long. Sepals 1.5–2 mm long. Fruit 20–30 mm long, c. 1 mm wide. Widespread. Waterways and low lying areas. Fl. spring–summer. *Eastern Bittercress* . . . . **C. microthrix** I.Thomps.
3 Pinnae without cilia. Terminal pinnae entire or 1–2-lobed or toothed 4

4 Plant stoloniferous. Basal leaves arising singly from horizontal stem. Flowering stems erect. Mature style 1–3 mm long. Petals 6–11 mm long white or pink outside. Fruit 20–30 mm long. Seeds 2 mm long. Blue Mts Uncommon. Damp places. Grassland. Fl. summer . . . .**C. astoniae** I.Thomps.
4 Plants not stoloniferous; basal rosette present 5

5 Seed more than 1.5 mm long. Petals white or pink, 7–11 mm long. Mature style up to 4 mm long. Fruit 10–50 mm long. Seed 1.5–3 mm long. Perennial herb up to 50 cm tall with taproot which branches with age. Blue Mts Damp places, usually in open forests. Fl. spring–autumn . . . . **C. lilacina** Hook.
5 Seed 1–1.2 mm long. 6

6 Cauline leaves 0–3, upper ones reduced, sometimes minutely papillose along margins, terminal pinna entire. Basal rosette usually persisting. Pedicels sometimes minutely papillose. Fruit 20–30 mm long, erect. Annual or perennial herb with taproot. Scattered in hilly areas. Fl. all year **C. papillata** I.Thomps.
6 Cauline leaves 3 or more, reducing only slightly towards top of stem. Leaf segments 1–4 pairs. Basal rosette usually not persisting. Fruit 10–30 mm long, usually with spreading stalk. Annual herb up to 30 cm high with tap and fibrous roots. Widespread in damp places on better soils. Fl. summer . . . . **C. paucijuga** Turcz.

11 Irenepharsus Hewson
3 species endemic Aust.; NSW, Vic., SA.

One species in the area
Annual, glabrous herb up to 1 m high. Stem leaves linear to elliptic, toothed, petiolate. Petals white, c. 4 mm long. Fruit linear, up to 20 mm long. Seeds sticky when wet. Coast and Illawarra Escarpment. Damp gullies. Endangered . . . . **I. trypherus** Hewson

12 Camelina Crantz
2 species naturalized Aust.; Qld, NSW, Vic., Tas., S.A., WA

One species in the area
Annual or biennial herb, slightly hirsute, up to 60 cm high. Leaves dentate to lobed. Petals yellow, 4–5 mm long. Fruiting racemes short and lax, with spreading fruits. Fruit c. 8 mm wide, compressed-globular, with several seeds. Lower Hunter River Valley. Weed of pastures. Introd. from Europe. *Stinking Flaxweed* . . . .*C. alyssum** (Mill.) Thell.

13 Lobularia Desv.
1 species naturalized Aust.; Qld, NSW, Vic., SA, WA.

One species in the area
Ascending, greyish, pubescent or glabrescent, perennial herb up to 40 cm high. Leaves linear to narrow-elliptic, acute, ± entire. Petals white pink or purple, 3–4 mm long. Fruit obovate, c. 3 mm wide, not notched, with a prominent style. Seed 1 per loculus. Widespread. Garden escape. Introd. from Europe. *Sweet Alison* . . . .*L. maritima** (L.) Desv.

14 Capsella Medik.

1 species naturalized Aust.; all states and territories

One species in the area

Erect, annual herb 8–40 cm high, glabrous or ± pubescent. Lower leaves in a rosette, pinnately lobed to entire; upper leaves variable but always clasping the stem with basal auricles. Flowers 2–3 mm diam., in terminal multiple racemes. Petals white. Fruits triangular, 6–9 mm long, with a broad emarginate upper end (Fig. 40 h). Pedicels longer than the fruits, spreading. Seeds 10 or more per loculus, brown or orange. Widespread. Weed of cultivated land and waste places. Cosmop.. Fl. most of the year. *Shepherd's Purse* . . *C. **bursa-pastoris** (L.) Medik.

15 Conringia Heist. ex Fabr.

1 species naturalized Aust.; Qld, NSW, Vic., S.A.

One species in the area

Annual, glabrous herb. Leaves obovate to elliptic, up to 10 cm long, entire, stem-clasping. Petals yellow or greenish, up to 15 mm long. Fruit 4-angled, up to 14 cm long, beaked, each valve with one vein. Coast and adjacent plateaus; Cumberland Plain. Weed. Introd. from Europe. *Hare's Ear* or *Treacle Mustard* . . . . *C. **orientalis** (L.) C.Presl

16 Lepidium L.

Cresses and *Peppercresses*

43 species in AUST, (35 endemic, 8 naturalized); all states and territories

Erect, annual or biennial herbs up to 1 m high. Leaves entire, serrate or pinnately divided and subdivided. Flowers small, in terminal racemes. Petals sometimes absent. Fruit short, compressed laterally, usually ovate, with a narrow septum. Seeds 1 per loculus.

Key to the species

1 Stamens 6. Stem leaves auriculate, entire or toothed. 2
1 Stamens usually 2, rarely 4 3

2 Petals yellow. Fruit elliptic, c. 4 mm wide. Annual herb up to 40 cm high. Blue Mts Weed. Introd. from Europe. . . . . *L. **perfoliatum** L.
2 Petals white. Fruit ovate, 5–6 mm long, covered with dry vesicles when mature. Annual herb up to 60 cm high. Widespread. Weed. Introd. from Europe. . . . . *L. **campestre** (L.) W.T.Aiton

3 Fruiting raceme not elongating, mostly 2 cm long or less. Fruit ovate, c. 1.5 mm wide. Lower leaves pinnatisect with linear dentate segments. Annual herb up to 40 cm high. Scattered . . . . **L. fasciculatum** Thell.
3 Fruiting raceme elongating, mostly more than 2 cm long. 4

4 Pedicels glabrous. Petals c. as long as sepals or absent. Fruit elliptic, c. 3 mm long, with a wide notch. Annual herb up to 60 cm high. Widespread. Weed of cultivation and waste places. *Peppercress* . . . . **L. pseudohyssopifolium** Hewson
4 Pedicels hairy. Petals shorter than sepals or absent` 5

5 Upper leaves lobed or dissected. Fruit ± orbicular, 2–4 mm diam., with a shallow notch. Annual up to 50 cm high. Widespread but not common. Weed of cultivation and waste places. Introd. from S. America . . . . *L. **bonariense** L.
5 Upper leaves entire or shortly toothed. 6

6 Hairs on the margins of the young leaves long, sometimes absent on the mature leaves. Fruit ovate to obovate, c. 3 mm long. Annual or perennial up to 70 cm high. Widespread. Weed of cultivation and waste places. Introd. from Africa. . . . . *L. **africanum** (Burm.f.) DC.

6 Hairs on the margins of the young leaves short, sometimes absent on the mature leaves. Petals reduced or absent. Fruit elliptic, c. 3 mm long, with a narrow notch. Annual herb up to 40 cm high. Coast. Uncommon . . . . **L. pseudotasmanicum** Thell.

17 Cardaria Desv.
1 species naturalized Aust.; all states and territories

One species in the area

Erect, rigid, hoary perennial. Upper leaves oblong-elliptic, dentate to sinuate, stem-clasping. Racemes up to 10 cm long, arranged in dense corymbs. Petals white, c. 5 mm long. Fruit ovate, c. 4 mm wide, not notched, indehiscent. Widespread. Weed of cultivation and waste places. *Hoary Cress*. . . . . ***C. draba** (L.) Desv.

18 Raphanus L.
3 species naturalized Aust.; all states and territories

Annual or biennial herbs. Lower leaves lyrate. Flowers in terminal multiple racemes. Petals white or pale yellow with violet veins. Fruit long, cylindrical, thick ± constricted between the seeds when dry, breaking into 1 seeded articles; beak long, conical. Seeds separated by a pithy substance in the fruit. Cotyledons conduplicate.

Key to the species

1 Fruit constricted between the seeds, ± moniliform, 2–5 cm long; articles bony. Pedicels c. 15 mm long. Tap-root slender, tough. Widespread. Weed of cultivation. Introd. from N. temp. regions. *Wild Radish*. . . . . ***R. raphanistrum** L.
1 Fruit scarcely constricted between the seeds, thick, inflated, spongy. Tap-root thick, fleshy. Garden escape near habitation. Introd. from N. temp. regions. *Radish* . . . . ***R. sativus** L.

19 Cakile Mill.
2 species naturalized Aust.; Qld, NSW, Vic., Tas., S.A., WA

Glabrous, sometimes glaucous, annual herbs. Leaves thick, fleshy. Flowers in racemes. Petals pinkish to purplish, lilac or white, c. 8 mm long. Fruit up to 30 mm long, separating at maturity into 2 articles different in shape from each other, each 1-seeded or the lower one seedless. The articles may be dispersed by sea currents.

Key to the species

1 Lower article of the fruit sagittate almost as wide as the upper one, with 2 blunt lateral horns at the top; upper article ovoid. Leaves pinnatisect with 2–4 pairs of lobes. Petals 8–14 mm long, white to purple. Sand dunes near the sea. Introd. from Europe. *European Sea Rocket* . . . . ***C. maritima** Scop.
1 Lower article of fruit scarcely sagittate, much narrower than the upper one, without lateral horns; upper article ellipsoid. Leaves lobed but not pinnatisect. Petals 4–8 mm long, white to lilac or pink. Sand dunes near the sea but becoming scarcer. Introd. from America. *American Sea Rocket*. . . . . ***C. edentula** (Bigelow) Hook.

20 Rapistrum Crantz
1 species naturalized Aust.; Qld, NSW, Vic., Tas.

One species in the area

Erect, annual herb often more than 1 m high, ± beset with stiff hairs. Lower leaves lyrate, 8–25 cm long. Flowers in multiple racemes. Petals pale yellow, 10 mm long. Fruit with 2 articles; upper one globular, 3–4 mm diam., ribbed, 1-seeded; lower one cylindrical, 1-seeded or seedless; style persistent forming a slender subulate beak (Fig. 40 l). Pedicels shorter than the fruit. Widespread. Weed of roadsides; railway enclosures. Introd. from Europe. *Turnip Weed*. . . . . ***R. rugosum** (L.) All.

21 **Coronopus** Zinn

3 species naturalized Aust.; all states and territories except NT

One species in the area

Annual herb with an unpleasant odour. Branches leafy, prostrate or ascending. Leaves pinnatisect, with narrow entire or incised segments. Petals shorter than the sepals, sometimes absent. Racemes at fruiting stage longer than the leaf; pedicels longer than the fruits. Fruit 2-lobed, 2 mm wide, separating at maturity into 2 ovoid articles (Fig. 40 m). Widespread. Garden weed. Introd. from S. America. Fl. most of the year. *Lesser Swinecress* . . . . *C. didymus (L.) Sm.

95 CAPPARACEAE

C. 45 gen., trop. and warm temp.

1 **Capparis** L.

18 species in Aust. (13 endemic) all states and territories except Vic. and Tas.

One species in the area

Minutely pubescent to glabrous, small tree or ± scrambling when young, often with pungent subulate prickles. Leaves ovate to elliptic, up to 12 cm long; those on the sterile branches often smaller and with a pungent tip. Buds tomentose. Sepals 4; the outer 2 connate in the bud and breaking irregularly. Petals 4, white. Stamens numerous. Fruit a hard skinned berry, up to 5 cm diam. Lower Hunter River Valley. Open Forests. Fl. summer. *Brush Caper Berry; Native Orange* . . . . **C. arborea** (F.Muell.) Maiden

96 GYROSTEMONACEAE

Trees or shrubs. Leaves alternate, simple, entire; stipules absent or minute. Flowers unisexual, often dioecious. Perianth segments connate into a lobed cup. Stamens 6 or more. Carpels superior, connate, with separate styles, each with a single ovule. 4 gen., Aust.

Key to the genera

1 Small tree. Carpels more than 2 . . . . **1 CODONOCARPUS**
1 Undershrub or herb usually less than 1 m high. Carpels 2 . . . . **2 GYROSTEMON**

1 **Codonocarpus** A.Cunn. ex Endl.

3 species endemic Aust.; all states and territories except Tas.

One species in the area

Small tree up to 15 m high with thin flexuose branches. Leaves narrow-obovate to narrow-elliptic, up to 10 cm long and 1.5 cm wide, acuminate, petiolate. Male flowers with 10–20 stamens. Female flowers small greenish; perianth segments connate into a sinuate-toothed tube; carpels numerous connate into a ring around a central disc. Fruit a campanulate schizocarp, 10–15 mm diam., breaking up into winged mericarps at maturity. Hunter River Valley; Mt Waring; Howes Valley. RF margins and open forests. *Bell-fruit Tree* . . . .**C. attenuatus** (Hook.) H.Walter

2 **Gyrostemon** Desf.

12 species endemic Aust.; all states and territories

One species in the area

Erect, dioecious herb slightly woody at the base, or undershrub, up to 60 cm high. Leaves linear to narrow-oblong, up to 4 cm long, glabrous. Flowers in spikes. Male flowers with 9–12 stamens. Female flowers with a 2-carpellary ovary. Nepean and George's Rivers area. Locally common after fire . . . .
. . . . **G. thesioides** (Hook.f.) A.S.George

97 RESEDACEAE

6 gen., mostly Mediterranean

One genus in the area

1 Reseda L.

5 species naturalized Aust.; all states and territories except NT

Annual or perennial herbs with simple or pinnatisect alternate leaves. Stems erect. Flowers few, in terminal spikes or racemes with small bracts. Sepals 4–6. Petals 4–6, free. Stamens 10-numerous, inserted on a basal disc. Ovary 1-locular, superior, with 3–4 parietal placentas; stigmas 3–4. 6 gen., mostly Mediterranean.

Key to the species

1 Leaves entire or with a few teeth towards the base, narrow-oblong to linear, 5–10 cm long, 5–15 mm wide. Flowers in a ± dense spike-like raceme. Sepals 4. Petals yellow, 4. Filaments persistent. Erect annual or biennial. S. Blue Mts Introd. from Europe. Fl. summer. *Weld* . . . . ***R. luteola** L.
1 Leaves mostly pinnatisect with a few lobes on either side, 5–10 cm long. Flowers in a terminal ± spreading raceme with caducous bracts. Sepals 6. Petals yellow, 6. Filaments caducous. Erect annual or biennial. S. Blue Mts Introd. from Europe. Fl. summer. *Cut-leaved Mignonette* . . . . ***R. lutea** L.

98 TROPAEOLACEAE

1 gen., South America

1 Tropaeolum L.

1 species naturalized Aust.; Qld, NSW, S.A., WA

One species in the area

Weak, decumbent, annual herb. Leaves orbicular, peltate, spirally arranged; stipules absent. Flowers irregular, bisexual. The anterior sepal spurred. Petals 5, imbricate, bright orange or yellow; the lateral pairs connate at the base. Stamens 8, hypogynous. Ovary 3-locular, superior; ovule 1 per loculus. Fruit a succulent green schizocarp. Garden escape, naturalized near Sydney. Introd. from S. America. Fl. spring–summer. *Nasturtium* . . . . ***T. majus** L.

99 MALVACEAE

Herbs, shrubs or small trees. Leaves alternate, simple or palmately lobed, often with stellate hairs; stipules present or absent. Epicalyx of free or connate bracteoles present or absent. Flowers bisexual unisexual or polygamous, regular. Sepals 5, usually connate, valvate. Petals 5; cohering at the base, minute or absent. Stamens 5–15, or numerous; free or ± connate; often alternating with staminodes. Ovary superior; carpels free or united; 2–many-locular; ovules 2–many; placentas axile or parietal. Style multi-fid.. Fruit a follicle, loculicidal capsule or schizocarp separating into 1- or 2-seeded mericarps. 243 gen., cosmop.

Key to the genera

1 Androgynophore present. Flowers unisexual, more than 10 mm long. Sepals petaloid, petals absent. Epicalyx absent. Anthers 2-locular. Trees. . . . . **1 BRACHYCHITON**
1 Androgynophore absent. Stamens fused into a column around the style, free or slightly connate at base. Flowers usually bisexual, if unisexual then petal present and less than 5 mm long. Epicalyx present or absent. Anthers 1 or 2-locular. 2

2 Stamens free or slightly connate at the base. 3
2 Stamens forming a tube around the ovary 7

3 Petals present, at least 2 mm long and well developed 4
3 Petals absent or reduced and scale like. (N.B: staminodes may be mistaken for petals) 5

4 Staminodes alternating with stamens, linear-lanceolate, petaloid. Fruit less than 12 mm diam. . . . . .
. . . . **2 RULINGIA**
4 Staminodes 3-lobed alternating with stamens; the central lobe the largest; lateral lobes filiform. Fruit c.
20 mm diam. . . . . **3 COMMERSONIA**

5 Staminodes absent. Petals scale like . . . . **6 LASIOPETALUM**
5 Staminodes present, subulate, often almost as long as the stamens or longer. Petals absent or scale like . 6

6 Calyx brownish. . . . . .4 SERINGIA
6 Calyx pale pink, drying bluish . . . . .5 KERAUDRENIA

7 Epicalyx present, caducous in *Lagunaria* 8
7 Epicalyx absent 15

8 Capsule 2–3 cm long, usually with 3–5 segments and covered with rigid hairs. Leaves with peltate
scales. Trees. . . . . .7 LAGUNARIA
8 Capsule usually with 5 or more segments. Leaves variously hairy but without peltate scales. Shrubs or
small trees. 9

9 Style branches twice as many as the ovary locules. Shrubs . . . . **14 PAVONIA**
9 Style branches as many as the ovary locules. Herbs, shrubs or small trees 10

10 Epicalyx of 5–12 bracteoles. Herbs, shrubs or small trees 11
10 Epicalyx of 3 bracteoles. Herbs 12

11 Stylar branches 5. Fruit a capsule . . . . .8 HIBISCUS
11 Stylar branches more than 15. Fruit a schizocarp. . . . . .9 ALCEA

12 Flowers pink, white or blue. Stigmatic surface longitudinal. 13
12 Flowers yellow, orange or red. Stigmatic surface capitate 14

13 Mericarps and styles 16–20. Mericarp not completely surrounding the seed . . . . . . **10 LAVATERA**
13 Mericarps and styles 6–12. Mericarp usually completely surrounding the seed . . . .**12 MALVA**

14 Mericarps and styles 16–22. Flowers solitary, pedunculate. . . . .**13 MODIOLA**
14 Mericarps and styles 5–18. Flowers in compact terminal spikes, sessile or nearly so. . . . .

. . . .**11 MALVASTRUM**

15 Leaves with peltate scales. Capsule 2–3 cm long, usually with 3–5 segments and covered with rigid
hairs. Trees. . . . . .7 LAGUNARIA
15 Leaves without peltate scales, variously hairy. Capsule less than 2 cm long, usually with 5 or more
segments (3 in *Howittia*). Shrubs or small trees. 16

16 Petals violet . . . . **15 HOWITTIA**
16 Petals yellow, orange or white 17

17 Stigmatic surface decurrent on the style . . . . **18 GYNATRIX**
17 Stigmatic surface capitate 18

18 Ovules 2 or more per loculus . . . . **16 ABUTILON**
18 Ovule 1 per loculus . . . .**17 SIDA**

1 **Brachychiton** Schott & Endl.
30 species in Aust.; all states and territories except Tas.

Trees. Leaves entire or lobed. Flowers in axillary panicles, unisexual or polygamous. Calyx petaloid,
tubular, 5-lobed. Petals absent. Stamens of male flowers 10–15, connate into a column. Female flowers
with 5 nearly or quite free carpels on a short gynophore, with 5 bundles of staminodes at the base. Styles
cohering when young but finally separating. Ovules numerous. Fruit of 5 (or fewer by abortion) stalked,
hard follicles opening along the inner suture.

Key to the species

1 Calyx bright coral-red, glabrous, 1–2 cm diam. Mature fruits c. 12 cm long. Leaves coriaceous, usually more than 10 cm long, palmately 5–7-lobed at juvenile stage, later becoming entire and ± ovate. Petioles stout. Coast and adjacent plateaus. RF. Fl. summer. *Flame Tree* . . . . **B. acerifolius** F.Muell.

1 Calyx creamy white, slightly downy pubescent outside, splashed with red and yellow inside, 1–2 cm diam. Fruits 4–7 cm long. Leaves thin; juvenile leaves often 3- or sometimes 5-lobed; mature leaves ovate to lanceolate or 3-lobed, c. 4 cm long, with slender petioles. Cumberland Plain; valleys of the Blue Mts e.g. Capertee; basalt areas in the drier parts of Hornsby Plateau. Fl. summer. *Kurrajong* . . . .
. . . . **B. populneus** (Schott & Endl.) R.Br. ssp. **populneus**

2 Rulingia R.Br.
20 species in Aust.; all states and territories except Tas.

Shrubs. Leaves simple, spirally arranged; stipules narrow, caducous. Flowers small, whitish, in cymes. Calyx 5-lobed. Petals 5, broad. Stamens 5, opposite the petals, alternating with 5 petal-like linear-lanceolate staminodes. Ovary sessile, 5-locular. Fruit a capsule, tomentose or beset with prickles or bristles.

Key to the species

1 Erect shrubs up to 4m high, softly pubescent except upper surface of leaves which may be scabrous . . 2
1 Prostrate shrubs, hirsute when young. Flowers 5–6 mm diam., in small cymes 3

2 Capsule globular, 5–7 mm diam., with rigid bristles which are glabrous except for tips. Calyx c. 2 mm long. Leaves ovate to lanceolate, mostly up to 7 cm long. Shrub to 4 m. Widespread. Ss., particularly in gullies. Fl. spring. *Kerrawang*. **R. dasyphylla** (Andrews) Sweet
2 Capsule globular, 8–15 mm diam., with rigid bristles which are softly pubescent. Calyx c. 4 mm long. Leaves linear to lanceolate, mostly 1–5 cm long. Shrub up to 2 m high. Coast and adjacent plateaus. Open forests. Fl. spring . . . . **R. rugosa** Steetz

3 Leaves lanceolate to oblong, rarely lobed at the base especially after fire, mostly 5–12 mm long but sometimes longer, almost sessile, wrinkled and glabrescent above, pubescent below; margins recurved, crenato-serrate. Capsule 4 mm diam., with short hirsute bristles up to 0.5 mm long. Coast and adjacent plateaus. Heath. Ss. Fl. spring. . . . . **R. hermanniifolia** (DC.) Endl.
3 Leaves ovate, mostly more than 15 mm long, only slightly wrinkled above, petiolate, pubescent underneath, glabrescent on upper surface; margins recurved, serrate. Capsule c. 6 mm diam., with long hirsute bristles (1–2 mm long towards the top of the capsule). Picton Lakes and southwards. Heath and DSF. Endangered. Fl. spring . . . . **R. prostrata** Maiden & Betche

3 Commersonia J.R.Forst. & G.Forst.
14 species in Aust. (12 endemic); all states and territories except Tas.

One species in the area

Tall, erect shrub with tomentose young branches. Leaves cordate to ovate-lanceolate, 7–15 cm long, often acuminate, glabrous or lightly pubescent on upper surface, white-tomentose underneath; margins irregularly dentate, rarely lobed. Flowers 8–10 mm diam., often numerous, whitish, in loose cymose panicles sometimes corymbose. Calyx lobes spreading, c. 4 mm long. Petals c. as long as the calyx lobes. Stamens 5. Staminodes 3-lobed as long as the petals. Ovary 5-locular. Capsule c. 2 cm diam., with soft hairy bristles. Widespread except higher Blue Mts River banks; near RF; on more fertile, cleared land. Fl. spring–summer. *Brown Kurrajong* or *Brush Kurrajong* . . . . **C. fraseri** J.Gay

4 Seringia J.Gay
1 species in Aust.; Qld, NSW

One species in the area

Tall shrub to 8 m high, with rusty tomentose young branches. Leaves ovate to ovate-lanceolate, mostly 4–10 cm long, often oblique at the base; upper surface glabrous or nearly so; undersurface densely stellate tomentose; margins coarsely toothed. Flowers bisexual, in short terminal or leaf-opposed cymes; bracteoles absent. Calyx angular in bud, tomentose, 4 mm long. Petals absent. Stamens 5, alternating with 5 small subulate staminodes. Carpels distinct in the fruit, tomentose, with a short broad vertical truncate wing on the back. Coast and adjacent plateaus; Blue Mts e.g. Yerranderie. Forests. Fl. spring . . . . **S. arborescens** (Aiton) Druce

5 Keraudrenia J.Gay
7 species in Aust.; Qld, NSW, NT, S.A., WA

One species in the area

Shrub c. 1 m high. Leaves ovate to lanceolate, up to 13 cm long, softly tomentose on undersurface but less so on upper surface, dentate. Flowers bisexual, in leaf-opposed cymes; bracteoles absent. Calyx tomentose, pink drying bluish, c. 5 mm long. Petals absent. Stamens 5, connate into a short tube with the 5 alternating staminodes. Colo River district. Open forests. Fl. spring–early summer. Endangered population in some LGAs . . . . **K. corollata** var. **denticulata** C.White

6 Lasiopetalum Sm.
35 species endemic Aust.; all states and territories except NT

Shrubs ± covered with stellate rust-coloured hairs. Leaves coriaceous, whitish or rust-coloured underneath, often glabrous on upper surface, without stipules. Flowers bisexual, in small axillary drooping or reflexed cymes which are loose or contracted almost into heads or apparently in simple racemes. Calyx 5-lobed nearly to the base, persistent, not much enlarging after flowering, usually with 3 bracteoles. Petals minute and scale-like or absent. Stamens 5, alternating with the calyx lobes. Staminodes absent. Ovary usually 3-locular, tomentose. Capsule shorter than the calyx.

Key to the species

1 Calyx lobes glabrous on the inside (except the margins), tomentose outside 2
1 Calyx lobes tomentose or pubescent both inside and outside. 3

2 Leaves usually linear to narrow-oblong, mostly 3–6 cm long, 3–7 mm wide, obtuse, smooth on upper surface. Calyx lobes 3–4 mm long; glabrous on the inside, whitish tomentose on the outside, Bracteoles almost linear. Shrub 0.5 m–1 m high. Cumberland Plain; Hornsby and Woronora plateaus; Burragorang. Usually on shales. Open forest or cleared areas. Fl. spring . . . . **L. parviflorum** Rudge
2 Leaves ovate-lanceolate to lanceolate, usually 5–10 cm long, 10–40 mm wide. Cymes short or almost capitate, reflexed, densely rusty tomentose. Calyx lobes 6–8 mm long, glabrous inside; rusty tomentose on the outside. Bracteoles lanceolate to ovate-lanceolate. Ovary sometimes 4-locular. Erect shrub 1–2 m high. Probably widespread but not common. Open forests. Fl. spring. . . . .**L. macrophyllum** Graham

3 Leaves narrow-oblong to lanceolate or cordate, 1–10 cm long 4
3 Leaves linear to narrow-oblong, mostly 3–6 cm long 5

4 Leaves narrow-oblong to lanceolate, 3–10 cm long, entire, sinuate or with short lobes, glabrous on upper surface. Cymes narrow or spreading, 5–10-flowered. Calyx lobes ovate, 5–8 mm long, acute, thick, deep rusty to light brown tomentose outside, whitish tomentose inside. Erect or diffuse shrub 1–3 m high but prostrate on exposed headlands near the sea. Widespread. Heath and DSF. Sandy soils. Fl. spring . . . .
. . . . **L. ferrugineum** Sm. var. **ferrugineum**
4 Leaves cordate at the base, ovate-lanceolate to lanceolate, 1–10 cm long. The leaves are often shorter than in var. *ferrugineum* but the leaf margins, inflorescence and flowers are similar. Erect or diffuse shrub up to 1 m high. Blue Mts Sandy soils. Fl. spring . . . . . . **L. ferrugineum** var. **cordatum** Benth.

5 Calyx lobes 8–10 mm long; bracteoles broader and closer to the calyx than in *L. rufum* but in all other respects similar. Erect shrub 50–150 cm high. Hornsby Plateau south of Hawkesbury River. Heath. Ss. Vulnerable. Fl. spring . . . . **L. joyceae** Blakely

5 Calyx lobes 3–5 mm long 6

6 Calyx lobes whitish tomentose inside, rusty brown tomentose outside. Cymes dense. Erect shrub up to 1 m high . . . . **L. ferrugineum** var. **ferrugineum** (see above)

6 Calyx lobes reddish inside, pubescent but glabrous towards the margins, c. 5 mm long; bracteoles linear-subulate, not close to the calyx. Cymes loose. Erect or diffuse shrub 50–150 cm high. Coast and adjacent plateaus; lower Blue Mts Ss. Fl. spring. **L. rufum** R.Br. ex Benth.

(In addition to variations within *L. ferrugineum*, intermediate forms between species, possible hybrids, will be found in some places.)

7 **Lagunaria** (DC.) Rchb.
1 species endemic Aust.; Lord Howe & Norfolk Islands

One species in the area
Tree up to 15 m high covered with scurfy scales. Leaves ovate to broad-lanceolate, C.10 cm long, 6 cm wide; margins entire;; petiole up to 2 cm long. Flowers solitary in leaf axil. Epicalyx 6 mm long, falling early. Calyx 10–25 mm long. Petals pink to mauve. Fruit a spiny capsule C. 3 cm long. Seeds red and fleshy. Garden escape. Widely cultivated and occasionally naturalized in coastal areas. *Norfolk Island Hibiscus*. . . . .*L. patersonii** (Andrews) G. Don.

8 **Hibiscus** L.
35 species native Aust.; all states and territories except Tas.

Herbs shrubs or small trees. Leaves often deeply divided. Flowers usually large. Epicalyx with 5 or more bracteoles. Calyx 5-lobed or 5-toothed. Petals usually marked with a deeper colour at the base. Staminal tube 5-lobed or truncate at the summit, with free filaments on the outside. Ovary 5-locular; stylar branches 5. Capsule 5-valved, coriaceous or membranous, dehiscing loculicidally.

Key to the species

1 Annual herb. Flowers solitary, on axillary pedicels. Leaves deeply 3–5-lobed, 3–8 cm long; the lobes often divided again. Calyx very shortly 5-lobed, membranous, with 20 raised veins, c. 12 mm long at flowering but much larger and inflated at fruiting stage. Petals pale yellow with dark purple bases. Capsule ovoid-globular, hirsute, enclosed in the calyx. Widespread. Garden weed. Probably introd. but origin unknown. Fl. spring–summer. *Bladder Ketmia* . . . .*H. trionum** L.

1 Shrubs or small trees. Calyx deeply lobed. 2

2 Leaves without glandular hairs on the midrib 3

2 Leaves with glandular hairs on the midrib 4

3 Petals pink, c. 7 cm long. Leaves entire, crenate or 3–5-lobed, 3–15 cm long, densely pubescent. Stems pubescent and with prickles. Pedicel articulate above the base. Capsule 20 mm long. Shrub or small tree up to 6 m high. Coast and adjacent plateaus. Open forest. Fl. spring–summer. . . . . **H. splendens** C.Fraser ex Graham

3 Petals purple, pink or white up to 2.5 cm long. Leaves ovate to lanceolate, margins crenate, 2 – 5 cm long, densely pubescent. Stems without prickles. Peduncles 2 – 5 cm long. Capsule 10 mm long. Shrub to 60 cm high, sometimes prostrate. Eucalypt woodland, rocky areas. *Hill Hibiscus* . . . . **H. sturtii** Hook. ssp. **sturtii**

4 Stems pubescent and with prickles. Leaves oblong to cordate or nearly orbicular, irregularly toothed, angular or ± 5-lobed, 5–7 cm wide, sparingly hispid. Flowers in a terminal raceme, with a bract under each pedicel. Calyx densely hispid with rigid hairs. Petals c. 5 cm long, pale yellow, darker below. Capsule acuminate, very hairy, c. 2 cm long. Spreading shrub. Coast. Swampy places. Fl. summer . . . . **H. diversifolius** Jacq.

4 Stems glabrous or glabrescent except for prickles. Leaves linear to narrow-elliptic, entire or deeply 3-lobed, up to 15 cm long, nearly glabrous. Flowers on short pedicels in the upper axils. Calyx 2 cm long, densely stellate tomentose. Petals up to 7 cm long, white pink or yellow with a dark spot near the base. Capsule ovoid-acute, c. as long as the calyx, densely stellate tomentose. Shrub or small tree up to 6 m high. Coast and adjacent plateaus. Sheltered places. Fl. summer. *Native Rosella* . . . .
. . . . **H. heterophyllus** Vent. ssp. **heterophyllus**

9 Alcea L.

1 species naturalized Aust.; NSW

One species in the area

Biennial herb up to 3 m high. Leaves c. 20 cm wide, 3–7-lobed, palmately veined, margins toothed. Upper leaves smaller than lower. Epicalyx with 6 connate bracteoles, 6–10 mm long. Sepals triangular 10 mm long. Petals white, pink or purple, 4 cm long. Fruit a schizocarp with more than 15 indehiscent mericarps. Mericarps with a central groove. An occasional garden escape near habitation. Introd. from Europe. *Hollyhock* . . . . ***A. rosea** L.

10 Lavatera L.

1 or 2 species naturalized Aust.; all states and territories except Tas.

One species in the area

Annual herbs to 50 cm high. usually covered with a stellate pubescence. Leaves ovate 2–6 cm long, ± lobed to unlobed, margins crenate. Petiole 1–7 cm long. Epicalyx with 3 broad connate bracteoles forming a lobed cup. Flowers usually solitary. Petals 8–30 mm long. Ovary 6–12-locular; stylar branches 6–12. Fruit a schizocarp, splitting into 1-seeded mericarps. Mericarps covered by a disc-like expansion of the central axis. An occasional garden escape. Introd. from the Mediterranean. Fl. spring–summer . . . . . .
. . . .* **L. trimestris** L.

11 Malvastrum A.Gray

3 species in Aust. (1 native, 2 introduced); all states and territories except Tas.

One species in the area

Herb or woody herb up to c. 1 m high. Leaves ovate, up to 6 cm long, hirsute-pubescent with stellate hairs, dentate, petiolate. Flowers in dense terminal spikes, c. 1 cm diam. Epicalyx with 3 lanceolate bracteoles. Calyx lobes longer than the tube, enclosing the fruit. Fruit a schizocarp with c. 10 mericarps. Introd. from America. *Spiked Malvastrum*. . . . .***M. americanum** (L.) Torrey

12 Malva L.

Mallow

C. 10 species in Aust. (1 endemic); all states and territories

Erect to prostrate annual herbs. Leaves palmately veined, with 5–7 short lobes. Epicalyx of 3 free or slightly connate bracteoles inserted near the base of the calyx. Petals emarginate or notched. Ovary 8–12-locular, with 1 ovule per loculus; stylar branches 8–12. Fruit an orbicular schizocarp separating into 1-seeded indehiscent mericarps.

Key to the species

1 Epicalyx segments ovate. 2
1 Epicalyx segments linear 5

2 Sepals half as long as epicalyx, 4–6 mm long. Flowers 2-7 in clusters. Petal 12–22 mm long, pale purplish with darker veins.. Mericarps c. 7, reticulate, ridged on the dorsal and lateral surfaces. Biennial with woody stems up to 3 m high. An occasional garden escape near habitation. Introd. from Europe. *Tree Mallow*. . . . . * **M. dendromorpha** M.F.Ray
2 Sepals nearly as long or longer than the epicalyx. 3

3 Sepals c.5 mm long, nearly as long as or just longer than epicalyx. Flowers in axillary clusters on short pedicels. Petals 5–12 mm long, pink to light purple or bluish twice as long as the sepals. Mericarps c. 10, reticulate. Mature calyx covering the fruit. Stem usually erect. Widespread. Common weed of waste ground. Introd. from the Mediterranean. Fl. spring–summer. . . . . *M. nicaeensis** All.

3 Sepals 6–16 mm long, Petals 12–22 mm long. 4

4 Petals 20 mm long or longer, lilac, pink or white. Leaves 20 cm long and 5-7-lobed, margins crenate or toothed, stellate hairy on both surfaces; petiole 10–15 mm long. Sepals enlarging and covering the fruit. Mericarps 12–15, dorsally reticulate. Fl. winter–summer (may be introduced to Sydney region from western NSW)n *Native Hollyhock* . . . . **M. australiana** M.F.Ray

4 Petals less than 18 mm long, bluish pink. Leaves broad-ovate, 3–10 mm long, sometimes to 20 mm; petiole up to 12 mm long. Sepals not inflated in the fruiting stage or only slightly so. Mericarps smooth or with a very few ridges with rounded angles between the dorsal and lateral faces. Weed of gardens and waste ground. Introd. from the Mediterranean. Fl. spring–summer . . . .*M. linnaei** M.F.Ray

5 Petals 6–12 mm long, lilac–pink, twice as long as sepals. Fruit furrowed on the rim between the carpels. Mericarps c. 15, smooth. Stems decumbent or prostrate. Widespread. Weed of waste ground. Introd. from the Mediterranean. Fl. spring–summer. *Dwarf Mallow.* . . . . *M. neglecta** Wallr.

5 Petals 4–6 mm long, as long as or just longer than the sepals. Fruit ridged on the rim between the carpels. Mericarps c. 11, strongly reticulate. Stems decumbent or prostrate. Widespread. Common weed of waste ground. Introd. from the Mediterranean. Fl. spring–summer. *Small-flowered Mallow* . . . . . .
. . . . *M. parviflora** L.

13 Modiola Moench

1 species naturalized Aust.; all states and territories except NT

One species in the area

Perennial herb with creeping or ascending stems rooting at the nodes. Leaves sub-orbicular, sometimes deeply palmatifid; margins crenate. Epicalyx with 3 free lanceolate bracteoles. Ovary and fruit villous at the summit, with 2 ovules per loculus. Fruit c. 8 mm diam., with c. 20 carpels. Widespread. Weed of cultivated and waste ground. Introd. from S. America. Fl. summer. *Red-flowered Mallow* . . . .
. . . . *M. caroliniana** (L.) G.Don

14 Pavonia Cav.

1 species naturalized Aust.; Qld, NSW

One species in the area

Small spreading shrub, covered with minute stellate hairs except the upper surfaces of the leaves which are scabrous. Leaves ovate-cordate to oblong-hastate, 1–3 cm long; margins crenate. Flowers axillary, solitary, on rather slender pedicels 1–3 cm long. Epicalyx with 5 free ovate bracteoles as long as the calyx. Calyx lobes c. 6 mm long. Petals reddish-purple with dark bases and veins, sometimes c. twice as long as the calyx but at other times short and closed over the stamens which are then reduced to 5. Ovary 5-locular. Fruit a schizocarp. Coast; Cumberland Plain; lower Blue Mts Open forest and partially cleared areas. Introd. from S. America. Fl. summer. . . . . **P. hastata** Cav.

15 Howittia F.Muell.

1 species endemic Aust.; NSW, Vic., S.A.

One species in the area

Erect shrub ± covered with stellate hairs, rusty on the younger parts. Leaves ovate-cordate to ovate-lanceolate, 2–10 cm long, yellowish underneath. Flowers axillary on slender pedicels. Epicalyx absent. Calyx lobes c. 5 mm long. Petals more than twice as long as the calyx, violet. Ovary 3-locular. Fruit a loculicidally dehiscing capsule, shorter than the calyx. Widespread. Open forests. Fl. spring–summer . . . .
. . . .**H. trilocularis** F.Muell.

16 Abutilon Mill.

30 species in Aust. (28 native, 2 naturalized); all states and territories except Tas.

Shrubs or herbs, usually softly pubescent with stellate hairs. Stipules subulate, caducous. Flowers solitary or 2 together on pedicels which articulate near the top. Epicalyx absent. Petals yellow. Ovary usually 5-locular or more. Fruit a schizocarpic capsule, with 2 seeds per loculus.

Key to the species

1 Calyx c. 20 mm long; the lobes longer than the fruiting carpels. Fruiting carpels 7–10, with erect tips. Petals yellow-orange, c. 3 cm long, adnate high up to the staminal tube. Shrub up to 1.5 m high. Shoalhaven River and Rylestone areas. Fl. spring . . . . **A. tubulosum** (A.Cunn. ex Hook.) Walp.
1 Calyx 4–10 mm long, shorter than the fruiting carpels. Fruiting carpels with divergent tips 2

2 Leaves broad-ovate to ovate. Calyx 8–10 mm long. Mericarps densely villous. Soft-wooded shrub up to 1.5 m high. Garden escape near habitation. Introd. from S. America. Fl. winter–spring . . . .
. . . . ***A. grandifolium** (Willd.) Sweet
2 Leaves ovate to narrow-ovate. Sepals 4–5 mm long. Stems with short patent hairs. Mericarps sparsely villous. Soft-wooded shrub up to 1.5 m high. Widespread but not common. Open forests and along creeks. Fl. summer–autumn. *Flannel Weed* . . . .**A. oxycarpum** (F.Muell.) F.Muell. ex Benth.

17 Sida L.

C. 40 species in Aust.; all states and territories except Tas.

Small shrubs with stellate hairs. Stipules subulate, caducous. Flowers on axillary peduncles (when the peduncles are 1-flowered they are often bent a short distance below the flower) Epicalyx absent. Petals yellow or orange. Ovary 5-locular. Fruit a schizocarp.

Key to the species

1 Plants prostrate to decumbent with a stout rootstock. Leaves orbicular to obovate, often cordate, serrato-crenate with large teeth. Calyx c. 4 mm long, broad and spreading under the ripe fruit. Petals longer than the calyx. Mericarps 6–10, without awns wrinkled on back. Widespread but not common
. . . .**S. corrugata** Lind.
1 Erect or ascending shrubs up to 2 m high. 2

2 Calyx 10-ribbed. Mericarp awned. 3
2 Calyx without prominent ribs. Mericarp not awned. 5

3 Peduncles more than 7 mm long. Mericarps 7–11. Flowers solitary. Calyx c. 5 mm long. Petals longer than the sepals. Leaves lanceolate to ovate, 2–5 cm long, deeply dentate, paler on undersurface. Widespread. Common weed of gardens and waste ground. Fl. spring–summer. *Common Sida* or *Paddy's Lucerne* . . . . **S. rhombifolia** L.
3 Peduncles less than 7 mm long. Mericarps 5–7. 4

4 Mericarps 5–7, glabrous. Plants glabrescent. Leaves narrow lanceolate 2–8.5 cm long. Flowers solitary sometimes crowded in leaf axils. Petals 6–9 mm long. Sepals 6 mm long. Fruit 3–5 mm diam., rounded on back. Shrub up to 1.5 m high. Occasional. Introduced from C. America. . . . . . . ***S. acuta** Burm.f.
4 Mericarps 5, pubescent. Plants pubescent. Leaves ovate to lanceolate, 1–4 cm long. Flowers solitary or 2–5 in leaf axils. Petals 5–7 mm long. Sepals 5 mm long. Fruit 4–6 mm diam., wrinkled on back. Shrub up to 1 m high. Disturbed sites . . . . **S. spinosa** L.

5 Peduncles 0–1 mm long. Mericarps 4 – 6, obtuse. Sepals 2–3 mm long. Petals only slightly longer than the sepals. Leaves lanceolate to narrow-oblong, up to 6 cm long, dentate, paler on undersurface. Hunter River Valley. . . . .**S. subspicata** F.Muell. ex Benth.
5 Peduncles more than 10 mm long, flowers solitary in leaf axils. Mericarps 6–9, wrinkled. Sepals 4–5 mm long. Petals longer than sepals. Leaves linear-oblong, 1–3 cm long. Northern part of region, Yengo National Park. Grows in clay soils . . . . **S. trichopoda** F.Muell.

18 Gynatrix Alef.

1 species endemic Aust.; NSW, Tas.

One species in the area

Dioecious shrub up to 2 m high with a variable stellate tomentum. Leaves ovate, up to 7 cm long, serrate, acuminate, cordate at the base. Flowers 3–4 mm diam., in axillary panicles. Epicalyx absent. Petals just exceeding the calyx. Ovary 5-locular. Fruit a schizocarp with 1 seed per loculus. Cumberland Plain and Blue Mts *Hempbush* . . . . **G. pulchella** (Willd.) Alef.

100 THYMELAEACEAE

Shrubs with a thin tough bark. Leaves alternate or opposite, simple, entire; stipules absent. Flowers regular, bisexual or unisexual. Floral tube cylindrical or campanulate. Sepals, which form the floral tube, 4, imbricate. Petals absent or rudimentary. Stamens 2–8, perigynous. Ovary superior, with a solitary pendulous ovule. Fruit a 1-seeded capsule or berry. 46 gen., mostly warm temp.

Key to the genera

1 Stamens 2. Petals absent . . . . **1 PIMELEA**
1 Stamens 8. Petals small and rudimentary . . . . **2 WIKSTROEMIA**

1 Pimelea Banks & Sol. ex Gaertn.

90 species endemic Aust.; all states and territories

Shrubs. Leaves simple, entire, alternate or opposite. Flowers bisexual or unisexual, variously arranged but often terminal and surrounded by an involucre of bracts sometimes different from the leaves. Floral tube cylindrical, usually circumsciss above the ovary; the upper part deciduous. Sepals 4. Stamens 2, inserted in the throat, Ovary 1-locular, enclosed in the floral tube but free from it; style elongated, simple. Fruit with a membranous epicarp, rarely succulent.

Key to the species

1 Flowers in spikes elongated and interrupted when fruiting. Leaves elliptic, 5–20 mm long. Flowers white to pinkish, 7–10 mm long. Decumbent to erect shrub up to 50 cm high. Coast and Cumberland Plain. Grasslands. Endangered . . . . **P. spicata** R.Br.
1 Flowers in heads 2

2 Flowers up to c. 20 mm long, white or pink to cream. Heads surrounded by bracts clearly different from the leaves 3
2 Flowers up to 10 mm long, greenish yellow or yellow to red. Heads not surrounded by bracts or surrounded by bracts similar to the leaves. 10

3 Inner bracts ciliate on margins 4
3 Inner bracts not ciliate on margins. 6

4 Floral tube glabrous at base. Leaves lanceolate, up to 20 mm long, bluish green. Heads with more than 7 flowers. Bracts 4, lanceolate to ovate, up to 15 mm long, glabrous outside. Open forests and grasslands. Widespread. Open forests and grasslands. . . . . **P. glauca** R.Br.
4 Floral tube hairy at base. Leaves oblanceolate to elliptic, up to 10 cm long, glabrous except when young. Heads with more than 15 flowers. Bracts 4–8, elliptic to ovate, 4–18 mm long. Shrubs up to 3 m high . 5

5 Peduncles glabrous or nearly so. Bracts glabrous inside, hairy outside. Shrub up to 1.5 m high. Widespread. Open forests. Fl. summer . . . . **P. ligustrina** Labill. ssp. **ligustrina**
5 Peduncles hairy just below the bracts. Bracts villous inside and outside. Shrub up to 2 m high. Coast and adjacent plateaus. Margins of WSF. Fl. summer. . . **P. ligustrina** ssp. **hypericina** (Cunn.) Threlfall

6 Stems hairy. Leaves elliptic to lanceolate, 5–15 mm long. Bracts 4, glabrous outside, hairy inside (sometimes only on midvein). Heads with up to c. 50 flowers. Straggling shrub up to 50 cm high. Robertson. DSF and heath . . . . **P. humilis** R.Br.

6 Stems glabrous except when very young. Leaves oblanceolate to narrow-elliptic, up to 35 mm long. Bracts 4–8, oblanceolate to ovate, up to 20 mm long. Heads with up to 60 flowers 7

7 Leaves and bracts with prominent raised intramarginal veins at least on undersurface, often drying to a bluish colour. Flowers usually silky. Erect shrub up to 1 m high. Widespread. Heath and forest, usually in damp situations. Fl. spring–summer . . . .**P. linifolia** ssp. **collina** (R.Br.) Threlfall

7 Leaves and bracts without prominent intramarginal veins. Flowers hairy but not silky 8

8 Leaves glaucous, narrow-elliptic, 5–10 mm long. Heads with usually less than 20 flowers. Prostrate or dense, low shrub up to 50 cm high. Widespread. Open forest and heath. Fl. spring–summer . . . .
. . . . **P. linifolia** ssp. **caesia** Threlfall

8 Leaves green. Heads with usually more than 20 flowers. Leaves narrow-oblong to narrow-elliptic, 5–35 mm long. 9

9 Luxuriant shrub up to 2.5 m high. Blue Mts Open forests and heaths. Fl. summer . . . .
. . . .**P. linifolia** ssp. **linoides** (A.Cunn.) Threlfall

9 Erect to prostrate shrub up to 1.5 m high. Very variable. Widespread. Open forests and heaths. Fl. spring–summer. . . . .**P. linifolia** Sm. ssp. **linifolia**

10 Leaves tomentose or villous on both surfaces. Erect shrubs up to 50 cm high 11

10 Leaves glabrous or nearly so on upper surface. Flowers greenish yellow to yellow or reddish. Heads with up to c. 20 flowers. 12

11 Leaves tomentose with spreading hairs, mostly narrow-elliptic, up to 20 mm long, 3–6 mm wide. Heads with 1–7 flowers. Shrub up to 2 m high. Coast to Blue Mts Open forests. Fl. spring–summer . .
. . . . **P. latifolia** R.Br. **ssp. hirsuta** (Meisn.) Threlfall

11 Leaves with antrorse-appressed hairs, mostly narrow-elliptic, up to 18 mm long, 3–6 mm wide. Heads with 6–20 flowers. Shrub up to 50 cm high. Hunter River Valley. Forests on rocky sites. Fl. spring–summer. . . . .**P. latifolia** ssp. **elliptifolia** Threlfall

12 Undersurface of leaves with sparse coarse appressed hairs. Leaves narrow-elliptic to oblanceolate, 5–10 mm long. Erect shrub up to 1.5 m high. Ss. Open forests. Coast and adjacent plateaus. DSF and heath. Fl. Vulnerable. spring–summer . . . .**P. curviflora** R.Br. var. **curviflora**

12 Undersurface of leaves with ± dense appressed or spreading hairs. 13

13 Undersurface of leaves with long spreading sparse hairs. Leaves narrow-elliptic, up to 22 mm long. Low shrub up to 50 cm high. Coast and adjacent plateau from Sydney northwards. Open forests. Fl. spring–summer . . . . **P. curviflora** var. **divergens** Threlfall

13 Undersurface of leaves with appressed hairs. 14

14 Undersurface of leaves with only a few very short coarse hairs. Leaves linear to elliptic or oblanceolate, 2 mm wide. Shrub to 30 cm high. Uncommon in scrub . . . . .**P. curviflora** var. **subglabrata** Threlfall

14 Undersurface of leaves moderate to densely hair. Hairs course or fine. 15

15 Pedicels with long antrorse hairs. Leaves narrow-elliptic to obovate, up to 20 mm long. Flowers bisexual and female. Erect shrub up to 1.5 m high. Widespread. Open forests. Fl. spring–summer . . .
. . . . **P. curviflora** var. **gracilis** (R.Br.)Threlfall

15 Pedicels with short appressed hairs. Leaves narrow-elliptic to elliptic, 6–20 mm long. Shrub to 30 cm high. Open forest. . . . .**P. curviflora** var. **sericea** Benth.

2 **Wikstroemia** Endl.

1 species native Aust.; Qld, NSW

One species in the area

Shrub up to 1.5 m high. Leaves mostly opposite, ovate to oblong or lanceolate, paler underneath, rather thin, glabrous, c. 3 cm long. Flowers in few-flowered terminal heads sometimes growing out into short spikes, bisexual, erect or slightly recurved. Floral tube cylindrical, shortly 4-lobed, greenish yellow, c. 8 mm long. Petals rudimentary. Stamens 8. Fruit a red drupe, 5–8 mm long. Coast and adjacent plateaus. RF margins; coastal headlands and other places near the sea. . . . .**W. indica** (L.) C.A.Meyer

101 ANACARDIACEAE

Trees with resin ducts in the bark. Leaves pinnately compound, alternate; stipules absent or minute. Flowers small, regular, unisexual, in panicles. Sepals 5, basally connate. Petals 5, white. Stamens 10, in 2 whorls, inserted around a perigynous disc. Ovary superior, usually 1-locular, with a solitary basal ovule; styles 3, free or basally connate. Fruit drupaceous. 70 gen., mostly trop., some temp.

Key to the genera

1 Leaflets 1 cm wide or less. Branchlets and leaves drooping . . . . **4 SCHINUS**
1 Leaflets more than 1.5 cm wide. Branchlets and leaves spreading 2

2 Flowers white to pink, fruit purple-black. Rainforest tree. . . . . **3 EUROSCHINUS**
2 Flowers yellow-green, fruit whitish or red and turning purple at maturity 3

3 Leaves imparipinnate, glabrous, lower surface of leaflets glaucous. Flowers with petals . . . .
. . . . **1 TOXICODENDRON**
3 Leaves paripinnate, bristles scattered along midrib, lower surface of leaflets not glaucous. Flowers without petals. . . . . **2 PISTACIA**

1 **Toxicodendron** Mill.

1 species naturalized AUST; NSW

One species in the area

Deciduous tree to 5 m high; branchlets glabrous; dioecious. Leaves, alternate, 10–30 cm long; petiole 3–10 cm long. Leaflets, opposite, 9–15, lanceolate to ovate 5–9 cm long, 15–30 mm wide; upper and lower surface glabrous. Inflorescence a panicle, flowers yellow-green, 5-merous. Stamens 5 inserted below disk. Ovary 1-locular. Fruit globose and laterally compressed 6–10 mm long, whitish. Occasionally naturalized. Introd. from E. Asia. *Rhus Tree* (a noxious weed in NSW, poisonous). . . . ***T. succedaneum** (L.) Kuntze

2 **Pistacia** L.

1 species naturalized in Aust.; NSW

One species in the area

Deciduous tree to 20 m high; branchlets hairy; dioecious. Leaves alternate, 10–25 cm long; petiole 4–10 cm long. Leaflets ± opposite, 8–14, lanceolate, 5–9 cm long, 10–20 mm wide. Inflorescence a panicle, flowers yellow-green, 5-merous. Stamens 5 inserted below disk. Ovary 1-locular. Fruit globose and laterally compressed, 5–6 mm long, red when immature, purple at maturity. Occasionally naturalized. Introd. from China. *Chinese Pistachio* . . . . ***P. chinensis** Bunge

3 **Euroschinus** Hook.f.

1 species endemic Aust.; Qld, NSW

One species in the area

Small to medium tree. Leaves alternate, pinnate; leaflets 4–10, ovate to lanceolate, very oblique, 5–7 cm long. Flowers unisexual, small, numerous, in panicles shorter than the leaves. Calyx 5-lobed, short. Petals

2 mm long. Stamens 10. Styles 3. Fruit ovate, 6–12 mm long. Coast, e.g. Gosford and Royal National Park southwards. In or near RF. *Ribbonwood*. . . . . **E. falcata** Hook.f. var. **falcata**

4 Schinus L.
2 species naturalized Aust.; Qld, NSW, Vic., S.A.

One species in the area

Tree with drooping branchlets. Leaves c. 20 cm long, pinnate; leaflets 3–17, narrow-lanceolate, 3–4 cm long, sessile, strongly aromatic. Flowers white, in a drooping panicle. Fruit red, c. 3 mm diam., containing aromatic oil. Naturalized in a few places, e.g. Razorback Range. Introd. from S. America. *Pepper Tree* . . . . ***S. areira** L.

102 MELIACEAE

Trees or large shrubs. Leaves pinnately compound, alternate; stipules absent. Flowers regular, bisexual, in axillary cymo-panicles. Calyx 4–6-lobed. Petals 4–6. Stamens hypogynous, as many as or twice as many as the petals; filaments connate into a tube or free. Ovary 3–5-locular, superior, often surrounded by an annular disc; style simple; placentas axile. Fruit drupaceous or a capsule. 52 gen., mostly trop.

Key to the genera

1 Leaves bipinnate. Filaments connate into a tube. Fruit drupaceous . . . . **1 MELIA**
1 Leaves once pinnate. Fruit a capsule 2

2 Leaves imparipinnate . . . . **3 SYNOUM**
2 Leaves paripinnate 3

3 Filaments free . . . . **2 TOONA**
3 Filaments connate . . . .**4 DYSOXYLUM**

1 Melia L.
1 species native Aust.; Qld, NSW, WA

One species in the area

Deciduous tree with spreading branches. Leaves bipinnate; leaflets ovate to narrow-ovate, acuminate, entire or coarsely dentate, 2–5 cm long. Panicle loose, slightly shorter than the leaves. Petals 8–10 mm long, lilac. Filaments connate into a dark-coloured (purplish) tube which is 10–12-toothed. Fruit drupaceous, ovoid, 12–20 mm long. Coast and Cumberland Plain. RF. often planted for shade. Fl. Oct.–Nov. *White Cedar*. . . . . **M. azedarach** L.

2 Toona (Endl.) M.Roem.
1 species native Aust.; Qld, NSW

One species in the area

Deciduous medium to very large tree with spreading branches. Leaves paripinnate; leaflets 11–17, obliquely ovate-lanceolate, acuminate, 7–12 cm long, with very small indistinct sunken glands (domatia) in the forks of several of the primary veins on the undersurface of the leaves (Fig. 30). Panicles large, pyramidal, many-flowered. Petals white or pinkish, 5–6 mm long. Stamens 5, free. Fruit an oblong capsule 2–3 cm long. Seeds winged. Widespread. RF. Fl. Nov.–Dec. *Red Cedar*. . . . . **T. ciliata** M.Roem.

3 Synoum A.Juss.
1 species endemic Aust.; Qld, NSW

One species in the area

Evergreen shrub or small tree. Leaves imparipinnate; leaflets mostly 5–9, elliptic to lanceolate, acuminate, entire, 3–10 cm long, with a distinct small hairy sunken gland (domatia) on the undersurface at the junction of several primary veins with the midvein. Panicles very short, rarely more than 25 mm long.

Petals 4 rarely 5, reddish on the outside, 4–5 mm long. Filaments connate into a broad tube. Ovary 3-locular. Fruit a reddish capsule, depressed-globular, 3-furrowed, 20–30 mm diam. Seeds with an aril. Widespread. RF and in sheltered situations; stabilized sand dunes. Fl. most of the year. *Bastard Rosewood* . . . . **S. glandulosum** (Sm.) A.Juss. ssp. **glandulosum**

4 Dysoxylum Blume
14 species in Aust. (3 endemic); Qld, NSW, NT, WA

One species in the area

Tree up to 30 m high; younger parts minutely pubescent. Leaves paripinnate; leaflets 5–9, oblong to elliptic, 7–15 cm long, acuminate. Flowers in axillary panicles. Calyx very short. Petals 4, c. 4 mm long, adnate to the staminal tube. Stamens 8; filaments connate. Ovary 3-locular, surrounded by and usually exceeding the tubular disc. Coast, Wyong northwards. RF. *Rosewood* . . **D. fraserianum** (A.Juss.) Benth.

103 RUTACEAE

Shrubs or trees. Leaves simple or compound, opposite or alternate, mostly entire, without stipules, dotted with obvious or sometimes obscure translucent oil glands. Flowers regular, bisexual (occasionally unisexual in *Melicope*). Sepals 4–5, free or cohering at the base or forming a cup. Petals equal in number to sepals, free (except *Correa* and one species of *Leionema*). Stamens equal in number to petals and alternating with them or double the number or rarely more numerous, free (except for 2 species of *Philotheca*). Ovary superior, surrounded by a perigynous disc, 4–5-locular, with 1–2 ovules per loculus; placentas axile or basal; style usually simple, gynobasic or terminal. Fruit a schizocarp-capsule, separating into mericarps and the endocarp separating explosively from the epicarp; or a berry; or an explosive capsule. 150 gen., temp. and subtrop., particularly Africa and Aust.

Key to the genera

1 Leaves opposite. Petals 4. 2
1 Leaves alternate. Petals 5 9

2 Leaves compound 3
2 Leaves simple 5

3 Stamens 8 . . . . **1 BORONIA**
3 Stamens 4 4

4 Leaves and stems glabrous and without prominent tubercles. . . . .**12 MELICOPE**
4 Leaves and stems pubescent, or, if glabrous then with prominent tubercles . . . .**2 ZIERIA**

5 Leaves stellate-tomentose at least underneath 6
5 Leaves glabrous or with simple hairs. 7

6 Petals pink, free. . . . . **1 BORONIA**
6 Petals white or green, sometimes reddish if coherent into a tube . . . .**3 CORREA**

7 Flowers unisexual. Plants dioecious. Buds globular or nearly so. Petals slightly imbricate . . . .
. . . .**14 SARCOMELICOPE**
7 Flowers bisexual, with buds longer than wide. Petals valvate. 8

8 Tree or shrub more than 2 m high . . . . **13 ACRONYCHIA**
8 Small shrub less than 2 m high . . . . **1 BORONIA**

9 Stamens 5 . . . . **11 GEIJERA**
9 Stamens 10 or more 10

10 Plants with axillary spines . . . .**15 CITRUS**
10 Plants without spines 11

11 Calyx inconspicuous, hidden beneath hairs. Stigma large . . . . **10 ASTEROLASIA**
11 Calyx conspicuous or if small, then not hidden beneath hairs. Stigma small 12

12 Petals yellow or white, 3–6 mm long, if more than 6 mm then petals connate forming tube 13
12 Petals pink or sometimes white, more than 6 mm long 15

13 Petals valvate, shrub with simple or stellate hairs. **6 LEIONEMA**
13 Petals imbricate, shrub or small tree with scurfy scales at least on branchlets 14

14 Anthers basifixed with a gland at apex, bracteoles attached at base of pedicel and insignificant . . . . .
. . . . **4 PHEBALIUM**
14 Anthers versatile without an appendage at the apex, bracteoles attached at or above the centre of the
pedicel . . . . **5 NEMATOLEPIS**

15 Stamens terminated by an appendage longer than the anther . . . . **7 CROWEA**
15 Stamens terminated by an appendage much shorter than anther or appendage obsolete 16

16 Leaves smooth, greyish, 3-veined from base; bracts prominent, imbricate, and rounded at tip. . . . . .
. . . . **8 ERIOSTEMON**
16 Leaves bullate, green, 1-viened from base; bracts not prominent.. . . . . **9 PHILOTHECA**

1 Boronia Sm.
c. 150 species endemic Aust.; all states and territories

Shrubs. Leaves opposite, simple or compound. Calyx 4-lobed. Petals 4, spreading. Stamens 8. Ovary
4-locular, deeply 4-lobed. Fruit an explosive capsule. Seeds 1 (2) per loculus.

Key to the species

(Intermediate forms occur between *B. floribunda* and *B. serrulata* and between *B. floribunda* and *B. microphylla*)

1 Lower surface of leaf or leaflet concealed by indumentum of dense hairs. Leaves simple or 3-foliolate,
sometimes with up to 7 leaflets, 1–4 cm long, linear to narrow-elliptic, glabrous above when mature;
margins recurved. Flowers solitary (rarely very few in a small cyme). Petals pink, 5–10 mm long. Shrub
30–150 cm high; young branches glandular, tomentose. Widespread. Heath and DSF. Ss. Fl. winter–
spring. *Ledum Boronia*. . . . . **B. ledifolia** (Vent.) DC.
1 Hairs if present not concealing the under surface of leaf 2

2 Stellate hairs present 3
2 Stellate hairs absent 5

3 Branches and petioles glabrous or nearly so. Stems 4-angled. Sepals broad-ovate. Petals pink, 6–8 mm
long. Otherwise similar to *B. mollis*; forms intermediate between these two species are included under
B. mollis. Coast and adjacent plateaus; lower Blue Mts Gullies on Ss. Fl. spring. *Fraser's Boronia* . . . . .
. . . . **B. fraseri** Hook.
3 Branches and petioles densely softly hirsute. Stems terete 4

4 Leaves tomentose underneath. Leaflets oblong-elliptic to obovate; the terminal one 15–40 mm long; the
others much shorter and broader in proportion. Sepals linear. Petals pink, 8–10 mm long. Shrub 1–2 m
high. Coast and Cumberland Plain. Ss. Fl. spring. *Soft Boronia* . . . . . . **B. mollis** A.Cunn. ex Lindl.
4 Leaves glabrous or nearly so. Leaflets elliptic to spathulate, 4–30 mm long, 3–12 mm wide. Petals pink,
6–10 mm long. Shrub up to 2 m high. Coast and adjacent plateaus. DSF. Ss. Fl. spring.. . . . .
. . . . **B. rubiginosa** A.Cunn. ex Endl.

5 Petal tips inflexed. 6
5 Petal tips not inflexed 11

6 Leaves simple 7

6 Leaves compound 8

7 Branchlets glabrous. Leaves very narrow-ovate to linear, 5–12 mm long. Flowers axillary, usually solitary. Sepals c. 1.5 mm long. Petals 4–6 mm long, reddish or pale pink. Anthers with a small recurved appendage. Shrub 8–20 cm high. Cumberland Plain; Hornsby Plateau. Usually on shale. Fl. spring. *Milkwort Boronia* . . . .**B. polygalifolia** Sm.
7 Branches and branchlets hispid. Otherwise similar to *B. polygalifolia*. Upper Blue Mts, Kanimbla Valley. Heath and DSF. Fl. spring–summer. *Dwarf Boronia* . . . . . . **B. nana** Hook. var. **hyssopifolia** Melville

8 Leaflets all entire, lanceolate to linear, 3–4 mm long, thick, grooved on the upper surface, often with recurved tips. Petals white to pale pink. Shrub rarely more than 40 cm high, with pubescent branches. Widespread. Heath and DSF. Ss. and stable dunes. Fl. spring. *Stiff Boronia* . . . . **B. rigens** Cheel
8 At least some of the leaflets divided into 3 or with 3 lobes or teeth towards the apex; oil glands distinct 9

9 Leaflets acute, leaves bipinnate, at least some of the leaflets further divided into 3 entire secondary leaflets, 3–15 mm long. Branchlets usually prominently 4–angled, glabrous. Flowers often more than 3 in each axillary cyme; the cymes shorter than the leaves. Petals pale pink to white, 3–5 mm long. Shrub up to 1 m high. Blue Mts Heath and DSF. Fl. winter–spring. *Narrow-leaved Boronia* . . . .**B. anethifolia** A.Cunn. ex Endl.
9 Leaflets 3-toothed, ± deeply 3-lobed or truncate. Branchlets shallowly 2–angles, glabrous to pilose. Petals pale pink to white, 4–5 mm long 10

10 Leaves hirsute usually with 3–5 leaflets. Leaflets linear-cuneate, thick, 3–8 mm long; terminal leaflet ± as long as lateral leaflets. Flowers solitary, in pairs or rarely in 3s in the upper axils. Petals persistent in fruit. Coast and Upper Blue Mts Shrub up to 80 cm high. Heath and DSF. Fl. spring. *Sticky Boronia* . . . . **B. anemonifolia** A.Cunn.ssp. **anemonifolia**
10 Leaves ± glabrous, pinnate or bipinnate with 3–5 leaflets. Leaflets broadly cuneate to oblanceolate, 3–13 mm long, terminal leaflet usually shorter than lateral leaflets. Flowers in groups of 6–9. Petals not persistent in fruit. Widespread in Blue Mts Shrub up to 80 cm high. Heath and DSF. Fl. spring. *Sticky Boronia* . . . . **B. anemonifolia** ssp. **variabilis** (Hook.) P.G.Neish

11 Leaves simple 12
11 Leaves compound 18

12 Leaves semi-terete, 2–12 mm long, obtuse or acute, smooth on the upper surface, ± warty underneath from the presence of oil glands. Flowers terminal solitary or 3 together (occasionally 2). A pair of leafy bracts present at the base of the peduncle. Petals white to deep pink, c. 5 mm long. Erect, glabrous shrub up to 1.5 m high, with very strong odour. 13
12 Leaves flat, 7–30 mm long 14

13 Apex of leaf obtuse. Lower surface of leaf slightly warty. Branches sparsely glandular-tuberculate. Blue Mts Clarence to Wolgan. Wet heath. Vulnerable. Fl. winter–spring. *Deane's Boronia* . . . . **B. deanei** Maiden & Betche ssp. **deanei**
13 Apex of leaf acute. Lower surface of leaf prominently warty. Branchlets with a dense row of warts along the decurrent leaf bases. Wingecarribee Reservoir. Wet heath. Vulnerable. Fl. spring. *Deane's Boronia* . . . . **B. deanei** ssp. **acutifolia** Duretto

14 Leaves rhomboidal or broad-ovate, serrulate, crowded, almost sessile, 7–15 mm long, coriaceous, veins very faint or not visible; acute. Sepals acute. Petals rose-pink, 8–10 mm long, broad-ovate, imbricate. Shrub 1–1.5 m high. Woronora and Hornsby plateaus. Sandy soil in wet heath. Fl. spring. . . . . **B. serrulata** Sm.
14 Leaves elliptic or very narrow-ovate 15

15 Petals 3–6 mm long, pink. Sepals persistent in fruit. Flowers solitary or few together, on pedicels c. 4 mm long. Leaves entire or finely serrulate, narrow-elliptic or very narrow-ovate, 7–15 mm long. Erect shrub 30–100 cm high. Woronora and Hornsby Plateaus. Wet sandy peat, often at the edges of swamps. Fl. most of the year. *Swamp Boronia* . . . . **B. parviflora** Sm.

15 Petals 6–11 mm long, pink to mauve. Sepals not persistent in fruit. Flowers mostly numerous, on pedicels 5–18 mm long. Leaves mostly serrulate sometimes entire, broad-elliptic lanceolate or oblanceolate, 15–30 mm long. Shrub, similar to *B. parviflora* but larger in all its parts. 16

16 Sepals 3.5–6 mm wide, ovate. Inner surface of petals glabrous. Pedicels 1.5–2 mm wide at apex. Leaves obovate to oblanceolate 12–23 mm long, 4–9 mm wide, margins toothed. Flowers 1–3 together. Sydney Harbour area south to Waterfall. Extinct. Fl. spring.. . . . . . **B. barkeriana** ssp. **gymnopetala** Duretto
16 Sepals 1.5–3 mm wide, triangular. Inner surface of petals glabrescent to minutely hairy. Pedicel 1 mm wide at apex. 17

17 Leaves elliptic to oblanceolate, margins prominently toothed, 16–30 mm long, 4–11 mm wide. Flowers in groups of 3–9. Petals 6–8 mm long. Wollomi National Park to Blue Mts Fl. spring–summer . . . . .
. . . . **B. barkeriana** F.Muell. ssp. **barkeriana**
17 Leaves narrow–elliptic to narrow–obovate, margins entire or slightly toothed, 15–33 mm long, 1.5–6.5 mm wide. Flowers in groups of 3–8 sometimes 12 flowers. Petals 5–10 mm long. Morton National Park to the coast. Fl. spring–summer . . . . **B. barkeriana** ssp. **angustifolia** Duretto

18 Stamens glabrous. Flowers solitary, terminal. Leaflets thick, obovate, 4–7 mm long. Sepals deltoid. Petals white to pink, c. 5 mm long. young branches slightly pubescent. Low, spreading shrub up to 1 m high;. Blue Mts DSF. Fl. spring–summer. . . . . **B. algida** F.Muell.
18 Stamens pilose. 19

19 Style obscured by globular stigma, almost as wide as the ovary. Petals mostly pale pink, sometimes deep pink or white, 6–10 mm long. Flowers c. 3 in each loose axillary cyme; the cymes scattered along the branches. Leaflets elliptic to narrow-oblong, 5–25 mm long. Shrub usually c. 1 m high. Widespread. Heath and DSF. Ss. Fl. Sept.–Oct. *Pale Pink Boronia*. . . . **B. floribunda** Sieber ex Rchb.
19 Style distinct, stigma not swollen, or if so, then scarcely half as wide as the ovary. 20

20 Stems obviously glandular tuberculate. Leaflet obovate or cuneate-oblong, mostly emarginate, thick, 4–6 mm long. Flowers 1–3 in the cyme. Petals 5–6 mm long, rarely up to 8 mm long, rose-pink to rose-purple. Low shrub up to c. 1 m high. Upper Blue Mts; Hilltop. Heath and DSF. Ss. Fl. Oct.–Jan. *Small-leaved Boronia*. . . . . **B. microphylla** Sieber ex Rchb.
20 Stems not obviously glandular tuberculate. Leaves narrow-elliptic or lanceolate to oblong-linear, mostly acute, 5–30 mm long. 21

21 Leaves 3-foliolate leaflets linear, thick, almost terete, up to 12 mm long. Flowers arranged in 1–3-flowered cymes in upper leaf axils. Petals pink 8–10 mm long. Undershrub with few stems, usually c. 50 cm high. Coast; Wet heath. Fl. spring (this description is based on only one specimen believed to have been collected in the Woy Woy district in 1927. It is otherwise recorded north of the Hunter River where it can have 3–5-foliate or simple leaves) . . . . **B. falcifolia** A.Cunn. ex Endl.
21 Leaves ≥5-pinnate. 22

22 Leaflets entire, 5–25 mm long, rather thick and coriaceous, mostly 5–9 per leaf. Flowers numerous in loose axillary or subterminal corymbose cymes. Petals rose-pink or rose-purple, 5–10 mm long. Sepals 2–3 mm long, acute. Shrub usually 1–1.5 m high. Widespread. Heath and DSF. Ss. Fl. Aug.–Oct. *Pinnate Boronia*. . . . . **B. pinnata** Sm.
22 Leaflets serrulate with distant serrations marked by a marginal oil gland, up to 30 mm long, thin, 9–15 per leaf. Flowers in loose axillary corymbose cymes. Petals deep pink. Style sometimes slightly dilated. Glabrous shrub 1–4 m high with a very strong odour. Woronora and Hornsby Plateaus; Blue Mts on King's Tableland. Creek banks and damp shady situations. Ss. Fl. spring–summer . . . .
. . . . **B. thujona** A.R.Penfold & M.B.Welch

2 Zieria Sm.

c. 60 species endemic Aust.; Qld, NSW, Vic., Tas., S.A.

Shrubs or small trees. Leaves opposite, 3-foliolate or rarely 1-foliolate. Flowers in axillary cymes, rarely solitary. Calyx 4-lobed. Petals 4, spreading, whitish or pink. Stamens 4. Ovary 4-locular, deeply 4-lobed. Fruit an explosive capsule. Seed 1 per loculus.

Key to the species

1 Young branches prominently ridged by the decurrent leaf bases. 2

1 Young branches not prominently ridged 6

2 Young branches stellate hairy or appearing glabrous 3

2 Young branches simple hairy. 5

3 Leaflets elliptic to narrow-elliptic or lanceolate, 5–10 cm long, hirsute underneath. Petiole 15–30 mm long. Calyx lobes shorter than the petals. Petals white, 3–7 mm long. Tall shrub or small tree. Widespread. WSF and margins of RF. Fl. spring–summer. *Stinkwood* . . . . **Z. arborescens** Sims ssp. **arborescens**

3 Leaflets linear, velvety on lower surface. Petiole ≤10 mm long 4

4 Staminal filaments hirsute with simple to stellate hairs towards apex; sepals >2 mm long. Inflorescence usually shorter than the leaves, commonly 3–flowered but up to 23. Petals very pale pink, 3–5 mm long. Leaflets linear, with revolute margins, 2–4 cm long, 1–2.5 mm wide, hairy underneath. Erect shrub to 1.5 m high. Widespread. Heath and DSF. Ss. Fl. late winter–spring . . . . **Z. laevigata** Bonpl.

4 Staminal filaments usually glabrous; sepals <2 mm long. Inflorescence usually longer than the leaves, commonly 9–flowered but up to 50. Petals very pale pink, c. 3.4 cm long. Leaflets linear, with revolute margins, 2–5 cm long, 1.5–4.5 mm wide, hairy underneath. Erect shrub up to 1.2 m high. North from Swansea. Coastal heath on sand or swampy sand further inland. Ss. Fl. late winter–spring. **Z. laxiflora** Domin

5 Leaflets with flat margins, obovate, 3–6 mm wide; petiole 2–8 mm long. Hairs on the stem arranged in 2 opposite rows through each internode. Petals pinkish to white, c. 2 mm long. Shrubs up to 2 m high, with prominent tubercles. Higher Blue Mts Heath. Fl. spring–summer. . . **Z. robusta** Maiden & Betche

5 Leaflets with margins revolute to the midrib, lanceolate to linear, up to 3 mm wide; petiole less than 2 mm long. Petals pale to deep pink, 4–6 mm long. Shrub up to 1 m high. Widespread. Heath and DSF. Ss. Fl. spring–summer . . . . **Z. aspalathoides** A.Cunn. ex Benth. ssp. **aspalathoides**

6 Young branches with mostly simple hairs. Calyx lobes c. as long as the petals. Petals white to pale pink, 3–5 mm long. Anthers with a minute terminal point. Divaricate shrub 30–100 cm high. Widespread. Heath; WSF and DSF. Ss. Fl. spring . . . . **Z. pilosa** Rudge

6 Young branches with mostly stellate hairs 7

7 Young branches prominently tuberculate. 8

7 Young branches not prominently tuberculate but oil glands may be present 11

8 Lower surface of leaflets glabrous or with scattered stellate hairs, petiole 8–20 mm long. Leaflets narrow-elliptic to ovate, 2–5 cm long, 4–7 mm wide, with flat or slightly recurved margins. Petals pink to white. Erect shrub 1–2 m high. Widespread on a variety of soils, usually in cleared areas. Fl. spring . . . . **Z. smithii** Jacks.

8 Lower surface of leaflets velvety with a dense indumentums of short stellate hairs 9

9 Upper surface of leaflets glabrous. Calyx lobes less than 1 mm long, ± triangular, ± glabrous. Leaflets linear, 2–4 cm long, 0.5–1.5 mm wide. with recurved to revolute margins. The whole plant covered with prominent yellowish glands. Petals white, valvate, 1.7–2.5 mm long. Erect, strongly aromatic shrub up to 2 m high. Dry rocky ridges. Coast in Kiama district. Open forests to RF margins. Endangered. Fl. spring–summer. . . . . **Z. granulata** (F.Muell.) C.Moore ex Benth.

9 Upper surface of leaflets densely pubescent. 10

10 Upper surface of leaflets not tuberculate. Calyx lobes more than 1 mm long. Stems and leaves with brownish hairy tubercles. Leaflets always 3, up to 8 cm long, 2–10 mm. Petals pale pink to white, 1–3 mm long. Softly hairy shrub up to 1 m high. Olney State Forest. DSF, WSF and RF. Fl. spring–summer . . . . **Z. furfuracea** R.Br. ex Benth. ssp. **furfuracea**

10 Upper surface of leaflets ± densely tuberculate. Petiole 4–8 mm long. Leaflets linear 2.5–5 cm long, 2–3.5 mm wide. Petals white. Shrub to 3.5 m high. Cambewarra Range. Exposed rock outcrops. Heath and DSF. Vulnerable. Fl. late winter–spring . . . . **Z. tuberculata** J.A.Armstr.

11 Lower surface of leaflets glabrous to slightly hirsute 12
11 Lower surface of leaflets densely pubescent 13

12 Primary inflorescences bracts large >10 mm long, deciduous. . . . .
. . . . **Z. arborescens** ssp. **arborescens** (see above)
12 Primary inflorescence bracts small <10 mm long, persistent . . . . **Z. smithii** (see above)

13 Apex of leaflets retuse or emarginate. Leaflets pubescent on both surfaces with dense, short stellate hairs. Flowers axillary in 7-flowered cymes. Calyx lobes triangular with stellate and simple hairs. Petals c. 3 mm long. Straggling open shrub to 0.8 m high. Nowra area. DSF. Ss. Endangered. Fl. autumn–spring . . . . **Z. baeuerlenii** J.A.Armstr.
13 Apex of leaflets acute or rounded 14

14 Upper surface of leaflets densely pubescent mostly with stellate hairs 15
14 Upper surface of leaflets glabrescent with stellate, bifurcate or simple hairs 16

15 Some leaves 3-foliolate and some simple. Flowers sessile or the pedicel less than 1 mm long. Leaflets oblong to elliptic, 2.5–4 cm long. Flowers dense and usually numerous, surrounded by ovate bracts nearly as long as the flower. Shrub, densely softly pubescent, without tubercles. Blue Mts, e.g. Springwood, The Valley. WSF. Endangered. Fl. spring . . . . **Z. involucrata** R.Br. ex Benth.
15 Leaves consistently 3-foliate. Pedicles usually ≥3.5 mm. Central leaflet obovate to elliptic, 1–3 cm long, pale and densely tomentose underneath, with recurved margins. Petals white to pink, 3–5 mm long. Erect, tomentose shrub up to 3 m high, without tubercles. Blue Mts Heath and DSF. Fl. winter–spring
. . . . **Z. cytisoides** Sm.

16 Upper surface of leaflets with predominately simple hairs. Filaments warty or not at apex 17
16 Upper surface of leaflets with stellate hairs. Filaments warty at apex 18

17 Filaments not warty at apex. Some leaves 3-foliolate and some simple, leaflets linear to oblong, 3–7 cm long. Inflorescence bracts deciduous. Peduncles 3–5 mm long. Petals pale pink to white, c. 2 mm long. Softly tomentose shrub up to 2 m high. Blue Mts DSF. Ss. Vulnerable. Fl. spring. . **Z. murphyi** Blakely
17 Filaments warty at apex. Leaves all 3-foliate, leaflets narrow-elliptic to oblong, mostly up to 3.5 cm long. Inflorescence bracts persistent. Peduncles 13–25 mm long. Calyx lobes broad-ovate, stellate hairy. Petals pale pink to white, c. 5 mm long. Coast and Blue Mts DSF. Ss . . . **Z. compacta** C.T.White

18 Young branches glabrous to glabrescent. Petiole 15–30 mm long. . . . .
. . . . **Z. arborescens** ssp. **arborescens** (see above)
18 Young branches densely stellate hairy. Petiole <15 mm long 19

19 Primary inflorescence bracts 8–12 mm long. Leaflet narrow elliptic to lanceolate, 4–10 cm long, 5–15 cm wide, margins entire. Flowers on pedicels 5–8 mm long. Petals 4.5 mm long, white. Tall shrub to small tree up to 6 m high. Escarpment ranges, south from Colo R. DSF, WSF and RF margins. Ss. Fl. autumn–spring . . . . **Z. caducibracteata** J.A.Armstr.
19 Primary inflorescence bract 6–8 mm long. Leaflets oblong to ovate 2.5–3.5 cm long, 7.5–10 mm wide, margins entire, recurved to revolute. Flowers ± sessile. Petals white to pink 6 mm long. Erect shrub to 2 m high. Narrow Neck Peninsula, Blue Mts Woodland. Endangered. Fl. summer . . . .
. . . . **Z. covenyi** J.A.Armstr.

3 Correa Andrews
12 species endemic Aust.; all states and territories except NT

Stellate-tomentose shrubs with opposite simple leaves. Calyx cup-shaped, persistent. Petals 4, valvate, tomentose outside, much larger than the calyx, cohering into a cylindrical tube, sometimes separating as the flower expands, spreading at the top. Stamens 8, free. Ovary 4-locular, each loculus with 2 ovules; style filiform, terminal. Fruit dehiscent.

Key to the species

1 Leaves 5–10 cm long, 3–6 cm wide, ovate to obovate. Calyx 7 mm long, sprinkled with stellate hairs. Corolla 20–25 mm long, greenish yellow, densely velvety hairy. Shrub. Patonga Creek area and Illawarra district. WSF. Fl. spring . . . . **C. lawrenciana** var. **macrocalyx** (Blakely) Paul G.Wilson
1 Leaves 2–4 cm long. 2

2 Petals separating when the flower expands, white, 10–15 mm long. Corolla campanulate and star-like, 11–13 mm long. Flowers 1–4 together, terminal. Leaves orbicular to ovate obovate or elliptic, 2–4 cm long, entire, thick, white with a close tomentum underneath. Compact, much branched shrub 1 m high with tomentose branches. Coast. Sandy and rocky situations near the sea. Fl. winter–early spring. *White Correa* . . . . **C. alba** Andrews var. **alba**
2 Petals cohering in a cylindrical tube for the greater part of their length, red or yellowish-green, 2–3 cm long. Flowers 1–3 together, terminal, erect to drooping. Leaves cordate-ovate to elliptic, 2–3 cm long. Erect shrub 0.5–1.5 m high. 3

3 Flowers drooping, clasped by 2 reflexed bracts, corolla green to yellow or red with green to pale lobes. Red flower coastal form, green flower inland form. Widespread. DSF and Heath. Sandy soils. Fl. spring. *Native Fuschia* . . . . **C. reflexa** (Labill.) Vent. var. **reflexa**
3 Flowers erect to drooping, not clasped by reflexed bracts, corolla red with green to pale lobes or pale yellow. Coastal areas on sand or Ss. DSF. Fl. winter–early spring. . . . .
. . . . **C. reflexa** var. **speciosa** (Don ex Andrews) Paul G.Wilson

4 **Phebalium** Vent.
25 species endemic Aust.; all states and territories except NT

Shrubs covered with silver to rusty scurfy scales. Leaves alternate, simple, entire or slightly toothed. Flowers in axillary or terminal corymbs umbels or short racemes. Calyx 5-cleft or 5-toothed. Petals 5, white to yellow, imbricate, covered with scales on outer surface. Stamens 10. Ovary 5-locular, rarely fewer. Fruit dehiscent. Seeds solitary.

Key to the species

1 Leaves strongly bilobed, 3.5–14 mm long, margins recurved. Petals cream to lemon yellow . Erect shrub to 1.5 m high. Capertee Valley. Heath or DSF. Endangered. Fl. spring . . . .
. . . . **P. bifidum** P.H.Weston & M.J.Turton
1 Leaves entire 2

2 Leaves mostly obovate less than 10 mm long, shining above, whitish below; the margins recurved. Flowers similar to the other ssp. Erect shrub c. 1 m high. Blue Mts, e.g. Mt York, Mt Victoria. Exposed situations on Ss. Fl. spring. . . . . **P. squamulosum** ssp. **ozothamnoides** (F.Muell.) Paul G.Wilson
2 Leaves linear to obovate or oblong, recurved or flat, mostly more than 10 mm long. Petals scaly outside, up to 4 mm long. Calyx short. Fruiting carpel scarcely beaked. Erect or diffuse shrubs 0.5–2 m high. . 3

3 Leaves linear, c. 1 mm wide, 10–25 mm long. Apex truncate to obtuse. Shoalhaven River district. Heath and DSF. Ss. Fl. spring . . . . **P. squamulosum** ssp. **lineare** Paul G.Wilson
3 Leaves narrow-oblong to elliptic or obovate, 2–10 mm wide 4

4 Leaves narrow-oblong, up to 50 mm long, obtuse or retuse or truncate, with brownish scales underneath, shining on the upper surface. Widespread. Chiefly in DSF. Ss. and granite. Fl. spring . . . .
. . . . **P. squamulosum** Vent. ssp. **squamulosum**
4 Leaves elliptic to obovate, up to 70 mm long, obtuse, with silvery scales underneath, dull on the upper surface. Widespread. Coastal heath. Ss. Fl. spring . . . **P. squamulosum** ssp. **argenteum** Paul G.Wilson

5 Nematolepis Turcz.

7 species endemic Aust.; Qld., NSW, Vic., Tas., WA

One species in the area

Tall shrub or small tree. Stems without warts. Leaves oblong-lanceolate to elliptic, 3–10 cm long, 10–22 mm wide, flat, silvery white underneath. Flowers mostly in axillary corymbs on short thick scaly peduncles. Calyx lobed, very small. Petals 5, imbricate, white 4–5 mm long. Fruiting carpel with a very short beak. Widespread. Gullies on Ss. Fl. spring . . . **N. squamea** (Labill.) Paul G.Wilson ssp. **squamea**

6 Leionema (F.Muell.) Paul G.Wilson

21 species endemic Aust.; Qld, NSW, Vic., Tas., S.A.

Shrub with simple or stellate hairs. Leaves alternate, simple, margins entire or rarely toothed. Flowers in axillary or terminal corymbs or flowers solitary. Petals 5, valvate, free or united. Carpels 5, styles fused. Fruit dehiscent. (some of the following species were previously included in *Phebalium*)

Key to the species

1 Petals connate into a tube 12–16 mm long. Stamens longer than or as long as the petals. Leaves narrow-elliptic to oblanceolate, 1.5–3.5 mm long, obtuse to retuse, with recurved margins. Shrubs up to 3 m high with glabrescent branches. Rylestone district. DSF. Ss. Vulnerable. Fl. spring . . . . **L. sympetalum** (Paul G.Wilson) Paul G.Wilson
1 Petals free from each other 2

2 Flowers in axillary, short-stalked umbels or solitary. 3
2 Flowers in terminal, umbellate or capitate clusters. Carpels usually beaked. Stamens exserted 4

3 Flowers solitary. Leaves almost terete, c. 8 mm long, with revolute margins. Carpels glabrous; the upper sterile beak two-thirds the length of the carpel. Shrub. Blue Mts, Blackheath. Rocky places. Endangered. Fl. winter–spring. . . . . **L. lachnaeoides** (A.Cunn.) Paul G.Wilson
3 Flowers in short, axillary, short-stalked umbels. Leaves linear, obtuse, 3–8 cm long, glabrous on the upper surface, hoary underneath; the margins recurved, often minutely toothed. Sepals very small. Petals c. 4 mm long. Stamens slightly exserted. Ovary glabrous. Fruiting carpel nearly orbicular, shortly beaked. Tall shrub with minute stellate hairs on the young branches. Coast and adjacent plateaus; lower Blue Mts Gullies on Ss. Fl. spring. . . . . **L. dentatum** (Sm.) Paul G.Wilson

4 Leaves obovate to orbicular, up to 6 mm long and 5 mm wide; the margins very slightly recurved, ± obtuse. Flowers in umbellate clusters. Sepals less than a quarter as long as the petals, pubescent. Petals cream-white, c. 3 mm long. Erect shrub up to 1 m high, with pubescent young branches. Blue Mts, Lithgow–Capertee. Heath. Fl. early summer . . . . **L. lamprophyllum** (F.Muell.) Paul G.Wilson ssp. **orbiculare** F.M.Anderson
4 Leaves linear to oblanceolate more than 6 mm long. 5

5 Stems glabrous and minutely glandular, Leaves glabrous, narrow elliptic to oblanceolate, 3–7 cm long, 10–15 mm wide, glossy above, apex acute margins finely toothed. Petals cream to pale yellow, 5 mm long. Shrub 1–3 m high. Avon Catchment south to Budawang Range. DSF on ridges and creek banks. Fl. spring–summer. . . . . **L. coxii** (F.Muell.) Paul G.Wilson
5 Stem pubescent with simple or stellate hairs 6

6 Leaves glabrous on lower surface, linear to oblanceolate, 15–50 mm long, 1–4 mm wide, dull above, apex acute to acuminate. Petals white and pale green on outside,4–5 mm long. Shrub to 2 m high. Colo River. Riparian Forest. Fl. autumn–winter . . . . **L. 'Colo River'** (P.H.Weston 2423)
6 Leaves hairy on lower surface. 7

7 Leaves linear, crowded, obtuse, 8–15 mm long, 1–5 mm wide, upper surface glabrous, pilose or scabrous, lower surface with hairs, apex obtuse, with very revolute margins. Sepals linear, erect, half as long as the petals, pubescent. Petals creamy yellow, c. 6 mm long. Erect shrub up to 1.5 m high, with hirsute branches. Widespread. Heath on Ss. Fl. spring. . . . . **L. diosmeum** (A.Juss.) Paul G.Wilson

7 Leaves narrow elliptic to oblanceolate, 25–65 mm long, 5–10 mm wide, glossy above, apex obtuse to marginate, margins entire to serrulate, flat to slightly recurved. Petals lemon yellow to greenish yellow, 5–8 mm long. Erect shrub to 4 m high. Restricted to Wollemi National Park. Ledges and clefts on Ss. Fl. winter–spring . . . .**L. scopulinum** B.M.Horton & Crayn

7 **Crowea** Sm.
3 species endemic Aust.; NSW, Vic., WA

Low shrubs with angular branches. Leaves alternate, simple, entire, thick, not prominently bullate-glandular. Flowers solitary, mostly axillary. Calyx 5-lobed. Petals 5, imbricate, bright pink inside, often greenish outside, persistent and imbricated around the young fruit. Stamens 10, free; anther terminated by a narrow bearded appendage longer than the pollen sacs; filaments hairy. Carpels 5; ovules 2 per carpel. Fruit a schizocarp.

Key to the species

1 Leaves mostly narrow-elliptic, 3–5 cm long, dark green, shining on the upper surface; midvein distinct. Branches conspicuously angular and glabrous. Sepals short and broad. Petals 12–18 mm long. Woronora and Hornsby plateaus; lower Blue Mts Sheltered situations on Ss. Fl. Jan.–June . . . **C. saligna** Andrews
1 Leaves linear to narrow-obovate or obcordate 1–4 cm long. Branches only slightly angular and with a few hairs. Petals 3.5–14 mm long. 2

2 Leaves obcordate, 1–11 mm long, 4–5.5 mm wide. Petals 7–9 mm long. Pedicels c. 3.5 mm long. Boyd Plateau. Endangered. Fl. throughout year . . . . **C. exalata** ssp. **obcordata** WAGebert
2 Leaves narrow-oblong to narrow-elliptic 3

3 Leaf apex obtuse to acute. Petals 5–8 mm long. Pedicels 0.5–1.5 mm long. Coast and tablelands south of Sydney. Heath and DSF. Sheltered situations. Fl. summer–autumn . . . **C. exalata** F.Muell. ssp. **exalata**
3 Leaf apex mucronate to apiculate. Petals 7–14 mm long. Pedicels 1.5–4 mm long. Coast and escarpment ranges. Heath and DSF. Sheltered situations. Fl. summer–autumn . . . . **C. exalata** ssp. **magnifolia** WAGebert

(There may be difficulty separating the above subspecies south of Sydney.)

8 **Eriostemon** Sm.
2 species endemic Aust.; Qld, NSW

One species in the area
Erect shrub up to 2 m high. Leaves alternate, simple, entire, linear to lanceolate, 2–8 cm long, 8–14 mm wide, rather thick, glabrous when mature, greyish-green, 3-veined from base. Flowers axillary. Sepals short, orbicular. Petals pink, 10–15 mm long, 5-veined from base. Stamens 10, free; anther appendage c. an eighth as long as the pollen sac, white, glabrous; filaments hairy. Fruit a schizocarp. Fruiting carpels 5 mm long, truncate. Widespread. Heath and DSF. Sandy soils. Fl. spring. *Wax Plant* . . . .**E. australasius** Pers.

9 **Philotheca** Rudge
45 species endemic Aust.; all states and territories except NT

Shrub glabrous or with simple hairs. Leaves alternate, simple, margins sometimes crenulate, 1–veined from base. Flowers in axillary or terminal clusters of solitary. Petals and sepals 5 occasionally 4. Petals mauve, pink or white. Stamens 10, filaments hairy. Fruit with 1–5 cocci.

(N.B. some of the following species were previously known as *Eriostemon*)

Key to the species

1 Staminal filaments cohering into a tube at the base 2
1 Staminal filaments free 3

2 Anthers glabrous. Heath-like undershrub 50–150 cm high, glabrous or with a minute pubescence. Leaves numerous, crowded, narrow-linear or almost terete, 10–20 mm long. Flowers terminal, solitary or 2–3 together. Sepals 5, broad-triangular. Petals 5, imbricate in the bud, pink to purple or bluish, 7–12 mm long. Stamens 10; filament hairs not concealing anthers. Carpels 5, nearly free except the style. Widespread. Heath and DSF. Sandy soils. Fl. spring–summer . . . . **P. salsolifolia** (Sm.) Druce ssp. **salsolifolia**

2 Anthers with a tuft of hairs at their apex. Heath-like undershrub 50–150 cm high, hispid. Leaves numerous, crowded, linear or almost terete, 4–12 mm long. Flowers terminal, solitary or 2–3 together. Sepals 5, broad-triangular. Petals 5, imbricate in the bud, pink to purple or bluish, 7–12 mm long. Stamens 10; filament hairs concealing anthers. Carpels 5, nearly free except the style. Sydney area. Heath on rock or sandy areas. Fl. spring–summer . . . . **P. reichenbachii** Sieber ex Spreng.

3 Leaves 3–10 cm long (very rarely less in *P. trachyphyllus*),. 4
3 Leaves less than 3 cm long. 5

4 Flowers usually in 2–4-flowered clusters on a stout peduncle c. 1 mm thick. Carpels free almost from the base, divergent, sharp pointed in the fruiting stage. Petals oblong-obovate, 8–12 mm long. Leaves 4–10 cm long, narrow-ovate to narrow-elliptic, greyish-green, mucronate; oil glands prominent. Erect, bushy shrub up to 2 m high Western and southern parts of the Cumberland Plain and adjacent plateaus; Blue Mts in valleys. Heath and DSF. Fl. spring. *Native Daphne* . . . . **P. myoporoides** (DC.) M.J. Bayly ssp. **myoporoides**

4 Flowers often solitary, but when clustered the slender pedicels arising from the leaf axil with no peduncle. Carpels connate to the midpoint or higher, slightly divergent, ± obtuse when ripe. Petals narrow-obovate, 5–9 mm long. Leaves, elliptic-obovate, 3–4 cm long, 5–8 mm wide, obtuse or minutely mucronate, with prominent oil glands. Shrub or small tree up to c. 6 m high. Coast in the south of the area; recorded once from Botany Bay. Fl. spring–early summer . . . . **P. trachyphylla** (F.Muell.) Paul G.Wilson

5 Leaves with 2 small black stipules at the base, elliptic to oblong, 5–7 mm long, 2–4 mm wide, crenate, thickened at the margin. Petals white, 4–5 mm long. Glandular-tuberculate shrub up to 1 m high. Singleton area. DSF. Fl. summer . . . . **P. difformis** ssp. **smithiana** (Benth.) Paul G.Wilson
5 Leaves without stipules 6

6 Leaves more than twice as long as broad. Flowers mostly solitary, axillary. Petals usually pale pink, 6–10 mm long. 7
6 Leaves rarely more than twice as long as broad. 9

7 Leaf margins revolute. Leaves linear or linear-spathulate, mucronate, with a straight or recurved point, 12–25 mm long; midvein prominent underneath; oil glands prominent. Erect shrub up to 1 m high. Hornsby Plateau; Blue Mts Heath and DSF. Fl. spring. . . . . **P. hispidula** (Spreng.) Paul G.Wilson
7 Leaf margins never revolute. Leaves convex underneath, flat or channelled on the upper surface or almost terete. Oil glands prominent. Divaricate shrubs up to c. 70 cm high 8

8 Leaves terete to linear lanceolate, c. 1–4 mm wide, 10–25 mm long. Coast and adjacent plateaus from Manly southwards. DSF and heath. Ss. Fl. spring . . . . . **P. scabra** (Paxton) Paul G.Wilson ssp. **scabra**
8 Leaves oblong-elliptic to ovate, 2–5 mm wide, 10–30 mm long. Southern Blue Mts DSF. Fl. spring. . . . . **P. scabra** ssp. **latifolia** (Paul G.Wilson) Paul G.Wilson

9 Leaves obovate to spathulate, narrowed at the base, often petiolate, very obtuse, thick; the midvein faint. Flowers axillary. Petals pink, 6–8 mm long. Small shrub up to 1 m high, with glabrous branches. Blue Mts Heath and DSF. Fl. spring–summer . . . . **P. obovalis** (A.Cunn.) Paul G.Wilson
9 Leaves orbicular to elliptic or obovate or ovate, mucronate, tuberculate, with the midvein prominent underneath; the base truncate or cordate. Flowers solitary, axillary. Petals pinkish, 8–10 mm long. Divaricate shrubs up to 1 m high, with pubescent branchlets 10

10 Leaves orbicular to elliptic, cordate at the base; margins entire. Coast and adjacent plateaus south of Port Jackson. Heath and DSF. Fl. spring . . . . **P. buxifolia** (Sm.) Paul G.Wilson ssp. **buxifolia**

10 Leaves obovate, truncate at the base; margins crenate-tuberculate. Coast and adjacent plateaus north of Port Jackson. Heath and DSF. Fl. spring . . . . . . **P. buxifolia** ssp. **obovata** (Don) Paul G.Wilson

(Intermediate forms occur where *P. buxifolia* and *P. scabra* adjoin south of Royal National Park and where *P. buxifolia* and *P. myoporoides* adjoin in the lower Hawkesbury area.)

10 Asterolasia F.Muell.
20 species endemic Aust.; NSW, Vic., S.A., WA

Erect shrubs, densely stellate-tomentose on all the parts except the older branches and upper surfaces of mature leaves. Flowers solitary or in axillary or terminal clusters. Petals valvate, cream-yellow or white, tomentose outside. Sepals minute, hidden beneath the stellate hairs. Stamens 10. Ovary 5-lobed, densely tomentose; stigma large, peltate. Fruiting carpels truncate, with an incurved beak.

Key to the species

1 Petals bright yellow. Flowers solitary or sometimes in clusters of up to 3 2
1 Petals white to cream. Flowers usually several in a cluster 4

2 Pedicels 0–2 mm long. Petals 5–6 mm long. Leaves ovate-spathulate, 3–10 mm long, 2–6 mm wide, upper surface with scattered stellate hairs, densely brownish tomentose underneath. Flowers solitary. Shrubs 1–2 m high. Penrose-Wingello district. Open forests. Fl. spring . . . .
. . . . **A. buckinghamii** (Blakely) Blakely
2 Pedicels 2–7 mm long. Petals 6–7 mm long. 3

3 Leaves narrow oblong to cunneate, 1–2.5 cm long, 2–3 mm wide. Upper surface with tubercles, densely brownish tomentose underneath. Flowers in clusters of 1–3. Shrub to 1.5 m high. Avon Dam to Buxton area. Growing along streams. Fl. spring. . . . .**A. rivularis** Paul G.Wilson
3 Leaves obovate, 0.5–1.8 cm long, 3–10 mm wide. Upper surface glabrous, stellate hairy underneath. Flowers solitary. Shrub to 1 m high. Bells Line of Road, Blue Mts Endangered. Known from only one collection. . . . . **A. buxifolia** Benth.

4 Petals c. 5 mm long. Leaves elliptic to lanceolate, 2–8 cm long, 8–20 mm wide, tomentose underneath. Shrubs 1–2 m high. Hornsby Plateau. Moist gullies. Ss. Fl. spring . . . . . . **A. correifolia** (Juss.) Benth.
4 Petals 8–14 mm long. Leaves lanceolate to elliptic, 4–13 cm long, acuminate, rusty tomentose underneath. Shrub up to 3 m high. Maroota. WSF. E. Fl. spring . . **A. elegans** L.McDougall & Porteners

11 Geijera Schott
5 species endemic Aust.; Qld, NSW, Vic., S.A., WA

One species in the area
Small tree. Leaves simple, alternate, coriaceous, ovate to lanceolate, 7–10 cm long. Flowers in a broad pyramidal terminal panicle shorter than the last leaves. Petals 5, yellowish white, 2 mm long, valvate. Ovary depressed, 4–5-lobed. Mericarps often only 1–2, obovoid, not beaked, 4–6 mm long. Coastal. RF. Fl. spring–summer. *Brush Wilga* . . . .**G. salicifolia** var. **latifolia** (Lindl.) Domin

12 Melicope J.R.Forst. & G.Forst.
14 species endemic Aust.; Qld, NSW, NT

One species in the area
Glabrous, tall shrub or small tree. Leaves digitately 3-foliolate, mostly opposite, usually clustered towards the ends of the branchlets; leaflets obovate to elliptic, 4–10 cm long, entire, obtuse. Flowers numerous in dense cymo-panicles on axillary peduncles. Petals 4, ovate, glabrous, 4–5 mm long, whitish. Stamens 4. Mericarps 4, c. 6 mm long. Coast and adjacent plateaus; lower Blue Mts RF. Fl. summer. *Doughwood* . . .
. . . . **M. micrococca** (F.Muell.) T.G.Hartley

13 Acronychia J.R.Forst. & G.Forst.

19 species in Aust.; (16 endemic) Qld, NSW, Vic.

Trees or shrubs. Leaves opposite or rarely alternate, glabrous, 1- or rarely 3-foliolate. Flowers in small axillary panicles or loose cymes, 4-merous. Stamens 8. Ovary 4–8-locular; ovules 2 per loculus. Fruit a truncate or globular berry.

Key to the species

1 Leaves 1–3 cm wide, 5–10 cm long, oblong to lanceolate or sometimes obovate. Petals narrow, 4–5 mm long, glabrous. Filaments rather thick, dilated, ciliate or hirsute above the base. Ovary 4-locular, hirsute around the base of the style but otherwise glabrous. Fruit whitish, globular, 6–12 mm diam., with a 4-angled endocarp. Shrub or tree up to c. 25 m high. Widespread. RF and sheltered situations. Fl. summer . . . . **A. oblongifolia** (A.Cunn. ex Hook.) Endl. ex Heynh.
1 Leaves 4–6 cm wide, 5–12 cm long, obovate to elliptic, coriaceous. Petals narrow-oblong, c. 5 mm long. Filaments villous towards the base. Ovary 4–8-locular, sometimes hirsute at the base of the style. Fruit a greenish yellow, globular berry. Shrub or small tree up to c. 10 m high. Coast from Gosford northwards. RF. Fl. autumn. *Silver Aspen* . . . .**A. wilcoxiana** (F.Muell.) T.G.Hartley

14 Sarcomelicope Engl.

1 species in Aust.; Qld, NSW

One species in the area

Dioecious shrub or tree. Leaves opposite, simple or lobed, petiolate, ovate elliptic or obovate, mostly 7–12 cm long. Flowers in axillary panicles, 4-merous. Petals slightly imbricate. Stamens 8; filaments ciliate. Ovary 4-locular; stigma broadly lobed. Minnamurra Falls; Royal National Park; Kurrajong. RF. Fl. autumn–winter . . . . **S. simplicifolia** (Endl.) T.G.Hartley ssp. **simplicifolia**

15 Citrus L.

8 species in Aust. (2 endemic); Qld, NSW, NT, WA

One species in the area

Shrubs or small trees with axillary spines. Leaves alternate, simple, articulate at the base, ± crenate. Flowers in axillary cymes. Petals usually 5. Stamens 10 or more. Ovary 4–8-locular. Fruit a rough orange berry. Naturalized on the margins of RF and around old habitations. Introd. from Asia and Europe. *Rough Lemon* . . . . *C. X taitensis** Risso

104 SAPINDACEAE

Trees shrubs or rarely climbers. Leaves opposite or alternate, frequently compound, without stipules. Flowers variously arranged, usually polygamous or unisexual, regular or irregular. Sepals 3–6, free or connate. Petals as many as the sepals or fewer or absent, imbricate in the bud. Disc usually present. Stamens usually 8, inserted between the ovary and the disc. Ovary mostly 1-4-locular, superior, often lobed; style simple; placentas axile. Ovules 1–2 per loculus. Fruit usually a capsule, 3-lobed or sometimes with 1 or 2 lobes aborted; or a 2-winged schizocarp ("samara"). Seeds frequently with a fleshy aril. 135 gen., mostly trop.

Key to the genera

1 Leaves opposite. . . . . **1 ACER**
1 Leaves alternate. 2

2 Climber or twiner . . . .**7 CARDIOSPERMUM**
2 Trees or shrubs 3

3 Leaves simple or pinnate and then with leaflets less than 12 mm long. Shrubs. . . . . . **2 DODONAEA**
3 Leaves pinnate with leaflets more than 20 mm long. Trees 4

4 Leaflets more than 20 cm long . . . . **3 DIPLOGLOTTIS**
4 Leaflets less than 10 cm long 5

5 Leaflets toothed. Fruit lobes globular, 2 or 1 by abortion . . . . **6 ALECTRYON**
5 Leaflets entire. Fruit a capsule opening loculicidally, 3-lobed or less by abortion 6

6 Leaflets glaucous underneath, 2–4. Lobes of the capsule compressed, thin-walled . . . . . . **4 GUIOA**
6 Leaflets not glaucous underneath, 6–10. Lobes of the capsule not compressed, thick-walled. . . . .
. . . . **5 CUPANIOPSIS**

1 Acer L.
2 species naturalized Aust.; NSW, Vic., Tas., S.A.

Trees or shrubs. Leaves opposite, compound or palmately lobed, usually deciduous; stipules absent. Flowers often unisexual, greenish or yellowish, arranged in corymbs or racemes, 5-merous but stamens often 8; disc usually prominent. Ovary 2-locular, superior. Fruit a 2-winged schizocarp ("samara").

Key to the species

1 Leaves compound, often pinnate, with 5–7 (but sometimes with only 3) leaflets. Flowers dioecious. Petals absent. Tree up to 6 m high. Naturalized along the Nepean River. Introd. from N. America. *Box Elder* . . . .*A. negundo* L.
1 Leaves simple, palmately lobed. Flowers monoecious. Petals present. Disc absent. Tree up to 5 m high. Naturalized at Jenolan Caves. Introd. from Europe. *Sycamore* . . . .*A. pseudoplatanus* L.

2 Dodonaea Mill.
61 species in Aust. (59 endemic); all states and territories

Shrubs. Leaves simple or pinnate, often ± viscid. Flowers usually unisexual (plants dioecious), rarely bisexual. Flowers terminal or apparently axillary by abortion of the flowering branch, solitary or clustered or in short racemes or panicles. Sepals 5 or sometimes fewer, valvate, usually deciduous. Petals absent. Disc inconspicuous. Stamens usually 8. Ovary superior, 3–4-locular, with 2 ovules per loculus. Fruit a membranous capsule; each valve with a dorsal angle often produced into a vertical wing.

Key to the species

1 Leaves simple 2
1 Leaves pinnately compound. 11

2 Wing of the capsule short reaching neither the style nor the base 3
2 Wing of the capsule longer than broad and reaching to the style and the base 4

3 Leaves elliptic to obovate 10–20 mm wide, usually dentate towards the top, usually pubescent; wing of the capsule less than 3 mm wide, often reduced to a point. Capsule usually 3-lobed. Erect shrub up to 2 m high. Howes Mt. DSF . . . . **D. triangularis** Lindl.
3 Leaves lanceolate to linear-lanceolate, 5–10 mm wide, 3–8 cm long, entire or obscurely sinuate-toothed. Capsule usually 4-lobed; the lobes with a wing more than 3 mm wide, truncate. Tall, erect, scarcely viscid shrub with angular branchlets. Blue Mts; Nepean River district. DSF . . **D. truncatiales** F.Muell.

4 Diffuse or prostrate shrub usually less than 50 cm high. Leaves broad at the base, sometimes almost stem-clasping; the margins and midrib decurrent on the stem as very prominent ridges, oblong, 1–3 cm long, 3–7 mm wide, sinuate, with slightly recurved margins. Fruit up to 1 cm long, often purplish. Seeds brown, dull. Widespread except higher Blue Mts DSF . . . . **D. camfieldii** Maiden & Betche
4 Erect shrubs, usually more than 1 m high. Leaves narrowing gradually towards the base or linear. . . 5

5 Leaves scarcely viscid 6
5 Leaves viscid. Branchlets usually ± angular. Seeds brown or black, dull. 7

6 Leaves acute to acuminate, narrow-ovate to narrow-elliptic, 4–10 cm long, entire. Sepals minute. Anthers linear, up to 3 mm long. Styles up to 12 mm long. Fruit 3-lobed, usually c. 1 cm long, often purplish. Seeds brown, shining. Erect, glabrous shrubs 2–3 m high; young branchlets angular or flattened. Widespread except higher Blue Mts Very common in DSF. *Hopbush* . . . .**D. triquetra** J.C.Wendl.

6 Leaves emarginate, with a short mucro . . . . **D. viscosa** ssp. **cuneata** (see below)

7 Leaves distinctly cuneate, 20–25 mm long, 7–11 mm wide, ± emarginate, with a small mucro continuing from the midvein. Fruit 3-lobed, c. 8 mm long and 10 mm wide, often purplish. Seeds brown, smooth. Erect or diffuse shrubs. Cumberland Plain and Singleton district. Open forests. . . . . **D. viscosa** ssp. **cuneata** (Sm.) J.G.West

7 Leaves linear, very narrow-obovate to narrow-oblong, obtuse or acute never cuneate 8

8 Leaves sessile; linear, concave or incurved, 0.5–1 mm wide, 15–40 mm long, with flat brownish glands on the surface. Fruit, 4-winged, c. 1.5 cm wide, sparsely pubescent. Erect shrub up to 1 m high. Hornsby Plateau. DSF. Ss. Uncommon . . . . **D. falcata** J.G.West

8 Leaves narrow-elliptic to narrow-oblong or oblanceolate, more than 1 mm wide, flat or with slightly recurved margins, often dentate. Fruit 3–4-winged, 9–20 mm long, 11–22 mm wide, glabrous 9

9 Leaves less than 7 times as long as broad; 4–10 mm wide, 5–8 cm long; petiole 0–10 mm long. Erect shrub up to 4 m high. Widespread. Open forests . . . . **D. viscosa** ssp. **spatulata** (Sm.) J.G.West

9 Leaves more than 7 times as long as broad. 10

10 Leaves with petiole 6–18 mm long, linear to lanceolate tapering at apex and base, 6–13 cm long, 5–10 mm wide. Compact shrub up to 5 m high. Often in rocky areas, Blue Mts Open DSF forests . . . .**D. viscosa** ssp. **angustifolia** (L.f.) J.G.West

10 Leaves sessile, linear to narrow oblong, up to 6 mm wide, 3–9 cm long; margins dentate, not narrowly tapering at apex. Erect shrub to 4m high. Bowral. Woodlands . . . . **D. viscosa** ssp. **angustissima** (DC.) J.G.West

11 Leaflets 5–9 (rarely more), all opposite, obtuse or truncate, usually toothed at the tip, 2–4 mm long. Branches and leaves slightly pubescent with short bent hairs. Much branched, often viscid shrub. Blue Mts Heath and DSF. . . . . **D. boroniifolia** G.Don

11 Leaflets 15–20 12

12 Leaflets c. 25 mm long and 4 mm wide, elliptic, acute, opposite or irregularly arranged; the rhachis winged, glabrous. Fruits on slender pedicels 2–3 cm long. Shrub or small tree up to c. 4 m high. Olney S.F. DSF to margins of RF . . . .**D. megazyga** (F.Muell.) F.Muell. ex Benth.

12 Leaflets up to 10 mm long 13

13 Branches, petioles, rhachis and both surfaces of the leaflets thickly beset with erect straight rather rigid hairs. Leaflets entire or slightly toothed at the tips, mostly 5–8 mm long, opposite or irregular; rhachis winged. Pedicels short. Shrub up to 1.5 m high. Coast and adjacent plateaus and lower Blue Mts . . . .**D. pinnata** Sm.

13 Branches and leaves scabrous-pubescent with short bent hairs or the upper surfaces of the leaflets nearly glabrous. Leaflets ± toothed at the tips, (2)3–6(10) mm long, opposite or alternate. Erect shrub up to 1.5 m high. Widespread. DSF. Sandy soils . . . .**D. multijuga** G.Don

3 Diploglottis Hook.f.
8 species endemic Aust.; Qld, NSW

One species in the area

Medium sized to large tree with dense soft rust-coloured tomentum on the young branches petioles and inflorescences. Leaves pinnate, sometimes exceeding 60 cm long. Leaflets 8–12, oblong-elliptic to narrow-ovate, usually c. 15 cm long but sometimes exceeding 30 cm. Flowers numerous in a large panicle. Calyx c. 3 mm long. Petals c. 3 mm long, orbicular, thin, ciliate. Capsule c. 12 mm diam., tomentose, 3-valved. Coast and adjacent plateaus. RF. *Native Tamarind* . . . . **D. australis** (G.Don) Radlk.

4 Guioa Cav.
5 species in Aust. (3 endemic); Qld, NSW

One species in the area
Small to medium sized tree. Leaves pinnate; leaflets 2–4 (rarely 6), elliptic to obovate or narrow-ovate, obtuse, glabrous, entire, 4–8 cm long, Pale and glaucous underneath. Flowers in axillary panicles shorter than the leaves. Sepals orbicular; the inner ones 2 mm diam. Petals shorter than the sepals, with 2 cuneate hairy scales as long as the petal. Stamens exserted. Capsule 8–12 mm diam., deeply divided with divaricate compressed lobes, thin-walled, loculicidally dehiscent. Seeds with a thin aril. Coast and adjacent plateaus. RF . . . .**G. semiglauca** (F.Muell.) Radlk.

5 Cupaniopsis Radlk.
11 species endemic Aust.; Qld, NSW, NT, WA

One species in the area
Medium sized tree, glabrous except the inflorescence. Leaves pinnate; leaflets 6–10 (usually 8), broad-ovate to obovate or elliptic-oblong, 6–12 cm long, entire, coriaceous, notched at the apex. Flowers in axillary panicles. Sepals orbicular; the inner ones 4 mm wide. Petals small, orbicular, with 2 very short hirsute scales at the base. Capsule 12–16 mm diam., acutely 3-lobed. Coast. Sheltered places near the sea. *Tuckeroo* . . . . **C. anacardioides** (A.Rich.) Radlk.

6 Alectryon Gaertn.
13 species in Aust. (12 endemic); Qld, NSW, WA

One species in the area
Small tree. Leaves pinnate; leaflets 4–6, narrowed into very short petiolules, oblong or elliptic to lanceolate, acuminate, 6–12 cm long; venation prominent on the undersurface; margins sometimes toothed. Flowers in a loose panicle. Calyx c. 2 mm diam., with very short broad lobes. Ovary 2–3-locular. Fruit with distinct globular lobes, 8–10 mm diam. Widespread. RF. *Native Quince* . . . . . . **A. subcinereus** (A.Gray) Radlk.

7 Cardiospermum L.
Balloon Vine
2 species naturalized Aust.; Qld, NSW, NT, WA

Herbaceous climbers. Leaves compound. Inflorescence branches frequently converted into tendrils. Leaflets 9, deeply dentate, up to 6 cm long. Flowers small, white. Fruit an inflated membranous capsule c. 25 mm diam. Seeds orbicular, black, with a heart-shaped spot.

Key to the species
1 Petals 6–10 mm long. Stems and leaves slightly pubescent especially on the undersurface. Terminal leaf segment ovate to narrow-ovate, deeply dentate. Widespread. Waste places. Introd. from trop. Asia, Africa or America . . . . *C. grandiflorum Sweet
1 Petals c. 3 mm long. Stems and leaves pubescent, especially the stems. Terminal leaf segment ovate to elliptic, deeply dentate. Widespread. Waste places. Introd. from trop. Asia, Africa or America . . . . . .
. . . . *C. halicacabum L. var. halicacabum

105 SIMAROUBACEAE
c. 25 gen., trop. to warm-temp.

1 Ailanthus Desf.
3 species in Aust. (2 native, 1 naturalized); Qld, NSW, Vic.

One species in the area
Tree up to 8 m high, suckering very freely and sometimes forming dense thickets. Leaves deciduous, imparipinnate, with 9–17 leaflets; leaflets ovate, 7–13 cm long, ± acuminate. Flowers polygamous, in

panicles, c. 6 mm diam. Sepals 5. Petals 5, narrow-oblong, white. Stamens 10 in the male flowers, fewer in the bisexual flowers. Ovary 5-locular. Fruit a schizocarp; each mericarp winged, c. 3 cm long. Common around old habitation and escaped in some places. Introd. from Asia. *Tree of Heaven* . . . .
. . . . *A. altissima** (Mill.) Swingle

106 BALSAMINACEAE
2 gen., trop. and subtrop.

One genus in the area

1 Impatiens L.
1 species naturalized Aust.; NSW

One species in the area

Glabrous, ± succulent herb to 1 m high. Leaves ovate, 4–2 mm long, margins toothed; petiole with scattered stalked glands. Flowers on long stalks, pink, white or red. Lower sepal petaloid and white, lateral 2 sepals small and green. Petals 5, 10–20 mm long, largest petal keeled, lateral petals slightly fused at base. Stamens 5. Style 5 lobed. Fruit a swollen capsule. Occasional garden escape. Introd. from S. Africa. Fl. all year. *Balsam, Busy Lizzie.* . . . . *I. walleriana** Hook.f.

107 EBENACEAE
4 gen., mostly trop.

1 Diospyros L.
15 species in Aust.; Qld, NSW, NT, WA

Trees or tall shrubs, dioecious. Leaves alternate, entire, without stipules. Flowers unisexual, regular; male flowers several together in axillary clusters; female flowers 1–3 together, larger than the male. Calyx and corolla 4–5-lobed; corolla lobes imbricate in the bud. Stamens 12–20, epipetalous or free. Ovary superior, 4–5-locular, with 2 ovules in each loculus; placenta axile. Fruit a berry. Seeds much compressed.

Key to the species

1 Calyx and corolla lobes 4. Leaves 3–10 cm long, elliptic, rounded or blunt at the apex, dark green on the upper surface, paler or yellowish underneath. Fruit ovoid, black, c. 15 mm long; the persistent calyx enlarged to 6–10 mm diam. Shrub or small tree. Coast and gullies in adjacent plateaus, e.g. Royal National Park and Illawarra region. RF. *Black Plum.* . . . . **D. australis** (R.Br.) Hiern
1 Calyx and corolla lobes 5. Leaves less than 5 cm long, elliptic, obtusely acuminate, shining on the upper surface, yellowish underneath. Fruit globular, c. 12 mm diam. Large tree. Coast and gullies in Illawarra region. RF . . . . **D. pentamera** (F.Muell.) F.Muell.

108 ERICACEAE
(now includes all genera from Epacridaceae)

Shrubs rarely small trees. Leaves alternate, whorled or spirally arranged, often crowded, rigid, often ericoid, entire or slightly denticulate, usually with parallel veins; stipules absent. Flowers bisexual; regular; axillary or terminal, solitary, or several together or in spikes, racemes or panicles. Sepals 4–5, much imbricate in the bud, often with several sepaloid bracts below the flower. Corolla tubular, urceolate or campanulate,4–5-lobed. Stamens hypogynous or epipetalous, up to twice as many as the corolla lobes; anthers opening apically by pores or longitudinal slits. Ovary superior, 1–5-locular; style simple; stigma undivided. Fruit a loculicidal dehiscent capsule or drupaceous. 157 gen., trop. to temp., cosmop. 36 gen. in Aust.

Key to the genera

1 Stamens 8–10, twice as many as corolla lobes. 2
1 Stamens 4–5, as many as corolla lobes or less. 5

2 Flowers ≥1.5 cm long. . . . . **3 RHODODENDRON**
2 Flower <1.5 cm long 3

3 Corolla lobes 4. Leaves whorled . . . . **2 ERICA**
3 Corolla lobes 5. Leaves alternate 4

4 Flowers ≤4 mm long. Stems with acicular bristles. Fruit a capsule enclosed by white or pink swollen
   calyx . . . . **1 GAULTHERIA**
4 Flowers ≥8 mm long. Stems with gland-tipped hairs. Fruit a berry, rough and orange at maturity . . . .
   . . . . **4 ARBUTUS**

5 Style terminal on the ovary. Fruit drupaceous; seeds solitary or few in each loculus 6
5 Style inserted in a ± deep pit in the ovary summit. Fruit a capsule; seeds numerous in each loculus . 16

6 Leaves at least 15 mm wide. Tree or rarely a shrub in RF . . . . **14 TROCHOCARPA**
6 Leaves less than 15 mm wide. Shrubs or rarely small trees, not in RF. 7

7 Anthers projecting beyond the corolla tube on long filaments . . . . **5 STYPHELIA**
7 Anthers wholly or partially enclosed within the corolla tube on short filaments 8

8 Corolla tube with tufts of hairs or scales inside below the middle .9
8 Corolla tube glabrous inside below the middle. 10

9 Corolla at least twice the length of the sepals, with 5 tufts of hairs below the middle. Filaments
   flattened . . . .**6 ASTROLOMA**
9 Corolla tube much less than twice as long as the sepals, with 5 glandular scales below the middle.
   Filaments not flattened . . . . **7 MELICHRUS**

10 Corolla tube c. 18 mm long . . . .**6 ASTROLOMA**
10 Corolla tube less than 6 mm long. 11

11 Throat of the corolla almost closed by reflexed hairs descending into the tube; lobes imbricate in bud
   . . . . **8 BRACHYLOMA**
11 Corolla throat with or without hairs, if hairs present, then they are not reflexed into the tube closing
   the throat; lobes valvate in bud 12

12 Corolla lobes densely woolly hairy on inside surface . . . . **12 LEUCOPOGON**
12 Corolla lobes not densely woolly inside; glabrous or with a few hairs scattered over the surface or at the
   tip. 13

13 Corolla with tufts of hair at the tips of the lobes and spreading across the tube throat . . . .
   . . . . **13 ACROTRICHE**
13 Corolla lobes glabrous or with a few scattered hairs 14

14 Flowers ± sessile. Ovary 1 or 2-locular. . . . . **11 MONOTOCA**
14 Flowers pedicellate. Ovary 5 or more–locular. 15

15 Corolla lobes glabrous. Stigma exserted from corolla tube . . . .**9 LISSANTHE**
15 Corolla lobes sparsely hairy. Stigma included in corolla tube. . . . .**10 AGIORTIA**

16 Leaves with a sheathing base which falls off with the leaf 17
16 Leaves petiolate sessile or stem-clasping but never sheathing the stem 18

17 Branches smooth, without annular scars . . . . **18 SPRENGELIA**
17 Branches with annular scars. . . . .**19 DRACOPHYLLUM**

18 Corolla lobes imbricate and contorted in the bud. . . . . **17 WOOLLSIA**

18 Corolla lobes valvate or imbricate but not contorted in the bud 19

19 Stamens inserted at the throat of the corolla tube. Anthers versatile, attached to filament above their middle . . . . **15 EPACRIS**

19 Stamens inserted at the base of the corolla tube. Anthers partially adnate to filament, connivent around the style but not cohering . . . .**16 RUPICOLA**

1 Gaultheria Kalm ex L.
4 species in Aust. (3 endemic); Qld, NSW, Vic., Tas.

One species in the area

Erect shrub up to 2 m high; stems with stiff ± appressed hairs. Leaves elliptic to ovate, up to 8 cm long, finely toothed; petiole 1–3 mm long. Raceme 3–11-flowered, towards the ends of the branches. Sepals becoming white and ± fleshy in the fruiting stage. Corolla white, up to 4 mm long. Stamens 10, enclosed within the corolla tube. Fruit a capsule enclosed in the persistent sepals. Blue Mts, e.g. Mt Werong. Open forests; RF margins; gullies. Fl. spring–summer. *White Waxberry.* . . . . **G. appressa** A.W.Hill

2 Erica L.
6 species naturalized Aust. NSW, Vic. Tas., S.A.

Shrub or small tree. Leaves in whorls linear or appearing terete, margins flat to strongly revolute. Flowers in axillary racemes or clusters. Sepals 4. Corolla 4-lobed, campanulate or urceolate. Stamens 8, epipetalous, enclosed within the tube. Ovary 4-locular. Fruit a loculicidal capsule, many-seeded.

Key to the species

1 Branchlets with simple hairs. Leaves 3–4 in each whorl, terete, grooved underneath, glabrous, 3–6 mm long, 0.5 mm wide. Flowers axillary, forming large pyramidal panicles. Calyx tubular, 4-lobed, much shorter than the corolla. Corolla narrowly campanulate, pink to white, 4–5 mm long. Erect shrub 1–2 m high Naturalized at places in Blue Mts and on the coastal plain. Introd. from W. Europe. Fl. autumn–spring . . . .***E. lusitanica** Rudolphi

1 Branchlets with branched hairs. Similar to *E. lusitanica* but leaves 0.7 mm wide, and floral parts slightly larger. Flowers white or cream. Naturalized in Blackheath area. Introd. from W. Europe. Fl. autumn– spring . . . . ***E. arborea** L.

(*E. glandulosa* has been recorded as naturalized at one site in Lawson, flowers are pink and white, 3 cm long.)

3 Rhododendron L.
3 species in Aust. (2 native, 1 naturalized); Qld, NSW

One species in the area

Shrub to 10 m high. Leaves elliptic, 10–17 cm long, 3.5–4.5 cm wide, margins entire. Inflorescence terminal with 5–20 flowers. Corolla pink-purple, campanulate, pubescent inside towards the base, 2.5–5 cm long, upper lobes with orange spots. Stamens filaments pink with white hairs. Ovary 5-locular, glabrous, style 3–4 cm long. Capsule 1.5–3 cm long. Blue Mts Introd. from Asia. Fl. spring . . . . ***R. ponticum** L.

4 Arbutus L.
1 species naturalized Aust.; NSW, Vic., S.A., Tas.

One species in the area

Tall shrub to 10 m high. Bark flaky, Stems with gland tipped hairs. Leaves alternate, ovate 3–9 cm long, 1.5–3.5 cm wide, dark green, thick, margins usually serrate. Inflorescence a raceme or panicle 2–9 cm long. Corolla 8–10 cm long, white or pink, lobes rounded and strongly recurved. Stamens wooly anthers with a narrow appendage. Fruit a berry, 1.5–2 cm diam., red-orange when mature. Blue Mts Introd. from Europe. Fl. winter . . . . ***A. unedo** L.

5 Styphelia Sm.

14 species endemic Aust.; all states and territories except NT

Erect or spreading shrubs. Leaves sessile or nearly so, rigid, pungent pointed. Flowers axillary, solitary (rarely 2–3 together). Corolla tube elongated, cylindrical; lobes linear, valvate in the bud, bearded inside, the lobes revolute exposing the stamens. Ovary 5-locular, with 1 ovule per loculus. Style filiform, longer than the corolla tube. Fruit drupaceous.

Key to the species

1 Young branches with long silky or hirsute hairs 2
1 Young branches glabrous or with minute hairs. 6

2 Leaves mostly convex on the upper surface, oblong-linear or obovate, 8–15 mm long, abruptly acuminate; margins recurved. Corolla red (rarely pale yellowish, green or whitish); tube slender, 20–25 mm long. Spreading shrub up to 1 m high. Coast and adjacent plateaus south of Sydney; lower Blue Mts DSF and heath. Sandy soils. Fl. autumn–winter . . . . **S. tubiflora** Sm.
2 Leaves mostly concave or flat on the upper surface 3

3 Leaves 25–50 mm long, 3–4 mm wide, very concave, narrow-lanceolate, tapering very gradually to a long fine point. Corolla greenish-yellow; the tube 18–25 mm long. Erect shrub 1–2 m high. Coast and adjacent plateaus. DSF. Sandy soils. Fl. autumn–winter . . . . **S. longifolia** R.Br.
3 Leaves mostly less than 25 mm long. Corolla yellow to red; tube up to 30 mm long. 4

4 Leaves 8–15 mm wide, shortly tapering to a fine point, broad-elliptic to ovate; petiole c. 1 mm long; margins ciliolate. Flowers usually spreading. Corolla tube 15–30 mm long. Erect shrub up to 2 m high. Hawkesbury River and Gosford districts. DSF. Sandy soils. Fl. mostly winter . . . .
. . . . **S. laeta** R.Br. ssp. **latifolia** (R.Br.) J.M.Powell
4 Leaves 3–8 mm wide. 5

5 Leaves mostly more than 6 mm wide, ovate to ovate-lanceolate or broad-oblong, shortly tapering to a fine point; petiole up to 0.5 mm long; margins minutely dentate. Flowers erect to horizontal. Corolla yellowish-green or red; tube 15–26 mm long. Erect shrub up to 2 m high. Coast and adjacent plateaus. DSF and heath. Sandy soils. Fl. autumn–winter . . . . **S. laeta** R.Br. ssp. **laeta**
5 Leaves mostly less than 6 mm wide, lanceolate, tapering gradually to a fine point; petiole up to 1 mm long; margins dentate. Flowers drooping. Corolla pale green; tube 15–20 mm long. Erect shrub up to 2 m high. Woronora Plateau; lower Blue Mts DSF and heath. Sandy soils. Fl. summer . . **S. angustifolia** DC.

6 Leaves entire, narrow- to oblong-lanceolate but often obovate at the base of the shoots, up to 35 mm long. Young branches glabrous or scabrous. Flowers often clustered at the base of a shoot, spreading or drooping. Corolla green or pink with yellow; tube 18–20 mm long. Erect shrub 1–2 m high. Widespread. Heath and DSF. Sandy soils. Fl. Aug.–Oct. on the coast, summer in the Blue Mts. . **S. triflora** Andrews
6 Leaves minutely fringed, broad-lanceolate to obovate-oblong, up to 30 mm long. Young branches minutely pubescent. Flowers almost erect to horizontal. Corolla green; tube 15–25 mm long. Erect shrub 1–2 m high. Coast and adjacent plateaus. Heath and DSF. Fl. winter–spring . . . .
. . . . **S. viridis** Andrews ssp. **viridis**

6 Astroloma R.Br.

18 species endemic Aust.; NSW, Vic., Tas., S.A., WA

Small shrubs. Leaves sessile or nearly so. Flowers axillary, solitary, on very short pedicels or almost sessile, surrounded by several bracts of which the inner ones are enlarged and embrace the calyx. Corolla tube elongated; the lobes straight or spreading at the top only. Filaments short, usually flattened, inserted in the throat. Ovary 5-locular; style filiform, as long as or longer than the corolla tube. Fruit drupaceous.

Key to the species

1 Leaves narrow-linear, crowded, 10–18 mm long; margins revolute, scabrous. Corolla red at the base, passing to yellow and green at the tip, c. 18 mm long; tube glabrous inside. Erect or diffuse shrubs 20–100 cm high. Coast and adjacent plateaus; lower Blue Mts Heath and DSF. Sandy soils. Fl. winter–spring. *Pine Heath*. . . . .**A. pinifolium** (R.Br.) Benth.

1 Leaves narrow-lanceolate to linear, flat or slightly convex or concave, 5–12 mm long; margins minutely ciliate-denticulate. Corolla tube red, 10–12 mm long, with 5 tufts of hair inside below the middle. Diffuse, much branched shrub 20–30 cm high. Widespread. Open forest and cleared land on Ss and WS, often on heavy soils. Fl. winter–spring. *Native Cranberry*. . . . .**A. humifusum** (Cav.) R.Br.

7 Melichrus R.Br.

4 species endemic Aust.; Qld, NSW, Vic.

Low shrubs with sessile crowded lanceolate or narrow-lanceolate leaves. Flowers axillary, solitary, with several bracts passing into sepals. Corolla tube short, broad, with 5 glandular scales inside alternating with stamens; lobes longer than the tube. Fruit drupaceous, usually nearly globular.

Key to the species

1 Leaves with scattered long hairs on lower surface, ciliate, flat, finely pointed but not pungent, 12–25 mm long. Calyx broadly campanulate, 10–12 mm diam. Corolla scarcely as long as the calyx; the tube very short and broad. Flowers facing the ground on the underside of the branches, producing large amounts of nectar. Coast and adjacent plateaus; lower Blue Mts Heath. Sandy soils. Fl. winter–spring. *Jam Tarts* . . . . **M. procumbens** (Cav.) Druce

1 Leaves glabrous or scabrous margins toothed, apex pungent. 2

2 Corolla white, cream or yellowish-green; tube urceolate 4–5 mm long, nearly as long as the lobes. Calyx ovoid, green to white. Leaves concave, spreading to reflexed; greyish-green; 5–25 mm long. Fruit white or purple. Erect shrub up to 1.5 m high, with a thick rootstock. Higher Blue Mts DSF. Fl. winter–spring **M. urceolatus** R.Br.

2 Corolla pink to red, tube 2.5–4 mm long. Calyx tinged with pink or red. Leaves held more or less erect, 11–30 mm long. Fruit red. Shrub to 1.2 m high. Capertee Valley and Newnes area. Fl. autumn–spring . . . . **M. erubescens** A.Cunn. ex DC.

8 Brachyloma Sond.

7 species endemic Aust.; all states and territories except NT

One species in the area

Erect, bushy shrub 30–100 cm high. Leaves sessile or nearly so, broad-lanceolate to oblong-elliptic or nearly ovate, obtuse or with a short callous point, flat or slightly concave; veins ± divergent from the base, 4–10 mm long. Corolla white; tube 3–4 mm long, with a ring of hairs reflexed inside the throat and descending into the tube between the stamens; lobes narrow, shorter than the tube. Ovary 3–5-locular; style short. Fruit drupaceous, globular, 3–4 mm diam. Widespread. Heath and DSF. Sandy soils, rarely clay soils. Fl. Aug.–Oct. on the coast; until Dec. in the Blue Mts. . . . . **B. daphnoides** (Sm.) Benth.

9 Lissanthe R.Br.

6 species in Aust.; all states and territories except NT

Erect shrubs. Leaves linear to narrow-lanceolate, very shortly petiolate, rigid, pungent pointed. Corolla tube short, hairy inside above the middle; lobes spreading, glabrous (or sparsely hairy) inside. Ovary 5–7-locular. Fruit drupaceous, depressed-globular.

Key to the species

1 Leaves lanceolate to linear, 4–17 mm long, 0.5–2 mm wide, rigid, tapering to a pungent point, very distinctly 3-ribbed on lower surface, varying in size and shape in different districts. Flowers numerous,

in short dense racemes. Corolla tube 2–3 mm long, pink to white. Fruit white or greenish, up to 3 mm long. Rigid, heath-like shrub c. 50 cm high. Widespread. Heath and DSF. Sandy soils; laterites. Fl. spring–summer. . . . . **L. strigosa** (Sm.) R.Br.

1 Leaves linear-oblong to narrow-lanceolate, 15–25 mm long, 2–5 mm wide, 7–8 ribbed or faintly striated. Flowers 1–4 in a loose raceme. Corolla tube 4–5 mm long, white. Leaves flat to convex, lower surface 7–8-ribbed. Fruit red, 8 mm diam. Erect shrub up to 1 m high with spreading branches. Blue Mts; Woronora Plateau. Open forest. Fl. winter–spring. . . . . **L. sapida** R.Br.

10 Agiortia Quinn
3 species endemic Aust.; Qld, NSW

One species in the area
Erect, open shrub to 1.5 m high. Leaves flat to concave, lower surface with faint striations; obtuse at the base, with a distinct petiole, oblong, 4–20 mm long, 2–5 mm wide, ± spreading, ± concave or flat towards the base. Flowers in groups of 4–9, white, corolla tube 1–2 mm long. Anthers not prominently exerted and without a sterile tip. Stigma inclosed within the corolla tube. Ovary 7–10-locular. DSF. Previously recorded as occurring at Woy Woy but this record cannot be confirmed. Fl. spring . . . .

. . . . **A. pleiosperma** (C.T.White) Quinn

11 Monotoca R.Br.
17 species endemic Aust.; all states and territories except NT

Small or tall shrubs. Leaves broad-linear to elliptic-oblong, with recurved margins or nearly flat, pale or whitish and finely veined underneath. Flowers small, often ± unisexual. Corolla almost campanulate, whitish; lobes spreading, glabrous; tube glabrous. Ovary usually 1-locular, with 1 ovule; style short. Fruit drupaceous.

Key to the species

1 Shrub up to 4 m high (but often very short in wind-swept coastal heaths). Flowers in terminal or axillary racemes which sometimes exceed the leaves or reduced to a few or a solitary flower. Bracts deciduous. Leaves elliptic-oblong to almost linear, 5–25 mm long, paler underneath. Fruit red or orange. Widespread, but frequently near the sea. Sandy soils. Fl. July–Sept . . . . **M. elliptica** (Sm.) R.Br.

1 Small, erect shrubs c. 30 cm or rarely more than 100 cm high. Flowers in clusters or very short racemes always shorter than the leaves, more rarely solitary or in pairs, usually axillary. 2

2 Leaves mucronate, usually oblong-linear, flat or with recurved margins, whitish underneath, 8–10 mm long; petiole greenish. Flowers in clusters or very short racemes, rarely solitary or in pairs. Widespread. Heath and DSF. Sandy soils. Fl. autumn–winter . . . .**M. scoparia** (Sm.) R.Br.

2 Leaves obtuse, scarcely paler underneath, oblong, 4–8 mm long, rather thick; petiole reddish. Flowers solitary or in pairs. Woronora Plateau; Blue Mts Uncommon. Fl. summer .**M. ledifolia** A.Cunn. ex DC.

12 Leucopogon R.Br.
Bearded Heaths
200 species endemic Aust.; all states and territories

Small or tall shrubs. Flowers small, in racemes or clusters. Corolla tube longer or shorter than the calyx; lobes densely bearded inside with white hairs; tube usually glabrous inside. Stamens enclosed in the corolla tube. Ovary 2–5-locular. Fruit drupaceous.

Key to the species

1 Leaves concave or with incurved margins. 2
1 Leaves convex or with recurved or revolute margins 5

2 Sepals acute. Ovary 3-locular. Leaves almost sessile, lanceolate, c. 6 mm long and 2 mm wide, acuminate-acute, often appressed to the stem, with 5 veins at the base. Small shrub up to 40 cm high, with wiry branches. Early record from Centennial Park . . . . **L. deformis** R.Br.

2 Sepals obtuse. Ovary 5-locular 3

3 Anthers obtuse, with a small white callous on the ± divergent tip. Leaves with a petiole c. 0.5 mm long, with 3 veins at the base, narrow-lanceolate, 5–15 mm long, tapering to a long but not hyaline point, often appressed to the stem. Small shrub up to 50 cm high, with wiry branches. Widespread. Heath and DSF. Ss and WS. Fl. spring–summer. . . . . **L. virgatus** (Labill.) R.Br.

3 Anthers emarginate, with no callous; tip scarcely divergent 4

4 Leaves with 3 central veins and a short petiole . . . . **L. neo-anglicus** (see below)

4 Leaves with numerous veins from the base, quite sessile on a broad pulvinus, lanceolate to ovate, 5–15 mm long, tapering to an hyaline point c. 1 mm long, appressed to the stem. Straggling shrub up to 60 cm high, with rigid wiry branches. Coast and adjacent plateaus; lower Blue Mts Heath and DSF. Sandy soils. Fl. summer–autumn . . . . **L. appressus** R.Br.

5 Leaves stem clasping, with rounded auricles, cordate-ovate, 12–25 mm long. Flowers in elongated spikes, much exceeding the leaves. Straggling shrub up to 80 cm high. Coast and adjacent plateaus. Sheltered places in DSF. Sandy soils. Fl. winter–spring . . . . **L. amplexicaulis** (Rudge) R.Br.

5 Leaves not stem-clasping 6

6 Anthers with a sterile callous tip 7

6 Anthers emarginate or obtuse, without a callous tip. 10

7 Leaves ovate to almost linear, usually 2–4 mm long but up to 6 mm on young plants, obtuse. Flowers in dense clusters. Sepals very acute. Erect or straggling shrub with wiry branches, up to 1 m high. Widespread. Heath and DSF. Sandy soils. Fl. most of the year . . . . **L. microphyllus** (Cav.) R.Br. var. **microphyllus**

7 Leaves narrow-elliptic-lanceolate or oblong, flat, obtuse or with a callous tip, more than 10 mm long. Sepals obtuse 8

8 Flowers in dense spikes mostly shorter than or equal to the leaves. Leaves oblanceolate to lanceolate, mostly 1–3 cm long, paler underneath. Fruit white. Erect shrub up to 5 m high but usually much less. Coast. Sandy soils near the sea. Heath. Fl. July–Aug. . . . .**L. parviflorus** (Andrews) Lindl.

8 Flowers in loose interrupted spikes as long as or longer than the leaves, 1–4 cm long. Leaves ± concolorous. Fruit red or yellowish. Bushy shrubs up to 2 m high 9

9 Young stems (but not inflorescence stems) glabrous or minutely pubescent. Leaves elliptic to ovate, 3–10 mm wide, mostly 1–4 cm long. Flowers c. 3 mm long when open. Widespread. Forests on most soil types. Fl. most of the year . . . .**L. lanceolatus** (Sm.) R.Br. var. **lanceolatus**

9 Young stems (but not inflorescence stems) villous-pubescent. Leaves narrow-ovate, 1–3 mm wide, mostly 1–3 cm long. Flowers c. 2 mm long when open. Hunter River Valley. Forests on sandy soils. Fl. most of the year. . . . . **L. lanceolatus** var. **gracilis** Benth.

10 Flowers pendulous. Corolla villous or glabrous outside 11

10 Flowers erect or spreading. Corolla glabrous outside. 14

11 Corolla villous outside. Leaves linear-lanceolate, 8–15 mm long, 1–1.5 mm wide, acuminate, with a pungent hyaline tip. Flowers on pedicels 2–4 mm long, solitary or 2–3 together in the leaf axils. Erect shrub up to 1 m high. Upper Georges River; Woronora River; Grose River. Ss. Vulnerable. Fl. spring . . . . **L. exolasius** (F.Muell.) F.Muell. ex Benth.

11 Corolla glabrous outside 12

12 Flowers 1–4 in the leaf axils on a peduncle 1–8 mm long. Corolla tube shorter than the sepals; the lobes as long as the tube. Leaves linear-lanceolate, 8–12 mm long, 2–4 mm wide, acuminate, with a pungent hyaline tip; margins recurved or flat. Spreading shrub up to 2 m high. Widespread. DSF. Sandy soils. Fl. spring. . . . .**L. setiger** R.Br.

12 Flowers solitary in the leaf axils on pedicels up to 1 mm long. Corolla tube slightly longer than the sepals; the lobes not quite as long as the tube. Leaves acuminate, with a pungent hyaline tip; margins revolute. 13

13 Shrubs to more than 1.5 m high. Branchlets with short rough hairs. Leaves up to 8 mm long, 1.5–2 mm wide. Flowers solitary. Sepals 3.5–4.5 mm long. Fruit obovoid, 4.5–5 mm long. Western Sydney and Blue Mts DSF. Sandy doils. Endangered. Fl. spring. . . **L. fletcheri** Maiden & Betche ssp. **fletcheri**
13 Shrub up to 1 m high. Branchlets with longer hairs. Leaves up to 8 mm long, 1.5–3 mm wide. Flowers 1–3. Sepals 2–3.5 mm long. Fruit ellipsoid, 3–3.5 mm long. Mt Werong. DSF. Sandy soils. Fl. spring . .
. . . . **L. fletcheri** ssp. **brevisepalus** J.M.Powell

14 Leaves obtuse or with a callous tip 15
14 Leaves with an acuminate pungent tip 17

15 Leaves revolute, slightly paler underneath, pubescent, narrow-oblong to linear, c. 10 mm long, 0.5–1 mm wide. Flowers few together, almost sessile in the leaf axils. Corolla c. 3 mm long. Erect shrub up to 70 cm high. Coast from Broken Bay northwards. Heath and DSF. Fl. spring–summer . . . .
. . . . **L. margarodes** R.Br.
15 Leaves oblanceolate or elliptic to oblong, flat convex or very slightly recurved 16

16 Leaves glaucous underneath, distinctly nerved, elliptic to oblong, 3–6 mm long, up to 2 mm wide, mostly spreading. Much branched undershrub up to 50 cm high. Barrington Tops. Heath and DSF. Fl. summer . . . .**L. hookeri** Sond.
16 Leaves greyish-green on both surfaces; the lower surface indistinctly nerved; tapering gradually into the petiole, oblanceolate to obovate or elliptic, 10–20 mm long, 2–4 mm wide, mostly erect or nearly so, convex at least at the base. Slender, erect, straggling shrub up to 2 m high. Widespread. Heath and DSF. Sandy soils. Fl. spring . . . . **L. muticus** R.Br.

17 Leaves recurved or convex (very rarely flat in *L. juniperinus*) 18
17 Leaves flat 20

18 Corolla tube twice as long as the sepals; the lobes very short. Leaves lanceolate to oblong-linear. Sepals 3–4 mm long, acute. Flowers solitary or few together in the leaf axils, often on the older branches. Rigid, divaricate shrub up to 1 m high. Coast and adjacent plateaus; lower Blue Mts Heath and DSF on a variety of soil types. Fl. winter–spring . . . .**L. juniperinus** R.Br.
18 Corolla tube less than twice as long as the sepals 19

19 Leaves spreading, oblong-linear, very shortly mucronate, 5–8 mm long, pungent pointed. Flowers 1–11 in short erect spikes not exceeding the leaves. Style longer than the ovary. Diffuse, heath-like shrub up to 1 m high, with wiry branches. Widespread. Heath and DSF. Sandy and clay soils. Fl. spring . . . . .
. . . . **L. ericoides** (Sm.) R.Br.
19 Leaves erect to ascending along the stem, oblong-linear, up to 7 mm long, 1–2 mm wide, acuminate. Flowers erect, 1–3 in the leaf axils. Style shorter than or equal to the ovary. Erect shrub up to c. 50 cm high. Widespread. DSF. Sandy soils. Fl. winter–summer . . . . **L. attenuatus** A.Cunn.

20 Leaves diverging almost horizontally, 5–11 mm long, 1–3 mm wide, paler underneath. Flowers 1–2 in the leaf axils. Corolla longer than the sepals; tube 4–5 mm long. Prostrate shrub up to 20 cm high, with bristly branches. Higher Blue Mts DSF. Sandy and granite soils. Fl. autumn . .**L. fraseri** A.Cunn.
20 Leaves erect or spreading upwards 21

21 Sepals 4 mm or more long. Leaves ovate to obovate, 5–10 mm long, 2–5 mm wide, paler underneath. Flower solitary in the leaf axils. Corolla longer than the sepals; tube 5–8 mm long. Erect shrub up to 1 m high. Widespread. Heath and DSF. Fl. winter–spring. . . . . . . **L. neo-anglicus** F.Muell. ex Benth.
21 Sepals less than 3 mm long. Leaves lanceolate, 10–20 mm long, 2–4 mm wide. Corolla as long as the sepals; tube 1–2 mm long. Compact, erect shrub up to 1 m high, with ± erect branches. Widespread. Heath and DSF. Sandy soils, often in damp situations. Fl. spring. . . . . **L. esquamatus** R.Br.

13 Acrotriche R.Br.
14 species endemic Aust.; all states and territories except NT

Rigid divaricate shrubs. Leaves rigid. Flowers very small, in bracteate clusters on very short spikes in the axils of the previous years' growth or on the stem below the leaves. Corolla greenish, equaling the calyx or longer; lobes spreading, with a tuft of long hairs on the inner surface at the end. Ovary 2–10-locular; style short. Fruit drupaceous, small, globular or depressed.

Key to the species

1 Leaves with strongly recurved margins, grooved below, spreading, lanceolate, 6–11 mm long, up to 2 mm wide, whitish, 3–5-veined. Corolla greenish; tube up to 3.5 mm long; lobes up to 2 mm long. Dense shrub up to 1.5 m high. Rylestone district. Heath and DSF. Fl. winter–spring . . **A. rigida** B.R.Paterson
1 Leaves flat or with very slightly recurved margins 2

2 Leaves up to 1.5 mm wide, c. 4 mm long, 3–5-veined with shallow grooves between the veins, pungent pointed. Ovary 5–7-locular, green. Corolla tube inflated, 2 mm long. Sepals 2–3 mm long. Low prostrate shrub often forming mats. Higher Blue Mts, e.g. Colong. DSF. Fl. spring. *Honeypots* . . . .
. . . . **A. serrulata** (Labill.) R.Br.
2 Leaves 3–7 mm wide, with more than 5 veins radiating from the petiole 3

3 Ovary 4–5-locular, green. Corolla tube c. 1 mm long. Sepals 1–2 mm long. Leaves oblong-elliptic to narrow-lanceolate, horizontal, flat or nearly so, 10–15 mm long, with pungent tips. Erect shrub 1–2 m high. Coast and adjacent plateaus; lower Blue Mts DSF. Fl. winter. . . . . **A. divaricata** R.Br.
3 Ovary 6–10-locular, with a band of red extending from the neck down both sides of the ovary. Corolla tube c. 3 mm long. Sepals 1–2 mm long. Leaves spreading, elliptic to obovate, up to 35 mm long and 7 mm wide. Erect shrub up to c. 1 m high. Blue Mts Ss and granite areas. Fl. spring . . **A. aggregata** R.Br.

14 Trochocarpa R.Br.
7 species endemic Aust.; Qld, NSW, Vic., Tas.

One species in the area
Glabrous tree or shrub. Leaves ovate to lanceolate, acuminate, flat, shining, 5–7-veined, up to 5 cm long. Flowers small, in terminal clustered or solitary spikes which are 2–4 cm long. Corolla white. Fruit drupaceous, dark bluish, 6–8 mm diam. Coast and adjacent plateaus. In or near RF. Fl. summer. *Tree Heath.* . . . . **T. laurina** R.Br.

15 Epacris Cav.
Heaths
38 species in Aust.; Qld, NSW, Vic., Tas., S.A.

Heath-like shrubs. Leaves sessile or petiolate, articulate on the stem, sometimes embracing it above the base but not sheathing. Flowers solitary, axillary, on short pedicels or almost sessile. Bracts numerous, passing gradually into the sepals and forming an involucre around them. Corolla tube cylindrical or campanulate; the lobes imbricate but not contorted in the bud. Filaments short; anthers wholly or partially enclosed in the corolla tube. Ovary 5-locular; style inserted in a tubular depression in the ovary. Fruit a small capsule, loculicidally dehiscent. Seeds numerous.

Key to the species

1 Corolla tube shorter than the calyx or almost as long 2
1 Corolla tube longer than the calyx 7

2 Leaves with an acute to acuminate apex, ... 3
2 Leaves with an obtuse apex 4

3 Leaves spreading to slightly reflexed, 1.5–4 mm long, apex acuminate, apiculate or with a short mucro, base cordate often wrapping around the stem, very concave and broad above the base Flowers usually

sessile. Corolla white or slightly tinged with pink. Sepals 1–2 mm long. Corolla tube broad; the lobes as long as the tube. Erect, wiry shrub usually 30–100 cm high. Widespread. Heath & DSF. Sandy soils. Fl. winter–spring. . . . . **E. microphylla** R.Br.

3 Leaves appressed, rhombic 2–4.5 mm long, 1.6–3 mm wide, apex with a callus or shortly apiculate, base obtuse or truncate, flat, margins slightly concave to convex. Corolla white, 1.5–2 mm diam. Barrington Tops and western ranges. Ranges. At higher altitudes. Fl. summer–autumn . . . .
. . . . **E. rhombifolia** (L.R.Fraser & Vickery) Crowden

4 Leaves distinctly but very shortly (1–3 mm) petiolate, obovate or broad-ovate, concave, smooth, 4–8 mm long, 3–8 mm wide. Sepals c. 2 mm long, rigid, obtuse. Corolla lobes as long as the tube. Procumbent to erect shrub up to 3 m high. Illawarra Ranges, e.g. Bulli Lookout. Ss cliffs. Fl. spring . . . .
. . . . **E. coriacea** A.Cunn. ex DC.
4 Leaves with petiole less then 1 mm long. 5

5 Leaves 8–14 mm long, 3–4 mm wide, 1 main vein, lateral veins not conspicuous. Petiole with sparse hairs on adaxial side. Branchlets glabrous. Peduncles 1–1.5 mm long. Flowers axillary clustered at end of new branches. Sepals 5–6 mm long, Corolla tube 5–7 mm long. Semi erect to erect branching shrub, to 50 cm high, occasionally to 1m. Mid to high altitudes on eastern slopes of ranges, south from Grose River valley. Grows on creek banks or in sandstone fissures. DSF. Fl. summer. . . . .
. . . . **E. pinoidea** Crowden & Menadue
5 Leaves 2–4 mm long, 0.5–3 mm wide, ovate or elliptic to lanceolate, very thick. 6

6 Branchlets pubescent. Peduncles mostly less than 2 mm long. Leaves with a prominent thick keel. Sepals 2–3 mm long, obtuse (rarely a few almost acute). Corolla tube as long as the sepals, with 5 transverse thickenings inside near the base; lobes longer than the tube, broad, obtuse. Erect, bushy shrub up to 1 m high. Blue Mts, e.g. Linden, Woodford. Open situations. Ss. Fl. spring . . . . **E. rigida** Sieber ex Spreng.
6 Branchlets glabrous. Peduncles 2–5 mm long. Thick keel on the leaves often scarcely prominent. Sepals less obtuse. Corolla tube shorter than the sepals to almost as long. Spreading or straggling shrub up to 30 cm high. Blue Mts, e.g. Mt. Wilson, Blackheath. On damp rocks in sheltered situations. Fl. summer. (not easily distinguished from *E. rigida*) . . . . **E. muelleri** Sond.

7 Leaves with an obtuse apex 8
7 Leaves with an acute to acuminate apex. 12

8 Stems with prominent ridged leaf-scars 9
8 Stems with indistinct leaf-scars. 11

9 Erect shrub to 30 cm with few stems. Branchlets hirsute. Leaves oblanceolate to narrow-elliptic, 6–14 mm long, margins thickened, 3 prominent main veins, intramarginal vein present, lateral veins not conspicuous. Flowers axillary, scattered along new branches. Sepals 3–4 mm long. Corolla tube 7–9 mm long. Coast and Blue Mts Growing on sandstone rock faces or at base of cliffs. DSF. Fl. summer. . .
. . . . **E. lithophila** Crowden & Menadue
9 Weak, trailing or ascending–shrub to 30 cm high, branchlets pubescent. Leaves spreading, obovate to oblanceolate, 4–12 mm long; margins thickened, sometimes ciliate, 1–3 main veins, lateral veins conspicuous. Flowers axillary, scattered along the branches but sometimes clustered at ends of branches 10

10 Corolla 3–5.5 mm diam., tube ≤7 mm long, rarely to 10 mm. Sepals 3–4 mm long, ½ the length of corolla tube. Anthers 1mm long. Style 3–5.5 mm long. Coast to Blue Mts Fl. summer . . . .
. . . . **E. crassifolia** R.Br. ssp. **crassifolia**
10 Corolla 6–10 mm diam., tube >10 mm long. Sepals 4–6.5 mm long ¼ to ⅓ as long as corolla tube. Anthers 2 mm long. Style 12–28 mm long. Blue Mts Fl. summer. . . . .
. . . . **E. crassifolia** ssp. **macroflora** Crowden & Menadue

11 Leaves less than 3 mm wide, erect or ascending, oblong-elliptic to lanceolate or broad-linear, 5–12 mm long. Flowers axillary, usually in long one-sided terminal leafy racemes. Corolla tube creamy white, slightly exceeding the sepals, 4–8 mm long, cylindrical or slightly campanulate. Erect shrub up to 1.50 m high. Widespread. Chiefly in swampy sandy soils. Fl. spring–summer . . . . **E. obtusifolia** Sm.

11 Leaves more than 3 mm wide, horizontal to reflexed, obovate to almost orbicular, 4–8 mm long, 3–7 mm wide. Flowers axillary in short leafy terminal inflorescences. Corolla tube white or cream, slightly exceeding the sepals. Erect shrub up to 1 m high. Jenolan Caves. Heath and DSF. Fl. spring . . . . **E. robusta** Benth.

12 Leaves broad- to narrow-lanceolate, sometimes pungent-pointed. Bracts and sepals acute. Corolla yellowish or white. 13

12 Leaves ovate, often cordate, sessile or nearly so. Corolla white to pink or red. 17

13 Leaves sparingly hairy, broad-lanceolate, 10 mm long. Corolla tube c. 6 mm long, 3–4 mm diam. Style longer than the tube. Flattened or matted shrub up to 1 m high, covered with soft whitish hairs, especially on the young stems. Blackheath. Damp situations. Ss. Endangered. Fl. winter–spring . . . . **E. hamiltonii** Maiden & Betche

13 Leaves glabrous 14

14 Corolla tube narrow, 4–6 mm long, 1–2 mm diam,; lobes obtuse, broad, spreading. Flowers usually crowded in the upper axils forming leafy spikes or heads. Sepals 4–5 mm long. Corolla white. Style c. as long as the corolla tube. Leaves narrow-lanceolate, 6–12 mm long. Erect shrub 1–2 m high. Woronora Plateau; Blue Mts Swampy places on Ss. Fl. spring . . . . **E. paludosa** R.Br.

14 Corolla tube more than 9 mm long. 15

15 Flowers few towards the ends of the branches; peduncle 3–5 mm long. Bracts acute. Sepals up to 4 mm long. Corolla tube 15–20 mm long. Erect shrub up to 1 m high with prominent cup-shaped leaf-scars. Grose River. Open forests amongst rocks. vulnerable. Fl. winter . . . . **E. sparsa** R.Br.

15 Flowers numerous in long inflorescences towards the ends of the branches. Corolla tube 2–4 mm diam, slightly swollen; the lobes narrow, acute. Style much longer than the corolla tube. Leaves tapering gradually towards the base, elliptic to narrow-ovate, 5–15 mm long, 1–3 mm wide, with a narrow subulate apex. Corolla tube more than 10 mm long. Style exserted from the corolla tube. Erect shrubs 1–2 m high. 16

16 Corolla tube white to yellow. Blue Mts; Woronora Plateau. DSF. Sandy soils. Fl. spring–summer . . . . **E. calvertiana** F.Muell. var. **calvertiana**

16 Corolla tube pink. Robertson and Mittagong districts. Open forests. Fl. spring–summer . . . .**E. calvertiana** var. **versicolor** Maiden & Betche

17 Corolla tube yellow or white. 18

17 Corolla tube red or pink 19

18 Leaves mostly 4–6 mm long. Rather diffuse shrub 1–2 m high with thin wiry branches. Sepals 3–4 mm long. Corolla lobes 2–3 mm long. Widespread. DSF and heath. Sandy soils. Fl. chiefly autumn on the coast and summer on the Blue Mts. . . . . **E. pulchella** Cav.

18 Leaves 8–12 mm long. Rigid shrubs 1–2 m high. Sepals 6 mm long. Corolla lobes 4–5 mm long; tube slightly longer than the sepals. Anthers exserted from the corolla tube. Blue Mts Wet and sheltered situations. Fl. spring–summer. . . . .**E. purpurascens** var. **onosmiflora** Maiden & Betche

19 Corolla tube less than twice as long as the sepals, 4–8 mm long; lobes 4–5 mm long. Leaves with an obtuse or cordate base, ovate, up to 20 mm long, 5–9 mm wide, acuminate. Anthers slightly exserted from the corolla tube. Style enclosed in the corolla tube. Diffuse shrub 1–2 m high with thin wiry branches. DSF and heath. Sandy soils. Fl. Vulnerable. spring–summer . . . . **E. purpurascens** R.Br. var. **purpurascens**

19 Corolla tube more than twice as long as the sepals. 20

20 Corolla tube 12–20 mm long, red; the lobes and sometimes part of the tube white. Sepals 4–5 mm long, acuminate, with long fine points. Leaves ovate, 5–12 mm long. Straggling shrub up to 2 m high. Coast and adjacent plateaus. Heath and DSF. Ss. Fl. most of the year . . . . **E. longiflora** Cav.

20 Corolla tube 8–12 mm long, usually uniformly red or pink. Sepals acute 2–3 mm long. Leaves ovate, 5–9 mm long. Small, weak shrub with straggling branches. Higher Blue Mts Sheltered rocks and gullies. Fl. winter–summer . . . . **E. reclinata** A.Cunn. ex Benth.

16 Rupicola Maiden & Betche
4 species endemic Aust.; NSW

Erect to decumbent shrubs with prominent leaf-scars. Leaves ovate to elliptic, 3–5-veined, shortly petiolate. Flowers solitary in the leaf axils forming leafy inflorescences towards the ends of the branches. Corolla glabrous; tube much shorter than the sepals; lobes spreading, imbricate in the bud. Filaments inserted at the base of the corolla tube; anthers dehiscing by a short apical slit. Style exserted from the corolla tube.

Key to the species

1 Leaves ovate, ± cordate and auriculate at the base, with a long callous tip, 3–10 mm long, horizontal or reflexed. Corolla lobes 4–6 mm diam.; tube up to 4 mm long. Slender, erect shrub with villous branchlets. Blue Mts Sandy soils. Fl. spring–summer . . . . **R. apiculata** (A.Cunn.) I.Telford
1 Leaves elliptic to narrow-elliptic, tapering towards the base 2

2 Erect shrub up to 150 cm high, with villous branchlets. Leaves ± erect, narrow-elliptic, 10–28 mm long, up to 3 mm wide, thickened and rounded at the tip. Corolla lobes 6–9 mm diam.; tube 4–7 mm long. Burragorang Valley. Ss. Fl. spring–summer. . . . . **R. sprengelioides** Maiden & Betche
2 Decumbent shrubs. Leaves spreading 3

3 Leaves more than 15 mm long, spreading to horizontal, elliptic to ovate, 12–20 mm long, 4–8 mm wide, with a small callous tip. Corolla lobes 5–9 mm long; tube c. 1 mm long. Style hairy towards the base. Glen Davis. DSF in rocky places. Ss. Fl. winter–summer . . . . **R. decumbens** I.Telford
3 Leaves less than 11 mm long, mostly horizontal, elliptic to ovate, 8–14 mm long, 3–4 mm wide, acute. Corolla lobes 5–6 mm long; tube c. 1.5 mm long. Style glabrous. Lower Blue Mts DSF in rocky places. Ss. Fl. spring–summer . . . . **R. ciliata** I.Telford

17 Woollsia F.Muell.
1 species endemic Aust.; NSW

One species in the area
Erect shrub up to 2 m high. Leaves crowded, ovate, acuminate, tapering into a rigid point, 6–12 mm long. Flowers axillary, sessile in the axils of the upper leaves. Bracts and sepals imbricate, acute, 8–10 mm long. Corolla tube white or purplish or pinkish, as long as or rather longer than the calyx; lobes contorted in the bud. Fruit a capsule. Widespread. Heath and DSF. Sandy soils. Fl. winter–summer . . . . **W. pungens** (Cav.) F.Muell.

18 Sprengelia Sm.
4 species endemic Aust.; Qld, NSW, Vic., Tas., S.A.

Erect or small and diffuse shrubs. Leaves with a sheathing often membranous base concealing the branches, very concave, often stem-clasping above the base (Fig. 30). Sheathing leaf-bases falling off leaving the stem smooth and without scars. Flowers terminal, solitary or grouped towards the ends of the branches resembling a terminal panicle. Corolla tube short, pink or white, sometimes almost completely divided into separate petals; the lobes very spreading. Anthers connivent or cohering around the style. Ovary 5-locular. Fruit a capsule.

Key to the species

1 Diffuse or procumbent, often very small shrub. Leaves narrow, lanceolate-subulate, spreading, 3–6 mm long. Flowers solitary. Sepals lanceolate, pale. Corolla 6–8 mm long, white or pinkish; the lobes broadly imbricate in the bud. Blue Mts Often on wet rock ledges. Ss. Fl. spring–summer . . . . **S. monticola** (A.Cunn. ex DC.) Druce
1 Erect shrubs. Leaves tapering from the broad base. 2

2 Sepals narrow, pale, not leaf-like. Leaves spreading or recurved, 5–20 mm long. Corolla pink or purplish; the lobes narrow, almost valvate in the bud. Flowers often numerous and appearing to form a terminal panicle. Shrub mostly 50–100 cm high. Widespread. Wet heath on swampy sandy soils. Fl. spring. *Bog Rose* . . . . **S. incarnata** Sm.

2 Sepals broad, green, much resembling the leaves. Leaves spreading or incurved, 3–5 mm long. Corolla whitish; lobes very broad, imbricate in the bud. Slender, erect, wiry shrub 30–100 cm high. Coast and adjacent plateaus. Wet heath in swampy places on sandy soils. Fl. spring. **S. sprengelioides** (R.Br.) Druce

19 Dracophyllum Labill.
4 species endemic Aust.; Qld, NSW, Tas.

One species in the area

Erect or procumbent shrub mostly 20–100 cm long, linear-lanceolate, spreading; leaf bases sheathing, falling off with the leaves and leaving annular scars on the bare branches. Flowers in a narrow terminal panicle or loose raceme. Sepals 6 mm long. Corolla white or pink; the tube cylindrical, 8–10 mm long; the lobes small, recurved. Fruit a small globular capsule. Widespread. On rock ledges in gullies. Ss. Fl. winter–spring on the coast, spring–summer in the Blue Mts . . . . **D. secundum** R.Br.

109 MYRSINACEAE

Trees, shrubs or herbs. Leaves opposite or alternate, simple, entire or toothed; stipules absent. Flowers regular, bisexual, solitary or in axillary or terminal umbels, racemes or clusters.. Calyx 4–5-lobed, usually imbricate in the bud. Corolla tubular or rotate, 4–5-lobed; lobes valvate or imbricate. Stamens 4–5, opposite the corolla lobes. Ovary superior, 1-locular, with free-central or basal placentas. Fruit a capsule, drupaceous or viviparous. 41 gen., mainly trop. to warm temp.

Key to the genera

1 Herbs. 2
1 Trees or shrubs 3

2 Corolla yellow. . . . . .2 **LYSIMACHIA**
2 Corolla red, pink, orange or blue . . . . **1 ANAGALLIS**

3 Corolla 8–12 mm long. Fruit elongated. Mangrove trees . . . . **5 AEGICERAS**
3 Corolla less than 4 mm long. Fruit globular. Forest trees or shrubs 4

4 Petals connate. Flowers in sessile axillary umbel-like clusters . . . . **3 MYRSINE**
4 Petals free to the base or cohering together. Flowers in short axillary racemes. . . . . **4 EMBELIA**

1 Anagallis L.
2 species naturalized Aust.; all states and territories

One species in the area

Small, annual herbs with weak quadrangular stems. Leaves cauline, opposite, sessile, ovate, 5–20 mm long. Flowers axillary, solitary, on filiform pedicels which are longer than the leaves and finally recurved. Calyx and corolla divided nearly to the base. Corolla c. 10 mm diam., red, pink, orange or blue. Capsule circumscissile. Widespread. Weed of cultivation and waste ground. Introd. from Europe. Fl. spring. *Pimpernel* . . . . ***A. arvensis** L.

2 Lysimachia L.
S species in Aust. (1 naturalized, 2 native); NSW, Vic.

One species in the area

Erect herb up to 1 m high. Leaves cauline, opposite, elliptic to oblanceolate, 4–8 cm long, sessile or shortly petiolate. Flowers in terminal panicles, up to 15 mm diam. Sepals with black glands near the margin.

Corolla yellow. Capsule ± 5-valved. Wingecarribee Swamp. Endangerred. Fl. summer. *Yellow Loosestrife*
. . . . **L. vulgaris** L. var. **davurica** (Lebed.) Kunth

3 Myrsine L.
15 species in Aust.; Qld, NSW, Vic.

Shrubs or small trees with buds covered in brownish or reddish-brown hairs. Flowers in axillary clusters mostly on old wood. Calyx and corolla 4–5-lobed. Corolla 3–4 mm long. Anthers erect, almost sessile. Style short. Fruit a small almost globular drupe (previously *Rapanea*).

Key to the species

1 Corolla divided to c. half its length. Flowers 4–5-merous. Leaves usually irregularly toothed, sometimes entire on part of the plant but rarely on the whole, oblanceolate, mostly 3–8 cm long, coriaceous, often shining on the upper surface; petioles 3–7 mm long. Shrub or small tree. Coast and adjacent plateaus. Forests on heavy soils and Ss gullies. Fl. spring–summer. *Muttonwood* . . . .
. . . . **M. variabilis** R.Br.

1 Corolla divided nearly to the base. Flowers 5-merous. Leaves always entire except the juveniles, elliptic to obovate or broad-oblanceolate, 5–10 cm long, rather thin; the veins further apart than in *M. variabilis*; petioles mostly more than 7 mm long. Shrub or small tree. Widespread. RF. Fl. spring. *Brush Muttonwood*. . . . .**M. howittiana** R.Br.

4 Embelia Burm.f.
2 species in Aust.; Qld, NSW

One species in the area

Shrub or small tree. Leaves ovate to elliptic, 5–10 cm long, 3–5 cm wide, entire, obtuse or acute to acuminate, glabrous, with many small yellow to brown resinous dots. Flowers in short racemes up to 1 cm long. Petals free or cohering near the base, c. 2 mm long, whitish. Fruit a drupe, globular 5–7 mm diam. Wyong district. RF. Fl. spring–summer.. . . . . **E. australiana** (F.Muell.) FM.Bailey

5 Aegiceras Gaertn.
1 species native Aust.; Qld, NSW, NT, WA

One species in the area

Glabrous shrub or small tree up to 4 m high with aerating knee-roots. Leaves obovate, very obtuse, 4–7 cm long, coriaceous. Flowers in umbels which are terminal or in the upper axils. Calyx 5-lobed; lobes stiff and much imbricate, 4–6 mm long. Corolla tube not exceeding the calyx; lobes about the same length, spreading or reflexed, very acute, white. Fruit viviparous, elongated, cylindrical, acute, incurved, horn-shaped, 3–4 cm long. Mangrove swamps, extending further upstream than *Avicennia marina*. *River Mangrove* . . . .**A. corniculatum** (L.) Blanco

110 POLEMONIACEAE
18 gen., trop. to warm temp., mostly America.

1 Cobaea Cav.
1 species naturalized Aust.; NSW

One species in the area

Herbaceous perennial climber of very rapid growth, climbing by means of leaf tendrils. Leaves pinnate, alternate; leaflets in 2–3 pairs; stipules absent. Flowers solitary on long peduncles. Sepals 5. Corolla 3–4 cm diam., tubular; lobes 5, spreading. Stamens 5, epipetalous. Ovary 3-locular, superior; placentas axile. Fruit a capsule. Garden escape, naturalized in a few places, e.g. Berowra Creek. Introd. from trop. America. Fl. spring–summer.. . . . . ***C. scandens** Cav.

111 SAPOTACEAE

53 gen., mostly trop.

1 Pouteria Aubl.

9 species in Aust.; Qld, NSW, NT, WA

One species in the area

Trees with a rough scaly bark, exuding latex when cut. Leaves alternate, coriaceous, rather prominently reticulate, elliptic to obovate, 5–8 cm long, tapering at the base into a short petiole 6–12 mm long; stipules absent. Flowers in axillary clusters of 2–6, bisexual, regular, c. 5mm long. Sepals 5. Corolla tubular, 5-lobed. Fertile stamens 5, epipetalous, opposite the corolla lobes and alternating with 5 slender staminodes. Ovary 5-locular, very hairy. Fruit a globular or ovoid berry, 1–5-seeded, black, 2–5 cm diam. Coast and gullies in adjacent plateaus. RF. *Black Apple* . . . . **P. australis** (R.Br.) Baehni

112 SYMPLOCACEAE

1 gen., trop. and subtrop., except Africa

1 Symplocos Jacq.

7 – 10 species in Aust.; Qld, NSW, Vic.

Trees or shrubs. Leaves alternate, coriaceous; stipules absent. Flowers sessile or shortly pedicellate, often numerous, in axillary spikes racemes or panicles. Calyx lobes 5, short, broad. Corolla tubular, deeply 5-lobed or the petals almost free, regular; lobes imbricate in the bud. Stamens numerous. Ovary inferior, 2-locular; placentas axile. Fruit drupaceous.

Key to the species

1 Petals 3 mm long. Flowers sessile. Inflorescence minutely pubescent. Leaves c. 10cm long, obovate or elliptic to lanceolate; margins entire or with very short teeth. Fruit ovate, c. 12 mm long, surmounted by the calyx lobes. Little Cattai Creek. . . . . **S. stawellii** F.Muell.

1 Petals 6 mm long. Flowers distinctly pedicellate. Inflorescence glabrous. Leaves c. 12 mm long, often more prominently and remotely toothed than *S. stawellii*, otherwise similar. Coast and gullies in adjacent plateaus. RF. . . . . **S. thwaitesii** F.Muell.

113 THEOPHRASTACEAE

6–9 gen., temp. trop.

One genus in the area

1 Samolus L.

4 species endemic Aust.; all states and territories

Perennial or annual, glabrous herbs. Stems erect or procumbent. Leaves basal or cauline; cauline leaves alternate. Flowers in terminal racemes. Calyx campanulate, 5-lobed; tube very short and broad. Stamens 5. Ovary and fruit half-inferior. Capsule opening in 5 valves.

Key to the species

1 Leaves rather thick; the lower ones ovate or oblong; the upper ones narrower. Corolla up to 10 mm diam, white or tinged with pink. Perennial with ± tufted rootstock. Stems usually warty or wrinkled, erect and bearing flowers or arched and stoloniferous or creeping and rooting at the nodes. Coast. Margins of saltmarshes and other sites near the sea. Fl. spring–summer. *Creeping Brookweed* . . . .**S. repens** (J.R.Forst. & G.Forst.) Pers.

1 Leaves very thin, basal leaves obovate, 3–4 cm long; cauline leaves narrower and smaller. Flowers on filiform pedicels in loose racemes. Corolla 8–10 mm diam., white. Glabrous, erect herb with flowering

stems up to 30 cm high. Widespread. Margins of swamps and rivers. Fl. spring-summer. *Common Brookweed* or *Water Pimpernel* . . . . **S. valerandi** L.

114 BORAGINACEAE

Herbs or shrubs. Whole plant usually covered with ± stiff hairs seated on large tubercles. Leaves alternate, sometimes basal, rarely a few opposite; stipules absent. Flowers usually in terminal or axillary cymes, which may be coiled in the bud and unrolling to resemble a one-sided raceme or spike. Flowers bisexual, regular or irregular. Sepals 5, free or basally connate, persistent. Corolla tubular or rotate, 5-lobed. Stamens 5, epipetalous, alternating with the corolla lobes. Ovary superior, 2-carpellary, 2-locular but becoming 4-locular at maturity; Style simple, terminal or gynobasic. Fruit schizocarpic, splitting into four 1-seeded articles or 2 mericarps, usually with a rugose spiny or sometimes glabrous epicarp; rarely the articles ± drupaceous. 156 gen., widespread, temp. to trop.

Key to the genera

1 Corolla irregular 2
1 Corolla regular 3

2 Stamens extending beyond corolla . . . . **9 ECHIUM**
2 Stamens not extending beyond corolla . . . . **10 ANCHUSA**

3 Tall shrub or tree . . . . **11 EHRETIA**
3 Herb or small shrub 4

4 Stamens forming a cone around the style 5
4 Stamens not forming a cone around the style 6

5 Coarse herb. Ovary lobed . . . . **5 BORAGO**
5 Shrub. Ovary entire . . . . **1 HALGANIA**

6 Corolla with scales in the throat 7
6 Corolla without scales in the throat 9

7 Articles of the fruit smooth, shining . . . . **4 MYOSOTIS**
7 Articles of the fruit beset with hooked bristles 8

8 Erect plants . . . . **2 CYNOGLOSSUM**
8 Decumbent straggling plants . . . . **3 AUSTROCYNOGLOSSUM**

9 Corolla yellow . . . . **7 AMSINCKIA**
9 Corolla white or blue 10

10 Stigma filiform, capitate. Flowers in loose interrupted monochasia and appearing to be in the axils of leaves . . . . **6 BUGLOSSOIDES**
10 Stigmas almost sessile, with a terminal conical appendage. Flowers in compact coiled monochasia . . . . **8 HELIOTROPIUM**

1 **Halgania** Gaudich.
18 species endemic Aust.; all states and territories except Tas.

One species in the area
Small shrub up to 60 cm high, scabrous, with appressed hairs. Leaves elliptic to oblanceolate, 5–15 mm wide, dentate, tapering into a petiole. Flowers pedicellate in irregular cymes. Stamens forming a cone around the style, with terminal appendages. Corolla deep blue, c. 8 mm diam. Ovary entire, with a terminal style. Coast and adjacent plateaus. DSF. Fl. summer . . . . **H. brachyrhyncha** P.G.Wilson

2 Cynoglossum L.

3 species in Aust. (2 endemic, 1 naturalized); all states and territories except NT

Erect, perennial herbs. Calyx deeply divided into 5 segments. Corolla white to blue, with a short broad tube and 5 spreading lobes; the throat closed with scales opposite the lobes. Ovary 4-lobed. Articles 4, covered with short hooked spines.

Key to the species

1 Racemes leafy at least in the lower half. Pedicels 1–2 cm long, longer than the calyx (sometimes twice as long), recurved after flowering. Articles convex. Erect perennial up to c. 50 cm high. Coast and Hornsby Plateau. Moist places. Fl. spring–summer . . . . **C. suaveolens** R.Br.
1 Racemes without bracts. Pedicels rarely more than 5 mm long, usually shorter than the calyx. Articles concave. Erect perennial up to 80 cm high. Widespread. DSF; coastal sand dunes. Fl. spring–summer .
. . . .**C. australe** R.Br.

3 Austrocynoglossum Popov ex R.R.Mill

1 species endemic Aust.; Qld, NSW, Vic.

One species in the area

Perennial, decumbent, scabrous herb with stems up to 1 m long. Leaves ovate to broad-ovate, 2–8 cm long. Flowers usually solitary, slightly displaced in the leaf axils, sometimes in bracteate monochasia, on slender recurved pedicels. Corolla white, up to 5 mm long. Mericarps densely prickly. Widespread. Moist places in open forests; margins of RF. Fl. spring–summer. . . . .**A. latifolium** (R.Br.) R.R.Mill.

4 Myosotis L.

Forget-Me-Not

7 species in Aust. (2 native, 5 naturalized); NSW, Vic., Tas., S.A., WA

Herbs, usually hispid. Flowers in simple or forked cymes without bracts. Calyx deeply 5-cleft. Corolla tube cylindrical, with 5 scales in the throat; limb spreading, 5-lobed. Ovary 4-lobed. Articles 4, smooth, shining, erect, fixed at the base only.

Key to the species

1 Stamens and style much exserted. Leaves elliptic to narrow-lanceolate, usually sessile or slightly decurrent. Corolla white to blue; lobes as long as the tube. Herb up to 50 cm high. Higher Blue Mts, e.g. Jenolan Caves. Rocky situations. Fl. spring–summer . . . . **M. exarrhena** F.Muell.
1 Stamens enclosed in the corolla tube or only the tips protruding. Corolla 2.5 – 4 mm long; lobes shorter than the tube 2

2 Hairs on leaves appressed; hairs on calyx hooked. Corolla pale blue. Basal leaves petiolate 2–8 cm long; cauline leaves ± stem clasping. Erect annual herb to 35 cm high. Blue Mts Occasionally naturalized weed. Introd. from Europe. Fl. spring–summer. . . . . .***M. laxa** Lehm. ssp. **caespitosa** (Schultz) Hyl. ex Nordh.
2 Hairs on leaves spreading; hairs on calyx not hooked 3

3 Corolla white or yellowish. Leaves obovate to spathulate, up to 20 cm long, sessile or shortly petiolate. Calyx lobes less than twice as long as broad in the fruiting stage. Herb up to 50 cm high. Blue Mts Rocky situations. Fl. spring–summer . . . . **M. australis** R.Br.
3 Corolla yellow to cream, becoming blue or violet with age. Leaves oblong to elliptic, sessile or shortly petiolate. Calyx lobes more than twice as long as broad in the fruiting stage. Herb up to 20 cm high. Blue Mts, e.g. Mt. Jellore. Moist places. Introd. from Europe. Fl. spring–summer . . . ***M. discolor** Pers.

5 Borago L.

1 species naturalized Aust.; NSW, Vic., S.A., WA

One species in the area

Annual, hispid herb up to 1.5 m high. Leaves elliptic to ovate, up to 20 cm long. Flowers in loose, bracteate monochasia. Sepals c. 1 cm long. Corolla blue, with scales in the throat, 15–30 mm diam. Articles rugose. Receptacle flat. Naturalized in a few places near Sydney. Introd. from Europe. Fl. spring. *Borage.* . . . . .
. . . . ***B. officinalis** L.

6 Buglossoides Moench

1 species naturalized Aust.; all states and territories

One species in the area

Erect, harshly hairy annual, mostly c. 30 cm high, ± hoary, with appressed hairs. Leaves narrow-lanceolate, 2–3 cm long. Flowers small, in the upper axils. Calyx deeply 5-cleft; lobes linear, 5 mm long increasing to 10 mm long in fruiting stage. Corolla white, with a cylindrical tube; throat not closed with scales; lobes spreading. Articles 4, erect, wrinkled. Widespread. Weed of cultivation and waste places. Introd. from Europe. Fl. spring. *Corn Gromwell* . . . . ***B. arvensis** (L.) I.M.Johnst.

7 Amsinckia Lehm.

4 species naturalized Aust.; all states and territories

Annual herbs covered with stiff hairs. Leaves basal and cauline. Flowers in terminal coiled bractless monochasia, sessile or shortly pedicellate. Corolla tubular, without scales in the throat. Stamens scarcely exserted. Articles papillose or wrinkled.

Key to the species

1 Corolla orange-yellow; lobes 3–4 mm long. Articles 3 mm long or longer, transversely wrinkled, grey to pale brown. Leaves lanceolate; basal ones up to 20 cm long. Annual up to 1 m high. Widespread. Weed of cultivation and waste places. Introd. from N. America. Fl. spring. *Fiddleneck* . . . .
. . . . ***A. intermedia** Fisch. & C.A.Mey.
1 Corolla pale yellow; lobes 1–2 mm long. Articles less than 3 mm long, not or very weakly transversely wrinkled, brown to black. Leaves oblanceolate to lanceolate, up to 10 cm long. Annual up to 50 cm high. Widespread. Weed of cultivation and waste places. Introd. from S. America. Fl. spring . . . .
. . . . ***A. calycina** (Moris) Chater

8 Heliotropium L.

78 species in Aust. (75 endemic, 3 naturalized); all states and territories except Tas.

Hairy, diffuse perennials up to 50 cm high. Leaves alternate, covered with rough hairs. Flowers sessile in dense bractless cymes. Corolla white or lilac, with a yellow tube. Fruit a globular schizocarp. Stigma almost sessile, with a terminal conical appendage.

Key to the species

1 Corolla white, c. 2 mm long. Leaves petiolate, ovate to elliptic, up to 40 mm long, 10–20 mm wide. Fruit splitting into 4 articles. Widespread. Weed of cultivation and waste places. Introd. from Europe. Fl. summer– autumn. *Heliotrope.* . . . . ***H. europaeum** L.
1 Corolla lilac with a yellow throat. Leaves tapering gradually towards the base and sessile, oblong-lanceolate, 2–8 cm long, mostly less than 10 mm wide. Fruit splitting into 2 articles. Widespread. Weed of cultivation and waste places. Introd. from S. America. Fl. spring–summer. *Blue Heliotrope* . . . . . .
. . . . ***H. amplexicaule** Vahl

9 Echium L.

4 species naturalized Aust.; all states and territories

Erect biennial herbs up to 1 m high, scabrous-hairy. Basal leaves petiolate; cauline leaves alternate, sessile. Flowers in monochasia arranged in a panicle. Calyx deeply 5-cleft. Corolla funnel-shaped, with a straight tube, purplish blue or rarely white; lobes 5, unequal. Stamens 5, unequal, at least some exserted. Articles 4, attached to the flat receptacle by their flat bases.

Key to the species

1 Two of the stamens long and exserted. Panicle loose. Corolla red changing to purple-blue, 20–30 mm long. Leaves sparingly scabrous; basal leaves ovate, up to 30 cm long, with prominent lateral veins; cauline leaves up to 9 cm long, oblong to lanceolate, cordate at the base. Coarse annual up to 1.5 m high. Widespread. Weed of cultivation, pastures and waste places. Introd. from the Mediterranean. Fl. winter– summer. *Paterson's Curse* or *Salvation Jane* . . . .***E. plantagineum** L.

1 Four stamens long and exserted. Corolla blue, 12– 15 mm long. Cymes short and dense. Leaves rough, harsh; basal leaves oblanceolate, up to 15 cm long, without prominent lateral veins; cauline leaves rounded at the base. Coarse annual up to 1 m high. Widespread. Weed of pastures and waste places. Introd. from Europe. Fl. spring–summer. *Viper's Bugloss*. . . . . ***E. vulgare** L.

10 Anchusa L.

3 species naturalized Aust.; NSW, Vic., Tas., S.A., WA

One species in the area

Erect herb to 60 cm high. Basal leaves up to 15 cm long and 15 mm wide. Cauline leaves shorter; margins dentate. Inflorescence terminal, flowers sessile. Sepals connate at base, 3.5–5 mm long, lengthening to 8 mm in fruiting stage. Corolla blue or white, tube curved, lobes unequal. Stamens inserted half way along tube. Fruit ovoid 1.5–2 mm long. Weed of disturbed areas. Introd., from Europe. Blue Mts Fl. winter. *Wild Bugloss*. . . . .***A. arvensis** (L.) Beib.

11 Ehretia P.Browne

6 species in Aust. (4 endemic); Qld, NSW, NT, WA

One species in the area

Tall shrub or small tree. Leaves alternate, glabrous, ovate to elliptic-lanceolate, 10–20 cm long, shortly acuminate, serrate, with callous teeth; petioles 1–3 cm long. Flowers sessile in a panicle much exceeding the leaves. Calyx segments nearly orbicular, very small. Corolla white, tubular; tube very short; lobes spreading, 4–6 mm diam. Fruit drupaceous. Kurrajong; Illawarra district. RF. Fl. summer. *Koda* . . . . . .
. . . . **E. acuminata** R.Br. var. **acuminata**

115 APOCYNACEAE

(includes *Asclepiadaceae*)

Shrubs, climbers or herbs, with white, yellow, orange or colourless latex. Leaves opposite, entire, without stipules. Flowers in cymes or solitary, regular, bisexual. Sepals 5, imbricate in the bud. Corolla connate, 5-lobed; the lobes usually contorted or almost valvate in the bud. Stamens 5, epipetalous, alternating with the corolla segments; anthers erect, usually connivent or stamens and style forming gynostemium and corona present. Ovary superior, 1–2-locular or consisting of 2 carpels united only by the styles; stigma usually dilated. Fruit usually 2 follicles or a berry. Seeds mostly pendulous, often with a tuft of long hairs (*coma*). 390 gen., widespread, mostly trop. and subtrop.

Key to the genera

1 Pollen granular; corona absent 2
1 Pollen consolidated into sac like pollinia, two pollina per anther; corona present. 7

2 Herbs, often trailing. Corolla blue . . . . **6 VINCA**
2 Climbers twiners or shrubs. Corolla white or yellow 3

3 Shrubs 4
3 Climbers or twiners 5

4 Leaves with spinose tips. Corolla tube less than 8 mm long. . . . . **3 ALYXIA**
4 Leaves without spinose tips. Corolla tube more than 8 mm long. . . . .**4 NERIUM**

5 Stamens exserted, cohering into a cone . . . .**1 PARSONSIA**
5 Stamens enclosed in the corolla tube, not cohering 6

6 Corolla yellow; the tube c. 4 mm long . . . . **2 MELODINUS**
6 Corolla white; the tube c. 20 mm long. . . . .**5 MANDEVILLA**

7 Corolla and corona blue . . . . **11 TWEEDIA**
7 Corolla and corona not blue. 8

8 Erect shrubs or herbs 9
8 Climbers. 11

9 Corolla not reflexed at anthesis; up to 6 mm diam. Follicle smooth ovoid to elongated. Leaves with
   glands at base of midvein . . . . **7 MARSDENIA**
9 Corolla reflexed at anthesis; more than 8 mm diam. Follicle smooth and elongated or ovoid and with
   spines. Leaves without glands at base of midvein. 10

10 Follicles ovoid and covered with weak green spines. Corolla c. 10 mm diam., white, corona white . . .
    . . . .**12 GOMPHOCARPUS**
10 Follicles smooth and elongated. Corolla red and corona yellow. . . . . **13 ASCLEPIAS**

11 Corolla tube 10–15 mm long. . . . .**10 ARAUJIA**
11 Corolla tube less than 10 mm long 12

12 Corolla dark purple. Corona depressed . . . .**9 TYLOPHORA**
12 Corolla whitish or yellowish. Corona compressed or erect 13

13 Pollinia held erect in each anther locule. Corona appendages erect; the upper end free . . . .
    . . . . **7 MARSDENIA**
13 Pollinia pendulous in each anther locule. Corona appendages pendulous; the lower end free . . . . . .
    . . . .**8 CYNANCHUM**

1 Parsonsia R.Br.

Silkpods
22 species in Aust. (20 endemic); all states and territories

Tall and woody, or slender climbers. Leaves opposite. Flowers in cymes. Corolla yellowish or whitish; the tube short; the lobes contorted or valvate in the bud. Anthers exserted, cohering into a cone around the stigma. Ovary 2-locular, surrounded by 5 scales or a 5-lobed disc; style filiform; stigma surrounded by a membranous expansion at the base, usually 2-lobed. Fruit separating into 2 follicles. Seed with a coma of long silky hairs.

Key to the species

1 Corolla glabrous inside; tube up to 2 mm long; lobes 3–4 mm long. Leaves lanceolate to broad-elliptic,
   up to 10 cm long and 5 cm wide, obtuse or shortly acute, reticulations very obscure. Flowers in compact
   cymes. Widespread. RF and WSF. Fl. summer–autumn . . . .**P. lanceolata** R.Br.
1 Corolla hairy at least in the throat or tube 2

2 Branchlets and petioles densely pubescent. Undersurface of leaves pubescent with soft brown hairs,
   sometimes glabrescent with age. Leaves ovate 7–19 cm long, 4–9 cm wide, cordate at the base. Juvenile

leaves 2-lobed at base. Flowers in axillary cymes. Corolla c. 4 mm long bearded inside at the base. Fruit 7–16 mm long, c. 10 mm diam. North from Gosford. Rainforest. Fl. spring–summer . **P. velutina** R.Br.

2 Branchlets and petioles glabrous to pubescent. Undersurface of leaves glabrous or minutely pubescent. . 3

3 Leaves 3–5 cm wide, mostly 7–12 cm long, strongly reticulate on both surfaces, oblong-elliptic to broad-lanceolate; main lateral veins making an angle of 45–50° with the midvein. Flowers in corymbose cymes, finally loose. Corolla c. 6 mm long, bearded inside to above the middle; lobes scarcely longer than the tube. Fruit hard, 15–20 cm long. Widespread. WSF and RF. Fl. most of the year *Common Silkpod* . . . . **P. straminea** (R.Br.) F.Muell.

3 Leaves mostly less than 3 cm wide, 5–10 cm long, lanceolate to ovate-lanceolate. Corolla c. 4 mm long. . 4

4 Leaves tapering towards both ends, paler underneath; main lateral veins making an angle of 30–45° with the midvein. Inflorescence rusty-tomentose. Corolla bearded inside at the base with reflexed hairs; lobes narrow, twice as long as the tube. Fruit 5–8 cm long, dividing into rather thin follicles. Blue Mts WSF. Fl. Summer. . . . . **P. brownii** (J.Britton) Pichon

4 Leaves rounded or cordate at the base, equally green on both surfaces; main veins making an angle of 45–50° with the midvein. Inflorescence glabrous. Corolla campanulate, bearded at the base inside; lobes broad, c. as long as the tube. Blue Mts; Illawarra region. WSF and RF. Fl. summer–autumn. *Black Silkpod* . . . . **P. purpurascens** J.B.Williams

2 Melodinus J.R.Forst. & G.Forst.
7 species endemic Aust.; Qld, NSW

One species in the area

Tall, woody climber. Leaves opposite, elliptic to oblong, often sessile, glabrous, 5–10 cm long. Calyx segments small, very obtuse. Corolla yellow; the tube c. 4 mm long; the lobes spreading, c. half as long as the tube. Anthers wholly enclosed in the corolla tube. Fruit a berry, ovoid, c. 5 cm long. Seeds enveloped in pulp, ovate, compressed, c. 6 mm long. Coast and adjacent plateaus. RF. Fl. summer–autumn. . . . . **M. australis** (F.Muell.) Pierre

3 Alyxia Banks. ex Benth.
9 species in Aust. (8 endemic); all states and territories

One species in the area

Erect, bushy shrub up to 3 m high. Leaves in whorls of 3–6, lanceolate to ovate, 1–6 cm long, 4–30 mm wide, with an acute or acuminate spiny tip, glossy, slightly recurved. Corolla tube 5–7 mm long. Anthers orange. Coast north of Wollongong. RF and WSF. Fl. spring–summer . . . . **A. ruscifolia** R.Br.

4 Nerium L.
1 species naturalized Aust.; NSW

One species in the area

Multistemmed shrub up to 4 m high. Leaves narrow-elliptic to lanceolate, up to 20 cm long, acute but not spiny. Corolla pink, red or white; tube up to 25 mm long; lobes c. 20 mm long. Occasionally naturalized in reserves near Sydney. Introd. from the Mediterranean. Fl. summer–autumn. *Oleander*. . . . . ***N. oleander** L.

5 Mandevilla Lindl.
1 species naturalized Aust.; NSW

One species in the area

Glabrous climber or scrambler, woody towards the base. Leaves opposite, ovate, c. 6 cm long and 3 cm wide, cordate, petiolate. Flowers fragrant, in terminal racemes. Calyx lobes c. 6 mm long, acute. Corolla white; the tube c. 20 mm long. Stamens enclosed in the corolla tube. Fruit consisting of 2 follicles cohering at the tips. Naturalized in a few places near Sydney. Fl. summer. *Chilean Jasmine*. . . . . ***M. laxa** (Ruíz Lopez & Pavón) Woodson

6 Vinca L.

1 species naturalized Aust.; NSW, Vic., Tas.

One species in the area

Perennial with long prostrate stems rooting and forming stolons. Leaves opposite, ovate, shining, 2–5 cm long, broad at the base, petiolate. Flowers solitary, 4–5 cm diam., axillary. Calyx 12–15 mm long, with 5 linear segments. Corolla blue, 4–5 cm diam.; tube slightly longer than the calyx; the lobes broad, spreading. Stamens enclosed in the corolla tube. Carpels 2, distinct. Widespread. Garden escape. Introd. from the Mediterranean. Fl. spring–summer. *Blue Periwinkle* . . . . ***V. major** L.

7 Marsdenia R.Br.

Milk Vines

20 species in Aust. (18 endemic); all states and territories except Tas.

Twiners or small erect shrubs. Leaves with some glands at the base of the lamina. Flowers fragrant, axillary, in simple or compound umbels. Corolla white to yellowish or greenish; the tube short; the limb spreading. Corona of 5 compressed segments, adnate to the base of the anthers; the upper end erect, free. Pollinia oblong, erect. Fruit of 1–2 turgid follicles.

Key to the species

1 Flowers in compound umbels often exceeding the leaves. Leaves elliptic to oblong-elliptic, obtuse or shortly acuminate, pale and slightly yellowish underneath, 4–10 cm long. Corolla nearly rotate, c. 3 mm diam., glabrous inside. Stigmatic head short, obtuse. Fruit slender, acuminate, 5 cm long. Tall twiner, ± softly pubescent except the upper surface of the leaves. Widespread. In or near RF. Fl. summer . . . . **M. flavescens** A.Cunn ex Hook.
1 Flowers in simple umbels on short peduncles. Corolla ± urceolate or campanulate 2

2 Leaves 6–15 times as long as broad, linear to narrow-lanceolate, 2–12 cm long, rather thick; midvein channelled above. Corolla campanulate, greenish to yellow, 3–4 mm diam. Fruit ovoid, 4–8 cm long. Razorback Range. Open forest and scrub. Fl. spring. Endangered population in some LGAs . . . .**M. viridiflora** R.Br. ssp. **viridiflora**
2 Leaves up to 5 times as long as broad 3

3 Stigmatic head long, rostrate. Leaves obtusely and usually abruptly acuminate, up to 13 cm long, 2–7 cm wide; ovate to almost orbicular, dark green above, slightly paler underneath; petioles 10–40 mm long. Flowers numerous in dense simple umbels up to 3 cm diam. Corolla c. 6 mm diam., bearded inside below the middle; lobes slightly longer than the tube. Fruit ovoid, acuminate, 5 cm long, 2–4 cm diam. Robust twiner, glabrous or the young growth tomentose. Widespread. In or near RF; also near the sea in sheltered places. Fl. spring–summer . . . . **M. rostrata** R.Br.
3 Stigmatic head narrow, conical. Leaves up to 7 cm long, 6–25 mm wide; oblong-elliptic to lanceolate or ovate, gradually acute or obtuse; petioles 1–4 mm long. Umbels less than 2 cm diam. Corolla lobes bearded inside from the base to the middle or higher. Fruit narrow, 5–10 cm long, up to 1.5 cm diam. Variable in habit but twining when in forests, often a small erect shrub in heath. Widespread. DSF; WSF; heath; RF. Fl. summer . . . .**M. suaveolens** R.Br.

8 Cynanchum L.

10 species in Aust. (9 endemic); Qld, NSW, NT, S.A, WA

One species in the area

Glabrous climber or twiner, slightly woody towards the base. Leaves opposite, broad-ovate to ovate, 2–3 cm long, 1.5–3 cm wide, truncate but scarcely cordate at the base. Flowers in cymo-umbels. Corolla white; the tube up to 4 mm long. Hunter River Valley; Cumberland Plain. RF; rocky places. Endangered. Fl. summer . . . .**C. elegans** (Benth.) Domin

9 Tylophora R.Br.

12 species in Aust. (11 endemic); Qld, NSW, Vic., NT, WA

Slender, woody climbers. Leaves opposite. Flowers in umbels or cymose-panicles. Corolla dark purplish, 5–6 mm diam. Pollinia ovoid to globular, erect. Fruit a ± turgid follicle tapering gradually towards the apex.

Key to the species

1 Flowers white to pale green in open cymose-panicles. Corolla 9–14 mm diam.; lobes linear to filiform. Leaves lanceolate to ovate, 3–9 cm long, up to 4.5 cm wide; fine hairs along veins and margins. Hairy climber. Uncommon. RF. Fl. spring-summer . . . . **T. paniculata** R.Br.

1 Flowers purple to red. Corolla 5–6 mm diam. 2

2 Corolla lobes glabrous, c. 2.5 mm long. Flowers in loose, irregular, axillary, cymose-panicles. Leaves ovate to broad-ovate, 4–6 cm long, 2.5–5 cm wide, cordate at the base. Corona segments broad, ± confluent to form a ring. Glabrous climber. Cumberland Plain. Open forest. Endangered. Fl. summer–autumn . . . . **T. woollsii** Benth.

2 Corolla lobes hairy, c. 2 mm long. Flowers in 1–2 short umbels in the leaf axils. Leaves ovate to narrow-ovate or elliptic, 2–5 cm long, 1–3 cm wide, obtuse at the base. Corona segments globular, discrete. Glabrous climber. Widespread. WSF. Fl. most of the year. . . . . **T. barbata** R.Br.

10 Araujia Brot.

1 species naturalized Aust.; Qld, NSW, Vic.

One species in the area

Twiner. Leaves oblong, acuminate, entire, glaucous on the upper surface, grey or whitish underneath, up to 8 cm long, 2–5 cm wide, on short petioles. Flowers several in a cyme on a short peduncle arising beside the petiole. Calyx lobes spreading. Fruit a pear-shaped follicle, at first green and fleshy, 5–8 cm long. Seeds with a long silky coma. Widespread. Weed of waste ground. Introd. from S. America. Fl. summer. *Moth Plant.* . . . . ***A. sericifera** Brot.

11 Tweedia Hook. & Arn.

1 species naturalized in Aust.; Qld, NSW

One species in the area

Herbaceous twiner, ± woody at the base, erect when young. Leaves oblong, cordate-hastate at the base, shortly petiolate. Peduncles axillary, 3–4-flowered. Corolla blue, c. 25 mm diam.; lobes spreading. Naturalized in the Helensburg district and lower Blue Mts Introd. from S. America. Fl. summer. . . . . ***T. coerulea** D.Don

12 Gomphocarpus R.Br.

3 species naturalized in Aust.; Qld, NSW, Vic., S.A, WA

Erect, somewhat herbaceous shrubs with pubescent branches. Flowers white, in loose umbels. Corolla deeply divided; the lobes reflexed. Corona segments compressed, the outer edge lower than the inner which terminates in 2 teeth. Follicle large, inflated, covered with soft green spines. Seeds with a long white coma.

Key to the species

1 Follicle ovoid-acuminate, 5 cm long, tapering into a very short curved beak. Teeth of the corona incurved. Shrub 1–2 m high. Leaves linear-lanceolate, 5–10 cm long, on short petioles. Widespread. Waste places and pastures. Introd. from S. Africa. Fl. spring–autumn. *Narrow-leaved Cotton Bush* . . . . ***G. fruticosus** (L.) W.T.Aiton.

1 Follicle ovoid-oblong, not tapering into the very short recurved beak. Teeth of the corona curved upwards. Otherwise similar to *G. fruticosus*. Widespread. Waste places and pastures. Introd. from S. Africa. Fl. spring–autumn. *Cotton Bush* . . . . ***G. physocarpus** E.Mey.

13 Asclepias L.

1 species naturalized Aust.; Qld, NSW, Vic., S.A, WA

One species in the area

Perennial herb or shrub to 1 m high. Stems finely ribbed. Leaves opposite, lanceolate to narrow elliptic, 5–15 cm long, 1–3 cm wide. Inflorescence terminal or axillary umbels. Calyx hairy. Corolla red and yellow; lobes 8 mm long. Stigma broad. Follicle smooth, held erect, 6–9 cm long, c. 1 cm wide. Disturbed sites. Introd. from Central America. Fl. spring–summer . . . . ***A. curassavica** L.

116 GELSEMIACEAE

2–3 gen. trop. to subtrop.

1 Gelsemium Juss.

1 species naturalized Aust.; NSW

One genus in the area

Glabrous, twining perennial. Leaves lanceolate, c. 3 cm long; the bases joined by a line. Flowers sweetly scented, in short axillary or terminal cymes. Calyx 5-partite. Corolla funnel-shaped, 5-lobed, c. 2 cm long, yellow. Near Hornsby. Garden escape. Introd. from N. and Central America . . . .***G. sempervirens** (L.) J.St Hil.

117 GENTIANACEAE

Glabrous herbs. Leaves simple, opposite; stipules absent. Flowers in cymes, regular, bisexual. Calyx tubular, sometimes divided nearly to the base, 4–5-lobed; lobes imbricate. Corolla tubular, 4–5-lobed; the lobes contorted in the bud. Stamens as many as the corolla lobes and alternating with them, epipetalous. Ovary superior, 1–2-locular, with parietal or subaxile placentas; style simple. Fruit a capsule. 80 gen., cosmop.

Key to the genera

1 Corolla red or pink. 2
1 Corolla white, yellow, blue or purple, sometimes greenish outside. 3

2 Stigmatic lobes reniform to shoe-shaped, fleshy; capsule linear . . . .**1 CENTAURIUM**
2 Stigmatic lobes rhombic to fan-shaped, not fleshy; capsule elliptic to ovate . . . . **2 SCHENKIA**

3 Corolla yellow. 4
3 Corolla blue, purple or white, sometimes greenish outside 5

4 Stigma peltate. Ovary 1-locular. . . . . **3 CICENDIA**
4 Stigma 2-lobed. Ovary 2-locular . . . . **4 SEBAEA**

5 Anthers versatile. Corolla white with purplish markings . . . . **6 CHIONOGENTIAS**
5 Anthers rigid. Corolla blue, greenish outside. . . . . **5 GENTIANA**

1 Centaurium Hill

Centauries

5 species in Aust. (1 endemic, 4 naturalized); all states and territories

Erect, glabrous, annual herbs up to 50 cm high, with slender quadrangular stems. Leaves sessile; the lowest often forming a basal rosette. Flowers in cymose panicles or corymbs. Calyx 4–5-lobed. Corolla red or pink, with a cylindrical tube and 5 spreading lobes. Stamens inserted in the throat of the corolla tube. Ovary 1-locular; stigma ± 2-lobed. Capsule narrow-oblong, enclosed in the persistent calyx or slightly exceeding it.

Key to the species

1 Corolla lobes 3–4 mm long, dark red; corolla tube slightly longer than the calyx. Leaves elliptic-oblong, 1–2 cm long. Erect herb up to 25 cm high. Widespread. Waste places, pastures and cleared land. Introd. from Europe. Fl. spring–summer . . . . *C. tenuiflorum (Hoffmanns. & Link) Fritsch ex Janch.

1 Corolla lobes 4–6 mm long, light red; corolla tube longer than the calyx. Leaves obovate to elliptic-oblong, 1–2 cm long. Erect herb up to 25 cm high. Widespread. Waste places, pastures and cleared land. Introd. from Europe. Fl. spring–summer *Common Centaury* . . . . *C. erythraea Rafn

2 Schenkia Griseb.

2 species in Aust. (2 native, 1? naturalized); all states and territories

One species in the area

Erect herb up to 35 cm high. Compound cymes solitary or few on the stem. Cyme branches spreading, spike-like after the first branching since one branch of each subsequent dichasial branching is stronger than the other; the ends of the branches monochasial. Basal rosette of leaves absent. Leaves sessile, elliptic 1–3 cm long, narrower towards the apex. Corolla pink; tube usually slightly longer than the calyx. Calyx 6–8 mm long, 4–5-toothed. Coast and adjacent plateaus; Cumberland Plain. Pastures and waste ground. Not common. Fl. spring–summer . . . . **S. spicata** (L.) Mans.

3 Cicendia Adans.

2 species naturalized in Aust.; NSW, Vic., Tas., S.A.

Erect, annual herbs up to c. 15 cm high, with a basal rosette of leaves and a few cauline ones. Flowers in terminal cymes or solitary. Calyx tube with 4 very short teeth. Corolla yellow, 4-lobed. Stamens not twisted. Ovary 1-locular with parietal placentas. Stigma peltate on the filiform style. Fruit a 2-valved capsule.

Key to the species

1 Cauline leaves linear to narrow-lanceolate up to 5 mm long. Calyx tubular, c. 4 mm long; lobes more than 1 mm long. Corolla tube c. 6 mm long. Coastal sites in Sydney. On damp sand. Introd. from Europe. Fl. spring . . . . *C. filiformis (L.) Delarbre

1 Cauline leaves ovate to lanceolate, up to 5 mm long. Calyx obovoid; lobes c. 0.5 mm long. Corolla tube 4–8 mm long. Widespread south of Sydney. Damp shady places. Introd. from Europe. Fl. spring. . . . . .
. . . . *C. quadrangularis (Lam.) Griseb.

4 Sebaea Sol. ex R.Br.

2 species in Aust. (1 endemic); all states and territories except NT

One species in the area

Glabrous, erect herb up to 50 cm high. Leaves in distant pairs, sessile, ovate or almost orbicular, 6–15 mm long. Flowers in a rather loose dichotomous cyme. Calyx divided nearly to the base. Corolla yellow; tube cylindrical; lobes spreading. Anthers tipped by a stipitate gland and finally recurved at the summit. Ovary 2-locular. Capsule oblong, enclosed in the calyx. Cumberland Plain; Blue Mts Damp places in open forests. Fl. spring. *Yellow Centaury* . . . . **S. ovata** (Labill.) R.Br.

5 Gentiana L.

4 species endemic Aust.; NSW

One species in the area

Erect, glabrous herb up to 10 cm high with reddish stems. Leaves sessile, oblong to ovate, 3–8 mm long, keeled. Flowers solitary, terminal. Corolla up to 2 mm long, greenish outside, blue inside. Stigmas 2, sessile. Capsule 2-valved, c. 5 mm long, with a gynophore 1–3 mm long. Wingecarribee Swamp. *Sphagnum* and sedge swamp. Endangered. Fl. spring. . . . . **G. wingecarribiensis** L.G.Adams

6 Chionogentias L.G.Adams
14 species endemic Aust.; NSW, Vic., Tas., S.A.

One species in the area
Erect or ascending, glabrous herb up to 50 cm high. Leaves linear to oblanceolate, up to 3 cm long, 3–4 mm wide. Flowers in cymes in the axils of the upper leaves. Calyx c. lobes 6–9 mm long; the lobes more than twice as long as the tube. Corolla campanulate lobes up to 15 mm long, white with purplish to blue markings. Stamens 5, anthers bluish black. Ovary 10 mm long, 1-locular, with parietal placentas; stigmas 2, 0.6–0.8 mm long, sessile on the ovary summit. Fruit a 2-valved capsule. Higher Blue Mts Swamps. Fl. summer . . . . **C. cunninghamii** L.G.Adams ssp. **cunninghamii**

118 LOGANIACEAE
Herbs or shrubs. Leaves opposite; stipules absent or rudimentary but the leaf bases often connected by a raised line. Flowers regular, bisexual or unisexual. Calyx tubular, 4–5-lobed or rarely 2-lobed or cleft. Corolla tubular 4–5-lobed, imbricate or contorted. Stamens 4–5, epipetalous, alternating with the lobes. Ovary superior, 2-locular; placentas axile; style simple but later splitting into 2 from the base in *Mitrasacme*. Fruit a small capsule. 22 gen., trop. to warm temp.

Key to the genera

1 Herbs. Style 2-fid . . . . **2 MITRASACME**
1 Herbs or shrubs. Style simple . . . . **1 LOGANIA**

1 Logania R.Br.
22 species endemic Aust.; all states and territories except Tas.

Herbs or shrubs. Leaves connected by a raised line or a short sheath. Flowers small, often unisexual, in axillary cymes or panicles or solitary. Calyx 5-cleft. Corolla white, 4–5-lobed; lobes imbricate in the bud. Stamens 5. Fruit a small capsule, almost separating into 2 mericarps at maturity.

Key to the species

1 Erect shrub 1–2 m high, glabrous or nearly so. Leaves mostly 2–5 cm long, lanceolate to linear, paler underneath, flat or with revolute margins; the midvein prominent. Flowers sweetly scented, in axillary cymes or panicles, but sometimes solitary. Corolla 2–3 mm long. Capsule narrow, 4–5 mm long. Widespread. Forests on a variety of soils. Fl. early spring. . . . . **L. albiflora** (Andrews & Jacks.) Druce

1 Herb or procumbent shrub c. 10 cm high. Leaves obovate to elliptic or oblong, obtuse, 5–10 mm long, with revolute margins. Flowers solitary, axillary, sessile or very shortly pedicellate. Calyx segments long and narrow. Corolla 5–6 mm long. Hornsby Plateau; Cumberland Plain. DSF. Fl. spring . . . . **L. pusilla** R.Br.

2 Mitrasacme Labill.
48 species in Aust. (41 endemic); all states and territories

Erect or prostrate herbs, usually small. Leaves opposite, without stipules or a raised line. Flowers solitary in the upper axils or in loose irregular terminal umbels. Calyx campanulate, 4-lobed. Corolla white, often yellowish in the throat, rarely all yellow; tube short and broad or elongated and cylindrical; lobes 4, spreading, valvate. Stamens 4. Style simple but splitting later from the base into 2. Fruit a small capsule.

Key to the species

1 Branches with coarse dense spreading to patent hairs. 2
1 Branches glabrous or with scattered minute ± appressed hairs. 3

2 Plants erect, up to 30 cm high, perennial, sometimes flowering in the first year, usually much branched near the base. Leaves 4–10 mm long, linear-lanceolate to oblong or elliptic; margins recurved. Calyx broadly campanulate, 2–3 mm long. Corolla tube broad, c. as long as the calyx; lobes spreading, as long as the tube. Peduncles terminal, simple or sparingly branched, bearing an irregular umbel of 3–5 flowers on filiform pedicels. Widespread. Heath. Sandy soils. Fl. spring–summer **M. polymorpha** R.Br.

2 Plants prostrate or procumbent or ascending from creeping stems. Leaves orbicular to ovate-lanceolate, often ciliate, obtuse, 3–6mm long. Stems ± hirsute with rigid hairs. Corolla lobes much shorter than the tube; tube bearded inside. Pedicels to 10 mm long, lengthening to 15 mm in fruit. Wet heath; rocks near waterfalls. Fl. spring–summer. (*M. pilosa* ssp. *stuartii* may occur in south of region. It has a pedicel up to 40 mm long) . . . . **M. pilosa** Labill. ssp. **pilosa**

3 Flowers solitary, terminal, almost sessile. Leaves ovate to oblong-lanceolate, obtuse, 2–4 mm long, glabrous or ciliate. Annual herb with filiform stems. Blue Mts; Woronora Plateau. Damp sandy soils. Fl. spring–summer. . . . . **M. serpyllifolia** R.Br.

3 Flowers in 3–5-flowered terminal umbels, mostly distinctly pedicellate 4

4 Corolla glabrous inside, up to 5 mm long. Leaves oblong-linear to linear-lanceolate, 3–8 mm long. Perennial herb. Coast and adjacent plateaus. Damp soils. Fl. spring–summer . . . . . **M. paludosa** R.Br.

4 Corolla hairy inside, usually more than 5 mm long. Leaves ovate to oblong, 2–8 mm long. Annual herb, sometimes very small, erect or almost trailing. Coast and adjacent plateaus; Cumberland Plain. Damp places. Fl. spring–summer . . . . **M. alsinoides** R.Br.

119 RUBIACEAE

Trees shrubs or herbs. Leaves opposite, simple, usually entire; stipules present, often interpetiolar, sometimes similar to the leaves so that apparent whorls of "leaves" are present. Domatia sometimes present on leaves. Flowers united into simple or compound heads or in cymes, regular, usually bisexual. Calyx 3–5-lobed or sometimes obsolete. Corolla tubular, often inserted around an epigynous disc, 4–5-lobed. Stamens as many as the corolla lobes and alternating with them, epipetalous. Ovary inferior, usually 2-locular; style simple or with as many branches as loculi. Fruit a capsule, berry or drupaceous. C. 650 gen., mostly trop.

Key to the genera

1 Flowers united by their calyces into simple or compound heads 2
1 Flowers not united with each other 4

2 Woody climber or scrambling shrub . . . . **1 MORINDA**
2 Herbs or small shrubs 3

3 Heads simple, arranged in an umbel. . . . . **2 POMAX**
3 Heads grouped together into a compound head. Plant usually foetid . . . . **3 OPERCULARIA**

4 Shrubs or small trees. 5
4 Herbs. 7

5 Leaves mostly less than 5 cm long. Style 2-lobed; the lobes very long . . . . **6 COPROSMA**
5 Leaves mostly more than 5 cm long. Style obsolete or very short. 6

6 Leaves glabrous. Flowers axillary. Stigma peltate. . . . . **4 CYCLOPHYLLUM**
6 Leaves hairy, tomentum often rusty when young. Flowers terminal. Style 2-lobed; the lobes short . . . . **5 PSYCHOTRIA**

7 Stipules much smaller than the leaves 8
7 Stipules leaf-like forming what appear to be whorls of "leaves" at the nodes 10

8 Stipules with several terete teeth . . . . **7 RICHARDIA**
8 Stipules entire. 9

9 Leaves ovate to orbicular . . . . **8 NERTERA**

9 Leaves linear to narrow-elliptic. . . . . **11 GALIUM**

10 Corolla rotate with a very short tube . . . . **11 GALIUM**

10 Corolla funnel-shaped, with a distinct tube at least in the male flowers 11

11 Calyx present, 6-toothed. . . . . **9 SHERARDIA**

11 Calyx absent . . . .**10 ASPERULA**

1 Morinda L.
6 species in Aust. (4 endemic); Qld, NSW, Vic.

One species in the area

Woody climber. Leaves oblong-lanceolate to almost ovate, acuminate, 3–17 cm long, glabrous. Peduncles slender, paired at the ends of the branches, each with a small head of usually 6–12 basally united flowers. Corolla white, 5–8 mm long; tube narrow, usually shorter than the lobes. Fruit a compound berry, irregularly globular, 6–14 mm diam., yellowish or reddish. Widespread. WSF and RF. Fl. spring–summer . . . . **M. jasminoides** A.Cunn.

2 Pomax Sol. ex DC.
1 species endemic Aust.; all states and territories except Tas.

One species in the area

Small, perennial herb or low shrub, much branched, sometimes suckering, erect or diffuse, usually ± hirsute, sometimes glabrous. Leaves ovate to lanceolate, 7–15 mm long. Flowers minute, united by their calyces into simple heads; heads arranged in a simple terminal umbel which is sessile within the uppermost leaves; umbel rays 4–12 mm long. Compound fruit a cluster of capsules. Widespread. Heath and DSF. Fl. spring . . . . **P. umbellata** (Gaertn.) Sol. ex A.Rich.

3 Opercularia Gaertn.
Stinkweeds
15 species endemic Aust.; all states and territories except NT

Perennial herbs, ± decumbent erect or diffuse, sometimes ± woody, often with a very foetid odour when fresh. Leaves with sheathing stipules. Flowers very small, united by their calyces into a globular compound or rarely a simple head. Corolla tube short, 3–5-lobed. Anthers exserted on long filaments. Ovary 1 or 2-locular. Fruit a compound capsule opening by a lid.

Key to the species

1 Plant hirsute, ± decumbent, the branches c. 10 cm long. Seeds transversely wrinkled, without prominent longitudinal lines. Leaves ovate to lanceolate, petiolate, 12–25 mm long. Stamens usually 1–2. Hawkesbury and Colo River district; Blue Mts Rocky places and along creek banks. Ss. Fl. spring–summer . . . .**O. hispida** Spreng.

1 Plants glabrous or scabrous-pubescent, with straggling branches 2

2 Seeds very much pitted, rugose on the inner face, without longitudinal ridges. Leaves linear to linear-lanceolate, up to 30 mm long. Stems c. 20 cm long. Widespread (uncommon in Blue Mts). Heath; DSF; grasslands. Fl. spring–summer . . . . **O. diphylla** Gaertn.

2 Seeds transversely wrinkled with 2 prominent smooth longitudinal ridges 3

3 Leaves lanceolate to ovate-lanceolate, usually 10–40 mm long, occasionally smaller. Whole plant scabrous-pubescent to almost glabrous. Variable shrub up to 2 m high. Widespread. Heath and DSF. Sandy and clay soils. Fl. spring–summer . . . . **O. aspera** Gaertn.

3 Leaves oblong-linear to lanceolate, rarely broader, usually less than 12 mm long but sometimes up to 18 mm. Whole plant often glabrous. Spreading shrubby plant up to 50 cm high. Widespread. Heath and DSF. Fl. winter–spring. . . . . **O. varia** Hook.f.

4 Cyclophyllum Hook.f.

9 species in Aust.; Qld, NSW, NT

One species in the area

Tall shrub or small tree, branches glabrous. Leaves mostly 5–10 cm long, elliptic-lanceolate to obovate; the veins distant and not prominent; domatia present. Stipules triangular, keeled. Flowers 1–4, on short pedicels in sessile axillary clusters. Corolla tube c. 8 mm long; lobes shorter than the tube, folded at apex. Style exserted; stigma broad, thick, peltate. Fruit up to 12 mm diam. North from Illawarra district. RF. Fl. summer–autumn . . . . **C. longipetalum** S.T.Reynolds & R.J.F.Hend.

5 Psychotria L.

13 species in Aust. (2 endemic); Qld, NSW, NT

One species in the area

Pubescent shrub or small tree, tomentum often rusty on young stems and leaves. Leaves ovate to elliptic or oblong, 6–10 cm long, narrowed into a petiole; stipules caducous. Flowers sessile in small clusters in a loose terminal corymb. Fruit drupaceous, yellowish white, c. 6 mm diam; endocarp longitudinally ribbed. Coast and adjacent plateaus; lower Blue Mts RF and WSF. Fl. summer . . . .
. . . . **P. loniceroides** Sieber ex DC.

6 Coprosma J.R.Forst. & G.Forst.

8 species in Aust. (6 endemic); NSW, Vic., Tas., S.A.

Erect, usually dioecious shrubs. Flowers solitary or clustered, axillary or terminal. Anthers exserted on long filaments. Style divided to the base into 2 long filiform branches, ± hairy, exserted. Fruit drupaceous.

Key to the genera

1 Leaves thin, glabrous, mostly 5–15 mm long, a few larger, narrow- to broad-lanceolate. Branches slender; branchlets often spinescent. Flowers solitary, terminating short axillary shoots. Corolla 4 mm long. Fruit bright red. Shrub 2–4 m high. Illawarra district; Blue Mts, e.g. Mt Tomah, Mt Wilson. WSF. Fl. summer. *Prickly Currant Bush.* . . . . **C. quadrifida** (Labill.) B.L.Rob.
1 Leaves thick, glossy on the upper surface, more than 10 mm long. Flowers several together 2

2 Leaves acuminate, broad-ovate to lanceolate, 1–7 cm long. Flowers few together in short pedunculate heads which are terminal but becoming lateral by the growth of the shoot. Corolla 6 mm long. Fruit red or brownish. Shrub up to 2 m high, usually minutely scabrous. Blue Mts, e.g. Blackheath. Open forests not usually on Ss. Fl. summer. . . . . **C. hirtella** Labill.
2 Leaves obtuse truncate or emarginate, broad-oblong to broad-ovate, 2–8 cm long. Domatia prominant on lower surface. Flowers numerous in compound axillary clusters. Corolla 5 mm long. Fruit orange-red. Shrub up to 8 m high but sometimes prostrate, pubescent on the young branches. Naturalized at places near the sea. Introd. from New Zealand. Fl. spring–summer. *Looking-glass* or *Mirror Bush* . . . .
. . . . ***C. repens** A.Rich.

7 Richardia Kunth

4 species naturalized Aust.; Qld, NSW

Prostrate herbs. Leaves lanceolate to elliptic, sessile or shortly petiolate; stipules sheathing, with terete bristles. Inflorescence terminal, dense, surrounded by 2 or 4 leaves. Flowers small, sessile. Calyx lobes 4–6. Corolla white, 3–6-lobed. Ovary usually 3–4-locular. Fruit separating into 1-seeded mericarps.

Key to the species

1 Calyx 6-lobed. Stems mostly more than 15 cm long. Mericarps 3 2
1 Calyx 4–5-lobed. Mericarps 4 3

2 Mericarps with large papillae; adaxial face with a raised ridge, 2–3 mm long. Leaves elliptic to ovate or obovate, up to 5 cm long. Flowers c. 20 or more per head. Corolla mostly 6-lobed; tube 3–8 mm long. Prostrate to ascending annual with stems up to 50 cm long. Coast. Weed of waste places and cultivation. Introd. from S. America. Fl. spring–autumn. *Mexican Clover* or *Brazil Weed* . . *R. brasiliensis Gomes

2 Mericarps with short papillae; adaxial face with a narrow groove, 2–4 mm long. Leaves elliptic to obovate up to 7 cm long. Flowers c. 20 or more per head. Corolla 6-lobed; tube 2–5 mm long. Decumbent to erect annual with stems up to c. 80 cm long. Coast. Weed. Uncommon. Introd. from S. America. Fl. spring–summer. . . . . *R. scabra L.

3 Floral leaves broad-obovate, mostly more than 5 mm wide, hairy on both surfaces. Leaves ovate to narrow-ovate, up to 2 cm long. Flowers up to 15 per head. Corolla 4-lobed; tube up to 2 mm long. Prostrate perennial with stems up to 20 cm long, sometimes forming dense mats. Hunter River Valley. Weed of cultivation. Introd. from S. America. Fl. spring–summer . . . . *R. humistrata (Cham. & Schldl.) Steud.

3 Floral leaves narrow-ovate, up to 5 mm wide, usually glabrous on the upper surface. Leaves narrow-ovate to ovate, up to 2.5 cm long. Flowers often few per head some of which are often cleistogamous. Corolla 4-lobed; tube 2–4 mm long. Prostrate perennial with stems up to 15 cm long. Widespread. Weed of cultivation and waste places. Introd. from S. America. Fl. spring–summer . . . . *R. stellaris (Cham. & Schldl.) Steud.

8 Nertera Banks & Sol. ex Gaertn.
1 species native Aust.; NSW, Vic., Tas., S.A.

One species in the area
Prostrate, creeping, perennial herb rooting at the nodes. Leaves ovate to orbicular, mostly 3–5 mm long, petiolate. Flowers minute, solitary, terminal, sessile. Fruit drupaceous, globular, red or yellow, 3–4 mm diam. Blue Mts, e.g. Govett's Leap. Wet moss on cliffs. Fl. spring–summer . . . . . . **N. granadensis** (Mutis) Druce

9 Sherardia L.
1 species naturalized Aust.; NSW, Vic., Tas., NT, S.A.

Monotypic genus
Small, annual herb with weak quadrangular stems. "Leaves" apparently in whorls of 6 (4–5), 3–7 mm long, mucronate-acute; the lower ones obovate; the upper ones lanceolate. Flowers 4–5 mm long, in terminal heads surrounded by an involucre of floral leaves. Corolla pale violet or pink; tube slender, longer than the lobes. Fruit dry, pubescent, indehiscent, 2-seeded, surmounted by the persistent calyx. Widespread. Weed of cultivation and waste places. Fl. spring–summer. *Field Madder* . . . . **S. arvensis** L.

10 Asperula L.
Woodruffs
17 species in Aust. (16 endemic, 1 naturalized); all states and territories except NT

Annual or perennial, rhizomatous herbs with 4-angular stems. Stipules leaf-like and, together with the leaves, forming a whorl of 2–9 "leaves" at the nodes. "Leaves" sessile or nearly so; the lower ones often reflexed. Flowers in terminal or axillary cymes. Calyx absent. Corolla tubular, 4-lobed. Anthers exserted. Style 2-lobed. Fruit usually schizocarpic.

Key to the species
1 "Leaves" mostly 4 in each whorl 2
1 "Leaves" 6 or more in each whorl. 3

2 "Leaves" in each whorl more or less equal; obovate to oblanceolate, up to 10 mm long and 3 mm wide, glabrous or nearly so, obtuse or apiculate. Flowers in terminal or axillary cymes. Corolla c. 3 mm long. Mericarps black, glabrous. Perennial, dioecious, ascending herb with stems up to 20 cm long. Blue Mts Open forests and grasslands. Fl. summer . . . . **A. gunnii** Hook.f.

2 "Leaves" in each whorl unequal in length. Leaves linear to linear-oblong, 2–5 mm long, c. 3 mm wide. Flowers in terminal cymes, 1–3-flowered. Corolla 2–3 mm long. Mericarps black. Perennial subshrub with stems up to 30 cm long. Damp rocky areas at higher altitudes. Capertee area. Fl. spring–summer . . . . **A. ambleia** Airy Shaw & Turrill

3 "Leaves" acuminate with a hyaline tip, linear, up to 10 mm long and 1 mm wide, ± recurved, ciliolate. Corolla c. 2 mm long. Mericarps rugose. Perennial, dioecious herb up to 15 cm high. Blue Mts Damp places in DSF and WSF. Fl. summer. *Prickly Woodruff* . . . . **A. scoparia** Hook.f.
3 "Leaves" obtuse or acute but without a hyaline tip 4

4 Corolla pinkish to white, c. 3 mm long. "Leaves" linear, 3–7 mm long; those on the lower part of the stem strongly reflexed, acute. Flowers in terminal or axillary dichasia, almost sessile. Mericarps reticulate. Weak, rhizomatous, dioecious herb up to 15 cm high. Widespread. Grasslands. Fl. spring–summer. *Common Woodruff* . . . . **A. conferta** Hook.f.
4 Corolla blue, c. 4 mm long. "Leaves" linear, obtuse, ciliate, up to 3 cm long. Flowers in terminal corymbose panicles. Fruit smooth, glabrous. Annual herb up to 30 cm high. Recorded from Penshurst but now probably extinct there. Introd. from Europe. *Blue Woodruff* . . . . ***A. arvensis** L.

11 Galium L.

Bedstraws
16 species in Aust. (11 native, 5 naturalized); all states and territories except NT

Perennial or annual herbs with quadrangular stems. Stipules leaf-like and with the leaves forming a whorl of 2–9 "leaves" at the nodes. "Leaves" sessile or subsessile. Flowers very small, in small axillary or terminal cymes. Calyx absent. Corolla white or yellow, rotate and with scarcely any tube; lobes 4, spreading. Anthers exserted. Style 2-lobed; stigma capitate. Fruit dry, indehiscent, 2-seeded, covered with bristles or hooks.

Key to the species

1 "Leaves" mostly more than 6 in each whorl; may be less on flowering stems. Annuals 2
1 "Leaves" 6 or less in each whorl. 4

2 "Leaves" with antrose hairs on the margins; 4–6 in each whorl, 3–7 mm long. Flowers in pairs, axillary. Fruit linear-oblong, c. 1 mm diam., covered with spreading hooked hairs. Weak, glabrous or scabrous annual up to 12 cm high. Coast and adjacent plateaus; Cumberland Plain. Damp places in open forests. Introd. from Europe. Fl. spring. . . . . ***G. murale** (L.) All.
2 "Leaves" with retrorse hairs on the margins; 6–8 in each whorl, 15–30 mm long. Flowers in axillary cymes 3

3 Pedicels strongly recurved at fruiting stage. Cymes 1–3-flowered, shorter than the subtending leaves. Corolla cream. Fruit granular with large papillae, 3–4 mm diam. Widespread. Weed of cultivation and waste places. Introd. from Europe. Fl. spring–summer . . . . ***G. tricornutum** Dandy
3 Pedicels straight. Cymes 3–5-flowered, longer than the subtending leaves. Fruits 4–6 mm diam., covered with hooked bristles with tuberculate bases. Corolla whitish. Widespread. Weed of cultivation and waste places. Introd. from Europe. Fl. spring–summer. *Cleavers* or *Goosegrass* . . . . . . ***G. aparine** L.

4 Stipules much smaller than the leaves ("leaves" of different sizes). Stems prostrate, weak, from a primary root and sometimes with a few nodal roots. Leaves linear-elliptic, up to 5 mm long; stipules c. 3 mm long. Peduncles conspicuous; pedicels up to 5 mm long. Fruit rugose or ribbed. Widespread. DSF and WSF. Fl. spring–summer . . . . **G. binifolium** N.A.Wakef.
4 Stipules and leaves the same size ("leaves" at each node similar). 5

5 Fruit and ovary with hooked hairs. 6
5 Fruit and ovary without hooked hairs. 7

6 Annual. Stems glabrous; week. Flowers in pairs on pedicels shorter than subtending leaves. . . . . **G. murale** (see above)

6 Perennial. Stems with short hairs; robust. Flowers 1–7 on pedicels longer than subtending leaves. Fruit reniform, 1.5 mm long. Uncommon in coastal forest. Endangered. Fl. summer . . . . . **G. australe** DC.

7 Fruit with 5–7 ribs. "Leaves" obovate to elliptic, membranous, petiolate, 6–20 mm long. Stems weak, prostrate from a primary root, seldom with nodal roots. Peduncles and pedicels c. 1 mm long. Coast to lower Blue Mts DSF and WSF. Fl. winter–spring . . . . **G. liratum** N.A.Wakef.
7 Fruit not ribbed. Peduncles and/or pedicels more than 1 mm long. Leaves not membranous 8

8 Stems weak and flexible to the base from an extensive branched underground rhizome, rooting at the nodes; primary root not usually evident. 9
8 Stems brittle towards the base from a conspicuous primary root. Rhizomes absent or very poorly developed 10

9 Upper surface of the leaves matt. Leaves elliptic to ovate, up to 15 mm long. Fruit rugose or pubescent. Stems up to 30 cm long. Probably widespread but mostly coastal. Open forests and grasslands. Fl. spring–summer. . . . .**G. propinquum** A.Cunn.
9 Upper surface of leaves glossy. Leaves ovate, 3–10 mm long. Fruit papillose. Stems up to 30 cm long. Blue Mts Open forests and grasslands. Fl. summer . . . . **G. ciliare** Hook.f.

10 Leavf margins very recurved especially towards the apex, irregularly curled when dried, , lanceolate to linear, 1–12 mm long. Pedicels 1–2 mm long. Fruit up to 1.5 mm long, rugose. Stems up to 30 cm long. Widespread. Drier sites in open forests and grasslands. Fl. winter–spring . . . . **G. gaudichaudii** DC.
10 Leaf margins flat or slightly recurved, drying almost flat, elliptic to ovate, up to 18 mm long. Pedicels usually more than 2 mm long, often persistent with the peduncles. Fruit c. 1.5 mm long, papillose to tuberculate or wrinkled or hairy. Widespread. Usually in damp places, often amongst rocks. Fl. spring–summer . . . . **G. migrans** Ehrend. & MacGillivray

120 ACANTHACEAE

Small trees or perennial herbs or climbers. Leaves simple, opposite, entire or nearly so; stipules absent. Flowers irregular, bisexual. Sepals 5 and almost free or calyx tubular, 5–15-lobed. Corolla tubular, sometimes 2-lipped, 4 or 5-lobed. Stamens 2–4, epipetalous. Ovary superior, 2-locular; placentas axile or free central. Fruit a loculicidal capsule, 2-valved. Seeds flat, attached to hooked processes from the septum; or solitary, without integuments, the embryo with 2 large cotyledons folded longitudinally. 250 gen., mostly trop.

Key to the genera

1 Small tree in salt marshes or on the shores of estuaries . . . . **8 AVICENNIA**
1 Plants not as above 2

2 Corolla yellow-orange with a dark throat; twiners or climbers. . . . .**1 THUNBERGIA**
2 Corolla colouring not as above; herbs or shrubs 3

3 Corolla lobes ± equal. 4
3 Corolla lobes very unequal appearing 2-lipped. 6

4 Corolla tube cylindrical, slender. Stamens 2, usually exserted . . . . **3 PSEUDERANTHEMUM**
4 Corolla tube gradually enlarging upwards. Stamens 4, enclosed in the corolla tube. 5

5 Corolla tube <1.5 cm long . . . . **2 BRUNONIELLA**
5 Corolla tube >1.5 cm long . . . .**9 STROBILANTHES**

6 Calyx of 4 lobes the upper lobe functioning as the upper lip of the corolla; corolla 3-lobed, 4–5 cm long . . . .**4 ACANTHUS**
6 Flower not as above; corolla up to 3 cm long 7

7 Corolla red . . . .**5 ODONTONEMA**
7 Corolla white, pink, mauve or purple 8

8 Stamens 2, leaves linear to lanceolate . . . . **6 HYPOESTES**

8 Stamens 4, leaves lanceolate to elliptic. . . . . .7 **HYGROPHILA**

1 Thunbergia Retz.

4 species in Aust. (1 native, 3 naturalized); Qld, NSW, NT, WA

One species in the area

Climber or twiner. Leaves opposite, ovate to broad-ovate, up to 5 cm long and 3 cm wide, acute to acuminate, pubescent, cordate, margins entire or toothed. Flowers axillary, on long peduncles, enclosed by 2 large ovate bracteoles. Calyx 10–15-lobed. Corolla campanulate, 2–3 cm diam., yellow-orange with a dark throat, 5-lobed. Coast. Garden escape near habitation. Introd. from Africa. Fl. most of the year. *Black-eyed Susan* . . . . *****T. alata** Bojer ex Sims

2 Brunoniella Bremek.

Blue Trumpets

6 species in Aust. (5 endemic); Qld, NSW, Vic., WA, NT

Perennial herbs. Leaves petiolate, obovate to lanceolate, 1–5 cm long. Flowers axillary, sessile or nearly so. Corolla blue; tube gradually enlarged upwards, exceeding the calyx; lobes broad, often longer than the tube. Stamens 4, enclosed in the corolla tube. Capsule oblong-linear.

Key to the species

1 Calyx segments linear, c. 0.5 mm wide, hirsute with long spreading hairs. Corolla c. 5 mm diam. Perennial herb 5–30 cm high. Widespread. Forests and cleared land, particularly on shales. Fl. spring–summer. . . . .**B. australis** (Cav.) Bremek.

1 Calyx segments narrow-oblong to lanceolate, c. 1 mm wide, pubescent with short hairs. Corolla c. 10 mm diam. Perennial herb usually c. 5 cm high. Coast and adjacent plateaus. Open forests often on Ss. Fl. spring–autumn . . . .**B. pumilio** (R.Br.) Bremek.

3 Pseuderanthemum Radlk.

3 species in Aust. (2 endemic); Qld, NSW, NT

One species in the area

Perennial herb with a creeping rhizome and erect stems 7–30 cm high. Leaves petiolate, ovate to lanceolate or linear, mostly 1–5 cm long. Flowers in a terminal narrow bracteate raceme. Calyx lobes linear-setaceous, 3–8 mm long. Corolla pale blue or pink to white; tube slender, straight, 7–16 mm long; lobes broad, spreading. Capsule narrow, 10–14 mm long; the lower half contracted, seedless. Widespread. Forests and cleared land. Fl. summer–autumn. *Pastel Flower* . . . . **P. variabile** (R.Br.) Radlk.

4 Acanthus L.

3 species in AUST, (2 native, 1 introduced); Qld, NSW, S.A., WA

One species in the area

Perennial herb. Leaves in a basal rosette, dark green, glabrous, to 50 cm long and 30 cm wide; margins dissected, the segments lobed. Petiole 20–30 cm long. Inflorescence up to 2 m high. Flowers purple and white, calyx of 4 lobes the upper lobe functioning as the upper lip of the corolla; corolla 3-lobed, 4–5 cm long. Capsule c. 2 mm long with 1–2 seeds, apex mucronate. Naturalized in disturbed areas of DSF. Introd. from Eurasia. Fl. summer. *Bear's Breeches* . . . . *****A. mollis** L.

5 Odontonema Nees

1 species introduced Aust.; Qld, NSW

One species in the area

Erect rhizomatous shrub to 2 m high. Stems 4 angles often with grooves. Leaves elliptic to 18 cm long, 8 cm wide; margins entire or crenulate, apex acuminate sometimes curved to one side. Inflorescence terminal, flowers in sessile groups of 5–7 in upper half of rachis. Calyx 2 mm long. Corolla red 2–3 cm long, lobes ciliate on margins. Stamens enclosed by corolla. Style 14–17 mm long. Occasional garden escape. Introd. from Central America . . . . ***O. tubaeforme** (Bertol.) Kuntze

6 Hypoestes Sol. ex R.Br.

3 species in Aust. (1 native, 1 endemic, 1 naturalized); Qld, NSW, WA, NT

One species in the area

Perennial herb or subshrub to 1 m high. Stems with hairs in vertical bands. Leaves ovate to lanceolate 3–13 cm long, 1.5–9 cm wide, margins entire; petiole to 6 cm long. Inflorescence axillary, flowers to 3.5 cm long, pink to purple, a white central band on upper lib with purple markings. Capsule 8–12 mm long, glabrous. Occasional garden escape. Cumberland Plain. Fl. winter . . . . ***H. aristata** (Vahl) Sol. ex Roem. & Schult.

7 Hygrophila R.Br.

2 species in Aust. (1 native, 1 naturalized); Qld, NSW, WA, NT

One species in the area

Herb to 2 m high. Stems often reddish, 4-angles, lower nodes swollen. Leaves lanceolate to elliptic, 3.5–18 cm long, up to 5 cm wide. Margins entire, undulate. Petiole to 3.5 cm long. Inflorescence axillary, flowers sessile in clusters of 3 or 4. Corolla white, 5–10 mm long. Capsule 8–13 mm long, glabrous. Waterways in disturbed areas. Introd. from Central and S. America. . . . . ***H. costata** Nees

8 Avicennia L.

2 species in Aust.; all states and territories except Tas.

One species in the area

Small trees. Leaves opposite, ovate-lanceolate to lanceolate, glabrous, shining on the upper surface, whitish below, 5–8 cm long. Flowers in small, dense cymes on angular peduncles in the upper axils or in terminal panicles. Calyx divided to the base into 5 segments. Corolla white becoming orangish; tube shorter than the sepals; lobes ovate, longer than the tube. Stamens 4, inserted in the throat. Fruit a compressed capsule, c. 3 cm diam. Seed solitary, without integuments; the embryo with 2 large cotyledons folded longitudinally, germinating before the fruit drops. Saltwater swamps and on estuarine muds. Fl. spring–summer. *Grey Mangrove* . . . .**A. marina** (Forssk.) Vierh. var. **australasica** (Walp.) J.Everett

9 Strobilanthes Blume

1 species naturalized Aust.; Qld, NSW

One species in the area

Shrub to 1.5 m high. Leaves opposite, glabrous, narrow ovate to lanceolate up to 11 cm long, c. 2 cm wide, discolorous, apex acuminate. Inflorescence a cymose-panicles in leaf axils. Sepals connate, glandular hairy , reddish. Corolla mauve, 25–35 mm long. Stamens 4, didymous. Fruit a capsule 7–10 mm long, glandular hairy. Blue Mts Introd. from India. Fl. spring–summer. *Goldfussia* . . . . *** S. anisophyllus** (G.Lodd.) T.Anderson

121 BIGNONIACEAE

Lianas trees or shrubs. Leaves opposite, imparipinnate-compound, entire or dentate, petiolate; stipules absent. Flowers irregular, bisexual. Calyx tubular. Corolla tubular, 5-lobed. Stamens usually 4, epipetalous. Ovary superior, 2-locular, with axile placentas. Fruit a loculicidal or septicidal capsule, 2-valved. Seeds often winged. C. 120 gen., trop. and subtrop. especially S. America

Key to the genera

1 Corolla white or cream with purplish markings in the throat . . . .**2 PANDOREA**
1 Corolla purple, yellow, or orange to red 2

2 Corolla purple with darker streaks. . . . .**4 CLYTOSTOMA**
2 Corolla not purple 3

3 Corolla yellow and densely pubescent on outside, creamy yellow inside . . . . **5 PITHECOCTENIUM**
3 Corolla yellow with orange markings in the throat or orange to red, not pubescent on outside 4

4 Leaflets regularly toothed . . . . **1 TECOMA**
4 Leaflets entire or nearly so . . . . **3 MACFADYENA**

1 Tecoma Juss.
2 species naturalized Aust.; Qld, NSW

One species in the area

Scrambling shrub up to 3 m high. Leaflets 5–9, elliptic to ovate, 10–40 mm long. Flowers in terminal racemes. Corolla 5–6 cm long, orange to red; tube curved; lobes c. 10 mm long. Capsule linear to narrow-oblong, up to 65 mm long. Coast. Garden escape naturalized in places. Introd. from S. Africa. Fl. spring–summer. *Cape Honeysuckle* . . . . *T. capensis* (Thunb.) Lind.

2 Pandorea (Endl.) Spach
4 species in Aust. (3 endemic); all states and territories

One species in the area

Woody, often tall, glabrous climber, with ± twining branches. Leaflets 5–9, glabrous, usually ovate-oblong to lanceolate, entire, 2–7 cm long (coarsely toothed and much larger in the juvenile stages). Flowers in loose terminal or axillary panicles. Corolla whitish, with purplish spots inside the throat; tube almost cylindrical, 10–20 mm long, c. 6 mm diam.; lobes broad, less than one third as long as the tube. Capsule oblong, 2-valved, 4–7 cm long. Seeds flat, winged all around. Widespread. In or near WSF and RF; often persisting in cleared areas. Fl. spring. *Wonga Wonga Vine.* . . . . **P. pandorana** (Andrews) Steenis

3 Macfadyena A.DC.
1 species naturalized Aust.; Qld, NSW

One species in the area

Woody, often tall, glabrous climber with twining branches. Leaves elliptic, up to 6 cm long and 2 cm wide, obscurely dentate to sinuate, often with 3-partite tendrils. Flowers in loose terminal or axillary panicles. Corolla yellow; the tube c. 5 cm long. Capsule narrow-oblong, up to 45 cm long. Seed winged on 2 sides. Coast. Weed in RF, WSF and gullies. Introd. from America. Fl. spring–summer. *Cat's Claw Creeper* . . .
. . . . **M. unguis-cati** (L.) A.H.Gentry

4 Clytostoma Miers ex Bureau
1 species naturalized Aust.; NSW

One species in the area

Climber with woody stems. Leaflets 2, 5–9 cm long, 2–5 cm wide, glabrous, margins entire, tendrils present. Flowers tubular to 8 cm long. Corolla pale purple with dark streaks. Capsule woody to 8 cm long, prickly. Eastwood area along creek lines. Introd. from S. America. Fl. spring. *Argentine Trumpet Vine* . .
. . . . **C. callistegioides** (Cham.) Baill.

5 Pithecoctenium Mart. ex DC.
1 species naturalized Aust.; NSW

One species in the area

Climber with woody, ribbed stems. Leaves 2.5–8 cm long, with 2–3 thin leaflets. Terminal leaflet often replaced with tendrils. Flowers in a terminal raceme or panicle, calyx truncate and cup shaped; corolla yellow and pubescent on outside, up to 5 cm long. Capsule woody, compressed, with dense bristles. Naturalized in bush. Introd. from S. America. Fl. summer. *White Trumpet Vine* . . **P. cynanchoides** DC.

122 GESNERIACEAE
120 gen., mainly trop. and subtrop.

1 Fieldia A.Cunn.
1 species endemic Aust.; Qld, NSW, Vic.

One species in the area

Scrambling or climbing, semi-epiphytic, slightly woody perennial. Young branches and leaves hirsute. Leaves ovate to elliptic or almost lanceolate, pale underneath, opposite; the pairs often very unequal in size; the larger 4–7 cm long; the smaller sometimes less than 1 cm long and almost sessile. Flowers axillary, solitary on pendulous peduncles. Bracteoles broad, up to 10 mm long. Calyx tubular, 5-lobed; lobes c. as long as the bracteoles. Corolla tubular, greenish yellow, ± irregular; the tube c. 3 cm long; the lobes very short. Stamens 4. Ovary 2-locular, superior; placentas axile. Fruit a berry, whitish. Widespread. RF, mostly on tree fern trunks. Fl. summer . . . . **F. australis** A.Cunn.

123 LAMIACEAE

Herbs or shrubs, usually strongly scented with aromatic oils. Stems nearly always quadrangular. Leaves opposite or whorled, without stipules. Inflorescence simple or compound, axillary or terminal. Flowers solitary or in opposite cymes which are frequently reduced to clusters (false whorls) in the axils of the leaves or floral bracts. Flowers usually irregular and bisexual. Calyx tubular, persistent, 5–10-toothed or 2-lipped. Corolla tubular, 4–5-lobed, often 2-lipped with the upper lip 2-lobed and the lower lip 3-lobed. Stamens 4 in pairs of unequal length or 2 reduced to staminodes or 2 only. Nectiferous disc often present between stamens and ovary. Ovary superior, entire or deeply 4-lobed, with basal placentas; style gynobasic or terminal 2–4-fid. Ovules 4. Fruit a drupe or schizocarp splitting into two 2-seeded or four 1-seeded articles (or fewer by abortion), often enclosed in the persistent calyx. 236 gen., cosmop.

Key to the genera

1 Calyx with 4–10 ± equal teeth 2
1 Calyx divided into 2 lips which may or may not be toothed 20

2 Calyx almost equally 4–5-toothed 3
2 Calyx with 8–10 teeth 19

3 Leaves in whorls of 3–4 4
3 Leaves opposite, in pairs. 5

4 Perfect stamens 2; staminodes 2. Corolla white to bluish . . . .**17 WESTRINGIA**
4 Perfect stamens usually 4. Corolla deep blue or purple . . . . **16 HEMIGENIA**

5 Corolla almost equally 4-lobed to indistinctly 2-lipped 6
5 Corolla lobes unequal; corolla distinctly 1- or 2-lipped 8

6 Flowers in dense oblong spikes . . . . **2 LAVANDULA**
6 Flowers not in dense oblong spikes. 7

7 Stamens 4. Fragrant herbs. . . . . **3 MENTHA**
7 Stamens 2; staminodes 2 . . . . **4 LYCOPUS**

8 Calyx campanulate c. 20 mm diam . . . . **11 MOLUCCELLA**
8 Calyx less than 20 mm diam 9

9 Corolla strongly one-sided with apparently 1 lip with up to 5 lobes 10
9 Corolla not strongly one-sided appearing 2-lipped and/or 5-lobed 11

10 Corolla blue to purple . . . . **19 AJUGA**
10 Corolla white . . . . **18 TEUCRIUM**

11 Style terminal on ovary. Trees or shrubs 12
11 Style gynobasic. Herbs or subshrubs 15

12 Corolla tube long and slender . . . . .20 CLERODENDRUM
12 Corolla tube long or short, if long then not slender 13

13 Trees, petioles >1 cm long, hairs simple . . . . 21 GMELINA
13 Shrubs, leaves decurrent or petioles <1 cm long, glabrous or with branched hairs 14

14 Leaves decurrent and wrinkled. Corolla 15–40 mm long . . . . 22 CHLOANTHES
14 Leaves not decurrent or wrinkled. Corolla <5 mm long . . . . 23 SPARTOTHAMNELLA

15 Flowers in terminal inflorescences 16
15 Flowers in axillary clusters 17

16 Leaves ovate to triangular, petiolate, base cordate . . . . 14 NEPETA
16 Leaves lanceolate, sessile . . . . .24 PHYSOSTEGIA

17 At least some leaves very deeply lobed . . . . 13 LEONURUS
17 Leaves crenate to dentate 18

18 Corolla with a ring of hairs in the throat . . . . 9 STACHYS
18 Corolla with scattered hairs in the corolla tube . . . . .10 LAMIUM

19 Herb up to 30 cm high, clothed with soft whitish hairs. Corolla white . . . . 8 MARRUBIUM
19 Shrub up to 2 m high, with very short hairs. Corolla orange to scarlet . . . . .12 LEONOTIS

20 Perfect stamens 2. Anthers separated by a long connective . . . . 5 SALVIA
20 Perfect stamens 4 21

21 One or both calyx lips toothed or lobed 22
21 Both calyx lips entire or nearly so 24

22 Upper calyx lip 3-toothed; lower lip 2-toothed . . . . 7 PRUNELLA
22 Upper calyx lip entire; lower lip acutely 4-toothed 23

23 Stamens free, exceeding corolla. Stigma capitiate . . . . 25 ORTHOSIPHON
23 Stamens basally connate, hardly exceeding corolla . . . . 1 PLECTRANTHUS

24 Herbs, upper calyx lip with a hollow protuberance . . . . .6 SCUTELLARIA
24 Subshrubs, shrubs or small trees, upper calyx lip with no protuberance . . . . 15 PROSTANTHERA

1 Plectranthus L'Hér.

28 species in Aust. (22 endemic, 6 naturalized); all states and territories except Tas.

Perennial herbs or shrubs, usually rather succulent, variously pubescent with glandular and simple hairs and with yellowish sessile glands. Leaves ovate to orbicular, serrate, obtuse or cordate at the base. Flowers in terminal leafless thyrses with 3–10 flowers per cyme. Calyx 2-lipped; upper lip broad, entire; lower lip with 4 lanceolate teeth. Corolla bluish purple; tube twice as long as the calyx; upper lip 4-lobed; lower lip entire, concave, enclosing the 4 stamens.

Key to the species

1 Bracts persistent at least until fruiting stage 2
1 Bracts caducous 3

2 Leaves and stems with reddish-purple simple hairs as well as glandular hairs. Leaves elliptic or ovate to almost orbicular, mostly up to 8 cm long, crenate. Corolla 12–15 mm long, whitish with purplish markings. Aromatic, straggling herb up to c. 50 cm high. Sydney district. Ss. Introd. from S. Africa. Fl. most of the year . . . .*P. ciliatus E.Mey. ex Benth.

2 Leaves and stems without reddish-purple hairs. Leaves fleshy, broad ovate to triangular, 3–7 cm long, upper surface densely hairy, 7–15 teeth on each side. Corolla 7–9 mm long, mauve or white, tube slightly curved at middle. Aromatic perennial herb to 1.5 m high, densely covered with hairs. Occasionally naturalized. Introd. from Africa. Fl. spring. *Allspice* . . . . ***P. amboinicus** (Lour.) Spreng.

3 Corolla 14–20 mm long, tube straight. Leaves ovate to oblong-elliptic, 6–17 cm long, 4–10 cm wide, pubescent to almost glabrous, 16–25 teeth on each side. Soft erect shrub to 3 m high sparsely hairy, with semi succulent base. Garden escape. Blue Mts Introd. from S. Africa. Fl. all year . . ***P. ecklonii** Benth.

3 Corolla less than 12 mm long, tube deflexed 4

4 Glandular hairs abundant amongst the simple hairs on the stems and leaves. Leaves ovate to orbicular 4–11 cm long, 3–7 cm wide; villous and glandular pubescent with 10–20 teeth on each side. Corolla 8–9 mm long, deflexed almost at a right angle. Weak shrub up to 1 m high without a tuberous base, heavily scented when bruised. Blue Mts Rocky outcrops. Fl. most of the year . . . . **P. graveolens** R.Br.

4 Glandular hairs absent or very few below the inflorescence. Leaves ovate to orbicular, 2–7 cm long, 2–4 cm wide; pubescent to almost glabrous, with up to 12 teeth on each side. Corolla 6–11 mm long, deflexed at less than a right angle. Perennial herb up to 1 m high with a fleshy tuberous base. Widespread. Rocky outcrops in gullies. Fl. most of the year . . . . **P. parviflorus** Willd.

Intermediates between *P. graveolens* and *P. parviflorus* appear to occur.

2 Lavandula L.
2 species naturalized Aust.; NSW, Vic, S.A., WA

One species in the area
Grey tomentose shrub 30–100 cm high. Leaves linear to oblong-linear, 1–2 cm long. Flowers in dense oblong-quadrangular, terminal spike-like thyrses surmounted by several large sterile violet coloured bracts. Camden and Dural districts. Roadsides. Introd. from the Mediterranean. Fl. spring–summer. *French Lavender* . . . . ***L. stoechas** L.

3 Mentha L.
10 species in Aust.; (4 endemic, 6 naturalized); all states and territories

Strongly scented herbs. Flowers small, in verticillate cymes. Calyx 5-toothed, regular, striate. Corolla white, lilac or pink; the tube not longer than the calyx; lobes almost equal, shorter than the tube. Stamens 4, equal, erect. Articles smooth.

Key to the species

1 Cymes in the axils of leaf-like bracts. 2
1 Flowers in cymes forming dense, terminal, cylindrical or globose, spike-like thyrses. Branches often purplish 5

2 Calyx lobes with short hairs but not villous inside. 3
2 Calyx lobes densely villous inside 4

3 Flowers numerous in leaf axils. Plants grey-green, villous with simple hairs. Leaves ovate to orbicular, 1–1.5 mm long; the floral ones smaller. Corolla lilac. Stoloniferous herb often rooting at the nodes. Coast. Escape from cultivation. Naturalized near habitation. Introd. from Europe. Fl. summer–autumn. *Pennyroyal.* . . . . ***M. pulegium** L.

3 Flowers in clusters of 2–6 in leaf axils, rarely more. Plants with multicellular or retrorse hairs. Leaves ovate 1–4.5 cm long. Corolla white, pink or mauve. Herb to 60 cm high, rooting at nodes. Damp or swampy areas on clay-rich soils. Widespread. Fl. summer–winter. *Forest Mint* . . . **M. laxiflora** Benth.

4 Branches almost glabrous. Leaves oblong to oblong-lanceolate, not broader towards the base, 5–30 mm long. Whorls 3-flowered in leaf axils. Calyx 3–4 mm long; teeth triangular, sometimes acuminate. Corolla white or pink. Dense, rhizomatous herb with ascending to prostrate branches. Widespread.

Open forests and grasslands on loams and clays. Fl. most of the year. *Native Pennyroyal* . . . .
. . . . **M. satureioides** R.Br.

4 Branches hairy. Leaves ovate to ovate-lanceolate, broader towards the base, 6–20 mm long. Whorls 3–8-flowered in leaf axils. Calyx 3–4 mm long; teeth lanceolate, acuminate. Corolla pale purple. Rhizomatous herb forming mats. Widespread. Open forests and grasslands, often on sandy soils. Fl. most of the year. . . . .**M. diemenica** Spreng.

5 Flowers in globose terminal heads, sometimes with some verticillate whorls distant from the terminal head. Leaves ovate, 2–8 cm long, serrate; petiole glabrous. Corolla dark lilac. Rhizomatous herb with characteristic lemon or eau-de-cologne scent. Coast. Escape from cultivation. Damp places. Introd. from Europe. Fl. summer–autumn. *Lemon Mint, Eau-de-Cologne Min* . . . .
. . . . ***M.** X **piperita** L. var. **citrata** (Ehrh.) Briq.
5 Flowers in elongated spike-like thyrses 6

6 Leaves with a slightly pubescent petiole mostly at least 4 mm long, lanceolate to narrow-elliptic, up to 6 cm long, serrate. Corolla dark lilac. Rhizomatous herb with characteristic peppermint scent. Coast. Damp places. Introd. from Europe. Fl. summer–autumn. *Peppermint* . .***M.** X **piperita** L. var. **piperita**
6 Leaves sessile, lanceolate to ovate, 2–9 mm long, serrate, glabrous or slightly pubescent. Corolla lilac to pink. Perennial herb with characteristic scent. Widespread. Damp places. Introd. from Europe. Fl. summer–autumn. *Spearmint* . . . . ***M.** X **spicata** L.

4 Lycopus L.

1 species endemic Aust.; Qld, NSW, Vic., Tas., S.A.

One species in the area

Erect herb up to 1 m high, glabrous or nearly so. Leaves lanceolate to elliptic, up to 15 cm long, serrate, acute to acuminate, tapering towards the base. Calyx with 5 lanceolate equal teeth. Corolla white, only slightly longer than the calyx or less; the lobes shorter than the tube. Stamens 2; staminodes 2. Coast and Cumberland Plain. Wet places. Fl. summer–autumn. *Gipsywort.* . . . . **L. australis** R.Br.

5 Salvia L.

Sages

6 species in Aust. (1 native, 5 naturalized); all states and territories except NT

Erect, branching, pubescent to hirsute, almost shrubby herbs. Leaves petiolate. Flowers arranged in terminal thyrses or racemes. Calyx 2-lipped, with acute teeth. Corolla 2-lipped; the upper lip convex, erect, 2-lobed; the lower lip spreading, partly reflexed, 3-lobed. Stamens 2; the connective elongated and the posterior cell of each anther absent.

Key to the species

1 Corolla bright red, 15–25 mm long. Leaves narrow-ovate, 2–5 cm long, serrate. Erect coarse herb up to 1 m high. Coast. Shaded places. Introd. from America. Fl. spring–autumn . ***S. coccinea** Juss. ex Murray
1 Corolla pale purple or blue, up to 15 mm long 2

2 Branches with long multicellular hairs. Leaf bases cordate to truncate; margins lobed and irregularly toothed; elliptic to ovate, 3–10 cm long. Corolla 8–15 mm long, bluish purple to lilac. Perennial herb to 70 cm high. Disturbed areas. Introd. from Europe. Fl. all year. *Wild Sage* . . . . ***S. verbenaca** L.
2 Branches with simple hairs which are not multicellular. Leaf bases cuneate to tapering 3

3 Corolla up to c. 5 mm long, pale purple. Leaves lanceolate to oblong, 2–7 cm long, serrate, minutely scabro-pubescent and sometimes glandular-viscid when young; bases usually cuneate. Perennial herb up to 1 m high. Blue Mts, e.g. Tuglow Caves. Often on limestone. Fl. spring–autumn . **S. plebeia** R.Br.
3 Corolla 7–10 mm long, pale blue. Leaves lanceolate to narrow-elliptic, mostly up to 5 cm long; margins with shallow distant teeth; bases tapering. Annual herb up to c. 50 cm high, sometimes woody towards the base. Coast and Cumberland Plain. Heavy and silty soils. Introd. from N. America. Fl. spring–autumn. *Mintweed* . . . . ***S. reflexa** Hornem.

6 Scutellaria L.

Skullcap

3 species in Aust. (2 native, 1 naturalized); Qld, NSW, Vic., Tas., S.A.

Perennial herbs with creeping rootstocks and ascending stems. Leaves petiolate. Flowers solitary in the axil of each floral leaf, all turned to one side of the rhachis. Calyx 2-lipped; both lips entire; upper lip with a hollow scale-like protuberance on the back. Corolla blue, with a rather long tube and small nearly closed lips; upper lip concave, emarginate; lower lip convex, spreading, 3-lobed. Stamens 4, in pairs, enclosed in the upper lip.

Key to the species

1 Plants glandular pubescent. Stems slightly branched, weak, ascending, mostly 30 cm long with acute angles. Leaves ovate or narrow-ovate, coarsely toothed, often more than 25 mm long. Corolla c. 10 mm long; the lower lip considerably longer than the upper. Coast and lower Blue Mts WSF; RF. Fl. spring–summer. . . . . **S. mollis** R.Br.
1 Plants glabrous or pubescent with minute white simple ± appressed hairs 2

2 Leaves ovate, often cordate at the base, crenate to lobed, rarely more than 12 mm long. Corolla c. 5 mm long; the lower lip slightly longer than the upper. Stoloniferous herbs with stems ascending up to c. 15 cm high, pubescent with minute white simple often appressed hairs or almost glabrous. Widespread. Damp sheltered places. Fl. spring–autumn . . . . **S. humilis** R.Br.
2 Leaves narrow-ovate to narrow-oblong, up to 2 cm long, hastate at the base, often with acuminate auricles, glabrous; margin shallowly sinuate to entire. Corolla c. 5 mm long, white with purple markings; the lower lip slightly longer than the upper. Weak, decumbent to ascending, glabrous, perennial herb. Cumberland Plain. Damp places. Introd. from S. America. Fl. spring–summer . . . . *S. racemosa Pers.

7 Prunella L.

2 species naturalized Aust.; all states and territories except NT

Perennial herbs with 4-angular branches. Leaves opposite, petiolate, toothed to lobed. Flowers in terminal ovoid to ellipsoid spike-like thyrses. Calyx 2-lipped; lower lip with deep lobes; upper lip with almost obsolete lobes. Corolla 2-lipped. Stamens 4.

Key to the species

1 Upper leaves deeply lobed, up to 5 cm long. Calyx 7–11 mm long. Corolla yellowish white or rarely pinkish to purplish, c. 15 mm long. Decumbent perennial with branches up to c. 30 cm long. Lower Blue Mts; Bowral district. Weed. Introd. from Europe. Fl. summer–autumn . . . . *P. laciniata (L.) L.
1 Upper leaves entire or toothed, 2–7 cm long. Calyx 8–12 mm long. Corolla violet, c. 12 mm long. Decumbent or shortly creeping perennial. Widespread. Weed of damp places; roadsides; pastures. Introd. from Europe. Fl. spring–autumn. *Self-heal* . . . . *P. vulgaris L.

8 Marrubium L.

1 species naturalized Aust.; all states and territories except NT

One species in the area

Perennial herb up 30–80 cm high. Stems thickly covered with white woolly hairs. Leaves petiolate, orbicular, 1–3 cm diam., irregularly crenate, soft, wrinkled, whitish with soft hairs. Flowers in dense axillary verticillate cymes. Calyx 4–5 mm long with 10 subulate recurved teeth. Corolla white, slightly longer than the calyx. Widespread. Weed of waste places. Introd. from Europe and W. Asia. Fl. spring–autumn. *Horehound* . . . . *M. vulgare L.

9 Stachys L.

2 species naturalized Aust.; all states and territories except NT

One species in the area

Weak, ascending annual with spreading hairs. Leaves ovate, petiolate, 15–40 mm long, regularly crenate. Flowers 2–6 in each verticillate cyme, forming a loose leafy spike-like thyrse. Calyx 5-toothed, 5–7 mm long; the teeth as long as the tube. Corolla pale purple, scarcely longer than the calyx. Stamens 4, in pairs, enclosed in the upper lip. Widespread. Weed of cultivation and waste places. Introd. from Europe and Asia. Fl. winter–spring. *Stagger Weed* . . . .*S. arvensis* (L.) L.

10 Lamium L.

3 species naturalized Aust.; Qld, NSW, Vic., Tas., S.A., WA

Annual or perennial herbs. Leaves decussate, simple, crenate to toothed, petiolate. Flowers in leafy thyrse-like cymes in leaf axils. Calyx 5-lobed. Corolla 2-lipped, lower lip hooded. Stamens 4. Mericarps 3-keeled.

Key to the species

1 Weak, ascending pubescent annual herb. Lower leaves petiolate; upper leaves sessile, orbicular-reniform, crenately lobed, 1–2 cm long, subtending and embracing the whorls of flowers. Calyx villous, 5 mm long; teeth 5 almost equal, subulate. Corolla purplish; tube long, narrow. Widespread. Weed of cultivation and waste places. Introd. from Europe and Asia. Fl. winter–spring. *Dead Nettle* . . . .
. . . .*L. amplexicaule* L.
1 Trailing perennial herb, rooting from lower nodes. Leaves petiolate, broadly ovate 2–7 cm long, variegated, margins crenate, base truncate to cuneate. Calyx sparsely hairy, 7–10 mm long. Corolla bright yellow with brown markings, tube hardly longer than calyx. Higher Blue Mts Garden escape. Introd. from Europe and Asia. Fl. spring. *Aluminium Plant*. . . . . *L. galeobdolon* (L.) L.

11 Moluccella L.

1 species naturalized Aust.; NSW, Vic., S.A., WA

One species in the area

Glabrous, erect or ascending annual up to 1 m high. Leaves on long petioles, broad-ovate to almost orbicular. Calyx c. 2 cm diam., enlarging to 4–5 cm after flowering, campanulate, oblique, membranous; the margin 5-angled. Sydney district; Cumberland Plain. Waste places, often on clay soils. Introd. from the Mediterranean. Fl. spring–summer. *Molucca Balm* . . . .*M. laevis* L.

12 Leonotis (Pers.) R.Br.

2 species naturalized Aust.; Qld, NSW, NT, WA

One species in the area

Tall shrub up to 2 m high. Leaves narrow-lanceolate, c. 7 cm long, irregularly toothed towards the apex, narrowed at the base into a short petiole. Flowers in dense verticillate cymes towards the ends of branches. Calyx 10-ribbed, 10-toothed; the tube 10–12 mm long at fruiting stage. Corolla orange to scarlet, 4–5 cm long, hairy; the upper lip long; the lower lip short, spreading. Coast. Garden escape naturalized near habitation. Introd. from S. Africa. Fl. spring–summer. *Lion's Tail*. . . . .*L. leonurus* R.Br.

13 Leonurus L.

1 species naturalized Aust.; NSW

One species in the area

Perennial herb up to 3 m high. Leaves ovate in outline, up to 7 cm long, deeply pinnately or palmately lobed; upper leaves sometimes linear. Flowers in distant or dense verticillate cymes. Calyx 4–5 mm long, 5-lobed. Corolla white to reddish or mauve, up 15–20 mm long, 2-lipped; lower lip lobed; upper lip entire. Coast. Weed; waste places. Introd. from Asia. Fl. winter–summer. . . . .*L. japonicus* Houtt.

14 Nepeta L.

1 species naturalized Aust.; NSW, Vic., Tas., S.A.

One species in the area

Perennial herb up to c. 1 m high, densely pubescent. Leaves ovate, 2–6 cm long, cordate at the base, serrate; bracts very reduced. Flowers in dense terminal spike-like thyrses. Calyx 5-toothed. Corolla c. 12 mm long, white with purple spots; tube curved and dilated at the middle. Coast. Near habitation. Introd. from Europe. Fl. summer. *Catmint.* . . . .*N. cataria* L.

15 Prostanthera Labill.

Mint Bushes

c. 100 species endemic Aust.; all states and territories

Shrubs with oil glands and often strongly scented. Flowers solitary and axillary, or in terminal racemes and/or thyrses. Calyx usually striate, 2-lipped; the lips usually entire. Corolla 2-lipped; tube short, dilated upwards; upper lip erect, broadly 2-lobed; lower lip spreading, 3-lobed, the middle lobe larger and usually emarginate. Stamens 4, in pairs; anthers 2-celled, the connective usually produced into one or two linear appendages. Style shortly 2-fid. Articles reticulate-rugose.

Key to the species

1 Leaves subtending the flowers or cymes much smaller than the lower leaves or reduced to deciduous bracts or absent 2
1 Leaves subtending flowers similar to the lower leaves or slightly smaller 18

2 Leaves with margins entire 3
2 Leaves with margins variously toothed 7

3 Branches glabrous except for nodes. Leaves linear to oblong-linear, obtuse, mostly 12–25 mm long, recurved. Flowers either axillary or the upper ones forming leafy or leafless racemes. Calyx glabrous or slightly pubescent. Corolla white, sometimes pinkish in the throat, 10–12 mm long. Erect shrub up to 3 m high. Widespread. Sheltered places. Fl. spring–summer . . . . **P. linearis** R.Br.
3 Branches sparsely to densely hairy 4

4 Leaves orbicular to broad-ovate, 3–20 mm long, 3–15 mm wide; petioles 2–8 mm long. Corolla 10–15 mm long, purple to pinkish. Erect, aromatic shrub up to 3 m high. Widespread. Sheltered situations in forests; margins of RF. Fl. spring . . . . **P. rotundifolia** R.Br.
4 Leaves lanceolate or narrow to broadly ovate, if ± circular then more than 15 mm long 5

5 Leaves mostly ovate (20)30–50 mm long, obtuse, entire or with a few short teeth; petioles 5–10 mm long. Calyx more than 5 mm long. Corolla white or bluish or rarely pinkish, 12–15 mm long. Flowers in simple terminal racemes; floral leaves reduced to broad obtuse concave bracts enclosing the calyx in the bud stage. Erect shrub 1–2 m high. Nepean River district and lower Blue Mts Open forests, often along creek banks. Fl. spring . . . . **P. prunelloides** R.Br.
5 Leaves lanceolate to ovate. Calyx 2–4 mm long. Corolla 6–12 mm long 6

6 Leaves ovate, up to 40 mm long but mostly less than 30 mm, 3–12 mm wide, entire to irregularly sinuate; petiole 1–6 mm long. Calyx ± glabrous. Corolla mauve or deep purple, rarely white. Erect shrub up to 4 m high, slightly to strongly aromatic. Coast and adjacent plateaus. Widely cultivated. DSF and WSF. Ss. Fl. spring (a number of as yet undefined taxa may occur within this species) . . . . . **P. ovalifolia** R.Br.
6 Leaves lanceolate, 15–30 mm long, 1–5 mm wide, entire to toothed; petiole up to 3 mm long. Calyx pubescent, 4–5 mm long. Corolla mauve to purple with darker spots, 7–10 mm long. Erect, strongly aromatic shrub up to 3 m high. Hunter River Valley. WSF. Fl. spring . . . . **P. discolor** R.T.Baker

7 Leaves flat 8
7 Leaves with recurved or revolute margins 15

8 Leaves mostly more than 25 mm long 9

8 Leaves mostly less than 25 mm long 11

9 Leaves 3–5 mm wide, linear to narrow ovate 28–40 cm long, margins toothed, petiole 2–5 mm long. Flowers in leafy terminal inflorescences, bracteoles 4–5 mm long. Corolla 10–12 mm long, white to purple with darker markings in the throat. Anther appendage present. Mostly glabrous, aromatic shrub up to 2 m high. Kangaroo Valley area. DSF. Fl. summer . . . . **P. sp. E** (*sensu* B.J.Conn (1992))

9 Leaves more than 5 mm wide 10

10 Both anther appendages shorter than the anther cell. Leaves ovate to lanceolate, slightly serrate, (25)30–50 mm long. Racemes simple or with 2 branches at the base; floral bracts ovate, concave, acuminate, falling from the very young bud. Corolla glabrous, 10–12 mm long, white to bluish. Erect, slightly aromatic shrub up to 3 m high. Coast and adjacent plateaus; Blue Mts Fl. spring . . . . **P. caerulea** R.Br.

10 One anther appendage much longer than the anther cell; the other very short or obsolete. Leaves lanceolate to oblong-lanceolate, serrate, 30–100 mm long, paler underneath. Racemes large, often thyrses; the leaves subtending the flowers much reduced or absent. Lower lip of the calyx somewhat smaller than the upper. Corolla pubescent inside and outside, white to bluish, often spotted, c. 12 mm long; the lobes very broad. Erect shrub or small tree up to 6m high. Blue Mts; Woronora plateau. RF; WSF; gullies. Fl. summer–autumn. *Victorian Christmas Bush* . . . . **P. lasianthos** Labill.

11 Leaves more or less regularly toothed or lobed 12

11 Leaves shallowly and irregularly toothed or lobed 13

12 Leaves deeply toothed, ovate, up to 25 mm long, covered with long hairs 1–1.5 mm long. Calyx 5–6 mm long. Corolla mauve, c. 8 mm long. Erect, strongly aromatic open shrub up to 1 m high. Ourimbah-Strickland SF district. In or near RF. Fl. spring . . . . **P. askania** B.J.Conn

12 Leaves ± shallowly toothed, glabrous or covered with hairs less than 1 mm long, ovate, up to 30 mm long; petiole 1–10 mm long. Calyx 3–5 mm long. Corolla mauve, 7–10 mm long. Erect strongly aromatic much branched shrub up to 3 m high. Widespread. WSF; margins of RF. Fl. spring . . . . **P. incisa** R.Br.

13 Leaves orbicular to broad-ovate . . . . **P. rotundifolia** (see above)

13 Leaves lanceolate to ovate 14

14 Calyx pubescent . . . . **P. discolor** (see above)

14 Calyx glabrous or slightly hairy . . . . **P. ovalifolia** (see above)

15 Leaves glabrous except for short rigid tuberculate hairs resembling minute prickles on the upper surface, ovate-lanceolate to narrow-linear, sessile or nearly so, 5–12 mm long, revolute, entire. Floral bracts ovate, acuminate, coloured. Calyx hirsute; lips broad, nearly equal. Corolla deep violet, 7–10 mm long. Straggling to decumbent shrub up to 1 m high. Frenchs Forest to Newport and Cowan. Open forests on Ss and shales. Fl. spring. . . . . **P. denticulata** R.Br.

15 Leaves hairy on both surfaces 16

16 Leaves with margins recurved, entire; linear to narrow-oblong, 8–20 mm long, scabrous-hirsute. Corolla violet with purple markings, c. 7 mm diam. Calyx lips nearly equal. Floral leaves obtuse. Strongly aromatic shrub up to 2 m high. Woronora Plateau; lower Blue Mts DSF; rocky places. Fl. spring. . . . . **P. hirtula** F.Muell. ex Benth.

16 Leaves with margins crenate. 17

17 Lower lip of the calyx slightly longer and narrower than the upper. Corolla mauve to bluish, 6–8 mm long. Calyx 4–5 mm long. Leaves only slightly recurved, orbicular to ovate, 2–6 mm long. Branches and leaves sprinkled with short stiff hairs. Aromatic, divaricate shrub up to 2 m high, with slender branches. Nepean River district and lower Blue Mts Open forests and RF; gullies. Fl. spring . . . . **P. violacea** R.Br.

17 Both lips of the calyx broad; the lower slightly exceeding the upper. Corolla 6–8 mm long, lilac. Leaves 3–20 mm long, with recurved margins, bullate-rugose. More densely hairy and more robust in all its parts than *P. violacea* with which some small-leaved forms of it may be confused. Erect, aromatic shrub to 2.5 m high. Nepean River and Blue Mts Open forests. Fl. spring . . . . **P. incana** A.Cunn. ex Benth.

18 Leaves with recurved or revolute margins 19
18 Leaves flat (sometimes slightly recurved in *P. hindii* and *P. junonis*) concave upwards or with incurved margins 27

19 Leaves, densely hairy, cordate to ovate-lanceolate, ± succulent, 10–15 mm long, almost sessile, strongly recurved, often crowded towards the end of branches. Branch hairs spreading. Calyx c. 5 mm long. Corolla mauve, 12–15 mm long. Erect, compact, aromatic shrub up to 2 m high. Coast. DSF and heath. Headlands and ranges near the coast. Fl. chiefly spring . . . . **P. densa** A.A.Ham.
19 Leaves not as above. Corolla mostly less than 12 mm long 20

20 Leaves lobed, wrinkled, ovate, up to 9 mm long, thick, often crowded towards the ends of the branches. Corolla whitish mauve, 10–13 mm long. Divaricate, aromatic shrub up to 2 m high. Moss Vale district. WSF and DSF. Fl. spring . . . . **P. rugosa** A.Cunn. ex Benth.
20 Leaves not lobed 21

21 Anther with distinct basal appendage 22
21 Anthers without distinct appendage 24

22 Upper surface of leaves glabrous . . . . **P. linearis** (see above)
22 upper surface of leaves densely hairy 23

23 Branches with retrorse hairs. Leaves more or less uniform in shape, ovate, recurved margins often give a triangular appearance, 4–10 mm long, 4–5 mm wide, sessile or shortly petiolate. Calyx ± hirsute, tube 1.5–2 mm long; lips both broad, nearly equal in length. Corolla purple to mauve, 7–10 mm long. Straggling, scarcely aromatic shrub up to 1 m high. Coastal in Sydney area. Ss. Open forests. Fl. spring (listed as Extinct but collections of this species have recently been made). . . . . . . **P. marifolia** R.Br.
23 Branches with antrorse, ± appressed hairs. Leaves often of varying shape narrow-elliptic to elliptic or narrow-obovate often appearing linear; 8–16 mm long, 1–6 mm wide, petiole up to 3 mm long. Calyx ± hirsute, tube 2–2.8 mm long, lips nearly equal in length. Corolla white to pale mauve, 8–12 mm long. Decumbent spreading shrub to 30 cm high. Somersby Plateau, Mangrove Mtn. Damp areas on sandstone. Fl. spring–summer . . . . **P. junonis** B.J.Conn

24 Upper surface of leaves densely hairy and often densely glandular. Leaves ovate, leaf margins revolute, 6–15 mm long, 2–5 mm wide. Calyx 5 mm long. Corolla 8–10 mm long, purple , mauve or occasionally white, darker markings on lower lip. Branches with short curled hairs. Spreading shrub to 1 m high, not aromatic. Heath and DSF in shallow soils. Fl. spring–summer. . . . **P. granitica** Maiden & Betche
24 Upper surface of leaves glabrous or nearly so 25

25 Branches with stiff, spreading, sometimes gland tipped hairs. Leaves glabrous or with a few stiff spreading hairs, orbicular to nearly rhomboidal, sometimes almost bullate, 4–6 mm long, entire. Calyx c. 4 mm long, shortly pubescent and very glandular; lips nearly equal in length; the upper one broader than the lower. Divaricate, strongly aromatic shrub up to 2 m high. Nepean River district. Open forests and margins of RF. Fl. spring . . . . **P. rhombea** R.Br.
25 Branches with short appressed or curled hairs 26

26 Some hairs on the branches curled. Upper surface of leaves almost glabrous. Calyx scabrous-hispid; lips broad, the lower much smaller than the upper. Corolla deep violet, c. 10 mm long. Leaves lanceolate, obtuse, 3–5 mm long. Erect, slender, spreading, scabrous, aromatic shrub up to 2 m high. Lane Cove River and Glenorie-Maroota districts. Open forests. Sandy soils. Fl. spring . . **P. howelliae** Blakely
26 Hairs short and appressed none curled. Calyx with a raised pubescent line inside at the base of the upper lip. Leaves acute, rarely more than 12 mm long. Corolla deep lilac, 8–10 mm long. Erect to decumbent shrub up to 3 m high. Woronora Plateau; Penrith-Windsor district. DSF on laterites and clays. Fl. spring . . . . **P. scutellarioides** (R.Br.) Briq.

27 Leaves crenate-lobed, oblong in outline, c. 10 mm long and 3 mm wide. Anther appendages ± equal. Flowers solitary in the upper axils. Calyx c. 10 mm long; the upper lip ± broader but not much longer than the lower. Corolla lilac, c. 15 mm long. Strongly aromatic, ± viscid, divaricate shrub up to 2 m high. Capertee district; Hunter River Valley. DSF. Fl. spring–summer . . . . **P. cryptandroides** A.Cunn. ex Benth. ssp. **cryptandroides**

27 Leaves entire. 28

28 Anther appendage absent but each cell with a basal swelling. Leaves ovate to narrow ovate, 10–25 mm long, 4–8 mm wide, glabrous, often with maroon midrib and margins. Calyx maroon, tube 2.5–3 mm long. Corolla 10–14 mm long, mauve with darker markings in throat. Erect shrub scarcely aromatic, up to 1 m high. Ranges. DSF. Fl. spring . . . . **P. hindii** B.J.Conn

28 Anther appendage present. 29

29 Decumbent spreading shrub to 30 cm high. Hairs on branches in opposite decusate bands. Anther appendage not twice as long as anther cells . . . . **P. junonis** (see above)

29 Shrubs up to 2 m high, hoary with appressed hairs or almost glabrous. Anthers with one appendage c. twice as long as the anther cell; the other appendage short. 30

30 Bracteoles 1.6–3 mm long. Leaves linear to lanceolate, 5–9 mm long, up to 1 mm wide, folded inwards, base cuneate. Corolla white to mauve, 8–10 mm long. Erect to prostrate shrub to 2 m high, spindly with branches intertwining. Blue Mts Heath and DSF. Fl. spring–summer . . . . **P. saxicola** var. **bracteolata** J.H.Willis

30 Bracteoles 0.5–2 mm long. Leaves 1.5–6 mm wide. Margins rarely folded inwards 31

31 Leaves elliptic to linear, entire, rather thick, usually up to 8 mm long and c. 2 mm wide. Flowers few in the upper leaf axils. Calyx 4–5 mm long. Corolla white, 8–10 mm long. Slender shrub up to 1 m high. Coast and adjacent plateaus. Heath and DSF. Fl. spring . . . . **P. saxicola** R.Br. var. **saxicola**

31 Leaves elliptic, thick, up to 15 mm long and c. 4 mm wide. Calyx 5–7 mm long. Corolla white, 10–12 mm long, decumbent shrub up to 1 m high. Blue Mts, e.g. Narrow Neck. DSF and heath. Ss. Fl. spring–summer . . . . **P. saxicola** var. **montana** A.A.Ham.

16 Hemigenia R.Br.
40 species endemic Aust.; Qld, NSW, WA

Shrubs. Leaves in whorls of 3–4. Flowers axillary, solitary, with a pair of bracteoles under the calyx. Calyx 5-toothed, Corolla purplish or bluish, with a dilated throat; the upper lip erect, ± concave, emarginate or 2-lobed; the lower lip longer than the upper, spreading, 3-lobed, the middle lobe larger and often itself 2-lobed. Stamens 4 or 2 with 2 staminodes. Articles reticulate-rugose.

Key to the species

1 Leaves linear-terete, channeled on the upper surface, 8–15 mm long, c. 1 mm wide. Flowers in the axils of the upper leaves, sessile or on short peduncles; bracteoles linear, shorter than the calyx. Calyx 4–5 mm long; the teeth longer than the tube. Corolla 8–10 mm long. Erect, slender shrub up to 2 m high. Widespread. DSF and heath. Sandy soils. Fl. spring–summer . . . . **H. purpurea** R.Br.

1 Leaves oblong-cuneate, 10–15 mm long, 2–4 mm wide, contracted into a long petiole. Calyx 4–6 mm long; teeth shorter than the tube. Corolla c. 8 mm long. Erect shrub up to c. 2 m high, Woronora Plateau; Hornsby Plateau; northern Blue Mts DSF. Sandy soils. Fl. spring–summer. . . . . **H. cuneifolia** Benth.

17 Westringia Sm.
25 species endemic Aust.; all states and territories except NT

Shrubs. Leaves in whorls of 3–4, entire, spreading. Flowers axillary, solitary, sessile or nearly so. Calyx campanulate, 5-toothed. Corolla white to bluish with a short tube and a dilated throat; upper lip erect, flat, broadly 2-lobed; lower lip spreading, 3-lobed. Fertile stamens 2; anthers 1-celled; the 2 lower stamens reduced to staminodes. Style shortly 2-fid. Articles reticulate-rugose.

Key to the species

1 Leaves 2.5 mm wide or wider; usually in whorls of 4, rarely 3 or 5; glabrous on the upper surface; the margins recurved or revolute. Branches undersurface of the leaves and the calyx hoary-tomentose with appressed hairs. Corolla 14 mm long. Compact shrub 1–2 m high. Near the sea; extensively planted as an ornamental. Fl. most of the year . . . . **W. fruticosa** (Willd.) Druce

1 Leaves less than 2.5 mm wide mostly in whorls of 3 2

2 Leaves linear, mostly 12–35 mm long, 1–2 mm wide, flat or slightly recurved, sparsely hairy on lower surface. Calyx with a few scattered hairs; the lobes deltoid, c. 2 mm long. Corolla c. 8 mm long. Shrub up to 3 m high. Cumberland Plain; Woronora Plateau; Blue Mts DSF, often near creeks. Fl. most of the year. . . . . **W. longifolia** R.Br.

2 Leaves linear to narrow-linear, 8–25 mm long, 1–2 mm wide, revolute, moderately to densely hairy on lower surface. Calyx densely hairy; the lobes deltoid, c. 2 mm long. Corolla 6–9 mm long. Shrubs 1–2 m high. Widespread. Open forests often near creeks. Fl. most of the year . **W. eremicola** A.Cunn. ex Benth.

18 Teucrium L.
Germander
13 species endemic Aust.; all states and territories

Herbs or shrubs with 4-angled stems. Simple hairs and/or sessile glands present. Leaves opposite, simple or 3-foliolate; margins toothed or entire. Calyx almost equally 5-toothed. Corolla white, 1- sided, 5-lobed,; the upper 4 lobes nearly equal; the middle and lower one twice as long. Stamens 4, in pairs, exserted. Style 2-fid. Fruit keeled or smooth.

Key to the species

1 Leaf bases truncate. Inflorescence unbranched. Leaves broad ovate to triangular, toothed or crenate, 1–5 cm long. Flowers 1 per axil. Calyx 4–7 mm long, corolla pink to purple, 8–10 mm long. Ovary glabrous. Erect to scrambling herb to 0.5 m high. Pasture & DSF. Fl. summer–autumn . . . . . **T. argutum** R.Br.

1 Leaf bases tapering. Inflorescence branched. Leaves thin, ovate to ovate-lanceolate, toothed or lobed, 2–11 cm long, petiolate. Flowers in loose axillary pedunculate cymes which grade into a narrow terminal panicle, up to 10 flowers per axil. Calyx 3–5 mm long. Corolla white, 8–12 mm long. Ovary hairy. Erect, perennial herb up to 1 m high. Widespread. Open forests in sheltered situations. Fl. summer. . . . . . .
 . . . . **T. corymbosum** R.Br.

19 Ajuga L.
2 species in Aust. (1 native, 1 naturalized); Qld, NSW, Vic., Tas., S.A.

Perennial or annual herbs. Leaves decussate, simple entire, toothed or lobed. Inflorescence leafy, thryse-like . Calyx 5-lobed, lobes equal. Corolla 2-lipped, the upper lip much reduced. Stamens 4, style gynobasic stigma shortly bifid.

Key to the species

1 Erect or ascending, perennial herb 4–30 cm high. Leaves basal and cauline, variously hairy; lower leaves obovate to oblong, contracted into a long petiole, bluntly and distantly toothed, 4–12 cm long, passing gradually into small sessile entire upper floral leaves. Flowers in false whorls of 6–20. Calyx 5-toothed, 4–7 mm long. Corolla blue or purple, variable in size up to 20 mm long; lower lip long and spreading, 3-lobed, the middle lobe larger and emarginate. Stamens 4, in pairs, exserted. Articles rugose. Widespread. Open forests. Mostly on shale. Fl. most of the year. *Australian Bugle.* . . . .
 . . . . **A. australis** R.Br.

1 Procumbent perennial herb. Leaves glabrescent, 1.5–5 cm long, 0.7–3 cm wide, braod-elliptic to oblong, margins undulate or crenate. Petiole 1–2 cm, long winged. Flowering stem densely hairy on 2 sides. Floral bracts similar to leaves but sessile and tinged with blue. Flowers are in whorls of c. 6. Calyx 4–5 mm long, purplish, acute with long white hairs. Corolla 13–17 mm long, blue with darker veins, the tube is longer than the calyx. The upper lip of the corolla is emarginate. Filaments are blue and glandular. Upper Blue Mts . Introd. from New Zealand. Fl. most of the year. *Bugle* . . . . ***A. reptans** L.

20 Clerodendrum L.
9 species in Aust. (7 endemic); Qld, NSW, NT, WA

One species in the area

Tall shrub or small tree. Leaves and inflorescences pubescent; older leaves sometimes glabrescent. Leaves opposite, ovate-elliptic to almost lanceolate, entire, 5–11 cm long, on rather long petioles. Flowers numerous in compact terminal corymbs or rarely in loose axillary cymes. Calyx campanulate, 5-lobed, c. 6 mm long but enlarging in the fruiting stage to c. 20 mm diam. and becoming bright red. Corolla tube c. 20 mm long; lobes 6–8 mm long, spreading, creamy-white. Stamens protruding from the corolla by as much as 2 cm, later rolled back. Fruit drupaceous, black, glossy. Coast. WSF; RF margins; often persisting in cleared areas. Fl. spring–summer . . . . **C. tomentosum** R.Br.

21 Gmelina L.
5 species in Aust. (2 endemic); Qld, NSW, NT

One species in the area

Tree, usually tall. Leaves opposite, entire, ovate, 7–14 cm long, glabrous on the upper surface, tomentose and with raised veins underneath. Flowers in a pyramidal terminal panicle. Calyx truncate, tomentose, enlarging under the fruit. Corolla white with purple markings; tube dilated, 8–10 mm long; lobes 5, spreading, shorter than the tube. Ovary 4-locular. Fruit nearly globular, drupaceous, blue, c. 25 mm diam. Coast, e.g. Minnamurra Falls, head of Patonga Creek. In or near RF. Fl. spring–summer. *White Beech* . . . . **G. leichhardtii** (F.Muell.) Benth.

22 Chloanthes R.Br.
4 species endemic Aust.; Qld, NSW, WA

Shrubs up to 1m high but usually less and sometimes straggling. Leaves opposite, decurrent, narrow-oblong to lanceolate, softly woolly underneath. Flowers irregular, solitary, axillary, almost sessile. Calyx deeply 5-lobed. Corolla bluish to greenish yellow, 2-lipped. Stamens 4. Ovary superior; placentas parietal, intruding into the loculus. Fruit a schizocarp with 2 mericarps.

Key to the species

1 Stamens and style shorter than the corolla tube. Corolla tube mauve or bluish yellow, with darker markings in the throat and long hairs on the lower lip and into the throat, usually less than 2.5 cm long. Leaves mostly 1–3 cm long and 2–4 mm wide, bullate, with deciduous stellate hairs. Shrub up to 1 m high. Rylestone-Singleton district; (also south of Nowra). DSF and heath. Fl. mostly spring. . . . .
. . . . **C. parviflora** Walp.

1 Stamens and style longer than the corolla tube. Corolla greenish yellow or greenish blue, usually more than 2 cm long, glabrous or nearly so on the lower lip and in the throat but with a ring of hairs near the base of the tube **2**

2 Leaves recurved, often only slightly so, lanceolate to linear-lanceolate, shortly hispid below, usually 5–7 cm long, 3–10 mm wide, with yellow-glandular hairs scattered over the surface as well as scabrid hairs. Shrub up to 1 m high. Central Blue Mts; also recorded for Port Jackson. WSF. Fl. spring–summer . . . .
. . . . **C. glandulosa** R.Br.

2 Leaves revolute, linear-lanceolate to terete with the midrib not visible underneath, usually 3–4 cm long and 3–4 mm wide, glandular hairs absent or infrequent, woolly villous underneath. Straggling shrub up to 1 m high. Widespread. DSF and heath. Fl. mostly winter . . . . **C. stoechadis** R.Br.

23 Spartothamnella Briq.

3 species endemic Aust.; all states and territories except Vic.

One species in the area

Glabrous, rigid or scrambling shrub or herb 1–3 m high with slender green 4-angled branches. Leaves opposite, reduced to small scales on mature stems; linear on juvenile stems. Flowers c. 3 mm diam., in short axillary cymes. Calyx divided almost to the base into 5 acute lobes, persistent and reflexed in fruiting stage. Corolla white. Fruit globular, drupaceous, orange, c. 3 mm diam., on a filiform reflexed pedicel. Coast and Cumberland Plain. WSF; margins of RF; coastal swamp forests. Fl. mostly summer . . . . **S. juncea** (A.Cunn. ex Walp.) Briq.

24 Physostegia Benth.

1 species naturalized Aust.; NSW

One species in the area

Perennial, rhizomatous, herb up to 1 m high. Leaves lanceolate, up to 12 cm long and 4 cm wide, sessile, margins toothed. Inflorescence a spike up to 25 cm long. Flowers white, lavender or purple. Corolla c. 3 cm long, 2 lipped. Upper lip hooded lower lip 3-lobed. Stamens 4 with purple anthers. Occasionally naturalized. Coastal. Introd. from N. America. *Obedient Plant* . . . . ***P. virginiana** Benth.

25 Orthosiphon Benth.

1 species native Aust.; Qld, NSW

One species in the area

Slender shrub to 50 cm high. Leaves, narrow-ovate to rhombic, 3–16 cm long, 2–5 cm wide, hairy and with sessile glands, margins toothed, petiole 1–6 cm long. Corolla 10–16 mm long, white to lilac, upper lip slightly 4-lobed and recurved. Stamens exserted. Style 5–6 cm long, stigma 2-fid. Gosford area. RF margins. Fl. autumn-winter. . . . . **O. aristatus** (Blume) Miq.

124 LENTIBULARIACEAE

3 gen., cosmop.

1 Utricularia L.

Bladderworts

c. 60 species in AUST, (42 endemic); all states and territories

Annual or perennial herbs, aquatic swamp or bog plants. Leaves alternate or basal, frequently dimorphic; submerged or subterranean leaves usually finely divided, bearing bladders which trap small aquatic animals; emerged basal leaves entire, sometimes absent at flowering time. Flowers solitary or in racemes, bisexual, irregular. Calyx 2-lipped. Corolla tubular, spurred at the base, 5-lobed, 2-lipped; upper lip erect, broad, 2-lobed or almost entire; lower lip 2–3-lobed or almost entire, with a convex palate almost closing the throat. Stamens 2, sometimes with 2 staminodes, epipetalous. Ovary superior, 1-locular; style 2-fid; placentas free-central. Fruit a globular capsule opening by 2 valves.

Key to the species

1 Floating aquatic plants. Flowers on axillary peduncles. Corolla yellow. Leaves submerged, divided into capillary segments interspersed with bladder-traps 2
1 Swamp plants, submerged or exposed or growing on wet ground. Flowers on erect scapes with or without scale leaves. Corolla blue to purple or white. Leaves basal, linear to linear-spathulate or ovate, often minute, usually absent at flowering time 3

2 Leaf segments few (up to c. 8). Stems very fine and often interwoven. Peduncles filiform, 2–20 cm long, 1–6-flowered. Calyx lobes 1–2 mm long when in flower. Corolla 4–6 mm across; spur almost as long as or longer than the lower lip. Coast and adjacent plateaus; Hunter River Valley. Slow moving water, ponds. Fl. spring–summer . . . . **U. gibba** L.

2 Leaf segments numerous (more than 15). Stems not interwoven. Corolla 12–20 mm long; spur shorter than the lower lip. Peduncle up to 30 cm long, 4–10-flowered. Calyx lobes 3–4 mm long. Coast and adjacent plateaus; Hunter River Valley; Cumberland Plain. Slow moving water, ponds. Fl. summer . . . . **U. australis** R.Br.

3 Upper corolla lip deeply 2-lobed, each lobe 3.5–4 mm long, white with lilac tinges. Spur curved forward, projecting passed lower corolla lobe. Leaves few, obovate, 10 mm long, 2–3 mm wide. Peduncle to 7 cm high, up to 4-flowered. Calyx lobes c. 2.3 mm long. Blue Mts Wet areas on sandstone walls. Introd. from S. Africa. Fl. summer . . . . *U. sandersonii** Oliv.
3 Upper corolla lip not deeply 2-lobed **4**

4 Corolla lips both 2-lobed; spur longer than the lower lip; whole corolla 6–8 mm long, dark blue. Flowers on slender pedicels in a simple or slightly branched raceme 15–30 cm high. Tending to blacken on drying. Coast and adjacent plateaus. Wet ground. Fl. summer . . . . **U. biloba** R.Br.
4 Corolla lips entire or obscurely lobed **5**

5 Scale leaves present on the scape below the lowest flower **6**
5 No scale leaves present on the scape below the lowest flower **7**

6 Corolla pale blue or whitish; upper lip obovate, shorter than the calyx or scarcely exceeding it, 3–30 mm long; lower lip somewhat longer, broader, very convex. Basal leaves usually pale green, linear to spathulate, 6–12 mm long. Scapes 5–8 cm high, rarely higher. Pedicels c. 4 mm long. Calyx segments acute. Coast and adjacent plateaus. Wet ground. Fl. spring–autumn . . . . **U. uliginosa** Vahl
6 Corolla purple; upper lip oblong to linear, slightly exceeding the calyx, c. 2 mm long; lower lip broad, 6–8 mm across. Leaves oblanceolate, up to 5 cm long. Scapes up to 25 cm high. Flowers almost sessile. Calyx segments obtuse. Widespread. Wet ground. Fl. most of the year . . . . **U. lateriflora** R.Br.

7 Corolla spur shorter than the lower lip. Flowers on slender pedicels in pairs or in whorls of 3–4 or solitary and terminal; scapes 10–30 cm high. Corolla purple to lilac or blue; upper lip small, broad-ovate; lower lip horizontal, broad-semicircular, 12–20 mm across. Widespread. Wet ground. Fl. most of the year. *Fairy Aprons* . . . . **U. dichotoma** Labill.
7 Corolla spur as long as the lower lip. Flowers solitary and terminal or the scape 2-flowered. Scapes up to 20 cm high. Corolla mauve to lilac; upper lip small, ovate; lower lip semicircular, c. 20 mm across. Widespread. Wet ground. Fl. spring–summer . . . . **U. uniflora** R.Br.

125 MARTYNIACEAE

Annual or perennial herbs covered with sticky glandular hairs. Leaves opposite, margins entire or toothed occasionally lobed, stipule absent. Inflorescence axillary or terminal racemes. Flowers irregular, bisexual 5-merous. Sepals 2, free or fused and split on one side. Petals tubular expanding towards the apex, 5-lobed and often 2-lipped. Stamens epipetalous fertile and staminodes present. Ovary superior, style slender, with 2 stigmatic flaps. Fruit a woody capsule with a fleshy outer layer, dehiscing longitudinally. Seeds many, compressed, black. 5 gen., cosmop.

Key to the species

1 Sepals fused . . . . **1 PROBOSCIDEA**
1 Sepals free . . . . **2 IBICELLA**

1 **Proboscidea** Zucc.

2 species naturalized Aust.; NSW, Vic., S.A., WA

One species in the area

Annual or perennial herb. Leaves opposite, reniform 10–16 cm long, 14–25 cm wide, margins entire to sinuate, petiole c. as long as lamina. cm diam. Inflorescence a terminal raceme 6–20 cm long. Calyx 10–20 mm long, cup-shaped, 5-lobed, split on lower side to base. Corolla 2–5 cm long, cream to reddish-purple, tubular, 2-lipped, unequally 5-lobed. Capsule 8–10 cm long, 1–2 cm wide, horns 10–25 cm long. Introd. from America. Noxious in some areas. Fl. summer–autumn . . . . *P. louisianica** (Mill.) Thell.

2 Ibicella (Stapf) Van Eselt.

1 species naturalized Aust.; NSW, Qld.

One species in the area

Annual her to 30 cm high. Leaves opposite, reniform 7–12 cm long, 7–19 cm wide, margins dentate, petiole 10–15 cm long. Inflorescences terminal, few-flowered. Calyx 10–20 mm long, lower 2 sepal broader than upper 3. Corolla spiny, c. 3 cm long, tubular, unequally 5-lobed, yellow with purple spots. Capsule 5–15 cm long, 2–3 cm wide, horns 5–12 cm long. Introd. from S. America. Noxious in some areas. Fl. summer–autumn . . . . *I. lutea (Lind.) Van. Eselt

126 OLEACEAE

Shrubs or trees. Leaves opposite, simple, entire, without stipules. Inflorescence racemose or cymose. Flowers regular, usually bisexual. Calyx tubular, 4-lobed. Corolla 4-lobed, connate into a short tube or almost free. Stamens 2, epipetalous. Ovary superior, 2-locular. Fruit drupaceous or a berry. 24 gen., cosmop.

Key to the genera

1 Corolla tube divided almost to the base into 4 petals which may be connected in pairs by the filaments
. . . . .3 NOTELAEA
1 Corolla with a distinct short tube 2

2 Flowers in axillary panicles shorter than the leaves . . . . 1 OLEA
2 Flowers in terminal leafless panicles . . . . 2 LIGUSTRUM

1 Olea L.

2 species in Aust. (1 native, 1 naturalized); Qld, NSW, Vic., S.A.

One species in the area

Small, evergreen tree, glabrous except the young tips which are scurfy. Branchlets drooping. Leaves lanceolate, 5–10 cm long, dark green and shining on the upper surface, paler underneath, acute, with a small recurved point. Flowers in axillary panicles shorter than the leaves. Calyx with 4 very short teeth. Corolla with 4 reflexed lobes 2–3 mm long; tube much shorter. Style short; stigma 2-lobed, almost capitate. Fruit globular, black, drupaceous, c. 6 mm diam. Coast and Cumberland Plain. Waste places. Fl. spring. *African Olive* . . . . *O. europaea ssp. cuspidata (Wall. ex Don) Cif.

2 Ligustrum L.

4 species in Aust. (1 endemic, 3 naturalized); Qld, NSW, S.A., Tas.

Tall, much branched shrubs or small trees, glabrous or the young branches pubescent. Flowers in terminal panicles, white, small, numerous. Calyx small, 4-toothed. Corolla funnel-shaped with a short tube and 4 spreading lobes. Style short. Fruit drupaceous. Seed solitary.

Key to the species

1 Quite glabrous. Leaves ovate to elliptic or narrow-ovate, up to 12 cm long. Tall shrubs up to 4 m high. Coast and adjacent plateaus. Naturalized in gullies. Introd. from E. Asia. Fl. summer. *Large-leaved Privet* . . . . *L. lucidum Aiton
1 Young branches pubescent 2

2 Leaves glabrous, lanceolate to elliptic, 2.5–6 cm long; petioles 1–3 mm long. Shrub up to 5 m high. Robertson district. Garden escape. Introd. from Europe. Fl. spring. *European Privet* . . . *L. vulgare L.
2 Leaves hairy at least underneath on the midrib, elliptic to narrow-ovate, up to 7 cm long. Tall shrub up to 3 m high. Coast and adjacent plateaus. Naturalized in gullies. Introd. from E. Asia. Fl. spring. *Small-leaved Privet* . . . . *L. sinense Lour.

3 Notelaea Vent.

9 species endemic Aust.; Qld, NSW, Vic., Tas.

Shrubs or small trees. Leaves coriaceous. Flowers small in short simple axillary racemes sometimes reduced to sessile clusters. Calyx small, 4-toothed. Petals 4, yellowish, small, connected in pairs by the filaments, induplicate-valvate in the bud. Ovary 2-locular with 2 ovules per loculus; style short. Fruit drupaceous. Seed solitary.

Key to the species

1 Venation of the leaf obscure on both upper and lower surfaces, only the primary veins distinctly visible. Leaves linear to narrow-elliptic or lanceolate, up to 10 cm long, usually 7–10 times as long as wide, ± pubescent when young, entire; the base acute. Corolla c. 2 mm long. Fruit c. 5 mm diam. Tall shrubs or small trees. Western Blue Mts DSF. Fl. spring . . . . **N. neglecta** P.S.Green
1 Venation of the leaf prominent on the upper surface, the reticulation clearly visible 2

2 Venation on the undersurface less prominent than on the upper surface, unevenly reticulate. Leaves lanceolate to linear, 2–12 cm long, entire, glabrous or glabrescent; the primary veins usually parallel to the margins. Small tree up to 10 m high, with a ± tessellated bark. Singleton district. Open forest. Fl. spring–summer. . . . . **N. microcarpa** R.Br. var. **microcarpa**
2 Venation on the undersurface as prominent as that on the upper surface. 3

3 Leaves mostly ovate, cordate to obtuse or tapering abruptly at the base into a short petiole, 3–12 cm long, crenulate; veins progressively smaller with each division; the primary veins not curving towards the apex. Inflorescences usually inserted above the axils. Fruit purple-black, c. 1.5 mm diam. Glabrous or pubescent shrub or small tree up to 8 m high, with flaky greyish bark. Widespread. Open forest or margins of RF. Fl. summer . . . . **N. ovata** R.Br.
3 Leaves narrow-ovate to linear, tapering gradually towards the petiole (at least in the Sydney region). Inflorescence inserted in the axil of leaf. 4

4 Venation unevenly reticulate since the veins become smaller with each division. Leaves narrow-ovate to lanceolate, 6–10 cm long, usually entire, velutinous to glabrous; the primary veins curving towards the apex and becoming divided. Shrub or small tree up to 8 m high, with a finely fissured greyish bark. Widespread. Usually in or near RF. Fl. winter–spring. *Mock Olive* . . . . **N. longifolia** Vent.
4 Venation evenly reticulate since the veins do not decrease in size after the primaries. Leaves narrow-ovate to linear, 6–14 cm long, glabrous or glabrescent, entire or obscurely crenulate. Shrub or small tree up to 8 m high, with a flaky greyish bark. Widespread. In or near RF. Fl. spring–summer. . . . .
. . . . **N. venosa** F. Muell.

127 PLANTAGINACEAE

3 gen., cosmop., particularly the Mediterranean and the Andes

1 Plantago L.

33 species in Aust.; all states and territories

Annual or perennial herbs. Leaves usually all basal (cauline in *P. scabra*); main veins parallel. Inflorescence a spike on a leafless scape. Flowers small, numerous, each subtended by a bract, regular, bisexual, protogynous, wind pollinated. Sepals 4. Corolla tubular, 4-lobed, green white or purplish. Stamens 4, epipetalous. Ovary superior, 2-locular or 4-locular by a false-septum. Fruit a circumscissile capsule enclosed in the withered corolla.

Key to the species

1 Anterior sepals connate. Bracts ovate-acuminate, longer than the sepals. Spikes ovoid to cylindrical, 1–7 cm long. Scapes usually much longer than the leaves, angular, furrowed. Leaves usually 5-veined, elliptic to narrow-ovate, up to 30 cm long. Widespread. Common weed of cultivation and waste places.

Introd. from Europe; Asia. Fl. spring–autumn. *Ribwort, Rib-Grass, Lamb's Tongue* or *Plantain.* . . . . .
. . . . ***P. lanceolata** L.
1 Anterior sepals free 2

2 Leaves opposite on a ± elongated stem, glandular hairy. Spike globular or shortly cylindrical. Seeds 2. Recorded once from the Sydney area. Introd. from Asia. Fl. summer. *Sand Plantain* . .***P. scabra** Moench
2 Leaves all basal 3

3 Ovules 6 or more per ovary. Leaves broad-ovate, 5–9-veined, 5–20 cm long. Scapes shorter or longer than the leaves. Spikes cylindrical, dense except at the base, 5–20 cm long. Bracts as long as the sepals or nearly so, ovate, obtuse. Capsule usually with 8–16 small seeds. Widespread. Weed of cultivation and waste places. Introd. from Europe and W. Asia. Fl. spring–summer. *Large Plantain* . . . . .***P. major** L.
3 Ovules 2–5 per ovary . . . . , ... 4

4 Corolla tube hairy outside. Leaves 1–3-veined, linear to narrow-ovate, deeply 1–2-pinnatifid with narrow lobes, 2–15 cm long. Capsule with 4 seeds, ± pubescent annual or biennial herbs 5
4 Corolla glabrous outside. 6

5 Scapes 0.5–1.8 mm diam., slender, mostly longer than the leaves. Spikes dense; the flowers ± spreading. Bracts longer than the sepals, acuminate. Widespread. Waste land. Introd. from Europe and W. Asia. Fl. spring–summer. *Buckshorn Plantain.* . . . .***P. coronopus** L.ssp. **coronopus**
5 Scapes 1.3–2.3 mm diam., rarely less, stout, shorter than the leaves or equalling them. Spikes very dense; the flowers appressed. Bracts mostly longer than or as long as the sepals, acute or obtuse. Coast. Waste land. Introd. from the Mediterranean. Fl. spring–summer . . . .
. . . . ***P. coronopus** ssp. **commutata** (Guss.) Pilger

6 Corolla lobes erect, stiff even after the flower has opened. Ovules 3 per ovary. Leaves elliptic-oblanceolate, pubescent, with 3 veins, entire or dentate. Scapes up to 40 cm long. Spikes 4–25 cm long. Seeds 3, greenish. Coast; Cumberland Plain. Open forests, grasslands and waste places. Introd. from S. America. Fl. spring–summer . . . . ***P. myosuros** Lam. ssp. **myosuros**
6 Corolla lobes spreading after the flower opens. Ovules 4–5. 7

7 Axillary hairs red-brown or golden-brown. Sepals 3–4 mm long 8
7 Axillary hairs whitish to yellow-brown. Sepals 1.5–3 mm long 9

8 Leaves 5–10 times as long as broad, narrow-elliptic, pubescent with soft hairs up to 1 mm long, up to 20 cm long, usually dentate. Scapes up to 25 cm long. Spikes 3–11 cm long. Seeds up to 5. Blue Mts; Cumberland Plain. Open forests and grasslands. Fl. spring–summer . . . . **P. varia** R.Br.
8 Leaves 15 times as long as broad, linear to narrow-oblong, glabrous or with long hairs. Scapes as long as the leaves. Spikes 4–20 cm long. Cumberland Plain; lower Blue Mts Open forests and grasslands. Fl. spring–summer . . . . **P. gaudichaudii** Barnéoud

9 Capsule with a cylindrical truncate beak which is minutely lobed. Leaves basal, pubescent, narrow elliptic 4–10 cm long, 8–25 mm wide, margins irregularly toothed. Scapes 8–20 cm long. Spikes cylindrical, 3–11 cm long, flowers crowded. Sepals 2.8–3.5 mm long. Occasionally naturalized in seasonally moist sites. Introd. from western divisions of NSW Fl. spring–summer . . . .
. . . . ***P. turrifera** B.G.Briggs, Carolin & Pulley
9 Capsule not as above. 10

10 Leaves mostly thin, oblanceolate to obovate rarely narrower, 3–15 cm long, pubescent, dentate to entire. Scapes up to 40 cm long. Spikes up to 10 cm long, with the flowers distant at the base. Sepals 1.5–2 mm rarely 3 mm long. Widespread. Open forests and grasslands. Fl. spring–summer . . . .
. . . . **P. debilis** R.Br.
10 Leaves narrow-oblong to oblanceolate, 2–16 cm long, usually pubescent but coastal forms often glabrous and thick. Scapes up to 30 cm long. Spikes up to 10 cm long. Sepals 2–3 mm long. Widespread. Open forests, often amongst rocks and frequently near the sea. Fl. spring–summer . . **P. hispida** R.Br.

128 BUDDLEJACEAE

7 gen. trop. to sub trop., temp.

1 Buddleja Houst. ex L.

6 species naturalized Aust.; Qld, NSW, Vic.

Scrambling shrub with densely tomentose young branches. Leaves opposite, margins entire, crenate or toothed. Stipules interpetiolar. Flowers numerous in terminal or axillary, panicles or spike-like. Sepals and petals connate, 4-merous. Stamens inserted on corolla tube. Fruit a capsule or berry. Seed numerous often winged.

Key to the species

1 Leaves narrow-ovate to elliptic, acuminate, 10–12 cm long, dark green on the upper surface, whitish- or yellowish-tomentose underneath, margins entire; petiole 15–25 mm long. Corolla yellow. Anthers inserted just below corolla mouth. Coast. Garden escape. Introd. from Madagascar . . . . *B. madagascariensis Lam.

1 Leaves narrow-ovate to oblong, acuminate; 4–20 cm long, upper surface dark green, white-tomentose underneath, margins crenate to minutely toothed; petiole 2–5 mm long. Corolla white to lilac. Anthers inserted at or below the middle of the corolla tube. Blue Mts Garden escape. Introd. from China and Asia. . . . . *B. davidii Franch.

129 CALLITRICHACEAE

1 gen., cosmop.

1 Callitriche L.

9 species in Aust. (6 native, 3 naturalized); all states except NT

Small, glabrous, monoecious herbs growing in water or on mud. Leaves opposite, entire, less than 1cm long; stipules absent. Flowers axillary, subsessile, minute, unisexual, either solitary or male and female together. Calyx and corolla absent. Male flower with 1 stamen on a long filament. Female flower with a 4-locular ovary; styles 2. Fruit a 4-lobed schizocarp separating into 4 compressed 1-seeded mericarps.

Key to the species

1 Leaves rhomboidal, apiculate, usually with a tooth on either edge. Styles less than 1 mm long. Mature fruit broader than long, brown, with a pale wing one third the width of the mericarp. Widespread . . . . C. muelleri Sond.

1 Leaves spathulate or ovate, blunt, not toothed. Styles 2–3 mm long. Mature fruit pale, not broader than long; the wing less than one third the width of the mericarp. Widespread. Introd. from N. hemisphere . . . . *C. stagnalis Scop.

130 SELAGINACEAE

Herbs or undershrubs. Leaves linear, spirally arranged. Flowers arranged in dense spikes, ± irregular. Calyx deeply divided into 3–5-lobes. Corolla 4–5-lobed, usually oblique. Stamens 4, usually arranged in pairs, inserted near the top of the corolla tube. Ovary superior, 2-locular, with 1 pendulous ovule per loculus. Style simple. Fruit usually schizocarpic, separating into 2 mericarps. 8 gen., Africa to W. Aust. 3 gen. in Aust.

Key to the genera

1 Corolla 5-lobed, indistinctly 2-lipped, 2–3 mm long . . . . 1 SELAGO

1 Corolla with a long 4-lobed upper lip; the lower lip absent or a rudimentary tooth c. 1 cm long. . . . . .2 HEBENSTRETIA

1 Selago L.

2 species naturalized Aust.; NSW

One species in the area

Erect, slightly pubescent undershrub up to 70 cm high. Flowers in numerous compact spikes at the apex of the branches giving a corymbose appearance. Leaves linear, 3–5 mm long, fasciculate. Corolla 5-lobed, slightly 2-lipped, white. Camden district. Grasslands. Introd. from S. Africa. Fl. winter. . *S. corymbosa L.

2 Hebenstretia L.

1 species naturalized Aust.; NSW

One species in the area

Erect, minutely pubescent undershrub up to 80 cm high. Flowers arranged in ± compact long spikes; spikes 1–few on each branch, not corymbose. Leaves linear, up to 15 mm long, sometimes fasciculate towards the top. Corolla with a long 4-lobed upper lip; the lower lip represented by a rudimentary tooth or absent, whitish with a dark orange throat. Coastal sand dunes near Newcastle. Introd. from Africa. Fl. most of the year. . . . . *H. dentata L.

131 MYOPORACEAE

Shrubs to small trees, often resinous on the surface. Leaves usually alternate, simple, usually sessile; stipules absent. Flowers irregular, usually bisexual. Sepals usually 5, connate. Corolla tubular, usually 5-lobed, usually 2-lipped. Stamens 4 (rarely 5), epipetalous. Ovary superior, 2-locular, with up to 4 ovules per loculus. Fruit usually drupaceous. 5 gen., trop. to warm temp.

Key to the genera

1 Calyx segments linear to oblong, 6–8 mm long . . . . 1 EREMOPHILA
1 Calyx segments narrow-triangular, acute, up to 4 mm long . . . . 2 MYOPORUM

1 Eremophila R.Br.

206 species in Aust. (205 endemic, 1 native); all states and territories

Shrubs or small trees. Leaves alternate, opposite or whorled, entire or toothed, sessile or petiolate. Flowers in clusters in leaf axils. Sepals free or fused at base. Corolla 2-lipped upper lip with 2–4 lobes, lower lip with 1–3 lobes. Stamens 4 or 5. Ovary 2 or 4-locular. Fruit woody or drupaceous.

Key to the species

1 Stamens 4. Prostrate to ascending shrub with a thick rootstock. Leaves elliptic to lanceolate, mostly 3–5 cm long, entire or with a few teeth. Flowers 1–2 together on short axillary pedicels. Corolla white to mauve, without spots, 4–10 mm long. Fruit slightly compressed, 6–8 mm long, reddish pink. Coast and adjacent plateaus; Cumberland Plain. Open forests; grazing land. Fl. spring–summer. *Winter Apple* or *Amulla* . . . . **E. debilis** (Andrews) Chinnock
1 Stamens 5. Shrub to 4 m high. Leaves linear to oblanceolate, 2–5 cm long, entire. Flowers 1–4 in leaf axils. Corolla cream, 4–7 mm long. Fruit ovoid 4–6 mm long, yellow to black. Not common. In west of area. Fl. winter–summer. *Turkey Bush* . . . . **E. desertii** (A.Cunn. ex Benth.) Chinnock

2 Myoporum Sol. ex G.Forst.

17 species in Aust. (16 endemic); all states and territories

Shrubs or small trees. Leave simple, alternate. Flowers axillary, bisexual, nearly regular. Sepals 5, usually basally connate. Corolla tubular, almost campanulate; lobes 4–6, spreading, nearly equal. Stamens 4 (rarely 5), adnate to the base of the corolla. Ovary superior, 2–5-locular, with 1 ovule per loculus. Fruit drupaceous.

Key to the species

1 Fruit dry and compressed, 2-locular 2

1 Fruit fleshy, ovoid, 3–5-locular 3

2 Leaves 2–11 cm long, 1–2 mm wide, linear to terete, pendent, serrate, concolorous. Corolla white to mauve, without spots, with prominent veins; lobes ovate, 1–1.5 mm wide. Flowers up to 5 in axillary clusters, pedicels 1–3 mm long. Stamens usually longer than the corolla lobes but sometimes only as long. Fruit compressed, obtuse, 2–3 mm long. Shrub up to 1.5 m high, with an unpleasant odour. Nepean River; Jenolan Caves; coast in south of the area. Fl. spring–summer . . . .
. . . . **M. floribundum** A.Cunn. ex Benth.

2 Leaves 5–15 cm long, 3–19 wide, linear-elliptic, finely toothed, tuberculate, discolorous. Flowers 3–10 in axillary clusters, pedicels 2–3 mm long. Corolla 4–6 mm diam., white to pale purple without spots. Stamens usually longer than the corolla lobes. Fruit compressed 2–3 mm long. Shrub to 4 m high, with a pleasant odour. South of area. Coastal ranges in DSF. Fl. spring–summer . . . . . **M. bateae** F.Muell.

3 Ovary and fruit deeply wrinkled. Leaves mostly 10 mm wide or wider, usually serrate. Flowers usually 6 or more in axillary clusters, pedicels 4–10 mm long. Corolla 6–10 mm diam., white, spotted with purple, pubescent on the inside of the lobes. Fruit, globular, purplish or bluish, 5–6 mm diam. Shrub up to 5 m high, with corky bark. Coast. Often near tidal waters; RF; WSF. Fl. spring . . . . **M. acuminatum** R.Br.

3 Ovary not as above 4

4 Upper leaf margins usually toothed. Sepals with prominent membranous margins. Leaves lanceolate to broad-ovate, 3–9 mm long, 7–20 mm wide, acute, concolorous. Flowers 3–8 in axillary clusters, pedicels 5–9 mm long. Corolla 6–8 mm diam., white with purple spots. Fruit 4–9 mm diam. black or purple. Shrub to 6 m high with corky bark. (N.B. South coast species recorded from Villawood area, may be garden escape.) Fl. spring–summer. . . . . **M. insulare** R.Br.

4 Leaf margins entire. Sepals without prominent membranous margins 5

5 D Leaves obovate to elliptic, mostly 3–7 cm long and 15–30 mm wide, mucronate. Flowers axillary, 2 or more together but sometimes solitary, pedicels 10–15 mm long. Corolla 4–6 mm diam., whitish, spotted with purple inside, densely pubescent on the inside of the lobes. Fruit globular, purplish blue, c. 6 mm diam. Shrub up to 1.5 m high. Sea cliffs; coastal dunes; coastal swamp forests. Fl. spring–summer. . . .
. . . . **M. boninense** Koidz. ssp. **australe** Chinnock

5 Leaves mostly less than 10 mm wide. Fruit succulent and smooth or dry, ± compressed. Corolla white with purple spots, 5–6 mm diam., sprinkled with hairs inside the lobes or glabrous. Leaves entire, linear to lanceolate or oblanceolate, concolorous. Fruit globular to ovoid, 5–7 mm diam., purplish. Shrub up to 3 m high, with an unpleasant odour. Blue Mts Forests. Fl. spring–summer. *Water Bush* . . . .
. . . . **M. montanum** R.Br.

132 SCROPHULARIACEAE

Herbs or rarely shrubs. Leaves alternate or opposite, without stipules. Flowers irregular or almost regular, bisexual. Calyx persistent, with 4–5 lobes or teeth. Corolla usually tubular, occasionally rotate, 2-lipped or nearly regular; the limb with 4–5 lobes, imbricate in the bud. Stamens 4 (in pairs, 2 long and 2 short), rarely 5 or 2. Ovary superior, 2-locular; placentas axile; style solitary. Fruit a 2-locular capsule opening by valves or pores. 220–250 gen. cosmop.

Key to the genera

1 Parasites with reduced scale-like leaves, without chlorophyll. . . . . **19 OROBANCHE**

1 Leaves not reduced, green 2

2 Bracteoles present on the pedicels 3

2 Bracteoles absent 5

3 Fertile stamens 2, staminodes 2 or absent . . . . **8 GRATIOLA**

3 Fertile stamens 4 4

4 Leaves basal . . . . **9 MAZUS**
4 Leaves cauline. . . . . **12 BACOPA**

5 Sepals fused for less than half their length 6
5 Sepals fused for half their length or more 15

6 Stamens 2 7
6 Stamens 4 9

7 Shrubs not rooting at the nodes . . . . **14 DERWENTIA**
7 Herbs, sometimes woody towards the base and rooting at the nodes 8

8 Herbs, sometimes rooting at the nodes . . . . **16 VERONICA**
8 Plants woody towards the base and rooting at the nodes . . . . **15 PARAHEBE**

9 Leaves alternate, at least on the upper parts of the stem. 10
9 Leaves opposite, at least on the upper parts of the stem 14

10 Corolla without a spur 11
10 Corolla spurred 12

11 Corolla tubular, white to pink-purple with dark purplish spots. . . . .**13 DIGITALIS**
11 Corolla rotate, yellow. . . . .**1 VERBASCUM**

12 Leaves pubescent . . . . **4 KICKXIA**
12 Leaves glabrous 13

13 Leaves reniform with 5–9 rounded or triangular lobes . . . .**3 CYMBALARIA**
13 Leaves linear to elliptic, not lobed . . . . **5 LINARIA**

14 Corolla pouched at the base . . . .**2 MISOPATES**
14 Corolla not pouched at the base. . . . . **11 SCROPHULARIA**

15 Calyx smooth with 3–4 unequal lobes. Stoloniferous herb with clustered leaves . **6 GLOSSOSTIGMA**
15 Calyx ribbed or angled with 3, 4 or 5 ± equal teeth 16

16 Leaves entire on long slender petioles, basal, or occasionally opposite on new growth. Flowers solitary
   in leaf axils at base of plant, bracts absent . . . . **7 LIMOSELLA**
16 Leaves entire or variously toothed, sessile or shortly petiolate. Flowers 1 or 2 or in racemes in the axils
   of leaf-like bracts 17

17 Flowers 1 or 2 in bract axils, calyx 5-toothed. Undersurface of leaves sometimes with scattered sessile
   glands. . . . . **10 MIMULUS**
17 Flowers in racemes, calyx 4-toothed. Under surface of leaves with patches of sessile glands 18

18 Upper corolla lobes recurved. Seeds with longitudinal ridges. . . . .**17 EUPHRASIA**
18 Upper corolla lobes not recurved. Seeds smooth or finely reticulate . . . . **18 PARENTUCELLIA**

1 Verbascum L.

5 species naturalized Aust.; all states and territories except NT

Herbs with a basal rosette of leaves and erect stems with alternate cauline leaves. Flowers in terminal
racemes or panicles, often with several flowers in the axil of each bract. Calyx deeply 5-lobed. Corolla
yellow, sometimes with purplish markings, rotate, 4-lobed, almost regular. Stamens 4–5. Fruit a capsule.

Key to the species

1 Hairs on the filaments white. Corolla up to 30 mm diam., in a large dense raceme. Cauline leaves
   obovate to elliptic, up to 15 cm long; the margins decurrent, forming conspicuous wings on the stem.

The whole plant covered with a dense white stellate tomentum. Widespread. Waste places. Introd. from Europe. Fl. spring–summer. *Aaron's Rod* . . . .*V. thapsus L. ssp. **thapsus**

1 Hairs on the filaments purplish. Glabrous to sparsely hairy plants except for the glandular hairy inflorescence. Leaf margins decurrent only as lines or low ridges on the stems 2

2 Pedicels longer than the calyx. Flowers in a raceme, usually solitary in the axil of each bract. Cauline leaves ovate to lanceolate or elliptic. Widespread. Waste places. Introd. from Europe. Fl. spring–summer. *Moth Mullein* . . . .*V. blattaria L.

2 Pedicels shorter than the calyx. Flowers in a loose spike-like raceme, mostly several in the axil of each bract. Cauline leaves broad-ovate to ovate. Widespread. Waste places. Introd. from Europe. Fl. spring–summer. *Twiggy Mullein* . . . .*V. virgatum Stokes

2 Misopates Raf.
1 species naturalized Aust.; Qld, NSW, Vic., S.A., WA

One species in the area
Slender, erect annual 20–50 cm high, glandular hairy in the upper part. Leaves lanceolate to linear, 25–50 mm long; the lower ones opposite; the upper ones alternate. Flowers subsessile, solitary in the axils of the upper leaves in a leafy raceme, Calyx segments 5, linear. Corolla pink, pouched at the base, 10–15 mm long. Coast. Naturalized in moist situations. Introd. from the Mediterranean. Fl. spring–summer. *Lesser Snapdragon* . . . .*M. orontium (L.) Raf.

3 Cymbalaria Hill
1 species naturalized Aust.; Qld, NSW, Vic., Tas., S.A., WA

One species in the area
Trailing creeping, glabrous, perennial herb. Leaves alternate, often opposite towards the base of the plant, petiolate, slightly succulent, broad-reniform, irregularly 5–9-lobed or angled, 10–12 mm diam. Corolla lilac, with a yellow palate; tube 8–10 mm long, spurred. Widespread. Walls; weed in gardens. Introd. from Europe. Fl. spring–summer. *Ivy-leaved Toadflax* . . .*C. muralis G.Gaertn. B.Mey. & Scherb. ssp. **muralis**

4 Kickxia Dumort.
Toadflax
3 species naturalized Aust.; all states and territories except NT

Prostrate to procumbent, hairy herbs or sometimes climbing. Leaves alternate, ± hastate, broad-elliptic to broad-ovate, up to 35 mm long. Flowers solitary in the leaf axils, irregular. Corolla yellowish to purplish, spurred, 2-lipped. Stamens 4, didynamous. Fruit a ± globular capsule, 3–4 mm diam.

Key to the sub-species

1 Pedicels glabrous except just beneath the sepals. Corolla 8–10 mm long (including the spur). Picnic Point near Sydney. Weed. Introd. from Europe and Asia. Fl. summer. . . . .
. . . .*K. elatine (L.) Dumort. ssp. **elatine**

1 Pedicels hairy. Corolla c. 10 mm long (including the spur). Blue Mts Weed. Introd. from Europe and Asia. Fl. summer . . . . * K. elatine ssp. **crinita** (Mabille) Greuter

5 Linaria Mill.
Toadflax
6 species naturalized Aust.; Qld, NSW, Vic., WA

Annual or perennial herbs. Non-flowering stems with whorled leaves. Flowering stems with alternate entire sessile leaves. Flowers in terminal spikes or racemes. Corolla 2-lipped with the palate usually closing the throat, spurred. Stamens 4, didynamous. Capsule globular.

Key to the species

1 Corolla yellow, with an orange palate, 9–12 mm long; spur up to 12 mm long. Leaves on the flowering stems narrow-elliptic to linear, up to 6 cm long. Perennial herb up to 70 cm high. Coast. Waste places, uncommon. Introd. from Europe. Fl. spring–summer . . . . *L. vulgaris* Mill.

1 Corolla purple to violet, 5–10 mm long; spur up to 10 mm long 2

2 Inflorescence glandular hairy, plant otherwise glabrous. Leaves on the flowering stems linear, up to 45 mm long. Corolla with a yellowish palate scarcely closing the mouth. Annual herb up to c. 50 cm high. Coast. Naturalized in a few scattered localities. Introd. from Europe. Fl. spring. . . . .
. . . . *L. incarnata* (Vent.) Spreng.

2 Inflorescence glabrous. Leaves on the flowering stems linear, up to 40 mm long. Corolla with a whitish palate closing the mouth. Annual herb up to 70 cm high. Coast. Waste places; pastures. Introd. from the Mediterranean. Fl. spring. . . . .*L. pelisseriana* (L.) Mill.

6 Glossostigma Wight & Arn. ex Arn.
Mudmats
6 species in Aust. (3 endemic); all states and territories

Aquatic or terrestrial stoloniferous herbs. Leaves opposite or appearing clustered on suppressed internodes, petiolate, margins entire. Flowers solitary in leaf axils, bracteoles absent. Calyx unequally 3 or 4-lobed, smooth. Corolla 2-lipped, upper lip usually 2-lobed lower lip usually 3-lobed. Stamens 2 or 4, anthers 1-locular. Stigma with 2 flaps, 1 much smaller than the other. Fruit a capsule with 2-locules.

Key to the species

1 Plants glabrous. Leaves oblanceolate to oblong, to 20 mm long and 3 mm wide. Pedicels c. 0.5 mm long. Calyx 3-lobes. Corolla rudimentary. Stamens 2. Stigma lobe narrow-elliptic. Ephemeral herbs sometimes mat forming. Uncommon. Creek beds, swamp margins, river flats. Fl. spring . . . .
. . . . .G. cleistanthum W.R.Barker

1 Plants sparsely hairy. Leaves obovate to elliptic, 6–20 mm long up to 3 mm wide. Pedicels 1–12 mm long. Calyx 4-lobed. Corolla c. 3 mm long, blue. Stamens 4. Stigma lobe broad. Perennial herbs forming mats. Uncommon. River flats or swamps margins. Fl. summer-autumn . . . .
. . . . G. elatinoides (Benth.) Benth. ex Hook.

7 Limosella L.
3 species native Aust.; all states and territories except NT

One species in the area
Tufted, aquatic herb. Leaves in basal tufts, erect, 15–40 mm long; blade shorter than the petiole or reduced to the terete petiole alone. Flowers 2–4 mm long. Peduncles shorter than the leaves. Corolla purple or pink to white. Widespread. Wet places. Fl. spring–summer. *Mudwort* . . . . **L. australis** R.Br.

8 Gratiola L.
c. 5 species native Aust.; all states and territories except NT

Perennial herbs with a creeping rootstock; the stems ascending occasionally up to 30 cm high. Leaves opposite, sessile, ovate to narrow-lanceolate, 10–25 mm long, serrate to almost entire. Flowers axillary, solitary. Calyx segments 5, occasionally 3 or 4, linear-lanceolate, 6 mm long, with 2 linear bracteoles at the base. Corolla 10–15 mm long, white or pale-coloured, tubular, shortly 2-lipped; upper lip 2-lobed or notched; lower lip 3-lobed, spreading. Fertile stamens 2, enclosed in the corolla tube. Capsule broad-ovoid, 4-valved, nearly as long as the calyx.

Key to the species

1 Flowers on peduncles more than 10 mm long. Staminodes absent. Leaves lanceolate to elliptic, up to 10 mm wide, dentate. Sepals ± glandular pubescent. Widespread. Wet places. Fl. spring–summer. . . . . **G. pedunculata** R.Br.

1 Flowers on short peduncles less than 5 mm long. Staminodes present 2

2 Calyx glandular pubescent. Leaves lanceolate, up to 6 mm wide, distinctly dentate. Widespread. Wet places. Fl. spring–summer. . . . . **G. pubescens** R.Br.

2 Calyx glabrous. Leaves ovate to elliptic, mostly more than 8 mm wide, entire or undulate or sometimes ± dentate. Widespread. Wet places. Fl. spring–summer . . . . **G. peruviana** L.

9 Mazus Lour.

1 species endemic Aust.; Qld, NSW, Vic., Tas., S.A.

One species in the area

Perennial, rhizomatous herb up to 15 cm high. Leaves in basal rosettes, obovate, up to 55 mm long and 20 mm wide, toothed to entire. Flowers few on an erect scape from the basal rosette. Corolla blue to purple, yellow or white in the throat, 8–12 mm long. Capsule ellipsoid, c. 7 mm long, tardily dehiscent. Coast. Wet places. Fl. spring–summer . . . . **M. pumilio** R.Br.

10 Mimulus L.

6 species in Aust. (4 native, 2 naturalized); all states and territories

Decumbent to prostrate herbs often rooting at the nodes, slightly succulent. Leaves opposite, sessile or shortly stalked, elliptic to ovate. Flowers pedunculate, solitary in the upper axils. Calyx tubular, 5-toothed. Corolla 5-lobed, slightly 2-lipped, tubular towards the base. Stamens 4. Stigmas 2, flap-like, closing together when touched. Fruit a capsule surrounded by the persistent calyx.

Key to the species

1 Corolla bluish or violet with a yellowish tube; the lobes broad. Leaves sessile, elliptic to ovate, 2–6 mm long. Glabrous, slightly succulent, much branched, prostrate herb. Coast. Wet ground, e.g. margins of coastal lagoons. Fl. spring–summer . . . . **M. repens** R.Br.

1 Corolla yellow. Leaves ovate, 2–5 mm long. Viscid, pubescent, decumbent to ascending herb with stems up to 50 cm long. Widespread but not common. Moist places. Introd. from N. America. Fl. spring–summer. *Musk Monkey Flower* . . . . *M. moschatus Douglas ex Lindl.

11 Scrophularia L.

2 species naturalized Aust.; NSW, Tas.

One species in the area

Erect herb up to 1 m high, with square stems. Leaves opposite, ovate, 6–15 cm long, dentate, shortly petiolate or almost sessile. Flowers in glandular-pubescent panicle-like cymes. Corolla reddish-brown, c. 10 mm long. Stamens 4, with a broad staminode, Capsule 6–10 mm long. Naturalized at some localities in the Illawarra district. Introd. from Europe. Fl. summer. *Figwort* . . . . *S. nodosa L.

12 Bacopa Aubl.

3 species in Aust. (2 native, 1 naturalized); Qld, NSW, NT, WA

Small, prostrate or creeping perennial herb, much branched, rooting at the nodes. glabrous. Leaves opposite, margins entire to sinuate, almost sessile. Flowers solitary or in pairs in leaf axils. Sepals 5, 3 broad, 2 narrow. Corolla weakly 2-lipped, upper lip 2-lobed lower lip with 3 segments. Stamens 4. Stigma 2-lobed. Fruit a capsule surrounded by the persistent calyx.

Key to the species

1 Branchlets glabrous. Leaves rather thick, cuneate-obovate to oblong, up to 12 mm long, base attenuate. Flowers with 2 small bracteoles c. 5 mm long, under the calyx. Sepals ovate to lanceolate, 4–5 mm long. Corolla 4.5–9 mm long, white or pale blue; lobes spreading, broad. Capsule c. 4 mm long. Coast. Wet ground. Fl. spring–summer. . . . . **B. monnieri** (L.) Pennell

1 Branchlets sparsely to densely hairy. Leaves ovate, up to 12 mm long, base cordate. Flowers with 2 inconspicuous bracteoles c. 0.5 mm long under the calyx. Sepals 6–11 mm long. Corolla 9–11 mm long, blue or purple; lobes spreading, rounded. Capsule ovoid 4–5 mm long, styles persistant. Margins of waterways or dams. Blue Mts Introd. from America. Fl. summer . . . . . ***B. caroliniana** (Walter) Rob.

13 **Digitalis** L.
1 species naturalized Aust.; NSW, Vic.

One species in the area

Annual herb up to 1.5 m high. Leaves ovate to narrow-ovate, alternate, often with a basal rosette, up to 30 cm long, pubescent on the upper surface and almost tomentose underneath; petiole winged. Flowers in a terminal raceme. Calyx deeply 5-lobed. Corolla tubular, narrowing abruptly near the base, 3–4.5 cm long, white to purplish pink with dark purple spots. Stamens 4. Fruit a capsule. Widespread. Waste places. Introd. from Europe. Fl. summer. *Foxglove* . . . . ***D. purpurea** L.

14 **Derwentia** Raf.
8 species endemic Aust.; Qld, NSW, Vic., Tas., S.A.

Suffruticose herbs or shrubs. Leaves sessile, opposite. Flowers in long, mostly axillary racemes. Calyx deeply divided into 4 segments. Corolla with a very short tube, rotate, 4-lobed with 1 lobe usually broader than the others. Stamens 2, epipetalous, exserted. Fruit a ± compressed capsule, at first dehiscing septifragally and then septicidally, 2-valved; each valve 2-fid.

Key to the species

1 Leaves 2–5 cm long, 1–3 mm wide, linear, entire, with a few teeth or with irregular lobes. corolla violet-blue, 7–10 mm long. Herb to 1 m high. Western Blue Mts DSF. Fl. spring–autumn . . . . **D. arenaria** (A.Cunn. ex Benth.) B.Briggs & Ehrend.

1 Leaves lanceolate to ovate, more than 10 mm wide 2

2 Leaves recurved and v-shaped in cross section, ovate to lanceolate, 8–18 pairs of teeth per side, glaucous. Corolla violet-blue, 6–7 mm long. Glabrous shrub or woody herb to 50 cm high. Western Blue Mts at high altitudes. Vulnerable. Fl. summer . . . . **D. blakelyi** B.G.Briggs & Ehrend.

2 Leaves not recurved or v-shaped 3

3 Leaves stem-clasping, often connate at the base, ovate to narrow-ovate, acuminate or acute, glaucous, usually entire. Corolla bluish, streaked with purple, 5–6 mm long. Herb up to 1.5 m high. Higher Blue Mts Open forests. Fl. summer. *Digger's Speedwell* . . . . **D. perfoliata** (R.Br.) Raf.

3 Leaves not stem-clasping, glaucous underneath, narrow-ovate to lanceolate, 5–10 cm long, acuminate, serrate, 30–80 pairs of teeth per side. Corolla white or blue, 5–6 mm diam. Herb up to 1 m high. Higher Blue Mts Fl. summer . . . . **D. derwentiana** (Andrews) B.G.Briggs & Ehrend. ssp. **subglauca** B.G.Briggs & Ehrend

15 **Parahebe** W.R.B.Oliv.
1 species native Aust.; NSW

One species in the area

Suffruticose herb, woody at the base and with a woody rootstock, Stems trailing up to 50 cm long, rooting at the nodes. Flowers in axillary racemes. Calyx 4-lobed. Corolla with a short tube; lobe 4, spreading, with 1 lobe usually broader than the others. Stamens 2, epipetalous, exserted. Fruit a compressed capsule,

dehiscing at first loculicidally. Higher Blue Mts Cliffs and rocky places. Fl. spring–summer. . . . .
. . . . **P. lithophila** B.G.Briggs & Ehrend.

16 Veronica L.

Speedwells

18 species in Aust. (9 naturalized); all states and territories

Herbs. Leaves opposite. Flowers axillary and solitary or in racemes. Calyx deeply divided into 4 segments. Corolla with an indistinct tube, usually rotate, 4-lobed with 1 lobe usually broader than the others and imbricate outside in the bud. Stamens 2, epipetalous, exserted. Fruit a compressed capsule at first dehiscing loculicidally and then septifragally and often subsequently septicidally.

Key to the species

1 Leaves mostly more than 30 mm long, usually more than 10 mm wide. Stems ascending usually more than 30 cm high from creeping or decumbent stolons. Flowers usually numerous, in axillary racemes. . 2
1 Leaves less than 25 mm long. Stems creeping and rooting at the nodes or, if erect, then up to 30 cm high 4

2 Leaves sessile, stem-clasping, entire to dentate, with fine teeth. Stems hollow. Flowers, 15–40 in racemes 30–80 cm long. Corolla 5–6 mm diam., pale blue. Widespread. Wet places. Introd. from Europe. Fl. spring–summer. *Water Speedwell* . . . . *V. anagallis-aquatica L.
2 Leaves petiolate, lanceolate to narrow-ovate, or triangular, coarsely serrate, acute. Stems solid. Racemes up to 17 cm long with up to 20 flowers. 3

3 Leaf bases cuneate. Stems with long spreading hairs. Corolla 4–8 mm long, white to bluish. Calyx lobes usually glabrous towards apex. Widespread. WSF and RF. Fl. summer . .**V. notabilis** F.Muell. ex Benth.
3 Leaf bases truncate. Stems with mostly short rigid hairs. Corolla 3–8 mm long, white becoming mauve with age. Calyx lobes usually finely ciliate on margins and apex. In north of region. WSF and RF. Fl. spring–autumn. . . . . **V. grosseserrata** B.G.Briggs & Ehrend.

4 Leaves entire, finely toothed or with 1–4 pairs of small irregular teeth 5
4 Leaves coarsely serrate, crenate or crenate-dentate. 9

5 Stems with scattered spreading glandular hairs. Leaves narrow-elliptic to spathulate, 3–10 mm long, 0.5–5 mm wide, ± sessile. Flowers solitary in the axils of leaf-like bracts. Corolla white to lavender, c. 1.5 mm long. Capsule truncate, glabrous 2–3 mm long. Herb to 20 cm high. Uncommon. Seasonally wet areas. Introd. from America. Fl. spring . . . . *V. peregrine L.
5 Stems with minute antrorse hairs or recurved to spreading hairs, rarely glandular hairs present 6

6 Leaves elliptic or obcordate to orbicular, up to 12 mm long, opposite or the floral leaves alternate. Racemes terminal, leafy. Flowers on short pedicels. Corolla pale blue with dark lines, c. 3 mm long. Coast, e.g. Minnamurra Falls, Manly. Weed of cultivation and waste places. Introd. from Europe. Fl. spring–summer. . . . .*V. serpyllifolia L.
6 Leaves ovate to narrow-ovate or linear, 10–25 mm long, all opposite. Stems arising from a creeping rootstock, ascending to erect, unbranched or slightly branched, usually 5–40 cm high 7

7 Fruiting calyx elliptic 2.5–4 mm long. Leaves ovate to lanceolate 8–20 mm wide. Racemes with 6–10 flowers. Corolla white tinged with mauve, 5–6 mm long. Rhizomatous herb to 40 cm high. Tablelands in north of region. Fl. spring–summer . . . . **V. sobolifera** B.G.Briggs & Ehrend.
7 Fruiting calyx lobes linear 0.8–2.5 mm long 8

8 Stem hairs c. 0.5 mm long. Leaves lanceolate 2–9 mm wide. Racemes with 1–6 flowers. Corolla mauve or blue, mostly 4–8 mm long. Capsule broader than long. Blue Mts Wet places. Fl. spring–summer . . .
. . . . **V. gracilis** R.Br
8 Stem hairs c. 0.1 mm long. Leaves linear, up to 3 mm wide. Racemes with 1–6 flowers. Corolla mauve or pale blue, 4–7 mm long. Capsule longer than broad. Widespread. Wet places. Fl. spring–summer . . . .
. . . . **V. subtilis** B.G.Briggs & Ehrend.

9 Flowers nearly sessile in terminal racemes. Calyx 2–3 mm long when in flower. Corolla minute. Leaves ovate, cordate, crenate-dentate, 5–8 mm long. Widespread. Weed of cultivation, waste places and grasslands. Introd. from Europe. Fl. spring–summer . . . . ***V. arvensis** L.

9 Flowers in axillary racemes or solitary in the axils of the upper leaves 10

10 Flowers solitary in the axils of the upper leaves. Calyx c. 4 mm long when in flower. Corolla light blue, longer than the calyx. Leaves ovate, crenate-dentate, 8–20 mm long. Capsule compressed, 2-lobed, 7–8 mm wide. Widespread. Weed of cultivation and waste places. Introd. from Europe and W. Asia. Fl. spring–summer . . . . ***V. persica** Poir.

10 Flowers in axillary, sometimes much reduced, racemes 11

11 Capsule oblong-orbicular. Stems ± evenly pubescent or the very short hairs tending to be denser in two broad indistinct longitudinal bands. Calyx segments acute, 4 mm long when in flower. Widespread. Common in many situations on heavier soils. Fl. spring–summer . . . . **V. plebeia** R.Br.

11 Capsule broad-obcordate to transverse-obcordate or truncate. Stems hairy 12

12 Stems villous-pubescent; the hairs ± patent and tending to be denser in two broad indistinct longitudinal bands. Leaves ovate to broad-ovate, crenate or dentate with obtuse teeth. Calyx segments obtuse, 4–5 mm long when in flower, enlarging in the fruit. Widespread. Open forests. Fl. spring–summer . . . . **V. calycina** R.Br.

12 Stems pubescent with short crisped hairs arranged in two narrow longitudinal bands with the stem surface between almost quite glabrous. Leaves ovate to lanceolate, serrate with acute teeth. Calyx segments ± obtuse, 4–5 mm long when in flower, enlarging slightly in the fruit. Valleys in the Blue Mts Open forests. Fl. spring–summer . . . . **V. brownii** Roem. & Schult.

17 Euphrasia L.

Eyebrights

20 species endemic Aust.; all states and territories except NT

Erect to trailing, perennial herbs branching from the base, parasitic on other plants. Leaves opposite, toothed or crenate, with sessile glands underneath. Flowers in terminal spike-like racemes, irregular. Calyx 4-lobed. Corolla yellow, white to purplish or blue, 5-lobed, 2-lipped; the upper lip recurved. Stamens 4, didynamous, dehiscing by pores, awned. Ovary 2-locular; stigma 2-lobed. Fruit a loculicidal capsule.

Two other species *E. scabra* (Endangered) *and E. orthocheila* are considered to possibly occur in the Sydney area. They can be distinguished by their yellow flowers.

Key to the species

1 Upper leaves entire or with 1 pair of teeth, elliptic to obovate, 4–8 mm long. Corolla mauve to violet with darker markings in throat, 7–11 mm long; tube with glandular hairs. Trailing to ascending herb. Higher Blue Mts Ss cliffs. Vulnerable. Fl. spring–summer. . . . . **E. bowdeniae** W.R.Barker

1 Upper leaves with 2 or more pairs of teeth. Leaves cuneate to spathulate, 5–15 mm long, obtuse, wrinkled. Erect, perennial herbs up to c. 50 cm high 2

2 Calyx hairy outside. Corolla blue to purple, without yellow markings on the lower lip. Widespread. Open forests and damp places. Fl. spring–summer . . **E. collina** R.Br. ssp. **speciosa** (R.Br.) W.R.Barker

2 Calyx glabrous outside except just below each cleft. Corolla white to purple or pink, often with yellow markings on the lower lip. Widespread. DSF, heath and damp places. Fl. spring–summer. . . . .

. . . . **E. collina** ssp. **paludosa** (R.Br.) W.R.Barker

18 Parentucellia Viv.

2 species naturalized Aust.; NSW, Vic., Tas., S.A., WA

Erect, annual herbs, often unbranched, parasitic on the roots of other plants. Leaves opposite or nearly so, sessile. Flowers in terminal spikes. Calyx tubular, 4-toothed. Corolla 2-lipped; the upper lip forming a hood; the lower lip 3-lobed. Stamens 4, didynamous. Fruit a capsule.

1 Corolla dark purplish red, c. 12 mm long. Calyx 7–10 mm long. Leaves ovate, 5–15 mm long, with 3–7 coarse teeth or lobes. Erect, glandular pubescent herb up to c. 40 cm high. Widespread. Open forests and grasslands. Introd. from the Mediterranean. Fl. spring–summer. . . . . . . ***P. latifolia** (L.) Caruel

1 Corolla yellow, 15–20 mm long. Calyx 12–15 mm long. Leaves lanceolate to narrow-oblong, 15–32 mm long, coarsely dentate with several teeth. Erect, glandular hairy, viscid herb up to c. 70 cm high. Blue Mts, e.g. Wentworth Falls. Weed of grasslands or cultivation. Introd. from the Mediterranean. Fl. spring–summer. . . . . ***P. viscosa** (L.) Caruel

19 Orobanche L.

Broomrapes

2 or 3 species in Aust. (1 native, 1 or 2 naturalized); all states and territories except NT

One species in the area

Erect, parasitic herb without chlorophyll, up to 50 cm high. Leaves pale, ovate, up to 15 mm long. Sepals 2-fid; the lobes linear. Corolla white to bluish, 9–18 mm long, covered with glandular hairs. Stamens 4, didynamous. Fruit a 1-locular capsule up to 8 mm long. Widespread. Weed of cultivation and waste places; parasitic on garden and pasture plants. Introd. from Europe. Fl. chiefly spring . . . ***O. minor** Sm.

133 VERBENACEAE

Herbs, shrubs or trees; stems often quadrangular. Leaves opposite, usually simple; stipules absent. Flowers bisexual, usually in racemes or heads. Calyx tubular, with 4–5 lobes or teeth, persistent. Corolla tubular, often irregular; lobes 4–5, imbricate, rarely 2-lipped. Stamens 4, adnate to the corolla tube. Ovary superior, 2-carpellary, entire or 4-lobed, 2–4-locular; style with 2 stigmatic lobes or simple. Fruit a capsule, drupaceous or schizocarpic and separating into four 1-seeded articles or 2 mericarps. 34 gen., widespread, mostly trop. with some temp.

Key to the genera

1 Shrubs often forming thickets . . . . **3 LANTANA**
1 Herbs or subshrubs. 2

2 Creeping herb often rooting at nodes . . . . **1 PHYLA**
2 Plants erect not rooting at nodes . . . . **2 VERBENA**

1 Phyla Lour.

1 species naturalized Aust.; Qld, NSW, Vic., SA., WA, NT

One species in the area

Prostrate, perennial herb, usually sparingly pubescent. Leaves opposite, obovate to linear-cuneate, toothed at the apex, 10–25 mm long. Flowers in cylindrical to ovoid axillary spikes on long peduncles. Bracts closely imbricate, longer than the calyx. Calyx 2-lipped. Corolla 4-lobed, white to pink or mauve, 2-lipped; lower lip twice as long as the upperlip; tube short. Stamens 4. Fruit separating into 2 mericarps. Coast and adjacent plateaus; Moist areas. Cumberland Plain; Hunter River Valley. Introd. from America. Fl. spring–summer. *Lippia* or *Carpet Weed* . . . . ***P. canescens** (Kunth) Greene

2 Verbena L.

7 species naturalized Aust. (one infraspecific taxa endemic); all states and territories

Herbs with quadrangular stems and branches. Leaves opposite. Flowers in bracteate spikes in terminal often corymbose panicles. Calyx minutely 5-toothed. Corolla with 5 spreading slightly unequal lobes; tube slightly bent. Stamens 4, enclosed in the corolla tube. Ovary 4-locular with 1 ovule per loculus. Fruit separating into four 1-seeded articles.

Key to the species

1 Leaves sessile, often stem-clasping, coarsely serrate, 3–18 cm long. Spikes dense both in flower and fruit. Scabrous to hairy perennials 2
1 Leaves shortly petiolate entire to deeply divided. Spikes dense at flowering but often elongating in fruit . 5

2 Bracts conspicuously longer than the calyx; corolla tube twice as long or longer than calyx 3
2 Bracts not or only slightly longer than the calyx; corolla tube twice as long as calyx 4

3 Corolla limb 7–8 mm diam. Calyx c. 4–6 mm long. Corolla tube nearly 3 times as long as the calyx. Perennial herb usually 20–40 cm high with creeping rhizome and erect or ascending stems. Coast; Cumberland Plain; lower Blue Mts Weed of waste ground. Introd. from S. America. Fl. spring–autumn . . . . *V. rigida Spreng. var. rigida
3 Corolla limb 1–1.5 mm diam. Calyx up to 3 mm long. Corolla tube scarcely twice as long as the calyx. Decumbent to ascending non stoloniferous herb up to 1 m high. Coast; Cumberland Plain; Hunter River Valley. Weed of cultivation and waste places. Not common. Introd. from S. America. Fl. most of the year . . . . *V. hispida Ruiz & Pav.

4 Spikes sub-cylindrical. Peduncles without glandular hairs. Corolla tube rarely twice as long as calyx. Stems scabrous pubescent. Leaves ovate-lanceolate. Perennial or annual herb to 2 m high. Widespread. Weed of cultivation and waste places. Introd. from S. America. Fl. spring–autumn. *Purple Top* . . . . *V. bonariensis L. var. bonariensis
4 Spikes in capitate clusters. Peduncles with glandular hairs. Corolla tube mostly twice as long as calyx. Leaves ovate-lanceolate. Perennial or annual herb to 2 m high. Widespread. Weed of cultivation and waste places. Introd. from S. America. Fl. spring–autumn. *Purple Top* . . . . *V. bonariensis var. conglomerata Briq.

5 Corolla limb 8–10 mm diam. Calyx 6–9 mm long. Upper and lower leaves c. 2–6 cm long, all deeply divided and subdivided. Prostrate and ascending annual up to 50 cm high. Coast. Weed of roadsides. Uncommon. Introd. from S. America. Fl. spring–summer . . . . *V. aristigera S.Moore
5 Corolla limb 2–5 mm diam.; the tube up to c. 4 mm long. Calyx c. 3 mm long 6

6 Leaves deeply pinnately lobed to pinnatifid, up to 7 cm long; upper ones often entire. Spikes elongating in the fruiting stage; the peduncles often becoming leafy just below the spike. Glandular pubescent also with some simple hairs. Corolla lilac, pink, to deep blue. Erect, annual herb up to 1 m high. 7
6 Leaves coarsely toothed. Peduncles usually naked for some distance below the spike. 8

7 Lower and mid stem leaves deeply divided or incised. Floral bracts usually as long as calyx. Widespread. Fl. spring–summer . . . . V. officinalis L. var. gaudichaudii Briq.
7 Lower and often mid stem leaves coarsely toothed or incised not deeply divided. Floral bracts about half the length of the calyx. Weed of cultivation and waste places. Introd. from the Mediterranean. Fl. spring–summer. *Common Verbena* . . . . *V. officinalis var. halei (Small) Munir

8 Spikes elongating in fruit. Hispid with simple hairs, becoming almost glabrous on the older parts. Corolla blue-purple. Leaves elliptic to lanceolate, 2–9 cm long. Erect, perennial herb up to c. 1 m high. Widespread. Weed of cultivation and waste places. Introd. from S. America. Fl. spring–summer . . . . *V. litoralis Kunth var. littoralis
8 Spikes not elongating in the fruit but sometimes slightly spreading at the base. Villous, becoming glabrous on the stems. Corolla white to purple. Leaves narrow-ovate, up to 5 cm long. Perennial herb up to 2 m high. Coast; Cumberland Plain. Weed of cultivation and waste places. Introd. from S. America. Fl. most of the year. . . . . *V. littoralis var. brasiliensis (Vell.) Briq.

3 Lantana L.

2 species naturalized Aust.; Qld, NSW, NT

Shrubs with weak scrambling scabrous branches. Leaves opposite, scabrous, ovate to lanceolate. Flowers in dense capitate axillary spikes; each flower sessile or nearly so in the axil of a small bract. Corolla 4-lobed. Stamens 4, enclosed within the corolla tube. Fruit drupaceous.

Key to the species

1 Large shrub. Corolla orange, yellow, red or pink. Branches often with soft recurved prickles. Leaves petiolate, ovate to slightly cordate, more than 25 mm long. Fruit black. Coast; Cumberland Plain. Weed of pastures, forests and waste places. Introd. from S. America. Fl. most of the year . . . . *L. camara L.

1 Low decumbent shrub with thin wiry branches. Corolla purplish. Leaves ovate-lanceolate, serrate, 8–25 mm long. Fruit purplish. Coast. Weed of waste places. Introd. from S. America. Fl. most of the year . . . . *L. montevidensis (Spreng.) Briq.

134 CONVOLVULACEAE

Annual or perennial herbs, sometimes woody, erect, creeping, trailing or twining, sometimes parasitic and leafless. Leaves alternate, without stipules. Flowers regular. Sepals 5, free or connate, persistent. Corolla campanulate or funnel-shaped, 5-angled or 5-lobed. Stamens 5, epipetalous, alternating with the corolla lobes. Ovary superior, 2–4-locular, sometimes incompletely 2-locular, rarely separated into 2 free carpels; stigmas capitate or 2–8-lobed; Placentas axile or basal. Fruit usually a loculicidal capsule, opening in valves and leaving the dissepiments attached to the axis. 55 gen., cosmop.

Key to the genera

1 Leafless parasitic twiners . . . . 8 CUSCUTA
1 Leafy plants 2

2 Calyx tubular, 5-toothed. Plants from saline areas. . . . . 7 WILSONIA
2 Sepals free 3

3 Styles 2, free 4
3 Style solitary, simple, bifid or 2–8-lobed. 5

4 Stigmas capitate . . . . .6 DICHONDRA
4 Stigmas 2-fid . . . . 3 EVOLVULUS

5 Stigma lobes 4–8-fid . . . . 5 POLYMERIA
5 Stigma capitate or 2-fid 6

6 Stigmatic lobes globular or the stigma capitate. . . . . .1 IPOMOEA
6 Stigmatic lobes elongated 7

7 Bracteoles minute, remote from the calyx. Capsule 2-locular . . . . .2 CONVOLVULUS
7 Bracteoles 2, enlarged, enclosing the calyx. Capsule 1 or incompletely 2-locular . . . .4 CALYSTEGIA

1 Ipomoea L.
50 species in Aust. (20 endemic, 12 naturalized); all states and territories

Twining, herbaceous perennials. Corolla campanulate or with a cylindrical tube; the limb usually entire, angular. Ovary 2–3-locular, with 2 ovules per loculus. Style filiform; capitate or 2-fid and lobes globular. Fruit a capsule, nearly globular.

Key to the species

1 Leaves 1–5 cm long, digitately divided into 5–7 lobes. Corolla purple pink or white, mostly 3–5 cm long, campanulate, contracted into a tube towards the base. Sepals obtuse. Ovary 2-locular. Seeds pubescent. Coast. Chiefly near the sea. Introd. from Asia. Fl. most of the year . . . . * I. cairica (L.) Sweet
1 Leaves 2–5-lobed or entire, not digitate 2

2 Leaf apex emarginate or obtusely 2-lobed, ovate to orbicular, glabrous, coriaceous, mostly 4–8 cm long. Corolla pink to purple, up to 65 mm long. Prostrate perennial with long trailing stems. Coast, e.g. Port Hacking, Broken Bay. Sandy beaches. Fl. spring–summer . . . . I. brasiliensis (L.) Sweet

2 Leaves 7–10 cm long; entire or 3-5-lobed; apex obtuse to acuminate, bases cordate cm. Stems twining, ± hirsute with reflexed hairs. 3

3 Corolla less than 3 cm long; white, each petal with a central band of hairs. Sepals ciliate. Leaves 3–8 cm long, 1–5 cm wide. Weed in suburban areas. Gullies. Lane Cove. Fl. all year . . . .**I. plebeia** R.Br.
3 Corolla more than 3 cm long; white to purple-blue; petals ± glabrous 4

4 Sepals 14–22 mm long, lanceolate, acuminate, with soft ± appressed hairs. Flowers mostly 3–12 in cymes on axillary peduncles. Corolla bluish-purple, 65–80 mm diam.; tube more than 50 mm long. Bracteoles more than 10 mm long. Coast. Weed in suburban areas; margins of RF; gullies. Introd. from trop. Asia. Fl. spring–autumn. *Blue Morning Glory* . . . . *I. indica** (Burman f.) Merr.
4 Sepals less than 15 mm long, lanceolate, scarcely acuminate. Flowers 1–3 together on short pedicels on axillary peduncles. Corolla purple blue to white, 20–40 mm diam.; tube 3–5 cm long. Bracteoles less than 7 mm long. Coast. Weed in suburban areas; margins of RF; gullies. Introd. from trop. Americ. Fl. most of the year. *Morning Glory* . . . . *I. purpurea** (L.) Roth

2 Convolvulus L.

7 species in Aust. (6 endemic, 1 naturalized); all states and territories

Twining or prostrate, perennial herbs. Leaves entire or toothed lobed or divided. Flowers solitary or in pairs on axillary peduncles. Bracteoles minute, distant from the calyx. Corolla campanulate, entire or angular. Style filiform, with 2 elongated stigmatic lobes. Ovary 2-locular with 2 ovules per loculus. Fruit a capsule, 2-locular, seed 3–4 mm long.

Key to the species

1 Outer sepals 4 mm or less long, apex rounded or retuse, glabrous to weakly hairy. Leaves similar in shape from base to top of stem, lanceolate-hastate or ovate-hastate, entire except for the basal lobe; glabrous or with a few crisped weakly spreading hairs; apex acute to rounded sometimes emarginate. Inflorescence solitary, axillary, a one sided dichasium with1–4 flowers, rarely 2 inflorescences per axil. Peduncles terete,ribbed. Pedicels recurved at fruiting stage. Seed surface with a regular pattern of tubercles. Perennial with a slender creeping rootstock. Widespread. Weed of cultivated and waste ground. Fl. spring–summer. *Field Bindweed* . . . . *C. arvensis** L.
1 Outer sepals more than 4 mm long, apex acute to rounded with recurved apiculum. Leaves extremely variable in shape from base to top of stem, sagittate-cordate to linear, slightly crenate to deeply incised 2

2 Pedicels straight or undulate at fruiting stage. Peduncles slightly ribbed. Outside of outer sepals moderately hairy to ± glabrous, hairs crisped and loosely appressed. Leaf apex obtuse to mucronulate. Leaves ranging from glabrous to moderately hairy above, moderately hairy below. Hairs crisped and appressed to loosely spreading. Margins undulate to lobed. Inflorescence solitary axillary, a one sided dichasium with1–3 flowers, occasionally 2 inflorescences per axil. Seed surface with small irregular tubercles. Widespread. Open forests and grasslands. Fl. spring–summer. *Australian Bindweed*. . . . . **C. erubescens** Sims
2 Pedicels recurved at fruiting stage. Peduncle terete. Outside of outer sepals moderately to sparsely hairy, hairs loosely ascending to spreading, a few appressed. Leaf apex acute to rounded sometimes emarginate. Margins crenulate to serrate. Leaves moderately hairy to ± glabrous with semi erect hairs on both surfaces. Inflorescence solitary axillary, with solitary flowers, rarely with 2 flowers or 2 inflorescence's per axil. Seed surface with reticulate ridges. Loam or clay soils. Open forests Fl. spring–autumn . . . . **C. angustissimus** R.Br. ssp. **angustissimus**

3 Evolvulus L.

1 species native Aust.; Qld, NSW, NT, S.A., WA

One species in the area

Erect or ascending perennial up to 50 cm high. Leaves entire, linear to linear-lanceolate, up to 30 mm long, hairy, hairs 2-fid. Flowers axillary, solitary or in few-flowered cymes. Corolla wide-funnel-shaped,

7–10 mm diam., blue. Styles 2, free, bifid. Capsule globular 3–4 mm diam., 4-valved. Coast. Open forests. Fl. spring–summer . . . .**E. alsinoides** var. **decumbens** (R.Br.) Ooststr.

4 Calystegia R.Br.
4 species in Aust. (3 native, 1 naturalized); all states and territories except NT

Twining or prostrate, perennial herbs. Leaves hastate or reniform, entire or irregularly toothed. Flowers solitary or in pairs. Bracteoles ovate to oblong, close to and enclosing the calyx. Corolla campanulate, 5-toothed. Ovary 1–2-locular; style filiform, 2-fid. Fruit a capsule, usually 1-locular or incompletely 2-locular.

Key to the species

1 Leaves reniform or rounded-cordate, 2–5 cm wide, glabrous. Corolla purplish, 30–50 mm long. Stems trailing or slightly twining. Coastal sand dunes. Fl. spring–summer . . . .**C. soldanella** (L.) R.Br.
1 Leaves not reniform 2

2 Corolla c. 2 cm long. Bracteoles ovate to orbicular-cordate. Leaves 3–5 cm long, narrow-lanceolate to deltoid, with acute divergent basal lobes. Widespread. Moist gullies and RF margins. Fl. spring–summer . . . . **C. marginata** R.Br.
2 Corolla 4–7 cm long 3

3 Bracteoles narrow-ovate to ovate, c. 1 cm wide, flat but very slightly keeled at the base, not inflated, not overlapping at flowering. Leaves ovate-cordate to lanceolate, hastate, acute-acuminate, mostly 5–10 cm long. Prostrate or twining herb. Coast. Weed of cultivation and waste places. Fl. spring–summer . . . .**C. sepium** (L.) R.Br.
3 Bracteoles broad-ovate, c. 2 cm long, inflated, one overlapping the other on both sides at flowering. Leaves ovate to broad-ovate, hastate, mostly 5–10 cm long. Prostrate or twining herb. Naturalized in Wahroonga and Bowral districts. Introd. from Europe. Fl. summer . . . . . ***C. silvatica** (Kit.) Griseb.

5 Polymeria R.Br.
7 species in Aust. (6 endemic); Qld, NSW, NT, WA

One species in the area

Prostrate or slightly twining herb, glabrous or slightly pubescent. Leaves 5–30 mm long, on thin petioles, cordate or rarely hastate at the base; lower leaves broad-ovate; upper ones oblong-linear. Peduncles 1-flowered, usually shorter than the leaves, with minute bracteoles at or below the middle. Sepals unequal, overlapping; the outer one very broad-ovate or cordate, c. 6 mm long; the inner one shorter, acuminate. Corolla 10–12 mm long, slightly pubescent outside, pink. Stigmatic lobes 4–8. Coast and adjacent plateaus; valleys in Blue Mts Grasslands and open swampy places. Fl. most of the year. . . . . .**P. calycina** R.Br.

6 Dichondra J.R.Forst. & G.Forst.
2 species native Aust.; all states and territories except NT

Perennial herbs with creeping stems rooting at the node. Leaves entire, reniform, on petioles longer than the blade, often clustered at the rooting nodes. Flowers solitary, axillary, small. Corolla greenish yellow, 5–6 mm long or less, with a short tube. Stamens 5. Carpels 2, almost free; styles 2; each with a capitate stigma. Fruit a 2-lobed irregularly 2-valved capsule.

Key to the species

1 Leaves similar on both surfaces, up to 25 mm long. Calyx more than 2 mm long and longer than the capsule. Pedicels more than 10 mm long. Widespread. Sheltered moist places in forests and grasslands; weed in lawns. Fl. most of the year. . . . .**D. repens** J.R.Forst. & G.Forst.
1 Leaves more densely and silky hairy underneath. Calyx up to 2 mm long and shorter than the capsule. Pedicels up to 10 mm long. Widespread. Sheltered moist places in forests and grasslands. Fl. spring–summer . . . .**D. species A** sensu Harden (1992)

7 Wilsonia R.Br.
3 species endemic Aust.; all states and territories except Qld & NT

Perennial herb or subshub. Leaves fleshy, entire, glabrous or with bifid hairs. Flowers solitary, axillary, erect. Calyx tubular, with 5 teeth. Corolla tubular with 5 lobes. Stamens 5. Style bifid. Ovary with 1 ovule per loculus. Fruit an indehiscent capsule.

Key to the species

1 Plants with spreading hairs. Leaves broadly ovate, apex acute to obtuse 1.5–4 mm long. Calyx c. 5 mm long, as long as the corolla. Corolla white, tube broad. Stamens and style slightly exserted. Herb with prostrate stems. Coastal. Royal National Park. Endangered. Fl. spring-summer. **W. rotundifolia** Hook.
1 Plants glabrous or nearly so, often forming dense dark green mats. Leaves linear to narrow-oblong, 6–15 mm long. Calyx 5–8 mm long; the teeth much shorter than the tube. Corolla tube slender, exceeding the calyx; the lobes reflexed, white. Stamens and style much exserted. Small procumbent perennial herb. Margins of saltmarshes, e.g. Parramatta River. Vulnerable. Fl. spring–summer . . . . **W. backhousei** Hook.f.

8 Cuscuta L.
Dodders
10 species in Aust. (4 endemic, 6 naturalized); all states and territories

Leafless, usually annual parasites; the small root usually dying away as soon as the filiform twining yellowish to reddish stems have attached themselves to the host by their haustoria. Flowers subglobular in spike-like or head-like cymes. Corolla white to reddish. Stamens epipetalous, with a scale below each. Ovary 2-locular. Fruit a circumcissile capsule.

Key to the species

1 Stigmas filiform; the style and stigma together c. as long as the ovary. Flowers in globular heads, 5–10 mm diam. Stamens exserted. Staminal scales oblanceolate, included in the corolla tube. Widespread, usually on leguminous plants. Introd. from Europe. Fl. spring–summer . . *C. epithymum (L.) Murray
1 Stigmas capitate 2

2 Staminal scales 2-fid. Corolla lobes ovate, obtuse. Styles shorter than the ovary. Capsule 3–4 mm diam. Chiefly coastal swampy areas. Parasitic on a variety of species but particularly *Polygonum* and *Persicaria*. Fl. summer. *Australian Dodder* . . . . **C. australis** R.Br.
2 Staminal scales fimbriate but not 2-fid. Corolla lobes deltoid, acute. Styles as long as the ovary. Capsule 2–3 mm diam. Widespread on leguminous plants. Introd. from Europe. Fl. summer. *Golden Dodder* . . . . *C. campestris Yunck.

135 SOLANACEAE
Herbs or shrubs. Leaves alternate, without stipules. Flowers solitary or in cymes, regular or sometimes slightly irregular, bisexual. Calyx persistent, usually with 5 teeth or lobes. Corolla rotate or tubular, 4–5-lobed. Stamens 4–5; if 4 present a staminode also usually present, epipetalous, alternating with the corolla lobes. Ovary superior, 2-locular, rarely becoming 4-locular by the growth of a secondary dissepiment; placentas axile; septum oblique; style simple. Fruit a berry or capsule. Seeds few to numerous, ± reniform; embryo curved in the endosperm. 102 gen., cosmop.

Key to the genera

1 Fruit enclosed in an enlarged bladdery calyx 2
1 Fruit not enclosed in an enlarged bladdery calyx 3

2 Corolla yellow or greenish yellow . . . . **1 PHYSALIS**
2 Corolla blue . . . . **2 NICANDRA**

3 Fruit a berry. **4**
3 Fruit a capsule **10**

4 Stamens fewer than the corolla lobes . . . . **9 DUBOISIA**
4 Stamens as many as the corolla lobes, normally 5 . . . . **5**

5 Flowers in lateral (extra axillary) or terminal cymes, very rarely solitary. **6**
5 Flowers solitary or in pairs, axillary **9**

6 Corolla violet or blue to white . . . . **3 SOLANUM**
6 Corolla pink, yellow to orange or greenish **7**

7 Corolla tube more than twice as long as the lobes . . . . **5 CESTRUM**
7 Corolla tube as long as or shorter than the lobes. **8**

8 Corolla pink. . . . . **7 CYPHOMANDRA**
8 Corolla yellow. . . . . **4 LYCOPERSICON**

9 Shrubs with ± thorny branches. . . . . **6 LYCIUM**
9 Herbs without thorns . . . . **8 SALPICHROA**

10 Fertile stamens 5. Herbs **11**
10 Fertile stamens 4 . . . . **12**

11 Fruit prickly . . . . **10 DATURA**
11 Fruit smooth. . . . . **11 NICOTIANA**

12 Shrubs. Corolla deeply lobed with narrow lobes . . . . **12 CYPHANTHERA**
12 Herbs. Corolla trumpet shaped; the lobes scarcely distinguished from each other. . . . **13 PETUNIA**

1 Physalis L.
8 species in Aust. (1 native, 7 naturalized); all states and territories

Ascending to decumbent, herbaceous perennials up to 1.5 m high. Leaves 1–2 per node, petiolate. Flowers solitary in the leaf axils. Calyx enlarged, inflated in the fruit, completely enclosing the globular edible berry. Corolla yellow with 5 coloured spots in the throat. Anthers 5 purplish.

Key to the species

1 Plants rhizomatous, perennials, up to 60 cm high, ± glabrescent. Leaves ovate, mostly up to 6 cm long. Corolla greenish yellow with dark spots, 10–12 mm long. Fruiting calyx 10-angled, 15–25 mm long. Coast. Uncommon weed. Introd. from N. America. Fl. summer. . . . . ***P. virginiana** Mill.
1 Plants not rhizomatous, annuals **2**

2 Densely pubescent, shrubby herb up to 1 m high. Leaves ovate, dentate to entire, up to 8 cm long. Corolla yellow with dark spots, 10–15 mm long. Fruiting calyx 10-angled, 3–4 cm long. Coast. Weed of cultivation and waste places. Introd. from S. America. Fl. summer. *Cape Gooseberry* . ***P. peruviana** L.
2 Glabrous plants or with scattered hairs **3**

3 Pedicel 20–25 mm long. Fruiting calyx 10-angled, 20–30 mm long. Leaves narrow-ovate to narrow-elliptic, irregularly serrato-dentate, up to 8 cm long. Corolla pale yellow with 5 brownish spots, 5–8 mm long. Shrubby herb up to 50 cm high. Coast. Banks of rivers and creeks; waste places. Fl. summer. *Wild Gooseberry* . . . . **P. minima** L.
3 Pedicel 5–15 mm long. Fruiting calyx not angled, circular in section, 15–30 mm long. Leaves narrow-ovate, up to 6 cm long. Corolla pale yellow with dark spots, 6–10 mm long. Herb up to 60 cm high. Coast. Weed of cultivation and waste places. Introd. from Central America. Fl. summer. *Ground Cherry* . . . . ***P. ixocarpa** Brot. ex Hornem.

2 Nicandra Adans.

1 species naturalized Aust.; all states and territories except NT

One species in the area

Erect, glabrous herb mostly 30–100 cm high. Leaves ovate-lanceolate, irregularly toothed or sinuate, 4–20 cm long. Flowers axillary. Corolla pale blue, c. 20 mm long. Ovary 4–5-locular. Fruit a rather dry berry enclosed in 5 enlarged cordate sepals. Widespread. Weed of cultivation and waste places. Introd. from S. America. Fl. summer. *Apple of Peru* . . . . *N. physalodes* (L.) Gaertn.

3 Solanum L.

Nightshades

c. 120 species in Aust. (native, endemic, naturalized); all states and territories

Herbs or shrubs. Leaves usually petiolate. Flowers in lateral (extra-axillary) cymes which may appear to be racemes or umbels or rarely solitary. Calyx 5- rarely 4-lobed, spreading. Corolla broad-campanulate or stellate to rotate, 4–5-lobed, with a short tube. Filaments very short; anthers basifixed, connivent, exserted, opening by terminal pores. Fruit a globular, usually 2-locular berry.

Key to the species

1 Prickles absent 2
1 Prickles present 18

2 Corolla mostly 15–50 mm diam., violet, pale blue or white. Flowers in loose pedunculate raceme-like inflorescences 3
2 Corolla less than 12 mm diam., white to pale violet. Mature berry globular 11

3 Leaves sessile, decurrent, usually lanceolate or narrow-lanceolate, undivided or with a few long lobes. Filaments as long as the anthers. Mature berry ovoid or globular, greenish or purple, 3–4 cm diam. Erect, spreading shrub up to 2 m high, glabrous except young parts. Widespread. Open forests. Fl. winter–spring. . . . . **S. vescum** F.Muell.
3 Leaves with petioles up to 3 cm long. 4

4 Leaves very variable in shape, usually pinnate-lobed (the lower ones often deeply so); the largest 20–30 cm long; upper ones entire 5
4 Leaves ovate to linear or narrow-elliptic, up to 10 cm long, only rarely longer, entire or undulate, rarely with shallow lobes 6

5 Climber or sprawling shrub. Leaves ovate in outline, up to 13 cm long, with hairs on the margins. Corolla mauve, 20–30 mm diam. Mature berry globular, red, 8–12 mm diam. Coast. Occasionally naturalized in moist gullies. Introd. from W. Indies. Fl. spring–autumn. *Brazilian Nightshade*. . . . . *S. seaforthianum* Andrews
5 Erect shrubs. Corolla up to 3.5 cm diam., broad-campanulate to rotate, pale purple; lobes acute, as long as the tube. Stems purplish or green. Glabrous, shrubby perennial up to 2 m high. Widespread. WSF and RF margins. Fl. summer. *Kangaroo Apple* . . . . **S. aviculare** G.Forst.

6 Sprawling or scrambling shrub, glabrous. Corolla star-like, with distinct lobes, white to violet, 20–25 mm diam. Leaves ovate to lanceolate, 10–20 mm wide, up to 50 mm long. Coast. Garden escape naturalized in gullies near habitation. Introd. from S. America. Fl. summer–autumn. *Potato Climber* or *Jasmine Nightshade*. . . . . *S. laxum* Spreng.
6 Erect, bushy shrubs 7

7 The entire leaves more than 10 times as long as broad, usually c. 5 cm long, linear to narrow-elliptic, glabrous or almost so. Corolla rotate, reddish violet, 35–45 mm diam. Mature berry globular, yellow with violet markings, 15–20 mm diam. Blue Mts Usually in exposed situations. Fl. spring–summer. . . . . **S. linearifolium** Geras.. ex Symon
7 The entire leaves less than 10 times as long as broad 8

8 Small leaves present in the axils, recurved, resembling stipules. Leaves hairy . . . .
. . . . **S. mauritianum** (see below)

8 No stipule-like small leaves present in the axils 9

9 Corolla white, 13–20 mm diam. Leaves ovate to elliptic, up to 25 cm long, densely hairy. Mature berry yellow, c. 10 mm diam. Shrub or small tree up to 6 m high. Coast. Margins of RF. Fl. autumn. *Potato Tree* or *Tobacco Tree* . . . . **S. erianthum** D.Don

9 Corolla violet or pale blue 10

10 Corolla violet 20–30 mm diam. Leaves elliptic to lanceolate, 3–8 cm long, glabrous. Mature berry green tinged with purple, 10–15 mm diam. Erect shrub up to 2 m high. Widespread. Disturbed areas. Fl. spring. *Oondoroo* . . . . **S. simile** F.Muell.

10 Corolla pale blue, 25–40 mm diam. Leaves hairy at least on undersurface . . . .S. brownii (see below)

11 Stellate hairs forming a dense soft tomentum on all parts. Leaves ovate to narrow-ovate, acuminate, velutinous, 8–20 cm long; the young axillary leaves semi-orbicular, recurved, resembling stipules. Flowers in dense pedunculate cymes. Corolla pale blue or whitish. Mature berry yellowish, c. 12 mm diam. Coast. Weed of waste places; margins of RF; WSF. Introd. from trop. Asia. Fl. autumn–spring. *Wild Tobacco Bush* . . . . *****S. mauritianum** Scop.

11 Hairs simple or glandular or absent 12

12 At least the lower leaves deeply cut or lobed 13

12 Leaves entire or pinnate toothed or sinuate, petiolate 14

13 Leaves ternate-lobed or ternate, up to 10 cm long and 6 cm wide; the lower lobes ovate-elliptic. Corolla white, up to 10 mm diam. Mature berry yellow-green, 4–5 mm diam. Ascending or sprawling herb with stems up to 1 m long. Woronora Plateau; Cumberland Plain; Hunter River Valley. Weed of cultivation and waste places. Introd. from S. America. Fl. summer . . . . *****S. radicans** L.

13 Leaves pinnately lacerate; the lobes linear-deltoid, 15–30 mm long. Flowers regularly in threes. Corolla white, 5–10 mm diam. Mature berry whitish green, marbled. Sprawling, branched herb. Widespread. Weed of cultivation and waste places. Introd. from N. America. Fl. summer. *Three-flowered Nightshade* . . . . *****S. triflorum** Nutt.

14 Mature berry 10–15 mm diam., orange-red, globular. Flowers solitary or 2–3 together on short lateral peduncles. Corolla c. 10 mm diam., white. Leaves lanceolate, entire, 2–10 cm long. Erect shrub 1–1.5 m high. Coast; Cumberland Plain. Weed of cultivation and waste places. Introd. from America. Fl. most of the year. *Madiera Winter Cherry*. . . . . *****S. pseudocapsicum** L.

14 Mature berry 5–7 mm diam., black or green. Leaves ovate to lanceolate, sinuate-toothed to entire, glabrous or hairy, mostly 2–5 cm long. Corolla white to purplish 15

15 Plants densely pubescent. Anthers 2–3.5 mm long 16

15 Plants glabrous or with scattered hairs. Anthers usually up to 2 mm long 17

16 Mature berry glossy, brownish green to dark brown, reticulate; the calyx membranous and c. half as long. Hairs on the stems and leaves mostly glandular. Corolla with triangular to narrow-triangular reflexed lobes. Infructescence raceme-like, with a few short internodes. Style equal to or slightly exceeding the anthers. Sprawling annual up to 70 cm high. Widespread. Weed of cultivation and waste places. Introd. from S. America. Fl. summer . . . .
. . . . *****S. physalifolium** Rusby var. **nitidibaccatum** (Bitter) Edmonds

16 Mature berry dull black; the calyx short, herbaceous. Hairs on the stems and leaves simple. Leaves whitish below. Corolla with narrow-oblong to narrow-elliptic lobes. Infructescence umbellate. Style exceeding the anthers by 1–2 mm. Sprawling perennial up to 1 m high. Widespread. Weed of cultivation and waste places. Introd. from S. America. Fl. summer . . . . . . *****S. chenopodioides** Lam.

17 Mature berry purplish to black, dull. Leaves concolorous and ± hairy. Hairs on stems and leaves simple and glandular. Corolla with ovate to linear lobes. Infructescence raceme-like, with a few short internodes near the base. Calyx spreading-recurved to appressed around the fruit. Style as long as or exceeding the anthers by 1 mm. Erect to sprawling herb up to 1 m high. Widespread. Weed of cultivation and waste places. Introd. from Europe. Fl. spring. *Black Nightshade* . . . . . ***S. nigrum** L.

17 Mature berry glossy with the flesh not aromatic. Calyx reflexed below the fruit. Hairs if present simple. Bushy, ascending to erect herb up to 1.5 m high. Widespread. Weed of cultivation and waste places. Possibly native. Fl. most of the year. . . . . **S. americanum** Mill.

18 Prickles present on the branches, sparse and occasionally 1–2 on the upper surfaces of the leaves or calyces 19
18 Prickles present on branches leaves and flowers. 23

19 Upper surface of leaves moderately to densely hairy 20
19 Upper surface of leaves glabrous to sparsely hairy (stellate hairs not overlapping) 21

20 Corolla purple, 15–18 mm diam. Style c. 7 mm long. Leaves elliptic to lanceolate, 5–12.5 cm long. Petiole 1–2 cm long. Mature berry 13–16 mm diam, pale green. Shrub to 2.5 m high. Wollongong to Nowra. RF. or WSF. Endangered. Fl. spring. . . . . **S. celatum** A.R.Bean
20 Corolla purple, 24–32 mm diam. Style 8–9 mm long. Leaves elliptic, 5–5.5 cm long, margins entire or sinuate, occasionally with up to 2 lobes. Petiole 0.5–1.2 cm long. Mature berry green, globular. Shrub to c. 2 m high. Yerranderie to Wombeyan Caves. DSF. Endangered. Fl. spring–summer . . . .
. . . . **S. armourense** A.R.Bean

21 Leaves oblong or narrow-elliptic, up to 4 cm long, c. 7 mm wide, with a dense yellow-grey pubescence underneath. Petiole 3–5 cm long. Corolla violet, 10–15 mm diam. Fruit subglobular, c. 5 mm diam. Mature berry 5–8 mm diam., red. Spreading shrub up to 1.5 m high. Berrima. Drier communities. Fl. spring. . . . . **S. parvifolium** R.Br.
21 Leaves lanceolate to ovate or elliptic, 4–12 cm long, more than 10 mm wide, with a loose brownish tomentum underneath 22

22 Corolla stellate, c. 10 mm diam.; the lobes narrow, often reflexed, pale violet. Petiole 5–20 cm long. Mature berry c. 5–12 mm diam., globular, red. Shrub up to 2 m high. Widespread. Forests. Fl. spring–summer. *Devil's Needles* . . . . **S. stelligerum** Sm.
22 Corolla rotate, 20–25 mm diam.; the lobes broad-deltoid, violet. Petiole to 1 cm long. Mature berry 10–15 mm diam., globular, white or green to yellow. Shrub up to 2 m high. Widespread. Drier forests. Fl. winter–spring . . . . **S. brownii** Dunal

23 Undersurfaces of leaves densely tomentose to hirsute and glandular pubescent 24
23 Leaves glabrous or sprinkled with stellate hairs. 26

24 Undersurfaces of leaves covered with a dense close whitish or greyish stellate tomentum. Branches and inflorescence with a similar stellate tomentum. Leaves ovate to ovate-lanceolate, acute, sinuate-pinnatifid, 5–14 cm long, green and almost glabrous on the upper surface. Flowers 2–4 in pedunculate cymes. Calyx prickly, 5–8 mm long. Corolla blue, 20–25 mm diam. Mature berry globular, 20–25 mm diam., yellowish. Shrub up to 1 m high. Widespread. DSF; WSF; RF. Fl. spring–autumn. *Narrawa Burr.* . . . . **S. cinereum** R.Br.
24 Leaves hirsute and with a brownish pubescence 25

25 Mature berry greenish yellow to black, globular 20–30 mm diam., with an enlarged prickly calyx. Prickles numerous on the stems. Leaves ovate to elliptic, 5–13 cm long; tomentum denser underneath where many stellate hairs also occur. Flowers few in lateral raceme-like cymes. Corolla blue to violet, c. 25 mm diam. Small, coarse shrub up to 1 m high. Cumberland Plain; Blue Mts DSF; WSF. Fl. most of the year . . . .**S. campanulatum** R.Br.
25 Mature berry red, 15–20 mm diam., with an enlarged ± prickly calyx. Prickles numerous on the stems. Leaves deeply lobed, ovate in outline, up to 14 cm long and 10 cm wide, ± similarly hairy on both surfaces. Corolla white to blue, stellate, 30–50 mm diam. Weed of waste land near Sydney. Introd. from America. Fl. spring–summer . . . . ***S. sisymbriifolium** Lam.

26 Prickles on the stems and leaves stout, mostly 1–4 mm wide at the base although some may be narrower. Mature berry orange or yellow when mature, more than 20 mm diam. 27

26 Prickles on stems and leaves all slender, with narrow bases less than 1 mm wide. Mature berry c. 12 mm diam., mottled white and green when mature 29

27 Mature berry bright orange, dry, with numerous winged seeds. Leaves ovate, up to 15 cm long, lobed; the lobes acute or shortly acuminate. Some prickles usually narrow and hair-like. Corolla stellate, yellow or white, 20–30 mm diam. Weak shrub up to 1 m high. Coast. Damp forests. Introd. from S. America. Fl. spring–summer. *Devil's Apple* . . . . ***S. capsicoides** All.

27 Mature berry yellow, sometimes changing to blackish, succulent, with compressed but not winged seeds 28

28 Flowers up to 6 together with a common peduncle, in lateral racemes. Mature berry yellow 20–30 mm diam., changing to blackish. Leaves elliptic in outline, deeply obtusely pinnately lobed, 7–15 cm long. Prickles almost all stout. Corolla violet, 25–30 mm diam. Herbaceous perennial with stout spreading branches up to 1 m high. Coast. Sheltered places in forests; roadsides; creek banks. Introd. from the Mediterranean. *Apple of Sodom.* . . . . ***S. linnaeanum** Hepper & P.-M.L.Jaeger

28 Flowers solitary or 2 together, each on a slender pedicel without a common peduncle. Mature berry yellow, 25–30 mm diam. Corolla 15–20 mm diam., pale purple. Leaves elliptic, deeply lobed, 5–8 cm long, slightly paler underneath. Widespread. Open forests; margins and clearings in RF. Fl. spring–summer . . . . **S. pungetium** R.Br.

29 Flowers mostly 2–3 in a loose lateral raceme; the common peduncle ± elongated beyond the lowest pedicel. Corolla 20–25 mm diam., violet. Mature berry c. 12 mm diam. Leaves ovate to broad-oblong-lanceolate, sinuate-lobed or pinnatifid, acute, thin, green on both surfaces, mostly up to 10 cm long. Widespread. DSF; WSF; margins of RF. Fl. most of the year . . . . **S. prinophyllum** Dunal

29 Flowers solitary or 2 together, each on a slender pedicel without a common peduncle . . . . **S. pungetium** (see above)

4 Lycopersicon Mill.
1 species naturalized Aust.; Qld, NSW

One species in the area

Ascending to decumbent, pubescent, annual herb up to 1.5 m high. Leaves ovate to oblong in outline, pinnate-lobed to deeply serrate. Flowers in few-flowered terminal cymes. Calyx deeply 5-lobed. Corolla yellow, c. 2 cm diam.; lobes acute-acuminate. Stamens connivent and coherent in a cone around the style. Fruit a large red berry. Garden escape near habitation and on rubbish tips. Introd. from S. America. Fl. spring–summer. *Tomato* . . . . ***L. esculentum** Mill.

5 Cestrum L.
4 species naturalized Aust.; Qld, NSW, Vic.

Shrubs, sometimes straggling and almost climbing. Leaves alternate, entire or obscurely toothed or undulate. Flowers in axillary or terminal cymes. Corolla trumpet-shaped to funnel-shaped, red, pink, orange, yellow-green to orange-yellow; the tube longer than the calyx. Stamens 5. Ovary 2-locular. Fruit a berry.

Key to the species

1 Flowers reddish orange to reddish-purple. Hairs on stems and inflorescences purple 2

1 Flowers greenish orange yellow to greenish yellow. Hairs when present not purple 3

2 Calyx and corolla densely pubescent on outside. Calyx 8–9 mm long. Corolla red or orange-red, tube 15–25 mm long. Leaves ovate, 5–13 cm long. Flowers in terminal panicles. Berry 15 mm diam., red. Shrub to 3 m high. Occasional garden escape. Wollongong area. Introd. from Mexico. Fl. winter-spring . . . . ***C. fasiculatum** (Schltdl.) Miers

2 Calyx and corolla ± glabrous. Calyx 4.5–6 mm long. Corolla reddish pink to reddish-purple, tube 15–23 mm long. Leaves lanceolate, 7–12 cm long. Flowers in terminal panicles. Berry 8–13 mm diam., red. Shrub to 5 m high. Occasional garden escape. Wollongong and southern ranges area. Introd. from Mexico. Fl. winter–spring . . . . *C. elegans (Brongn. ex Neumann) Schltdl.

3 Corolla orange-yellow, up to 2 cm long. Leaves elliptic, up to 10 cm long, with a few obscure teeth or entire. Straggling shrub up to 2 m high. Poisonous to stock. Garden escape hear habitation. Introd. from Central America. Fl. most of the year . . . . *C. aurantiacum Lindl.

3 Corolla greenish yellow, up to 2 cm long. Leaves narrow-elliptic to lanceolate, up to 7 cm long, obscurely undulate to entire. Ascending shrub up to 2 m high. Flowers very fragrant at night. Poisonous to stock. Garden escape near habitation. Introd. from Central America. Fl. most of the year. *Green Poisonberry* . . . . *C. parqui L'Hér

6 Lycium L.
4 species in Aust. (1 endemic, 3 naturalized); all states and territories

Shrubs, usually glabrous, with greyish usually spiny branches. Corolla funnel-shaped; lobes spreading, lilac. Stamens 5, ± unequal, exserted. Ovary 2-locular. Fruit a berry.

Key to the species

1 Lateral branches leafy, mostly more than 10 mm long, ending in a stout spine. Leaves thick, oblanceolate to obovate, 15–40 mm long. Corolla 8–12 mm long; tube slightly exceeding the calyx. Berry globular, orange-red, 10–12 mm diam. Widespread. Weed of pastures and disturbed sites. Introd. from Africa. Fl. most of the year. *African Box-thorn* . . . . *L. ferocissimum Miers

1 Lateral branches reduced to leafless spines less than 10 mm long. Leaves thin, ovate to elliptic, 8–20(70) mm long. Corolla 10–12 mm long. Berry red, ellipsoid, 3–4 mm diam. Coast. Waste places. Introd. from China. Fl. summer–autumn. *Chinese Box-thorn.* . . . . *L. barbarum L.

7 Cyphomandra Mart. ex Sendtn.
1 species naturalized Aust., Qld, NSW.

One species in the area

Soft-wooded, pubescent shrub or small tree up to 3 m high with an unpleasant odour. Leaves ovate, mostly up to 12 cm long, cordate at the base. Flowers in cymes often resembling racemes. Corolla stellate, pink, c. 20 mm diam. Anthers dehiscing by apical pores. Ovary 2-locular. Fruit an ovoid berry 5–7 cm long, red. Coast. Naturalized in sheltered situations. Introd. from S. America. Fl. summer. *Tamarillo or Tree Tomato* . . . . *C. betacea (Cav.) Sendtn.

8 Salpichroa Miers
1 species naturalized Aust.; all states and territories except NT

One species in the area

Minutely hairy, perennial herb 30–60 cm high, with weak straggling branches. Leaves thin, petiolate, ovate to rhomboid, 6–20 mm long. Flowers axillary, solitary on slender reflexed pedicels. Corolla tubular, white, 5–6 mm long. Stamens inserted above the middle of the tube; filaments inserted at the back of each anther. Berry yellow, ovoid, up to 18 mm long. Widespread. Weed in vacant land near habitation. Introd. from S. America. Fl. spring–summer. *Pampas Lily-of-the-Valley*. . . . . *S. origanifolia (Lam.) Baill.

9 Duboisia R.Br.
3 species in Aust., (2 endemic); all states and territories except Tas.

One species in the area

Erect, glabrous shrub or small tree. Leaves alternate, obovate to oblanceolate, 3–10 cm long, entire, tapering into a petiole. Flowers in a loose terminal panicle. Calyx broadly campanulate, with obtuse teeth. Corolla campanulate, c. 4 mm long, white to pale lilac; lobes rather shorter than the tube. Fruit a small globular berry 5–8 mm diam., purplish black. Coast; Illawarra ranges; Kurrajong district. RF. and WSF. Fl. spring. *Corkwood* . . . . .D. myoporoides R.Br.

10 Datura L.

Thornapples or *Jimson Weeds*

6 species naturalized Aust.; all states and territories

Robust, annual herbs up to 1.5 m high, with an unpleasant odour. Stems green to purplish. Leaves alternate, usually opposite towards the top, petiolate, glabrous or nearly so. Flowers apparently solitary in the forks of the branches; pedicels elongating in the fruiting stage. Calyx 5-angled, acutely 5-toothed. Corolla funnel-shaped, white or purplish. Stamens 5, enclosed in the corolla tube. Ovary 2-locular. Capsule ovoid, 4-locular by a secondary dissepiment, armed with pickles.

Key to the species

1 Prickles on the capsule stout, mostly more than 3 mm diam at the base, 10–15 mm long or longer. Corolla white, 2–3 cm diam. Leaves irregularly sinuate, ovate to oblong in outline. Robust herb up to 1.5 m high. Poisonous weed. Widespread. Weed of cultivation, stock yards and waste places. Introd. from America. Fl. summer–autumn. . . . .***D. ferox** L.

1 Prickles on the capsule mostly 2 mm wide at the base or less, up to 12 mm long. Corolla white or purplish, 3–4 cm diam. Leaves sinuate-dentate, ovate in outline. Robust herb up to 1.5 m high. Poisonous weed. Widespread. Weed of cultivation, stock yards and waste places. Introd. from America. Fl. summer–autumn . . . . ***D. stramonium** L.

11 Nicotiana L.

17 species in Aust. (16 endemic, 1 naturalized); all states and territories

Herbs or shrubs. Leaves alternate. Calyx acutely 5-toothed. Corolla tubular, with 5 short lobes. Stamens 5, enclosed in the corolla tube. Ovary 2-locular; stigma capitate. Fruit a capsule enclosed in the persistent calyx, opening septicidally in 2 bifid valves.

Key to the species

1 Glaucous shrub 2–4 m high. Leaves ovate to ovate-lanceolate, 4–10 cm long, on slender petioles. Flowers in short loose panicles at the ends of branches. Calyx 10 mm long. Corolla pale yellow, 3–4 cm long, tubular, swollen below the short limb. Widespread. Waste places. Introd. from S. America. Fl. chiefly spring. *Tobacco Bush*. . . . . ***N. glauca** Graham

1 Erect herbs, not glaucous 2

2 All stamens inserted in the lower half of the corolla tube. Cauline leaves sessile, ± stem-clasping, auriculate, narrow-elliptic to linear, up to 20 cm long. Stem viscid in the upper part with ellipsoid-headed hairs. Hawkesbury River district. Fl. summer. . . . . **N. forsteri** Roem. & Schult.

2 Four stamens inserted near the mouth of the corolla tube; one stamen usually inserted in the lower half of the corolla tube, its anther always 4–5 mm below the other anthers. Cauline leaves mostly petiolate, ovate to elliptic, up to 15 cm long. Stems not viscid, usually only slightly pubescent. Widespread. Sheltered situations; creek banks. Fl. spring–summer. *Native Tobacco* . . . . **N. suaveolens** Lehm.

12 Cyphanthera Miers

9 species endemic AUST,; all states and territories except NT

Shrubs. Leaves entire, alternate. Flowers in panicles or cymes or solitary, on filiform or stout peduncles. Calyx 5-toothed. Corolla campanulate; tube shortly contracted at the base with 5 ± unequal lobes. Stamens 4, unequal, inserted at the base of the corolla tube, sometimes with an extra staminode. Fruit a small globular capsule.

Key to the species

1 Flowers on pedicels longer than the subtending leaves. Leaves pubescent with short stellate hairs. Corolla white, 5–10 mm long. Shrub up to 2 m high. Valleys of the lower Blue Mts, e.g. Bilpin, Mt Wilson. DSF and WSF. Fl. spring. . . . .**C. scabrella** (Benth.) Miers

1 Flowers almost sessile; pedicels shorter than the subtending leaves or obsolete. Leaves and young stems tomentose to densely pubescent, with long white often stellate hairs. Corolla white with purple markings. Shrub up to 2 m high. Cumberland Plain; valleys of western Blue Mts DSF. Fl. spring–summer . . . **C. albicans** (Cunn.) Miers ssp. **albicans**

13 Petunia Juss.
1 or 2 species naturalized Aust.; Qld, NSW, WA

One species in the area

Ascending to decumbent, softly glandular hairy herb up to 80 cm high. Leaves alternate, elliptic to obovate or narrow-obovate, up to 7 cm long. Flowers solitary on peduncles longer than the subtending leaves. Corolla white, 4–6 cm diam.; tube longer than the calyx. Stamens 4 with 1 staminode. Ovary 2-locular. Fruit a 2-valved capsule 8–12 mm long. Coast. Waste places. Introd. from S. America. Fl. spring–summer. . . . . ***P. axillaris** (Lam.) Britton, Sterns & Poggenb

136 POLYOSMACEAE
1 gen., trop. subtrop. temp.; Asia, to New Caledonia

One genus in the area

1 Polyosma Blume
6 species endemic Aust.; Qld, NSW

One species in the area

Small tree. Leaves opposite, ovate-elliptic to narrow-ovate, acuminate, glabrous, paler underneath, 5–10 cm long, with irregular coarse callous teeth. Flowers in racemes shorter than the leaves. Floral tube ovoid. Calyx teeth 4, persistent. Petals 4, c. 10 mm long, cohering into a tube for half their length, spreading at the tips. Ovary inferior. Berry ovoid, 12 mm long. Coast and adjacent plateaus. RF. Fl. spring–summer. . . . . **P. cunninghamii** Bennett

137 QUINTINIACEAE
1 gen., trop. subtrop. temp., N. Guinea, N. Zealand, New Caledonia.

One genus in the area

1 Quintinia A.DC.
4 species endemic Aust.; Qld, NSW

One species in the area

Glabrous tree which frequently begins life on the trunks of tree ferns where the seed germinates. Leaves alternate, broad- to narrow-elliptic, 5–10 cm long, thick, entire. Flowers numerous in a terminal racemose panicle. Floral tube obconical. Calyx teeth 5, persistent. Petals 5, white, 4 mm long. Stamens 5. Ovary inferior, 3–5-locular; style 3–5-furrowed, persistent; stigma 3–5-lobed. Fruit an inferior capsule; the persistent style separating up to the stigmas. Widespread. RF. Fl. spring. *Possum Wood* **Q. sieberi** A.DC.

138 APIACEAE
Herbs or shrubs. Leaves usually alternate; often highly dissected or compound, often basal; in the case of herbs the petiole often dilated into a sheathing base (Fig. 30). Stipules absent except in *Centella*. Flowers in simple or compound umbels or heads often subtended by an involucre of bracts, usually bisexual, regular. Sepals 5, small or sometimes obsolete. Petals 5, often bent inwards. Stamens 5, alternating with the petals. Ovary inferior, 1–2-locular, with 1 ovule per loculus; styles 2, free, swollen at the base into a fleshy epigynous nectiferous disc. Fruit a schizocarp, usually separating at maturity into two 1-seeded mericarps (except *Actinotus*) often suspended individually on the divided filiform central axis (*carpophore*);

each carpel usually marked with longitudinal ribs and frequently by linear oil ducts or glands (*vittae*) which can be observed only in transverse section. 250 gen., mostly N. hemisphere.

N.B. recent research has resulted in the transfer of the genera *Hydrocotyle* and *Trachymene* to the family Araliaceae.

Key to the genera

1 Leaves and bracts pungent pointed. Flowers in dense heads or short cylindrical spikes . . **14 ERYNGIUM**
1 Leaves and bracts never pungent pointed 2

2 Umbels simple 3
2 Umbels compound 9

3 Flowers numerous, crowded, surrounded by radiating involucral bracts exceeding the flowers, the whole resembling a head. Fruit undivided . . . . **1 ACTINOTUS**
3 Flowers not arranged as in *Actinotus* with the bracts seldom exceeding the flowers. Fruit a schizocarp . . 4

4 Leaves septate, simple, linear-terete (Fig. 28) . . . . **4 LILAEOPSIS**
4 Leaves not septate, compound or simple 5

5 Stipules present 6
5 Stipules absent 7

6 Involucral bracts conspicuous. Petals imbricate . . . . **3 CENTELLA**
6 Involucral bracts very small or absent. Petals valvate (*Hydrocotyle*) . . . . **see ARALIACEAE**

7 Calyx lobes conspicuous . . . . **2 XANTHOSIA**
7 Calyx lobes minute or absent 8

8 Fruit ovate, slightly compressed . . . . **5 OREOMYRRHIS**
8 Fruit orbicular, compressed (*Trachymene*) . . . . **see ARALIACEAE**

9 Fruit and ovary densely prickly . . . . **6 DAUCUS**
9 Fruit and ovary without prickles 10

10 Fruit and ovary with a long beak . . . . **8 SCANDIX**
10 Fruit and ovary without a long beak 11

11 Calyx with conspicuous teeth . . . . **2 XANTHOSIA**
11 Calyx without conspicuous teeth 12

12 Leaves entire or lobed 13
12 Leaves compound or very deeply dissected 15

13 Leaves peltate (*Hydrocotyle*) . . . . **see ARALIACEAE**
13 Leaves not peltate 14

14 Leaves not stem clasping . . . . **12 PLATYSACE**
14 Leaves stem clasping . . . . **15 BUPLEURUM**

15 Stems and petioles with purplish blotches . . . . **13 CONIUM**
15 Stems and petioles without purplish blotches 16

16 Flowers yellow. Leaves with a strong odour of aniseed when crushed . . . . **11 FOENICULUM**
16 Flowers white. Leaves without an aniseed odour 17

17 Involucral bracts present at least at the base of the pedicels . . . . **7 AMMI**
17 Involucral bracts absent 18

18 Leaves filiform, c. 0.5 mm wide . . . . **10 CICLOSPERMUM**
18 Leaves not filiform, 1 mm or more wide . . . . **9 APIUM**

1 Actinotus Labill.

14 species endemic Aust.; all states and territories except S.A.

Herbs, perennial or appearing annual by flowering in the first season, sometimes becoming woody. Leaves ternately divided. Umbels simple, surrounded by a radiating involucre of white or slightly coloured often very woolly or tomentose bracts, the whole having the appearance of a composite (Asteraceae) flower. Flowers very numerous, crowded; the outer ones often male. Calyx lobed or truncate. Petals absent or 5. Ovary 1-locular, with 2 styles. Fruit ovate, 1-seeded, compressed, without vittae, 5-ribbed.

Key to the species

1 Involucre 5–8 cm diam., densely white-tomentose, sometimes tipped with green. Erect annual or perennial 30–150 cm high, usually slender when in open forest but stouter and more woody when in exposed situations, covered with a soft dense woolly white or brownish tomentum. Leaves twice 3-partite; segments linear or oblong-linear, almost obtuse, sometimes further divided. Umbels on long peduncles. Flowers on filiform pedicels 3–4 mm long. Calyx lobes very small, linear. Fruit c. 4 mm long, covered with silky hairs. Widespread. DSF and heath. Sandy soils and coastal headlands. Fl. spring–summer. *Flannel Flower* . . . . **A. helianthi** Labill.

1 Involucre less than 3 cm diam. Plants diffuse, with wiry branches, glabrous or silky hairy. 2

2 Peduncles less than 1 cm long. Leaves glabrous or slightly hairy, up to 15 mm long. Involucre 3–8 mm diam. Flowers pinkish. Ascending or decumbent annual or perennials with stems up to 50 cm long. Western Blue Mts DSF and heath. Sandy soils. Fl. spring–summer . . . . **A. gibbonsii** F.Muell.

2 Peduncles more than 2 cm long. 3

3 Involucre more than 15 mm diam.; bracts white to pink, silky hairy. Petals small, spathulate, with slender claws. Leaves 3–5-partite; segments much incised, with acute lobes. Calyx truncate or sinuate. Loose decumbent herb with stems up to 50 cm long. Blue Mts DSF and heath. Fl. spring–autumn. . . . . **A. forsythii** Maiden & Betche

3 Involucre rarely exceeding 12 mm diam.; bracts white, tomentose. Petals absent. Leaves 3-partite; segments entire or lobed, acute, glabrous on the upper surface, silky or white-tomentose underneath. Calyx lobes acute. Sprawling herb with stems up to 50 cm long. Widespread. Heath and DSF. Sandy soils. Fl. most of the year. *Lesser Flannel Flower* . . . . **A. minor** (Sm.) DC.

2 Xanthosia Rudge

20 species endemic Aust.; all states and territories except NT

Herbs or small shrubs. Leaves toothed to lobed or dissected or rarely entire. Umbels usually compound, sometimes simple or reduced to a few or a single flower. Compound umbels usually with 3–4 rays and as many bracts; umbellules with 2–3 bracts and several almost sessile flowers. Calyx lobes broad, orbicular to lanceolate. Petals usually with a much inflexed apex. Fruit without vittae.

Key to the species

1 Leaves cauline, simple or 3-foliolate; petioles shorter than the blades 2

1 Leaves mostly basal, ternately dissected; petioles up to 12 cm long, usually longer than the blades . . . 5

2 Leaves simple but nearly equally 3-toothed at the tip, cuneate, mostly less than 12 mm long, glabrous or white-tomentose underneath. Umbels mostly 1–3-flowered; peduncles short mostly reflexed. Calyx lobes acute, peltate. Diffuse perennial c. 30 cm high, hirsute or nearly glabrous. Widespread. Heath. Fl. winter . . . . **X. tridentata** DC.

2 Leaves simple or compound, oblong to ovate in outline. Calyx not peltate 3

3 Older branches of plants with flaking bark. Upper surface of leaves grey-green. Young stems and leaves with stellate hairs. Juvenile leaves trifoliolate; adult leaves simple. Inflorescence with 2–3 rays, 1–4 flowers per ray. Fruit 2 mm long. Small compact shrub with many branches to 20 cm high. Jamison Valley. Sandstone cliffs. Vulnerable. Fl. summer . . . . **X. scopulicola** J.M.Hart & Henwood

3 Older branches without flaking bark. Upper surface of leaves green 4

4 All leaves compound, terminal leaflet ± equal to lateral leaflets and similar in shape. Inflorescence with 1–2 rays, 1–2 flowers per ray. Fruit 3 mm long. Plants sparsely stellate hairy. Small erect to procumbent shrub to 20 cm high. Blue Mts DSF. Ss. Fl. mainly summer. . . . . . . **X. stellata** J.M.Hart & Henwood

4 Leaves simple or compound. If compound then the terminal leaflet is at least 1.5 times longer than lateral leaflets. Inflorescence with 1–4 rays, 1–4, sometimes 9 flowers per ray. Fruit 2–3 mm long. Plants glabrous to densely covered with dendritic hairs. Erect or scrambling shrub 10–50 cm high. Widespread. DSF and heath. Fl. all year . . . . **X. pilosa** Rudge

5 Leaves up to 15 mm long, finely dissected, usually twice tripartite, glabrous; the ultimate segments acute. Peduncles slender, axillary, much shorter than the leaves. Bracteoles green. Perennial herb up to 15 cm high with a woody rootstock. Blue Mts; Illawarra district. Wet heath; swamps. Fl. spring–summer . . . . **X. dissecta** Hook.f.

5 Leaves more than 20 mm long, 3-partite; segments cuneate, 3-partite, acute, thin, wiry. Peduncles stout, up to 18 cm long. Flowering stems wiry, almost leafless. Bracteoles white. Erect perennial herb up to 60 cm high with a woody rootstock. Blue Mts DSF. Fl. spring–summer . . . . . . **X. atkinsoniana** F.Muell.

3 Centella L.
2 species native Aust.; all states and territories except NT

Perennial herb. Stems creeping, rooting at the nodes. Leaves broad-cordate or orbicular to reniform, entire, sinuate or crenate, glabrous or pubescent.; stipules broad, usually entire. Flowers 3–4 in small pedunculate or sessile heads or umbels. Petals broad, much imbricate in the bud. Fruit laterally compressed, c. 3mm diam., mericarps with 3–4 ribs on each side with reticulations between.

Key to the species

1 Leaves orbicular to reniform, entire, sinuate or crenate, glabrous with strongly crenate margins, 1–4 cm wide. Fruit strongly reticulate. Sepals absent. Petals pink to purple or white. Coast. Margins of swamps; damp places in gardens and by roadsides . . . . **C. asiatica** (L.) Urb.

1 Leaves longer than broad, 1–4 cm long, with entire to weakly crenate margins, usually glabrescent. Fruit slightly reticulate, glabrous or with hairs along the margin. Sepals a minutely lobed persistent ring. Petals purple. Damp places . . . . **C. cordifolia** (Hook.f.) Nannf.

4 Lilaeopsis Greene
3 species endemic Aust.; NSW, Vic., Tas.

One species in the area
Weak herb, creeping, rooting at the nodes. Leaves (reduced to phyllodes) 3–30 cm long, cylindrical, septate. Umbels simple, 2–4-flowered, on slender peduncles arising at the nodes. Involucral bracts few, minute. Pedicels filiform, 2–4 mm long. Fruit ovoid-oblong, contracted towards the base, 2–3 mm long. Widespread. Margins or lagoons and streams. Fl. summer . . . . **L. polyantha** (Gand.) H.Eichler

5 Oreomyrrhis Endl.
7 species endemic Aust.; NSW, Vic., Tas., S.A.

One species in the area
Tufted, perennial herb. Leaves once to thrice pinnatisect, usually 2–5 cm long on long petioles. Umbels simple, often on stiff erect peduncles 15–30 cm long; pedicels short. Involucral bracts short, entire or toothed. Calyx inconspicuous. Petals minute, white tinged with red. Fruit narrow, tapering towards the summit, 4–6 mm long, slightly compressed; mericarps bluntly 5-ribbed, with 1 vitta under each furrow and 2 at the commissure. Higher Blue Mts Grasslands and open forests. Fl. summer. . . . .
. . . . **O. eriopoda** (DC.) Hook.f.

6 Daucus L.

3 species in Aust. (1 native, 2 or 3 naturalized); all states and territories

Erect, annual or biennial herbs. Leaves 2–3-pinnate, usually slightly hairy. Umbels compound. Flowers small, white to cream. Fruit with barbed prickles and vittae.

Key to the species

1 Umbellules 1–6-flowered; umbels with 2–5 unequal rays, terminal or in whorls along the stem. Mature fruit 4–5 mm long. Erect annual up to 20 cm high. Widespread in a variety of communities. Fl. spring–summer. *Native Carrot* . . . . **D. glochidiatus** (Labill.) Fischer, C.A.Mey. & Avé-Lall.
1 Umbellules more than 6-flowered; umbels usually with more than 6 regular rays, terminal. Mature fruit 3–4 mm long. Coarse, erect herb up to 150 cm high. Coast and Cumberland Plain. Waste places. Introd. from Europe. Fl. spring–autumn. *Wild Carrot* . . . .* **D. carota** L.

7 Ammi L.

2 species naturalized Aust.; QLD, NSW, Vic., S.A., WA

One species in the area

Glabrous annual or biennial 60–130 cm high. Leaves pinnately or twice pinnately divided; segments 1–3 cm long; those of the basal leaves oblong-ovate; those of the cauline leaves narrow- or linear-lanceolate; all serrulate. Umbels terminal and 3–6 cm diam. or lateral and smaller. Involucral bracts c. half as long as the rays, mostly pinnatifid with linear segments. Bracteoles subulate, c. equalling the pedicels. Flowers white, numerous, c. 3 mm diam. Fruit oblong to ovoid, c. 2 mm long; the ridges pale, prominent. Widespread. Weed of cultivated and vacant land. Introd. from Europe, Asia and N. Africa. Fl. summer. *Bishop's Weed* . . . . *****A. majus** L.

8 Scandix L.

1 species naturalized Aust.; NSW, Vic., Tas., S.A.

One species in the area

Nearly glabrous annual up to 50 cm high. Leaves 2–3-pinnate, ovate in outline. Umbels mostly 2-rayed, with bracts only at the base of the pedicels. Fruit with a long beak up to 7 mm long. Weed of waste land, not recorded since 1915. Introd. from Europe. *Shepherd's Needle* . . . . *****S. pecten-veneris** L.

9 Apium L.

4 species in Aust. (1 naturalized, 3 native); all states and territories

Glabrous herbs with hollow grooved stems and divided leaves. Umbels compound, terminal or leaf-opposed, without involucral bracts or bracteoles. Flowers white. Sepals obsolete. Petals with inflexed apices. Fruit subglobular, laterally compressed; mericarps with 5 prominent ribs and 1 vitta under each furrow.

Key to the species

1 Erect, aromatic biennial c. 1 m high, stems ribbed. Leaves pinnatisect; segments 3–5, ovate-cuneate, ± deeply 3-lobed. Umbels with 4–8 unequal rays. Fruit c. 3 mm long; mericarps with 5 whitish ribs. Coast and Cumberland Plain. Escape from cultivation. Introd. from Europe and Asia. Fl. spring–summer. *Garden Celery*. . . . . *****A. graveolens** L.
1 Stems prostrate or ascending or rarely erect. Leaves once or twice pinnatisect, often rather thick; the segments tripartite. Compound rays 3–6, each with an umbel of rather numerous small white flowers. Fruit c. 3 mm long; mericarps with 5 thick ribs. Small perennial with a thick rootstock. *Sea Celery* . . 2

2 Leaflets divided, primary segments 2–3 times as long as broad. Widespread. Coastal dunes and headlands. Fl. spring–autumn . . . . **A. prostratum** var. **filiforme** (A.Rich.) Kirk
2 Leaflets divided or entire, primary segments 6–15 times as long as broad. Near inland creeks and brackish water. Fl. spring–summer . . . . **A. prostratum** Labill. ex Vent. var. **prostratum**

10 Ciclospermum Lag.

1 species naturalized Aust.; Qld, NSW, Vic., NT, S.A.

One species in the area

Slender ascending annuals up to 70 cm high. Leaves thrice pinnatisect, glabrous, up to c. 7 cm long; segments linear-filiform, c. 0.5 mm wide. Umbels leaf-opposed with 2–4 rays; umbellules 6–20-flowered. Flowers white. Fruit 1–3 mm long. Widespread. Weed of cultivation and waste places. Introd. from America. Fl. most of the year. *Slender Celery* or *Carrot Weed* . . . .*C. **leptophyllum** (Pers.) Sprague

11 Foeniculum Mill.

1 species naturalized Aust.; Qld, NSW, Vic., Tas., S.A.

One species in the area

Erect, glabrous biennial 2m or more high, strongly aromatic (aniseed). Leaves finely divided with numerous filiform segments; petioles hollow, with sheathing bases. Flowers yellow, numerous, in compound umbels without involucral bracts or bracteoles; rays 10–30, on long peduncles. Calyx truncate. Petals rolled inwards. Fruit oblong, 5 mm long, slightly compressed; mericarps with 5 prominent ribs; vittae 1 under each furrow. Widespread. Common weed in vacant land. Introd. from Europe, W. Asia. *Fennel* . . . .*F. **vulgare** Mill.

12 Platysace Bunge

25 species endemic Aust.; all states and territories except Tas.

Shrubs. Leaves entire or lobed. Flowers small, white, in terminal compound umbels with small involucral bracts. Calyx teeth small, usually deciduous. Petals induplicate-valvate in bud with inflexed tips. Fruit slightly compressed laterally, ± rough, without vittae.

Key to the species

1 Leaves narrow-linear or subulate. 2
1 Leaves linear-lanceolate to orbicular, entire to dentate or lobed .3

2 Leaves mostly 5–10 mm long, linear to subulate, more spreading and usually firmer than in *P. linearifolia*. Stems short, diffuse, ± glandular pubescent. Umbels small, compact. Low shrub, often less than 30 cm high. Widespread. Heath and DSF. Sandy soils. Fl. spring–summer. . . . . **P. ericoides** (Sieber ex Spreng.) C.Norman
2 Leaves 10–25 mm long, narrow-linear to subulate, much more slender and less spreading than *P. ericoides*. Stems erect or ascending, glabrous. Umbels with slender rays. Erect or spreading shrub up to 150 cm high. Widespread. Heath and DSF. Sandy soils. Fl. spring–summer . . . . **P. linearifolia** (Cav.) C.Norman

3 Leaves all entire, linear-lanceolate to elliptic to ovate or orbicular, mostly less than 15 mm long when broad and obtuse, often twice as long when narrow, sessile or very shortly petiolate. Diffuse or erect shrub 60–150 cm high. Very variable in leaf form. Widespread. Heath and DSF. Sandy soils. Fl. summer . . . . **P. lanceolata** (Labill.) Druce
3 At least the lower leaves lobed or dentate 4

4 Lower or all the leaves deeply divided into 3 spreading lobes; the outer lobes sometimes themselves 2-lobed; upper leaves sometimes undivided, lanceolate to narrow-lanceolate. All leaves rigid, very acute, almost pungent pointed. Stems ± pubescent. Mericarps glabrous, bullate-rugose. Widespread. Heath and DSF. Fl. summer. (*P. stephensonii* hybribizes with *P. lanceolata*)**P. stephensonii** (Turcz.) C.Norman
4 All the leaves dentate at the apex with usually more than 3 acute teeth, obovate to orbicular, up to 10 mm long and 10 mm wide. Stems pubescent, with some villous hairs. Mericarps ± smooth, pubescent. Hornsby Plateau. DSF. Fl. spring–summer . . . .**P. clelandii** (Maiden & Betche) L. Johnson

13 Conium L.

1 species naturalized Aust.; Qld, NSW, Vic., Tas., S.A.

One species in the area

Erect, glabrous biennial 1.5–2 m high. Leaves much dissected, up to 30 cm long and wide, with an unpleasant odour; petioles hollow. Stems and petioles marked with purple spots. Umbels compound, with 10–15 rays and with c. 5 small reflexed involucral bracts; umbellules many-flowered, with 3 unilateral bracteoles. Sepals absent. Petals white with an incurved point. Fruit broad-ovoid, c. 3 mm long; mericarps with 5 prominent undulate ribs; vittae absent. Naturalized in places as a garden escape. Very poisonous. Introd. from Europe and W. Asia. *Hemlock* . . . . ***C. maculatum** L.

14 Eryngium L.

7 species in Aust. (4 native, 3 naturalized); all states and territories except NT

Glabrous herbs. Leaves usually with rigid spiny lobes, alternate near the base, opposite or whorled under the flowering branches or peduncles. Flowers sessile, in compound heads or short spikes, with a bract under each flower; the outer bracts, and sometimes the inner, much longer than the flowers, rigid, pungent pointed. Calyx lobes rigid, acute or pungent. Petals with a long inflexed point. Fruit ovoid, scarcely compressed, covered with bladdery scales. Plants resembling thistles.

Key to the species

1 Plants 1–2 m high. Basal leaves linear, c. 1 m long; margins with distinct fine prickles. Heads (umbels) in a tall cymo-panicle. Coast. Margins of watercourses. Introd. from S. America. Fl. summer–autumn
. . . . ***E. pandanifolium** Cham. & Schltdl.
1 Plants less than 1 m high 2

2 Main stems very short; branches long, prostrate, like stolons but not rooting at the nodes. Basal leaves lanceolate to broad-linear, 4–15 cm long, with spreading spiny teeth. Valleys of the western Blue Mts Fl. summer–autumn. . . . . **E. vesiculosum** Labill.
2 Main stems erect, branched, rigid, ± ribbed, 15–50 cm high 3

3 Basal leaves pinnatisect, 10–20 cm long; segments linear, rigid, pungent pointed; rhachis about as broad as the short petiole. Whole plant with a bluish tinge. Valleys of the western Blue Mts Open forests; pastures; waste places. Fl. summer. *Blue Devil* . . . . **E. ovinum** A.Cunn.
3 Basal leaves lobed or dentate, broad-ovate in outline; each lobe or tooth terminating in a rigid pungent point; petiole short, broad. Plant without a bluish tinge. Budgewoi. Coastal sand-dunes. Fl. summer–autumn . . . . **E. maritimum** L.

15 Bupleurum L.

3 species naturalized Aust.; Qld, NSW, Vic., S.A., WA

One species in the area

Annual herb 3–70 cm high. Leaves stem clasping, ovate to broad-lanceolate, 1.5–4.5 cm long, apex with mucronate tip; veins on lamina dark and conspicuous. Flowers yellow-green, in compound umbel with 3 or more rays. Bracts absent. Bracteoles circular. Sepals usually absent. Petals circular. Fruit with 5 ridges, 3–5 mm long, tuberculate. Occasional weed. Introd. from Europe. Fl. spring-summer . . . .
. . . . ***B. lanicifolium** Hornem.

139 ARALIACEAE

Trees, shrubs, herbs or rarely climbers or scramblers. Leaves alternate, simple or compound; stipules, small, inconspicuous or absent. Flowers bisexual or rarely polygamous, in paniculate umbels, heads or racemes. Calyx very small or truncate. Petals 4–5, valvate. Stamens 4–5, alternating with the petals. Ovary inferior, surmounted by an entire epigynous disc, usually 2-locular, with 1 ovule per loculus. Fruit drupaceous, with 2 hard flattened 1-seeded pyrenes, or a schizocarp. 80 gen., trop. to a few temp., particularly S.E. Asia and S. America.

Key to the genera

1 Hydrocotyle L.

Pennyworts

55 species in Aust. (most endemic); all states and territories

Small herbs, creeping and rooting at the nodes or sometimes ascending or erect. Leaves petiolate, orbicular to cordate in outline, entire or deeply divided; stipules scarious ± adnate to the petiole. Flowers minute, in simple axillary umbels or heads. Involucral bracts inconspicuous or absent. Sepals inconspicuous. Petals straight at the apex, acute, white, pink or yellowish. Fruit compressed laterally; mericarps 5-ribbed.

Key to the species

1 Leaves peltate, orbicular, crenulate, 1–6 cm or more diam., glabrous, on long petioles. Stems stoloniferous, rooting at the nodes 2
1 Leaves not peltate. 3

2 Umbels compound, 1–6 cm diam. Sandy soil near sea beaches. Introd. from S. America. Fl. summer. *Beach Pennywort* . . . . ***H. bonariensis** Lam.
2 Umbels simple, 2–5-flowered or sometimes the peduncle with successive whorls of flowers. Georges River. Freshwater swamps. Fl. summer. . . . . **H. verticillata** Thunb.

3 Leaves divided to the base (or nearly so) into 3–7 segments. 4
3 Leaves undivided and with crenate margins, or slightly lobed (rarely more deeply than to the middle) 5

4 Leaf segments large, lanceolate to narrow-lanceolate, acute, tapering gradually to the apex, toothed or lobed; leaf bases fringed; the central segment up to 5 cm long. Flowers white, rather numerous in each umbel; peduncles slender; pedicels filiform, 1–3 mm long. Stems lax, diffuse, up to 50 cm long, rooting at the nodes near the base. Widespread. Sheltered places. Fl. summer . . . . . **H. geraniifolia** F.Muell.
4 Leaf segments rarely more than 6 cm and sometimes less than 3 cm long, cuneate, entire to toothed or lobed, glabrous or sprinkled with a few hairs; leaf bases entire. Peduncles shorter than the leaves; flowers few, nearly sessile, greenish or yellowish. Stems creeping, rooting at the nodes. Widespread. Margins of streams; sheltered places; sometimes a weed in lawns. Fl. summer . . . . **H. tripartita** R.Br. ex A.Rich.

5 Fruit more than 3 mm wide. Leaves 1–5 cm wide.; reniform to orbicular-cordate; margins crenate; unbroken or with 3 to 11 shallow lobes. Flowers 5–20 per head ± sessile. Fruit smooth with thick dorsal ribs, base emarginate. Stems creeping, rooting at nodes. Moist places in open forest or RF. Fl. spring–autumn . . . . **H. pterocarpa** F.Muell.
5 Fruit less than 3 mm wide 6

6 Flowers usually unisexual, 30–50 in open umbels. Male flowers distinctly pedicellate, pedicels 6–8 mm long, those of the females much shorter. Leaves orbicular-cordate, shortly and broadly 5–11-lobed, crenate, 1–3 cm diam. Flowering branches and both surfaces of the leaves hirsute. Stems creeping, rooting at the nodes; flowering stems erect or ascending, up to 12 cm high. Widespread. Moist places. Fl. spring–summer . . . . **H. laxiflora** DC.

6 Flowers bisexual, sessile or very shortly pedicellate, up to 40 in small dense umbels 7

7 Fruits and flowers 15–40 per head, ± sessile. Leaves hairy; reniform to orbicular-cordate, 1–4 cm wide; margins dentate to crenate; lobes obtuse, shallow to deeply 5–7-lobed. Stipules ciliate with red markings. Petals yellow may have red markings, as wide as fruit at time of flowering. Fruit flattened with area between ribs convex often with red markings. Stems prostrate. Widespread. Moist shady places. Fl. summer . . . . **H. hirta** A.Rich.

7 Fruits and flowers less than 15 per head, occasionally with 15 but then pedicels to 1.5 mm long. Petals wider than fruit at time of flowering 8

8 Leaf lobes rounded, lamina reniform-cordate, 1–5 cm wide, upper surface glabrous and shiny lower surface hispid. Stipules entire and overlapping. Petiole with long hairs just below lamina. Petals off white often with purple markings. Fruit smooth to tuberculate; sibs prominent. Stems creeping, rooting at the nodes. Widespread. Sheltered places. Fl. summer . . . . **H. sibthorpioides** Lam.

8 Leaf lobes acute, peduncles always shorter than petiole; lamina orbicular to reniform-cordate; margins crenate; lobes acute, shallow to deeply 3–7-lobed. Stipules entire. Petals yellow, green or purple. Fruit smooth, olive to brown, ribs inconspicuous. Fl. summer . . . . **H. acutiloba** (F.Muell.) Wakef.

2 **Trachymene** Rudge
36 species in Aust. (34 endemic); all states and territories

Herbs with a perennial rootstock. Leaves dissected, 3–5-partite; the segments often twice tri-fid with acute lobes. Flowers very small, white. Umbels simple, usually c. 15 mm diam., on terminal or axillary peduncles. Calyx teeth minute. Fruit laterally compressed, usually flat, notched at the base, without vittae.

Key to the species

1 Leaves much divided with acute lobes, reduced in size or absent from the base of the peduncle. Taproot tuberose. Coast and adjacent plateaus. Heath and DSF. Sandy soils. Fl. summer . . . . **T. incisa** Rudge var. **incisa**

1 Leaves 3-partite 2

2 Leaves cut to the base into 3 primary segments, orbicular to broad-ovate in outline, up to 10 cm long. Umbels 2–3 together, 50–60-flowered. Widespread. Swamps and other moist places. Fl. most of the year . . . . **T. composita** (Domin) B.L.Burtt var. **composita**

2 Leaves cut to c. halfway into 3–5 primary segments, orbicular, up to 4 cm long. Umbels 20–30-flowered. Jenolan Caves. Fl. summer . . . . **T. scapigera** (Domin) B.L.Burtt

3 **Astrotricha** DC.
25 species endemic Aust.; Qld, NSW, Vic., WA

Shrubs, often tall, ± clothed with a stellate tomentum. Leaves simple, entire, petiolate, without stipules. Flowers c. 5 mm diam., articulate on the pedicels, in umbels which are arranged in a terminal often large panicle. Calyx teeth minute. Petals 5, pubescent outside. Fruit a laterally compressed schizocarp tardily separating into 2 mericarps.

Key to the species

1 Leaves more than 20 mm wide, acuminate to acute, flat. Panicles 30–50 cm long. Shrubs up to 3 m high 2

1 Leaves mostly less than 20 mm wide. 3

2 Petals white to cream, spreading to erect. Leaves with loose floccose hairs underneath, elliptic to narrow-ovate, 7–27 cm long; petioles usually less than 25 mm long. Coast and adjacent plateaus north of Port Jackson; lower Blue Mts DSF. Ss. Fl. spring–summer . . . . **A. floccosa** DC.

2 Petals greenish yellow, reflexed. Leaves with a thin even tomentum underneath, oblong-ovate to elliptic; petioles usually 40–80 mm long. Widespread. DSF and sheltered places on Ss. Fl. spring–summer. . . .
. . . . **A. latifolia** Benth.

3 Leaves ± acute, flat or nearly so. Tomentum fine and even 4

3 Leaves very obtuse or rarely emarginate, linear to oblong-obovate, with revolute or recurved margins 7

4 Leaves 10–20 mm wide, narrow-lanceolate, coriaceous, shining on the upper surface. Stout shrub. . . .
. . . . **A. crassifolia** (see below)

4 Leaves mostly less than 10 mm wide. 5

5 Leaves linear, mostly less than 3 mm wide . . . . **A. linearis** (see below)

5 Leaves more than 3 mm wide. 6

6 Spreading many-branched shrub. Stems glabrescent and often purplish. Leaves narrow-lanceolate, 0.3–1 cm wide, thin, flat or nearly so, rather dull on the upper surface and lateral veins obscure. Inflorescence up to 60 cm long, axes purplish and glabrescent. Widespread on coast. DSF and gullies. Ss. Fl. spring–summer. (Individuals from the Sydney Basin have tessellate fruit and might represent a distinct taxon.)
. . . . **A. longifolia** Benth. (coastal form)

6 Erect, few branched shrub. Stems densely hairy, rarely glabrescent. Leaves lanceolate, 0.8–2.5 cm wide, lateral veins conspicuous on upper surface. Inflorescence up to 25 cm long, axes pale and densely hairy. DFS. Fl. summer . . . . **A. longifolia** Benth. (inland form)

7 Inflorescence up to 6 cm long, with few branches and few flowers on each branch. Petals greenish yellow. Leaves oblong to obovate, 1–3 cm long, up to 10 mm wide, with a dense tomentum underneath; upper surface smooth or slightly scabrous. Erect or spreading shrub up to 1 m high, sometimes suckering. Hornsby Plateau in St. Albans district northwards; Hunter River Valley. DSF. Sandy soils. Fl. summer
. . . .**A. obovata** Makinson

7 Inflorescence more than 6 cm long, much branched and with many flowers. Petals white to cream or rarely cream-green or purplish 8

8 Upper surface of the leaves smooth, shining. Leaves linear 2–6 cm long, 2–4 mm wide, thick, with revolute margins, with a loose very dense tomentum underneath, very shortly petiolate or sessile. Petals white to cream. Erect shrub up to 2 m high, suckering. Woy Woy district; Woronora Plateau. DSF. Ss. Fl. spring . . . . **A. crassifolia** Blakely

8 Upper surface of leaves minutely scabrous 9

9 Leaves narrow-oblong to oblong-elliptic, 3–7 mm wide, up to 45 mm long, with an even tomentum underneath; petiole 1–2 mm long. Petals white to cream-green or purplish. Erect shrub up to 2 m high. Blue Mts DSF. Fl. summer. . . . . **A. ledifolia** DC.

9 Leaves linear, mostly less than 2 mm wide, 2–6.5 cm long, with a woolly tomentum underneath; petiole up to 5 mm long. Petals white to greenish. Erect shrub up to 2 m high. Bungonia district. DSF on a variety of soils. Fl. spring–summer . . . . **A. linearis** A.Cunn. ex Benth.

4 Hedera L.
1–2 species naturalized Aust.; NSW, Vic.

One species in the area

Woody climber, climbing by roots but sometimes creeping along the ground. Leaves simple, coriaceous, glabrous, dark green on the upper surface, pale underneath, palmately 3–5-lobed, 4–10 cm long. Corolla greenish. Fruit succulent, blue, 5–8 mm diam. Naturalized at places in Blue Mts Introd. from Europe. *English Ivy* . . . .*H. helix L.

5 Polyscias J.R.Forst. & G.Forst.

10 species in Aust. (9 endemic, 1 naturalized); Qld, NSW, Vic., NT

Tall trees or erect shrubs, glabrous or nearly so. Leaves pinnately compound, imparipinnate; rhachis articulate. Flowers bisexual or sometimes polygamous, articulate on the pedicels, in umbels or racemes arranged in panicles. Calyx truncate or slightly toothed. Petals 5. Stamens 5. Ovary 2-locular. Fruit ± succulent, laterally compressed.

Key to the species

1 Flowers in racemes arranged in a large panicle. Leaves bipinnate (or pinnate on some flowering shoots), up to 1 m long; leaflets opposite, ovate, acuminate, coriaceous, 5–10 cm long, entire; rhachis articulate. Tree up to 30 m high with few branches. Coast and gullies in adjacent plateaus. RF. Fl. winter. *Celery Wood* or *Silver Basswood* . . . . **P. elegans** (C.Moore & F.Muell.) Harms
1 Flowers in umbels which are arranged in panicles 2

2 Tree up to 25 m high with few branches. Leaves pinnate, up to 1.3 m long; leaflets opposite, 11–17, oblong-elliptic, 5–20 cm long, 30–90 mm wide, often toothed. Widespread. RF. Fl. autumn. *Pencil Cedar* or *Umbrella Tree.* . . . . **P. murrayi** (F.Muell.) Harms
2 Suckering usually much branched shrub up to 5 m high. Leaves 10–60 cm long 3

3 Leaves 1-pinnate; leaflets 5–11, 20–60 mm wide, margins ± toothed. DSF, WSF or RF margins. Widespread. Fl. spring–summer *Elderberry Panax* . . . . **P. sambucifolia** (Sieber ex DC.) Harms ssp. **A**
3 Leaves twice or thrice pinnate; leaflets 5–15 mm wide; margins entire to deeply pinnatifid. Primary or secondary rachis maybe winged. Widespread. DSF; WSF; RF margins. Fl. summer. *Ferny Panax* . . . .
. . . .**P. sambucifolia** (Sieber ex DC.) Harms ssp. **C**

6 Cephalaralia Harms

1 species endemic Aust.; Qld, NSW

One species in the area

Climber, often quite tall, or scrambling shrub, glabrous except the inflorescences and young growth. Leaves usually 3-foliolate, on slender petioles; leaflets on slender petiolules, oblong to lanceolate, acuminate, entire, 5–10 cm long. Flowers sessile, in small pedunculate heads arranged in short racemes or panicles. Minnamurra Falls; Blue Mts in gorges. RF. Fl. spring–autumn . . . .
. . . .**C. cephalobotrys** (F.Muell.) Harms

7 Tetrapanax (K.Koch) K.Koch

1 species naturalized Aust.; NSW

One species in the area

Shrub up to 4 m high. Leaves ovate to broad-ovate in outline, up to 27 cm long and 25 cm wide, palmately 7–9-lobed, dentate, stellato-pubescent, paler underneath; each lobe itself often lobed. Flowers in umbels arranged in racemes. Petals 4. Stamens 4. Styles 2 distinct, slender. Naturalized in a few places in the Illawarra district. Fl. spring–summer. *Rice-paper Plant* . . . . **T. papyrifer** (Hook.f.) K.Koch

140 PENNANTIACEAE

1 gen., trop. to temp.

1 Pennantia J.R.Forst. & G.Forst

1 species native Aust.; Qld, NSW

One species in the area

Small, glabrous tree, sometimes with weak branches and resembling a climber. Leaves alternate, elliptic or ovate to narrow-ovate, entire, acuminate, thin, 8–14 cm long, marked underneath by domatia in the forks of lateral veins. Flowers numerous on short pedicels, in broad rather dense panicles, either terminal or in the upper axils, bisexual or unisexual. Calyx absent or rudimentary. Fruit an ovoid drupe c. 12 mm long. Widespread. RF. Fl. summer. *Brown Beech*. . . . .**P. cunninghamii** Miers

141 PITTOSPORACEAE

Trees, shrubs or twiners. Leaves alternate, without stipules. Flowers regular, bisexual. Sepals and petals 5, imbricate. Stamens 5, alternating with the petals, free, hypogynous. Ovary superior, 2–5-locular; placentas axile or parietal; style simple. Fruit a capsule or berry. Seeds several per loculus. 7 gen., all Australia except *Pittosporum* which extends to Africa, Asia and Oceania.

Key to the genera

1 Anthers opening through terminal often confluent pores or short slits. Flowers blue, held erect . . . . .
. . . .**6 CHEIRANTHERA**
1 Anthers opening through longitudinal slits. Flowers white, cream, yellow, green or blue and pendant. 2

2 Scandent shrubs or twiners with pendant blue or yellow-green flowers and fruit a green or black berry
. . . .**5 BILLARDIERA**
2 Trees or shrubs sometimes prostrate but never twinning. Fruit a 2 or 3-valved capsule 3

3 Flowers in terminal lateral panicles; petals spreading from base of flower. Ovary prominently stipitate. Capsule strongly flattened. . . . . **3 BURSARIA**
3 Flowers solitary or in terminal or axillary clusters; petals tubular in lower half. Ovary sessile or scarcely stipitate 4

4 Prostrate to procumbent dwarf shrubs. Leaves usually 3 mm or less wide, entire with an apical tooth. Flowers 4–6 mm long. . . . .**4 RHYTIDOSPORUM**
4 Shrubs or trees. Leaves more than 30 mm wide and entire or; 3.5–7 mm wide, toothed in upper half, and branches spinose 5

5 Petals 30–40 mm long with hairs on outer surface. . . . . **2 HYMENOSPORUM**
5 Petals less than 25 mm long, glabrous.. . . . . **1 PITTOSPORUM**

1 Pittosporum Banks ex Gaertn.
20 species in Aust. (mostly endemic); all states and territories

Trees or shrubs. Petals white or yellow, usually cohering into a tube at the base; the limb spreading. Ovary almost sessile, incompletely 2-locular; style short. Capsule with leathery or thick hard valves. Seeds very viscid, often angular.

Key to the species

1 Shrub with spinose branches. Leaves nearly sessile, ovate to orbicular or obovate to broad-cuneate, 4–12 mm long, entire or with a few prickly teeth. Flowers axillary, solitary, not numerous. Petals white, 4 mm long. Ovary 1-locular. Berry globular, 4–10 mm diam., orange. Rigid, much branched shrub up to 3 m high; Widespread. In or near RF and in shaded places on shales. Fl. spring–summer. *Orange Thorn* . . .
. . . . **P. multiflorum** (A.Cunn. ex Loudon) L.Cayzer, Crisp & I.Telford
1 Shrub or small tree without spines. 2

2 Petioles and peduncles densely clothed in white hairs. Petals dark red c. 10 mm long. Leaves 7.5–12.5 mm long, 2.5–5.5 cm wide, elliptic to obovate. Fruit 3-valved, pubescent. Shrub to 4 m tall. Blue Mts Introd. from New Zealand . . . . ***P. ralphii** Kirk
2 Petioles and peduncles not densely hairy or clothed with rusty hairs. Petals white or yellow 3

3 Leaves rusty-tomentose underneath at least when young, ovate- to oblong-elliptic, 4–10 cm long; margins slightly revolute. Petals yellow, c. 12 mm long. Capsule oblong, rough, orange. Seeds bright red. Shrub up to 3 m high. Widespread. Forests. Fl. spring . . . . . **P. revolutum Dryand. ex A. T.Aiton**
3 Leaves glabrous, narrow-ovate to elliptic or ovate-oblong, flat or undulate. 4

4 Leaves dark green, shining, mostly 5–12 cm long. Flowers in terminal compound clusters shorter than the leaves. Sepals 5–10 mm long. Petals white, 10–12 mm long. Capsule smooth, nearly globular before opening, c. 10 mm diam. Seeds numerous, red-brown. Small tree. Widespread. Sheltered situations and RF. Fl. early spring. *Pittosporum* . . . . **P. undulatum** Vent.

4 Leaves glossy, dark green sometimes variegated, 5–15 cm long. Flowers in paniculate inflorescences. Sepals c. 2 mm long. Petals creamy, c. 7 mm long. Capsule 5–6 mm long, finely granulate. Small tree to 12 m high. Blue Mts Introd. from New Zealand . . . . ***P. eugenioides** A.Cunn.

2 Hymenosporum R.Br. ex F.Muell.
1 species native Aust.; Qld, NSW

One species in the area

Small tree. Leaves alternate, obovate to oblanceolate, acuminate, glabrous, 8–16 cm long. Flowers in a terminal loose panicle. Petals cream changing to yellow, tomentose inside, 3–4 cm long. Fruit an ovate compressed capsule 2–3 cm long. Seeds numerous, flat, winged. Coast and adjacent plateaus. RF. Fl. spring . . . . **H. flavum** (Hook.) F.Muell.

3 Bursaria Cav.
7 species endemic Aust.; Qld, NSW, Vic., WA, NT

Rigid much-branched shrubs up to 3 m high, usually with thorny branches. Leaves entire. Flowers small, often numerous, in panicles. Sepals deciduous or persistent. Capsule compressed, obcordate or reniform, thin-walled.

Key to the species

1 Leaves variable in size and shape. Flowers in terminal pyramidal panicles, often numerous. Sepals less than 2 mm long. Petals narrow, to 6 mm long, creamy white. Leaves obovate oblong or cuneate, 10–35 mm long, truncate or emarginate, narrowed at the base. Capsule 4–6 mm long. Widespread. Shrub to small tree. Tall forests particularly on shales; weed of cultivated land. Fl. spring–summer. *Blackthorn* 2

1 Leaves consistent in size and shape. Sepals prominent. Petals 6–8 mm long. 3

2 Tall shrub or tree 5–10 m high. Indumentum on young shoots and undersurface of leaves not usually persisting with age. Petals 4–6 mm long . . . . **B. spinosa** Cav. ssp. **spinosa**

2 Multi-stemmed shrub under 5 m high. Indumentum of dense appressed hairs on young shoots and undersurface of leaves persisting with age. Petals to 4 mm long. Higher altitudes. . . . . **B. spinosa** ssp. **lasiophylla** (E.M.Benn.) L.Cayzer, Crisp & I.Telford

3 Branchlets scabrous with erect, rigid hair bases. Other parts of plant glabrous. Mature style longer than ovary. Fruit with 2 valves, not woody. Flowers in lateral panicles or clusters rarely longer than the spines, fewer than in *B. spinosa*. Mature leaves up to 6–9 mm long, 1–3 mm wide crowded along the branches. Spines numerous. Often procumbent shrub. Blue Mts Open forests. Fl. spring–summer. . . . . **B. longisepala** Domin

3 Whole plant with soft appressed hairs. Mature style shorter than ovary. Fruit with 2 or 3 valves often woody. Inflorescence usually terminal. Mature leaves 8–12 mm long, 4 mm wide. Erect shrub. Wombeyan Caves area. Fl. spring–summer. . . . . **B. calcicola** L.Cayzer, Crisp & I.Telford

4 Rhytidosporum F.Muell.
5 species endemic Aust.; Qld, NSW, Vic., Tas.

Prostrate to erect shrubs, sometimes rhizomatous. Leaves up to 2 cm long. Flowers in terminal corymbs or solitary. Petals white and purplish outside, 3–5 mm long. Ovary 1-locular. Fruit a compressed orbicular capsule with few to several seeds.

Key to the species

1 Inflorescence ± sessile umbel with 4–8 flowers, pedicels less than 4 mm long. Leaves linear to elliptic, 10–20 mm long, 1.5–2 mm wide; apex with prominent mucro. Coastal sandstone plateau. Heath and woodland. Fl. spring . . . .**R. diosmoides** (Putt.) L.Cayzer, Crisp & I.Telford

1 Inflorescence pedunculate, flowers on pedicels 6–15 mm long. Leaves obovate usually less than 12 mm long, 2–3 mm wide, tri-lobed. 2

2 Stems much branched. Flowers 1–2 together on pedicels up to 8 mm long rarely longer. Leaves clustered, stem-clasping. Undershrub. Widespread. Heath and forests. Fl. spring–summer . . . . **R. procumbens** (Hook.) F.Muell.

2 Stems unbranched or with very few branches, up to 100 cm long. Flowers 1–6 together on slender pedicels more than 8 mm long. Leaves alternate, 2-ranked. Prostrate to trailing shrub. Illawarra Ranges; southern Blue Mts Heath and open forests in damp places. Fl. spring. . . . **R. prostratum** McGillivray

5 **Billardiera** Sm.
23 species endemic Aust.; all states and territories except NT

Small or medium sized scandent shrub or twiner with slender woody stems, villous or velutinous to nearly glabrous. Leaves narrow-ovate to linear or rarely ovate, entire or undulate. Petals cohering or free. Ovary 1–2-locular. Fruit an oblong to ovoid berry.

Key to the species

1 Petals free, blue. Leaves 30–50 mm long. Inflorescence with up to 10 flowers. Berry oblong-cylindrical, glabrous, green-purple. Introd. from Western Australia. Fl. summer. *Bluebell Creeper* . . . .*B. heterophylla** (Lindl.) L.Cayzer & Crisp

1 Petals spreading from the middle to form a narrow campanulate corolla, greenish or pale yellow often tinged with purple, 6–24 mm long. Leaves mostly 15–35 mm long. Flowers 1 or 2, pendulous on slender terminal peduncles. Ovary glabrous or pubescent. Berry cylindrical or ovoid-oblong, greenish, 1–3 cm long. Widespread. Heath and open forest. Ss. and WS. Fl. spring–summer. *Appleberry* or *Snotberry*. . . . . **B. scandens** Sm.

6 **Cheiranthera** Brongn.
5 species endemic Aust.; all states and territories except NT and Tas.

One species in the area

Erect, glabrous shrub or undershrub up to 50 cm high. Leaves linear, up to 5 cm long, acute-obtuse, entire or denticulate; margins ± incurved to involute. Flowers solitary or in short terminal racemes, 2–3 cm diam. Petals blue. Anthers opening through 2 confluent terminal pores or short slits. Capsule oblong, hard, 2-valved. Essentially a species of the Western Slopes but reported from Lithgow, Kurrajong and Ebenezer. DSF. Fl. summer–autumn. *Finger Flower* . . . .**C. cyanea** Brongn.

142 CARDIOPTERIDACEAE
7 gen., trop. to temp.

1 **Citronella** D.Don
2 species endemic Aust.; Qld, NSW

One species in the area

Tall, dioecious tree with a fluted trunk, glabrous except the inflorescence. Leaves ovate-lanceolate to oblong, 3–10 cm long; small domatia present on the under surface along the midrib in the vein axils. Flowers in narrow panicles, unisexual. Petals 5, Fruit an ovoid black drupe, 20–25 mm long. Coast and adjacent plateaus. RF. Fl. winter–spring. *Churnwood* . . . . .**C. moorei** (F.Muell. ex Benth.) R.A.Howard

143 ASTERACEAE

Herbs, shrubs, rarely climbers; some with latex. Leaves usually alternate, sometimes opposite, often basal, simple dissected or compound; stipules usually absent. Inflorescence a head, solitary on a scape or several together in compound inflorescences (corymbs etc.) or sometimes the heads aggregated into compound heads; each head surrounded by an involucre of 1–several rows of bracts; involucre green or coloured, chaffy, papery or fleshy, sometimes with spines. Florets bisexual or unisexual (the plants then usually monoecious); each flower sometimes subtended by a bract. Calyx represented by a pappus of fine bristles, hairs, scales, barbs, thorns, or spines; or absent. Corolla segments 5, either regular and connate into a tube (tubular florets or if very slender, filiform florets); or irregular and connate into a very short basal tube terminating in a unilateral often strap-shaped limb 3–5-lobed at the apex or entire (ligulate florets); or rarely 2-lipped (e.g *Podolepis*). Within a single head, florets may be: (1) all tubular); (2) tubular and filiform; (3) tubular and ligulate; (4) all ligulate; the former at the centre are called disc florets, the latter at the periphery are called ray florets or rays, and are often female or sterile by the abortion of the styles. Stamens 5, epipetalous, alternating with the corolla lobes; anthers linear, with 2 pollen sacs, connate around the style, rarely free. Ovary inferior, 2-carpellary, 1-locular; ovule solitary, basal; style usually 2-fid with 2 stigmas. Fruit a *cypsela* surmounted by the pappus or pappus absent. c. 1500 gen., cosmop.

Key to the genera

1 Involucral bracts of the female heads or of the outer florets enclosing the fruit(s) at maturity and forming a burr (Fig. 42) . . . . **GROUP 1**
1 Burr not formed around the fruits at maturity 2

2 Florets all tubular or filiform and tubular (Fig. 41) or; florets with ligule present but the ligule less than 3mm 3
2 Florets ligulate and tubular, or all ligulate; ligule more than 3 mm long 5

3 Pappus of several to many soft simple or plumose hairs or bristles equal to or exceeding the length of the cypselas (Fig. 41) . . . . **GROUP 2**
3 Pappus absent or consisting of a few short spines, or of rigid bristles or scales, or reduced to a short rim, or the cypselas surmounted by a hardened spiny style (Fig. 41) 4

4 Cypselas without a pappus or hardened spiny style . . . . **GROUP 3**
4 Cypselas surmounted by a spiny hardened style or a pappus of spines, rigid bristles or scales. . **GROUP 4**

5 All florets ligulate . . . . **GROUP 5**
5 Ray florets ligulate; disc florets tubular 6

6 Pappus absent or minute (less than 0.5 mm long) . . . . **GROUP 6**
6 Pappus present and usually conspicuous (maybe hidden by woolly hairs) 7

7 Pappus of scales, barbs, spines or awns, (may be obscured by cypsela hairs) or; the cypselas covered by woolly hairs . . . . **GROUP 7**
7 Pappus of soft capillary hairs . . . . **GROUP 8**

Group 1
Fruits forming a burr

1 Burrs without hooked appendages, each burr containing 1 fruit (Fig. 41) . . . . **30 AMBROSIA**
1 Burrs with hooked appendages. 2

2 Burrs 5–8, whorled, each burr derived from a single ray floret enclosed in a bract . . . . **31 ACANTHOSPERMUM**
2 Burrs solitary or few together in the axils of alternate leaves, each burr containing 2 fruits . . . . **32 XANTHIUM**

FIGURES 41 & 42

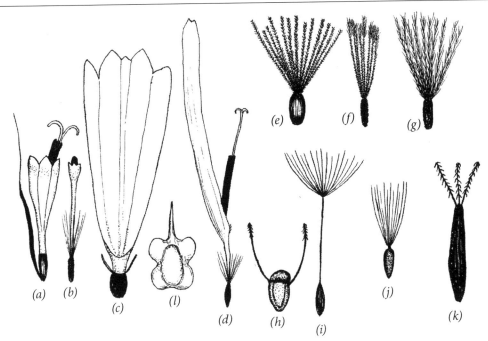

Fig. 41 ASTERACEAE: diagram (a) of a tubular flower and subtending bract; (b) of a filiform flower; (c) of a sterile female ray flower (style and stigmas lacking); (d) of a bisexual ligulate flower; (e) cypsella and pappus of *Xerochrysum bracteatum*; (f) cypsella and pappus of *Chrysocephalum semipapposum*; (g) cypsella and pappus of *Leucochrysum albicans*; (h) cypsella of *Calotis cuneifolia* surmounted by 2 barbed awns and 2 scales; (i) beaked cypsella of *Lactuca serriola*; (j) cypsella and pappus of *Leptorhynchus squamatus*, with a short beak; (k) cypsella and barbed pappus of *Bidens pilosa*; (l) winged cypsella of *Soliva sessilis* wi th persistent, spiny style. e–l x 5.

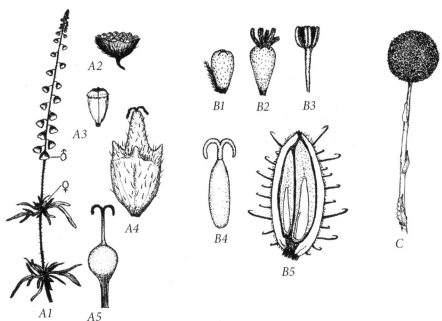

Fig. 42 ASTERACEAE: A, *Ambrosia tenuifolia*: A1, terminal shoot bearing male and female inflorescences (nat. size); A2, male inflorescence with 16 flowers (X5); A3, single male flower (X15); A4, female involucre with emergent styles (X10); A5, female flower (X10). B, *Xanthium spinosum*: B1, single male flower-bud and involucral bract (X5); B2, mature male flower (X5); B3, androecium (X5); B4, female flower (X10); B5, longitudinal section through the almost mature burr showing 2 young fruits (X4). C, inflorescence of *Craspedia sp.*

Group 2
Florets all tubular or tubular and filiform; pappus of hairs

1 Involucral bracts and/or leaves with spinescent tips (thistles) 2
1 Involucral bracts and leaves without spinescent tips. 7

2 Leaves mottled with white veins . . . . **.87 SILYBUM**
2 Leaves green, not mottled 3

3 Receptacle without scales or bristles . . . . **88 ONOPORDUM**
3 Receptacle with scales or bristles. 4

4 Receptacle with scales. Pappus bristles plumose . . . . **.84 CIRSIUM**
4 Receptacle with bristles. Pappus bristles simple or plumose 5

5 Involucral bracts not ending in spines. **83 STEMMACANTHA**
5 Involucral bracts ending in spines 6

6 Pappus of simple bristles . . . . **.85 CARDUUS**
6 Pappus of plumose bristles . . . . **.86 CYNARA**

7 Inflorescence solitary on a long scape, consisting of groups of 3–12 florets arranged in a tight globular
head with a common receptacle . . . . **.82 CRASPEDIA**
7 Inflorescence not as in *Craspedia*. 8

8 All florets on individual plants either functionally male or female. Shrubs . . . . . . **13 BACCHARIS**
8 Herbs, subshrubs or shrubs. Flower heads with at least some bisexual florets; or male and female florets
on same plant 9

9 Leaves opposite. 10
9 Leaves alternate or basal. 11

10 Florets white. . . . . **2 AGERATINA**
10 Florets mauve to purple . . . . **3 AGERATUM**

11 Involucral bracts in 1 whorl, but sometimes with a few much smaller bracts at the base 12
11 Involucre bracts in 2 or more whorls. 16

12 Pappus mauve to reddish . . . . **.49 ERECHTITES**
12 Pappus whitish 13

13 Florets orange, brownish or red. . . . . **51 CRASSOCEPHALUM**
13 Florets yellow 14

14 Disc florets without stigmas . . . . **50 ARRHENECHTHITES**
14 Disc florets with stigmas 15

15 Climbers or scramblers. Stipules present, 5–10 mm wide . . . . **.47 DELAIREA**
15 Erect to ascending or rarely scrambling plants. Stipules usually absent . . . . **.45 SENECIO**

16 Leaves 1–2 cm long, crowded, spathulate, stem clasping; apex emarginate . . . . **67 FACELIS**
16 Leaves not as above 17

17 Shrubs with persistent, woody stems. 18
17 Herbs; or subshrubs with short woody base. 19

18 Receptacle with chaffy scales or; when florets 3 or less per receptacle then scales absent . **68 CASSINIA**
18 Receptacle without scales or scales 0–3 and many less than florets. . . . . **72 OZOTHAMNUS**

19 At least outer involucre bracts with ciliate margins 20
19 Involucre bracts lacking ciliate margins, outer ones sometimes woolly or lacerate towards base. . . 22

434 ❀

20 Heads sessile surrounded by a cluster of leaves . . . . **71 TRIPTILODISCUS**
20 Heads on long peduncles not surrounded by a cluster of leaves. 21

21 Corollas longer than involucre. Cypsela ribbed . . . . **78 LEPTORHYNCHOS**
21 Corollas shorter than involucre. Cypsela not ribbed . . . . **73 CHRYSOCEPHALUM**

22 Involucral bracts herbaceous or with membranous or scarious apex or margins. 23
22 Involucral bracts with at least the base papery, white or coloured, opaque 30

23 Outer filiform florets in 2 to several whorls 24
23 Filiform florets few or absent 28

24 Flower heads pedunculate; plants not woolly . . . . **12 CONYZA**
24 Flower heads ± sessile; plants woolly 25

25 Inner involucre bracts with scarious claws. Heads at least as wide as long . . . .
. . . . **62 PSEUDOGNAPHALIUM**
25 Inner involucre bracts without scarious claws. Heads campanulate to urceolate longer than wide . 26

26 Heads in dense, terminal, clusters surrounded by a number of leaf-like bracts. Stoloniferous herbs
(except *E. sphaericus*). . . . . **65 EUCHITON**
26 Heads arranged in elongated often leafy spikes, or in large corymbs of spikes. Erect to prostrate herbs,
not stoloniferous 27

27 Cypsela with papillae. Involucre bracts with membranous tips. Leaves, stems and heads covered by
loose woolly hairs. Weak, erect to decumbent herbs . . . . **64 GNAPHALIUM**
27 Cypsela with globose hairs. Involucre bracts with firm ± leathery tips. Leaves, stems and heads with
tangled appressed hairs. Robust, erect or prostate herbs. . . . . **66 GAMOCHAETA**

28 Pappus in 2 whorls outer bristles shorter than inner, not fused. . . . . **1 VERNONIA**
28 Pappus in 1 whorl, shortly fused at the base. 29

29 Pappus plumose for entire length. . . . . **70 RHODANTHE**
29 Pappus barbate, or plumose at tip. . . . . **78 LEPTORHYNCHOS**

30 Heads crowded into dense clusters, sessile or almost so . . . . **63 VELLEREOPHYTON**
30 Heads on distinct peduncles, usually not crowded but if so then the heads separable 31

31 Pappus hairs plumose from the base 32
31 Pappus hairs simple, barbellate or sometimes ± plumose towards the tips 33

32 Leaves basal, woolly on both surfaces . . . . **69 LEUCOCHRYSUM**
32 Leaves cauline, glabrous or hairy but not woolly . . . . **70 RHODANTHE**

33 At least some of the involucre bracts with an elongated claw . . . . **74 HELICHRYSUM**
33 Involucral bracts with claw reduced . . . . **75 XEROCHRYSUM**

Group 3
Florets all tubular or tubular and filiform; pappus absent

1 Involucral bracts terminating in spines . . . . **89 CENTAUREA**
1 Involucral bracts not spiny 2

2 Heads numerous in a terminal panicle 3
2 Heads few together or solitary 6

3 Involucre 1–2 mm diam. Leaves entire, green on both surfaces . . . . **79 CALOMERIA**
3 Involucre more than 3 mm diam. 4

4 Involucre 12–20 mm diam. Leaves opposite . . . . **5 GYMNOCORONIS**

4 Involucre 3–7 mm diam. Leaves alternate 5

5 Involucre 3 mm diam. Leaves silvery . . . . **44 ARTEMISIA**
5 Involucre 6–7 mm diam. Leaves green. . . . .**43 TANACETUM**

6 Heads sessile or on peduncles less than 1 cm long 7
6 Heads on peduncles longer than 1 cm 8

7 Involucral bracts with scarious margins, acute, in 2 rows. . . . .**37 CENTIPEDA**
7 Involucral bracts herbaceous, very obtuse, in 3–4 rows . . . . **61 EPALTES**

8 Leaves basal. 9
8 Leaves cauline. 10

9 Ray more than 1 mm long. Cypselas beaked . . . .**4 LAGENOPHORA**
9 Ray less than 1 mm long. Cypselas not beaked . . . . **6 SOLENOGYNE**

10 Style 2-fid. Disc florets fertile . . . .**35 COTULA**
10 Style with a terminal disc. Disc florets sterile . . . . **36 LEPTINELLA**

Group 4
Florets all tubular or tubular and filiform; pappus of spines, rigid bristles or scales or style a persistent spine

1 Involucral bracts terminating in spines 2
1 Involucral bracts not spiny 3

2 Involucral bracts all similar. . . . . **89 CENTAUREA**
2 Outer involucral bracts larger than the inner and resembling cauline leaves. . . . . **90 CARTHAMUS**

3 Cypselas surmounted by a single unbranched spine which is the rigid persistent style (Fig. 41 l) . . . . .
. . . .**38 SOLIVA**
3 Spine absent. 4

4 Pappus of 2–4 barbellate awns 5
4 Pappus a membranous cup or crown or, whitish scales 6

5 Most leaves basal . . . . **22 GLOSSOGYNE**
5 Most leaves cauline. . . . .**24 BIDENS**

6 Pappus of simple scales 7
6 Pappus a cup or crown or absent 8

7 Leaves pinnate with filiform lobes . . . . **34 SCHKUHRIA**
7 Leaves entire, sessile . . . . **77 RUTIDOSIS**

8 Stems winged. Receptacle with scales . . . . **76 AMMOBIUM**
8 Stems not winged. Receptacle naked . . . .**43 TANACETUM**

Group 5
Florets all ligulate

1 Florets blue to violet 2
1 Florets yellow to brown 3

2 Pappus of minute scales . . . . **92 CICHORIUM**
2 Pappus of capillary hairs on a long beak. . . . . **93 TRAGOPOGON**

3 Heads solitary on unbranched scapes arising from a rootstock 4

3 Inflorescence branched or; the heads solitary or a few together arising from leaf axils 7

4 Pappus of scales. 5
4 At least the inner cypselas with a pappus of plumose hairs 6

5 Involucral bracts in 2–3 rows. Pappus of flat scales tapering into bristles . . . . **91 MICROSERIS**
5 Involucral bracts in 1 row, hardened at maturity and embracing the outer cypselas. Pappus crown-shaped . . . . **94 HEDYPNOIS**

6 All cypselas with a pappus of plumose hairs . . . . **95 TARAXACUM**
6 Outer cypselas with a pappus of scales; inner cypselas with a pappus of plumose hairs. . . . .
. . . .**96 LEONTODON**

7 Cypselas beaked (Fig. 41) 8
7 Cypselas not beaked 12

8 Pappus hairs simple 9
8 Pappus hairs plumose 10

9 Involucral bracts in 2 rows. Heads axillary, sessile or almost so. Cypselas subcylindrical, rough towards the summit . . . . **97 CHONDRILLA**
9 Involucral bracts in 4–5 rows. Heads terminal or terminating branches. Cypselas flattened, ribbed longitudinally. . . . . **98 LACTUCA**

10 Leaves all basal in a flat rosette. Rhachis of the inflorescence with scale-leaves only. Receptacle with chaffy scales . . . . **99 HYPOCHAERIS**
10 Leaves basal and cauline, the latter smaller. Receptacle naked 11

11 Cypsela narrowing gradually into the beak. Leaf hairs 2-fid . . . . **100 PICRIS**
11 Cypsela narrowing abruptly into a beak. Leaf hairs 3–5-fid . . . . **101 HELMINTHOTHECA**

12 Pappus of 4–10 hairs with a crown of minute scales at their bases; outer cypselas with scales only. Outer florets yellow; inner florets brownish . . . . **105 TOLPIS**
12 Pappus of numerous silky hairs. Florets all yellow 13

13 Cypselas sub-cylindrical, or sometimes outer ones dorsiventrally compressed. 14
13 All cypselas laterally compressed 16

14 Pappus hairs plumose . . . . **100 PICRIS**
14 Pappus hairs simple. 15

15 Upper peduncle and involucre bracts with glandular hairs, all cypselas sub-cylindrical . **104 CREPIS**
15 Upper peduncle and involucre bracts without glandular hairs, outer cypselas compressed dorsiventrally. . . . . **106 YOUNGIA**

16 Cypselas brown, with narrow wings or not winged. Involucres 5–6 mm diam.. . . . . **102 SONCHUS**
16 Cypselas straw-coloured, winged. Involucres c. 10 mm diam. . . . . **103 ACTITES**

Group 6
Ray florets ligulate; disc florets tubular; pappus absent or minute

1 At least lower leaves opposite 2
1 Leaves alternate or mostly basal 10

2 Stems decumbent. 3
2 Stems ascending to erect. 4

3 Heads on peduncles 1–2 cm long. Rays bilobed. Terrestrial. . . . . **18 ECLIPTA**
3 Heads sessile. Rays 3-toothed. Usually aquatic . . . .**23 ENYDRA**

4 Rays more than 5 mm long 5
4 Rays up to 5 mm long 9

5 Rays white, bluish or pinkish . . . . **28 COSMOS**
5 Rays yellow 6

6 Leaves sessile, stem-clasping, lamina entire, toothed or shallowly lobed . . . . **26 GUIZOTIA**
6 Leaves petiolate or almost sessile but not stem-clasping. 7

7 Cypselas of disc florets prominently winged . . . . **29 COREOPSIS**
7 Cypselas not prominently winged 8

8 Heads 10 mm diam., mostly in 3s . . . . **20 MELANTHERA**
8 Heads 10–15 mm diam., solitary . . . . **19 WEDELIA**

9 Heads bearing viscid glandular hairs . . . . **17 SIGESBECKIA**
9 Heads without viscid glandular hairs . . . . **25 GALINSOGA**

10 Leaves all or mostly cauline 11
10 Leaves all or mostly basal 22

11 Heads in a dense terminal corymb 12
11 Heads solitary or few together. Rays not regularly 3-lobed 13

12 Rays 3-lobed . . . . **39 ACHILLEA**
12 Rays entire . . . . **43 TANACETUM**

13 Rays white 14
13 Rays yellow to orange. 18

14 Receptacle with chaffy scales (floral bracts) . . . . **40 ANTHEMIS**
14 Receptacle naked, smooth or pitted. 15

15 Rays purple underneath 16
15 Rays not purple underneath 17

16 All florets producing cypselas . . . . **54 DIMORPHOTHECA**
16 Only outer florets producing cypselas . . . . **55 OSTEOSPERMUM**

17 Leaves entire to pinnate-lobed. . . . . **42 LEUCANTHEMUM**
17 Leaves 3-pinnatisect . . . . **41 TRIPLEUROSPERMUM**

18 Cypselas with 2 broad pale-coloured wings . . . . **27 VERBESINA**
18 Cypselas without wings 19

19 Leaves pinnatisect. . . . . **52 EURYOPS**
19 Leaves entire or toothed 20

20 Cypselas tuberculate to spiny, herb. . . . . **53 CALENDULA**
20 Cypselas neither tuberculate nor spiny, shrubs or herbs 21

21 Cypsela ovoid, drupaceous, finally hard and bony. Shrub up to 2 m high **56 CHRYSANTHEMOIDES**
21 Cypselas compressed, finely pubescent. Small shrubs or herbs . . . . **21 HELIANTHUS**

22 Rays yellow or orange . . . . **58 CYMBONOTUS**
22 Rays white, mauve, pink or blue. 23

23 Cypselas beaked or contracted at the summit. . . . . **4 LAGENOPHORA**
23 Cypselas not beaked, obtuse or acute. 24

24 Rays more than 4 mm long . . . . **7 BRACHYSCOME**
24 Rays less than 3 mm long . . . . **6 SOLENOGYNE**

Group 7

Ray florets ligulate; disc florets tubular; pappus of barbs, scales or spines

1 Cypselas enveloped in woolly hairs . . . . .59 ARCTOTHECA
1 Cypselas glabrous or with short to long, sometimes silky hairs 2

2 Involucral bracts connate into a tube which surrounds the florets. Plants with a disagreeable odour . . .
. . . . .33 TAGETES
2 Involucral bracts not connate into a tube (may be basally connate in *Gazania*) 3

3 Leaves opposite 4
3 Leaves alternate or basal. 6

4 Leaves entire. Heads axillary . . . . 25 GALINSOGA
4 Leaves toothed or pinnatisect. 5

5 Pappus of 2–4 retrorsely barbed awns . . . . .24 BIDENS
5 Pappus a short irregularly toothed crown of minute scales or awns . . . . 19 WEDELIA

6 Pappus of rigid awns with recurved barbs, sometimes also with scales 7
6 Pappus of fine bristles or thin scales 8

7 Involucre cylindrical. Cypselas linear, flattened, c. 4 times longer than broad, with 2 barbellate awns . .
. . . . 22 GLOSSOGYNE
7 Involucre usually hemispherical. Cypselas angular, sometimes winged, mostly as long as broad.
   Barbellate awns 2–several . . . . 8 CALOTIS

8 Rays white to mauve . . . . .57 ARCTOTIS
8 Rays yellow to orange 9

9 Cypsela covered with long hairs. Rays usually black at the base with or without a white spot . . . .
. . . . 60 GAZANIA
9 Cypsela glabrous or with short hairs. Rays without a white spot at the base . . . . . 21 HELIANTHUS

Group 8

Ray florets ligulate; disc florets tubular; pappus of hairs

1 Rays white, blue or mauve. 2
1 Rays yellow 7

2 Shrubs . . . . 9 OLEARIA
2 Herbs, sometimes woody at the base. 3

3 Leaves basal. Heads solitary on scapes exceeding the leaves. Involucre more than 1 cm diam. . . . . . .
. . . . 10 CELMISIA
3 Leaves cauline (a few only basal). Heads in compound inflorescences, or solitary and the involucre c. 5
   mm diam. 4

4 Heads solitary or few. 5
4 Heads in large compound inflorescences 6

5 Ray florets in 1 row. . . . . 11 VITTADINIA
5 Ray florets in several rows. . . . . 14 ERIGERON

6 Leaves scabrous to hairy. Rays minute. . . . . .12 CONYZA
6 Leaves glabrous. Rays conspicuous. . . . . 15 ASTER

7 Involucral bracts yellowish, membranous. . . . . .80 PODOLEPIS
7 Involucral bracts green, herbaceous 8

1 Vernonia Schreb.

1 species native Aust.; Qld, NSW, NT

One species in the area

Perennial herb up to 50 cm high, nearly glabrous to woolly (very variable). Leaves alternate, ovate-oblong to lanceolate, irregularly toothed lobed or almost entire. Heads on slender peduncles in a terminal leafless cymo-panicle. Involucre hemispherical, 5 mm diam.; the bracts linear, acute. Florets all tubular, purple or white. Cypselas cylindrical, hairy, 2 mm long. Pappus of fine bristles, 5–7 mm long. Widespread. Chiefly clay soils. Fl. summer–autumn . . . . **V. cinerea** (L.) Less. var. **cinerea**

2 Ageratina Spach

3 species naturalized Aust.; Qld, NSW, S.A.

Herbaceous perennials. Leaves opposite. Heads numerous in terminal corymbs. Involucre cylindrical; the bracts in 2 or more rows. Receptacle without scales. Florets numerous, white, all tubular, bisexual. Cypselas 5-angled. Pappus a single row of capillary hairs 3–4 mm long.

Key to the species

1 Leaves rhomboidal, glandular hairy; blades almost as broad as long, hastate, serrate, conspicuously veined, up to 10 cm long, with long petioles. Cypselas glabrous. Perennial herb 1–2 m high. Widespread. Weed in moist places. Introd. from America. Fl. spring. *Crofton Weed.* . . . .
. . . . ***A. adenophora** (Spreng.) R.King & H.Rob.
1 Leaves lanceolate, hairy but not glandular hairy, much longer than broad, serrate, up to 10 cm long, petiolate. Cypselas hairy. Perennial herb up to 1 m high. Widespread. Weed in moist places. Introd. from America. Fl. spring. *Mist Flower* . . . . ***A. riparia** (Regel) R.M.King & H.Rob.

3 Ageratum L.

2 species naturalized Aust.; Qld, NSW, NT

One species in the area

Annual, decumbent to erect herb, with many stems up to 50 cm long. Leaves opposite, ovate to triangular; the blades 2–3 cm long, 2–3 cm wide, crenate-serrate; petioles 10–15 mm long. Heads terminal in clusters. Involucre 3–5 mm diam.; the bracts green, in 2–3 rows. Florets all tubular, mauve. Cypselas angled, dark brown to black, c. 2 mm long. Pappus of subulate scales c. 2 mm long. Coast. Waste places. Introd. from Central America. Fl. most of the year . . . . ***A. houstonianum** Mill.

4 Lagenophora Cass.

4 species native Aust.; all states and territories except NT

Perennial herbs. Leaves in a basal rosette, usually toothed. Heads solitary on leafless unbranched scapes. Involucre with numerous bracts in 3–4 rows. Ray florets female, white to purple, in 3–4 rows. Disc florets bisexual or sterile. Pappus absent.

Key to the species

1 Plants with stolons. Roots fibrous. Leaves obovate, 1–15 cm long, hirsute. Scapes up to 18 cm long. Heads 4–8 mm diam. Ligule of ray florets more than 2 mm long. Widespread. Open forests and grasslands. Fl. spring–summer. *Blue Bottle-daisy* . . . . **L. stipitata** (Labill.) Druce

1 Plants without stolons. Roots fleshy. Leaves obovate, up to 8 cm long, hirsute or glabrous. Scapes up to 35 cm long. Heads 3–4 mm diam. Ligule of ray florets up to 2 mm long. Widespread. Moist places in open forests. Fl. spring–summer *Slender Lagenophora* . . . .**L. gracilis** Steetz

5 Gymnocoronis DC.

1 species naturalized Aust.; NSW

One species in the area

Erect to decumbent, glabrous, perennial herb. Up to 2 m high with ribbed, hollow stems. Leaves elliptic, lanceolate or ovate 4–20 cm long; apex acuminate; margins toothed. Heads 12–20 mm diam. in dense terminal corymbs. Involucre bracts 4–5 mm long in 2 rows. Florets bisexual, white, c. 4 mm long with glandular hairs. Cypsela 0.5 mm long, compressed, 5-angled. Pappus absent. Noxious weed growing in wet areas. Introd. from S. America. Fl. summer–autumn . . . . ***G. spilanthoides** (D.Don) DC.

6 Solenogyne Cass.

3 species endemic Aust.; Qld, NSW, Vic., Tas.

Perennial herbs with a short rootstock and fleshy roots. Leaves in a basal rosette, toothed. Heads solitary on leafless unbranched scapes, 4–7 mm diam. Involucre with numerous imbricate bracts in 3–4 rows. Ray florets female, in 3–4 rows; ligule very small. Disc florets bisexual. Pappus absent.

Key to the species

1 Scapes c. 0.5 mm diam. Involucral bracts reflexed in the fruiting stage. Leaves obovate 20–70 mm long. Scapes up to 20 cm long. Ray florets white to bluish. Widespread. Open forests and grasslands. Fl. spring–summer. . . . . **S. bellioides** Cass.

1 Scapes c. 0.5–1 mm diam. Involucral bracts erect in the fruiting stage. Ray florets pinkish 2

2 Plants glabrous or with very fine hairs. Leaves usually with more than 14 teeth on each margin, oblanceolate, up to 65 mm long, paler underneath. Widespread. Open forests and grasslands. Fl. spring–summer. . . . . **S. dominii** L.G.Adams

2 Plants hirsute with coarse hairs. Leaves usually with less than 14 teeth on each margin, oblanceolate to obovate, up to 110 mm long, similar colour on both surfaces. Higher Blue Mts Open forests and grasslands. Fl. spring–summer . . . . **S. gunnii** (Hook.f.) Cabrera

7 Brachyscome Cass.

60 species endemic Aust.; all states and territories

Herbs. Heads terminal. Involucre hemispherical; bracts in c. 2 rows, nearly equal, with scarious often purplish margins. Ray florets female, mostly in 1 row. Disc florets usually numerous, bisexual. Cypselas obtuse at the summit, often winged. Pappus of minute bristles or absent.

Key to the species

1 Cypselas not winged 2

1 Cypselas winged 8

2 Cypselas either tuberculate or longitudinally furrowed 3

2 Cypselas smooth, compressed 6

3 Cypselas tuberculate, 2 mm long, dark brown to black. Leaves up to 7 cm long, once or twice pinnatisect into 7–10 segments of various shapes. Heads solitary. Involucre 4 mm diam; bracts oblanceolate.

Rays 7–10 mm long, mauve pink or white. Pappus minute, white. Glabrous annual up to 40 cm high. Widespread. Open forest and grassland, usually on clay soils. Fl. spring–summer 4

3 Cypselas longitudinally furrowed, 2–3 mm long, quadrangular, light to very dark reddish-brown. Leaves basal and cauline, up to 10 cm long. Involucre 10–20 mm diam. Rays c. 10 mm long, white. Pappus white, c. 0.5 mm long. Perennial up to 40 cm high. Blue Mts Open forests. Fl. summer–winter 5

4 Leaf segments narrow-linear, subulate. . . . . **B. multifida** DC. var. **multifida**
4 Leaf segments broad-linear oblanceolate or cuneate. . . . . **B. multifida** var. **dilatata** Benth.

5 Leaves once pinnatisect . . . . **B. diversifolia** (Graham) Fisch. & C.A.Mey. var. **diversifolia**
5 Leaves twice pinnatisect. . . . . **B. diversifolia** var. **dissecta** G.L.Davis

6 Leaves mostly pinnatifid, with short linear lobes. Rays c. 3 mm long, pink. Cypselas c. 1 mm long, with longitudinal ribs and 2 wings. Pappus c. 0.2 mm long. Glabrous annual 5–10 cm high. Higher Blue Mts Moist places. Fl. spring–summer. . . . . **B. ptychocarpa** F.Muell.
6 Leaves entire 7

7 Leaves oblanceolate, 6–15 cm long, in a basal rosette. Scapes 12–30 cm long, usually unbranched, with a solitary head. Ray florets c. 40, white or mauve. Cypselas cuneate with rib-like margins. Pappus minute. Tufted perennial without stolons. Blue Mts Open forests. Fl. summer–autumn . . . .
. . . . **B. scapigera** (Sieber ex Spreng.) DC.
7 Leaves cauline, linear to oblanceolate, up to 14 cm long and 7 mm wide. Scapes up to 70 cm high, branched with terminal heads. Ray florets 9 mm long, blue or violet. Cypselas 2 mm long. Pappus minute. Stoloniferous perennial. Widespread. Wet places. Fl. spring–summer . . . .
. . . . **B. graminea** (Labill.) F.Muell.

8 Leaves with pungent tips, pinnatisect, up to 2 cm long. Involucre 8–10 mm diam; bracts broad, obtuse, with torn-ciliate margins. Ray florets 8 mm long, blue. Cypselas 2–3 mm long, flat, brown, with entire or dissected narrow wings. Western fringe of the Blue Mts Well-drained situations amongst rocks. Fl. summer . . . . **B. rigidula** (DC.) G.L.Davis
8 Leaves without pungent tips 9

9 Cypselas tuberculate. 10
9 Cypselas smooth 12

10 Plants without stolons. Leaves entire or sometimes pinnatisect, cuneate, up to 50 mm long with 3 acute linear lobes at the top. Heads solitary, on long scapes. Involucre c. 10 mm diam. Rays up to 9 mm long, white to bluish. Cypselas with toothed wings. Pappus c. 0.7 mm long. Erect perennial up to 50 cm high. Blue Mts Usually in wet situations. Fl. spring–summer . . . . **B. dentata** Gaudich.
10 Plants stoloniferous. Leaves entire or sometimes pinnatifid, oblanceolate to elliptic, up to 50 mm long. Heads c. 5 mm diam. Rays blue to pink, 6–10 mm long. Cypselas with crenate wings. Pappus c. 0.5 mm long. Perennial up to 35 cm high. Widespread. DSF. Sandy soils. Fl. summer–autumn 11

11 Leaves entire. **B. angustifolia** A.Cunn. ex DC. var. **angustifolia**
11 Leaves pinnatisect or deeply toothed . . . . **B. angustifolia** var **heterophylla** (Benth.) G.L.Davis

12 Scapes up to 50 cm long, leafless or with reduced leaves in the lower third. Leaves mainly basal, 5–10 cm long, spathulate, crenate. Involucre 10–15 mm diam. Rays 5–6 mm long, mauve. Cypselas 3–4 mm long, winged. Pappus up to 1 mm long. Erect perennial up to 50 cm high. Blue Mts Forests on clay soils. Fl. spring–summer. . . . . **B. spathulata** Gaudich.
12 Scapes leafy. Leaves mainly cauline. Lower leaves petiolate, up to 10 cm long. Involucre 10–20 mm diam. Rays c. 8 mm long, white, lilac or blue. Cypselas 3–4 mm long. Pappus minute. Perennial up to 60 cm high. Widespread. Open forests. Fl. spring–summer . . . . **B. aculeata** (Labill.) Less.

8 Calotis R.Br.

26 species in Aust. (23 endemic, 3 native); all states and territories except Tas.

Annual or perennial herbs, occasionally shrubs. Leaves alternate, entire or dissected. Heads solitary or clustered, pedunculate. Involucral bracts green or with a scarious margin, in 3–4 rows. Ray florets female, in 1–several rows; rays white pink mauve blue or yellow. Disc florets usually male, tubular, 5-toothed, yellow. Cypselas flattened, cuneate. Pappus of rigid awns with recurved barbs, sometimes with scales alternating with the awns.

Key to the species

1 Pappus of rigid awns alternating with an equal number of scales 2
1 Pappus of awns only; scales absent 5

2 Awns very finely but densely barbed along their whole length. Cypselas hairy. Cauline leaves cuneate, distally 5-toothed or occasionally entire, up to 2 cm long. Heads numerous in a dense cymo-panicle. Involucre hemispherical; bracts lanceolate to spathulate, entire, hairy. Rays c. 1 mm long, filiform, yellow. Cypselas very dark brown, cuneate, flattened, c. 2 mm long, hairy on each face and woolly towards the top. Pappus of 5–6 awns, c. 2 mm long. Prostrate or ascending annual up to 25 cm long, with white septate hairs. Widespread. Grasslands, open forests and roadsides in dry situations. Fl. chiefly winter. *Bogan Flea* . . . . **C. hispidula** (F.Muell.) F.Muell.
2 Awns barbed only distally. Cypselas glabrous 3

3 Pappus scales longer than broad. Body of the cypsela minutely but densely tuberculate, cuneate, flattened, 1.5–2.5 mm long; awns 4–5, unequally barbed at the tip. Lower leaves cuneate, c. 3 cm long and 8 mm wide, with 5–7 acute lobes near the tips; upper leaves narrower, sessile. Heads c. 2 cm diam. Rays blue, 7 mm long. Perennial up to 30 cm high. Higher Blue Mts Open forests and grasslands. Vulnerable. . . . . **C. glandulosa** F.Muell.
3 Pappus scales broader than long. Cypsela with few large tubercles 4

4 Leaves linear-lanceolate or oblong; basal leaves up to 8 cm long, distally serrate, occasionally pinnatifid; cauline leaves entire. Heads numerous. Involucre hemispherical, 5–10 mm diam., on axillary peduncles; bracts lanceolate to ovate, entire. Rays c. 10 mm long, white. Cypselas reddish-brown, c. 2 mm long, broad, flattened; pappus scales 2, fringed at their free edges; awns usually 2 but sometimes 1 or 3. Hairy perennial up to 80 cm high. Widespread. Open forests grasslands and roadsides. Usually clay or loam soils. Fl. all the year. *White Daisy Burr* . . . . .C. dentex R.Br.
4 Leaves cuneate or spathulate, distally toothed. Basal leaves with slender petioles; cauline leaves up to 4 cm long and 20 mm wide, narrow-ovate to linear. Rays numerous, up to 10 mm long, white to lilac. Cypselas reddish-brown, up to 1.5 mm long, cuneate, flattened; awns usually 2 but occasionally 3–4, c. 3 mm long; pappus scales infolded distally so as to appear entire. Erect to prostrate perennial up to 60 cm high, with stiff hairs. Widespread. Open forests grasslands and roadsides. Fl. most of the year . . . . **C. cuneifolia** R.Br.

5 Pappus awns equal in length, 4–6, 1–4 mm long. Leaves mostly basal, up to 25 cm long, linear to linear-lanceolate, entire or sparsely toothed, almost glabrous. Heads on unbranched peduncles exceeding the basal leaves and bearing a few bract-like leaves. Involucre hemispherical, 5–10 mm diam., bracts oblanceolate, obtuse, entire, c. 3 mm long. Ray florets numerous; rays 3–5 mm long, white or lavender. Cypselas brown; the body cuneate, flattened, glabrous, c. 2 mm long. Stoloniferous perennial up to 40 cm high. Coast and Cumberland Plain. Damp clay soils. Fl. most of the year . . . . **C. scapigera** Hook.
5 Pappus awns of unequal length. 6

6 Major pappus awns 2, erect; secondary awns 4 or more, horizontal; awns barbed distally, with hairs at the base. Basal leaves up to 6 cm long, cuneate, toothed or pinnatifid, sessile, up to 25 mm long. Involucre hemispherical, 3–6 mm diam.; bracts numerous, 2–3 mm long. Ray florets numerous; rays c. 3 mm long, yellow. Cypselas flattened, minutely tuberculate, c. 1.5 mm long. Widespread. Open forests and grasslands on heavy clay soils. Fl. all the year. *Yellow Daisy Burr* . . . . **C. lappulacea** Benth.

6 Major pappus awns 5–6, up to 4 mm long; secondary awns when present, c. 1 mm long, hairy. Leaves up to 18 cm long, linear to elliptic, toothed, pinnatifid or pinnatipartite; cauline leaves sessile, entire or toothed. Involucre hemispherical, 6–10 mm diam.; bracts numerous, up to 5 mm long, elliptic, acuminate. Rays white or mauve. Cypselas reddish-brown, 3–4 mm long, broad, cuneate, flattened. Stoloniferous, hairy perennial up to 40 cm high. Blue Mts Open forests and grasslands, often on heavy clay soils. Fl. most of the year . . . **C. scabiosifolia** Sond. & F.Muell. var. **integrifolia** F.Muell. ex Benth.

9 Olearia Moench.
Daisy Bushes
130 species endemic Aust.; all states and territories

Shrubs. Leaves alternate or opposite. Heads solitary or in compound inflorescences. Involucre broad-hemispherical to narrow-ovate; bracts imbricate in several rows. Receptacle pitted, without scales. Ray florets female, in a single row, white or blue. Disc florets bisexual tubular. Cypselas, terete or slightly compressed. Pappus of numerous, usually unequal capillary bristles.

Key to the species

1 Upper surface of the leaves viscid 2
1 Upper surface of the leaves not viscid 3

2 Leaves mostly opposite, linear, 4–7 mm wide, up to 10 cm long, entire, white-tomentose underneath. Heads on slender peduncles, solitary or few together in a loose corymb. Involucre 5–6 mm diam. Ray florets 8–10, white. Shrub up to 2 m high. Widespread. DSF and Woodland. Fl. chiefly spring . . . . . .
. . . . **O. viscidula** (F.Muell.) Benth.
2 Leaves alternate, 20–30 mm wide, 3–8 cm long, elliptic oblong or lanceolate, with glandular dots; glabrous underneath. Heads numerous, pedunculate. Involucre 6–8 mm diam. Ray florets 6–8, white. Shrub up to 2 m high. Widespread. Wet situations. Fl. spring–summer. *Sticky Daisy Bush* . . . .
. . . . **O. elliptica** DC.

3 Leaves opposite 4
3 Leaves alternate 5

4 Leaves sessile, linear, 12–90 mm long, up to 4 mm wide, revolute, entire, grey tomentose underneath. Heads on axillary peduncles forming a terminal leafy panicle. Ray florets 6–8, white. Cypselas glandular-papillose; pappus hairs unequal. Shrub up to c. 1.5 m high. Western parts of the area. Open forests. Fl. spring–summer . . . . **O. rosmarinifolia** (DC.) Benth.
4 Leaves petiolate, elliptic, 15–115 mm long, 4–38 mm wide, flat, irregularly toothed to entire, brown-tomentose underneath. Heads pedunculate, in terminal panicles. Ray florets 4–7, white. Cypselas glabrous; pappus in 2 rows. Shrub up to 3 m high. Blue Mts Open forests. Fl. summer . . . .
. . . . **O. chrysophylla** (DC.) Benth.

5 Leaves glabrous or nearly so, linear, 3–4 cm long, c. 1 mm wide, bluntly denticulate, blistered. Heads numerous pedunculate, in terminal compound corymbs. Ray florets 15–22, white. Cypselas silky. Shrub up to 2 m high. Blue Mts Swamps; creek banks. Fl. summer–winter . . . **O. glandulosa** (Labill.) Benth.
5 Leaves hairy at least underneath 6

6 Leaves densely glandular hairy, linear to narrow-lanceolate, 10–20 mm long, 2–8 mm wide, cordate at the base. Heads pedunculate, few or solitary on axillary branches. Ray florets 10–18, deep mauve. Cypselas silky. Shrub up to 2 m high. Hornsby Plateau, Wiseman's Ferry northwards. DSF. Sandy soils. Vulnerable. Fl. summer–autumn. . . . . .O. cordata Lander
6 Leaves not densely glandular hairy but often with other types of hair, rarely with few glandular hairs on the upper surface 7

7 Leaves mostly more than 30 mm wide. 8
7 Leaves mostly less than 30 mm wide. 10

8 Ray florets more than 10, white. Leaves elliptic to ovate, 5–150 mm long, 5–30 mm wide, entire or toothed, flat, brownish tomentose underneath, hairy or glabrous on the upper surface. Heads pedunculate, in terminal leafy panicles. Ray florets white. Cypselas silky. Shrub up to 3 m high. Illawarra district. Open forests. Fl. spring–summer . . . . **O. stellulata** (Labill.) DC.

8 Ray florets less than 10. 9

9 Heads sessile, in dense terminal corymbs, 9–18 mm diam. Ray florets 1–2, white. Leaves broad-elliptic, up to 65 mm long and 40 mm wide, silvery tomentose underneath. Blue Mts Extinct . . . .
. . . . **O. oliganthema** F.Muell. ex Benth.

9 Heads pedunculate, in terminal corymbs, 13–27 mm diam. Ray florets 3–8, white. Leaves broad-elliptic, 15–185 mm long, entire or toothed, silvery tomentose underneath. Shrub or small tree up to 10 m high, with a musky odour. Widespread. WSF and RF margins. Fl. spring–summer. *Native Musk* or *Silver Bush* . . . . **O. argophylla** (Labill.) F.Muell. ex Benth.

10 Hairs on the undersurface of the leaves stellate 11
10 Hairs on the undersurface of the leaves T-shaped or simple. 14

11 Leaves with acuminate tips, 40–60 mm long, 8–20 mm wide, sinuately lobed or dentate. Heads pedunculate, numerous, in mainly terminal corymbs. Involucre 3–4 mm diam. Ray florets 9–20, white. Shrub up to 2 m high. Coast, Gosford northwards. Open forests. Fl. winter–spring . . . .
. . . . **O. nernstii** (F. Muell.) F.Muell. ex Benth.

11 Leaves with blunt or rounded tips. 12

12 Leaves glabrous or scabrous on the upper surface, sinuately lobed, 25–35 mm long, 10–15 mm wide. Heads pedunculate, mostly solitary and axillary towards the ends of branches. Involucre 6–7 mm diam. Ray florets 7–15, white. Shrub up to 2 m high. Blue Mts Swampy places. Fl. spring–summer . . .
. . . . **O. quercifolia** Sieber ex DC.

12 Leaves with hairs on both surfaces 13

13 Leaves 2–4 mm wide, 12–30 mm long, sinuately lobed, hairy usually on both surfaces. Heads pedunculate or sessile, few, terminal and solitary on axillary branches. Involucre 6–8 mm diam. Ray florets 12–20, white. Shrub up to 50 cm high. Blue Mts Heath and DSF. Fl. spring–summer . . . .
. . . . **O. asterotricha** (F. Muell.) F.Muell. ex Benth.

13 Leaves 4–10 mm wide, mostly 30–50 mm long, sinuate or entire, grey-tomentose underneath. Heads solitary and terminal on axillary branches. Involucre 4–5 mm diam. Ray florets 16–30, white. Shrub up to 1 m high. Higher Blue Mts Heath and DSF. Fl. spring–summer . . **O. phlogopappa** (Labill.) DC.

14 Disc florets pink. Leaves linear, up to 24 mm long and 1 mm wide, glabrous or upper surface with a few glandular hairs. Heads pedunculate, in terminal leafy panicles, up to 1.5 mm diam. Ray florets 14–20, white to pink. Shrub up to 70 cm high. Capertee–Wallerawang. Fl. summer–autumn . . . . . .
. . . .**O. suffruticosa** D.A.Cooke

14 Disc florets white to yellowish. 15

15 Leaves sessile 16
15 Leaves pedunculate. 19

16 Leaves not scabrous on the upper surface, sometimes with scattered hairs, elliptic to obovate, 3–13 mm long, 1–4 mm wide, entire, slightly revolute, hairy at least underneath. Heads terminal, solitary or in loose corymbs, 10–16 mm diam. Ray florets 8–11, white. Shrub up to 2 m high. Burragorang to Picton. Open forests. Fl. winter–summer. . . . . **O. burgessii** Lander

16 Leaves scabrous on the upper surface 17

17 Heads more than 20 mm diam., solitary at the ends of branches forming a leafy panicle. Leaves elliptic to triangular, up to 5 mm long, grey tomentose underneath, slightly revolute. Ray florets 12–15, blue to mauve. Disc florets blue or yellowish. Shrub up to 1.5 m high. Shoalhaven Gorge area. Open forests. Fl. spring. . . . . **O. ramosissima** (DC.) Benth.

17 Heads less than 20 mm diam 18

18 Leaves obovate to spathulate, 2–7 mm long, 1–3 mm wide, very revolute; grey tomentose underneath. Heads terminal, solitary, sessile. Ray florets 6–8, white. Shrub up to 1.5 m high. Widespread. DSF and heath. Fl. spring . . . . **O. microphylla** (Vent.) Maiden & Betche

18 Leaves linear to narrow-obovate, 2–8 mm long, 1–2.5 mm wide, revolute, grey tomentose underneath. Heads axillary or terminal on short lateral branches, sessile or pedunculate. Ray florets 2–13, blue or sometimes white. Shrub up to 1.5 m high. Widespread. DSF. Fl. spring–summer . . . .
. . . . **O. ramulosa** (Labill.) Benth.

19 Involucre 2–3 mm diam., on peduncles up to 1 cm long. Leaves obovate, 1–2 cm long, 4–7 mm wide, entire or denticulate, grey tomentose underneath. Ray florets 2–4, white. Shrub up to 1.5 m high. Blue Mts Open forests and grasslands. Fl. summer–autumn . . **O. myrsinoides** (Labill.) F.Muell. ex Benth.

19 Involucre 6–10 mm diam. on peduncles up to 4 cm long 20

20 Leaves scabrous on the upper surface, broad-ovate, 20–60 mm long, 10–25 mm wide, sinuately lobed, brownish or grey tomentose underneath. Ray florets 13–29, blue to white. Shrub up to 2 m high. Coast and adjacent plateaus; lower Blue Mts Heath and DSF, particularly near the coast. Fl. spring–summer . . . . **O. tomentosa** (J.C.Wendl.) DC.

20 Leaves not scabrous on the upper surface, 20–40 mm long, 7–20 mm wide; silvery tomentose underneath. Heads 3–5 together on lateral branches. Ray florets 4–7, white. Shrub up to 2 m high. Widespread. DSF. Fl. spring–summer . . . . **O. erubescens** (Sieber ex DC.) Dippel

10 Celmisia Cass.
10 species endemic Aust.; NSW, Vic., Tas.

One species in the area

Perennial herb. Leaves basal, linear to linear-lanceolate, entire, up to 20 cm long, with a broad sheathing base, softly mucronate or obtuse, revolute, densely white tomentose underneath, silvery tomentose or glabrescent on the upper surface. Heads solitary on scapes exceeding the leaves; peduncles with linear or lanceolate bracts up to 4 cm long. Involucre broad-turbinate or hemispherical, woolly or glabrous when old, 2–3 cm diam.; bracts imbricate, in several rows; the margins dry or scarious. Ray florets numerous, in a single row, spreading, female; rays 1–2 cm long, white or pink. Disc florets numerous, c. as long as the involucre, bisexual, yellow. Cypselas slightly compressed, with 2–3 prominent veins on each side, silky pubescent, 6 mm long. Pappus of numerous slightly unequal capillary hairs up to 7 mm long. Higher Blue Mts Wet ground. Fl. summer. *Snow Daisy* . . . . **C. longifolia** Cass.

11 Vittadinia A.Rich.
27 species in Aust. (25 endemic); all states and territories

Perennial herbs up to 40 cm high, usually with a woody base. Leaves alternate. Heads solitary on leafy peduncles. Involucre cylindrical or campanulate; bracts in 3–5 unequal rows, with scarious margins. Ray florets female, in 2 or more rows; rays narrow. Disc florets fewer, tubular, bisexual. Cypselas narrow, compressed. Pappus of numerous capillary bristles.

Key to the species

1 Leaves linear or 3-lobed, up to 0.5 mm wide 2

1 Leaves broader than linear, 3-toothed if 3-lobed then more than 0.5 mm wide 4

2 Leaves almost or quite glabrous, 10–20 mm long. Involucre 4–5 mm long. Cypselas 3–4 mm long; the lower half glabrous apart from short appressed hairs; the upper part with sparse slender spreading hairs with 2-fid apices. Coast. Fertile soils. Fl. spring . . . . **V. tenuissima** (Benth.) J.M.Black

2 Leaves hairy 3

3 Cypselas ribbed on the surface, 4–5 mm long; the lower part with slender appressed 2-fid hairs. Leaves with scanty septate hairs on the margins and midrib on the lower surface, 10–50 mm long. Involucre 6–9 mm long. Ray blue to purple. Leaves entire or with 2 narrow spreading lateral lobes slightly above the middle. Leaves almost lacking from the peduncles. Widespread. Open forests and grasslands. Fl. spring–summer. . . . . **V. muelleri** NTBurb.

3 Cypselas not ribbed on the surface but with marginal ribs . . . . **V. hispidula** (see below)

4 Leaves mostly deeply 3-lobed, obovate in outline, 10–40 mm long, with hispid hairs; the lobes themselves usually lobed or toothed. Involucre 4–6 mm long. Rays white to blue. Cypselas 4–5 mm long, ribbed on the surface. Coast. Open forests and grasslands. Fl. most of the year. . . . .
. . . . **V. dissecta** (Benth.) NTBurb. **dissecta**

4 Leaves entire or 3-toothed. 5

5 Plant covered with soft white woolly hairs. Leaves linear to obovate, 10–40 mm long. Involucre 7–8 mm long. Rays purplish. Cypselas 4–5 mm long, ribbed on the surface, glandular pubescent. Widespread. Open forests, grasslands and roadsides. Fl. most of the year . . . . **V. gracilis** (Hook.f.) NTBurb.

5 Plant with ± hispid hairs 6

6 Cypselas not ribbed on the surface but sometimes with marginal ribs 7
6 Cypselas ribbed on the surface 8

7 Cypselas with 1 or more pustules at the base, without prominent marginal ribs, ± 4-angled, 2–4 mm long. Leaves obovate, 5–20 mm long. Involucre 5–6 mm long. Coast and Cumberland Plain. Grasslands and waste places. Fl. spring . . . .**V. pustulata** NTBurb.

7 Cypselas without pustules at the base, prominently ribbed on the margins, 2–3 mm long. Leaves oblong to obovate, 10–60 mm long. Involucre 6–7 mm long. Rays white to purplish. Coast and adjacent plateaus . . . . **V. hispidula** F.Muell. ex A.Gray

8 Cypselas without glandular hairs, 5–8 mm long. Leaves more than 5 mm wide, obovate, 8–50 mm wide, flat. Involucre 7–9 mm long. Rays purple. Coast and Cumberland Plain. Waste places, creek banks. Fl. spring . . . . **V. sulcata** NTBurb.

8 Cypselas with glandular hairs 9

9 Cypselas cuneate, 4–7 mm long, ribs continuing to the summit. Leaves up to 5 mm wide, oblanceolate to obovate, 10–25 mm long, ± folded along the midrib. Involucre 5–8 mm long. Rays blue to mauve. Widespread. Open forests, grasslands and waste places. Fl. most of the year   **V. cuneata** DC. var. **cuneata**

9 Cypselas oblanceolate, 6–10 mm long, , ribs converging below the summit. Leaves to 10 mm wide, oblanceolate, obovate to spathate, 10–30 mm long, often folded along middle. Involucre 8–12 mm long. Ray florets purple to white. DSF and grasslands. Fl. All year.. . . . . **V. cervicularis** NTBurb.

12 Conyza Less.
9 species naturalized Aust.; all states and territories

Annual herbs. Leaves basal and cauline, alternate. Heads in corymbs or panicles. Involucre ovoid to hemispherical; bracts numerous, nearly equal, imbricate in several rows. Florets ligulate and tubular or filiform and tubular. Cypselas flattened; the margins thickened. Pappus of numerous capillary, nearly equal hairs.

Key to the species

1 Outer florets with white ligules c. 1 mm long. Involucre c. 2 mm diam, glabrous or with a few hairs . . 2
1 Outer florets filiform. Involucre more than 2 mm diam. 3

2 Stems glabrous or sparsely hairy. Leaves linear to oblong mostly up to 5 cm long, entire or crenate. Heads in spike-like panicles. Inner surface of the involucral bracts whitish, often with a red spot at the apex. Pappus cream. Erect annual up to 50 cm high. Coast. Weed on sandy soils. Introd. from S. America. Fl. summer–autumn . . . . ***C. parva** Cronq.

2 Stems conspicuously hairy. Leaves oblong to linear, up to 10 cm long, lower ones toothed. Heads in spreading panicles. Inner surface of the involucral bracts brownish, without a red spot at the apex. Pappus cream. Erect annual up to 1.5 m high. Widespread. Weed of cultivation and waste places. Introd. from N. America. Fl. most of the year. *Canadian Fleabane.* . . . .***C. canadensis** (L.) Cronq. var.**canadensis**

3 Involucral bracts glabrous, red-brown. Pappus yellowish-brown. 2–3 mm long. Heads numerous in panicles. Erect annual up to 2 m high, with glabrescent or sparsely hairy stems. Widespread. Weed of cultivation and waste places. Introd. from S. America. Fl. most of the year . . . . *C. bilbaoana J.Remy
3 Involucral bracts hairy. Stems hispid 4

4 Pappus white or pink, c. 3 mm long. Involucral bracts pale inside. Heads numerous in a pyramidal or corymb-like panicle. Erect perennial up to c. 1 m high. Widespread. Weed of cultivation and waste places. Introd. from N. America. Fl. most of the year. *Flaxleaf Fleabane* . . *C. bonariensis (L.) Cronq.
4 Pappus straw-coloured, c. 3 mm long. Involucral bracts reddish-brown inside. Heads in an extensive panicle. Erect perennial up to c. 2 m high. Widespread. Weed of cultivation and waste places. Introd. from N. America. Fl. most of the year. *Tall Fleabane* . . . . *C. sumatrensis (Retz.) E.Walker

13 Baccharis L.
1 species naturalized Aust.; Qld, NSW

One species in the area
Shrub up to 6m high, Leaves elliptic to obovate, 3–7 cm long, toothed, petiolate, ± viscid. Heads in clusters on axillary and terminal peduncles, functionally unisexual. Involucre 3–5 mm diam.; bracts scarious or slightly herbaceous. Ray florets absent. Disc florets yellowish. Cypselas up to 2 mm long, glabrous. Pappus white, 10–12 mm long on the fruit. Wollongong. Swamps near the sea. Introd. from N. America. Fl. spring–summer. *Groundsel Bush* . . . . *B. halimifolia L.

14 Erigeron L.
8 species in Aust. (5 endemic, 3 naturalized); NSW, Vic., Tas., S.A., NT

One species in the area
Perennial herb up to c. 30 cm high, much branched from the base. Lower leaves 3-lobed or coarsely 3–5-toothed at the apex, 1–3 cm long; upper leaves linear-lanceolate. Heads few, on long peduncles, 8–15 mm diam. Involucral bracts linear, acute, hairy, in several rows. Ray florets in several rows, pale purple turning white and finally pink, 6–8 mm long. Cypselas compressed. Pappus of fine whitish hairs longer than the cypsela. Widespread. Garden escape into moist areas. Introd. from Central America. Fl. spring–summer. *Fleabane* . . . . *E. karvinskianus DC.

15 Aster L.
2 species naturalized Aust.; Qld, NSW, Vic., S.A., WA

Mostly perennial herbs, sometimes with rhizomes. Leaves cauline, alternate. Heads clustered or solitary. Involucral bracts in few to several rows, herbaceous, unequal, imbricate. Receptacle pitted. Ray florets in 1–2 rows; ray white blue red or purple. Disc florets yellow. Cypselas usually compressed. Pappus of numerous bristles in 1–2 rows.

Key to the species

1 Involucre 2–4 mm diam. Leaves 10 mm wide, linear-lanceolate, up to 12 cm long, sessile. Heads in a large loose leafy panicle. Ray florets female; rays white pink or blue. Cypselas 2 mm long. Pappus bristles 5–7 mm long. Erect, almost glabrous herb up to 2 m high. Widespread. Weed of cultivation and waste places, often in wet ground; margins of saltmarshes. Introd. from America. Fl. chiefly winter. *Bushy Starwort* . . . . *A. subulatus Michx.
1 Involucre 5–7 mm diam. Leaves 12–20 mm wide, 6–8 cm long, sessile, stem-clasping, lanceolate, usually with a few small teeth. Rays blue-purple. Rhizomatous perennial up to 30 cm high. Widespread. Weed near habitation. Introd. from Europe. *Michaelmas Daisy* or *Easter Daisy* . . . . *A. novi-belgii L.

16 Solidago L.

1 species naturalized Aust.; Qld, NSW, S.A.

One species in the area

Perennial herb up to 1.5 m high, with stolons and rhizomes. Leaves alternate, simple, sessile, glabrous or pubescent, serrate, 3-veined, 5–10 cm long. Heads numerous on one side of the rhachises, in a terminal pyramidal panicle. Involucre 2–3 mm diam. Ray florets in 1 row; rays 1–2 mm long, yellow. Disc florets tubular, yellow. Cypselas terete, 1 mm long. Pappus of fine hairs c. 2 mm long. Garden escape near habitation. Introd. from N. America. Fl. summer. *Golden Rod.* . . . . ***S. canadensis** L. var. **scabra** Tor. & A.Gray

17 Sigesbeckia L.

2 species in Aust. (1 native, 1 endemic); all states and territories

One species in the area

Annual herb up to 2 m high. Leaves opposite, toothed, broad-ovate-triangular to lanceolate, 4–7 cm long; petioles usually dilated upward but not at the base. Heads pedunculate, in a dichotomous leafy panicle. Involucre 6–12 mm diam; bracts in 2 rows, the outer ones c. 10 mm long and covered with glandular hairs. Ray florets female, shortly ligulate or irregularly 2–3-lobed, yellow. Disc florets tubular, yellow. Cypselas c. 2 mm long, dark brown, usually curved. Pappus absent. Widespread. Open forests; margins of RF; roadsides. Usually on better soils, often abundant after fire. Fl. spring–autumn. *Indian Weed.* . . . .
. . . . **S. orientalis** L. ssp. **orientalis**

18 Eclipta L.

2 species endemic in Aust.; all states and territories except Tas.

One species in the area

Perennial herb, usually prostrate. Stems up to 50 cm long. Leaves opposite, sessile or shortly petiolate, narrow-lanceolate, with short appressed hairs. Involucre 2–3 mm diam., on axillary peduncles 1–2 cm long; bracts herbaceous, in 2 rows. Ray florets female or sterile; ray 1–2 mm long, yellow, 2-lobed. Disc florets tubular, yellow. Cypselas cylindrical, tuberculate, light brown, 1–2 mm long. Pappus absent. Coast and Cumberland Plain. Swamps. Fl. summer. *Yellow Twin-heads* . . . . **E. platyglossa** F.Muell.

19 Wedelia Jacq.

8 species in Aust. (1 naturalized); all states and territories except Tas.

One species in the area

Perennial herb up to c. 60 cm high. Leaves opposite, 6–10 cm long, 5–7 mm wide, often irregularly toothed. Involucre 10–15 mm diam; bracts herbaceous, in 2 rows. Ray florets yellow, female. Disc florets tubular, bisexual. Cypselas 4 mm long. Pappus a short irregularly toothed crown of minute scales or awns Coast and Hunter River Valley. Weed of cultivation and waste places. Introd. from S. America. Fl. autumn. *Pascalia Weed* . . . . ***W glauca** (Ortega) S.F.Blake

20 Melanthera Rohr

1 species native Aust.; Qld, NSW, NT

One species in the area

Perennial, straggling herb up to 1 m high, usually with white appressed hairs. Leaves opposite, 8–10 cm long, 2–4 cm wide, petiolate, 3-veined, toothed. Heads usually 3 together on a leafless peduncle. Involucre 10 mm diam.; bracts herbaceous, in 2 rows. Ray florets 10–12, ligulate, female, white. Disc florets tubular, bisexual. Cypselas 2–4 mm long, slightly flattened. Pappus absent or consisting of 1–3 caducous awns 1 mm long. Coast. Dunes and headlands. Fl. summer . . . . **M. biflora** (L.) Wild

21 Helianthus L.

3 species naturalized Aust.; all states and territories except Tas.

Annual or perennial herbs. Stems erect. Leaves opposite or alternate. Heads terminal and solitary or numerous in a terminal corymb. Involucral bracts in 2–several rows. Ray florets sterile. Disc florets bisexual. Receptacle flat or convex, with chaffy bracts. Cypselas flattened. Pappus of 2 scale-like awns.

Key to the species

1 Leaves mostly alternate, mostly ovate and serrate, up to 20 cm long and 10 cm wide. Stems often mottled. Heads 3–10 cm diam. (larger in cultivated plants). Rays yellow. Disc florets brownish purple. Cypselas 1–2 cm long. Erect annual up to 3 m high. Commonly cultivated. Widespread. Garden escape near habitation; roadsides. Introd. from America. Fl. summer. *Sunflower* . . . . ***H. annus** L.

1 Leaves mostly opposite, ovate to oblong, serrato-dentate, up to 16 cm long, 5–8 cm wide; petioles winged. Rays yellow. Disc florets yellow. Heads numerous in a corymb, mostly 5–8 cm diam. Erect perennial up to 3 m high, with edible root tubers. Occasional near habitation. Introd. from America. Fl. summer. *Jerusalem Artichoke* . . . . ***H. tuberosus** L.

22 Glossogyne Cass.

2 species in Aust. (1 endemic); all states and territories except Tas.

One species in the area

Erect, perennial herb up to 60 cm high. Stems often almost leafless or decumbent and leafy at the base. Leaves mostly basal; the lowest cuneate and 3-lobed; all the others pinnatipartite into 5 or 7 stiff almost linear segments which are either entire or 2–3-lobed. Heads on slender terminal peduncles. Involucre 5–8 mm diam.; bracts in 2 rows, narrow, nearly equal. Ray florets female; rays up to 10 mm long, yellow, sometimes absent. Disc florets tubular, bisexual. Cypselas linear, c. 8 mm long, flattened, striate. Pappus of 2 erect barbellate awns 4 mm long. Widespread. Grasslands and tall forests. Usually on clay soils. Fl. summer. *Cobbler's Tack* . . . . **G. tannensis** (Spreng.) Garn.-Jones

Some botanists now refer to this taxon as *Glossocardia bidens* (Retz.) Valdkemp)

23 Enydra Lour.

1 species endemic Aust.; Qld, NSW

One species in the area

Perennial herb, aquatic or growing in swamps. Stems creeping, rooting at the nodes. Leaves opposite, oblong or lanceolate, petiolate or sessile and stem-clasping, 4–6 cm long, 6–10 mm wide, entire sinuate or toothed. Heads sessile in the forks of the stem or leaf axils. Involucral bracts 4; the 2 outer ones 3–6 mm long; the inner ones shorter; all longer than the florets. Ray florets in several rows, female, with very short 3-toothed rays. Disc florets tubular, bisexual. Coast near Sydney. Swamps. Fl. spring–summer . . . . **E. fluctuans** Lour.

24 Bidens L.

5 species naturalized Aust.; all states and territories except Tas.

Herbs with opposite mostly compound or pinnatisect leaves. Heads on terminal peduncles. Involucral bracts few, in 2–3 rows. Ray florets sterile, ligulate or sometimes absent. Disc florets tubular, bisexual. Cypselas slender, 4-angled. Pappus of 2–4 rigid barbellate awns.

Key to the species

1 Leaves divided pinnatisectly into 3 segments; terminal segment 6–8 cm long, 2 cm wide; lateral segments half as large; all coarsely toothed. Herb up to 1 m high, with a single erect stem with few or no branches. Heads terminal and axillary. Involucral bracts 1–2 cm diam. Ray yellow. Cypselas 6–8 mm long. Pappus awns 2–4 mm long, finely barbed. Coast. Near habitation. Introd. from America. Fl. autumn–winter. *Burr Marigold* . . . . ***B. tripartita** L.

1 Leaves pinnately divided into 3–5 segments or pinnatisectly into 5 or more segments; the lower ones sometimes simple. Cypselas up to 12 mm long, 4-angled. Pappus awns up to 3 mm long 2

2 Leaflets serrate or dentate very rarely lobed, stalked, usually 3 but sometimes 5, ovate to lanceolate, 3–6 cm long. Ray white, up to 1 cm long, sometimes absent. Heads few, mostly terminal on slender peduncles. Involucre cylindrical, c. 6 mm diam.; outer bracts herbaceous; inner ones bordered with a white thinnish margin. Erect, glabrous or hairy herb up to 2 m high, sometimes woody towards the base, with angular branches. Widespread. Weed of cultivation and waste places. Introd. from S. America. Fl. spring–summer. *Cobbler's Pegs*. Benson (2007) indicates that this species may be native to Australia. . . . . ***B. pilosa** L.

2 Leaflets deeply divided or the leaves bipinnate, 5–7-lobed linear to lanceolate, 4–11 cm long. Ray yellow, up to 1 cm long. Involucral bracts herbaceous. Erect ± glabrous herb up to 2 m high. Coast. Weed of cultivation and waste places. Introd. from S. America. Fl. spring–summer. *Beggars Ticks* . . . . ***B. subalternans** DC.

25 Galinsoga Ruiz & Pav.
1 species naturalized Aust.; all states and territories

One species in the area

Erect annual up to 60 cm high, glabrous or silky hairy. Leaves opposite, petiolate, ovate to ovate-lanceolate, usually toothed or lobed, up to 5 cm long. Heads axillary, on slender peduncles. Involucre hemispherical, 4 mm diam. Bracts 5, broad, nearly equal. Ray florets c. 5, female; rays white. 1–3 mm long, 3-lobed. Disc florets tubular, yellow. Cypselas angular, slightly flattened, slightly hairy, 1–2 mm long. Pappus of the ray cypselas reduced to a few minute bristles or absent; that of the disc cypselas of 12–20 chaffy scales, less than 1 mm long, ± plumose-ciliate. Widespread. Weed of cultivation and waste places. Introd. from America. Fl. summer. *Potato weed* . . . . ***G. parviflora** Cav.

26 Guizotia Cass.
1 species naturalized Aust.; QLD, NSW

One species in the area

Annual herb up to 2 m high. Lower leaves opposite, ovate to obovate, up to 10 cm long, toothed, sessile and stem-clasping. Heads c. 15 mm diam. Involucral bracts in 2 rows. Ray florets female; rays yellow, 10–15 mm long, 3-lobed. Cypselas 3–4-angled. Pappus absent. Weed near Sydney. Introd. from E. Africa. Fl. winter. *Rantil* . . . . ***G. abyssinica** (L.f.) Cass.

27 Verbesina L.
1 species naturalized Aust.; Qld, NSW, Vic., NT, S.A.

One species in the area

Annual herb up to 60 cm high. Leaves alternate, up to 10 cm long, toothed to deeply lobed, with broad stem-clasping auricles. Heads solitary, on peduncles up to 25 cm long. Involucre broad-hemispherical, 2 cm diam.; bracts in 2 rows; the outer ones green. Ray florets female or sterile; rays yellow, 1–2 cm long, deeply 3-lobed. Cypselas flat, dark brown, 5–7 mm long. Pappus of 2 bristles, caducous. Weed in drier areas. Introd. from America. Fl. summer. *Crownbeard*. . . . .***V. encelioides** (Cav.) A.Gray ssp.**encelioides**

28 Cosmos Cav.
2 species naturalized Aust.; Qld, NSW

One species in the area

Erect, annual herb up to 3 m high. Leaves opposite, glabrous or pubescent, up to 12 cm long, bipinnately segmented into linear segments. Heads solitary, on long peduncles. Involucre 10–15 mm diam.; bracts herbaceous, in 2 rows; the outer ones connate at the base. Ray florets female but without styles; ray 2–4 cm long, 10–15 mm wide, white pink red or purple. Cypselas flattened, oblong, glabrous, 8–10 mm long, beaked. Pappus absent. Widespread. Garden escape near habitation; roadsides. Introd. from Mexico. Fl. summer. *Cosmos* . . . . ***C. bipinnatus** Cav.

29 Coreopsis L.

1 species naturalized Aust.; Qld, NSW, Vic.

One species in the area

Perennial herb up to 1 m high, hirsute. Leaves basal and cauline, up to 30 cm long, spathulate, on long petioles; uppermost cauline leaves sessile, c. 2 cm long; remaining leaves pinnatisect, almost pinnate. Head solitary on axillary peduncles 25–30 cm long. Involucre 1–2 cm diam.; bracts broad, green, in 3 rows; the outermost ones narrowest. Ray florets female; ray yellow, 3–4 cm long, toothed. Disc florets tubular, yellow. Cypselas compressed, black, 2–3 mm long, with 2 broad wings. Pappus absent. Widespread. Weed along railway lines and roadsides. Introd. from N. America. Fl. summer . . . .*C. lanceolata L.

30 Ambrosia L.

Ragweeds

4 species naturalized Aust.; NSW, Vic., S.A., WA

Monoecious, annual or perennial herbs, frequently becoming woody towards the base. Leaves opposite or alternate, Male heads numerous, in terminal racemes, each several–many-flowered; receptacle with scales between the florets; corolla 5-toothed; stamens 5. Female heads solitary or few together in the leaf axils below the male heads, 1-flowered; Florets without corollas, surrounded by several connate bracts; style 2-fid, protruding beyond the involucre. Pappus absent. Cypsela surrounded by the persistent involucre.

Key to the species

1 Fruiting head with a whorl of 4–6 spines surrounding the head near the summit, glabrous; head 2.5–4 mm diam. Leaves pinnatipartite, lanceolate to ovate, mostly up to 10 cm long; the upper ones alternate; the lower ones almost opposite. Annual herb up to 1 m high. Coast and Cumberland Plain. Weed of cultivation and waste places. Introd. from N. America. Fl. summer–autumn . . . .*A. artemisiifolia L.
1 Fruiting head without spines but with teeth, hairy. Perennials. 2

2 Leaves pinnatifid, lanceolate to ovate, 10 cm long, c. 3 cm wide. Male involucre 3 mm diam. Perennial herb up to 2 m high with creeping rootstock. Widespread. Weed of roadsides and waste places. Introd. from America. Fl. spring–summer. . . . . *A. psilostachya DC.
2 Leaves 2–4-pinnate, linear to oblong, mostly up to 10 cm long. Perennial herb with long suckering roots. Widespread. Weed of roadsides and waste places. Introd. from America. Fl. spring–summer . . .
. . . .*A. tenuifolia Spreng.

31 Acanthospermum Schrank

2 species naturalized Aust.; Qld, NSW, NT

One species in the area

Prostrate herb with stems up to 60 cm long. Leaves opposite, petiolate, ovate to rhomboidal, 15–20 mm long, 1 cm wide, serrate or crenate towards the tip. Heads solitary on short axillary leafy branches. Involucral bracts in 2 rows; the outer of 5 (up to 8) simple bracts 1–2 mm long; the inner row of 6 (up to 8) bracts up to 10 mm long, enclosing the fruits of the ray florets at maturity; the bracts becoming spinescent with hooked spines. Ray florets bisexual. Disc florets sterile. Cypsela enclosed by a spiny bract. Coast. Weed of waste places. Introd. from America. Fl. summer. . . . . *A. australe (Loefl.) Kuntze

32 Xanthium L.

6 species naturalized Aust.; all states and territories

Monoeciouis, annual herbs. Leaves alternate. Heads in axillary clusters or short racemes. Male heads almost globular, surrounded by an involucre of free bracts in one row; corolla 5-toothed; stamens 5. Female head ovoid, 2-flowered, with an involucre in several rows; the lowest row remaining free; the upper consolidated into a hard burr, the free tips of the bracts becoming hooked spines; burr 2-beaked at the summit, 2-seeded; corolla absent. Cypselas ovoid, completely enclosed in the burr. Pappus absent.

Key to the species

1 Stems with 3-branched yellow spines at the base of the leaves. Leaves narrow-elliptic to lanceolate, up to 8 cm long, entire or 3-lobed, white tomentose underneath. Burrs 10–12 mm long, pubescent, spines numerous, c. 3 mm long. Herb up to 1 m high. Widespread. Weed of cultivation, pastures and waste places. Introd. from S. America. Fl. most of the year. *Bathurst Burr* . . . . ***X. spinosum** L.

1 Stems without spines. 2

2 Leaves usually 3–5-lobed, ovate, 5–15 cm long, cordate at the base, petiole and veins purplish. Burrs ellipsoid, 16–18 mm long, with 2 larger terminal straight spines. Herb up to 2 m high with purplish stems. Widespread. Weed of cultivation and waste places, particularly along creek flats. Fl. summer–autumn. *Noogoora Burr* or *Cockle Burr* . . . . ***X. occidentale** Bertol.

2 Leaves very shallowly and unevenly lobed, ovate, up to 12 cm long. Burrs ellipsoid, 15–30 mm long, with 2 larger terminal straight spines 6–8 mm long. Herb up to 1 m high. Weed of cultivation and waste places, particularly along creek flats. Fl. summer–autumn. *South American Burr* . . . .
. . . . ***X. cavanillesii** Schouw

33 Tagetes L.

1 species naturalized Aust.; Qld, NSW, S.A., WA

One species in the area

Glabrous annual with a strong odour, 2–3 m high. Leaves usually opposite, pinnate, up to 10 cm long, with linear-lanceolate segments. Heads numerous, in dense terminal panicles. Involucre 1 cm diam; bracts connate into a tube. Florets 6–12, rarely exceeding the involucre. Ray florets c. 3, female; rays yellow, 1–2 mm long. Disc florets tubular. Cypselas linear, black, c. 8 mm long. Pappus of 5–6 bristles, one much longer than the others. Widespread. Weed of cultivation and waste places. Introd. from America. Fl. summer. *Stinking Roger* . . . .***T. minuta** L.

34 Schkuhria Roth

1 species naturalized Aust.; Qld, NSW, Vic..

One species in the area

Herb up to c. 50 cm high, much branched. Leaves alternate, pinnate; the segments filiform, 2–3 cm long. Heads numerous, terminal on branches, pedunculate. Involucre tubular to conical, 7–8 mm long, 3–4 mm diam.; bracts few, elliptic, with hyaline tips. Ray floret solitary, female; rays yellowish, scarcely exceeding the involucre. Disc florets c. 6, bisexual. Cypselas c. 4 mm long, flat, pappus of several pink scales c. 1 mm long. Weed of waste places in the drier parts of the area, usually on clay. Introd. from S. America. Fl. summer. *Dwarf Marigold* . . . .***S. pinnata** (Lam.) Cabrera var. **abrotanoides** (Roth) Cabrera

35 Cotula L.

7 species in Aust. (4 endemic); all states and territories except NT

Small, decumbent herbs, often rooting at the nodes. Leaves alternate, entire lobed or dissected. Heads small, pedunculate. Involucre hemispherical or campanulate, with few nearly equal bracts in 2 rows. Outer florets female, without corolla; style 2-fid. Disc florets tubular, bisexual. Cypselas flattened, sometimes winged, hairy. Pappus absent.

Key to the species

1 Peduncles hollow, scarcely exceeding the leaves at flowering. Leaves oblong, mostly 2–4 cm long, pinnatifid. Heads c. 4 mm diam. Perennial, decumbent herb. Lithgow. Wet places; bogs. Fl. summer . .
. . . .**C. alpina** (Hook.f.) Hook.f.

1 Peduncles solid, exceeding the leaves at flowering 2

2 Leaves entire toothed or coarsely lobed, oblong-lanceolate to broad-oblong, 2–6 cm long, stem-clasping. Heads on solitary terminal peduncles. Involucre 6–10 mm diam.; bracts oblong. Outer female florets, in a single row. Disc florets very numerous, yellow. Cypselas 1–2 mm long; those of the female florets

broadly winged, on distinct pedicels; those of the disc florets with a narrow wing, on short pedicels. Glabrous perennial with weak succulent stems. Widespread. Wet situations, fresh or brackish water. Fl. winter–spring. *Waterbuttons* . . . . **C. coronopifolia** L.

2 Leaves once or twice pinnatisect, obovate to oblanceolate, c. 2 cm long. Heads on solitary terminal peduncles. Involucre 3–5 mm diam.; bracts oblong to oblanceolate. Outer female florets in 3–4 rows. Disc florets yellowish white to greenish. Cypselas pedicellate; the outer ones winged; the inner ones wingless. Prostrate to ascending, ± hairy herb. Widespread. Weed of cultivation and waste places, pastures and open forests. Fl. most of the year. *Common Cotula* or *Carrot Weed* . . . .
. . . . **C. australis** (Sieber ex Spreng.) Hook.f.

36 Leptinella Cass.
2 species endemic Aust.; Qld, NSW, Vic., Tas.

One species in the area

Usually prostrate, stoloniferous, perennial herbs. Leaves alternate, mostly more than 5 cm long, pinnate, with entire toothed or deeply lobed segments; petiole up to 10 cm long. Heads terminal, solitary on peduncles exceeding the leaves. Involucre c. 6 mm diam.; bracts orbicular, in 1 row; receptacle without scales. Outer florets female, corolla present; style with a terminal disc. Disc florets usually sterile. Cypselas sessile, globular to compressed, up to 2 mm long, with thickened margins. Pappus absent. Coastal swamps. Fl. all the year. *Creeping Cotula* . . . . **L. longipes** Hook.f.

37 Centipeda Lour.
5 species in Aust. (3 endemic); all states and territories

Small, usually decumbent herbs. Leaves alternate, toothed. Heads sessile or shortly pedunculate, axillary or terminal. Involucral bracts in 2 rows, almost equal, with scarious margins. Florets all tubular. Outer florets female, in several rows, 2–3-toothed. Inner florets campanulate, 4-toothed, bisexual. Cypselas with 4 obtuse longitudinal hairy ribs. Pappus absent.

Key to the species

1 Inner bisexual florets 0.3–0.4 mm long. Heads 3–4 mm diam., sessile or on short axillary peduncles; 10–20 bisexual florets per head. Leaves oblanceolate to oblong-cuneate, petiolate, 5–15 mm long. Cypselas c. 1 mm long, 0.3 mm wide; area between ribs flat or convex. Aromatic annual, glabrous pubescent or woolly, with usually prostrate stems up to 20 cm long. Widespread. Usually in wet places. Fl. summer. *Spreading Sneeze Weed* . . . . **C. minima** (L.) A.Braun & Asch. ssp. **minima**

1 Inner bisexual florets 0.6–0.8 mm long 2

2 Cypselas less than 1 mm long; 0.4–0.5 mm wide; area between ribs concave; trichomes arranged in rows. Heads 2.5–5 mm diam, sessile; 9–11 bisexual florets per head. Leaves ± spathulate, 3–12 mm long. Cottony annual with stems to 15 mm long. Uncommon. Wet places. Clay soils . . . .
. . . .**C. nidiformis** N.G.Walsh

2 Cypselas 1.5–2.5 mm long; trichomes scattered. Heads 4–9 mm diam., sessile. 20–50 bisexual florets per head. Leaves sessile, stem-clasping, 10–12 mm long, 2–3 mm wide, with sinuate margins. Erect to decumbent aromatic perennial up to 20 cm high, glabrous to woolly. Widespread. Wet places. Fl. summer. *Sneeze Weed* . . . . **C. cunninghamii** (DC.) A.Braun & Asch.

38 Soliva Ruiz & Pav.
3 species naturalized Aust.; all states and territories except NT

Small, diffuse herbs. Leaves basal and cauline; cauline leaves alternate, finely dissected. Heads sessile. Involucral bracts in 2 rows. Outer florets female, in 2–3 rows; corolla absent with thick wings tapering to the persistent style. Inner florets bisexual but sterile. Cypselas compressed; style persistent as a spine. Pappus absent.

Key to the species

1 Stolon-like branches absent or rare. Cypselas with wings or marginal ribs narrower than the body of the cypsela. Leaves up to 10 cm long, forming a dense tuft, twice or thrice pinnatisect, with linear lobes, glabrous or with soft hairs. Heads usually clustered amongst the leaves. Involucre ± globular, up to 10 mm diam. at fruiting stage. Cypselas 3 mm long; persistent style shorter than the cypsela. Widespread. Weed in lawns. Introd. from America. Fl. summer. *Jo-jo* . . . . ***S. anthemifolia** (Juss.) Sweet.

1 Stolon-like branches usually present. Cypselas with wings broader than the body of the cypsela. Leaves up to 5 cm long, forming a dense tuft, once or twice pinnatisect, with linear lobes, glabrous or with soft hairs. Heads not usually clustered. Involucre 3–5 mm diam. Cypselas 3 mm long; persistent style c. as long as the cypsela. Widespread. Weed in lawns. Introd. from America. Fl. summer. *Jo-jo* or *Bindyi*. . . . . ***S. sessilis** Ruiz Lopez & Pavón

39 Achillea L.

2 species naturalized Aust.; all states and territories except NT

One species in the area

Perennial herb up to 1 m high. Leaves basal and cauline, alternate, narrow- lanceolate in outline, thrice pinnatisect, up to 20 cm long. Heads in a dense broad corymb. Involucre ovoid to elliptic, 3–4 mm diam. Ray florets 5; ray white or pink rarely red, 3-lobed, 1–3 mm long. Disc florets yellow. Cypselas compressed, 1.5 mm long. Pappus absent. Blue Mts, Illawarra district. Weed in moist situations. Introd. from Europe. Fl. spring–summer. *Milfoil* or *Yarrow* . . . . ***A. millefolium** L.

40 Anthemis L.

3 species naturalized in Aust.; all states and territories except NT

Annual biennial or perennial herbs, with sparsely hairy stems. Leaves once to thrice pinnatisect. Heads solitary, on slender terminal peduncles. Involucral bracts in several rows. Ray florets in 1 row, usually female. Disc florets numerous, tubular, bisexual. Cypselas ribbed. Pappus a short crown or absent.

Key to the species

1 Erect annual up to 60 cm high with an unpleasant scent when crushed. Leaves 2–3-pinnatisect, 3–4 cm long; the segments linear. Involucre 6–10 mm diam. Rays up to 15, white, 10–12 mm long. Cypselas 2 mm long, tuberculate-ribbed. Widespread. Weed of cultivation and waste places. Introd. from Europe. Fl. summer. *Stinking Mayweed* . . . . ***A. cotula** L.

1 Erect annual up to 60 cm high without an unpleasant scent when crushed. Leaves 1–3-pinnatisect, 2–5 cm long; the segments linear. Involucre up to 10 mm diam. Rays white, usually more than 15, 10–12 mm long. Cypselas c. 2 mm long, ribbed with smooth ribs. Western parts of the area. Weed of pastures. Fl. summer. *Corn Chamomile* . . . . ***A. arvensis** L.

41 Tripleurospermum Sch.Bip.

1 species naturalized in Aust.; NSW, Tas.

One species in the area

Annual or biennial herb up to 1 m high, glabrous or hairy. Lower leaves alternate, 3-pinnatisect, up to 15 cm long; upper leaves smaller and less cut. Heads few, in corymbs, 3–5 cm diam. Involucral bracts in 2–several rows. Ray florets female, c. 10, in 1 row; rays white, up to 20 mm long. Disc florets yellow, tubular, bisexual. Cypselas c. 2 mm long, dark brown, wrinkled. Pappus a truncate rim. Sydney area. Weed of waste places and roadsides. Introd., from Europe. Fl. spring–summer. *Scentless Mayweed* . . . . ***T. inodorum** Sch.Bip.

42 Leucanthemum Mill.

2 species naturalized in Aust.; NSW, Vic., Tas., S.A.

One species in the area

Rhizomatous, perennial herb up to 60 cm high. Leaves alternate, simple, dentate or lobed; the lowest ones broad-spathulate, up to 5 cm long. Heads 2–3 together or solitary on leafless peduncles 5–15 cm long. Ray florets c. 18, usually female: rays white, 1–2 cm long. Disc florets yellow. Cypselas 1 mm long, with ± mucilaginous ribs. Pappus an irregular rim. Higher Blue Mts Weed in pastures. Introd. from Europe. Fl. summer. *Ox-eye Daisy* . . . .*L. vulgare Lam.

43 Tanacetum L.

2 species naturalized in Aust.; NSW, Vic., Tas., S.A., WA

Perennial herbs. Leaves alternate, entire or dissected. Heads in terminal corymbs. Involucre hemispherical. Involucral bracts in several rows. Outer row of florets female; the ray sometimes scarcely developed. Disc florets bisexual, tubular. Cypselas 5-ribbed or 3–5-angled. Pappus a short membranous crown or absent.

Key to the species

1 Leaves pinnatisect, c. 10 cm long and 6 cm wide; ultimate segments c. 2 mm wide, often toothed. Heads numerous, 6–7 mm diam. Florets yellow. Ray scarcely developed. Cypselas 1–2 mm long. Strongly scented perennial up to 1 m high. Widespread. Garden escape near habitation. Introd. from the Mediterranean. Fl. summer. *Tansy.* . . . . *T. vulgare L.
1 Leaves pinnate; the segments toothed or pinnatifid; largest leaves (lowest) up to 7 cm long, almost glabrous. Heads numerous, massed together. Ray white, 10–15 mm long. Disc florets yellow. Cypselas 1 mm long. Scented perennial up to c. 70 cm high. Sydney; Blue Mts Garden escape near habitation; roadsides. Fl. summer. *Pyrthrum or Feverfew.* . . . .*T. parthenium (L.) Sch.Bip.

44 Artemisia L.

5 species naturalized in Aust.; all states and territories except Tas. and NT

One species in the area

Erect herb with a ± woody base, up to 1 m high; stems sometimes purplish. Leaves alternate, 6–10 cm long, scented, divided into lanceolate lobes, whitish below; the upper leaves entire. Heads in a terminal panicle. Involucre c. 3 mm diam.; the bracts in several rows. Florets all tubular, yellow; the inner ones bisexual; the outer ones female. Cypselas 1–2 mm long. Pappus absent. Widespread. Usually near habitation. Introd. from Europe and Asia. Fl. summer. *Mugwort* or *Chinese Wormwood.* . . . .
. . . .*A. verlotiorum Lamotte

45 Senecio L.

60 species in Aust. (10 naturalized, 50 native); all states and territories

Herbs or shrubs, glabrous to woolly. Leaves alternate, entire or divided. Heads terminal, solitary, in corymbs or panicles. Florets yellow. Involucre cylindrical or campanulate; the longest bracts in a single row; scale-like bracts at the base of the involucre. Receptacle naked or pitted. Ray florets (when present) female or rarely neuter. Disc florets tubular. Cypselas striate or angular. Pappus of numerous capillary hairs.

Key to the species

Unless stated otherwise, leaf measurements and characters apply to middle stem leaves. A leaf is considered dissected if depth of incision is >30% of the distance from leaf outline to midrib. Coarse-dentate =30–50%; lobate =50–75%; pinnatisect = >75%. When pressed the diameter of floral heads can increase by up to 50%, measurements given here are for unpressed heads.

1 Scrambler or climber. 2
1 Erect or prostrate herbs or small shrubs. 3

2 Stems up to 2 m long. Leaves broad-triangular, c. 7 cm long and wide, usually with a few triangular lobes. Venation palmate. Petiole 2–5 cm long. Involucre bracts 5–7. Ray florets, yellow, 3–6; ray c. 10 mm long. Disc florets 10–12. Cypselas c. 2 mm long, glabrous. Garden escape near Sydney. Moist waste places; gullies. Introd. from S. Africa. Fl. winter . . . . ***S. tamoides** DC.

2 Stems to 5 m long. Leaves 3–5 cm long, 1–5 cm wide, rhombic to ovate, toothed or shallowly lobed. Venation not palmate. Petiole 1–4 cm long. Involucre bracts 7–9. Ray florets 4–6, yellow, ray c. 6–9 mm long. Disc florets 10–15. Cypsela 2.5 mm long with scattered hairs. Coastal dunes and near urban areas. Introd. from S. Africa. Fl. autumn–winter . . . . ***S. angulatus** L.f.

3 Florets all filiform and tubular 4
3 Florets ligulate and tubular 19

4 Leaves deeply lobate to bipinnatisect. 5
4 Leaves not deeply dissected; or if pinnatisect then only so in lower half and upper stem leaves usually not pinnatisect. 6

5 Stems glabrous to slightly hairy. Leaves twice pinnatisect, upper surface ± glabrous, lower surface ± sparsely hairy, 10–15 cm long. Heads few to very many (50–200), in a terminal corymb. Involucre 5–7 mm long, bracts 8–10. Florets 15–25. Fruiting head 2.5 mm diam., involucre bracts ± reflexed. Cypselas 2 mm long, minutely granular. Pappus 4–5 mm long. Erect herb 1–2 m high. Widespread. Fl. most of the year . . . . **S. bipinnatisectus** Belcher

5 Stems sparsely to densely hairy. Leaves scabrous or hispid on both surfaces, 7–15 cm long. Heads few to many (50–200); in terminal corymbs. Involucre 4–6.5 mm long, bracts mostly 11–13. Florets 20–35. Fruiting head 3–4 mm diam., involucre bracts ± erect. Cypselas 1.5–2 mm long. Pappus 3–4 mm long. Erect herb to 1.5 m. Widespread in poorer soils. Fl. spring–summer . . **S. bathurstianus** (DC.) Sch.Bip.

6 Flower heads with less than 25 florets and 7–13 involucre bracts. 7
6 Flower heads with more than 25 florets and 11–25 involucre bracts.. 14

7 Plant surface densely covered with matted whitish hairs without a multicellular base. Middle leaves linear to lanceolate, 6–20 cm long, 2–8 mm wide. Heads numerous (50–200), in a terminal corymb. Involucre 1–2 mm diam.; bracts 10–13, cobwebby at base or glabrous. Florets 18–50. Fruiting head 2–3 mm diam., involucre bracts spreading. Cypselas 2.5–4 mm long. Pappus 6–8 mm long. Erect herb up to 1 m high. Widespread. Natural and disturbed areas. Fl. most of the year. . . .**S. quadridentatus** Labill.

7 Plant with course multicellular hairs on stems and leaves, not covered with matted hairs.. 8

8 Surfaces of leaves glabrous (some marginal hairs may be present). Leaves oblanceolate to linear, 10–18 cm long, 15–25 mm wide, coarsely dentate, sometimes irregularly toothed, often auriculate at the base. Heads numerous (50–200) in terminal corymbs. Involucre 1–1.5 mm diam.; bracts 8–10. Florets 12–25. Fruiting head 2.5 mm diam., involucre bracts erect. Cypsela 2–2.5 mm long. Pappus 4–5 mm long. Erect herb up to 1 m high. Widespread. Moist places. Fl. summer–autumn . . **S. diaschides** D.G.Drury

8 Leaves with hairs on one or both surfaces. 9

9 Leaves not dissected; upper leaves usually without auricle but if present then not toothed or lobed. . 10
9 Leaves dissected or margins with regular crowded teeth (*S. minimus*); upper leaves with auricle present which is usually lobed or toothed 12

10 Lower stem bearing coarse multicellular hairs. Cypselas >2.8 mm long . **S. prenanthoides** (see below)
10 Lower stem not developing coarse hairs, cypselas <2.8 mm long. 11

11 Leaves never dissected, margins entire or with scattered teeth; narrow-oblanceolate to linear, 5–10 mm long; both surfaces scabrous, lower surface purplish. Heads 20–60 in terminal corymbs. Involucre 1–1.5 mm diam., bracts 7–13. Florets 12–25. Fruiting head 1.5 mm diam., involucre bracts usually reflexed. Cypsela 2–2.5 mm long. Pappus 5–6 mm long. Erect herb to 50 cm high. DSF. South of Sydney. Fl. spring–summer . . . . **S. microbasis** I.Thomps.

11 Leaves lobate to dissected, oblanceolate, 6–15 cm long, 5–15 mm wide; upper surface hispid, lower surface of leaves green or purplish usually with a matt of multicellular hairs with wispy tips. Heads numerous (20–60), in a terminal corymb. Involucre 1–1.5 mm diam.; bracts 9–13. Florets 15–35.

Fruiting head 2–3 mm diam., involucre bracts usually reflexed. Cypsela 2–2.5 mm long, slightly curved, glabrous or with some papillae in grooves. Pappus 5–6 mm long. Erect herb up to 50 cm high. Open forests and woodlands. Fl. most of the year . . . . **S. tenuiflorus** (DC.) Sieber ex Sch.Bip.

12 Stems glabrous above sparsely and minutely hairy below middle. Leaves lanceolate to narrow-elliptic, mostly 5–20 cm long and 10–30 mm wide, regularly and finely toothed, distinctly auriculate and stem-clasping at the base; upper surface glabrous or finely scabrous, lower surface green, glabrous or with matted wispy hairs. Heads numerous, in terminal corymbs. Involucre 1–1.5 mm diam.; bracts 7–10 Florets 12–25. Fruiting heads 1.5–2 mm diam., involucre bracts reflexed. Cypselas 1.5–2 mm long, ± lustrous. Pappus 5–6 mm long. Erect, herb up to 1.5 m high. Widespread. Moist places, Fl. summer–autumn . . . . **S. minimus** Poir.
12 Plant moderately to densely covered with course multicellular hairs; leaves dissected; 7–13 involucre bracts 13

13 Involucre with 7–10 bracts, rarely 12. Leaves narrow-oblong to narrow-elliptic, 6–12 cm long, 2.5–4 cm wide, teeth or lobes more developed towards top of leaf; base attenuate to auriculate. Upper leaves narrowly elliptic and widest away from the auricle. Heads numerous in terminal corymbs. Involucre 5–7 mm long. Florets 12–20. Fruiting heads 2 mm diam., involucre bracts reflexed. Cypsela 2–2.5 mm long. Pappus 5–6 mm long. Erect herb to 1.5 m high. Higher Blue Mts Forests over altitudes of 800 m. Fl. summer–autumn . . . . **S. distalilobatus** I.Thomps.
13 Involucre with 11–13 bracts, rarely 9. Leaves narrow-ovate to lanceolate, 7.5–15 cm long, 5–25 mm wide; teeth or lobes more developed towards base of leaf; base auriculate. Upper leaves lanceolate to linear and widest at base. Head numerous in terminal corymbs. Involucre 3–5 mm long. Florets 18–35. Fruiting heads 3–3.5 mm diam., involucre bracts erect. Cypsela 1.5–2.2 mm long. Pappus 4–6 mm long. Erect herb up to 1 m high. Widespread. Forest and disturbed sites. Fl. spring–summer . . . . . .
. . . . **S. hispidulus** A.Rich.

14 Stems with course, multicellular hairs.. 15
14 Stems with fine hairs which are not multicellular. 17

15 Cypselas more than 2.8 mm long. Leaves narrow oblanceolate 6–17 cm long, 2–6 mm wide, attenuate at base, both surfaces with coarse multicellular hairs; lower surface often purple. Heads numerous on terminal corymbs. Involucre 6–9 mm long, 1.4–1.5 mm diam., bracts 8–13. Florets 15–35. Fruiting heads 2–2.5 mm diam., involucre bracts reflexed. Pappus 6–8 mm long, erect herb to 50 cm high. Widespread. Forest. Fl. spring–autumn.. . . . .**S. prenanthoides** A.Rich.
15 Cypselas less than 2.2 mm long. 16

16 Florets 50–70 per head. Involucre bracts 7 mm long. Leaves glabrous or with soft white ± matted hairs, usually once pinnatifid, with sinuately toothed lobes; upper leaves auriculate and stem-clasping. Heads few in terminal corymbose panicles. Involucre c. 4 mm diam.; bracts 14—23. Cypselas 2 mm long. Pappus 6–8 mm long. Erect annual up to 50 cm high. Widespread. Weed of cultivation and waste places. Introd. from Europe. Fl. most of the year. *Common Groundsel.* . . . . ***S. vulgaris** L.
16 Florets 18–35 per head. Involucre bracts 3–5 mm long. . . . . **S. hispidulus** (see above)

17 Cypselas oblong to elliptic, less than 2.5 mm long, glabrous or with some papillae in grooves . . . . . .
. . . .**S. tenuiflorus** (see above)
17 Cypselas bottle shaped, more than 2.5 mm long. 18

18 Leaves linear to lanceolate, 6–20 cm long, 2–8 mm wide. Lower surface without coarse hairs . . . . . .
. . . . **S. quadridentatus** (see above)
18 Leaves narrow-elliptic more than 15 mm wide. Lower surface with coarse multicellular hairs. Heads numerous in terminal corymbs. Involucre 2–3 mm diam, bracts 12–14, cobwebby or occasionally glabrous. Florets 30–50. Fruiting heads 3–3.5 mm diam., involucre bracts reflexed. Cypselas 4–6 mm long, with long neck. Pappus 6–7 mm long. Erect herb to 1.5 m high. Uncommon. Moist areas. Northern Blue Mts Fl. summer–autumn. . . . . **S. longicollaris** I.Thomps.

19 Rays pink to purple. Stems ribbed and woody, slightly hairy. Leaves10–15 cm long, 3–5 cm wide, coarsely toothed, prickly towards base, stem clasping. Inflorescence loose corymbs with several

hundred heads per plant. Bracts 3–5.5 mm long. Involucre bracts 19–23, 6–9 mm long. Ray florets 12–22. Cypsela c. 2.5 mm long, pappus 7–9 mm long. Perennial shrub to 1.5 m high. Coastal on sand. Introd. from S. Africa. Fl. spring . . . . *S. glastifolius L.

19 Rays yellow to whitish 20

20 Plant greyish throughout. Leaves silvery white, with copious hairs, spathulate, 6–8 cm long, sessile, toothed towards the apex. Involucre 1–2 cm diam.; the bracts silvery white, c. 20. Ray florets 15–22; rays 20–30 mm long. Cypselas up to 5 mm long. Pappus 14–15 mm long. Decumbent herb up to 50 cm high. Coast. Near habitation. Introd. from S. America. Fl. most of the year . . . . . *S. crassiflorus (Poir.) DC.

20 Plant not greyish throughout. 21

21 Leaves cordate with broad rounded auricles, glabrous, glaucous; stem-clasping. Leaves ovate-oblong or lanceolate, up to 20 cm long, entire or coarsely dentate. Heads 5–80 in a terminal corymb. Involucre cylindrical to campanulate, 3–5 mm diam.; bracts 12–16, 5–7 mm long. Ray florets c. 10; ray yellow, c. 10 mm long. Cypselas 2–4 mm long, brown. Pappus 3–5 mm long. Erect perennial up to 1.5 m high, sometimes ± shrubby. Widespread. Open forests; RF. Fl. spring–autumn. . . . . S. velleioides A.Cunn. ex DC.

21 Leaves without broad auricles; petiolate or sessile, sometimes stem-clasping. 22

22 Involucre up to 3 mm diam. and 5.5 mm long; 4–8 bracteoles subtending the involucre; 4–8 ligulate florets, ligules 5–8 mm long.. 23

22 Plants without this combination of characters 24

23 Plants glaucous. New leaves not woolly. Leaves linear to lanceolate, 6–10 cm long, 8–15 mm wide, with a cuneate base, under surface with scattered spreading hairs sometimes with wispy extensions, serrate or sinuate on the margin. Heads numerous, in a terminal corymb. Pappus 6–9 mm long. Shrub up to 1 m high. Widespread from Barrington Tops the Saddleback Mountain. Open forests. Fl. most of the year . . . . .S. linearifolius var. macrodontus (DC.) I.Thomps.

23 Plants not glaucous. New leaves densely woolly. Leaves linear to lanceolate, 6–10 cm long, 8–15 mm wide, mostly with an attenuate and petiolate base, usually cottony underneath, serrate or sinuate on the margin. Heads numerous, in a terminal corymb. Pappus 6–9 mm long. Shrub up to 1 m high. Disjunct populations south from Bulli. Open forests and coastal scrub. Fl. most of the year . . . . S. linearifolius var. arachnoideus I.Thomps.

24 Cypselas of ray florets glabrous, cypselas of disc florets with rows of short appressed hairs. Pappus of the ray florets deciduous. Plants biennial with basal rosette of leaves in first year. Lower leaves pinnatifid to lyrate, with ragged irregular lobes, mostly 6–12 cm long. Involucral bracts 5–6 mm long, c. 13. Heads in a terminal corymb. Involucre campanulate, 5–7 mm diam. Ray florets 12–15; ligules 6–10 mm long. Cypselas c. 2 mm long. Pappus 5–6 mm long. Perennial, rhizomatous, glabrescent herb up to 1 m high. Bowral district. Weed of pastures and waste places. Introd. from Europe. Fl. autumn. *Ragwort* . . . . *S. jacobaea L.

24 Cypselas not as above; leaves not forming basal rosette, simple or variously pinnatisect, lobes ± regular 25

25 Lower surface of leaves whitish grey with closely appressed hairs. Bracteoles subtending the involucre 14–20. Leaves decurrent as a wing-like extension on the stem, lanceolate to narrow-elliptic, mostly up to 10 cm long, toothed or rarely entire, whitish underneath. Heads numerous. Involucre campanulate, c. 4 mm diam.; bracts 18–22, c. 5 mm long. Cypselas c. 2 mm long, brown. Shrub up to 1.5 m high. Garden escape in a few places. Introd. from S. Africa. Fl. summer. . . . . *S. pterophorus DC.

25 Lower surface of leaves glabrous or with only a few hairs 26

26 Leaves with obvious delineation between lamina and petiole; never divide, ovate-lanceolate, 5–10 cm long, 10–30 mm wide, dentate. Involucre narrow-campanulate, c. 10 mm diam.; bracts 12–14, 8–10 mm long. Ray florets up to 10; rays 10–13 mm long. Cypselas 4–6 mm long. Pappus pinkish, c. 10 mm long. Erect perennial up to 1.5 m high. Widespread. WSF. Fl. most of the year. . . . . .S. amygdalifolius F.Muell.

26 Leaves with delineation between lamina and petiole indistinct; often divided 27

27 Leaves in middle of stem ovate in outline with a short petiole-like base. Bracteoles subtending the involucre more than 5 mm long. Ligules consistently 7-nerved c. 10 mm long. Involucral bracts 13–15. Lower leaves pinnatisect with ± entire lobes, mostly up to 12 cm long. Involucre campanulate, 6–8 mm diam. Ray florets 6–9. Cypselas c. 5 mm long, brown. Pappus 7–10 mm long. Erect perennial up to 1 m high. Widespread. Moist places in open forests and gullies. Fl. spring–summer . . . . **S. vagus** F.Muell. ssp. **eglandulosus** Ali

27 Leaves in middle of stem not as above. Bracteoles subtending the involucre less than 5 mm long. Ligules not consistently 7-nerved. Involucre bracts glabrous 28

28 Leaves ± linear, glabrous, 4–8 cm long, 1–3 mm wide with strap like basal segments. Ligules 1–1.5 cm long. Involucre c. 10 mm diam. Involucre bracts 18–22. Ray florets 8–13. Cypselas up to 5 mm long, glabrous. Pappus up to 10 mm long. Erect perennial up to 1 m high. Widespread. Moist places in open forest. Fl. spring. . . . . **S. macranthus** A.Rich.

28 Leaves not linear or if so then involucre less than 8 mm long. 29

29 Majority of heads with c. 20 involucre bracts and c. 13 ray florets. Bracts always great in number than ray florets 30

29 Majority of heads with c. 13 involucre bracts and c. 13 ray florets. Involucre bracts ± equal in number to ray florets 31

30 Leaves generally undivided occasionally with one or two lobes on either side; not fleshy and drying thin; margins with frequent small teeth15–25 per side. Rays 5–10 mm long. Inflorescence with 2–20 heads. Cypsela, 1.5–2 mm long, 0.3–0.5 mm diam. Pappus 3–5 mm long, not persistent. Perennial, glabrous or sparsely hairy herb up to 50 cm high. Widespread. Weed of cultivation, waste places and pastures. Introd. from S. Africa. Fl. most of the year. *Fireweed* . . . . ***S. madagascariensis** Poir.

30 Leaves divided or if undivided then margins with few small teeth per side; fleshy or not. Cypsela 2–4 mm long, 0.5–0.8 mm diam. . . . . **S. pinnatifolius** (see below)

31 Leaves undivided, attenuate, up to 5 cm long; very fleshy, drying thick; 0-15 small teeth per side. Inflorescence with 1–5 heads. Heads 6–10 mm long. Rays 10–20 mm long. Cypsela 4–6 mm long, 0.5–0.7 mm diam. Pappus 5–7 mm long, persistent. Coastal sand dunes. Endangered. Fl. most of the year . . . . **S. spathulatus** A.Rich var. **attenuatus** I.Thomps.

31 Leaves divided or undivided, up to 10 cm long; thin or fleshy; margins of undivided leaves entire or with 0–20 teeth per side. Divided leaves with 1–5 segments per side. Inflorescence with 1–20 heads. Rays 6–20 mm long. Heads 3–7.5 mm long. Cypsela 2–4 mm long, 0.5–0.8 mm diam. Pappus 3–6 mm long not persisting. Widespread. Fl. winter–summer . . . **S. pinnatifolius** A.Rich. var. **pinnatifolius**

46 Cineraria L.
1 species naturalized Aust.; NSW

One species in the area

Annual herb, glabrous or glandular hairy. Leaves lobed, lyrate, oblong, 2–8 cm long; margins toothed; lower leaves petiolate upper leaves sessile. Heads 4–8 mm diam. Involucre bracts 12–14, 4–5 mm long. Florets yellow. Cypsela c. 2.5 mm long with pale wings. Introd. from Africa. Not common. Fl. summer. *African Marigold* . . . . **C. lyratiformis** Cron.

47 Delairea Lem.
1 species naturalized Aust.; NSW. Vic., Tas., S.A.

One species in the area

Climbing or scrambling perennial with stems several metres long. Heads in terminal corymbs. Leaves alternate, broad-ovate to orbicular, palmately 3–7-lobed or angled, cordate-hastate, 4–8 cm long; stipules reniform. Involucre cylindrical 2–3 mm diam. Large bracts 8, in one row. Florets 8–12, all tubular. Cypselas glabrous, cylindrical, 1–2 mm long. Pappus 3–4 mm long. Coast and gullies in adjacent plateaus. Weed in moist waste places in forests, gullies and margins of RF. Introd. from Africa. Fl. winter–spring. *Cape Ivy* . . . . ***D. odorata** Lem.

48 Roldana La Llave
1 species naturalized AUST;. Qld, NSW

One species in the area
Perennial, hairy herb up to 2 m high, sometimes woody at the base. Leaves broad-ovate to orbicular, up to 20 cm long, palmately 9–13-lobed, petiolate. Heads numerous in a terminal panicle. Involucre 9–11 mm long; bracts in one row. Ray florets 4–6; rays bright yellow, 10–12 mm long. Disc florets 10–15, yellow. Cypselas subcylindrical, glabrous. Pappus up to 10 mm long. Coast. Garden escape, usually near habitation. Introd. from Mexico. Fl. spring . . . .*R. petasitis (Sims) H.Robinson & Brettell

49 Erechtites Raf.
1 species naturalized Aust.; Qld, NSW

One species in the area
Erect annual up to 2 m high, usually glabrous. Lower leaves petiolate, lanceolate or ovate-lanceolate, entire serrate or irregularly dentate; upper leaves pinnately lobed or pinnatisect, up to 14 cm long; the petiole sometimes narrowly winged. Inflorescences terminal or axillary, ± corymbose. Heads sessile or on stalks up to 1 cm long. Involucre tubular, c. 1 cm long and 3 mm diam.; bracts 12–14, in 1 row, with some short basal bracts. Marginal and disc florets all bisexual. Cypselas 2.5–3.5 mm long, ribbed. Pappus of fine hairs, mauve or reddish, 7–8 mm long. Widespread. Weed in open forests, margins of RF and roadsides. Introd. from S. America. Fl. summer . . . . *E. valerianifolia (Wolf) DC.

50 Arrhenechthites Mattf.
1 species endemic AUST;. NSW, Vic.

One species in the area
Perennial herb up to c. 1 m high, ± woody at the base, sparingly branched. Basal leaves broad-lanceolate to ovate, with winged petioles; upper leaves lobed to pinnatisect, up to 10 cm long, glabrous or cottony, sessile or petiolate, stem-clasping. Inflorescence a terminal cymose corymb, much branched, lax. Heads solitary on a stalk up to 5 cm long. Involucre 6–8 mm diam., 10–12 mm long; bracts 5–8, ± in 2 rows. Outer florets female, filiform, slightly exceeding the involucre. Disk florets 2–3, functionally male because of the absence of stigmas. Cypselas 7–8 mm long. Pappus of white hairs, 10–12 mm long. Blue Mts Moist places; gullies. Fl. summer . . . . A. mixta (A.Rich.) Belcher

51 Crassocephalum Moench
1 species naturalized AUST;. Qld, NSW

One species in the area
Erect herb up to 75 cm high, thinly downy throughout. Stems prominently ribbed. Leaves alternate, ovoid to obovate, usually with 2 (sometimes more) prominent lobes near the base of the blade, irregularly and sharply toothed throughout, up to 8 cm long and 4 cm wide; petiole c. 1 cm long. Inflorescence terminal, loosely branched. Heads mostly long-stalked, sometimes 2–3 in a cluster. Involucral bracts numerous, narrow, up to 12 mm long, with a few basal bracts. Florets tubular, slightly exceeding the bracts. Corollas orange brownish or red. Cypselas ribbed. Pappus of numerous fine white silky hairs, up to 20 mm long. Sydney district. Weed of cultivation and waste places. Introd. from Africa. Fl. summer. *Thickhead* . . . .
. . . .*C. crepidioides (Benth.) S.Moore

52 Euryops (Cass.) Cass.
1 species naturalized Aust.; NSW

One species in the area
Shrub up to 1.5 m high. Leaves alternate, pinnatisect, elliptic to obovate in outline, up to 8 cm long and 2 cm wide, glabrous except sometimes for tufts of hair in the axils. Heads solitary in the leaf axils. Involucre 3–5 mm diam.; bracts fused for the lower third, c. 5 mm long. Ray florets female; rays yellow, 14–18 mm long. Cypselas up to 0.5 mm long. Pappus absent. Coast. Weed on roadsides. Introd. from S. Africa. Fl. winter–spring. . . . .*E. chrysanthemoides (DC.) B.Nord.

53 Calendula L.
3 species naturalized Aust.; all states and territories except NT and WA

One species in the area
Erect to ascending, annual herb up to 50 cm high, glandular pubescent, clammy. Leaves lanceolate to oblong, up to 6 cm long and 20 mm wide, entire or with fine teeth, stem-clasping. Heads solitary. Involucre 5–15 mm diam.; bracts numerous, in 2 rows, up to 8 mm long. Ray florets in 2–3 rows; rays orange to yellow. Cypselas up to 1.5 mm long; the outer ones with a spiny crest; inner ones curved. Pappus absent. Widespread but not common. Weed of cultivation and waste places. Introd. from the Mediterranean. Fl. spring. *Field Marigold.* . . . . ***C. arvensis** L.

54 Dimorphotheca Moench
1 species naturalized AUST;. NSW

One species in the area
Ascending, annual herb up to 20 cm high. Leaves oblanceolate, 1–6 cm long, up to 20 mm wide, entire to sinuate, pubescent, sessile. Heads solitary. Involucre 10–15 mm diam.; bracts 12–18, in 1 row. Ray florets 12–18; rays up to 20 mm long, white, purple underneath. Disc florets yellow, purplish towards the top. Outer cypselas tuberculate; inner ones winged, up to 8 mm long. Garden escape near habitation. Introd. from S. Africa. Fl. spring–summer. *Cape Marigold* . . . . ***D. pluvialis** (L.) Moench

55 Osteospermum L.
5 species naturalized Aust.; NSW, Vic., S.A., WA

One species in the area
Shrub up to 1 m high, glandular pubescent, slightly clammy. Lower leaves elliptic to oblanceolate, 5–8 cm long, coarsely toothed. Heads solitary or in few-flowered corymbs. Involucre 15–20 mm diam.; bracts in 1–several rows, up to 15 mm long. Ray florets female, up to 20, in 1 row; rays white, purplish underneath. 15–30 mm long. Disc florets blue, cypselas aborting. Cypselas c. 7 mm long, with acute narrow wings. Garden escape near habitation. Introd. from S. Africa. Fl. spring–summer . . . ***O. ecklonis** (DC.) Norl.

56 Chrysanthemoides Fabr.
1 species naturalized Aust.; Qld, NSW, Vic., S.A., WA

Shrubs 1–2 m high. Leaves alternate, petiolate. Heads in a terminal corymb. Involucre campanulate, 9–10 mm diam. Ray florets 5–13; rays yellow, 9–10 mm long. Disc florets tubular. Cypselas of the ray ovoid to globular, 6–8 mm diam., purplish black, with a hard bony endocarp. Pappus absent.

Key to the species

1 Leaves obovate to elliptic, up to 6 cm long and 3 cm wide, attenuate towards the base and scarcely petiolate, toothed especially towards the summit. Inner involucral bracts broad-ovate to broad-lanceolate, usually woolly on the margin. Cypselas globular or subglobular. Erect shrub. Widespread. Disturbed sites. Introd. from S. Africa. Fl. spring. *Boneseed* . . ***C. monilifera** (L.) Norl. ssp. **monilifera**
1 Leaves obovate to almost orbicular, up to 7 cm long and 5 cm wide, abruptly narrowing near the base into a ± distinct petiole, ± entire or with obscure teeth. Inner involucral bracts broad-ovate to orbicular, almost glabrous on the margins. Cypselas ellipsoid to obovoid. Sprawling shrub with decumbent branches. Coastal sand dunes. Introd. from S. Africa. Fl. spring. *Bitou Bush* . . . .
. . . .***C. monilifera** ssp. **rotundata** (DC.) Norl.

57 Arctotis L.
1 species naturalized Aust.; Qld, NSW, Vic., Tas., S.A., WA

One species in the area
Usually prostrate, softly white-tomentose, perennial herb, sometimes forming mats; stems mostly up to 40 cm long. Leaves alternate, sessile, up to 10 cm long, 2–4 cm wide; the lower ones lyrate; the upper

ones obovate-cuneate, sometimes lobed or toothed. Heads terminal, solitary. Involucral bracts in several rows. Ray florets female, in 1 row; rays white, often violet below, 2–3 cm long. Disc florets tubular, violet. Cypselas silky, 5-ribbed; the 2 lateral ribs winged and incurved. Pappus of 7–8 pink oblong scales. Coast. Garden escape near habitation. Introd. from S. Africa. Fl. spring–summer ***A. stoechadifolia** P.J.Bergius

58 Cymbonotus Cass.
3 species endemic Aust.; all states and territories except NT

Perennial herbs. Leaves basal, on long petioles, ovate, coarsely toothed, green on the upper surface, cottony-white underneath. Peduncles shorter than the leaves. Involucre 10 mm diam. Ray florets yellow. Disc florets yellow, shorter than the bracts. Cypselas dark brown, oblong, ribbed, 2–3 mm long. Pappus absent.

Key to the species

1 Cypselas 2.5 mm long, strongly curved, pubescent and almost smooth on the convex outer face. Leaf blades 5–8 cm long, sometimes pinnatifid. Widespread. Open forests, especially on clay soils. Fl. most of the year. *Bear's Ear* . . . .**C. lawsonianus** Gaudich.

1 Cypselas 3 mm long, straight or slightly curved, glabrous, rugose. Leaf blades 5–10 cm long, lobed or almost lyrate. Jenolan Caves district. DSF. Fl. most of the year. . . . . **C. preissianus** Steetz

59 Arctotheca J.C.Wendl.
3 species naturalized in Aust.; all states and territories

Annual or perennial herbs with white tomentum. Leaves basal or alternate. Heads campanulate to hemispherical, solitary, pedunculate. Involucral bracts in 3–5 rows. Ray florets in 1 row, neuter, ligulate. Disc florets tubular, bisexual. Cypselas ribbed. Pappus of scales, sometimes minute.

Key to the species

1 Leaves lobed, lyrate, toothed, mostly basal, oblanceolate, up to 25 cm long and 6 cm wide, toothed; upper surface glabrous or pubescent but not white-tomentose. Rays yellow, purplish to greenish below, 15–25 mm long. Disc florets very dark purplish. Annual herb with a rosette of leaves; peduncles up to 30 cm high. Widespread. Weed of cultivation, waste places and pastures. Introd. from S. Africa. Fl. spring–summer. *Capeweed* or *Cape Dandelion*. . . . . ***A. calendula** (L.) Levyns

1 Leaves entire or slightly toothed, white tomentose on both surfaces; lamina ovate, up to 6 cm long and 5 cm wide; petiole up to 8 cm long. Rays yellow, 5–7 mm long. Disc florets yellow. Perennial herb up to 30 cm high, with decumbent branches. Coastal sands. Introd. from S. Africa. Fl. most of the year. *Beach Daisy*. . . . .***A. populifolia** (P.J.Bergius) Norl.

60 Gazania Gaertn.
2 species naturalized Aust.; NSW, Vic., S.A.

One species in the area

Decumbent, perennial herb up to 50 cm high. Leaves cauline, elliptic to narrow-oblanceolate, pinnatisect, up to 10 cm long and 50 mm wide, white tomentose underneath. Heads solitary on axillary peduncles up to 20 cm long. Involucre 10–15 mm diam.; bracts in 2–3 rows. Ray florets in 1 row, neuter; rays orange to yellow, usually black towards the base with or without a white dot, 30–50 mm long. Disc florets tubular, bisexual, orange to yellow. Cypselas hairy, c. 5 mm long. Pappus of scales. Coast. Roadsides and coastal dunes. Introd. from S. Africa. Fl. most of the year . . . . ***G. rigens** (L.) Gaertn.

61 Epaltes Cass.

5 species endemic Aust.; all states and territories except Tas.

One species in the area

Much branched, perennial herb, diffuse or prostrate, rarely more than 25 cm high, glabrous or hirsute. Leaves petiolate, ovate or cuneate-oblong, entire or deeply lobed, 1–4 cm long. Heads lateral, sessile or shortly pedunculate. Involucre depressed-hemispherical, 4–6 mm diam.; bracts very obtuse. Florets all tubular, yellow; the outer ones female; the inner ones usually bisexual. Cypselas 1 mm long, striate. Pappus absent. Coast and Cumberland Plain. Heavy soils in damp places. Fl. summer. *Spreading Nutheads.* . . . .**E. australis** Less.

62 Pseudognaphalium Kirp.

1 species native AUST;. all states and territories

One species in the area

Annual or biennial herb up to 50 cm high, woolly white. Leaves alternate; lower leaves obovate to oblong or spathulate, up to 7 cm long; upper leaves linear to lanceolate. Heads in a dense terminal cluster. Involucre c. 4 mm diam.; bracts scarious, brown or straw-coloured, with obtuse scarious tips, not spreading. Outer florets filiform, female; inner florets tubular, bisexual. Cypselas c. 0.5 mm long, hairy. Pappus of 8–12 bristles. Widespread. Weed in almost all situations. Fl. spring–summer. *Jersey Cudweed* . . . . ***P. luteoalbum** (L.) Hilliard & B.L.Burtt

63 Vellereophyton Hilliard & B.L.Burtt

1 species naturalized Aust.; NSW, VIC, Tas., S.A.

One species in the area

Ascending, annual herb up to 50 cm high often forming mats. Leaves alternate oblanceolate to spathulate, mostly up to 30 mm long and 5 mm wide. Heads in terminal globular woolly clusters. Involucral bracts in 3–5 rows; the claws purplish; the spreading limb white, obovate, c. 0.5 mm long. Achenes c. 0.5 mm long. Pappus of bristles. Coastal swamps near Sydney, e.g. Maroubra. Introd. from S. Africa. Fl. most of the year. *White Cudweed*. . . . . ***V. dealbatum** (Thunb.) Hilliard & B.L.Burtt

64 Gnaphalium L.

3 species in Aust.; All states and territories

One species in the area

Erect to ascending annual up to 15 cm high. Stems grey-green and cottony. Leaves oblanceolate, up to 5 cm long and 10 mm wide, cottony. Heads in axillary clusters arranged in a panicle Involucral bracts with broad scarious tips; involucre 2–3 mm diam. Outer florets filiform, female, in 2 or more rows; inner florets tubular. Cypselas oblong to obovate with imbricate papillae. Pappus of capillary hairs Western Blue Mts Moist places; creek banks. Fl. spring–summer . . . . **G. polycaulon** Pers.

65 Euchiton Cass.

14 species in Aust.; all states and territories

Perennial or occasionally annual herbs. Stolons present (absent in *E. sphaericus*). Leaves alternate, sessile, entire. Heads solitary or in dense terminal cluster surrounded by a number of leaf like bracts. Outer florets female, filiform, more in number than inner bisexual florets. Cypselas small, oblong. Pappus of free barbate hairs.

Key to the species

1 One bisexual floret per head. Plants annual without stolons. Heads 1–1.5 mm diam., in clusters surrounded by 3–8 subtending leaves.. Basal leaves 5–7 cm long, 2–15 mm wide, finally glabrous on the upper surface. Erect herb up to 50 cm high. Widespread, in a variety of situations. Fl. most of the year . . . . **E. sphaericus** (Willd.) Anderb.

1 More than one bisexual floret per head. Plants stoloniferous perennials. Heads 2–4 mm diam.. 2

2 Basal rosette of leaves alive at flowering; basal leaves mostly 10–18 cm long, 4–10 mm wide, acute, glabrous above. Clusters of heads with 1–2 subtending leaves. Stoloniferous perennial with leafy stems, c. 40 cm high. Widespread. DSF. Fl. most of the year. *Creeping Cudweed* . . . .
. . . . **E. gymnocephalus** (DC.) Anderb.

2 Basal rosette of leaves dead at flowering time; basal leaves 7–18 cm long, 4–11 mm wide, oblong to linear. Cluster of heads with 3–5 subtending leaves. Erect perennial up to 70 cm high. Widespread. Open forests and grasslands; sometimes a weed in gardens. Fl. most of the year. *Star Cudweed*. . . . . .
. . . . **E. involucratus** (G.Forst.) Anderb.

66 Gamochaeta Wedd.
4 species naturalized Aust.; all states and territories

Robust annual or perennial herbs. Stolons absent. Leaves alternate, sessile, entire. Heads small and arranged in elongated often leafy spikes, or in large corymbs of spikes. Involucre bracts with firm ± leathery tips. Leaves, stems and heads with tangled but appressed hairs. Outer florets female, filiform greater in number than inner bisexual florets. Cypsela oblong with globose hairs. Pappus of barbate hairs.

Key to the species

1 Leaf surfaces contrasting in colour and/or hairiness 2
1 Leaf surfaces ± similar, densely woolly 4

2 Heads almost hidden amongst woolly hairs, with 3–4 subtending leafy bracts, in terminal spikes. Leaves darker and with fewer hairs on the upper surface, oblanceolate to spathulate, with obtuse mucronate tips, 25–80 mm long, 4–18 mm wide. Involucre 3–4 mm diam.; bracts green to straw-coloured. Procumbent to erect, woolly annual up to 50 cm high. Coast. Open forests on sandy soils. Introd. from America. Fl. spring. . . . . ***G. pensylvanica** (Willd.) Cabrera
2 Heads not hidden amongst woolly hairs. 3

3 Midrib of leaf similar in colour to the surface. Basal leaves spathulate to obovate, 10–12 cm long, 8–20 mm wide. Heads in axillary clusters in a dense spike-like panicle. Involucre 1.5–3 mm diam.; bracts straw-coloured, scarious, hyaline towards the tip. Erect herb up to 40 cm high. Sydney district. Disturbed sites. Introd. from Central and S. America. Fl. spring–summer . . ***G. purpurea** (L.) Cabrera
3 Leaves with distinctly paler usually furrowed midrib on the upper surface, oblanceolate to spathulate, up to 7 cm long and 15 mm wide. Heads in axillary clusters in a dense spike-like panicle. Involucre 2–3 mm diam.; bracts scarious towards the top purplish. Annual or short-lived perennial herb up to 50 cm high. Widespread. Weed of cultivation and waste places. Introd. from America. Fl. summer. *Purple Cudweed*. . . . . ***G. coarctata** (Willd.) Kerguelen

4 Heads 2–3 mm diam. Heads initially in a cylindrical uninterrupted Inflorescence, the main axis obscured by heads at least distally. Involucre bracts in 3–4 rows apices broadly acute, not tinged with purple. Introd. from S. America . . . . ***G. antillana** (Urb.) Anderb.
4 Heads 3–3.5 mm diam. Heads in an interrupted inflorescences the main axis visible from the terminal head down. Involucre bracts in 5–7 rows, apices sharply acute, commonly tinged with purple. Leaves narrow-elliptic to spathulate, up to 3 cm long and 8 mm wide. Erect to procumbent herb up to 20 cm high. Coast. Weed of cultivation and waste places. Introd. from S. America. Fl. spring–summer. . . . .
. . . . ***G. calviceps** (Fernald) Cabrera

67 Facelis Cass.
1 species naturalized Aust.; NSW

One species in the area

Perennial, decumbent herb. Leaves crowded, spathulate, emarginate, 1–2 cm long. Stem-clasping, with recurved margins, glabrous on the upper surface, woolly white underneath. Heads in short terminal racemes. Involucre cylindrical, acuminate, 4–7 mm diam.; bracts silvery white, scarious, not radiating.

Florets all tubular, numerous. Cypselas silky hairy, 1–1.5 mm long. Pappus of numerous capillary hairs, 10–15 mm long, each bearing numerous fine barbellate interlacing branches. Coast. Weed near habitation. Introd. from S. America. Fl. spring–summer . . . . ***F. retusa** (Lam.) Sch.Bip.

68 Cassinia R.Br.
c. 18 species endemic Aust.; all states and territories except NT

Shrubs. Leaves entire, alternate. Heads small, numerous, in terminal corymbs or panicles. Involucre narrow-ovoid to cylindrical; bracts imbricate, in several rows, scarious or coloured, inner bracts with connivent tips. Receptacle with chaffy scales between the florets only 1–3 fewer than florets or; when florets 3 or less per receptacle then scales absent. Florets all tubular, few, bisexual. Cypselas angular or nearly terete, usually papillose. Pappus of several simple entire or minutely denticulate capillary hairs in a single row and slightly connate into a ring at the base.

Key to the species

1 Inner involucral bracts bright yellow at the tip. Involucre c. 1 mm diam. Heads in a dense corymb up to 12 cm diam. Leaves linear to elliptic, up to 7 cm long and 5 mm wide, entire; upper surface dark green, ± glabrous. Florets 5–6. Shrub up to 2 m high. Widespread. Open forests and heath. Fl. spring–summer . . . . **C. aureonitens** N.A.Wakef.
1 Involucral bracts white or straw-coloured, sometimes tinged with pink 2

2 Heads in a terminal panicle 3
2 Heads in a dense terminal corymb 4

3 Leaves linear, rarely more than 1 cm long, up to 2 mm wide, obtuse or with recurved tips, revolute. Heads numerous in a long loose terminal panicle. Involucre cylindrical, often curved, 2–3 mm long, 1 mm diam. Florets usually 2–3. Shrub up to 2 m high with white tomentose branches. Widespread in a variety of open habitats. Fl. chiefly summer. *Chinese Shrub* or *Sifton Bush* . . . . **C. arcuata** R.Br.
3 Leaves up to 4 cm long, linear, up to 2 mm wide, revolute. Heads numerous in pyramidal panicles. Involucre narrow-oblong, 4 mm long, 1 mm diam.; bracts narrow, obtuse, straw-coloured. Florets c. 5. Glabrous ± viscid shrub 2–3 m high. Western parts of the area. Open forests. Fl. summer. . . . . **C. quinquefaria** R.Br.

4 Leaves 5–7 cm long, 2–3 mm wide, scabrous, linear, revolute. Heads crowded, very numerous. Involucre 5–6 mm long, 2 mm diam.; bracts scarious, straw-coloured. Scales often absent. Florets 2–3. Hunter River. DSF. Fl. summer . . . . **C. leptocephala** F.Muell. ssp. **leptocephala**
4 Leaves 1–4 cm long. 5

5 Leaves 1–2 mm wide 6
5 Leaves ≥2 mm, with flat or recurved margins 8

6 Involucral bracts of 2 kinds; the inner white or pinkish, shiny; the outer brownish, dull, thinner. Leaves crowded, linear-terete, 1–4 cm long, margins recurved, tips mucronate and sometimes slightly recurved. Involucre 1–3 mm diam. Florets 6–12. Shrub 1–2 m high, with rusty to yellowish tomentose-pubescent branches. Widespread. Open forests. Fl. chiefly summer. *Dolly Bush* . . . . . . **C. aculeata** (Labill.) R.Br.
6 Involucral bracts all straw-coloured and similar in texture 7

7 Involucre cylindrical, deep straw-coloured throughout; bracts in longitudinal parallel rows. Leaves linear, 20-35 mm long. Involucre 3–5 mm diam. Scales often absent. Florets 3–4. Shrub up to 2 m high, with white woolly hairs on the stems. Widespread. DSF. Fl. most of the year . . **C. cunninghamii** DC.
7 Involucre campanulate, upper inner paler than outer bracts; bracts alternating and not in parallel longitudinal rows. Leaves narrow-linear, 12–30 mm long. Involucre 1–2 mm diam. Florets usually 4–6. Usually viscid shrub up to 2 m high. Drier parts of the area. Open forests. Fl. spring–summer. *Cough Bush* . . . . **C. laevis** R.Br.

8 Leaves with a midvein and 2 smaller longitudinal veins near the margins especially near the base, glabrous or glandular-scabrous above, narrow-lanceolate, 6–8 cm long, 3–6 mm wide. Involucre c.

1 mm diam.; bracts with white tips. Florets 3–4. Shrub or small tree up to 8 m high, with yellowish branches. Widespread. WSF and margins of RF. Fl. summer–autumn . . . . . . **C. trinerva** N.A.Wakef.

8 Leaves with a single longitudinal vein 9

9 Leaves stem-clasping, 10–25 mm long, oblong to broad-lanceolate, glabrous on the upper surface, hoary or rusty underneath. Involucre 2–4 mm diam.; bracts straw-coloured. Florets 10–12. Shrub up to 2 m high, with yellowish stems. Widespread. DSF. Sandy soils. Fl. spring–summer . . . **C. denticulata** R.Br.

9 Leaves not stem-clasping 10

10 Involucral bracts with a white, opaque limb. Leaves 3–8 cm long, 2–6 mm wide, linear to lanceolate, glabrous on the upper surface, often white tomentose underneath, aromatic, viscid. Involucre oblong, 1 mm diam. Florets 6–8. Shrub 2–3 m high with yellowish stems. Widespread. Tall forests and margins of RF. Fl. summer–autumn . . . . **C. longifolia** R.Br.

10 Involucral bracts with a translucent limb. Leaves 3–8 cm long, 1–3 mm wide, linear, viscid, scabrous or glabrous on the upper surface, glandular hairy underneath. Involucre oblong, c. 2 mm diam. Florets 5–6. Shrub up to 3 m high with yellowish stems. Hawksebury/Colo River areas. DSF. Sandy and rocky soils. Fl. spring–summer . . . . **C. accipitrum** Orchard

69 Leucochrysum (DC.) Paul G.Wilson
5 species endemic Aust.; all states and territories

Tufted, perennial herbs, ± woody towards the base. Leaves towards the base, linear to obovate, up to 10 cm long. Heads solitary on leafless scapes. Involucre hemispherical, 3–4 cm diam.; bracts in several rows, all petaloid, spreading, 8–10 mm long, pure white, or tinged with pink-brown or yellow; inner bracts with a claw. Florets all tubular. Receptacle without scales. Cypselas glabrous, 2 mm long. Pappus of hairs 5–6 mm long.

Key to the species

1 Leaves glabrescent on the upper surface, filiform, up to 7 cm long. Involucral bracts narrow-elliptic; outer ones pale brown; inner ones yellow. Perennial up to 20 cm high, with a woody rootstock. Mostly western parts of the Blue Mts DSF and heath. Sandy soils. Fl. summer . . . . **L. graminifolium** (Paul G.Wilson) Paul G.Wilson

1 Leaves woolly, linear to obovate, up to 10 mm long. Involucral bracts ovate to narrow-elliptic. Perennials up to 30 cm high, with a woody rootstock. Widespread. In a variety of situations. Fl. spring–summer.

2 Inner involucral bracts yellow . . . . . **L. albicans** (A.Cunn.) Paul G.Wilson ssp. **albicans** var. **albicans**
2 Inner involucral bracts white . . . . **L. albicans** ssp. **albicans** var. **tricolor** (DC.) Paul G.Wilson

70 Rhodanthe Lindl.
46 species endemic Aust.; all states and territories

Annual or perennial herb. Leaves entire, sessile, alternate. Heads cylindrical to subglobular, solitary or in clusters. Involucre bracts multi-serriate, scarious with flattened claw. Florets 1 to many usually bisexual or the outer ones female. Cypselas with plumose bristles.

Key to the species

1 Branches woolly and leaves without sessile glands. Herb up to 6 cm high. Leaves linear and triangular, 5–10 mm long. Heads sessile, solitary or clustered. Involucre 7 mm long, light brown. Florets 4–5, corolla 3 mm long. Cypselas silky hairy, 2.5 mm long. Pappus of hairs c. 6 mm long. Grows in seasonally wet areas. Fl. spring . . . . **R. pygmaea** (DC.) Paul G.Wilson

1 Branches glabrous and leaves with sessile glands. Herb up to 30 cm high. Leaves linear, crowded, 10 mm long, glabrous. Heads solitary on bracteate scapes. Involucre c. 5 mm long.; outer bracts scarious, tinged with brown with purple mid rib; inner bracts white, spreading, 6–10 mm long. Receptacle without scales. Florets many, corolla 6 mm long. Cypselas densely silky hairy, 1–2 mm long. Pappus of hairs c. 4

mm long. Western parts of the area. Grasslands and open forests. Fl. spring–summer . . . .
. . . . **R. anthemoides** (Spreng.) Paul G.Wilson

71 **Triptilodiscus** Turcz.
1 species endemic Aust.; Qld, NSW, Vic., S.A., WA

One species in the area
Decumbent to erect annual up to 10 cm high. Leaves linear, 1–3 cm long, with a few hairs. Heads sessile terminal or lateral. Involucre broad-ovoid, 4–6 mm diam.; bracts lanceolate, straw-coloured, scarious, fringed with long cilia, not radiating. Florets all tubular. Cypselas glabrous, 1–2 mm long. Pappus of 3–4 shortly plumose bristles c. 2 mm long, sometimes absent. Widespread. Grasslands and tall forests on clay soils. Fl. spring–summer. . . . . **T. pygmaeus** Turcz.

72 **Ozothamnus** R.Br.
44 species endemic Aust.; Qld, NSW, Vic., Tas., S.A.

Shrubs. Leaves alternate, usually entire, usually sessile. Heads in clusters or large umbellate corymbs. Involucre usually small; bracts papery or scarious, inner bracts with spreading tips. Florets all tubular or a few filiform. Receptacle without scales or scales 0–3 and many less than florets. Cypselas oblong, glabrous hairy or papillose. Pappus of barbellate hairs.

Key to the species

1 Leaves less than 3 cm long (rarely 4cm) 2
1 Leaves 3–7 cm long. 5

2 Florets c. 20 or more 3
2 Florets 14 or fewer 4

3 Leaves 4–15 mm long with margins flat or recurved; shortly tomentose underneath. Heads in small compact corymbs. Involucre 3–5 mm diam.; bracts scarious, whitish to brown. Uncommon. Rainforest margins. Fl. spring . . . . **O. rufescens** DC.
3 Leaves linear, 1–2 cm long, with revolute margins, glabrous or hoary tomentose underneath. Heads in a dense terminal corymb. Involucre globular or broadly campanulate, 2–3 mm diam.; bracts broad, opaque, white or tinged with pink, without a spreading limb. Cypselas glabrous or papillose, 1 mm long. Pappus hairs 2–3 mm long. Shrub up to 5 m high. Widespread. Tall forests, especially on shales. Fl. chiefly spring . . . . **O. diosmifolius** (Vent.) DC.

4 Leaves with decurrent margins, in prominent lines, linear, 2–8 mm long. Heads in dense terminal corymbs. Involucre 3–4 mm diam.; bracts obtuse, white or straw-coloured, without a spreading limb. Florets 10–12. Cypselas glabrous or papillose, less than 1 mm long. Pappus hairs few, 2 mm long. Heath-like shrub 1–2 m high. Woronora Plateau, Hurstville–Waterfall; Blue Mts Open forests on sandy soils. Fl. spring . . . . **O. adnatus** DC.
4 Leaves not decurrent, linear, mostly obtuse, 5–50 mm long (rarely 40 mm). Heads numerous in dense corymbs. Involucre 3 mm diam.; inner bracts with a white erect limb. Florets 6–14, white. Cypselas ribbed. Pappus hairs clavate at the tip. Shrub up to 3 m high. Blue Mts Open forests, creek banks. Fl. summer . . . . **O. rosmarinifolius** (Labill.) Sweet

5 Leaves 6–10 mm wide, 4–7 cm long, lanceolate, greyish white or rusty underneath, ± viscid on the upper surface. Involucre 2 mm diam.; inner bracts with a broad white sometimes spreading limb. Heads numerous. Florets c. 14. Aromatic viscid shrub up to 3 m high. Coast. Forests and margins of RF. Fl. spring–summer . . . . **O. argophyllus** (A.Cunn. ex DC.) Andrb.
5 Leaves 2–3 mm wide, 3–5 cm long, narrow-lanceolate, white tomentose underneath. Involucre 2–4 mm diam.; inner bracts with a white erect or spreading limb. Heads numerous. Florets 4–7. Tall shrub or small tree up to 5 m high. Widespread. Forests. Fl. summer. *Tree Everlasting* . . . .
. . . . **O. ferrugineus** (Labill.) Sweet

73 Chrysocephalum Walp.

Yellow Buttons

8 species endemic Aust.; All states and territories

Perennial herbs often with a woody rootstock. Leaves alternate, glandular hairy, sessile. Heads clustered. Involucre turbinate to globular; bracts membranous, ± transparent, yellow to brown, ciliate at the margin. Florets mostly tubular. Receptacle without scales. Cypselas oblong, papillose. Pappus of barbellate bristles, sometimes plumose at the top.

Key to the species

Intermediates between *C. apiculatum* and *C. semipapposum* can occur

1 Leaves usually oblong-cuneate to lanceolate (cauline leaves sometimes linear), up to 5 cm long, 10–25 mm wide, usually with a silvery tomentum. Involucre broadly turbinate to globular, 3–10 mm diam.; bracts numerous, bright yellow, usually radiating. Florets as long as the involucre. Cypselas glabrous, 1 mm long. Pappus hairs few, 4–5 mm long. Perennial up to 50 cm high. Widespread. Grasslands and forests on better soils. Fl. spring–summer. . . . . **C. apiculatum** (Labill.) Steetz

1 Leaves usually linear, without a silvery tomentum, up to 5 cm long, 1–2 mm wide. Involucre 6–7 mm diam.; bracts yellow, usually radiating. Florets as long as the involucre. Cypselas glabrous or papillose. Pappus bristles c. 4 mm long. Aromatic perennial up to 60 cm high. Widespread. Grasslands and forests on better soils. Fl. spring–summer . . . .**C. semipapposum** (Labill.) Steetz

74 Helichrysum Mill.

32 species endemic Aust.; all states and territories

Perennial herbs, often ± clothed with cottony hairs. Leaves alternate, or the lower ones very rarely opposite, entire. Heads solitary or in compound inflorescences. Involucres mostly ± hemispherical; bracts imbricate in several rows, papery or scarious, sometimes with a basal claw. Florets numerous, bisexual or some sterile. Receptacle without scales. Cypselas glabrous papillose or hairy. Pappus of barbellate hairs.

A recent revision places all the following species except for *H. calvertianum* and *H. leucopsideum* in a new Genus *Coronidium* Paul G.Wilson.

Key to the species

1 Involucral bracts white often tinged with pink 2
1 Involucral bracts straw-coloured to yellow, sometimes tinged with brown 5

2 Leaves 20–40 mm wide, up to 10 cm long, lanceolate to ovate-lanceolate; the uppermost ones glabrous or with sparse tufts of wool. Heads solitary or loosely paniculate. Involucre hemispherical, 3–4 cm diam.; bracts narrow, petal-like. Florets very numerous, less than half as long as the involucre. Cypselas glabrous or papillose, 2–3 mm long. Pappus hairs 6–10 mm long, shortly cohering at the base. Perennial herb or small shrub up to 2 m high, clothed with cottony hairs. Widespread. Tall forests especially on better soils. Fl. spring. *White Paper Daisy.* . . . .**H. elatum** A.Cunn. ex DC.

2 Leaves c. 5 mm wide or less 3

3 Stems and branches scabrous with glandular hairs at least when young. Leaves scabrous or glandular hairy, linear 2–5 cm long, 1–3 mm wide. Involucre up to 45 mm diam. Cypselas oblong, pitted. Erect annual up to 50 cm high. Widespread. DSF. Fl. summer . . . . **H. adenophorum** F.Muell. var. **waddelliae** J.H.Willis

3 Stems and branches with soft woolly hairs 4

4 Leaves 20–50 mm long, with ± flat or slightly recurved margins, cottony hairy underneath, narrow-lanceolate to linear, 1–5 mm wide. Involucre 25–35 mm diam. Cypselas 4-ribbed, glabrous. Perennial herb up to 50 cm high. Widespread. Sandy soils. Fl. chiefly summer. *Satin Everlasting.* . . . .**H. leucopsideum** DC.

4 Leaves up to 10 mm long, revolute or recurved with the lower surface concealed, linear, up to 1 mm wide. Involucre up to 15 mm diam. Cypselas terete, glandular. Undershrub up to 30 cm high. Fitzroy Falls to Berrima. DSF. Fl. spring–summer . . . . **H. calvertianum** F.Muell.

5 Involucral bracts subulate, straw-coloured to yellow. Involucre usually 2–3 cm diam., broadly hemispherical. Leaves narrow-lanceolate, up to 10 cm long. Heads solitary on almost leafless peduncles. Cypselas glabrous, 2–3 mm long. Pappus hairs 5 mm long. ± woody plant up to 80 cm high. Widespread. DSF and WSF, often on heavy soils. Fl. spring . . . . **H. collinum** DC.

5 Involucral bracts obtuse. Involucre 10 mm diam., rarely more than 20 mm 6

6 Heads solitary on almost leafless peduncles. Leaves oblong, lanceolate to linear, acute or mucronate, stem-clasping, up to 5 cm long. Involucre hemispherical, c. 10 mm diam; bracts in many rows. Cypselas glabrous, 1 mm long. Pappus hairs 3–4 mm long. Perennial usually decumbent hairy or woolly. herb. Widespread. Grasslands, WSF, margins of RF. Fl. most of the year . . . . **H. rutidolepis** DC.

6 Heads solitary on leafy peduncles. Leaves basal and cauline, up to 6 cm long, oblong-spathulate to linear, glabrous to woolly on the upper surface, woolly underneath. Involucre hemispherical 10–20 mm or more diam.; bracts numerous spreading. Cypselas glabrous, 1–2 mm long. Perennial herb, often decumbent, up to 30 cm high, woolly. Widespread. Grasslands and forests, usually on heavy soils. Fl. spring–summer. . . . . **H. scorpioides** Labill.

75 **Xerochrysum** Tzvelev
7 species endemic Aust.; all states and territories

Annual or short-lived perennial herbs. Leaves basal or cauline and alternate, flat, sessile. Heads terminal, usually solitary. Involucre hemispherical to globular; bracts numerous in several rows, papery or scarious, yellow brownish or rarely white. Florets mostly tubular, bisexual; a few filiform and female, less than half as long as the involucre. Cypselas glabrous or papillose. Pappus of barbellate bristles 6–8 mm long.

Key to the species

1 Leaves viscid and minutely scabrous, narrow, linear to elliptic, 3–9 cm long. Involucre 20–30 mm diam.; bracts bright yellow to brownish. Cypselas glabrous, c. 2 mm long. Annual or short-lived viscid perennial herb up to 1 m high. Widespread. Open forests. Usually on sandy soils. Fl. spring . . . .
. . . . **X. viscosum** (DC.) R.J.Bayer

1 Leaves not viscid, glabrous or with a few hairs 2

2 Annual or short-lived perennial herb up to 1 m high with glandular hairs and a taperoot. Erect and branched. Leaves up to 10 cm long, linear to oblong-lanceolate; the lower ones obovate-oblong. Involucre 20–40 mm diam.; bracts petal-like, bright yellow to reddish-brown. Cypselas glabrous or slightly papillose, 2–3 mm long. Widespread in a variety of situations. Fl. spring. *Golden Everlasting* . .
. . . .**X. bracteatum** (Vent.) Tzvelev

2 Perennial herb with cobwebby hairs or glabrescent, rhizomatous. Erect usually with single stem up to 45 cm high. Leaves up to 10 cm long, linear-elliptic. Involucre solitary, 2.5–5 cm diam.; bracts golden-yellow. Cypsela 3 mm long. Uncommon. Swamps. Fl. summer. *Swamp Everlasting.* . . . .
. . . .**X. palustre** (Flann) R.J.Bayer

76 **Ammobium** R.Br.
2 species endemic Aust.; Qld, NSW, S.A.

One species in the area

Perennial herb. Leaves basal, entire, lanceolate to spathulate, 10–20 cm long, tapering into a winged petiole, green-glaucous above, silvery white with simple hairs below. Aerial stems up to 1 m long, silvery white with 4 wings 1–3 mm wide running from top to bottom. Heads 6 to c. 10 in a terminal corymb. Heads mostly 2 cm diam. Involucral bracts in several rows, white, papery; the uppermost large with expanded laminas. Florets all tubular, yellow, becoming darker with age, bisexual, each subtended by a scale. Cypselas 4-angled, c. 4 mm long, 1 mm diam. Pappus a membranous cup c. 0.5 mm high. Mainly in cooler parts of the area. Often common in grasslands and woodlands. Fl. summer . . **A. alatum** R.Br.

77 **Rutidosis** DC.

6 species endemic Aust.; all states and territories

One species in the area

Perennial herb. Leaves alternate, linear, to 3.5 cm long, c. 2 mm wide; margins revolute. Heads solitary, c. 20 mm diam. Involucre bracts in several rows, 3–10 mm long, outer bracts shortest. Receptacle naked. Florets mostly bisexual, outer ones female. Corolla yellow, 5 lobed and the tube longer than involucre. Cypsela papillose, 1–1.5 mm long. Pappus of ciliate, elliptic scales 1.5–2 mm long. Northern parts of the area. Coastal heath. Vulnerable. . . . . **R. heterogama** Philipson

78 **Leptorhynchos** Less.

10 endemic species Aust.; all states and territories except NT

Annual or perennial herbs. Leaves alternate, entire. Heads pedunculate, usually hemispherical. Involucral bracts in several rows; the outer scarious and sometimes descending along the peduncle. Receptacle flat, without scales. Florets all tubular; the outer ones often female, 3–4-toothed. Disc florets bisexual, 5-toothed. Cypselas contracted at the top or beaked, 2–3 mm long. Pappus of several capillary bristles which are barbellate or almost plumose towards the ends.

Key to the species

1 Involucral bracts with cottony-woolly ciliate margins; involucre c. 18–15 mm diam. Pappus 5–6 mm long, connate into a ring at the base. Leaves lanceolate to linear, up to 4 cm long, glabrous on the upper surface, cottony underneath. Perennial herb up to 25 cm high. Higher altitudes in the west of the area. Grasslands and forests. Fl. summer–autumn . . . .**L. squamatus** (Labill.) Less. ssp. **squamatus**
1 Involucral bracts entire or lacerated, not ciliate 2

2 Involucral bracts acute to acuminate, descending down the peduncle and grading into the upper leaves; involucre 20–30 mm diam. Pappus distinctly shorter than the corolla. Leaves oblong to narrow-oblanceolate, 2–4 cm long, ± pubescent on both surfaces. Erect to ascending, perennial herb up to 40 cm high. Western parts of the area. Grasslands and forests. Fl. spring–summer. . . . **L. elongatus** DC.
2 Involucral bracts obtuse, descending a short way down the peduncle but not intergrading with the leaves; involucre c. 10 mm diam. Pappus c. as long as the corolla. Leaves linear 2–4 cm long, glabrescent. Stoloniferous herb up to 25 cm high. Widespread south of Sydney. Grasslands, open forests, roadsides. Fl. summer–autumn . . . .**L. nitidulus** DC.

79 **Calomeria** Vent.

1 species endemic Aust.; NSW, Vic.

One species in the area

Biennial up to 2 m high, glandular pubescent to almost glabrous, strongly scented. Leaves alternate entire; lower leaves ovate-lanceolate or oblong, acuminate, stem-clasping or decurrent at the base, 12–25 cm long, rugose and scabrous-pubescent, green on both surfaces; upper leaves smaller. Heads numerous in a large loose terminal panicle. Involucre 6 mm long, 1–2 mm diam.; bracts thin, scarious, red-brown to pink. Florets 3–4, tubular. Cypsela glabrous, less than 1 mm long. Pappus absent. Coast; Blue Mts Margins of RF and moist sheltered situations. Fl. summer. *Incense Plant*. . . . . **C. amaranthoides** Vent.

80 **Podolepis** Labill.

Copperwire Daisies

20 species endemic Aust.; all states and territories

Perennial or annual herbs. Leaves basal and cauline, alternate, entire. Heads terminal solitary or few together. Involucre ovoid to hemispherical; bracts unequal, in several rows, with hyaline non-radiating laminas. Outer florets female, ligulate or tubular, with 2–4 lobes or teeth. Disc florets tubular, 5-toothed. Cypselas usually terete. Pappus of barbellate or capillary bristles, often united at the base.

Key to the species

1 Laminas of intermediate involucral bracts as long as or longer than the claws. Heads usually solitary (rarely 2–8 on the same scape). Leaves basal and cauline; the lowest largest, up to 20 cm long and 2 cm wide. Involucres up to 3 cm diam. Florets yellow. Ray florets 30–40; rays up to 25 mm long, c. 2.5 mm wide. Pappus c. 10 mm long. Perennial up to 80 cm high. Widespread. Open forests and grasslands. Fl. summer . . . . **P. jaceoides** (Sims) Voss

1 Laminas of intermediate involucral bracts shorter than the claws. Heads in clusters of 3–20. Leaves basal and cauline; the lowest largest, up to 16 cm long, 25 mm wide. Involucres 15–20 mm diam. Florets yellow. Ray florets 15–20; rays 15–18 mm long, 2 mm wide. Perennial up to 80 cm high. Blue Mts Open forests and grasslands. Fl. summer. . . . . **P. hieracioides** F.Muell.

81 **Dittrichia** Greuter
1 species naturalized Aust.; all states and territories except NT

One species in the area
Annual herb, glandular pubescent, up to 50 cm high. Leaves alternate, narrow-lanceolate to linear, obscurely dentate; the longest 5–6 cm long. Heads numerous, on short peduncles in the leaf axils. Involucre campanulate, 5–7 mm diam.; bracts linear-lanceolate, in several rows. Ray florets female; rays 1–2 mm long, yellow. Disc florets tubular. Cypselas light brown, 2 mm long, surmounted by a cup supporting the pappus. Pappus of capillary hairs 5–7 mm long. Widespread. Weed of cultivation and waste places. Introd. from the Mediterranean. Fl. summer. *Stinkwort* . . . . . ***D. graveolens** (L.) Greuter

82 **Craspedia** G.Forst.
15 species endemic Aust.; Qld, NSW, Vic., Tas.

Erect, annual or perennial herbs with 1–several scapes. Leaves basal (often in a rosette) and sometimes alternate and cauline, entire. Heads grouped into a compound head on a common receptacle and terminal on a scape; each head with a subtending ± herbaceous bract and surrounded by scarious involucral bracts; the compound head globular to hemispherical, up to 25 mm diam., surrounded by a leaf-like involucre. Florets tubular, bisexual. Cypselas silky hairy. Pappus of plumose bristles.

Key to the species

1 Leaves covered with fine silvery hairs, glandular hairs often present but small and hidden by the silky hairs, greyish-green, up to 25 cm long and 15 mm wide. Partial heads with 5–9 florets, yellow. Cypselas 1–1.5 mm long. Pappus up to 4 mm long. Herb up to 50 cm high. Widespread. Open forests and grasslands. Fl. spring–summer . . . . **C. canens** J.Everett & Doust

1 Leaves covered with coarse multicellular hairs and glandular hairs, a few fine silky ones sometimes present, narrow-oblanceolate, up to 13 cm long and 13 mm wide. Partial heads with 7–12 florets. Cypselas 1.5–2.5 mm long. Pappus up to 6 mm long. Herb up to 50 cm high. Widespread. Open forests and grasslands. Fl. spring–summer . . . . **C. variabilis** J.Everett & Doust

83 **Stemmacantha** Cass.
1 species endemic Aust.; Qld, NSW., Vic.

One species in the area
Herb to 60 cm high. Stems slightly woolly. Leaves oblanceolate, toothed to deeply pinnatifid, to 18 cm long reducing in size up the stem; petiolate below ± sessile above. Heads terminal, 3–6 cm diam. Outer Involucre bracts with lobed appendage, inner bracts lanceolate. Corolla 25–50 mm long, purple. Cypsela striated, 7–8 mm long. Pappus c. 20 mm long. Heavy soils. Western Blue Mts Extinct. Fl. spring–autumn. *Austral Cornflower* . . . . **S. australis** (Gaudich.) Dittrich.

84 Cirsium Mill.

4 species naturalized Aust.; all states and territories except NT

One species in the area

Biennial herb up to 1 m high. Leaves basal and cauline, pinnatipartite, with spiny lobes, scabrous on the upper surface, white-woolly underneath. Heads in a terminal corymb. Involucre 2–3 cm diam.; bracts linear-lanceolate, green. Florets all tubular, purple. Cypselas oblong, glabrous, 5 mm long. Pappus of numerous plumose hairs 10–15 mm long. Widespread. Weed of pastures and waste places. Introd. from Europe; Asia and Africa. Fl. spring–summer. *Spear Thistle*. . . . . *C. vulgare* (Savi) Ten.

85 Carduus L.

4 species naturalized Aust.; all states and territories except NT

Erect, annual or biennial herbs. Stems winged; the wings terminating in spines. Leaves basal and cauline, decurrent, pinnatifid, with sinuate lobes terminating in spines. Heads sessile. Involucre cylindrical to spherical; bracts in several unequal rows, ending in spines. Florets all tubular, bisexual. Cypselas oblong, 5 mm long, glabrous. Pappus of numerous simple bristles c. 10 mm long.

Key to the species

1 Flower heads solitary, spherical, 4–5 cm diam. Florets red-purple. Biennial up to 1.5 m high. Widespread. Weed of cultivation and waste places. Introd. from Europe. Fl. spring–summer. *Musk* or *Nodding Thistle*. . . . . *C. nutans* L. ssp. **nutans**
1 Flower heads in terminal clusters 2

2 Stems discontinuously spinous-winged, naked just below the terminal flower heads. Heads ovoid, 2–3 cm diam. Florets pink or white. Widespread. Weed of cultivation, pastures and waste places. Introd. from Europe. Fl. spring–summer. *Shore* or *Slender Thistle* . . . . *C. pycnocephalus* L.
2 Stems continuously spinous-winged. Heads cylindrical, c. 2 cm long. Florets purple. Annual up to 1 m high. Widespread. Weed of cultivation, pastures and waste places. Introd. from Europe. Fl. spring–summer. *Winged Slender Thistle* . . . . *C. tenuiflorus* Curtis

86 Cynara L.

2 species naturalized Aust.; all states and territories except NT, Tas.

One species in the area

Coarse, tufted perennial up to 2 m high. Leaves basal and cauline, pinnatisect with spinose tips to the lobes, up to 80 cm long. Heads globular, usually solitary, up to 15 cm diam.; bracts in several rows often with short spinose tips. Florets numerous, purplish blue. Cypselas c. 6 mm long. Pappus of unequal brown hairs 3–4 cm long. Naturalized in a few places throughout the area. Introd. from the Mediterranean. Fl. summer. *Globe Artichoke* . . . . *C. scolymus* L.

87 Silybum Adans.

1 species naturalized Aust.; all states and territories except NT

One species in the area

Erect, glabrous biennial up to 3 m high. Leaves basal and cauline, glossy, mottled with white veins, ± pinnatifid, with spiny margins; the uppermost stem-clasping. Heads solitary on leafy peduncles. Involucre subglobular, 4 cm diam.; bracts broad, rigid, with a rounded appendage terminating in a spine. Florets all tubular, purple. Cypselas black, glossy, glabrous, 5–6 mm long. Pappus of numerous simple unequal hairs, 12–20 mm long. Widespread. Weed of cultivation, pastures and waste places especially around stockyards. Poisonous to stock. Introd. from the Mediterranean. Fl. chiefly spring. *Variegated Thistle*. . . . . *S. marianum* (L.) Gaertn.

88 Onopordum L.

4 species naturalized Aust.; all states and territories except Qld, NT

One species in the area

Biennial herb up to 1 m high, with woolly stems. Leaves mostly cauline, lanceolate, up to 35 cm long; margins with spine-tipped teeth, decurrent on the stem. Heads solitary or 2–3 together. Involucre 35 mm diam.; bracts subulate, up to 25 mm long with a yellowish terminal spine. Florets purple. Cypselas 4–5 mm long. Pappus bristles barbellate up to 10 mm long. Widespread. Weed of cultivation, pastures and waste places. Introd. from Europe. Fl. spring–summer. *Scotch Thistle* . . . . . ***O. acanthium** L. ssp. **acanthium**

89 Centaurea L.

13 species naturalized Aust.; all states and territories except NT

Erect annuals. Leaves basal and cauline, alternate. Heads solitary or clustered. Involucre globular or ovoid; bracts imbricate in several rows, ending in a prickle or toothed appendage. Florets all tubular; the outer ones often neuter. Receptacle with numerous bristles between the florets. Cypselas glabrous. Pappus of short simple bristles or scales or absent.

Key to the species

1 Corollas purple. Pappus absent. Involucral bracts horny, with a spreading spine 1–2 mm long. Leaves pinnatifid, with a few long linear or lanceolate lobes. Heads sessile in the forks or within the last leaves of the branches. Cypselas 2 mm long. Erect annual up to 25 cm high, usually with a cottony down. Widespread. Weed of cultivation and waste places. Introd. from the Mediterranean. Fl. summer. *Star Thistle* . . . . ***C. calcitrapa** L.
1 Corollas yellow 2

2 Heads solitary. Involucral bracts with a spine 2–3 cm long, spinules palmately arranged. Basal leaves lyrate, up to 10 cm long; cauline leaves linear, decurrent, entire. Involucre urn-shaped, 1 cm diam. Cypselas compressed, brown, 2–3 mm long. Pappus bristles 5 mm long. Erect, white-tomentose annual up to 60 cm high. Widespread. Weed of cultivation, pastures and waste places. Introd. from the Mediterranean. Fl. chiefly summer. *St Barnaby's Thistle* . . . . ***C. solstitialis** L.
2 Heads sometimes clustered. Bracts with a spine less than 1 cm long, spinules pinnately arranged. Basal leaves pinnately divided; cauline leaves narrow, decurrent, entire or slightly toothed. Involucre more than 1 cm diam. Cypselas 3–4 mm long. Pappus of bristles up to 5 mm long. Erect annual up to 50 cm high, woolly or glabrous. Widespread. Weed of cultivation, pastures and waste places. Introd. from the Mediterranean. Fl. summer. *Maltese Cockspur* . . . . ***C. melitensis** L.

90 Carthamus L.

5 species naturalized Aust.; all states and territories

Annual herbs. Leaves basal and cauline, alternate. Heads terminal, solitary or corymbose. Involucre with spreading leafy outer bracts and spiny inner ones. Receptacle chaffy. Florets all tubular; the corolla 5-toothed. Cypselas glabrous, mostly 4-ribbed. Pappus scale-like or absent.

Key to the species

1 Longest involucral bracts 30–40 mm long, with spinescent lobes or spines up to 10 mm long. Leaves deeply lobed or pinnatifid; the lobes ending in spines. Involucre 2–3 cm diam. Corolla yellow. Cypselas 6 mm long. Pappus of scales. Annual herb up to 80 cm high, sometimes slightly woolly. Widespread. Weed of pastures and waste places. Introd. from Europe; Africa and Asia. Fl. summer. *Saffron Thistle* . . . . ***C. lanatus** L.
1 Longest involucral bracts c. 15 mm long, with spines c. 2 mm long. Leaves oblong or ovate to lanceolate, with minute spines; the upper ones stem-clasping. Involucre 2–4 cm diam. Corollas orange-yellow. Cypselas 6 mm long. Weed near habitation, Cultivated for seed-oil. Introd. from Europe and Asia. Fl. summer–autumn. *Safflower* . . . . ***C. tinctorius** L.

91 Microseris D.Don
1 species endemic Aust.; all states and territories except NT

One species in the area

Perennial herb with a fleshy taproot. Leaves basal, glabrous, linear-lanceolate, 10–20 cm long, entire or with a few distant teeth or lobes. Heads solitary on leafless scapes 10–40 cm high. Involucre green, cylindrical, 20–25 mm long; the bracts in 2 rows. Florets all ligulate; rays 10–15 mm long. Cypselas 8–10 mm long, with 10 obtuse, longitudinal ribs. Pappus of 10–12 thin white scales, c. 15 mm long, tapering into barbellate awns. Western Blue Mts Open forests and grasslands. Fl. summer . . . . . .**M. lanceolata** (Walp.) Sch.Bip.

92 Cichorium L.
2 species naturalized Aust.; all states and territories except NT

One species in the area

Much branched perennial up to 1 m high. Basal leaves pinnatifid, 10–20 cm long; cauline leaves lanceolate, stem-clasping. Heads solitary or 2–3 together, terminal or sessile and axillary. Involucre oblong, 5 mm diam.; bracts in 2 rows. Florets all ligulate; rays blue, 1 cm long. Cypselas compressed, striate, 2 mm long. Pappus of minute scales. Widespread. Weed of roadsides and waste places. Introd. from Europe; Asia and Africa. Fl. summer. *Chicory* . . . . ***C. intybus** L.

93 Tragopogon L.
2 species naturalized Aust.; all states and territories except NT

One species in the area

Glabrous biennial up to 1 m high, with a long taproot. Leaves linear, stem-clasping, subulate, up to 30 cm long. Heads solitary on naked peduncles which are swollen and hollow towards the summit. Involucre 3–4 cm diam.; bracts herbaceous, in 1 row. Florets all ligulate; rays violet. Cypselas fusiform. 10–12 mm long, with a beak longer than the body of the cypsela. Pappus of plumose hairs 2 cm long with soft interwoven barbs. Widespread. Weed of roadsides and waste places. Introd. from Europe. Fl. spring–summer. *Salsify* . . . . ***T. porrifolius** L.

94 Hedypnois Mill.
1 species naturalized Aust.; all states and territories except NT

One species in the area

Procumbent or erect annual; stems up to 30 cm long. Basal leaves in a rosette, up to 15 cm long, ± spathulate, mostly lobed. Cauline leaves alternate, stem-clasping, toothed or entire, up to 5 cm long. Heads solitary on leafless peduncles which are hollow and swollen towards the summit. Involucre 7–9 mm long; bracts c. 12, in 1 row, with a few smaller ones at the base. Florets all ligulate, yellow. Cypselas terete, 6–10 mm long, curved, ribbed longitudinally. Pappus of 4–5 scales, c. 3 mm long, forming a crown, sometimes double on the inner cypselas. Widespread. Weed of cultivation, pastures and waste places. Introd. from the Mediterranean. Fl. spring–summer. *Cretan Weed* . . . . ***H. rhagadioloides** (L.) Schmidt ssp. **cretica** (L.) Hayek

95 Taraxacum Weber
Currently under revision. At least 2 species native, c. 7 introduced Aust.; all states and territories except NT

One species in the area

The following description follows (Willis 1973) who includes all introduced species in *T. officinale*. A number of species may be separated with further revision. See Flora of Victoria Vol. 4.

Perennial herb with thick taproot. Leaves basal, linear-lanceolate, usually deeply pinnatifid but sometimes entire, up to 20 cm long. Heads solitary on leafless hollow peduncles up to 15 cm long. Involucre 10–15 mm diam.; bracts in several rows; the outer ones imbricate. Florets numerous, all ligulate; rays yellow, 1–2 cm long. Cypselas scarcely compressed, striate, with a beak c. 1 cm long. Pappus of numerous simple hairs 5–7 mm long. Widespread. Weed of cultivation, pastures and waste places. Introd. from Europe and Asia. Fl. most of the year. *Dandelion* . . . .*T. officinale* Weber

96 Leontodon L.

1 species naturalized Aust.; all states and territories except NT

One species in the area

Biennial or perennial herb. Leaves basal, oblong, slightly and distantly lobed or sinuate-pinnatifid, often with triangular lobes. Heads solitary, on leafless solid scapes 10–20 cm long. Involucre glabrous or bristly; bracts c. 12, equal, herbaceous, with smaller bracts at the base. Florets numerous, all ligulate, exceeding the involucre; rays yellow, 3–7 mm long. Cypselas striate, wrinkled transversely; outer ones thick, curved, with a short pappus of small ± united scales; inner ones tapering at the summit, with a pappus of long plumose bristles and an outer row of barbellate bristles. Widespread. Weed of cultivation, pastures and waste places. Introd. from Europe. Fl. most of the year. *Lesser Hawkbit*. . . . .
. . . . *L. taraxacoides* (Vill.) Mérat ssp. **taraxacoides**

97 Chondrilla L.

1 species naturalized Aust.; Qld, NSW, Vic., SA, WA

One species in the area

Perennial herb with a long taproot. Leaves mostly basal, mostly c. 10 cm long, toothed or acutely lobed; cauline leaves linear, c. 2 cm long. Scapes up to 1 m high, leafless at maturity. Heads solitary or 2–3 together on the upper branches. Involucre cylindrical, 10–13 mm long; bracts c. 8. Florets 9–15, all ligulate; rays yellow. Cypselas 4 mm long, 5-angled and ribbed, surmounted by a 5-toothed crown and a beak 4–5 mm long, with numerous white pappus hairs 6–7 mm long. Widespread. Weed of cultivation, pastures and waste places. Introd. from Europe and Asia. Fl. summer. *Skeleton Weed* . . . . *C. juncea* L.

98 Lactuca L.

3 species naturalized Aust.; all states and territories

Erect, biennial herbs. Involucre narrow-cylindrical; bracts unequal, herbaceous, imbricate, in 5 rows. Florets all ligulate; rays yellow. Cypselas flattened, obovate-cuneate, striate on both faces, terminating in a long filiform beak. Pappus of numerous silky hairs.

Key to the species

1 Cauline leaves linear-lanceolate, entire, glabrous, stem-clasping; basal leaves pinnatisect, with few lobes, up to 20 cm long. Heads subsessile in the leaf axils. Involucre 3 mm diam. Cypselas glabrous, black, 2–3 mm long, with a beak 3–5 mm long. Pappus hairs 3–4 mm long. Glabrous herb up to 1 m high. Widespread. Weed of cultivation, pastures and waste places. Introd. from Europe; Asia and Africa. Fl. summer. *Willow-leaved Lettuce* . . . .*L. saligna* L.
1 Cauline leaves broad-oblong, pinnatifid; the longest up to 10 cm long, mostly twisted at the base so as to be vertical, spinulose on the margins and often on the midrib underneath; upper leaves stem-clasping. Heads in a terminal panicle. Involucre 3–4 mm diam. Cypselas grey, ciliate towards the summit, 3 mm long; the beak as long or longer. Pappus hairs 5 mm long. Herb up to 2 m high. Widespread. Weed of cultivation, pastures and waste places. Introd. from Europe; Central Asia. Fl. summer. *Prickly Lettuce* or *Compass Plant*. . . . . *L. serriola* L.

99 Hypochaeris L.

3 species naturalized Aust.; all states and territories

Herbs with basal leaves in a rosette. Heads on solid scapes with a few scale-like or herbaceous leaves. Involucre cylindrical to conical; bracts herbaceous in 2–5 rows. Florets all ligulate; rays white or yellow. Receptacle with a few linear chaffy scales. Cypselas usually striate. Pappus of 1–2 rows of plumose bristles.

Key to the species

1 Rays white. Hairs of the pappus in 1 row, all plumose, 6–7 mm long. Leaves basal and cauline, 5–15 cm long; basal leaves sinuately lobed, 15–25 mm wide; cauline leaves narrower, with a few teeth. Scapes up to 25 cm high, usually branched several times. Heads solitary. Involucre 10–12 mm long. Cypselas 5 mm long; beak 5 mm long. Perennial with a fleshy taproot. Coast and Cumberland Plain. Weed of cultivation and waste places. Introd. from Europe; Asia and Africa. Fl. summer. *White Flatweed* . . . .***H. microcephala** (Sch.-Bip.) Cabrera var. **albiflora** (Kuntze) Cabrera
1 Rays yellow. Hairs of the pappus in 2 rows; the inner long and plumose; the outer short and simple. Leaves in a basal rosette. Heads on branched stems up to 30 cm high. 2

2 Outer cypselas without a beak. Leaves up to 15 cm long, toothed, sinuate to pinnatifid, glabrous or scabrous. Heads, up to 10 mm diam. Cypselas 3–4 mm long; inner ones with a beak c. 7 mm long. Pappus hairs 8–10 mm long. Widespread. Weed of cultivation, pastures and waste places. Annual herb. Introd. from Europe; Asia and Africa. Fl. summer. *Smooth Catsear*. . . . .***H. glabra** L.
2 All cypselas beaked. Leaves oblanceolate, up to 20 cm long, toothed, sinuate to pinnatifid, hairy with hispid hairs. Heads, up to 15 mm diam. Cypselas 4-7 mm long, with a beak 7–10 mm long. Pappus hairs 8–14 mm long. Perennial herb. Widespread. Weed of cultivation, pastures and waste places. Introd. from Europe; Asia and Africa. Fl. summer. *Catsear* or *Flatweed*. . . . .***H. radicata** L.

100 Picris L.

12 species in Aust. (2 naturalized); all states and territories except NT

Annual, biennial or perennial herbs with taproot. Hairs are rigid and spinose, or have 2–4-fid tips which are straight or hooked. Leaves simple, entire to dentate; basal leave in rosette; cauline leaves alternate, stem clasping becoming smaller towards top of stem. Heads terminal or few to many in corymbose or cymose panicles. Involucral bracts 20–50 in 1–4 outer rows and 2 inner rows. Florets 20–90, all ligulate; rays yellow. Stylar branches yellow. Cypselas tapering to narrow tip, 5-ribbed; outer ones curved, inner ones ± straight. Pappus cream-white, plumose, in two rows, outer row smaller than inner.

Key to the species

1 Outer involucral bracts almost as long as or at least 2/3 the length of longest inner bracts; in 2–4 rows; straight to spreading or recurved but becoming straight towards inner rows. Peduncles 1–6 cm long, sparingly to densely hairy with short 2 hooked hairs 0.3–0.6 mm long. Peduncle bracts 0–3 near head. Heads with 30 to 40 florets; corolla 9–13 mm long. Annual 50–120 cm high. DSF & RF. Hunter Valley. Uncommon. Fl. spring–summer . . . . **P. burbidgei** S.Holzapfel
1 Outer involucral bracts usually shorter than longest inner bract by 1/3 or more; in 2–3 row; straight and appressed or outer rows spreading slightly. Peduncles 1–20 cm long, glabrous to densely hairy with 2 hooked hairs and a few 2-fid hairs, 3 hooked hairs sometimes present; hairs 0.5–2 mm long. Peduncle bracts 0–6 near head or rarely spread further apar.t Heads with 20–60 florets; corolla 10–17 mm long. Annual or perennial 7–140 cm high 2

2 Flowering heads 1–14 mm long. Achenes 4.5–11 mm long. Coast and ranges. Uncommon. Fl. summer–autumn . . . .**P. angustifolia** DC. ssp. **angustifolia**
2 Flowering heads 8–10 mm long. Achene 3–4.5 mm long. Coast and ranges. Occasional. Fl. all year . . . . **P. angustifolia** ssp. **carolorum-henricorum** (Lack) S.Holzapfel

101 Helminthotheca Zinn

1 species naturalized Aust.; all states and territories except NT

One species in the area

Prostrate to erect annual or biennial herb up to 1 m high. Basal leaves elliptic to oblanceolate, up to cm 75 cm long, ± petiolate; upper cauline leaves stem-clasping. Heads 10–15 mm diam., in clusters forming a panicle. Involucral bracts herbaceous, in 2 rows. Florets all ligulate; rays yellow, sometimes marked with red. Stylar branches black. Cypselas c. 3 mm long, with a beak c. 3 mm long. Pappus pure white. Coast.

Weed of cultivation and waste places. Introd. from Europe; Asia and Africa. Fl. spring–summer. *Ox-tongue* . . . . ***H. echioides** (L.) Holub

102 Sonchus L.

5 species in Aust. (1 native, 4 naturalized); all states and territories

Herbs with milky latex and hollow stems. Leaves basal and cauline; cauline leaves stem-clasping. Heads in irregular corymbs. Involucre ovoid in bud, later conical; bracts numerous, herbaceous, unequal, in several rows. Florets all ligulate, yellow, exceeding the involucre. Cypselas flattened. Pappus of numerous white silky hairs up to 8 mm long.

Key to the species

1 Leaves soft, thin, basal and cauline, pinnatifid, irregularly toothed, up to 20 cm long. Heads 5–6 mm diam. Cypselas wrinkled, winged, scabrous on the margins. Annual herb up to 2 m high, glabrous or glandular hairy towards the summit. Widespread. Weed of cultivation and waste places. Introd. from Europe. Fl. most of the year. *Common Sowthistle* or *Milk Thistle*. . . . . ***S. oleraceus** L.
1 Leaves leathery, stiff. Cypselas ribbed 2

2 Leaves with spinose margins, pinnatifid to pinnatisect, oblanceolate to lanceolate, up to 20 cm long and 7 cm wide. Heads 5–6 mm diam. Annual or biennial herb up to 1 m high. Widespread. Weed of cultivation and waste places. Introd. from Europe. Fl. most of the year. *Prickly Sowthistle* . . . . ***S. asper** (L.) Hill ssp. **glaucescens** (Jord.) Ball
2 Leaves without spinose margins, shallowly sinuate-pinnatifid to entire, oblanceolate, up to 30 cm long and 3 cm wide. Heads 10–20 mm diam. Annual or biennial herb up to 1 m high. Coast. Margins of lakes and creeks. Fl. most of the year. *Native Sowthistle*. . . . . **S. hydrophilus** Boulos

103 Actites Lander

1 species endemic Aust.; all states and territories except NT

One species in the area

Perennial, rhizomatous herb up to 50 cm high, glabrous or scabrous. Leaves basal and cauline, pinnatisect to entire, oblanceolate to lanceolate, up to 17 cm long and 45 mm wide, petiolate or sometimes cordate at the base, sinuate-toothed. Heads solitary, 10–20 mm diam. Involucral bracts in 3–5 rows. Florets all ligulate. Rays yellow. Cypselas glabrous, 4–8 mm long. Pappus of unequal barbellate bristles in several rows. Coastal dunes. Fl. most of the year. *Dune Thistle* . . . . **A. megalocarpa** (Hook.f.) Lander

104 Crepis L.

6 species naturalized Aust.; all states and territories except NT

Annual or perennial herbs. Leaves basal and forming a rosette or cauline and alternate. Heads terminal, pedunculate, in loose corymbose panicles. Involucral bracts herbaceous with scarious margins, in 3 rows. Florets all ligulate; rays yellow, sometimes marked with red. Cypselas terete, ribbed, sometimes beaked.

Key to the species

1 Plants glabrous or nearly so except for the finely hairy involucre. Basal leaves oblanceolate to lanceolate, pinnatifid, mostly 10–25 cm long; cauline leaves smaller, stem-clasping, auriculate. Heads 5–6 mm diam. Herb up to 80 cm high. Cypselas without beaks. Herb up to 80 cm high. Widespread. Weed of roadsides and waste places. Introd. from Europe. Fl. summer. *Smooth Hawkesbeard*. . . . . ***C. capillaris** (L.) Wallr.
1 Plants hairy. At least the inner cypselas beaked 2

2 Involucral bracts pubescent with soft glandular and simple hairs. Basal leaves oblanceolate, up to 13 cm long and 3 cm wide, pinnatifid to toothed; cauline leaves smaller. Heads 10–15 mm diam. Cypselas more than 7 mm long; only the inner ones distinctly beaked. Herb up to 50 cm high with an unpleasant scent. Widespread. Weed of roadsides and waste places. Introd. from Europe. Fl. summer. *Stinking Hawkebeard*. . . . . *C. foetida* L. ssp. **foetida**

2 Involucral bracts with yellow bristles. Basal leaves oblanceolate, up to 20 cm long and 5 cm wide, pinnatifid; cauline leaves smaller, stem-clasping. Heads 6–10 mm diam. Cypselas less than 7 mm long, all beaked. Annual herb up to 80 cm high. Recorded from Hornsby only once. Introd, from Europe. Fl. summer . . . . *C. setosa* Haller f.

105 Tolpis Adans.
1 species naturalized Aust.; NSW, Vic., S.A.

One species in the area

Annual herb up to 40 cm high. Leaves chiefly basal, lanceolate to linear, up to 8 cm long; the lower leaves denticulate; the upper ones entire. Heads solitary on long peduncles bearing small leaves from the axils of which smaller heads proliferate, thus producing cymose compound inflorescences. Involucre 4–6 mm diam.; bracts herbaceous, subulate, in 2–3 rows. Florets all ligulate; outer rays yellow; inner rays brownish. Cypselas less than 1 mm long. Pappus of 2–4 capillary hairs, 2–3 mm long. Widespread. Grasslands and open forests on heavy soils. Introd. from the Mediterranean. Fl. chiefly summer. *Yellow Hawkweed*. . . . .*T. barbata* Bertol.

106 Youngia Cass.
1 species native Aust.; Qld, NSW

One species in the area

Annual herb. to 60 cm high. Leaves 1.5–12 cm long; basal and cauline, oblong-ovate, lyrate, pinnatifid; margins finely toothed. Heads numerous 2–3 mm diam. Outer involucre bracts small inner bracts 5 mm long. Florets all ligulate, yellow. Cypsela 1.5–2 mm long with 11–13 ribs. Pappus bristles united at base, 3 mm long. Moist sheltered positions. Fl. most of the year . . . .*Y. japonica* (L.) DC.

144 CAMPANULACEAE

Prostrate or erect herbs, often with latex. Leaves alternate or opposite sometimes rosetted; stipules absent. Flowers in racemes or solitary in the leaf axils, bisexual or sometimes unisexual and the plants dioecious. Sepals 5. Corolla tubular, often with 5 irregular lobes, usually slit to the base or notched anteriorly although the pedicel is twisted to bring the slit to the posterior position. Stamens 5; anthers (and often the upper parts of the filaments) connate around the style (Fig. 43), all or only the anterior ones hairy at the apex; filaments often adnate to the corolla tube basally. Ovary inferior, 2–5-locular; placentas axile. Fruit a capsule or indehiscent and ± fleshy. 79 gen., mostly temp.

Key to the genera

1 Flowers regular, funnel-shaped or rotate. Stamens free . . . . **1 WAHLENBERGIA**
1 Flowers irregular, ± 2-lipped or fan-shaped. Stamens connate (Fig. 43). 2

2 Corolla tube entire, notched or slit half-way to the base anteriorly . . . . **5 ISOTOMA**
2 Corolla tube slit open to the base. 3

3 Only two anterior anthers hairy at the apex, each with one bristle longer than the other. Fruit indehiscent, ± fleshy . . . . **2 PRATIA**
3 Either; all the anthers hairy at the apex; or, if only the anterior pair hairy at the apex, then the back of the posterior ones hairy 4

4 Corolla red . . . . **4 MONOPSIS**
4 Corolla blue, white, purple or pink. . . . . **3 LOBELIA**

FIGURE 43

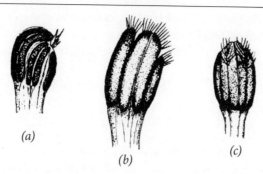

Fig. 43 Campanulaceae: staminal tubes: (a) *Pratia purpurascens*, lateral view, (b) *Lobelia dentata*, lateral view, (c) *Lobelia anceps*, back view. All X10.

1 Wahlenbergia Schrad. ex Roth.
Native Bluebells
22 species in Aust. (1 naturalized); all states and territories

Annual or perennial herbs with fleshy or almost woody rootstocks or rhizomes. Leaves alternate or opposite, obovate to linear. Flowers solitary on long peduncles or in loose cymes. Sepals 4–5. Corolla tubular to rotate, regular, 3–5-lobed, blue to white. Stamens 5, free; pollen deposited on hairs on the upper part of the style before the flower opens. Ovary inferior, 2–3-locular; style 2–3-fid. Fruit a capsule dehiscing through terminal valves.

Key to the species

1 Perennial herb with spreading rhizomes; glabrous or ± hirsute on lower stems. Stems and inflorescences unbranched or with 1–2 branches. Flowers solitary or 2–3 per stem. Leaves sessile, 5–50 mm long, obovate to narrow elliptic becoming linear to lanceolate on upper stem; margins entire. Sepals narrow triangular, 2.5–6 mm long, glabrous. Corolla blue or white, lobes acute, 10–23 mm long, tube 3–7.5 mm long. Capsule obconic 5–10 mm long, glabrous. At high altitudes in dense vegetation. Fl. summer–autumn . . . . **W. ceracea** Lothian

1 Tufted perennial herbs without rhizomes the stems and inflorescences becoming much branched . . . 2

2 Style constricted in the region of the pollen collecting hairs either close to the stigmatic branches (when the unopened stigmatic lobes appear almost globular) or c. half-way down (when the upper part of the style is distinctly thicker than the lower part) 3

2 Style not constricted, ± the same diameter throughout once the pollen has been removed. 7

3 Style constricted about half-way down. Corolla campanulate; lobes less than 4 times as long as the tube . 4

3 Style constricted close to the stigmatic lobes. Corolla ± rotate; lobes more than 4 times as long as the tube 6

4 Leaves opposite and linear may become alternate towards top of stem; sessile, 4–60 mm long, 1–4 mm wide. Capsule at least 3 times as long as broad. Leaf margins entire or with a few small teeth. Flowers blue inside whitish outside often with a yellow/brown ting. Corolla lobes acute 6–14 mm long, tube 2–5 mm long. Capsule 5–12 mm long, glabrous. Herb with stems to 80 cm long. Ranges, with a few early collections from coastal areas. Woodland and grasslands. Fl. all year. . . . . **W. luteola** P.J.Sm.

4 Lower leaves alternate or, if opposite then broader than linear and capsule not 3 times as long as broad 5

5 Plants glabrous or nearly so. Lower leaves linear or narrow-oblanceolate, 4–60 mm long. Sepals 2–4 mm long. Corolla blue, lobes 5–9 mm long; tube 2–3 mm long. Capsule obconic 3–8 mm long, glabrous. Tufted perennial herb up to 80 cm high, with few stems from the base but branching above. Widespread. Open forests, grasslands and heath. Fl. most of the year . . . . **W. littoricola** P.J.Sm.

5 Plants hirsute towards the base. Lower leaves obovate to oblanceolate, hirsute, up to 80 mm long. Sepals 2–6 mm long. Corolla lobes 6–19 mm long, tube 2–6 mm long. Capsule obconic 3–10 mm long, glabrous or hairy. Tufted perennial up to 80 cm high, with few stems from the base but branching above. Blue Mts Open forests. Fl. most of the year . . . . **W. graniticola** Carolin

6 Capsule more than twice as long as broad, elongated-obconic, 4–12.5 mm long. Leaves obovate to oblanceolate, up to 80 mm long, 1–6 mm wide, mostly acute, glabrous or slightly hirsute. Corolla lobes 2–9 mm long. Tufted perennial up to 80 cm high, with few stems from the base but branching above. Widespread. Open forests and grasslands. Fl. most of the year. Endangered population in some LGAs . . . . **W. multicaulis** Benth.

6 Capsule less than twice as long as broad, obconic, 3–8 mm long. Leaves obovate to oblanceolate at the base of the stem, 4–95 mm long, 1–11 mm wide, obtuse to acute, glabrous to hirsute. Corolla lobes 6–14 mm long. Tufted perennial up to 70 cm high, with few stems from the base but branching above. Widespread. Open forest and grasslands. Fl. most of the year . . . **W. planiflora** P.J.Sm. ssp. **planiflora**

7 Corolla lobes 1.5–6 mm long, tube c. as long as the ovary, 1–5 mm long . Sepals up to 3 mm long. Capsule obconic 2.5–7 mm long. Tufted perennial up to 80 cm high, often ± straggling, with many branches. Widespread. Many communities and often a weed of cultivation. Fl. most of the year . . . . **W. gracilis** (G.Forst.) A.DC.

7 Corolla lobes 6–20 mm long, tube distinctly longer than the ovary 8

8 At least the lowermost leaves opposite and obovate or oblanceolate, becoming alternate and linear up the stem, 5–70 mm long, usually hirsute, margins typically undulate. Sepals 3–16 mm long. Corolla tube 4–11 mm long; lobes 6–20 mm long. Capsule ellipsoid to globular, 3–10 mm long, glabrous or hirsute. Tufted perennial up to 90 cm high, with few stems from the base and few branches. Widespread. Open forests and grasslands, occasionally in heath. Fl. most of the year . . **W. stricta** (R.Br.) Sweet ssp. **stricta**

8 Lowermost leaves alternate, linear or the lowermost sometimes oblanceolate, 4–80 mm long, glabrous to hirsute, margins typically flat. Sepals up to 6 mm long. Corolla tube 4–9 mm long; lobes 6–13 mm long. Capsule elongated-obconic, 4–9 mm long, glabrous. Tufted perennial up to 80 cm high, with numerous stems from the base. Widespread. Open forests and grasslands and as a roadside weed. Fl. most of the year. . . . . **W. communis** Carolin

2 Pratia Gaudich.
7 species endemic Aust.; all states and territories

Annual or perennial, often prostrate herbs. Leaves alternate, entire or dentate, usually sessile. Flowers solitary in the leaf axils, bisexual or unisexual and the plants dioecious. Corolla white to pale purple; tube slit to the base. Stamens free from the corolla; two anterior anthers bearing a number of short hairs and one bristle (Fig. 43). Ovary 3-locular. Fruit indehiscent, ± fleshy, sometimes tardily dehiscent.

Key to the species

1 Leaves obovate to spathulate, margins entire or with scattered indentations. Leaves distichous, petiole 0–5 mm long. Corolla 5–11 mm long, white or tinged with blue or violet. High altitudes. Damp places. Fl. summer. *Mud Pratia* . . . . **P. surrepens** (Hook.f.) E.Wimm.

1 Leaves not as above, margins toothed 2

2 Leaves purplish underneath, shortly petiolate, ovate to elliptic, irregularly dentate, 10–25 mm long. Corolla 8–10 mm long, pale pink-purple. Capsule tardily dehiscent. Widespread. Damp places and WSF. Fl. spring–summer. *White-root* . . . . **P. purpurascens** (R.Br.) E.Wimm.

2 Leaves green underneath 3

3 Leaves oblong to lanceolate, 12–25 mm long, up to 10 mm wide, dentate, sessile, regularly distichous. Male flowers with blue-black anthers; female flowers with paler abortive anthers. Corolla white to pinkish or purplish, 6–8 mm long. Cumberland Plain. Damp places. Uncommon. Fl. summer . . . . **P. concolor** (R.Br.) Druce

3 Leaves ovate-elliptic to almost orbicular. Male flowers with black to deep brown anthers. Female flowers with abortive yellow-brown anthers 4

4 Peduncles 20–100 mm long. Corolla pinkish white to purplish, 6–8 mm long. Leaves 5–12 mm long, 3–6 mm wide, dentate, sessile. Widespread. Damp places and WSF. Fl. summer . . . .
. . . . **P. pedunculata** (R.Br.)Benth.

4 Peduncles 6–17 mm long. Corolla whitish, 4–6 mm long. Leaves 3–6 mm long, 2–4 mm wide, entire or obscurely dentate, sessile. Blue Mts Damp places. Fl. spring–summer . . . . **P. puberula** Benth.

3 Lobelia L.
22 species in Aust. (18 endemic, 2 naturalized); all states and territories

Glabrous, erect or procumbent herbs. Leaves alternate. Flowers in racemes spikes or solitary in the leaf axils, bisexual. Corolla blue purple or white, slit to the base, irregular. Stamens free from the corolla or adnate basally to it; anthers all hairy at the apex or only the two anterior ones hairy at the apex and the posterior ones hairy on the back (Fig. 43). Ovary 2-locular. Fruit a capsule.

Key to the species

1 Flowers solitary in the leaf axils. Posterior anthers hairy on the back, only the anterior anthers with apical bristles. Corolla pale blue to white or pink. Peduncles shorter or only slightly longer than the leaves, 5–7 mm long, rarely longer. Leaves very variable, obovate to cuneate or almost linear, distantly dentate to entire. Capsule elongate-obconic, 6–10 mm long. Procumbent to ascending herb. Coast. Damp places or damp rocks near the sea. Fl. summer–autumn. (previously *L. alata*). . . .**L. anceps** L.f.

1 Flowers in terminal usually ± unilateral racemes. Anthers all with bristles at the apex of the anther tube. Corolla dark blue 2

2 Weak prostrate or decumbent herb. Leaves ovate, cordate at the base, mostly 15–35 mm wide, crenate, petiolate. Corolla c. 10 mm long. Hornsby Plateau. WSF and RF. Fl. summer–autumn. . . . .
. . . .**L. trigonocaulis** F.Muell.

2 Erect herbs. Leaves linear to ovate, deeply incised or dentate, tapering towards the base. Capsule oblique, subglobular 3

3 Three anterior corolla lobes broad-obovate, c. 4 mm wide. Corolla 10–12 mm long. Flowers on slender pedicels. Stems often branched. Capsule c. 4 mm diam. Widespread. Open forests and heath. Ss. Fl. spring–summer. . . . . **L. gracilis** Andrews

3 Three anterior corolla lobes oblong to narrow-oblong or narrow-obovate, up to 6 mm wide. Corolla 15–25 mm long. Stems simple to slightly branched. Capsule more than 5 mm diam 4

4 Lower leaves ovate, deeply cut, up to 3 cm long and 4 mm wide. Pedicels c. 2 cm long. The middle anterior corolla lobe narrow-oblong. Widespread. Open forests and heath. Ss. Fl. winter–spring. . . . . .
. . . .**L. dentata** Cav.

4 Lower leaves linear, entire or nearly so, up to 5 cm long and 4 mm wide. Pedicels c. 1 cm long. The middle anterior corolla lobe oblong. Widespread. Open forests and heath. Ss. Fl. spring–autumn . . . .
. . . . **L. gibbosa** Labill.

4 Monopsis Salisb.
1 species naturalized Aust.; NSW, Vic., S.A., WA

One species in the area

Procumbent or ascending herb. Leaves sessile, narrow-oblanceolate to almost linear, up to 15 mm long and 5 mm wide, serrate, with narrow acute teeth. Corolla red, slit to the base anteriorly, irregular. All anthers with a tuft of hairs at the apex. Stigmatic lobes filiform. Fruit a capsule, up to 9 mm long and 3 mm diam. Naturalized in places near Sydney. Damp places. Introd. from S. Afric. Fl. spring–summer . .
. . . . ***M. debilis** (L.f.) C.Presl

5 Isotoma Lindl.

c. 15 species in Aust. (1 native, 14 endemic); all states and territories

Glabrous to minutely pubescent herbs. Leaves alternate. Flowers in racemes or solitary in the leaf axils, sessile or on long peduncles. Corolla tubular, entire or ± notched anteriorly, with almost equal lobes. Stamens adnate to the corolla tube; anterior anthers each bearing a long bristle which may cohere together and appear as one. Ovary 2-locular. Fruit an indehiscent or dehiscent capsule.

Key to the species

1 Corolla pale blue; the tube 2–3 cm long. Leaves deeply pinnate-lobed, narrow-oblong to narrow-elliptic in outline, 4–6.5 cm long. Capsule ellipsoid, c. 8 mm diam., on peduncles up to 13 cm long. Acrid, tufted herbs up to 50 cm high. Widespread. Stony sites. Fl. spring–autumn . . . . . . **I. axillaris** Lindl.
1 Corolla tube less than 1cm. Leaves entire or toothed less than 2 cm long. 2

2 Leaves glabrous, obovate-oblong, 2–4 mm long, 1–2.5 mm wide, obtuse, entire or nearly so; margin undulate. Fruit indehiscent, ± fleshy. Flowers solitary in the leaf axils, sessile or nearly so. Corolla white to pale purple, 4–5 mm long; tube entire or notched, ± regular. Small, prostrate herb. Cumberland Plain. Damp places. Endangered. Fl. spring (previously *Hypsela sessiliflora* and may be immature form of *I. fluviatilis*) . . . . **I. sessiliflora** (E.Wimm.) Lammers
2 Leaves pubescent or sometimes glabrous, entire or dentate, ovate to linear, 4–10 mm long. Fruit dehiscent. Flowers with peduncles 5–30 mm long. Corolla pale blue, 4–12 mm long. Capsule obconic, up to 6 mm long and 3 mm diam. Procumbent to ascending herbs 3

3 Flowers unisexual, dioecious. Corolla glabrous inside. Coast and Cumberland Plain. Damp places; sometimes a weed in lawns. Fl. spring–summer . . **I. fluviatilis** (R.Br.) F.Muell. ex Benth.ssp. **fluviatilis**
3 Flowers bisexual. Corolla pubescent inside. Blue Mts Damp places. Fl. spring–summer . . . .
. . . . **I. fluviatilis** ssp. **borealis** McComb

145 GOODENIACEAE

Annual or perennial herbs or shrubs. Leaves basal or alternate; stipules absent. Flowers in cymes, racemes, spikes or solitary or few in the leaf axils, bisexual, irregular. Sepals 3–5. Corolla tubular often slit open to the base posteriorly; lobes 5, valvate, usually with induplicate winged margins, often hairy on the back except the glabrous wings. Stamens 5, hypogynous or epigynous; anthers free or connate around the style. Ovary superior or inferior, 1–2-locular; stigmatic surface enclosed within a cup-like indusium (Fig. 25), growing out when mature. Fruit a capsule nut or drupe. 11 gen., mostly temp. Australia, few throughout trop.

Key to the genera

1 Ovary superior 2
1 Ovary inferior 3

2 Corolla yellow. . . . . .1 VELLEIA
2 Corolla blue . . . . .7 BRUNONIA

3 Anthers connate into a tube around the style. . . . . 6 DAMPIERA
3 Anthers free. 4

4 Corolla yellow. . . . . .3 GOODENIA
4 Corolla blue to purplish, whitish or brownish 5

5 Corolla lobes without broad marginal wings. Prostrate glabrous herb . . . . .4 SELLIERA
5 Corolla lobes with broad marginal wings. Plants usually hairy 6

6 Corolla lobes ± equal, spreading like a fan, no stellate hairs present. . . . . 5 SCAEVOLA
6 Corolla lobes unequal, not spreading like a fan, stellate hairs present but sometimes sparse. . . . .
. . . . .2 COOPERNOOKIA

1 **Velleia** Sm.

21 species in Aust. (1 native, 20 endemic); all states and territories

Herbs with short thick rootstocks. Leaves mostly basal. Flowers in an erect or ascending dichotomous cyme with opposite bracts which are free or connate around the forks. Sepals 3–5. Corolla yellow, oblique, slit on the upper side and with a small protuberance or spur near the base. Stamens free. Ovary incompletely 2-locular, superior; indusium with stiff hairs.

Key to the species

1 Sepals 5; the outer one longer and broader than the others, up to 12 mm long. Corolla 12–18 mm long. Flowering stems 15–45 cm long. Bracts free; the lower ones lobed. Basal leaves petiolate, broad-ovate to narrow-oblong, 5–10 cm long, dentate or entire. Cumberland Plain. Open forests and grasslands on WS. Fl. spring. . . . . **V. paradoxa** R.Br.
1 Sepals 3 2

2 Bracts connate, large, entire or toothed. Basal leaves petiolate, obovate or oblong, toothed or entire. Corolla c. 12 mm long. Flowering stems mostly up to 50 cm long. Hawkesbury River district. Endangered. Fl. spring. . . . . **V. perfoliata** R.Br.
2 Bracts almost or quite free from one another 3

3 Flowering stems shorter than the leaves. Corolla 8–10 mm long. Leaves 3–4 cm long, spathulate, entire or nearly so, rather thick. Blue Mts, e.g. Blackheath. Open communities. Fl. spring . . . .
 . . . . **V. montana** Hook.f.
3 Flowering stems longer than the leaves 4

4 Sepals broad-ovate-cordate. Flowering stems erect or ascending, up to 30 cm long, Leaves lyrate-pinnatifid to dentate towards the base, up to 10 cm long. Corolla 10–12 mm long. Coast and adjacent plateaus. Moist sandy soil. Fl. spring–summer . . . . **V. lyrata** R.Br.
4 Sepals lanceolate to nearly ovate or oblong-lanceolate. Flowering stems weak, procumbent to ascending, 3–20 cm long. Corolla 8–10 mm long. Leaves spathulate, entire or with distant teeth, 2–12 cm long. Coast and Hornsby Plateau. Moist soil. Fl. summer . . . . **V. spathulata** R.Br.

2 **Coopernookia** Carolin

6 species endemic Aust.; Qld, NSW, Vic., S.A., WA

Herbs, ± shrubby towards the base. Flowers mostly solitary between 2 bracteoles in terminal leafy racemes. Sepals 5. Corolla pinkish, purple, bluish or white; the superior lobes arching around the indusium; lobes winged. Stamens free. Ovary inferior, incompletely 2-locular with 2 ovules. Fruit a capsule. Seed ovoid, with a caruncle.

Key to the species

1 Leaves linear to linear-lanceolate, glandular hairy, with a few stellate hairs, up to 30 mm long and 5 mm wide, ± revolute. Corolla bluish to pink-purple, 11–14 mm long. Erect, shrubby herb up to 60 cm high. Widespread on lower Blue Mts DSF. Fl. most of the year . . . . **C. barbata** (R.Br.) Carolin
1 Leaves narrow-elliptic to elliptic, 40–90 mm long, 10–25 mm wide, stellate hairs obscuring the glandular hairs; margins almost flat. Erect, shrubby herb up to 80 cm high. Hunter River Valley. WSF. Fl. spring–summer. . . . . **C. chisholmii** (Blakely) Carolin

3 **Goodenia** Sm.

178 species endemic Aust.; all states and territories

Herbs or sometimes shrubs. Leaves basal or alternate. Flowers axillary or in terminal thyrses racemes or spikes. Sepals 5. Corolla yellow or rarely purplish; the 3 anterior lobes sub-erect or reflexed and the 2 posterior ones arching over the indusium. Stamens free. Ovary semi- or wholly inferior, incompletely 2-locular; lip of the indusium covered with stiff hairs (Fig. 25). Fruit a capsule. Seeds compressed.

Key to the species

1 Stellate hairs present on the outer surface of the corolla. 2

1 Stellate hairs not present on the outer surface of the corolla .7

2 Leaves mostly basal, often in a distinct rosette 3

2 Leaves mostly cauline 5

3 Flowers in a spike or spike-like thyrse. Leaves linear to narrow-oblanceolate, entire or distantly and irregularly toothed, thick, 1–15 cm long. Corolla 14–16 mm long. Ovary villous. Capsule ovoid-oblong, 6–8 mm long. Perennial herb. Widespread. Wet soil. Fl. spring . . . . **G. stelligera** R.Br.

3 Flowers in a spreading thyrse or raceme. Capsule linear-oblong, c. 10 mm long. Corolla 12–14 mm long . . 4

4 Leaves oblong to ovate-spathulate, 2–5 cm long, glabrous. Blue Mts; Hornsby Plateau S. of Gosford. Damp sandy soil. Fl. spring–summer . . . . **G. dimorpha** Maiden & Betche var. **dimorpha**

4 Leaves linear to linear-lanceolate, entire or obscurely dentate, 2–4 cm long, glabrous. Woronora Plateau. Damp sandy soil. Fl. spring–summer . . . . **G. dimorpha** var. **angustifolia** Maiden & Betche

5 Leaves decurrent on the stem, oblong, glabrous, 5–10 cm long, coarsely serrate. Flowers in a terminal bracteate thyrse or raceme on the stiff erect stems. Corolla 14–18 mm long. Erect shrubby herb up to 80 cm high. Blue Mts; lower Hawkesbury River district. DSF. Ss cliffs or amongst rocks. Fl. spring–summer . . . . **G. decurrens** R.Br.

5 Leaves not decurrent on the stem 6

6 Ovary glabrous or nearly so. Leaves obovate to narrow-obovate, 3.5–7 cm long, usually obscurely dentate, obtuse, glabrous. Corolla 14–18 mm long. Flowers in a loose raceme or thyrse. Ascending or decumbent, straggling shrubby herb. Blue Mts, Leura to Lawson. Damp south facing Ss cliffs. Fl. spring–summer. . . . . **G. rostrivalvis** Domin

6 Ovary villous. Leaves obovate to narrow-obovate, up to 8 cm long, distinctly dentate, acute, villous at least in the young stages. Corolla up to 20 mm long. Flowers in a terminal compact thyrse or raceme. Perennial with stems up to 60 cm high. Southern Blue Mts DSF, often amongst Ss rocks in seepages. Fl. spring–summer. . . . . **G. glomerata** Maiden & Betche

7 Bracteoles very close under the ovary; peduncles long but pedicel (above the bracteoles) 1 mm long or obsolete. Leaves orbicular, ovate-elliptic or obovate, 8–20 mm long, 5–20 mm wide, petiolate, dentate to crenate, pubescent to hirsute with mostly simple hairs. Prostrate to decumbent herb. Hunter River Valley. Open forests. Fl. spring–summer . . . . **G. rotundifolia** R.Br.

7 Bracteoles well below the ovary or absent 8

8 Bracteoles absent. Corolla glabrous outside but with a short beard inside, 10–14 mm long; lobes often reflexed. Leaves obovate to oblanceolate, 3–8 cm long, crenate-dentate or lobed, obtuse, pubescent or glabrescent. Erect herb up to 50 cm high. Glen Davis. Grasslands and open forests. Fl. spring–summer . . . .**G. pinnatifida** Schldl.

8 Bracteoles present 9

9 Leaves mostly basal **10**
9 Leaves mostly cauline **11**

10 Inflorescence narrow and spike-like, villous. Leaves obovate to oblong-spathulate, tapering at the base into a petiole, irregularly dentate or entire, 1–8 cm long. Flowers sessile or nearly so. Corolla 10–12 mm long. Capsule ovoid, 4–5 mm long. Perennial herb up to 70 cm high. Widespread. Heath and open forests. Fl. most of the year . . . . **G. bellidifolia** Sm. ssp. **bellidifolia**
10 Inflorescence spreading and panicle-like, glandular hairy. Corolla 8–10 mm long. Leaves obovate to narrow-spathulate, 2–10 cm long, irregularly dentate, pubescent to almost glabrous. Capsule ellipsoid or ovoid, 3–5 mm long. Erect herb up to 50 cm high. Widespread. Wet soil. Fl. summer . . . .
. . . . **G. paniculata** Sm.

11 Prostrate herb. Leaves obovate to orbicular, entire or toothed, 1–2 cm long, with webby hairs underneath at least when young. Flowers in a leafy thyrse or raceme. Corolla 10–15 cm long. Widespread. DSF. Fl. spring–summer . . . . **G. hederacea** Sm. ssp. **hederacea**
11 Erect to ascending herbs or shrubs **12**

12 Leaves apparently glabrous (*G. ovata* has minute peltate hairs invisible to the naked eye), either decurrent on the stem or viscid **13**
12 Leaves variously hairy (very rarely glabrous in *G. heterophylla* but then neither decurrent on the stem nor viscid) **14**

13 Leaves decurrent on the stem . . . . **G. decurrens** (see above)
13 Leaves not decurrent on the stem, ovate to narrow-ovate, dentate, 2–5 cm long. Flowers in leafy thyrses or racemes; the partial inflorescences shorter than the leaves. Corolla 10–15 mm long. Capsule narrow-cylindrical, 10–12 mm long. Erect shrubs up to 2 m high, usually viscid on the young parts. Widespread. Usually heavier soils in open forests, cliffs and headlands. Fl. spring–summer . . . .
. . . . **G. ovata** Sm.

14 Bracteoles close to the axil; peduncle below the bracteoles up to 3 mm long or obsolete; pedicel above the bracteoles long. Leaves covered with short glandular hairs, ovate to narrow-ovate, 2–5 cm long. Flowers in terminal leafy thyrses. Corolla 10–15 mm long, yellow streaked with purple. Capsule ovoid-oblong, 10–12 mm long. Shrubby herb up to 1.5 m high. Lower Blue Mts DSF amongst rocks. Uncommon. Fl. summer . . . . **G. grandiflora** Sims
14 Bracteoles inserted away from the axil; pedicel c. as long as the peduncle. **15**

15 Corolla pouch distinct, Leaves narrow-ovate to lanceolate or linear, almost sessile, irregularly dentate, 3–5 cm long. Flowers in terminal leafy racemes. Erect herb, ± woody at the base. Gosford district; Hunter River Valley. DSF. Fl. summer . . . . **G. stephensonii** F.Muell.
15 Corolla pouch not distinct. Shrubby herbs up to c. 60 cm high. **16**

16 Herbs with simple and glandular hairs. Leaves usually ovate sometimes linear, up to 3 cm long, 3–8 mm wide; margins toothed or lobed with acute scarcely acuminate teeth, usually with 2 larger lobes or teeth near the base very rarely entire, sometimes recurved. Widespread. Open forests. Soils derived from sandstone and sometimes in clay soils. Fl. spring–summer. . . . .
. . . . **G. heterophylla** Sm. ssp. **heterophylla**
16 Herbs with simple hairs, sometimes glabrescent **17**

17 Leaves covered with soft simple hairs, ovate, up to 3 cm long, 3–8 mm wide; margins toothed, each tooth mucronate, recurved, sometimes with 2 basal lobes. Open forests. Coast. Soils derived from deep sands or Permian conglomerates. Fl. spring–summer . . . . **G. heterophylla** ssp. **eglandulosa** Carolin
17 Leaves with villous to cottony hairs when young, scabrous with short simple hairs when mature, linear to narrow-oblong, 1.5–2.5 cm long, mostly 2–3 mm wide; margins entire or nearly so, revolute. Blue Mts Open forests. Fl. spring–summer . . . . **G. heterophylla** ssp. **montana** Carolin

4 Selliera Cav.

1 species native Aust.; NSW, Vic., Tas., S.A.

One species in the area

Glabrous, creeping, perennial herb rooting at the nodes. Leaves linear-spathulate to ovate, entire, shining, thick, mostly 2–8 cm long. Flowers solitary or several together on short axillary peduncles. Sepals 5. Corolla white, brownish outside, 6–8 mm long, slit to the base on the posterior side; the lobes ± erect, without a membranous wing or almost so. Stamens free. Ovary inferior, 2-locular. Fruit an ovoid berry, 4–5 mm long. Seeds very sticky in water. Coast. Margins of saltmarshes. Fl. summer . . **S. radicans** Cav.

5 Scaevola L.

71 species in Aust. (70 native); all states and territories except Tas.

Herbs, scarcely shrubby. Flowers sessile or pedunculate between 2 bracteoles, in leafy thyrses or racemes. Sepals 5. Corolla oblique, slit to the base on the posterior side and open like a fan; tube ± hairy inside. Stamens free. Ovary inferior, 2-locular with 1 ovule per loculus or 1-locular with 2 ovules; indusium lip covered with stiff hairs. Fruit drupaceous, often rather dry.

Key to the species

1 Flowers conspicuously pedunculate, in racemes or thyrses 2
1 Flowers sessile or nearly so, in spikes 3

2 Corolla 18–25 mm long, purplish to blue; the lobes with broad marginal wings. Peduncles sometimes with more than 1 flower; bracteoles linear. Ascending, rather straggling plants, scabrous, with few branches. Leaves linear to lanceolate, sessile, 20–50 mm long. Widespread. Heath and DSF. Sandy soils. Fl. most of the year . . . . **S. ramosissima** (Sm.) K.Krause
2 Corolla 3–5 mm long, blue to whitish. Peduncles 1-flowered. Bracteoles elliptic to oblong, 4–6 mm long. Leaves obovate to lanceolate or spathulate, irregularly toothed, 5–12 mm long. Prostrate, creeping, hirsute perennial rooting at the nodes. Blue Mts Damp places on sandy soils. Fl. summer. . . . .
. . . . **S. hookeri** (de Vriese) F.Muell. ex Hook.f.

3 Prostrate or decumbent perennial, often forming large mats. Leaves entire or with a few teeth towards the apex, thick, mostly 25–40 mm long, obovate to oblong-spathulate. Corolla bright blue, 15 mm long. Fruit a purplish fleshy drupe, 5 mm diam. Coastal sand dunes. Fl. most of the year . . . .
. . . . **S. calendulacea** (Andrews) Druce
3 Ascending or decumbent perennials. Leaves coarsely toothed. Corolla blue. Fruit a 1- or 2-seeded dry drupe up to 3 mm diam 4

4 Indusium with a number of longer often purplish hairs at the base. Leaves 10–50 mm long, ovate to elliptic. Mostly coast in the area. DSF. Fl. spring–summer . . . . **S. aemula** R.Br.
4 Indusium ± uniformly hairy with short silvery hairs. Leaves 3–35 mm long, ovate to narrow-lanceolate. Widespread. Open forests. Fl. spring–summer . . . . **S. albida** (Sm.) Druce ssp. **albida**

6 Dampiera R.Br.

66 species endemic Aust.; all states and territories

Small, erect shrubs often almost herbaceous. Flowers solitary or few together on axillary or terminal peduncles. Sepals almost obsolete. Corolla blue or purplish rarely pink or white; posterior lobes with auricles which enclose the indusium. Anthers connate. Ovary 1-locular with 1 erect ovule; lips of the indusium glabrous (Fig. 25). Fruit a small nut.

Key to the species

1 Branches triangular, glabrous or nearly so. Leaves elliptic to almost linear, entire or with a few coarse teeth, 1–3 cm long, glabrous. Corolla 10–14 mm long, with short appressed rusty hairs outside. Widespread. Heath and DSF. Sandy soils. Fl. winter–summer . . . . **D. stricta** (Sm.) R.Br.

1 Branches terete, sometimes ribbed. 2

2 Corolla 12–15 mm, long covered with dense dark grey woolly hairs outside. Hairs on the short flowering branches usually brown. Leaves stellate tomentose underneath, becoming scabrous on the upper surface, elliptic to nearly orbicular, 1–6 cm long, entire or toothed. Shrub with several stems, up to 1 m high. Widespread. Open forests. Sandy soils. Fl. spring–summer . . . . **D. purpurea** R.Br.

2 Corolla up to 12 mm long covered with fine silvery hairs outside. Hairs on the short flowering stems pale grey to silvery. Leaves with pale grey hairs underneath, glabrous on the upper surface, linear to elliptic, 9–30 mm long, entire or with a few teeth. Shrub with several stems, up to 60 cm high. Hornsby Plateau south of the Hawkesbury River. Uncommon. Open forests and heath. Sandy soils. Fl. spring–summer
. . . . **D. scottiana** F.Muell.

7 Brunonia Sm. ex R.Br.
1 species endemic Aust.; all states and territories

Monotypic genus

Tufted, perennial herb up to 50 cm high, clothed with long silky hairs. Leaves mostly basal, obovate to linear-cuneate, entire, mucronate, mostly 5–10 cm long. Flowers numerous, intermixed with bracts, in a dense hemispherical head on a leafless unbranched scape surrounded by an involucre of bracts. Sepals 5, ciliate. Corolla tubular, ± regular, blue; lobes 5, 3–4 mm long, oblong, nearly as long as the tube. Ovary 1-locular, superior. Fruit a nut enclosed within the persistent hairy calyx. Blue Mts, e.g. Blackheath, Bell. Open forests. Fl. summer . . . . **B. australis** Sm. ex R.Br.

146 MENYANTHACEAE

Aquatic or marsh herbs. Leaves simple, basal or alternate except sometimes on the flowering stems; stipules absent. Flowers regular, bisexual, in cymes. Calyx tubular; lobes 5, imbricate. Corolla tubular; lobes 5, valvate, winged. Stamens 5, alternating with the corolla lobes, epipetalous. Ovary superior or half-inferior, 1-locular, with parietal placentas; style simple. Fruit a capsule; seeds numerous. 5 gen., temp. to trop.

Key to the genera

1 Erect herb with at least some emergent leaves . . . . **1 VILLARSIA**

1 Leaves floating or plant prostrate on mud. . . . . **2 NYMPHOIDES**

1 Villarsia Vent.
13 species endemic Aust.; all states and territories except NT

Perennial aquatic usually stoloniferous herbs with emergent leaves. Leaves basal or alternate, on long petioles, ovate to reniform or orbicular, entire or nearly so. Flowers in cymo-panicles. Corolla yellow; lobes spreading, copiously bearded or fringed inside at the base. Ovary half-inferior or almost superior. Capsule 4-valved.

Key to the species

1 Basal leaves erect, never floating, longer than broad, ovate to broad-ovate but when broad-ovate then not cordate at the base, rounded attenuate or slightly cordate at the base, 4–12 cm long 2–8 cm wide; upper surface matt, slightly darker than the undersurface or the same colour. Flowers heterostylous. Capsule adnate to the calyx lobes in lower half or third. Tufted, perennial herb up to 1.5 m high. Coast and Cumberland Plain. Swamps or pools or slow moving water. Fl. spring–summer . . . . **V. exaltata** (Sol. ex Sims) G.Don

1 Basal leaves usually floating, at least in deepest water, ± as long as broad, broad-ovate to orbicular or depressed-ovate (reniform), deeply cordate at the base; upper surface darker than undersurface, glossy. Flowers homostylous. Capsule adnate to the calyx lobes only at the very base. Perennial herb up to 1 m high. Widespread south of Sydney. Swamps or pools or slow moving water. Fl. spring–summer . . . . **V. reniformis** R.Br.

2 Nymphoides Ség.
19 species in Aust. (17 endemic); all states and territories

Aquatic herbs. Leaves floating or on mud, broad-reniform or orbicular, entire or crenate. Flowers on long pedicels arising from weak peduncles, in pairs or clusters. Corolla yellow or white, campanulate or almost rotate, deeply 5-lobed; lobes with broad membranous margins (wings). Ovary 1-locular. Fruit often indehiscent.

Key to the species

1 Corolla white, sometimes yellow or orange in the tube, 25–35 mm diam., bearded in the throat; lobes hairy inside, fringed on the margin. Leaves solitary or two on each stem, orbicular-cordate, up to 40 cm wide. Mostly coastal. Uncommon in scattered localities. Ponds or slow moving streams. Fl. spring–autumn. *Water Snowflake*. . . . . **N. indica** (L.) Kuntze

1 Corolla yellow. 2

2 Corolla lobes crested along the middle to a broad fringed membrane, fringed on the margins and bearded inside at the base with a few long hairs. Leaves crenate, glandular-dotted underneath. Uncommon. Coast and adjacent plateaus. Ponds and margins of rivers. Fl. spring–autumn. . . . . **N. crenata** (F.Muell.) Kuntze

2 Corolla lobes without a crested membrane along the middle. Leaves entire or very slightly crenate . . 3

3 Corolla lobes fringed on the margins, with a transverse fringe at the base. Seeds smooth, ellipsoid, compressed. Southern Blue Mts Ponds. Fl. spring–summer . . . . **N. montana** Aston

3 Corolla lobes scarcely fringed on the margin or near the base. Seeds ± globular, sculptured. Coast and adjacent plateaus. Ponds and margins of rivers. Fl. spring–autumn . . . . . **N. geminata** (R.Br.) Kuntze

147 ROUSSEACEAE
4 gen., trop, subtrop, temp., Mauritius and scattered from New Guinea to New Zealand

One genus in the area

1 Abrophyllum Hook.f. ex Benth.
1 species endemic Aust.; Qld, NSW

One species in the area

Spreading shrub or tree. Leaves alternate, thin, elliptic or ovate to lanceolate, 10–20 cm long, 4–10 cm wide, acuminate, with a few short teeth in the upper part. Flowers in terminal or axillary corymbose panicles which are irregularly dichotomous and much shorter than the leaves. Disc very short, adnate to the broad base of the ovary. Calyx lobes 5, spreading, deciduous. Petals 5, yellowish, 4 mm long, valvate, deciduous. Stamens 5. Ovary superior, 5-locular. Style 5-lobed. Berry globular, 6–8 mm diam. Widespread. RF. and sheltered gullies. Fl. summer. *Native Hydranger* . . . .

. . . . **A. ornans** (F.Muell.) Hook.f. ex Benth.

148 STYLIDIACEAE

Herbs or small shrubs. Leaves basal or cauline, fasciculate or spirally arranged; stipules absent. Flowers usually irregular, in terminal racemes or cymes. Calyx 5-lobed, variously connate into 2 lips. Petals 5, connate into a short tube, imbricate, the lower lobes usually smaller. Stamens 2, adnate to the style forming a column (*gynostemium*); anther cells finally divaricate and exposing the stigma between them. Ovary inferior, incompletely 2-locular; placentas axile or free-central. Fruit a 2-valved capsule. 5 gen., mostly temp. S. hemisphere.

Key to the genera

1 Column almost erect, not sensitive. Corolla whitish. . . . . **1 LEVENHOOKIA**
1 Column bent, with a trigger action when touched. Corolla usually red or pink . . . . . **2 STYLIDIUM**

1 Levenhookia R.Br.
Styleworts
8 species endemic Aust.; NSW, Vic., Tas., S.A., WA

One species in the area

Small, annual, glandular-hairy herb 2–6 cm high. Leaves obovate to elliptic, 3–4 mm long, spirally arranged. Flowers 2–3 mm diam., in short terminal corymbs. Sepals narrow-triangular, free. Petals white with a yellowish throat, one ± smaller than the others and forming a hood over the ± erect column. Ovary globular; floral tube glandular hairy. Wolgan and Capertee valleys. Damp ground in open communities. Fl. spring–summer . . . . **L. dubia** Sond.

2 Stylidium Sw.
Trigger Plants or *Springbacks*
180 species in Aust., all states and territories

Herbs or small shrubs. Leaves basal tufted or cauline and spirally arranged, linear to obovate. Calyx lobes 5, ± united into 2 lips. Corolla tubular, deeply 5-lobed; the lower lobe much smaller and usually turned down. Column elongate and bent, sensitive with a trigger-like action when touched.

Key to the species

1 Leaves spathulate to obovate, 12–40 mm long, 4–8 mm wide, basal or mostly tufted just below the scape(s). Corolla c. 4 mm long, pink. Calyx lobes free or the two posterior ones ± connate. Ovary linear-obconic. Flowers in a raceme. Short-lived herb with scapes up to 40 cm high. Coast and adjacent plateaus. Wet ground. Fl. spring–summer . . . . **S. debile** F.Muell. var. **debile**
1 Leaves linear to linear-lanceolate. 2

2 Leaves tufted, basal or at the top of a short stem or scattered and in tufts along a sprawling stem. . . . 3
2 Leaves cauline, spirally arranged. 5

3 Leaves linear-subulate, c. 1 mm wide, 2–5 cm long, often incurved. Corolla 5–8 mm long, pink to deep pink. Ovary narrow-cylindrical to narrow-obovoid. Perennial herb with reddish scapes up to 20 cm high. Widespread. Heath and DSF. Fl. spring–summer . . . . **S. lineare** Sw. ex Willd.

3 Leaves linear to narrow-lanceolate, almost flat, mostly more than 2 mm wide, 8–20 cm long, entire or denticulate. Corolla 5–10 mm long. Ovary ovoid to subglobular. 4

4 Stems short; the scapes arising up to 1 m high from a basal tuft of leaves. Leaves without a prolongation at the base; stomata in bands on both ad- and abaxial surfaces. Labellum narrowing to an obtuse tip. Widespread. Heath and DSF. Fl. spring–summer . . . . **S. graminifolium** Sw. ex Willd.

4 Stems sprawling and elongating up to 1 m long with leaves scattered along the stem but tufted at intervals particularly below the scape, with a basal abaxial prolongation; stomata in two bands on the abaxial surface only. Labellum broad-obovate. Widespread. Heath and DSF. Only on sandy soils. Fl. summer . . . . **S. productum** Hind. & Blaxell

5 Leaves less than 4 mm long and 1 mm wide, thick, narrow-lanceolate, few (usually less than 15) on the stem. Flowers in a raceme. Corolla c. 2 mm long, white to pale pink. Ovary linear. Annual herb up to 15 cm high, ± fleshy. Coast. Damp places. Fl. spring–summer. . . . . **S. despectum** R.Br.

5 Leaves more than 10 mm long, c. 1 mm wide, linear, with many on the stem. Flowers in a lax thyrse or raceme. Corolla pale pink, 10–15 mm long. Ovary ovoid to subglobular. Shrubby perennial with stems up to 50 cm high. Widespread. Open forests. Fl. spring. . . . . **S. laricifolium** Rich.

149 ADOXACEAE

Small trees or shrubs, branches often with lenticels. Leave opposite, simple or pinnate. Flowers in terminal inflorescences, bisexual, regular or irregular. Calyx tubular, 3–5-lobed. Corolla tubular, 3–5-lobed. Stamens inserted in corolla tube, sessile. Style terminal. Ovary inferior, 3–5-locular. Fruit a drupe. 5 gen, temp. and trop.

One genus in the area

1 Sambucus L.
3 species in Aust. (2 endemic, 1 naturalized); Qld, NSW, Vic., Tas., S.A.

Large shrubs or small trees. Leaves imparipinnate, the lowest pair of leaflets sometimes resembling stipules. Flowers small, numerous in large terminal corymbose cymes. Calyx 3–5-toothed. Corolla white or yellow, with a very short tube and 3–5-equally spreading lobes. Ovary 3–5-locular; stigma sessile, 3–5-lobed. Fruit berry-like.

Key to the species

1 Lowest pair of leaflets of each leaf close to the stem, short and resembling leafy stipules. Leaflets 5–11, coarsely and acutely toothed, ovate or ovate-lanceolate, 5–12 cm long. Corolla white. Berry white. Large shrub with herbaceous branches arising from a perennial rootstock. Widespread. WSF and damp places in other open forests. Fl. spring–summer. *White Elderberry* . . . .**S. gaudichaudiana** DC.

1 Leaves without stipule-like leaflets at the base; the lowest pair distant from the branch. Leaflets mostly 3–5 2

2 Flowers mostly 3-merous. Berry yellow. Leaflets narrow-elliptic to oblanceolate, 3–10 cm long, entire serrate or again divided. Corolla yellowish. Shrub or small tree up to 4 m high. Widespread. In or near RF. *Native Elder* . . . .**S. australasica** (Lindl.) Fritsch

2 Flowers mostly 5-merous. Berry black. Leaflets ovate to elliptic, mostly up to 8 cm long, toothed, coriacous. Corolla white. Deciduous shrub up to 5 m high. Blue Mts Occasionally naturalized in moist places. Introd. from Europe. Fl. spring–summer. *Elder* . . . . **S. nigra** L

150 CAPRIFOLIACEAE

Shrubs or woody climbers. Leaves opposite, simple, without stipules. Flowers bisexual, regular or irregular. Calyx tubular, 4–5-lobed. Corolla tubular, 4–5-lobed; lobes spreading, nearly equal or the corolla 2-lipped. Stamens epipetalous, as many as the corolla lobes and alternating with them. Ovary inferior, 2–5-locular; placentas axile or basal. Fruit a berry or drupaceous. 11 gen., mainly N. hemisphere.

Key to the genera

1 Corolla distinctly 2-lipped. Twiner or climber . . . . **1 LONICERA**
1 Corolla ± equally 5-lobed. Erect shrub . . . . **2 LEYCESTERIA**

1 Lonicera L.
2 species naturalized Aust.; Qld, NSW, Vic., SA, WA

One species in the area

Woody twiner, sometimes running along the ground and rooting at the nodes. Branches pubescent when young. Adult leaves ovate or elliptic-oblong to broad-lanceolate, 3–7 cm long. Flowers axillary, in pairs on a short common peduncle, each flower subtended by a leafy bract. Corolla pale yellow tinged with red inside, c. 4 cm long, 2-lipped; upper lip broad, shortly 4-lobed; lower lip narrow; the lips rather shorter than the narrow tube. Stamens 5, exserted. Style exserted. Widespread. Weed in a number of native communities. Introd. from China and Japan. *Japanese Honeysuckle* . . . . ***L. japonica** Thunb.

2 Leycesteria Wall.
1 species naturalized Aust.; NSW, Vic., Tas.

One species in the area

Shrub up to 2 m high with hollow stems. Leaves opposite, ovate to broad-ovate, up to 15 cm long and 8cm wide, acuminate, petiolate, minutely pubescent. Flowers in compact cymes with conspicuous often purplish bracts. Corolla c. 2 cm long, nearly equally 5-lobed, swollen at the base. Fruit a dark red berry. Blue Mts Occasionally naturalized in moist places. Introd. from Asia. Fl. summer. *Himalayan Honeysuckle* . . . . *L. **formosa** Wall. in Roxb.

151 DIPSACACEAE

Perennial or annual herbs or low shrubs. Leaves opposite or whorled; stipules absent. Flowers bisexual, irregular, usually arranged in cymose heads with involucral bracts at the base of the head and each flower subtended by a receptacle bract. Petals connate, imbricate. Stamens 2–4, alternating with the petals, epipetalous. Ovary inferior, with a solitary pendulous ovule. Fruit an inferior nut. 11 gen., cosmop.

Key to the genera

1 Stems armed with prickles . . . .**1 DIPSACUS**
1 Stems without prickles. . . . . **2 SCABIOSA**

1 Dipsacus L.
2 naturalized Aust.; NSW, Vic., Tas., S.A.

One species in the area

Erect herbs up to 2 m high with a thick tap-root and stem armed with prickles. Leaves basal and cauline. Cauline leaves lanceolate, opposite, connate at the base, entire or crenate or dentate. Flower heads conical, pungent. Receptacle bracts acute-acuminate, pungent. Corolla purplish to white. Southern Highlands, e.g. Berrima, Moss Vale. Introd. from Europe. *Teazel* . . . . *D. **fullonum** L. ssp. **fullonum**

2 Scabiosa L.

2 species naturalized Aust.; all states except Qld, NT

One species in the area

Annual, erect herb up to 50 cm high. Leaves lyrato-pinnatifid to coarsely dentate, elliptic to oblanceolate in outline, up to 7cm long and 2 cm wide. Flower heads hemispherical on naked scapes usually up to 20 cm long. Receptacle bracts linear-lanceolate, not pungent. Corolla dark purple to white. Blue Mts Occasionally naturalized along roadsides. Introd. from the Mediterranean. *Pincushion*. . . . .
. . . . *S. atropurpurea* L.

152 VALERIANACEAE

17 gen. trop., Mediterranean.

One genus in the area

1 Centranthus Neck. ex Lam. & DC.

2 species naturalized Aust.; all states and territories

One species in the area

Annual or perennial herbs, glabrous, young leaves glaucous. Leaves 3–12 mm long, opposite, narrow elliptic to oblanceolate, petiolate or sessile and stem clasping. Flowers pink to red or sometimes white. Calyx with 10 to 12 lobes. Corolla tube 7–11 mm long. Fruit compressed topped with feather like calyx. Occasional garden escape. Introd. from Europe and Asia. Fl. spring-summer. *Red Valerian* . . . .
. . . . *C. ruber* (L.) DC.

153 ALISMATACEAE

Erect, aquatic or marsh herbs with perennial rootstock. Leaves basal; petioles elongated, sheathing but open at the base; principal veins of the lamina parallel with the margins but converging at the apex. Inflorescence a much branched panicle or raceme. Flowers bisexual or unisexual, regular. Sepals 3, green. Petals 3, white or pinkish. Stamens 6–many. Carpels 6–numerous, free, superior. Fruit an achene. 12 gen., temp. and trop., mostly N. hemisphere.

Key to the genera

1 Flowers unisexual . . . . **3 SAGITTARIA**
1 Flowers bisexual 2

2 Carpels c. 20, narrowed at the base, scarcely beaked, each with 1 ovule . . . . **1 ALISMA**
2 Carpels 6–9, broad at the base, beaked, each with 2 ovules . . . . **2 DAMASONIUM**

1 Alisma L.

2 species in Aust. (1 native, 1 naturalized); NSW, Vic., S.A.

One species in the area

Erect, aquatic herb with a perennial rootstock, Leaves basal, oblong to cordate-ovate, 7–15 cm long, 5–9-veined, on long petioles. Flowers bisexual, c. 5 mm diam., in a large loose erect panicle up to 150 cm high with whorled branches and pedicels. Corolla white or pinkish. Carpels very small, c. 20, rounded on the back and summit. Coast and Cumberland Plain. Along watercourses and on margins of pools. Fl. summer. Cosmop.. *Water Plantain*. . . . . **A. plantago-aquatica** L.

2 Damasonium Mill.

1 species endemic Aust.; all states and territories

One species in the area

Erect, aquatic herb with a perennial rootstock. Leaves lanceolate to cordate-ovate, 2–5 cm long, 3–5-veined. Flowers bisexual, c. 3 mm diam., in a loose erect panicle 20–50 cm high. Carpels 6–9, each tapering to a beak, radiating like a star. Widespread. Wet pastures and margins of pools. Fl. summer. *Star Fruit* . . . .**D. minus** (R.Br.) Buchenau

3 Sagittaria L.

4 species naturalized Aust.; Qld, N.SW., Vic., S.A.

Tufted, aquatic herbs, usually perennial. Flowers unisexual, monoecious, in whorls; the males in the upper whorls; the females in the lower whorls. Stamens numerous. Carpels numerous, spirally arranged, compressed. Fruit an achene.

Key to the species

1 Leaves saggitate, up to 20 cm long; the sheath with obtuse auricles. Flowers 10–15 mm diam. Erect emergent aquatic herb up to 1 m high. Centennial Park. Introd. from S. America. Fl. spring–summer . . . .***S. montevidensis** Cham. & Schltdl.

1 Leaves elliptic, up to 20 cm long; the sheath acuminate towards the top. Flowers 10–15 mm diam. Erect emergent aquatic up to 50 cm high. Georges River. Introd. from N. America. Fl. spring–summer . . . .***S. platyphylla** (Engelm.) J.G.Sm.

154 APONOGETONACEAE

1 gen., trop. to temp.

1 Aponogeton L.

5 species in Aust. (4 native, 1 naturalized); all states and territories except Tas.

One species in the area

Aquatic herb with roots and leaves arising from the top of a tuberous rootstock. Leaves floating with long petioles; laminas oblong, up to 30 cm long, with prominent cross-veins. Flowers bisexual, irregular, in a forked, spike-like cyme enclosed at first in a spathe. Perianth segment solitary, up to 1.5 cm long, white to pinkish. Stamens 15–20. Carpels 3–6, free, superior. Fruit a follicle. Still water. Introd. from S. Africa. Fl. winter . . . . * **A. distachyos** L.f.

155 ARACEAE

Perennial glabrous herbs usually with an acrid watery or milky juice. Rhizomes tuberous or elongated. Woody flowering stems present only in *Gymnostachys*. Leaves solitary or few. Flowers minute, crowded into a dense cylindrical spadix which is usually enclosed in a spathe, bisexual or unisexual, usually monoecious with the males in the upper part of the spadix and the females below, sometimes separated by sterile flowers or a bare interval. Perianth segments present in bisexual flowers, mostly absent from the others. Stamens 6 or less. Ovary superior or embedded in the spadix, 1- or 3-locular. Fruit a berry. 110 gen., mostly trop., some temp.

Key to the genera

1 Plants floating. Lettuce-like. . . . . .7 PISTIA
1 Plants rooted in substrate 2

2 Perianth present. Leaves and flowering stems fibrous, tough . . . . .1 GYMNOSTACHYS
2 Perianth absent. Leaves and flowering stems soft, not fibrous .3

3 Small veins between the secondary veins of the leaf parallel. Spathe whitish. . . . **2 ZANTEDESCHIA**
3 Small veins between the secondary veins of the leaf reticulate or confluent 4

4 Male and female zones of the spadix separated by a region of sterile flowers only; no sterile flowers above the males. 5
4 Male and female zones of the spadix separated by a region of sterile flowers and a bare interval, or sterile flowers above and below the males. 6

5 Leaves peltate. Spathe yellow . . . . **4 COLOCASIA**

5 Leaves not peltate (except when immature). Spathe green . . . . **5 ALOCASIA**

6 Spathe pale green, sometimes flecked with purple inside . . . . **6 ARUM**

6 Spathe dark purple inside . . . . **3 TYPHONIUM**

1 Gymnostachys R.Br.

1 species endemic Aust.; Qld, NSW

One species in the area

Erect perennial with tuberous roots. Leaves tough, fibrous, linear, often 1–2 m long, basal. Flowering stems 1–2 m high, flattened or 4-angled, with acute often scariose edges. Spadices with or without a minute spathe, 2–6 clustered together in the axils of leafy bracts in the upper portion of the flowering stem, slender, up to c. 15 cm long. Flowers bisexual, small, sessile, not closely packed. Perianth segments 4, scale-like, in 2 whorls. Stamens 4, opposite the perianth segments. Ovary oblong, 1-locular, with 1 ovule; stigma sessile. Fruit a berry, ovoid or globular, 5–8 mm long. Sheltered gullies and RF. *Settler's Flax.* . . . . **G. anceps** R.Br.

2 Zantedeschia Spreng.

1 species naturalized Aust.; NSW, Vic., Tas., S.A., WA

One species in the area

Succulent, robust, bright green, perennial herb c. 1 m high, with a short stout rhizome. Leaves cordate-saggitate, 15–45 cm long, 10–25 cm wide, pointed at the apex, with obtuse lobes at the base. Peduncles up to 1 m long. Spathe white, yellowish at the base, 10–25 cm long; the blade slightly recurved. Spadix orange-yellow; the female portion c. one quarter as long as the male. Fruit a berry, greenish or yellowish, ovoid, with a persistent stylar vestige. Coast. Wet ground. Introd. from S. Africa. Fl. spring–summer. *Arum Lily* . . . . ***Z. aethiopica** (L.) Spreng.

3 Typhonium Schott

13 species in Aust. (10 endemic, 1 naturalized); Qld, NSW, WA, NT

Perennial, erect, glabrous herbs with a tuberous rhizome. Leaves all basal, 3-lobed to deeply hastate; the small veins between the secondaries reticulate. Peduncle shorter than the petioles. Spathe constricted above the ovoid base; the lamina lanceolate to narrow-lanceolate, greenish outside, deep purplish inside. Spadix with a sterile terminal appendage. Female flowers at the base followed by sterile flowers, a bare interval, and male flowers. Fruit a berry.

Key to the species

1 Sterile terminal appendage of the spadix ± conical, obliquely truncate at the base, 2–5 cm long, black, finally putrescent. Spathe lamina ovate, 5–15 cm long, acuminate. Leaf lobes lanceolate to narrow-ovate, 5–20 cm long. Hunter River valley; Capertee valley. Forests. Fl. spring–summer . . . **T. brownii** Schott

1 Sterile terminal appendage of the spadix terete, not dilated or truncate at the base, 5–15 cm long, purplish. Spathe lamina lanceolate, acuminate. Leaf lobes linear to lanceolate, 5–20 cm long. Coast and adjacent plateaus. Sheltered places. Uncommon. Fl. spring–summer . . . . **T. eliosurum** (F.Muell. ex Benth.) O.D.Evans

4 Colocasia Schott

1 species native Aust.; Qld, WA, NT

One species in the area

Perennial herb up to 1.5 m high, with a stout tuberous rhizome. Leaves ovate, ± cordate, peltate, up to 60 cm long; the principal veins pinnate; the basal veins pedate; the smaller veins reticulate, confluent. Petioles often purplish. Base of the spathe persistent, confluent, oblong, c. 4 cm long; the lamina c. 15 cm long or more. Spadix much shorter than the spathe. Ovary 1-locular, with numerous ovules. Wet places.

Introd. from Oceania and northern Aust. Cultivated in tropical countries for the edible rhizome. *Taro* *C. esculenta* (L.) Schott

5 Alocasia (Schott) G.Don
2 species in Aust. (1 endemic); Qld, NSW

One species in the area

Large, erect, perennial herb with a stout tuberous rhizome which often later becomes an erect stem. Leaves basal, hastate-cordate, up to 90 cm long. Peduncle thick, spongy, 30–100 cm long. Spathe spoon-shaped, 15–20 cm long, greenish. Female flowers at the base of the spadix separated from the males by rudimentary flowers. Stamens usually 2. Ovary 1-locular, with few ovules. Berry ovoid, 4–8 mm long, with 3–4 seeds. Inflorescence strongly and sweetly scented. All parts of the plant contain an intensely acrid poisonous substance; the leaves especially causing severe pain and sometimes death. Coast. Moist forest; often in gardens. *Cunjevoi* or *Spoon "Lily"* . . . . **A. brisbanensis** (F.M.Bailey) Domin

6 Arum L.
1 species naturalized Aust.; NSW, S.A.

One species in the area

Perennial, erect, glabrous herb up to 50 cm high, with small tubers. Leaves hastate to saggitate, 12–20 cm long, veined with white; the smaller veins between the secondary veins reticulate. Peduncle shorter than the petioles. Spathes white-green with purple flecks, constricted above the ovoid base. Spadix with a thick yellow terminal appendage: female flowers at the base followed by sterile flowers, often a bare interval, male flowers, and a further zone of sterile flowers. Perianth absent. Naturalized in forests near Gosford. Introd. from Europe. Fl. spring. . . . . *A. italicum** Mill.

7 Pistia L.
1 species native Aust.; Qld, NT

One species in the area

Floating herb with leaves arranged in a rosette. Leaves alternate, broad-ovate, up to 13 cm long and 8 cm wide, hairy and ribbed. Spathe tubular at base. Spadix 1.5 cm long, white-green. Male zone a whorl of stamens. Female flowers partly fused to spathe. Perianth absent. Fruit a berry. Grows on still or slow moving water. Introd. from N. Aust. *Water Lettuce.* . . . . *P. stratiotes** L.

156 LEMNACEAE

Small, floating or submerged, monoecious, aquatic herbs, consisting of a flat or subglobular, green, often leaf-like cladode. Reproduction is usually vegetative; in the posterior portion, on either side or centrally, is a groove or pouch under the edge, in which arise branches which later become detached and form new plants. Flowers unisexual, very rarely observed. Perianth absent. Stamens 1–2. Ovary 1-locular; style simple. 5 gen., temp. and trop.; fresh water.

Key to the genera

1 Roots absent. . . . . **4 WOLFFIA**
1 Roots present on the undersurface 2

2 Roots never more than 1 on each cladode . . . . **3 LEMNA**
2 Roots variable in number on each cladode, often 2–5, only occasionally fewer or more. 3

3 Cladodes elliptic to obovate, 3–10 mm long, with 7–16 veins; length/breadth ration 1–1.5 . . . .
. . . .**1 SPIRODELA**
3 Cladodes ± round or broad-ovate, 2–5 mm long, with 3–7 veins; length/breadth ration 1.5–2. . . . . . .
. . . .**2 LANDOLTIA**

<h1 style="text-align:center">1 Spirodela Schleid.</h1>

1 species native Aust.; Qld, NSW, Vic., WA, NT

One species in the area

Small, floating, aquatic herbs frequently forming a green carpet on the surface of still waters. Cladodes elliptic to obovate. Roots 5–18 per cladode. Inflorescence minute, in a pocket on the margin of the cladode, with 1 female flower and 2 males enclosed in a sheath. Widespread in still water, ditches, ponds and creeks . . . . **S. polyrhiza** (L.) Schleid.

2 **Landoltia** Les & D.J.Crawford

1 species native Aust.; Qld, NSW, Vic., S.A., WA

One species in the area

Small, floating, aquatic herbs frequently forming a green carpet on the surface of still waters. Cladodes ± round or broad-ovate. Roots 2–11 (occasionally 1) per cladode. Inflorescence minute, in a pocket on the margin of the cladode, with 1 female flower and 2 males enclosed in a sheath. Widespread in still water, ditches, ponds and creeks . . . . **L. punctata** (G.Mey.) Les & D.J.Crawford

3 **Lemna** L.

Duckweeds

5 species in Aust. (4 native, 1 naturalized); all states and territories

Small, floating, aquatic herbs similar to *Spirodela*. but with one root or none on each cladode.

Key to the species

1 Floating on the surface. Cladodes thick, opaque, broad-ovate to elliptic, 1–4 mm long, stalkless. In still water. Widespread but uncommon in the area. Fl. summer . . . . **L. disperma** Hegelm.
1 Floating beneath the surface. Cladodes thin, translucent, narrow-oblong to lanceolate, 2–8 mm long, stalked when mature. Picton Lakes; Eastlakes. In still water . . . . **L. trisulca** L.

4 **Wolffia** Horkel ex Schleid.

3 species in Aust. (3 native, 1 endemic); all states and territories

Very small, floating herbs usually less than 1 mm long. Roots absent.

Key to the species

1 Cladode widest just below the water surface, globular to ovoid, less than twice as deep as wide, up to 0.7 mm long, green at the edge and paler in the centre, usually with less than 25 stomata on the upper surface, flat near the apex, acute at the edges. Widespread. Still water . . . . . **W. globosa** (Roxb.) Hartog & Plas
1 Cladode widest at the water surface, at least twice as deep as wide 2

2 Cladodes green, shining on the upper surface, up to 1 mm long, ± convex, rounded on the edges, with 50–80 stomata on the upper surface. Widespread. Still water, usually in grazing land . . . . **W. australiana** (Benth.) Hartog & Plas
2 Cladodes whitish green in the centre of the upper surface, green towards the edges, up to 0.8 mm long, flat near the apex, acute on the edges, with up to 25 stomata on the upper surface. Coast and Cumberland Plain . . . . **W. angusta** Landolt

157 HYDROCHARITACEAE

Fresh or salt water herbs, partly or wholly submerged. Leaves crowded at the base, or alternate or whorled or opposite and dispersed along elongated stems. Flowers bisexual or unisexual, regular, arranged within a bifid spathe or within 2 opposite bracts; male flowers several together or solitary. Spathes sessile or on long peduncles, sometimes spirally twisted. Perianth segments free when present, in 1 or 2 whorls, 3 in each whorl; the outer ones green, valvate; the inner ones usually petaloid, imbricate. Stamens-1, or

2–numerous. Ovary inferior or superior, 1-locular, with parietal placentas bearing numerous ovules or 1-seeded. Staminodes sometimes present in female flowers and a rudimentary ovary in the males. 18 gen., temp. and trop.

Key to the genera

1 Ovary superior; male flowers with 1 stamen . . . . **7 NAJAS**
1 Ovary inferior; male flowers with more than 1 stamen 2

2 Marine submerged plants with creeping stems. . . . . **1 HALOPHILA**
2 Fresh water plants 3

3 Leaves all basal arising from a rootstock growing on the bottom of a stream or pond 4
3 Leaves cauline, opposite or whorled, on elongated occasionally branched submerged stems with adventitious roots. Flowers unisexual, axillary. 5

4 Leaves linear, ribbon-like. Flowers unisexual; petals in female flowers rudimentary . **2 VALLISNERIA**
4 Leaves elliptic. Flowers bisexual; petals larger than the sepals . . . . **3 OTTELIA**

5 Leaves serrulate, often crisped on the margins. Styles undivided, free to the base. . . . . **6 HYDRILLA**
5 Leaves with entire or minutely denticulate margins. Styles in female flowers lobed or notched 6

6 Leaves in whorls of 3 very rarely 4 . . . . **4 ELODEA**
6 Leaves in whorls of 4–6 . . . . **5 EGERIA**

1 **Halophila** Thouars
Sea Wrack
5 species native Aust.; all states and territories

Submerged marine herbs with a creeping stem. Leaves opposite, oblong to elliptic or oblong-lanceolate, nearly transparent, 2–7 cm long with one central vein an intramarginal vein and very oblique lateral veins; stipules broad. Flowers small, unisexual, within a 2-leaved sessile axillary spathe. Male flowers pedicellate, emerging from the spathe; the female flower sessile, enclosed. Fruit ovoid, membranous, c. 10 mm long. Reproductive parts rarely seen.

Key to the species

1 Leaves glabrous, with 10–14 pairs of lateral veins. Plants dioecious. Spathe 1-flowered. On sand and mud in protected marine habitats . . . . **H. ovalis** (R.Br.) Hook.f.
1 Leaves with fine hairs on the lower surface, with 6–9 pairs of lateral veins. Plants monoecious. Spathe with 1 male and 1 female flower. On sand in protected marine habitats . . . . . . **H. decipiens** Ostenf.

2 **Vallisneria** L.
5 species native Aust.; all states and territories

One species in the area
Submerged, fresh water, stoloniferous, dioecious herb. Leaves basal, tufted, linear, long and ribbon-like, 1–2 cm wide; margin obscurely dentate; nerves up to 9. Male flowers minute, numerous, enclosed in an ovoid spathe, detaching and floating to the surface; petals absent; stamens 2–3. Female flowers solitary, sessile in a tubular spathe on a long peduncle which is spirally coiled unwinding to allow the flower to reach the surface for pollination; sepals 3; petals obsolete. Widespread in fresh water. Fl. summer. *Eel Grass* . . . . **V. gigantea** Graeb

3 Ottelia Pers.

2 species native Aust.; all states and territories except Tas.

One species in the area

Fresh water aquatic herb with a perennial rootstock. Leaves radical; petioles long, dilated at the base; laminas mostly ovate or elliptic, 4–10 cm long, floating on the surface. Flowers bisexual, 3–4 cm diam., solitary, sessile, within a 2-lobed tubular spathe on a long peduncle. Corolla white or pale yellow. Stamens 6–12; anthers linear. Styles 6, 2-lobed. Coast. In ponds. Fl. summer. *Swamp Lily* . . . . **O. ovalifolia** (R.Br.) Rich. ssp. **ovalifolia**

4 Egeria Planch.

1 species naturalized Aust.; Qld, NSW, Vic., NT, WA

One species in the area

Submerged, dioecious herb. Stems elongated, occasionally branching, with adventitious roots. Leaves linear, sessile, in whorls of 4–6, up to 2 cm long, densely crowded towards the apex; margins entire or minutely denticulate. Flowers unisexual, solitary in tubular spathes in the leaf axils. Male flowers pedicellate; petals 3, white; stamens 3–9. Female flowers sessile; styles 3, each 2-lobed or notched, free to the base. Coast and Cumberland Plain. Plants form dense masses in still water. Fl. spring–summer . . . . **E. densa** Planch.

5 Elodea Michx.

1 species naturalized Aust.; NSW, Vic., Tas., WA

One species in the area

Submerged, aquatic herb. Leaves sessile in whorls of 3 or rarely 4, dark green, oblong-lanceolate, up to 10 mm long, c. 3 mm wide, serrulate. Flowers c. 5 mm diam., floating, green-purple, dioecious. Found in deep water. Introd. from N. America. *Canadian Pondweed* . . . . ***E. canadensis** Rich.

6 Hydrilla Rich.

1 species native Aust.; all states and territories except Tas.

One species in the area

Submerged, dioecious herb. Stems elongated, branching, with adventitous roots. Leaves serrulate, often crisped, sessile in whorls of 3–4 or opposite. Flowers unisexual, sessile in the leaf axils, solitary within a spathe. Perianth segments 6, in 2 rows. Stamens 3. Ovary 1-locular, with 2–3 parietal placentas; style 3- or rarely 2-fid. Coast and Cumberland Plain. Plants form dense masses in still water. Fl. spring–summer. *Water Thyme* . . . . **H. verticillata** (L.f.) Royle

7 Najas L.

Water Nymphs

8 species native Aust.; all states and territories except Tas.

Submerged plants growing in fresh or brackish water; dioecious or monoecious. Leaves opposite or apparently whorled, linear, ± toothed. Flowers small, unisexual, solitary, axillary, sessile or nearly so. Flowers unisexual. Male flowers; anther solitary, with 1–4 cells, enclosed in a membranous ± tubular bract or perianth irregularly rolled back. Female flowers with 1 carpel, sessile; style 2–4-branched. Fruit an achene.

Key to the species

1 Leaves prominently sharply toothed so as to appear almost pinnatifid, 12–15 mm long, c. 5 mm wide, dilated at the base into a short sheath without auricles. Carpels mostly with 3 stigmas. Dioecious annual. Coast. Lagoons, in fresh to slightly brackish water. Fl. spring–autumn . . . . **N. marina** L. ssp. **latior** (K.Schum.) Triest

1 Leaves inconspicuously toothed, with auriculate sheaths, up to 5 cm long, 1–3 mm wide. Carpels with 2 or 3 stigmas 2

2 Leaf auricles obtuse up to 0.5 mm long. Seed less than 1.7 mm long. Monoecious annual or short-lived perennial. Port Jackson. Fl. spring–autumn . . . . **N. browniana** Rendle

2 Leaf auricles acuminate, 0.5–1.5 mm long. Seed 1.7 mm or more long. Monoecious or dioecious annual or short-lived perennial. Coast. Fl. spring–autumn . . . . **N. tenuifolia** R.Br.

158 JUNCAGINACEAE

Annual or perennial herbs, aquatic or growing in swamps or marshes. Rhizome sometimes with tuberous roots. Leaves mostly basal, linear, sheathing at the base, sometimes floating. Flowers bisexual, regular, small, on scapes in spikes or racemes; bracts absent. Perianth segments 6, in 2 whorls, greenish or reddish, deciduous. Stamens 3–6, subsessile. Carpels 2–6, free or basally connate or cohering. Fruit cylindrical or obovoid, consisting of free or connate follicles. 4 gen., mainly temp. S. hemisphere.

Key to the genera

1 Carpels 3 or 6, cohering at the apices only. Fruit a follicle. . . . . **1 TRIGLOCHIN**

1 Carpels usually 2 or 4, cohering for their full length. Fruit ± succulent. . . . . **2 MAUNDIA**

1 Triglochin L.

20 species native Aust.; all states and territories

Fresh water or saltmarsh plants, glabrous. Leaves basal, linear. Flowers in spikes or spike-like racemes on a simple leafless scape. Perianth segments usually 6, in 2 whorls, caducous. Anthers 3–6, sessile or nearly so. Stigmas prominent and recurved after flowering. Carpels 3–6. Fruit a cluster of follicles.

Key to the species

1 Rhizomes slender, tuberous roots absent. Basal leaf forming an obtuse ligule. Fertile carpels 3, alternating with 3 persistent barren ones. Leaves 1–3 mm wide, up to 30 cm long. Scape slender, shorter or longer than the leaves, with few to many small flowers. Fertile carpels streaked on the rounded back. Perennial with a creeping rhizome. Coast. Margins of salt marshes. Fl. spring–summer. *Arrow Grass*. .
. . . . **T. striatum** Ruiz & Pav.

1 Rhizomes thick, tuberous roots present. Fertile carpels 3–6 not alternating with barren ones. Basal leaf without ligule 2

2 Tubers 4–13 mm long, globose, clustered closely beneath the rhizomes on short roots. Dorsal surface of mature carpels flat, never keeled. Leaf sheath gradually inrolled touching or overlapping on margins. Leaf erect, thickening and spongy towards base. Terminal raceme 15–24 mm wide, fruit touching, 7–9 per cm of rhachis; fruit base star-shaped. Robust perennial with creeping rhizome. Still or slow flowing water. Fl. spring–autumn . . . . **T. microtuberosum** Aston

2 Tubers elongated, more than 13 mm long, distributed at a distance from the rhizome on long roots. Dorsal surface of mature carpels keeled or ridged 3

3 Leaves submerged, thin, similar on both surfaces, (maybe glossy on stranded specimens), sometimes spiral or margins undulate. Leaf sheath narrow never meeting across the blade, margins tightly inrolled. Terminal raceme 15–30 mm wide, fruit mostly touching, 4–9 per cm of rhachis. Fruit straight or slightly twisted. Robust perennial. Swiftly flowing clear water. Fl. spring–summer. . . . . **T. rheophilum** Aston

3 Leaves emergent or floating, spongy, upper surface glossy. Leaf sheath broad sometimes meeting across the blade, margins gradually inrolled 4

4 Fruit straight, small, globular 3–5 mm long, crowded 14–27 per cm of rhachis. Rhachis maroon. Robust perennial. Irrigation channels, swamps and steams. Fl. spring–autumn . . . . . **T. multifructum** Aston

4 Fruit straight or twisted, 7–15 mm long, globular to ellipsoid, not crowded 3–11 per cm of rhachis. Robust perennial. Widespread. Swamps and slow flowing streams, sometimes in deeper water. Fl. spring–summer. *Water Ribbons* . . . . **T. procerum** R.Br.

2 **Maundia** F.Muell.

1 species endemic Aust.; Qld, NSW

Monotypic genus

Robust, aquatic perennial. Leaves linear, flat, ± succulent, resembling those of *Triglochin procera*. Carpels usually 2–4, cohering for their full length; apex truncate. Ovules pendulous. Fruit cylindrical, ± succulent. Coast. Swamps. Vulnerable . . . . **M. triglochinoides** F.Muell.

159 POSIDONIACEAE

1 gen., widespread on the coasts of warmer seas

1 **Posidonia** K.D.Koenig.

8 species endemic Aust.; NSW, Vic., Tas., S.A., WA

One species in the area

Submerged, marine plant with a creeping rhizome. Leaves linear, up to 1 cm wide, up to 1 m long; leaf bases sheathing, persistent, fibrous, covering the stem. Flowers bisexual, in a compound spike with a long floral leaf at the base of the whole and shorter floral leaves at the bases of each single spike. Perianth absent. Stamens 3; anther lobes dark coloured, separated by an extended ± awned green connective. Ovary superior, 1-locular, sessile; stigma 3–4-lobed. Fruit c. 2 cm long, obliquely ovoid-acuminate, fleshy, indehiscent. Coast, in bays and estuaries below low tide mark. Fl. spring–summer . . **P. australis** Hook.f.

160 POTAMOGETONACEAE

Aquatic perennials or annuals. Stems floating or creeping. Leaves simple, opposite or alternate. Inflorescence a spike; flowers regular, bisexual, perianth absent or 4, sepaloid appendages present. Stamens 2–4. Ovary superior, carpels 2–19. Fruit nut-like or drupe-like. 3 gen., cosmop., fresh water herbs.

Key to the genera

1 Inflorescence always 2-flowered. Carpels 2–19. Mature fruit usually stalked . . . . **1 RUPPIA**
1 Inflorescence usually more than 2-flowered. Carpels 4. Mature fruit ± sessile . . . **2 POTAMOGETON**

1 **Ruppia** L.

4 species native Aust.; all states and territories

Submerged, grass-like plants, sometimes occurring abundantly in brackish water. Leaves opposite or alternate, linear or filiform, mostly up to 15 cm long, dilated and sheathing at the base. Flowers bisexual, very small, few in a short spike or cluster enclosed when young in the sheathing bases of the leaves but finally elongated. Perianth absent. Stamens 2, opposite each other, sessile or nearly so, with 2-lobed anthers. Carpels up to 16 but usually 4, free, becoming stalked. Fruit an achene, often beaked.

Key to the species

1 Peduncle at fruiting stage not coiled or loosely coiled 1–2 times. Flowers submerged. Fruit black, 2–3 mm long, asymmetrical, with stalks c. 1 cm long. Leaves linear, up to 5 cm long, c. 0.5 mm wide, with an acute or obtuse tip. Coast. Estuaries in brackish to very salt water. Fl. spring–autumn . . . .
. . . . **.R. maritima** L.
1 Peduncle at fruiting stage coiled several times. Flowers floating or close to the water surface. Fruit submerged 2

2 Leaves truncate or emarginate, 5–15 cm long, 0.5–0.8 mm wide. Carpels 2–6 usually 4. Peduncle at flowering more than 15 cm long. Fruit brown-black, asymmetric, 3–4 mm long; fruit stalk often reflexed. Coast. Lagoons and estuaries in brackish to very salt water. Fl. spring–autumn . . . .
. . . . **.R. megacarpa** R.Mason

2 Leaves obtuse or acute, 5–10 cm long, usually less than 0.5 mm wide. Carpels 4–12 but usually 6–8. Peduncle at fruiting stage usually more than 10 cm long. Fruit slightly asymmetric, 2–3 mm long. Coast. Lakes and lagoons in brackish to very salt water and usually in water less than 1 mdeep. Fl. spring–autumn . . . . **R. polycarpa** R.Mason

2 Potamogeton L.
Pond Weeds
8 species native Aust.; all states and territories

Freshwater plants submerged or floating, with long stems. Leaves alternate (opposite when under a branch or peduncle), thin when submerged, often leathery when above the water. Flowers bisexual, regular, in dense simple spikes on axillary peduncles. Perianth segments apparently 4, sometimes regarded as appendages of the stamens. Stamens 4, almost sessile, inserted at the base of the perianth segments. Carpels 4, superior, free, ± drupe-like when ripe.

Key to the species

1 Leaves distinctly petiolate; the upper ones with a floating lamina; a few of the lower ones submerged and of a different shape 2
1 Leaves sessile or nearly so, all alike, submerged, membranous 3

2 Floating leaves narrow-oblong to elliptic or lanceolate, 2–3 cm long, 5 mm wide; submerged leaves linear, 3–10 cm long. Spikes narrow, 2–3 cm long, on slender peduncles. Coast. Uncommon. Fl. spring–summer . . . . **P. javanicus** Hassk.
2 Floating leaves cordate to elliptic-ovate, 5–10 cm long. Fruits c. 3 mm long, with 3 dorsal ridges. Spikes dense, 25–50 mm long, on rather stout peduncles. Floating leaves with 15–25 veins; submerged leaves narrow-lanceolate. Widespread. Fl. spring–autumn . . . **P. tricarinatus** F.Muell. & A.Benn. ex A.Benn.

3 Leaves narrow- to broad-ovate or almost orbicular, 1–4 cm long, cordate and stem-clasping, with 5–7 strong longitudinal veins and fainter intermediate veins. Spikes 1–2 cm long; peduncles longer than the leaves. Fruiting carpels scarcely beaked. Nepean River. Fl. spring–summer . . . . **P. perfoliatus** L.
3 Leaves lanceolate or oblong to linear. Spikes ± interrupted. Flowers few. Fruiting carpels distinctly beaked 4

4 Leaves lanceolate to linear-lanceolate, usually 3–6 cm long, with serrate and usually undulate or crisped margins, 3–5-veined, without faint intermediate veins; ligule obtuse. Fruit with a beak c. as long as the fruit. Spikes c. 1 cm long, few-flowered. Widespread. Fl. spring–summer . . . . **P. crispus** L.
4 Leaves almost or quite flat, with entire margins, linear to narrow-oblong. Fruit with a beak shorter than the fruit. 5

5 Leaves up to 1 mm wide; ligule long, obtuse. Spikes interrupted, with the flowers in few-flowered globular groups, 2–5 cm long. Widespread. Fl. spring–aurumn . . . . **P. pectinatus** L.
5 Leaves 3 mm wide or wider; ligule acuminate. Spikes dense, 2–3 cm long. Widespread. Fl. spring–summer . . . . **P. ochreatus** Raoul

161 ZANNICHELLIACEAE
4 gen., cosmop., brackish water or marine plants

1 Zannichellia L.
1 species native Aust.; NSW, S.A.

One species in the area

Monoecious, submerged herb in brackish water. Leaves linear, up to 5 cm long, with a basal stipular sheath; distal leaves apparently in whorls of 3. Roots single or paired at the nodes of the rhizome. Flowers unisexual, usually 1 male and 1 female at each leaf whorl; male flowers with a single stamen; female flowers usually with 4--5 carpels surrounded by a sheath; stigma funnel-shaped. Fruit an achene. Ash Island, Hunter River. Fl. summer. . . . . **Z. palustris** L.

162 ZOSTERACEAE

Submerged, marine, rhizomatous perennials. Leaves distichous, linear with an auriculate ligulate sheath. Flowers unisexual; the males and females in alternate rows forming a flat spike enclosed in a sheath near the base of a floral leaf. Perianth absent. Stamens 2. Carpel 1, with 1 ovule; stigma 2-fid. Fruit an achene.

2 gen., cosmop.

Key to the genera

1 Erect flowering and vegetative stems both with elongated internodes. Rhizome and stem with 4 or more cortical bundles near the middle (in cross-section) . . . . **1 HETEROZOSTERA**
1 Only the erect flowering stems with elongated internodes; no such vegetative stems present. Rhizome and stem with 2 cortical bundles near the middle (in cross-section) . . . . **2 ZOSTERA**

1 Heterozostera (Setch.) Hartog
1 species native Aust.; NSW, Vic., S.A.

Monotypic genus

Submerged rhizomatous marine plant with erect flowering and vegetative stems up to 30 cm high which are deciduous in winter. Leaves with laminas linear, up to 30 cm long; apex obtuse, with a notch; sheaths 2–4 cm long. Marine habitats. Fl. spring–summer . . . . **H. tasmanica** (Martens ex Asch.) Hartog

2 Zostera L.
3 species native Aust.; Qld, NSW, Vic., Tas., S.A.

One species in the area

Submerged marine rhizomatous plants with erect flowering stems up to 50 cm high. Leaves with laminas up to 30 cm long; apex entire, truncate; sheath up to 10 cm long. Marine habitats. Fl. summer. *Eel Grass*
. . . . **Z. capricorni** Asch.

163 ALLIACEAE

Herbs with underground bulbs or rhizome. Flowers regular, bisexual, in terminal cymose umbels with a scariose spathe, often 2-valved. Perianth segments 6, free or connate. Stamens 6, adnate to the perianth at the base. Ovary superior, 3-locular; style simple, capitate or 3-lobed. Fruit a capsule. c. 13 gen., mostly N. temp.

Key to the genera

1 Perianth segments not connate . . . . **1 ALLIUM**
1 Perianth segments connate . . . . **2 NOTHOSCORDUM**

1 Allium L.
8 species naturalized Aust.; NSW; all states and territories except Qld and NT

Herbs. Leaves terete or linear, crowded at the base of a 3-angular or terete stem. Bulbils often present amongst the flowers. Perianth segments 6, free, pale blue to white. Inner filaments sometimes 3-fid. Ovary imperfectly 3-locular, with 2 ovules per loculus. Capsule almost membranous.

Key to the species

1 Scape terete. Stamens exserted; inner filaments 3-fid, with the central branch bearing the anther and c. half as long as laterals. Perianth segments c. 5 mm long. Leaves cylindrical, grooved, hollow, c. 25 cm

long. Umbels bearing both flowers and bulbils or bulbils alone. Spathe usually 1-valved. Waste ground around Sydney but not recently recorded. Introd. from Europe. *Crow Garlic* . . . .*A. **vineale** L.

1 Scape 3-angular. Stamens not exserted; inner filaments entire. Perianth segments 12–18 mm long. Leaves linear, flat, c. 25 cm long. Umbels with flowers only. Spathe 2-valved. Waste ground around Sydney. Introd. from western the Mediterranean. *Triquetrous Garlic* . . . . *A. **triquetrum** L.

2 Nothoscordum Kunth

1 species naturalized Aust.; NSW, Vic., Tas.

One species in the area

Herb with underground bulbils surrounding the bulb. Leaves basal, linear, 20–80 cm long. Umbels with flowers only. Spathe 2-valved. Perianth segments connate into a short tube, white, green at the base, with a dull red midrib, 8–14 mm long. Filaments entire. Ovary 3-locular, with 6–12 ovules per loculus; style filiform, 3-fid. Seeds numerous, black. Garden weed and in waste places. Introd. from N. America. *Onion Weed* . . . . * **N. borbonicum** Kunth

164 AGAPANTHACEAE

Robust rhizomatous perennial herbs. Roots fleshy. Leaves linear, basal, distichous with sheathing bases. Inflorescence an umbel with 2 basal bracts. Flowers bisexual, perianth segments connate at base. Stamens adnate to perianth at base, anthers dorsifixed. Ovary superior, 3–locular, ovules numerous. Stigma capitate. Friut a capsule. 1 gen., Africa.

1 Agapanthus L'Hér.

1 species naturalized Aust.; NSW, Vic., WA

One species in the area

Herb with branching rhizome to 1.2 m high. Leaves 40–80 cm long. Umbel 10–20 cm wide, many-flowered. Flowers blue or white, c. 4 cm long. Capsule 25–35 mm long. Seeds black, winged. Garden escape. Katoomba area. Introd. from Africa. Fl. summer . . . .*A. **praecox** Willd. ssp. **orientalis** (F.M.Leight.) F.M.Leight.

165 ASPARAGACEAE

2 gen., Africa, Europe and Asia

1 Asparagus L.

9 species naturalized Aust.; all states and territories

Rhizomatous shrubs, climbers or scramblers. Stems with prickles or smooth. Leaves reduced to membranous scales subtending flattened or terete cladodes. Flowers regular, bisexual or unisexual, small, solitary, clustered or in racemes. Perianth segments 6, ± equal. Stamens 6. Ovary 3-locular, superior. Fruit a berry.

Key to the species

1 Cladodes terete, c. 5 mm diam. 2
1 Cladodes flat, most more than 1 mm wide 4

2 Flowers unisexual. Stems erect, not spiny. Leaves reduced to membranous scales each subtending 3 unequal filiform cladodes. Cladodes 5–30 mm long. Flowers solitary in the scale axils, greenish white, 4–8 mm diam., on filiform pedicels 6–15 mm long. Fruit a red berry. Introd. from Europe. Garden escape in waste ground near towns. Fl. spring–summer. *Asparagus* . . . .*A. **officinalis** L.
2 Flowers bisexual 3

3 3–6 cladodes at each node. Stems erect, not spiny. Cladodes 3–15 mm long, terete. Flowers axillary, solitary, pedicels 7–12 mm long. Berry orange. Occasionally naturalized. Introd. from S. Africa. Fl. spring–summer. . . . .*A. virgatus Baker

3 8–15 cladodes at each node. Stems scrambling, spines present on older sections. Cladodes 4–7 mm long, terete. Scales ± spine-like. Flowers single or paired. Pedicels 1–2 mm long. Berry black. A weed in bushland. Introd. from S. Africa. *Climbing Asparagus Fern* . . . .*A. plumosus Baker

4 Tepals free. Scales spine-like or with spines in the axils. Cladodes linear, flattened, 15–25 mm long, 2–3 mm wide. Scales with axillary spines. Flowers in racemes. Roots with tubers. Berry orange to red. A weed in bushland near towns. Introd. from S. Africa. Fl. summer. *Asparagus Fern* . . *A. aethiopicus L.

4 Tepals shortly united. Scales membranous, without spines in the axils 5

5 Cladodes single in each axil, ovate to lanceolate, 1–7 cm long, up to 25 mm wide. Anthers orange-red. Berry 6–10 mm diam. Garden escape near towns, sometimes common on coastal dunes. Introd. from S. Africa. Fl. summer . . . . *A. asparagoides (L.) Druce

5 Cladodes 3 in each axil, linear-lanceolate, 5–15 mm long, up to 2 mm wide. Anthers yellow. Berry 5–7 mm diam. Garden escape near towns. Introd. from S. Africa. Fl. summer . . . . . *A. scandens Thunb.

166 AGAVACEAE

Shrubby or coarse, xerophytic herbs. Leaves in tufts, large, basal or at the ends of branches, broad-linear, sometimes fleshy and fibrous. Flowers bisexual, regular, in racemes or panicles which may be head-like. Perianth segments 6, connate into a tube. Stamens 6, adnate to the perianth basally. Ovary superior or inferior, 3-locular, with numerous ovules on axile placentas; style 3-fid. Fruit a capsule. 23 gen., mainly drier trop. and subtrop., few temp.

Key to the genera

1 Stamens longer than perianth segments. . . . .**1 AGAVE**

1 Stamens shorter than perianth segments . . . .**2 FURCRAEA**

1 Agave L.

3 species naturalized Aust.; Qld, NSW

One species in the area

Coarse herb up to 5 m high. Leaves succulent, fibrous, glaucous, up to 1.5 m long, Usually more than 15 cm wide; margins with large spines. Inflorescence a spreading loose panicle on a scape up to 5 m long. Perianth segments white or yellowish. Stamens much longer than the perianth segments. Ovary with a short beak. Usually near habitation. Waste places. Introd. from Mexico and neighbouring countries. *Century Plant* . . . . *A. americana L.

2 Furcraea Vent.

2 species naturalized Aust.; Qld, NSW

One species in the area

Plants to 50 cm high. Leaves fleshy, lanceolate, 1–1.5 m long, 7–16 cm wide, margins entire or toothed, apex with brown spine. Inflorescence 2–6 m high, flowers greenish, white on inside, 3–4 cm long. Bulbils developing on inflorescences branches after flowering. Garden escape on Sydney foreshores. Intod. from America. Fl. autumn . . . .*F. foetida (L.) Haw.

167 ANTHERICACEAE

Herbs with short rhizomes. Leaves mostly basal. Flowers regular, bisexual, in terminal racemes panicles or umbels. Perianth segments 6, free or connate at the base. Stamens 6, 3 sometimes as staminodes, free. Ovary 3-locular; style simple or 3-lobed. Fruit a capsule. c. 20 gen., temp., particularly Australia.

Key to the genera

1 Perianth persistent and twisted after flowering. 2
1 Perianth deciduous or if persistent then not twisted. 4

2 Inner perianth segments fringed . . . . .2 THYSANOTUS
2 Inner perianth segments not fringed. 3

3 Filaments bearded . . . . 6 TRICORYNE
3 Filaments not bearded . . . . 5 CAESIA

4 Flowers in loose panicles or racemes. 5
4 Flowers in compact heads or umbels. 7

5 Filaments bearded . . . . 3 ARTHROPODIUM
5 Filaments not bearded 6

6 Filaments with two small appendages just below the anther . . . . 4 DICHOPOGON
6 Filaments without an appendage. . . . . 1 CHLOROPHYTUM

7 Leaves basal. Functional stamens 3 . . . . .9 SOWERBAEA
7 Leaves cauline. Functional stamens 6 8

8 Inflorescence a terminal head-like umbel. Flowers sessile or shortly pedicellate. . . . 7 LAXMANNIA
8 Inflorescence an axillary condensed raceme. Flowers distinctly pedicellate . . . . 8 ALANIA

1 Chlorophytum Ker Gawl.
2 species in Aust. (1 native, 1 naturalized); Qld, NSW, NT, WA

One species in the area
Erect or lax, perennial herbs c. 60 cm high with short rhizomes. Leaves linear, up to 50 cm long, 18 mm wide. Flowers in panicles. Perianth segments 6, free, white. Stamens 6, free, shorter than the perianth segments. Ovary distinctly 3-angular, 3-locular; style simple. Fruit a 3-angled capsule, truncate, with few (often 1) seeds per loculus. Garden escape into grasslands and Ss. gullies. Introd. from S. Africa. Fl. summer . . . . *C. comosum (Thunb.) Jacques

2 Thysanotus R.Br.
Fringe Lilies
49 species in Aust. (47 endemic, 2 native); all states and territories

Erect or trailing herbs with fibrous or tuberous roots. Leaves basal, linear. Flowers in umbels or panicles, rarely solitary. Perianth segments 6, blue or purple, free, persistent and spirally twisted after flowering; inner segments fringed and broader than the outer ones. Stamens 6, 3 somewhat longer than the others. Ovary 3-locular, with several seeds per loculus; style simple. Fruit a capsule. Seeds black.

Key to the species

1 Roots tuberous. Basal leaves 10–20 cm long. Cauline leaves few, almost scarious, usually subtending a branch or pedicel. Inflorescence much branched, paniculate. Widespread in a variety of habitats. Fl. spring–summer. . . . . T. tuberosus R.Br. ssp. tuberosus
1 Roots not tuberous. Basal leaves absent or short, generally withered. Inflorescences branched once or twice 2

2 At least the upper parts of the stems glabrous. Cauline leaves reduced but herbaceous, often numerous; lower leaves not subtending branches or pedicels. Outer perianth segments 5-nerved, 1.5–2 mm wide. Widespread. Heath and DSF. Chiefly Ss. Fl. spring–summer . T. juncifolius (Salisb.) J.H.Willis & Court
2 Stems densely hirsute on the ridges. Cauline leaves reduced, often numerous. Outer perianth segments 6–7-nerved, 3–4 mm wide. Royal National Park. Heath. Lateritic gravels. Fl. spring–summer . . . . . .
. . . . T. virgatus Brittan

3 Arthropodium R.Br.

7 species endemic Aust.; all states and territories except NT

Erect herbs. Leaves basal, linear, flat. Flowers 2–3 together or solitary on each pedicel, arranged in panicles. Perianth segments free, purplish white, spreading, persistent, not spirally twisted after flowering. Stamens 6; filaments bearded. Ovary 3-locular; style simple. Fruit a capsule.

Key to the species

1 Flowers usually solitary on pedicels lacking an articulation. Leaves 5–35 cm long. Anthers green to white. Filaments bearded almost to the base. Widespread. Fl. spring–summer . . . . . . **A. minus** R.Br.

1 Flowers 2–3 together on pedicels with an articulation. Filaments bearded above the middle 2

2 Anthers white to green. Hairs on filament white. Leaves never glaucous, 3–40 cm long. Perianth 4–6 mm long, pink tinged with green. Widespread in a variety of habitats. Fl. spring–summer . . . . **A.** sp. **B** (*sensu* Harden1993)

2 Anthers purple. Hairs on filament purple. Leaves often glaucous, 3–60 cm long. Perianth 7–10 mm long, white, pink or pale blue. Widespread in a variety of habitats. Fl. spring–summer . . . . **A. milleflorum** (DC.) J.F.Macbr.

4 Dichopogon Kunth

6 species in Aust. (5 endemic, 1 native); all states and territories except NT

Erect herbs up to 1 m high. Leaves flat, basal, 1–3 cm long, 3–5 mm wide. Flowers in loose racemes or panicles, chocolate-scented. Perianth segments 6, blue-violet (rarely white), free, spreading, persistent but not spirally twisted after flowering. Stamens 6, with 2 small appendages immediately below the anthers; filaments glabrous. Ovary 3-locular; style simple. Fruit a globular capsule.

Key to the species

1 Flowers 3–4 together (or occasionally solitary) in the axils of floral bracts on spreading pedicels. Capsules erect. Widespread. Grassland and open forest. Fl. spring–summer. *Chocolate Lily* . . . . **D. fimbriatus** (R.Br.) J.F.Macbr.

1 Flowers solitary in the axils of floral bracts on erect pedicels. Capsules reflexed. Widespread. Grasslands and open forests. Fl. spring–summer . . . . **D. strictus** (R.Br.) Baker

5 Caesia R.Br.

8 species in Aust. (7 endemic , 1 native); all states and territories

Tufted herbs with fibrous, often tuberous roots. Leaves basal, linear. Flowers arranged in irregular clusters on a scape up to 50 cm high. Perianth segments 6, connate basally, blue to white, spirally twisted over the ovary immediately after flowering; inner segments broader and thinner than the outer, ones. Stamens 6, adnate to the perianth basally. Ovary 3-locular with 2 superposed ovules per loculus; style simple. Fruit a truncate 3-angled capsule.

Key to the species

1 Plants usually less than 20 cm high. Leaves 2 mm or less wide. Inflorescence branches spreading. Perianth less than 5 mm long, mostly white, sometimes blue or purple. Uncommon. Moist sites. Endangered. Fl. spring–summer . . . . **C. parviflora** var. **minor** R.J.F.Hend.

1 Plants usually more than 20 cm high. Leaves up to 8 mm wide. Inflorescence branches ascending. Perianth up to 8 mm long, white to blue. 2

2 Perianth white or sometimes pale blue or rarely pink.Filaments greenish white. Leaves to 5 mm wide. Capsule less than 2 mm diam. Widespread in a variety of habitats. Fl. spring–summer. *Pale Grass Lily* . . . . **C. parviflora** R.Br. var. **parviflora**

2 Perianth blue. Filaments blue often with bands of white. Leaves up to 8 mm wide. Capsule 2–3 mm diam. Widespread in a variety of habitats. Fl. spring–summer . . . .**C. parviflora** var. **vittata** (R.Br.) Hend.

6 Tricoryne R.Br.
7 species in Aust. (1 native, 6 endemic); Qld, NSW, Vic., S.A., WA

Herbs. Leaves mostly basal. Flowers in terminal umbels with small scarious bracts of which 2 are often long and leafy. Perianth segments 6, subequal, free, persistent and spirally twisted after flowering. Stamens 6; filaments with a tuft of hairs just below the anthers. Ovary 3-locular; style simple. Fruit a schizocarp separating into three 1-seeded mericarps.

Key to the species

1 Stems with numerous clustered branches, usually over 30 cm high. Leaves shorter than the stems. Umbels 2–8-flowered, numerous on each stem. Perianth yellow to white, c. 2 cm diam. Widespread. Various habitats. Fl. summer . . . . **T. elatior** R.Br.
1 Stems simple or sometimes lightly branched, usually less than 30 cm high. Umbels solitary or 2–3 together, 6–25-flowered. Perianth white to yellow, 2 cm diam. Coast. Various habitats. Fl. summer . . . . **T. simplex** R.Br.

7 Laxmannia R.Br.
13 species endemic Aust.; all states and territories

Perennial herb with fibrous roots. Leaves linear, with broadened leaf bases, usually tufted leaving leafless intervals on the stems. Flowers almost sessile in pedunculate heads surrounded by scariose bracts; inner bracts margins jagged to finely divided. Perianth segments 6, in 2 whorls; outer whorl free, scarious; inner whorl freee or connate basally. white to pink. Stamens 6; inner whorl adnate to perianth. Ovary apparently stipitate; ovules 2–4 per loculus. Fruit a capsule. Flowers nocturnally.

Key to the species

1 Plants erect to 40 cm high. Outer perianth usually shorter than inner. Anthers 0.5–1 mm long. Widespread. DSF. Fl. autumn–spring . . . . **L. gracilis** R.Br.
1 Plants prostrate to 10 cm high. Outer perianth usually equal to or longer than inner perianth. Anthers c. 0.25 mm long. Widespread. Heath. Fl. spring . . . . **L. compacta** Conran & P.I.Forst.

8 Alania Endl.
1 species endemic Aust.; NSW

One species in the area
Perennial herb with branching stems woody at the base, roots fibrous. Leaves dense, linear-filiform, up to 12 cm long, 0.5 mm wide, with broadened scarious bases. Inflorescences umbellate when mature, on slender peduncles. Flowers subtended by scarious bracts, with slender pedicels. Perianth segments 6, cream, free, c. 3 mm long, scarious, persistent but not spirally twisted after flowering. Stamens 6, free, with filiform filaments. Ovary 3-locular; ovules several per loculus. Fruit a globular capsule. Blue Mts; Hornsby Plateau. Gullies on moist rocks. Fl. summer . . . . **A. endlicheri** Kunth

9 Sowerbaea Sm.
5 species endemic Aust.; all states and territories except S.A.

One species in the area
Tufted perennial with fibrous roots. Leaves radical, linear, 20–50 cm long, 1–1.8 mm wide. Flowers in dense terminal umbellate clusters. Perianth segments 6, ± scarious, pink-lilac, free, persistent but not spirally twisted after flowering. Stamens 3; anther lobes linear; 3 staminodes alternating with stamens.

Ovary 3-locular, with 3 ovules per loculus. Fruit a capsule. Widespread. Wet heath. Fl. spring . . . .
. . . . .S. juncea Andrews

168 AMARYLLIDACEAE

Herbs with bulbs. Leaves linear, flat or terete, mostly basal. Flowers in cymose umbels or solitary surrounded by a spathe or spathes, regular, bisexual. Perianth segments 6, in 2 whorls, free or connate. Stamens 6, adnate to the perianth. Ovary 3-locular, inferior; placentas axile. Fruit a capsule. c. 59 gen., trop., mostly cool to warm temp.

Key to the genera

1 Perianth tube c. 60 mm long . . . . .1 CRINUM
1 Perianth tube up to 4 mm long or absent 2

2 Anthers dehiscing through apical pores. Leaves flat .2 LEUCOJUM
2 Anthers dehiscing through slits. Leaves terete at the base, compressed above . . .3 ZEPHYRANTHES

1 Crinum L.

5 species in Aust. (4 endemic, 1 naturalized); all states and territories except Tas.

Perennial or deciduous herbs with underground bulb. Leaves usually basal, linear to broad linear. Inflorescence a cymose umbel. Flowers pedicellate or sessile, with 6 perianth segments variously connate. Stamens 6. Style filiform. Ovary inferior, 3-locular. Fruit globular, indehiscent.

Key to the species

1 Deciduous herb to 1.4 m high neck of underground bulb extending up to 30 cm aboveground. Leaves ≤1 m long, c. 20 cm wide, dying back after flowering. Flowers 5–10 per inflorescence, pink or sometimes white, funnel-shaped, lobes broadly lanceolate c. 10 cm long, c. 4 cm wide. Garden escape. Lower Blue Mts Introd. from S. Africa. *Natal Lily* . . . . *C. moorei Hook.f.
1 Tall erect perennial with underground bulb extending to 45 cm above ground. Leaves broad-linear, up to 75 cm long, 20 cm wide. Flowers 10–40 per inflorescence on pedicels usually longer than the ovary, white; trumpet shaped, lobes narrow 4–8 cm long. Coast. Swampy situations. Fl. summer. *Swamp Lily*
. . . . .C. pedunculatum R.Br.

2 Leucojum L.

1 species naturalized AUST,; NSW, Vic., S.A.

One species in the area

Perennial with underground bulb. Leaves linear, 30–40 cm long. Flowers in cymose umbels. Perianth segments 10–16 mm long, white, tipped with green, connate basally, ovate; tube up to 3 mm long. Waste places. Introd. from the Mediterranean. Fl. winter. *Snowflake* . . . . * L. aestivum L.

3 Zephyranthes Herb.

2 species naturalized Aust.; Qld, NSW

One species in the area

Perennial with underground bulbs. Leaves basal, linear, terete at base, becoming flattened above, up to 35 mm long, 2–4 mm wide. Flowers solitary on erect stems c. 25 cm long. Perianth segments 30–50 mm long, white sometimes tinged with pink, free. Stamens 6, equal. Style with three stigmatic lobes. Fruit a capsule. Coast, e.g. Berry. Damp grasslands . . . . *Z. candida (Herb.) Herb.

169 ASPHODELACEAE

Herbs usually with short rhizomes. Leaves mostly basal. Flowers regular, bisexual, in racemes or panicles. Perianth segments 6, free or nearly so. Stamens 6, free. Ovary 3-locular; style simple. Fruit a capsule. c. 15 gen., warm temp., particularly Africa.

Key to the genera

1 Perianth yellow or red **2**
1 Perianth white or pink sometimes with purplish-brownish or green lines and sometimes with a yellow
spot. **3**

2 Perianth less than 25 mm long . . . . **3 BULBINE**
2 Perianth more than 25 mm long . . . . **4 KNIPHOFIA**

3 Tubular membranous scales present at base of leaves . . . . **2 TRACHYANDRA**
3 Scales not present at base of leaves . . . . **1 ASPHODELUS**

1 Asphodelus L.

1 species naturalized Aust.; all states and territories except NT

One species in the area

Glabrous annual up to 50 cm high. Leaves basal, subulate, up to 30 cm long, striate. Flowers in terminal racemes. Perianth segments pinkish or white, up to 10 mm long with a distinct brownish mid-vein. Anthers versatile, pitted where the filament is inserted. Fruit a capsule. Naturalized in Hunter River Valley and in a few places on the coast. Introd. from the Mediterranean. Fl. summer . . . . *A. fistulosus L.

2 Trachyandra Kunth

1 species naturalized Aust.; NSW, S.A., WA

One species in the area

Perennials up to 70 cm high. Leaves basal, linear, flat, up to 60 cm long and 8 mm wide, glabrous with tubular membranous scales at the base. Flowers in racemes or panicles with membranous bracts. Perianth segments up to 14 mm long, white, with a green line, and a yellow spot near the base, free. Stamens 6, ± unequal. Style filiform; stigma capitate. Fruit a globular capsule, c. 6 mm diam. Coastal sand dunes, e.g. Wollongong. Introd. from S. Africa. Fl. summer . . . . *T. divaricata (Jacq.) Kunth

3 Bulbine Wolf

5 species endemic Aust.; all states and territories

Herbs up to 60 cm high. Leaves succulent, basal, up to 40 cm long. Flowers in racemes, pedicellate. Perianth segments 6, subequal, yellow. Stamens 6; filaments bearded above the middle. Ovary 3-locular; style filiform, simple. Fruit a capsule.

Key to the species

1 Perianth segments usually 5 mm long. 3 outer stamens without bearded filaments. Leaves linear, up
to 15 cm long. Pedicels c. 5 mm long. Herbs with fibrous roots. Illawarra. Damp places, often amongst
rocks. Fl. spring. . . . . **B. semibarbata** (R.Br.) Haw.
1 Perianth segments 8–15 mm long. All stamens with bearded filaments. Leaves linear, up to 40 cm long.
Pedicels 2–4 cm long. Herbs with tuberous roots. Cumberland Plain. Grasslands. Fl. spring . . . .
. . . .**B. bulbosa** (R.Br.) Haw.

4 Kniphofia Moench
Red-hot Pokers
1 species naturalized Aust.; NSW, Vic.

One species in the area

Erect herb, roots rhizomatous. Leaves basal 3, 40–90 cm long, 5–25 mm wide. Inflorescence up to 1 m high, with >50 spreading to pendulous flowers in a dense raceme. Perianth green to orange-red, 3–4 cm long, lobes 2–3 mm long. Stamens 6, free, anthers yellow becoming black. Ovary superior, style filiform, stigma capitate. Capsule 7–14 mm long, seeds numerous. Garden escape. Introd. from Africa. Fl. all year . . . . **K. uvaria** (L.) Hook.f.

170 ASTELIACEAE
5 gen. mostly S. hemisphere

1 Cordyline Comm. ex R.Br.
8 species in Aust. (7 endemic); Qld, NSW

One species in the area

Shrub up to 2 m high with annular scars on the stems. Leaves crowded towards the top of the stems, linear, up to 80 cm long. Flowers in a panicle up to 80 cm long. Perianth segments white to purplish; the outer ones c. 5 mm long, shorter than the inner. Stamens adnate to the perianth segments near the base. Fruit ± succulent when young but becoming dry, tardily dehiscent. Coast north of the Hawkesbury River; Kurrajong. RF. . . . . **C. stricta** (Sims) Endl.

171 BLANDFORDIACEAE
1 gen., Aust.

1 Blandfordia Sm.
Christmas Bells
4 species endemic Aust.; Qld, NSW, Tas.

Erect, perennial herbs with fibrous roots. Leaves mostly crowded at the base of the stem, linear. Flowers in racemes with small bracts and 2 bracteoles at the base of the pedicels. Perianth segments connate into a tube much longer than the 6 lobes, red with yellowish lobes or all yellow. Anthers 6, adnate to the perianth; filaments filiform. Anthers versatile. Ovary 3-locular, superior; stigma 3-grooved, scarcely branched. Ovules numerous. Fruit a capsule with numerous seeds.

Key to the species

1 Leaf margins quite smooth. Flowers 7–18 per raceme. Perianth tube 30–45 mm long, 18–30 mm diam., narrowing near the base. Stamens attached below the middle of the perianth tube. Leaves 4–8 mm wide, up to 100 cm long. Blue Mts Damp places. Fl. summer. . . . . **B. cunninghamii** Lindl.

1 Leaf margins crenulate, rough. Flowers 3–10 per raceme. Perianth lobes obtuse-mucronate. Stamens usually attached to the middle of the perianth tube. Leaves usually less than 5 mm wide, up to 60 cm long. 2

2 Perianth tube ± cylindrical, narrowing abruptly 5–10 mm above the base, 20–35 mm long, 5–10 mm diam., often ± constricted just below the lobes. Widespread south of Gosford. Damp places. Sandy soils. Fl. summer . . . . **B. nobilis** Sm.

2 Perianth tube funnel-shaped, narrowing gradually towards the base, 35–45 mm long, 25–40 mm diam. Coast north of Gosford, sandy soil. Damp places. Fl. summer . . . . **B. grandiflora** R.Br.

Intermediates between the last two species occur where their ranges overlap.

172 DORYANTHACEAE

1 gen. endemic Aust.

1 Doryanthes Corrêa

2 species endemic Aust.; Qld, NSW

One species in the area

Coarse herb up to 5 m high. Leaves fibrous, c. 1 m long, up to 10 cm wide, entire. Flowers in a compact head-like panicle on a leafy scape up to 5 m high. Perianth segments red, c. 10 cm long. Ovary not beaked. Coast and adjacent plateaus. Fl. spring–summer. *Giant Lily, Flame Lily, Illawarra Lily* or *Gymea Lily* . . .
. . . . **D. excelsa** Corrêa

173 HEMEROCALLIDACEAE

Rhizomatous herbs, sometimes ± woody. Leaves distichous. Flowers regular, bisexual, in panicles, cymo-panicles or umbels. Perianth segments 6, free or nearly so. Stamens 6, free, often with thickened filaments; anthers often dehiscing by terminal pores. Ovary 3-locular; style simple. Fruit a berry or capsule. c. 19 gen., S. hemisphere.

Key to the genera

1 Filaments not bearded . . . . **3 DIANELLA**
1 Filaments bearded 2

2 Flowers nodding . . . . **1 STYPANDRA**
2 Flowers erect . . . . **2 THELIONEMA**

1 Stypandra R.Br.

1 species endemic Aust.; Qld, NSW, S.A., WA

One species in the area

Tufted, rhizomatous perennial with tough aerial stems up to 1.5 m high. Leaves linear, up to 20 cm long, 5–15 mm wide, sheathing the stem. Flowers bisexual, nodding, in a terminal cymo-panicle. Perianth segments 6, free, deep blue to white, elliptic, 8–16 mm long. Stamens 6; filaments bent, woolly with yellow hairs towards the top; anthers dehiscing through slits. Ovary 3-locular with several ovules per loculus; style filiform. Fruit a capsule. Widespread. DSF. Fl. summer. *Nodding Blue Lily, Candyup Poison* or *Blind Grass* . . . . **S. glauca** R.Br.

2 Thelionema R.J.F.Hend.

3 species endemic Aust.; Qld, NSW, Vic., Tas.

Tufted, rhizomatous perennials without elongated aerial stems. Leaves linear, sheath closed at the top. Flowers in terminal cymo-panicles, bisexual. Perianth segments 6, free. Stamens 6; filaments papillose; anthers dehiscing through slits. Ovary 3-locular with several ovules per loculus. Fruit a capsule.

Key to the species

1 Inflorescence spreading, cymose, with 2–many branches, 60 cm high, usually exceeding the leaves; the scape flattened towards the base. Perianth segments 8–15 mm long, blue or cream. Leaves linear, 10–35 cm long, 5–8 mm wide. Widespread. DSF and heath. Chiefly Ss. Fl. summer . . . .
. . . . **T. caespitosum** (R.Br.) R.J.F.Hend.
1 Inflorescence umbel-like, with 1–3 branches, usually less than 30 cm long, shorter than or just exceeding the leaves; the scape compressed or terete towards the base. Perianth segments 5–7 mm long, cream. Leaves linear, 10–30 cm long, c. 5 mm wide. Widespread. DSF and heath. Chiefly Ss. Fl. summer . . . .
. . . . **T. umbellatum** (R.Br.) R.J.F.Hend.

Intermediates between these species are known to occur in the Southern Blue Mts

3 Dianella Lam.

48 species in Aust. (endemic, native); all states and territories

Glabrous, perennial herbs often with rhizomes which turn upwards and become erect aerial stems at the end. Leaves distichous, sheathing at the base, linear; the adaxial surface folded and the surfaces on either side of the midrib ± fused together ("occluded") just above the sheath. Flowers in terminal panicles. Perianth segments blue to greenish white, 6, subequal, free, persistent but not spirally twisted after flowering. Stamens 6; filaments with swelling below anther glabrous; anthers erect, opening by terminal or ± elongated pores. Ovary 3-locular, with several ovules per loculus; style filiform, simple or with a grooved stigma. Fruit a subglobular berry, usually blue. (Thrips may cause considerable distortion to the flowers.)

Key to the species

1 Tussock forming herb c. 10 cm wide at base and up to 50 cm high. Leaves narrow mostly 3–4 mm wide. Leaf sheath with uniform crimson markings, ± fully occluded at summit. Inflorescences much shorter than the leaves with 3–10 flowers. Perianth very pale purple, filaments violet, anthers pale to deep purple with a pale yellow swelling beneath. Berry ovoid, obscurely 3-lobed, violet. Restricted to Narrabeen sandstones of Blue Mts between altitude 800–1200 m . . . . **D. tenuissima** G.W.Carr
1 Plants without the above combination of characters. 2

2 Aerial stems less than 15 cm long, usually without scale-like leaves 3
2 Aerial stems more than 15 cm long, usually with scale-like leaves below the green leaves 10

3 Cross-section of occluded region of leaf just above the sheath V- or Y-shaped 4
3 Cross-section of occluded region of leaf just above the sheath biconvex with two short projections on the adaxial edge 8

4 Anthers yellowish-brown. Filaments swelling yellow. Perianth segments blue to dark blue, 6–10 mm long. Tufted, solitary perennial up to 50 cm high. Widespread. Heath and DSF. Ss. Fl. spring. *Blueberry Lily* or *Blue Flax-lily* . . . . **D. caerulea** Sims var. **caerulea**
4 Anthers yellow. Filaments swelling orange 5

5 Leaves strongly decurved, i.e., curving outwards. Perianth segments blue to dark blue. Tufted, solitary perennial up to 80 cm high. Sandy soil near the sea. Heath and DSF . . . . . . **D. crinoides** R.J.F.Hend.
5 Leaves erect or flaccid and slightly drooping, often greyish-green. Perianth segments up to 10 mm long. Tufted perennials up to 1.5 m high. 6

6 Leaves less than 4 mm wide. Roots with tubers. Perianth segments pale blue to nearly white. Probably widespread in heath and DSF. . . . . **D. longifolia** var. **stenophylla** Domin
6 Leaves more than 4 mm wide. Roots without tubers. 7

7 Perennials with small tufts up to 5 cm diam. Perianth segments pale blue. Widespread in a variety of habitats . . . . **D. longifolia** R.Br. var. **longifolia**
7 Perennials with large tufts up to 60 cm diam. Perianth segments greenish white to pale blue. Hornsby Plateau, but probably widespread. DSF . . . . **D. longifolia** var. **grandis** R.J.F.Hend.

8 Leaf sheath distinctly different from the lamina, with an acute keel; lamina sometimes glaucous, 4–12 mm wide. Perianth segments dark blue to violet. Filaments swelling yellow. Anthers brown to black. Tufted or mat-forming perennial up to 80 cm high. Widespread in a variety of situations but not in RF
. . . .
. . . .**D. revoluta** R.Br. var. **revoluta**
8 Leaf sheath not distinctly different from lamina 9

9 Anthers up to the same length as the swollen part of the anther, pale yellow. Filaments swelling golden yellow. Mature berry longer than wide. Leaves 14–32 mm wide, dark green. Perianth segments lavender to violet. Tufted, solitary perennial up to 1 m high with a golden-yellow rhizome. Widespread. Forests
. . . .**D. tasmanica** Hook.f.

9 Anthers longer than the swollen part of the filament, yellowish-brown. Filaments swelling yellow. Mature berry depressed-globular or globular. Leaves up to 25 mm wide, greyish-green. Perianth greenish white to pale blue. Tufted perennial sometimes growing in gregarious colonies, with a brown rhizome. Blue Mts Forests. . . . .**D. caerulea** var. **cinerascens** R.J.F.Hend.

10 Cross-section of occluded region of leaf just above the sheath V- or Y-shaped. Perianth greenish white to blue. Filaments yellow. Anthers brownish yellow. Tufted, solitary perennial up to 1.3 m high, with an elongated stem, sometimes branching in the leaf sheaths. Blue Mts; Hornsby Plateau. DSF. . . . . **D. caerulea** var. **producta** R.J.F.Hend.

10 Cross-section of occluded region of leaf just above the sheath biconvex with two short projections on the adaxial edge. 11

11 Leaves glaucous, 8–38 mm wide. Perianth blue to violet. Filaments swelling yellow. Anthers brownish yellow to blue. Tufted, solitary perennial up to 2 m high. Widespread. DSF. . .**D. prunina** R.J.F.Hend.

11 Leaves green to yellowish-green. 12

12 Inflorescence not exceeding the leaves, decurving and congested. Leaves 10–15 mm wide, often yellowish-green. Perianth blue. Filaments swelling yellow. Anthers yellowish-brown. Perennial up to 1 m high forming mats in gregarious colonies which are often extensive. Coast. Stabilized sand dunes . . . . **D. congesta** R.Br.

12 Inflorescence usually exceeding the leaves, neither decurving nor congested (except when attacked by thrips) 13

13 Stems covered with scales for most of their length. Perianth blue, streaked with green. Tufted, solitary perennial up to 2 m high, often branching from the sheaths. Blue Mts DSF; WSF; RF margins . . . . **D. caerulea** var. **assera** R.J.F.Hend.

13 Stems covered with leaves for most of their length 14

14 Aerial shoots touching or very close together. Perianth pale blue. Tufted, solitary perennial, up to 50 cm high. Blue Mts Open forest . . . . **D. caerulea** var. **protensa** R.J.F.Hend.

14 Aerial shoots close together or well separated. Perianth blue or very pale blue. Tufted, solitary perennial up to 1.3 m high. Blue Mts Open forest. . . . . **D. caerulea** var. **vannata** R.J.F.Hend.

174 HYPOXIDACEAE
4 gen., S.E. Asia and Aust.

1 Hypoxis L.
Golden-weather Grass
10 species endemic Aust.; all states and territories

Erect herbs with corms. Leaves basal, linear, up to 40 cm long. Flowers bisexual, regular, solitary or in short racemes. Perianth segments 6, white to yellow, 7–12 mm long, united into a short tube, persistent. Stamens 6. Ovary inferior, 3-locular, with numerous ovules on axile placentas; style 3-fid. Fruit a capsule up to 6 mm long.

Key to the species

1 Filaments alternate long and short. Flowers usually 1–3 on each stem. Leaves up to 20 cm long, 1–4 mm wide 2

1 Filaments ± equal. Flowers usually 4–5 on each stem. Leaves up to 40 cm long, 1–4 mm wide 4

2 Flowers usually 2 on each stem. Sepals hairy outside. Leaves up to 2 mm wide. Blue Mts Heath. Fl. spring–autumn . . . . **H. hygrometrica** var. **villosisepala** R.J.F.Hend.

2 Flowers usually solitary on each stem. Sepals glabrous or glabrescent. 3

3 Anthers versatile, tapering upwards, emarginate. Leaves up to 1.5 mm wide. Style slender; stigmas much shorter than the style. Widespread. Grasslands and Heath. Fl. spring–autumn . . . . **H. hygrometrica** Labill. var. **hygrometrica**

3 Anthers erect, oblong, forked at the top. Leaves usually more than 1.5 mm wide. Style thick; stigmas ± as long as the style. Higher Blue Mts Heath and DSF. Fl. spring–autumn. . . . .**H. hygrometrica** var. **splendida** Hend.

4 Anthers up to 2 mm long. Seeds not tuberculate. Widespread. Grasslands. Fl. summer–autumn . . . . **H. pratensis** R.Br. var. **pratensis**

4 Anthers more than 2 mm long. Seeds tuberculate. Widespread. Grasslands. Fl. summer–autumn . . . . **H. pratensis** var. **tuberculata** Hend.

175 IRIDACEAE

Erect, perennial herbs with rhizomes or corms. Leaves linear or almost so; sheathing, often distichous. Flowers regular or irregular, bisexual, enclosed individually or severally in a cyme within spathes, in spikes racemes or thyrses or solitary. Perianth segments 6, in 2 similar or subsimilar whorls, connate basally into a tube or free. Stamens 3, often adnate to the perianth; anthers dehiscing by longitudinal slits. Ovary inferior, 3-locular, with numerous ovules on axile placentas. Style filiform, 3-fid; branches entire or 2-fid, filiform dilated or petaloid. Fruit a capsule with numerous seeds. 67 gen., cool temp. to drier subtrop., particularly S. Africa.

Key to the genera

1 Perianth segments free or connate basally into a tube less than 3 mm long. 2
1 Perianth segments connate basally into a tube more than 3 mm long 7

2 Filaments connate into a tube. 3
2 Filaments not connate 5

3 Perianth blue; outer whorl larger than inner. Stylar branches distinctly 2-fid . . . . . . **2 HERBERTIA**
3 Perianth not blue; whorls subequal. Stylar branches entire or 2-toothed at the apex 4

4 Perianth pink, orange, or yellow. Plants with only 1 or 2 leaves from the base of the stem; cauline leaves sometimes present . . . . **3 MORAEA**
4 Perianth white. Plants with more than 2 leaves from the base of the stem . . . . **5 LIBERTIA**

5 Stylar branches winged, petaloid. . . . . **1 IRIS**
5 Stylar branches neither winged nor petaloid 6

6 Perianth pink. Flowers solitary . . . . **4 ROMULEA**
6 Perianth yellow or rarely blue. . . . . **6 SISYRINCHIUM**

7 Either flowers regular or perianth tube curved and the stamens not on the upper side of the flower . . 8
7 Flowers irregular; perianth tube curved and the stamens on the upper side of the flower 14

8 Stylar branches entire 9
8 Stylar branches 2-fid 12

9 Outer perianth segments conspicuously larger than inner. Flowers in a terminal cymose cluster . . . . **7 PATERSONIA**
9 Perianth segments all similar, c. the same size. Flowers in loose sometimes compound cymes 10

10 Flowers nodding . . . .**8 DIERAMA**
10 Flowers erect. 11

11 Perianth tube green, lobes bright blue. Rhizome short and thick. . . . .**19 ARISTEA**
11 Perianth tube not green, lobes white, mauve, pink. Corm globose . . . . **9 IXIA**

12 Rhachis of the inflorescence straight . . . . **12 WATSONIA**

1 Iris L.

3 species naturalized Aust.; NSW, Vic., Tas., S.A., WA

One species in the area

Perennial herb with rhizomes. Leaves linear, distichous, up to 80 cm long, up to 5 cm wide. Flowers in spikes; spathes scarious above. Perianth whorls dissimilar; outer segments large, deflexed, purple above, yellowish below, with bronze veins, or white; inner segments erect, almost equalling the outer ones, pale purple; perianth tube longer than the ovary. Stylar branches petaloid. Garden escape in waste ground around towns. Introd. from the Mediterranean. Fl. spring. *Garden Flag* . . . . * **I. germanica** L.

2 Herbertia Sweet

1 species naturalized Aust.; NSW

One species in the area

Erect, perennial herb with corms. Leaves linear, mostly radical, up to 8 cm long and 3 mm wide. Flowers solitary or 2–3 together on long peduncles with a sheathing spathe. Perianth blue; outer whorl longer than the inner and scarcely connate at the base. Stamens adnate to the perianth below; filaments connate into a tube; anthers sessile on the tube. Style branches filiform, 2-fid. Grasslands around Sydney. Introd. from S. America . . . .*H. lahue** (Molina) Goldblatt ssp. **caerulea** (Herb.) Goldblatt

3 Moraea Mill.

11 species naturalized Aust.; NSW, Vic., Tas., S.A., WA

Erect, perennial herbs with corms. Leaves 1–2 linear from the base of the stem up to 80 cm long and 10 mm wide, and some much shorter cauline leaves. Perianth whorls subequal, connate basally into a tube. Filaments connate into a tube; anthers sessile or almost so on the filament tube. Stylar branches entire or 2-dentate, scarcely dilated. Capsule linear. Cormils often present at the nodes. Noxious weeds.

Key to the species

1 Two well developed leaves from the base of the stem. Flowers pink, yellow or white with triangular yellow nectary guides dotted with green in centre. Staminal tube swollen and pubescent at base; anthers c. 2 mm long. Pasture, roadsides and disturbed areas. Introd. from S. Africa. Fl. summer. *Two-leaved Cape Tulip* . . . .*M. miniata** (Andrews) Sweet
1 Only one well developed leaf from the base of the stem 2

2 Style crest minute and obtuse or absent. Bracts on flowering stems 6–7 cm long. Flowers yellow, sometimes with orange centre or all orange. Staminal tube glabrous, anthers 5–8 mm long. Pasture, roadsides and disturbed areas. Introd. from S. Africa. Fl. spring . . . . *M. ochroleuca* Salisb.

2 Style crest erect, c. 1 mm long. Bracts on flowering stems 2–6 cm long 3

3 Anthers 8–11 mm long. Flowers orange with deep yellow nectary guides in centre. Staminal tube smooth to sparsely papillate at base. Pasture, roadsides and disturbed areas. Introd. from S. Africa. Fl. summer. *One-leaved Cape Tulip.* . . . . *M. flaccida* Sweet

3 Anther 5–6 mm long. Flowers pale yellow or pink deep yellow nectory guides with green edge present or absent. Staminal tube sparely pubescent in lower half. Occasionally naturalized in disturbed areas. Introd. from S. Africa. Fl. spring. *Cape Tulip* . . . . *M. collina* (Thunb.) Salisb.

4 Romulea Maratti

4 species naturalized Aust.; all states and territories except NT

One species in the area

Erect perennial with corms. Leaves narrow-linear, up to 25 cm long and 3 mm wide. Flowers solitary, surrounded by 2 scarious spathes. Perianth segments pink, subequal, connate basally into a tube. Anthers coherent around the style. Stylar branches filiform, 2-fid. Fruit a membranous capsule with numerous seeds. Widespread. Grasslands and waste places. Introd. from S. Africa. Fl. spring–summer. *Onion Grass* . . . . * **R. rosea** (L.) Eckl. var. **australis** (Ewart) M.P.deVos.

5 Libertia Spreng.

2 species native Aust.; Qld, NSW, Vic., Tas.

Perennial with rhizomes. Aerial stems erect or lax. Leaves linear, distichous. Flowers often clustered, in irregular cymes. Perianth segments white, subequal, connate basally for a short distance only. Stamens almost or quite free from the perianth; filaments free. Stylar branches filiform, entire. Capsule subglobular.

Key to the species

1 Inner perianth segments elliptic, 3–5 mm long. Stems less than 20 cm high. Leaves up to 25 cm long and 5 mm wide. Flower clusters few, scarcely spreading. Blue Mts Damp places; rock ledges. Fl. summer . . . . **L. pulchella** (R.Br.) Spreng.

1 Inner perianth segments obovate, 8–15 mm long. Stems usually more than 20 cm high. Leaves up to 60 cm long and 10 mm wide. Flower clusters numerous in irregular spreading cymes. Widespread. Margins of RF. Fl. summer . . . . **L. paniculata** (R.Br.) Spreng.

6 Sisyrinchium L.

2 species naturalized Aust.; Qld, NSW, Vic., Tas., S.A.

Erect herbs with rhizomes. Leaves linear, basal and cauline. Flowers on slender pedicels, clustered within spathes. Perianth segments yellow or blue to purple, connate basally for a short distance only, subequal. Filaments connate into a membranous tube. Stylar branches filiform, entire. Capsule subglobular.

Key to the species

1 Perennial or annual with compressed or winged stems, up to 35 cm high. Perianth 10–20 mm diam., yellow or blue to purple. Capsule c. 6 mm diam. Sporadic in a variety of habitats. Introd. from America. Fl. spring–summer. *Blue Pigroot* . . . . *S. iridifolium* Kunth

1 Annual with only slightly compressed stems, up to 15 cm high. Perianth 5–7 mm diam., yellow. Capsule c. 4 mm diam. Widespread in a variety of habitats. Introd. from S. America. Fl. spring. (Previously included in *S. micranthum* and may be a subspecies of *S. iridifolium*.) *Scour Weed* or *Yellow Rush Lily.* . . . . *S. sp. **A.** (*sensu* D.A.Cooke)

7 Patersonia R.Br.

Native Iris

17 species endemic Aust.; all states and territories

Erect, rhizomatous herbs sometimes with woody aerial stems. Leaves linear, distichous. Flowers regular, in a terminal cluster (cymes) enclosed by two large striate spathes. Outer perianth segments connate into a filiform tube with 3 broad spreading lobes; inner segments smaller, erect. Filaments connate below. Stylar branches entire, dilated. Capsule narrow, 3-angular.

Key to the species

1 Plants glabrous. Leaves rigid, 15–40 cm long. Spathes green to pale brown, herbaceous in texture. Scapes leafless, usually shorter than the leaves. Lobes of the outer perianth segments 10 mm long. Widespread. Damp soil. Ss. Fl. spring. . . . . **P. fragilis** (Labill.) Asch. & Graebn.
1 Plants hairy. Spathes brown to black. Scapes and leaves 30 cm high 2

2 Leaves cauline on erect stems. Leaf margins minutely hairy at the base. Scape glabrous. Spathes dark brown, villous. Widespread. DSF and heath. Fl. spring . . . . **P. glabrata** R.Br.
2 Leaves basal, linear. Stems very short. Spathes dark brown to black 3

3 Leaves deeply grooved, hairy on the margin never regularly inflexed 2–6 mm wide, usually erect but sometimes lax towards the tip. Spathes 35–60 mm long. Widespread. DSF and heath. Fl. spring . . . .**P. sericea** R.Br.
3 Leaves smooth, with inflexed hairs on the margin, 1–2 mm wide, lax, trailing. Spathes 20–35 mm long. Coast and adjacent plateaus. DSF and heath. Fl. spring . . . . **P. longifolia** R.Br.

8 Dierama K.Koch

2 species naturalized Aust.; NSW, S.A.

One species in the area

Erect, perennial herb c. 1 m high, with corms. Leaves linear, c. 5 mm wide; basal leaves up to 70 cm long; cauline leaves shorter. Flowers regular, nodding, in a loose elongated series of cymes; spathes scarious-hyaline, toothed at the apex. Perianth segments white to purple, connate into a funnel-shaped tube. Stamens adnate to the perianth tube. Stylar branches entire, ± dilated at the apex. Capsule subglobular. Garden escape. Blue Mts Introd. from S. Africa . . . . ***D. pendulum** (Hook.f.) Baker

9 Ixia L.

c. 5 species naturalized Aust.; all states and territories except NT

Key to the species

Erect annual herbs c. 30 cm high, with corms. Leaves basal. Flowers regular, erect, in spikes; spathes scarious-membranous, 2-fid. Perianth connate basally into a short tube. Filaments free or connate basally into a very short tube. Stylar branches entire, filiform. Capsule oblong or subglobular. Seeds numerous, glossy.

1 Perianth tube 4–7 mm long. Spathe bracts tinged with pink. Leaves 10–25 cm long, c. 10 mm wide. Widespread. Garden escape in waste ground near towns. Introd. from S. Africa. Fl. spring–summer . . . .***I. flexuosa** L.
1 Perianth tube 6–14 mm long. Spathe bracts white. Leaves 20–60 cm long, 2–6 mm wide. Shaded and moist site. Introd. from S. Africa. Fl. summer . . . . ***I. polystachya** L.

10 Freesia Klatt
1 naturalized Aust.; NSW, Vic., S.A., WA

One species in the area

Weak, erect, perennial herb with reticulate corms. Leaves linear, mostly basal, c. 25 cm long. Flowers almost regular, on branched stems in few-flowered one-sided spikes with a flexuose rhachis; spathes scarious-membranous, ovate, toothed, 5–7 mm long. Perianth segments yellow or cream, subequal, connate into a tube longer than the lobes; lobes c. 15 mm long and 10 mm wide. Filaments adnate to the perianth tube. Stylar branches 2-fid, ± dilated. Capsule ovoid. Garden escape in waste ground near towns. Introd. from S. Africa. Fl. spring . . . . *F. alba (G.L.Mey.) Gumbl. x F. leichtlinii Klatt

11 Anomatheca Ker Gawl.
1 species naturalized Aust.; Qld, NSW, Vic.

One species in the area

Erect, perennial herb with corms. Leaves linear, basal and cauline, up to 30 cm long, 3–6 mm wide. Flowers regular, in one-sided spikes with a flexuose rhachis; spathes sub-herbaceous, ovate, entire, c. 5 mm long. Perianth segments blue-purple, connate into a narrow tube c. 15 mm long; lobes lanceolate to narrow-oblong, c. 7 mm long. Stylar branches 2-fid, filiform. Capsule subglobular. Frenchs Forest; Avalon. Garden escape in waste ground near habitation. Introd. from S. Africa. Fl. summer . . . .
. . . . *A. laxa (Thunb.) Goldblatt

12 Watsonia Mill.
6 species naturalized Aust.; all states and territories except NT

Erect, perennial herb with corms. Leaves basal and cauline, linear, fibrous. Flowers in a series of cymes with or without cormils in the position of flowers; spathe 2-fid. Perianth segments connate basally into a curved tube. Stamens arranged onto the upper side of the flower. Stylar branches 2-fid, filiform. Capsule woody, globose to cylindrical.

Key to the species

1 Flowers usually orange to red, 3–4 cm apart. Cormils often in the position of flowers in the lower spathes. Leaves linear, 50–80 cm long and up to 5 cm wide. Spathe bracts triangular to oblong. Perianth tube c. 4 cm long; lobes c. 2 cm long. Style longer than the perianth, branches c. 0.5 cm long. Capsule subglobular. Herb up to 2 m high Garden escape in waste ground near habitation. Introd. from S. Africa. Fl. summer . . . . *W. meriana (L.) Mill.

1 Flowers usually pink to mauve, 2–2.5 cm apart. Cormils absent from cymes. Leaves linear, 20–100 cm long, c. 4 cm wide. Spathes lanceolate to oblong. Perianth tube 3–4 cm long, lobes c. 3 cm long. Style shorter than the perianth, branches c. 0.5 cm long. Capsule ovoid-cylindrical. Herb to 2 m high. Garden Escape. Introd. from S. Africa. Fl. spring . . . .
. . . . *W. borbonica (Pourr.) Goldblatt ssp. ardernei (Sander) Goldblatt

13 Babiana Ker Gawl.
1 species naturalized Aust.; NSW, Vic., S.A., WA

One species in the area

Erect, pubescent, perennial herb c. 50 cm high, with corms. Leaves linear, 25 cm long, c. 15 mm wide, distichous. Flowers irregular, in a branched spike on a hairy scape; spathe subherbaceous, striate, hairy. Perianth segments red-purple, subequal, connate into a curved funnel-shaped tube. Stamens arranged onto the upper side of the flower. Stylar branches entire, ± dilated at the apex. Capsule ± inflated. Garden escape in waste ground near habitation. Introd. from S. Africa. Fl. summer . . *B. stricta (Sol.) Ker Gawl.

14 Crocosmia Planch.

1 species naturalized Aust.; NSW, Vic., Tas., S.A., WA

One species in the area

Erect, perennial herbs c. 60 cm high, with corms. Leaves basal and cauline. Flowers sessile in scarious spathes, arranged in compound cymes or spike-like inflorescences. Perianth segments subequal, red to orange-yellow, connate into a curved cylindrical tube. Stamens arranged onto the upper side of the flower. Stylar branches entire, often with a ± dilated denticulate tip. Capsule globular. Garden hybrid escape into waste ground near habitation. Introd. from S. Africa. Fl. summer. . . . . *C. X **crocosmiiflora** (Lemoine ex E.Morren) N.E.Br.

15 Tritonia Ker Gawl.

3 species naturalized Aust.; NSW, Vic., Tas., S.A., WA

One species in the area

Erect, perennial herb with corms. Leaves linear, basal, up to 30 cm long, c. 15 mm wide. Flowers irregular, in spikes with membranous to scarious spathes. Perianth segments cream with dark veins, subequal, connate into a curved cylindrical tube c. 13 mm long; lobes c. 15 mm long. Stamens arranged onto the upper side of the flower, exserted. Stylar branches entire, filiform or very slightly dilated. Capsule oblong, with many seeds. Garden escape in waste ground near habitation. Introd. from S. Africa. Fl. summer. . . . .* **T. lineata** (Salisb.) Ker Gawl.

16 Sparaxis Ker Gawl.

2 species naturalized Aust.; NSW, Vic., Tas., S.A., WA

Perennial herbs, with corms. Leaves basal, linear, soft, veins prominent. Flowers few, few within scarious-membranous spathes, regular to slightly irregular. Perianth connate, lobes spreading to erect longer than the tube. Style 3-fid, branches entire. Fruit an ovoid, membranous capsule. Seeds numerous and smooth.

Key to the species

1 Stamens arranged symmetrically around the style; anthers yellow. Bracts entire or slightly lacerate. Perianth tube bright yellow c. 8 mm long. Style branches cuneate, 3–5 mm long. Bulbils few in lower leaf axils. Garden escape in disturbed areas. Introd. from S. Africa. Fl. winter–spring. *Tricolour Harlequin Flower* . . . .*S. tricolor (Schneev.) Ker Gawl.
1 Stamens arranged on one side of the style; anthers white. Bracts deeply lacerate. Perianth tube pale yellow, c. 15 mm long. Style branches filiform, 7–10 mm long. Bulbils numerous in leaf axils. Garden escape in disturbed areas. Introd. from S. Africa. Fl. spring. *Harlequin Flower* . . . .*S. bulbifera (L.) Ker Gawl.

17 Gladiolus L.

12 species naturalized Aust.; all states and territories except NT

Perennial herbs with corms. Leaves annual mostly basal, fibrous. Flowers irregular, in distichous or one-sided spikes on erect scapes. Perianth segments connate into a straight or curved, funnel-shaped tube. Stamens arranged onto the upper side of the flower. Style branches entire, widened at the apex.

Key to the species

1 Perianth c. 3 cm long, mauve with a purple stripe on each lobe. Leaves linear, up to 75 cm long, thick, succulent, much longer than the scape. Flowers in a one-sided spike. Coastal sand dunes. Introd. from S. Africa . . . . *G. gueinzii Kunze
1 Perianth more than 3 cm long. Leaves not succulent and scarcely exceeding the scapes 2

2 Perianth tube c. twice as long as the lobes. Leaves linear, up to 60 cm long, 5–8 mm wide. Flowers in a one-sided spike. Perianth cream to pink, 7–10 cm long. Garden escape near habitation. Introd. from S. Africa . . . . *G. angustus L.

2 Perianth tube c. as long as the lobes . . . . ³

3 Perianth lobes with tapering undulate tips. Leaves linear, up to 75 cm long, 5–15 mm wide. Flowers in a distichous spike. Perianth greenish white to cream, 9–12 cm long. Garden escape near habitation. Introd. from S. Africa . . . .*G. undulatus L.

3 Perianth lobes acute without tapering tips. Leaves linear, up to 60 cm long, 6–20 mm wide. Flowers in a distichous spike. Perianth white to pink or lavender, 5–8 cm long. Garden escape near habitation. Introd. from S. Africa . . . .*G. carneus D.Delaroche

18 Chasmanthe N.E.Br.
1 species naturalized Aust.; NSW, Vic., Tas., S.A., WA

One species in the area
Erect, perennial herb with corms. Leaves linear, c. 90 cm long and 3 cm wide. Flowers irregular, oblique, in spikes c. 25 cm long; spathes scarious. Perianth segments red, connate into a curved tube c. 4 cm long, narrow below, widening abruptly c. 1 cm above the ovary; inner lobes somewhat larger than outer ones; posterior lobe largest. Stamens arranged onto the upper side of the flower. Stylar branches entire, only slightly dilated. Capsule ovoid. Garden escape near habitation. Introd. from S. Africa. Fl. summer . . . .
. . . .*C. floribunda (Salisb.) N.E.Br.

19 Aristea Sol. ex Aiton
1 species naturalized Aust.; NSW, Vic.

One species in the area
Erect herb to 70 cm high. Leaves linear 10–60 cm long, 5–11 mm wide, bases sometimes red. Inflorescence a lose panicle, spathe bracts brown acuminate 5–14 mm long. Flowers on pedicels 2–10 mm long, opening for only a few hours. Perianth bright blue tube short and green. Stamens exserted, filamens free. Style with 3 branches. Fruit an oblong capsule c. 20 mm long. Occasional garden escape northern suburbs. Introd. from Africa. . . . .*A. ecklonii Baker

176 ORCHIDACEAE

Perennial herbs; epiphytic or terrestrial; some terrestrial species saprophytic. Epiphytes with creeping stems or rhizomes and fibrous roots with *velamen*, or with erect or pendulous stems and without rhizomes; *pseudobulbs* sometimes present. Terrestrial or saprophytic species with fleshy rhizomes or tubers (Fig. 44). Leaves usually alternate or basal, sometimes reduced to scales. Inflorescence a spike, raceme or panicle or the flower solitary. Flowers bisexual, usually very irregular, often twisted through 180° (*resupinate*). Sepals 3, often similar to the lateral petals but usually larger. Petals 3, alternating with the sepals; one of them (the *labellum*, originally nearest the floral axis but often remote from the axis as a result of resupination) usually different in size, shape and colour from the other 2. Stamens, stigma and style fused into a single, central structure (the *column*) (Fig. 44). Anther 1 (or 2 in genera not in the area), usually with 2 pollen sacs, sunken in the apex of the column; pollen waxy, granular, rarely mealy, aggregated into 1, 2 or 4 pairs of pollen masses (*pollinia*) (pollinia absent in some genera not in the area); pollinia usually lying free in the apical cavity of the column and sometimes held together by a thin thread which later elongates into a stalk (*caudicle*). Stigma convex or concave, in front of the column, viscid and in the upper margin usually produced into a platform-like appendage (the *rostellum*). Ovary inferior, 3-carpellary, 1-locular; placentas 3, parietal. Fruit a capsule dehiscing longitudinally. Seeds minute, extremely numerous. c. 788 gen.; worldwide, the epiphytes chiefly trop. and subtrop.

FIGURE 44

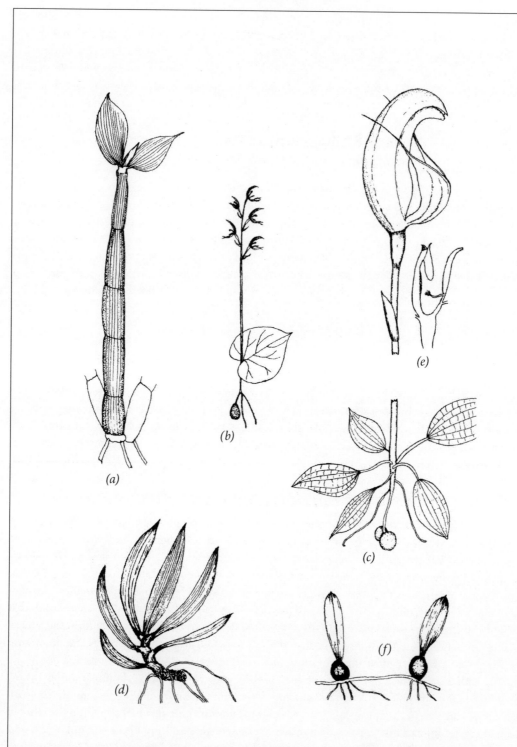

(a)

(b)

(c)

(d)

(e)

(f)

Fig. 44   ORCHIDACEAE: (a) *Dendrobium aemulum*, showing a jointed pseudobulb;
(b) *Acianthus exsertus*, with a single cordate leaf and smaller tuber; (c) *Pterostylis
baptistii*, with several leaves and 2 tubers; (d) *Sarcochilus falcatus*; (e) *Pterostylis* sp.
flower, insert column and labellum, the former with pendulous wings, the latter with a
brush-like appendate; (f) *Bulbophyllum exiguum*.

FIGURE 45

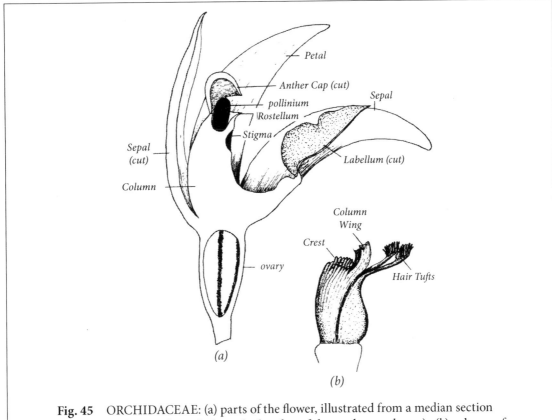

Fig. 45 ORCHIDACEAE: (a) parts of the flower, illustrated from a median section through the flower of *Dendrobium* (only a few of the ovules are shown) ; (b) column of *Thelymitra ixioides* from the outside (X8) (anther and stigma not visible).

Colour forms, hybrids, and some ill-defined varieties are not included in the key below.

Key to the genera

1 Plants epiphytic on trees, logs, rocks. . . . . .GROUP 1
1 Plants terrestrial with tubers or rhizomes. Leaves basal and/or cauline, large or scale-like. 2

2 Saprophytes. Leaves scale-like . . . . .GROUP 2
2 Autotrophs with green leaves and stems; stems sometimes coloured. Leaves sometimes absent at
   flowering time 3

3 Leaves 2 or more. Stem-bracts present or absent . . . . .GROUP 3
3 Leaves 1 or none or leaves absent at flowering time. Stem-bracts present or absent . . . . . . .GROUP 4

Group 1
Epiphytes

1 Leaves short, usually no more than 5 times longer than broad 2
1 Leaves linear, many times longer than broad 9

2 Pseudobulbs usually present. Lateral sepals dilated at the base and united with the projecting foot of the
   column to form a spur. Labellum not spurred 3
2 Pseudobulbs absent 4

3 Pseudobulbs ± globular (Fig. 44). Labellum entire . . . .**34 BULBOPHYLLUM**
3 Pseudobulbs elongate, jointed (Fig. 44). Labellum 3-lobed . . . .**33 DENDROBIUM**

4 Saprophytes with scale-leaves only. Stems climbing . . . . .27 ERYTHRORCHIS
4 Autotrophs with green leaves 5

5 Leaves leathery 6
5 Leaves membranous or thick and/or succulent and ribbed or warted 8

6 Labellum articulate on the column foot .37 SARCOCHILUS
6 Labellum adnate to the column foot 7

7 Callus absent on posterior of labellum spur . . . . .39 PAPILLILABIUM
7 Callus present on posterior of labellum spur . . . . .38 PLECTORRHIZA

8 Leaves membranous . . . . 15 RIMACOLA
8 Leaves thick and/or succulent, ribbed or warted . . . . .33 DENDROBIUM

9 Leaves not basal, terete . . . . .33 DENDROBIUM
9 Leaves basal, flat or folded. Plants tufted 10

10 Column elongate, incurved, shortly winged at the top . . . . 31 CESTICHIS
10 Column erect or slightly incurved at the top, semi-terete . . . . 36 CYMBIDIUM

Group 2
Saprophytes

1 Climbing plants with short sucker-like aerial roots . . . . .27 ERYTHRORCHIS
1 Plants not climbing 2

2 Plants flowering underground . . . . 29 RHIZANTHELLA
2 Plants flowering above the ground 3

3 Sepals and petals whitish or brown, fused together into a 5-lobed tube . . . . 28 GASTRODIA
3 Sepals and petals free 4

4 Petals reddish to pink . . . . 35 DIPODIUM
4 Petals yellow . . . . 25 CRYPTOSTYLIS

Group 3
Terrestrial autotrophs with 2 or more leaves

1 Leaves large and cauline, leathery, spreading, 1.5–12 cm long . . . . 40 EPIDENDRUM
1 Leaves not as above 2

2 Dorsal sepal and lateral petals fused to form a hood (Fig. 44) . . . . 26 PTEROSTYLIS
2 Hood absent 3

3 Labellum 3-lobed near the base 4
3 Labellum not 3-lobed near the base 5

4 Petals well developed, spreading or recurved . . . . .3 DIURIS
4 Petals minute, incurved over the column . . . . .4 ORTHOCERAS

5 Flower solitary . . . . 11 CHILOGLOTTIS
5 Flowers several to many 6

6 Flowers not resupinate . . . . 25 CRYPTOSTYLIS
6 Flowers resupinate 7

7 Leaves prostrate on the ground, usually absent at flowering time . . . . 10 ARTHROCHILUS
7 Leaves not prostrate 8

8 Basal leaves petiolate; cauline leaves sheathing the stem . . . . **16 RIMACOLA**
8 Petioles absent 9

9 Flowers 2–5. Perianth reddish to purplish-brown outside, white inside. . . . . **19 BURNETTIA**
9 Flowers many in a spike. Perianth pink. Labellum white . . . .**30 SPIRANTHES**

Group 4
Terrestrial autotrophs with 1 leaf or none

1 Leaves hairy. Labellum without hairs, or with calli 2
1 Leaves glabrous, or absent at flowering time 4

2 Labellum sitting at ± right angle to the column. Labellum sessile, with calli at the base only . . . .
. . . .**23 GLOSSODIA**
2 Labellum standing erect against the column, shortly clawed, with calli along the lamina 3

3 Flowers blue. . . . . **20 CYANICULA**
3 Flowers not blue . . . . **21 CALADENIA**

4 Leaves absent at flowering time. 5
4 Leaves present at flowering time, glabrous 6

5 Flowers hooded. . . . . **26 PTEROSTYLIS**
5 Flowers not hooded. Petals and sepals reflexed. . . . . **10 ARTHROCHILUS**

6 Stems glandular-pubescent. Labellum covered with long hairs. . . . . **14 ERIOCHILUS**
6 Stems glabrous 7

7 Flower solitary 8
7 Flowers 2 or more 10

8 Leaf linear. . . . .**8 CALEANA**
8 Leaf ovate orbicular or broad-lanceolate 9

9 Dorsal sepal and labellum greatly developed; the outer perianth segments inconspicuous . **24 CORYBAS**
9 Perianth segments unequal but all well developed. . . . . **22 ADENOCHILUS**

10 Leaf orbicular to broad-lanceolate (Fig. 44) 11
10 Leaf linear to narrow-lanceolate 15

11 Stem-bracts absent 12
11 Stem-bracts 1–3 14

12 Lateral petals more or less as long as sepals. Leaf basal, lower surface green . . . . **14 CYRTOSTYLIS**
12 Lateral petals much shorter than sepals. Leaf cauline, lower surface reddish-purple or green 13

13 Labellum with 2 well developed basal glands . . . . **12 ACIANTHUS**
13 Labellum without well developed basal glands . . . . **13 ACIANTHELLA**

14 Leaf ovate-cordate. Dorsal sepal hooded. . . . .**18 PYRORCHIS**
14 Leaf ovate-lanceolate. Dorsal sepal concave . . . . **19 BURNETTIA**

15 Labellum 2–3 times as long as the other perianth segments, covered with long hairs . **2 CALOCHILUS**
15 Labellum not conspicuously longer than the other perianth segments. 16

16 Perianth segments similar in shape and size; perianth almost radially symmetrical . . **1 THELYMITRA**
16 Flower conspicuously irregular 17

17 Flowers few (usually 3–5) 18
17 Flowers numerous (usually more than 10). 20

18 Dorsal sepal conspicuously hooded. . . . . **17 LYPERANTHUS**
18 Dorsal sepal erect, incurved 19

19 Labellum smooth, shining. . . . . **.8 CALEANA**
19 Labellum tuberculate. . . . . **.9 PARACALEANA**

20 Labellum 3-lobed; the central lobe bifid . . . . **32 CALANTHE**
20 Labellum otherwise. 21

21 Flowers resupinate. Dorsal sepal hooded . . . . **5 MICROTIS**
21 Flowers not resupinate. Dorsal sepal concave 22

22 Free part of leaf terete. Labellum not articulate on a claw . . . . **6 PRASOPHYLLUM**
22 Free part of leaf flattened. Labellum articulate on a claw . . . . **7 GENOPLESIUM**

1 Thelymitra J.R.Forst. & G.Forst.
Sun Orchids
c. 37 species in Aust. (c. 23 endemic); all states and territories except NT

Terrestrial herbs usually glabrous, with ovoid tubers. Leaf solitary, narrow, channelled, of varying length, stem-clasping in its lower portion, with 1–2 foliaceous bracts above it. Flowers in a terminal raceme, rarely solitary, usually expanding only on bright sunny days. Perianth almost regular; the labellum differing only slightly from the paired petals; sepals petaloid. Column rather short and stout, prominently winged (Fig. 45 b); the wings united at the base in the front; then either expanding upward behind the anther, or extending only on one side of the anther; in all cases each wing is furnished with a lateral lobe, which in many species terminates in a conspicuous tuft of hairs; the apex of the column wings is sometimes referred to as a *hood*. Anther erect or bent forward between the lobes of the wings. Pollinia granular.

Key to the species

1 Lateral lobes of the column-wings terminating in hair tufts (Fig. 45 b) 2
1 Column-wings without hair tufts 15

2 Hair tufts yellow or cream. 3
2 Hair tuffs white, pink or blue 5

3 Perianth bright rose-pink, c. 25 mm diam. Hair tufts on column wings yellowish. Column between the lateral arms crested. Plant up to 40 cm high. Flowers 2–4. Coast. Fl. Sept.–Oct. (A hybrid between *T. ixioides* and *T. carnea*). . . . . **T. X irregularis** Nicholls
3 Perianth blue to purplish occasionally white 4

4 Column between the lateral arms hooded, deeply notched. Hair tuffs on column arms sparse, 2. Perianth deep blue to mauve. Hair tufts on lateral arms sparse. Uncommon. Bundanoon. Fl. Nov.–Dec. . . . . **.T. holmesii** Nicholls
4 Column between the lateral arms not hooded, warty or toothed with fringed lobes. Hair tufts on column arms dense, 3; the middle one facing the labellum; the hairs yellow. Perianth pale or dark lilac. Plant up to 60 cm high. Leaf leathery, keeled. Flowers 3–10. Blue Mts Damp situations. Fl. Dec.–Feb. . . . . **T. circumsepta** Fitzg.

5 Column between the lateral arms tuberculate, dentate or fringed with glands. 6
5 Column between the lateral arms entire, emarginate or deeply v-cleft 8

6 Perianth with spots. Flowers 30–40 mm diam. Plants c. 70 cm high. Middle lobe of the column papillose. The upper 3 segments with darker spots; segments up to c. 20 mm long. Column up to 5 mm long. Plants c. 70 cm high. Flowers few to numerous. Widespread. DSF and heath. Fl. Sept.–Nov. *Dotted Sun Orchid* (hybrids occur between this species, *T. nuda* and *T. pauciflora* giving rise to *T. X merraniae* Nicholls and *T. X truncata* R.S.Rogers respectively. Both have blue–purple flowers but show variation in the characters of the column.) . . . . **T. ixioides** Sw. var. **ixioides**

6 Perianth without spots. 7

7 Leaf 5–10 mm wide. Perianth pale to dark blue, c. 20 mm diam., 1–6 flowers per inflorescence. Perianth segments 8–12 mm long, 4.5–5.5 mm wide. Plants to 40 cm high. Coastal. Fl. Oct.–Dec. . . . . **T. longiloba** D.L.Jones

7 Leaf 10–20 mm wide. Perianth blue or purple, c. 30 mm diam, 7–30 flowers per inflorescence. Perianth segments 12–17 mm long, 6–10 mm wide. Plant up to 90 cm high. Mainly coastal to lower Blue Mts DSF and heath. Fl. Oct.–Jan. *Tall Sun Orchid* . . . .**T. media** R.Br. var. **media**

8 Perianth segments less than 11 mm long. 9

8 Perianth segments 15 mm long or longer 13

9 Plants glaucous. Lower sterile bracts leaf-like, up to 4 in number. Post-anther lobe semi-cylindrical and widely open on the lower side. Plants to 45 cm high. Uncommon. Heath and grasslands. Fl. Oct.–Nov. . . . . **T. planicola** Jeanes

9 Plants without the above combination of characters. Lower sterile bracts not leaf-like 10

10 Leaf lanceolate, flat, thin, less than half the length of the inflorescence, often with purplish markings. Usually 3–10-flowered. Perianth purple, rarely pink or white. Column between lateral arms emarginate to deeply and irregularly split into lobes, brown or reddish orange. Plant to 60 cm high. Widespread. Fl. Sept.–Nov. . . . . **T. brevifolia** Jeanes

10 Leaf linear to lanceolate, fleshy or thin, usually more than half the length of the inflorescence. Column between lateral arms not irregularly split, usually brown to black with a yellow tip. Hair tufts on the column pointing upward. Perianth white, blue, pink or purple. 11

11 Leaf thin, ≥75% the length of the inflorescence. Flowers 2–10, purplish-blue. Lateral lobes of the column 1–1.5 mm long. Column between the lateral arms shallowly bi-lobed at apex. WSF. Coastal. Fl. June – Oct.. . . . . **T. angustifolia** R.Br.

11 Leaf leathery or fleshy, <75% the length of the inflorescence 12

12 Leaf leathery. Flowers 4–10, mauve or deep purplish-blue. Lateral lobes of the column ≥1.2 mm long. Column between lateral arms deeply bi-lobed at apex. Widespread. Fl. Sept.–Nov. . . . .**T. peniculata** Jeanes

12 Leaf fleshy. Flowers 1–3, pale blue, occasionally pink or white. Lateral arms of the column ≤1 mm long. Column between lateral arms entire to emarginate. Widespread. Fl. Sept.–Nov. *Slender Sun Orchid* **T. pauciflora** R.Br.

13 Hairs on lateral arms of the column forming a ciliate margin to the arm, horizontal. Plant up to 45 cm high. Flowers 3–15. Perianth segments usually blue, up to 20 mm long. Widespread. Open forest and heath. Fl. Sept.–Nov. . . . .**T. nuda** R.Br.

13 Hairs on lateral arms of the column in dense tufts 14

14 Hairs on lateral arms of the column white to yellowish. Column between the lateral arms emarginate. Flowers fragrant, 3–8. Perianth usually mauve, up to c. 16 mm long. Plant up to 30 cm high. Widespread in the north of the area. DSF and heath. Fl. Sept.–Nov.. . . . .**T. aristata** Lindl.

14 Hairs on lateral arms of the column mauve. Column between the lateral arms notched. Flowers fragrant, 2–20. Perianth segments blue or white to deep violet, to 20 mm long. Plant to 70 cm high. Coast and adjacent plateaus. DSF in wet sites and wet heath. Fl. Nov.–Jan. . . . . **T. malvina** M.A.Clem., D.L.Jones & Molloy

15 Perianth segments deep pink to cream or yellow 16

15 Perianth segments blue with darker veins or rarely white to pale pink. 17

16 Perianth usually more than 8 mm long. Outer perianth segments acute ending in a fine point. Column to 5 mm long. Lateral arms of column with long, warty teeth. Plant up to 40 cm high. Flowers 1–4. Uncommon in the area. Heath and DSF. Fl. Sept.–Nov. *Red Sun Orchid* . . . . **T. rubra** Fitzg.

16 Perianth up to 7 mm long. Outer perianth segments obtuse or with very short points. Column up to 4 mm long. Lateral arms of column without warty teeth. Plant up to 45 cm high. Flowers 1–4. Widespread. Open communities, often in disturbed sites. Fl. Sept.–Nov. *Pink Sun Orchid* . . . . . . **T. carnea** R.Br.

17 Perianth segments 16–25 mm long. Column up to 8 mm long; lateral arms, not lobed, twisted or tightly coiled. Plant up to 70 cm high. Flowers 1–6. Blue Mts on Ss. Damp situations. Fl. Oct.–Dec. *Veined Sun Orchid* . . . . **T. venosa** R.Br.

17 Perianth segments up to 16 mm long. Column up to 6 mm long' lateral arms often lobed, twisted or loosely coiled. Plant up to 50 cm high. Flowers 1–6. Blue Mts Damp situations at high altitudes. Fl. Nov.–Mar. . . . . **T. cyanea** R.Br.

2 Calochilus R.Br.
Beardies
c. 28 species in Aust. (25 endemic, 3 native); all states and territories

Terrestrial, glabrous herbs with ovoid tubers. Leaf basal, usually long and narrow, conspicuously channelled; loose stem-bracts 1–3. Flowers few–c. 15, in a terminal raceme. Perianth green with red or purplish striae. Dorsal sepal broad, ± hooded; lateral sepals narrower, acute, spreading. Petals much shorter, broad, usually erect. Labellum longer than the sepals, sessile, undivided, with an oblong basal portion and triangular or almost trapezoid lamina most frequently ending in a ribbony filament and with long brilliant purplish or red hairs. Column short and broad, with wide wings. Anther terminal, incumbent or horizontal, with an obtuse beak, with 2 pollen sacs. Pollinia 2 or 4, deeply bilobed: pollen granular.

Key to the species

1 Column without basal glands. Plants 16–90 cm high. Leaf 10–27 cm long. Stem-bracts 2–4. Flowers 2–15. Perianth green; labellum 2 cm long, beset with red hairs. Widespread. Swamps or dry land. Fl. Sept.–Oct. . . . . **C. paludosus** R.Br.

1 Column with a dark gland on either side at the base. 2

2 Column glands connected by a prominent coloured ridge. Plants 18–36 cm high. Leaf c. 20 cm long. Stem-bracts 2–4. Flowers 2–8. Labellum 2 cm long, beset with purple hairs. Open forest. Fl. Sept.–Nov. (Occasionally a specimen occurs which has a labellum without calli or hairs. Here it is considered to be a variant of this species but is recognized as *C.imberbis* R.S.Rogers by some) . . . **C. robertsonii** Benth.

2 Column glands not connected by a coloured ridge or band or band present but obscure. 3

3 Flowers large. Labellum to 45 mm long, tip elongated and glabrous. Perianth pale yellow-green to golden bronze. Leaf to 50 cm long and 5 mm wide. Plants often spindly, to 60 cm high. Coastal heath. North from Gosford. Fl. winter–spring . . . . **C. grandiflorus** (Benth.) Domin

3 Labellum to 35 mm long, clothed with calli and hairs for nearly all its length. Perianth green with purple striae. 4

4 Labellum c. 20 mm long, narrow, becoming filiform towards the apex, densely beset with long reddish-purple hairs almost to the tip. Plants 20–35 cm high. Stem-bracts 2. Flowers 2–8. Woy Woy; Mt Irvine. Open forest. Fl. midsummer . . . . **C. gracillimus** Rupp

4 Labellum c. 15 mm long, two smooth blue–purple plates at top; almost lanceolate, or with a short glabrous ribbon-like prolongation of the apex, rather sparsely beset with coppery red or reddish blue hairs. Leaf erect, 12–30 cm long, sometimes absent. Stem-bracts 1–3, 5–8 cm long. Plant up to 60 cm high. Flowers 7–15. Perianth yellowish-green with reddish-brown or purplish markings. Widespread. WSF. Fl. Sept.–Nov. . . . . **C. campestris** R.Br.

Calochilus therophilis D.L.Jones and *C. montanus* D.L.Jones as described by D.Jones (2006) may occur in the region. They are found in montane habitats above 700 m.

3 Diuris Sm.

45 species endemic Aust.; all states and territories except NT

Terrestrial glabrous herbs with ovoid tubers. Leaves few or occasionally numerous, linear, basal, passing into sheathing bracts above. Flowers several in a terminal raceme, usually with long pedicels and floral bracts. Dorsal sepal erect or reflexed, broad, its basal portion embracing the column. Lateral sepals ± deflexed, usually linear, most frequently green, but occasionally petaloid in shape and colour, sometimes very long. Petals broad, usually on linear claws. Labellum deeply 3-lobed; the middle lobe often much longer and broader than the lateral lobes. Column consisting of a separate stamen and style. Stamen with 2 wings arising from a very short filament, and an anther bearing 2 pollinia facing the back of the style. Style broad, with a viscid disc behind on which the pollinia stick; stigma and rostellum facing the labellum.

Key to the species

1 Lateral lobes of the labellum c. equal to or longer than the middle lobe. 2
1 Lateral lobes of the labellum smaller than the middle lobe 4

2 Flowers orange to golden, 2–5 per spike, 25 mm diam., usually without dark markings, c Lateral sepals 8–13 mm long, c. as long as the petals. Plant up to 50 cm high. Blue Mts; Cumberland Plain. Open forest. Endangered. Fl. spring–summer. *Buttercup Double-tail* . . . . . . **D. aequalis** F.Muell. ex Fitzg.
2 Flowers yellow, with darker markings 3

3 Perianth spotted with brown chiefly on the back. Lateral sepals 14–16 mm long; dorsal sepal 7–10 mm long. Flowers 2–8. Plant up to 30 cm high. Leaves 2–3. Widespread. Open forest. Fl. July–Sept. *Leopard Orchid* or *Spotted Double-tail*. . . . . **D. maculata** Sm.
3 Perianth heavily diffused and blotched with darker colour. Lateral sepals 8–18 mm long; dorsal sepal 5–12 mm long. Flowers 1–10. Plant up to 50 cm high. Leaves 2–3. Widespread. Open forest. Fl. spring–early summer. *Leopard Orchid* . . . . **D. pardina** Lindl.

4 Perianth white to purple. 5
4 Perianth yellow to orange sometimes with purplish markings 7

5 Flowers more than 40 mm diam. Perianth purplish to lilac. Lateral sepals 30–90 mm long, more than twice as long as the petals. Petals 10–30 mm long. Plants up to 35 cm high, with 2–3 leaves. Flowers 2–10. Widespread. Open forest. *Purple Donkey Orchid* . . . . **D. punctata** Sm. var. **punctata**
5 Flowers up to 40 mm diam.. 6

6 Perianth mauve with darker markings. Lateral sepals 35–65 mm long, usually more than twice as long as the sepals. Petals 6–12 mm long. Plants up to 40 cm high, with 1–2 leaves. Blue Mts Open forest. Fl. spring–summer. . . . . **D. dendrobioides** Fitzg.
6 Perianth white with purple markings. Lateral sepals 20–40 mm long. Petals 10–22 mm long. Plants up to 40 cm high, with 1–3 leaves. Coast in north of the area. Open forest. . . . . **D. alba** R.Br.

7 Lateral sepals much longer than the petals, c. 30 mm long, green. Petals 17–18 mm long. Perianth yellow with purple splashes about the centre; petal claws purple. Plant up to 42 cm high, with 2–3 leaves. Flowers 3–6. Blue Mts Open forest. Vulnerable. Fl. Oct.–Nov.. . . . . **D. tricolor** Fitzg.
7 Lateral sepals c. as long as the petals 8

8 Lateral sepals crossed beneath the labellum 9
8 Lateral sepals parallel or divergent. 11

9 Peduncles with c. 5 leafy bracts 2–5 cm long. Perianth yellow, with a few dark brown spots; the dorsal sepal with 2 brown blotches at the base; the lateral sepals brown with green at the ends, c. 10 mm long. Gladesville in 1889. Endangered. Fl. Aug.–Sept. . . . . **D. bracteata** Fitzg.
9 Peduncles with 2–3 bracts which are not leafy 10

10 Perianth orange to golden yellow. Dorsal sepal broad-ovate, 7–14 mm wide. Plant 15–60 cm high. Leaves 2. Flowers 2–5. Widespread. Open forest and heath. Fl. Aug.–Oct. . . . . **D. aurea** Sm.

10 Perianth lemon yellow. Dorsal sepal ovate, c. 5 mm wide. Plant up to 60 cm high. Leaves 2. Flowers 2–8. Blue Mts Open forest. Fl. Oct . . . . **D. platichila** Fitzg.

11 Petals erect or divergent 12
11 Petals drooping 13

12 Labellum with a single callus ridge. Plant up to 70 cm high, with 2–3 leaves. Flowers 3–6. Perianth yellow with brown markings; the lateral sepals green; dorsal sepal c. 20 mm long. Petals 20–30 mm long. Widespread. Open forest. Fl. Aug.–Nov. *Tiger Orchid* . . . . **D. sulphurea** R.Br.
12 Labellum with two callus ridges. Plant up to 40 cm high, with 2–3 leaves. Flowers 6–10. Perianth yellow with a few darker markings. Lateral sepals 12–15 mm long; dorsal sepal 9–11 mm long. Petals 8–12 mm long. Coast in north of the area. Open forest. Vulnerable. Fl. July–Sept . . . .**D. praecox** D.L.Jones

13 Leaves 2. Perianth clear yellow, with orange labellum, with green lateral sepals. Lateral sepals 12–20 mm long, directed upwards. Plant up to 20 cm high. Flowers 1–2.. Recorded from Hawkesbury River area. Open forest. Endangered. Fl. spring–summer . . . . **D. pedunculata** R.Br.
13 Leaves 3–9. Perianth clear yellow. Lateral sepals 12–25 mm long, brownish-green, directed upwards. Petals 8–20 mm long. Plant up to 40 cm high. Flowers 1–4. Widespread. Open forest. Fl. winter–spring. *Snake Orchid*. . . . .**D. chryseopsis** D.L.Jones

HYBRIDS: Many named hybrids between species of *Diuris* occur. *D.* X *polymorpha* is a hybrid between *D. chryseopsis* and *D. platichila* it has a light yellow perianth with dark marks on the sepals and labellum and occurs in the Blue Mts *D.* X *nebulosa* is a hybrid between *D. punctata* and *D. aurea* it has a pink and yellow perianth and labellum and occurs across the region.

4 Orthoceras R.Br.
1 species native Aust.; Qld, NSW, Vic., Tas., S.A.

Monotypic genus

Terrestrial, glabrous herb 20–60 cm high, with ovoid tubers. Leaves basal, linear, 2–5, channelled below. Flowers 1–9, with large floral bracts, almost sessile, in a terminal raceme, or rarely solitary. Dorsal sepal erect, incurved, hooded, 10–12 mm long; lateral sepals linear, long and straight, erect or spreading, brown or green. Petals very short, blunt, erect. Labellum sessile, 9–12 mm long, 3-lobed, recurved; lateral lobes much shorter than the middle lobe. Column short, with 2 wings which are not as high as the anther. Anther incurved, tapering to an obtuse point, with 2 pollen sacs; pollinia 2, bilobed, mealy, without a caudicle. Stigma erect in front of the anther. Chiefly sandy soils. *Horned Orchid, Bird's-mouth Orchid* . . . . **O. strictum** R.Br.

5 Microtis R.Br.
Onion Orchids
c. 12 species in Aust.; all states and territories except NT

Terrestrial, glabrous herbs with globular tubers. Leaf solitary, sheathing from the base of the stem to half or two-thirds of the distance to the base of the inflorescence, then extending into a free terete lamina. Flowers very small, numerous in a terminal spike. Perianth green. Dorsal sepal ± hooded, 2–3 mm long. Lateral sepals lanceolate to oblong, not appreciably longer than the dorsal one, spreading or much reflexed. Petals narrower and not much shorter than the sepals, spreading or incurved. Labellum sessile, not lobed, very obtuse, occasionally emarginate, usually oblong, deflexed or recurved, with raised callosities on the upper surface. Column very short and broad with 2 wings or auricles which are not as high as the apex of the anther. Anther erect, with 2 pollen sacs; pollinia 2, bilobed, granular; caudicle usually present.

Key to the species

1 Upper surface of the labellum with 2 calli at the base and none at the apex. Margin of labellum entire or ± crenulate, tip apiculate; lamina granulated near the tip, with 2 cushion-like elevations towards the base and 2 inconspicuous horizontal glands at the extreme base (these may be absent). Plant 25–60 cm high. Widespread. Fl. spring – summer . . . . **M. parviflora** R.Br.

1 Upper surface of the labellum with 2 callosities at the base and 1 at the apex 2

2 Petals spreading below dorsal sepal. Labellum at least ⅔ as long as ovary, 3–5 mm long, emarginate, margins wavy. Flowers well spaced, yellowish-green to bright green. Petals more than 2 mm long. Lateral sepals spreading, recurved or tightly revolute. Plant up to 50 cm high (usually much less). Widespread. Fl. Oct.–Nov. *Scented Onion Orchid* . . . . **M. rara** R.Br.
2 Petals erect held within the dorsal sepal. Labellum less than half as long as ovary. Flowers crowded, green. 3

3 Calli at the base of the labellum not united. Petals 2 mm long. Labellum 1–3 mm long, emarginate. Lateral sepals reflexed or rolled. Ovary Plant 15–60 cm high. Widespread. Chiefly in grassland. Fl. Sept.–Nov. *Common Onion Orchid*. . . . . **M. unifolia** (G.Forst.) Rchb.f.
3 Calli at the base of labellum united. Petals less than 2 mm long. Labellum 1.6–1.8 mm long, emarginate, papillose on margin. Plant to 60 cm high. Known only from Mona Vale. Shale. Woodland. Endangered. Fl. May–Oct. . . . . **M. angusii** D.L. Jones

6 **Prasophyllum** R.Br.
Leek Orchids
58 species in Aust. (57 endemic, 1 native); all states and territories except NT

Terrestrial glabrous herbs, usually with ovoid tubers. Leaf solitary, closely sheathing the stem for half to the whole of the distance to the base of the inflorescence, then produced into a terete lamina. Flowers not resupinate, usually numerous but occasionally few, in a terminal spike. Dorsal sepal ± lanceolate, concave on the inner side, often recurved; lateral sepals usually narrower but quite as long as the dorsal one, free or ± connate. Petals shorter and narrower than the sepals. Labellum sessile, undivided, oblong to lanceolate-acuminate; margins crisped, denticulate, fringed or quite smooth; upper surface with a longitudinal callus of varying form, often bordered at least in part by a membrane; basal portion ± erect; the lamina often variously curved. Column short, with a variable lateral wing on either side. Anther with 2 pollen sacs, erect behind the stigma; pollinia 2, bilobed; caudicles present.

Key to the species

1 Lateral sepals free 2
1 Lateral sepals connate for at least part of their length 7

2 Labellum reflexed back on itself, margins undulate; callus bright green, extending beyond the bend. Perianth white with yellow-green or pink, 13–17 mm long. Dorsal sepal 9–11 mm long. Plant 24–70 cm high. Widespread. Fl. Aug–Jan . . . . **P. odoratum** R.S.Rogers
2 Labellum variously bent but not reflexed back on itself. Perianth predominantly green or brown but white and green in *P. patens* 3

3 Labellum margins undulate, crenate or pleated 4
3 Labellum margins not as above, ± entire 5

4 Callus slightly raised, not extending past the bend. Labellum pale or white. Flowers 8–11 mm long. Dorsal sepal 5–6 mm long. Plant up to 30 cm high. Widespread. Fl. Sept.–Dec. . . . . . **P. patens** R.Br
4 Callus distinctly raised, extending beyond the bend to tip of labellum 5

5 Flowers reddish-brown. Labellum sharply recurved in middle and extending into tapering tail. Dorsal sepal c. 9 mm long. Plants 30–45 cm high. Georges River. Fl. Oct.–Dec. . . . . **P. fuscum** R.Br.
5 Flowers pink or green-white. Labellum sharply recurved and constricted near the middle. Dorsal sepal c. 10 mm long.; Plant 14–40 cm high. Blue Mts Fl. spring–summer . . . . **P. pallens** D.L.Jones

6 Callus well developed before the bend and extending to ¾ labellum length. Column wings pink. Plant 30–45 cm high. Swampy areas. Moss Vale area. Endangered. Fl. early summer . . **P. uroglossum** Rupp
6 Callus developed at bend and extending past it. Column wings dull red. Heath or DSF in damp places. Fl. early summer . . . . **P. appendiculatum** Nicholls

7 Perianth yellow. Plant 25–65 cm high, tubers on rhizomes distant from base of stem. Petals and sepals 10 mm long. Widespread. Fl. Nov.–Jan. . . . .**P. flavum** R.Br.

7 Perianth not yellow. Plants with tubers at base of stem 8

8 Labellum green to purplish. Ovary ovoid, turgid, not appressed to the stem. Flowers very strongly scented, c. 40. Perianth segments up to 12 mm long. Plants up to 50 cm high. Coast and adjacent plateaus. Dry heath. Endangered. Fl. summer . . . . **P. affine** Lindl.

8 Labellum white. Ovary appressed to stem or not. 9

9 Ovary narrow-ovoid almost appressed to stem. Perianth brown or purple. Plant up to 40 cm high. Dorsal sepal 7–11 mm long. Labellum reflexed at an angle greater than 90°. Widespread. Fl. Aug. –Jan.
. . . . **P. brevilabre** (Lindl.) Hook.f.

9 Ovary cylindrical, appressed to the stem 10

10 Plant 6–30 cm high. Petals and sepals striate green and white. Dorsal sepal triangular, acute, 10 mm long. Labellum with narrow claw. Widespread. Fl. April–June . . . .**P. striatum** R.Br.

10 Plants usually more than 30 cm and up to 120 cm high. Dorsal sepal ovate to lanceolate. 11

11 Labellum gibbous at the base, reflexed at an angle greater than 90°. Dorsal sepal 8 mm long. Plant 30–90 cm high. Perianth green white and brown, striate. Widespread. Fl. Nov.–Dec. . . **P. australe** R.Br.

11 Labellum not gibbous at the base, reflexed at an angle less than 90°. Dorsal sepal up to 11 mm long. Plant up to 120 cm high. Perianth green yellow-green or brown. Widespread. Fl. Aug.–Oct.. . . . . . .
. . . .**P. elatum** R.Br.

7 **Genoplesium** D.L.Jones
Midge Orchids
40 species in Aust. (37 endemic, 3 native); all states and territories except NT

Glabrous terrestrial herbs with paired tubers. Leaf solitary, sheathing the stem for most of its length, terete below, flattened above the sheath. Flowers not resupinate, usually numerous in a terminal spike. Dorsal sepal ± lanceolate, free, concave on the inner side; lateral sepals usually narrower and slightly longer, free or ± connate. Petals shorter and narrower than the dorsal sepal. Labellum clawed, usually articulate to the base of the column, with a single longitudinal callous; margin fringed to smooth. Column short, with a variable lateral wing on either side; anther with 2 pollen sacs, erect behind the stigma; pollinia 2, bilobed; caudicles present.

D.L.Jones (2006) places most of the following species in a new genus *Corunastylis*

Key to the species

1 Yellowish-green to purplish saprophytic plant up to 15 cm high. Flowers c. 15 mm diam., green to reddish. Coast and adjacent plateaus. DSF and heath. Vulnerable. Fl. summer–autumn **G. baueri** R.Br.

1 Green autotrophic plants 2

2 Labellum entire to dentate but never ciliate. 3

2 Labellum ciliate. 6

3 Bract emerging just below the flowers and just touching them; the tip sometimes exceeding the inflorescence. Flowers c. 3 mm diam. Labellum ± toothed. Plant up to 18 cm high. Widespread. DSF and heath. Ss. Fl. summer . . . . **G. nudiscapum** (Hook.f.) D.L.Jones & M.A.Clem.

3 Bract emerging well below the flowers, usually appressed to the stem. 4

4 Labellum acuminate. Flowers 3–4 mm diam., dark purple to greenish purple. Labellum entire or toothed. Plant up to 25 cm high. Coast in south of area and Illawarra range. DSF and heath. Fl. summer–autumn. **G. despectans** (Hook.f.) D.L.Jones & M.A.Clem.

4 Labellum acute or obtuse 5

5 Perianth red to reddish-brown, 3–4 mm diam. Labellum entire to toothed. Plant up to 25 cm high. Widespread. DSF and heath. Fl. summer–autumn. . . . . . . **G. rufum** (R.Br.) D.L.Jones & M.A.Clem.

5 Perianth green to yellowish-green, sometimes with red markings. Flowers c. 3 mm diam. Plant up to 20 cm high. Coast to lower Blue Mts DSF and heath. Fl.summer–autumn . . . .
. . . . **G. pumilum** (Hook.f.) D.L.Jones & M.A.Clem.

6 Labellum ciliate; other perianth segments not ciliate 7
6 Labellum, and at least one other perianth segment, ciliate 11

7 Flowers 2–3 mm diam., green and red to purplish. Dorsal sepal entire. Plant up to 30 cm high. Widespread. DSF in damp areas. Fl. summer–autumn . . **G. nudum** (Hook.f.) D.L.Jones & M.A.Clem.
7 Flowers 4–6 mm diam. 8

8 Flowers purplish 9
8 Flowers green to brownish 10

9 Flowers c. 4 mm diam. Dorsal sepal entire. Petals ovate, 3–3.5 mm long. Plant up to 20 cm high. Widespread. Open forest. Fl. summer–autumn . . . . . . **G. archeri** (Hook.f.) D.L.Jones & M.A.Clem.
9 Flowers c. 5 mm diam. Dorsal sepal entire. Petals lanceolate–narrow ovate, 4.5–5 mm long. Plant to 18 cm tall. Heath on sand. Wyong area. Endangered. Fl. spring. . . . . **G. insignis** D.L.Jones

10 Lateral sepals more than 7 mm long. Labellum oblong, c. 5 mm long. Dorsal sepal entire. Plant up to 20 cm high. Coast and adjacent plateaus south of Sydney. DSF. Ss. Endangered. Fl. summer–autumn
. . . .**G. plumosum** (Rupp) D.L.Jones & M.A.Clem.
10 Lateral sepals up to 6 mm long. Labellum obovate, c. 4 mm long. Dorsal sepal entire. Plant up to 20 cm high. Widespread. DSF and heath. Fl. summer–autumn . . . .
. . . . **G. sagittiferum** (Rupp) D.L.Jones & M.A.Clem.

11 Petals and labellum ciliate 12
11 Petals, dorsal sepal and labellum ciliate 13

12 Lateral sepals deflexed. Flowers purplish, 5–6 mm diam. Petals coarsely ciliate. Plant up to 25 cm high. Kanangra-Boyd National Park. Open forest. Fl. summer . . . . **G. morinum** D.L.Jones
12 Lateral sepals erect. Flowers c. 4 mm diam., purplish . . . .**G. archeri** (see above)

13 Cilia usually glandular, very short 14
13 Cilia not glandular, long (c. 0.5 mm) 17

14 Lateral sepals spreading. Labellum thin. Flowers c. 5 mm diam., greenish to purple. Plants up to 40 cm high. Widespread north of Sydney. DSF and heath. Fl. summer–autumn. . . . .
. . . .**G. filiforme** (Fitzg.) D.L.Jones & M.A.Clem.
14 Lateral sepals deflexed. Labellum fleshy 15

15 Flowers less than 4 mm diam., purple to brownish purple. Plants up to 20 cm high. Mt Wilson. DSF and heath. Fl. summer . . . . **G. eriochilum** (Fitzg.) D.L.Jones & M.A.Clem.
15 Flowers 4–6 mm diam.. 16

16 Flowers greenish with purple labellum. Plants to 15 cm high. In north of region. Swampy to heath areas. Fl. summer–autumn . . . . **G. ruppii** (R.S.Rogers) D.L.Jones
16 Flowers purple with darker markings. Plants up to 40 cm high. Widespread. Open forest and heath. Fl. autumn. . . . .**G. woollsii** (F.Muell.) D.L.Jones & M.A.Clem.

17 Labellum with red to pink cilia 18
17 Labellum with brown or dark purplish cilia. 19

18 Labellum hinged and movable on the hinge. Flowers c. 10 mm diam., lemon-scented. Plant up to 30 cm high. Widespread. DSF and heath. Fl. summer–autumn . . . .
. . . . **G. fimbriatum** (R.Br.) D.L.Jones & M.A.Clem.
18 Labellum not hinged and not movable. Flowers c. 5 mm diam., sometimes remaining closed. Plant up to 40 cm high. Coast and adjacent plateaus; lower Blue Mts Open forest and heath. Fl. summer–autumn. . . . .**G. apostasioides** (Fitzg.) D.L.Jones & M.A.Clem.

19 Flowers c. 5 mm diam., very dark purple, lemon-scented. Plants up to 40 cm high. Blue Mts DSF in moist sites. Fl. summer–autumn . . . . **G. citriodorum** D.L.Jones & M.A.Clem.

19 Flowers 7–8 mm diam., very dark purple, without a strong lemon-scent. Plants up to 50 cm high. Blue Mts; Illawarra ranges. DSF and heath. Fl. summer–autumn . . . . **G. simulans** D.L.Jones

8 Caleana R.Br.
One species in the area

Terrestrial herb 15–50 cm high, with underground tubers. Leaf solitary, basal, oblong to lanceolate, 6–10 cm long. Flowers 1–4, not resupinate. Sepals and petals linear, red-brown, rarely green, 20 mm long. Labellum smooth. Column without an elongated foot, appressed to the ovary. Widespread. Fl. spring–summer. *Bee Orchid* or *Duck Orchid* . . . . **C. major** R.Br.

9 Paracaleana Blaxell
One species in the area

Terrestrial herb 6–17 cm high, with tubers. Leaf solitary, linear, 4–12 cm long. Flowers 1–4, not resupinate. Perianth green or dull red-brown, c. 8 mm long. Labellum tuberculate. Column with an elongated foot; the wings decurrent on the column. Coast; Blue Mts Fl. spring–summer. . . . . . **P. minor** (R.Br.) Blaxell

10 Arthrochilus F.Muell.
10 species in Aust. (8 endemic); all states and territories except WA

Terrestrial herbs with small tubers. Leaves 2–5 (absent in one species), basal, oblong to lanceolate, prostrate, usually but not always separate from the flowering stem, sometimes absent at flowering time. Flowers in a terminal raceme, sometimes numerous. Sepals and petals very slender, somewhat similar. Dorsal sepal erect; lateral sepals and petals spreading or deflexed. Labellum articulate by a long slender claw with the column foot; lamina narrow, peltate, hammer-shaped or insectiform; upper lobe either emarginate or separated into 2 long divergent filiform tails. Column elongate, very slender, incurved or erect or reflexed towards the ovary, with 2 wing-like auricles on either side of its upper part; foot either absent or rudimentary or long. Anther erect, obtuse, with 2 pollen sacs. Rostellum almost obsolete.

Key to the species

1 Basal leaves absent. A few loose bracts present on the stem. Sepals and petals 5 mm long, green with darker tints. Plant 6–15 cm high. Flowers 3–5. Blue Mts Uncommon. Fl. summer. *Elbow Orchid*. . . . .
. . . . **A. huntianus** (F.Muell.) Blaxell

1 Basal leaves prostrate, 2–5. Dorsal sepal 12–14 mm long; lateral sepals and petals shorter, green with darker tints. Plants 5–37 cm high. Flowers 3–20, in a loose spike. Coast. Fl. chiefly summer . . . .
. . . . **A. prolixus** D.L.Jones

11 Chiloglottis R.Br.
18 species in Aust. (15 endemic); Qld, NSW, Vic., Tas., S.A.

Small terrestrial herbs with ovoid or globular tubers. Leaves 2, basal ± prostrate. Flowers usually solitary, on a comparatively short stem which usually elongates considerably after fertilization of the ovary. Dorsal sepal erect or incurved, ± contracted at or towards the base. Lateral sepals linear. Petals narrow-lanceolate. Labellum undivided, sessile, but the lamina contracted into a long or short horizontal claw towards the base, beyond this expanding widely and provided on the upper surface of the expansion with variously shaped and arranged prominent calli. Column elongate, slender, rather narrowly winged. Anther erect, higher than the rostellum; pollinia granular.

Key to the species

1 Petals diverging, sometimes curving upwards 2
1 Petals reflexed against the ovary 3

2 Perianth green to yellowish. Petals c. 15 mm long. Labellum with c. 12 calli not forming a flange towards the base with the margin. Plant up to c. 15 cm high. Coast and adjacent plateaus. Damp, sheltered places. Fl. spring–summer . . . . **C. chlorantha** D.L.Jones

2 Perianth reddish to purplish-brown. Petals 15–18 mm long. Labellum with c. 12 calli forming a flange towards the base with the margin. Plant up to c. 15 cm high. Blue Mts Open forest. Fl. summer . . . . .
. . . . **C. pluricallata** D.L.Jones

3 Lateral sepals with osmophores less than 1 mm long. Flowers 12–16 mm across. Labellum with a shiny black callous 4

3 Lateral sepals with osmophores more than 2 mm long 5

4 Labellum with a callus at the base only. Perianth brownish. Plant up to c. 10 cm high. Widespread. Open forest. Moist places. Fl. spring . . . . **C. trapeziformis** Fitzg.

4 Labellum with a callus reaching to the apex. Perianth greenish brown. Plant up to c. 10 cm high. Widespread. Open forest. Moist places. Fl. spring. *Ant Orchid* . . . . **C. formicifera** Fitzg.

5 Osmophores yellowish. 6

5 Osmophores reddish 7

6 Flower more than 28 mm long, greenish brown. Lateral sepals ± parallel with osmophores more than 5 mm long. Plant up to c. 12 mm high. Coast and adjacent plateaus. Open forest. Moist places. Fl. summer–autumn. . . . . **C. diphylla** R.Br.

6 Flower up to c. 25 mm long, greenish pink. Lateral sepals divergent, with osmophores up to 3 mm long. Widespread. Open forest and RF. Moist places. Fl. summer–autumn . . . .
. . . . **C. sylvestris** D.L.Jones & M.A.Clem.

7 Labellum with a black callus on the lower two-thirds, obtuse. Lateral sepals decurved. Petals reflexed. Flowers green to reddish, 10–14 mm long. Plants up to c. 15 cm high. Blue Mts Open forest in moist sheltered sites. Fl. summer–autumn . . . . **C. seminuda** D.L.Jones

7 Labellum with a black callus almost to the apex 8

8 Labellum mucronate to almost obtuse. Lateral sepals reflexed and incurved beneath the labellum. Petals reflexed. Flowers greenish to reddish, 25–30 mm long. Plants up to c. 15 cm high. Widespread. Open forest in moist places. Fl. summer–autumn. . . . . **C. reflexa** (Labill.) Druce

8 Labellum usually with a short tail. Lateral sepals recurved but not incurved beneath the labellum. Petals reflexed. Flowers greenish brown, 30–35 mm long. Plants up to c. 15 cm high. Widespread. Open forest in moist places. Fl. summer–autumn . . . . **C. trilabra** Fitzg.

12 Acianthus R.Br.
13 species in Aust. (12 endemic); Qld, NSW, Vic., Tas., S.A.

Small terrestrial glabrous herbs with globular tubers. Leaf solitary, often close to the base of the stem, cordate to ovate or reniform, lower surface reddish-purple (Fig. 44). Flowers in a terminal raceme, rarely solitary, usually very small. Dorsal sepal erect or incurved, occasionally hooded, often produced into a fine point. Lateral sepals narrower than the dorsal sepal, usually spreading. Petals often shorter, reflexed or spreading. Labellum undivided, spreading, with 2 basal calli. Column long, rather slender, almost terete, incurved, seldom winged. Anther broad, erect, with 2 pollen sacs, valvate; pollinia granular or mealy; caudicle absent.

Key to the species

1 Sepals long-filiform; the dorsal sepal up to 40 mm long. Stems 6–16 cm high. Flowers 1–6, reddish-purple. Chiefly coast. Sandy soils. Fl. July–Sept. *Mayfly Orchids* . . . . **A. caudatus** R.Br.

1 Sepals not long-filiform, up to 20 mm long. Perianth brown, green or pinkish 2

2 Petals spreading. Column covered by the dorsal sepal. 3

2 Petals reflexed against the ovary. Column prominently exserted. 4

3 Labellum shortly acuminate, callus cordate. Plant 5–22 cm high. Flowers 2–15. Perianth greenish-pink 7–10 mm long. Widespread in coastal areas. Fl. winter. *Pixie Caps* . . . . **A. fornicatus** R.Br.

3 Labellum with a long acuminate tip; callus narrow-cordate. Plant to 20 cm high. Flowers 2–9. Perianth pink, 9–12 mm long. Moist forest on tablelands. Fl. autumn . . . . **A. apprimus** D.L.Jones

4 Flowers 8–12 mm long, green to pink. Labellum up to 5 mm long. Plant up to 20 cm high. Widespread. Open forest and margins of RF. Fl. autumn–winter. *Gnat Orchid* . . . . **A. pusillus** D.L.Jones

4 Flowers 12–16 mm long, purplish. Labellum 5 mm long or longer. Plant 6–30 cm high. Flowers 1–10. Perianth usually pale reddish-brown, sometimes green, up to 20 mm long. Widespread. Fl. May–Sept. *Mosquito Orchid* . . . . .A. exsertus R.Br.

13 Acianthella D.L.Jones & M.A.Clem.
2 species endemic Aust.; Qld, NSW

One species in the area
Plant 5–9 cm high. Leaf broad, entire or lobed, green on both surfaces. Flowers green, 2–20, c. 3 mm wide. Sepals and petals narrow, 2–3 mm long, spreading to decurved. Labellum oblong, c. 2 mm long with 3–5 course teeth. Callus extending to half the length of the labellum. Uncommon. Coastal scrub and littoral RF. Fl. April–May . . . . **A. amplexicaulis** (F.M.Bailey) D.L.Jones & M.A.Clem.

14 Cyrtostylis R.Br.
5 species in Aust. (4 endemic, 1 native); all states and territories except NT

One species in the area
Plant 5–27 cm high. Leaf solitary, reniform to orbicular, 15–40 mm long; upper surface greyish; lower surface crystalline. Flowers 1–8, sessile, reddish-brown or rarely yellowish-green, 9–15 mm long. Sepals and petals almost equal in length. Labellum entire to denticulate, with 2 basal calli and 2 longitudinal ridges. Petals spreading or curving downwards. Column slender, c. 5 mm long. Widespread. Chiefly sandy soils. Fl. June–Oct. *Mosquito Orchid* . . . . **C. reniformis** R.Br.

15 Eriochilus R.Br.
6 species endemic Aust.; all states and territories except NT

Tubers globular. Leaf solitary, basal, ovate to broad-lanceolate, glabrous, often purple beneath, cordate at the base. Flowers 1–5, terminal. Perianth pink or white. Stem bractless above the leaf. Dorsal sepal erect, slightly concave, sometimes with undulate edges; lateral sepals on a slender claw; the lamina lanceolate. Petals linear almost as long as the dorsal sepal. Labellum equal to or shorter than the petals, on an erect claw embracing the column; the basal margins often produced into small lateral lobes; middle lobe of lamina much recurved, convex above. Column slightly curved or erect, sometimes ciliate in front, narrowly winged. Stigma concave. Anther erect, blunt, valvate, with 2 pollen sacs; pollinia 8, 4 pyriform masses in each pollen sac, waxy or granular.

Key to the species

1 Upper surface of the leaf hairy, with prominent veins, ovate to orbicular, up to 15 mm long. Flowers 1–3, white to pale pink. Lateral sepals up to 10 mm long, linear to filiform. Plants up to c. 30 cm high. Coast to lower Blue Mts Heath in moist places. Fl. autumn–winter . . . **E. petricola** D.L.Jones & M.A.Clem.

1 Upper surface of leaf glabrous, without prominent veins, ovate, up to 35 mm long. Flowers 1–5, white to deep pink. Lateral sepals 10–17 mm long. Plants up to c. 25 cm high. Widespread. DSF and heath. Moist places. Fl. Jan.–May. *Parson's Bands* . . . . **E. cucullatus** (Labill.) Rchb.f.

16 Rimacola Rupp
1 species endemic Aust.; NSW

Monotypic genus
Semi-terrestrial plant with succulent terete rhizomes. Basal leaves several, elliptic-lanceolate, petiolate, 3–5 cm long; cauline leaves 1–3, ovate, loosely stem-clasping. Flowers 6–18, rather crowded in a weak terminal raceme, often drooping. Perianth green. Sepals acuminate; the dorsal one ± hooded but straight or slightly curved towards the tip; lateral sepals spreading, hardly divergent. Petals shorter, obtuse, slightly deflexed. Labellum on a slender claw, undivided, white, with dark red stripes; calli rather obscure, sometimes obsolete. Column slender, elongate, very narrowly winged above. Anther acute; pollinia 4 in 2 pairs, granular. Stigma conspicuous. Coast to Blue Mts Clay crevices or wet dripping ledges of sandstone cliffs. Fl. summer, occasionally at other times. *Green Beaks* . . . . **R. elliptica** (R.Br.) Rupp

17 Lyperanthus R.Br.
3 species endemic Aust.; all states and territories except NT

One species in the area
Terrestrial herbs 18–44 cm high, with ovoid or globular tubers. Basal leaf solitary, linear, erect, coriaceous, 12–26 cm long. Plant Cauline bracts 1–3. Flowers in a terminal raceme, 2–6, with large bracts. Perianth red-brown to yellow-green. Dorsal sepal 20 mm long, incurved and hooded, ± deflexed towards the tip. Lateral sepals c. as long as but narrower than the dorsal sepal, erect or spreading; petals similar. Labellum much shorter, sessile or almost so, 3-lobed near the base, ± papillose; the lamina with calli, sometimes ridged. Middle lobe of the labellum entire. Column incurved, c. as long as the labellum, sometimes obscurely winged. Anther erect with 2 sacs; pollinia 2 or 4, granular or mealy. Coast. Fl. Aug.–Oct. *Brown Beaks* . . . . **L. suaveolens** R.Br.

18 Pyrorchis D.L.Jones and M.A.Clem.
1 species endemic Aust.; all states and territories except Qld, & NT

Monotypic genus
Terrestrial herbs with ovoid or globular tubers. Basal leaf solitary; cauline bracts usually 2. Flowers in a terminal raceme, rarely solitary, usually 2–8. Dorsal sepal incurved and hooded, ± deflexed towards the tip; lateral sepals c. as long as but narrower than the dorsal sepal, erect or spreading; petals similar. Labellum much shorter, sessile or almost so, 3-lobed near the base, ± papillose; the lamina with calli, sometimes ridged. Column incurved, c. as long as the labellum, sometimes obscurely winged. Anther erect with 2 sacs; pollinia 2 or 4, granular or mealy.

One species in the area
Leaf broad-ovate, prostrate, fleshy, up to 12 cm long, 6 cm wide. Middle lobe of the labellum fimbriate or dentate. Plant 10–30 cm high, jet black when dry. Cauline bracts 2, erect, loosely sheathing. Flowers 1–8. Perianth whitish with purplish red striae. Dorsal sepal 20 mm long. Coast. Fl. Sept.–Oct. *Red Beaks* or *Undertaker Orchid* . . . . **P. nigricans** (R.Br.) D.L.Jones & M.A.Clem.

19 Burnettia Lindl.
1 species endemic Aust.; NSW, Vic., Tas.

Monotypic genus
Dwarf, terrestrial herb with small globular tubers. Leaves 1–2, ovate-lanceolate, basal, usually absent at flowering time. Cauline bracts 1–2, loosely clasping. Plant 5–10 cm high. Flowers 2–5, large for the size of the plant. Perianth reddish or purplish-brown outside, white inside, sometimes with conspicuous dark veins. Dorsal sepal 12–15 mm long, broad-lanceolate, concave, hardly hooded; lateral sepals and petals as long as dorsal sepal, not very widely expanding; petals narrower. Labellum sessile, short, undivided, erect at the base, recurved towards the tip, with 2 longitudinal ridges or raised plates broken up into calli above the middle. Column c. 7 mm long, incurved, winged. Anther erect, with 2 pollen sacs; pollinia 2, granular. Usually sandy soils. Uncommon. Fl. autumn. *Lizard Orchid* . . . . **B. cuneata** Lindl.

20 Cyanicula Hopper & A.P.Br.
9 species endemic Aust.; Qld, NSW, Vic., WA

One species in the area
Terrestrial herb to 20 cm high, usually less. Leaf linear 7 cm long, sparsely hairy. Inflorescence 1-flowered. Perianth blue (rarely white) up to 17 mm long; outer surface with sparse blue hairs. Labellum cuneate 3-lobed. Calli in 2 rows, golden yellow. Widespread. Open forests. Fl. spring. . . . .
. . . . **C. caerulea** (R.Br.) Hopper & A.P.Br.

21 Caladenia R.Br.
c. 150–250 species in Aust.; all states and territories except NT

Terrestrial herbs usually ± hirsute on the stem ovary and leaf, with small globular tubers. Leaf solitary, basal, linear to lanceolate. Flowers sometimes large, solitary or in small terminal racemes of 2–6 flowers. Cauline bract 1, small. Dorsal sepal erect, usually ± incurved behind the column sometimes deeply so, usually rather narrow; lateral sepals similar, sometimes smaller, spreading or reflexed. Petals narrower and often shorter than the sepals, erect spreading or reflexed. Labellum usually on a moveable claw, entire or 3-lobed, erect at the base; anterior portion recurved; margins often fringed or variously toothed; lateral lobes erect; middle lobe with stalked or sessile calli of various forms in longitudinal rows or more rarely distributed irregularly. Column ± incurved, rather long, winged above. Anther terminal, oblique, pointed, with 2 pollen sacs, valvate; pollinia 4, without a caudicle. Stigma discoid.

Key to the species

1 Sepals finely acuminate, often produced into long tails, sometimes club-shaped at the tip. Petals similar but usually narrower and shorter, not club-shaped at the tip. Sepals and petals more than twice as long as the labellum. *Spider Orchids* 2
1 Sepals and petals acute, c. as long as or less than twice as long as the labellum 8

2 Sepal tails filiform to the end 3
2 Sepals clubbed at the end 6

3 Base of column without yellowish glands. Perianth crimson or reddish-purple. Labellum entire. Sepals 4–7 cm long. Plants up to 20 cm high. Widespread. Fl. spring. *Daddy Long-legs*. . **C. filamentosa** R.Br.
3 Base of column with yellowish glands. Perianth in shades of red, brown, green, yellow-green, or white. . 4

4 Sepals up to 25 mm long. Calli crowded and overlapping towards base of labellum. Sepals neither caudate nor clavate. Plant 10–26 cm high. Widespread. Endangered. Fl. Sept.–Oct. . **C. tesselata** Fitzg.
4 Sepals 4–8 cm long. Calli not crowded 5

5 Labellum margins entire or minutely denticulate on the red-brown middle lobe. Plant up to 30 cm high, very hairy. Perianth red and green, tails with sparse short hairs. Sepals up to 5 cm long, tipped with a gland. Blue Mts Fl. Sept.–Nov. . . . . **C. clavigera** A.Cunn. ex Lindl.
5 Mid lobe of labellum toothed. Tail of perianth segments covered with long, dark hairs. Calli stout but not conspicuously overlapping. Perianth up to 8 cm long, creamy to yellowish. Plants up to 30 cm high. Blue Mts Endangered. Fl. Sept.–Oct . . . . **C. arenaria** Fitzg.

6 Labellum lateral lobes deeply fringed, teeth c. 8 mm long; middle lobe denticulate or crenate, maroon in the front. Sepals up to 70 mm long, deflexed. Plants up to 30 cm high. Perianth yellow-green. Sepals up to 5 cm long. Widespread. Fl. July–Oct. . . . . **C. tentaculata** Schltdl.
6 Labellum lateral lobes not deeply fringed 7

7 Labellum almost as long as wide, greenish, reddish at the tip; lateral margins entire, or minutely denticulate towards labellum tip. Sepals up to c. 50 mm long, shortly club-shaped . . . .
. . . .**C. clavigera** (see above)
7 Labellum ovate to triangular, yellowish, maroon at tip, margins ± fringed Sepals c. 45 mm long ending in a club c. 15 mm long, spreading. Labellum ovate. Plants to 30 cm high. Montane areas. Fl. Aug.–Nov. *Spider Orchid* . . . .**C. fitzgeraldii** Rupp

8 Perianth white to pink or greenish outside 9
8 Perianth brownish to purplish or reddish outside 17

9 Labellum with a pair of flat, irregular, yellowish teeth either side of the middle lobe 10
9 Labellum without large flat teeth at the base of the middle lobe or, if teeth are present, then ± the same size as others on the margin 11

10 Leaves up to 6 cm long. Flowers less than 15 mm diam.. Perianth segments white with green stripes on the back, up to 7 mm long. Plants up to c. 10 cm high. Coast and adjacent plateaus. DSF and heath. Moist places. Fl. spring. *Fairy Orchid*. . . . .**C. alata** R.Br.
10 Leaves 6–12 cm long. Flowers more than 20 mm diam. Perianth segments pink with red blotches, with a dark strip on the back, 12–18 mm long. Plants to 25 cm high. DSF. Coastal. Fl. Aug.–Oct. . . . . **C. hillmannii** D.L.Jones

11 Labellum with red transverse bars 12
11 Labellum without red transverse bars 14

12 Labellum with 4–6 rows of club-shaped yellowish calli down the centre; margin of the middle lobe with stalked teeth. Flowers 1–2. Perianth segments up to 20 mm long, greenish pink outside with glandular hairs, pink inside. Plants up to 20 cm high. Coast in north of the region. DSF. Fl. spring. . . . .**C. quadrifaria** D.L.Jones
12 Labellum with 2 rows of calli down the centre 13

13 Calli at base of labellum orange, the rest white, becoming larger towards tip of labellum. Marginal calli flat. Labellum white, pink or purplish, tip orange. Perianth bright pink to dark pink inside, greenish white or purplish-brown outside. Dorsal and lateral sepals 14–20 mm long, bright to dark pink inside, greenish white to purplish-brown outside without a central dark band. Flowers solitary. Plants to 35 cm tall. Restricted to coastal areas. Heath and woodland. Endangered. Fl. Sept.–Oct. . . . . **C. porphyrea** D.L.Jones
13 Labellum with 2 rows of club-shaped yellowish calli down the centre; margin of the middle lobe with unstalked teeth. Dorsal and lateral sepals ± equal, 8–15 mm long, greenish pink outside with glandular hairs, greenish pink to greenish white inside. Flowers 1–3, usually musk scented. Plants up to 22 cm high. Widespread. Open forest and heath. Fl. Aug.–Oct. *Pink Fairy* or *Pink Fingers* . . . .**C. carnea** R.Br. var. **carnea**

14 Labellum with a long narrow tip. Calli on middle lobe crowded, shiny and black; lateral lobes falcate, pink. Perianth segments c. 20 mm long, bright pink, dark hairs on outside. Flowers 1–3. Dorsal sepal forming a cap over the column. Plants up to 60 cm high. Montane areas. Fl. Oct.–Dec. *Black Tongue Caladenia* . . . . **C. congesta** R.Br.
14 Labellum not as above 15

15 Calli in 4 rows. Perianth white inside, outside greenish with scattered hairs. Dorsal sepal forming a cap over the column. Labellum white, sometimes with a dark spot on the tip; upper section margins with slender teeth. Plants to 30 cm high. DSF. Blue Mts . . . . **C. dimorpha** Fitzg.
15 Calli in 2 rows. 16

16 Middle lobe of the labellum with long, acute, undulating teeth giving it the appearance of a comb, with a yellow-orange tip. Perianth segments up to 20 mm long, shining white. Column greenish. Flowers usually 1–2. Plant up to 30 cm high. Coast and adjacent plateaus. DSF and heath. Fl. winter–spring. *White Fingers* . . . . **C. catenata** (Sm.) Druce
16 Middle lobe of the labellum with short blunt teeth and a yellowish tip. Perianth segments up to 20 mm long, white with pale pink tips. Column with a red spot. Flowers 1–2. Plant up to 15 cm high. Coast. Open forest mostly on clay soils. Fl. winter . . . . **C. picta** (Nicholls) M.A.Clem. & D.L.Jones

17 Inner surface of perianth greenish yellow. 18
17 Inner surface or perianth white or pinkish 19

18 Lateral lobes of labellum entire. Plant up to 20 cm high. Widespread. DSF. Fl. Oct.–Nov. . . . .
. . . . **C. transitoria** D.L.Jones
18 Lateral lobes of the labellum toothed. Plants up to 20 cm high. Widespread. DSF and heath. Fl. spring–
summer . . . .**C. testacea** R.Br.

19 Dorsal sepal hooded over the column 20
19 Dorsal sepal not hooded over the column 21

20 Labellum with coarse teeth, white, purplish towards the base and tip; teeth ± club-shaped, with rough
stalks. Flowers up to 6. Perianth segments up to 15 mm long, greenish to brownish outside, white
inside. Plant up to 25 cm high. Blue Mts DSF. Fl. spring–summer . . . .**C. cucullata** Fitzg.
20 Labellum with slender teeth, white, purplish towards the tip; teeth with slender heads and smooth
stalks. Flowers up to 6, with musky scent. Plant up to 40 cm high. Blue Mts DSF. Fl. spring–summer .
. . . . **C. gracilis** R.Br.

21 Calli at base of labellum orange, the rest white . . . . **C. porphyrea** (see above)
21 All central calli yellow 22

22 Lateral sepals up to 7 mm long, covered with reddish glandular hairs outside, white to pink inside.
Flower usually solitary. Labellum toothed on the middle lobe, white to pink. Plant up to 15 cm high.
Widespread. Open forest. Fl. spring . . . . **C. fuscata** (Richb.f.) M.A.Clem. & D.L.Jones
22 Lateral sepals more than 7–16 mm long, with lines of green glandular hairs down the centre outside,
white or cream to pink inside. Flower usually solitary. Labellum with broad flat teeth on the middle
lobe. Plant up to 15 cm high. Coast and adjacent plateaus. DSF and heath. Ss. Fl. spring . . . .
. . . . **C. curtisepala** D.L.Jones

22 Adenochilus Hook.f.
1 species endemic Aust.; NSW

One species in the area

Small, glabrous, terrestrial herb 10–20 cm high, with fleshy rhizomes. Leaf solitary, ovate to cordate,
either sessile on the flowering stem or on a long petiole arising from the rhizome. Flower solitary, with
a subtending sheathing bract well below the ovary; the bract with a conspicuous filament between its
margins at the base, spathulate at the tip. Perianth white, 1–2 cm diam. Dorsal sepal broad, acute, ±
hooded. Lateral sepals narrower and a little longer than the dorsal sepal, acuminate. Petals narrower and
shorter than the sepals. Labellum on a claw, 3-lobed; the surface covered with yellow calli except at the
tip. Column erect, widely winged; the wings higher than the anther. Labellum and column spotted with
red. Blue Mts Rock crevices or in *Sphagnum* at elevations above 900 m. Fl. spring–summer. . . . .
. . . . **A. nortonii** Fitzg.

23 Glossodia R.Br.
Waxlip Orchids
2 species endemic Aust.; Qld, NSW, Vic., Tas., S.A.

Terrestrial herbs with small tubers. Leaf solitary, basal, oblong to lanceolate, prostrate or nearly so.
Flowers 1–2, rarely more. Sepals and petals C. equal, all spreading. Labellum very much shorter, sessile,
undivided, with entire margins, without calli on the lamina; two tall linear clavate calli (sometimes
fused) at the base of the lamina standing erect against the column. Column hardly as long as the labellum,
widely winged above. Anther erect, with 2 pollen sacs; the outer valves broad; the inner valves much
smaller; the connective produced into a small point; pollinia 4, granular.

Key to the species

1 Calli at the base of the labellum fused, with a bilobed head. Flowers usually 3–4 cm diam. (up to 6 cm),
1–2 (rarely 3) together. Perianth purplish or white. Plants up to 30 cm high. Leaf oblong to lanceolate,
4–15 cm long, hairy. Widespread. Fl. spring . . . . **G. major** R.Br.

1 Calli at the base of the labellum not fused, clavate. Flowers usually c. 2 cm diam., 1–2 together. Perianth violet-blue rarely white. Plants up to 16 cm high. Leaf broad-lanceolate, 2–4 cm long. Widespread. Fl. spring . . . . **G. minor** R.Br.

24 Corybas Salisb.
Helmet Orchids
20 species in Aust. (19 endemic, 1 native); all states and territories except NT

Dwarf, terrestrial herbs with small globular tubers. Leaf solitary, cordate, ovate or nearly orbicular, usually flat on the ground. Flower solitary, almost sessile or very shortly stalked from the base of the leaf, occasionally rather large relative to the size of the plant. Dorsal sepal and labellum greatly developed; the other segments very small and inconspicuous. Dorsal sepal narrow at the base, becoming broad and hooded towards the apex. Lateral sepals linear, small. Labellum ± tubular towards the base, expanding into a broad concave lamina with variously fringed or entire margins. Column short and stout, occasionally winged; pollinia 4, granular or mealy. Stem often extensively elongating after fertilization of the ovary.

Key to the species

1 Margins of the labellum fringed or serrulate 2
1 Margins of the labellum entire 5

2 Margins of the labellum minutely and irregularly serrulate. Labellum tube 2-spurred. Leaf cordate, reddish-grey below, 7–20 mm long. Perianth dark purplish red, 10–15 mm long. Frenchs Forest. Fl. May–July . . . . **C. undulatus** (A.Cunn.) Rupp
2 Margins of the labellum fringed. Leaf green on both surfaces 3

3 Labellum with conspicuous white protuberance which is notched at apex. Flowers 3–35 mm long. Perianth reddish-purple. Petals auriculate at base. Leaf 20–35 mm wide. Sheltered slopes in Blue Mts Fl. summer–autumn. . . . . **C. hispidus** D.L.Jones
3 Labellum protuberance inconspicuous and not notched at apex; reddish-purple or whitish 4

4 Flowers 2–3 cm long. Perianth reddish-purple. Petals auriculate at base. Leaf orbicular-cordate, 20–40 mm wide. Widespread. Fl. May–June . . . . **C. fimbriatus** (R.Br.) Rchb.f.
4 Flowers less than 2 cm long. Perianth purple. Petals often 2-fid at apex. Leaf 10–20 mm wide. Widespread. Fl. May–June . . . . **C. pruinosus** (R.Br.) Rchb.f.

5 Labellum longer than the dorsal sepal. Leaf cordate. Perianth reddish-purple. Coast. Fl. June–July . . . . **C. unguiculatus** (R.Br.) Rchb.f.
5 Labellum shorter than the dorsal sepal 6

6 Perianth white to pink. Leaf cordate, dull green on the upper surface, purplish underneath. Hornsby Plateau. Open forest. Fl. winter. . . . . **C. barbarae** D.L.Jones
6 Perianth reddish-purple. 7

7 Leaf green on both surfaces, ovate-cordate. Flower less than 15 mm long. Perianth reddish-purple. Wentworth Falls. Fl. July–Aug . . . . **C. fordhamii** Rupp
7 Leaf usually reddish below, orbicular-cordate 8

8 Flowers 26–32 mm long. Dorsal sepal 6–10 mm wide. Perianth dark reddish-purple. Coastal area south of Newcastle. Fl. June–Aug. . . . . **C. dowlingii** D.L.Jones
8 Flower 15–24 mm long. Dorsal sepal 10–12 mm wide. Perianth reddish-purple rarely white. Widespread. Fl. April–June. . . . . **C. aconitiflorus** Salisb.

25 Cryptostylis R.Br.

Tongue Orchids

5 species in Aust. (4 endemic, 1 native); all states and territories except NT

Terrestrial, glabrous herbs with fleshy rhizomes. Leafless or with a few erect lanceolate to ovate, basal leaves on long petioles. Flowers not resupinate, several in a terminal raceme. Perianth green or yellowish except for the red-brown labellum. Sepals and petals linear; the sepals almost equal in length; the petals smaller. Labellum undivided, sessile; the base enclosing the column; the lamina sometimes very broad, concave on the upper surface or convex through reflexion of the margins, longitudinally ridged or with sessile calli. Column very short and broad, winged; the wings forming auricles or produced posteriorly into a glandular or membranous process with fringed margins behind the anther.

Key to the species

1 Plant leafless, 15–45 cm high, saprophytic. Cauline bracts 6–8. Flowers 3–10, sessile, reversed. Sepals 2 cm long. Petals 1 cm long, subulate, yellow. Labellum pubescent, narrow-oblong, convex, 3 cm long, yellowish-green below, with red markings; upper portion light green merging into black. Kuring-gai area. Sandy soils. Vulnerable. Fl. Aug.–Feb . . . . **C. hunteriana** Nicholls
1 Leaves 1–3, broad- to narrow-lanceolate, 5–10 cm long 2

2 Labellum very concave, forming a hood over the rest of the flower, broad, erect, striate, reticulate, with a broad median vertical ridge. Flowers 2–10. Coast to Blue Mts Sandy soils. Fl. Nov.–April . **C. erecta** R.Br.
2 Labellum not forming a concave hood, horizontal, straight 3

3 Labellum broad but the margins much reflexed and thus appearing narrow; lamina smooth, glossy, with a bilobed glandular process near the apex. Sepals and petals linear. Flowers 2–14. Plant 15–80 cm high. Coast to Blue Mts Sandy soils. Fl. Oct.–March. . . . . **C. subulata** (Labill.) Rchb.f.
3 Labellum narrow, very dark, pubescent, recurved at the apex; the margins inturned forming a thick linear channelled lamina. Sepals and petals filiform, Flowers 4–15. Plant rarely more than 40 cm high. Chiefly Blue Mts, also coast. Sandy soils. Fl. Nov.–March. . . . . **C. leptochila** F.Muell. ex Benth.

26 Pterostylis R.Br.

Greenhoods

Terrestrial herbs with small globular tubers and succulent slender roots. Leaves often in a rosette and ± ovate, or cauline and lanceolate to linear; the rosette in many species dissociated from the flowering stem, and often absent at flowering time. Flowers solitary, or from few to many in a terminal raceme. Perianth usually green but often red with red-brown tints. Dorsal sepal very concave, usually much incurved; lateral margins dove-tailed into or connate with those of the petals to form a hood over the column (Fig. 44). Lateral sepals ± connate in their lower portions to form a lip in front of the column, or deflexed against the ovary, or rarely spreading; lower portions divergent and often tapering into fine points. Labellum usually small, ovate to linear or rarely filiform, attached to the projecting foot of the column by a moveable claw which is usually ± irritable and occasionally furnished with long translucent hairs, usually undivided, frequently with a curved brush-like appendage below its junction with the column foot. Column elongate inside the hood and ± attached to the median line of its wall; the upper part with transluscent wings on either side of the rostellum; the base prolonged into a nearly horizontal foot. Anther very blunt, with 2 pollen sacs; pollinia 4, powdery. Stigma often prominent, near the middle of the column; rostellum higher up, just below the pollinia; caudicle absent.

Key to the species

1 Basal leaves in a cluster or rosette encircling the base of the stem, always present at flowering time. Cauline leaves reduced to sheathing bracts or well developed and scattered . . . .**GROUP 1**
1 Basal leaves in a cluster or rosette usually separated from the flowering stem, sometimes attached to its base by a short scape, but very rarely encircling it and often present at flowering time. Cauline leaves alternate, reduced to scales near the base, usually well developed above but occasionally bract-like . . . . .
. . . .**GROUP 2**

Group 1
Basal leaves in a cluster or rosette encircling the base of the stem

15 Labellum red-brown brownish or black 16
15 Labellum green 20

16 Lateral sepals reflexed against the ovary, joined section flat, margins strongly reflexed, free points reflexed. Perianth green, with a dark brown to black labellum. Plants 15–40 cm high. Leaves ovate to orbicular. Coast in the south of the area. Endangered. Fl. Aug.–Nov. . . . . **P. gibbosa** R.Br.
16 Lateral sepals deflexed but not against the ovary. Perianth dark reddish-brown or green with dark markings. 17

17 Labellum obovate 4.5–5 mm long. Leaves obovate. Lateral sepals with joined section shallowly concave, margins strongly incurved, free points curved forward. Plants to 25 cm high. Western Sydney. Endangered. Fl. Sept.–Nov. . . . . **P. saxicola** D.L.Jones & M.A.Clem.
17 Labellum oblong, c. 4 mm long 18

18 Flowers nodding on slender pedicels, with brown, grey or red-brown marks. Lateral sepals subulate, free points curved backwards. Labellum with 8–10 pairs of marginal hairs c. 1.6 mm long. Widespread. Fl. Sept.–Dec. . . . . **P. rufa** R.Br.
18 Flowers ± erect 19

19 Lateral sepals folded inwards at the margin only. Flowers c. 1.5 cm long, green, with pale rusty brown marks. Apices of sepals produced into needle-like points. Labellum with 5–10 pairs of marginal hairs 0.3–2.5 mm long. Blue Mts Fl. Sept.–Dec. . . . . . . **P. aciculiformis** (Nicholls) M.A.Clem. & D.L.Jones
19 Outside third of lateral sepals folded inwards, edges nearly touching beneath labellum. Flowers c. 1 cm long, with green and red markings. Labellum with 2–4 pairs of marginal hairs c. 0.6 mm long. Blue Mts Fl. Sept.–Nov.. . . . . **P. pusilla** R.S.Rogers

20 Appendage at the base of the labellum pointing inwards. Perianth 6–9 mm long. Plant up to 30 cm high. Leaves narrow-ovate to lanceolate. Widespread. Fl. Sept.–Dec. . . . . **P. mutica** R.Br.
20 Appendage at the base of the labellum pointing outwards 21

21 Labellum appendage dark green; middle lobe long, pointed. Flower c. 8 mm long. Plant up to 20 cm high. Widespread. Open forest and grassland. Fl. spring–summer. . . . . **P. cycnocephala** Fitzg.
21 Labellum appendage blackish; middle lobe short, blunt. Flower c. 10 mm long. Plant up to 40 cm high. Widespread. Open forest and grassland. Fl. spring–summer . . . . . **P. bicolor** M.A.Clem. & D.L.Jones

22 Labellum narrow ovate, brown. Lower sepals erect but inclined forward. Lower leaves petiolate, lanceolate-ovate, 4–8 cm long, passing into stem bracts. Flower solitary. Perianth 4–6 cm long, green with tinges of brownish-red at the tips of the segments. Plant up to c. 20 cm high. Southern parts of the Blue Mts Fl. Nov.–Jan. *Sickle Orchid* . . . . **P. falcata** R.S.Rogers
22 Labellum filiform, beset with long yellow hairs almost to the apex. Plant 12–20 cm high. Leaves numerous, lanceolate, 2–3 cm long, crowded near the base of the stem, tipped with red-brown. Perianth c. 4 cm long, pale green, with fine darker green veins 23

23 Flowers leaning forward. Coast. Endangered. Fl. Aug.–Nov. . . . **P.** sp. **'Botany Bay'**(Bishop J221/1-13)
23 Flowers held ± erect. Blue Mts Fl. Aug.–Oct. . . . . **P. plumosa** Cady

Group 2
Basal leaves in a cluster or rosette usually separated from the flowering stem

1 Flowers solitary or rarely 2 2
1 Flowers several in racemes 13

2 Labellum usually notched at the tip, occasionally obtuse. Plant 10–15 cm high. Leaves linear to broad-lanceolate. Perianth green with red-brown, 2–3 cm long. Blue Mts and Illawarra Escarp. Vulnerable. Fl. April–May. . . . . **P. pulchella** Messmer
2 Labellum always entire 3

3 Labellum narrow-oblong for half its length then becoming filiform and terminating in a gland at its apex. Plant 14–24 cm high. Leaves ovate. Perianth green with red-brown, up to 4 cm long. Petal margins widely flared, red-brown. Widespread. Fl. April–Aug. . . . . **P. grandiflora** R.Br.

3 Labellum acuminate or obtuse 4

4 Labellum acuminate 5

4 Labellum always very obtuse 11

5 Dorsal sepal apex acuminate, tip to 3 mm long, Petals bluntly truncate, margins decurved. Leaves 2–5, lanceolate to ovate. Perianth 35–50 mm long, translucent green, longitudinally striped with darker green or red. Labellum sometimes blunt at the apex. Plant 5–16 cm high. Coast. Fl. April–July. . . . . **P. truncata** Fitzg.

5 Dorsal sepal apex filiform tip longer than 3 mm 6

6 Perianth mainly white with green or brown bands. Labellum brown to red-brown, 10–13 mm long. Plants 15–45 cm high. Basal leaves ovate to hastate. Blue Mts . . . . **P. laxa** Blackmore

6 Perianth green and red or red-brown 7

7 Sinus between the lower sepals broad, flat or raised when viewed from front, may have small central notch. Petal margins flared or decurved. 8

7 Sinus between the lower sepals narrow, V-shaped when viewed from front. Petal margins decurved. . 9

8 Flowers less than 20 mm long, green and white, shining. Petal margins decurved. Plant up to c. 15 cm high. Coast and adjacent plateaus. Moist places. Fl. autumn–winter . . . . **P. alveata** Garnet

8 Flowers more than 20 mm long. Petals acute, margins flared. Plant 12–22 cm high. Basal leaves usually absent at flowering time. Perianth c. 4 cm long, mainly bright red with some green. Blue Mts Fl. Jan.– March . . . . **P. coccina** Fitzg.

9 Flower erect. Lower part of the hood not as long as the upper curved part. Plant 7–24 cm high. Basal leaves ovate; cauline leaves linear to narrow-lanceolate. Perianth green with darker bands, 2–3 cm long. Widespread. Fl. March–May. *Dainty Greenhood* . . . . **P. reflexa** R.Br.

9 Flower nodding. Lower part of the hood longer than the upper curved part 10

10 Perianth green and white. Flower up to 30 mm long. Plant 15 cm high. Widespread. Fl. Feb.–June . . . . **P. revoluta** R.Br.

10 Perianth green with brownish markings. Flower up to 25 mm long. Plant up to 25 cm high. Coast and adjacent plateaus; lower Blue Mts DSF and heath in moist places. Fl. winter . . . . **P. longipetala** Rupp

11 Cauline leaves c. 15 mm long. Basal leaves usually 3, c. 15 mm long. Plant 15–20 cm high. Perianth green streaked with red, 25 cm long excluding the filiform ends (3 cm) of the sepals. Widespread in colder parts. Fl. winter . . . . **P. fischii** Nicholls

11 Cauline leaves 3–4 cm long 12

12 Cauline and basal leaves similar in shape and size; the margins often crenate. Plant 12–25 cm high. Perianth green with red near the tips of the segments, c. 25 mm long excluding the filiform ends (2 cm) of the sepals. Widespread. Fl. Feb.–June . . . . **P. obtusa** R.Br.

12 Cauline leaves c. half as wide as the basal leaves. Plant 10–30 cm high. Perianth green streaked with red, c. 3 cm long excluding the filiform ends (2–3 cm) of the sepals. Colder parts of the area. Uncommon. Fl. summer. *Summer Greenhood* . . . . **P. decurva** R.S.Rogers

13 Lower sepals erect. 14

13 Lower sepals spreading or deflexed 15

14 Perianth shine, green and white, apex darker green, 6–7.5 mm long. Scape and ovary fleshy. Labellum not visible. Plant to 15 cm high. Damp places. Uncommon. Fl. summer . . . . **P. uliginosa** D.L.Jones

14 Perianth green and white, apex often red-brown, 7–10 mm long. Labellum just visible. Plant 5–35 cm high. Widespread. Fl. chiefly autumn. *Baby Greenhood* . . . . **P. parviflora** R.Br.

15 Labellum with 2 conspicuous lateral lobes. Leaves less than 3 cm long, arising laterally from the base of the stem. Perianth green and white, 6–8 mm long. Plant 12–30 cm high. Widespread. Fl. April–July . . . . **P. daintreana** Benth.

15 Labellum without 2 conspicuous lateral lobes. Leaves arising only on non-flowering plants 16

16 Labellum green with a blackish stripe down the centre, more than 2 mm wide. Flowers c. 15 mm long. Leaves 3–6 cm long (rarely longer). Plants 12–90 cm high. Widespread. Open forest and heath. Moist places. Fl. April–Oct. . . . . **P. longifolia** R.Br.

16 Labellum dark brown, less than 2 mm wide. Flowers c. 10 mm long. Plants 15–50 cm high. Woronora Plateau; Illawarra ranges. Open forest. Moist places. Fl. winter . . **P. tunstallii** D.L.Jones & M.A.Clem.

27 Erythrorchis Blume
1 species endemic Aust.; Qld, NSW

One species in the area

Terrestrial, leafless saprophytes with thick fleshy rhizomes. Stems usually numerous, branching, up to 6m long, climbing on tree trunks by means of short sucker-like aerial roots. Flowers very numerous in terminal panicles; pedicels 8–10 mm long. Perianth brown and yellow. Bracts at the base of the branches and panicles ± stem clasping. Sepals and petals almost equal but the petals narrower. Labellum sessile, undivided, 12–14 mm long, whitish with brown or red streaks, anteriorly undulate-crisped. Column long, erect, not winged or very obscurely so, shorter than the labellum. Anther with 2 pollen sacs, operculate, with a broad flat or convex dorsal appendage; pollinia 2, deeply bilobed, without a caudicle; pollen granular-farinaceous or waxy. Seeds winged, in large irregularly terete capsules. Coast to Blue Mts Shady places. Fl. spring–summer. *Climbing Orchid.* . . . . **E. cassythoides** (A.Cunn. ex Lindl.) Garay

28 Gastrodia R.Br.
7 species in Aust. (6 endemic, 1 native); all states and territories except NT

Leafless, terrestrial saprophyte with tuberous rhizomes. Flowers in a terminal raceme. Perianth segments connate into a 5-lobed tube very gibbous at the base under the labellum. Labellum shorter than the perianth and connate with it at the base; margins crisped or undulate; lamina with two longitudinal ridges converging and uniting towards the front or occasionally with only one ridge. Column erect, elongate, slightly winged. Anther almost hemispherical; pollinia 2, bilobed, granular; caudicle absent.

Key to the species

1 Racemes with up to 20 flowers. Perianth whitish; tube 10–12 mm long. Sepals 2–3 mm long. Labellum yellow obscurely 3-lobed. Stems up to 70 cm long, with 3–6 short scaly bracts. Capsule 10–15 mm long. Coast to Blue Mts Fl. late winter–summer. *Potato Orchid.* . . . . **G. sesamoides** R.Br.

1 Racemes with up to 75 flowers. Perianth brown with white tips on outside, white on inside; tube 15–20 mm long. Sepals c. 5 mm long. Labellum orange prominently 3-lobed. Stems up to 90 cm high with 6–8 short scaly bracts. Capsule 25–30 mm long. Blue Mts Fl. spring–summer . . . . . **G. procera** G.W.Carr

29 Rhizanthella R.S.Rogers
2 species endemic Aust.; Qld, NSW, WA

Monotypic endemic genus

Subterranean saprophyte with fleshy rhizomes up to 15 cm long, clothed with imbricate scaly bracts. Flowers 15–30 in terminal heads. Perianth segments free. Dorsal sepal 5–8 mm long, concave. Lateral sepals 8–12 mm long, acuminate. Petals acute, 4–6 mm long. Labellum ovate, papillose, dark reddish-purple. Column 4–6 mm long, with 3-fid appendages level with the rostellum. Flowers pollinated and fertilized below the ground. The fruits are pushed to the surface where the seeds are dispersed. Coast; Blue Mts Vulnerable. Fl. Oct.–Nov. *Underground Orchid* . . . . **R. slateri** (Rupp) M.A.Clem. & P.J.Cribb

30 Spiranthes Rich.

1 species native Aust.; Qld, NSW, Vic., Tas.

One species in the area

Terrestrial herb 16–50 cm high, with oblong tubers or thick fibrous roots. Leaves 3–10, chiefly basal, linear, 4–10 cm long. A few bracts present on the stem. Flowers small, spirally arranged in a terminal spike. Perianth bright pink; segments almost equal, 4–5 mm long. Dorsal sepal ovate, concave. Lateral sepals free, erect or spreading. Petals truncate, forming with the dorsal sepal a hood sheltering the column. Labellum undivided, mostly white, on a short claw; the margins at least partially fringed or denticulate; the tip recurved; lamina with 2 ovoid glands at the base. Column shorter than the labellum, erect, contracted basally. Anther usually blunt with 2 pollen sacs. Column wings membranous. Stigma large, U-shaped, rostellum higher than the anther. Pollinia in 2 pairs, granular; caudicles present. Wet places throughout the area. Fl. summer. *Ladies Tresses* . . . .**S. australis** (R.Br.) Lindl.

31 Cestichis Thouars & Pfitzer

8 species endemic Aust.; Qld, NSW

Terrestrial or epiphytic herbs. Stems sometimes thickened into pseudobulbs. Leaves at or near the apex of the stem, basal or nearly so. Flowers in terminal pedunculate racemes. Sepals and petals all free spreading or reflexed, c. equal in length but the dorsal sepal and petals often narrower than the lateral sepals. Labellum shortly embracing or united with the column at the base, erect or ascending, entire. Column elongated, incurved, winged at the top. Anther terminal, operculate; pollinia 4, obovoid.

Key to the species

1 Peduncle winged. Leaves 4–8 cm long. Flowers pale translucent green with orange tints; odour unpleasant or absent. Labellum cuneate, 5–6 mm long, 5 mm wide; the front margin irregularly denticulate, not reflexed; the shorter narrower base embracing the column. Column slender, erect, slightly to moderately incurved, c. 4 mm long. Epiphyte on tree trunks. Fl. Aug.–Feb. . . . .**C. coelogynoides** (F.Muell.) M.A.Clem. & D.L.Jones
1 Peduncle not winged. Leaves up to 30 cm long. Flowers yellowish-green, rarely reddish-purple, odour unpleasant, sickly. Labellum sessile, entire, reflexed about the middle. Column incurved, 5–6 mm long. Widespread. Rock ledges and crevices. Fl. March–May. *Yellow Rock Orchid.* . . . .**C. reflexa** (R.Br.) M.A.Clem. & D.L.Jones

32 Calanthe R.Br.

1 species native Aust.; Qld, NSW

One species in the area

Tall, terrestrial herb with fleshy rhizomes. Stems reduced to short pseudobulbs. Leaves up to 90 cm long, pale green. Scapes up to 1.5 m long with sheathing bracts. Flowers numerous in a terminal raceme, up to 3 cm diam. Perianth white; segments nearly equal, free, spreading. Labellum spurred, adnate to the base of the column; lamina 3-lobed, with several calli. Column short. Anther operculate; pollinia 8, tapering below, fixed to a divisible gland. Coast. Deep shaded gullies. Fl. summer. *Scrub Lily* . . . .**C. triplicata** (Will.) Ames

33 Dendrobium Sw.

71 species in Aust. (52 endemic, 19 native); all states and territories except S.A.

Epiphytes or lithophytes; rhizomes tufted or creeping. Stems erect, creeping or pendulous, stout and rigid or slender and lax, often reduced to short pseudobulbs about the base. Leaves variously shaped. Flowers usually in racemes rarely solitary. Sepals nearly equal; the lateral pair obliquely dilated at the base, adnate to the column foot to form a spur under the labellum. Petals often as long as or longer than the dorsal sepal. Labellum articulate at the end of the column foot, erect, concave near the base; the lamina with longitudinal ridges; the margins expanding into 2 lateral lobes which usually embrace the column; the middle lobe recurved or spreading, acute or less commonly obtuse, occasionally the 3 lobes confluent into an undivided lamina. Column usually shorter than the labellum, winged or toothed above. Anther operculate; pollinia 4, in pairs, waxy.

Key to the species

1 Leaves long, terete succulent circular in cross section or ± flattened. Flowers not resupinate (i.e. labellum uppermost in flower). 2

1 Leaves not terete thin to leathery, conduplicate, Stems swollen for at least part of their length. Flowers usually resupinate ie labellum lower most in flower 4

2 Leaves obscurely corrugated, usually curved, 6–11 cm long. Rhizomes creeping on rocks. Stems branching, pendulous, 25–50 cm long. Flowers solitary or 2 together, 25–30 mm diam. Coast to Blue Mts On rocks. Fl. Sept.–Nov. . . . . **D. striolatum** Rchb.f.

2 Leaves smooth, pendulous, 10–70 cm long. Stems up to 2 m long. Racemes lateral or from the bases of the leaves, many-flowered. *Pencil Orchids*. 3

3 Perianth white cream or yellowish, 5–6 cm diam; the labellum sometimes dotted or striped with red. Leaves mostly c. 5 mm diam., 10–50 cm long. On trees (usually Casuarinaceae). Fl. July–Oct. *Rats-tail Orchid* . . . . **D. teretifolium** R.Br.

3 Perianth greenish or the base striated with red or purple-brown, 6–10 cm diam. Leaves mostly 4 mm diam., 10–70 cm long. Blue Mts; lower Hunter River Valley. RF. Fl. Aug.–Oct. . . . .
. . . . **D. fairfaxii** F.Muell. & Fitzg.

4 Stems swollen, not branched. Leaves clustered 5

4 Stems not swollen, usually branched. Leaves not clustered 10

5 Stems swollen from close to the base and thickest near the base. Plants 12–80 cm high. Leaves 2–5 on each stem, thick, leathery, 8–24 cm long. Racemes almost terminal, up to 45 cm long, many-flowered. Perianth cream white or yellow, c. 4 cm diam. Coast to Blue Mts On trees and rocks. Fl. Aug.–Oct. *Rock Lily* . . . . **D. speciosum** Sm.

5 Stems thickest above the base 6

6 Upper part of stems swollen, square in cross-section; lower part thin and wiry. Perianth green to deep yellow 7

6 Upper part of stem circular in cross-section; lower part not thin and wiry. 8

7 Dorsal sepal up to 30 mm long. Leaves 2–5, broad-lanceolate, 3–7 cm long. Plants 6–45 cm long. Coast. Humid places; on rocks and trees. Fl. chiefly spring. *Spider Orchid* . . . . . . **D. tetragonum** A.Cunn.

7 Dorsal sepal more than 35 mm long. Leaves 2–4, elliptic 4–9 cm long. Plants up to 50 cm long. Coast. On rocks and trees, particularly *Melaleuca styphelioides*. Endangered. Fl. winter–spring . . . .
. . . . **D. melaleucaphilum** M.A.Clem. & D.L.Jones

8 Racemes 20 cm long or longer. Perianth white to cream. Dorsal sepal 15–30 mm long. Labellum with purplish spots and stripes. Some roots erect. Leaves elliptic to oblong, up to 25 cm long. Plant up to 100 cm high. Coast and adjacent plateau north of the Hawkesbury River. RF and open forest. Growing on trees or rocks. Fl. winter–spring (hybridizes with *D. gracilicaule*) . . **D. tarberi** M.A.Clem. & D.L.Jones

8 Racemes up to 20 cm long. 9

9 Perianth white to pale yellow or pale pink. Leaves 2–4, at the apex of each stem, ovate, 3–7 cm long. Racemes 5–10 cm long. Plants 10–30 cm long. Coast. Open forests or RF; on trees. Fl. Aug.–Oct. *White Feather Orchid* or *Ironbark Orchid* . . . . **D. aemulum** R.Br.

9 Perianth deep yellow with dense red markings on the outside. Leaves 3–6 at the apex of each stem, elliptic, up to 13 cm long. Racemes 5–12 cm long. Plants up to 90 cm long. Coast. Epiphyte in RF and in open forest; sometimes on rocks. Fl. winter–spring . . . . **D. gracilicaule** F.Muell.

10 Leaves tapering to a dagger-like point, 2–5 cm long, 1–2 cm wide. Stems chiefly aerial, draped from the branches of trees, up to 1 m long. Flowers solitary or 2–3 together. Perianth light green with purple or red. Coast to Blue Mts On trees. Fl. Oct.–Nov. *Dagger Orchid*. (hybridizes with *D. linguiforme* and *D. striolatum*) . . . . **D. pugioniforme** A.Cunn.

10 Leaves with obtuse tips. Stems creeping 11

11 Leaves corrugated above, ovate, 2–3 cm long. Perianth white with faint purple. Flowers numerous in racemes 6–12 cm long. Coast to Blue Mts On rocks. Fl. Aug.–Nov. *Tongue Orchid* or *Button Orchid* . . . . **D. linguiforme** Sw.

11 Leaves tuberculate, resembling small gherkins. Perianth yellowish white or greenish white with red, 3 cm diam. Burragorang Valley. Usually on trees. Fl. chiefly late summer. *Cucumber Orchid* . . . . **D. cucumerinum** Macleay & Lindl.

34 Bulbophyllum Thouars
32 species in Aust. (28 endemic, 4 native); Qld, NSW

Small, epiphytic herbs with ± extensively creeping stems usually covered with thin scarious sheathing bracts. Pseudobulbs from the creeping stems usually minute, bearing solitary or paired leaves at the top (Fig. 44). Flowers usually small, solitary or in racemes; pedicels arising in the axils of the sheathing bracts of the creeping stems. Sepals erect, free, nearly equal; the lateral pair dilated at the base, adnate to the column foot to form a short blunt spur. Petals usually much shorter than the sepals. Labellum articulate at the end of the column foot, usually clawed and undivided. Column very short; the apex bidentate. Anther operculate, terminal; pollinia 4, waxy, without a caudicle.

Key to the species

1 Flowers solitary, often numerous 2
1 Flowers in racemes, few, scattered 3

2 Flowers crowded, on pedicels 5–9 mm long. Perianth whitish. Leaf thick, succulent, solitary, 2–4 cm long. Creeping stems extensive, rather intricate. Bracts conspicuous. Pseudobulbs minute. Widespread. RF and humid areas. Fl. chiefly spring. *Wheat-leaved Orchid* . . . . . **B. shepherdii** (F.Muell.) F.Muell.
2 Flowers scattered. Perianth red. Leaf thin, coriaceous, minute, often withering early. Stems creeping. Pseudobulbs orbicular, 1–2 mm diam. Humid areas north and south of Sydney. On tree trunks and branches. Fl. Oct.–Nov. . . . .**B. minutissimum** F.Muell.

3 Pseudobulbs 4–10 mm diam., ovoid-globular. Leaf solitary, 1–5 cm long. Perianth whitish or pale green. On trees and rocks in humid areas. Fl. chiefly autumn . . . . **B. exiguum** F.Muell.
3 Pseudobulbs 11-13 mm diam., ovoid, tubercular. Leaves 1–2, 5–10 cm long. Perianth usually green or yellow with dark red or purplish labellum. Coast; Hunter River Valley; Blue Mts Fl. May–Nov. *Pineapple Orchid* . . . .**B. elisae** (F.Muell.) Benth.

35 Dipodium R.Br.
Hyacinth Orchids
11 species in Aust. (10 endemic, 1 native); all states and territories

Terrestrial, saprophytic herb leafless, but the stems with loosely imbricate sheathing bracts at the base. Flowers in a terminal raceme on a long scape, often numerous and showy, usually spotted. Sepals and petals nearly equal, free, spreading. Labellum sessile, semi-terete, adnate to the column at its base and then ± gibbous, 3-lobed; the lateral lobes rather small; the middle lobe longer, usually oblong-ovate, pubescent near the apex. Column erect, semi-cylindrical, with sinuate or dentate membranous margins. Anther operculate; pollinia 2; deeply bilobed or in 4 pairs; caudicle present.

Key to the species

1 Pedicel and ovary with reddish spots, together 12–25 mm long. Perianth segments up to 20 mm long, cream to pale pink with red spots. Labellum mauve to deep pink. Widespread. DSF; WSF; heath. Fl. spring–summer. . . . .**D. variegatum** M.A.Clem. & D.L.Jones
1 Pedicel and ovary without reddish spots 2

2 Perianth segments dull to bright greenish yellow with deep pink spots, up to c. 25 cm long. Labellum white with pink and yellowish markings. Stems green, spotted, up to 60 cm high. Flowers 6–15 per raceme. Coast. Uncommon. Fl. summer . . . . **D. hamiltonianum** F.M.Bailey

2 Perianth segments white to pink with reddish markings. Stems usually reddish **3**

3 Perianth segments dark pink with heavy darker markings; the tips straight or slightly recurved, up to 18 mm long. Plant up to c. 50 cm high. Widespread. DSF and WSF. Fl. summer–autumn . . . .
. . . . **D. punctatum** (Sm.) R.Br.

3 Perianth segments pale pink with light darker markings; the tips distinctly recurved, up to 20 mm long. Plant up to c. 50 cm high. Widespread. DSF and WSF. Fl. summer–autumn. . . . .
. . . . **D. roseum** D.L.Jones & M.A.Clem.

36 Cymbidium Sw.
3 species endemic Aust.; Qld, NSW, WA, NT

Epiphytes on trees. Stems usually short and reduced to pseudobulbs covered with sheathing leaf bases, but occasionally longer and not pseudobulbous. Leaves long, narrow, channelled or at least slightly concave above. Flowers numerous, in often long racemes pedunculate in the lower leaf axils; peduncles usually with rigid sheathing bracts at the base. Sepals and petals nearly equal, free, spreading. Labellum sessile at the base of the column, slightly concave above, 3-lobed or sometimes undivided. Column erect or slightly incurved, semi-terete. Anther operculate, concave; pollinia 2, or 4 united in pairs.

Key to the species

1 Stems not pseudobulbous, up to 30 cm long. Racemes c. 20 cm long. Leaves flaccid, shining green, 18–35 cm long, less than 2 cm wide. Flowers numerous. Perianth golden green to light green, sometimes with indistinct red blotches, c. 3 cm diam. Coast to Blue Mts Epiphytic on hollow logs and branches. Fl. spring–early summer. *Snake Orchid* . . . . **C. suave** R.Br.

1 Stems pseudobulbous. Racemes 14–40 cm long. Leaves rigid, dull green, 12–40 cm long, 2–4 cm wide, deeply channelled above. Flowers numerous. Perianth variable in colour combining green brown and purple, spotted with purplish-brown or red, 3–4 cm diam. Gosford District. Fl. spring–early summer. *Tiger Orchid*. . . . . **C. canaliculatum** R.Br.

37 Sarcochilus R.Br.
16 species endemic Aust.; Qld, NSW, Vic., Tas.

Epiphytic herbs usually with rather short stems covered by the persistent bases of the older leaves. Leaves broad-lanceolate to linear, sometimes channelled, often ± falcate, Racemes axillary, with few or many flowers. Sepals and petals nearly equal, free, spreading; the lateral sepals ± dilated at the base, adnate to the column foot, Labellum articulate at the end of the column foot, without a basal spur, 3-lobed; lateral lobes erect; middle lobe small, with a large fleshy dorsal protuberance or spur at its base; lamina between the lateral lobes with various forms of calli. Column short, erect, with a prominent foot. Anther operculate; pollinia 4, in pairs; caudicle present.

Key to the species

1 Leaves linear, often spotted. Perianth white or pink. Labellum with lateral lobes shorter than the middle lobe. Leaves 3–6 cm long. Racemes with 2–6 flowers. Mainly coast. On trees in humid places. Fl. summer . . . . **S. hillii** (F.Muell.) F.Muell.

1 Leaves narrow-oblong to lanceolate. Perianth greenish or white. **2**

2 Sepals and petals narrow-oblong or sometimes almost falcate. Stem 1–6 cm long. Leaves dark green, 5–14 cm long, thick, falcate. Racemes with 2–11 flowers. Perianth 20–35 mm diam., olive-green or golden. Coast. On trees, usually RF. Fl. Oct.–Nov.. . . . . **S. parviflorus** Lindl.

2 Sepals and petals broad-ovate to broad-lanceolate. **3**

3 Perianth white except for coloured markings on the labellum, 20–35 mm diam. Leaves pale green. Widespread. On trees, usually RF. Fl. chiefly Oct. *Orange-blossom Orchid* . . . . **S. falcatus** R.Br.

3 Perianth greenish brown, 15–30 mm diam. Labellum white with coloured markings. Leaves nearly straight, lanceolate, 3–7 cm long, slightly twisted. Flowers 5–14 per raceme. Widespread. On trees in humid places. *Butterfly Orchid* . . . . **S. australis** (Lindl.) Rchb.f.

38 Plectorrhiza Dockrill
3 species endemic Aust.; Qld, NSW, Lord Howe Isd.

One species in the area
Epiphytic herbs with stems up to 30 cm long. Roots aerial, often tangled. Leaves 4–10 cm long, linear to lanceolate. Flowers in racemes, either in alternating rows or facing all ways, usually facing upwards, 3–10 per raceme. Perianth 10–12 mm diam.; segments similar, free, short or rather long and narrow, brown and green or light and dark green; labellum immovable, joined to the column foot, saccate with a callus within the front of the sac, 3-lobed; lateral lobes ± erect; middle lobe short or long, usually fleshy, sometimes whitish. Column short, with a distinct broad foot not sharply defined from the base of the labellum; pollinia 4, unequal, united in pairs. Coast to Blue Mts Fl. Sept.–Jan. *Tangle Orchid*. . . . .
. . . . **P. tridentata** (Lindl.) Dockrill

39 Papillilabium Dockrill
1 species endemic Aust.; Qld, NSW

One species in the area
Epiphytic herbs with stems less than 5 cm long. Roots creeping on branches. Leaves 2–4, linear, 2–5 cm long. Racemes with 3–7 flowers. Perianth c. 10 mm diam.; segments similar, free, dull green. Labellum 3-lobed, spurred, shortly adnate by the posterior margins of the lateral lobes to the column foot; spur with a ± conspicuous fleshy callus on the posterior wall at its entrance, sometimes septate but the septum often reduced to a midrib; lateral lobes often triangular; middle lobe curved forwards and downwards or sometimes incurved. Column short; the foot also usually short. Anther hooded, rostrate, obtuse; pollinia 4, united in pairs; caudicle present. Gosford district. Fl. Sept.–Dec. . . . .
. . . . **P. beckleri** (F.Muell. ex Benth.) Dockrill

40 Epidendrum L.
1 species naturalized Aust.; Qld, NSW

One species in the area
Erect terrestrial or lithophytic herb. Stems terete to 1 m long. Leaves cauline, leathery, spreading, 1.5–12 cm long. Flowers red, orange or yellow in a terminal inflorescence. Sepals spreading 10–22 mm long. Petal with undulate margin, 12–20 mm long. Labellum deeply 3-lobed, mid-lobed retuse to deeply lobed. Callus 3-horned. Fruit 2.5–4 cm long. Occasionally naturalized in coastal areas. Introd. from C. & S. America. Fl. all year . . . . ***E. radicans** Lindl.x **secundum** Jacq.hybrid complex

177 LOMANDRACEAE
c. 4 gen., Australia, New Guinea and New Caledonia

1 Lomandra Labill.
52 species in Aust. (49 endemic, 3 native); all states and territories

Erect and tufted or decumbent perennials with rhizomes or stolons. Leaves narrow-linear to narrow-oblong, basal or cauline. Flowers unisexual, often clustered, in spikes or panicles or semi-globular heads. Perianth segments 6, arranged in 2 whorls. Male flowers with 6 stamens, 3 attached to the inner perianth segments, the other 3 free; anthers versatile, deeply lobed; ovary rudimentary or absent. Female flowers similar, with a persistent perianth; staminodes present or absent; ovary 3-locular with 1 erect ovule per loculus, superior. Capsule subglobular, glossy.

Key to the species

1 Flowers separate on the inflorescence, well spaced or 2–3 together; the male flowers distinctly, and the female flowers very shortly, pedicellate 2

1 Flowers arranged in clusters on the inflorescence; or, if separate, then sessile 8

2 Flowers purplish or reddish (some other species may have purplish flowers but they lack the following characteristics); male flowers 2 mm diam., open-campanulate; female flowers c. 5 mm long: perianth segments ± thin and similar, lanceolate to ovate. Leaves narrow-linear, 30–40 cm long, 1–2 mm wide, quadrangular to terete; bases with purplish-brown to black margins but scarcely shredded. Male inflorescence a broad panicle; female inflorescence a narrower panicle or raceme. Widespread. DSF. Fl. autumn–winter . . . . **L. micrantha** (Endl.) Ewart. ssp. **tuberculata** J.Everett

2 Flowers yellow to cream; male flowers closed-campanulate. Inner perianth segments often ± thicker than the outer ones 3

3 Leaves terete or nearly so or plano-convex, margins not inrolled 4

3 Leaves flat or with inrolled margins 5

4 Leaves cylindrical to semi-cylindrical, usually twisted above, 15–60 cm long, 1–2 mm wide, with a few shreds at the base, often only 2 per shoot with several prominent scales below them. Male inflorescence a panicle, up to 15 cm long. Female inflorescence similar or a raceme. Flowers 1–3 mm long. Widespread. DSF . . . . **L. cylindrica** A.T.Lee

4 Leaves plano-convex, scarcely twisted above, up to 60 cm long, 1–2 mm wide, several to many per shoot, with numerous dark brown shreds at the base, and some inconspicuous leaf scales below. Male inflorescence a panicle Usually more than 20 cm long. Female inflorescence a smaller panicle, Flowers 1–3 mm long. Widespread. DSF . . . . **L. gracilis** (R.Br.) A.T.Lee

5 Densely tufted plants up to 15 cm high. Leaves involute, narrow-linear, 5–23 cm long, c. 1 mm wide with an entire apex, purplish near the base above the whitish shreds. Male inflorescence a narrow panicle up to 9 cm long with smooth axis. Coast and adjacent Plateaus. Damp places .**L. brevis** A.T.Lee

5 Tufted plants usually taller than 15 cm. Leaves narrow-linear with a minutely toothed apex or, if obtuse then, with flat margins, mostly up to 3 mm wide and 30 cm long, with numerous dark brown shreds at the base. Male flowers 1–2.5 mm long. Female flowers 2–3 mm long 6

6 Leaves flat or nearly so, leathery and firm. Male inflorescences up to half as long as the leaves. Female inflorescence smaller. Southern part of the area. Often on soils not derived from sandstones. DSF . . . . **L. filiformis** ssp. **coriacea** A.T.Lee

6 Leaves involute or incurved 7

7 Male inflorescence a raceme or panicle almost half as long as the leaves. Female inflorescence slightly smaller. Leaves firm. Widespread in all but southern and extreme western parts of the area. DSF. Ss . . . . **L. filiformis** (Thunb.) Britten ssp. **filiformis**

7 Male inflorescence a panicle much less than half as long as the leaves; scape shorter than rhachis; bracts at branches of rhachis larger. Female inflorescence much smaller. Hornsby Plateau, Blue Mts DSF. Ss . . . . **L. filiformis** ssp. **flavior** A.T.Lee

8 Leaves 8–14 mm wide, very rarely less, 2–3-toothed at the apex, up to 80 cm long, flat, glossy, with few or no shreds at the base. Male inflorescence a panicle of clusters. Female inflorescence ± similar. Male flowers ± cylindrical, often purplish, 3–4 mm long; inner perianth segments connate. Female flowers slightly shorter. Tufted plants up to 1 m high. Widespread in a variety of situations particularly on sandy soils. . . . . **L. longifolia** Labill.

8 Leaves up to 4 mm wide 9

9 Leaves conspicuously twisted or curled backwards towards the stem, 2–8 cm long (rarely longer), 1–2 mm wide, flat or concave, green or glaucous, with a few white shreds at the base or, more usually, none. Stems up to 50 cm long with the leaves arranged regularly and distichously along them. Male inflorescence a spike or small panicle of clusters. Female inflorescence smaller. Widespread. DSF. Ss . . . . **L. obliqua** (Thunb.) J.F.Macbr.

9 Leaves not as in *L. obliqua*. **10**

10 Leaf sheaths white to pale brown at the margins, disintegrating into a lattice. Male flowers with inner perianth segments connate above the middle, 2 mm long. Female flowers with almost free segments. Leaves flat to concave, 5–30 cm long, 1–2 mm wide, glaucous. Male inflorescence a small panicle of clusters, 2–12 cm long; the scape shorter. Female inflorescence a solitary head; scape up to 3 cm long, concealed in the leaf bases. Widespread. Heath and DSF. Sandy soils . . . . . . **L. glauca** (R.Br.) Ewart

10 Leaf sheaths shredding or not but not disintegrating into a lattice. Male flowers with free perianth segments **11**

11 Leaf bases ± intact **12**
11 Leaf margins shredding **13**

12 Scape flattened . . . . **L. confertifolia** ssp. **pallida** (see below)
12 Scape not flattened. Male flowers (when mature) on pedicels 8–10 mm long, 2–3 mm long, constricted above the base. Female flowers sessile, 3–4 mm long. Male inflorescence a spike or panicle of clusters, 5–30 cm long, usually longer than the scape. Female inflorescence slightly smaller. Leaves flat to concave, dark greyish-green, 40–75 cm long, 2–4 mm wide, obtuse; sheath white to brown. Widespread. DSF . . . . **L. multiflora** (R.Br.) Britten ssp. **multiflora**

13 Leaf shreds white or pale brown. Leaves flat to channelled, 30–70 cm long, 1–3 mm wide. Male inflorescence a spike or panicle of clusters, usually c. a quarter as long as the leaves; scape shorter than or equal to the rhachis. Female inflorescence a spike, smaller; scape concealed amongst the leaf bases. Hornsby Plateau and N. Blue Mts DSF. Sandy soils (intermediates between this and *L. confertifolia* ssp. *rubiginosa* can occur). . . . . **L. confertifolia** (F.M.Bailey) Fahn ssp. **pallida** A.T.Lee
13 Leaf shreds dark brown or reddish-brown. **14**

14 Stems long, decumbent, up to 50 cm long. Leaves ± incurved, 0.5–2 mm wide, 15–40 cm long, with reddish-brown shreds at the base. Male inflorescence a narrow panicle or spike of clusters, shorter than the scape. Female inflorescence smaller. Widespread. DSF . . . . **L. confertifolia** ssp. **rubiginosa** A.T.Lee
14 Stems short, tufted ascending **15**

15 Male inflorescence a narrow panicle of clusters, 5–17 cm long, usually shorter than the scape. Female inflorescence slightly smaller. Leaves 25–50 cm long, 1–3 mm wide, concave to inrolled, irregularly toothed at the apex, with brown shining shreds at the base. Male flowers 3–4 mm long. Female flowers 2–3 mm long. Coast and adjacent Plateaus. Creek beds . . . . **L. fluviatilis** (R.Br.) A.T.Lee
15 Male inflorescence a spike of clusters, 5–8 cm long, longer than the flattened scape. Female inflorescence slightly smaller. Leaves 35–50 cm long, 2–4 mm wide, flat to concave, irregularly toothed at the apex; basal shreds brown, becoming dull. Male flowers 2–3 mm long. Blue Mts Near creeks . . . . **L. montana** (R.Br.) L.R.Fraser & Vickery

178 XANTHORRHOEACEAE
1 gen., Aust.

1 Xanthorrhoea Sm.
28 species endemic Aust.; all states and territories

Perennials with woody secondarily thickened aerial or subterranean stems densely covered with leaf bases. Leaves narrow-linear, entire, in a terminal crown. Flowers arranged in numerous monochasial cymes packed together into a large cylindrical spike-like or a head-like inflorescence on a leafless scape; bracts numerous with often larger cluster bracts subtending groups of flowers and even more numerous packing bracts giving the spike its characteristic appearance. Perianth segments free, ± membranous. Stamens 6, exserted, white. Ovary 3-locular, with several ovules per loculus, superior; style simple. Fruit a capsule. The species with aerial stems are often known as *Grass Trees* or *Blackboys*.

Key to the species

1 Scape c. 10 times as long as the spike, up to 2 m long. Spike brush-like with prominent inner perianth segments 8–10 mm long and stamens c. 10 mm long. Leaves up to 1.5 m long, 2–4 mm wide, glossy, triangular in cross-section, scabrid on the margins. Stem buried. Coast and adjacent plateaus. Fl. summer . . . .**X. macronema** F.Muell. ex Benth.

1 Scape up to 5 times as long as the spike 2

2 Leaves soft and spongy, rhombic in cross-section, 2–4 mm wide (rarely more), not glaucous. Scape 130–200 cm long, c. as long as the spike, 20–30 mm diam. Stem usually aerial, 2–6 m high, often branched. Coast and adjacent Hornsby Plateau north from Gosford. WSF and RF margins . . . .
. . . . **X. malacophylla** D.J.Bedford

2 Leaves hard, not soft and spongy. 3

3 Packing bracts densely hairy on their abaxial surface making the whole spike velvety 4

3 Packing bracts glabrous or ± hairy but the spike not velvety 6

4 Spike dark brown when flowering. Leaves rhombic in cross-section, 2–4 mm wide, almost erect, glaucous. Scape up to 2m long, c. as long as the spike, 10–30 mm diam. Stem buried or up to 60 cm high not usually branched. Widespread south of Sydney. Heath and DSF. Sandy soils . . . . .**X. resinosa** Pers.

4 Spike cream to light brown when flowering. Leaves concave in cross-section or with a median channel or ± flat on adaxial surface. Stems buried, branched; the clusters of leaves thus often appearing to grow in colonies. 5

5 Leaves ± concave in cross-section, 3–6 mm wide, spreading to almost erect, glaucous. Scape 50–250 cm long, c. twice as long as the spike, c. 10 mm diam. Spike light brown when flowering. Widespread south of Sydney. Heath and DSF, often in wet sites . . . . **X. concava** (A.T.Lee) D.J.Bedford

5 Leaves triangular in cross-section, up to 3 mm (rarely up to 3.5mm) wide, ± erect, glaucous. Scape 20–160 cm long, 2–3 times as long as the spike, 5–20 mm diam. Spike very pale brown to cream when flowering. Coast and Hornsby Plateau north of Gosford. Wet heath on sand. . . . .
. . . . **X. fulva** (A.T.Lee) D.J.Bedford

6 Leaves glaucous. 7

6 Leaves not glaucous 8

7 Leaves up to 3 mm wide, rhombic in cross-section. Scape 18–40 mm diam. Spike up to 1.6m long. Cluster bracts prominent, hairy on the margins. Stem aerial, 1–5 m high, sometimes branched. Blue Mts Open forest. . . . . **X. glauca** D.J.Bedford ssp. **angustifolia** D.J.Bedford

7 Leaves 5–7 mm wide, ± flat on adaxial surface or concave . . . . **X. arborea** (see below)

8 Scape and spike together usually less than 75 cm long. Leaves erect, ± flat on adaxial surface or concave, 2–4 mm wide. Scape 2–5 times as long as the spike, 7–20 mm diam. Fruit often curved with the tip pointing upwards. Stems buried, branched; the clusters of leaves thus often appearing to grow in colonies. Widespread. Damp sites, often on clay soils (Some specimens of other species may have the same scape:spike ratio as *X. minor*. When identifying *Xanthorrhoea* the characteristics of the population from which the specimen has been taken should be borne in mind.) . . . . . **X. minor** R.Br. ssp. **minor**

8 Scape and spike together usually more than 90 cm long 9

9 Leaves up to 3 mm wide. Old leaves strongly reflexed 10

9 Leaves more than 3 mm wide, spreading to erect 11

10 Young leaves in upright tufts. Stems aerial up to 5 m long. Scape 1–2 m long, usually the same length as the spike, 7–20 mm diam. North from Singleton. DSF and Heath . . . . **X. johnsonii** A.T.Lee

10 Leaves in a hemispherical crown. Stem aerial and up to 60 cm high or buried, usually unbranched. Scape 1–2 m long, usually twice as long as or rarely equalling the spike, 5–11 mm diam. Widespread. DSF. Ss . . . .**X. media** R.Br.

11 Scape less than twice as long as the spike, 130–170 cm long (rarely longer), 12–16 mm diam. Stem aerial, often branched, up to 2 m high. Widespread north of Heathcote. DSF and WSF. Ss and sands in sheltered sites . . . . **X. arborea** R.Br.

11 Scape twice as long as or equalling the spike or longer, 1–2 m long, 10–16 mm diam. Stem aerial and up to 2 m high or buried, sometimes branched. DSF. Sandy soils . . . . **X. latifolia** (A.T.Lee) D.J.Bedford

179 BURMANNIACEAE
20 gen., mostly trop. and subtrop.

1 Burmannia L.
3 species native Aust.; all states and territories except S.A.

One species in the area
Erect herb. Basal leaves linear-oblong, 5–9 cm long, 5–10 mm wide, acute, entire. Scapes leafy, usually solitary, up to 60 cm high. Flowers bisexual, arranged in a pair of monochasia. Perianth segments 6, in 2 rows; the outer ones c. 2 mm long. Stamens 3. Ovary 3-locular, inferior; the floral tube blue, petaloid, winged, 12–15 mm long. Fruit a capsule. Coast N. of Gosford. Swamps. Fl. summer . . . . . **B. disticha** L.

180 DIOSCOREACEAE
4 gen., warm temp. and trop.

1 Dioscorea L.
5 species in Aust. (2 endemic, 1 naturalized); Qld, NSW, NT, WA

One species in the area
Glabrous twiner often several metres long, with underground tubers. Leaves alternate or subopposite, hastate or ovate-cordate, acuminate, with 5–7 prominent longitudinal veins, 6–10 cm long. Male flowers in interrupted spikes 3–5 cm long; the spikes clustered in axillary panicles. Female flowers in simple interrupted racemes 5–10 cm long. Perianth segments minute. Stamens 6. Ovary 3-locular, with 2 ovules per loculus, inferior. Style 2-lobed. Fruit a 3-angled capsule, 2 cm long, 3 cm wide. Coast. RF. Fl. spring–summer. *Native Yam* . . . . **D. transversa** R.Br.

181 ALSTROEMERIACEAE
3 gen., temp. to trop., Central and S. America.

1 Alstroemeria L.
2 species naturalized Aust.; NSW, Vic.

One species in the area
Weak herb. Leaves cauline, mostly on the upper part of the stem, broad-lanceolate to elliptic, 4–7 cm long, 15–30 mm wide, acute, on slender twisted petioles 1–6 cm long. Flowers in irregular umbels with a few leafy bracts at the base; scapes up to 70 cm high. Perianth red and green; posterior lobe ± larger than the others, spotted. Stamens 6, inserted on an annular nectary at the base of the perianth. Ovary 3-locular; style obtusely 3-angular, 3-fid. Capsule ovoid. Escaped from cultivation into shady places near Hawkesbury River and Blue Mts Introd. from trop. S. America . . . . * **A. pulchella** L.f.

182 COLCHICACEAE
Herbs with corms. Leaves basal or cauline, often few. Flowers bisexual or unisexual, regular, in cymes or umbels. Perianth segments 6, free or connate at the base. Stamens 6, free. Ovary 3-locular; styles 3, free or connate towards the base. Fruit a capsule. c. 19 gen., mostly Africa.

Key to the genera

1 Leaves ending in a tendril . . . . **3 GLORIOSA**
1 Leaves not ending in a tendril 2

2 Leaves ≥7 mm wide, ovate to lanceolate . . . . **4 SCHELHAMMERA**
2 Leaves <7 mm wide 3

3 Flowers in monochasial cymes . . . . **1 WURMBEA**
3 Flowers in umbels . . . .**2 BURCHARDIA**

1 Wurmbea Thunb.

19 species endemic Aust.; all states and territories

Erect herbs up to 30 cm high, dioecious polygamous or hermaphrodite, with corms. Leaves 3, linear, up to 30 cm long; upper ones broader and shorter. Flowers in monochasial cymes of up to 10 flowers. Perianth segments 6, white, with a purplish swollen band or two pinkish nectaries near the base, persistent but not spirally twisted after flowering. Stamens 6. Ovary 3-locular, with numerous ovules per loculus. Fruit a capsule.

Key to the species

1 Nectary a single, purplish, swollen band across each perianth segment, occasionally with a slight break in the middle. Plants with 1–10 flowers all female or all male or the lowest 1–2 hermaphrodite; occasionally all hermaphrodite. Widespread. Grasslands. Fl. summer. *Early Nancy* . . . .
. . . . **W. dioica** (R.Br.) F.Muell. ssp. **dioica**
1 Nectaries 2, thickened, pinkish, shelf-like or shallowly cupped structures at the margins of each perianth segment, clasping the filaments. Plants with 1–6 hermaphrodite flowers or sometimes the uppermost flower male. Cumberland Plain. Grasslands. . . . . . .**W. biglandulosa** (R.Br.) T.D.Macfarl.

2 Burchardia R.Br.

5 species endemic Aust.; all states and territories except NT

One species in the area

Erect herbs with fibrous roots. Leaves few, linear, surrounded by the dead bases of old leaves. Flowers in terminal umbels. Perianth segments 6, free, pale pink or white, convolute around the stamens in the bud, flattened and spreading in flower, deciduous. Stamens 6; filaments flattened basally. Ovary 3-angular, 3-locular, with numerous ovules per loculus. Fruit a capsule. Widespread. Heath and DSF. *Milkmaids* . .
. . . . **B. umbellata** R.Br.

3 Gloriosa L.

1 species naturalized Aust.; Qld, NSW

One species in the area

Glabrous, climbing herb, stems to 4 m long. Leaves 6–20 cm long, alternate, lanceolate, sessile, ending in a tendril. Flowers solitary in leaf axils, yellow to red, pedicels up to 20 cm long. Perianth c. 6 cm long, strongly reflexed. Stamens 6, free. Ovary 3-locular. Style 3-fid bent near base. Capsule 4–10 cm long, 10–20 mm wide. Seeds red. Occasionally naturalized. Introd. from Africa and Asia. Fl. spring–autumn .
. . . . ***G. superba** L.

4 Schelhammera R.Br.

2 species in Aust. (1 endemic); Qld, NSW, Vic.

One species in the area

Erect or diffuse herbs with stems up to 20 cm high. Leaves sessile, undulate, broad-lanceolate, 3–4 cm long, 7–18 mm wide, with numerous parallel veins . Flowers usually solitary on slender pedicels. Perianth 15 mm diam., pale lilac; segments 6, subequal, free, convolute around the stamens in bud. Stamens

6. Ovary 3-locular, with several ovules per loculus; style 3-fid. Fruit a globular almost fleshy capsule. Widespread. Shady places. Fl. spring. . . . . **S. undulata** R.Br.

183 LILIACEAE
c. 16 gen., mostly N. temp.

1 Lilium L.
1 species naturalized Aust.; NSW, Vic.

One species in the area
Erect, bulbous herb up to 1 m high. Leaves linear, up to 16 cm long, sessile, scattered along the stem. Flowers in very short terminal racemes. Perianth segments white, with purplish markings on the back, c. 13 cm long. Stamens 6; filaments filiform. Ovary 3-locular; style 3-lobed. Introd. from Asia. Garden escape, naturalized and widespread . . . .*L. formosanum Wallace

184 PHILESIACEAE
Shrubs or climbers, sometimes with rhizomes. Leaves alternate, linear to ovate or oblong, with prominent parallel veins. Flowers clustered or in paniculate cymes in the leaf axils, regular, bisexual. Perianth segments 6, free, in 2 whorls. Stamens 6; filaments connate into a tube or free. Ovary superior, 3-locular, with axile placentas; style capitate or 3-lobed. Fruit a berry. 7 gen., temp. S. hemisphere.

Key to the genera

1 Inner perianth segments fringed. Filaments connate into a membranous tube . . . . **1 EUSTREPHUS**
1 Inner perianth segments not fringed. Filaments free . . . .**2 GEITONOPLESIUM**

1 Eustrephus R.Br.
1 species native Aust.; Qld, NSW, Vic.

One species in the area
Glabrous much branched climber several metres long. Leaves usually ovate-lanceolate, sometimes linear, 5–10 cm long. Flowers clustered in the upper leaf axils. Perianth segments c. 6 mm long, white to pale pink or purple; inner segments fringed. Filaments connate into a membranous tube. Fruit a globular orange berry c. 1 cm diam. Seeds black. Widespread. Moist communities. Fl. spring. *Wombat Berry* . . . . **E. latifolius** R.Br.

2 Geitonoplesium A.Cunn. ex R.Br.
1 species native Aust.; Qld, NSW, Vic.

One species in the area
Climber with tangled branches several metres long. Leaves usually linear, sometimes narrow-ovate, alternate, 5–8 cm long. Flowers in paniculate cymes. Perianth purplish green to white, 6–8 mm long. Stamens free. Berry subglobular, purple-black, c. 6 mm diam. Seeds black. Widespread. Moist communities. Fl. spring . . . . **G. cymosum** (R.Br.) A.Cunn. ex R.Br.

185 RIPOGONACEAE
1 gen., temp. and trop.

1 Ripogonum J.R.Forst. & G.Forst.
5 species native Aust.; Qld, NSW

Shrubby climber with or without prickles on the stems. Leaves sub-opposite, 3–5-veined, coriaceous; petioles without tendrils. Flowers bisexual, in axillary racemes or spikes. Perianth segments free, greenish white, reflexed. Anthers sagittate. Berry red, globular.

Key to the species

1 Branches pubescent; prickles absent. Leaves ovate to narrow-ovate, 8–20 cm long, cordate at the base, on short petioles or almost sessile, acuminate. Coast and adjacent plateaus. WSF and RF . . . .
. . . . **R. fawcettianum** F.Muell. ex Benth.
1 Branches glabrous, the larger ones with prickles. Leaves ovate to narrow-ovate, 8–15 cm long, narrowing towards the base, with a twisted petiole, shortly acuminate. Widespread. RF . . . . **R. album** R.Br.

186 SMILACACEAE
1 gen., temp. and trop.

1 Smilax L.
7 species in Aust. (3 endemic); Qld, NSW, Vic., NT, WA

Branched, shrubby, dioecious climbers. Leaves alternate, with reticulate venation and longitudinal veins, broad-lanceolate to ovate, coriaceous; stipular tendrils usually present at base of each leaf. Flowers unisexual, in axillary umbels. Perianth segments free, greenish white, spreading. Anthers oblong. Berry black, globular, with 1–3 hard glossy seeds.

Key to the species

1 Leaves lanceolate to ovate-lanceolate, 4–6 cm long, with 3 longitudinal veins, glaucous below. Prickles absent. Widespread. Most common in gullies and humid places. Fl. summer. *Sarsparilla* . . . .
. . . .**S. glycophylla** Sm.
1 Leaves oblong to almost orbicular, 5–10 cm long, with 5 longitudinal veins, green on both surfaces. Prickles present on the stem. Fruit 10–15 mm diam. Widespread. Usually in or near RF. Fl. summer . .
. . . .**S. australis** R.Br.

187 ARECACEAE
Trees. Leaves large, pinnately or palmately divided; the petiole with or without a sheathing base. Flowers bisexual or unisexual, regular, numerous in a large panicle which is enclosed in a large deciduous spathe when young. Perianth segments 6, in 2 whorls. Stamens usually 6, in 2 whorls, hypogynous. Ovary superior, 1–3-locular. Fruit a berry or drupaceous. 187 gen., trop. to warm temp.

Key to the genera

1 Leaves palmately divided, fan-shaped . . . . .2 LIVISTONA
1 Leaves pinnately divided 2

2 Leaf petiole without sheathing base; lower pinnae spine-like. . . . . .3 PHOENIX
2 Leaf petiole with sheathing base; lower pinnae not spine-like . . . . .1 ARCHONTOPHOENIX

1 Archontophoenix H.W.Endl. & Drude
3 species endemic Aust.; Qld, NSW

One species in the area
Tall, erect, monoecious tree with pinnately divided leaves 3–4 m long. Flowers pale lilac, unisexual, usually 2 male and 1 female together, sessile in the notches of the numerous branches of the panicle which arises from below the base of the leaf sheaths. Fruit ovoid-globular, c. 12 mm diam, red, drupaceous. Coast. RF. *Bangalow Palm*. . . . .**A. cunninghamiana** (H.Wendl.) H.Wendl. & Drude

2 Livistona R.Br.

c. 12 species native Aust.; Qld., NSW, Vic., NT, WA

One species in the area

Tall, erect tree with palmately divided fan-shaped, plicate leaves 1 m or more across. Flowers small, yellowish, numerous, bisexual, sessile, solitary or clustered along the slender branches of the panicle which arises below the leaf bases. Fruit globular, 12–20 mm diam., blackish, drupaceous. Widespread. RF and sheltered places. *Cabbage-tree Palm*. . . . . **L. australis** (R.Br.) Mart.

3 Phoenix L.

2 species naturalized in Aust.; NSW, Vic., S.A., WA, NT

One species in the area

Tall, erect tree with thick trunk to 1 m or more diam. Leaves up to 6 m long, pinnate with stiff pinnae which become spine-like towards base of leaf. Petioles short, bases remaining as stubs on trunk. Flowers bisexual, yellowish, sessile in axil of bracts. Fruit an elliptic berry, 15–20 mm long, yellow to red. Garden escape near habitation. Introd. from the Mediterranean. *Canary Island Date Palm* . . . .***P. canariensis** Chabaud.

188 HAEMODORACEAE

Perennial herbs. Leaves mostly basal with sheathing leaves. Inflorescence with smaller leaves or leaves absent. Tepals 6 free or connate. Stamens 3 to 6, anthers basifixed. Ovary inferior to half inferior, 3-locular. Stigma 1. Fruit a capsule. 17 gen., temp. to trop., S. hemisphere.

Key to the genera

1 Stamens 3 . . . . .1 HAEMODORUM
1 Stamens 6 . . . . .2 ANIGOZANTHOS

1 Haemodorum Sm.

Bloodroot
20 species in Aust. (19 endemic); Qld, NSW, Tas., NT, WA

Erect herbs with short erect rhizomes. Leaves linear, mostly basal. Inflorescence paniculate or corymbose. Perianth segments 6, purplish black, in 2 whorls; the outer whorl slightly shorter than the inner, often convolute around the base of the stamens. Stamens 3. Ovary half-inferior, 3-locular; placentas axile. Fruit a capsule, 8–12 mm diam.; seeds 2 per loculus.

Key to the species

1 Leaves flat, 2–6 mm wide; the lowest 30–40 cm long. Widespread. Heath and DSF. Ss. Fl. summer . . . . **H. planifolium** R.Br.
1 Leaves terete, less than 2 mm wide; the lowest up to 50 cm long. Coast and adjacent plateaus. Heath and DSF. Fl. summer . . . .**H. corymbosum** Vahl

2 Anigozanthos Labill.

Kangaroo Paw
25 species native Aust.; WA, NSW

Erect rhizomatous herb. Leaves linear, sheathing, mostly basal. Inflorescence a raceme, scape to 3 m long. Perianth 30–45 mm forming a long tube with 6 short lobes, covered with short yellow-green hairs. Stamens 6, anthers orange, 2–4 mm long, outer filaments longer than inner filaments. Style 30–40 mm long. Ovary inferior. Fruit 3-celled. Royal National Park roadside and on eastern boundary between park and Bundeena. Introd. from WA Fl. summer. . . . . ***A. flavidus** DC.

189 COMMELINACEAE

Perennial herbs. Leaves alternate, with a basal closed sheath. Flowers regular or slightly irregular, bisexual. Sepals 3, usually green. Petals 3, free or rarely connate. Stamens 6 or sometimes 2–3 with 3–4 staminodes, free, hypogynous. Ovary 3-locular; placentas axile; style filiform. Fruit a capsule or nut. 40 gen., trop. to warm temp.

Key to the genera

1 Fertile stamens 6; filaments bearded. 2
1 Fertile stamens 2–3, filaments glabrous or bearded 3

2 Petals >4 mm long . . . . **2 TRADESCANTIA**
2 Petals ≤4 mm long . . . . **6 GIBASIS**

3 Inflorescence enclosed in a folded leafy spathe. Filaments glabrous . . . . **1 COMMELINA**
3 Inflorescence not enclosed in a folded leafy spathe but often with 2 leaf-like bracts at the base 4

4 Leaves 10–15 cm long. Fruit indehiscent, 6 mm diam., with a bluish dry brittle pericarp. Stems creeping, rooting at the base but finally erect. Filaments glabrous. . . . . **5 POLLIA**
4 Leaves less than 10 cm long. Fruit dehiscent, small, inconspicuous 5

5 Filaments all glabrous. Ovary with 2 perfect loculi. **3 ANEILEMA**
5 Some or all the filaments bearded. Ovary 3-locular . . . . **4 MURDANNIA**

1 Commelina L.
10 species in Aust. (2 naturalized); Qld, NSW, NT, S.A., WA

One species in the area

Weak, procumbent, ascending herb, with stems rooting at the nodes, glabrous or nearly so.Leaves with sheathing base. Flowers enclosed in a folded green spathe; spathe cordate at the base, with rounded auricles, obliquely ovate-acuminate in side view. Peduncles in pairs in the spathe; one usually with a single flower; the other with 2–3 flowers.

Key to the species

1 Leaves ovate-lanceolate, 3–8 cm long, 5–15 mm wide, sheathing base with short hairs at mouth. Petals blue, nearly 12 mm long. Coast and adjacent plateaus. Forests; persistent weed in gardens. Fl. spring . . . . **C. cyanea** R.Br.
1 Leaves ovate-lanceolate, 2.5–5 cm long, 10–15 mm wide, sheathing base with long hairs at mouth. Petals yellow. Garden escape. Greenwich area. Introd. from S. Africa. Fl. spring–autumn. . . . .*C. africana** L.

2 Tradescantia L.
3 species naturalized Aust.; Qld, NSW, Vic., Tas.

Perennial herbs, stems rooting at the nodes. Leaves sheathing at the base, glabrous or ciliate. Inflorescences axillary or terminal. Flowers regular, bisexual in a sessile, terminal cluster subtended by 2 unequal leaf-like bracts. Sepals free or fused, petals free or fused at base. Stamens 6, filaments bearded. Ovary 3-locular. Fruit a papery capsule.

Key to the species

1 Leaves purplish-red underneath, silvery stripes on the upper surface, lanceolate-ovate, c. 5 cm long. Leaf sheaths with long ciliate hairs. Petals pink. Filament hairs purple. Weak, procumbent or ascending, glabrous, perennial herb. Garden plant occasionally naturalized. Introd. from Mexico. Fl. spring–summer . . . . **T. zebrina** Bosse
1 Leaves green on lower surface, ovate-acute, c. 5 cm long, shining, ciliate at the base, without a distinct petiole. Petals white. Filament hairs white. Weak procumbent or ascending, perennial, herb. A common

weed in shaded places. Introd. from S. America. Fl. spring–summer *Wandering Jew* . . . .
. . . . **T. fluminensis** Vell.

(**T. pallida* (Rose) D.R.Hunt (leaves deep purple on both surfaces) and **T. blossfeldiana* Mildbr. (leaves green on top purple below) have recently been recorded as garden escapes in the Blue Mts area)

3 **Aneilema** R.Br.
3 species endemic Aust.; Qld, NSW, NT

One species in the area

Weak herbs; stems creeping, or ascending from prostrate stems. Flowers small. Inflorescence a loose terminal panicle or with 2 short branches rarely exceeding the highest leaf. Petals white. Fertile stamens 2–3; staminodes 2–4; filaments all glabrous. Ovary with 2 perfect loculi.

Key to the species

1 Inflorescence a narrow panicle with filiform branches, usually much exceeding the last leaves. Petals 4–6 mm long. Leaves 3–8 cm long, lanceolate, acuminate. Capsule oblong, c. 4 mm long, obtuse, much flattened. Coast and adjacent plateaus. In or near RF . . . .**A. acuminatum** R.Br.
1 Inflorescence usually consisting of 2 short branches each bearing 2 flowers, rarely exceeding the highest leaves. Flowers smaller than in *A. acuminatum*. Stems weak, procumbent. Leaves lanceolate, 2–5 cm long, acuminate. Damp shaded, places . . . . **A. biflorum** R.Br.

4 **Murdannia** Royle
5 species in Aust. (4 endemic, 1 naturalized); Qld, NSW, NT

Annual or perennial herbs with erect or prostrate stems. Leaves glabrous, sessile, distichous or spirally arranged. Inflorescence terminal and/or axillary. Flowers bisexual, regular to slightly irregular. Sepals and petals free. Fertile stamens 2 or 3, staminodes 3 or absent; filaments bearded. Ovary 3-locular. Fruit a capsule.

Key to the species

1 Herb with creeping stems. Internodes with a line of dense hairs. Leaves linear-lanceolate, 2–8 cm long, sometimes slightly folded, apex acuminate. Flower in a terminal mostly 1-flowered inflorescence. Petals pale pink to purple, obovate. Garden escape. Damp places. Introd. from E. Asia. Fl. autumn . . . .
. . . . **M. keisak** (Hassk.) Hand.-Mazz.
1 Weak, perennial herb. Stems erect, 10–60 cm high. Leaves linear, usually 6–10 cm long, sometimes much longer, basal with a few cauline leaves. Flowers in a terminal irregularly branched panicle. Petals usually mauve or pink, ovate, c. 10 mm long. Coast. Uncommon . . .**M. graminea** (R.Br.) G.A.Brückn.

5 **Pollia** Thunb.
2 species endemic Aust.; Qld, NSW

One species in the area

Glabrous, herbaceous perennial. Stems erect or ascending, 30–100 cm high, creeping, rooting at the nodes. Leaves lanceolate, 10–15 cm long, tapering towards the sheathing base, often with a crisped margin near the base. Flowers in a terminal cymose bracteate panicle. Sepals much imbricate. Petals 4–6 mm long, usually blue. Fertile stamens 3; staminodes 3. Fruit an ovoid, blue, shining nut c. 6 mm long; pericarp brittle. Widespread. In or near RF. . . . .**P. crispata** (R.Br.) Benth.

6 **Gibasis** Raf.
1 species naturalized Aust.; NSW

One species in the area

Perennial, decumbent, herb, rooting at nodes. Leaves distichous, oblong to lanceolate 4–7 cm. Inflorescences terminal and/or axillary. Flowers in pedunculate cymes. Sepals keeled, 2–3 mm long, petals white, 4 mm long. Stamens 6, filaments bearded, Fruit ovoid, 2.5 mm long. Garden escape. Blue Mts Introd. from Mexico. *Tahitian bridal-veil* . . . . . . ***G. pellucida** (M.Martens & Galeotti) D.R.Hunt

190 PHILYDRACEAE
4 gen., S.E. Asia and Aust.

1 Philydrum Banks ex Gaertn.
1 species native Aust.; Qld, NSW, Vic., NT, WA

One species in the area

Erect, perennial, aquatic herb, slightly woolly-hairy, up to 1 m high. Leaves sword-shaped, 20–60 cm long, mostly basal but passing gradually upward into smaller floral leaves and bracts. Flowers in a simple or branched spike, sessile in the axil of a leafy acuminate bract. Perianth segments 4, yellow, free, in 2 whorls; the outer ones c. 12 mm long; the inner ones shorter. Stamen 1. Ovary superior, 3-locular; placentas axile. Fruit an oblong capsule, 3-valved, c. 12 mm long. Widespread. Near water. Fl. summer. *Frogmouth*. . . . .**P. lanuginosum** (Banks & Sol.) Gaertn.

191 PONTEDERIACEAE
9 gen., temp. to trop., aquatics.

1 Eichhornia Kunth
1 species naturalized Aust.; Qld, NSW, Vic., WA

One species in the area

Floating, aquatic herb with numerous roots, spreading vegetatively by stolons. Leaves tufted; petioles swollen and spongy, 5–50 cm long; blades flat, elliptic to orbicular, c. 10 cm long. Flowers in erect racemes with c. 8 flowers. Flowers irregular, bisexual. Perianth pale violet with some blue and yellow on the lower lobe, 6-lobed, c. 5 cm diam., tubular. Stamens 6. Ovary 3-locular, superior. Fruit a capsule. Widespread. Streams and ponds. Regarded as a noxious weed because it blocks watercourses. Introd. from Central and S. America. *Water Hyacinth*. . . . . *** E. crassipes** (Mart.) Solms

192 CENTROLEPIDACEAE
4 gen., temp., mainly Aust.

1 Centrolepis Labill.
20 species in Aust. (19 endemic); all states and territories

Small tufted annuals. Leaves crowded, linear, filiform, basal, 1–5 cm long. Flowering culms leafless, rather rigid, usually several times longer than the leaves but sometimes only slightly exceeding them, with a single short terminal spikelet within 2 green often striate sheathing bracts. Flowers usually bisexual. Stamen 1. Ovary 3- or more-carpellate, superposed in 2 rows to one side of a linear receptacle. Styles 1 to each carpel, with a linear filiform stigma, free or united basally. Fruit a capsule.

Key to the species

1 Outer floral bracts each with an awn usually longer than the broader basal portion. Leaves obtuse, with or without a minute blunt hyaline tip. Widespread. Damp or wet sandy soil . . . . **C. fascicularis** Labill.
1 Outer floral bracts with a very short blunt tip, never even half as long as the broader basal portion and often almost obsolete. Leaves terminated by a fine hair-like hyaline point. Widespread. Damp or wet sandy soil . . . .**C. strigosa** (R. Br.) Roem. & Schult.

193 CYPERACEAE
Herbs usually with solid or (rarely) ± hollow stiff culms. When perennial the rhizomes often produce short stolons covered with short sheathing scales. Leaves chiefly basal, usually with closed sheaths, or

leaves reduced to sheathing scales. Flowers bi- or unisexual, in spikelets, subtended by scale-like bracts (glumes) which are distichous or spirally imbricate; lower glumes often sterile. Spikelets either solitary or clustered in terminal or apparently lateral simple or compound spikes panicles or umbels. Compound inflorescence and its branches subtended by bracts the lower of which often closely resemble cauline leaves; the upper bracts increasingly glumaceous. Perianth absent or present as bristles or scales. Stamens 1–3, rarely more; filaments free, filiform or slightly flattened; anthers usually exserted from the spikelets, attached by their bases, oblong or linear, with 2 parallel cells opening by slits. Ovary 1-locular, with 1 erect ovule; style terminal, filiform or thickened at the base, 2–3(4)-fid. Fruit a small nut, flattened when style is 2-fid, ± 3-angular when style is 3-fid. c. 98 gen., cosmop.

Key to the genera

1 Flowers unisexual, female flowers enclosed in a flask-shaped utricle (Fig. 46 e,g) or in separate spikelets . . . . .GROUP 1
1 Flowers bisexual or unisexual but never enclosed in a utricle and never apparently separated into unisexual spikelets 2

2 Spikelets with 1–2 perfect flowers, other flowers male or sterile . . . . .GROUP 2
2 Spikelets with more than 2 perfect flowers . . . . .GROUP 3

Group 1
Female flowers enclosed in a flask-shaped utricle or in separate spikelets

1 Spikelets dioecious. Fruit not enclosed in a utricle . . . . 26 SCLERIA
1 Spikelets monoecious. Fruit enclosed in a utricle 2

2 Hooked bristle protruding from the utricle . . . . 27 UNCINIA
2 Bristle absent from female flower or short and not exserted . . . . 28 CAREX

Group 2
Female flowers never enclosed in a utricle and never apparently separated into unisexual spikelets; spikelets with 1–2 perfect flowers

1 Perianth scales or bristles present 2
1 Perianth scales or bristles absent 9

2 Style 2-fid 3
2 Style 3-fid 4

3 Sterile glumes 3–4. Stamens 3. Perianth bristles 6–7 . . . . 14 RHYNCHOSPORA
3 Sterile glumes 1–2. Stamens usually 2. Perianth bristles usually 4 . . . . 15 CYATHOCHAETA

4 Inflorescence globular . . . . .17 GYMNOSCHOENUS
4 Inflorescence paniculate, spicate or capitate 5

5 Bracts surrounding the head of spikelets with broad hyalo-membranous margins 1.5–3 mm wide . . . . 18 PTILOTHRIX
5 Bracts of the inflorescence, which is not usually a head, with no hyalo-membranous margin or a very narrow one less than 1 mm wide 6

6 Perianth present as scales 7
6 Perianth present as bristles 8

7 Spikelet solitary, terminal . . . . 16 SCHOENUS
7 Spikelets numerous in a terminal panicle . . . . .20 LEPIDOSPERMA

8 Glumes distichous (Fig. 46 i) . . . . 16 SCHOENUS
8 Glumes spirally imbricate (Fig. 46 a,m) . . . . .19 TRICOSTULARIA

FIGURE 46

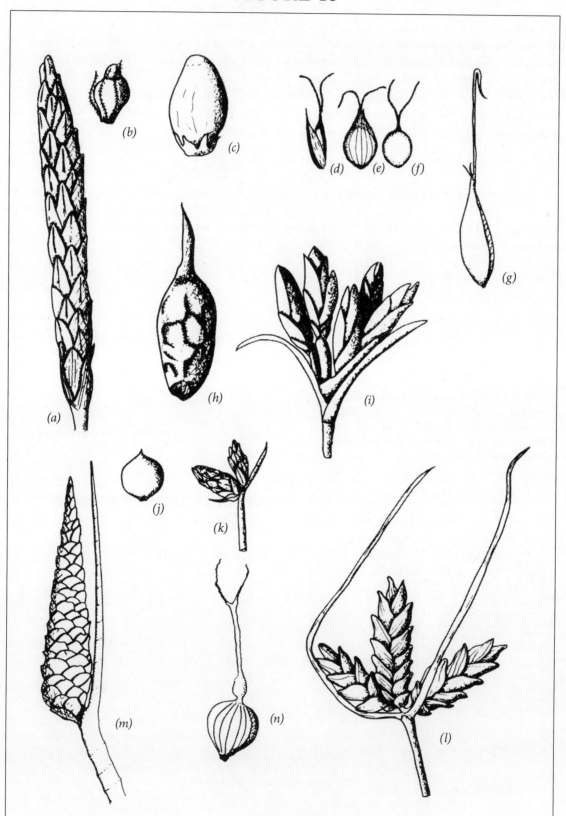

Fig. 46 CYPERACEAE: (a) *Eleocharis acuta*, spikelet, ✕4; (b) *Eleocharis sphacelata*, nut, ✕5; (c) *Lepidosperma laterale*, nut, ✕5; (d) *Carex inversa*. female flower, ✕5; (e) *Carex inversa*, mature utricle, ✕5; (f) *Carex inversa*, nut, ✕5; (g) *Uncinia tenella*, mature utricle, ✕5; (h) *Caustis flexuosa*, nut, ✕5; (i) *Schoenus ericetorum*, spikelets, ✕4; (j) *Isolepis prolifera*, nut, x S; (k) *Isolepis hookeriana*, spikelets, ✕5; (l) *Cyperus gracilis*, spikelets, ✕4; (m) *Lepironia articulata*, spikelet, ✕2; (n) *Fimbristylis dichotoma*, nut with style attached, ✕5.

9 Glumes distichous (Fig. 46 i,l) 10
9 Glumes spirally imbricate (Fig. 46 a,m) 13

10 Spikelets solitary or in compact heads or umbels. 11
10 Spikelets in narrow panicles or racemes 12

11 Culms distinctly 3-angular . . . . **1 CYPERUS**
11 Culms terete ridged or compressed. . . . . **16 SCHOENUS**

12 Nut surmounted by a conical beak as long as the body. Uppermost flower perfect. . . **25 TETRARIA**
12 Nut surmounted by the thickened stylar base scarcely half as long as the body or almost obsolete.
Lowermost flower perfect . . . .**22 BAUMEA**

13 Leaves reduced to brown scales sheathing the culm . . . . **24 CAUSTIS**
13 Leaves with distinct laminas 14

14 Leaves without a ligule. Upper flower abortive or male or both flowers perfect. . . . . . **21 CLADIUM**
14 Leaves with a ligule. Lower flower abortive or male . . . . **23 GAHNIA**

Group 3
Female flowers never enclosed in a utricle and never apparently separated into unisexual spikelets; spikelets with more than 2 perfect flowers

1 Perianth scales often numerous but always with 2 lateral ones keeled or at least concave; usually a
number of flat ones inside these lateral scales 2
1 Perianth scales very rarely more than 6 or absent; lateral scales not keeled and scarcely concave 3

2 Spikelet solitary, globular or subglobular. Nut with 8 prominent ribs. Glumes loose; perianth scales ±
exserted from the glumes . . . . **13 CHORIZANDRA**
2 Spikelet solitary, conical. Nut smooth. Glumes closely imbricate; perianth scales not exserted . . . . . .
. . . . **12 LEPIRONIA**

3 Base of style dilated 4
3 Base of style not dilated 7

4 Stylar base deciduous, not persistent on the nut 5
4 Stylar base persistent on the nut 6

5 Glumes distichous at least towards base. . . . . **4 ABILDGAARDIA**
5 Glumes imbricate around rhachis . . . .**3 FIMBRISTYLIS**

6 Leaves reduced to scales sheathing the base of the culms . . . . **2 ELEOCHARIS**
6 Leaves with well developed laminas . . . . **5 BULBOSTYLIS**

7 Glumes distichous (Fig. 46 i,l) 8
7 Glumes spirally imbricate (Fig. 46 a,m) 9

8 Inflorescence generally umbellate or corymbose, occasionally clustered. Rhachis of spikelet ± straight
. . . . **1 CYPERUS**
8 Inflorescence generally spicate or paniculate. Rhachis of spikelet zig-zag. . . . . **16 SCHOENUS**

9 Perianth scales 2, enclosing the nut . . . . **11 LIPOCARPHA**
9 Perianth scales or bristles more than 2 or absent. 10

10 Perianth absent or present as a disc at the base of the nut 11
10 Perianth present as bristles or scales 13

11 Nut rugose . . . .**10 SCHOENOPLECTUS**
11 Nut smooth or reticulate, often with swollen ribs. 12

12 Plants with thick long, rhizome; hypogynous disc present at base of nut . . . . **8 FICINIA**
12 Plants in tufts or with a filiform rhizome; nut without hypogynous disc at base . . . . . . **7 ISOLEPIS**

13 Culms with nodes and cauline leaves. . . . .**6 SCIRPUS**
13 Culms with basal leaves or sheathing scales, without nodes except right at the base. 14

14 Leaves reduced to sheathing scales or with very short laminas less than half as long as the culm . . . .
. . . .**10 SCHOENOPLECTUS**
14 Leaves with distinct long laminas often as long as the culms . . . . **9 BOLBOSCHOENUS**

1 Cyperus L.
150 species in Aust. (50 endemic); all states and territories

Tufted or creeping rhizomatous perennials or annuals. Culms 3-angular or terete. Leaves few or reduced to scales at the base of the culms. Involucral bracts 1 to several at the base of the inflorescence. Inflorescence usually either capitate or umbellate; rays simple or compound, commonly bearing divaricate spikes or clusters of spikelets. Spikelets flat to subterete, 1–many-flowered (Fig. 46). Rhachilla either not winged or bearing at each node a pair of wings which are the decurrent bases of the next distal glume. Glumes distichous, concave or keeled; the lowest 1 or 2 sterile. Perianth absent. Stamens 1–3. Style undilated at the base, deciduous, 3–2-fid or rarely entire. Nut equally 3-angular or lenticular, naked or clasped by the wings of the rhachilla.

Key to the species

1 Style 2-fid. Nut compressed, biconvex or plano- or concavo-convex. . . . .**GROUP 1**
1 Style 3-fid. Nut almost equally 3-angular 2

2 Spikelets arranged spicately on the rhachis of the ultimate branches of the inflorescence in the form of very short or ± elongated spikes. Rhachilla of spikelet bordered by hyaline wings, but these are sometimes deciduous . . . .**GROUP 2**
2 Spikelets ± digitate or fasciculate in clusters on a very reduced rhachis, sometimes very few in number or solitary and then sometimes appearing lateral when the lowest involucral bract is erect. Rhachilla winged or not . . . .**GROUP 3**

Group 1
Style 2-fid

1 Spikelets 6 or more flowered. Flowers falling from spikelet leaving persistent rhachilla, glumes falling away at maturity 2
1 Spikelets 1-4 flowered. Rhachilla articulate above the 2 lowest sterile glumes, spikelet falling away as a unit at maturity leaving a knob or protuberance on the rhachis. Inflorescence a dense globular to ovoid or cylindrical capitate spike. 6

2 Inflorescence apparently lateral. Bracts 2, the upper glume-like, the lower erect and appearing to be a continuation of the culm. Nut dorsally compressed, plano-convex, with the flat side against the rhachilla. Culms erect, 3-angular or terete, up to 40 cm high. Spikelets 1–16 or more, clustered, sessile, spreading, subterete, subacute. Perennial with short or long creeping rhizome. Coast. In margins of lagoons or other wet places. Not common in the area . . . . **C. laevigatus** L.
2 Inflorescence terminal umbellate or capitate. Nut laterally compressed with one edge against the rhachilla 3

3 Spikelets narrow-oblong, 2–3 mm wide, 5–20 mm long, divergent. Glumes dark brown, red or black with pale margins. Loosely distichous, 1–3 mm wide. Inflorescence a simple umbel of 3–5 short rays or reduced to a head. Nut obovoid to globular, less than half as long as the glume. Perennial or apparently annual. Culms tufted or solitary. Widespread. Grassland and waste places . . **C. sanguinolentus** Vahl
3 Spikelets narrow-linear, 1–2 mm wide. Glumes pale yellow to brown, green when immature, closely distichous. Tufted perennials or apparently annuals. Rhizome, when present, very short 4

4 Spikelets mostly erect to suberect, usually densely crowded, either in a compound head or in an umbel with short rays. Leaves flat, keeled, 1–3 mm wide. Nut oblong, c. half as long as the glume. Coast and adjacent plateaus. Damp places. . . . . **C. polystachyos** Rottb.

4 Spikelets radiating, not very crowded. Leaves 1–2 mm wide, sometimes almost filiform 5

5 Glumes yellowish to yellowish-brown. Spikelets c. 1.5 mm wide. Nut obovoid to globular, with minute wrinkles. Tufted herb up to 30 cm high. Somersby. Damp places. Uncommon. Introd. from Europe. . . . . ***C. flavescens** L.

5 Glumes reddish-brown. Spikelets 1.5–2 mm wide. Nut obovoid to globular, smooth, matt, less than half as long as the glume. Tufted herb up to 50 cm high. Cumberland Plain, banks of Nepean River. Damp places. Uncommon. . . . . **C. flavidus** Retz.

6 Tufted annual. Rhizome absent or very short. Spike whitish or very pale green, ovoid to oblong-cylindrical, 6–12 mm long, often with 2 smaller ones at its base. Nut finally black. Culms up to 30 cm high. Widespread. Grasslands. Introduced status uncertain. . . . ***C. sesquiflorus** (Torr.) Mattf. & Kük.

6 Perennials with creeping rhizomes. 7

7 Glumes purplish-brown. 2–4 flowers per spikelet. Stamens 3. Inflorescence obovate, with several densely clustered heads of spikes, purple to black in colour. Culms triangular, involucral bracts 3- 5 up to 24 cm long, often reflexed. Spikelets 3–5 mm long. Glumes with smooth or toothed midrib, sides with several nerves. Nut about half as long as glumes. Herb with long stolons. Possible serious environmental weed. Blue Mts Creeklines. Introd. from Africa. Fl. summer . . . . ***C. teneristolon** Mattf. & Kuek.

7 Glumes light to pale brown. 1 or 2 flowers per spikelet. Stamens 1, 2 or 3. 8

8 Spikelets numerous, often c. 100, greenish, in a ± ovoid head. Glumes usually with a denticulate keel. Stamens 1. Widespread. Wet places. Weed in lawns, gardens and grasslands. Introduced status uncertain. *Mullumbimby Couch* . . . . ***C. brevifolius** (Rottb.) Hassk.

8 Spikelets not usually more than 50, in a globular head, yellowish. Glumes only rarely with a denticulate keel. Stamens 2–3. Widespread. Wet places. Weed in lawns and gardens . . . . **C. sphaeroideus** L.A.S.Johnson & O.D.Evans

Group 2
Style 3-fid; spikelets arranged spicately

1 Flowers falling from spikelet leaving persistent rhachilla; glumes usually falling away individually at maturity 2

1 Spikelet usually falling away as unit at maturity leaving the two empty glumes and a knob or protuberance on the rhachis; glumes usually persistent 9

2 Spikelets very numerous, more than 20 per spike 3

2 Spikelets less than 20 per spike. 7

3 Glumes 1–1.5 mm long, mucronate. Nut less than half as long as the glume. Stout tufted perennial with a very short rhizome. Coast. Wet places. . . . . **C. exaltatus** Retz.

3 Glumes 2–4 mm long 4

4 Glumes red-brown with white margin on upper edges, 2.5–3 mm long, sides c. 3-nerved, obtuse in side view. Spikelets 2.5–3.5 mm wide, up to 36-flowered. Perennial herb with long rhizome. Swamps and seasonally wet areas. Uncommon . . . . **C. procerus** Rottb.

4 Glumes without white margins, acute to mucronate in side view 5

5 Leaves and bracts thin flat or with revolute margins, never internally septate. Umbel with a few rays or often with only a single compound cluster. Glumes c. 4 mm long, sides 3–4-nerved, acute. Keel green. Spikelets 1–2 mm wide, 8–24- flowered. Tufted perennial, often apparently annual. Coast and adjacent plateaus. Weed in waste places . . . . ***C. congestus** Vahl

5 Leaves and bracts septate or not, leathery, flat or folded, scabrous on midrib and margins. 6

6 Glumes 2.5–4 mm long acute. . . . . **C. ihotskyanus** (see below)

6 Glumes 1.5–2.5 mm long, obtuse but may have short mucro . . . . **C.gunnii** (see below)

7 Spikelets 1–1.5 mm wide. . . . . **C. subulatus** (see below)
7 Spikelets >1.5 mm wide 8

8 Glumes dark brown, 3–3.5 mm long. Spikelets 10–30 mm long. Rhizomes dark brown to purplish. Tubers purplish. Culms up to 40 cm high. Widespread. Weed of cultivation and waste places. Introd. from Africa. *Nut Grass.* . . . . ***C. rotundus** L.
8 Glumes yellowish-green to yellowish-brown, only rarely tinged with red. Spikelets up to 15 mm long. Rhizomes pale brown. Tubers tomentose when mature. Culms up to 40 cm high. Uncommon. Introd. from Africa. *Tigernut* or *Yellow Sedge* . . . . ***C. esculentus** L.

9 Spikelets less than 20 per spike, linear-terete or subulate. Glumes distant, not or scarcely keeled. Nut c. two-thirds as long as the glume, brown or black 10
9 Spikelets more than 20 per spike. Spikes ± dense. 11

10 Glumes striate all over, tawny on the sides, green on the back. Spikelets 1–3-flowered . . . .
. . . . **C. leiocaulon** (see below)
10 Glumes partly or wholly smooth and nerveless on the sides, reddish towards the margin, greenish and 3–7-nerved on the back. Leaves 1–2 mm wide. Uncommon in the area . . . .**C. subulatus** R.Br.

11 Leaves septate-nodulose. Umbels compound or decompound, with 5–13 unequal rigid rays. Glumes with 2–4 nerves per side, sometimes shortly mucronate, Robust perennials 1 m or more high. . . . 12
11 Leaves smooth or only obscurely septate-nodulose. 14

12 Glumes pale brown, 2.5–4 mm long. Spikes cylindrical up to 2 cm long. Tufted perennial up to 1.5 m high. Parks and reserves in Sydney. Introd. from Africa. . . . .***C. vorsteri** K.L.Wilson
12 Glumes dark to reddish-brown 13

13 Glumes 4 mm long, subacute. Terminal spike on the rays ovate-cylindrical, 2–3 cm long; laterals smaller, divergent or deflexed. Widespread . . . . **C. lucidus** R.Br.
13 Glumes 2–2.5 mm long, obtuse but sometimes shortly mucronate by extension of the keel. Spikelets in dense subglobular spikes 1–2 cm diam., usually with more than 8 flowers. Coast. Recorded a few times from waste land . . . . **C. gunnii** Hook.f. ssp. **gunnii**

14 Glumes closely appressed to rhachis and remote not or scarcely overlapping the next glume. 15
14 Glumes not as above 16

15 Inflorescence compound. Spikelet separating into individual units of nut and glume. Glumes 1–2.5 mm long, yellow to red-brown, 2–4 nerves per side. Spikelets 4–16-flowered. Coastal areas. Creek banks and swamps . . . . **C. odoratus** L.
15 Inflorescence simple. Spikelet falling as unit. Glumes 2–3 mm long tawny on the side, striate, with 2–5 nerves on either side of the green keel. Spikelets 1–3-flowered. Coast and adjacent plateaus. Waste places . . . . **C. leiocaulon** Benth.

16 Spikelets with 1–3 flowers 17
16 Spikelets with 7–20 flowers. 18

17 Spikes oblong-cylindrical to ovoid, 1–6, sessile or nearly so at the apex of the culm. Spikelets oblong-elliptic, with setaceous bracteoles, Glumes 3–4 mm long, greenish or brown, striate all over. Culms sharply 3-angular. Leaves flat, smooth. Bracts divergent or reflexed. Widespread. Weed in waste places. Introd. from N. America. . . . . ***C. aggregatus** (Willd.) Endl.
17 Spikes globular to ovoid on the ends of slender rays. Spikelets linear-terete, 5–8 mm long. Glumes narrow, closely appressed, striate. Nut brown, closely embraced by the wings of the rhachilla Glumes 3–3.5 mm long, green on the keel and yellowish-green on the sides, with 4–5 nerves on either side. Spikes c. 15–40 mm long. Herbs with short rhizomes. Woronora Plateau . . **C. cyperoides** (L.) Kuntze

18 Glumes 3–3.5 mm long, acute, greenish to pale brown; sides 2–3 nerved; margin hyaline. Umbel 4–7-radiate, simple or semicompound, rays up to 10 cm long. Spikes ovate, 1.5–2.5 cm long. Nut dark reddish-brown. Culms acutely 3-angular. Perennial with a short woody rhizome the apex of which is enclosed in sheaths which persist on the young culm up to 20 cm above the base. Blue Mts Roadside near Lawson. Introd. from S. America . . . . *C. rigens C.Presl

18 Glumes mucronate, red-brown . . . . C. ihotskyanus (see Group 3)

Group 3
Style 3-fid; spikelets ± digitate or fasciculate

1 Spikelet usually falling away as unit at maturity leaving the two empty glumes and a knob or protuberance on the rhachis; glumes usually persistent. Spikelets in globular to hemispherical clusters. 2

1 Flowers falling from spikelet leaving persistent rhachilla glumes falling away from the base of the spikelet at maturity. 6

2 Spikelets with 6 flowers or less 3

2 Spikelets with 6 flowers or more 4

3 Glumes acute. Spikelet 1–3 flowered. Bases of culm shortly swollen. . . . *C. aggregatus (see Group 2)

3 Glumes obtuse, 2.5–4 mm long, 7–9 nerves per side, light to straw-coloured. Spikelet 3–6 flowered. Bases of culms enlarged by leaf sheaths. Tufted perennial with a short rhizome. Disturbed areas. Introd. from Africa and Asia. Fl. spring–summer . . . . *C. dubius Rottb.

4 Glumes 3–3.5 mm long, acute in side view, with a distinct straight or reflexed mucro; sides reddish-brown, with 3–4 prominent nerves. Tufted perennial with a short woody sometimes creeping rhizome. Blue Mts River banks . . . . C. lhotskyanus Boeck.

4 Glumes 1–2.5 mm long, obtuse but sometimes shortly mucronate by extension of the keel 5

5 Glumes dark reddish-brown. Spikelets 8–14-flowered, in dense clusters 1–2 cm diam. Leaves strongly septate-nodulose. Culms up to 1 m high or more. Densely tufted perennial . . . C. gunnii (see Group 2)

5 Glumes pale, yellowish-brown, green when immature, loosely imbricate, at length spreading. Spikelets in dense clusters 6–10 mm diam. Leaves ± septate-nodulose. Culms up to 40 cm high. Tufted perennial. Coast and Cumberland Plain. Disturbed pastures . . . . C. fulvus R.Br.

6 Culms leafless at the base 7

6 Culms leafy at the base 10

7 Culms 3-angular for all or most of their length, smooth or scabrous 8

7 Culms terete for all or most of their length, firm to rigid, smooth 9

8 Involucral bracts 12–22. Plants 50–150 cm tall. Spikelets oblong to elliptic, 1.5–3 mm wide. Garden escape. Damp places. Introd. from Africa. Fl. summer . . . . *C. involucratus Poir.

8 Involucral bracts 1, 2 or 3. Plants weak, 20–40 cm high. Spikelets linear to lanceolate, 1–1.5 mm wide. . . . . C. haspan ssp. haspan (see below)

9 Bracts 2–4, mostly less than 5 cm long, rigid, pungent with incurved margins; shorter than or just exceeding inflorescence. Umbels head-like or with 2–3 branches Spikelets 2–3 mm wide. Banks of creeks and lakes. Inland species possibly introduced to area . . . . C. gymnocaulos Steud.

9 Bracts mostly 5–8, flat, 5–15 cm long, acute, firm, all exceeding the inflorescence. Umbel simple or compound, with 4–12 branches, sometimes reduced to a head. Spikelets oblong, 2–2.75 mm wide, crowded. Capertee Valley, Hunter River Valley. Creek banks. . . . . C. vaginatus R.Br.

10 Spikelets pale, yellowish, greenish, pallid or whitish when dry 11

10 Spikelets brown or reddish-brown or darker at maturity 22

11 Leaves and involucral bracts up to 15 mm wide, thin, flat, with 3 prominent faintly whitish principal nerves. Sheaths reddish. Umbel compound, with 8–24 filiform rays. Spikelets often solitary, sometimes 2–7. Rhachilla winged. Coast in south of area. Garden escape. Introd. from S. Africa. . . . . *C. albostriatus Schrad.

11 Leaves and bracts not as above 12

12 Glumes minutely cellulose-reticulate near the keel, yellowish-green, 1-nerved, with short acute spreading or recurved tips; keel acute. Spikelets ovate changing to oblong or oblong-linear, evenly and sharply serrate, up to 20-flowered. Low growing tufted perennial or apparently annual. Coast and Cumberland Plain. Damp pasture . . . . **C. flaccidus** R.Br.

12 Glumes not as in *C. flaccidus* 13

13 Leaves flat or with recurved margins. Glumes with 1 prominent median nerve on each side or nearly nerveless, especially when fresh. 14

13 Leaves filiform or nearly so, rarely more than 0.5 mm wide. 18

14 Spikelets very numerous in each compact subglobular cluster (100s per culm), oblong, flat, 2.5–3 mm wide. Glumes 2–2.5 mm long, acute, keeled, greenish yellow or pale brown. Leaves 3–4 mm wide. Tufted perennial 20–100 cm high. Widespread. Waste ground. Introd. from S. America. *Umbrella Sedge* . . . . ***C. eragrostis** Lam.

14 Spikelets less than 20 in each cluster. Inflorescence often reflexed. Leaves linear, up to 3 mm wide. Culms up to 50 cm long. Tufted perennials 15

15 Bracts of the inflorescence mostly 2. Nut obtuse but apiculate, trigonous, with concave faces, two-thirds as long as the glume. Spikelets in a sessile cluster or a few short rays may develop, sometimes with apparently pedicellate spikelets. Glumes with 1 nerve on either side of the keel 17

15 Bracts of the inflorescence mostly 3 or more. Glumes green to yellow, usually with 2–6 nerves on either side of the keel at least towards the base. Spikelets sessile in a cluster but usually with a few rays bearing a terminal cluster of 2–4(5) spikelets 16

16 Most clusters of spikelets with some pedicellate spikelets. Glumes shortly acuminate and recurved at the tip, with a prominent green keel, 1.6–1.7 mm long. Tufted herb with more slender habit than *C. laevis*. Coast and adjacent plateaus; Cumberland Plain. RF and sheltered situations **C. imbecillis** R.Br.

16 All spikelets in a cluster sessile or occasionally rays developed bearing a cluster of 2–3 terminal spikelets. Glumes mostly straight at the tip, 1.7–2 mm long. Tufted herb. Coast and adjacent plateaus. Sheltered places . . . . **C. laevis** R.Br.

17 Spikelets 1–1.5 mm wide. Glumes with 2–3 nerves on either side of the keel, 1.5–2 mm long. Nut pointed at both ends, trigonous with obtuse angles, convex on the face. Widespread. Sheltered situations. . . . . **C. trinervis** R.Br.

17 Spikelets 2 mm wide. Glumes with 4–6 nerves on either side of the keel, 2.5–4 mm long. Nut truncate but minutely apiculate and flat to slightly convex on the face. Pyrmont. Probably introd. from N. Australia . . . . ***C. compressus** L.

18 Annual with fine fibrous roots. Culms solitary or tufted, erect, not exceeding 100 cm long. Inflorescence apparently lateral . . . . **C. tenellus** (see below)

18 Perennials, densely tufted, with very short ± erect rhizomes. Culms erect or drooping, up to 40 cm long. Glumes, when dry, with 3–4 nerves on each side or 1–2 close to the keel 19

19 Glumes ≥2 mm long (excluding mucro) with 3–4 nerves on each side; keel green, curved from the base, slightly recurved near the apex and produced into a short point. Nut smooth to minutely rough, broad, obovate to obcordate, truncate, apiculate, c. three-quarters as long as the glume, dark, external cells minute. Coast and adjacent plateaus; Blue Mts Damp places, sheltered lawns and gardens, uncommon along creeks . . . . **C. gracilis** R.Br.

19 Glumes ≤2 mm long (excluding mucro) with 1–3 nerves on each side often close to the keel. 20

20 Nut longitudinally striate, tuberculate, external cells transversely elongated in 8–10 rows, finally dark, c. half as long as the glume. Glume 1–3 nerved on each side, slightly recurved near the apex and produced into a short point, white to pale brown. Coast. Banks of creeks and drains, sometimes in sheltered lawns . . . . **C. mirus** C.B.Clarke

20 Nut smooth, not much shorter than the glume, obovate, truncate, abruptly and shortly apiculate; sides concave. 21

21 Nut ellipsoid without a swollen stylar base, yellow-brown. Style divided to base. Glumes with 2–3 nerves on each side. Coast. Littoral RF. . . . . **C. eglobosus** K.L.Wilson

21 Nut trigonous with a swollen stylar base, red-brown. Coast. RF. WSF. . . . . **C. enervis** R.Br.

22 Inflorescence apparently lateral, with 1–2 setaceous bracts; the lower erect, appearing to continue the culm. Culms filiform, 1–8 cm high. Spikelets 1–4, sessile, oblong at maturity, 4–8 mm long, 2–2.5 mm wide, very flat. Glumes green changing to reddish-brown; keel green. Tufted annual with filiform leaves. Widespread. Wet places and damp pastures. Introd. from Africa. . . . . ***C. tenellus** L.f.

22 Inflorescence umbellate or capitate. 23

23 Spikelets mostly erect or suberect on the rhachis, 1–6 on each umbel ray. Glumes dark red or blackish, contrasting with the whitish nut which is about as long as the glume but spreading beyond the margins of the glume. Umbel rays up to 10, simple, unequal. Bracts 3–6, leafy. Culms up to 60 cm high. Widespread. RF . . . . **C. tetraphyllus** R.Br.

23 Spikelets radially divergent on the rhachis. 24

24 Spikelets obtuse, very numerous in dense globular clusters on each umbel ray, only slightly compressed, 4–8 mm long, 1 mm wide, 10–40-flowered. Glumes 0.8 mm long, very obtuse. Nut c. as long as the glume. Culms up to 50 cm high. Annual. Widespread. Weed in damp place throughout the warmer parts of the world . . . . **C. difformis** L.

24 Spikelets acute to subacute. Glumes 1–1.75 mm long. 25

25 Inflorescence compact, capitate (in NSW). Culms slender, erect, rigid. Leaves 1–2 mm wide. Bracts 3–4. Spikelets ovate to linear-oblong, in globular heads. Glumes reddish, 1-nerved on the sides; keel greenish or pallid, cellulose-reticulate; apex acute, excurrent. Nut less than half as long as the glume. Coast. Uncommon. Introd. from S. America . . . . ***C. reflexus** Vahl

25 Inflorescence loose 26

26 Spikelets c. 1 mm wide, linear to lanceolate, brown. Bracts 1–3. Nut trigonous, shortly apiculate, c. two-fifths as long as the glume. Perennial or annual. Coast. Uncommon . . . **C. haspan** L. ssp. **haspan**

26 Spikelets c. 2 mm wide, ovate to oblong, very dark reddish-brown. Bracts 2–4. Nut trigonous, c. half as long as the glume. Perennial up to 60 cm high. Capertee. Open forest. . . . . **C. concinnus** R.Br.

2 Eleocharis R.Br.

Spike Rushes

30 species in Aust. (10 endemic); all states and territories

Annual or perennial herbs, frequently producing slender stolons or sometimes with tubers or descending or ascending rhizomes. Leaves reduced to sheathing scales. Spikelets solitary, terminal, erect, without bracts, few–many-flowered. Glumes spirally imbricate, rarely ± distichous. Flowers bisexual. Perianth up to 10 bristles or absent. Stamens 1–3. Style 2–3-fid. Nut surmounted by the persistent dilated stylar base. (Fig. 46).

Key to the species

1 Culms longitudinally and transversely septate; 4–5 angled. Spikelets 2–3 mm wide. Glumes leathery, many-nerved, greenish to straw-coloured, keeled, margin hyline. Nut surface deeply pitted. Plants to 50 cm high. Swampy areas. Uncommon. Usually North coast. Fl. spring–summer. . . . . **E. philippinensis** Svenson

1 Culms transversely septate only or not septate. Glumes with a few nerves only or nerveless either side of midrib 2

2 Culms transversely septate 3

2 Culms not transversely septate. Spikelet c. 5 mm wide or less. Glumes membranous, with a distinct midvein, usually distinctly keeled. Spikelets much wider than the culms unless the latter is flat 4

3 Glumes without side veins. Spikelets cylindrical, up to c. 3 cm long, 3–4 mm wide. Wyong. Ponds and swamps . . . . **E. equisetina** C.Presl

3 Glumes hardened on the back, with a distinct side vein. Spikelets cylindrical, 3–5 cm long, acute. Nut lenticular. Plants up to 2 m high. Widespread. Ponds and swamps. *Tall Spike Rush*.   **E. sphacelata** R.Br.

4 Style 2-fid. Culms mostly 5–15 cm high, angular, 0.4–0.6 mm wide. Sheaths purplish at base. Spikelets ovoid-oblong or ovoid, dark coloured, 3–7 mm long, c. 2 mm wide. Perennial, sometimes with short stolons or rhizomes. Coast. Swamps, sometimes wholly submerged except the inflorescences. . . . .
. . . . **E. minuta** Boeck.
4 Style 3-fid 5

5 Nut obscurely 3-angular or terete, vertically ribbed 6
5 Nut 3-angular or lenticular, finely reticulate 8

6 Nut pyriform with c. 8 ribs on each face. Spikelets cylindrical, 5–18 mm long. Glumes to 3.5 mm long. Plant to 75 cm high. Wet areas on clay. Uncommon. Fl. spring–summer . . **E. macbarronii** K.L.Wilson
6 Nut obovoid, with 3–5 ribs on each face. 7

7 Plants neither proliferating through spikelets nor bearing tubers. Spikelet ovate to linear, 2–7 mm long, 1.5–1.7 mm wide. Glumes 1.7–2.5 mm long, 2–2.2 mm wide. Culms tufted, up to 25 cm (usually 2–15cm) high, rarely up to 0.5 mm wide. Rhizome slender. Widespread. Swampy ground. . . . .
. . . . **E. pusilla** R.Br.
7 Plants with stolons c. 0.7 mm wide bearing ovoid tubers and often proliferating through spikelets. Spikelet lanceolate to linear, 10–20 mm long, 2–3 mm wide, brown. Glumes 3 mm long or longer. Probably widespread. Swampy ground . . . . **E. atricha** R.Br.

8 Nut 3-angular; dorsal angle prominently ridged 9
8 Nut 3-angular or lenticular; dorsal angle, when present, not ridged. 10

9 Leaf sheath oblique at the orifice. Culms ± 4-angular, deeply striate, up to 20 cm high, less than 0.8 mm wide. Spikelet ovoid to oblong or lanceolate, 5–9 mm long, 2–4 mm wide. Rhizome creeping, 2–3 mm wide. Widespread, swampy ground . . . . **E. gracilis** R.Br.
9 Leaf sheath truncate, mucronate. Culms with 6–9 ribs, up to 30 cm long, 0.7–1 mm wide. Sheaths purplish, membranous. Spikelet ovoid or oblong, 6–9 mm long, 2–3 mm wide. Coast. Wet ground . . .
. . . . **E. dietrichiana** Boeck.

10 Leaf sheath oblique at the orifice, not or only minutely mucronate. 11
10 Leaf sheath truncate and mucronate at the orifice 12

11 Culms terete, to 5 cm high. Spikelets 2–4 mm long, 1–2 mm wide. Bristles 6, longer than nut. Matt forming herb. Uncommon in south. Saline flats. Introd. from America and Asia . . . .
. . . . ***E. parvula** (Roem. & Schult.) Link ex Bluff
11 Culms 4-angular, tufted, up to 20 cm high. Spikelet lanceolate ellipsoid or ovoid, 4–7 mm long, 2–3 mm wide, dark brown. Bristles 4, less than or only just equal to nut. Swamps. Introd. from S. America
. . . . ***E. pachycarpa** E.Desv.

12 Rhizome creeping Margins of the nut not swollen or ribbed 13
12 Rhizome very short. Margins of the nut swollen or ribbed. Culms densely tufted, 0.5–1.7 mm wide 14

13 Culms in distant tufts, terete, 1–3 mm wide. Spikelet dark brown or variegated, 13–30 mm long, 3–7 mm wide. Glumes tardily deciduous. Widespread. In or near water . . . . **E. acuta** R.Br.
13 Culms flattened, 2–4 mm wide. Spikelet straw-coloured to dark red-brown, 10–30 mm long, c. 2.5 mm wide. Mostly inland species. Uncommon. Moist areas. Fl. spring–summer . . . . . **E. plana** S.T.Blake

14 Glumes acute. Spikelet acute, pallid or brownish, 1–2 cm long, c. 2 mm wide. Anther appendage 0.1–1.5 mm long. Culms 0.5–1 mm wide, shining, up to 50 cm high. Cumberland Plain. In or near water . . .
. . . . **E. pallens** S.T.Blake
14 Glumes obtuse. Spikelet obtuse, pallid or pale brown rarely dark brown, linear-cylindrical, 2.5–3 cm long, 1–2 mm wide. Widespread. In or near water . . . . **E. cylindrostachys** Boeck.

3 **Fimbristylis** Vahl

85 species in Aust. (25 endemic); all states and territories except Tas.

Tufted herbs with narrow often filiform leaves. Spikelets with numerous usually bisexual flowers, solitary, irregularly umbellate or clustered, with unequal narrow involucral bracts some often exceeding the inflorescence. Glumes spirally imbricate; the lowest 1–4 empty. Stamens 1–3. Style 2–3-fid (Fig. 46).

Key to the species

1 Spikelet solitary (occasionally 2–3 together). Glumes brownish. Spikelet ovoid, not compressed, mostly 6–10 mm long. Leaves reduced to oblique sheaths. Culms up to 50 cm high. Cumberland Plain. Uncommon . . . .**F. nutans** (Retz.) Vahl
1 Spikelets in umbels or clusters 2

2 Nut distinctly transversely ribbed, yellowish. Spikelets c. 5 mm long. Leaves few; laminas up to 40 cm long, 1–2 mm wide. Perennial, usually more than 25 cm high. Coast. Wet places . **F. dichotoma** (L.) Vahl
2 Nut smooth or very finely striate, brown 3

3 Spikelets more than 6 mm long, c. 5 mm wide, brown. Stylar base without long spreading hairs. Stamens 3. Nut biconvex but much flattened. Leaves few; laminas up to 3 cm long, c. 1 mm wide. Perennial, usually more than 25 cm high. Coast. Saline or brackish swamps . . . . **F. ferruginea** (L.) Vahl
3 Spikelets less than 6 mm long, 1–2 mm wide, pale brown. Stylar base with long hairs spreading or reflexed over the ovary. Stamen 1. Nut biconvex. Leaves numerous; laminas up to 8 cm long, less than 1 mm wide. Annual, usually less than 25 cm high. Coast. Moist places . . . . **F. velata** R.Br.

4 **Abildgaardia** Vahl

7 species in Aust. (6 endemic); Qld, NSW, NT

One species in the area

Tufted herbs with filiform rigid leaves shorter than culms. Culms up to 35 cm high. Inflorescence usually a single, compressed, spikelet ovate or narrow-ovate, to 3 cm long, solitary or with 1 or 2 extra spikelets on branchlets. Glumes pale coloured (green or yellowish), leathery, distichous at least at the base but often becoming spiral towards the apex. Stamens 2 or 3. Style 3-fid, 2–5 mm long. Nut 2–3 mm long, tuberculate. Coast. Uncommon, often on headlands. . . . . **A. ovata** (Burm.f.) Kral

5 **Bulbostylis** Kunth

7 species in Aust. (3 endemic); all states and territories except Vic. and Tas.

Tufted annual herbs. Culms and leaves capillary, up to 0.5 mm wide. Leaf sheath usually bearded. Inflorescence a head or branched. Involucre bracts leaflike. Spikelets angular several to many-flowered. Rachilla winged. Glumes spirally imbricate, keeled side nerves absent. Flowers bisexual. Perianth absent. Stamens 1–3. Style 3-fid, with a dilated base which is persistent on the nut. Nut 3-angled.

Key to the species

1 Spikelets with several flowers, in a terminal head. Stamen usually 1. Small tufted herb up to 30 cm high. Agnes Banks. Sandy soils . . . .**B. barbata** (Rottb.) C.B.Clarke
1 Spikelets in branched inflorescence occasionally reduced to 1 spikelet. Stamens 2. Small tufted herb to 40 cm high. Uncommon. In north of region. Style 3-fid, with a dilated base which is persistent on the nut . . . . **B. densa** (Wall.) Hand.-Mazz.

6 Scirpus L.

2 species in Aust. (1 endemic, 1 native); Qld, NSW, Vic.

One species in the area

Tufted perennial up to 1 m high. Spikelets 4–7 mm long, greyish-brown, numerous along the rays in terminal clusters with several basal bracts, several-flowered. Perianth bristles 2–3 times as long as the pale nut. Stamens 3. Styles mostly 3-fid; base not dilated. Blue Mts Wet places . **S. polystachyus** F.Muell.

7 Isolepis R.Br.

28 species in Aust. (14 endemic); all states and territories

Rhizomatous, stoloniferous or tufted annuals or aquatic perennials. Culms mostly without nodes. Spikelets solitary or clustered, apparently lateral since a bract appears to continue the culm, several-flowered, sometimes proliferating (Fig. 46). Perianth bristles absent. Stamens 1–3. Style 2–3-fid; base not dilated. Nut variously trigonous, sometimes with swollen ridges, smooth reticulate or granular.

Key to the species

1 Style 2-fid 2
1 Style 3-fid 5

2 Stamens mostly 1 or 2, rarely 3. Spikelets usually with a continuing bract 3
2 Stamens mostly 3. Spikelets solitary without a continuing bract or only with short ones. Culms weak, floating or creeping 4

3 Spikelet solitary, 3–8 mm long. Nut greyish, c. 1.5 mm long. Lowest glume empty, only rarely produced into a linear lamina longer than the spikelet. Stamens 2, rarely 3. Usually aquatic, often with long narrow leaves rhizomes and culms. When on mud, the leaves are up to 1 mm wide and the culms more erect. Widespread. In water or on wet mud . . . . **I. fluitans** (L.) R.Br.
3 Spikelets usually more than 1 in each cluster; basal bract apparently continuing the culm, 2–3(5) mm long. Stamens 1 rarely 2. Nut yellowish or brownish, c. 1 mm long. Leaves linear, less than 1 mm wide; innermost leaf ensheathing each culm and usually reduced to a sheath alone or with a mucro. Erect herb with rhizome, often proliferating. Widespread. Wet places . . . . **I. inundata** R.Br.

4 Glumes reddish-purple at least on the margins. Spikelets 2–3 mm long, 1–1.5 mm wide. Anthers 1.5–3 mm long. Trailing aquatic plants with long culms and floating linear leaves. Widespread. Streams . . .
. . . . **I. producta** (C.B.Clarke) K.L.Wilson
4 Glumes greenish or brownish. Spikelets 5–8 mm long, 3–4 mm wide. Anthers 0.9–1.2 mm long. Usually in swamps or on mud but sometimes aquatic in streams. More robust than *I. producta*. Widespread. Wet places . . . . **I. crassiuscula** Hook.f.

5 Inflorescence proliferating 6
5 Inflorescence not proliferating 7

6 Inflorescence apparently terminal. Stamens usually 3. Spikelets often numerous, only very rarely reduced to 1. Nut acutely 3-angular, smooth or minutely granular. Leaves reduced to a sheath with a very short lamina. Plants up to 40 cm high, weak, rhizomatous. Widespread. Wet places. Introd. from S. Africa . . . . ***I. prolifera** (Rottb.) R.Br.
6 Inflorescence apparently lateral with basal bract apparently continuing the culm. Stamen 1, rarely 2 . .
. . . . **I. inundata** (see above)

7 Plant with a creeping rhizome. Spikelet solitary or rarely 2 together, 1–4 mm long, apparently lateral. Glumes keeled. Stamens usually 1, sometimes 2–3 in lower flowers. Blue Mts Wet places . . . .
. . . . **I. subtilissima** Boeck.
7 Plants tufted or with a short ascending rhizome 8

8 Stamens 1–2 9
8 Stamens 3 14

9 Glumes obtuse 10

9 Glumes acute, distinctly keeled straw-coloured and marked with reddish-brown or dark reddish-brown on either side of the keel. Invcolucre bract exceeding spikelets. 11

10 Nut 0.6–0.8 mm long, straw-coloured. Glumes with mucro, indistinctly nerved, white to straw-coloured. Involucre bract at least one third as long as culm. Stamen 1. Tufted plant to 15 cm high. Widespread. Damp places . . . . **I. gaudichaudiana** Kunth

10 Nut 0.8–1 mm long yellow-brown to grey-brown. Glume scarcely keeled but concave, green or mostly green or greenish yellow, occasionally some glumes on a plant reddish-brown in places on either side of the centre line but not regularly so. Involucre bract usually only shortly exceeding spikelets. Stamens 3 . . . .
I. cernua (see below)

11 Glume 1.3–1.8 mm long. Nut 0.5–0.8 mm diam. 12

11 Glumes less than 1.3 mm long. Nut c. 0.3 mm diam. 13

12 Glumes 3–4-nerved on either side of keel, straw-coloured and tinged with red-brown or uniformly red-brown. Leaves to 12 cm long. Stamens 2 or rarely 3. Nut ± equally trigonous, 0.6–0.8 wide. Plant with short rhizome to 30 cm high, occasionally proliferating. Damp places . . **I. habra** (Edgar) Soják

12 Glumes 1–3 nerved on either side of keel, straw-coloured, often marked with red-brown. Leaves to 2.5 cm long. Stamens 1. Nut unequally trigonous 0.5–0.6 wide. Tufted plant up to 10 cm high. Widespread. Damp places. Possibly introd. from S. Australia . . . . **I. stellata** (C.B.Clarke) K.L.Wilson

13 Glumes c. 1 mm long, nerveless or with one nerve close to midrib, margins hyaline. Nut obovoid, yellowish-brown to red-brown . . . . **I. australiensis** (see below)

13 Glumes 0.5–1 mm long, 3–4 nerved on either side of the green keel, margins hyaline to pale brown. Nut ellipsoid, dark grey to black. Spikelets 1–2(4) in each cluster, 2–3 mm long. Leaves filiform. Tufted plants up to 15 cm high. Coast and adjacent plateaus; Cumberland Plain. Damp places. Introd. from S. Africa. . . . . ***I. sepulcralis** Steud.

14 Nut globular, with 10–12 longitudinal striations, stipitate, whitish. Spikelets usually (1)2–3 per cluster. Culms 5–15 cm high. Leaves with short filiform laminas or reduced to a mucronate sheath. Widespread. Damp places . . . . **I. hookeriana** Boeck.

14 Nut variously triangular, reticulate or rugose 15

15 Glumes obtuse 16

15 Glumes ± acute, distinctly keeled, generally striate on either side of the keel. Nut acutely 3-angular 17

16 Glumes 1–2 mm long. Nut obscurely 3-angular or flattened on one side, pale yellow to grey-brown. Spikelet usually solitary. Leaves usually reduced to sheaths with short laminas. Involucre bract usually only shortly exceeding Spikelets. Plants to 20 cm high with shortly rhizomes. Widespread. Damp places. . . . . **I. cernua** (Vahl) Roem. & Schult.

16 Glumes 1–1.5 mm long Spikelets usually solitary. Nut 0.5–0.8 mm long; the angles well defined, swollen or ribbed, brown–black. Tufted plant up to 20 cm high. Widespread. Damp places . . . .
. . . . **.I. platycarpa** (S.T.Blake) Soják

17 Nut 0.4–0.6 mm long, up c. 0.3 mm wide. Glumes nerveless or with 1 on either side of the keel, very thin, membranous except for the keel. Spikelets 1–3. Leaf laminas to 3 cm long. Plants up to 15 cm high. Widespread. Damp places. . . . . **I. australiensis** (Maiden & Betche) K.L.Wilson

17 Nut more than 0.6 mm long 18

18 Nut 0.6–0.8 mm wide. Glumes dull, 3–4 nerved either side of midvein. Spikelets 2 or 3. Leaves to 12 cm long. Perennial plant with rhizome . . . . **I. habra** (see above)

18 Nut 0.5–0.7 mm wide. Glumes shiny, 4–6-nerved either side of midvein. Spikelets mostly 3–6, rarely 1. Leaves with very short to 4 cm long, filiform laminas or reduced to sheaths. Tufted annual to 8 cm high. Coast. Damp places. Possibly introd. from S. Africa. . . . . **I. marginata** (Thunb.) A.Deitr.

8 **Ficinia** Schrad.

1 species in Aust.; NSW, Vic., Tas., S.A., WA

One species in the area

Rhizomatous perennials with long, thick and woody rhizome. Leaves reduced to an obtuse sheath or with a minute mucro. Spikelets numerous in a compact apparently lateral globular cluster. Glumes concave, obtuse or mucronate, brown, closely imbricate. Hypogynous disc present at the base of the smooth plano-convex nut. Stamens 3. Tufted plants up to 70 cm high. Coast. Damp places, especially near the sea on dunes and cliffs . . . . **F. nodosa** (Rottb.) Goetgh., Muasya & D.A.Simpson

9 **Bolboschoenus** (Asch.) Palla

4 or 5 species in Aust., all states and territories

Rhizomatous perennials with a tuber at the base of the culms, often aquatics or in swamps. Culms leafy, with nodes. Leaves with well developed laminas. Spikelets in irregular umbels. Perianth present. Stamens 1–3. Style 2–3-fid. Nut trigonous.

Key to the species

1 Style 2-fid. Nut pale brown, often greyish below, minutely reticulate, obovoid, compressed and flat on the abaxial surface. Longest perianth bristles up to half as long as the nut. Up to 1.5 m high. Coast. Near permanent water . . . . **B. caldwellii** (V.J.Cook) Soják
1 Styles mostly 3-fid (some flowers in each spikelet usually with 2-fid styles but the majority of flowers in each spikelet have 3-fid styles) **2**

2 Leaves 5–8 mm wide. Nut glossy black; dorsal angle rounded or obtuse, ± lenticular. Perianth bristles c. half as long as the nut. Plant up to 2 m high. Coast. Lake margins, drainage ditches etc . . . .
. . . . **B. medianus** (V.J.Cook) Soják
2 Leaves 7–11 mm wide. Nut trigonous; dorsal angle acute. Perianth bristles c. as long as the nut. Plant up to 2 m high. Coast and Cumberland Plain. Near permanent water . . . . . . **B. fluviatilis** (Torr.) Soják

10 **Schoenoplectus** (Rchb.) Palla

11 species in Aust. (2 endemic); all states and territories

Rhizomatous perennials. Culms not nodose. Leaves basal, reduced to the sheaths. Spikelets in irregular umbels or apparently lateral clusters, with several flowers. Perianth present or absent. Stamens 1–3. Nut trigonous.

Key to the species

1 Inflorescence apparently lateral. Spikelets sessile. **2**
1 Inflorescence a terminal compound irregular umbel. Spikelets mostly pedicellate. Culms often more than 1 m high **5**

2 Styles mostly 3-fid. Culms usually more than 2 mm wide, sharply 3-angled **3**
2 Styles 2-fid. Culms up to 1 mm diam., terete or with 3 rounded angles **4**

3 Leaf blades absent, plants 50–100 cm high. Culms yellow-green. Perianth bristles as long as or longer than the nut. Spikelet 7–10 mm long. Glumes acute. Nut plano-convex to trigonous. Widespread. Damp places. Uncommon. . . . . **S. mucronatus** (L.) Palla ex Kern.
3 Leaf blades to 20 cm long, plants 20–60 cm high. Culms grey-green. Perianth bristles as long as or shorter than the nut. Spikelets 6–10 mm long. Glumes retuse. Nut plano-convex. Uncommon. Damp places. . . . . .S. pungens (Vahl) Palla

4 Perianth bristles absent or shorter than the nut. Culms terete up to 50 cm high. Spikelets 5–10 mm long, mostly 2–3 together in lateral clusters. Glumes reddish-brown towards the top. Nut plano-convex, rugose. Widespread. Damp places . . . . .S. erectus (Poir.) Palla ex J.Raynal

4 Perianth bristles at least as long as the nut, with reflexed hairs. Culms subterete, usually less than 30 cm high. Spikelets solitary, lateral, c. 4 mm long. Glumes pale brown. Nut plano-convex, smooth. Rarely collected. Nepean River and Manly Dam. Damp places. Probably introd. from Japan . . . .
. . . . ***S. lineolatus** (Franch. & Sav.) T.Koyama

5 Perianth bristles with reflexed hairs. Involucral bracts usually shorter than the inflorescence. Widespread. Damp places. . . . . **S. validus** (Vahl) A. & D.Löve

5 Perianth bristles often flattened or folded, plumose. Involucral bracts usually as long as the inflorescence or longer. Coast. Swampy places . . . . **S. subulatus** (Vahl) Lye

11 Lipocarpha R.Br.
2 species native Aust.; all states and territories except Tas.

One species in the area
Annual herbs with basal leaves and subterete culms up to 30 cm high. Spikelets 1–5 (usually 3), sessile, in compact terminal clusters, with numerous perfect flowers. Glumes spirally imbricate; lower 1–2 sterile. Perianth scales 2, narrow, hyaline, longer than the nut. Stamens 1–3. Style 2-fid. Nut compressed. Coast. Swampy places . . . . **L. microcephala** (R.Br.) Kunth

12 Lepironia Rich.
1 species native Aust.; Qld, NSW, NT

One species in the area
Rhizomatous perennials up to 1 m high. Leaves reduced to basal sheathing scales. Culms terete, with transverse septa. Spikelet solitary, ovoid or oblong-fusiform, 1–3 cm long, lateral, with numerous flowers. Glumes spirally imbricate, obtuse. Perianth scales numerous; 2 outer lateral ones with ciliate keels. Stamens 8. Style 2-fid. Nut compressed, smooth. Coastal swamps; lower Blue Mts, e.g. Glenbrook Lagoon. (Fig. 46) . . . . **L. articulata** (Retz.) Domin

13 Chorizandra R.Br.
5 species in Aust. (4 endemic): all states except NT

Rhizomatous perennial herbs with terete culms and a few basal leaves. Culms and leaves often transversely septate, up to 80 cm high. Spikelet solitary, apparently lateral, compact, globular to ovoid-globular, with numerous flowers, with one leafy involucral bract apparently continuing the culm. Glumes spirally imbricate; some lower ones empty. Perianth scales numerous; 2 outer lateral ones keeled; others in several rows. Stamens 6–12 rarely more. Style deeply 2- or 4-fid. Nut ovoid-globular, with c. 8 longitudinal ribs.

Key to the species

1 Inflorescence ovoid-globular, almost sheathed by the elongated base of the bract. Leaves up to 80 cm long, c. 2 mm diam.. Culms often shorter. Flowers not flattened. Glumes obtuse. Perianth scales all oblong-spathulate, ± concave. Coast. Swampy places . . . . **C. cymbaria** R.Br.

1 Inflorescence globular; base of bract scarcely enlarged and not sheathing the spikelet. Leaves up to 60 cm long, c. 2 mm diam. Flowers flat. Glumes acute-acuminate. 2 outer perianth segments keeled; inner ones spathulate, acuminate, flat. Coast. Swampy places . . . . **C. sphaerocephala** R.Br.

14 Rhynchospora Vahl
17 species in Aust. (5 endemic); Qld, NSW, Vic., NT, WA

One species in the area
Perennial herb. Leaves flat or concave, narrow, 3-angular. Culms slender, c. 30 cm high. Spikelets clustered in loose corymbs, with 1–2 perfect flowers and 1–3 male flowers above them. Glumes spirally imbricate; 3–4 lowest ones sterile. Perianth bristles 6–7, longer than the nut. Stamens 1–3. Style 2-fid, with a dilated persistent base. Nut globular, wrinkled. Coast and Cumberland Plain. Heath and DSF . . . .
. . . . **R. brownii** Roem. & Schult.

15 Cyathochaeta Nees
3 species endemic Aust.; NSW, Vic., WA

One species in the area

Rhizomatous perennial. Culms terete or angular, up to 50 cm high. Leaves rigid, channelled, mostly basal, often twisted, usually as long as the culm. Inflorescence paniculate. Spikelets brown, c. 15 mm long, several within each bract, distant, with 1 perfect flower and a male or sterile one below it. Glumes spirally imbricate; 2 lowest ones sterile. Perianth bristles usually 4, long, rigid. Stamens usually 2. Style 2-fid, with a dilated persistent base. Nut oblong. Widespread. Heath. Ss . . . . . .**C. diandra** (R.Br.) Nees

16 Schoenus L.
90 species in Aust. (85 endemic); all states and territories

Perennial herbs. Culms terete or obscurely 3-angular, seldom leafy. Leaves narrow, usually basal, sometimes reduced to sheathing scales. Spikelets with 2–6 flowers, in compact or spreading clusters or solitary and terminal, all perfect or uppermost sterile. Glumes distinctly distichous, 2–several lowest ones empty and often with another empty one above the fertile ones. Perianth bristles or scales up to 6 or absent. Stamens 1–6 usually 3. Style 3-fid, rarely ± thickened at the base. Nut oblong-globular, ± trigonous (Fig. 46).

Key to the species

1 Perianth bristles or scales absent 2
1 Perianth bristles or scales present 7

2 Leaves with long laminas. Inflorescence an erect interrupted panicle. Spikelets clustered, compact . . 3
2 Leaves with very small laminas or reduced to sheaths. 4

3 Sheaths membranous and not bearded at the orifice. Inflorescence 2–4 cm long. Glumes pale, membranous at the margin, not bearded; 1–2 lowest ones empty. Culms c. 15 cm high; leaf laminas as long, c. 1 mm wide. Widespread. Damp sandy soil. Uncommon. . . . .**S. moorei** Benth.
3 Sheaths bearded at the orifice with long hairs. Inflorescence usually 4–20 cm long but sometimes as short as 2 cm. Glumes lanceolate, bearded at least at the margin with long hairs; 3–4 lowest ones empty. Culms up to 60 cm high; leaf laminas c. 20 cm long, 1–2 mm wide. Widespread. Damp sandy soil. . . . .
. . . . **S. villosus** R.Br.

4 Inflorescence capitate; spikelets in a compact terminal cluster. Culms slender but rigid, c. 1 mm wide, c. 30 cm high 5
4 Inflorescence an interrupted panicle. Culms usually more than 1 mm wide 6

5 Leaf sheaths bearded at the orifice with long hairs. At least lowest glumes bearded at the margins, black to brown; 4–7 lowest ones empty. Spikelet with 2 perfect flowers. Widespread. Sandy soils and laterite. Heath and DSF . . . . **S. ericetorum** R.Br.
5 Leaf sheaths glabrous at the orifice. Glumes ciliate but scarcely bearded at the margins; 3–4 lowest ones empty. Spikelet with 2 perfect flowers. Widespread. Sandy soils and laterite. Heath and DSF . . . .
. . . . **S. imberbis** R.Br.

6 Spikelets brown, 9–12 mm long, numerous, on pedicels 1–2 mm long, 3–5-flowered. Nut smooth. Leaf sheaths bearded at the orifice. Culms up to 70 cm high, almost pungent at the tip. Widespread. Sandy soils. Damp places in heath and DSF . . . . **S. brevifolius** R.Br.
6 Spikelets black, 5–8 mm long, scarcely numerous, on filiform pedicels 3–5 mm long or longer. Nut tuberculate, 3-ribbed. Leaf sheaths ± bearded at the orifice. Culms up to 1 m high but usually less, not pungent at the tip. Widespread. Damp places, usually on creek banks . . . . . . **S. melanostachys** R.Br.

7 Leaves with very small laminas or reduced to sheaths alone 8
7 Leaves with distinct long laminas 9

8 Plants more than 30 cm high. Spikelets more than 1, black, in a loose panicle. Perianth present as bristles or absent . . . . **S. melanostachys** (see above)

8 Culms up to 30 cm high. Spikelets solitary, terminal. Perianth present as scales elongating in the fruiting stage. Nut smooth. Widespread. Damp places . . . .
. . . .**S. lepidosperma** (F.Muell.) K.L.Wilson ssp. **pachylepis** (S.T.Blake) K.L.Wilson

9 Spikelets in a compact terminal cluster which may appear lateral since a bract appears to continue the culm. Leaves channelled to appear terete, filiform, less than 1 mm wide 10

9 Spikelets not in terminal clusters. Leaves flat to concave 12

10 Involucral bracts more than 2. Spikelets in a terminal head, 1-flowered. 3–4 lowest glumes empty. Perianth bristles longer than the nut. Nut 3-ribbed, rugose, brown or grey. Culms up to 40 cm high. Widespread. Sandy soils. Heath and DSF . . . . **S. turbinatus** (R.Br.) Roem. & Schult.

10 Involucral bract 1, apparently continuing the culm 11

11 Perianth bristles plumose slightly longer than the nut. Inflorescence apparently a terminal cluster, 2–10 mm diam. Spikelets few, 2–3-flowered. Nut trigonous, smooth. Culms up to 30 cm high, 0.5–1 mm diam. Coast and adjacent plateaus. Damp places . . . . **S. nitens** (R.Br.) Roem. & Schult.

11 Perianth bristles not plumose, about as long as nut. Inflorescence 1–4 cm long. Spikelets 1–4 in 1 or 2 clusters. Nut trigonous, 3-ribbed and minutely reticulate. Culms up to 12 cm long, 0.2–0.4 mm diam. Blue Mts Ss. Damp, sheltered places. . . . . **S. evansianus** K.L.Wilson

12 Plant robust. Spikelets distant, in long loose panicles, pedicellate, with 1 perfect flower. Glumes acute; usually lowest 4 empty. Nut brown, with 3 ribs. Culms c. 30 cm high. Basal leaves up to 8 cm long, 1–2 mm wide. Widespread. Ss. Wet places in heath and DSF . . . . . **S. paludosus** (R.Br.) Roem. & Schult.

12 Plants small, delicate. Spikelets in numerous terminal and axillary clusters or solitary in the axils of leafy bracts. Nut white to grey. 13

13 Spikelets in loose terminal umbels and pedicellate axillary compound umbels. Spikelets 4–5 mm long, 2-flowered. 3–4 lowest glumes empty. Perianth bristles longer than the nut. Culms 10–25 cm high; leaves usually as long, c. 1 mm wide. Widespread. Damp places. *Fluke Bogrush* . . . .
. . . . **S. apogon** Roem. & Schult.

13 Spikelets 2–3 in the axils of leafy bracts, sessile or nearly so, 2–4 mm long, with 1 perfect flower, 2 or rarely 3 lowest glumes empty. Perianth bristles as long as the nut. Culms 10–25 cm long; leaves c. 1 mm wide or less. Widespread. Sandy soils. Damp places . . . .**S. maschalinus** Roem. & Schult.

17 Gymnoschoenus Nees
2 species endemic Aust.; NSW, Vic., Tas., S.A., WA

One species in the area

Perennial forming large dense tufts. Culms subterete or compressed, often 1 m or more high. Leaves linear, up to 50 cm long, 1–2 mm wide, rigid, flat or concave, with a dark brown bearded sheath. Spikelets numerous, with 1 perfect flower and a male one below it, clustered in compact solitary globular terminal spikes c. 15 mm diam., with 3 broad bracts at the base which are usually only half as long as the spike. Glumes 7–8, obscurely distichous, ovate to orbicular; the lowest few empty. Perianth bristles 3, stiff. Stamens 3. Style 3-fid with broad lobes. Nut obovoid, surmounted when young by the ± persistent stylar base. Widespread. Sandy soils. Swamps in heath. *Button Grass* . . . .**G. sphaerocephalus** (R.Br.) Hook.f.

18 Ptilothrix K.L.Wilson
1 species endemic Aust.; Qld, NSW

One species in the area

Erect, rigid, perennial herb. Culms compressed, up to 40 cm high. Leaf lamina concave or channelled, up to 30 cm long, often less than 1 mm wide. Spikelets numerous, 8–10 mm long or longer, in solitary compact terminal clusters. Involucral bracts with black to brown sheaths hyalo-membranous at the margins, the enlarged sheaths as long as the spikelets. Glumes obscurely distichous, lanceolate, with hyaline margins; 4–5 lowest ones empty. Perianth bristles 3, ciliate below. Stamens 3. Style 3-fid. Nut ovoid, obtusely trigonous. Widespread. Sandy soils. Heath and DSF . . . . . **P. deusta** (R.Br.) K.L.Wilson

19 Tricostularia Nees ex Lehm.
5 species in Aust. (3 endemic); all states and territories

One species in the area

Tufted perennial herb. Culms terete, striate, c. 30 cm high. Leaves basal, sometimes with very small laminas. Spikelets solitary or 2–3 in a terminal cluster, brown, each with an uppermost perfect flower and lower 1 or 2 male or sterile. Glumes spirally imbricate; 2–3 lowest ones empty and a small empty one above the flowers. Perianth bristles up to 6. Stamens 3. Style 3-fid. Nut ovoid. Widespread. Sandy soils. Swampy places in heath and DSF. . . . . **T. pauciflora** (F.Muell.) Benth.

20 Lepidosperma Labill.
55 species in Aust. (53 endemic); all states and territories except NT

Perennial herbs. Leaves basal, flat with an edge against the culm, or terete. Spikelets with 1 perfect flower and 1 or more male or abortive ones below it, scarcely flattened, usually sessile, in a terminal panicle or spike. Glumes often obscurely distichous, several lower ones empty. Perianth scales 6 or less. Stamens usually 3. Style 3-fid. Nut obtusely 3-angular.

Key to the species

1 Culms flattened, biconvex or ribbon-like 2
1 Culms terete or nearly so or with 4 angles 12

2 Culms more than 3 mm wide. 3
2 Culms 1–3 mm wide, up to 60 cm high. Leaves usually shorter than the culms (except *L. curtisiae*), 2–3 mm wide. 9

3 Culms hollow or with a very loose pith, distinctly convex on both sides, usually 1–1.5 m high. Leaves shorter than the culm, c. 1 cm wide. Panicle loose, 12–25 cm long; the secondary rhachis not usually obscured by the spikelets. Glumes lanceolate. Nut shining, brownish, surrounded by the lanceolate to narrow-deltoid perianth scales. Widespread. Swamps. . . . .**L. longitudinale** Labill.
3 Culms solid with a dense pith. 4

4 Culm margins viscid, scabrous, often ciliate, up to 60 cm high, 7–10 mm wide. Leaves as long as the culms, 3–6 mm wide. Panicle 7–15 cm long, oblong to ovate, erect. Rhizomes long and loosely rooted. Widespread. Sandy soils. . . . . **L. viscidum** R.Br.
4 Culms not viscid at the margins 5

5 Spikelets numerous and densely arranged in the inflorescence; secondary rhachis obscured 6
5 Spikelets loosely arranged in the inflorescence and/or inflorescences small with few spikelets; secondary rhachis visible between the spikelets. 8

6 Panicle 7–15 cm long, mostly 1–2 cm wide rarely to 3 cm narrow-oblong, erect, branches ± appressed to central rhachis. Culms convex on both sides, often more than 1 m high but sometimes less. Leaves often as long as the culms, c. 5 mm wide. Glumes lanceolate. Nut greyish. Outer perianth scales lanceolate, c. as long as the inner ones but both whorls often irregular in length. Widespread. Swamps. Usually on sandstone platforms . . . . **L. limicola** N.A.Wakef.
6 Panicle 2–7 cm diam., ± ovate, branches spreading. Culms flat on one side, to 60 cm high. 7

7 Culm margins scabrous, sharp. Leaves usually shorter than culms, 3–6 mm wide. Panicle 4–15 cm long. Glumes 6–8, red-brown to grey-brown, minutely pubescent. Nut shining, brown. Outer perianth scales narrow-oblong, shorter than the obovate-acuminate inner ones. Mostly coastal extending to Blue Mts Sandy soils. . . . .**L. concavum** R.Br.
7 Culm margins smooth, cutting, becoming irregular with age. Leaves shorter than culms, 2.5–4 mm wide. Panicle 5–12 cm long. Glumes 5 or 6, red-brown with blackish apex; glabrous or minutely

pubescent. Nut grey to red-brown, shining. Rhizomes short and firmly rooted. Mostly in Blue Mts extending to coast. Sandy soils . . . . **L. latens** K.L.Wilson

8 Panicles ovate to oblong, lax, drooping, very compound, 25–40 cm long. Spikelets 4–5 mm long. Nut shining greyish or reddish-brown. Perianth scales ± equal, lanceolate to narrow-oblong or narrow-elliptic. Culms 1–2 m high. Damp places. Coast and adjacent plateaus. Near RF . . . . **L. elatius** Labill.

8 Panicles linear, erect to spreading, not very compound, 6–20 cm long. Spikelets c. 5 mm long. Nut shining, greyish. Perianth scales lanceolate to ovate-acuminate, all ± the same length but the inner ones ± broader. Culms up to 100 cm high. Leaves often as long as the culms, 3–5 mm wide. Widespread. Sandy soils. . . . . **L. laterale** R.Br.

9 Leaves always longer than the culms. Panicle up to 5 cm long. Spikelets reddish-brown, 5–6 mm long. Nut ribbed. Blue Mts DSF and heath. Sandy soils . . . .**L. curtisiae** K.L.Wilson & D.I.Morris

9 Leaves shorter than the culms 10

10 Panicle ± linear. Spikelets usually several, clustered in a panicle. Nut brownish. Perianth scales oblong-elliptic, outer ones often narrower than the inner ones. Widespread. Sandy soils . **L. gunnii** Boeckeler

10 Panicle ± ovate. Spikelets few (usually 3–6), spreading, the upper one (when more than 3) twisted downwards below the lower ones. Nut greyish-brown. Perianth scales oblong-elliptic; outer ones often slightly narrower than the inner ones. Blue Mts Sandy soils . . . . **L. tortuosum** F.Muell.

12 Inflorescence spreading; the rhachis zig zag. 13

12 Inflorescences not spreading; the rhachis straight 14

13 Perianth scales 3, oblong-ovate, ± equal. Culms less than 1 m high, c. 1 mm diam. Mature plants with usually more than 20 culms. Spikelets up to 11 mm long, c. 1 mm wide, narrow-lanceolate. Glumes acute-acuminate. Nut greyish-brown. Panicle less than 5 cm long. Leaves much shorter than the culms, c. 1 mm wide; sheath reddish or pale yellow. Coast and adjacent plateaus. Sandy soils. Damp places . . . .**L. filiforme** Labill.

13 Perianth scales 5 or 6, oblong-ovate, ± equal. Culms often more than 1 m high, 1.5-2 mm wide. Mature tussocks usually with less than 20 culms. Spikelets 11–15 mm long, c. 2 mm wide, lanceolate. Glumes mucronate. Nut greyish-brown. Panicles 4–7 cm long. Leaves much shorter than the culms, often reduced to the sheaths alone; sheath dark brown. Wet places. Coast and Woronora Plateau . . . . **L. forsythii** A.A.Ham.

14 Spikelets in dense clusters appressed to rhachis. 15

14 Spikelets loosely arranged 17

15 Culms quadrangular, up to 80 cm high. Leaves 2–3 mm wide, often as long as the culms or longer. Spikelets 5–7 mm long. Inflorescence a spike-like panicle, up to 5 cm long. Coast and adjacent plateaus. Wet heath and swamp forest. . . . . **L. quadrangulatum** A.A.Ham.

15 Culms terete or oval in cross-section. 16

16 Stylar base on nut rounded. Culms terete, 1–2 mm wide, 1 or sometimes 2-grooved. Leaves c. 1 mm wide, usually shorter than the culms. Panicles 2–7 cm long. Spikelets few, sessile on an almost straight rhachis which is hidden by them. Perianth scales very small, broad-obovate, tending to be connate. Widespread. Damp, sandy soils. . . . . **L. neesii** Kunth

16 Stylar base on nut elongated, half as long to equaling nut. Culms 1–1.5 mm wide, oval in cross section with a groove on each surface. Leaves to 1.5 mm wide, slightly shorter than culms. Panicles 5–7 cm long. Spikelets ± numerous. Perianth scales c. three quarters the length of nut, including bristle at apex. Damp areas. Ss. cliffs. Blue Mts Vulnerable. . . . .**L. evansianum** K.L.Wilson

17 Leaf sheaths pale yellow to reddish . . . . **L. filiforme** (see above)

17 Leaf sheath black to dark red-brown. Inflorescence erect, narrow, 4–12 cm long; the rhachis straight. Spikelets 7–9 cm long. Nut greyish. Perianth scales ovate-elliptic, ± equal. Culms erect, straight, up to 1 m high. Widespread. Sandy soils . . . .**L. urophorum** N.A.Wakef.

21 Cladium P.Browne

1 species native Aust.; Qld, NSW, Vic., S.A., NT

One species in the area

Coarse robust stoloniferous or rhizomatous perennials. Culms ± terete, 1–2 m high. Leaves flat, 3-ranked, scabrous on the margin and keel. Spikelets in dense corymbose clusters, usually with 2 flowers but usually only the lowest perfect. Glumes spirally imbricate, obtuse; c. the lowest 4 empty but without empty ones above the perfect flowers. Perianth absent. Stamens 2. Style 2-fid. Nut ± drupe-like, with an inconspicuous corky base. Coast. Swamps and margins of lakes. . . . . **C. procerum** S.T.Blake

22 Baumea Gaudich.

5 species in Aust. (9 endemic); all states and territories

Rhizomatous perennials. Culms terete or compressed, pithy and/or septate. Leaves laterally compressed or terete, 2-ranked (distichous), entire. Spikelets arranged in ± erect panicles, 2- or sometimes 3-flowered rarely 1-flowered; the lowest or middle one perfect. Glumes distichous; lowest and uppermost ones empty. Perianth absent. Stamens usually 3. Style 3-fid. Nut surrounded at the base by a rudimentary disc, surmounted by a relatively inconspicuous stylar base.

Key to the species

1 Either leaves 4-angled or flat and 1–4 mm wide 2
1 Leaves terete or compressed and more than 2 mm wide or even reduced to the sheaths 6

2 Leaves with a raised midrib on either side and thus often 4-angular. Nut pitted or covered with small ridges, surmounted by the persistent stylar base. Inflorescence 40–100 cm long. Blue Mts; Woronora Plateau. Damp places . . . . **B. tetragona** (Labill.) S.T.Blake
2 Leaves flattened, striate but with raised midribs, up to 150 cm long 3

3 Leaves and culms smooth. Leaves with yellowish-brown sheaths 4
3 Leaves and culms scabrous. Leaf sheaths brown, red-brown or black 5

4 Lowest cluster of spikelets on each culm shorter than the next internode above on the culm. Inflorescence lax, usually 8–20 cm long. Nut pale red-brown, c. 3 mm long. Coast. Damp, sandy soils
. . . . **B. muelleri** (C.B.Clarke) S.T.Blake
4 Lowest cluster of spikelets on each culm equal to or shorter than the next internode above on the culm. Inflorescence compact, usually 2–4 cm long. Nut red-brown, c. 2 mm long. Coast. Damp, sandy soils. .
. . . . **B. acuta** (Labill.) Palla

5 Glumes <4 mm long. Leaves to 6 mm wide, with brown sheaths, midrib strongly raised. Inflorescence to 14 cm long. Nut pale to dark brown, 1.5–2 mm long. Blue Mts Swamps and streams. Sandy soil . . . . .
. . . . **B. planifolia** (Benth.) K.L.Wilson
5 Glume >6 mm long. Leaves 2–3 mm wide, with reddish sheaths, without a raised midrib. Inflorescence up to 50 cm long. Nut red-brown to black, 2.5–3.5 mm long, with a short stalk. Blue Mts Damp, sheltered situations . . . . **B. johnsonii** K.L.Wilson

6 Leaves all reduced to tightly sheathing scales on the culm below the inflorescence bracts. Culms slender, rigid, up to 70 cm high. Inflorescence 15–80 mm long. Spikelets few, sessile, on short branches. Glumes acute, with ciliate keels. Nut ovoid, obtuse. Widespread. Swamps and lake sides . .**B. juncea** (R.Br.) Palla
6 Either leaves with distinct long laminas or reduced to sheathing scales at the base of the culm 7

7 Nut whitish, acutely 3-angular. Panicle ± spreading, reddish-brown, c. 30 cm long. Leaves terete or slightly compressed, septate but air spaces between septa filled with looser tissue, usually as long as the culms. Glumes with or without cilia. Rhizomatous perennial up to 1 m high. Coast. Swamps . . . . . .
. . . . **B. arthrophylla** (Nees) Boeckeler
7 Nut brown, obtusely 3-angular 8

8 Leaves articulate with broad white septa separating large air spaces, often as long as the culms, 4–12 mm wide. Panicle spreading, lax, grey-brown, 20–50 cm long. Usually lower or middle flower perfect, sometimes upper 2 perfect. Glumes without cilia. Rhizomatous perennial 1–2 m high. Coast. Wet places. . . . .
B. articulata (R.Br.) S.T.Blake

8 Either leaves not septate or if septate then the spaces filled with looser tissue (subseptate) **9**

9 Spikelets at maturity scarcely clustered, arranged on a zigzag ultimate rhachis which is usually visible between them. **10**

9 Spikelets at maturity clustered and obscuring the ultimate rhachis **11**

10 Leaves with laminas c. 0.5 mm diam., usually exceeding the inflorescence. Nut c. 2 mm long. Slender rhizomatous perennial up to 1 m high. Coast. Damp places . . . . **B. nuda** (Steud.) S.T.Blake

10 Leaves usually reduced to sheaths alone; laminas if present c. 1.5 mm diam., not exceeding the inflorescence. Nut 2.5–3.5 mm long. Stout rhizomatous perennial up to 1 m high. Widespread. Swamps . . . . **B. gunnii** (Hook.f.) S.T.Blake

11 Spikelets in broad-ovoid to subglobular clusters on arcuate to ascending peduncles, greyish-brown to reddish-brown, usually with 3 flowers but sometimes reduced to 2. Nut smooth. Leaves sub-terete to compressed, sometimes sub-septate. Glumes usually ciliate. Rhizomatous perennial up to 1 m high. Widespread. Damp places . . . . **B. rubiginosa** (Spreng.) Boeck.

11 Spikelets in ovoid to narrow-ovoid clusters on erect peduncles, black to reddish-brown, often with 2 flowers but sometimes reduced to 1. Nut deeply wrinkled. Leaves usually terete but sometimes compressed, ± subseptate. Rhizomatous perennial up to 80 cm high. Widespread. Damp places . . . . **B. teretifolia** (R.Br.) Palla

23 Gahnia R.J.Forst. & G.Forst.

22 species in Aust. (20 endemic); all states and territories except NT

Tufted perennials up to 2 m high. Culms terete. Leaves long, linear, terete or inrolled, usually very scabrous. Inflorescence paniculate, each branch subtended by a ± leafy bract. Glumes dark brown to black, spirally imbricate; at least the lowest 4 empty. Perianth absent. Stamens 3–6, Style 3-fid or more. Nut ovoid-fusiform to globular, obtusely 3-angular to terete.

Key to the species

1 Spikelets with 1 or 2 flowers; an apparently terminal bisexual flower (very rarely with an imperfect one below it). Inflorescence erect or narrowly spreading. **2**

1 Spikelets with 3 flowers; an apparently terminal bisexual one with a fertile and a male flower below it . . **3**

2 Spikelets erect in dense terminal clusters. Inflorescence ± interrupted. Flowering glumes orbicular, very obtuse, c. 5 mm diam., brown. Nut smooth, brownish-red, with a distinct mucro, c. 5 mm long. Leaves scabrous on the back, often c. 80 cm long. Culms usually 40–50 cm high. Widespread. Damp places . . . . **G. aspera** (R.Br.) Spreng.

2 Spikelets in narrow panicles. Glume of the bisexual flower orbicular, obtuse, c. 3 mm diam. Nut black to very dark brown, 3–4 mm long, without a distinct mucro. Culm usually c. 1 m high. Leaves smooth on the back or almost so, as long as the culms. Widespread. Damp places.. . . . . **G. melanocarpa** R.Br.

3 Basal sterile glumes more than 10; the lowest ones less than half as long as the spikelet **4**

3 Sterile glumes fewer than 10, the lowest ones mostly more than half as long as the spikelet **6**

4 Nut 4–5 mm long, obtuse or truncate, usually bright reddish-brown. Glumes acute to acuminate, black or dark brown; upper ones obtuse. Spikelets 7–8 mm long. Culms up to 2 m high. Leaves usually as long as the culms, c. 1 cm wide. Coast and adjacent plateaus. Damp hillsides on Ss. . . . **G. erythrocarpa** R.Br.

4 Nut less than 4 mm long. **5**

5 Staminal filaments to 20 mm long. Nut reddish-brown, ovoid, obtuse, 2.5–3 mm long; when ripe it is *stuck* to the persistent filaments of the flowers and hangs out of the glumes. Culms up to 2 m high.

Leaves usually as long as the culms. Coast and adjacent plateaus. Creek banks and margins of swamps
. . . . **G. clarkei** Benl
5 Staminal filaments to 8 mm long. Nut brown, oblong, acute, 3–4.5 mm long; when ripe it hangs out
of the glumes on the *intertwining* filaments of the flower. Culms up to 2 m high or sometimes higher.
Leaves usually as long as the culms. Blue Mts Damp places . . . . **G. grandis** (Labill.) S.T.Blake

6 Bracts subtending the lower spikelets long-awned; lower bracts of the cluster 3–4 times as long as the
cluster. Clusters of spikelets erect, compact; spikelets almost sessile. Stamens 4. Nut oblong. Culms up
to 1 m high. Leaves usually as long as the culm, 1–2 mm wide, channelled. Blue Mts Damp places. . . .
. . . . **G. filifolia** (C.Presl) Kük ex Benl
6 Bracts subtending the spikelets short-awned or acute to acuminate; the lower ones less than half as long
as the cluster 7

7 Glumes brown or pale brown in fruiting stage. Spikelets almost sessile, 3–4 mm long, in ± erect
clusters. Nut brown. Leaves usually as long as the culm, 1–2 mm wide, channelled. Culms c. 80 cm
high. Widespread. Drier places. . . . . **G. microstachya** Benth.
7 Glumes very dark brown to black. Spikelets 5–8 mm long 8

8 Nut dark brown to black, ± trigonous, c. 2 mm long. Mature spikelets 1–2 mm wide. Leaves up to 8 mm
wide. Culms up to 10 mm wide towards the base. Spikelets erect, very shortly pedicellate. Coast and
adjacent plateaus. DSF in drier places . . . . **G. radula** (R.Br.) Benth.
8 Nut reddish-brown, c. 4 mm long. Mature spikelets 3–4 mm wide. Leaves 8–20 mm wide. 9

9 Culms hollow, with c. 6–11 nodes. Glumes smooth or obscurely scabrous; lower and upper ones usually
much shorter than the middle ones. Panicle very spreading; each partial panicle dividing 4–5 times.
Coarse tufted herbs up to 2 m high. Widespread in a variety of situations . . . . . **G. sieberiana** Kunth
9 Culms solid, with 2–4 nodes. Glumes scabrous; lower and upper ones not much shorter than the middle
ones. Panicle spreading to ascending; each partial panicle dividing up to 2 times. Coarse tufted herbs
up to 1.5 m high. Widespread in a variety of situations . . . . **G. subaequiglumis** S.T.Blake

24 Caustis R.Br.

7 species endemic Aust.; all states and territories except NT

Rhizomatous perennials. Culms terete, striate. Leaves reduced to sheathing scales on the culms and
at their bases. Spikelets with 1 bisexual or female flower, often with a male one below it. Glumes 3–6,
spirally imbricate; outer ones shorter and empty. Perianth absent. Stamens 3–6. Style slender, 3-fid, with a
thick hard persistent base. Nut obovoid or oblong (Fig. 46).

Key to the species

1 Ultimate branches straight, erect or pendulous. Culms often more than 1 m high. Spikelets in pairs,
12–16 mm long. Widespread. Sandy soils. Heath and DSF . . . . **C. pentandra** R.Br.
1 Ultimate branches flexuose, curved into spirals. Culms up to 60 cm high 2

2 Stamens 3. Spikelets functionally bisexual. Flower-bearing branches usually all flexuose. Widespread.
Ss. Heath and DSF . . . .**C. flexuosa** R.Br.
2 Stamens 5–6. Spikelets functionally unisexual with males and females on separate culms. Branches
bearing male spikelets considerably less flexuose than those bearing female spikelets 3

3 Glumes and rhachis glabrous at maturity. Spikelet c. 2 mm wide. Coast. Heath and DSF. Deep sands . .
. . . . **C. recurvata** Spreng. var. **recurvata**
3 Glumes and rhachis pubescent at maturity. Spikelet 3–4 mm wide. Coast south of Port Jackson. Heath
and DSF. Deep sands. . . . . **C. recurvata** var. **hirsuta** Kük

25 Tetraria P.Beauv.

4 species in Aust. (3 endemic); all states and territories except NT

One species in the area

Rhizomatous perennial herb. Culms 20–40 cm high, up to 1 mm wide, terete or ridged. Leaves with filiform laminas up to 5 mm long but frequently reduced to a sheath alone. Spikelets few, in a short panicle or raceme, pedicellate, narrow-lanceolate, c. 4 mm long, with 2 flowers or rarely 1-flowered; only the upper flower perfect. Glumes distichous. Perianth bristles absent or minute. Stylar base persistent, conical, as long as the body of the nut. Coast and adjacent plateaus. Sandy soils. . . . . .T. capillaries (F.Muell.) J.M.Black

26 Scleria P.J.Bergius

Razor Grass

23 species in Aust. (3 endemic); Qld, NSW, NT, WA

One species in the area

Perennial rhizomatous herb, glabrous. Culms 3-angular, up to 1 m high but usually much less. Leaves flat, linear, usually as long as the culms, scabrous. Spikelets unisexual, monoecious, in several clusters towards the end of the culm, c. 2 mm long; male spikelets with several flowers; stamens 3; female spikelets with a solitary flower; style 3-fid; perianth absent but possibly represented by the swollen disc at the base of the ovary. Nut whitish, tuberculate, with the base of the style persistent as a minute mucro. Razorback Range. Open forest . . . . .S. mackaviensis Boeck.

27 Uncinia Pers.

11 species in Aust. (6 endemic); NSW, Vic., Tas.

One species in the area

Densely tufted delicate perennial herb. Leaves filiform. Spikelets 1-flowered (see *Carex*), unisexual, arranged in a loose oblong spike 15–20 mm long. Female spikes enclosed in a flask-shaped 2-fid utricle with a hooked bristle exserted from the mouth of the utricle (Fig. 46). Male spikelets in upper part of the spike. Perianth absent. Stamens 3. Style 3-fid. Nut narrow-oblong, 3-angular, as long as the utricle. Blue Mts Damp shaded places. . . . . .U. tenella R.Br.

28 Carex L.

60 species in Aust. (20 endemic); all states and territories except NT

Perennial herbs. Leaves flat. Culms 3-angular or terete. Spikelets unisexual, often in separate spikes. Pseudo-glumes spirally imbricate. Perianth absent. Female flower enclosed in a flask-shaped utricle (see below) with a straight bristle (Fig.46). Stamens usually 3. Style 2- or 3-fid. Nut flattened or trigonous, enclosed in the persistent utricle.

The inflorescence of *Carex* and *Uncinia* is interpreted thus: The spikelet is reduced to a single flower in both male and female spikelets. Thus, the glumes of many authors, including Bentham, are really the bracts subtending the spikelets, called here 'pseudoglumes'; the true glumes are absent in the male, and modified to form the utricle in the female. The bristle, a reduced perianth segment according to Bentham, is the sterile upper part of the spikelet rhachilla. Thus, the inflorescence of these two genera is a collection of spikelets arranged in spikes which, in their turn, may be paniculate, spicate, umbellate, or solitary (Fig. 46).

Key to the species

1 Style 3-fid, nut trigonous 2

1 Style 2-fid, nut not trigonous 7

2 Spikes sessile or nearly so, erect to ascending in the fruiting stage. Utricle may be corky, beak 1 mm or less long 3

2 Spikes with long pedicel, usually pendulous in the fruiting stage. Utricle never corky but with a prominent beak 2 mm long or longer 5

3 Utricle corky, obscurely veined but not ridged, smooth, 2–3.5 mm diam.; the beak c. 1 mm long, notched. Female spikes ovoid, 1.5–4 cm long, greenish, Leaves 15–30 mm long, up to 6 mm wide. Perennial with a creeping rhizome. Coast near the sea. Sand dunes . . . . **C. pumila** Thunb.
3 Utricle not corky, with several minutely scabrous ridges, 1–1.5 mm diam., beak c. 0.5 mm long 4

4 Female spikes ovoid to oblong, 0.5–1.8 cm long, green. Utricle hairy. Leaves usually more than 20 cm long, up to 3 mm wide. Culms mostly up to 8 cm high. Rhizomatous perennial. Widespread. Usually in grassland . . . . **C. breviculmis** R.Br.
4 Female spikes oblong, (1)1.5–4 cm long, usually red-brownish. Utricle papillose. Leaves usually up to 30 cm long, c. 3 mm wide. Culms mostly more than 10 cm high. Tufted perennial up to 40 cm high. Coast in north of area. Damp places . . . .**C. maculata** Boott

5 Spikes solitary at nodes, 2–6 per culm, lanceolate-oblong, 3–6 cm long, 8–10 mm wide. Pseudoglumes aristate. Utricle 1–2 mm diam., prominently nerved with strong lateral nerves; beak patent when mature, 2–3 mm long, deeply notched. Leaves c. 30 cm long, up to 6 mm wide. Tufted perennial up to 1 m high. Widespread. Swamps and creek banks . . . . **C. fascicularis** Sol. ex Boott
5 Spikes clustered at nodes, 6–20 per culm 6

6 Female pseudoglumes yellow-brown to red-brown, 1.5–2 mm wide; acuminate, ascending. Female spike 3–5 mm diam at maturity. Utricle c. 1.5 mm diam., with prominent veins, 3-angular; beak ascending to erect, 2–2.5 mm long, very slightly notched to entire. Culms 40–80 cm high. Leaves 2–4 mm wide. Widespread. Swamps and creek banks . . . . **C. longebrachiata** Boeck.
6 Female pseudoglumes pale yellow-brown, 2.5–4 mm wide, aristate, ascending. Female spike 5–8 mm diam at maturity. Utricle c. 1.5 mm wide with prominent veins, 3-angular; beak ascending to erect, 1–1.5 mm long, apex notched. Culms 20–100 cm high. Leaves 3–6 mm wide. Blue Mts Swamps and creek banks . . . . **C. iynx** Nelmes

7 Lowermost spikes, or collection of spikes, on each culm, subtended by a leafy bract as long as or longer than the spike 8
7 Lowermost spikes, or collection of spikes, on each culm, subtended by a short glumaceous bract . . 15

8 Spikes ovoid, 3–10 mm long, yellowish-green or green, often all clustered together into a ± ovoid head or racemose 9
8 Spikes oblong, more than 10 mm long, brownish to blackish-green 12

9 Lowermost spike on each culm subtended by a bract more than twice as long as the spike. Utricle with a prominent beak, c. 1.5 mm diam., rugose, strongly, laterally ribbed. Male spikelets at the base of the individual spikes (occasionally 2 at the summit). Utricle with a very short entire beak. Widespread. Grassland and open forest . . . . **C. inversa** R.Br.
9 Lowermost spike on each culm subtended by a bract as long as or up to twice as long as the spike . . 10

10 Utricle >5 mm long . . . . **C. klaphakei** (see below)
10 Utricle <5 mm long 11

11 Female glumes 3–4 mm long, yellow-brown to red-brown. Utricle with a very short beak, nerveless not rugose or costulate . . . . **C. chlorantha** (see below)
11 Female glumes 1.5–2.5 mm long, red-brown, midrib green, margins hyline. Utricle beak prominent, c. 1 mm long, faintly nerved. Upper spikes with male and female spikelets, low spikes with all female spikelets. Uncommon at high altitudes. Alpine habitat . . . . **C. echinata** Murray

12 Lower spikes pedicellate, drooping to ascending 13
12 Lower spikes sessile to subsessile, erect to spreading 14

13 Utricle beak prominent, 1 mm long, 2-fid at the apex. Utricle brown to dark brown, minutely white-pubescent, c. 1.5 mm diam., prominently nerved. Spikes narrow-oblong, 1–4 cm long, c. 3 mm wide, usually clustered at nodes. Tufted perennial up to 80 cm high. Widespread. RF . . **C. brunnea** Thunb.

13 Utricle beak short, c. 0.3 mm long, truncate at the apex. Utricle c. 1.5 mm diam., minutely papillose, red-brown. Spikes 2–6 cm long, solitary at nodes. Loosely tufted herb to 90 cm high. Blue Mts Swamps. . . . . **C. lobolepis** F.Muell.

14 Female spike narrow-oblong, 5–14 cm long, up to 4 mm wide in the fruiting stage. Pseudoglumes dark brown to almost black. Utricles green to brown. Leaves up to 80 cm long and 1 cm wide. Rhizomatous perennial. Blue Mts Swamps and creek banks. **C. polyantha** F.Muell.

14 Female spikes oblong to elliptic, 1–4 cm long, 4–6 mm wide in fruiting stage. Pseudoglumes dark brown to almost black. Utricles green. Leaves up to 30 cm long, c. 5 mm wide. Rhizomatous perennial. Blue Mts Swamps and creek banks . . . . **C. gaudichaudiana** Kunth

15 Inflorescence with up to 12 spikes clustered into a head or very short raceme usually up to 3 cm long at the top of the culm. 16

15 Inflorescence of more than 12 spikes in a raceme or panicle usually more than 5 cm long at the top of the culm (rarely less in *C. tereticaulis*) 18

16 Female spikelets above male spikelets. Utricles winged on upper margins, 3.5–5 mm long, pale brown; beak 1.5 mm long. Leaves shorter than culms. Inflorescence 2–4 cm long. Female glumes yellow-brown, 3–4.5 mm long. Densely tufted herb. Tablelands. Damp places. Introd. from Europe . . . . ***C. leporina** L.

16 Male spikelets above female spikelets. Utricles without wings 17

17 Spike 1–3. Utricle >5 mm long. 1.7–2 mm diam., with 8–10 nerves. pseudoglumes mucronate, margins not or only slightly hyline. Inflorescence 0.8–1.5 cm high. Culms slender, 80–160 cm high. Rhizomatous perennial with leaves to 7 cm long, 1.5 mm wide. Blue Mts Swamps. Endangered. . . . . **C. klaphakei** K.L.Wilson

17 Spikes 4–12. Utricle <5 mm long, c. 1.5 mm diam., with a very short slightly notched beak. Pseudoglumes acute, margins hyline. Inflorescence 1–2.5 cm long. Culms 8–35 cm high. Rhizomatous perennial. Uncommon at higher altitudes (a few old records for coastal areas . . . **C. chlorantha** R.Br.

18 Utricle glabrous, margins not winged, distinctly 3-angular or plano-convex; beak c. 2 mm long, notched, smooth. Tufted herb with often septate-nodulose, sharply scabrous leaves which are shorter than culms. Widespread. RF and WSF. . . . . **C. declinata** Boott

18 Utricles hispid along slightly winged margins 19

19 Culm grey-green, terete or obtusely angular. Leaves reduced to basal sheathing scales or with laminas up to half as long as the culm. Glumes orange-brown with hyline margins. Utricle c. 1.5 mm diam., thickened at base, with a very short slightly notched beak. Tufted perennial. Wet places. Uncommon . . . . **C. tereticaulis** F.Muell.

19 Culms ± to acutely 3-angular, at least in upper half 20

20 Utricle dark yellow-brown when mature; base truncate, thickened; compressed, c. 1.5 mm diam; beak 1 mm long, slightly notched, scabro-pubescent. Inflorescence usually more than 12 cm high. Leaves shorter than to as long as culms with antrorse marginal prickles. Densely tufted herb forming large tussocks. Widespread. Damp places . . . . **C. appressa** R.Br.

20 Utricle black at maturity, base without basal thickening, ellipsoid c. 1.5 mm diam., beak 0.3 mm long, apex notched. Inflorescence to 12 cm high. Leaves shorter than to equaling culms with retrorse marginal prickles, (may be antrorse towards apex). Herb with slender tufts spread loosely along a rhizome. Blue Mts Grassy woodland and forests . . . . **C. incomitata** K.R.Thiele

194 ERIOCAULACEAE
13 gen., mainly trop.

1 Eriocaulon L.
20 species native Aust.; all states and territories except Tas.

One species in the area
Tufted monoecious herb. Leaves linear, radical, 2–8 cm long, with broad sheathing bases; lamina flat, tapering gradually to the apex; margins entire. Flowers very small, crowded into a solitary terminal hoary head 6–8 mm diam. on a peduncle 5–50 cm long which exceeds the leaves. Flowers unisexual, concealed within imbricate usually scarious bracts; the outer rows mostly female; the inner ones males. Perianth segments membranous, up to 6, in 2 whorls. Stamens 6. Ovary 2–3-locular, superior. Fruit a capsule. Coast. Wet ground in the open . . . . **E. scariosum** Sm.

195 FLAGELLARIACEAE
1 gen., trop. to subtrop.

1 Flagellaria L.
1 species native Aust.; Qld, NSW, NT, WA

One species in the area
Glabrous, climbing perennial up to 10 m high. Leaves narrow-lanceolate, 10–30 cm long, with a spirally coiled tip and a closed sheath around the stem. Flowers small, in a terminal panicle, regular, bisexual. Perianth segments 6, free, in 2 whorls. Stamens 6. Ovary superior, 3-locular; style 3-lobed. Fruit drupaceous, small, nearly globular. Coast. RF. . . . . **F. indica** L.

196 JUNCACEAE
Annual or perennial herbs with fibrous roots, often with rhizomes. Leaves mostly tufted at the base of the culms, flat or terete or reduced to sheaths at the base of the culms. Flowers regular, bisexual, arranged in loose compound cymes or compact heads or solitary in the leaf-axils. Perianth segments 6, free, arranged in two whorls, glumaceous subherbaceous or coriaceous, green or brown to black. Stamens 3–6, free. Ovary superior, 1–3-locular; style 3-fid; ovules inserted on axile parietal or basal placentas. Fruit a capsule. 8 gen., mostly temp.

Key to the genera

1 Leaf sheaths open; seeds numerous . . . . **1 JUNCUS**
1 Leaf sheaths closed (i.e., the margins fused); seeds 3 . . . . **2 LUZULA**

1 Juncus L.
Rushes
68 species in Aust. (31 endemic, 21 naturalized); all states and territories

Annual or perennial herbs. Leaves flat to terete, often crowded at the base of the culm or reduced to sheaths. Flowers arranged in loose compound cymes or heads or solitary in the leaf axils. Perianth segments green to black, subequal. Stamens 3–6; filaments filiform or dilated at the base; anthers linear-oblong, erect. Ovary 1–3-locular with numerous ovules inserted on parietal axile or basal placentas. Fruit a capsule. (Colour of culms described for fresh material. Check for other culm characters in the middle of the culm)

Key to the species

1 Leaves flat and channelled, or if terete then less than 1 mm wide and not septate . . . . **GROUP 1**
1 Leaves terete (sometimes compressed) or reduced to sheaths with the lamina as a small acute tip to the sheaths 2

2 Lamina well developed, terete or ± compressed . . . . .GROUP 2

2 Lamina reduced to an acute or mucronate tip on the sheaths . . . . .GROUP 3

Group 1
Leaves flat and channelled, or if terete then less than 1 mm wide

1 Each flower with a prophyll (bracteole) beneath it. Leaves with or without auricles. 2

1 Each flower without a prophyll beneath it. Leaves without auricles, sheaths tapering above 7

2 Leaf sheaths tapering above. Flowers solitary on an elongated rachis or cymose or clustered in the leaf axils. Perianth segments lanceolate, acute, 3–4 mm long, as long as the capsule. Leaves up to 10 cm long, but often less than 5 cm. Annual. Widespread. Damp places. *Toad Rush.* . . . . . . **J. bufonius** L.

2 Leaf sheaths auriculate or truncate. 3

3 Inflorescence divaricate; flowers clustered into groups of 3 or more. Perianth segments acutely acuminate, 4–5 mm long, as long as the capsule. Leaves usually channelled, usually c. 8 cm long, but up to 15 cm. Widespread. Grassland and open forest . . . . **J. homalocaulis** F.Muell. ex Benth.

3 Inflorescence erect or slightly spreading; flowers never clustered into distinct groups. Perianth segments 3–4 mm long 4

4 Several bracts at the base of the inflorescence longer than it. Perianth segments acute, as long as the capsule. Leaves generally flat or concave, c. 10 cm long. Tufted plants. Introd. from temp. N. America. Well established around Sydney in waste places . . . . ***J. tenuis** Willd.

4 Only one bract at the base of the inflorescence longer than it, or all shorter 5

5 Leaf laminas terete with a narrow channel or line on the posterior side, up to 0.5 mm wide. Capsule 2–3 mm long, only slightly longer than the perianth segments. Flowers few in a single head-like cluster on each scape. Tufted plants up to 20 cm high. Coast and Cumberland Plain. Grasslands and open ground. Introd. from S. America. . . . . ***J. capillaceus** Lam.

5 Leaf laminas semi-terete with a wide channel on the posterior side, up to 1 mm wide. Capsule 3–5 mm long, glossy. Flowers in ascending cymes 6

6 Capsule longer than the perianth segments. Tufted plant up to 40 cm high. Coast. Waste places. Introd. from temp. S. America. . . . . ***J. imbricatus** Laharpe

6 Capsule shorter than, or just as long as, the perianth segments. Tufted plant up to 40 cm high. Widespread. Waste places. Introd. from temp. S. America . . . . ***J. cognatus** Kunth

7 Stamens 3 8

7 Stamens 6 9

8 Clusters up to 10-flowered, generally solitary or rarely 2–3 on a scape. Perianth segments green, distinctly acuminate, longer than the capsule. Leaves c. 1 mm wide, 3–5 mm long. Plants up to 8 cm high. Annual. Widespread. Seasonally damp places. Introd. from Europe . . . . . . ***J. capitatus** Weigel

8 Clusters up to 8-flowered, arranged in a ± elongated or branched cyme. Perianth segments brown, acute or almost obtuse, as long as the capsule. Leaves more than 1 mm wide. Plants perennial over 10 cm high, rarely less. Widespread. Damp places. . . . . **J. planifolius** R.Br.

9 Perianth segments acuminate, 4–5 mm long, the outer ones longer than the inner. Flower clusters usually more than 8, in a terminal cyme. Stigmas well exserted from the flower. Leaves c. 8 cm long. Plants perennial. Introd. from S. Africa. Waste land near Sydney . . . .
. . . . ***J. capensis** var. **macranthus** Adamson

9 Perianth segments acute (inner ones almost obtuse), the outer equal or shorter than the inner. Flower clusters less than 8 in an irregular terminal cyme. Style short. Leaves c. 8 cm long. Plants perennial Coast, Botany Bay, Port Hacking. Near the sea. Uncommon . . . . **J. caespiticius** E.Mey.

Group 2
Leaves terete with well developed laminas

1 Leaves not jointed with internal septa, stiff and hard; plants robust; inflorescence apparently lateral. . 2

1 Leaves jointed with internal septa (incompletely so in *J. prismatocarpus*); inflorescence rather loose, apparently terminal, with distinct clusters of flowers or individual flowers. 3

2 Perianth segments oblong-elliptic, obtuse or emarginate, c. half as long as the capsule. Capsule c. 2.5 mm diam. Inflorescence up to 7 cm long. Leaves up to 50 cm long, very pungent. Coast and Cumberland Plain. Damp places. Introd. from Europe . . . .*J. acutus** L. ssp. **acutus**

2 Perianth segments lanceolate, acute or acuminate, as long as or slightly shorter than the capsule. Capsule c. 1 mm diam. Inflorescence more than 10 cm long. Leaves c. 10 cm long. Coast. Damp, usually saline, places near the sea . . . . **J. kraussii** ssp. **australiensis** (Buchenau) Snogerup

3 Leaves less than 1 mm wide. Flowers solitary or 2–4 per cluster. Perianth segments green. Plants up to 10 cm high, rarely more, tufted. 4

3 Leaves 1 mm wide or wider, usually more than 3 flowers per cluster. Tufted plants usually more than 10 cm high 5

4 Plants tufted and bulbous, sometimes stoloniferous. Leaves filiform. Up to 4 clusters of flowers in an irregular cyme. Perianth segments green or tinted with brown, 2–3.5 mm long. Upper Blue Mts Damp places. Introd. from Europe. . . . .*J. bulbosus** L.

4 Plants with underground rhizomes. Leaves linear-terete. Up to 4 clusters of flowers in an irregular cyme. Perianth segments green, 2–3 mm long. Upper Blue Mts Damp places . . **J. sandwithii** Lourteig

5 Cross septa incomplete, sometimes obscure. Leaves compressed. Perianth segments acute or acuminate, 3–4 mm long, shorter than the capsule. Stamens 3. Leaves up to 25 cm long. Widespread. Damp places . . . . **J. prismatocarpus** R.Br.

5 Cross septa complete. Leaves terete 6

6 Seeds with a white tail. Leaves 1–2 mm wide. Flowers more than 8 per cluster. Perianth segments straw-coloured. Stamens 3–6. Capsule as long as or slightly longer than the perianth. Erect tufted plants up to 40 cm high. Blue Mts Damp places. Introd. from N. America . . . . . .*J. canadensis** J.Gay ex Laharpe

6 Seeds without a tail. 7

7 Stamens 3. Leaves 2–3 mm wide, as long as or shorter than the culm. Flowers solitary or up to 4 per cluster, in an irregular cyme. Perianth segments green to straw-coloured, as long as the capsule. Coast e.g. Royal National Park. Uncommon. Damp places near streams. Introd. from N. America . . . .*J. acuminatus** Michx.

7 Stamens 6 8

8 Capsule dark purplish-brown. Perianth segments reddish-brown, 2–3.5 mm long, shorter than the capsule. Leaves 5–20 cm long. Widespread. Damp places. Introd. from Europe . . . . .*J. articulatus** L.

8 Capsule light brown. Perianth segments straw-coloured to pale brown. 9

9 Leaves mostly shorter than the scapes, 2–4 mm wide. Flowers mostly up to 4 per cluster 10

9 Leaves mostly as long as or longer than the scapes, 1–3 mm wide 11

10 Perianth segments 3–4 mm long, with a broad membranous margin, c. as long as the capsule. Coast and adjacent plateaus. Damp places. Introd. from S. America . . . .*J. microcephalus** Kunth

10 Perianth segments 2–2.5 mm long, with a narrow membranous margin, c. as long as the capsule. Wentworth Falls. Swamps. Introd. from Europe . . . . *J. acutiflorus** Ehrh. ex Hoffm.

11 Capsule shorter than the perianth segments or slightly longer, suddenly contracted towards the top. Widespread. Grasslands and open forests . . . . **J. holoschoenus** R.Br.

11 Capsule usually longer than the perianth segments, gradually tapering towards the top. Widespread. Grasslands and open forests. . . . .**J. fockei** Buchenau

Group 3
Leaves reduced to an acute or mucronate tip on the sheath

1 Pith of the culm continuous throughout 2
1 Pith interrupted by transverse spaces at least in part of the culm 10

2 Tepals, culms and inflorescence branches finely scabrous. Culms, soft, 0.7–2 mm diam. Inflorescence diffuse, 4–12 cm long, flowers solitary or in loose clusters. Slender rhizomatous herb to 65 cm high. Lithgow, Mt Blaxland. Woodland and grassland. Seasonally damp places. May be introduced from western regions . . . . **J. radula** Buchenau
2 Plants not scabrous 3

3 Leaf sheaths dark brown to reddish-brown 4
3 Leaf sheaths yellowish-green to yellowish-brown 7

4 Capsule longer than the perianth segments 5
4 Capsule shorter than the perianth segments 6

5 Stamens 6. Culms soft, 0.5–2.5 mm diam. Inflorescence very loose with the branches clearly visible between the flower clusters. Tufted rhizomatous herb with culms up to c. 1 m high. Uncommon in the area. Wet places, often in forests . . . . **J. pauciflorus** R. Br
5 Stamens mostly 3, sometimes up to 5. Culms hard, 2.0–3.0 mm diam. Inflorescence up to 10 cm long, with flowers clustered towards the ends of the branches. Culms with coarse longitudinal striations. Higher Blue Mts Wet places . . . . **J. phaeanthus** L.A.S.Johnson

6 Culms spongy, easily crushed. Inflorescence ± loose with the tips curving downwards, up to 5 cm long. Tufted rhizomatous herbs up to 1 m high. Widespread. Wet places. Introd. from Europe . ***J. effusus** L.
6 Culms hard, not easily crushed. Inflorescence very loose with branches curving downwards, up to 10 cm long. Rhizomatous herb up to 1 m high. North western Blue Mts Wet places on fertile soils . . . . .
. . . . **J. alexandri** ssp. **melanobasis** L.A.S.Johnson

7 Striations on culm coarse, c. 3 per mm across the culm. Inflorescence large, spreading, up to 15 cm long; flower clusters subglobular. Tufted rhizomatous herb up to 1.5 m high, 4–8 mm wide, with soft spongy culms. Widespread. Damp places . . . . **J. vaginatus** R.Br.
7 Striations on culm fine, c. 6 per mm across the culm. Inflorescence up to 10 cm long, ± ascending; flower clusters irregular and indefinite 8

8 Stamens 6. Perianth segments 2.5–3.5 mm long. Tufted rhizomatous herbs up to 2 m high, with hard pale green culms. Damp places near sea . . . . **J. pallidus** R.Br.
8 Stamens mostly 3 9

9 Inflorescence with ± erect branches; bracts at the base often more than 3 times as long as the inflorescence. Culms ± easily crushed. Widespread. Sandy soils . . . . **J. continuus** L.A.S.Johnson
9 Inflorescence with some branches curved downwards; bracts at the base usually less than 3 times as long as the inflorescence. culms hard, not easily crushed, pith continuous or occasionally slightly interrupted. Basalt and alluvial soils. Illawarra region . . . . **J. laeviusculus** ssp. **illawarrensis** L.A.S.Johnson

10 Culms blue-green 11
10 Culms bright green to yellowish-green 17

11 Spaces in pith very small (less than 0.5 mm long), occurring rather irregularly. Inflorescence usually a single globular head. Leaf sheaths dull, yellow to brown. culms up to 1.5 mm wide. Perianth segments equal to or longer than the capsule. Stamens 3. Tufted rhizomatous herb up to 60 cm high. Widespread. Grassy places . . . . **J. filicaulis** Buchenau
11 Spaces in pith mostly c. 1 mm long. Inflorescence not a head. Stamens 3–6 12

12 Stomata superficial. Leaf sheaths straw-coloured to pale brown 13

12 Stomata sunken into pits or depressions. Leaf sheaths chestnut to dark brown. Culms 2 mm wide, hard. Stamens 3 16

13 Capsule shorter than or equaling outer perianth segments 14
13 Capsule longer than outer perianth segments 15

14 Culms hard, 1–2 mm wide striations usually 15–30. Leaf sheaths straw-coloured to dark yellow-brown at base. Tufted rhizomatous herb up to 70 cm high. Widespread. Damp places. . . . .
. . . . **J. subsecundus** N.A.Wakef.
14 Culms soft, 0.9–2.0 mm wide striations on culm 25-60. Short, rhizomatous herb to 70 cm high. Leaf sheaths yellow to dark yellow-brown at base. Widespread, Temporarily wet situations. . . . .
. . . . **J. remotiflorus** L.A.S.Johnson

15 Striations on culm 20–35. Culms 1–2.5 mm wide, ± soft. Tufted rhizomatous herb up to 70 cm high. Leaf sheaths yellow brown to red brown at base. Widespread but uncommon. Damp places . . . .
. . . . **J. aridicola** L.A.S.Johnson
15 Striations on culm 50–60. Culms 2–3 mm wide, soft. Tufted rhizomatous herb up to 80 cm high. Leaf sheaths golden-brown to dark pink-brown at base Coast and adjacent plateaus. Uncommon. Damp places . . . . **J. polyanthemus** Buchenau

16 Flowers in clusters at tips of branchlets of the inflorescence. Capsule as long as the perianth segments. Tufted rhizomatous herb up to 1 m high. Blue Mts Damp places . . . . **J. australis** Hook.f.
16 Flowers arranged along the ultimate branches of the inflorescence. Capsule shorter than the perianth segments. Blue Mts Damp places . . . . **J. sarophorus** L.A.S.Johnson

17 Culms bright green, glossy especially near the base, 1–3 mm wide. Leaf sheaths dark brown inside and outside; the uppermost one usually loose. Inflorescence very variable. Stamens 3. Perianth segments as long as the capsule. Widespread. Damp places . . . . **J. gregiflorus** L.A.S.Johnson
17 Culms green to yellowish-green, dull. 18

18 Flowers clustered at the tips of the branchlets of the inflorescence 19
18 Flowers arranged along the ultimate branches of the inflorescence. 23

19 Culms grey-green to blue-green. 20
19 Culms yellow-green. 21

20 Culms hard, 1.4–2.3 mm diam. with 35–50 longitudinal striations. Leaf sheaths loose, dark yellow brown to red brown. Inflorescence 4–10 (15) cm long. Robust perennial up to 1 m high. Hunter Valley.
. . . . **J. subglaucus** L.A.S.Johnson
20 Culms soft, 1–2.5 mm diam. with 30–60 longitudinal striations. Leaf sheaths loose, yellow-brown to dark red-brown. Inflorescence 1–5 (10) cm long. Perennial, up to 1 m high. Introduced to Sydney region from western NSW. . . . . ***J. amabilis** Edgar

21 Culms soft, 4–10 mm diam. with 45–140 longitudinal striations. Leaf sheaths loose. Robust perennial, to 1.5 m high. Southern Blue Mts Uncommon. Wet places . . . . **J. procerus** E.Mey.
21 Culms hard, 3 mm or less diam. 22

22 Pith finely interrupted for entire length of culm with 20–52 longitudinal striations. Leaf sheaths loose, pale yellow to golden-brown. Inflorescence 6–15 cm long. Stamens 4–6, rarely 3. Robust perennial, to 60 cm high. Wollombi area . . . . **J. ochrocoleus** L.A.S.Johnson
22 Pith interrupted below middle of culm only with 35–80 longitudinal striations. Stamens 3. Leaf sheaths loose, dark red-brown to black . . . . **J. continuus** (see above)

23 Stomata on culms deeply sunken. Capsule usually shorter than or sometimes as long as the perianth segments. Culms 1–3 mm wide, yellowish-green with 38–90 longitudinal striations. Leaf sheaths golden-brown and glossy at the base. Tufted rhizomatous herb up to 60 cm high. Widespread but uncommon in the area . . . . **J. flavidus** L.A.S.Johnson
23 Stomata on culms superficial. Capsule longer than the (outer) perianth segments. Culms green . . 24

24 Leaf sheaths lax around culms. Perianth segments with narrow hyaline margins. Culms 2–5 mm diam., hard., green with 25–90 longitudinal striations. Leaf sheaths yellow–brown or red–brown. Capsule 1.5–2.5 mm long. Blue Mts. . . . . **J. laeviusculus** L.A.S.Johnson ssp. **laeviusculus**

24 Leaf sheaths tightly enclosing culms. Perianth segments with broad hyaline margins 25

25 Culms 1–2 mm wide, with less than 35 longitudinal striations, pith dense. Leaf sheaths reddish-brown and glossy near the base. Tufted rhizomatous herb up to 60 cm high. Widespread. Damp places . . . .
. . . . **J. usitatus** L.A.S.Johnson

25 Culms mostly more than 2 mm wide, with more than 35 longitudinal striations, pith very open. Leaf sheaths yellowish or pinkish-brown to black near the base. Lowest involucre bract longer than inflorescence. Rhizomatous herb up to 1.5 m high. Coast. Swamp forests and wet places in pastures . .
. . . . **J. mollis** L.A.S.Johnson

2 Luzula DC.

Wood Rushes

15 species in Aust. (12 endemic); all states and territories except NT

Perennial herbs up to 40 cm high. Leaves linear, flat (or channelled in species not found in the area), with at least a few white hairs. Flowers clustered in irregular loose compound cymes or in a single head. Stamens 6, 3 of them adnate to the inner perianth segments; filaments filiform. Ovary 1-locular with 3 or fewer basal ovules. Fruit a capsule.

Key to the species

1 Flowers in a single head or in 2–3 clusters on each scape; occasionally the clusters stalked 2

1 Flowers in more than 3 stalked heads or clusters which are arranged in irregular cymes on each scape. . 4

2 Plants with long rhizomes. Flowers in oblong heads or clusters. Leaves linear, up to 14 cm long and 5 mm wide. Capsule pale brown at the base, reddish-brown above. Upper Blue Mts Wet heath and swamps . . . . **L. modesta** Buchenau

2 Plants tufted or bulbous, without rhizomes. 3

3 Plants not bulbous. Flowers in obovate to globular heads or clusters. Leaves linear, up to 12 cm long and 4 mm wide. Capsule purplish-brown almost to the base. Southern Blue Mts Swamps . . **L. ovata** Edgar

3 Plants bulbous at the base. . . . . **L. densiflora** (see below)

4 Plants tufted, not bulbous at the base. Leaves linear, up to 18 cm long. Capsule usually creamish-brown. Perianth segments brown, paler at the margin. Widespread. Damp forests. **L. flaccida** (Buchenau) Edgar

4 Plants bulbous at the base. 5

5 Anthers 0.4–0.5 mm long. Bracts of the flower clusters quite acute. Capsule dark brown. Perianth segments brown and paler at the margin, occasionally cream. Widespread. Wet heath and swamps . . .
. . . . **L. densiflora** (H.Nordensk.) Edgar

5 Anthers 0.7–1.6 mm long. Bracts of the flower clusters with an obtuse tip. Capsule purplish-brown. Perianth segments dark brown. Widespread. Damp places . . . . **L. meridionalis** H.Nordensk.

197 POACEAE

Annual or perennial herbs, or rarely with woody stems, often with rhizomes or stolons, mostly tufted with basal leaves; sometimes creeping; rarely climbing, rarely amphibious or aquatic. Aerial stems (*culms*) usually hollow except at the nodes. Leaves alternate, solitary at the nodes, with the base modified to form an open sheath which encloses the stem; blade usually with parallel venation, rarely with a distinct midrib, linear or lanceolate, tapering towards the tip, flat or inrolled: *ligule* present at the junction of the sheath and blade and consisting either of a membranous flap which is sometimes cleft or ciliate, or of a few to several hairs or papillae or absent (Fig. 47 a–b). The basic unit of the inflorescence is the *spikelet* which is analogous to the flower of other inflorescences. Spikelets pedicellate or sessile, arranged in spikes racemes panicles or irregular clusters (Fig. 47 e–h) and with 2 (rarely 1) basal sterile *glumes* and 1–many florets arranged on a narrow axis or *rhachilla* which is sometimes prolonged at the tip of the spikelet.

Florets consisting of 2 bracts: the (fertile) *lemma* (outer and lower) and the *palea* which surround the flower (Fig. 48 a-c). Flowers usually bisexual (rarely unisexual when the plants are either monoecious or dioecious). Perianth represented by 1–3 (usually 2) *lodicules* or absent. Stamens usually 3, hypogynous; filaments long; anthers versatile. Ovary 1-locular, with 1 laterally inserted ovule; styles 2, with feathery stigmas (rarely the stigmas connate). Flowers sometimes abortive or male, when the subtending lemmas are referred to as *sterile lemmas* or *staminate* (or *male*) *lemmas* respectively. Glumes, lemmas, sterile lemmas and rarely paleas sometimes bear branched or unbranched *awns* attached terminally, dorsally, basally or in the base of a terminal notch (Fig. 48 d). At maturity the spikelets break up in one of two ways; either the rhachilla disarticulates above the glumes which remain at the tip of the pedicel after the florets have fallen (sometimes also disarticulating between the lemmas as well), or the disarticulation occurs below the glumes and the spikelets fall whole with the glumes attached (Fig. 48 c). Fruit a *caryopsis* often referred to as a *grain*) usually enclosed by the lemma, rarely free. Endosperm large. C. 668 gen., cosmop.

Unless specifically stated in the text:

(a) the lengths of the glumes and lemmas do not include the awns; awn lengths are stated separately;

(b) the lengths of the spikelets are measured from the base of the glumes to the tips of the awns;

(c) the lengths of the inflorescence are measured from the lowermost branch of the inflorescence (or the base of the lowermost spikelet for unbranched inflorescences) to the tip of the inflorescence.

FIGURE 47

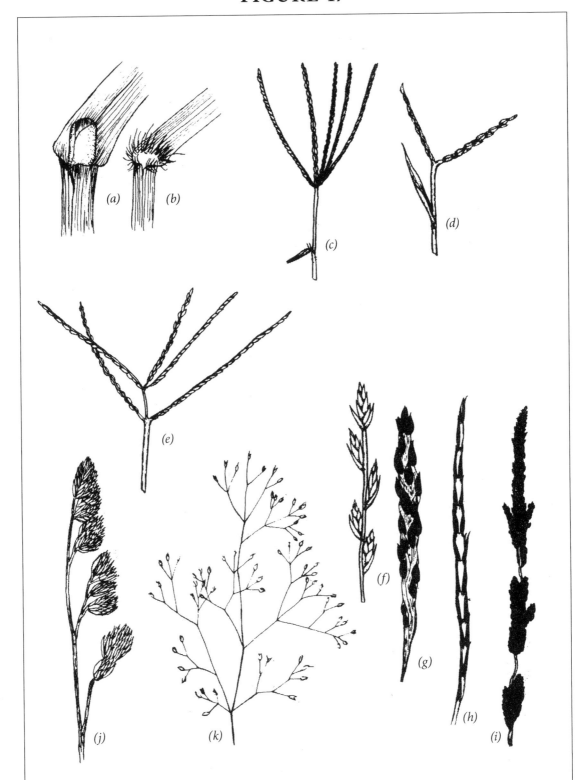

Fig. 47 Poaceae, ligules and inflorescences: (a) large membranous ligule; (b) ligule of hairs; (c) digitate panicle with 5 branches (*Cynodon dactylon*); (d) digitate panicle with 2 branches (*Paspalum distichum*); (e) subdigitate panicle (*Digitaria sanguinalis*); (f) spike, the spikelets with several florets (*Lolium temulentum*); (g) one-sided spike with flattened rhachis, the spikelets with 1 floret (*Stenotaphrum secundatum*); (h) spike with spikelets half embedded in the rhachis (*Hemarthria uncinata*); (i) spike-like, interrupted panicle (*Sporobolus elongatus*); (j) one-sided panicle (*Dactylis glomerata*); (k) open panicle (*Aira caryophyllea*). (c)–(k) ✕1.

FIGURE 48

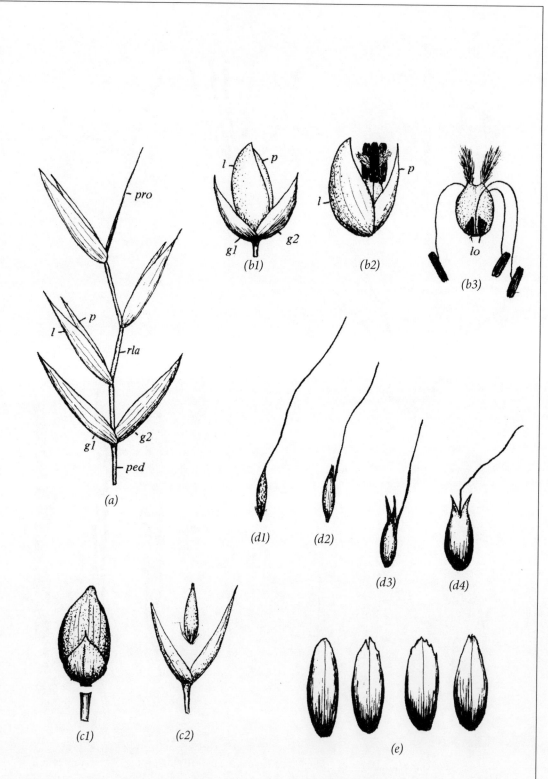

Fig. 48   Poaceae: (a) diagram of a spikelet with 3 florets, with the rhachilla elongated; (b1) diagram of a spikelet with 1 floret; (b2) floret taken from the same spikelet; (b3) flower of same; (c1) spikelet disarticulating below the glumes; (c2) spikelet disarticulating above the glumes; (d1) lemma with a terminal awn; (d2) lemma with a dorsal awn; (d3) lemma with a bifid apex and dorsal awn; (d4) lemma with a bifid apex and awn arising from the base of the notch; (e) lemmas with ragged apices (*Diplachne fusca*). Symbols: *ped* pedicel, *rla* rhachilla. *pro* prolongation of the rhachilla, *g1* lower glume, *g2* upper glume, *l* lemma, *p* palea, *lo* lodicule.

FIGURE 49

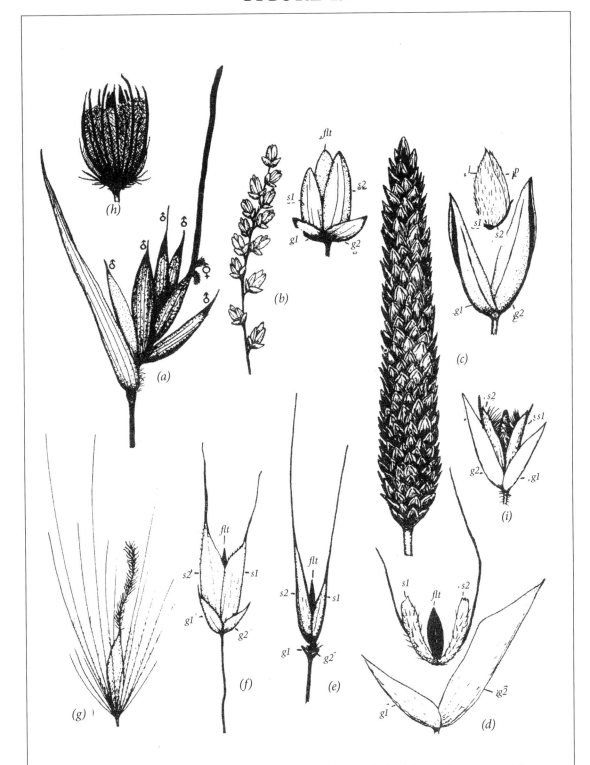

Fig. 49 Poaceae: (a) cluster of spikelets of *Themeda australis* (✕2) (part of awn removed);
(b) *Tetrarrhena juncea*, inflorescence (✕1) and a single spikelet; (c) *Phalaris aquatica*
inflorescence (✕1) and a single spikelet (floret and sterile lemmas separated from the glumes);
(d) spikelet (✕3) of *Anthoxanthum odoratum* (floret and sterile lemmas separated from the
glumes); (e) spikelet (✕3) of *Microlaena stipoides*; (f) spikelet (✕3) of *Ehrharta longiflora*;
(g) spikelet (✕2) of *Pennisetum villosum* with surrounding bristles (note connate styles); (h) burr
(✕2) of *Cenchrus caliculatus* (tips of the twc· spikelets visible); (i) *Hierochloe rariflora*, spikelet
(✕4). Symbols: *g1* lower glume, *g2* upper glume, *sl* lower sterile lemma, *s2* upper sterile lemma, *flt*
floret, *l* lemma, *p* palea.

FIGURE 50

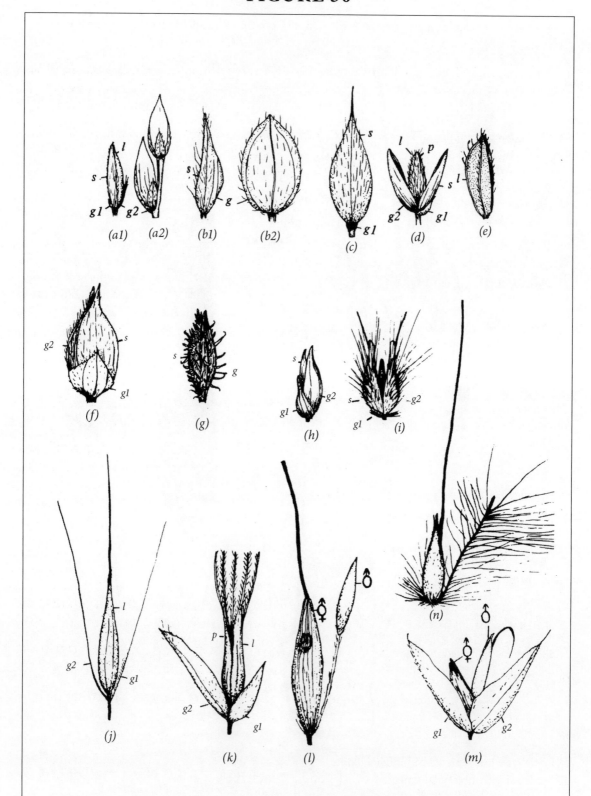

Fig. 50 Poaceae spikelets: (a1) *Digitaria sanguinalis*; (a2) paired spikelets of *D. sanguinalis*; (b1) *Paspalum dilatatum*, edge-on; (b2) the same face-on; (c) *Eriochloa pseudoacrotricha*; (d) *Entolasia marginata*, with upper glume and sterile lemma separated to show the lemma; (e) *Leersia hexandra*; (f) *Echinochloa crus-galli*; (g) *Tragus australianus*; (h) *Sacciolepis indica*; (i) *Melinis repens*; (j) *Hordeum leporinum*; (k) *Amphipogon strictus*; (l) *Bothriochloa macra*; (m) *Holcus lanatus*; (n) *Andropogon virginicus*. Symbols: *g* solitary glume; *g1* lower glume; *g2* upper glume; *s* sterile lemma; *l* lemma; *p* palea. (j) X2½, the others X5.

Key to groups and genera

Phyllostachys aurea and *P. nigra* (Bamboo), introduced from China, are naturalized in some areas where they have escaped from cultivation. They have been declared noxious weeds.

1 Spikelets unisexual, male and female markedly different and in different inflorescences; plants monoecious or dioecious **2**
1 Spikelets bisexual (rarely unisexual, and if so, the spikelets similar and in the same inflorescence). . . **4**

2 Female flowers arranged in a globular head 20–30 cm diam. Plants dioecious. Stoloniferous perennials on coastal sand dunes. . . . . **GROUP 12**
2 Female flowers not arranged in a globular head **3**

3 Female flowers surrounded by a bony utricle. Plants monoecious . . . . **111 COIX**
3 Female flowers not surrounded by a bony utricle. Plants dioecious . . . . **GROUP 5**

4 Spikelets in pairs (or rarely triplets), usually one sessile and awned, the other pedicellate and awnless (Fig. 50 l). The sessile spikelet can be sunken into the thick rhachis (Fig. 47 h) or; the two spikelets are similar, awnless and surrounded by silky hairs or; the pedicellate spikelet is reduced to the pedicel (Fig. 50 n). Spathe present just below the inflorescence in some genera.. . . . . **GROUP 13**
4 Spikelets not as in Group 13. . . . . **5**

5 Inflorescence a solitary spike or raceme. **6**
5 Inflorescence branched **12**

6 Spikelets sunken in the rhachis and scarcely protruding beyond the contour of the rhachis until they open **7**
6 Spikelets not sunken in the rhachis, although the bases alone may be slightly sunken **8**

7 Spikelets with a single floret. . . . . **GROUP 3**
7 Spikelets with two florets; upper one bisexual; lower one male or reduced to a lemma . . . . **GROUP 13**

8 Glume 1, except 2 in the terminal spikelet . . . . **GROUP 1**
8 Glumes 2 . . . . **9**

9 Spikelets sessile or if in groups of 3 then fertile central spikelet sessile and lateral male spikelets shortly pedicellate, Spikelets mostly c. 1 cm long or more . . . . **GROUP 2**
9 Spikelets pedicellate to sub sessile; spikelets usually less than 1 cm long **10**

10 Either; rhachis flattened, broad, corky; or spikelets surrounded by hairs and bristles . . . . **GROUP 12**
10 Neither the rhachis flattened broad and corky; nor spikelets surrounded by bristles. **11**

11 Spikelets with 2 or more florets, with or without an awn or mucro. . . . . **GROUP 6**
11 Spikelets with 1 floret, awnless . . . . **GROUP 11**

12 Silky hairs longer than the lemmas, attached to the pedicel or the rhachilla or the backs of the lemmas around the bisexual or female florets. **13**
12 Hairs around the florets much shorter than the lemmas or absent. **14**

13 Spikelets with more than 1 bisexual floret. . . . . **GROUP 5**
13 Spikelets with 1 bisexual floret . . . . **GROUP 13**

14 Inflorescence digitate (Fig. 47). **15**
14 Inflorescence not digitate **16**

15 Spikelets falling whole, disarticulating below the glumes . . . . **GROUP 12**
15 Spikelets disarticulating above the glumes, usually the glumes remaining on the pedicel . **GROUP 10**

16 Either; spikelets with 2 or more bisexual florets; or with 1 bisexual and 1 male or sterile lemma above **17**

16 Either; spikelets with 1 bisexual floret only; or with 1 bisexual and 1–2 male or sterile lemmas below (the spikelets may appear to have 1 floret from the outside; sometimes sterile spikelets are present as well as those containing a bisexual floret) 23

17 Lemma with 9 awns . . . .**74 ENNEAPOGON**
17 Lemma with up to 3 awns 18

18 Glumes almost as long as the florets or longer; awns sometimes protruding beyond the glumes . . 19
18 Glumes shorter than the lowest floret; the upper florets prominently exserted above the glumes. Lemmas not notched (except *Bromus*), usually awnless or with a terminal awn 21

19 Lemmas awnless . . . .**GROUP 12**
19 Lemmas awned from the back or notched at the tip 20

20 Lemmas awned from the back or awnless; the apex entire or notched. Ligule membranous   **GROUP 4**
20 Lemmas awned from the base of a notch in the bifid apex or ± equally 3-toothed at the apex. Ligule of hairs or papillae, or very short and ciliate (Fig. 47) . . . .**GROUP 6**

21 Lemmas 1–3-nerved; nerves prominent, sometimes hairy. . . . .**GROUP 9**
21 Lemma usually 5–many-nerved; nerves often obscure. 22

22 Florets 2; ligule of cilia . . . .**GROUP 6**
22 Florets more than 2 or if 2 the ligule membranous . . . .**GROUP 1**

23 Glumes absent, reduced to a hyaline rim. Stamens 6. Semi-aquatic. . . . . **51 LEERSIA**
23 Glumes 1–2. Stamens usually 3 24

24 Bisexual spikelets mixed with sterile spikelets 25
24 Bisexual spikelets not mixed with sterile spikelets 26

25 All spikelets drooping. Inflorescence one-sided. . . . .**GROUP 1**
25 Spikelets ± erect. Inflorescence not one-sided . . . .**GROUP 4**

26 Spikelets with 1 bisexual floret and 1–2 male or sterile lemmas below. 27
26 Spikelets with 1 bisexual floret only, usually disarticulating above the glumes which remain on the pedicel 31

27 1 male or sterile lemma present. 28
27 2 male or sterile lemmas present 30

28 Spikelet disarticulating below the glumes; falling whole. Lemmas awnless or with a straight awn . . .
. . . .**GROUP 12**
28 Spikelet usually disarticulating above the glumes. Lemmas with a twisted bent awn 29

29 Lemma awned from the tip . . . . **39 ARUNDINELLA**
29 Lemma awned from the back . . . .**GROUP 4**

30 Glumes as long as or longer than the floret . . . .**GROUP 4**
30 Glumes shorter than the floret . . . .**GROUP 7**

31 Spikelets 1–2 mm long, dull green-grey, crowded in narrow spike-like panicles, awnless . . . .
. . . .**70 SPOROBOLUS**
31 Either; spikelets more than 2 mm long; or with an awn or mucro; or in an open panicle 32

32 Lemma awnless or with a dorsal awn, hyaline or membranous at maturity. . . . .**GROUP 4**
32 Lemma with 1–3 terminal awns, hardened and rigid at maturity, tightly enveloping the fruit 33

33 Lemma with 3 awns or awn 3-branched . . . . **73 ARISTIDA**
33 Lemma with 1 awn 34

34 Glumes awnless. . . . .**GROUP 8**

34 Glumes awned. . . . . .GROUP 4

Group 1

1 Fertile spikelets with one floret; bisexual and sterile spikelets mixed, pendulous in a 1-sided inflorescence . . . . **13 LAMARCKIA**
1 Spikelets with 2 or more florets. 2

2 Fertile and sterile spikelets together in the same inflorescence; sterile spikelets with numerous rigidly awned lemmas . . . . **9 CYNOSURUS**
2 Spikelets all bisexual. 3

3 Glumes 1, except 2 on the terminal spikelet. Spikelets sessile, alternate on a flattened rhachis. . . . .
. . . . **14 LOLIUM**
3 Glumes 2 4

4 Panicle one-sided; the spikelets crowded in one-sided clusters at the ends of stiff branches. .**8 DACTYLIS**
4 Panicle not one-sided or if so then the spikelets not in clusters. 5

5 Lemmas cordate at the base. Glumes and lemmas inflated. Florets becoming progressively smaller upwards . . . . **10 BRIZA**
5 Lemmas and glumes otherwise. 6

6 Ovary with a terminal hairy appendage; the hairy stigmas attached beside the appendage. . . .**1 BROMUS**
6 Ovary without a hairy appendage in addition to the stigmas. 7

7 Lemmas laterally compressed and keeled 8
7 Lemmas rounded on the back, sometimes slightly compressed towards the top. 9

8 Leaves convolute. Plants found adjacent to the coast. . . . .**6 POA**
8 Leaves flat or folded. Plants found in various habitats. 9

9 Lemmas with nerves confluent towards the tip, apex very acute to shortly awned often finely 2-toothed.
. . . . **4 HOOKEROCHLOA**
9 Lemmas nerves not confluent towards the tip, apex obtuse to acute. . . . .**6 POA**

10 Lemmas awned 11
10 Lemmas awnless or with a short mucro 13

11 Glumes very unequal. . . . . **5 VULPIA**
11 Glumes subequal 12

12 Inflorescence a panicle . . . . **3 FESTUCA**
12 Inflorescence a raceme . . . . **2 BRACHYPODIUM**

13 Annual . . . . **7 DESMAZERIA**
13 Perennials 14

14 Lemmas with nerves (often obscure) converging towards the apex. Usually not in wet places . . . . . .
. . . . **3 FESTUCA**
14 Lemma usually with prominent nerves which are usually parallel and scarcely converging towards the apex. Usually in water or wet places 15

15 Spikelets 2–3.5 mm long . . . . **11 GLYCERIA**
15 Spikelets c. 9 mm long . . . . **12 DRYOPOA**

Group 2

1 Spikelets with 1 floret, in clusters of 3 at each node. Glumes linear, awn-like (Fig. 50) . **15 HORDEUM**

1 Spikelets with more than 1 floret 2

2 Spikelets with 2 florets; the rhachilla usually prolonged above them. Glumes very narrow, rigid, acuminate or subulate . . . .**16 SECALE**
2 Spikelets with more than 2 florets (if only 2, then the glumes broad) 3

3 Glumes bulging on the back, 4–6 mm wide. Lemmas abruptly pointed or awned. Spikelets with 2–5 florets. Annual . . . .**17 TRITICUM**
3 Glumes not bulging on the back, up to 3 mm wide. Lemmas tapering into an awn or awnless. Spikelets usually with 3 or more florets. Perennials . . . . **18 ELYMUS**

Group 3

1 Glume 1 (2 in the terminal spikelet) . . . . **19 HAINARDIA**
1 Glumes 2, standing side-by-side in the spikelet . . . .**20 PARAPHOLIS**

Group 4

1 Some florets male or sterile 2
1 All florets bisexual or only one bisexual floret present 6

2 Lowest floret bisexual 3
2 Lowest floret male 4

3 Spikelets pendulous; bisexual floret usually awned . . . . **23 AVENA**
3 Spikelets erect; bisexual floret awnless (Fig. 50m) . . . . **25 HOLCUS**

4 Male or sterile lemma awnless or with a terminal awn . . . .**22 PHALARIS**
4 Male or sterile lemma with a large dorsal awn 5

5 Spikelets with 2 florets; the lowest 1 male . . . . **26 ARRHENATHERUM**
5 Spikelets with 3 florets; the lowest 2 sterile (Fig. 49d) . . . . **21 ANTHOXANTHUM**

6 Spikelets with >1 floret 7
6 Spikelets with 1 floret 11

7 Inflorescence a spike-like panicle. Spikelets with 2–4 florets . . . . **28 ROSTRARIA**
7 Inflorescence an open panicle 8

8 Spikelets with 4–10 florets . . . .**27 AMPHIBROMUS**
8 Spikelets with 2–3 florets 9

9 Spikelets 2–3 mm long, with 2 florets. Pedicels and branches thread-like and delicate . . . . . **24 AIRA**
9 Spikelets more than 3 mm long 10

10 Leaves flat or folded . . . . **23 AVENA**
10 Leaves inrolled . . . . **31 DEYEUXIA**

11 Glumes covered with long hairs giving the spikelets and inflorescence (a dense ovoid spike) a woolly appearance . . . .**34 LAGURUS**
11 Glumes not covered with long woolly hairs 12

12 Both glumes awned 13
12 Glumes awnless 14

13 Glumes truncate, with a rigid awn 1–2 mm long . . . .**36 PHLEUM**
13 Glumes tapering into an awn or awn arising from a notch; awn 4–7 mm long . . . . **35 POLYPOGON**

14 Glumes swollen at base . . . . **37 GASTRIDIUM**

14 Glumes not swollen at base 15

15 Glumes rigidly ciliate on the keels (rough to the touch). Lemma 5–11-nerved. Panicle spike-like . . . .
. . . . **32 ECHINOPOGON**
15 Glumes not rigidly ciliate on the keels 16

16 Spikelets 9–16 mm long. Panicle spike-like. Coarse perennial on coastal dunes . . .**38 AMMOPHILA**
16 Spikelets less than 9 mm long 17

17 Margins of the 2 glumes united near the base; glumes with keels fringed with fine hairs (no longer
considered to occur in the Sydney region) . . . . **ALOPECURUS**
17 Margins of the glumes free at the base; keels not fringed with fine hairs 18

18 Lemma with an awn attached near the top at least twice as long as the lemma . . **33 DICHELACHNE**
18 Lemma awnless or with an awn less than twice as long as the lemma 19

19 Lemma papery or hardened, partially scabrous . . . . **31 DEYEUXIA**
19 Lemma membranous, smooth or softly hairy 20

20 Palea absent or up to half as long as the lemma. Lemma usually glabrous and awnless   **29 AGROSTIS**
20 Palea more than half the length of the lemma. Lemma usually hairy and awned . . . .
. . . .**30 LACHNAGROSTIS**

Group 5

1 Leaves crowded near the base of the plant, at least 1 m long. Inflorescence up to 1 m long, drooping. . .
. . . . **40 CORTADERIA**
1 Leaves distributed along the upright stems. Inflorescence pyramidal, erect 2

2 Lemmas glabrous. Rhachilla joints hairy . . . . **41 PHRAGMITES**
2 Lemmas hairy. Rhachilla joints glabrous . . . . **42 ARUNDO**

Group 6

1 Spikelets awnless . . . . **49 ERIACHNE**
1 Spikelets awned 2

2 Palea with 2 awns or deeply bifid 3
2 Palea entire or very shortly bifid 4

3 Central awn of the lemma much longer than the lateral ones . . . . **43 ANISOPOGON**
3 Central awn of the lemma ± the same length as the lateral ones . . . .**44 AMPHIPOGON**

4 Ligule a row of papillae . . . .**50 NOTOCHLOE**
4 Ligule a row of hairs, or short, membranous and ciliate. 5

5 Lemma with a row of long hairs immediately below the apical sinus (notch) or hairs tufted or lemma
almost glabrous 6
5 Lemma without a row of conspicuously longer hairs below the apical sinus; hairs on the rest of the
lemma scattered 7

6 Lemma with hairs even in length on back, hairs below sinus longer and in a row of tufts . . . .
. . . . **48 NOTODANTHONIA**
6 Hairs on lemma not arranged as above, varying in length and density, arranged in tufts or rows. . . . .
. . . . **47 AUSTRODANTHONIA**

7 Lemma with an awn up to 6 mm long. . . . . **46 PLINTHANTHESIS**
7 Lemma with an awn 7 mm long or longer. . . . . **45 JOYCEA**

Group 7

1 Inflorescence a simple spike (Fig. 49 b) . . . . **55 TETRARRHENA**
1 Inflorescence a spike-like or open panicle. 2

2 Glumes very small, remote from the sterile lemmas, with a tuft of hairs between them (Fig. 49 e) . . . .
. . . . **54 MICROLAENA**
2 Glumes overlapping the sterile or staminate lemmas 3

3 Two sterile lemmas present below the terminal bisexual floret, usually awned terminally (Fig. 49f) . . .
. . . . **53 EHRHARTA**
3 Two staminate lemmas present below the terminal bisexual floret (Fig. 49i). Awns absent . . . .
. . . . **52 HIEROCHLOE**

Group 8

1 Lemma elliptic lanceolate or ovate; awn delicate, straight, deciduous, c. as long or slightly longer than
the lemma, ... **58 PIPTATHERUM**
1 Lemma cylindrical or linear-oblong; awn stout, usually twisted, several times longer than the lemma . . 2

2 Awn on the lemma deciduous . . . .**57 NASSELLA**
2 Awn on the lemma persistent. . . . . **56 AUSTROSTIPA**

Group 9

1 Inflorescence digitate 2
1 Inflorescence not digitate 3

2 Axes of the spikes projecting beyond the uppermost spikelet as a ± sharp point. . . . .
. . . . **65 DACTYLOCTENIUM**
2 Axes of the spikelets terminated by a spikelet. . . . . **64 ELEUSINE**

3 Spikelets with 2–3 florets . . . . **63 LEPTOCHLOA**
3 Spikelets with more than 3 florets 4

4 Ligule membranous . . . . **61 DIPLACHNE**
4 Ligule a row of hairs or absent 5

5 Leaves pungent-pointed . . . .**60 TRIODIA**
5 Leaves not pungent-pointed. 6

6 Lemmas awnless . . . . **59 ERAGROSTIS**
6 Lemmas awned . . . . **62 TRIRAPHIS**

Group 10

1 Spikelets with 1 bisexual floret; the pedicel sometimes extended and bearing a vestigial lemma 2
1 Spikelets with 1 or more bisexual florets and with 1 or more reduced sterile florets above 3

2 Glumes shorter than the floret . . . . **68 CYNODON**
2 Glumes longer than the floret. . . . . **69 BRACHYACHNE**

3 Fertile floret compressed dorsally . . . . **66 CHLORIS**
3 Fertile floret compressed laterally . . . . **67 ENTEROPOGON**

Group 11

1 Spikelets sessile or shortly pedicellate in a terminal spike or spike-like raceme. Lower glume absent; upper glume smooth. Perennials with rhizomes or stolons . . . .**71 ZOYSIA**

1 Spikelets pedicellate, borne in clusters of 2 in a simple raceme. Lower glume absent; upper glume with hooked spines on the nerves (Fig. 50g). Annuals . . . . **72 TRAGUS**

Group 12

1 Spikelets unisexual, markedly different. Female heads 20–30 cm diam. Male spikelets in sessile clusters. Plants dioecious. Stoloniferous perennial on coastal dunes. . . . . **95 SPINIFEX**

1 Spikelets bisexual or, if unisexual, then male and female very similar and in the same inflorescence. . 2

2 Rhachis flattened, corky; the spikelets sunken in the rhachis. . . . . **76 STENOTAPHRUM**

2 Rhachis not flattened and corky; spikelets not sunken in the rhachis 3

3 Spikelets or groups of spikelets subtended or surrounded by 1–several bristles which are simple or fused at the base to form a burr 4

3 No long hairs and bristles surrounding the base of the spikelet 6

4 Burr present surrounding the spikelet (Fig. 50g) . . . . **93 CENCHRUS**

4 Bristles present at base of spikelet (Fig. 49) 5

5 Bristles falling with the spikelet . . . . **92 PENNISETUM**

5 Bristles left on the rhachis when spikelets fall. . . . .**91 SETARIA**

6 Inflorescence digitate or subdigitate (2 or more branches) 7

6 Inflorescence not digitate 9

7 Spikelets borne singly, sessile or nearly so. . . . . **78 AXONOPUS**

7 Spikelets paired or in 3s, one pedicellate, the others almost sessile 8

8 Glume 1; the 1 sterile lemma similar in size and shape to the glume (Fig. 50 b) . . . . . **80 PASPALUM**

8 Glumes 2 with 1 sterile lemma (Fig. 50 a) . . . . **75 DIGITARIA**

9 Spikelets with 2 florets of C. equal size, usually both bisexual . . . . **96 ISACHNE**

9 Spikelets with 1 bisexual floret 10

10 Upper glume and sterile lemma with very long silky hairs (Fig. 50i) . . . . **94 MELINIS**

10 Upper glume and sterile lemma glabrous or with short or hispid hairs 11

11 Lower glume rudimentary forming a swollen annular callus at the base of the upper glume . . . .
. . . . **77 ERIOCHLOA**

11 Callus at base of upper glume absent 12

12 Lower glume awned from the back of the entire apex or from the base of a small apical notch. . . . . .
. . . .**86 OPLISMENUS**

12 Lower glume otherwise 13

13 Glume and sterile lemma beset with spreading hooked or curved hairs with swollen bases. Rigid perennial, almost shrubby or climbing. . . . . **89 ANCISTRACHNE**

13 Glume and sterile lemma not beset with hooked or curved hairs. 14

14 Rhachis of the inflorescence branches prolonged into a bristle beyond the uppermost spikelet . . . 15

14 Rhachis of the inflorescence branches terminated by a spikelet. 16

15 Bristle terminating the inflorescence branches less than 8 mm long. Spikelets obtuse. . . . .
. . . . **81 PASPALIDIUM**

15 Bristle terminating the inflorescence branches 10 mm or more long. Spikelets pointed . . . .
. . . . **83 PSEUDORAPHIS**

16 Spikelets arranged in rows or clusters on one side of the rhachis 17
16 Spikelets not arranged on one side of the rhachis 20

17 Glumes stiff-hispid; spikelets ± loosely arranged on one side of the rhachis. . . . **90 ECHINOCHLOA**
17 Glumes not hispid; spikelets ± neatly arranged in 2 rows 18

18 Glumes 2; the lower one usually shorter and different from the sterile lemma . . . . .**79 UROCHLOA**
18 Glumes 1, equalling and similar to the sterile lemma (Fig. 50 b) 19

19 Spikelets of 2 kinds, some opening, others not, oblong-elliptic, much longer than broad . . . .
. . . . **88 CLEISTOCHLOA**
19 Spikelets all similar, all opening, ± as long as broad . . . . **80 PASPALUM**

20 Fertile lemma densely pubescent with fine white hairs (Fig. 50d) . . . . **82 ENTOLASIA**
20 Fertile lemma glabrous. 21

21 Inflorescence a spike-like panicle, less than 5 mm diam. . . . . **85 SACCIOLEPIS**
21 Inflorescence not a spike-like panicle 22

22 Glumes equal or sub-equal, shorter than the spikelet. Plants decumbent . . . . . . **87 OTTOCHLOA**
22 Glumes very unequal; the upper one equalling the sterile lemma; the lower one sometimes absent in
Cleistochloa 23

23 Spikelets all similar, globular or ellipsoid. Inflorescence usually broad and/or the spikelets distant.
Leaves usually not xeromorphic. . . . . **84 PANICUM**
23 Spikelets of 2 kinds, some opening, others not, oblong-elliptic, much longer than broad. Inflorescence
narrow. Leaves xeromorphic. . . . . **88 CLEISTOCHLOA**

Group 13

1 Inflorescence rhachis thickened. Spikelets closely appressed and sometimes embedded in the rhachis
(Fig. 47 h) 2
1 Inflorescence rhachis not markedly thickened, spikelets not embedded in the rhachis 4

2 Spikelet pedicel and rhachis internodes not fused . . . . **100 ISCHAEMUM**
2 Spikelet pedicel and rhachis internodes fused 3

3 Leaves glabrous. . . . . **101 HEMARTHRIA**
3 Leaves hairy. . . . . **102 ROTTBOELLIA**

4 Inflorescence spatheate 5
4 Inflorescence not spatheate 8

5 Pedicellate spikelet reduced to the pedicel alone (Fig. 50n) . . . .**107 ANDROPOGON**
5 Spikelets all developed. 6

6 Spikelets clustered, each cluster composed of 1 bisexual spikelet with an awn c. 5 cm long, and 4–6
awnless male or sterile spikelets (Fig. 49a) . . . . **110 THEMEDA**
6 Spikelets in pairs 7

7 Lower glume of sessile floret with 2 distinct keels. Awn inconspicuous or absent **109 CYMBOPOGON**
7 Lower glume of sessile floret rounded on back. Awns conspicuous, column hairy **108 HYPARRHENIA**

8 Paired spikelets similar in shape and sex 9
8 Paired spikelets dissimilar in shape and/or sex. 11

9 One spikelet of a pair sessile . . . . **99 MICROSTEGIUM**

9 Both spikelets pedicellate 10

10 Inflorescence a spike-like panicle. Spikelets awnless . . . . **97 IMPERATA**
10 Inflorescence an open panicle. Spikelets awned . . . . **98 MISCANTHUS**

11 Joints of the rhachis and pedicels with a translucent middle line between the thickened margins (Fig. 50 l) 12
11 Joints of the rhachis and pedicels without a translucent middle line 13

12 Inflorescence branched once or twice, usually digitate . . . . **105 BOTHRIOCHLOA**
12 Inflorescence branched many times . . . .**104 CAPILLIPEDIUM**

13 Inflorescence branches whorled, branched several times . . . . **103 SORGHUM**
13 Inflorescence digitate . . . . **106 DICHANTHIUM**

1 Bromus L.

23 species in Aust. (1 native, 22 naturalized); all states and territories

Annual or biennial, tufted herbs. Spikelets with several florets, large, pedicellate, disarticulating above the glumes. Glumes unequal or subequal, acute or shortly awned. Lemmas convex on the back or keeled, 2-toothed at the apex, awned terminally or from a little below the notch in the apex or awnless. Inflorescence an open or spike-like panicle.

Key to the species

1 Spikelets more than 4 cm long (including the awns) 2
1 Spikelets less than 3.5 cm long (including the awns if present) 4

2 Spikelets 6–10 cm long (including the long awns), with 6–10 florets, scabrous. Panicle up to 25 cm long. Annual up to 60 cm high. Widespread. Pastures; weed. Introd. from Europe and Asia. Fl. winter–spring. *Great Brome* . . . .***B. diandrus** Roth
2 Spikelets 4–5 cm long (including the awns if present), with 4–6 florets. Panicle 5–10 cm long. Erect annuals up to 40 cm high 3

3 Culms pubescent below the panicle. Lemmas scabrous to pubescent; awn ultimately bent upwards or almost spreading. Widespread. Pastures; weed. Introd. from the Mediterranean. Fl. winter–spring. *Red Brome* . . . .***B. rubens** L.
3 Culms glabrous below the panicle; lemmas scabrous or glabrous; awn remaining straight and erect. Widespread. Pastures; weed. Introd. from the Mediterranean. Fl. winter–spring. *Madrid Brome* . . . .***B. madritensis** L.

4 Spikelets mostly 1–2 cm long (including the awns if present), c. 5 mm wide, pale green, often ± whitish, mostly erect; the lowermost sometimes slightly drooping. Longest of the pedicels c. 2 cm long but most spikelets on pedicels c. 5 mm long or less. Awns, if present, up to 3 mm long. Panicle 6–20 cm long, usually compact, especially near the top. Annual or biennial, sometimes with rhizomes, up to 60 cm high. Widespread. Weed. Introd. from S. America. Fl. winter–spring. . . . .***B. brevis** Steud.
4 Spikelets mostly 2–3 cm long (including the awns), 5–10 mm wide. Awns, if present, up to 7 mm long 5

5 Awns on the lemmas 6–7 mm long. Panicle compact, 5–10 cm long. Spikelets contracted at the top, with 8–10 florets. Lemma 7–8 mm long. Annual up to 80 cm high. Widespread. Pastures; weed. Introd. from Europe. Fl. winter–spring. *Soft Brome* . . . .***B. molliformis** J.Lloyd
5 Awns on the lemmas less than 5 mm long. Panicle open, 10–30 cm long. Spikelets bright green, sometimes tinged with purple, often drooping. Longest of the pedicels 2–4 cm long; most spikelets on pedicels 5–10 mm long. Annual up to 1 m high, sometimes biennial. Widespread. Pastures. Introd. from America. Fl. winter–spring. *Prairie Grass* . . . . ***B. catharticus** Vahl

2 Brachypodium P.Beauv.
2 species naturalized Aust.; NSW, Vic., Tas., S.A., WA

One species in the area
Erect, tufted annual up to 50 cm high. Culms geniculate, hairy at the nodes. Leaf blades up to 3 mm wide. Spikelets 1–6, in a raceme up to 4 cm long, 20–40 mm long (including the awn); rhachilla disarticulating above the glumes. Glume unequal. Lemmas with a straight or curved terminal awn. Camden district. Waste places. Introd. from Europe. Fl. spring. *False Brome Grass* . . . . **B. distachyon** (L.) P.Beauv.

3 Festuca L.
10 species in Aust. (5 native, 5 naturalized); all states and territories except NT

Tufted perennials. Leaves flat or folded. Spikelets with several florets, pedicellate, disarticulating above the glumes. Glumes unequal, narrow. Lemmas rounded on the back, bifid or with a short terminal awn. Inflorescence a loose spreading panicle, ± one-sided.

Key to the species

1 Plants c. 1 m high. Leaf blades 3–10 mm wide. Spikelets 8–12 mm long. Lemmas 4–9 mm long, with or without awns 0.5–2 mm long. Panicles 15–30 cm long; the branches paired but unequal. Fodder grass in pastures. Introd. from Europe. Fl. spring. *Tall Fescue* . . . . **F. pratensis** Huds.
1 Leaf blades 1–2 mm wide 2

2 Leaf scabrous, 5–30 cm long. Leaf sheath always open to base. Spikelets 5–23 mm long. Lemma 5–8 mm long. Panicles 10–20 cm long. Tufted plant with rhizome. Higher altitudes. Fl. spring–summer . . . . .
. . . . **F. asperula** Vickery
2 Leaf smooth. Leaf sheath connate on young leaves 3

3 Plants rhizomatous, less than 50 cm high. Spikelets 7–9 mm long. Lemmas 4–5 mm long, with awns 2–3 mm long. Panicle compact. Lawn grass; weed; pastures. Introd. from Europe. Fl. summer. *Red Fescue* .
. . . . **F. rubra** L. ssp. **rubra**
3 Plants to 1 m tall, rhizome absent. Spikelets 7–12 mm long. Lemmas 4–5 mm long, with awns 1–2 mm long. Panicle open. Introd. from Europe. Fl. spring . . . . **F. nigrescens** Lam.

4 Hookerochloa E.B.Alexeev
2 species native Aust.; NSW, Vic., Tas., S.A., WA

Tufted perennials. Leaf blades flat or folded. Spikelets with several flowers, disarticulating above the glumes. Glumes unequal narrow, smooth or scabrous. Lemmas keeled, acute or shortly awned, sometimes bifid, slightly scabrous; nerves confluent towards the tip. Inflorescence a narrow or spreading panicle.

Key to the species

1 Leaves 2–4 mm wide, 5–40 cm long. Panicle c. 30 cm long, spreading. Spikelets 8–12 mm long. Plant to 1.2 m high. Wet places. Bundanoon area. Fl. spring . . **H. hookeriana** (F.Muell. ex Hook.f.) E.B.Alexeev
1 Leaves less than 1 mm wide, 20 cm long. Panicle spreading; the branches 3–6 cm long. Spikelets c. 15 mm long. Densely tufted plants up to 80 cm high. Higher Blue Mts Damp places. Fl. summer . . . . . .
. . . . **H. eriopoda** (Vickery) E.B.Alexeev

5 Vulpia C.C.Gmel.
5 species naturalized Aust.; all states and territories except NT

Erect annuals mostly less than 40 cm high. Ligule membranous, truncate. Spikelets with 3–8 flowers, disarticulating above the glumes, mostly c. 3 cm long; pedicels filiform. Glume acute, awnless. Lemmas subulate, rounded on the back, terminating in a straight subulate awn 20–25 mm long. Inflorescence a narrow one-sided panicle up to 20 cm long.

Key to the species

1 Lemma smooth or scaberulous, with a few to many silky hairs on the margins usually near the summit below the awn. Widespread. Weed. Introd. from America. Fl. spring–summer . . . .
. . . . ***V. myuros** f. **megalura** (Nutt.) Rydb.
1 Lemma smooth or scaberulous, but without silky hairs on upper margin 2

2 Upper glume twice as long as the lower glume (in the lateral spikelets of the branches only). Lower glume 3–6 mm long. Widespread. Weed. Introd. from Europe. *Squirrel's Tail Fescue.* . . . .
. . . . ***V. bromoides** (L.) A.Gray
2 Upper glume 3–4 times as long as the lower glume (in the lateral spikelets of the branches only). Lower glume up to 2 mm long 3

3 Panicle partly enclosed by leaf sheath. Upper glume 4–7 mm long; lower glume 1–2 mm long. Disturbed areas. Weed. Introd. from Europe. Fl. spring . . . . **V. muralis** (Kunth) Nees
3 Panicle well exserted from leaf sheath. Upper glume 3–5 mm long; lower glume 0.5–2 mm long. Widespread. Weed. Introd. from Europe. Fl. spring–summer. *Rat's Tail Fescue* . . . .
. . . . ***V. myuros** (L.) C.C.Gmel. f. **myuros**

6 Poa L.

41 species in Aust. (35 native, 6 naturalized); all states and territories

Tufted annuals or perennials. Ligule membranous or sometimes reduced to cilia. Leaf blade, flat, folded, involute or convolute. Spikelets with 2–10 florets, disarticulating above the glumes. Inflorescence a panicle, loose and spreading or rarely narrow and spike-like. Glumes keeled, awnless, unequal, usually acute. Lemmas keeled, obtuse or acute; nerves not confluent towards the tip. Callus often with loose woolly hairs.

Key to the species

1 Annual up to 30 cm high. Panicle ovate to triangular in outline, spreading. Leaves bright green; blades up to 14 cm long, c. 5 mm wide. Spikelets 3–10 mm long with 3–10 florets. Widespread. Weed in lawns and pastures. Introd. from Europe. Fl. winter. *Winter Grass* . . . . ***P. annua** L.
1 Perennials 2

2 Basal internodes bulbous or thickened; the thickening composed of the enlarged and fleshy inner basal leaf sheaths. Culms usually erect, up to 40 cm high, 2–4-noded. Leaves glabrous; blades up to 10 cm long, 1–2.5 mm wide; ligule membranous, 2–8 mm long. Panicle up to 6 cm long, dense, ovate or oblong in outline. Spikelets ovate to oblong, with 3–6 florets, compressed, 3–5 mm long. Blue Mts Pastures; weed. Introd. from Europe. Fl. spring. *Bulbous Poa* . . . . ***P. bulbosa** L.
2 Basal internodes not bulbous or thickened 3

3 Internodes of the culms strongly compressed, 2-sided. Stiff perennial up to 60 cm high, spreading by wiry rhizomes. Leaves glabrous; blades up to 12 cm long, 1–4 mm wide; ligule membranous, obtuse, 0.5–3 mm long. Panicle stiff, usually contracted, up to 10 cm long, c. 3 cm wide. Spikelets with 3–10 florets, 3–8 mm long, green or purplish. Widespread. Pastures; weed. Introd. from Europe. Fl. summer. *Canada Blue Grass* . . . . ***P. compressa** L.
3 Internodes terete or slightly compressed 4

4 Leaves convolute, rigid. Plants growing on coastal dunes, foreshores, estuaries or cliff faces 5
4 Leaves not convolute but may be folded or inrolled. Plants growing in heath, forested areas or as lawn weeds 6

5 Glumes 3–5 mm long. Callus with loose, woolly hairs. Culms much exceeding the leaves. Panicle usually contracted, ± linear in outline, 8–30 cm long, 1–6 cm wide. If some of the branches spread, then some of these bear spikelets almost to the base. Spikelets 6–10 mm long with 3–7 florets, mostly lanceolate. Lemma 3–6 mm long, sub-acute or truncate, hairy on the keel. Leaf blades usually very smooth on the lower (outer) surface, rarely faintly scabrous, convolute-terete or angular-terete, 0.25–1.5 mm diam (up to 2 mm when open). Coast. Fl. summer . . . . **P. poiformis** (Labill.) Druce var. **poiformis**

5 Glumes 8–9 mm long. Callus glabrous or with hairs. Culms usually as long as the leaves or shorter. Panicle 8–12 cm long, narrow; the branches c. 3 cm long. Spikelets 8–10 mm long, awnless or nearly so. Lemma 7–11 mm long, obtuse or slightly 3-fid. Leaves mostly c. 50 cm long, 1–15 mm wide, rigid, sharply acute, densely scabrous on inner surface. Large dense tussocks up to 80 cm high. Confined to coastal dunes. Fl. winter–spring. *Beach Fescue* . . . .**P. billardierei** (Spreng.) St.-Yves

6 Leaf blades hairy; the hairs of 2 lengths; the longer ones spreading; the shorter ones stiff and scabrous; blades 10–25 cm long, loosely inrolled, 0.7–0.8 mm diam.; ligule membranous to papery, obtuse or truncate, c. 1 mm long. Panicle 4–18 cm long, narrow to pyramidal, often contracted and one-sided, on a long peduncle. Tufted perennial up to 90 cm high. Blue Mts Usually in *Eucalyptus* forests. Fl. summer. . . . . **P. induta** Vickery

6 Leaf blades glabrous or scabrous 7

7 Ligule thinly membranous, usually 2 mm long or longer, usually not minutely pubescent or scabrous on the back 8

7 Ligule firmly membranous, 0.1–2 mm long, minutely pubescent or scabrous on the back and/or ciliolate at the apex, or reduced to cilia 10

8 Leaf blades 4–15 mm wide, flat, up to 40 cm long; ligule membranous, thin, 2–4 mm long, truncate, often laciniate. Panicle 12–30 cm long, up to 16 cm wide. Spikelets 3–5 mm long, with 2–4 florets. Loosely tufted perennial up to 150 cm high. Mainly coastal. Margins of RF. Fl. late summer. . . . . **P. queenslandica** C.E.Hubb.

8 Leaf blades up to 6 mm wide 9

9 Ligule acute to acuminate, 4–10 mm long. Loosely tufted perennial up to 1 m high with creeping leafy stolons. Leaves green or purplish, glabrous; blades flat or folded, up to 20 cm long, 2–6 mm wide. Panicle ovate to oblong, 3–30 cm long, 15 cm wide. Spikelets with 2–4 florets, 3–4 mm long. Weed in lawns. Introd. from Europe, Asia and Africa. Fl. spring–summer. *Rough Meadow Grass*. ***P. trivialis** L.

9 Ligule obtuse or truncate, 1–3 mm long. Perennial up to 80 cm high, with creeping scaly rhizomes. Leaves usually dark green; blades flat or folded, up to 30 cm long, mostly 1–4 mm wide. Panicle ovate pyramidal or oblong, erect or ± drooping. Spikelets with 2–5 florets, 2.5–6 mm long, usually green. Widespread. Lawn grass; pastures. Introd. from Europe and Asia. Fl. mainly summer. *Kentucky Blue Grass* . . . .***P. pratensis** L.

10 Some or all of the leaf blades flat and usually expanded at least when living, or folded and the margins ± inrolled on drying, either broad or narrow 11

10 Leaf blades closely folded and/or usually the margins also rolled and overlapping so that the blade is ± angular-terete, narrow 15

11 Stems often leafy and branching from the aerial nodes, or stoloniferous 12
11 Leaves and shoots arising mostly from the base, plants tufted 13

12 Leaf blades 1–5 mm wide when expanded, firm, 10–30 cm long; ligule 0.5–2 mm long, truncate, finely laciniate to minutely ciliolate. Plants up to 120 cm high with glabrous nodes. Panicle up to 18 cm long. Spikelets with 2–7 florets. Widespread but mainly coast. Heath and DSF. Fl. early summer . . . . **P. affinis** R.Br.

12 Leaf blades 1–1.5 mm wide when expanded, soft, ± filiform, up to 20 cm long; ligule 0.5–1.5 mm long, truncate, often scabrous on the back. Green flaccid trailing perennial or sometimes forming tussocks, with long stolons. Panicle 2–12 cm long. Spikelets compressed, with 2–4 florets, green rarely purplish. Blue Mts Shady moist places. Fl. early summer.. . . . .**P. tenera** F.Muell. ex Hook.f.

13 Plants tufted, not normally developing rhizomes which initiate shoots remote from the parent tussock. Panicle often widely spreading; branches devoid of spikelets in the lower part. Coarse tufted green-glaucous or greyish perennial, up to 120 cm high. Leaf blades flat, up to 80 cm long, up to 3.5 mm wide, inrolled, scabrous below; ligule 0.5 mm long, minutely ciliolate at the apex. Panicle 10–25 cm long; lowest branches up to 12 cm long, devoid of spikelets for c. half their length. Spikelets with 3–4

florets (rarely up to 8), greenish to purplish, strongly compressed. Widespread. Open forests and grasslands. Fl. summer. . . . . **P. labillardieri** Steud. var. **labillardieri**

13 Plants commonly developing horizontal rhizomes which initiate new shoots ± remote from the parent tussock. Scales usually present on new shoots. 14

14 Leaf blades herbaceous, often dark green. Rhizomes slender, with thin pointed scales . . . .
. . . . **P. pratensis** (see above)

14 Leaf blades relatively rigid, green, 8–25 mm long, 1–4 mm wide; ligule 1 mm long, truncate. Rhizomes short and thick, with firm obtuse scales. Perennials up to 1 m high. Panicle 8–25 cm long. Spikelets with 3–6 florets, green or purplish. Widespread but mainly coast. DSF on Ss. Fl. early summer. . . . .
. . . . **P. cheelii** Vickery

15 Culms either branching to form fascicles of aerial shoots or stoloniferous and trailing 16
15 Stolons absent; aerial shoots rarely produced 17

16 Tussocks soft and flaccid; the aerial shoots branching to form stolons. Leaves green . . . .
. . . .**P. tenera** (see above)

16 Tussocks not flaccid. Leaves often bluish, 3–10 cm long; ligules 1–2 mm long. Lemmas finely pubescent on the lower back. Blue Mts In a variety of communities. Fl. summer. . . . .
. . . . **P. sieberiana** var. **cyanophylla** Vickery

17 Lemmas hairy in the lower part on the keels and lateral nerves, but usually not between the nerves 18
17 Lemmas hairy between the nerves on the lower back, with or without longer and/or denser hairs on the keels and lateral nerves 19

18 Tussocks large and coarse. Leaf blades up to 80 cm long, up to 3.5 mm wide . **P. labillardieri** (see above)
18 Tussocks small. Leaf blades mostly 5–15(30) cm long, green to greyish-green, soft, inrolled, c. 0.3 mm diam. Panicle up to 10 cm long. Spikelets green or purplish, with 3–5 florets. Coast and eastern slopes of Blue Mts Open forests. Fl. summer . . . . **P. meionectes** Vickery

19 Leaves mostly green to greyish-green or glaucous-green; blades inrolled-terete or angular-terete, up to 60 cm long, 0.2–0.7 mm wide; ligule 0.1–1 mm long, truncate, minutely ciliolate at the apex. Panicle 3–20 cm long, pyramidal at maturity; branches devoid of spikelets in the lower half. Spikelets compressed, often purplish or greenish rarely yellowish, with 2–7 florets. Widespread in a variety of communities. Fl. summer . . . . **P. sieberiana** Spreng.var. **sieberiana**
19 Leaves bluish, 3–10 cm long . . . .**P. sieberiana** var. **cyanophylla** (see above)

7 Desmazeria L.
1 species naturalized Aust.; all states and territories except Qld, & NT

One species in the area
Spikelets with usually 4–7 florets, compressed laterally, disarticulating above the glumes, 4–5 mm long, scabrous. Glumes subequal, keeled, shorter than the florets. Lemmas 2 mm long, narrow, rigid, sharp-pointed. Inflorescence a narrow panicle 4–8 cm long, somewhat one-sided. Annual up to 20 cm high; stems often geniculate at the base. Leaf blades flat or involute, scabrous; ligule membranous, torn. Widespread. Pastures; lawns; roadsides. Introd. from the Mediterranean. Fl. spring–summer. *Rigid Fescue* . . . . *D. rigidum (L.) C.E.Hubb.

8 Dactylis L.
1 species naturalized Aust.; all states and territories except NT

One species in the area
Spikelets with 3–several flowers, almost sessile, compressed, disarticulating above the glumes. Glumes equal, membranous, 5–6 mm long, subulate or with an awn c. 1 mm long. Lemmas similar to the glumes. Inflorescence paniculate, often one-sided, up to 25 cm long; the spikelets crowded on the branches which are distantly spaced. Erect tufted perennial up to 1 m high. Leaves scabrous; blades c. 25 cm long and 5

mm wide; ligule membranous, c. 1 cm long. Widespread. Pastures; roadsides; weed. Introd. from Europe. Fl. summer. *Cocksfoot* . . . .*D. glomerata* L.

9 Cynosurus L.

2 species naturalized Aust.; NSW, Vic., Tas., S.A., WA

Tufted annual or perennial. Culms 1–3-noded, erect to spreading, smooth. Ligule membranous. Leaf narrow, flat. Inflorescence a dense one-sided panicle with short branches. Spikelets of 2 kinds, sterile and fertile mixed together in clusters; the sterile ones pedicellate, almost concealing the sessile, fertile spikelets. Fertile spikelets oblong or wedge-shaped. Rachilla of spikelets disarticulating above glumes. Glumes keeled. Lemma rounded on the back mucronate to awned. Sterile spikelets 4–6 mm long, persistent, composed of up to 18 narrow finely pointed bracts.

Key to the species

1 Lemmas 3–5 mm long; awns inconspicuous. Fertile spikelets with 3–5 florets, 3–6 mm long. Panicles up to 14 cm long. Leaf blades up to 15 cm log, 1–4 mm wide; ligule blunt 0.5–1.5 mm long. Plant up to 70 cm high. Waste and disturbed areas. Introd. from Europe. Fl. spring. . . . .*C. cristatus* L.
1 Lemmas 5–7 mm long, awns conspicuous. Fertile spikelets with 1–5 florets, 3–6 mm long. Panicles up to 14 cm long. Leaf blades up to 20 cm long, 2–10 mm wide; ligule blunt to 10 mm long. Plant up to 1 m high. Waste and disturbed areas. Introd. from Europe. Fl. spring. *Rough Dog's Tail* . . *C. echinatus* L.

10 Briza L.

3 species naturalized Aust.; all states and territories except NT

Spikelets with several florets, compressed, disarticulating above the glumes, pedicellate. Glumes subequal, rounded on the backs, awnless. Lemmas rounded on the back, becoming successively smaller towards the apex of the spikelets. Ligule long, scarious or hyaline. Glabrous annuals.

Key to the species

1 Spikelets 10–20 mm long, with 9–17 florets, on drooping pedicels c. 2 cm long. Inflorescence with less than 10 spikelets. Plants mostly 20–40 cm high. Widespread. Weed of cultivation and waste places. Introd. from the the Mediterranean. Fl. late winter–spring. *Quaking Grass* . . . .*B. maxima* L.
1 Spikelets 4–5 mm long. 2

2 Pedicels 10–15 mm long. Panicle very open, symmetrical, 4–8 cm long, composed of 10–20 spikelets. Plants usually c. 20 cm high. Widespread. Weed of cultivation and waste places. Introd. from Europe and Asia. Fl. winter–spring. *Shivery Grass* . . . .*B. minor* L.
2 Pedicels c. 2 mm long. Panicle dense, one-sided below, composed of numerous spikelets. Plants up to 60 cm high. Coast, Illawarra region to Sydney. Weed in poor soils. Introd. from S. America. Fl. spring–early summer . . . .*B. subaristata* Lamk

11 Glyceria R.Br.

Sweet Grasses

5 species in Aust. (2 native, 3 naturalized); all states and territories except NT

Perennial rhizomatous herbs. Culms smooth. Ligule membranous. Leaf blade flat with rough margins. Inflorescence a contracted or open panicle. Spikelets slightly compressed, disarticulating above the glumes; florets 4–20, disarticulating above the lemmas. Glumes unequal, persistent, membranous, 1–3 nerved. Lemmas thin, rounded on back, overlapping but exserted from glumes. Palea 2-keeled.

Key to the species

1 Lemmas >6 mm long. Spikelets mostly 2–3 cm long, with 8–12 florets, oblong. Glumes 5–6 mm long, acuminate. Pedicels 1–2 cm long. Panicle very open, up to 25 cm long. Leaves glabrous; blades 7–8 mm

wide Erect perennial 2–3 m high. Usually in swamps and on river banks. Fl. spring–summer . . . . . . L.
. . . . **G. australis** C.E.Hubb.

1 Lemmas <6 mm long. 2

2 Lemmas 2–3 mm long, obtuse, apex not toothed. Spikelets 5–10 mm long with 4–8 florets. Palea slightly
exceeding the lemma. Plant to 2.5 m high. Waste land. Riparian. Introd. from Europe. Fl. spring–
summer . . . . *****G. maxima** (Hartm.) Holmb.
2 Lemmas 4–6 mm long, toothed at tip. Spikelets 12–25 mm long with 8–15 florets. Palea not exceeding
the lemma. Plant to 80 cm high. Damp areas. Introd. from Europe and America.. . *****G. declinata** Bréb.

12 **Dryopoa** Vickery
1 species endemic Aust.; N.SW., Vic., Tas.

Monotypic genus

Spikelets with 3–7 florets, c. 9 mm long, greenish or purplish, laterally compressed, disarticulating above
the glumes. Glumes unequal, compressed, with scabrous keels; the lower glume 3–5.5 mm long; the upper
one 4–7 mm long. Lemma 4–5.5 mm long, very firm, turgid, scarcely keeled, coarsely scabrous, with a few
hairs at the base. Panicle broadly pyramidal, 20–50 cm long, c. 50 cm wide; pedicels long. Erect perennial
1.5–3 m high; culms hollow, c. 1 cm diam., smooth, 5-noded. Leaf blades up to 50 cm long, 7–18 mm
wide, flat or folded; ligules firmly membranous to chaffy, obtuse, laciniate, 8–20 mm long. Wet places in
tall forests from Woronora Plateau southwards. WSF. Fl. spring–summer. *Giant Mountain Grass* . . . . .
D. dives (F.Muell.) Vickery

13 **Lamarckia** L.
1 species naturalized Aust.; Qld, NSW, Vic., Tas., S.A., WA

One species in the area

Spikelets with 1 fertile floret, or sterile; fertile florets intermixed with sterile ones in little clusters on the
very short branches of a unilateral spike-like panicle, or the sterile ones lying above the fertile panicle.
Rhachilla of the fertile spikelets prolonged above the floret, bearing a narrow empty awn-like glume, and
sometimes a second rudimentary one above it. Sterile spikelets consisting of truncate awnless glumes
only. Glumes of fertile spikelets subulate, subequal, 1–2 mm long; lemmas 2–3 mm long with a dorsal awn
6–8 mm long. Glabrous annual up to 20 cm high. Leaf blades flat, up to 10 cm long; ligule membranous,
glabrous, 5–8 mm long. Widespread. Grasslands; roadsides. Introd. from the Mediterranean. Fl. summer.
Golden Top . . . .*****L. aurea** (L.) Moench

14 **Lolium** L.
5 species naturalized Aust.; all states and territories

Spikelets with 3–20 florets, placed edgeways to the rhachis, sessile or almost so, distichous in alternate
notches on the angular rhachis, oblong fusiform-elliptic or cuneate in outline, less than 2 cm long
(excluding the awns), appressed to the rhachis, so that the lemmas touch the rhachis. Glume 1 except in
the terminal spikelet which has 2 glumes. Lemmas rounded on the back. Leaf blades flat; ligule short,
membranous, truncate. Plants glabrous throughout.

Key to the species

1 Glume shorter than the spikelet (less than half to two-thirds the length). 2
1 Glume c. as long as, or longer than the spikelet. 3

2 Lemmas with a fine awn as long as the lemma (5–6 mm). Spikes 15–25 cm long. Spikelets with 10–20
florets. Annual up to c. 50 cm high. Widespread. Pastures; garden weed. Introd. from the Mediterranean.
Fl. spring. *Italian Ryegrass.* . . . . *****L. multiflorum** Lam.
2 Lemmas awnless. Spikes 7–25 cm long. Spikelets with 3–12 florets. Perennial up to c. 50 cm high.
Widespread. Pastures; garden weed. Introd. from Europe and Asia. Fl. winter–spring (Hybrids between
this species and *L. rigidum* occur). *Perennial Ryegrass.* . . . . *****L. perenne** L.

3 Spikelets oblong-cuneate, much broader than the rhachis, with 3–9 florets, swollen at the fruiting stage. Glume consistently longer than the spikelet. Lemma awnless or with a dorsal awn 1–2 mm long. Spike 15–20 cm long. Rhachis conspicuously zig zagged. Annual up to 50 cm high. Widespread. Pastures; uncommon garden weed. Introd. from Europe, Asia and Africa. *Darnel* . . . . ***L. temulentum** L.

3 Spikelets linear to narrow-lanceolate, as broad as, or a little broader than the rhachis, not swollen at the fruiting stage, deeply embedded in the rhachis. Rhachis faintly zig zagged 4

4 Rhachis roughly square in section. Spikelets with 3–5 florets. Otherwise similar to *L. temulentum*. *Wimmera Ryegrass* . . . . ***L. rigidum** Gaudin

4 Rhachis much flattened, rectangular in section. Uncommon. *Rigid Ryegrass* . . . . ***L. loliaceum** (Bory & Chaub.) Hand.-Mazz.

15 Hordeum L.

7 species naturalized Aust.; all states and territories

Spikelets 3 together, sessile or nearly so; central spikelet with a bisexual flower; lateral spikelets usually male or sterile. Each spikelet with 1 floret disarticulating above the glumes. Glumes awned or awn-like from the base. Lemmas convolute, with a straight terminal awn. Inflorescence dense, cylindrical, spike-like.

Key to the species

1 Rhachis persistent, not fragmenting at maturity. Annual up to 1 m high 2
1 Rhachis fragmenting at maturity. 3

2 All spikelets in triplet fertile and awned. Inflorescence 5–7 cm long. Awns up to 10 cm long. Cultivated; weed or in pastures. Introd. from Europe and Asia. Fl. summer. *Six row Barley* . . . . . ***H. vulgare** L.

2 Central spikelet of triplet fertile, lateral male. Male spikelets awnless. Awn on female spikelet to 15 cm long. Inflorescence not including awns, 6–10 cm long. Cultivated; weed or in pastures. Introd. from Europe and Asia. Fl. summer. *Two-row Barley* . . . . ***H. distichon** L.

3 Leaves without auricles. Inflorescence 2–4 cm long. Spikelets spreading, especially when dry, less than 2 cm long (including the awns). Glumes of the lateral spikelets bristle-like. Annual up to 20 cm high. Widespread. Weed of cultivation, pastures and waste places. Introd. from Europe. Fl. winter. *Barley Grass* . . . . ***H. hystrix** Roth

3 Leaves with auricles 4

4 Anthers of central spikelet 0.7–1.4 mm long, pale and exerted. Inflorescence 3–10 cm long, cylindrical. Spikelets 4–6 cm long (including the awns). Annual up to 30 cm high. Widespread. Weed of cultivation, pastures and waste places. Introd. from Europe, Asia and N. Africa. Fl. winter–spring. *Barley Grass* . . . . ***H. leporinum** Link

4 Anthers of central spikelet 0.2–0.5 mm long, usually black and not exerted. Inflorescence 2–8 cm long. Spikelets 4–6 mm long (including the awns). Annual to 40 cm tall. Widespread. Introd. from Europe and Asia. Fl. spring. . . . . ***H. glaucum** Steud.

16 Secale L.

1 species naturalized Aust.; NSW, Vic., S.A., WA

One species in the area

Spikelets with 2 florets, sessile, appressed against the rhachis, disarticulating above the glumes. Glumes linear, subulate. Lemmas 1 cm long, with a terminal awn as long or up to 3 times as long as the lemma. Inflorescence a dense narrow cylindrical spike 10–15 cm long, 15 mm wide (excluding the awns). Annual 1–2 m high. Coast and Cumberland Plain. Cultivated and escaped into pastures and waste places. Introd. from Europe. Fl. summer. *Rye* . . . . ***S. cereale** L.

17 Triticum Pers.

1 species naturalized Aust.; all states and territories except NT

One species in the area

Spikelets with 3–5 florets, sessile and alternate in excavations of a dense spike, disarticulating above the glumes. Glumes shorter than the florets, equal, awned. Inflorescence a spike, 4-angled, 7–12 cm long. Annual up to c. 1 m high. Leaves up to 2 cm wide; ligule membranous, truncate. Widespread. Cultivated; roadsides; weed. Introd. but origin uncertain. Fl. winter. *Wheat.* . . . . *T. aestivum* L.

18 Elymus L.

6 species in Aust. (5 native, 1 introduced); all states and territories except NT

Tufted or rhizomatous perennials. Ligule membranous, truncate. Leaf blade flat, rarely involute, scabrous on upper surface. Inflorescence an erect to drooping spike. Spikelets solitary, or in groups of 2–3, alternate, compressed, disarticulating above the glumes. Lemmas acute to awned, rounded on the back.

Key to the species

1 Plants with long yellowish rhizomes. Lemmas awnless, 10 mm long. Spikelets 10–15 mm long, oblong, with 3–6 florets, sessile or almost so. Leaf blades stiff, flat, finely pointed, 2–7 mm wide. Perennial up to 1 m high. Usually on clay soils. Introd. from temperate N. hemisphere. *English Couch* . . . .
. . . . *E. repens* (L.) Gould
1 Plants tufted, some lemmas with awns 2

2 Spikelets 5–7 cm long, cuneate, with 6–12 florets, sessile or almost so, awns slender 15 mm long or more. Rhachilla distinctly hairy. Glumes unequal. Lemmas 8–20 mm long, with awns 2–4 cm long. Tufted perennial c. 1.5 m high. Widespread. Chiefly on clay soils. Fl. summer. *Common Wheat Grass.* .
. . . .*E. scaber* (R.Br.) A.Löve
2 Spikelets 1.5–4 cm long, with 6–12 florets. Awns less than 15 mm long. Rhachilla glabrescent. Glumes subequal. Lemmas 7–12 mm long, mucronate or with awn to 14 mm long. Tufted perennial to 80 cm high. Coastal dunes and riparian habitats. *Short-awned Wheatgrass* . . . .
. . . . *E. multiflorus* (Banks & Sol. ex Hook.f.) A.Löve & Connor

19 Hainardia Greuter

1 species naturalized Aust.; NSW, Vic., Tas., S.A., WA

Monotypic genus

Spikelets with 1 floret, sessile, distichous, half embedded in the rhachis. Glume 1 (2 in the terminal spikelet), c. 7 mm long. Lemma shorter than the glume, awnless, hyaline, with its back to the rhachis. Palea hyaline. Inflorescence a cylindrical spike c. 2 mm diam., up to 20 cm long, straight or slightly curved, fragmenting at maturity below each spikelet. Plant up to 50 cm high. Leaf blades flat or inrolled, 5–10 cm long; ligule short, truncate. Introd. from the Mediterranean. Fl. summer. *Common Barb Grass* .
H. cylindrica (Willd.) Greuter

20 Parapholis C.E.Hubb.

2 species naturalized Aust.; NSW, Vic., Tas., S.A., WA

Monotypic genus

Spikelets with 1 floret, sessile, alternate, half embedded in the rhachis and appearing to form part of it, separating from the rhachis and becoming obvious only after flowering. Glumes stiff, lanceolate, 6–10 mm long, side by side on the outer edge of the spikelet except in the terminal spikelet. Lemmas shorter than the glumes, with the back to the rhachis, awnless. Inflorescence a cylindrical spike, 1.5–2.5 mm diam. Leaf blades glabrous, usually 2–4 cm long but up to 10 cm; ligule minute. Tufted annual with branching decumbent culms up to 50 cm long, turning upwards and inwards. Coast. Salt marshes and other saline places. Introd. from the Mediterranean. Fl. summer. *Coast Barb Grass* . . . .
* . . . . *P. incurva* (L.) C.E.Hubb.

21 Anthoxanthum L.

1 species naturalized Aust.; all states and territories except NT

One species in the area

Spikelets with 1 terminal bisexual floret and 2 sterile lemmas below it, subsessile, 5–7 mm long, disarticulating above the glumes. Glumes membranous subulate; the longer (upper) one 8 mm long. Sterile lemmas with golden hairs, both awned dorsally, 5–6 mm long (including the awn) (Fig. 49 d); fertile lemmas c. 2 mm long. Inflorescence a spike-like panicle 4–7 cm long, 12–20 mm wide. Erect perennial up to 80 cm high. Leaf blades flat, glabrous or hairy, 3–4 mm wide; ligule membranous, bilobed, 3 mm long. Widespread. Pastures; roadsides; garden weed. Introd. from Europe. *Sweet-scented Vernal Grass* . . . . .
. . . . ***A. odoratum** L.

22 Phalaris L.

9 species naturalized Aust.; all states and territories except NT

Spikelets with 1 bisexual floret, and 1 or 2 minute sterile lemmas below it (except *P. paradoxa*), flat, densely crowded into ovoid or cylindrical spike-like panicles, disarticulating above the glumes. Glumes equal, complicate, enclosing the florets, usually with green stripes along the nerves, keeled; the keel usually bordered by a scarious wing (Fig. 49 c).

Key to the species

1 Spikelets in groups of 7; fertile spikelet surrounded by 6 sterile spikelets. Plants up to 1 m high. Panicle spike-like, 4–7 cm long. Widespread. Pasture grass. Introd. from the Mediterranean. Fl. spring. *Paradoxa Grass* . . . . ***P. paradoxa** L.
1 Spikelets not in groups; all with 1 bisexual floret. 2

2 Panicle ovoid, c. twice as long as broad, 20–25 mm long. Plant up to 40 cm high. Cultivated for grain. Widespread. Naturalized in disturbed areas on sandy soils. Fl. spring. Introd. from the Mediterranean. *Canary Grass* . . . . ***P. canariensis** L.
2 Panicle cylindrical, more than twice as long as broad 3

3 Panicle lobed, with conspicuous branches, usually interrupted, 5–25 cm long. 4
3 Panicle cylindrical, neither lobed nor interrupted, 5–40 cm long. Spikelets densely crowded, 3.5–7.5 mm long. Robust perennial to 2 m high 5

4 Leaf blades green, with a pale midrib. Widespread. Pasture plant. Floodways. Introd. from Europe and Asia. Fl. spring–summer. *Reed Canary Grass* . . . .***P. arundinacea** L. var. **arundinacea**
4 Leaf blades striped green and cream. Garden escape near habitation. Introd. from Europe. *Ribbon Grass* . . . . ***P. arundinacea** var. **picta** L.

5 Panicle usually c. 5 mm diam. (rarely up to 10 mm), up to 6 cm long. Glumes ± wingless, blunt, usually finely toothed on the margins. Annual, usually c. 1 m high. Coast. Floodways. Introd. from America. Fl. spring. . . . .***P. angusta** Nees ex Trin.
5 Panicle more than 1 cm diam. Glumes winged, usually acuminate 6

6 Stems slightly to much swollen at the bases of the lower internodes. Erect tufted perennial 1–2 m high with short creeping rhizomes. Sterile lemmas 2, unequal (Fig. 49 c). Glumes winged, 6–7 mm long. Panicle 5–12 cm long, 1–2 cm diam. Blue Mts Pasture grass. Introd. from the Mediterranean. Fl. spring–summer. *Phalaris* . . . . ***P. aquatica** L.
6 Stems not swollen near the base. Erect annual usually c. 1 m high. Sterile lemma 1. Glumes winged, 6–7 mm long. Panicle 3–8 cm long, 1–2 cm diam. Widespread. Pasture grass; weed. Fl. spring. *Lesser Canary Grass* . . . . ***P. minor** Retz.

23 Avena L.

8 species naturalized Aust.; all states and territories

Spikelets with few florets, in a loose panicle, disarticulating above the glumes; rhachilla hairy below the lemmas. Glumes scarious (at least at the top), tapering. Lemmas smaller than the glumes, 2-cleft at the top, usually with a long twisted dorsal awn; terminal lemma often small and empty or rudimentary.

Key to the species

1 Lemmas scabrous or almost glabrous. Spikelets mostly 2-flowered; florets not readily separating from the glumes; awns usually straight, sometimes absent. Annual up to 70 cm high. Widespread. Cultivated; roadsides. Introd. from Europe. Fl. mainly spring. *Common Oat* . . . . ***A. sativa** L.

1 Lemmas with red or brown or white hairs near the base 2

2 Terminal teeth of the lemmas extended as bristles 2–4 mm long. Spikelets mostly 2-flowered; pedicels usually with a single kink. Lemmas with awns slightly bent near the base. Annual up to 70 cm high. Widespread. Weed of cultivation; roadsides. Introd. from the Mediterranean. Fl. mainly spring . . . . .
. . . . ***A. barbata** Pott. ex Link

2 Terminal teeth of the lemmas acute, c. 1 mm long. Spikelets mostly 3-flowered. Hairs on the lemmas brown or white 3

3 Rhachilla readily disarticulating between all the florets at maturity. Glumes c. 25 mm long. Lemmas 2 cm long; awns 3–4 cm long, bent near but below the middle. Rhachilla and lower part of the lemmas usually with brown or white hairs. Annual up to 70 cm high. Widespread. Weed of pastures and cultivation. Fl. mainly spring. Introd. from Europe. *Wild Oat* . . . . ***A. fatua** L.

3 Rhachilla continuous and tough between the florets which do not separate at maturity 4

4 Abscission scar at the base of the lemmas elongated. Glumes mostly 3–5 cm long. Lowest lemma 2.5–4 mm long. Otherwise similar to *A. fatua*. Introd. from Europe. . . . . ***A. sterilis** L.

4 Abscission scar at the base of the lemmas circular. Glumes mostly 2–3 cm long. Otherwise similar to *A. fatua*. Introd. from Europe. . . . . ***A. ludoviciana** Duricu

24 Aira L.

5 species naturalized Aust.; all states and territories

Spikelets with 2 florets, disarticulating above the glumes. Glumes subequal, acute. Lemmas shorter than the glumes, finely pointed or shortly bifid, with a twisted awn attached dorsally below the middle. Inflorescence a loose or rarely contracted panicle with capillary branches.

Key to the species

1 Panicle spike-like, 1–2 cm long. Spikelets c. 3 mm long; awn 1–1.5 mm long. Pedicels usually shorter than the spikelets. Erect annual up to c. 12 cm high. Sandy soils. Introd. from Europe and Asia. Fl. spring. *Early Hair Grass* . . . . ***A. praecox** L.

1 Panicle open. Pedicels as long as or longer than the spikelets. 2

2 Pedicels up to twice as long as spikelets 3

2 Pedicels twice as long or longer than the spikelets 4

3 Spikelets mostly 2 mm long but up to 2.5 mm, mostly with 1 exserted awn but sometimes 2; glumes obtuse. Erect annual up to 25 cm high. Widespread. Chiefly grasslands. Introd. from the Mediterranean. Fl. spring. *Silvery Hair Grass* . . . . ***A. cupaniana** Guss.

3 Spikelets 2.5–3 mm long, mostly with 2 exserted awns but sometimes only 1. Glumes acute. Erect annual up to 30 cm high. Widespread. Grasslands; roadsides. Introd. from Europe. Fl. summer. *Silvery Hair Grass* . . . . ***A. caryophyllea** L.

4 Spikelets c. 3.5 mm long, mostly with 1 exserted awn. Pedicels mostly 4–6 times as long as the spikelets with clavate swelling below the spikelets. Erect annual up to 35 cm high. Blue Mts Introd. from Europe. Fl. summer . . . . *A. **provincialis** Jord.

4 Spikelets 1–2.5 mm long, with one or 2 exserted awns. Pedicels 2–5 times as long as spikelets, without clavate swelling. Erect annual to 50 cm high. Blue Mts Introd. from Europe. Fl. summer . . . .

. . . . *A. **elegantissima** Schur

25 Holcus L.

4 species naturalized Aust.; all states and territories except NT

Spikelets with 2 florets, falling whole; lower floret bisexual, awnless; upper floret male, with a dorsally awned lemma. Glumes subequal, awnless. Lemmas shorter than the glumes (Fig. 50m).

Key to the species

1 Rhizomes absent. Panicle compact, 8–15 cm long, 2–3 cm wide, pubescent, with a pink or violet tinge. Spikelets 4–5 cm long; the awn on the upper lemma becoming hooked when dry. Villous tufted perennial up to 1 m high. Leaf blades flat, soft, 7–8 mm wide; ligule membranous. Widespread. Better soils, usually near habitation. Introd. from Europe. Fl. spring–summer. *Yorkshire Fog.* . *H. **lanatus** L.

1 Rhizomes present. Awn on the upper lemma remaining straight when dry and clearly projecting beyond the tips of the glumes. Otherwise similar to *H. lanatus.* Blue Mts *Creeping Fog* . . . . *H. **mollis** L.

26 Arrhenatherum P.Beauv.

1 species naturalized Aust.; all states and territories except NT

One species in the area

Spikelets with 2 florets, disarticulating above the glumes. Glumes subequal, membranous; the long one c. 8 mm long. Lower floret male; lemma 7–8 mm long, with a bent dorsal awn c. 15 mm long. Upper floret bisexual; lemma with a short straight or bent awn. Rhachilla ending in a fine bristle. Panicle 12–30 cm long, usually open. Erect, almost glabrous perennial mostly c. 1 m high.

Key to the varieties

1 Basal internodes not swollen. Nodes usually glabrous. Introd. from temp. N. hemisphere. Widespread but not Blue Mts Weed. Fl. summer. *False Oat Grass* *A. **elatius** (L.) P.Beauv. ex J.Presl & C.Presl var. **elatius**

1 Basal internodes swollen, globose. Nodes often hairy. Introd. from temp. N. hemisphere. Widespread. Weed. Fl. summer. *Bulbous Oat Grass.* . . . . *A. **elatius** var. **bulbosum** (Willd.) Spenn.

27 Amphibromus Nees

10 species native Aust.; all states and territories except NT

Spikelets with 3–10 florets, disarticulating above the glumes, pedicellate; the uppermost floret sometimes male. Glumes unequal to subequal. Lemmas 2–4-fid, with a bent ± twisted awn from the middle of the back. Inflorescence an erect loose panicle with ± erect branches.

Key to the species

1 Ligule 2–5 mm long. Panicle up to 25 cm long. Spikelets 8–15 mm long. Lemma swollen. Culms with 2–3 nodes. Tufted perennial up to 1 m high. Blue Mts and valleys. Swampy areas. Fl. spring–summer. .

. . . . **A. pithogastrus** S.W.L.Jacobs & Lapinpuro

1 Ligule 10–20 mm long. Panicle up to 40 cm long. Spikelets 10–16 mm long. Lemma not swollen. Culms with 2–5 nodes. Tufted perennial up to 1.5 m high. Widespread. Flood plains and swamp areas. Fl. spring–summer. *Swamp Wallaby Grass* . . . . **A. nervosus** (Hook.f.) Baill.

28 Rostraria Trin.

2 species naturalized Aust.; all states and territories except NT

One species in the area

Spikelets with 2–6 florets, compressed, 2–3 mm long, on pedicels 1 mm long, disarticulating above the glumes. Glumes membranous, 2–3 mm long, awnless. Lemmas similar to the glumes but with a fine awn 1 mm long. Upper florets smaller or sterile or reduced to a bristle. Inflorescence a cylindrical dense, spike-like panicle up to 15 cm long, sometimes lobed, 6–20 mm diam.; the spikelets in clusters. Leaves hairy; blades up to 10 cm long, 1–3 mm wide; ligule of bristles. Erect annual. Pastures; roadsides; weed. Introd. from the Mediterranean. Fl. spring–summer. *Annual Catstail* . . . .***R. cristata** (L.) Tzvelev

29 Agrostis Desf.

15 species in Aust. (4–5 naturalized); all states and territories except NT

Inflorescence a narrow panicle, not detaching as a whole unit. Rhachilla extension <20% of lemma length. Spikelets with 1 floret, disarticulating above or below the glumes. Glumes keeled, acute, awnless. Lemmas shorter than the glumes, broad, glabrous, usually awnless or with a dorsal awn attached below the middle. Palea very thin, shorter than the lemma or absent.

Key to the species

1 Palea absent or minute. Plants tufted, without stolons, up to 80 cm high. Leaf blades up to 15 cm long, 2–4 mm wide; ligule membranous, jagged. Panicle 10–30 cm long. Blue Mts Usually in damp or swampy areas. Fl. summer. . . . .**A. bettyae** S.W.L.Jacobs

1 Palea c. 50% the length of lemma. 2

2 Panicle open. 3

2 Panicle dense 5

3 Ligule shorter than wide, up to 2 mm long. Plants up to 40 cm high, with rhizomes. Leaf blades 5–10 cm long, 1–3 mm wide. Panicle 5–10 cm long. Lawn grass; pastures; weed. Introd. from Europe, Asia and N. America. Fl. spring–summer. *Brown Bent*. . . . .***A. capillaris** L.

3 Ligule longer than wide 4

4 Ligule truncate, 2–6 mm long, glabrous. Erect perennial up to 1 m high, with decumbent culms rooting at the nodes or with rhizomes. Leaf blades 5–20 cm long, 2–8 mm wide. Panicle up to 25 cm long. Widespread. Pastures; waste land. Introd. from Europe, Asia and N. America. Fl. summer. *Redtop Bent* . . . . ***A. gigantea** Roth

4 Ligule acute or obtuse but not truncate, 1–6 mm long. Tufted perennial up to 40 cm high, rhizomes absent but stolons present. Leaf blades up to 10 cm long, 1–5 mm wide. Panicle up to 13 cm long. Widespread. Lawns; waste land. Introd. from Europe, Asia and N. America. Fl. summer. *Creeping Bent* . . . . ***A. stolonifera** L.

5 Ligule acute or obtuse. Glumes not falling with the floret. . . . . ***A. stolonifera** (see above)

5 Ligule truncate, 2–3 mm long. Spikelets falling whole. Plants up to 60 cm high. Leaf blades up to 18 cm long, 2–10 mm wide. Panicle up to 15 cm long. Near or in water. Introd. from S. Europe and Asia. Fl. summer. *Water Bent* . . . . ***A. viridis** Gouan

30 Lachnagrostis Trin.

10 species native Aust.; all states and territories except NT

Inflorescence open often detaching as a whole unit at maturity. Rhachilla extension >30% of lemma length. Spikelets usually with 1-floret. Glumes keeled, acute, awnless. Lemmas shorter than the glumes, hairy or occasionally glabrous, awned. Palea at least more than 50% the length of the lemma.

Key to the species

1 Lemma glabrous. Erect annual up to 45 cm high. Culms 3-noded. Leaf blades 5–40 (usually c. 20) cm long, 2–8 (usually 4–6) cm wide; ligule membranous, 4–8 mm long, obtuse, laciniate. Panicle spreading. Spikelets c. 6 mm long, greenish purplish or straw-coloured, laterally compressed. Lemma 3–4 mm long, awned from c. the middle. Usually on sandy soils near the coast. Fl. summer. *Coastal Blown Grass*
. . . .**L. billardierei** (R.Br.) Trin.
1 Lemma pubescent 2

2 Pedicels of the spikelet usually diverging from the rhachis, some pedicels c. 1 cm long. Spikelets c. 5 mm long, tinged with purple. Lemma 2–3 mm long, awned from c. the middle. Erect annual up to 60 cm high. Culms 2–4-noded. Leaf blades 8–20 cm long, mostly 3–4 mm wide. Panicle spreading. Widespread. Usually in damp places. Fl. summer. *Blown Grass* . . . . . .**L. aemula** (R.Br.) S.W.L.Jacobs
2 Pedicel of spikelets usually close to the rhachis, less than 5 mm long. Spikelets 2–4 mm long, usually light green or straw-coloured sometimes tinged with purple. Lemma 2 mm long or less. Erect annual up to 70 cm high. Culms 4-noded. Leaf blades 8–25 cm long, usually c. 2 mm wide. Panicle spreading. Widespread. Usually in damp places. Fl. summer. *Blown Grass* . . . .**L. filiformis** (G.Forst.) Trin.

31 **Deyeuxia** Clarion ex P.Beauv.

29 species native Aust.; all states and territories except NT

Spikelets with 1 floret, usually with the rhachilla prolonged into a short hairy or glabrous bristle. Glumes keeled, equal or unequal, awnless. Lemma membranous, notched at the summit; awn fine and may be easily dislodged, dorsal, straight or bent, occasionally twisted in the lower part; callus often with a tuft of hairs which surrounds (but is much shorter) than the lemma. Palea nearly as long as the lemma.

Key to the species

1 Lemma awnless even in the young stages 2
1 Lemma awned at least in the young stages 3

2 Leaf blades 2–4 cm long, upper surface glabrous along ribs. Panicle narrow, elliptic to linear in outline, 2–8 cm long. Glumes subequal. Erect perennial to 25 cm tall. Blue Mts Swamps and creeklines. Fl. spring–summer. . . . . **D. innominata** D.I.Morris
2 Leaf blades 4–10 cm long, 1–2 mm wide, upper surface scabrous along ribs. Panicle loose, with few spikelets, lanceolate in outline, 3–20 cm long. Glumes unequal. Culms with c. 6 nodes. Ascending perennial up to 70 cm high. Cumberland Plain; Woronora Plateau. Shady places. Fl. summer . . . . . .
. . . . **D. nudiflora** Vickery

3 Lemma shorter than, or almost equal, to the upper glume, rarely very slightly exceeding it 4
3 Lemma distinctly and regularly longer than both glumes 15

4 Lower glume usually longer than the upper. 5
4 Lower glume shorter than the upper. 6

5 Awn attached near the base of, or at least in the lower third, of the lemma, exserted beyond the glumes. Panicle dense, usually more than 8 cm long, cylindrical in outline, sometimes lobed. Leaf blades flat or slightly inrolled. Perennial up to 120 cm high. Widespread and very variable. In a variety of habitats. Fl. spring–summer. *Reed Bent Grass*. **D. quadriseta** (Labill.) Benth.
5 Awn attached above the lower third of the lemma. Spikelet c. 3 mm long; awn only slightly exceeding the lemma. Panicle dense, 15–25 cm long; the branches slightly spreading. Leaf blades linear, flat. Tufted perennial up to 80 cm high. Widespread. Swamps and river banks. Fl. spring . . . .
. . . .**D. mesathera** Stapf ex Vickery

6 Panicle dense or spike-like or, if loose, then the shorter branches of each cluster bearing spikelets almost from the base 7
6 Panicle ± loose, with filiform ± spreading branches which are naked for some distance from the base . 11

7 Awn attached in the lower third of the lemma. Spikelet 4.5–6 mm long, greenish or very slightly purplish. Leaf blades strongly inrolled. Panicle linear to linear-lanceolate, 5–15 cm long. Tufted, subglabrous perennial 10–70 cm high. Higher altitudes. Wet places. Fl. spring–summer. . . . . **D. monticola** (Roem. & Schult.) Vickery var. **monticola**

7 Awn attached above the lower third of the lemma **8**

8 Rhachilla not produced beyond the floret. Awn more than 1 mm long. Spikelet less than 4 mm long. Panicle dense, linear to narrow-lanceolate, 20–30 cm long; the branches obliquely spreading. Erect perennial up to 1 m high. Coast and Cumberland Plain. Wet places. Endangered. Fl. spring–summer. . . . .**D. appressa** Vickery

8 Rhachilla produced beyond the floret **9**

9 Rhachilla produced into a densely hairy bristle. Awn less than 1 mm long. Spikelet 3.5–4 mm long. Panicle 3–10 cm long. Tufted, subglabrous perennial 40–60 cm high. Higher altitudes. Forests. Fl. spring . . . . **D. microseta** Vickery

9 Rhachilla produced into a glabrous bristle **10**

10 Spikelet less than 4 mm long. Awn shorter than or c. as long as the lemma. Panicle up to 15 cm long. Erect perennial up to c. 1 m high. Leaf blades up to 4.5 mm wide. Blue Mts Usually on granite. Fl. spring. . . . . **D. imbricata** Vickery

10 Spikelet more than 4 mm long **11**

11 Callus at the base of the floret naked. Awn very short, not exceeding the lemma. Panicle sparsely branched, 3–9 cm long, 4–6 mm wide. Glumes purplish. Tufted perennial up to 70 cm high. Leaf blades inrolled. Higher altitudes. Swamps. Fl. spring–summer . . . . **D. angustifolia** Vickery

11 Callus at the base of the floret clothed with short hairs. Awn about as long as the lemma or exceeding it, stiff, reflexed. Panicle loose and sometimes spreading, 8–17 cm long, 8–20 mm wide. Glumes often purplish. Tufted perennial 60–120 cm high. Leaf blades ± flat, up to 7 mm wide. Higher altitudes. Wet places. Fl. spring. . . . . **D. brachyathera** (Stapf) Vickery

12 Lemma distinctly and closely scaberulous. **13**

12 Lemma sparsely scaberulous or smooth **14**

13 Panicle 12–25 cm long, with lax branches. Spikelet 3–3.2 mm long; rhachilla produced into a bristle 0.8 mm long, usually hairy. Perennial 60–100 cm high. Higher altitudes. Forests, often amongst rocks. Fl. spring. . . . . **D. scaberula** Vickery

13 Panicle 6–9 cm long, with stiff contracted branches. Spikelet 3.5 mm long; rhachilla produced into a short densely hairy bristle. Glumes unequal. Lemma scaberulous, equal to the lower glume. . . . . **D. microseta** (see above)

14 Awn exceeding the lemma by more than half the length of the lemma, reflexed. Panicle dense, 8–13 cm long. Spikelet 5–5.5 mm long . . . . **D. brachyathera** (see above)

14 Awn minute, very fine, not or scarcely exceeding the lemma. Panicle loose, 12–30 cm long, with lax or nodding filiform branches. Erect, glabrous perennial up to 1 m high. Spikelets 2–3 mm long, green or purplish, racemosely arranged on short branches. Widespread. Sandy soil. Fl. spring. . . . . **D. decipiens** (R.Br.) Vickery

15 Panicle up to 9 cm long. Spikelet c. 2 mm long. Lemma truncate; awn c. 0.5 mm long; callus with short bristles. Erect to ascending perennial up to 50 cm high. Leaf blades filiform, less than 1 mm wide. Higher altitudes. Wet places. Fl. spring–summer. . . . .**D. gunniana** (Nees) Benth.

15 Panicle more than 10 cm long **16**

16 Spikelet 3–4 mm long; rhacilla produced into a bristle 0.2–1.2 mm long. Tufted subglabrous perennial up to 120 cm high. Panicle 20–40 cm long, lax, spreading. Higher altitudes. Forests. Fl. spring–summer . . . . **D. mckiei** Vickery

16 Spikelet 1.5–2.5 mm long; rhachilla produced into a bristle 0.6–0.8 mm long. Erect, subglabrous perennial 60–120 mm high. Panicle loose, spreading, 12–20 cm long. Higher altitudes. WSF and RF. Shady places. Fl. spring. **17**

17 Glumes unequal. Lemma exceeding glumes. . . . .**D. parviseta** Vickery var. **parviseta**
17 Glumes almost equal. Lemma equal to or slightly exceeding glumes . . . .
. . . .**D. parviseta** var. **boormanii** Vickery

32 Echinopogon P.Beauv.
7 species native Aust.; all states and territories except NT

Spikelets with 1 floret, disarticulating above the glumes, nearly sessile in a dense globular or ovoid spike-like panicle; rhachilla produced into a short bristle above the floret. Peduncle often scabrous. Glumes acute, keeled. Lemma 3-lobed; the lateral lobes awnless; the central one produced into a fine straight awn. Palea narrow.

Key to the species

1 Lemma bilobed with lobes at least 2 mm long. Spikelets mostly 5 mm long or longer. Leaf blades up to 10 mm wide. Panicle erect, dense, ovoid to oblong, up to 70 mm long and 40 mm wide. Loosely tufted perennial up to 1 m high, with slender rhizomes. Coast and adjacent plateaus. Open forest. Fl. spring–summer. *Erect Hedgehog Grass* . . . . **E. intermedius** C.E.Hubb.
1 Lemma entire or bilobed with lobes up to 1.5 mm long. Spikelets mostly less than 5 mm long 2

2 Loosely tufted perennial up to 1.2 m high with a slender often elongated rhizome. Culms often bent at a basal node or nodes. Lobes of the lemma minute or lemma entire. Leaf blades 2–8 mm wide. Panicle ovoid to oblong, up to 50 mm long and 25 mm wide. Widespread. Usually in WSF and damper situations in DSF. Fl. spring. *Forest Hedgehog Grass* . . . . **E. ovatus** (G.Forst.) P.Beauv.
2 Tufted perennial up to 1.5 m high, without rhizomes. Culms not usually bent at a basal node or nodes. Lobes of the lemma up to 1.5 mm long. Leaf blades 2–5 mm wide. Panicle ovoid to oblong, sometimes interrupted towards the base, up to 10 cm long and 5 cm wide. Widespread. Open forest and grasslands. Fl. spring–summer . . . . **E. caespitosus** C.E.Hubb. var. **caespitosus**

33 Dichelachne Endl.
5 species native Aust.; all states and territories except NT

Erect, tufted, shallow-rooted perennials. Spikelets with 1 floret, disarticulating above the glumes, in a dense narrow panicle, often tinged with purple. Glumes subequal, hyaline, subulate. Lemma with a short hairy callus and a dorsal curved or bent usually twisted awn affixed a little below the 2-lobed summit. Leaves glabrous or slightly hirsute. Ligule membranous, lobed or torn-ciliate at the summit.

Key to the species

1 Awn 2.5 cm or more long, 4–6 times as long as the lemma. Panicle very dense, spike-like, up to 30 cm long; the awns concealing the spikelets. Tufted herb up to 1.5 m high. Widespread. Open forest or cleared areas. Sandy soils. Fl. spring. *Long-hair Plume Grass* . . . . **D. crinita** (L.) Hook.f.
1 Awns less than 2.5 cm long, 2–5 times as long as the lemma 2

2 Lemma less than 4 mm long. Stamens 1–3 3
2 Lemma 4 mm or more long. Stamens 3 4

3 Lemma (excluding the awn) shorter than the glumes. Stamen 1. Panicle dense, up to 23 cm long; pedicels of spikelets ± concealed by the awns; branches not flexuous. Tufted herb up to 1.2 m high. Widespread. Open forests often on heavy textured soils. Fl. spring–summer. *Short-hair Plume Grass* . . . .
. . . . **D. micrantha** (Cav.) Domin
3 Lemma (excluding the awn) shorter than the lower glume. Stamens 3. Panicle dense, pedicels of spikelets visible between the awns, up to 28 cm long; branches flexuous. Tufted herb up to 80 cm high. Widespread. Forests in damp places on a variety of soils. Fl. summer. . . . .**D. parva** B.K.Simon

4 Lemma (excluding the awn) more than 6 mm long (rarely 6 mm long), as long as or shorter than the lower glume; awn 15–26 mm long. Panicle dense, up to 40 cm long, with stiff branches. Densely tufted herb up to 1.3 m high. Higher Blue Mts Woodlands. Fl. summer . . . .**D. hirtella** N.G.Walsh

4 Lemma 6 mm long or less 5

5 Lemma (excluding the awn) longer than the lower glume; awn 12–20 mm long. Panicle ± dense, up to 30 cm long. Tufted herb up to 1 m high. Widespread. Open forests on better soils. Fl. summer . . . . **D. inaequiglumis** (Hack. ex Cheeseman) Edgar & Connor

5 Lemma (excluding the awn) ± as long as the lower glume. 6

6 Culms hairy or scabrous. Lemma usually scabrous, c. as long as the lower glume; awn 12–16 mm long. Panicle ± dense, up to 20 cm long. Tufted herb up to 1 m high. Widespread. Open forests on better soils. Fl. summer . . . . **D. sieberiana** Trin. & Rupr.

6 Culms glabrous. Lemma usually smooth, shorter than or as long as the lower glume; awn up to 17 cm long. Panicle dense or loose, up to 30 cm long; branches sometimes flexuous. Tufted perennial up to 1.2 m high. Widespread. Open forests on better soils. Fl. summer. . . . . **D. rara** (R.Br.) Vickery

34 Lagurus L.

1 species naturalized Aust.; NSW, Vic., S.A., WA

One species in the area

Spikelets with 1 floret, disarticulating above the glumes. Glumes densely villous or plumose, 1 cm long. Lemma with 3 awns; the terminal ones 10–12 mm long; the dorsal one 15–20 mm long, with a bend below the centre. Rhachilla prolonged into a bristle almost as long as the shorter awns. Inflorescence a dense ovoid spike. Leaf blades villous or pubescent; ligule hairy. Annual. Sea coasts on sand dunes. Introd. from the Mediterranean. Fl. summer. *Harestail Grass* . . . . ***L. ovatus** L.

35 Polypogon Desf.

5 species naturalized Aust.; Qld, NSW, Vic., Tas., S.A., WA

Spikelets with 1 floret, compressed, falling whole, in a dense spike-like panicle 3–10 cm long. Glumes equal 2–3 mm long, notched, with a fine awn arising in the notch. Lemmas smaller than the glumes, hyaline, often with a short awn. Ligule subulate, membranous. Loosely tufted annuals or perennials.

Key to the species

1 Awns on the glumes up to 3 mm long. Lemmas sometimes awnless. Panicles loose, 2–10 cm long, with the branches not hidden by the spikelets. Leaf blades flat, 5–20 cm long, up to 10 mm wide; ligule 3–5 mm long. Tufted perennial up to 50 cm high. Coast. Damp disturbed areas. Introd. from Europe. Fl. spring. *Perennial Beardgrass* . . . .***P. littoralis** Sm.

1 Awns on the glumes 4–9 mm long. Lemmas with an awn 1–3 mm long. Panicles dense, 2–11 cm long, branches hidden by the spikelets. Leaf blades flat, 5–10 cm long, up to 6 mm wide; ligule 5–10 mm long. Tufted annual up to 50 cm high. Widespread. Damp disturbed areas; edges of swamps. Introd. from the Mediterranean. Fl. spring. *Annual Beardgrass* . . . . ***P. monspeliensis** (L.) Desf.

36 Phleum L.
One species in the area

Spikelets with 1 floret, disarticulating above the glumes, laterally compressed, 4–5 mm long. Glumes truncate, with awns 1 mm long, ciliate along the keels. Lemma much shorter than the glumes, awnless. Inflorescence a dense spike-like cylindrical panicle 7–12 cm long, up to 1 cm diam. Erect, tufted perennial up to 1.5 m high. Widespread on better soils. Pastures; roadsides; weed. Introd. from Europe and Asia. Fl. spring. *Timothy*. . . . .***P. pratense** L.

37 Gastridium P.Beauv.

1 species naturalized Aust.; NSW, Vic., S.A., WA

One species in the area

Glabrous, tufted annual herb to 50 cm high. Leaf blade to 4 mm wide. Inflorescence compact. Spikelets c. 5 mm long, with 1 bisexual floret. Glumes swollen near base. Lemmas toothed, silky, awned. 3 bundles of hairs on callus. Disturbed areas. Introd. from Europe. Fl. summer . . . . ***G. phleoides** (Nees & Meyen) C.E.Hubb.

38 Ammophila Host

1 species naturalized Aust.; all states and territories except NT

One species in the area

Spikelets with 1 floret, pedicellate, disarticulating above the glumes. Glumes subequal, 10–12 mm long, awnless. Lemma equal to the glumes, awnless, toothed near the apex. Inflorescence a dense spike-like panicle, 10–25 cm long. Stoloniferous perennial up to c. 1 m high. Leaf blades inrolled, with a sharp point; ligule hyaline, 20–30 mm long. Planted as a sand-binder on coastal dunes. Introd. from the Mediterranean. *Marram Grass* . . . . ***A. arenaria** (L.) Link

39 Arundinella Raddi

4 species native Aust.; Qld, NSW, WA, NT

One species in the area

Spikelets disarticulating above the glumes; with 1 terminal bisexual floret and a male one below it. Glumes subequal, 3–4 mm long, 5-nerved. Fertile lemma 3–4 mm long, with a fine awn 3–4 mm long which is twisted in the lower part and bent back at or below the middle. Inflorescence an open panicle 20–30 cm long. Erect perennial rarely more than 1 m high, with short rhizomes. Leaf blades 5–7 mm wide, 20–30 cm long; ligule membranous, less than 1 mm long. Cumberland Plain; Hornsby Plateau. In drainage lines or on creek banks. Fl. spring–summer. *Reedgrass* . . . . **A. nepalensis** Trin.

40 Cortaderia Stapf

3 species naturalized Aust.; all states and territories except NT

Large coarse dioecious or gynodioecious tussocky perennials. Leaves mostly towards the base of the culms. Inflorescence a large silvery silky panicle sometimes pinkish or purplish. Spikelets usually unisexual, disarticulating below the glumes; males with a usually sterile ovary. Glumes white or pinkish, papery. Lemmas with a slender terminal awn, glabrous in the male, silky hairy in the female.

Key to the species

1 Inflorescence white or purplish, slightly exceeding the basal leaves. Leaf blades dull green, ± folded at the base, arching away from the stem and thus drooping well above the sheath. Large, tussocky plant up to 4 m high. Naturalized in wetter disturbed situations, particularly in urban bushland. Fl. summer–autumn. *Pampas Grass* . . . . ***C. selloana** (Schult. & Schult.f.) Asch. & Graebn.
1 Inflorescence always pinkish when young, fading to white, usually well exceeding the basal leaves. Leaf blades bright green, flat, drooping from immediately above the sheath. Only female plants found. Large, tussocky plant up to 4.5 m high. Naturalized in wetter disturbed situations, particularly in urban bushland. Fl. summer–autumn. *Pink Pampas Grass* . . . . ***C. jubata** (Lem.) Stapf

41 Phragmites Adans.

2 species native Aust.; all states and territories

One species in the area

Spikelets with several florets; the lowest floret male or rudimentary; rhachilla clothed in long silky hairs. Glumes thin, unequal. Lemma narrow, subulate, c. 1 cm long. Inflorescence a dense much branched panicle, up to 30 cm long, drooping, often with a purplish-brown tinge. Perennial up to 3 m high, with rhizomes. Leaf blades 1–3 cm wide. Widespread but especially near the coast. Amphibious in brackish and fresh water. Fl. summer. *Common Reed* . . . . **P. australis** (Cav.) Trin. ex Steud.

42 Arundo L.

1 species naturalized Aust.; Qld, NSW, Vic., S.A., WA

One species in the area

Spikelets pedicellate, c. 12 mm long, disarticulating above the glumes and between the lemmas. Glumes equal. Lemmas hairy on the back, usually only the lowest containing a bisexual flower. Panicles up to 60 cm long. Perennials with thick horizontal rhizomes and crowded almost woody culms 2–6 m high. Leaf blades up to 7 cm long. Along River banks, e.g., Macdonald River, or in damp places. Introd. from Asia. *Giant Reed.* . . . . ***A. donax** L.

43 Anisopogon R.Br.

1 species endemic Aust.; Qld, NSW, Vic.

One species in the area

Spikelets with 1 floret, pendulous on pedicels 1–2 cm long, disarticulating above the glumes; rhachilla prolonged as a bristle 2–4 mm long. Glumes 3–5 cm long. Lemma stalked, 10–18 mm long, cylindrical, densely hairy, with 3 rigid awns; the central awn 5–8 cm long, twisted, bent; the outer awns straight or twisted, 15–25 mm long. Panicle open. Leaves convolute, with subulate tips; ligules truncate, ciliate, less than 1 mm long. Erect glabrous perennial up to 1 m high. Widespread. DSF and heath. Ss. Fl. spring. *Oat Spear Grass* . . . . **A. avenaceus** R.Br.

44 Amphipogon R.Br.

7 species endemic Aust.; all states and territories except Tas.

One species in the area

Spikelets with 1 floret, subsessile, disarticulating above the glumes. Glumes subequal, 4–5 mm long, awnless. Lemmas smaller than the glumes, but with 3 rigid awns exceeding the spikelet (Fig. 50k). Inflorescence a cylindrical spike-like panicle, 2–5 cm long, 1 cm diam. Tufted perennial up to 50 cm high, with short rhizomes. Leaves glabrous, filiform-cylindrical; ligule of hairs. Widespread. Usually on ridges among rocks. Fl. spring–summer, particularly after fire. *Grey-beard Grass*. **A. strictus** R.Br. var. **strictus**

45 Joycea H.P.Linder

3 species endemic Aust.; NSW, Vic., S.A.

One species in the area

Spikelets with 2–6 florets, 8–20 mm long, flattened, disarticulating above the glumes. Glumes equal, longer than the lemmas. Lemmas 2-lobed with the lobes shortly awned, usually with scattered hairs on the back, with a basal hairy callous; central awn between the lobes, bent and twisted near the base, longer than the lateral awns. Densely tufted perennials up to 1.2 m high, sometimes purplish. Panicle loose, spreading, lanceolate in outline, 8–35 (usually 10–20) cm long. Chiefly Blue Mts Poor, usually sandy soils. Fl. summer. *Red-anther Wallaby Grass* . . . . **J. pallida** (R.Br.) H.P.Linder

46 Plinthanthesis Steud.

3 species endemic Aust.; NSW, Vic.

Spikelets with 3–5 florets, disarticulating above the glumes; the uppermost floret sometimes male or sterile. Glumes equal, longer than the lemmas. Lemmas 2-fid with acute or subulate lobes; awn from between the lobes, bent, with a minute callous, short or absent. Leaf blades inrolled, up to 40 cm long. Panicle open.

Key to the species

1 Central awn c. 1 mm long, flattened, reflexed; lobes of the lemma less than half as long as the body of the lemma. Spikelets mostly 5–9 mm long, with 3–5 florets, greenish purple or straw-coloured. Erect, glabrous, tufted perennial up to 70 cm high. Panicle 7–20 cm long, lanceolate or pyramidal in outline. Coast. Moist situations on sandy or peaty soils. Fl. summer . . . . **P. paradoxa** (R.Br.) S.T.Blake

1 Central awn more than 2 mm long, reflexed or twisted at the base; lobes of the lemma half as long as the body of the lemma. Spikelets 6–10 mm long, mostly with 3 florets. Erect, tufted perennial up to 60 cm high. Panicle up to 20 cm long, ovate to broad-pyramidal. Blue Mts; Woronora Plateau. Fl. summer. . .
. . . .**P. urvillei** Steud.

47 **Austrodanthonia** H.P.Linder

Wallaby Grasses
26 species native Aust.; all states and territories except NT

Spikelets with 2–several florets, pedicellate or rarely almost sessile, in a loose panicle or raceme. Spikelets disarticulating above the glumes. Glumes keeled, awnless, usually as long as the spikelet. Lemma usually with a thickening (callous) below the terminal sinus, variously hairy or rarely glabrous, convex with 2 terminal lobes and a usually twisted and bent awn between them. Palea broad, conspicuous.

Key to the species

1 Body of the lemma with abundant hairs scattered over the back. 2

1 Either; lemma with hairs in distinct tufts or transverse rows of tufts and otherwise smooth; or the tufts of hairs reduced or absent 8

2 Lateral lobes of the lemma usually 2–3 times as long as the body (2–3 mm) of the lemma. Tufted perennial up to 45 cm high. Panicle lanceolate to ovate in outline, contracted or reduced almost to a raceme, with 5–15 spikelets. Spikelets with 4–6 florets; the florets much shorter than the glumes. Glumes 8–11 mm long. Chiefly Blue Mts; also Hornsby–Parramatta. Usually on poor soils. Fl. spring–summer . . . . **A. monticola** (Vickery) H.P.Linder

2 Either; the lateral lobes of the lemma not twice as long as the body of the lemma; or the body 3 mm long or longer. 3

3 Lateral lobes of the lemma shorter than the body of the lemma. Spikelets 7–10 mm long, with 3–5 florets. Lemma 4–6 mm long; lateral lobes minutely awned or awnless; central awn reflexed. Tufted perennial up to 50 cm high. Leaf blades fine, usually inrolled. Panicle ovate, up to 4 cm long, compact. Cumberland Plain. Grassland and open forest on clay soils. Fl. spring–summer . . . .
. . . . **A. carphoides** (Benth.) H.P.Linder

3 Lateral lobes of the lemma as long as the body of the lemma or longer 4

4 Hairs scattered on the backs of the lemma, uniformly short, of 2 different kinds, mostly c. 0.5 mm long. Lateral lobes of the lemma with capillary bristles 4–6 mm long, usually equalling or exceeding the glumes. Tufted perennial 15–60 cm high. Panicle 3–9 cm long. Spikelets with 4–10 florets. Glumes 8–14 mm long. Cumberland Plain. Clay soils. Fl. spring–summer. *Small-flowered Wallaby Grass.* . . . .
. . . . **A. setacea** (R.Br.) H.P.Linder

4 Shorter dorsal hairs of the lemma usually more than 1 mm long and grading upwards into the uppermost band of longer hairs. Body of the lemma 3 mm long or longer 5

5 Central awn of the lemma reflexed or loosely twisted once at the base 6

5 Central awn strongly twisted at the base 7

6 Palea lanceolate to elliptic, acute, distinctly exceeding the sinus of the lemma. Lateral lobes of the lemma often twice as long as the body. Central awn of the lemma scarcely twisted. Spikelets with c. 6 florets. Glumes subequal 8–13 mm long. Densely tufted perennial up to 70 cm high. Cumberland Plain. Grasslands and open forests. Fl. summer–autumn . . . .**A. bipartita** (Link) H.P.Linder

6 Palea broad-ovate, obtuse, usually only shortly exceeding the sinus of the lemma. Lateral lobes of the lemma less than twice as long as the body. Central awn slightly twisted and reflexed. Glumes subequal, 10–15 mm long. Densely tufted perennial up to 1 m high. Cumberland Plain (but probably now extinct there); Awaba. Grasslands and open forests. Fl. spring–summer . . . . **A. richardsonii** (Cashmore) H.P.Linder

7 Central awn equal to the lateral lobes or exceeding them by up to 5 mm. Glumes 12–17 mm long. Panicle often elongated and loose. Tufted perennial up to 1 m high. Cumberland Plain. Grasslands and open forests. Fl. spring–summer . . . . **A. fulva** (Vickery) H.P.Linder
7 Central awn exceeding the lateral lobes by 6–9 mm. Tufted perennial 75–120 cm high. Panicle linear-lanceolate, 9–18 cm long. Spikelets many, with c. 3 florets. Glumes 14–18 mm long, distinctly longer than the florets. Blue Mts Open forests on sandstone hillsides. Fl. spring–summer . . . . **A. induta** (Vickery) H.P.Linder

8 Lemma with a complete transverse row of hair-tufts just below the sinus 9
8 Lemma with dorsal or marginal tufts of hairs below the sinus, not or rarely forming a ± interrupted row across the back, or the tufts of hairs wholly or partly suppressed 15

9 Lemma without a complete transverse row of hairs above the thickening but with one below the sinus, shiny and glabrous below the hair tufts. Spikelets with 4–6 florets. Glumes 11–18 mm long. Panicles c. 4 cm long. Leaf blades filiform, inrolled. Tufted perennial up to 60 cm high. Colo Vale. Open forest. Fl. spring–summer. . . . .**A. laevis** (Vickery) H.P.Linder
9 Lemma with a complete transverse row of hair tufts across the back just above the thickening 10

10 Lateral lobes of the lemma with membranous lateral margins ending abruptly in small triangular points. Spikelets with 4–6 flowers. Glumes 9–16 mm long. Panicles c. 4 cm long. Leaf blades filiform, inrolled. Tufted perennial up to 50 cm high. Colo Vale. Open forest. Fl. spring–summer. . . . .**A. auriculata** (J.M.Black) H.P.Linder
10 Lateral lobes of the lemma with membranous lateral margins narrowing gradually into a bristle-like awn 11

11 Body of the lemma 1.3–3 mm long, rarely slightly longer in some forms of *A. setacea*. 12
11 Body of the lemma 3 mm long or longer 13

12 Lateral lobes of the lemma with bristles 2–4 mm long. Central awn only shortly exceeding the lateral lobes, not or scarcely exceeding the glumes . . . . **A. monticola** (see above)
12 Lateral lobes of the lemma with capillary bristles 4–6 mm long. Central awn exceeding the bristles by 3–6 mm, exceeding the glumes . . . .**A. setacea** (see above)

13 Lateral lobes of the lemma with bristles c. 2 mm (rarely 4mm) long, shorter than the flat portion. Tufted perennials 30–90 cm high. Panicles lanceolate to lanceolate-ovate, 5–20 cm long. Glumes 10–17 mm long. Widespread. Usually on sandy soils. Fl. spring–summer . . .**A. tenuior** (Steud.) H.P.Linder
13 Lateral lobes of the lemma with bristles c. as long as the flat portion. 14

14 Palea broad-obovate to oblanceolate, obtuse or 2-fid, only slightly exceeding the sinus of the lemma. Spikelets with 4–8 florets. Glumes 11–16 mm long. Leaf blades filiform, c. 1 mm wide, inrolled, usually hirsute. Densely tufted perennial up to 70 cm high. Woronora Plateau. Heath and DSF. Fl. spring–summer. . . . . **A. eriantha** (Lindl.) H.P.Linder
14 Palea lanceolate, acute or truncate or 2-fid, usually much exceeding the sinus of the lemma. Spikelets with 4–9 florets. Glumes 14–25 mm long. Leaf blades up to 2 mm wide. Tufted perennial up to 1 m high. Widespread. Open forests. Fl. chiefly summer . . . . **A. caespitosa** (Gaudich.) H.P.Linder

15 Panicle 4–5 cm long, with crowded spikelets. Glumes straight, slightly longer than the group of florets. Tufted perennials up to 50 cm high. Spikelets with 6–10 florets, usually crowded and overlapping. Glumes 9–13 mm long. Widespread. Various habitats. Fl. summer. *Smooth-flowered Wallaby Grass* . . . . **A. pilosa** (R.Br.) H.P.Linder
15 Panicle more than 5 cm long. Spikelets usually not crowded; the group of florets usually exceeding the glumes. 16

16 Lateral lobes of the lemma narrowing abruptly into bristles. Thickening on the lemma narrow, distinctly differentiated from the rest of the lemma. Spikelets with 6–10 florets. Glumes 7–16 mm long. Panicle up to 15 cm long. Tufted perennials 20–60 cm high. Widespread. Open forests. Chiefly shales. Fl. summer. . . . . **A. racemosa** (R.Br.) H.P.Linder var. **racemosa**

16 Lateral lobes of the lemma tapering gradually into the bristles. Thickening of the lemma less distinctly differentiated from the rest of the lemma. Spikelets with 4–7 florets. Glumes 8–13 mm long. Panicle up to 7 cm long. Tufted perennial up to 80 cm high. Blue Mts, Wentworth Falls. Forests. Fl. spring–summer . . . .**A. penicillata** (Labill.) H.P.Linder

48 Notodanthonia Zotov
2 species native Aust.; all states and territories except NT

Spikelets with 2–several florets, pedicellate or rarely almost sessile, in a loose panicle or raceme. Spikelets disarticulating above the glumes. Glumes keeled, awnless, usually as long as the spikelet. Lemma usually with a thickening (callous) below the terminal sinus, short hairs evenly covering back and longer hairs in a row of tufts below sinus convex with 2 terminal lobes and a usually twisted and bent awn between them. Palea broad, conspicuous.

Key to the species

1 Culms c. 3-noded. Leaf blades not flexuose, involute, firm; ligule of hairs c. 0.5 mm long. Tufted perennials 20–45 cm high. Panicle 5–8 cm long. Glumes 7–11 mm long. Cumberland Plain. Usually on clay soils. Fl. spring–summer. *Tasmanian Wallaby Grass*. . . . . **N. semiannularis** (Labill.) Zotov

1 Culms usually 5–many-noded. Leaf blades long, ± flexuose, becoming inrolled-filiform; ligule of hairs up to 0.2 mm long. Tufted perennials up to 75 cm high. Panicle 5–15 cm long. Spikelets with 5–6 florets. Glumes 8–13 mm long. Coast and adjacent plateaus. Open forest on sandy soils. Fl. spring–summer . . . . **N. longifolia** (R.Br.) Veldkamp

49 Eriachne R.Br.
40 species native Aust.; all states and territories except Tas.

One species in the area
Tufted perennial up to 60 cm high. Leaf blades inrolled, up to 3 mm wide; ligule a row of hairs. Spikelets with 2 florets, awnless, gaping, in a panicle 3–5 cm long and c. 1 cm wide; rhachilla disarticulating above the glumes. Coast near Sydney; Royal National Park. Damp sandy soils. Fl. summer. . . . .**E. glabrata** (Maiden) W.Hartley

50 Notochloe Domin
1 species endemic Aust.; NSW

Monotypic genus
Spikelets with 8–14 bisexual florets, disarticulating above the glumes, 2 cm long, 3–5 mm wide. Glumes awnless, glabrous. Lemmas like the glumes but slightly larger and with 3 minute terminal awns, the longest c. 1 mm long. Inflorescence a short panicle 6–10 cm long, usually with 5–6 spikelets. Tufted perennial 30–50 cm high. Leaf blades flat or inrolled, glabrous, c. 30 cm long, 2–3 mm wide; ligule of minute papillae. Blue Mts Swamps. . . . . **N. microdon** (Benth.) Domin

51 Leersia Sw.
2 species in Aust. (1 native, 1 naturalized): Qld, NSW, Vic., NT

One species in the area
Spikelets with 1 floret, flattened, sessile or pedicellate, falling whole. Glumes absent. Lemma slightly hardening at the fruiting stage, 3–4 mm long. Palea similar to the lemma but smaller. Stamens 6. Panicle 5–10 cm long; the branches filiform, zig-zag. Perennial usually with trailing stems sometimes more than 1 m long and often rooting at the nodes. Leaf blades 3–6 mm wide, 8–10 cm long; ligule of short hairs. Coast and Cumberland Plain. In or near water. Fl. all the year. *Swamp Ricegrass* . . . . . **L. hexandra** Sw.

52 Hierochloe R.Br.

4 species native Aust.; NSW, Qld., Vic., Tas.

One species in the area

Spikelets with 1 terminal bisexual floret and 2 males below it, disarticulating above the glumes, on fine pedicels. Glumes 5–6 mm long, subequal, enclosing the florets, awnless. Sterile lemmas 4–5 mm long, finely awned from the back close to the tip; fertile lemma 4–5 mm long, awnless. Inflorescence a loose panicle. Perennial up to c. 1 m high. Leaf blades 3–8 mm wide; ligule membranous. Moist places at higher altitudes. Fl. winter–spring. *Scented Holy Grass* . . . .**H. rariflora** Hook.f.

53 Ehrharta Thunb.

6 species naturalized Aust.; all states and territories except NT

Spikelet with 1 terminal bisexual floret and 2 sterile lemmas below it, pedicellate, in a terminal panicle, disarticulating above the glumes. Glumes awnless, closely appressed to 2 larger, awned, sterile lemmas. Fertile lemmas awnless. Palea absent. Lodicules large, thin. Stamens 6.

Key to the species

1 Sterile lemmas with long or short silky hairs 2
1 Sterile lemmas glabrous or scabrous 3

2 Spikelets 12–15 mm long. Perennial up to 1 m high. Leaf blades 4–7 cm long. Panicle 6–9 cm long. Coast. Pastures; weed. Introd. from S. Africa. Fl. spring–summer. *Pypgrass* . . . .
. . . .***E. villosa** (L.f) Schult.f. ex Schult. & Schult.f.
2 Spikelets less than 6 mm long. Perennial up to 60 cm high. Leaf blades 2–9 cm long. Panicle 7–22 cm long. Widespread. Pastures; weed. Introd. from S. Africa. Fl. spring–summer. *Perennial Veldtgrass* . . .
. . . .***E. calycina** Sm.

3 Sterile lemmas with stiff subulate awn-like points. Annual up to 60 cm high. Leaf blades 8–15 cm long. Panicle up to 15 cm long. Spikelets 7 mm long. Coast and Cumberland Plain. Pastures; weed. Introd. from S. Africa. Fl. spring–summer. *Annual Veldtgrass* . . . .***E. longiflora** Sm.
3 Sterile lemmas obtuse, glabrous. Perennial up to 45 cm high. Leaf blades 5–15 cm long. Panicle 5–20 cm long. Coast and Cumberland Plain. Pastures; weed. Introd. from S. Africa. Fl. spring–summer. *Panic Veldtgrass* . . . .***E. erecta** Lam.

54 Microlaena R.Br.

2 species native Aust.; all states and territories except NT

One species in the area

Spikelets with 1 terminal bisexual floret and 2 sterile lemmas below it, up to 3 cm long, disarticulating above the glumes, on filiform pedicels, in a loose panicle up to 20 cm long (Fig. 49 e). Glumes less than 1 mm long, persistent, separated from the sterile lemmas and subtending tufts of hairs. Sterile lemmas rigid, with subulate terminal awns; fertile lemmas less than 1 cm long. Palea shorter than the fertile lemmas. Stamens 4. Erect, tufted perennials up to 1 m high. Leaf blades flat, glabrous or slightly hairy; ligule membranous. *Ricegrass.*

Key to the varieties

1 Awns of the lower sterile lemmas as long as, or up to twice as long as, the lemmas. Spikelets green. Widespread. Forests in shady places. Fl. spring–summer . . . **M. stipoides** (Labill.) R.Br. var. **stipoides**
1 Awns of the lower sterile lemmas much shorter than the lemmas. Spikelets usually purplish. Ranges. Forests in shady places. Fl. spring–summer. . . . .**M. stipoides** var. **breviseta** Vickery

55 Tetrarrhena R.Br.
6 species endemic Aust.; all states and territories except NT

Spikelets with 1 terminal bisexual floret and 2 sterile lemmas below it, sessile or almost so, in a simple spike or spike-like panicle. Glumes unequal. Sterile lemmas subequal, awnless, keeled. Fertile lemmas 2–3 mm long, membranous. Stamens 4. Leaf blades flat or inrolled; ligule a ciliate rim.

Key to the species

1 Leaf blades flat, sometimes scabrous, 1–7 mm wide. Inflorescence mostly 7–9 cm long. Tufted, scrambling perennial with much-branched wiry stems up to 4 m long. Widespread. Damp places in heath. Ss. Fl. spring–summer. *Wiry Ricegrass* . . . . **T. juncea** R.Br.
1 Leaf blades tightly involute around the culm, smooth, up to 1 mm wide. Inflorescence mostly less than 3 cm long. Densely tufted rhizomatous perennial up to 1.3 m high. Widespread. Swamps and along creek banks. Fl. spring–summer . . . . **T. turfosa** N.G.Walsh

56 Austrostipa S.W.L.Jacobs & J.Everett
Spear Grasses
60 species in Aust. (58 native, 2 naturalized); all states and territories

Spikelets with 1 floret disarticulating above the glumes, on filiform pedicels or nearly sessile in a terminal panicle. Glumes awnless. Lemmas narrow, rolled around the palea, with a terminal undivided often bent awn twisted below the bend. Palea thin. Lodicules usually large. Fruit enclosed in the lemma.

Key to the species

1 Lemma variously hairy. Culms not branched at the nodes 2
1 Lemma glabrous or with a few scattered white hairs. Inflorescence branches whorled. Culms branching freely at the nodes 13

2 Awn hairy with ± spreading white hairs up to 2 mm long, especially towards the base 3
2 Awn glabrous or pubescent with appressed hairs less than 1 mm long 5

3 Hairs on the awn more than 1 mm long. Lemma 6–8 mm long; awn 7–9 cm long, with 2 bends; the column 1 cm long, twisted. Panicle dense, 20–25 cm long. Tufted perennials up to 80 cm high. Leaf blades inrolled, 20–30 cm long. Coast. Sandy soils and rocky areas. Fl. summer. *Soft Spear Grass* . . . .
. . . . **A. mollis** (R.Br.) S.W.L.Jacobs & J.Everett
3 Hairs on the awn less than 1 mm long. 4

4 Glumes glabrous except on the nerves. Lemma 9–12 mm long; awn with 2 bends, 7–11 mm long; the column of the awn 3–4 cm long. Panicle up to 30 cm long. Tufted perennials up to 1.5 m high. Leaf blade 1–7 mm wide, flat to inrolled. Berrima district. DSF and heath. Poor, often sandy soils. Fl. spring–summer . . . . **A. semibarbata** (R.Br.) S.W.L.Jacobs & J.Everett
4 Glumes hairy. Lemma 5–7 mm long; awn with 2 bends, 3–5 cm long; the column of the awn 13–20 mm long. Tufted perennial up to c. 1.5 m high. Leaf blades up to 5 mm wide, ± rolled. The Oaks. Open forests on poor soils. Fl. spring–summer . . . . **A. densiflora** (Hughes) S.W.L.Jacobs & J.Everett

5 Both glumes acutely acuminate 6
5 Lower glume (or both) truncate, sometimes with 1–3 fine points irregularly placed 12

6 Awn with 2 bends 7
6 Awn falcate; the terminal bristle curved 9

7 Palea with a deep central groove. Ligule 2–8 mm long. Awn 2–4 cm long; the column of the awn c. 10 mm long, tightly twisted; the terminal bristle straight. Lemma 5 mm long, 1 mm diam. at the widest point near the summit, dark coloured, with white hairs. Glumes unequal, lower 10–12 mm, upper 7–8 mm long, hyaline, subulate. Panicle narrow, 10–20 cm long. Leaf blades terete, 16–18 cm long. Plants up to 80 cm high. Not common. Clay soils. Fl. most of the year . . . . **A. setacea** (R.Br.) S.W.L.Jacobs & J.Everett

7 Palea convex or with slight depressions between nerves. Ligule less than 3 mm long mostly less than 2 mm 8

8 Ligule papery, usually 0.5 mm long or less, auricle glabrous. Callus 0.4–1.5 mm long. Culms ribbed. Leaves 3–6 mm wide. Glumes green, unequal 9–14 mm long. Panicle 25–45 cm long. Plants to 2 m high. Clay soils. West of Blue Mts Fl. spring–summer . . .**A. aristiglumis** (F.Muell.) S.W.L.Jacobs & J.Everett

8 Ligule leathery, 0.8 mm long or more; auricle usually with tufts of 1 mm long hairs. Callus 1–2.5 mm long . Culms smooth or slightly ribbed. Leaves c. 3 mm wide. Glumes unequal, purplish 12–18 mm long. Panicle 40–50 cm long. Plants to 1 m high. Woodland. Ranges. Fl. summer–summer . . . . **A. bigeniculata** (Hughes) S.W.L.Jacobs & J.Everett

9 Leaves linear, more than 1 mm wide, usually ± spreading 10

9 Leaves filiform, mostly up to 1 mm wide, usually erect. Lemma narrow-lanceolate in outline, c. 5 mm long, less than 1 mm diam., pubescent; awn with 2 bends, 4–7 cm long; the column of the awn 10–15 mm long, twisted. Panicle 25 cm long. Tufted perennial up to 60 cm high. Leaf blades terete. *Rough Spear Grass* 11

10 Panicle dense, base partly enclosed by upper leaf sheath. Glumes green and shiny. New shoots emerging from within the axis of the leaf sheath. Awn 4.5–7 cm long. Lemma 4–6 mm long, ± scabrous. Tufted perennial to 1 m high. Leaf blades flat or inrolled, to 40 cm long. Sandy soils. Fl. in response to rain . . . . **A. nitida** (Summerh. & C.E.Hubb.) S.W.L.Jacobs & J.Everett

10 Panicle spreading, its base exserted from the upper leaf sheath. Glumes purplish. New shoots breaking through the leaf sheath. Awn 5–10 cm long. Lemma 4–7 mm long, 1–3 mm diam., ± scabrous. Tufted perennial up to 1.5 m high. Leaf blades flat or inrolled, to 30 cm long. Hornsby . . . . **A. nodosa** (S.T.Blake) S.W.L.Jacobs & J.Everett

11 Panicle narrow. Upper leaf auricular lobe to 4 mm long, usually glabrous. Leaves often more than 15 cm long. Widespread. Grasslands, usually on clay soils. Fl. in response to rain . . . . **A. scabra** (Lindl.) S.W.L.Jacobs & J.Everett ssp. **scabra**

11 Panicle spreading. Upper leaf auricular lobe less than 1 mm long, usually with a row of hairs. Leaves often less than 15 cm long. Woodland. Ranges. Fl. spring–summer . . . .**A. scabra** ssp. **falcata** (Hughes) S.W.L.Jacobs & J.Everett

12 Glumes 15–25 mm long, subequal. Lemma 10–12 mm long; awn 6–9 cm long, with 1 or 2 bends; the column of the awn 3–5 cm long. Panicle up to 30 cm long, spreading. Leaf blades 20–30 cm long, inrolled. Tufted perennial up to 1.5 m high. Widespread. Usually on sandy or stony soils. Fl. summer. *Tall Spear Grass* . . . . **A. pubescens** (R.Br.) S.W.L.Jacobs & J.Everett

12 Lower glume 9–15 mm long; the upper shorter. Lemma 6–9 mm long; awn 2–7 cm long with 1–2 bends. Lower glume 9–15 mm long. Panicle spreading, 20–40 cm long. Leaf blades up to 40 cm long, inrolled. Tufted perennial up to 1 m high. Widespread. Usually on sandy soils. Fl. most of the year . . . . **A. rudis** (Spreng.) S.W.L.Jacobs & J.Everett

13 Lemma glabrous, dark coloured, c. 2 mm long; awn c. 3 cm long, strongly bent once. Panicle mostly c. 50 cm long, with whorled branches. Tufted perennial 1–2 m high. Leaf blades up to 10 mm wide. Widespread. Wet places. Fl. spring–summer. *Stout Bamboo Grass* . . . .**A. ramosissima** (Trin.) S.W.L.Jacobs & J.Everett

13 Lemma with scattered fine white hairs, 2.5–4 mm long, pale coloured but darkening with maturity; awn 3–5 cm long, with 1 or 2 weak bends. Similar to *S. ramosissima*. Widespread. *Slender Bamboo Grass* . . . . **A. verticillata** (Nees ex Spreng.) S.W.L.Jacobs & J.Everett

57 **Nassella** Desv.
6 species naturalized Aust.; NSW, Vic., Tas.

Spikelets with 1 floret, disarticulating above the glumes. Glumes subequal, longer than the floret, acuminate. Lemmas with a terminal loosely twisted slightly bent deciduous awn and often with a cup-like corona at the base of the awn. Inflorescence a loose or contracted panicle. Tufted perennials usually with branching culms. Leaf blades flat or inrolled; ligule small.

Key to the species

1 Lemma c. 2 mm long, c. as long as broad, without a corona; awn eccentric, c. 3 cm long. Spikelets 5–8 mm long. Panicle up to 25 cm long, often purplish; the whole panicle detaching at maturity. Leaf blades inrolled, bristly, c. 0.5 mm wide. Densely tufted perennial up to 70 cm high. Widespread in drier pastures, a declared noxious weed. Introd. from S. America. Fl. spring. *Serrated Tussock* . . . . ***N. trichotoma** (Nees) Hack. ex Arech.

1 Lemma 6–10 mm long, much longer than broad, with a fringed corona around the base of the awn; awn bent 2–3 times, up to 9 mm long. Spikelets 14–18 mm long. Panicle up to 40 cm long; the branches breaking up at maturity. Leaf blades flat to slightly inrolled, 2–2.5 mm wide, scabrous. Tufted perennial up to 80 cm high. Widespread. Roadsides. Introd. from S. America. . . . . ***N. neesiana** (Trin. & Rupr.) Barkworth

58 Piptatherum P.Beauv.
1 species naturalized Aust.; NSW, Vic., Tas., S.A., WA

One species in the area

Spikelets with 1 floret, 3–4 mm long, disarticulating above the glumes. Glumes subequal, 3–4 mm long, membranous, subulate. Lemma coriaceous, shorter than the glumes, with a fine caducous awn. Inflorescence a very open panicle 30–40 cm long. Glabrous perennial c. 1 m high, with rigid culms. Leaf blades 5–7 mm wide, c. 30 cm long; ligule membranous, lobed. Roadsides; cultivated as an ornamental. Introd. from the Mediterranean. Fl. summer. *Rice Millet* . . . . ***P. miliacea** (L.) Coss.

59 Eragrostis Wolf
69 species in Aust. (14 naturalized); all states and territories

Spikelets with 3–many florets, mostly bisexual, usually laterally compressed. Rhachilla disarticulating above the glumes and between the florets. Glumes persistent, usually shorter than the lowest lemma. Lemmas 1–3-nerved, entire, emarginate or 2–4-lobed at the tip, awnless or with a straight awn from the tip or the sinus. Inflorescence usually an open panicle, sometimes digitate. Annuals or perennials. (The following key has been adapted from Lazarides [1997]).

Key to the species

1 Glumes subequal 2
1 Glumes unequal 9

2 Plants with glands on leaf nerves and/or margins 3
2 Plants without glands on leaves 5

3 Leaf blade 2.5 mm or less wide, becoming involute on drying. Rhachilla glabrous. Panicle up to 40 cm long and 30 cm wide; the main branches divided once or twice. Pedicels of non-terminal spikelets 2–11 (usually 4–8) mm long, stiff, spreading. Spikelets dark green or leaden, shining, mostly 7–13 mm long, 1.5–2 mm wide, flattened, with 6–22 (usually 12–14) florets. Lemmas with slightly spreading apices. Fruit elongated. Perennial with culms up to 80 cm high. Coast; Cumberland Plain; lower Blue Mts Fl. summer . . . . **E. alveiformis** Lazarides
3 Leaf blade more than 2.5 mm wide 4

4 Leaves with glands on margin and blade. Leaf blades 2–7 mm wide. Spikelets 5–15 mm long, 2–3 mm wide, with 10–40 florets (often fewer than 10 in stunted specimens), dark grey-green or purple. Panicle erect, 5–20 cm long, branches often glandular. Tufted annual up to 60 cm high. Widespread. Weed; roadsides; waste places. Fl. summer. Introd. from the Mediterranean. *Stink Grass* . . . . ***E. cilianensis** (All.) Vignolo ex Janch.
4 Leaves with pitted glands only on blade . . . . ***E. mexicana** (see below)

5 Leaf blade 5–10 mm (occasionally less) wide. Spikelets up to 10 mm long, 1.5–2 mm wide, mostly with 5–9 florets, green or purplish. Panicle 12–15 cm long, 6–10 cm wide. Tufted annual up to 1.2 m high. Widespread. Weed; waste places. Introd. from America. Fl. summer. *Mexican Lovegrass* . . . . ***E. mexicana** (Hornem.) Link

5 Leaf blade 3 mm or less wide **6**

6 Spikelets pedicellate 0.7–1 mm wide. Panicles open. Lemmas 1–1.5 mm long. Spikelets terete or compressed, 3–14 mm long. Panicle 4–16 cm long, 2–8 cm wide. Tufted perennial up to 50 cm high. Leaf blades up to 2 mm wide, usually inrolled. Cumberland Plain. Fl. spring–autumn. *Purple Lovegrass* . . . . **E. lacunaria** Lazarides

6 Spikelets subsessile or shortly pedicellate, 1.5 mm or more wide, compressed. Lemmas 1.5 mm or more long. Panicles open or spikelike **7**

7 Stamens 2 or 3, spikelets shortly pedicellate. Panicles open. Glumes persistant . **E. brownii** (see below)

7 Stamens always 2, Spikelets sessile to subsessile. Panicles spikelike. Glumes deciduous **8**

8 Palea smooth. Lemmas 2–3 mm long, inflated. Spikelets 6–12 mm long, 2.5 mm wide. Keels of the palea with a broad membranous connective between them at the apex. Similar to *E. brownii*. Cumberland Plain; lower Blue Mts Grasslands and open forests, usually on sandy soils. Fl. all year . . . . . **E. sororia** Domin

8 Palea granular. Lemmas 1.5–2 mm long not inflated. Spikelets clustered on the primary branches, 3–6 mm long, 2 mm wide, with 6–14 florets, pale coloured, sessile. Panicle spike-like, often interrupted, 5–12 cm long. Tufted perennial up to 80 cm high. Widespread. Usually on clay soils. Fl. summer. *Clustered Lovegrass* . . . . **E. elongata** (Willd.) J.Jacq.

9 Glandular ring below nodes on culms and usually on pedicels. Pedicels of non-terminal spikelets mostly as long as or longer than the spikelet, 3–7 mm long. Panicle up to 25 cm long and 16 cm wide; the main branches divided usually only once. Spikelets dark green or leaden, mostly 4–7 mm long, 1.5–2 mm wide, flattened, with 6–12 (mostly 8–10) florets. Lemmas with straight apices. Perennials with culms up to 60 cm high. Leaf blades c. 2 mm wide, becoming involute when dry. Widespread. Grasslands, open forests and woodlands. Fl. late summer–autumn. *Paddock Lovegrass*. . . **E. leptostachya** (R.Br.) Steud.

9 Culms and pedicels without glandular rings **10**

10 Rhachilla of spikelet distinctly zigzag **11**

10 Rhachilla straight or wavy **12**

11 Palea entire. Panicle open or contracted, tufts of hair not present in axils. Spikelets shortly pedicellate, 4–12 mm long, 1.5–3 mm wide, with 8–24 florets. Leaf blades flat to inrolled. Perennial with culms up to 60 cm high. Widespread. Woodlands and grasslands. Fl. summer (includes *E. benthamii* Mattei). *Brown's Lovegrass* . . . . **E. brownii** (Kunth) Nees ex Wight

11 Palea lobed. Panicle usually contracted becoming open with maturity, branches and longer pedicels with tufts of hairs in axils. Culms sometimes with glandular striations below nodes. Tufted annual to 60 cm high. Otherwise similar to *E. parvifolia*. Introd. from the Mediterranean. *Soft Lovegrass*. . . . . ***E. pilosa** (L.) P.Beauv.

12 Palea apex entire or divided into 2 parts **13**

12 Palea apex with 3 lobes or teeth **14**

13 Lemma 1.5 mm long. Spikelet rhachilla without hairs. Lower leaf sheaths glabrous at base. Tufted annual with culms up to 110 cm high. Spikelets 2–10 mm long, 1–1.5 mm wide , usually with 10–20 florets (often fewer), dark green or leaden. Panicle usually 20–60 cm long, 10–20 cm wide. Leaf blades 1–3 mm wide. Widespread. Heavy soils. Fl. summer. *Weeping Lovegrass* . . **E. parviflora** (R.Br.) Trin.

13 Lemma >2 mm long. Spikelet rhachilla shortly hairy. Lower leaf sheaths hairy at base. Tufted perennial up to 110 cm high. Spikelets, dark olive green, 8–10 mm long, c. 1.5 mm wide, usually with 8–10 florets. Panicle 20–30 cm long. Leaf blades 2–3 mm wide, inrolled. Widespread. Weed; roadsides; pastures. Fl. most of the year. Introd. from S. Africa. *African Lovegrass* . . . . ***E. curvula** (Schrad.) Nees

14 Palea 3-lobed. Seed strongly compress, smooth. Panicle 12–25 cm long, 4–12 cm wide, axils of the branches and spikelet pedicels hairy. Spikelets 7–12 mm long, 1.5–2 mm wide, flattened, with 6–14

florets; pedicels of the spikelets up to 12 mm long. Tufted perennial with culms up to 70 cm high. Leaf blades flat, up to 3 mm wide. Coast. Trampled sites in public recreation areas. Introd. from America and Africa. Fl. summer. *Elastic grass* . . . .*E. tenuifolia* (A.Rich) Hochst. ex Steud.

14 Palea 3-toothed. Seed globose, honeycomb. Panicle spreading, 20–30 cm long, 10–27 cm wide, axils of lower primary branches sparsely hairy. Spikelets, 2–4 mm long, c. 1 mm wide with 3–5 florets; pedicels 2–6 times longer than the spikelets. Tufted perennial with culms up to 45 cm high. Leaf blades flat to inrolled, 2–3 mm wide. Widespread. Open forests. Fl. summer . . . . **E. trachycarpa** (Benth.) Domin

60 Triodia R.Br.
45 species endemic Aust.; all states and territories

One species in the area

Perennial forming hummocks or rings. Spikelets with 4–7 florets, loosely oblong, when young the association with others gives a narrow-lanceolate or cylindrical outline, 10–15 mm long, 5–7 mm wide; pedicels 10–15 mm long. Glumes stiff, rarely more than half as long as the spikelet, 5.5–9 mm long. Lemmas broad-lanceolate, obtuse; lower ones 6–7 mm long; the apex ragged subtruncate or emarginate or with the nerve extended as a mucro. Inflorescence a narrow panicle, 15–30 cm long. Leaf blades rigid, with a pungent point; ligule of dense short cilia. Capertee Valley. *Porcupine Grass* . . . .
. . . . .T. **scariosa** NTBurb. ssp. **scariosa**

61 Diplachne P.Beauv.
4 species in Aust. (3 native, 1 naturalized); all states and territories except Tas.

Tufted annuals or perennials. Culms erect with prominent nodes. Ligule membranous, sometimes laciniate. Spikelets solitary, shortly pedicellate, dorsally compressed, with numerous bisexual florets, disarticulating above the glumes and between the lemmas, in spike-like racemes arranged along the upper stem. Glumes unequal, membranous, 1-nerved, shorter than the lemmas. Lemmas membranous, 2-fid, notched or truncate, 3-nerved, shortly awned below the apex or from the notch.

Key to the species

1 Lemma 3.5 mm or more long. Upper glume more than 3 mm long. Spikelets usually dark coloured, usually with more than 6 florets, in a dense branched panicle up to 50 cm long. Erect, tussocky herb up to 1.2 m high. Widespread. Usually in or near water. Fl. after flooding. *Brown Beetle Grass* . . . .
. . . . **D. fusca** (L.) P.Beauv.

1 Lemma up to 2.5 mm long. Upper glume less than 2 mm long. Spikelets usually with 6 florets or fewer, in a panicle up to 30 cm long. Tufted herb up to 70 cm high. Homebush Bay. Ditches and drains. Introd. from America. Fl. summer . . . .*D. **uninervia** (J.Presl) Parodi

62 Triraphis R.Br.
1 species in Aust.; Qld, NSW, S.A., WA, NT

One species in the area

Annual or perennial herb to 80 cm high. Nodes on culms constricted, purplish. Ligule ciliate with rigid hairs c. 3 mm long at each end. Leaf blade rolled and bristle-like. Inflorescence spike-like. Spikelets bisexual, solitary, with 3–9 florets, disarticulating above the glumes and between the lemmas, in dense spike-like panicles. Glumes membranous, unequal, keel and margin scabrous, upper glume c. 4 mm long. lower glume c. 5 mm long. Lemmas longer than glumes, keeled, 3-nerved, ciliate on nerves. Central nerve extending to an awn 5–7 mm long. Lateral awns 6–7 mm long. Callus with short hairs. Sandy soils. Fl. summer. *Purple Needlegrass.* . . . . **T. mollis** R.Br.

63 Leptochloa P.Beauv.
9 species in AUST, (7 native, 2 naturalized); all states and territories except Vic. and Tas.

Inflorescence a much branched panicle or subdigitate. Spikelets with 1 to several florets, disarticulating above the glumes, compressed. Glumes subequal , shorter than lemmas, awnless or mucronate, 1-nerved. Lemmas 3-nerved, lateral nerves usually hairy. Palea shorter than lemma, keels ciliolate.

Key to the species

1 Seed compressed. Spikelets with 2–6 florets, pedicellate to sessile, 2–5 mm long, awnless. Glumes subequal , 0.5–2 mm long. Lemmas 1.5–2 mm long. Inflorescence 7–25 cm long. Tufted or tussock forming perennial to 1 m high. Leaf blades up to 9 cm long. Moist sheltered sites. Fl. summer . . . . . .
. . . . **L. decipiens** (R.Br.) Stapf ex Maiden

1 Seed trigonous. Spikelets with 2–5 florets, pedicellate to sessile, 1.5–3 mm long, awnless. Glumes subequal, 0.7–1.5 mm long. Lemma c. 1.7 mm long. Inflorescence 10–30 cm long. Erect, tufted, glabrous perennial up to 1 m high. Leaf blades up to 20 cm long, inrolled when dry. Camden district. Heavy soils. Fl. summer. . . . .**L. ciliolata** (Jedwabn.) S.T.Blake

64 Eleusine Gaertn.
3 species naturalized Aust.; all states and territories except Tas.

Spikelets with several florets, disarticulating above the glumes, flat, imbricate in 2 rows along one side of the digitate or subdigitate branches of the rhachis. Glumes spreading, unequal, acute or mucronate. Lemmas less acute, shorter; the terminal one empty or rudimentary. Palea folded.

Key to the species

1 Digitate branches of the inflorescence 2–4, equal or unequal, 1–3 cm long. Spikelets densely crowded. Erect, glabrous perennial up to 20 cm high or forming a mat. Widespread. Weed; roadsides; usually around habitation. Introd. from S. Africa. Fl. summer. *Goose Grass* or *Crabgrass.* . . . .
. . . . *E. tristachya** (Lam.) Lam.

1 Digitate branches of the inflorescence 6–8, equal or nearly so, 5–8 cm long. Spikelets overlapping. Erect, glabrous perennial, occasionally forming mats, up to 50 cm high. Widespread. Weed; roadsides; usually around habitation. Cosmop. in tropics and subtropics. *Crowsfoot Grass* . . . . . *E. indica** (L.) Gaertn.

65 Dactyloctenium Willd.
4 species in Aust.; (1 native, 3 naturalized); all states and territories except Tas.

Spikelets with 2–7 florets, disarticulating above the glumes, compressed, sessile, crowded into 2 rows along the rhachis. Inflorescence digitate, with 2–10 branches. Glumes unequal. Rhachilla prolonged. Leaf blades scabrous, sometimes ciliate.

Key to the species

1 Perennial with wiry stolons, up to 50 cm high. Inflorescence with 2–5 branches, each 1–5 mm long. Spikelets 3–4 mm long. Coast and Cumberland Plain. Lawn grass naturalized near habitation. Introd. from S. Africa. *Durban Grass*. . . . . *D. australe** Steud.

1 Tufted annuals or short-lived perennials, not stoloniferous 2

2 Spikelets crowded and touching those of adjacent spikes, c. 5 mm long. Inflorescence with 3–10 branches, each 5–15 mm long. Tufted annual up to 20 cm high. Coast. Fl. spring–summer. *Button Grass* or *Finger Grass* . . . . **D. radulans** (R.Br.) P.Beauv.

2 Only the lowest spikelets touching those of the adjacent branches; branches of the inflorescence more spreading than those of *D. radulans*. Inflorescence with 2–4 branches, each 1–5 cm long. Tufted annual or short lived perennial up to 70 cm high. Sandy soils mostly near the coast where it has been used as a stabilizer. Introd. from Africa. *Coastal Button Grass* . . . . *D. aegyptium** (L.) P.Beauv.

66 Chloris Sw.
13 species in Aust. (7 native, 6 naturalized); all states and territories except Tas.

Spikelets with 2–3 florets, disarticulating above the glumes, sessile, in 2 rows on one side of simple spikes which are digitate at the ends of the peduncle; the lower floret bisexual. Glumes awnless. Lemmas entire, usually with a toothed lobe and with a single fine straight awn, with or without a short awn on either side

⚘ 635

of the terminal one. Rhachilla sometimes prolonged behind the palea and bearing 1 or more empty or staminate lemmas, all awned and usually with their ends on a level with that of the fertile lemma.

Key to the species

1 Lemmas truncate or rounded at the summit 2
1 Lemmas acute at the summit 3

2 Lemma truncate at the summit. Glabrous perennial up to 50 cm high. Leaf blades folded or flat, 2–4 mm wide. Spikes 6–9, each 5–14 cm long. Glumes 2–3 mm long, hyaline. Lemmas c. 3 mm long, dark brown to black; awns 8–15 mm long. Widespread. Usually in grasslands on heavy soils. Fl. summer. *Windmill Grass* . . . . **C. truncata** R.Br.
2 Lemma rounded at the summit. Glabrous tufted perennial up to 1 m high. Leaf blades flat, 2–3 mm wide. Spikes 3–6, each 5–8 cm long. Glumes 2–3 mm long, hyaline. Lemmas 3–4 mm long, pale brown or purplish; awns c. 4 mm long. Grasslands or forests on clay soils. Widespread. Fl. summer. *Tall Chloris* . . . . **C. ventricosa** R.Br.

3 Spikelets with 2 terminal imperfect florets; at least one of them male. Lemma shortly bearded. Erect perennial up to c. 1 m high, with stolons. Leaf blades 3–5 mm wide. Spikes in 1–2 whorls, usually c. 10–15, each 5–10 cm long. Lemmas c. 3 mm long, pale coloured; awns 1–5 mm long. Widespread. Cultivated pasture grass; roadsides; weed. Introd. from S. Africa. Fl. summer. *Rhodes Grass* . . . . ***C. gayana** Kunth
3 Spikelets with 1 terminal sterile floret 4

4 Lemma with a long beard at the summit, c. 3 mm long; awns 5–10 mm long. Glabrous annual up to 60 cm high. Leaf blades flat, 2–5 mm wide. Spikes 5–8, each 2–8 cm long, feathery or silky. Grasslands; weed. Introd. from trop. America. Fl. summer. *Feathertop Rhodes Grass* . . . . ***C. virgata** Sw.
4 Lemmas beardless, scabrous, 2–3 mm long. Mostly glabrous annual up to 50 cm high. Leaf blades flat, up to 4 mm wide. Spikes 3–8, widely spreading, each 7–20 cm long. Grasslands. Fl. summer. *Slender Chloris* . . . . **C. divaricata** R.Br. var. **divaricata**

67 **Enteropogon** Nees
5 species native Aust.; all states and territories except Tas.

One species in the area
Spikelets solitary, with 1 bisexual floret and 1–2 sterile or male ones above it, disarticulating above the glumes. Glumes unequal, membranous. Lower lemma up to 9 mm long, with an awn 10–20 mm long. Paleas usually absent. Inflorescence with 7–22 digitate spikes, each 8–17 cm long. Leaf blades up to 4 mm wide. Culms usually branched. Tufted perennials up to 60 cm high. Grasslands. Cumberland Plain, possibly introd. from arid parts of the state. Fl. summer. *Windmill Grass* . . . . . . **E. acicularis** (Lindl.) Lazarides

68 **Cynodon** Rich.
3 species in Aust. (1 native, 2 naturalized); all states and territories

Spikelets with 1 floret, disarticulating above the glumes, almost sessile in 2 rows along one side of a flattened rhachis. Glumes narrow, acute, shorter than the florets. Lemmas awnless, folded, keeled. Inflorescence digitate. Perennials with rhizomes and/or stolons.

Key to the species

1 Ligule of hairs. Rhachilla produced beyond the lemma. Leaves distichous; blades 2–5 cm long. Inflorescence branches 2–5 cm long, turning reddish or purplish-brown after flowering. Widespread. Common lawn grass; weed; pastures. Possibly introd. Fl. spring–summer. *Common Couch* or *Bermuda Grass* . . . . **C. dactylon** (L.) Pers.
1 Ligule membranous, fringed with very short hairs. Rhachilla not produced beyond the lemma. Otherwise similar to *C. dactylon*. Uncommon. Introd. from S. Africa. *Blue Couch* . . . . ***C. incompletus** Nees

69 Brachyachne (Benth.) Stapf

5 species native Aust.; Qld, NSW, S.A., WA, NT

One species in the area

Annual herb to 50 cm high. Culms branched, geniculate near base. Ligule a ciliate membranous rim. Leaf blade 2–3 mm wide. Spikelets bisexual, 3–4 mm long, 1-flowered, disarticulating above the glumes, in 3–5 digitately arranged spikes. Glumes 1-nerved, 3–5 mm long, acute to mucronate. Lemmas c. 2 mm long, densely ciliate on keel and margins. Palea equal to lemma with 2, densely ciliate nerves which end in a point. Toxic to stock. Occasionally flooded areas. Fl. in response to rain. *Native Couch*. . . . .
. . . . **B. convergens** (F.Muell.) Stapf

70 Sporobolus R.Br.

19 species in Aust. (17 native, 2 naturalized); all states and territories

Spikelets with 1 floret, disarticulating above the glumes, dark coloured, 1–2 mm long, nearly sessile or pedicellate, in a narrow spike-like or loose panicle. Glumes awnless, unequal. Lemmas as long as or longer than the glumes, awnless. Leaf blades folded or rolled; ligule a ciliate rim. Fruit free, readily falling from the lemma.

Key to the species

1 Panicle very open, lowest branches of inflorescence whorled. Spikelets 1–2 mm long. Upper glume half the spikelet length. Palea entire. Anther 3. Seed c. 0.7 mm long. Erect, annual herb to 75 cm high. Moist areas, grassland. Fl. all year . . . . **S. caroli** Mez
1 Lowest branches of inflorescence not whorled (1 or 2 branches only) 2

2 Upper glume ± as long as lemma. Rhizomatous or stoloniferous perennials. Leaves often apparently in pairs; blades 4–7 cm long. Inflorescence 3–13 cm long, spike-like, dense. Spikelets 2–4 mm long. . . . 3
2 Upper glume shorter than lemma 4

3 Leaves more than 1 mm wide. Coastal dunes in wet places. Fl. summer. *Sand Couch*. . . . .
. . . . **S. virginicus** (L.) Kunth var. **virginicus**
3 Leaves 1 mm wide or less. Coast. Salt marshes. Fl. summer. *Marine Couch* . . . .
. . . . **S. virginicus** var. **minor** F.M.Bailey

4 Inflorescence dense, narrow and spike-like; main axis hidden by the appressed branches except sometimes in the lower third.. 5
4 Inflorescence loose, or narrow and interrupted and ± spike-like with the main axis visible between the branches in the upper two-thirds 6

5 Spikelets 2.1–2.5 mm long. Inflorescence 10–18 cm long; the branches stiff, appressed to the main axis. Tufted perennial up to 90 cm high. Coast and adjacent plateaus; Cumberland Plain. Weed in gardens and pastures; waste places. Introd. from S. Africa. Fl. summer. *Parramatta Grass*. . . *S. africanus (L.) R.Br.
5 Spikelets 1.6–2 mm long. Inflorescence 25–45 cm long with the branches usually lax at maturity. Tufted perennial up to 1.6 m high. Coast. Weed in pastures and waste places. Introd. from Asia. Fl. summer. *Giant Parramatta Grass* . . . . *S. fertilis** (Steud.) Clayton

6 Inflorescence loose, branches spreading, most of the axis in the upper half visible at flowering stage. Spikelets 1.5–2 mm long, placed loosely along branches. Leaf blades up to 3 mm wide, usually with ciliate margins. Inflorescence an open panicle up to 50 cm long. Erect tufted perennial up to 1.4 m high. Coast. Open forests. Fl. summer . . . . **S. sessilis** B.K.Simon
6 Inflorescence narrow, and interrupted. Spikelets contacted along the primary axis. 7

7 Inflorescence with stiff appressed branches, interrupted in the upper two-thirds, up to 40 cm long. Spikelets 1.3–1.8 mm long. Fruit c. two-thirds as long as the spikelet. Leaf blades up to 45 cm long and to 3 mm wide. Erect tufted perennial 40–140 cm long. Hornsby Plateau; Cumberland Plain. Grasslands and open forests on sandy or loamy soils. *Slender Rat's Tail Grass*. . . . . **S. creber** De Nardi

7 Most of the axis in the upper half of the inflorescence hidden by the appressed branches. Leaf blades up to 35 cm long and up to 3 mm wide, often ciliate on the margin. Inflorescence spike-like with the axis visible between the appressed branches in the lower half, up to 30 cm long. Spikelets 1.5–2.3 mm long. Fruit c. half the length of the spikelet. Erect tufted perennial up to c. 1 m high. Widespread. Pastures and open woodlands. Fl. summer. *Slender Rat's Tail Grass* . . . . **S. elongatus** R.Br.

71 Zoysia Willd.

2 species in Aust. (1 native, 1 endemic); Qld, NSW, Vic., Tas., S.A.

One species in the area

Spikelets with 1 floret, sessile or almost so, appressed to the notched rhachis, falling whole. Glume 1, keeled, coriaceous, dark brown, c. 5 mm long. Lemma hyaline. Inflorescence a narrow spike, 3–4 mm diam., 3–5 cm long. Perennial with rhizomes and erect aerial stems up to 30 cm high. Leaves glabrous, flat or convolute, 4–8 cm long, with a subulate often pungent pointed tip; ligule of fine hairs. Brackish swamps, dunes or cliffs near the sea coast. Fl. summer. *Prickly Couch* . . . . . **Z. macrantha** Desv. var. **macrantha**

72 Tragus Hallerf.

1 species endemic Aust.; all states and territories except Tas.

One species in the area

Spikelets with 1 floret, awnless, 2 or rarely 3 together on very short pedicels in a simple spike-like panicle 5–10 cm long. Lower glume absent; upper glume and sterile lemma subequal, 3–4 mm long, each with rows of rigid hooked bristles on the back (Fig. 50g). Annual up to 30 cm high. Leaf blades flat, bordered by rigid cilia, 2–5 cm long; ligule ciliate. Grasslands; roadsides; usually on clay soils. Fl. summer. *Small Burr Grass* . . . .**T. australianus** S.T.Blake

73 Aristida L.

Three-awned Spear Grasses or *Wire grasses*
58 species in Aust. (52 endemic, 6 native); all states and territories

Spikelet with 1 floret on filiform pedicels or nearly sessile in a terminal panicle, disarticulating above the glumes. Glumes awnless. Lemma rolled around the floret, with a trifid awn. Palea small, thin. Fruit enclosed in the lemma.

Key to the species

1 Awn with a twisted column 10–15 mm long below the 3 branches; awn branches unequal, 15–30 mm long. Lemma convolute, without a longitudinal groove; callus at base of lemma hairy. Tufted perennial up to 80 cm high. Widespread on sandy soils and rocky ridges. Fl. summer . . . . . . **A. warburgii** Mez
1 Column absent 2

2 Margins of lemma involute, producing a longitudinal furrow on the inner surface 3
2 Margins of the lemma overlapping, lemma therefore circular in section without a longitudinal furrow. . 6

3 Inflorescence a much branched panicle 20–30 cm long. Lemma 5–6 mm long; awns 8–10 mm long. Tufted perennials up to 80 cm high. Widespread. Sandy soils. Fl. summer. . . . .**A. benthamii** Henrard var. **spinulifera** B.K.Simon
3 Panicle spike-like. 4

4 Lateral awns shorter than the central one by more than 2 mm. Panicle 5–8 cm long. Tufted perennial up to 2 m high. Culms very branched. Widespread. DSF and heath. Sandy soils. Fl. summer . . . .**A. calycina** R.Br. var. **calycina**
4 Lateral awns shorter than the central one by up to 2 mm. Panicle 4–27 cm long. Compact tufted perennials up to 1 m high. Culms simple or sometimes branched at the base. 5

5 Furrow of the lemma without tubercles. Widespread. DSF and heath. Fl. summer . . . . **A. jerichoensis** (Domin) Henrard var. **jerichoensis**

5 Furrow of the lemma with tubercles. Coast; Hornsby Plateau. Fl. summer . . . .
. . . . **A. jerichoensis** var. **subspinulifera** Henrard

6 Lemma distinctly longer than both glumes 7
6 Lemma not longer than glumes 8

7 Lower glume 4.5 mm long or less. Panicle finally becoming divaricately branched, up to 18 cm long. Lemma 8–12 mm long. Widespread. Sandy soils. Fl. summer–autumn . . . . **A. vagans** Cav.
7 Lower glume more than 4.5 mm long. Panicle spike-like or with some ± erect branches, up to 18 cm long. Lemma 6–10 mm long; awn branches 10–15 mm long. Widespread. Sandy soils. Fl. summer–autumn (*A. vagans* and *A. ramosa* are known to hybridize in areas where they co-occur. The hybrid has a lower glume >4.5 mm long and a lemma 10–14 mm long.) . . . . **A. ramosa** R.Br.

8 Lemma tuberculate 6–11 mm long. Panicle spike-like, branches appressed, to 15 cm long. Fl. all year . .
. . . . **A. enchinata** Henrard
8 Lemma smooth or scabrous on margins, 7–12 mm long. Panicle ± open, branches loosely appressed, to 30 cm long. Widespread. Fl. all year . . . . **A. personata** Henrard

74 Enneapogon Desv. ex P.Beauv.
16 species in Aust. (15 endemic, 1 native); all states and territories except Tas.

One species in the area
Tufted annual or short-lived perennial up to 70 cm high. Culms usually much-branched. Leaf blades convolute, up to 3 mm wide. Spikelets with 3–6 flowers with 1 fertile, in spike-like panicles up to 10 cm long; rhacilla disarticulating above the glumes. Glumes ± unequal. Lemmas with 9 plumose awns up to 9 mm long. Barralier. Open forests. Fl. spring–summer. *Slender Nine-Awn* . . . **E. gracilis** (R.Br.) P.Beauv.

75 Digitaria Haller f.
37 species in Aust. (10 naturalized); all states and territories

Spikelets with 1 bisexual floret and a sterile lemma below it, in pairs (rarely 3s), one of them pedicellate the other sessile or almost so, arranged alternately in 2 rows on the upper side of a frequently winged rhachis. Lower glume small or absent; upper glume larger or equalling the sterile lemma. Inflorescence digitate or racemosely arranged spike-like racemes.

Key to the species

1 Rhachis devoid of spikelets for at least 2 cm at the base 2
1 Rhachis with spikelets to the base or devoid of spikelets up to 1 cm from the base 3

2 Racemes numerous, up to 35 cm long. Leaves glabrous or loosely pubescent. Spikelets 3 mm long, ± woolly with silky hairs, distant on the branches of the inflorescence. Erect, tufted perennial up to 80 cm high. Coast and Cumberland Plain. Forests. Fl. summer. *Umbrella Grass* . . . .
. . . . **D. divaricatissima** (R.Br.) Hughes
2 Racemes up to 11, up to 20 cm long. Leaves glabrous except towards the summit. Spikelets c. 3 mm long, covered with silky brownish or purplish hairs. Erect perennial up to 80 cm high, with short rhizomes. Hunter River Valley. Fl. summer. *Cotton Panic Grass* . . . . **D. brownii** (Roem. & Schult.) Hughes

3 Spikelets with long villous hairs . . . . **D. brownii** (see above)
3 Spikelets hirsute to glabrous 4

4 Plants with long prostrate stolons and tufted culms 5
4 Plants tufted. Culms sometimes bent near the base and rooting at the nodes but never with prostrate stolons 6

5 Spikelets pubescent with short fine hairs with verrucose walls, 1.3–1.8 mm long. Lower glume absent or minute; upper glume equalling the sterile lemma and both as long as the spikelet. Racemes usually 2–3, subdigitate, 3–8 cm long. Upper leaf sheaths glabrous. Stoloniferous annual with culms up to 50 cm high. Coast. Fl. summer . . . . **D. longiflora** (Retz.) Pers.

5 Spikelet with fine white hairs with ± smooth walls, 2–3 mm long. Lower glume minute, ovate; upper glume two-thirds to three-quarters as long as the spikelet. Sterile lemma as long as the lower glume. Racemes usually 2–3, digitate, 2–7 cm long. Leaf sheaths pilose. Tufted, slender, stoloniferous perennial up to 40 cm high. Coast. Fl. summer. *Queensland Blue Couch* . . . . **D. didactyla** Willd.

6 Hairs on the spikelets crinkled. Spikelets pallid to purple 7
6 Hairs on the spikelets not crinkled 8

7 Spikelets with short fine hairs with verrucose walls, purplish black, c. 1.5 mm long. Lower glume absent or minute; upper glume slightly shorter than the spikelet. Racemes usually 4–6, 5–10 cm long, on a short common axis. Ascending annual up to 50 cm high, rooting at the lower nodes. Coast. Weed. Introd. from trop. America and Asia. Fl. summer . . . . *D. violascens** Link
7 Spikelets with short fine hairs with smooth walls, pallid to purplish, 2–2.5 mm long. Lower glume absent or minute; upper glume slightly shorter than the spikelet. Sterile lemma as long as the spikelet. Racemes 2–6, 3–10 cm long, on a short common axis. Ascending or decumbent annual up to 40 cm high. Widespread. Weed. Fl. summer–autumn. Introd. from Europe and Asia. *Smooth Summer Grass* .
. . . . **D. ischaemum** (Schreb.) Schreb. ex Muhl.

8 Mature florets with a green, deep purple ,deep brown or black fertile lemma. Spikelets loosely and irregularly arranged, 1.5–2 mm long 9
8 Mature florets with a pallid, pale grey or pale purple fertile lemma. Spikelets densely arranged, c. 3 mm long. 12

9 Decumbent perennial up to 40 cm high. Culms rooting at the lower nodes. Spikelets 1.5–2 mm long. Lower glume up to a quarter as long as the spikelet; upper glume c. as long as the spikelet. Fertile lemmas dark brown to almost black. Racemes 2–7, 3–7 cm long, on a common axis 3–5 cm long. Widespread. Open forests and woodlands. Fl. summer. . . . .**D. diffusa** Vickery
9 Loosely tufted perennials up to 70 cm high. 10

10 Upper glume as long as the spikelet or nearly so. Spikelet c. 2 mm long. Lower glume up to half as long as the spikelet. Fertile lemma dark brown. Racemes 6–15 cm long, arranged racemosely 6–20 together on a common axis 6–10 cm long. Coast; upper Colo River. Open forests and woodlands. Fl. summer. *Small-flowered Fingergrass*. . . . .**D. parviflora** (R.Br.) Hughes
10 Upper glume distinctly shorter than the spikelet. 11

11 Upper glume a half to three-quarters as long as the spikelet. Spikelet c. 2 mm long. Lower glume short. Fertile lemma dark brown to black. Racemes 3–10, 3–13 cm long, racemosely arranged on a common axis 3–10 cm long. Widespread. Dry forests. Fl. summer . . . . **D. ramularis** (Trin.) Henrard
11 Upper glume less than half as long as the spikelet Spikelets c. 1.5 mm long. Lower glume minute. Fertile lemma brownish-black. Racemes 2–10, 2–10 cm long, racemosely arranged on a common axis 1–8 cm long. Widespread. Open forests and woodlands. Fl. summer . **D. breviglumis** (Domin) Henr.

12 Lower glume absent; upper glume equalling the sterile lemma and both slightly longer than the floret. Racemes 3–5, 5–8 cm long, digitate or nearly so. Erect or ascending annual up to 40 cm high, sometimes rooting at the nodes. Coast. Swampy ground. Introd. from S. America. Fl. summer . . . . .
. . . . *D. aequiglumis** (Hack. & Arech.) Parodi
12 Lower glume present; upper glume shorter and narrower than the sterile lemma. Racemes 4–9, 3–30 cm long, digitate or racemosely arranged on a common axis up to 5 cm long 13

13 Sterile lemma scabrous on the lateral nerves. Spikelets acute. Upper glume half to two-thirds as long as the spikelet. Ascending annual up to 70 cm high, usually rooting at the lower nodes. Widespread. Weed. Probably introd. Fl. summer. *Summer Grass* or *Crab Grass* . . . . . . *D. sanguinalis** (L.) Scop.
13 Sterile lemma smooth on the lateral nerves. Spikelets acuminate. Upper glume half to three-quarters as long as the spikelet. Ascending annual up to 70 cm high. Widespread. Introd. from tropics. Fl. summer. *Summer Grass* . . . .*D. ciliaris** (Retz.) Koeler

76 Stenotaphrum Trin.

2 species in Aust. (1 native, 1 naturalized); all states and territories except NT

One species in the area

Spikelets with 1 bisexual floret and a sterile or staminate lemma below it, dorsally compressed, c. 5 mm long, 2–5 together, half-embedded in hollows of the corky rhachis which finally disarticulates between the notches. Spike one-sided, flattened, c. 5 cm long, 8 mm wide. Glabrous perennial with long stolons often forming a spongy turf. Leaves keeled; blades up to 10 cm long, c. 5 mm wide. Lawn grass; near habitation; hind dunes near coast. Introd. from America and Asia. Fl. winter–spring. *Buffalo Grass.* . . .
. . . .
. . . . ***S. secundatum** (Walter) Kuntze

77 Eriochloa Kunth

Early Spring Grass

7 species in Aust. (6 native, 1 naturalized); all states and territories except Tas.

Spikelets with 1 bisexual floret and a sterile or staminate lemma below it, disarticulating below the glumes, arranged in 1 or 2 rows along one side of the branches of a simple panicle. Glume 1, with a callus supposed to represent the lower glume (Fig. 50c). Sterile lemma 1, c. as long as the glume.

Key to the species

1 Sterile lemma subulate, with an awn 1–2 mm long. Panicle 5–18 cm long; the branches 3–5 cm long. Spikelets 4.5–6 mm long, silky hairy. Light green, tufted, erect or ascending perennial to 1 m high. Nodes finely pubescent. Coast in drier parts. Open forest. Fl. summer . . . .
. . . . **E. pseudoacrotricha** (Stapf ex Thell.) J.M.Black
1 Sterile lemma with an awn less than 1 mm long or awnless. Panicle 6–15 cm long. Spikelets 3–4 mm long, scabrous or with a few silky hairs. Erect or ascending annual or perennial up to 80 cm high. Coast. Forests and grasslands. Fl. summer . . . .**E. procera** (Retz.) C.E.Hubb.

78 Axonopus P.Beauv.

2 species naturalized Aust.; Qld, NSW, W.A

One species in the area

Spikelets with 1 bisexual floret and a sterile lemma below it, sessile or subsessile, awnless, 2 mm long, alternate in 2 rows on one side of a 3-angled rhachis. Glume 1, similar to the sterile lemma. Inflorescence branched into 2, 3 or 4 spike-like racemes each 5–8 cm long. Stoloniferous perennial forming a mat. Leaf blades 3–5 cm long, 5–6 mm wide; ligule hyaline. Coast. Weed; sandy soils; roadsides. Introd. from West Indies. *Narrow-leaved Carpet Grass* . . . .***A. fissifolius** (Raddi) Kuhlm.

79 Urochloa P.Beauv.

28 species in Aust. (18 native, 10 naturalized); Qld, NSW, SA, NT, WA

Spikelets with 1 bisexual floret and a sterile lemma below it, disarticulating below the glumes, solitary, shortly pedicellate to nearly sessile, usually in 2 rows on one side of a 3-angled rhachis. Lower glume well developed, adaxial (facing the rhachis); upper glume equalling the sterile lemma. Ligule of fine hairs.

Key to the species

1 Tops of the pedicels with fine hairs c. 1 mm long (hairs best seen when the spikelets have fallen). . . . 2
1 Hairs absent from the tops of the pedicels 3

2 Spikelets pubescent, 5–6 mm long. Erect to decumbent annual with culms up to 100 cm long, usually softly pubescent especially at the nodes. Leaf blades 8–20 cm long, 7–15 mm wide; the margins undulate or crimped. Panicle up to 20 cm long; branches 2–4 cm long, usually appressed to the main axis; axis of the rhachis pubescent. Coast. Introd. from N. America. Fl. summer. *Texas Millet* . . . .
. . . . ***U. texana** (Buckley) R.D.Webster

2 Spikelets glabrous, 4–5 mm long. Decumbent to stoloniferous annual up to 70 cm high, ± pubescent. Leaf blades up to c. 20 cm long, 3–15 mm wide. Racemes usually 1–6 cm long. Cumberland Plain. Grasslands. Introd. from Africa. *Liverseed Grass*. . . . .*U. panicoides P.Beauv.

3 Lower lemma without a palea. Ascending annual up to 1 m high. Leaf blade up to 9 mm wide, hairy or glabrous. Spikelets 3–5 mm long. Racemes up to 5, 2–6 cm long. Cumberland Plain. Grasslands. Fl. summer . . . . **U. piligera** (F.Muell. ex Benth.) R.D.Webster

3 Lower lemma with a palea. Decumbent perennial. Leaf blades 4–18 mm wide, pubescent. Racemes up to 10, 4–9 cm long. Spikelets 5–7 mm long, distant. Coast. Usually near RF. Fl. summer. *Leafy Panic* . . . . **U. foliosa** (R.Br.) R.D.Webster

80 Paspalum L.

21 species in Aust. (5 native, 16 naturalized); all states and territories

Spikelets with 1 bisexual floret and a sterile lemma below it, awnless, compressed, solitary or in pairs in 2–4 rows on one side of the rhachis. Lower glume usually absent; upper glume and sterile lemma equal. Fertile lemma and palea coriaceous. Inflorescence a panicle; the branches (racemes) arranged digitately, subdigitately or racemosely. Ligule membranous or scarious.

Key to the species

1 Inflorescence bifurcating at the summit into 2 (3) equal racemes. Plants with creeping rhizomes and/or stolons 2

1 Inflorescence with several branches. Plants tufted or with short rhizomes 4

2 Glume and sterile lemma pubescent. Spikelets 2–3 mm long. Racemes 2–7 cm long. Plants up to 50 cm high, decumbent or floating, with stolons and rhizomes. Widespread. Wet places and in water. Fl. summer. *Water Couch* . . . . **P. distichum** L.

2 Glume and sterile lemma glabrous 3

3 Plants stoloniferous but without rhizomes, up to 60 cm high. Leaf blades up to 7 mm wide. Spikelets flattened, 3–4 mm long, not shiny. Racemes 2–8 cm long. Coast. In wet sandy salty areas. Fl. summer. *Salt-water Couch* . . . . **P. vaginatum** Sw.

3 Plants rhizomatous or tufted but without stolons, up to 80 cm high. Leaf blades up to 10 mm wide. Spikelets 3–4 mm long, shiny. Racemes 5–12 cm long. Coast. Roadside weed. Introd. from S. America. Fl. summer. *Bahia Grass*. . . . . *P. notatum Flugge

4 Racemes 3–7 5

4 Racemes usually more than 10 7

5 Spikelets glabrous, sessile, elliptic, 2–3 mm long, slightly acuminate. Racemes 2–7 cm long. Plants usually less than 50 cm high, tufted. Coast. Wet places. Fl. summer . . . . **P. orbiculare** J.R.Forst.

5 Spikelets with long hairs 6

6 Spikelets 3–4 mm long, on pedicels 1 mm long, acuminate. Hairs present on the margins of the glume, sterile lemma, pedicel and rhachis. Racemes 4–7 cm long. Plant up to 1 m high, with short rhizomes and arising from a 'crown' at ground level. Introd. from S. America. Fl. summer. *Paspalum* . . . . *P. dilatatum Poir.

6 Spikelets up to c. 2.5 mm long. Plants coarsely tufted . . . . **P. urvillei** (see below)

7 Racemes less than 5 cm long, very numerous. Panicle narrow-elliptic in outline. Spikelets 2–3 mm long, with few short hairs. Leaf blades 5–8 mm wide, glabrous. Plants 1–2 m high, tufted. Roadsides. Introd. from S. America. Fl. spring–autumn. *Tussock Paspalum*. . . . . *P. quadrifarium Lam.

7 Racemes up to 12 cm long. Spikelets 3–4 mm long, with long hairs so that the raceme appears cottony. Leaf blades 3–15 mm wide, hirsute at the base but glabrous above. Plants 1–2 m high, densely tufted. Coast. Disturbed sites. Introd. from S. America. Fl. summer. *Vasey Grass* . . . . *P. urvillei Steud.

81 Paspalidium Stapf

23 species native Aust.; all states and territories except Tas.

Spikelets with 1 bisexual floret and a sterile lemma below it, falling whole. Lower glume and lemma abaxial (away from the rhachis). Inflorescence a once-branched panicle; the branches spikes or spike-like, arranged alternately, with the spikelets in 1 or 2 rows on one side of the rhachis. Rhachis of the branches prolonged into a short bristle beyond the uppermost spikelet. Lower glume much smaller than the upper; upper glume similar to the sterile lemma.

Key to the species

1 Spikelets usually in 2 rows, usually many to each branch. Spikelets close together on the rhachis and usually touching 2
1 Spikelets in 1 row or irregularly arranged. 4

2 Leaf blades 3–8 mm wide. Lower glume inflated at the base. Spikelets mostly 2.5 mm long, bending away from the rhachis. Leaf blade scabrous or glabrous. Ascending perennial, usually >1 m high, sometimes rooting at the nodes. Hornsby Plateau and Cumberland Plain. Clay soils. Fl. summer . . . .
P. aversum J.Vickery
2 Leaf blades 1–3 mm wide. Spikelets mostly 1.5–2 mm long, sessile, overlapping. Lower glume not inflated. 3

3 Main rhachis and branches with soft spreading hairs. Leaf blades up to 15 cm long, 1–4 mm wide. Inflorescence 8–15 cm long; branches 1–2 cm long. Spikelets purplish. Kurrajong to Parramatta and northwards to Hunter River Valley. Woodlands and open forests. Fl. mainly summer . . . .
. . . . **P. albovillosum** S.T.Blake
3 Rhachies glabrous. Leaf blades up to 18 cm long, 1–6 mm wide. Inflorescence 6–15 cm long; branches 1–3 cm long. Spikelets often purplish. Mainly coast. Woodlands and open forests. Fl. summer . . . . . .
. . . .**P. distans** (Trin.) Hughes

4 Spikelets elliptic to obovate-elliptic. Ligule ciliate, c. 1–1.25 mm long altogether. Leaf blades ± convolute, up to 14 cm long and 3.5 mm wide when expanded. Inflorescence exserted; the main axis with up to 10 short racemes slightly shorter than the internodes or much reduced with a few few-flowered branches. Wiry perennial 10–80 cm high. Widespread. Open forests. *Slender Panic* . . . **P. gracile** (R.Br.) Hughes
4 Spikelets elliptic-lanceolate to oblong-lanceolate. Ligule ciliate, up to 0.5 mm long altogether. Leaf blades up to 10 cm long and 2.5 mm wide, often loosely involute. Inflorescence 2.5–12 cm long, with 4–6 appressed racemes, often branched below. Glabrous perennial, 10–60 cm high. Widespread. Woodlands on shales. Fl. summer . . . . **P. criniforme** S.T.Blake

82 Entolasia Stapf

3 species native Aust.; Qld, NSW, Vic.

Spikelets with 1 bisexual floret and a sterile lemma below it, disarticulating below the glumes, slightly compressed, solitary or the lowest paired, in spiciform or paniculately or racemosely arranged racemes. Glumes unequal; the lower one smaller than the upper. Upper lemma papery to coriaceous, finely and densely pubescent, obtuse. Palea equalling the lemma. Ligule a ciliate rim.

Key to the species

1 Upper glume and sterile lemma exceeding the floret by 0.25–1 mm. Spikelets ovate to narrow-ovate, 2.5–4 mm long, acute. Leaf blades flat, usually with 1 to 3 veins more prominent than the others. Ascending to decumbent perennial with a short rhizome, up to 60 cm high. Culms ± rigid, with slightly pubescent nodes. Coast and lower Blue Mts Damp areas on sandy soils. Fl. spring–summer . . . .
. . . . **E. marginata** (R.Br.) Hughes
1 Upper glume and sterile lemma c. equal to the floret. Leaf blades incurved, all veins equally prominent
. . . . 2

2 Spikelets 2–4 mm long, obtuse or acute. Racemes up to 7 cm long but often much shorter. Erect or ascending perennial, tufted but with a short rhizome. Culms rigid, up to c. 80 cm high. Widespread. Drier sandy soils. Fl. spring–summer . . . . **E. stricta** (R.Br.) Hughes

2 Spikelets 4–6 mm long, obtuse to acute. Racemes up to 3 cm long. Erect or ascending perennial, with a short rhizome. Culms rigid, up to c. 80 cm high. Coast. Damp areas on sandy soils. Fl. summer . . . . .

. . . . **E. whiteana** C.E.Hubb.

83 Pseudoraphis Griff.

Mudgrasses

2 species in Aust. (1 endemic, 1 native); Qld, NSW, S.A., WA, NT

Spikelets with 1 bisexual floret and a male floret below it, solitary and distant along the simple branches of a small panicle, with an erect bristle below the terminal spikelet. Lower glume minute, abaxial; upper glume almost equalling the sterile lemma. Rhachis produced into an awn-like point exceeding the spikelet.

Key to the species

1 Panicle spike-like but loose. Spikelets often 2 together on the lower branches, acute. Lower glume 1–1.5 mm long. Stems creeping, forming a mat. Leaf blades usually c. 5 cm long. Chiefly coast. Usually near water. Fl. summer . . . . **P. paradoxa** (R.Br.) Pilg.

1 Panicle spreading, 5–9 cm long. Spikelets distant on filiform branches, obtuse. Lower glume less than 1 mm long. Stems creeping or floating in water. Leaf blades flat, 4–5 cm long. In or near water. Fl. summer. *Floating Couch* or *Mudgrass* . . . . **P. spinescens** (R.Br.) Vickery

84 Panicum L.

33 species in Aust. (25 native, 8 naturalized); all states and territories

Spikelet with 1 bisexual floret and a male or sterile lemma below it, disarticulating below the glumes. Glumes unequal; the lower one sometimes minute; the upper one equalling the sterile lemma. Lemmas obtuse, inrolled over the palea; awns usually absent. Inflorescence a simple or compound panicle, usually very wide.

Key to the species

1 Lower glume 75% to 95% the length of the spikelet. Spikelets 5–8 mm long, often purplish. Sterile lemma and glumes with long acuminate tips. Panicle ± symmetrical, very open; the lowermost branches whorled. Perennial 40–50 cm high. Leaf blades 10–20 cm long, 2–5 mm wide, inrolled, glabrous. Hunter River Valley. Fl. summer . . . .**P. queenslandicum** Domin var. **acuminatum** Vickery

1 Lower glume 60% or less the length of the spikelet 2

2 Lower glume oblong or truncate 3

2 Lower glume acute 10

3 Either; plants decumbent with panicles under 10 cm long; or plants erect or decumbent and less than 40 cm high with inflorescences with 2–3 branches and few spikelets 4

3 Plants erect with much branched panicles mostly more than 10 cm long 5

4 Plants erect or slightly decumbent, up to 40 cm high. Leaf blades 5–8 cm long, c. 3 mm wide. Panicle with 2–3 spreading branches each with a few spikelets arranged close together. Spikelets 2–3 mm long. In or near water. Fl. summer. *White Water Panic* . . . . **P. obseptum** Trin.

4 Plants decumbent, rooting at the nodes. Leaf blades 3–5 cm long, 4–6 mm wide. Spikelets c. 1.5 mm long, distant on the rhachis. Panicle 4–10 cm long, usually with several branches. In or near RF or humid places. Fl. winter–summer. *Dwarf Panic* . . . . **P. pygmaeum** R.Br.

5 Swelling present at the base of the stem. Plants c. 1 m high. Panicle c. 25 cm long. Spikelets 5 mm long. Leaf blades 5–6 mm wide. Richmond district. Escaped from cultivation. Introd. from Mexico. Fl. summer. *Bulbous Panic* . . . . ***P. bulbosum** Kunth

5 No swelling present at the base of the stem **6**

6 Panicle with few (3–6) alternate branches, 5–12 cm long; the lower branches ± drooping and one-sided. Rhizomes long, creeping, often forming mats, with leaves reduced to sheaths. Leaf sheaths and blades usually softly hairy. Spikelets 2–3 mm long. Perennial up to 1 m high. Wet and waterlogged soils. Lower Hunter River Valley. Introd. from tropical coastal areas. Fl. summer . . . . ***P. repens** L.

6 Branches of the inflorescence numerous (usually more than 6); lowermost ones usually whorled. Rhizomes short or absent; plant tufted **7**

7 Silky hairs 5–7 mm long present at the junction of the leaf blade and sheath on the margins and the back, with fewer hairs on the inner surface particularly around the ligule. Perennial up to 1 m high; nodes often hairy. Leaf blades 1–3 cm wide. Panicle 20–50 cm long. Spikelets c. 3 mm long, usually clustered **8**

7 Hairs absent (rarely a few hairs present near the junction of the leaf blade and sheath margins). Nodes always glabrous **9**

8 Glumes and sterile lemma glabrous. Sydney area. Pastures; roadsides and waste places. Introd. from Africa. Fl. summer. *Guinea Grass* . . . . ***P. maximum** Jacq. var. **maximum**

8 Glumes and sterile lemma pubescent. Sydney area. Pastures; roadsides and waste places. Introd. from Africa. Fl. summer. *Green Panic* . . . . ***P. maximum** var. **trichoglume** A.Robyns

9 Ascending tufted annual up to 80 cm high, with soft compressible culms. Leaf blades up to 10 mm wide. Panicle 15–30 cm long. Spikelets c. 2.5 mm long. Illawarra region; Blue Mts Pastures and waste places. Introd. from Africa. Fl. summer. . . . . ***P. schinzii** Hack.

9 Perennial rarely more than 60 cm high, with hard stems. Leaf blades up to 12 mm wide. Panicle mostly 20–25 cm long. Spikelets c. 3 mm long, usually distant. Widespread on heavy soils, particularly in drier areas. Fl. summer. *Native Millet* . . . . **P. decompositum** R.Br.

10 Aquatic or semi-aquatic glabrous perennial with branching stems floating on water or rooting at the nodes in mud. Spikelets lanceolate-elliptic, acute, 2.5–3 mm long. Panicle loose; the flowering culms and panicle 20–80 cm high. Leaf blades 2–15 cm long, 4–13 mm wide. Grows along streams. Introd. from Asia. Fl. summer. *Black-seeded Panic* . . . . ***P. bisulcatum** Thunb.

10 Tufted grasses, not aquatic or semi-aquatic **11**

11 Leaf sheaths hispid; leaves usually more than 10 mm wide **12**

11 Either; leaf sheaths glabrous; or, if hispid, then leaf blade c. 5 mm wide or less. **13**

12 Spikelets 5 mm long, ovate, acuminate. Panicle 10–30 cm long. Annual up to 100 cm high. Leaf blades 15–20 mm wide, hairy to almost glabrous. Widespread. Weed in cultivated areas. Introd. from Europe and Asia. Fl. summer. *Millet Panic* . . . . ***P. miliaceum** L.

12 Spikelets 2–3 mm long. Panicle dense, up to 40 cm long. Annual up to 80 cm high. Leaf blades 5–15 mm wide, hispid. Pastures; roadsides. Introd. from N. America. Fl. summer. *Witchgrass* . . . .
 . . . . ***P. capillare** L. var. **capillare**

13 Lower florets usually male. Leaf blades 4–5 mm wide. Panicle dense, 15–20 cm long. Perennial up to 100 cm high, with rhizomes. Spikelets 2–3 mm long. Pastures; waste places. Introd. from India. Fl. summer. *Giant Panic* . . . . ***P. antidotale** Retz.

13 Lower florets sterile. Leaf blades 2–4 mm wide. Panicle very open, with distant spikelets **14**

14 Lower glume 25–35% the length of the spikelet. Panicle partly exserted, branches slender, ± 3-angled at base; primary branches spreading, secondary branches appressed. Spikelets 2.5–3.5 mm long. Leaf blade 3–8 mm wide. Erect to decumbent annual, often tinged with purple, to 70 cm long. Introd. from Africa. Fl. summer . . . . ***P. gilvum** Launert

14 Lower glume c. 50% the length of the spikelet **15**

15 Panicle 20 cm long, as broad as long, pyramidal in outline with the main branches mostly horizontal. Spikelets 2 mm long. Leaf blades 2–6 mm wide. Perennial up to 50 cm high. Widespread. Usually on sandy soils. Fl. spring–summer. *Poison Panic* or *Hairy Panic* . . . . **P. effusum** R.Br.

15 Panicle rarely more than 10 cm long, longer than broad; the branches ± erect. Spikelets 2–3 mm long. Plants up to 40 cm high. Leaf blades c. 3 mm wide. Perennial up to 70 cm high. Widespread. Many situations. Fl. summer. *Two-colour Panic* . . . . **P. simile** Domin

85 Sacciolepis Nash
2 species native Aust.; Qld., NSW, NT, WA

One species in the area

Spikelets with 1 bisexual floret and a sterile lemma below it, 2–3 mm long, awnless, disarticulating below the glumes. Glumes hard, rounded; the lower one c. half the length of the spikelet; the upper glume equalling the sterile lemma, slightly inflated (Fig. 50h). Inflorescence a narrow spike-like panicle, 1–6 cm long, 3–4 mm diam., on a long peduncle. Annual up to 60 cm high. Leaf blades 3–4 mm wide; ligule membranous, very short. Coast. Wet places. Fl. summer. *Indian Cupscale Grass* . . . **S. indica** (L.) Chase

86 Oplismenus P.Beauv.
4 species native Aust.; Qld, NSW, Vic., NT

Spikelets with 1 terminal bisexual floret and a rudimentary one below it, in one-sided clusters in a panicle, disarticulating below the glumes. Glumes awned; the lower one with the longer awn. Lemmas awnless. Palea hardened around the fruit. Inflorescence a one-sided panicle up to 10 cm long. Decumbent perennials much branched, rooting at the nodes. Ligule of hairs.

Key to the species

1 Lower branches of the inflorescence 3–5 cm long. Leaf blades ovate to lanceolate, less than 7 times as long as wide, 4–18 mm wide, softly pubescent. Widespread. Shaded places. Fl. summer–autumn . . . . **O. aemulus** (R.Br.) Roem. & Schult.

1 Lower branches of the inflorescence mostly up to 1.5 cm long. Leaf blades more than 10 times as long as wide, up to 7 mm wide, pubescent. Coast and adjacent plateaus. Shaded places. Fl. summer–autumn . . . . **O. imbecillis** (R.Br.) Roem. & Schult.

87 Ottochloa Dandy
2 species native Aust.; Qld, NSW

One species in the area

Spikelets with 1 bisexual floret, with either a sterile lemma or a male floret below it. Spikelets dorsally compressed, 2 mm long, disarticulating below the glumes. Glumes thinly membranous, usually glabrous, narrowly hyaline on the margins, equal or subequal, much shorter than the spikelet. Lower lemma usually with a shallow longitudinal depression on the back. Panicle 3–5 cm long , with 2–4 primary branches; the spikelets distant on the branches. Stems decumbent, rooting at the lower nodes. Culms up to 15 cm long; internodes 25–37 mm long. Leaf sheaths glabrous or rarely hispid; leaf blades c. 3.5 cm long and 4 mm wide. North from Gosford area. WSF. Fl. summer . . . . **O. gracillima** C.E.Hubb.

88 Cleistochloa C.E.Hubb.
3 species endemic Aust.; Qld, NSW

One species in the area

Spikelets of 2 kinds, one kind chasmogamous (opening); the other cleistogamous (not opening). Both kinds of spikelets oblong-elliptic, acuminate or mucronate, dorsally compressed, disarticulating above the glumes; both types with 1 bisexual floret and a sterile lemma below it. Chasmogamous spikelets borne singly in spike-like racemes; spikelets c. 5 mm long, c. 1.5 mm wide; lower glume hyaline, up to 0.35 mm long or absent; upper glume a little shorter than the spikelet. Cleistogamous spikelets solitary at the apex of short densely leafy branches of the culms; spikelets 5–7 mm long, 1.5–2 mm wide; lower glume absent;

upper glume slightly shorter than the spikelet. Densely tufted perennial with short rhizomes, up to 1 m high. Culms 7–10-noded, robust, densely or loosely hirsute; nodes bearded. Leaf blades 2–8 cm long, 2–3 mm wide, xeromorphic, c. 3 times as long as the sheaths. Widespread. Dry sandstone ridges. Fl. summer . . . . **C. rigida** S.T.Blake

89 Ancistrachne S.T.Blake
2 species native Aust.; Qld, NSW

One species in the area

Spikelets with 1 bisexual floret and a sterile lemma below it, disarticulating below the glumes, 2 mm long, arranged in a spike-like panicle. Lower glume very small or abortive; upper glume and lower lemma covered with tubercle-based hairs; fertile lemma with a hyaline margin. Hawkesbury River district. Shaded places. Vulnerable. Fl. autumn . . . . **A. maidenii** (A.A.Ham.) Vickery

90 Echinochloa P.Beauv.
17 species in Aust. (9 native, 8 naturalized); all states and territories

Spikelets with 1 bisexual floret, sometimes with a male floret below it, 2–3 mm long, sometimes awned, usually in 3–4 rows, disarticulating below the glumes, subsessile in dense spike-like racemes placed alternately on the angular rhachis. Lower glume c. one-third the length of the spikelet (Fig. 50f). Annuals with erect geniculate culms.

Key to the species

1 Spikelets awnless 2
1 Spikelets awned 5

2 Fruit 2–2.5 mm long, 1.8–2.25 mm wide. Panicle very compact, densely crowded 3
2 Fruit 1.4–2 mm long, 1.1–1.6 mm wide. Panicle less compact, if densely crowded then more openly spreading 4

3 Spikelets ± strongly purplish to blackish brown (rarely pallid when bleached), shortly acute, 3–4 mm long, 2–2.5 mm wide. Fruit brownish. Culms up to 1 m high. Widespread. Cultivated. Roadsides. Introd. from E. Asia. Fl. summer. *Japanese Millet* or *Ankee Millet*. . . *E. esculenta** (A.Braun) H.Scholz
3 Spikelets always pallid, shortly acute when young, usually very obtuse at maturity, 3–5 mm long, c. 2.3 mm wide. Fruit whitish. Culms up to 1 m high. Coast. Cultivated. Roadsides. Introd. from India. Fl. summer. *Siberian Millet* . . . . *E. frumentacea** Link

4 Spikelets 2–3 mm long, usually violet tinged and conspicuously hispid, awned or awnless. Hairs present at the base of the inflorescence branches. Culms up to 90 cm high. Widespread. Garden weed; pastures. Introd. from tropics. Fl. summer. *Barnyard Grass* . . . . *E. crus-galli** (L.) P.Beauv.
4 Spikelets c. 3 mm long, green, awnless. Bases of inflorescence branches pubescent or with a few bristles. Culms up to 60 cm high. Coast and Cumberland Plain. Garden weed; pastures, particularly on clay soils and floodways. Fl. summer. *Awnless Barnyard Grass* . . . . **E. colona** (L.) Link

5 Panicle nodding. Spikelets elliptic to broad-elliptic, 3.5 mm long. Panicles up to 18 cm long. Culms up to 1.3 m high. Widespread but occasional. Weed of cultivation. Introd. from S. America. Fl. summer . . . . *E. crus-pavonis** (Kunth) Schult.
5 Panicle erect or very slightly nodding 6

6 Culms up to 1.8 m high. Spikelets narrow-elliptic, 3–4.2 mm long. Panicle 20–35 cm long. Coast. In or near water. Fl. summer. *Swamp Barnyard Grass* . . . . *E. telmatophila** P.W.Michael & Vickery
6 Culms 25–90 cm high. Spikelets ovate-elliptic, 2.5–4 mm long . . . . *E. crus-galli** (see above)

91 Setaria P.Beauv.
17 species in Aust. (7 native, 10 naturalized); all states an territories

Spikelets with 1 terminal bisexual floret and sometimes a male floret below it, surrounded by numerous bristles, disarticulating below the glumes; awns absent. Inflorescence a cylindrical, dense or interrupted spike-like (rarely spreading) panicle.

Key to the species

1 Inflorescence up to 35 cm long, with long spreading branches. Leaf blades 4–6 cm wide. Perennial up to over 1 m high. Garden escape and weed near habitation. Introd. from India. Fl. summer. *Palm Grass* . . . . *S. palmifolia* (Koenig) Stapf
1 Inflorescence spike-like, up to 15 cm long. Leaf blades less than 2 cm wide 2

2 Spikelets seated directly on the main axis, or apparently so 3
2 Spikelets on short lateral branches. Bristles 1–2 below each spikelet, retrorsely scabrous 6

3 Bristles 1–2 (rarely 3) below each spikelet. Inflorescence 2–5 cm long, 12–15 mm diam. (including bristles), green. Annual up to 40 cm high. Widespread on ranges. Weed. Introd. from Europe. Fl. summer. *Green Pigeon Grass* . . . . *S. viridis* (L.) P.Beauv.
3 Bristles several below each spikelet 4

4 Inflorescence (including bristles) 10–15 mm diam., 2–8 cm long. Bristles 6 mm long. Annual up to 1 m high. Widespread. Weed. Introd. from Europe, temperate Asia. *Pale Pigeon Grass* . . . . *S. pumila* (Poir.) Roem. & Schult.
4 Inflorescence (including bristles) less than 10 mm diam. 5

5 Bristles 6–8 below each spikelet, 2–4 mm long, green. Inflorescence 2–10 cm long, 4–5 mm diam. (including bristles). Perennial up to 50 cm high. Widespread. Weed. Introd. from America. Fl. summer. *Slender Pigeon Grass* . . . . *S. parviflora* (Poir.) Kerguelen
5 Bristles 8–12 below each spikelet, 4–5 mm long, golden-brown. Inflorescence 4–6 cm long, 8–10 mm diam. (including bristles). Annual up to 40 cm high. Cumberland Plain. Waste places. Introd. from Africa. Fl. summer . . . . *S. sphacelata* (Schum.) Stapf & C.E.Hubb.

6 Inflorescence mostly 6–8 mm diam., 6–8 cm long. Annual up to 60 cm high. Widespread. Weed. Introd. from Europe and trop. America. Fl. summer. *Whorled Pigeon Grass* . . . . *S. verticillata* (L.) P.Beauv.
6 Inflorescence mostly 10 (–20) mm diam., 6–20 cm long, sometimes interrupted near the base. Annual up to 1.5 cm high. Leaf blades up to 20 mm wide. Upper floret falling from the spikelet at maturity. Regarded as a cultivated form of *S. viridis*. Near habitation. Cultivated for bird seed. Fl. summer. *Foxtail Millet* or *Indian Millet* . . . . *S. italica* (L.) P.Beauv.

92 Pennisetum Rich. ex Pers.
12 species in Aust. (3 native, 9 naturalized); all states and territories

Spikelets with 1 bisexual floret and a sterile lemma below it, awnless, solitary or 2–3 together, sessile or nearly so, each surrounded by an involucre of bristles which are united at the base; involucre falling off with the spikelet (Fig. 49.g). Lower glumes smaller than the upper which is almost as long as the sterile lemma. Stigmas connate or the tips only free.

Key to the species

1 Plants with long stolons or rhizomes often several metres long. Leaves bright green; blades mostly 5–8 mm wide. Spikelets 2–4 together enclosed in the upper leaf sheaths. Bristles half to one-third the length of the spikelet. Chiefly coast. Lawns and pastures; weed; waste places. Introd. from E. Africa. Fl. summer. *Kikuyu Grass* . . . . *P. clandestinum* Hochst. ex Chiov.
1 Plants upright, tufted. Spikelets in spike-like panicles. 2

2 Plants 1.5–3 m high (rarely higher). Bristles unequal, numerous, scabrous or rarely glabrous, mostly 10–16 mm long. Spikelets 4–7 mm long. Panicle 8–30 cm long, 14–30 mm diam. Coast. Cattle fodder; roadsides. Introd. from trop. Africa. Fl. summer. *Elephant Grass* or *Napier Fodder.* . . . . ***P. purpureum** Schumach.

2 Plants less than 1.5 m high 3

3 Bristles scarcely exceeding the spikelets which are 4–5 mm long. Panicle 10–30 cm long. Plants up to c. 1.5 m high. Coast. Cultivated as an ornamental; weed near habitation. Introd. from S. Africa. *African Feather Grass* . . . . ***P. macrourum** Trin.

3 Bristles much longer than the spikelets 4

4 Bristles almost equal, c. 30–40 mm long, with some shorter. Panicle 4–8 cm long, 3–5 cm diam., pale-coloured. Plants 40–50 cm high. Chiefly coast. Pastures; roadsides. Introd. from East Africa. Fl. summer. *Feather Grass* . . . . ***P. villosum** R.Br.

4 Bristles very unequal. 5

5 Inner bristles plumose, mostly up to 26 mm long, scaberulous. Spikelets 4.5–6.5 mm long. Panicle 10–25 cm long, 12–16 mm diam. Plants 25–90 cm high. Cultivated as an ornamental; weed near habitation. Introd. from the Middle East. Fl. summer. *Fountain Grass* . . . . ***P. setaceum** (Forssk.) Chiov.

5 Inner bristles not plumose, scabrous; the longest 30–40 mm long, but mostly 10–20 mm long. Panicle 8–14 cm long, 4–5 cm diam., usually purplish. Plants up to 1 m high. Widespread. Floodways; wet places. Fl. summer. *Swamp Foxtail.* . . . . ***P. alopecuroides** (L.) Spreng.

93 Cenchrus L.
10 species in Aust. (4 native, 6 naturalized); all states and territories except Tas.

Spikelets with 1 terminal bisexual floret and a male or reduced lemma below it, usually in clusters of 2–8 surrounded by a "burr" of rigid spikes or bristles which fall with the spikelet, disarticulating above the glumes (Fig. 49h). Lower lemma similar to the glumes. Inflorescence a spike-like panicle. Ligule of fine hairs.

Key to the species

1 Inner involucral spines flattened, connate for at least half their length 2

1 Inner involucral spines connate only at the base for up to a quarter of their length 3

2 Burr 4–10 mm long, 3–6 mm diam., often purplish at the tips but pallid below, subtended by an outer ring of bristles. Annual herb with culms up to 90 cm long. Coast. Sands. Introd. from America. Fl. summer. *Mossman River Grass.* . . . . ***C. echinatus** L.

2 Burr 7–10 mm long, outer ring of bristles absent. Annual herb to 80 cm high. Sandy disturbed areas. Noxious weed. Introd. from America. Fl. summer–autumn . . . . ***C. incertus** M.A.Curtis

3 Burr up to 7 mm long, usually purplish black when mature; bristles retrorsely barbed. Spikelets usually 1 but up to 3 per burr. Tall or scrambling perennial usually with a knotted rootstock. Culms up to 2 m long or longer. Widespread. Open forests on poor soils. Fl. summer. *Burr Grass* . . **C. caliculatus** Cav.

3 Burr c. 12 mm long; bristles antrorsely barbed, one distinctly longer the others. Spikelets 2–5 per burr. Plants tufted, up to 80 cm high. Coast. Deep sands. Introd. from Africa and Asia. Fl. summer. *Buffel Grass.* . . . . ***C. pennisetiformis** Hochst. & Steud. ex Steud.

94 Melinis P.Beauv.
2 species naturalized Aust.; Qld, NSW, Vic., WA

One species in the area
Spikelets with 1 bisexual floret and a sterile lemma below it, pedicellate, solitary, disarticulating below the glumes. Lower glume c. 1 mm long, hairy. Upper glume and sterile lemma equal, 5 mm long, purplish red, covered with long hairs, shortly awned, completely enclosing the bisexual floret. Panicle open, 7–15 cm long. Leaf blade flat, 2–5 mm wide. Tufted, short-lived perennial or annual up to 1 m high. Roadsides; weed. Introd. from S. Africa. Fl. summer. *Natal Red Grass* or *Natal Red Top.* . . ***M. repens** (Willd.) Zizka

95 Spinifex L.

3 species native Aust.; all states and territories

One species in the area

Plants dioecious. Spikelets unisexual, sessile, disarticulating below the glumes. Male spikelets 2-flowered, in spikes which are clustered into groups of 4 or 5 or 6; glumes 5–6 mm long; lemmas c. 8 mm long. Male spikes 2–4 cm long, subtended by bracts 5–6 cm long (including the subulate tips). Female spikelets 12–15 mm long, very numerous, in a large dense globular head; each spikelet solitary at the base of the spine-like rhachis which protrudes beyond the spikelets and is 10–12 cm long; subtending bract 2–3 cm long. Hirsute or scabrous perennial with stout rhizomes and stolons extending for several metres over sand. Leaves silvery-silky; ligules of hairs. Coast. Sand dunes. Fl. summer. *Hairy Spinifex* . . . **S. sericeus** R.Br.

96 Isachne R.Br.

4 species native Aust.; Qld, NSW, Vic., NT

One species in the area

Spikelets with 2 bisexual florets or with 1 bisexual floret and a male floret below it, awnless, disarticulating above the glumes which fall soon after the florets fall so that the pedicel soon becomes naked. Glumes equal, orbicular, 1–2 mm long. Lemma and palea each 1–2 mm long. Inflorescence an open panicle 8–10 cm long. Perennial with decumbent stems up to 50 cm long, rooting at the lower nodes. Leaf blades narrow-lanceolate, 6–8 cm long, usually c. 5 mm wide, scabrous; ligule of fine hairs. Widespread. In or near water or wet places. Fl. summer. *Swamp Millet* . . . .**I. globosa** (Thunb.) Kuntze

97 Imperata Cirillo

1 species native Aust.; all states and territories

One species in the area

Spikelets with 1 bisexual floret and a sterile lemma below it, all alike, in pairs, disarticulating below the glumes; one shortly pedicellate; the other on a long pedicel. Glumes thin, 3–4 mm long, subulate, with long silky hairs arising from the base. Lemma hyaline, awnless. Stamens 1–2. Style dark-coloured, conspicuous. Inflorescence a dense spike-like panicle 8–20 cm long, densely silky hairy with long hairs surrounding the spikelets. Stiff, erect perennial with deeply buried pungent pointed wiry rhizomes. Usually glabrous. Leaf blades 50–80 cm long, 5–10 mm wide, scabrous; ligule of hairs. Usually on deep soils (sands or clays); often common after fire. Fl. spring–summer. *Blady Grass* . . . **I. cylindrica** P.Beauv.

98 Miscanthus Andresson

1 species naturalized Aust.; NSW, WA

One species in the area

Spikelets with 1 bisexual floret and a sterile lemma below it, in pairs, unequally pedicellate, with a tuft of silky hairs at the base as long as or up to twice as long as the glumes. Glumes equal, c. 3 mm long, membranous. Fertile lemma with a bent flexuose awn 6–8 mm long. Plants up to more than 2 m high. Leaves mostly basal; blades flat, 50–100 cm long, 1 cm wide. Panicle fan-shaped, 10–20 cm long. Introd. from E. Asia.

Key to the varieties

1 Leaves green throughout. Cultivated. Coast and Blue Mts Roadsides. Fl. summer . . . .
. . . . ***M. sinensis** Andress.var. **sinensis**
1 Leaves green and white striped 2

2 Leaves transversely banded. Cultivated. Coast. Roadsides. Fl. spring–summer . . . .
. . . . ***M. sinensis** var. **zebrinus** Beal.
2 Leaves longitudinally banded. Cultivated. Coast. Roadsides. Fl. spring–summer . . . .
. . . . ***M. sinensis** var. **variegatus** Beal.

99 Microstegium Nees
1 species native Aust.; Qld, NSW

One species in the area

Spikelets equal, 1 sessile and 1 pedicellate, with a basal callus of short hairs. Florets 2; lower one usually sterile; upper one bisexual. Upper lemma with a geniculate awn. Inflorescence a panicle of 3–6 racemes, subdigitate. Decumbent annual or short-lived perennial up to 50 cm high. Leaf blades lanceolate, 4–8 mm wide; ligule of stiff hairs. Hornsby. Fl. summer–autumn . . . . **M. nudum** (Trin.) A.Camus

100 Ischaemum L.
10 species in Aust. (9 native, 1 naturalized); Qld, NSW, NT, WA

One species in the area

Erect to decumbent perennial herb to 1.2 m high. Ligule membranous 2 mm long with ciliate margin. Blade 3–5 mm wide. Inflorescence digitate sometimes appearing spike like, with 2 or more appressed racemes held erect. Spikelets in pairs 6–10 mm long, 1 sessile the other pedicellate. Sessile spikelet with glumes subequal, lower one shiny and 7-nerved; 1–2 florets the lower one male, upper one bisexual or female, awn on fertile lemma 8–10 mm long and slightly twisted near base. Pedicellate spikelet similar to sessile spikelet or male with a shorter awn. Coastal. Deporporate soils or swampy areas. Fl. summer . . . . **I. australe** R.Br. var. **australe**

101 Hemarthria R.Br.
1 species native Aust.; all sates and territories except NT

One species in the area

Spikelets half embedded in the rhachis, in pairs, the "pedicellate" spikelet becoming sessile through the fusion of its pedicel with the rhachis (terminal spikelet truly pedicellate), "pedicellate" spikelet above the sessile spikelet on the same side of the rhachis. Spikelets with 1 bisexual floret and a sterile lemma below it (Fig. 47 h). Glumes 7–8 mm long, awnless those on the pedicellate spikelet narrower than those on the sessile spikelet. Lemmas smaller than the glumes, membranous. Inflorescence spike-like, up to 15 cm long, cylindrical in outline, with the spikelets opposite or alternate; spathe usually present. Decumbent perennial with long thin rhizomes. Culms ascending up to 40 cm high. Leaves glabrous; blades up to 15 cm long, 2–4 mm wide; ligule of short membranous hairs. Coasts of Hawkesbury River; Blue Mts; Hornsby Plateau. Wet places. Fl. summer. *Mat Grass* . . . . **H. uncinata** R.Br.var. **uncinata**

102 Rottboellia (L.) L.f.
1 species native Aust.; Qld, NSW, NT

One species in the area

Spikelets with 2 florets, paired, unequal, 1 sessile and 1 pedicellate with the pedicel fused to the rhachis, half embedded in the rhachis; lower floret male; upper floret bisexual. Glumes subequal, obtuse or truncate. Inflorescence a solitary spike-like raceme up to 12 cm long. Erect annual up to 3 m high with proproots. Leaf blades flat, up to 3 cm wide, with stiff hairs particularly on the sheath; ligule membranous, truncate. Mangrove Mt. Grassland. Probably introd. from N. Australia . . . . ***R. cochinchinensis** (Lour.) W.D.Clayton

103 Sorghum Moench
24 species in Aust. (17 native, 7 naturalized); all states and territories except Tas.

Spikelets sessile and pedicellate, paired except at the ends of the branches where one sessile spikelet lies between 2 pedicellate spikelets; sessile spikelets with 1 bisexual floret and a sterile lemma below it; pedicellate spikelets with 1 male or sterile floret. Glumes awnless, hard. Lemma of the bisexual floret with an awn arising between the lobes, twisted in the upper half, bent above the middle. Male (and sterile) spikelets awnless. Palea very small or none.

Key to the species

1 Nodes densely hairy. Glumes hairy. Leaf sheaths hairy. Tufted perennial c. 1 m high. Grasslands. Fl. summer. *Wild Sorghum* . . . . **S. leiocladum** (Hack.) C.E.Hubb.

1 Nodes smooth. Glumes usually smooth, shiny 2

2 Perennial c. 2 m high, with long rhizomes. Pastures; weed; roadsides. Fl. most of the year. Introd. from the Mediterranean. *Johnson Grass* . . . . ***S. halepense** (L.) Pers.

2 Annual or perennial without a rhizome. 3

3 Fruit exposed between the glumes at maturity. Cultivated; roadsides. Introd. from Africa, America. Fl. summer. *Forage Sorghum* . . . . **S. bicolor** (L.) Moench ssp. **bicolor**

3 Fruit not exposed, enclosed within the glumes at maturity. Cultivated; roadsides. Introd. from Africa. Fl. summer. *Sudan Grass.* . . . . ***S. bicolor** ssp. **drummondii** (Steud.) de Wet

104 Capillipedium Stapf
Scented Top
2 species native Aust.; Qld, NSW, NT

Spikelets 2 together, one sessile, the other pedicellate, falling whole. Sessile spikelet with 1 bisexual floret and a male or empty lemma below it Pedicellate spikelet male or neuter. Glumes awnless. Lemma of the sessile spikelet awned, that of the pedicellate spikelet awnless. Inflorescence a loose panicle, built up from 1–8-jointed racemes at the ends of the panicle branches. Ligule truncate, ciliate.

Key to the species

1 Racemes 1–2-jointed (rarely 3). Lower glume of the sessile spikelet 6-nerved including 2 on the keel, prominently concave. Panicle very narrow-ovate to oblong, 8–25 cm long. Erect, tufted annual up to 1.5 m high. Coast. Uncommon. Open forests and woodlands. Fl. summer . . . **C. parviflorum** (R.Br.) Stapf

1 Racemes 3–8-jointed. Lower glume of the sessile spikelet 8–9-nerved including 4–5 on the keel, not prominently concave. Panicle ovate to very narrow-ovate, 11–25 cm long. Erect, tufted annual up to 2 m high. Coast. Forests. Fl. summer . . . . **C. spicigerum** S.T.Blake

105 Bothriochloa Kuntze
9 species in Aust. (7 native, 2 naturalized); all states and territories except Tas.

Spikelets in pairs (terminal spikelets in threes), one sessile, the other pedicellate (Fig. 50l). Sessile spikelet with one bisexual floret and a sterile lemma below it; glumes subequal, the lower glabrous with scabrous edges and often a pit near the centre, upper glume hairy; sterile lemmas shorter than the glumes and enclosed in the upper glume; fertile lemma with a minute body, with a long twisted awn much longer than the glumes. Pedicellate spikelet sterile or male, awnless. Inflorescence a digitate or subdigitate panicle with 2–several branches. Pedicels and articles of the rhachis silky hairy and with a longitudinal transparent furrow.

Key to the species

1 Anthers 3. Pedicellate spikelet lanceolate, reduced to 2 glumes. Sessile spikelet 5–7 mm long; lemma 3 mm long, with an awn 17–20 mm long. Erect to decumbent tufted perennial 40–80 cm high, reddish-purple, glabrous except the inflorescence. Panicle 4–8 cm long. Rhachis with silky hairs at the joints. Chiefly grasslands and open woodlands. Common. Fl. spring–summer. *Red Leg* or *Red Grass* . . . . . .
. . . . **B. macra** (Steud.) S.T.Blake

1 Pedicellate spikelet subulate, reduced to 1 glume. Anther 1. Sessile spikelet c. 5 mm long. Otherwise similar to *B. macra*. Uncommon. *Pitted Blue Grass* . . . . **B. decipiens** (Hack.) C.E.Hubb.

106 Dichanthium Willemet
10 species in Aust. (7 native, 3 naturalized); all states and territories

One species in the area
Spikelets 1-flowered, in pairs; the sessile one fertile; the pedicellate one sterile or male. Outer glume obtuse, 4–5 mm long, hairy. Lemma of the fertile floret with a terminal bent awn 2–3 cm long. Pedicellate spikelets without awns. Inflorescence a panicle with 2–7 spike-like racemose branches digitately arranged and each 3–5 cm long. Pedicels and articles of the rhachis opaque. Tufted perennial up to 70 cm high; nodes usually hairy. Leaves flat; ligule of hairs. Grasslands and open forests on clay soils. Fl. summer. *Queensland Blue Grass.* . . . . **D. sericeum** S.T.Blake ssp. **sericeum**

107 Andropogon L.
3 species naturalized Aust.; Qld, NSW, WA

One species in the area
Sessile spikelets with 1 bisexual floret and a sterile lemma below it, 3 mm long; lemma with a straight awn 15–20 mm long. Pedicellate spikelet represented by the pedicel only (Fig. 50 n). Joints and pedicels with long fine hairs up to 1 cm long. Inflorescence usually digitate, composed of 2–4 racemes each 2–3 cm long, initially enclosed in a spathe 4–5 cm long. Coarse, tufted perennial up to 1 m high, often tinged with purple. Nodes glabrous. Leaf blades 10–20 cm long, 2–4 mm wide. Ligule very short, papery, ciliolate at the apex. Widespread. Waste places. Introd. from America. Fl. summer. *Whisky Grass****A. virginicus** L.

108 Hyparrhenia Fourn.
3 species in Aust. (1 native, 2 naturalized); all states and territories except Tas.

One species in the area
Tufted perennial herb to 1.2 m high. Ligule scarious and minutely toothed. Blade 2–3 mm wide, ± glaucous. Racemes enclosed by spathes. Spikelets in pairs. Fertile spikelet sessile, 4–6 mm long; lower glume 9–11-nerved, upper glume 3-nerved; lemmas 2, fertile lemma 2.5–4.5 mm long with 2 short lobes, awn geniculate 15–35 mm long; palea absent or minute. Pedicellate spikelet male, 4–7 mm long, hairy. Roadsides, woodland or grassland. Introd. Medit. Fl. summer. . . . . **H. hirta** (L.) Stapf

109 Cymbopogon Spreng.
10 species in Aust. (9 native, 1 naturalized); all states and territories except Tas.

Spikelets in pairs, one pedicellate, the other sessile, overlapping on the rhachis and closely appressed to it. Sessile spikelet with 1 bisexual floret and a sterile lemma below it. Pedicellate spikelet male or sterile. Spikelets arranged in 2 racemes together subtended by a small spathe; racemes initially closely appressed, arranged in a narrow panicle, finally separating and becoming reflexed.

Key to the species

1 Racemes slightly hairy to almost glabrous, 15–20 mm long, sharply reflexed and divergent at maturity. Lemma of the sessile spikelet awnless or with a fine awn. Erect, tufted perennial with a purplish tinge, up to 1 m high. Leaf blades up to 25 cm long, inrolled, 2–3 mm wide; ligule membranous, sinuate. Widespread. Woodlands, forests, pastures and sand dunes. Fl. spring–summer. *Barbwire Grass* . . . . **C. refractus** (R.Br.) A.Camus

1 Racemes ± densely villous hairy, 15–25 mm long, reflexed at maturity. Lemma of the sessile spikelet with an awn from between the lobes up to 6 mm long or sometimes absent. Aromatic, tufted perennial up to 1 m high. Leaf blades up to 35 cm long and 2.5 mm long, folded or with revolute margins. Drier western parts of the region in woodlands . . . . **C. obtectus** S.T.Blake

110 Themeda Forssk.
5 species in Aust. (3 native, 2 naturalized); all states and territories

One species in the area
Spikelets clustered in groups of usually 5–7, of which 1 is bisexual and the others male or neuter. Glumes of the bisexual spikelet rigid, 1 cm long, enclosing a sterile lemma and the bisexual floret; lemma entirely awn-like 4–6 cm long; palea minute; lodicules cuneate. Male and neuter florets 1 cm long, enclosed in shortly (less than 1mm) awned glumes. Clusters of spikelets in groups of 2–5, subtended by a sheathing bract so that the cluster resembles a single spikelet; compound clusters solitary or 2–3 together, each on a peduncle 3–5 cm long arising from a leaf axil. Inflorescence a panicle. Erect, glabrous perennial often more than 1 m high. Leaf blades up to 25 cm long and 4 mm wide; ligule ciliate, less than 1 mm long. Stems leaves and glumes tinged with brown and purple. Widespread. Open forests, grasslands, sand dunes. Fl. spring–summer. *Kangaroo Grass*. . . . . **T. australis** (R.Br.) Stapf

111 Coix L.
2 species in Aust.(1 native, 1 naturalized); Qld, NSW, WA

One species in the area
Erect annual or perennial with branched culms up to 2 m high, with prop roots. Leaf blades 2–5 cm wide, up to 45 cm long; ligule membranous. Male and female inflorescences numerous, produced on the one peduncle in the upper axils. Female inflorescence with 1 sessile fertile spikelet and 2 pedicellate sterile spikelets surrounded by a hard bead-like globular utricle 6–12 mm diam. Fertile spikelets with 2 florets. Male inflorescences with c. 10 spikelets, 7–8 mm long in slender racemes protruding from the utricle. Cumberland Plain. Waste places; roadsides. Introd. from Asia. Fl. summer. *Job's Tears* . . . .
. . . . ***C. lachryma-jobi** L.

198 RESTIONACEAE
Perennial herbs, dioecious or rarely monoecious, tough and wiry or rush-like, with creeping or tufted rhizomes. Leaves reduced to sheaths which are dark to pale brown, distichous, split on one side and not tubular as in Cyperaceae, imbricate at the base of the culm, distant or rarely absent on the upper portion. Flowers small, unisexual, rarely bisexual, each subtended by a glume, arranged in spikelets in lax or dense inflorescences. Perianth segments 6 or fewer, in 2 whorls, glumaceous; female flowers sometimes very few in the spikelet, rarely solitary; male flowers usually with 3 stamens opposite the perianth segments; anthers usually 1-celled. Ovary 3-carpellary, 1-3-locular, superior; styles 3 or 3-fid or simple. Fruit a nut or capsule. 40 gen., temp. to trop., mostly S. hemisphere, particularly S. Africa and Aust.

Key to the genera

1 Culms flattened. . . . . **4 EURYOCHORDA**
1 Culms terete. 2

2 Culms branched 3
2 Culms mostly unbranched 8

3 Mid culms with sterile branches . . . . **3 BALOSKION**
3 Culms without sterile branches. 4

4 At least the upper culm sheaths with a reflexed lamina . . . . **6 EMPODISMA**
4 Culm sheaths without a reflexed lamina. 5

5 Culms hoary . . . . **8 HYPOLAENA**
5 Culms green, not hoary 6

6 Flowers without bracteoles . . . . **5 CHORDIFEX**
6 Flowers with 1-2 bracteoles 7

7 Culms with angular internal air cavities. Perianth segments very unequal. . . . . **2 SPORADANTHUS**

7 Culms with circular internal air cavities. Perianth segments equal . . . . **1 LEPYRODIA**

8 Culms greyish . . . . **7 LEPTOCARPUS**
8 Culms green or yellow-green 9

9 Flowers without bracteoles . . . . **3 BALOSKION**
9 Flowers with 1-2 bracteoles 10

10 Culms with angular internal air cavities. Perianth segments very unequal . . . . **2 SPORADANTHUS**
10 Culms with circular internal air cavities. Perianth segments equal . . . . **1 LEPYRODIA**

1 Lepyrodia R.Br.
22 species endemic Aust.; all states and territories except NT

Perennial herb with a creeping scaly rhizome and erect culms up to 1 m high; dioecious or monoecious. Flowers mostly unisexual in a narrow panicle or in apparent spikelets. Glumes loose, not closely imbricate. Bracteoles usually 2 under the flower, within the glume. Perianth segments 6, glume-like or hyaline. Stamens 3, usually reduced to staminodes in the female flowers. Ovary 3-angled, 3-locular; stylar branches 3. Capsule opening at the angles. Male and female inflorescences not very different.

Key to the species

1 Culms with neither leaf sheath nor node except at the base (except rarely one leaf sheath), very slender, unbranched. Flowers in a spike-like panicle c. 2 cm long. Widespread. Heath and DSF . . . .
. . . . **L. anarthria** F.Muell.
1 Culms with leaf sheaths. Flowers in a narrow panicle 2

2 Leaf sheaths loose, often more than 2 cm long. Floral bracts with long points. Inflorescence pale brown, 2–5 cm long or sometimes longer. Perianth segments 3–4 mm long. Widespread. Damp ground. Heath and open forest. Ss. Fl. summer. . . . . **L. scariosa** R.Br.
2 Leaf sheaths appressed. Floral bracts obtuse or shortly acuminate. Perianth segments 2–3 mm long. Plants often monoecious. Coast and adjacent plateaus. Wet ground in the open. Fl. summer . . . .
. . . . **L. muelleri** Benth.

2 Sporadanthus F.Muell.
6 species in Aust.; Qld, NSW, Vic., Tas., WA

One species in the area
Perennial monoecious or dioecious herb with creeping rhizome culms 30–150 cm high. Male and female flowers similar, arranged in sparse terminal panicles. Inflorescence bracts inconspicuous soon falling. Outer tepals shorter than inner not exceeding 2.3 mm long, inner tepals to 207 mm long. Ovary 3-angled , 3-locular; stylar branches 3. Fruit a capsule 1.7-3.5 mm long opening at the angles. Grows in wet groud in open. Widespread . . . . **S. gracilis** (R.Br.) B.G.Briggs & L.A.S.Johnson

3 Baloskion Raf.
8 species endemic Aust.; Qld, NSW, Vic., Tas., S.A.

Perennial, dioecious herbs with creeping or partly erect often woolly rhizomes. Culms erect, simple or branched, with persistent leaf sheaths. Flowers unisexual, arranged in spikelets with imbricate bracts; the spikelets usually in a panicle. Lower part of flower stalk attached to the glume. Perianth segments 4–6, glume like or the inner ones almost hyaline. Stamens 3, usually shortly exserted. Ovary 2–3-locular; styles or stylar branches 2–3. Capsule flat and 2-locular or 3-angled and 3-locular, usually opening at the angles. Male and female inflorescences similar.

Key to the species

1 Culms much branched 100–150 cm high, with dense clusters of filiform dichotomous green barren branches in the axils of the leaf sheaths. Rhizome creeping, woolly. Spikelets in a long narrow loose panicle mostly c. 30 cm long. Male spikelets narrow-ovate to nearly globular, 4–6 mm long; female spikelets usually narrower and longer. Perianth segments 6 in the male flowers, 4 in the female. Coast and adjacent plateaus. Wet ground. . . . . **B. tetraphyllum** (Labill.) B.G.Briggs & L.A.S.Johnson

1 Culms unbranched, 30–100 cm high. Spikelets in a narrow terminal panicle or raceme. Perianth segments 6 in the male flowers, 4 in the female. 2

2 Sheaths on the upper portion of the culm mostly with an apical tuft of fine hairs 1–4 mm long. Glumes mostly fringed with fine hairs. Culms erect, wiry, 20–80 cm high. Spikelets ovate to globular, 4–7 mm long. North-west Hornsby Plateau; Blue Mts Damp sandy soils on swamp margins . . . .

. . . . **B. fimbriatum** (L.A.S.Johnson & O.D.Evans) B.G.Briggs & L.A.S.Johnson

2 Apex of all the leaf sheaths glabrous or fringed with hairs less than 0.25 mm long. Glumes glabrous to ciliolate 3

3 Upper leaf sheaths loose; the lower ones appressed. Culms ± rugose, 50–100 cm high, 1.5–3 mm wide; stomata sunken in irregular depressions or pits. Spikelets ovoid to oblong-cylindrical, 8–15 mm long. Blue Mts Wet peaty, sandy, or gravelly soils. . . . . **B. australe** (R.Br.) B.G.Briggs & L.A.S.Johnson

3 Leaf sheaths all closely appressed. Culms smooth to striate 4

4 Non-overlapping part of the lowest bract gradually tapering to an acute or obtuse apex, longer than the sheathing base. Culms 0.5-1.5 mm wide. Female spikelets oblong-cylindrical, 6–16 mm long; male spikelets ellipsoid to globular, 5–10 mm long. Coast and adjacent plateaus. Wet or damp sandy soils . .

. . . .**B. gracile** (R.Br.) B.G.Briggs & L.A.S.Johnson

4 Non-overlapping part of the lowest bract abruptly tapered, shorter than the sheathing base. Culms 1.5–4 mm wide.. 5

5 Spikelets all shortly pedicellate to sessile. Female spikelets ellipsoid to globular, 4–6 mm long, glumes 2.5–4 mm long, male spikelets similar. Coast and Cumberland Plain. Deep sandy soils . . . .

. . . . **B. pallens** (R.Br.) B.G.Briggs & L.A.S.Johnson

5 Spikelets on lower part of inflorescence borne on long pedicels or branches. Female spikelets ovate–elliptic 8.5–9 mm long, glumes 4–6 mm long; male spikelets ovate 5–8 mm long. Blue Mts Swamps and damp areas on alluvium. Vulnerable. . **B. longipes** (L.A.S.Johnson & O.D.Evans) L.A.S.Johnson & B.G.Briggs

4 **Euryochorda** B.G.Briggs & L.A.S.Johnson
1 species endemic Aust.; Qld, NSW, Vic., Tas.

One species in the area

Culms much flattened, unbranched, 30–100 cm high; leaf sheaths obtuse, thin, appressed, 12–20 mm long. Rhizome short, almost erect. Male spikelets narrow-ovate to ovate, 5–7 mm long, on filiform pedicels. Female spikelets on very short pedicels, narrower than the males and sometimes longer. Perianth segments 4. Stamens and styles 2; ovary 2-locular, Widespread. Wet ground in the open . . . .

. . . . **E. complanatus** (R.Br.) B.G.Briggs & L.A.S.Johnson

5 **Chordifex** B.G.Briggs & L.A.S.Johnson
20 species in Aust.; NSW, Tas., WA

Key to the species

Perennial dioecious herbs with creeping or partly erect often woolly rhizomes. Culms erect, simple or branched, with persistent leaf sheaths. Flowers unisexual, arranged in spikelets with imbricate bracts; the spikelets usually in a panicle. Perianth segments 5–6, glume like or the inner ones almost hyaline. Stamens 3, usually shortly exserted. Ovary 2-locular; styles or stylar branches 2. Capsule flat and firm-walled. Male and female inflorescences similar or very different.

1 Culms divided into numerous, erect, uniformly, straight, slender flowering branches. Leaf sheaths appressed. Male spikelets very narrow, c. 4 mm long. Female spikelets resembling the males but pedunculate; the bracts more mucronate, empty except for a single terminal flower. Widespread. Damp ground in the open. Ss . . . .**C. fastigiatus** (R.Br.) B.G.Briggs

1 Culms divided into numerous often flexuose flowering branches. Leaf sheaths very loose, spreading upwards, obtuse. Male spikelets numerous, ovate, c. 4 mm long. Female spikelets fewer, sessile. Coast and Woronora Plateau. Damp ground in heath. Ss. . . . . **C. dimorphus** (R.Br.) B.G.Briggs

6 Empodisma L.A.S.Johnson & D.Cutler
2 species endemic Aust.; all states and territories

One species in the area

Weak perennial, dioecious herb. Culms slender, very much branched, often flexuose, sometimes climbing to nearly 2 m high but may be quite short and nearly erect. Leaf sheaths loose, green or light coloured; the upper ones and the floral bracts, or sometimes nearly all of them, with short subulate usually reflexed tips. Flowers in axillary solitary sessile spikelets 4–6 mm long; male spikelets several-flowered; female spikelets 1-flowered. Glumes lanceolate, acute, rigid. Perianth segments 6, narrow in the male flowers; short broad and almost hyaline in the female. Stamens 3. Ovary 1-locular; stylar branches 3. Fruit a nut. Widespread. Wet places . . . . **E. minus** (Hookf.) L.A.S.Johnson & D.Cutler

7 Leptocarpus R.Br.
3 species endemic Aust.; all states and territories except Tas..

One species in the area

Perennial dioecious herb up to 1 m high. Culms slender, wiry, erect, straight, mostly unbranched, glabrous except the rhachis of the inflorescence. Rhizome creeping, glabrous. Leaf sheaths dark, closely appressed, 8 mm long or less, without spreading tips. Flowers in terminal spikelets. Male spikelets often very numerous, c. 4 mm long, narrow-ovate, dark brown, pendulous on filiform pedicels, in a panicle; female spikelets few, up to 12 mm long, erect, sessile or shortly pedicellate, in a shortly contracted panicle; bracts and glumes acutely acuminate, rich dark brown. Perianth segments 6. Stamens 3. Stylar branches usually 3. Ovary 1-locular. Fruit an angular capsule. Widespread. Damp ground . **L. tenax** (Labill.) R.Br.

8 Hypolaena R.Br.
8 species endemic Aust.; all states and territories except NT

One species in the area

Perennial dioecious herb 30–60 cm high. Culms slender, erect to ascending, flexuose, much branched, striate, greyish or hoary, sometimes glabrous. Leaf sheaths dark, often exceeding 10 mm long, appressed, without spreading tips. Rhizomes creeping. Flowers in terminal spikelets. Male spikelets narrow-ovate, 4–7 mm long, rich brown, pendulous on filiform pedicels, in a graceful panicle; female spikelets sessile, 1–3 together at the tips of the branches. Perianth segments 6, narrow in the male flowers; short broad and almost hyaline in the female. Stamens 3. Ovary 1-locular; stylar branches 3. Fruit a nut. Widespread. Damp sandy heath . . . . **H. fastigiata** R.Br.

199 SPARGANIACEAE
1 gen., mostly N. hemisphere.

1 Sparganium L.
2 species in Aust. (1 native, 1 introduced); Qld, NSW, Vic.

One species in the area

Aquatic perennial with a short creeping rhizome. Stem erect, less than 1 m high. Leaves cauline, long, linear, 3–6 mm wide. Flowers unisexual, regular, numerous in globular clusters which are distant from each other along the upper stem; upper clusters with male flowers; lower clusters with female flowers. Bracts and perianth scale-like. Stamens usually 3. Ovary nearly sessile. Fruits indehiscent, sessile, crowded, with a drupaceous pericarp. Coast. Margins of fresh water ponds and streams. . . . .
. . . . **S. subglobosum** Morong

200 TYPHACEAE
1 gen., temp. and trop.

1 Typha L.
Bull-rush or *Cumbungi*
3 species in Aust. (2 native, 1 naturalized); all states and territories

Tall, erect reed-like emergent aquatic perennials 1–3 m high, with stout rhizomes, monoecious. Leaves distichous, linear, 5–20 mm wide, thick spongy, semicylindrical to flat, with long sheaths; the upper ones often auriculate. Flowers unisexual, regular, crowded into dense cylindrical spike-like inflorescences; upper part of the spike with male flowers; lower part with female flowers often separated from the males by a short bare interval. Male flowers with 3 stamens, interspersed with hair-like bracts. Female flowers fertile or sterile, solitary or several on very short lateral branches; fertile female flowers with a solitary carpel surrounded by hairs, at first sessile but later raised on a gynophore, usually subtended by a small delicate bract. Fruit a small 1-seeded follicle surrounded by silky hairs.

Key to the species

1 Female inflorescence black-brown or dark red-brown, usually continuous with the male inflorescence, bracts absent. Stigmas narrow-obovate to obovate, club-shaped. Leaves up to 2 cm wide, ± glaucous, nearly flat. Introd. from Europe. Faulconbridge. *Catstail* or *Great Reedmace* . . . . ***T. latifolia** L.
1 Female spikes pale brown to chestnut brown 2

2 Female inflorescence pale brown due to the exserted tips of the numerous bracts, interspersed with dark brown stigmas, usually separated from the male inflorescence by 2–5 cm of bare axis. Stigmas linear. Leaves green, biconvex. Erect perennial 1–2 m high. Widespread in standing water. Sometimes a pest . . . . **T. domingensis** Pers.
2 Female inflorescence chestnut brown due to the almost complete absence of bracts at least at the surface, sometimes separated from the male inflorescence by 2 cm of bare axis or more (or continuous). Stigmas narrow-obovate. Leaves ± blue-green, biconvex. Erect perennial, usually 2–3 m high. Widespread in standing water. more common than *T. domingensis* in the area, and sometimes a pest . . . . **T. orientalis** C.Presl

201 XYRIDACEAE
5 gen., trop. to temp.

1 Xyris L.
19 species in Aust.; all states and territories

Tufted, perennial herbs. Leaves narrow, basal, sheathing at the base. Flowers bisexual, slightly irregular, sessile with broad imbricate rigid bracts, in a terminal globular or ovoid head. Sepals 3; the lateral 2 exterior, glumaceous; the third one interior, petaloid or membranous. Petals 3, connate at the base, yellow, equal, spreading. Stamens 3, alternating with 3 staminodes. Ovary superior, 1-locular or imperfectly 3-locular; placentas parietal; style 3-branched. Fruit a 1-locular capsule.

Key to the species

1 Leaf margins pale, often scabrous. Leaves flat. Ovary and fruit not thickened towards the apex 2
1 Leaf margin not pale, smooth. Leaves terete or ± flattened at the apex 5

2 Lateral sepals with narrow membranous margins; apex glabrous. Style branched for half its length. Placentas short 3
2 Lateral sepals with wide membranous margins; apex ciliate. Style lobed. Placentas long 4

3 Leaves narrow-linear, less than 2 mm wide. Peduncles 0.5–1 mm diam. Flower heads ovoid to ellipsoid, 6–8 mm long, 3–5 mm diam. Bracts 6–14; the basal ones closely appressed. Flowers 1–4 per head. Widespread in damp soil in the open. Fl. spring–summer . . . . **X. gracilis** R.Br.

3 Leaves linear, 2–5 mm wide. Peduncles 1–2 mm diam., up to 60 cm long. Flower heads subglobular, 10 mm long, 7–8 mm diam. Bracts 18–30; lower ones lax, ± spreading. Flowers 8–12 per head. Coast and adjacent plateaus. Damp soil in the open. Fl. spring–summer . . . . **X. bracteata** R.Br.

4 Peduncle flattened, 1–2.5 mm diam., 20–60 cm long. Leaves linear, 5–30 cm long 1–4 mm wide. Flower heads ovoid to cylindrical, 1–2.5 cm long, 7–10 mm diam. Widespread. Damp places. Fl. spring–summer . . . . **X. complanata** R.Br.

4 Peduncle filiform, terete or very slightly angular. Flower heads globular . . . . . . **X. juncea** (see below)

5 Ovary and fruit not thickened at the apex. Flower heads globular, 4–8 mm diam. Peduncles filiform, ± terete, 7–30 cm long, c. 1 mm wide. Leaves ± terete or linear, 2–12 mm long, c. 1 mm wide. Widespread. On damp soil. Fl. spring–summer . . . .**X. juncea** R.Br.

5 Ovary and fruit thickened at the apex. Flower heads ± ovoid, 8–20 mm long, 5–15 mm diam. **6**

6 Lower bracts arranged in 5 rows, closely imbricate. Flower heads tapering gradually towards the base. Peduncles terete, 30–100 cm long. Leaves terete or nearly, 20–60 cm long, 1 mm wide. Widespread. In wet soil. Fl. spring–summer. . . . . **X. operculata** Labill.

6 Lower bracts not arranged in 5 rows, lax, often spreading. Flower heads rounded at the base. Peduncle terete or slightly angular, 50–150 cm long, 2 mm diam. Blue Mts and Woronora Plateau. Damp soil. Fl. spring–summer. . . . . **X. ustulata** L.A.Nilsson

202 CANNACEAE
1 gen., mainly trop. and subtrop. S. America

1 Canna L.
1 species naturalized Aust.; Qld, NSW

One species in the area
Erect, herbaceous perennial 1–1.5 m high with a fleshy rhizome. Leaves broad, pinnately veined, 2-ranked, c. 60 cm long, without ligules or pulvini; base sheathing. Flowers in a terminal raceme or narrow panicle, bisexual, irregular. Sepals 3, free, herbaceous. Petals 3, yellow and red, connate at the base. Stamen 1, petaloid, bearing half an anther on one edge, surrounded by four or fewer petaloid staminodes. Ovary 3-locular, inferior. Fruit a capsule. Seeds black, globular. Introd. from W. Indies, Central and S. America. Fl. spring–summer. *Canna* or *Indian Shot* . . . . ***C. indica** L.

203 ZINGIBERACEAE
Rhizomatous herbs. Leaves distichous, sheathing stems. Flowers in terminal spike or thyrse, bisexual, irregular, 3-merous. Stamen 1. Staminodes 3; 2 of them short and narrow; the other broadened into a petaloid labellum. Ovary inferior 1 or 3-locular. Fruit a capsule or berry. Seeds with an aril. c. 46 gen., trop., mostly Indo-Malaysia

Key to the genera

1 Inflorescence appearing unbranched. Stamens exserted; lateral staminodes petaloid. **1 HEDYCHIUM**

1 Inflorescence branched. Stamens not exserted, lateral staminodes reduced to teeth or absent. . . . . . .
. . . .**2 ALPINIA**

1 Hedychium J.Koenig
2 species naturalized AUST,; Qld, NSW

One species in the area
Perennial herb to 2 m high. Leaves lanceolate, up to 40 cm long, 15 cm wide; ligule membranous, 25 mm long, 20 mm wide. Inflorescence a spike to 35 cm long with 1–2 flowers in each bract. Flowers yellow.

Staminodes twice as long as the labellum, filaments bright red. Capsule walls thin, orange on inner surface. Aril red. North from Sydney. Introd. from India. Fl. summer-autumn . . . .
. . . . ***H. gardnerianum** Ker Gawl.

2 **Alpinia** Roxb.
5 species in Aust. (4 endemic); Qld, NSW

Perennial herbs. Leaves with a prominent midrib. Inflorescence terminal on shoots, flowers yellow. Outer perianth fused into a shallowly cleft tube, tube of inner perianth usually shorter than outer. Stamen 1. Staminodes 3; 2 of them short and narrow; the other broadened into a petaloid labellum. Ovary inferior, 3-locular. Fruit a capsule, sometimes fleshy.

Key to the species

1 Inflorescence 2–8 cm long. Ligule 3–4 mm long. Leaves lanceolate, 12–25 cm long, 2–4 cm wide. Outer perianth 3–4 mm long: inner perianth tube 3–4 mm long. Labellum reniform. Lateral staminodes minute. Ovary 2 mm long, glabrous. Fruit blue-black. Herbs with erect leafy stems up to 1.5 m high. North from Wyong. WSF & RF. Fl. spring – summer *Native Ginger*. . . . .
. . . . **A. arundelliana** (F.M.Bailey) K.Schum.
1 Inflorescence 12–30 cm long. Ligule usually 10–15 mm long. Leaves oblong-lanceolate to 40 cm long, 3–10 cm wide. Outer perianth c. 10 mm long; inner perianth tube c. 12 mm long, lobes c. 7 mm long. Labellum circular. Lateral staminodes c. 1 mm long. Ovary to 6 mm long, usually hairy. Fruit globular, c. 10 mm diam., blue, with a brittle pericarp. Herb to 3 m high. North from Gosford. Coastal RF. Fl. spring . . . . **A. caerulea** (R.Br.) Benth.

BIBLIOGRAPHY

GENERAL (references relevant to multiple families)

Angiosperm Phylogeny Group (2003). An update of the Angiosperm phylogeny group classification for the orders and families of flowering plants: APGII. *Botanical Journal of the Linnean Society* 141, 399–36.

Australia's Virtual Herbarium. Centre for Plant Biodiversity Research, Council of Heads of Australian Herbaria. [Online] Available: www.cpbr.gov.au/cgi-bin/avh.cgi

Australian Biological Resources Study (1998–). *Flora of Australia*. Canberra: Australian Government Publishing Service and CSIRO Publishing. Also [Online] Available: www.environment.gov.au/biodiversity/abrs/online-resources/flora/main/

Australian Government Department of the Environment, Water, Heritage and the Arts (2006). Weed Identification & Information. National Weeds Strategy. [Online] Available: www.weeds.gov.au/index.html

Australian Plant Census, IBIS database, Centre for Plant Biodiversity Research, Council of Heads of Australian Herbaria. [Online] Available: www.chah.gov.au/apc/index.html

Australian Plant Name Index, IBIS database, Centre for Plant Biodiversity Research, Australian Government, Canberra. [Online] Available: www.cpbr.gov.au/cgi-bin/apni

Beadle NCW, Evans OD & Carolin RC (1962). *Handbook of the vascular plants of the Sydney District and Blue Mountains*. Armidale: Brown Gem Print.

Beadle NCW, Evans OD & Carolin RC (1972). *Flora of the Sydney Region*. 2nd Edition. Sydney: Reed.

Beadle NCW, Evans OD & Carolin RC (1982). *Flora of the Sydney Region*. 3nd Edition. Frenchs Forest: Reed.

Carolin RC & Tindale MD (1994). *Flora of the Sydney Region*. 4th Edition. Chatswood: Reed.

Botanic Gardens Trust (2005–2009). PlantNet – The Plant Information Network System of the Botanic Gardens Trust Version 2.0. [Online] Available: www.plantnet.rbgsyd.nsw.gov.au

Cronquist A (1988). *The evolution and classification of flowering plants*. 2nd Edition. Bronx, NY: New York Botanical Garden.

Harden GJ (1990). *Flora of New South Wales*. Volume 1. Kensington: UNSW Press.

Harden GJ (1992). *Flora of New South Wales*. Volume 3. Kensington: UNSW Press.

Harden GJ (1993). *Flora of New South Wales*. Volume 4. Kensington: UNSW Press.

Harden GJ (2000). *Flora of New South Wales*. Volume 2, Revised Edition. Kensington: UNSW Press.

Harden GJ & Murray LJ (2000). *Supplement to Flora of New South Wales*. Volume 1. Kensington: UNSW Press.

Hosking JR, Conn B, Lepischi J & Barker CH (2007). Plant species first recognised as naturalised for New South Wales in 2002 and 2003, with additional comments on species recognised as naturalised in 2000–2001. *Cunninghamia* 10, 139–66.

Hosking JR, Conn BJ & Lepischi J (2003). Plant species first recognised as naturalised for New South Wales over the period 2000–2001. *Cunninghamia* 8, 175–87.

International Plant Name Index (2005). [Online] Available: www.ipni.org

Kubitzki K (Ed) (1998). *Flowering plants, Monocotyledons: Alismatanae and Commelinanae (except Gramineae)*. New York: Springer.

Kubitzki K (Ed) (1998). *Flowering plants, Monocotyledons: Lilianae (except Orchidaceae)*. Berlin and New York: Springer.

Kubitzki K (Ed) (2004). *Flowering plants, Dicotyledons: Celastrales, Oxalidales, Rosales, Cornales, Ericales*. Berlin and New York: Springer.

Kubitzki K & Bayer C (Eds) (2003). *Flowering plants, Dicotyledons: Malvales, Capparales, and non-betalain Caryophyllales*. Berlin and New York: Springer.

Kubitzki K, Rohwer JG & Bittrich V(Eds) (1993). *Flowering plants, Dicotyledons: Magnoliid, Hamamelid, and Caryophyllid families*. Berlin and New York: Springer.

Moore C & Betche E (1893). *Handbook of the Flora of New South Wales: a description of the flowering plants and ferns indigenous to New South Wales*. Sydney: Government Printer.

Morley BD & Toelken HR (Eds) (1983). *Flowering plants in Australia*. Adelaide: Rigby.

NSW Department of Environment and Climate Change (2005–2009). Threatened Species Information Network. [Online] Available: www.nationalparks.nsw.gov.au/npws.nsf/

Stanley TD & Ross EM (1983). *Flora of south-eastern Queensland*. Brisbane: Queensland Department of Primary Industries.

Stevens PF (2001–). Angiosperm Phylogeny Website. Versions 8 and 9. [Online] Available: www.mobot.org/MOBOT/research/APweb/

Walsh NG & Entwisle TJ (1994). *Flora of Victoria: Ferns and Allied Plants, Conifers and Monocotyledons*. Melbourne: Inkata Press.

Walsh NG & Entwisle TJ (1996). *Flora of Victoria: Dicotyledons, Winteraceae to Myrtaceae*. Melbourne: Inkata Press.

Walsh NG & Entwisle TJ (1999). *Flora of Victoria: Dicotyledons, Cornaceae to Asteraceae*. Melbourne: Inkata Press.

Western Australian Herbarium (1998–). FloraBase – The Western Australian Flora. Department of Environment and Conservation. [Online] Available: florabase.dec.wa.gov.au/

LYCOPODIOPHYTA, MONILOPHYTA, CYCADOPHYTA, PINOPHYTA

Andrews SB (1990). *Ferns of Queensland*. Brisbane: Queensland Department of Primary Industries.

Australian Biological Resources Study (1998). Ferns, Gymnosperms and allied groups (1998). In *Flora of Australia*. Volume 48. Canberra: Australian Government Publishing Service and CSIRO Publishing.

Jones WG, Hill KD & Allen JM (1995). *Wollemia nobilis*, a new living Australian genus and species in the Araucariaceae. *Telopea* 3, 173–76.

Kramer KU & Green PS (1990). *The families and genera of vascular plants*. Volume 1. Pteridophytes and Gymnosperms. Berlin: Springer.

Kramer KU & Tindale MD (1976). The Lindsaeoid ferns of the Old World VII. Australia and New Zealand. *Telopea* 1, 91–128.

Øllgaard B (1987). A revised classification of the Lycopodiaceae s. lat. *Opera Botanica* 92, 153–78.

Tindale MD (1956). The Cyatheaceae in Australia. *Contributions from the New South Wales National Herbarium* 2, 327–61.

Tindale MD (1960). Contributions to the Flora of New South Wales: new species and combinations in *Acacia* and *Blechnum*. *Proceedings of the Linnean Society of New South Wales* 85, 248–55.

Tindale MD (1961). Pteridophyta of South Eastern Australia. *Contributions from the New South Wales National Herbarium* 208–11, 1–78.

Tindale MD (1963). Pteridophyta of South Eastern Australia. *Contributions from the New South Wales National Herbarium* 201, 1–49.

Tindale MD (1965). A monograph of the genus *Lastreopsis* Ching. *Contributions from the New South Wales National Herbarium* 3, 249–339.

MAGNOLIOPHYTA (by Family)

Acanthaceae

Heenan PB, de Lange PJ, Cameron EK, Ogle CC & Champion PD (2004). Checklist of dicotyledons, gymnosperms and pteridophytes naturalized or casual in New Zealand: additional records, 2001–2003. *New Zealand Journal of Botany* 42, 797–814.

Anthericaceae

Stringer S & Conran JG (1991). Stamen and seed cuticle morphology in some *Arthropodium* and *Dichopogon* species (Anthericaceae). *Australian Journal of Botany* 39, 129–35.

Apiaceae

Hart JM & Henwood MJ (2000). Systematics of the *Xanthosia pilosa* complex (Apiaceae: Hydrocotyloideae). *Australian Systematic Botany* 13, 245–66.

Henwood MJ & Hart JM (2001). Towards an understanding of the phyogenetic relationships of Australian Hydrocotyloideae (Apiaceae). *Edinburgh Journal of Botany* 58, 269–89.

Asteraceae

Andenberg AA (1991). Taxonomy and phylogeny of the tribe Gnaphalieae (Asteraceae). *Opera Botanica* 104, 5–195.

Benson D & Eldershaw G (2007). Backdrop to encounter: the 1770 landscape of Botany Bay, the plants collected by Banks and Solander and rehabilitation of natural vegetation at Kurnell. *Cunninghamia* 10, 113–38.

Flann C, Ladiges PL & Walsh NG (2002). Morphological variation in *Leptorhynchos squamatus* (Gnaphalieae: Asteraceae). *Australian Systematic Botany* 15, 205–19.

Holland AE & Funk VA (2006). A revision of *Cymbonotus* (Compositae: Arctotideae, Arctotidinae). *Telopea* 11, 266–75.

Holzapfel S (1994). A revision of the genus *Picris* (Asteraceae, Lactucaceae) s.l. in Australia. *Willdenowia* 24, 97–218.

Orchard EA (2004). A revision of *Cassinia* (Asteraceae: Gnaphalieae) in Australia. 1. Introduction and generic infrageneric considerations. *Australian Systematic Botany* 17, 469–81.

Orchard EA (2004). A revision of *Cassinia* (Asteraceae: Gnaphalieae) in Australia. 2. Sections *Complanatae* and *Venustae*. *Australian Systematic Botany* 17, 505–33.

Orchard EA (2004). A revision of *Cassinia* (Asteraceae: Gnaphalieae) in Australia. 3. Section *Leptocephalae*. *Australian Systematic Botany* 17, 535–65.

Orchard EA (2005). A revision of *Cassinia* (Asteraceae: Gnaphalieae) in Australia. 4. Section *Costatae*. *Australian Systematic Botany* 18, 455–71.

Thompson IR (2004). Taxonomic studies of Australian *Senecio* (Asteraceae): 2. The shrubby discid species and the allied radiate species *Senecio linearifolius*. *Muelleria* 20, 67–110.

Thompson IR (2004). Taxonomic studies of Australian *Senecio* (Asteraceae): 1. The disciform species. *Muelleria* 19, 101–214.

Thompson IR (2004). Taxonomic studies of Australian *Senecio* (Asteraceae): 3. Radiate, arid region species allied to *S. magnificus* and the radiate, alpine species *S. pectinatus*. *Muelleria* 20, 111–38.

Thompson IR (2005). Taxonomic studies of Australian *Senecio* (Asteraceae): 5. The *S. pinnatifolius/S. lautus* complex. *Muelleria* 21, 23–76.

Walsh NG (2001). A revision of *Centipeda* (Asteraceae). *Muelleria* 15, 33–64.

Wilson PG (2008). *Coronidium*, a new Australian genus in the Gnaphalieae (Asteraceae). *Nuytsia* 18, 295–328.

Brassicaceae

Thompson IR (1996). A revision of the *Cardamine paucijuga* complex (Brassicaceae). *Muelleria* 9, 161–73.

Thompson IR & Ladiges PY (1996). A revision of the *Cardamine gunnii – lilacina* complex (Brassicaceae). *Muelleria* 9, 145–59.

Campanulaceae

Smith PJ (1992). A revision of the genus *Wahlenbergia* (Campanulaceae) in Australia. *Telopea* 5, 91–175.

Caryophyllaceae

Smissen RD & Garnock-Jones (2002). Relationships, classification and evolution of *Scleranthus* (Caryophyllaceae) as inferred from analysis of morphological characters. *Botanical Journal of the Linnean Society* 140, 15–29.

Chenopodiaceae

Rilke S (1999). Species diversity and polymorphism in *Salsola* sect. *Salsola* sensu lato (Chenopodiaceae). *Systematics and Geography of Plants* 68, 305–14.

Sheperd KA & Wilson PG (2007). Incorporation of the Australian genera *Halosarcia*, *Pachycornia*, *Sclerostegia* and *Tegicornia* into *Tecticornia* (Salicornioideae, Chenopodiaceae). *Australian Systematic Botany* 20, 319–31.

Wilson PG (1984). Chenopodiaceae. In ABRS. *Flora of Australia*. Volume 4. pp. 81–318. Canberra: Australian Government Publishing Service.

Convolvulaceae

Johnson RW (2001). A taxonomic revision of *Convolvulus* L. (Convolvulaceae) in Australia. *Austrobaileya* 6, 1–39.

Cyperaceae

Johnson LAS & Evans OD (1973). *Cyperus brevifolius* and an allied species in Eastern Australia. *Contributions from the New South Wales National Herbarium* 4, 378–79.

Klaphake V (2003). *Key to the commoner species of sedges and rushes of Sydney and the Blue Mountains*. Sydney: Van Klapake.

Muasya AM, Simpson DA & Goetghebeur P (2000). New combinations in *Trichophorum*, *Scirpoides* and *Ficinia* (Cyperaceae). *Novon* 10, 132–33.

Wilson KL (1981). A synopsis of the genus *Scirpus sens. lat.* (Cyperaceae) in Australia. *Telopea* 2, 153–76.

Wilson KL (1991). Systematic studies in *Cyperus* section *Pinnati* (Cyperaceae). *Telopea* 4, 361–96.

Wilson KL (1996). A new Australian species of *Carex* (Cyperaceae) and notes on two other species. *Telopea* 6, 569–77.

Dilleniaceae

Toelken HR (1995). Notes on *Hibbertia*. 1. New taxa from south-eastern Australia. *Journal of the Adelaide Botanic Gardens* 16, 59–72.

Toelken HR (1998) *Hibbertia aspera* group (Dilleniaceae). *Journal of the Adelaide Botanic Gardens* 18, 107–60.

Toelken HR (2000) Notes on *Hibbertia* (Dilleniaceae) 3. *H. sericea* and associated species. *Journal of the Adelaide Botanic Gardens* 19, 1–54.

Ericaceae

Crowden RK & Menadue Y (1996). *Epacris crassifolia* R. Br. (Epacridaceae) – A reappraisal. *Annals of Botany* 77, 333–39.

Judd WS & Kron KA (1993). Circumscription of Ericaceae (Ericales) as determined by preliminary cladistic analyses based on morphological, anatomical and embryological features. *Brittonia* 45, 99–114.

Powell JM, Robertson G, Wiecek BM & Scott JA (1992). Studies in Australian Epacridaceae: changes to *Styphelia*. *Telopea* 5, 207–28.

Quinn CJ, Brown EA, Heslewood MM & Crayn DM (2005). Generic concepts in Styphelieae (Ericaceae): the *Cyathodes* group. *Australian Systematic Botany* 18, 439–54.

Euphorbiaceae

Esser H (1996). Proposal to conserve the name *Homalanthus* (Euphorbiaceae) with a conserved spelling. *Taxon* 45, 555–56.

Hunter JT & Bruhl JJ (1997). Two new species of *Phyllanthus* and notes on *Phyllanthus* and *Sauropus* (Euphorbiaceae: Phyllantheae) in New South Wales. *Telopea* 7, 149–65.

Hassall DC (1976). Numerical and cytotaxonomic evidence for generic delimitation in Australian Euphorbieae. *Australian Journal of Botany* 24, 633–40.

Hassall DC (1977). The genus *Euphorbia* in Australia. *Australian Journal of Botany* 25, 429–53.

Webster GL (1994). Synopsis of the genera and suprageneric taxa of Euphorbiaceae. *Annals of the Missouri Botanical Gardens* 81, 33–144.

Fabaceae

de Kok RPJ & West JG (2002). A revision of *Pultenaea* (Fabaceae) 1. Species with ovaries glabrous and/or with tufted hairs. *Australian Systematic Botany* 15, 81–113.

de Kok RPJ & West JG (2003). A revision of the genus *Pultenaea* (Fabaceae) 2. Eastern Australian species with velutinous ovaries and incurved leaves. *Australian Systematic Botany* 16, 229–73.

de Kok RPJ & West JG (2004). A revision of the genus *Pultenaea* (Fabaceae). 3. The eastern species with recurved leaves. *Australian Systematic Botany* 17, 273–326.

Pfeil BE, Tindale MD & Craven LA (2001). A review of the *Glycine clandestina* species complex (Fabaceae: Phaseolae) reveals two new species. *Australian Systematic Botany* 14, 891–900.

Thompson IR (2001). Morphometric analysis and revision of Eastern Australian *Hovea* (Brongnairtieae-Fabaceae). *Australian Systematic Botany* 14, 1–99.

Gentianaceae

Adams LG (1995). *Chionogentias* (Gentianaceae): a new generic name for the Australasian 'snow-gentians' and a revision of the Australian species. *Australian Systematic Botany* 8, 935–1011.

Mansion G (2004). A new classification of the polyphyletic genus *Centuarium* Hill (Chironiinae – Gentianaceae): description of the New World endemic Zeltnera, and reinstatement of *Gyrandra* Griseb. and *Schenkia* Griseb. *Taxon* 53, 719–40.

Haloragaceae

Wilson PG & Moody ML (2006). *Haloragodendron gibsonii* (Haloragaceae): a new species from the Blue Mountains, New South Wales. *Telopea* 11, 141–46.

Hemerocallidaceae

Carr GW (2006). *Dianella tenuissima* (Hemerocallidaceae): a remarkable new species from the Blue Mountains, New South Wales, Australia. *Telopea* 11, 300–06.

Iridaceae

Goldblatt P (1980). Systematics and biology of *Homeria* (Iridaceae). *Annals of the Missouri Botanical Gardens* 68, 413–503.

Juncaceae

Johnson LAS (1991). New Australian taxa in *Juncus* (Juncaceae). In MR Banks et al. (Eds). *Aspects of Tasmanian botany – a tribute to Winifred Curtis,* (pp. 35–46). Hobart: Royal Society of Tasmania.

Johnson LAS (1993). New species of *Juncus* (Juncaceae) in eastern Australia. *Telopea* 5, 309–18.

Juncaginaceae

Aston HI (1995). A revision of the tuberous-rooted species of *Triglochin* L. (Juncaginaceae) in Australia. *Muelleria* 8, 331–64.

Lamiaceae

Conn B (1997). Four rare and/or threatened new species of *Prostanthera* Section *Prostanthera* (Labiatae) from New South Wales. *Telopea* 7, 231–44.

Conn B (1999). The *Prostanthera crypatandroides – P. euphrasioides – P. adoratissima* complex (Labiatae). *Telopea* 8, 266.

Kadereit JW (2004). *Flowering plants, Dicotyledons: Lamiales (except Acanthaceae including Avicenniaceae)*. Berlin and New York: Springer.

Webb CJ, Sykes WR & Garnock-Jones PJ (1988). *Flora of New Zealand*. Christchurch: Botany Division DSIR.

Lemnaceae

Les HL & Crawford DJ (1999). *Landoltia* (Lemnaceae), a new genus of Duckweeds. *Novon* 9, 530–33.

Les HL, Crawford DJ, Landolt E, Gabel JD & Kimball RT (2002). Phylogeny and systematics of Lemnaceae, the Duckweed Family. *Systematic Botany* 27, 221–40.

Rothwell GA, Van Atta MR, Ballard HE & Stockey RA (2004). Molecular phylogenetic relationships among Lemnaceae and Aracaceae using the chloroplast *trnL – trnF* intergenic spacer. *Molecular Phylogentics and Evolution* 30, 378–85.

Malvaceae

Barker RM (1998). *Sida* section *Sida* in Australia. *Journal of the Adelaide Botanic Gardens* 18, 33–41.

Pfeil BE & Crisp MD (2005). What to do with *Hibiscus*? A proposed nomenclatural resolution for a large and well known genus of Malvaceae and comments on paraphyly. *Australian Systematic Botany* 18, 49–60.

Ray MF (1995). Systematics of *Lavatera* and *Malva* (Malvaceae, Malveae) – a new perspective. *Plant Systematics and Evolution* 198, 29–53.

Ray MF (1998). New combinations in *Malva* (Malvaceae: Malveae). *Novon* 8, 288–95.

Myrsinaceae

Jackes BR (2005). A revision of *Myrsine* (Myrsinaceae) in Australia. *Australian Systematic Botany* 18, 399–438.

Myrtaceae

Bean AR (1997). Reinstatement of the genus *Babingtonia* Lindl. (Myrtaceae, Leptospermoideae). *Austrobaileya* 4, 627–45.

Bean AR (1999). A revision of the *Babingtonia virgata* (J.R.Forst. & G.Forst.) F.Muell. complex (Myrtaceae) in Australia. *Austrobaileya* 5, 157–71.

Craven LA & Lepsci BJ (1999). Enumeration of the species and infraspecific taxa of *Melaleuca* (Myrtaceae) occurring in Australia and Tasmania. *Australian Systematic Botany* 12, 819–927.

Wilson PG, Heslewood MM & Quinn CJ (2007). Re-evaluation of the genus *Babingtonia* (Myrtaceae) in eastern Australia and New Caledonia. *Australian Systematic Botany* 20, 302–18.

Orchidaceae

Bell S, Branwhite B & Driscoll C (2004). *Thelymitra adorata* Jeanes ms (Orchidaceae): population size and habitat of a highly restricted terrestrial orchid from Central Coast of New South Wales. *The Orchadian* 15, 6–77.

Bishop T (2000). *Field guide to the orchids of New South Wales and Victoria*. Kensington: UNSW Press.

Cameron KM (2005). Leave it to the leaves: a molecular phylogenetic study of Malaxideae (Epidendroideae, Orchidaceae). *American Journal of Botany* 92, 1025–32.

Clements MA & Jones DL (2002). Nomenclatural changes in the Dendrobieae (Orchidaceae) 1. The Australian Region. *The Orchadian* 13, 485–97.

Clements MA & Jones DL (2006). Australian Orchid Name Index (27/4/2006). [Online] Available: www.anbg.gov.au/cpbr/cd-keys/orchidkey/html/AustralianOrchidNameIndex.pdf

Clements MA & Jones DL (2007). Australian Orchidaceae: current genera and species list (31/1/2007). [Online] Available: www.anbg.gov.au/cpbr/program/ha/ha-orchidaceae.html

Clements MA, Sharma IK & Mackenzie AM (2001). A new classification of *Caladenia* R.B.r (Orchidaceae). *The Orchadian* 13, 389–429.

Jeanes JA (2004). A revision of the *Thelymitra pauciflora* R.Br. (Orchidaceae) complex in Australia. *Muelleria* 19, 19–29.

Jones DL (1996). *Microtis angusii*, a new species of Orchidaceae from Australia. *The Orchadian* 12, 10–12.

Jones DL (1998). A taxonomic review of *Prasophyllum* in Tasmania. *Australian Orchid Research* 3, 118–19.

Jones DL (1999). Eight new species of *Caladenia* R.Br. (Orchidaceae) from Eastern Australia. *The Orchadian* 13, 5–24.

Jones DL (2000). Four new names in *Caladenia* R.Br. (Orchidaceae), and a note on *C. carnea* var. *subulata* Nicholls. *The Orchadian* 13, 255–57.

Jones DL (2000). Ten new species of *Prasophyllum* R.Br. (Orchidaceae) from south-eastern Australia. *The Orchadian* 13, 149–73.

Jones DL (2001). Six new species and a new combination in *Genoplesium* R.Br. (Orchidaceae) from eastern Australia. *The Orchadian* 13, 293–307.

Jones DL (2006). *A complete guide to the native orchids of Australia: including the island territories*. Sydney: New Holland.

Jones DL (2006). *Speculanthus vernalis* (Orchidaceae). A critically endangered new species from south-eastern New South Wales. *The Orchadian* 15, 277–81.

Jones DL & Clements MA (2004). Miscellaneous new species, new genera, reinstated genera and new combinations in Australian Orchidaceae. *The Orchadian*, Supplement 14, i–xvi.

Jones DL & Clements MA (2005). Miscellaneous nomenclatural notes and changes in Australian, New Guinea and New Zealand Orchidaceae. *The Orchadian* 15, 33–42.

Jones DL & Clements MA (2006). The infrageneric taxonomy or *Diuris* Smith (Orchidaceae). *The Orchadian* 15, 203–06.

Jones DL, Clements MA, Sharma IK, Mackenzie AM & Molloy PJ (2002). Nomenclatural notes arising from studies into the tribe Diurideae (Orchidaceae). *The Orchadian* 13, 437–67.

Kores PJ, Molvray M, Weston PH, Hopper SD, Brown AP, Cameron KM & Chase MW (2001). A phylogenetic analysis of Diurideae (Orchidaceae) based on Plastid DNA sequence data. *American Journal of Botany* 88, 1903–14.

NSW Department of Environment and Climate Changes Threatened Species Information Network. Final Determination *Corybas dowlingii*. [Online] Available: www.nationalparks.nsw.gov.au/npws.nsf/

Rupp HMR (1947). A review of the species *Caladenia carnea* R.Br. (Orchidaceae). *Proceedings of the Linnean Society of New South Wales* 71, 280.

Pennantiaceae

Karehed J (2001). Multiple origins of the tropical forest tree family Icacinaceae. *American Journal of Botany* 88, 2259–74.

Karehed J (2003). The family Pennantiaceae and its relationships to Apiales. *Botanical Journal of the Linnean Society* 141, 1–24.

Pittosporaceae

Allan HH (1961). *Flora of New Zealand*. Wellington: Government Printer.

Cayzer LW, Crisp MD & Telford IRH (1999). *Bursaria* (Pittosporaceae): a Morphometric analysis and revision. *Australian Systematic Botany* 12, 117–43.

Cayzer LW, Crisp MD & Telford IRH (1999). Revision of *Rhytidosporum* (Pittosporaceae). *Australian Journal of Botany* 12, 689–708.

Cayzer LW, Crisp MD & Telford IRH (2000). Revision of *Pittosporum* (Pittosperaceae) in Australia. *Australian Systematic Botany* 13, 845–902.

Cayzer LW, Crisp MD & Telford IRH (2004). Cladistic analysis and revision of *Billardiera* (Pittosporaceae). *Australian Journal of Botany* 17, 83–125.

Plantaginaceae

Briggs BG & Ehrendorfer F (2006). New species and typifications in *Veronica* sens. lat. (Plantaginaceae). *Telopea* 11, 276–92.

Poaceae

Australian Biological Resources Study (2005). Poaceae. In *Flora of Australia*. Volume 44b. Canberra: Australian Government Publishing Service and CSIRO Publishing.

Hsiao C, Jacobs SWL, Barer NP & Chatterton NJ (1998). A molecular phylogeny of the subfamily Arundinoideae (Poaceae) based on sequences of rDNA. *Australian Systematic Botany* 11, 14–52.

Hsiao C, Jacobs SWL, Chatterton NJ & Asay KH (1999). A molecular phylogeny of the grass family (Poaceae) based on the sequences of nuclear ribosomal DNA (ITS). *Australian Systematic Botany* 11, 667–88.

Jacobs SWL (2001). Four new species of *Agrostis* (Gramineae) for Australia. *Telopea* 9, 679–84.

Jacobs SWL (2001). The genus *Lachnagrostis* (Gramineae) in Australia. *Telopea* 9, 439–48.

Jacobs SWL (2004). The tribe Triodieae (Chloridoideae:Gramineae). *Telopea* 10, 701–03.

Jacobs, SWL, Gillespie, LJ & Soreng, RJ (2008). New combinations in *Hookerochloa* and *Poa* (Gramineae). *Telopea* 12, 273–78.

Jacobs SWL, Whalley RDB & Wheeler DJB (2008). *Grasses of New South Wales*. 4th Edition. Armidale: University of New England.

Klaphake V (2002). *Key to the grasses of Sydney*. Sydney: Van Klapake.

Lazarides M (1997). A revision of *Eragrostis* (Eragrostideae, Eleusininae, Poaceae) in Australia. *Australian Systematic Botany* 10, 77–187.

Simons BK & Jacobs SWL (1999). A revision of the Genus *Sporobolus*. *Australian Systematic Botany* 12, 375–448.

Polygonaceae

Rechinger KH (1984). *Rumex* (Polygonaceae) in Australia: a reconsideration. *Nuytsia* 5, 75–122.

Wilson KL (1996). Nomenclatural notes on Polygonaceae in Australia. *Telopea* 7, 83–94.

Restionaceae

Briggs BG & Johnson LAS (1998). New genera and species of Australian Restionaceae (Poales). *Telopea* 7, 345–73.

Briggs BG & Johnson LAS (2004). New combinations in *Chordifex* (Restionaceae) from eastern Australia and new species from western Australia. *Telopea* 10, 683–700.

Klaphake V (2003). *Key to the commoner species of sedges and rushes of Sydney and the Blue Mountains*. Sydney: Van Klapake.

Meney KA & Pate JS (Eds) (1999). *Australian rushes biology: identification and conservation of Restionaceae and allied families*. Nedlands: University of Western Australia Press.

Rhamnaceae

Millott J & McDougall (2005). A new species of *Pomaderris* (Rhamnaceae) from the Central Tablelands of New South Wales. *Telopea* 11, 79–86.

Theile KR & West JG (2004). *Spyridium burragorang* (Rhamnaceae), a new species from New South Wales, with new combinations for *Spyridium buxifolium* and *Spyridium scortchinii*. *Telopea* 10, 823–29.

Walsh NG & Coates F (1997). New taxa, new combinations and an infrageneric classification in *Pomaderris* (Rhamnaceae). *Muelleria* 10, 27–56.

Rosaceae

Bean AR (1997). A revision of *Rubus* subg. Malachobatus (Focke) Focke and *Rubus* subg. Diemenicus A.R. Bean (Rosaceae) in Australia. *Austrobaileya* 5, 47–48.

Bean AR (1999). Queensland rasperries. *"Bulletin" Newsletter of the Queensland Region of SGAP*. Australian Plants. [Online] Available: http://asgap.org.au/

Evans KJ, Symon DE, Whalen MA, Hosking JR, Barker RM & Oliver JA (2007). Systematics of the *Rubus fruticosus* aggregate (Rosaceae) and other exotic *Rubus* taxa in Australia. *Australian Systematic Botany* 20, 187–251.

Rousseaceae

Lundberb J (2001). *Phylogeneric studies in the Euasterids II with particular reference to Asterales and Escalloniaceae*. Comprehensive Summaries of Uppsala Dissertations from the Faculty of Science and Technology, 676. Uppsala: Uppsala Universitet.

Rubiaceae

Reynolds ST & Henderson RJF (2001). Vanguerieae A. Rich. ex Dum. (Rubiaceae) in Australia, 2. *Cyclophyllum* Hook.f. *Austrobaileya* 6, 41–66.

Rutaceae

Armstrong JA (2002). *Zieria* (Rutaceae) a systematic and evolutionary study. *Australian Systematic Botany* 15, 277–463.

Duretto MF (2003). Notes on *Boronia* (Rutaceae) in eastern and northern Australia. *Muelleria* 17, 19–135.

Duretto MF & Forster PI (2007). A taxonomic revision of the genus *Zieria* Sm. (Rutaceae) in Queensland. *Austrobaileya* 7, 451–472.

Gebert WA & Duretto MF (2008). Geographic variation in Crowea exalata (Rutaceae) and the recognition of two new subspecies. *Telopea* 12, 193–213

Horton BM, Crayn DM, Clarke SW & Washington H (2004). *Leionema scopulinum* (Rutaceae) a new species from Wollemi National Park. *Telopea* 10, 815–22.

Mole BJ, Udovicic F, Ladiges PY & Duretto MF (2004). Molecular phylogeny of *Phebalium* (Rutaceae: Boronieae) and related genera based on the nrDNA regions ITS 1 + 2. *Plant Systematics and Evolution* 149, 197–212.

Weston PH & Turton M (2004). *Phebalium bifidum* (Rutaceae) a new species from the Capertee Valley, New South Wales. *Telopea* 10, 787–92.

Wilson PG (1970). A taxonomic revision of the genera, *Crowea, Eriostemon* and *Phebalium* (Rutaceae). *Nuytsia* 1, 5–155.

Wilson PG (1998). New species and nomenclatural changes in *Phebalium* and related genera (Rutaceae). *Nuytsia* 12, 267–88.

Wilson PG (1998). A new species and nomenclatural changes in *Phebalium* and related Genera (Rutaceae). *Nuytsia* 12, 167–288.

Solanaceae

Bean AR (2001). A revision of *Solanum brownii* Dunal (Solanaceae) and its allies. *Telopea* 9, 639–61.

Verbenaceae

Michael PW (1997). Notes on *Verbena officinalis* sensu stricto and *V macrostachya* (Verbenaceae) with new combinations in two closely related taxa. *Telopea* 7, 177–302.

Munir AA (2002). A taxonomic revision of the Genus *Verbena* L. (Verbenaceae) in Australia. *Journal of the Adelaide Botanic Gardens* 20, 21–103.

Violaceae

Theile KR & Prober SM (2003). New species and a new hybrid in the *Viola hederacea* species complex, with notes on *Viola hederacea* Labill. *Muelleria* 18, 7–25.

Thiele KR & Prober SM (2006). *Viola silicestris,* a new species in *Viola* section Erpetion from Australia. *Telopea* 11, 99–104.

INDEX

687

Manufactured by Amazon.com.au
Sydney, New South Wales, Australia

16623780R00415